BOOK LOAN

Please RETURN or RENEW it no later
than the last date shown below

CANCELLED		
20 DEC 1996 M		
CANCELLED		
2 CANCELLED		
18 DEC 1998 K		
CANCELLED		
1 5 DEC 2000		
1 4 DEC 2001		
2 0 DEC 2002		

Environmental problems in the humid tropical regions, where the focus is on the fate and management of the surviving rainforest and climate change, are attracting increasing international attention. The distribution of tropical rainfall is highly variable, and in many regions the supply of potable water is inadequate. By the end of the century one-third of the world's population will be living in the humid tropics. This book considers all aspects of hydrology in the humid tropics.

The first four parts of the book cover the physical basis of hydrology in the humid tropics: climatology, meteorology, process hydrology, sedimentation, water quality and freshwater ecology. This is followed by extensive treatment of the human and societal issues: land-use changes, water resource management, and rural and urban water supply in the tropical regions. The book is a uniquely integrated summary of hydrology in the tropics.

Hydrology and water management in the humid tropics

Hydrological research issues and strategies for water management

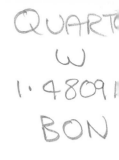

INTERNATIONAL HYDROLOGY SERIES

Hydrology and Water Management in the Humid Tropics

Hydrological research issues and strategies for water management

Edited by

Michael Bonell *(Department of Geography, James Cook University of North Queensland, Australia)*

Maynard M. Hufschmidt *(Program on Environment, East-West Center, Honolulu)*

John S. Gladwell *(Hydro Tech International Consultants, Vancouver and formerly UNESCO, Division of Water Sciences, Paris)*

Published by the Press Syndicate of the University of Cambridge
The Pitt Building, Trumpington Street, Cambridge CB2 1RP
40 West 20th Street, New York, NY 10011-4211, USA
10 Stamford Road, Oakleigh, Melbourne 3166, Australia
and United Nations Educational, Scientific and Cultural Organisation
7 Place de Fontenoy, Paris, France

First published 1993

Printed in Great Britain at the University Press, Cambridge

A catalogue record for this book is available from the British Library

Library of Congress cataloguing in publication data applied for

Cambridge University Press ISBN 0 521 45268 6 hardback
UNESCO ISBN 92 3 102854 5

The designations employed and the presentation of material throughout this publication
do not imply the expression of any opinion whatsoever on the part of the UNESCO
Secretariat concerning the legal status of any country, territory, city or area or of its
authorities, or the delimitations of its frontiers or boundaries.

Contents

Contents

List of Authors

DR. ARTHUR J. ASKEW
Hydrology and Water Resources
Department
World Meteorological Organization
Case postale No. 2300
CH-1211 Geneva 2
Switzerland

NII BOI AYIBOTELE
Water Resources Research Institute
Council for Scientific & Industrial
Research
P.O. Box M32
Accra
Ghana

DR. J. BALEK
ENEX - Environmental
Engineering Consultancy
Kopeckeho 8
Prague 6
169 00 Czech Republic

DR. MIKE BONELL
Department of Geography
and Institute for Tropical
Rainforest Studies
James Cook University of
North Queensland
Townsville, Qld 4811
Australia
Present address:
UNESCO
Division of Water Sciences
1, rue Miollis
75732 Paris Cedex 15
France

DR. J. P. BRUNEL
Institut Francais de Recherche
Scientifique pour le Développement
en Coopération (ORSTOM)
911 Avenue Agropolis
34032 Montpellier Cedex 1,
France

DR. ANDY BULLOCK
Institute of Hydrology
Crowmarsh Gifford
Wallingford, Oxon, OX10 8BB
United Kingdom,

DR. MARY J. BURGIS
City of London Polytechnic
Old Castle Street
London El 7NT
United Kingdom

J-H. CHANG
Chinese Culture University
Hwa Kang
Yang Ming Shan
Taiwan

JOHN CHILTON
Groundwater and Geotechnical Survey
Division (Hydrogeology Group)
British Geological Survey
Wallingford
Oxon OX10 8BB
United Kingdom

TONY FALKLAND
Hydrology and Water Resources
Branch
ACT Electricity and Water
GPO BOX 366
Canberra, ACT 2601,
Australia

P. M. (MICK) FLEMING
CSIRO
Division of Water Resources
PO Box 1666
Canberra, ACT. 2610
Australia

DR. STEPHEN FOSTER
Groundwater & Geotechnical
Surveys Division
British Geological Survey
Keyworth, Nottingham NG12 5GG
United Kingdom

DR. J. S. GLADWELL
Hydro Tech International,
Consultants
P.O. Box 40504
Waterfront Center
#11 - 200 Burrard Street
Vancouver, B.C. V6C 3L0
Canada
Previous address:
Division of Water Sciences
UNESCO
1, rue Miollis
75732 Paris Cedex 15
France.

DR. H. L. GOLTERMAN
Station Biologique
de la Tour du Valat
Le Sambuc
13200 Arles
France

B. GRIESINGER
Brazilian National Committee for the
International Hydrological Programme,
IBAMA/SEMA
SQN. 316 Block G Apto. 604
70.3775 Brasilia D. F.
Brazil

DR. MAYNARD M. HUFSCHMIDT
Program on the Environment
East-West Center
1777 East-West Road
Honolulu, Hawaii 96848
U.S.A

DR. V. KLEMEŠ
National Hydrology Research Institute
Environment Canada
Saskatoon, Saskatchewan, Canada
Present address:
3460 Fulton Road
Victoria, B.C. V9C 3N2
Canada

DR.-ING. U. KÜFFNER
The World Bank (Water Resources)
1818 H. Street, N.W.
Washington, D.C. 20433
U.S.A

PROF. R. LAL
Department of Agronomy
Ohio State University
2021 Coffey Road
Columbus, Ohio 43210
U.S.A

PROF. L. STEPHEN LAU
Water Resources Research Center
University of Hawaii at Manoa
2540 Dole Street
Honolulu, Hawaii 96822
U.S.A

DR. YOK-SHIU LEE
Program on Environment and
Program on Population
East-West Center
1777 East-West Road
Honolulu, Hawaii 96848
U.S.A.

DR. J. LEMOALLE
ORSTOM
B.P. 5045
34032 Montpellier Cedex 1
France

DR. G. J. M. LE MOIGNE
The World Bank (Water Resources)
1818 H Street, N.W.
Washington, D.C. 20433
U.S.A.

PROF. GUNNAR LINDH
Department of Water Resources Eng.
Lund University
Box 118
S-221 00 Lund
Sweden

PROF. KWAI-SIM LOW
Department of Geography/
Institute of Advanced Studies
Universiti of Malaya
Pantai Valley
59100 Kuala Lumpur
West Malaysia

R. E. MANLEY
Consultant in Engineering Hydrology
78 Huntingdon Road
Cambridge CB3 0HH
United Kingdom
formerly:
Chief Technical Advisor
WMO Project INS/83/029
Bandung, Indonesia

DR. M. J. MANTON
Bureau of Meteorology Research
Centre
GPO Box 1289K
Melbourne, Vic. 3001
Australia

DR. DAVID S. McCAULEY
Office of Environment and
Natural Resources
U.S. Agency for International
Development
Washington, D.C. 20523-1812
U.S.A.
Previous address:
U.S. Agency for International
Development
Jakarta
Indonesia

LUIZ CARLOS MOLION
Foundation for Advanced Studies in the
Humid Tropics (UNITROP)
Rua Floriano Peixoto, 287
69.003 Manaus, Amazonas
Brazil
Permanent address:
Instituto Nacional de Pesquisas
Espaciais
(INPE)
C. Postal 515
12.201 Sao Jose dos Campos
Sao Paulo
Brazil

DR. J. NIEMCZYNOWICZ
Department of Water Resources
Engineering
University of Lund
Box 118
S-22100 Lund
Sweden

DR. VINCENTE NOGUEIRA
Centro de Ciencias do Ambiente
Universidade do Amazonas
69.000 Manaus-Amazonas
Brazil

DR. A. PROST
World Health Organization
Avenue Appia
CH-1211 Geneva 27
Switzerland

DR. MICHEL-ALAIN ROCHE
Centre de Recherche en Ecologie
Marine
et Aquaculture (CREMA) de
l'Houmeau
B.P. 5
17137 L'Houmeau
France
Previous address:
ORSTOM
BP 5045
34032 Montpellier
France

PROF. CALVIN ROSE
Faculty of Environmental Sciences
Griffith University
Brisbane
Queensland
Australia

BRUCE STEWART
Hydrology Branch
Bureau of Meteorology
GPO BOX 1289K
Melbourne, VIC. 3001
Australia,

DR. J. F. TALLING
Freshwater Biological Association
The Ferry House
Ambleside, Cumbria LA22 0LP
United Kingdom

DR. K. G. TEJWANI
Center for Natural Resources and
Enviroment Management (CENREM)
25/31 Old Rajinder Nagar
New Delhi 110060
India

E. VAN BEEK
Delft Hydraulics
Water Resources & Environment Div.
P.O. Box 177
2600 MH Delft
The Netherlands

DR. PETER WURZEL
UNICEF
Water and Sanitation Section
Caixa postal 4713
Maputo
Mozambique

Foreword

It is widely recognized that water is going to be one of the major issues confronting humanity at the turn of the century and beyond. We are already facing a crisis as regards the quantity and quality of water supply, but we have yet to experience the full social and political impact of that crisis. Water is the life-blood of living organisms, and a very potent factor in human behaviour. Where plentiful, water is accepted unreflectively as a gift of nature. But when scarcity makes it a precious commodity, it can become a source of dispute and even conflict between its users. We know that civilization has always been crucially dependent on water. It is essential to remember that the converse is also true: vital water resources depend upon civilization, or more precisely on the "civilized" use of a finite and vulnerable resource. Culture and tradition are therefore important dimensions to be taken into account in water conservation.

The interdependence of water and civilization highlights the need for close co-operation among all players in the water game, extending from the local through the regional right up to the global level. At the international level, one of the most important needs is for co-ordinated efforts to understand the processes occurring in the water cycle, to assess surface and ground water resources, and to promote attitudes conducive to maintaining the quality and quantity of water resources for generations to come. Recognition of the importance of these objectives led to the launching of the International Hydrological Decade (IHD) in 1964, the first truly international scientific and educational effort ever made in hydrology.

IHD proved a truly remarkable example of international co-operation and, despite the formidable political and ideological barriers that existed at the time, yielded unique results in assessing the world's water resources. Yet important gaps remained, particularly in the application of scientific advances to the solution of practical problems. Economic and social activities tend to aggravate difficulties arising from the natural fluctuations in the hydrological regime, and the human impact on water resources increases with population growth and the spread of urbanization. Such problems, it was felt, could best be tackled by further strengthening international and regional co-operative efforts. In 1974, UNESCO decided to set up the long-term International Hydrological Programme (IHP) with the aim of finding solutions to the specific problems of countries with different geographical and climatic conditions and at various levels of technological and economic development. IHP was thus launched as an intergovernmental programme and today constitutes, together with the Man and the Biosphere (MAB) programme, the Intergovernmental Oceanographic Commission (IOC) and the International Geological Correlation Programme (IGCP), the scientific backbone of UNESCO's environmental activities.

The general objective of the IHD, and later of the IHP, was to improve the scientific and technological basis and the human resource base for the rational development and management of water resources, including the protection of the environment. The pursuit of this objective is closely bound up with the search for solutions to such crucial problems as the lack of reliable water supplies and sanitation, shortage of food and fibre, inadequate supplies of electrical energy, pollution of surface and ground waters, erosion and sedimentation, floods, drought and desertification. Attaining the general objective presupposes advances in knowledge and their effective application in a number of fields. In particular, a better understanding is required of the hydrological cycle as affected by man in terms of both quantity and quality under various climatic conditions.

The water resource situation in the developing countries will present a serious challenge to water management well into the twenty-first century. The challenge for water users, educators, planners, managers, policy-makers and politicians is how to devise policies that will be environmentally sound yet will contribute effectively to meeting social and economic goals. World leaders meeting at the Earth Summit in Rio de Janeiro in June 1992 acknowledged that protecting our rapidly changing and increasingly vulnerable environment will call for the deployment of our best legal, organizational and scientific efforts. A continued focus on satisfying short-term needs is a recipe for disaster. Greater attention must be given among other things to providing a scientifically valid basis for integrated resource development which takes into consideration both socio-cultural factors and the safeguarding of our natural heritage for future generations. The need for hydrological science as a foundation for integrated water management can thus only increase, and this is what the International Hydrological Programme of UNESCO has set out to satisfy, with the overriding goal of helping nations to help themselves.

It is great pleasure to launch the International Hydrology Series as a joint undertaking of UNESCO's International Hydrological Programme and Cambridge University Press, and I hope that this first volume concerned with the humid tropics will prove useful in the context of addressing the hydrological and water resources problems with which our planet is confronted.

Federico Mayor
Director-General of UNESCO

May 1993

Preface

The humid tropics are a treasure-house of natural resources. Beside making up twenty-two per cent of the globe's land area or 29.4 million km^2, these warm and humid areas hold most of the world's uncut forest, most of the unharnessed hydroelectric power and most of the world's genetic riches among their estimated 30 milllion species of plants and animals. They may also contain vast, untapped supplies of minerals.

A significant proportion of the world's developing countries lie within the humid tropics and adjacent areas, and estimates suggest that by the year 2000 some 33% of the global population or more than two billion people will inhabit the humid regions.

The environmental and social problems found in the humid tropics are particularly complementary – and nearly all are related to water in some way. They are a result of population and land-use pressures and the failure to consider water resource management adequately within the context of general development plans for the region.

Until recently, little coordinated attention was paid to studying the hydrology of the humid tropics. With plentiful rainfall, apparently endless forests, and major rivers – like the Mekong, the Irrawaddy, the Ganges–Brahmaputra of East and South Asia, the Congo and Niger of Africa, and the Amazon and its South American tributaries – dominating the landscapes, it did not seem that water could be a problem.

UNESCO's International Hydrological Programme (IHP) has for several years realized the importance of the humid tropics and has organized projects to study various aspects of its hydrology and managing its water. In April 1987, the IHP-III Working Group connected with Project 4.2 (Hydrology of Humid Tropical Areas) met at the University of Hawaii's Water Resources Research Center. In recognition of the escalation in socio-economic problems of the humid tropics and the urgency of the need to address the associated water management issues, it was proposed that a more comprehensive programme needed to be developed under the auspices of the IHP. An international meeting was suggested to formulate the directions of this programme.

In October 1987, the planning process for such a meeting was started at UNESCO headquarters (Paris) by Dr. John S. Gladwell and Dr. Michael Bonell. This included identifying from a diverse range of literature what was perceived to be the most critical water management issues and the formulation of a proposed structure for the meeting. An Organizing Committee was later established and met at the Environment and Policy Institute, East-West Center, Honolulu, Hawaii in February 1988 to consider existing planning proposals and the terms of reference for the international meeting. The Committee also addressed the scope of a major programme for action.

The result was a decision to hold an international meeting during July 1989 under the title of the "International Colloquium on the Development of Hydrologic and Water Management Strategies in the Humid Tropics." The offer of the James Cook University of North Queensland, Townsville, Australia to serve as the site of the Colloquium was also accepted. Dr. Michael Bonell agreed to serve as Chairman of the Organizing and subsequently Co-ordinating Committees, and to work closely with the Project Officer, Dr. John S. Gladwell, in the preparations for the colloquium.

Plans were then made for some 70 experts from around the world to be invited to the colloquium, with a number to prepare papers for discussion. Four workshops were scheduled to follow the general sessions at which recommendations for follow-up actions would be developed. It was made clear that the important output of the colloquium would be the recommendations for action – not just what needed to be done, but equally important, how needed action should be implemented. (Many of the workshop recommendations are now in the process of being put into action.)

Following the colloquium, it was decided that the papers that had been produced, and the workshop results, should be considered for publication. On the other hand, simply publishing a proceedings of the colloquium would not suffice. As a result, an Editorial Committee was constituted in February 1990. The members of that group carefully reviewed the existing material and concluded that the basis for a book existed, but that considerable editorial work would be needed. All authors who contributed to the colloquium were requested to revise (in some cases very extensively) their manuscripts over the 1990–92 period. It was also decided that several additional topics should be included and invitations to additional authors were initiated.

The result of these actions and the contributions of many people are represented in this book. However, while this book is a very important contribution to the humid tropics programme of the IHP, it is necessary that the total programme be understood. In addressing the water resource problems in the humid tropics, the International Hydrological Programme has chosen a three-pronged approach. First, there is the scientific component stressing hydrological research as the basis for all other actions. Then there is a management component, by which water managers, technicians and policy planners are exposed to the demands – and the rewards – of taking a broader view of natural resources and development. Finally, there is the information and training component, which diffuses knowledge by building networks for the exchange of information. A major effort to produce documents suitable for the higher level decision-makers and for the public is also underway.

In order to accomplish the various activities foreseen as needed (and this book explores many, if not most, of them), networks of water and water-related experts and research organizations involved in warm humid region hydrology and water management studies are being established. Three regional centres (Latin America and the Caribbean, Africa, Asia) and a fourth on the special problems of small islands of the warm humid tropics are being established. The centres are intended to serve within each of their assigned areas as focal points for networks of research programmes, knowledge and technology transfer activities, including the development of literature, symposia, seminars and workshops. In addition, they will serve for the coming together of established and younger scientists for short periods of study on specific subjects.

The International Hydrological Programme, of which the Humid Tropics programme is a part, also works through a number of networks, one of which is that of the national committees for the IHP. These networks will be used in implementing regionally-coordinated cooperative studies and capacity-building that is deemed to be essential for the development and maintenance of programmes leading to the rational management of water resources.

This book, which the IHP is very pleased to present to the scientific and water management community of the humid tropics, will serve in defining the future of its own programme. It is considered that this is one of the few monographs to integrate scientific and management aspects in a comprehensive way such that the problems and water-related issues of the humid tropics might be better understood. As such, it is hoped that it will be accepted by the professional community as a contribution to the solution of the many difficulties that have been identified in the region.

We wish to express our appreciation and indebtedness to the many individuals – authors, participants in the colloquium, editors, and other contributors – for their major effort that went into the preparation of this volume.

A. Szollosi-Nagy
Director
Division of Water Sciences
UNESCO
Paris

All cartographic work in this volume was prepared
by John Ngai and Edward Rowe,
Cartographic Centre, Department of Geography,
James Cook University of North Queensland,
Townsville, Australia.

Acknowledgements

The preparation of this book could not have been undertaken without the support from our institutions and the many individuals within them. We would like to acknowledge them as follows:

James Cook University of North Queensland, Townsville, Australia

Executive: Professor R. Golding (Vice-Chancellor) who immediately offered the university facilities for the Colloquium, following agreement within the Organizing Committee that Australia should be the venue. His subsequent interest in the development of the IHP Humid Tropics programme is also appreciated.

Department of Geography: Professor Richard Jackson and teaching staff colleagues for their strong interest, patience and support in this exercise. Rosie Aziz for her highly efficient secretarial work. She carried the burden of secretarial responsibility, both in the preparation for the Colloquium and in the subsequent editorial work undertaken by M. Bonell. Cathy Everitt (secretarial) cheerfully spent many hours at the copying machine replicating documents. Clive Grant (technician) and Andrew Howard (data processing) are thanked for helping to maintain the Babinda tropical rainforest hydrology project on the occasions when Bonell was diverted into this editing work. Finally, the interest taken by many of the undergraduate students was appreciated. Special mention is made of the 1991 Tropical Climatology optional group: J. Albert, D. Boon, W. Browne, M. Cameron, G. Fischer, F. Mullins, A. Squire, and A. West who co-operated in the teaching schedule so that Bonell could spend some time at the UNESCO headquarters in Paris to initiate this editorial exercise and to help plan the implementaion of the Humid Tropics Programme.

Institute for Tropical Rainforest Studies: Special thanks to Nicola Goudberg (Assistant to the Director) who arrived just in time to provide vital assistance in the Colloquium and subsequently, along with her successor Moya Tomlinson, provided support and interest in the protracted period of editing.

East-West Center, Environment and Policy Institute, Honolulu, Hawaii, USA
Mrs. Joyce Kim for her unfailing and cheerful secretarial support for Maynard Hufschmidt.

UNESCO, Division of Water Sciences
Martine Bastide provided much needed secretarial support to J. Gladwell for the Colloquium and during the early stages of the editing of this book. Subsequently, Eloise Loh cheerfully and efficiently carried on this responsibility until and even beyond the retirement of Gladwell from UNESCO in August 1992. She continued to provide assistance to Michael Bonell during the concluding stages of this book editing.

We would also like to give mention to other individuals outside our institutions:
Text Editing: David Flaherty of David C. Flaherty and Associates, Pullman, Washington State, USA, is especially thanked for meticulously screening the work for overlooked grammatical errors and compiling all the manuscripts onto computer disk to conform with the requirements of Cambridge University Press.
Cambridge University Press: The following persons are thanked for clearing their desks to enable rapid publication of this book once all materials were received: Adam Black, Sheila Champney, Pauline Ireland and Brenda Youngman.

In the event we have omitted any persons or institutions who should be recognized, please accept our apologies and thanks.

M. Bonell
M. Hufschmidt
J. Gladwell
Editors

I. Introduction

1: The Objectives and Structure of this Book

M. Bonell
M. M. Hufschmidt
J. S. Gladwell

Editors

During the last decade, increasing international attention has been directed towards the environmental problems of the humid tropics, with the focus on the fate and management of the surviving tropical rain forest. In terms of water management (also termed water resource management in this volume), several rigorous reviews concerning the hydrological effects of tropical forest conversion to other land uses have emerged (Hamilton with King, 1983; Hamilton, 1988; Bruijnzeel, 1989, 1990) with particular emphasis on the changes in water yield and related land management issues connected with soil and nutrient loss.

The possible impacts of large-scale conversion of forests on climate have also been elevated in research priority within the Amazon Basin. Of particular interest are the changes in the land surface's energy balance and water vapour transfer. This new attention has led to the establishment of more ambitious field experiments in the Basin in conjunction with remote sensing. The data obtained from both sources are being used to calibrate General Circulation Models (GCMs) (Shuttleworth, *et al.*, 1991).

These developments, however, only cover a relatively narrow part of the spectrum of water management issues. Within the developed world, the remaining problems have received less publicity, based on the false perception that being the humid tropics, the supply of potable water is more than adequate to meet the needs of the region. As a later presentation by Manton & Bonell (this volume) will outline, the distribution of tropical rainfall can still be highly variable over space and time, arising from various mechanisms that disrupt the atmospheric pressure patterns. Consequently, droughts are not unknown.

The cores of the region's environmental problems, however, are essentially socio-economic, resulting from the escalation in population across the humid tropics. By the year 2000, it is anticipated that 33% of the world's population will be living in this region and that this percentage will continue to rise dramatically well into the Twenty-First Century. The resulting biophysical impacts already are causing a series of water and related land management problems. A consequence is also a major reduction in the apparent plentiful supply of potable water.

Gradwohl & Greenberg (1988) provided a succinct review of the socio-economic causes inducing large-scale clearance of tropical forests. The "equation" is a complex one with over 30 variables involved. One key factor is the increase in demands on the water and land resources of the region for food production to meet the requirements of the escalating population. A consequence is poor land management over large areas which is resulting in degradation in the quality of both groundwater and surface water. Erosion is a prime cause but the concentration of various water-borne chemicals through irrigation and even poor waste management from rural communities also aggravate the overall water quality. There is an inevitable decline in agricultural productivity when erosion removes a large proportion of the plant-available nutrients stored in the top 20 cm of soil. The future of food production therefore, must depend to a great extent on better knowledge of intensified production and not simply on the continued expansion of production into previously unused (and more marginal) areas. Improved management in soil and water conservation and erosion controls are vital to the success of this area.

With the possible exception of Africa, a consequence of the population increase has been the rapid expansion in urbanization, most notably in South East and South Asia and Latin America, and the totally inadequate facilities to service the water management requirements of such urban centres. This includes storm drainage and both surface and ground water pollution induced by rapid industrialization and inadequate solid waste management. Urban development, with all of its social, economic and physical implications, may well be one of the most serious problems the region has to face.

Tropical islands also require special attention because their small area and limited water resources result in an exacerbation of the various management problems associated with both urban and rural areas.

Finally, a consequence of the contracting supply of potable water is that the linkage between water supply and health stands out as *the* most critical issue facing both the rural and urban areas.

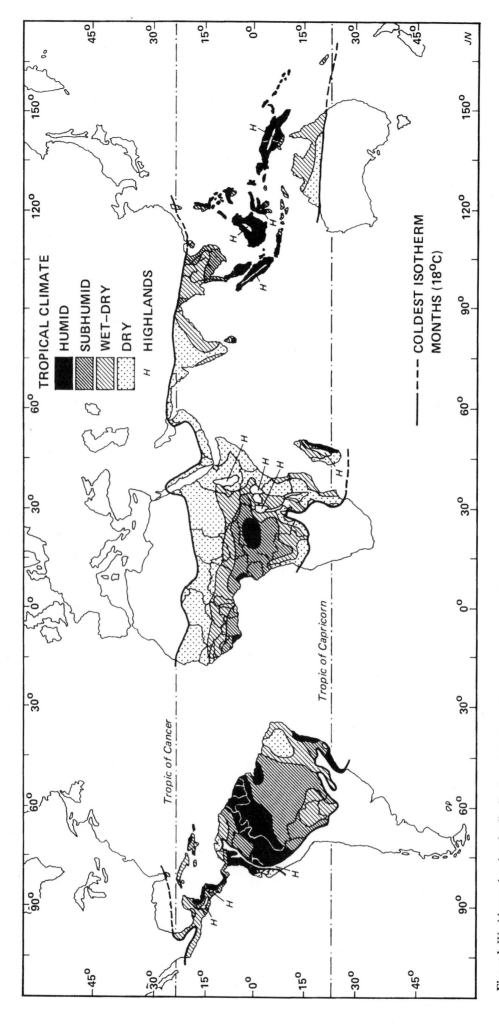

Figure 1: World map showing the distribution of the three climatic sub-types (humid, subhumid, wet-dry) of the humid tropics. Also shown is the dry tropical region.

THIS BOOK

This book is divided into seven Sections, including this Introduction. The initial sections of the book provide the physical basis of the humid tropics through a consideration of various systematic areas. A holistic approach is also adopted by the inclusion of the regional hydrology of the continents and islands. The interface between physical processes and human issues is later described as a basis for evaluating the complex water management issues of the region.

The Koppen and Thornthwaite methods of climatic classification are those most commonly used in the literature. We have followed, however, the definition of the humid tropics as developed by Chang and Lau (1983) as part of the UNESCO International Hydrological Programme's Projects IHP-II Projects A1.l0 (Hydrology of Humid Tropical Regions) and IHP-III Project 4.2 (b) (Hydrology of Humid Tropical Areas). The definition is summarized in Appendix A.

In that definition, the region is subdivided into three climatic sub-types on the basis of the number of their "wet" months, viz, Humid Tropical (9.5 to l2 months), Subhumid Tropical (7 to 9.5 months) and Wet-Dry Tropical (4.5 to 7 months). Figure 1 presents the world map of these climatic subregions.

It is significant that the Humid Tropical sub-type is only the most geographically extensive across the "maritime continent" (Ramage, 1968) of the Indonesian archipelago, and is particularly limited in distribution within Africa. Each of these three climatic sub-types not only serve to differentiate vegetation and agricultural land-use patterns, but also have implications for hydrological research. Each sub-type may be considered as a possible climate analogue where at least some of the hydrological characteristics are sufficiently similar to permit the successful transfer of methodology from one area to another.

SECTION II: HUMID TROPICS SETTING

A summary by Manton & Bonell initially presents the various meteorological systems which control the climate and rainfall variability across the humid tropics. Several points emerge from this chapter. Rain-producing synoptic systems are different between the monsoon and non-monsoon regions, and it important to define each system rigorously through an understanding of the atmospheric circulation. Of great interest to hydrologists is the description of various mechanisms producing rainfall at a mesoscale, and various oscillations (30–60 day, Southern Oscillations, Quasi-Biennial) which cause rainfall to be more variable in time and space than commonly credited, especially within hydrological literature. An underlying theme of this chapter is the consequences of high net radiant energy available which produce more severe storms, e.g., tropical cyclones, higher rainfall intensities and a higher frequency of devastating floods in comparison with the temperate areas.

A major problem for both meteorological and hydrological research, and water resource management is the relative absence of comprehensive datasets within the humid tropics. Manton & Bonell introduce this issue, but the following chapter by Manley & Askew provides a detailed assessment of the operational problems involved with data collection. Their contribution provides various environmental and socio-economic reasons why gaining good quality data is difficult in most humid tropical countries, in contrast to developed countries, where such databases are taken for granted. The location of this chapter near the front of the book serves to bring to the reader's immediate attention the basic hydrometric needs of the region – a subject of increasing neglect in the increasingly adverse socio-economic climate and yet highly significant if we are to improve our understanding of the climatology and hydrology. The chapter also serves as mandatory reading both for economists and politicians who are responsible for decision-making. It is also appropriate for the emerging generation of younger scientists, who were recently criticized by Philip (1991) for placing too much faith in computer modelling at the expense of supporting field data collection and experimentation to validate such models.

The concluding chapter of Section II by Klemeš provides a personal philosophy of the current inadequacies of existing hydrological methodology with its roots in hydraulic engineering and the challenges in producing an improved methodological framework for the humid tropics. He particularly focuses on three areas. First, the need for greater consideration of non-stationarity over time in the major geo- and bio-physical processes to more appropriately address the hydrological impacts of the population explosion within the region, through increasing urbanization and conversion of forests into agricultural land. Second, he examines the development of quantitative eco-hydrology, given the importance of vegetation in the hydrology of the humid tropics. A thorough understanding of the quantitative relationships across the soil-vegetation-atmosphere spectrum is particularly highlighted by Klemes, including adequate parameterization of the land-surface processes for modelling the dynamics of water and energy fluxes. His third point logically follows, with the call for a dynamically sound understanding of macrohydrology to address the needs of atmospheric General Circulation Models (GCMs), and the problems of scale within many existing methodologies associated with process hydrology.

Klemes particularly stresses the inadequacies of many water resource engineering techniques, both from a philosophical standpoint and as to their suitability for application to water management problems of the humid tropics. He calls for a new avenue in hydrological education with alternative roots in the earth sciences to produce specialists who can meet the water-related geo- and bio-physical problems of the humid tropics.

SECTION III: REGIONAL HYDROLOGY

The six chapters of the Regional Hydrology section attempt to present a summary of the hydrology and water resources of the world's major humid tropic regions. Chang's chapter dealing with the Asian region presents information concerning the variability of rainfall, including the reasons for the presence of droughts in what is normally considered to be a very moist area. Chang notes that in this region, studies of evaporation in mountain areas also are scarce, and that, as the temperature and psychrometric parameters decrease with elevation, the energy term rapidly loses its importance as the aerodynamic term assumes a dominant role in high mountains. Furthermore, although the monsoon atmospheric circulation governs the variability of the river flows in this region, there are still significant regional variations. From the standpoint of river regimes, four general regions may be recognized: (1) typhoon-prone, (2) equatorial, (3) tropical, with a distinct and prolonged dry season, and (4) areas affected by intense tropical cyclones (bordering the Bay of Bengal). In addition, heavy soil erosion poses a serious problem in this region, as evidenced by the exceedingly high sediment loads in many rivers. Besides the very high rainfall erosivity, rapid conversion of forests in recent years has been a major contributing cause to such sediment loads.

The chapter by Stewart looks specifically at the hydrology and water resources of the Wet-Dry Tropical region of northern Australia and makes contrasts with Papua New Guinea, which is mostly in the Humid Tropical Region, except for the Highlands and parts of the southern coastal strip. Stewart notes that while northern Australia's rainfall is markedly seasonal, occurring during the summer monsoon, and can be highly variable, Papua New Guinea, by comparison, receives very reliable rainfall from the four major rainfall-producing mechanisms (convection, convergence, orographic and cyclonic). An inverse association between evaporation in northern Australia and median annual rainfall is also noted. In Papua New Guinea, evaporation measurements have been made only in limited areas and for limited time periods so that extrapolation to other parts of the country are questionable. The mean annual runoff in Papua New Guinea has been estimated at 2,100 millimetres, with the runoff coefficients ranging from 0.25 to 0.75. The variability of the annual runoff in northern Australia is greater than other humid tropical regions, with the annual values ranging from over 1,000 millimetres on the east coast to under 25 millimetres inland.

Griesinger and Gladwell discuss the hydrology and water resources of the Latin America and Caribbean region. They note that this region is probably the world's most humid of the world, although the Humid Tropical sub-type is only the most extensive in the northern part of the Amazon Basin ,with smaller patches elsewhere. South America has some of the highest rainfall and runoff amounts of any continent. And because most of the rivers of the region are rainfed, the seasonal river regimes are mostly directly related to the rainfall distribution. In the Northern hemisphere, the incidence of tropical hurricanes can confuse the regionality of the rainfall amounts because of the random occurrence with which the islands and mainland are impacted by these cyclonic systems. Throughout the mainland of Latin America, the major physiographic factor affecting the water resources is the mountain chain that runs from Mexico to Chile. This causes the rivers generally to be longer and larger on the Atlantic side, and shorter and with much higher gradients on the Pacific side. Notable exceptions are to be found in Brazil, and in northern Bolivia and Paraguay where several rivers have their headwaters in the Central Plateau of Brazil.

Up to the present, the various human impacts on water resource uses have been mainly concentrated along the coastal areas of South America. Lesser impacts have been observed on the larger river systems, except primarily the response to regulation of the rivers for irrigation and hydroelectric power development. Nevertheless, the major land-use changes now occurring within the Amazon Basin suggest that, in the future, the various impacts on the surface waters will intensify and become of more serious concern.

The chapter by Molion looks specifically at the Amazon region. He notes that the hydrological processes in Amazonia vary widely from year to year due to natural changes in the atmospheric conditions. The chapter reviews the dynamic mechanisms that produce rainfall in the region. Subsequently, the characteristics and fluctuations of rainfall arising from the inter-annual variability of the large-scale atmospheric circulations associated with the El Niño - Southern Oscillation (ENSO) phenomenon are discussed along with the effects of blocking patterns within the atmospheric circulation, both of which cause variability in space as well as time. The effects of large-scale deforestation on local hydrology as well as the possible impacts on global climate are also considered.

Ayibotele identifies data availability and accuracy as major constraints in defining the hydrology and water resources of Africa. He also notes that except for their physical characteristics, information on the African lakes is also inadequate. Likewise, while a generalized knowledge about the sediment yield for the African region has been put forward, it is only tentative because of the data constraints. These conclusions reinforce the earlier operational problems described by Manley & Askew.

The high rates of population growth are also increasing the areal extent of land degradation within Africa – in response to the basic food requirements. As a result, Ayibotele discusses the impact of this land-use change on the water balance components and linkages with severe floods and droughts, increasing erosion and the deterioration of water quality.

Falkland and Brunel describe the hydrology and water resources of tropical islands. They note that island water resources (especially of the smaller ones) are often very limited. Many have no surface water resources and rely on limited groundwater resources in the form of thin fresh water lenses. They examine some of the major hydrological and water resource issues of small islands, with the main topics discussed being water resource assessment, water use, and water resource development and management. Approaches to resolving some of the major problems are also provided.

SECTION IV: PHYSICAL PROCESSES

Section IV provides the scientific basis of the book through an examination of the physical processes related to the hydrological cycle, erosion and sedimentation, water quality and freshwater biology.

Bonell with Balek systematically reviews process research connected with the water balance components: rainfall, evaporation, water movement in the unsaturated zone, groundwater and runoff generation. The broad field of the subject matter requires a more extensive coverage than found in other contributions of this Section.

The application of process hydrology within the context of various land-use conversions and their impact on the water balance are then considered. This includes some consideration of "physically-based" modelling techniques which have potential for application in land management issues. Throughout this review, several gaps in research are identified. Consequently, a large proportion of this chapter considers the technology-transfer of various methodologies and research findings from mostly temperate latitudes which need further testing in different humid tropical environments. Bonell with Balek conclude by describing the early results from macrohydrology projects connected mostly with the Amazon Basin. In accordance with an underlying theme of this chapter, this final section particularly highlights the meteorological and climatological linkages with process hydrology at different scales.

As indicated in Section V by Wurzel, groundwater is an important source of potable water (with its chemical and microbiological quality) for reducing the transmission of water-borne diseases. The modest pumping and reticulation costs, and the minimal treatment required are causing an increasing interest in the development of groundwater resources in the humid tropics. A separate chapter devoted to the physical basis of groundwater is therefore presented by Foster and Chilton. The principal features of five main types of groundwater system are described, viz, major alluvial formations, basement regoliths, intermontane valley-fill, active volcanic areas, and karstic limestones. Furthermore, the problems which arise in their exploitation and management are also highlighted. Foster and Chilton appropriately conclude on the vulnerability of shallow groundwater to pollution from a range of human activities.

Natural freshwater wetlands and lakes have a broad range of management applications such as flood control, groundwater discharge and water supply. Bullock initially presents material pertaining to natural lakes such as long-term changes in water levels, aquatic weeds and modelling of lake/river interactions. Greater emphasis, however, is placed on the status of tropical wetland research – drawing on the author's own experience in southern Africa. Different modelling strategies, empirical water balance studies and wetland influences upon downstream flows are considered. Linkages between hillslope hydrology processes (Bonell with Balek, this volume) and the hydrology of wetlands are also clearly demonstrated by Bullock. The author concludes by addressing the appropriate strategies for optimal wetland management and associated research needs.

Erosion and sedimentation by water-borne processes is a key element in land degradation following forest conversion. The chapter of Rose initially outlines the scientific issues involved in erosion and sedimentation, and subsequently presents arguments for a major change in direction concerning soil erosion modelling methodology beyond the commonly used Universal Soil Loss Equation. In this way, our present understanding of the physical processes are better represented in modelling for application at a hillslope scale. There is an acceptable bias in this account towards methodologies used in the development of the author's own model, which also has the advantage of being currently tested in the humid tropics of South East Asia. Nevertheless, how such theoretical developments within the Rose model relate to alternative methodologies are still discussed, with particular attention given to the Water Erosion Prediction Program (WEPP) in the USA. Evident from this review is that the field application of these new soil erosion models are at an earlier stage of development in contrast to hillslope hydrology models. Specific parameters, e.g. the J parameter of the Rose model, still require continued research effort under controlled experimental conditions to obtain a comprehensive understanding of their physical interpretability. In addition, at the present stage of technology, some of the parameters are difficult to measure during storms under field conditions. Rose concludes by considering erosion processes and modelling at a catchment scale, including the adaptation of physically-based hillslope hydrology models based on digital terrain models for predicting areas of erosion and deposition. The author also highlights developments using isotopic tracer methods for evaluating soil erosion and deposition, and recent work in the sediment behaviour and management within humid tropical catchments.

The standard of water quality has direct implications for health, agriculture and the environment. Roche isolates the main problem areas for scientific research and the related aspects of institutions and adverse management which have caused them. Water quality problems considered include the impact of industrialization through the introduction of microtoxics; organic (faecal) waste water from both urban and rural communities; and changes in salt and nutrient cycles plus the over-use of pesticides due to adverse land and water resource management. The impact on human health of water-borne pollution and aquatic vectors and larvae responsible for water-borne endemics is also one of the core messages of Roche's account. The problem of setting the appropriate water quality standards and even more important, the provision of adequate laboratory facilities to monitor such standards is strongly highlighted. A brief survey is also made of current experiences in water quality modelling within the humid tropics.

The final chapter of this Section by Golterman *et al.*, describes the water chemistry and freshwater biology of tropical surface waters. Initially, the chemical composition of river water and its variability, both in space and time, are considered. The utilization of freshwater nutrients for plant growth as well as photosynthetic activity are then considered. Various reactions on the water chemistry by plants are also reviewed in terms of dissolved oxygen, carbon dioxide and the recycling of plant decomposition products. The concluding part of this chapter concentrates on several aspects of the consumer's food web. They describe the web components, the invertebrate vectors of human disease and the various environmental factors controlling the food chain. Special consideration is also given to freshwater fisheries.

SECTION V: PHYSICAL PROCESSES – HUMAN USES – THE INTERFACE

This Section deals with the physical processes involved in the human uses of the water resources of the humid tropics, and their interactions. In the chapter by Lal, attention is primarily on the agricultural and forest hydrology aspects. He focuses primarily on the Humid Tropical and parts of the Subhumid Tropical regions associated with the narrow geographic band of latitudes 5° to 7° north and south of the equator, where tropical cyclonic influences are absent. The gaps in research knowledge related to the water balance, erosion and sedimentation and change in land-use are addressed. He also makes brief reference to the applications of remote sensing and geographic information systems, and to the problem of scale.

In the following chapter, Fleming describes the limited Australian experience in changes in land-use hydrology within the Wet-Dry Tropics, and also devotes some attention to temperate Australian work, e.g., the hydrology of eucalyptus trees, so as to evaluate the potential for technology transfer. He concludes that experimental data and design methods should be reviewed for their relevance to problems in tropical hydrology, using the simple balance equations for water, energy, salt and sediment. He stresses that an attempt should be made to develop predictive models that are parsimonious with respect to the detailed specification of environmental boundary conditions, but at the same time are realistic with respect to relevant processes. This, he suggests, should allow the extension of data from experimental sites concerned with manipulation of a limited suite of external variables.

The chapter by Gladwell reviews the situation currently to be found in the urban areas of the humid tropics. He argues that if the critical water management problems of these areas are to be successfully addressed, the water resources will have to be viewed as a continuum, rather than in a piecemeal manner – as is too often now done. The chapter reviews the various regions of the humid tropics, and presents the problems that have accumulated, their causes, and suggests directions in which the solutions will be found. Those directions include technical as well as nontechnical approaches. He concludes that unless both are considered in an integrated fashion, the

outlook for the urban areas of the humid tropics is not good. Decision-makers, planners, engineers and the public need to be aware that the results of narrowly focused and limited approaches to the solution of the various water-related problems have consequences far beyond the obvious limits of the urban area.

In the chapter by Prost, it is argued that innovative approaches will be needed to overcome the professional barriers between health specialists, engineers and decision-makers. The main health benefits accrue, it is argued, from the availability of unlimited quantities of water, whatever its quality; and unnecessarily stringent quality standards may be counterproductive since they may reduce the quantities available, delay the supply or increase its cost. It is pointed out that despite the momentum created by the International Drinking Water Supply and Sanitation Decade, coverage of the world population with adequate supply services is far from satisfactory. Progress can hardly keep pace with the increases in population. The majority of rural people in poor tropical countries are not served and there is little prospect of achieving universal coverage in the foreseeable future. There is concern expressed that a shift in resources from other vital sectors such as nutrition and health could offset the benefits of a safe water supply. It is argued that engineering techniques and non-medical interventions in water management can be shown to be the most cost-effective measures for controlling diseases of economic importance.

Wurzel's chapter concentrates largely on rural water supplies and sanitation as they relate to health in the humid tropics. It is argued that by interrupting the transmission cycle of water-borne and water-washed diseases, these interventions can contribute significantly to improved health in the developing world. The issues, trends, strategies and philosophy for the future are explored, with the accent on low-cost and appropriate technologies – a number of which are evaluated. Particular emphasis is given to groundwater as the primary water resource. As in the chapter by Prost, emphasis is placed on the fact that the installation of drinking water and sanitation facilities has hardly kept pace with the increasing population. The unfortunate paradox is that it is the most affluent and accessible populations which have been the first to receive attention. Wurzel observes that those remaining will be progressively more difficult to reach.

SECTION VI: MANAGEMENT ISSUES

The special nature of the humid tropics and the serious problems arising from a rapidly expanding population and economic activity have important consequences for water resource management (alternatively termed water management in some chapters). As defined by Hufschmidt in the opening chapter, water resource management is a process through which water resources are put to beneficial uses of humans and actions are taken to reduce detrimental effects of pollution and natural hazards on humans and natural systems. The management process consists of the key stages of assessment, planning, and implementation, each of which has its own problems and issues, and, taken together, provide the structure for developing the management strategy, as presented by Hufschmidt.

Water resource management issues in the rural resource-related context are analyzed in depth by Tejwani with a focus on watershed management. A historical perspective is used to examine the programmes and policies promoting water management. Selected management issues involving population pressure, integrated planning, research and financing are discussed. The urban context of water resource managment is treated in considerable detail in three additional chapters. Initially, Low examines urbanization trends and the accompanying water problems of the ASEAN region, primarily of Thailand, Malaysia, Indonesia and the Philippines. Water pollution, sedimentation, solid waste accumulation and flooding are important and growing problems of cities in this region. Lindh and Niemczynowicz subsequently emphasize the many obstacles to achieving effective and environmentally-sound management solutions. They call attention to innovative technical options involving recycling, wastewater reuse, and low-cost non-structural alternatives that are based upon sound ecological principles. The final urban contribution of Lee presents a revised urban water supply and sanitation strategy based on four lessons learned from the International Water Decade: technology alone is not enough, appropriate, low-cost technology is required, innovative cost recovery is a must, and community participation is a key element. The concluding chapter of this Section by LeMoigne and Kuffner emphasizes the particular issues of the humid tropics associated with excess water and flooding, the unique potential for hydroelectric power and the traditional emphasis on rice irrigation. Further details of the content of this Section are presented in the opening summary chapter on water resource management by Hufschmidt.

SECTION VII: APPENDICES

Section VII consists of four appendices. Appendix A contains the document prepared by Chang and Lau which provides the basis for the definition of the humid tropics used in this book.

The report was originally presented at an IAHS Symposium in Hamburg, Germany in 1983, and later included in a May 1986 UNESCO Report entitled *Hydrology of Humid Tropical Regions.*

In Appendix B there are four reports compiled by the rapporteurs from the working groups developed during the International Colloquium at James Cook University of North Queensland, Townsville, Australia in July 1989. Each report deals respectively with the following topics: hydrological processes (Bullock), erosion and sedimentation (Nogueira), water quality (Stewart) and water management (McCauley and Van Beek).

Finally, Appendix C contains lists of the members of the Organizing and Co-ordinating Committees for the Townsville International Colloquium and Appendix D acknowledges the institutions who sponsored the Colloquium.

REFERENCES

Bruijnzeel, L. A. (1989) Review Paper (De)forestation and dry season flow in the tropics: A closer look. *J, Trop. Forest Sci. 1*:229–243.

Bruijnzeel, L. A. (1990) *Hydrology of Moist Tropical Forests and Effects of Conversion: A State of Knowledge Review.* UNESCO IHP, Humid Tropics Programme, Paris.

Gradwohl, J. & Greenberg, R. (1988) *Saving the Tropical Forests.* Earthscan Publications.

Hamilton, L. S. (1988) The environmental influence of forests and forestry in enhancing food production and food security. Presented at *Expert Consultation on Forestry and Food Production/Security,* Bangalore, India Feb. 1988. FAO, Rome (1987).

Hamilton, L. S. with King, P. N. (1983) *Tropical Forested Watersheds: Hydrologic and Soils Response to Major Uses or Conversions.* Boulder, Westview Press.

Philip, J. R. (1991) Soils, natural science and models. *Soil Sci. 151:* 91–98.

Ramage, C. S. (1968) Role of a tropical "maritime continent" in the atmospheric circulation. *Mon. Weath. Rev.* 96:365–370.

Shuttleworth, W. J., Gash, J. H. C., Roberts, J. M., Nobre, C. A., Molion, L. C. B. & DeNazare Goes Ribeiro M. (1991) Post-deforestation Amazonian climate: Anglo-Brazilian research to improve prediction. In: Braga B. P. F.& Fernandez-Jauregui, C. A. (eds.) *Water Management of the Amazon Basin Symp.* Proc. Manaus Symp., Aug. 1990. UNESCO (ROSTLAC), Montevideo, Uruguay, 275–288.

II. Humid Tropics Setting

2: Climate and Rainfall Variability in the Humid Tropics

M.J. Manton

Bureau of Meteorology Research Centre, Melbourne, Australia

M. Bonell

Department of Geography and Institute for Tropical Rain Forest Studies, James Cook University of North Queensland, Townsville, Australia.

Present address: UNESCO, Division of Water Sciences, 1 rue Miollis, 75732 Paris Cedex 15, France

ABSTRACT

Although the absolute variability of weather features in the tropics is less than that at higher latitudes, tropical weather and climate are rich in variation on time scales from the diurnal to decadal. Moreover, the climate of the tropics is closely linked to the behaviour of the upper oceans, which provide inertia to the climate system and which act as a moisture source for convective heating of the atmosphere. Latent heating from convection is a common factor in all aspects of variability in the tropical climate. The links between the major causes of variability are reviewed in the present work.

INTRODUCTION

Weather at the mid-latitudes is clearly associated with baroclinic fronts and other synoptic features that are controlled by the earth's rotation. It is, therefore, natural to assume that the tropics, where these Coriolis effects are small, should have weather with a distinctive lack of structure. Charney & Shukla (1981) quantify this effect by showing that the temperature variance increases away from the equator. This apparent lack of structure is supported by the observation that persistence is a useful weather forecast in the tropics. Indeed, operational short-range weather forecasting is generally no more accurate than persistence (Holland *et al.*, 1987). But this result suggests that the basically quiescent weather is punctuated by intermittent and unpredictable events. In this paper, these events, their interconnections and progress on improving our understanding of them are reviewed.

One manifestation of the apparent lack of features in tropical weather is the observation that the temperature generally does not vary greatly, even following the passage of a weather system. This lack of change occurs because there is normally a local balance between diabatic heating from condensation and adiabatic cooling from lifting as the air flows out through the weather system. Such a localized vertical response is restricted by the earth's rotation at higher latitudes.

Significant weather features are described in the present work in order of increasing time-scale. They are diurnal convection, easterly disturbances, tropical cyclones, the 30–60 day oscillation, monsoons, quasi-biennial oscillations, El Niño-Southern Oscillation (ENSO) events and the greenhouse effect. The time scales of tropical weather events stretch from hours to decades. The longer-term variations tend to modulate the local response of shorter-term features. Thus the Australian monsoon is weaker in an El Niño year and tropical cyclones are more likely to form when the environment is preconditioned by convective disturbances, such as the 30-60 day oscillation.

At all scales, the atmospheric behaviour is closely coupled to the behaviour of the upper ocean. The tropical oceans are the source of the moisture that condenses and so heats the atmosphere to drive the global circulation. Similarly, the wind stress and latent heating (or cooling) at the ocean surface due to the atmospheric fluxes drive the upper ocean circulations. Any explanation of the tropical weather and climate must account for this coupling between the ocean and atmosphere. The importance of this link is recognized in the establishment of such projects as the Tropical Ocean Global Atmosphere (TOGA) project by the World Climate Research Programme (WCRP) in 1985.

ATMOSPHERIC CIRCULATION IN THE TROPICS

A common component of all the weather features in the tropics is convection and associated rainfall. Because the tropical atmosphere is almost invariably near a state of conditional instability, the overall dynamics is essentially a balance between latent heating and radiative cooling. The background large-scale flow for the tropics is the Hadley circulation, which

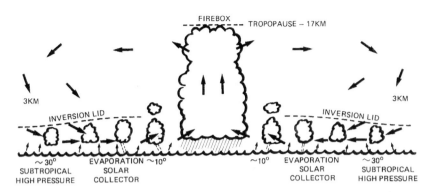

Figure 1: Schematic diagram of the Hadley circulation (from a seminar by J. Simpson, unpublished).

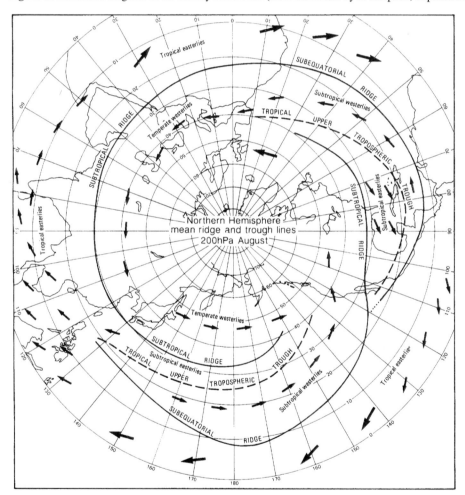

Figure 2a: The location of troughs, ridges, and major currents at 200 hPa during August (after Sadler, 1975a).

drives the low-level, easterly trade winds and in which heat is transported from the tropics to higher latitudes by the meridional return flow associated with the upper westerlies.

This meridional circulation involves a vigorous upward branch in the tropics, fed by low-level convergence of moist air flowing over the warm sea and driven by the latent heat released from the "hot towers" of cumulo-nimbus clouds (Fig. 1). The flow from these cumulus towers can extend to the

lower stratosphere, whence there is upper level divergence as the air streams towards the sub-tropics. Under the action of radiative cooling to space, this air sinks in the region of the sub-tropical high pressure systems, thus completing the overall circulation. A simple analytical model of the Hadley circulation is presented by Houghton (1986), which shows how the meridional pressure gradients lead to the low-level easterlies of the trade winds.

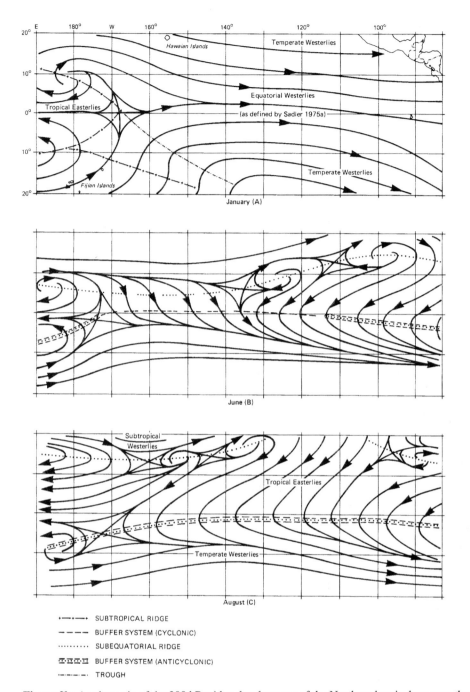

January (A)

June (B)

August (C)

```
·—·—·—·  SUBTROPICAL RIDGE
— — — —  BUFFER SYSTEM (CYCLONIC)
··········  SUBEQUATORIAL RIDGE
☐·☐·☐·☐  BUFFER SYSTEM (ANTICYCLONIC)
·—·—·—·  TROUGH
```

Figure 2b: A schematic of the 200 hPa ridge development of the Northern hemisphere over the eastern Pacific and the concurrent establishment of a buffer system in the equatorial region as part of the Hadley circulation (after Sadler, 1975a).

While the Hadley circulation provides a useful theoretical framework for understanding the nature of the large-scale flow, the actual circulation in the tropics involves substantial zonal and regional variations. Examination of Sadler's (1975a) upper tropospheric flow charts shows a rather complex pattern over the tropics and subtropics, especially in the Northern hemisphere summer. The principal channels summarized in Fig. 2a are direct entry into the *temperate westerlies* from the *subtropical ridge* or at lower latitudes via the *subtropical westerlies* which subsequently join the temperate westerlies. The *tropical easterlies* are also important in the Hadley circulation by eventually backing through a wind-turning zone close to the equator between two independently driven wind systems,

termed a *buffer system*. The resulting counter-clockwise rotation is anticyclonic when just south of the equator and cyclonic when north of the equator. Alternatively, the strong cross-equatorial outflow from the South Asia monsoon can displace the subtropical ridge poleward (to near 15°S at 80°E in midsummer). This system similarly allows the return flow to back anticyclonically through it. In both cases, there is a strong return flow to the temperate westerlies of the winter hemisphere. Figure 2b summarizes the development of a buffer system and changes in the 200 hPa ridge development in the Northern hemisphere over the eastern Pacific, as part of changes in the upper Hadley circulation. Mean wind streamline maps for the surface (Sadler *et al.*, 1987), and the 300 hPa

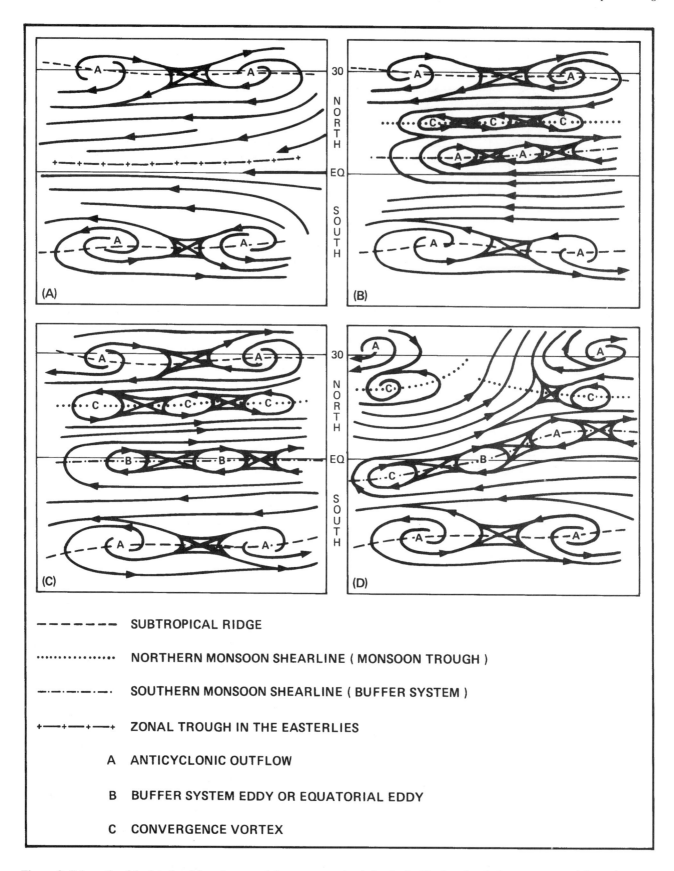

Figure 3: Schematic of the low-level flow for an evolving monsoon circulation in the Northern hemisphere summer and the various synoptic systems. Part D is typical of the low-level flow pattern between the longitudes of India and the Philippines (adapted from Sadler & Harris, 1970).

and 200 hPa levels (Sadler, 1975a) across the global tropics, provide further details of monthly changes in the Hadley circulation.

The axis of the Hadley circulation is commonly known as the Inter-Tropical Convergence Zone (ITCZ). This is a general term which embodies more than one synoptic-scale feature (Sadler & Harris, 1970; Sadler, 1975b). In addition, confusion arises when ITCZ is used to refer to the *monsoon trough* (Sadler & Harris, 1970) in monsoon regions. As Sadler (1975b) notes, satellite imagery from the GATE area (GARP Atlantic Tropical Experiment, where GARP represents the Global Atmospheric Research Programme) indicated that the *maximum cloud zone* (known as MCZ) was always equatorwards of the monsoon trough, and associated with the low-level equatorial westerlies, not the monsoon trough. Similar characteristics have been observed during the north Australian and Indian summer monsoons (Davidson *et al.,* 1983; McBride & Keenan, 1982; Gadgil, 1988). If ITCZ is understood to be the zone of maximum convergence (and therefore cloud persistence) then it is inappropriate to use this term for the monsoon trough (Sadler & Harris, 1970; Sadler, 1975b).

Figure 3 shows schematically the different surface wind streamline patterns that commonly occur in the tropics. Also shown are the various synoptic systems and the terms commonly used for them. The basic pattern is the tangential convergence of the northeast and southeast trade winds (Fig. 3a) into a feature termed the *zonal trough in the easterlies* (ZTE) by Sadler (1975b), with the word "zonal" purposely included to avoid confusion with meridional "easterly waves or troughs" embedded in the trade wind belt. Other writers refer to this feature as the *near-equatorial trough* (Riehl, 1979) or *near-equatorial convergence zone* (Ramage *et al.,* 1979) because of its geographic location being close to the equator. During anti-ENSO (El Niño-Southern Oscillation) events, this feature persists in the central Pacific Ocean, east of the International Date Line, during the Southern hemisphere summer (see Sadler *et al.,* 1987; Bonell *et al.,* 1991).

Further west of the GATE study area, Sadler (1975b) noted the same system in the western Atlantic Ocean associated with Mesoscale Cloud Complexes (Maddox, 1980) or Mesoscale Cloud Systems (Houze, 1982) on satellite imagery. (Spatial and temporal definitions of mesoscale meteorological phenomena are provided by Orlanski, 1975). The precise location of the MCZ in relation to the ZTE can, however, be difficult to determine because of the problems of specifying the position of the trough line within the very broad and weak-gradient pressure zone. During Sadler's (1975b) study, the August mean position of the MCZ in the west Atlantic appeared to be north of the ZTE, west of 40°W, possibly being influenced by the South American land mass. On the other hand, aircraft soundings between 150-158°W in the central Pacific by Ramage *et al.,* (1979) showed that the edge of the ZTE could be easily delineated by an abrupt discontinuity in moisture and temperature. The maximum in cloudiness and rainfall also was coincident with the ZTE in this area which is anchored between 4°N and 8°N throughout the year. A secondary cloudiness maximum can occur along 5°S when the ZTE is closest to the equator in March and April (Sadler *et al.,* 1976).

Ramage *et al.,* (1979) also established that the structure of the ZTE was more complex in the central Pacific than previously thought and was characterized by alternate strips of convergence and divergence rather than a single strip of convergence. Maximum convergence and rainfall were commonly measured at the edge of the ZTE coinciding with the discontinuity in low-level atmospheric moisture and temperature. Divergence and the presence of a dry, midtropospheric layer occurred somewhere in between. The penetration of the trade wind inversion to the latitude of the ZTE without any significant change in height suggested that the inversion may even extend across the ZTE, thus inhibiting deep convection in the form of "hot towers" of cumulonimbus commonly associated with the monsoon regions. For the most part, Ramage *et al.,* (1979) considered that activity within the ZTE was an "orographic" phenomenon (drawing an analogy with the passage of the trade winds over the Hawaiian Islands) where mostly shallow uplift resulted from the horizontal convergence of moisture. Weak, stationary and ephemeral cyclonic vortices occasionally may develop within the ZTE in this area.

Figures 3b to 3d show patterns of surface flow for the monsoon regions that can occur using the Northern hemisphere summer as an example. Two monsoon shearlines separating the respective trade wind easterlies from the opposing equatorial westerlies are evident. The monsoon shearline in the summer hemisphere, otherwise known as the *monsoon trough* (Sadler & Harris, 1970), is a thermally-induced low pressure belt. It is the more active of the two shearlines and is the seat of tropical vortices, some of which develop into tropical cyclones over oceans where atmospheric conditions are favourable. Satellite imagery shows that the cloudiness is not necessarily persistent within and south of the monsoon trough (referring to the Northern hemisphere summer), especially when well-organized tropical vortices occur. Large cloud clusters are often separated by relatively clear areas (Sadler, 1975b; Davidson *et al.,* 1983). The term *buffer system* (Sadler & Harris, 1970) was reserved for the less active monsoon shearline of the winter hemisphere because it is the same wind-turning mechanism previously described for the upper circulation. In the northwestern Pacific, however, "out of season" tropical cyclones can occur on the buffer system when it is positioned north of 5°N during the winter months (Sadler, 1967a). This allows typhoons to continue well into January. Hendon *et al.,* (1989) discussed their influence on the Australian summer monsoon. During the transition months (April, May, November and December) both shearlines can be equally active in the Indian Ocean and the western Pacific, as evidenced by the occasional pairing of tropical cyclones astride the equator (Lander, 1987). Consequently, the terms *northern monsoon shearline* and *southern monsoon shearline*

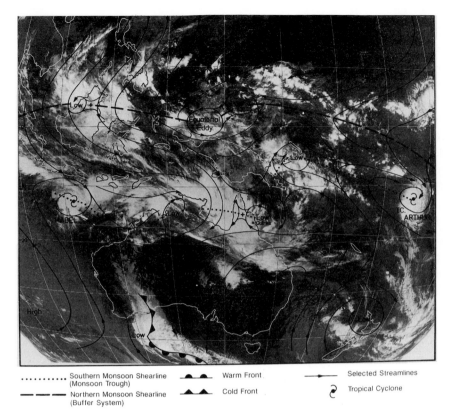

.... Southern Monsoon Shearline   ⌒⌒⌒ Warm Front   ——→ Selected Streamlines
              (Monsoon Trough)

— — — — Northern Monsoon Shearline   ▲▲▲ Cold Front   ⌇ Tropical Cyclone
              (Buffer System)

Figure 4: Japanese geostationary meteorological satellite (GMS) visible imagery taken at 1000 hr Australian Eastern Standard Time (0000 GMT), 13 January 1981 and associated simplified wind streamlines and synoptic systems (originally presented in Bonell *et al.*, 1991).

are preferred (Bonell *et al.*, (1991), following R. Falls in McAlpine *et al.*, (1983)).

Figures 3c and 3d also show that weak rotational eddies can occur along the less active monsoon shearline, despite its proximity to the equator where the Coriolis force is weak. As with the upper circulation, these eddies can be cyclonic or anticyclonic, depending on whether they are positioned in the respective Northern and Southern hemispheres. When they occur astride the equator, either the terms *equatorial* or *buffer eddy* (Sadler & Harris, 1970) are often used to denote them.

The deep convection of the MCZ, associated with the equatorial westerlies in monsoon regions, arises partly from the convergence of inter-hemispheric airstreams on passing through the respective monsoon shearlines (McBride & Keenan, 1982). Deep convection associated with Mesoscale Cloud Systems (MCSs) waxes and wanes, coinciding with the varying strength of surges of trade wind airstreams into the equatorial regions (Davidson, 1984; Love, 1985a, b).

Mean surface wind maps (Sadler, 1975a; Sadler *et al.*, 1987) show that on progressing eastwards, the ZTE splits into the two monsoon shearlines, either in the central or eastern Atlantic, depending on the season. The monsoon pattern is then maintained across Africa and through the Indian Ocean and the "maritime continent" of the Indonesian Archipelago (Ramage, 1968). The two monsoon shearlines then recombine into the ZTE in the mid-Pacific Ocean (Sadler *et al.*, 1987). During the transition months, the equatorial westerlies are confined to a narrow belt, or are even occasionally absent, with

daily streamline patterns often showing a temporary reversion to the Fig. 3a model. For example, this phenomenon is commonly observed in certain longitudes of the "maritime continent." In the Northern hemisphere summer, the eastern Pacific Ocean also develops a monsoon pattern (Sadler *et al.*, 1987) as far as about 130°W.

Many of the features described for the monsoon regions are shown in Fig. 4. The synoptic situation shows a near-stationary low pressure system embedded within a very active southern monsoon shearline across northern Australia. Rainfall totals up to 2,600 mm were recorded in coastal stations of northeastern Queensland over 13 consecutive wet days associated with this atmospheric circulation, including a few days with daily rainfalls far exceeding 250 mm (Bonell *et al.*, 1991). These included 818 mm at Cape Tribulation (north of Cairns) and maximum daily falls ranging between 426 mm and 565 mm elsewhere along the coast. Callaghan (1985) provided a detailed meteorological description of this period. The hydrological consequences of these rainfalls at one location (Babinda) will be later summarized in Bonell with Balek (this volume). Also highlighted are relatively cloud-free areas between the well-organized vortices along the southern monsoon shearline. In contrast there is less activity along the northern monsoon shearline. The MCSs within the equatorial westerlies are particularly evident over the Indonesian Archipelago. The cloud signatures also indicate the effect of the upper equatorial or *tropical easterlies* (Sadler, 1975a) steering the outflow westwards as eventual return flow to the

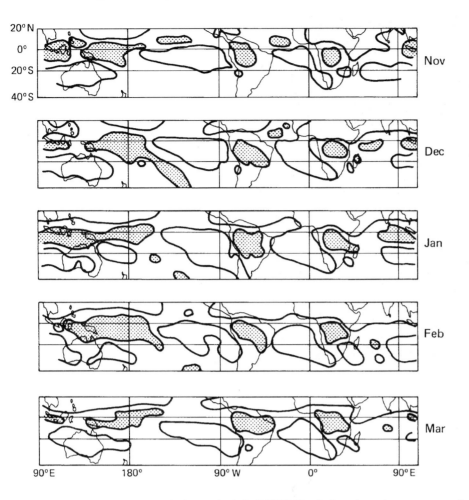

Figure 5: Outgoing long wave radiation (NOAA-9 AVHRR IR window channel measurements by NESDIS/ESL). Data are accumulated and averaged over 2.5° areas and interpolated to a 5° Mercator grid for display. Contours are at 220 and 260 W m⁻². Areas less than 220 W m⁻² are shaded. (a) Nov. 1986, (b) Dec. 1986, (c) Jan. 1987, (d) Feb. 1987, (e) Mar. 1987. Figures taken from Climate Diagnostics Bulletins, Nov. 1986-Mar. 1987 (Climate Analysis Center/NMC 1986, 1987) (after Gunn *et al.*, 1989).

higher latitudes of both hemispheres to complete the Hadley circulation.

A primary cause of regional variations in the tropical circulation is the distribution of land around the equator. Satellite imagery (see Fig. 5) clearly shows that there are three regions of maximum convection: the "maritime continent" of the Indonesian Archipelago, the Amazon Basin in South America and the Congo Basin in Africa. These variations in surface forcing give rise to regional zonal circulations around the equator, and the most important of these is the Walker circulation which involves rising motion over the maritime continent and sinking over the eastern Pacific (Fig. 6). The Walker circulation is associated with a coupled ocean-atmosphere phenomenon called the El Niño-Southern Oscillation (ENSO) (Troup, 1965; Bjerknes, 1969), which has a significant influence on the inter-annual variability of climate in some parts of the tropics and subtropics.

DIURNAL CONVECTION

Even from casual observation it is apparent that there is a strong diurnal cycle in convection in the humid tropics. The

GATE study of 1974 provided a comprehensive dataset on convection in the tropical east Atlantic. Convective activity, determined from satellite imagery, and rainfall from raingauges, were observed over both the ocean and West Africa, and so the nature of these diurnal variations could be clarified (McGarry & Reed, 1978). The large diurnal variation in surface heating over land leads to maxima in cloud cover and rainfall in the evening or night. On the other hand, the peak rainfall over the ocean occurs in the morning and early afternoon.

A contrast between land and ocean convection was also found in the observations from the Australian Monsoon Experiment (AMEX), where diurnal variations in convection were closely linked to corresponding variations in the larger-scale vertical velocity in the troposphere (McBride & Holland, 1989). The AMEX was held in 1986–87 during the transition and summer monsoon seasons in northern Australia. Six-hourly upper-air observations were taken at a dozen sites across northern Australia for 30 days. These measurements were supplemented by weather radar data from four stations.

Both satellite and radar data demonstrate very strong diurnal variations in convection over the land during the monsoon sea-

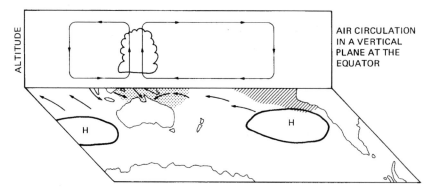

Typical patterns in a La Nina or strong Walker Circulation year.

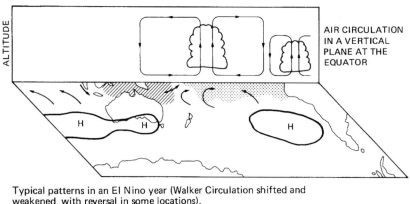

Typical patterns in an El Nino year (Walker Circulation shifted and weakened, with reversal in some locations).

Warmer sea temperatures

Cooler sea temperatures

(H)   Typical summer positions of high pressure systems

Surface winds

Figure 6: Tropical patterns associated with the Walker circulation during opposite phases of the Southern Oscillation (after Australian Bureau of Meteorology, Seasonal Outlook Service).

son, with the peak occurring at about 1530 local time. The afternoon maximum in convection is followed by an increase in stratiform cloud, which peaks at about 2030 local time when the vertical velocity in the middle troposphere is also at a maximum (Keenan *et al.*, 1989a). The diurnal cycle over the ocean is discernible but much less in magnitude than that over the land. Moreover, the phasing is different, with a gradual build up of convection throughout the night and a morning peak at about 0930 local time.

These observations for northern Australia are consistent with the results of a comprehensive study of convective activity in the tropics of the American region by Meisner & Arkin (1987). The latter writers showed that the diurnal variation over the ocean is generally less marked than that over land, and that the first harmonic is the dominant component of the diurnal variation.

It is clear that the diurnal behaviour of convection is strongly modulated by mesoscale convergence zones caused by the interaction of the prevailing large-scale flow and locally generated circulations. For example, squall lines that propagate across the Gulf of Carpentaria in northern Australia usually form in the convergence zone of the afternoon sea-breeze on the west coast of Cape York Peninsula. The resulting deep convective cells lead to a squall line that propagates across the Gulf in the prevailing easterly flow in the transition season before the onset of the summer monsoon (Drosdowsky *et al.*, 1989). Houze *et al.*, (1981) described – from the WMONEX study (Winter Monsoon Experiment) – the development of mesoscale squall lines off the north coast of Borneo. Such disturbances were the result of convergence between the nocturnal land breeze with the prevailing northeast monsoon over the South China Sea.

The most active region of convection on the globe is the maritime continent of the Indonesian Archipelago. A major contribution to the overall heating in this region is through the action of thunderstorms that develop diurnally over tropical islands. Such storms are caused by diurnally-oscillating boundary layer convergence patterns established by land and sea breeze circulations so that peak activity over the Indonesian islands occurs at 1800 hrs local time and over the adjacent waters at 0900 local time (Johnson & Houze, 1987). Other mechanisms, such as differential radiative heating both within and between MCSs over the open oceans, have been put forward to account for variations in diurnal cloudiness (e.g., Webster & Stephens, 1980). A recent study of island thunderstorms has been carried out near Darwin, Australia (Keenan *et al.*, 1989b). The storms are found to occur in the afternoon on about 65% of days during the transition season before the summer monsoon. They can have cloud tops of up to 20 km above the surface.

Convection in the tropics occurs through a variety of mesoscale processes, some of which have already been indicated in this section. These include continental thunderstorms (Murakami, 1983; Keenan *et al.*, 1989a); propagating squall lines (Drosdowsky *et al.*, 1989), with their origin from the convergence of nocturnal land breezes with synoptic scale winds, for example in Borneo (Houze *et al.*, 1981) and in the Amazon Basin (Molion, this volume); island thunderstorms (Murakami, 1983; Keenan *et al.*, 1989b) and oceanic cloud clusters (Kraus, 1963; Murakami, 1979, 1983). An important finding from GATE (Cheng & Houze, 1979) was that about 40% of the total precipitation from tropical systems is associated with stratiform cloud in the mature stage of development of the systems. This result has been confirmed in other regions (Johnson & Houze, 1987; Houze, 1989) and the precipitation is maintained by mesoscale updrafts of the order of 20 cm s^{-1} in the upper part of the cloud layer. The heating from stratiform clouds, therefore, plays a significant role in the overall balance between latent and radiative heating, as described by Johnson & Houze (1987). This has led Houze (1989) to refine his model (Houze, 1982) of MCSs, based on the structure of convective and stratiform precipitation. The hydrological implications are addressed by Bonell with Balek (this volume).

While the coupling between the ocean and atmosphere is clearly significant for long-term climate variations, there is also evidence of short-term feedbacks between tropical rainfall and sea-surface temperature (SST). In the western Caribbean, Malkus (1957) noted that cloud groups generally formed over and slightly downwind of warm ocean spots within the trade winds. Clouds were not found in the absence of warm spots, although warm spots were occasionally encountered without clouds. A lead and lag correlation between rainfall and SST has also been established by Greenhut (1978), which can be attributed to a warm ocean enhancing rainfall and cloud processes reducing the SST.

In general, however, there is an inadequate understanding of the dynamics of mesoscale atmosphere-ocean-topography interactions in terms of diurnal cloud and rainfall within the humid tropics (for application in hydrology and water management studies) compared with recent progress in higher lati-

tudes. See, for example, various contributions in Blumen (1990). Some of the existing knowledge and outstanding issues pertaining to the dynamics of tropical MCSs were reviewed by Johnson & Houze (1987), based particularly on research experience in GATE and both the Winter and Summer MONEX of South and Southeast Asia.

An interesting orographic mechanism is the development of precipitating MCSs within the southwest monsoon over the Arabian Sea, prior to their passage over the Western Ghats. Aircraft data showed that airflow up to a depth of 300 m decelerated as it approached the Western Ghats and began to rise at a rate of 2–3 cm s^{-1} about 200 km from land, thus releasing conditional instability (Grossman & Durran, 1984). The fact that deep convection is organized so far offshore implies that the dynamics of airflow over topographic barriers are significant over a more extensive area than usually considered. The internal structure, (viz, convective, stratiform) of these orographically-triggered cloud systems is still not understood (Johnson & Houze, 1987).

One of the few tropical areas that has received close attention is the Hawaiian Islands and surrounding waters. The remote sensing study of Larson (1978) provided an initial step by defining different cloud patterns embedded within the northeast trades for different synoptic situations, and their modification on passage over individual islands – in the context of the distribution of summer rainfall. Later, the application of wave theory concerning orographic airflow over mountains, commonly used in temperate studies (Tripoli & Cotton, 1989a, b; Weissbluth & Cotton, 1989) was applied to a Hawaiian study as part of an understanding of the spatial variations in rain patterns (Takehashi, 1988; Takehashi & Yoneyama, 1989).

The need for similar studies in other high rainfall areas was indicated in the context of diurnal rainfall-hillslope hydrology response of tropical rain forest of northeastern Queensland, especially in relation to the southeast trades (Bonell & Gilmour, 1980; Bonell & Jaycock, 1990, unpublished). The wet tropical coast of northeastern Queensland is typical of many coastal, humid tropic situations and the mesoscale processes are not understood. The environment presents a variety of research issues such as possible atmospheric-ocean interactions with thermal oceanic fronts (Burrage *et al.*, 1991) in the Great Barrier Reef lagoon and Coral Sea, the processes causing nocturnal enhancement of convective cloud (Kraus, 1963; Houze, 1982) and the fluid dynamics of the southeast trade flow during their passage over the coastal mountains in enhancing orographic cloud and rainfall (*cf.* Grossman & Durran (1984) for the Western Ghats). During the course of describing the January 1981 monsoon, Callaghan (1985) noted that the heaviest rain occurred just to the south of the monsoon trough (southern monsoon shearline). This is a common pattern along the northeastern coast of Queensland. Radar and satellite imagery showed that convective cells developed over and adjacent to the coast (Fig. 7), and were attributed by

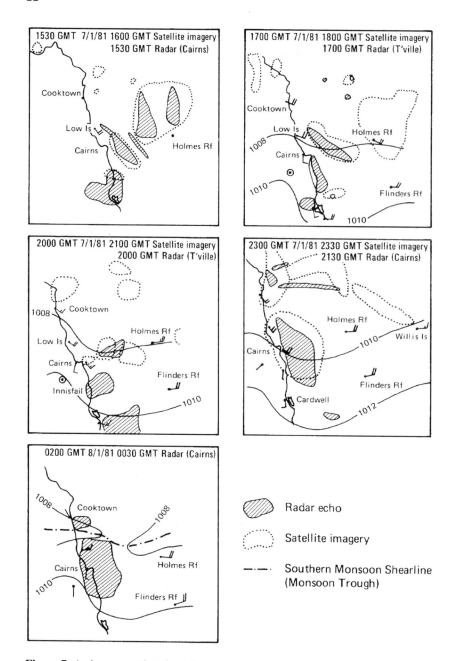

Figure 7: Active convective cloud from satellite imagery and radar rain echoes along the northeastern Queensland coast 1530 GMT 7 January 1981 to 0200 GMT 8 January 1981. Also shown are the isobaric patterns and surface wind observations (after Callaghan, 1985).

Callaghan (1985) to the process of differential shear stress described elsewhere by Ramage (1971). In Fig. 7, coastal stations, such as Cairns and Innisfail, show a greater cross-isobaric flow compared with the reef (Holmes, Flinders) or island (Fitzroy, immediately east of Cairns; Low Isles) stations. Such flow is thought to initiate a frictionally-induced convergence zone along the coast just south of the monsoon trough (Callaghan, 1985). Subsequent upward motion backing with height would transport moist air into the upper northwesterlies, causing the zone of heaviest rain. The development of a frictionally-induced convergence zone in combination with orographic effects may enhance mesoscale precipitation at other times along this coast, especially when the southeast trade wind flow is strong.

EASTERLY DISTURBANCES

In the summer period from May to October, a dominant feature of the weather in the tropical Atlantic is the steady sequence of meridional disturbances in the trade wind easterlies, known as *easterly waves*. Originally described by Riehl (1954) and later confirmed by the same author (Riehl, 1979), they are considered to be a persistent feature of the weather in the Caribbean and are even tracked across to the eastern Pacific. These waves have a length scale of 2,500 km, a period of 3 to 5 days, and a propagation speed of about 8 ms[-1] (Burpee, 1980). Such disturbances are thought to originate from more than one source area. For example, in the case of the Atlantic Ocean, Reed *et al.*, (1988, in their Fig. 2) pre-

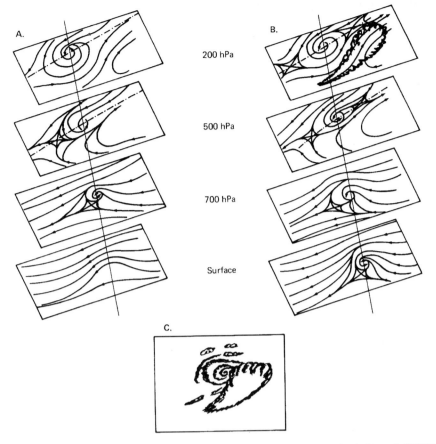

Figure 8: Three-dimensional structure of the Tropical Upper Tropospheric Trough (TUTT) systems and the plan view of a commonly observed cloud system. In Fig. 8a the vortex has penetrated through the 700 hPa level and shows in the surface level as an induced trough, whereas in Fig. 8b the penetration is to the surface as a vortex in the trade wind easterlies. Figure 8c is a schematic plan view of the cloud system of a moderate-to-strong cell in the central Pacific having a southeast slope with decreasing altitude. Whilst the view is under Fig. 8b, it is equally appropriate to Fig. 8a (after Sadler, 1967a).

sented evidence of two main source regions over western Africa. The first is from relatively cloud-free low-level circulation in the desert region of northern Mali, southern Algeria and eastern Mauritania, triggered by orographic uplift over the Hogger Mountains. The second, Reed *et al.*,(1988) identified with the monsoon region from Lake Chad across Nigeria to northern Ghana where the low-level circulation is associated with deep convection in proximity to the upper easterly zonal jet (Burpee, 1980) which forms part of the summer monsoon circulation. Earlier, the generation of easterly waves was initially described by the theory of Charney & Stern (1962) for instabilities of a baroclinic jet, such as that identified over in Africa in summer. Subsequently, Speth & Sperling (1989) considered that there were two types of waves, viz, barotropic waves south of the axis of the low-level jet and baroclinic waves north of the axis. Irrespective of their origin, these systems then continue westwards into the Atlantic, propagated essentially as Rossby waves on the mean easterly flow with a maximum amplitude at the 700 hPa level, and driven by latent heating from convection. Similar wave disturbances have been described for the eastern Pacific, although their vertical structure is though to be different from the Atlantic (Shapiro, 1986). Johnson & Houze (1987) also reported the presence of easterly

waves in the western Pacific and the South China Sea, but there is little evidence of any connection between these waves and those with origins in Africa. The generally cool waters of the eastern Pacific tend to suppress convection and so help the dissipation of any long-lived easterly wave from the Atlantic.

A series of papers by Sadler (1967a, b; 1976a, b; 1978) suggested alternative mechanisms for the described disturbances in the easterlies, and with the use of upper air data, showed in one case study for the western Pacific that the vertical structure of such disturbances does not fit the Riehl model (Sadler, 1967a). A principal source of these disturbances was ascribed to the Tropical Upper Troposphere Trough (TUTT) at 200 to 300 hPa levels (Sadler, 1976a, 1978) which forms during the summer months in both the North Pacific and North Atlantic (Sadler, 1975a). High-level convergence (which has no cloud signature as shown in Sadler, 1976a) extends towards the surface along a slope whose axis is displaced southeasterly with decreasing elevation. The surface circulation and cloud cover are dependent upon the areal extent, intensity and downward penetration of the upper cell. When penetration is moderate to strong (as shown in the schematic of Fig. 8), this often induces maximum amplitude at the 700 hPa level so that the vortex is underneath the divergent region of the sub-equatorial upper

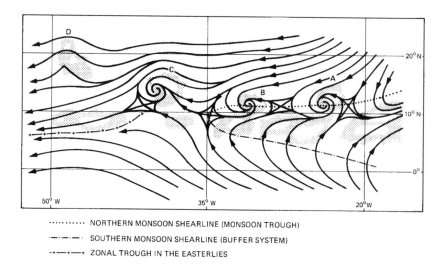

Figure 9: A schematic model of the low-level cyclones in the north Atlantic. The model depicts either a chain of cyclones (A to D) or the life history of one cyclone. The major satellite observed cloud systems are stippled and open areas within the stippling represents the deeper, more convective cloud groups. Note the wind streamlines have been modified near 50°W to conform with later information presented in Fig. 1 of Sadler (1975b) (modified from Sadler, 1967).

ridge (SER) because of the slope. The resulting cloud signature can then be displaced up to 500 km southeast from the upper tropospheric disturbance. In addition, TUTT provides an ideal tropospheric outflow channel to a large-scale westerly flow which complements the normally available channel to the upper tropical easterlies. Figure 8 shows that the 700 hPa vortex can either induce a trough at the surface or even a vortex within the trade easterlies. According to Sadler (1967a), subsequent westward movement of these TUTT-induced disturbances can be mistakenly identified as easterly waves. They are also an alternative mechanism for tropical cyclogenesis in both the western Pacific and north Atlantic (Sadler, 1976a, 1978). In the case of western Pacific, the proximity of the northern monsoon shearline and TUTT often makes it difficult to ascribe which system is responsible for tropical cyclogenesis (Sadler, 1967a).

Many of the disturbances regarded as "easterly waves" in the Caribbean are also considered to be former vortices originating from the summer monsoon over West Africa which are now in the process of decay. A conceptual presentation of the resulting cloud signature changes is given in Fig. 9 (Sadler, 1967a). A subsequent paper (Fig. 6 in Sadler, 1975b) showed tracks of vortices in August 1963 originating from the northern monsoon shearline which subsequently lost their structure on moving into the ZTE in the central Atlantic. If the premise is that all tropical vortices develop initially from a shear zone between two opposing airstreams (Sadler, 1967a), then the collapse of the West African systems can be attributed to their movement out of the monsoon shear zone into the "decay-producing" easterlies of the ZTE (Sadler, 1967a). A few of the eastern North Atlantic vortices intensify to tropical cyclones before leaving the monsoon shear zone. Others are able to maintain their vortex structure in the lower levels (but not necessarily at the surface) right across to the western Atlantic or Caribbean before decaying. According to Sadler (1967a,

1975b), many of these systems are mistaken for "easterly waves", following Riehl's model. On rare occasions, some of these disturbances reintensify. During late summer, a complex upper wind pattern occurs over the western North Atlantic (Sadler, 1975a), but favourable conditions for reintensification near the Lesser Antilles may occur when TUTT extends into the Hispaniola area of the Caribbean.

Surprisingly, there has been little reappraisal of these dichotomous ideas of easterly disturbances since the early balanced study of Merritt (1964), in connection with the Caribbean. Even from the limited database then available, Merritt (1964, p. 379) observed that the classic Riehl "waves in the easterlies" model was "... smaller in scale than indicated in previous studies and has a smaller frequency of occurrence." More appropriately, Merritt preferred the expression "easterly perturbations" and identified five distinctly different cloud distributions, with those most frequently observed being related to a closed cyclonic circulation in the mid-troposphere. There was little in that study which disputed Sadler's later ideas. Merritt (1964, p. 367) also made other pertinent observations, such as that "...distortion of Riehl's original concepts has been the prime source of this controversy and confusion" (see Merritt, 1964, pp 367–371 for historical background to easterly waves). In addition, the same writer noted that part of the problem would seem to be many meteorologists "forcing" all easterly disturbances into the classic easterly wave model in their "... analysis whenever any perturbation less intense than a tropical cyclone is detected in the tropical easterlies" (Merritt, 1964, p. 370).

Elsewhere, McBride (1983) did not detect any "easterly wave" phenomena during WMONEX in the south-west Pacific during a study of time-longitude cloud strips. Following Sadler (1975a), the absence of a persistent TUTT feature in that area may partly explain this observation.

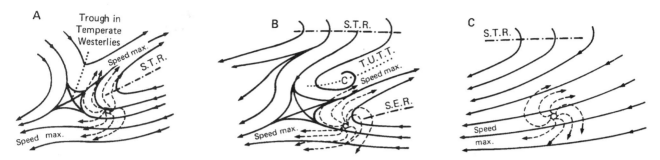

Figure 10: Schematic of storm outflow interaction (dash lines) with the larger scale upper tropospheric circulation (solid lines). STR is the subtropical ridge; SER, the subequatorial ridge. TUTT, the tropical upper tropospheric trough (adapted from Sadler, 1976a).

It is clear from this discussion that our understanding of disturbances in the easterlies is far from complete, and that further research is urgently required. As considerable areas of the humid tropics are dominated by the low-level easterlies, such research would have significant applications in related process hydrology and water management studies.

TROPICAL CYCLONES

Tropical cyclones are a major contribution to natural disasters in the humid tropics. The average rainfall within the inner 200 km of a tropical cyclone can be about 10 cm day^{-1} (Frank, 1977). Their impact is felt mostly in coastal regions where they cause damage not only directly from flooding, rains and strong winds but also from storm surges and high seas. Because of their devastating impact on human activities over a large fraction of the globe, there has been continuing work on the analysis and prediction of tropical cyclones. A series of international workshops on tropical cyclones has been initiated by the World Meteorological Organization (WMO), and a comprehensive review of our knowledge of tropical cyclones was one result of the first workshop in 1985 (Elsberry *et al.*, 1987).

The generation and development of tropical cyclones are closely linked to the transfer of moisture across the air-sea interface. Thus, Miller (1958) calculated an upper bound on the intensity of a tropical cyclone as a function of the underlying SST. A basis for the coupling between the flux of moisture from the sea surface and the release of latent heat in the core of a tropical cyclone is given by the theory of Conditional Instability of the Second Kind (CISK), introduced by Charney & Eliassen (1964) and Ooyama (1969). This theory suggests that the low-level convergence of moisture, induced by surface friction, causes an enhancement of cumulus convection in the core, which, in turn, increases the large-scale convergence. Many refinements and interpretations have been made to the theory of CISK over the last 20 years (Emanuel, 1987), but convective processes are still seen to be fundamental to the development of tropical cyclones.

In a series of papers, Gray (1968, 1988) has described the environmental conditions under which tropical cyclones develop over the globe. It is found that the seasonal frequency of tropical cyclones can be related to six factors: the Coriolis parameter, the low-level relative vorticity, the vertical wind shear, the SST, the vertical gradient of equivalent potential temperature and the mid-level relative humidity.

Most tropical cyclones form from two different systems, viz, the monsoon shearline (Sadler, 1967a; McBride & Keenan, 1982) or the Tropical Upper Tropospheric Trough (TUTT) (Sadler, 1976a; 1978). The latter feature is restricted to the North Pacific, North Atlantic and southeast Pacific (Tahiti area) during the summer months (Sadler, 1975a). In a series of papers, Sadler described the possible mechanisms of tropical cyclonic formation from TUTT (Sadler, 1967a, b, 1976a, b; 1978) and further suggested that about 10% of the North Atlantic and North Pacific tropical cyclones originate from this system. The initial disturbance from which a tropical cyclone grows is generally a cloud cluster of only about 200 km radius, sustained by a synoptic-scale feature such as the penetration of an upper trough from the mid-latitude westerlies (Hendon *et al.*, 1989; Sadler, 1976a; Davidson *et al.*, 1983) interacting with a monsoon shearline vortex or TUTT, both of which enhance outflow channels from the cyclone. Such multidirectional outflow channels are shown in Figs. 10a and 10b for the Northern hemisphere. The opposing winds effectively vent the upper anticyclonic outflow from the tropical storm. In the case of Fig. 10a, the interaction between an upper, mid-latitude westerly trough and the diffluent axis of the subtropical ridge over the monsoon shearline accentuates the existing westerly channel. Mention has already been made of the addition of a westerly channel associated with TUTT (Fig. 10b), complementing the persistent upper tropical easterlies. Figure 10c shows that a unidirectional airstream is less favourable for venting of the upper anticyclonic flow of a vortex, and only favours the southern sector. In the northern sector, anticyclonic outflow is impeded over much of the area. This upper tropospheric pattern is typical of the mean conditions in the western North Pacific, west of TUTT (Sadler, 1975a) (see Fig. 2a).

Holland (1984a, b, c) provided a comprehensive summary of the nature of tropical cyclones in the Australian region, including reference to various upper atmospheric circulation features influencing their genesis, subsequent development and decay of these systems. The Southern hemisphere equiva-

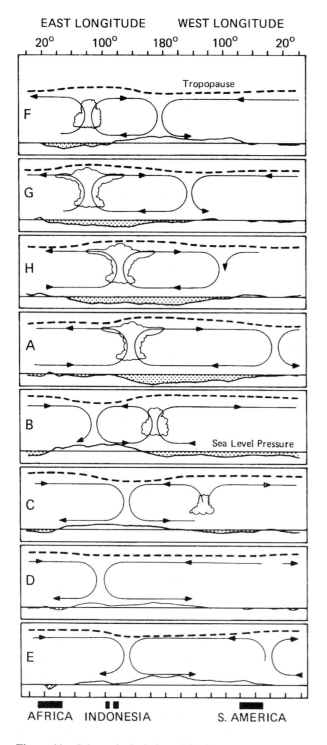

EAST LONGITUDE   WEST LONGITUDE

Figure 11a: Schematic depiction of the time and space (zonal plane) variations of the disturbance associated with the 30-60 day oscillation. Dates are indicated symbolically at the left of each chart and correspond to dates associated with Canton Island's (3°S, 172°W) surface atmospheric pressure indicated in Fig. 11b. The mean pressure disturbance is plotted at the bottom of each chart with negative anomalies being shaded. The circulation cells are based on the mean zonal wind disturbance previously presented in Fig. 13 of Madden & Julian (1972). Enhanced large-scale convection is schematically indicated by cumulus and cumulonimbus clouds. The relative tropopause height is indicated at the top of each chart (after Madden & Julian, 1972).

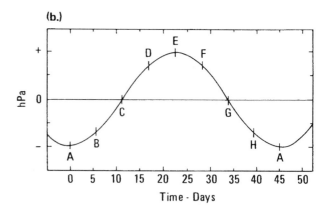

Figure 11b: Identical depiction of simple surface pressure wave in the Canton Island (3°S, 172°W) time series with the symbolic dates shown linked with Fig. 12a (after Madden & Julian, 1972).

lent of Fig. 10a seemed particularly important for tropical storm development in the Australian region. In contrast, a frequent cause of decay (especially over the Coral Sea) was the movement of such storms entirely beneath upper mid-latitude westerlies, thus presenting the same problems of effective venting as described in Fig. 10c.

30–60 DAY OSCILLATION

From a careful spectral analysis of zonal wind observations at Canton Island in the tropical Pacific, Madden & Julian (1971) found evidence of a quasi-periodic oscillation with a period of order of 30–60 days and a magnitude of about 5 m s^{-1}. Subsequently, Madden & Julian (1972) extended their analysis across the global tropics, and noted from cross spectral analysis that the observations at Canton Island (3°S, 172°W) were mostly linked with stations in the Indian and western Pacific Oceans. More detailed studies of composited numerical analyses (Knutson & Weickmann, 1987) show that the oscillation propagates continually around the equator with a phase speed of about 10 m s^{-1} and with a strong relationship between the divergence of the wind field and the outgoing long wave radiation (i.e. cumulus convection).

Because the feature is found primarily in the zonal wind component and it extends over the full troposphere, it appears to be a Kelvin wave; i.e. a trapped wave that propagates eastward along and decays away from the equator. Further theoretical studies (Lau & Peng, 1987) support the concept of the oscillation being a Kelvin wave, forced by deep convection arising from the atmospheric circulation over the monsoon regions in combination with high SST (Murakami & Nakazawa, 1985). The dependence upon the monsoon atmospheric circulation and SST is apparent from the observation that the region of convection is most active in the Indian and western Pacific Oceans, and that there is no convection where there are entirely different atmospheric circulation systems over cooler waters, such as the central east Pacific.

During an analysis of the 1979 Northern hemisphere summer, Murakami & Nakazawa (1985) suggested possible mechanisms which could act as a trigger in activating the Kelvin wave east of the African continent. Following Madden & Julian (1972), Fig. 11a schematically summarizes the growth and decay of convection arising from the eastward translation of a Kelvin wave, linked with the symbolic dates shown for the Canton Island time series (Fig. 11b). It is significant that this feature has minimal impact across the Atlantic Ocean and most of Latin America.

A 30-60 day oscillation is apparent in data on tropical cyclone activity, such that there are alternating periods of cyclone activity and inactivity on a global scale (Gray, 1988). It is suggested that these periods are influenced by the waxing and waning of favourable large-scale circulations for the generation of tropical cyclones.

The major regions of active convection associated with the 30-60 day oscillation have a scale of several thousand kilometres, and they are called super clusters. A super cluster is made up of a group of smaller cloud clusters with scales of a few hundred kilometres. Analysis of satellite data (Nakazawa, 1988) shows that the individual cloud clusters have a life cycle of several days and they propagate westward through the super cluster. Thus, the diurnally varying cloud clusters have a Rossby wave behaviour.

Super clusters and their related warm SST anomalies act as source regions for Rossby waves that propagate from the tropics to higher latitudes, producing the so-called teleconnections between the tropics and the extra-tropics. Horel & Wallace (1981) identify the Pacific-North America (PNA) pattern in the geopotential height field which can link the winter weather in North America with convection in the west Pacific. It is also found that the tropical west Atlantic is a source region for Rossby waves that can influence the weather over the North Atlantic. The relationship between these tropics-extra-tropics teleconnections and the 30–60 day oscillation is explored by Puri (1988) who uses a general circulation model to show that both phenomena are supported by low-frequency gravity waves.

While waves propagate to the extra-tropics from regions of enhanced convection in the tropics, it is also found that disturbances from the extra-tropics can lead to enhanced convection in the tropics. For example, Palmer (1988) uses principal component analysis to demonstrate that extra-tropical disturbances move equatorward from the Asian continent to enhance convection in the west Pacific. The importance of cold surges moving from Asia over the South China Sea has long been recognized (Ninomiya & Akiyama, 1976) and they are often related to enhanced convection in the Southern hemisphere with the onset of the Australian summer monsoon (Davidson *et al.*, 1983).

MONSOON

The summer monsoon in many regions of the tropics is the prime source of rainfall to support local agricultural and other activities. Because of its direct impact on communities in the humid tropics, the monsoon has been studied and documented for centuries. Halley (1686) pointed out the role of differential heating between land and sea on the evolution of the monsoon, and Hadley (1735) showed the need for a tropical circulation that conserves angular momentum (i.e. that accounts for the Coriolis forces). The simplest model of the summer monsoon is that it represents the seasonal migration away from equatorial latitudes of the appropriate monsoon shearline for each hemisphere (Sadler *et al.*, 1987). Current reviews of the physics and impacts of the monsoon are presented in Chang & Krishnamurti (1987) and Fein & Stephens (1987).

The monsoon has an annual cycle which tends to dominate all aspects of large-scale variability in the tropics. Meehl (1987) analyses outgoing long wave radiation (OLR), cloud, rainfall and sea-level pressure data to show that the region of maximum convection moves southeastward from the Indian sector as the annual cycle proceeds from northern summer to winter. Rainfall peaks occur near 90°E in the Indian monsoon season and near 130°E during the Australian monsoon. The return journey of the convection region from east to west is much less well-defined.

This annual cycle is also evident in the tropical SST, particularly in the eastern sectors of the Pacific and Atlantic Oceans. However, Meehl (1988) uses different general circulation simulations to show that the eastward migration of the region of maximum convection is not dependent upon the detailed SST distribution. The dominance of the annual cycle in the tropics suggests that the phase and strength of the monsoon circulation could be monitored by indices such as the mean meridional wind component near the equator (Ropelewski *et al.*, 1988).

The monsoon has significant inter-annual and intra-annual fluctuations. The latter are induced by the higher frequency weather features previously discussed. A major feature of the summer monsoon is the fluctuation between "active" and "break" periods of convection. While the 30–60 day oscillation plays a role in the determination of these fluctuations, other synoptic scale influences as well as local disturbances can also be important (Davidson, 1984; Love, 1985a; McBride, 1987; Gadgil, 1988; Hendon *et al.*, 1989). Murakami & Nakazawa (1985) provided a detailed analysis of the "active" and "break" phases of the Indian monsoon during the First GARP Global Experiment (FGGE), including the influence of the 30–60 day oscillation. They also give some consideration to determining the favourable channels for interhemispheric energy transfer.

QUASI-BIENNIAL OSCILLATION

Because of an apparent correlation between the 11-year solar cycle and the Arctic temperature in the stratosphere during the westerly phase of the stratospheric quasi-biennial oscillation (QBO), there has recently been increased interest in the nature of the QBO throughout the atmosphere (Labitzke & van Loon, 1988). On the other hand, the existence of a strong QBO signal in the tropical atmosphere has been recognized in its own right for some time. For example, Gray (1988) gives an explanation for the observation that the intensity of TCs varies with the phase of the stratospheric QBO. However, the present discussion will be limited to tropospheric and surface QBOs.

A QBO is found in the tropical surface pressure (Trenberth, 1975), monsoon rainfall (Bhalme & Jadhav, 1984), SSTs and tropospheric winds (Kawamura, 1988). Ropelewski *et al.,* (1987) point out that the QBO explains only 10 to 20% of the total variance in any of these variables. However, they demonstrate that the QBO signal is statistically significant, particularly in the eastern and central Pacific SSTs.

One reason for a QBO in tropical weather variables is that there is a tendency for a strong monsoon season to be followed by a weak one. Meehl (1987) analyses the annual migration of the convection maximum in the Indian monsoon over many years, and shows that the warm SST anomalies just ahead of the convection maximum tend to "over-shoot" and move into the eastern Pacific at the end of a strong monsoon season. Thus, conditions are established for a weak monsoon in the following season.

A simple model for the generation of a QBO in the tropics is given by Nicholls (1978, 1979). The model represents the seasonal behaviour of the SST and sea-level pressure (SLP) in the vicinity of Darwin. Observations show that a warm SST in this region enhances large-scale convergence, and so causes the SLP to decrease. On the other hand, the response of the SST to changes in SLP varies with the season. This result occurs because the SST is reduced by an increase in wind speed but the wind speed is controlled by the meridional (geostrophic) pressure gradient. Thus an increase in the local SLP in summer tends to reduce the pressure gradient and so to reduce the westerly wind speed and increase the SST. The opposite situation occurs in winter when there are prevailing easterlies.

Nicholls' model leads to a clear QBO signal in the SST and SLP, with a coherence that is consistent with observations. In particular, the extrema in the variables are found to be phase-locked to the annual cycle. Further work is required to extend the theory to account for the spatial variations in the QBO.

ENSO EVENTS AND OTHER ANOMALIES

It was stated above that the main feature of the global tropical flow is the Hadley circulation, with its axis marked by convection in the MCZ of the monsoon regions or close to the ZTE. However, the dominance of the convection maximum in the region of the Indonesian Archipelago leads to another large-scale but zonal circulation near the equator, called the Walker cell (Bjerknes, 1969). The Walker cell spans the Pacific Ocean, with warm moist air rising in the west and cool dry air sinking in the east. This circulation is named after Sir Gilbert Walker who first documented the global scale correlations

between seasonal variations in tropical weather features. The main zonal feature of the tropics is a strong anti-correlation between the SLP in the Indonesian Archipelago and that in the eastern Pacific, which makes up the Southern Oscillation (SO) and which is quantified by parameters such as the Southern Oscillation Index (SOI) (Troup, 1967). The SOI is defined by

$$\text{SOI} = 10 \times [\text{dP (Tahiti)} - \text{dP (Darwin)}] / \text{SD} \qquad (1)$$

where

dP (Tahiti) = Tahiti monthly pressure anomaly (monthly mean minus 1882–1985 mean, averaging 3-hourly observations);

dP (Darwin) = Darwin monthly pressure anomaly (monthly mean minus 1882–1985 mean, averaging 0900 hr, 1500 hr observations);

SD   = monthly standard deviation of the difference.

Chen (1982) and Lockwood (1984) provided an assessment of various other indices of the Southern Oscillation. Variations in the SO reflect fluctuations in the Walker circulation that are associated with marked seasonal anomalies in the climate of many parts of the global tropics and extra-tropics. The history and physics of the SO are well reviewed by Rasmusson (1985).

The Walker circulation is closely coupled with the SST distribution in the Pacific, with relatively cool water in the east and a warm pool in the west. The cool water of the east Pacific is maintained by upwelling driven by the prevailing easterly trade winds. Because of its impact on the fishing industry of South America, it has long been noted that the waters of the eastern Pacific occasionally become anomalously warm. Such El Niño events – which last for about a year – occur irregularly, with a recurrence interval of two to seven years. In describing the Walker circulation, Bjerknes (1969) shows the link between El Niño and variations in the Southern Oscillation. Thus, El Niño-Southern Oscillation (ENSO) events occur when the SO is weak and the waters of the east Pacific are warm. Such conditions produce negative phases of the SOI.

Regions of widespread convection are well marked by an underlying SST of at least 28°C in the tropics. The waters of the central Pacific are generally near this threshold temperature, with the pressure gradient being extremely small. Thus a relatively small SST anomaly can lead to a significant change in the pressure pattern and the distribution of convection in the Pacific. ENSO events occur when the convection maximum of the monsoon moves further east than normal and the SST in the Pacific is anomalously warm in the east and cool in the west. Similarly, anti-ENSO events occur when the convection maximum in the Indonesian Archipelago is anomalously strong. Because of the tendency of the convection maximum to generate Rossby waves that propagate into the extra-tropics (Das, 1986; Hendon *et al.*, 1989), ENSO and anti-ENSO events produce effects far from the tropics.

These events are also synchronized with the biennial oscil-

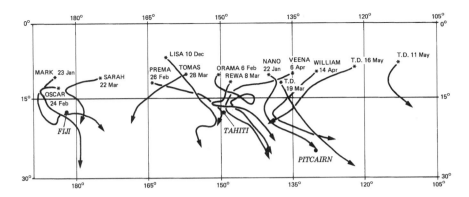

Figure 12: The tracks of the Southern hemisphere tropical cyclones, east of 180° from December 1982 to May 1983 during the strong negative phase of the SOI of 1982-83 (after Sadler, 1984).

lation in the monsoon, so that ENSO events tend to coincide with extrema in the biennial component of variables such as the SLP at Darwin (Ropelewski *et al.*, 1987). Thus, ENSOs and anti-ENSOs are a major source of the inter-annual variability in rainfall, particularly that associated with the monsoon. Nicholls (1988) shows that the inter-annual variability of rainfall in regions influenced by the SO is relatively greater than that in other regions of the globe, and these fluctuations could, over a long period, affect the evolution of local flora and fauna. It is also found that tropical cyclone activity is affected by the SO (Gray, 1988), and this relationship is expected to occur primarily through the readjustment of the atmospheric circulation patterns coupled with higher SSTs. The displacement eastwards of anomalously high SSTs in the Pacific Ocean results in a corresponding eastward penetration of the equatorial westerlies out from the traditional monsoon areas of the western North and South Pacific. Such changes replace the ZTE in the central Pacific (beyond the International Dateline) by the paired monsoon shearlines which, in turn, enhances tropical cyclonic activity. For example, Hastings (1990) presented the relative distributions of tropical cyclone occurrence in the South Pacific region for five ENSO and three anti-ENSO seasons. During an ENSO year, Hastings showed that there is a major shift eastwards of storm activity into the central Pacific. In contrast, there is a more concentrated clustering of storm tracks within the Coral Sea area during an anti-ENSO event.

Elsewhere, Sadler (1984) described the changes in the synoptic circulation patterns and the tracks of tropical cyclones east of 180°, resulting from the strong negative phase of the SOI in 1982–1983. During the transition season of October-November, three tropical cyclones developed near the International Dateline in the Northern hemisphere, with Hurricane "Iwa" forming far to the east at 167°W. An early commencement of the Southern hemisphere tropical cyclone season also occurred in similar longitudes over the same period. As noted by Sadler (1984, p. 55) however, the 1982–1983 tropical cyclone season of the Southern hemisphere "... has no counterpart in recorded history in terms of number of tropical cyclones, extended season, and eastward longitudes of

formation." A summary of the storm tracks is presented in Fig. 12. Five storms traversed the Society Islands within 500 km of Tahiti, for example, when only nine storms had previously affected this area between 1939 and 1982. Furthermore, the most easterly occurrence on record of a cyclonic storm took place at 8°S, 113°W.

The intensive research on the links between the SO and SST has led to studies of the climatic impact of the global distribution of SST. For example, there are two main principal components of winter rainfall for the Australian region and each of these is correlated with a different SST pattern. While one SST pattern is clearly related to ENSO events, the other links the SST gradient between the Indian Ocean and the Indonesian region to rainfall extending from the northwest to the southeast of Australia (Nicholls, 1989a). Similarly, up to 40% of the variance in seasonal rainfall in East Africa can be explained by correlations with the global SST pattern (Ogallo *et al.*, 1988).

Because of the large heat capacity of water, SST anomalies tend to persist and so ENSO and anti-ENSO events tend to run their natural course once they are initiated. This persistence suggests that seasonal forecasts of weather variables, such as rainfall, should be feasible if based on the evolution of the SO through an ENSO or anti-ENSO event. There has been some success in using both direct statistical correlation methods and physically-based numerical climate models to predict ENSO events (Barnett *et al.*, 1988).

Similar techniques have also been applied to the prediction of rainfall in areas such as the Sahel, where the SO is not dominant but where there is strong statistical relationship between rainfall and the global SST distribution (Owen & Ward, 1989; reviewed in Mott McDonald/BCEOM/SOGREAH/ORSTOM, 1992). Since the mid-1960s a marked warming has occurred in the Southern hemisphere oceans (plus the Indian Ocean) relative to the Northern hemisphere, although the causal factors are not fully understood. The SST anomaly difference between these two ocean groups against rainfall anomalies for the west African Sahel shows a strong negative correlation (Folland *et al.*, 1991). Thus on a global scale, the location of the boundary between the Wet-Dry Tropical sub-type and the Dry Tropical regions (Chapter 1, Fig. 1) probably shows the greatest oscillation in Africa (north of the equator) in response to the high temporal and spatial variability of rainfall (Hulme, 1992).

GREENHOUSE EFFECT

There is now some evidence of long-term variations in the global climate consistent with an enhanced greenhouse effect induced by human activities (Ramanathan, 1988). However, there is no clear picture of any regional trends due to the greenhouse effect, particularly in the tropics where the inter-annual variability is naturally high. That is, any climatic trend due to the greenhouse effect is a small signal superimposed on a background of large noise, due to the wealth of natural fluctuations in the tropics. For example, Nicholson (1989) reviews the variability of African rainfall over palaeontological time scales to suggest that the recent period of abnormally low rainfall in the Sahel is not necessarily associated with any long-term trend.

The major climate features that impact on human activity in the humid tropics are tropical cyclones (TCs) and the monsoon. Both of these suffer large inter-annual variability, especially from the SO. The possible consequences of the greenhouse effect on TCs and ENSO events are considered by Nicholls (1989b). It has been suggested that, because there is a relationship between TC frequency and SST, the frequency and regions of occurrence of TCs will increase with the greenhouse effect (Pearman, 1988). However, as pointed out above, TC frequency depends upon a number of parameters and it is not clear how the greenhouse effect will influence the other parameters such as vertical and horizontal shear. Moreover, there is not a universally positive correlation between SST and TC frequency. The frequency of TCs has not increased globally as the global SST has increased over the last few decades.

Similar uncertainties arise with predictions of possible changes in TC intensity with an enhanced greenhouse effect. The work of Emanuel (1987) relating TC intensity to SST implies that TC intensity may be increased by about 10% for each 1°C increase in global SST. On the other hand, Gray (1988) shows that the TC intensity in both the Atlantic and the north-west Pacific has decreased in the last 18 years, in comparison with previous decades.

The greenhouse effect on the monsoon will be related to the response of the SO. However, it is not clear that the greenhouse effects can be extracted from natural fluctuations in the SO. Over the last 15 years there have been a number of major ENSO events that can explain about half of the observed increases in the global SST (Nicholls, 1989b). It cannot be determined at present whether this occurrence of more ENSO than anti-ENSO events is a natural fluctuation in the SO or a consequence of the greenhouse effect.

Reliable estimates for the global climate in the coming decades will be obtained from sophisticated numerical models of the global climate system. Thus, there is a focused international effort to refine the current general circulation models for this purpose. Although the current models are reasonably consistent in their predictions of global trends – for example, a doubling of CO_2 concentration is expected to lead to an increase in global surface temperature of about 3°C and an increase in precipitation of about 10% (Schlesinger & Mitchell, 1987) – their predictions of regional effects are quite confused. Moreover, the capability of current general circulation models to simulate the present behaviour of the monsoon and ENSO events is minimal.

CONCLUSIONS AND THE FUTURE

The tropics constitute a vital part of the global climate system. They are the prime regional source for the heat engine that drives the general circulation, and the strong coupling between the ocean and the atmosphere gives rise to a richness in natural variability, extending from the diurnal to the decadal. Although there has been good progress in improving our understanding of the climatic processes peculiar to the tropics, much remains to be learned and increased knowledge is required for us to be able to satisfy the needs of the global community for assessments of possible changes in the global climate. Of immediate interest to hydrology and water management is an urgent need to intensify research into the dynamics of cloud development and precipitation structure at a mesoscale across the humid tropics. Such findings would assist related developments in rainfall hydrology concerned with the spatial and temporal distribution of precipitation at a mesoscale, e.g., stochastic and scale invariant rainfall models (see Bonell with Balek, this volume). Furthermore, the role of MCSs in inducing general upward motion within the broader synoptic-scale circulation, such as the MCSs occurring in the maritime continent, are also of great meteorological interest. In addition, a greater understanding of disturbances within the low-level easterlies at both the mesoscale and synoptic-scale deserves more attention.

In the 1960s and 1970s, our understanding of tropical weather features was greatly improved from the results of projects such as GATE in the Global Atmospheric Research Programme (GARP). Over the last decade and into the future, the initial thrust of GARP has been extended by the World Climate Research Programme (WCRP), which aims to determine the extent to which climate can be predicted and the extent of man's influence on climate. A number of projects have been established to focus research on particular aspects of the climate system. Two of these projects with special significance to the tropics are TOGA, which is relevant to the monsoon and ENSO events, and the Global Energy and Water Cycle Experiment (GEWEX), which is initially focused on the short-term aspects of the water cycle.

A problem with carrying out research on the tropical climate is the relative absence of comprehensive datasets. This problem will gradually be overcome as appropriate satellite-based instruments are developed to observe the full three-dimensional structure of weather systems in the tropics. The two out-standing measurements needed at present are the wind velocity and precipitation. Current plans for satellites in the 1990s should begin to satisfy these needs, with precipitation observations coming before wind measurements. In particular, a joint USA-Japan project called the Tropical Rainfall Measuring Mission (TRMM) is planned to launch an active weather radar in the mid-1990s. TRMM will produce comprehensive data on rainfall over the oceans for the first time, and so it should make a substantial contribution to GEWEX (Simpson *et al.*, 1988).

With the progress that has been made over the last three decades and the current plans for the next decade, we should be optimistic about prospects for our ability to predict the many fluctuations that occur in tropical weather and climate.

REFERENCES

Barnett, T., Graham, N., Cane, M., Zebiak, S., Dolan, S., O'Brien, J. & Legler, D. (1988) On the prediction of the El Niño of 1986–1987. *Science 241:* 192–196.

Bhalme, H. N. & Jadhav, S. K. (1984) The Southern Oscillation and its relation to the monsoon rainfall. *J. Climatol. 4:* 509–520.

Bjerknes, J. (1969) Atmospheric teleconnections from the equatorial Pacific. *Mon.Weath. Rev.*, 97: 163–172.

Blumen, W. (Ed.) (1990) *Atmospheric Processes over Complex Terrain*, Amer. Met. Soc., Boston, MA.

Bonell, M. & Gilmour, D. A. (1980) Variations in short-term rainfall intensity in relation to synoptic climatological aspects of the humid tropical north-east Queensland coast. *Singapore J. Trop. Geog. 1*(2): 16–30.

Bonell, M., Gilmour, D. A. & Cassells, D. S. (1991) The links between synoptic climatology and the runoff response of rain forest catchments on the wet tropical coast of north-eastern Queensland. In: P. A. Kershaw & G. Werran, (eds.) *Australian National Rain Forests Study Report 2:* 27–62, Australian Heritage Commission, Canberra, Australia.

Burpee, R. (1980) The structure of easterly waves during GATE. *Proc. Seminar on the impact of GATE on large-scale numerical modeling of the atmosphere and ocean.* National Research Council. Woods Hole, MA.

Burrage, D. M., Steinberg, C.R. & Kleypas, J.A. (1991) Mesoscale circulation features of the GBR region inferred from NOAA satellite imagery (in preparation). Available on request from the Australian Institute of Marine Science, Townsville, Australia.

Callaghan, J. (1985) *The North Queensland Monsoon of January 1981.* Australian Bureau of Meteorology, Meteorological Note 163.

Chang, C. P. & Krishnamurti, T. N. (eds.) (1987) *Monsoon Meteorology.* Oxford University Press, New York.

Charney, J. G. & Eliassen, A. (1964) On the growth of the hurricane depression. *J. Atmos. Sci., 21:* 68–75.

Charney, J. G. & Shukla, J. (1981) Predictability of monsoons. In: M. J.Lighthill & R. P. Pearce (eds.) ,*Monsoon Dynamics.* 99–109.

Charney, J. G. & Stern, M. E. (1962) On the stability of internal baroclinic jets in a rotating atmosphere. *J. Atmos. Sci., 19:* 159–172.

Chen, W. Y. (1982) Assessment of Southern Oscillation sea-level pressure indices. *Mon. Weath. Rev. 110:* 800–807.

Cheng, C.-P. & Houze, R. A. (1979) The distribution of convective and mesoscale precipitation in GATE radar echo patterns. *Mon. Weath. Rev., 107:* 1370–1381.

Climate Analysis Center/NMC (1986, 1987) Climate Diagnostics Bulletins, Nov. 1986, Dec. 1986, Jan. 1987, Feb. 1987, Mar. 1987. National Weather Service, National Oceanic and Atmospheric Administration, Washington, DC 20233.

Das, P. K. (1986) *Monsoons.* Fifth WMO Lecture, WMO No. 613, World Meteorological Organization.

Davidson, N. E. (1984) Short-term fluctuations in the Australian monsoon during winter MONEX. *Mon. Weath. Rev., 112:* 1697–1708.

Davidson, N. E., McBride, J. L. & McAvaney, B. J. (1983) The onset of the Australian monsoon during winter MONEX: Synoptic aspects. *Mon. Weath. Rev.: 111,* 496–516.

Drosdowsky, W., Holland, G. J. & Smith, R. K. (1989) Structure and evolution of North Australian cloud lines observed during AMEX Phase I. *Mon. Weath. Rev. 117:* 1181–1192.

Elsberry, R. L., Frank, W. M., Holland, G. J., Jarrell, J. D. & Southern, R. L. (1987) *A global view of tropical cyclones.* USA Office of Naval Research.

Emanuel, K. A. (1987) The dependence of hurricane intensity on climate. *Nature, 326:* 483–485.

Fein, J. S. & Stephens, P. L. (eds.) (1987) *Monsoons.* John Wiley & Sons, New York.

Folland, C., Owen, J., Ward, M. N. & Colman, A. (1991) Prediction of seasonal rainfall in the Sahel region using empirical and dynamical methods. *J. Forecasting 10:* 21–56.

Frank, W. M. (1977) The structure and energetics of the tropical cyclone. Part I: Storm structure. *Mon. Weath. Rev. 105:* 1119–1135.

Gadgil, S. (1988) Recent advances in monsoon research with particular reference to the Indian monsoon. *Aust. Met. Mag., 36:* 193–204.

Gray, W. M. (1968) Global view of the origin of tropical disturbances and storms. *Mon. Weath. Rev.: 96,* 55–73.

Gray, W. M. (1988) Environmental influences on tropical cyclones. *Aust. Met. Mag. 36:* 127–139.

Greenhut, G. K. (1978) Correlations between rainfall and sea surface temperature during GATE. *J. Phys. Ocean., 8:* 1135–1138.

Grossman, R. L. & Durran, D. R. (1984) Interaction of low-level flow with the Western Ghat Mountains and offshore convection in the summer monsoon. *Mon. Weath. Rev., 112,* 652–672.

Gunn, B. W., McBride, J. L., Holland, G. J., Keenan, T. D., Davidson, N. E. & Hendon, H. H. (1989) The Australian Summer Monsoon Circulation during AMEX Phase II. *Mon. Weath. Rev. 117:* 2554–2574.

Hadley, G. (1735) Concerning the cause of the general trade winds. *Phil. Trans. Roy. Soc. 39.*

Halley, E. (1686) An historical account of the trade winds and monsoons observable in the seas between and near the tropics, with an attempt to assign the physical cause of the winds. *Phil. Trans. Roy. Soc.,* 153–168.

Hastings, P. A. (1990) Southern Oscillation influences on tropical cyclone activity in the Australian/South-west Pacific region. *J. Climatol. 10:* 291–298.

Hendon, H., Davidson, N. E. & Gunn, B. (1989) Australian summer monsoon onset during AMEX 1987. *Mon. Weath. Rev. 117:* 370–390.

Holland, G. J. (1984) On the climatology and structure of tropical cyclones in the Australian/south-west pacific region: (a) Data and tropical storms. *Aust. Met. Mag. 32:* 1–15; (b) Hurricanes, 17-31, (c) Major hurricanes, 33–46.

Holland, G. J., Leslie, L. M., Fraedrich, K. & Love, G.B. (1988) The challenge of very short range forecasting in the tropics. *Proc. Symp. Mesoscale Analysis & Forecasting*, Vancouver, Canada, August 1987, 287–295.

Horel, J. D. & Wallace, J. M. (1981) Planetary-scale atmospheric phenomena associated with the Southern Oscillation. *Mon. Weath. Rev., 109*: 813–829.

Houghton, J. T. (1986) *The Physics of Atmospheres*, 2nd ed., Cambridge University Press, Cambridge.

Houze, R. A. (1982) Cloud clusters and large-scale vertical motions in the tropics. *J. Met. Soc. Japan, 60*: 396–410.

Houze, R. A. (1989) Observed structure of mesoscale convective systems and implications for large-scale heating. *Q.J.R. Met. Soc., 115*: 425–461.

Houze, R. A. Geotis, S. G. Marks, F. D. and West, A. K. (1981) Winter monsoon convections in the vicinity of North Borneo – Part 1: Structure and time variation of the clouds and precipitation. *Mon. Weath. Rev., 109(8)*: 1595–1614.

Hulme, M. (1992) Rainfall changes in Africa: 1931–60 to 1961–90. *Int. J. Climatol. 12*: 685–700.

Johnson, R. H. & Houze, R. A. (1987) Precipitating cloud systems of the Asian monsoon. In: C. P.Chang & T. N. Krishnamurti (eds.),*Monsoon Meteorology*, 298–353.

Kawamura, R. (1988) Quasi-biennial oscillation modes appearing in the tropical sea water temperature and 700 mb zonal wind. *J. Met. Soc. Japan, 66:* 955–965.

Keenan, T. D., McBride, J., Holland, G., Davidson, N. & Gunn, B. (1989a) Diurnal variations occurring during the Australian Monsoon Experiment (AMEX) Phase II. *Mon. Weath. Rev. 117*: 2535–2552.

Keenan, T. D., Morton, B. R., Manton, M. J. & Holland, G.J. (1989b) The Island Thunderstorm Experiment (ITEX) – a study of tropical thunderstorms in the maritime continent. *Bull. Am. Met. Soc., 70:* 152–159.

Knutson, T. R. & Weickmann, K. M. (1987) 30–60 day atmospheric oscillations: composite life cycles of convection and circulation anomalies. *Mon. Weath. Rev. 115:* 1407–1436.

Kraus, E. B. (1963) The diurnal precipitation change over the sea. *J. Atmosph. Sci. 20*: 546–551.

Labitzke, K. & van Loon, H. (1988) Association between the 11-year solar cycle, the QBO and the atmosphere. Part I: The troposphere and stratosphere in the Northern hemisphere winter, *J. Atmos. Terr. Phys. 50*: 197–206.

Lander, M. A. (1987) An investigation of the large-scale changes of the wind, sea level pressure and clouds associated with tropical cyclone twins symmetrical about the equator in the western Pacific. Preprints, 17th Conf. on Hurricanes and Tropical Meteorology, Miami, Fl, April 7–10, 1987, *Am. Met. Soc.,* 212–214.

Larson, R. N. (1978) Summer trade wind rainfall in the Hawaiian Islands. Thesis and partial contribution to MSc, Dept. Meteorology, University of Hawaii.

Lau, K. M. & Peng, L. (1987) Origin of low-frequency (intraseasonal) oscillations in the tropical atmosphere. Part I: Basic theory. *J. Atmos. Sci., 44*: 950–972.

Lockwood, J. G. (1984) The Southern Oscillation and El Niño. *Progress in Physical Geography 8(1)*: 102–110.

Love, G. (1985a) Cross-equatorial influence of winter hemisphere subtropical cold surges. *Mon. Weath. Rev. 113(9)*: 1487–1498.

Love, G. (1985b) Cross-equatorial interactions during tropical cyclogenesis. *Mon. Weath. Rev. 113(9)*: 1499–1509.

Madden, R. A. & Julian, P. R. (1971) Detection of a 40–50 day oscillation in the zonal wind in the tropical Pacific. *J. Atmos. Sci. 28*: 702–708.

Madden, R. A. & Julian, P. R. (1972) Description of global-scale circulation cells in the tropics with a 40-50 day period. *J. Atmos. Sci. 29*: 1109–1123.

Maddox, R.A. (1980) Mesoscale convective complexes. *Bull. Am. Met. Soc. 6l:* 1374–1387.

Malkus, J. S. (1957) Trade Cumulus Cloud Groups: Some observations suggesting a mechanism of their origin. *Tellus IX*, 33–44.

McAlpine, J, Keig, G. & Falls, R. (1983) *Climate of Papua New Guinea*, CSIRO-ANU Press, 11–38.

McBride, J. L. (1983) Satellite observation of the southern hemisphere monsoon during winter MONEX, *Tellus, 35A(3)*: 189–197.

McBride, J. L. (1987) The Australian summer monsoon. In: C. P. Chang & T. N. Krishnamurti (eds.) *Monsoon Meteorology*. Oxford University Press, New York, 203–231.

McBride, J. L. & Holland, G. J. (1989) The Australian Monsoon Experiment (AMEX): early results. *Aust. Met. Mag. 37,*: 23–35.

McBride, J. L. & Keenan, T. D. (1982) Climatology of tropical cyclone genesis in the Australian region. *J. Climatol. 2*: 13–33.

McGarry, M. M. & Reed, R. J. (1978) Diurnal variations in convective activity and precipitation during Phases II and III of GATE. *Mon. Weath. Rev. 106:* 101–113.

Meehl, G. A. (1987) The annual cycle and inter-annual variability in the tropical Pacific and Indian Ocean regions. *Mon. Weath. Rev. 115*: 27–50.

Meehl, G. A. (1988) Tropical-mid-latitude interactions in the Indian and Pacific sectors of the southern hemisphere. *Mon. Weath. Rev. 116:* 472–484.

Meisner, B. N. & Arkin, P. A. (1987) Spatial and annual variations in the diurnal cycle of large-scale tropical convective cloudiness and precipitation. *Mon. Weath. Rev. 115:* 2009–2032.

Merritt, E. S. (1964) Easterly waves and perturbations, a reappraisal. *J. Applied Met. 4:* 367–382.

Miller, B. I. (1958) On the maximum intensity of hurricanes. *J. Met., 15*: 184–195.

Mott MacDonald International/BCEOM/SOGREAH/ORSTOM (1992) Sub-Saharan Africa Hydrological Assessment West African Countries, Regional Report, December 1992. Available from The World Bank.

Murakami, T. (1983) Analysis of the deep convective activity over the western Pacific and southeast Asia. Part I: Diurnal variation. *J. Met. Soc. Japan 61*: 60–76.

Murakami, T. & Nakazawa, T. (1985) Tropical 45 day oscillations during the 1979 northern hemisphere summer. *J. Atmos. Sci. 42:* 1107–1122.

Nakazawa, N. (1988) Tropical super clusters within intraseasonal variations over the western Pacific. *J. Met. Soc. Japan 66:* 823–839.

Nicholls, N. (1978) Air-sea interaction and the quasi-biennial oscillation. *Mon. Weath. Rev. 106:* 1505–1508.

Nicholls, N. (1979) A simple air-sea interaction model. *Quart. J. Roy. Met. Soc. 105:* 93–105.

Nicholls, N. (1988) El Niño-Southern Oscillation and rainfall variability. *J. Climate 1:* 418–421.

Nicholls, N. (1989a) Sea surface temperatures and Australian winter rainfall. *J. Climate 2:* 965–973.

Nicholls, N. (1989b) Global warming, tropical cyclones and ENSO. *Proc. of Conf. on Responding to the threat of global warming: options for the Pacific and Asia, Hawaii.*

Nicholson, S. E. (1989) Long-term changes in African rainfall. *Weather 44*: 46–56.

Ninomiya, K. & Akiyama, T. (1976) Structure and heat energy budget of mixed layer capped by inversion during the period of polar out-break over Kuroshio region. *J. Met. Soc. Japan 54*: 160–174.

Ogallo, L. J., Janowiak, J. E. & Halpert, M. S. (1988) Teleconnections between seasonal rainfall over east Africa and global sea surface temperature anomalies. *J. Met. Soc. Japan 66*: 807–821.

Ooyama, K. V. (1969) Numerical simulation of the life cycle of tropical cyclones. *J. Atmos. Sci. 26*: 3–40.

Orlanski, I. (1975) A rational subdivision of scales for atmospheric processes. *Bull. Am. Met. Soc. 56*: 527–530.

Owen, J. A. & Ward, M. N. (1989) Forecasting Sahel rainfall. *Weather 44*: 57–64.

Palmer, T. N. (1988) Large-scale tropical, extra-tropical interactions on time-scales of a few days to a season. *Aust. Met. Mag. 36*: 107–125.

Pearman, G. I. (ed.) (1988) *Greenhouse: Planning for climate change*, CSIRO Publications, Australia.

Puri, K. (1988) On the importance of low-frequency gravity modes for the evolution of large-scale flow in a general circulation model. *J. Atmos. Sci. 45*: 2523–2544.

Ramage, C. S. (1968) Role of a tropical "maritime continent" in the atmospheric circulation. *Mon. Weath. Rev. 96*: 365–370.

Ramage, C. S. (1971) *Monsoon Meteorology*. Academic Press, New York.

Ramage, C. S., Khalsa, S. J. S. & Meisner, B. N. (1979) *The Central Pacific Near-Equatorial Convergence Zone*, UHMET 79–11, Department of Meteorology, University of Hawaii.

Ramanathan, V. (1989) The greenhouse theory of climate change: a test by an inadvertent global experiment. *Science 240*: 293–299.

Reed, R J. ; Hollingsworth, A. ; Heckley, W. A. & Delsol, F. (1988) An evaluation of the performance of the ECMWF operational system in analysing and forecasting easterly wave disturbances over Africa and the tropical Atlantic. *Mon. Weath. Rev. 116*: 824–865.

Rasmusson, E. M. (1985) El Niño and variations in climate. *Am. Sci. 73*: 168–177.

Riehl, H. (1954) *Tropical Meteorology*. McGraw-Hill, New York.

Riehl, H. (1979) *Climate and Weather in the Tropics*. Academic Press.

Ropelewski, C. F., Halpert, M. S. & Rasmusson, E. M. (1987) *Tropospheric biennial variability and its relationship to the Southern Oscillation*. 12th Annual Climate Diagnostics Workshop, Salt Lake City, Utah, USA.

Ropelewski, C. F., Halpert, M. S. & Rasmusson, E. M. (1988) *Climatological aspects of the tropical annual cycle*. 13th Annual Climate Diagnostics Workshop, Cambridge, Mass.

Sadler, J. C. (1967a) On the origin of tropical vortices. *Proc. of working panel on tropical dynamic meteorology*, Monterey, California, Navy Weather Research Facility, Report 12–1167–132, 39–75. Available from the Department of Meteorology, University of Hawaii.

Sadler, J. C. (1967b) *The tropical upper tropospheric trough as a secondary source of typhoons and a primary source of trade wind disturbances*, Hawaii Institute of Geophysics. Available from the Department of Meteorology, University of Hawaii.

Sadler, J. C. (1975a) *The upper tropospheric circulation over the global tropics*, Dept. of Meteorology, University of Hawaii.

Sadler, J. C. (1975b) The monsoon circulation and cloudiness over the GATE area. *Mon. Weath. Rev. 103(5)*: 369–387.

Sadler, J. C. (1976a) A role of the tropical upper tropospheric trough in early season typhoon development. *Mon. Weath. Rev. 104(10)*: 1266–1278.

Sadler, J. C. (1976b) *Tropical cyclone initiation by the tropical upper tropospheric trough*. Naval Environmental Prediction Research Facility, Monterey, California 93940. Available from the Department of Meteorology, University of Hawaii.

Sadler, J. C. (1978) Mid-season typhoon development and intensity changes and the tropical upper tropospheric trough. *Mon. Weath. Rev. 106*: 1139–1152.

Sadler, J. C. (1984) The anomalous tropical cyclones in the Pacific during the 1982–83 El Niño. Postprints volume, *15th Conf. Hurricanes and Tropical Meteorology*, January 9–13, 1984, Miami, Florida, American Meteorological Society, Boston, Mass., 51–55.

Sadler, J. C. & Harris, B. E. (1970) *The mean tropospheric circulation and cloudiness over southeast Asia and neighbouring areas*, Hawaii Institute of Geophysics, University of Hawaii. Available from the Department of Meteorology, University of Hawaii.

Sadler, J. C., Oda, L. & Kilonsky, B. J. (1976) *Pacific Ocean Cloudiness from Satellite Observations*. UHMET 76–01, Department of Meteorology, University of Hawaii.

Sadler, J. C., Lander, M. A., Hori, A. M. & Oda, L. K. (1987) *Tropical Marine Climatic Atlas, Volume I* (Indian Ocean and Atlantic Ocean) and *Volume II* (Pacific Ocean), Department of Meteorology, University of Hawaii.

Schlesinger, M. E. & Mitchell, J. F. B. (1987) Climate model simulations of the equilibrium climate response to increased carbon dioxide. *Rev. Geophys. 25*: 760–798.

Shapiro, L. J. (1986) The three-dimensional structure of synoptic-scale disturbances over the tropical Atlantic. *Mon. Weath. Rev. 114*: 1876–1891.

Simpson, J., Adler, R. F. & North, G .R. (1988) A proposed Tropical Rainfall Measuring Mission Satellite. *Bull. Amer. Met. Soc. 69*: 278–295.

Speth, P. & Sperling, T. (1989) Easterly waves over Africa during FGGE, In: 18th Conference on Hurricanes and Tropical Meteorology. May 16–19, 1989 San Diego, California, USA, Amer. Met Soc. 24–27.

Takehashi, T. (1988) Long-lasting trade wind showers in a three-dimensional model. *J. Atmos. Sci. 45(22)*: 3333–3353.

Takehashi, T. & Yoneyama, K. (1989) Rain duration in Hawaiian trade-wind rainbands aircraft observations. *J. Atmos. Sci. 46(7)*: 937–955.

Trenberth, K. E. (1975) A quasi-biennial standing wave in the southern hemisphere and interrelations with sea surface temperature. *Quart. J. Roy. Met. Soc. 101*: 55–74.

Tripoli, G. H. & Cotton, W. R. (1989a) A numerical study of an observed orogenic mesoscale convective system. Part I: Simulated genesis and comparison with observations. *Mon. Weath. Rev. 117*: 273–304

Tripoli, G. J. & Cotton, W. R. (1989b) A numerical study of an observed orogenic mesoscale convective system. Part II: Analysis of governing dynamics. *Mon. Weath. Rev. 117*: 305–328.

Troup, A. J. (1965) The southern oscillation. *Quart. J. Roy. Met. Soc. 91*: 490–506.

Troup, A. J. 1967) Opposition of anomalies of upper tropospheric winds at Singapore and Canton Island. *Aust. Met. Mag. 15*: 32–37.

Webster, P.J. & Stephens, G. L. (1980) Tropical upper-tropospheric extended clouds: Inferences from Winter MONEX. *J. Atmos. Sci. 37*: 1521–1541.

Weissbluth, M. J. & Cotton, W. R. (1989) Radiative and nonlinear influences on orographic gravity wave drag. *Mon. Weath. Rev. 117*: 2518–2534.

3: Operational Hydrology Problems in the Humid Tropics

R. E. Manley

Chief Technical Adviser, WMO Project INS/83/029, Bandung, Indonesia
Permanent address: 78 Huntingdon Road, Cambridge CB3 0HH, United Kingdom

A. J. Askew

Chief, Water Resources Division, WMO Secretariat, Geneva, Switzerland

ABSTRACT

Each of the major fields of activity within operational hydrology is considered. A number of the particular problems faced in the humid tropics are identified and discussed from a practical point of view. Possible solutions are reviewed and the overall conclusion is drawn that the principal barrier to progress is the low priority assigned by governments to water-resource assessment.

INTRODUCTION

Operational hydrology has a pivotal role in the development of water management for without the quantitative data it provides, no analysis of problems or estimation of the effectiveness of solutions are possible. In this paper, the authors look at many aspects of operational hydrology in the humid tropics. They indicate those areas where they feel there is scope for advancement and indicate what improvements they consider to be appropriate.

In preparing this paper, the authors have chosen to highlight typical problems as they exist rather than to attempt a comprehensive but general overview. Such an overview could never be truly comprehensive and would not stimulate consideration of the real practical difficulties faced in the field. Many other examples of problems and solutions, therefore, can be added to those cited below. The reader is invited to expand the list to suit specific circumstances.

The scope of the paper covers operational hydrology from the measurement of parameters in the field through to the preparation of data for analysis. Given that many of the countries in the humid tropics are classed as developing countries, the authors give more weight to such countries.

The authors cite a number of examples from their own experience to give anecdotal evidence in support of their arguments and proposals. They have not, however, indicated the countries involved as they to not wish a particular country to feel it was being singled out. Furthermore, the examples are illustrative of problems which are found in many countries.

Texts issued by commercial publishers are also valuable sources of information, particularly where efforts are made to update them from time to time, as for example, the handbook being prepared by Maidment (1990) and his collaborators to update the work of Chow (1964).

In the international arena, the World Meteorological Organization (WMO) is responsible for all aspects of operational hydrology and has published a large number of reports on the subject. Where relevant, these make reference to the specific problems encountered in tropical regions and in developing countries. However, to avoid burdening this paper with a long list of references, only the most pertinent are mentioned. Of these, three are of particular importance: WMO's basic reference on operational practice, the Guide to Hydrological Practices, WMO (1981; 1983); the manual which describes the mechanism developed by the Organization to facilitate the transfer of technology, namely HOMS (WMO, 1988); and a report on the very subject of this paper (WMO, 1987).

ANALYSIS OF THE PRESENT SITUATION

Parameters in operational hydrology

Operational hydrology deals with the measurement of parameters related to the quantitative and the qualitative aspects of the rainfall/runoff process. At one end of the hydrological cycle, there is an overlap with meteorology and the parameters of interest include rainfall and other climatic elements, particularly those related to evapotranspiration. At the other end of the cycle, the overlap is with oceanography, where problems of salinity in estuaries and coastal groundwater are quantified. Between these two extremes, and central to operational hydrology, is the measurement of flow in rivers and streams, or rather its estimation, as direct measurement of flow is not possible. It can only be inferred from other observable parameters such as water level. Other important parameters are groundwater levels and those related to water quality.

Many of the phenomena which operational hydrology tries to quantify occupy more than one dimension of the space-time continuum. For example, rainfall over an area can only be deduced from point measurements at a number of locations, or the variation in a water quality related variable over time can only be estimated from samples taken at a particular instant.

Regions concerned

For the purposes of the study reported in WMO (1987), it was necessary to establish a definition of the humid tropics. Chosen were those regions, located principally between the Tropics of Cancer and Capricorn which have a mean annual precipitation of at least 1,000 mm and a mean monthly temperature in all months of at least 20° C. By and large, these regions coincide with the areas defined by Köppen in Trewartha (1954) as "tropical constantly humid" and "tropical humid and dry." While this definition differs from that chosen for the present publication, it is interesting to note the close alignment of the areas so defined, when Trewartha (1954) is consulted. It should also be noted that in an era of rapid climatological change, such boundaries cannot be considered immutable.

Planning of activities in operational hydrology within the humid tropics should take account of the important sub-divisions based on:

(a) whether or not a subregion is affected by tropical cyclones;

(b) whether a subregion is continuously wet or is subject to clearly defined wet and dry seasons; and

(c) the types of meteorological systems which give rise to the major rainstorms.

The differences between these subregions have important implications for both the collection and analysis of hydrological data.

Difficulties particular to the humid tropics

Many of the difficulties of operational hydrology are the same the world over but some of those specific to, or of greater importance in, the humid topics are:

(a) *Access* Lack of metalled roads and difficult surface conditions during heavy rain often make access to the sites of measurement stations very difficult. In certain cases (heavily vegetated slopes of mountains), access may be almost impossible. In some countries the only convenient means of access is by helicopter.

(b) *Distance* Because of low population densities the distance between stations is often large. This poses problems for maintenance visits and flow gauging.

(c) *Absolute size of phenomenon* Most of the largest river basins in the world are in the humid tropics, thus the scale of phenomena to be measured is often much larger

than elsewhere. Rainfall intensities also can be much higher.

(d) *Climate* The continuous high levels of humidity can have a deleterious effect on measuring equipment.

(e) *Insects* Termites, in particular, can reduce a wooden structure to powder in a short time.

(f) *Personnel* In many developing countries it is not easy to find staff to work in government agencies. The salaries and conditions may not be sufficiently attractive and there are very few who have the appropriate training. Even when training is provided, there are often too few candidates with a good secondary education. Once trained, they frequently and rapidly find more attractive jobs in the private sector.

(g) *Funds* The shortage of local and, above all, hard currency is a major if not the chief impediment to progress.

(h) *Language* While many countries of the humid tropics use one of the major European languages for technical purposes, this is not always the case. Even where it is, the staff responsible for the day-to-day running of the stations require instruction and documentation in a national (or even local) language.

Instrumentation

The Guide to Hydrological Practices (WMO, 1981) offers guidance as to the density and location of networks of hydrological instruments. This guidance recognizes the differences between arid and humid regions and between tropical and extra-tropical regions.

Most instruments for hydrological data collection were developed in industrialized countries and, therefore, are not always well-adapted to the needs of the humid tropics. This section looks at different instrumentation for the collection of hydrological data and how appropriate they are. Meteorological instruments are considered to be outside the scope of this paper and are not discussed.

The principal factors of importance are:

(a) the great temporal variability of the phenomena, which calls for the use of automatic recorders;

(b) the difficulty of access, which calls for the teletransmission of data even where they are not required in real time; and

(c) the extreme climatic and hydrologic conditions under which the instruments need to operate, which suggest the installation of duplicate instruments at critical sites.

Precipitation The main emphasis of the paper is on water level and discharge measurements. Many of the more general remarks made below on that subject also apply to precipitation measurement. It is worth noting that the very high intensities and depths of rainfall experienced in the humid tropics can give problems for instruments designed for more temperate climates; tipping buckets may not be able to respond rapidly enough and storage gauges may overflow.

Figure 1: A telemetering raingauge which is part of a flood forecasting system installed in Jamaica under a WMO/UNDP project.

Where the funds are available, radar can provide very valuable information on the areal distribution of precipitation but requires highly-trained personnel and adequate ground-truthing. The latter requirement does, however, make radar inappropriate in much of the humid tropics. Most rain is derived from relatively small convective storm cells and, therefore, a very dense and costly network of telemetering raingauges is required for real time calibration (Fig 1). The use of satellite-based techniques also offers possible advantages. It is interesting to note that these techniques work best in estimating precipitation from isolated convective storms and that these storms are found more often in the tropics than in temperate regions.

River level The most basic method of river level measurement is the staff gauge. This method is acceptable where the catchment area is large and therefore the variations in level occur slowly, but in small catchments it is not suitable. This is particularly the case where most of the rain is convective, because

clouds build up during the morning and rainfall occurs in the afternoon or evening. A river in such an area is most likely to be in flood during the hours of darkness when observations are not possible. A lack of observers calls for the installation of automatic level recorders, but this, in turn, leads to a need for more highly trained staff to maintain the equipment.

High rates of both sedimentation and erosion are common in the humid tropics, which can make it difficult to locate a stable cross-section at which to measure the stage. All too often the staff gauges are either swept downstream or left high and dry on a new bank of silt.

The most common method of automatic level measurement is the float gauge. Two types of construction are used. The first consists of a "well" dug in the river bank from which a small pipe leads to the bed of the river. The siltation problem referred to above frequently leads to problems with blocked inlet pipes. The second consists of a large diameter pipe set in the bed of the river; the float rises and falls within the pipe and the instrument is set on top of the pipe. These types of instruments are generally considered to be the most accurate with a precision of better than 3 mm. However, they are not without their problems. First, in the case of a wide river or one with a rocky bank, the construction of a well with a pipe to the centre of the river is almost impossible. Secondly, the use of a vertical pipe is difficult unless it can be fixed to a nearby bridge to give stability and allow access. Finally, the instrument is usually a precision clockwork-driven device which requires regular visits to change the chart, wind the mechanism and check that there are no other anomalies.

Another type of gauge which overcomes some of the problems is the bubble gauge. This has many advantages over the float type in that there is no need to dig a well in the bank or to lay a pipe from the low point of the well to the bed of the river. Although a long trench may be required, it can be used successfully in wide rivers. A further advantage is that a single bottle of compressed gas can last for a year or more so regular visits are not essential. This type of gauge can still operate with a thin layer of sediment above the outlet, though at reduced sensitivity. The disadvantages are, firstly, that compressed gas is not always widely available and secondly, the mechanism which measures pressure is quite delicate. Leaving the gauge unattended for a long time can result in long periods of missing data if something goes wrong with the instrument.

An additional type of gauge introduced in the last few years uses a pressure transducer. The pressure transducer itself is temperature-sensitive so most of the latest instruments of this type measure the temperature and compensate for its effects. The only connection which is needed for this type of gauge is a cable from the station to the bed of the river. However, as the transducer needs air at atmospheric pressure for a reference, the cable also has to contain a capillary pipe. This type of device is the easiest of all to install but as it has been introduced fairly recently it is too early to be precise about its long-term future, although it does look promising. This device also can operate with a layer of sediment above the sensor, but some countries have reported problems of clogging in rivers with very fine silt.

Wildlife and the local population are other factors to consider with all instrumentation to be placed out of doors in the tropics. There is the problem of spiders, frogs, lizards and a host of other creatures interfering with instruments, even to the extent of "setting up home" inside the instrument shelter or the instrument itself. Vandalism, sadly, is a world-wide problem, but in its "pure" form of wanton damage it is not as frequent in developing countries. This is, to some extent, counteracted by the fact that a station costing thousands of dollars may be damaged by the theft of cables worth a few cents.

Discharge Water level data are of limited value if they cannot be converted to estimates of river discharge. However, such conversion requires the measurement of river velocity and the establishment of stage-discharge relationships. These make considerable demands in equipment and personnel which are not easily satisfied in many developing countries. It is quite common to find a well-maintained network of water level stations, plus associated data archives, with no means of converting these data to measures of discharge.

An ideal situation is reached in some countries where a team of technicians with a full set of equipment is stationed permanently at each major gauging site so that velocities can be measured over the full rise and fall of each flood. This is rare indeed. More common is the availability of a single team to cover a vast region with only limited transport facilities and a near impossibility of reaching any gauging site in times of flood because the roads are under water.

The above problems are compounded by the difficulty of finding gauging sites where the stage-discharge relationships will be stable for at least a few months and preferably years.

While it is not directly related to discharge measurement, it is worth making a reference to the need to clearly and unambiguously define the major river network. Standard topographic maps are frequently in error in this regard, particularly in heavily forested regions. Earth resource satellites can be of great assistance in these circumstances.

Sediment The standard method of measurement is to use a specially designed bottle to take a sample of the water containing the sediment and then carry out the analysis in the laboratory. This method is very accurate, but only gives an indication of the amount of sediment in suspension at the particular location and at the moment of taking the sample.

To obtain continuous measurements, a turbidity meter is used. This measures the transmission (or dispersion) of light by the solids suspended in the water. Although a turbidity meter can give a continuous reading, it is not as accurate as manual sampling. Therefore, the continuous method is most effective when undertaken in conjunction with routine sampling. It is important to note that the turbidity meter requires regular cleaning – at least once a week and more frequently if there is a large concentration of silt.

Water quality The measurement of many different parameters will be considered under this one heading. The traditional method, as with sediment measurement, is to take samples for later analysis in the laboratory. However, certain parameters which are related to water quality can be measured on-site. These include:

(a) Conductivity – related to the quantity of dissolved solids,
(b) Dissolved oxygen – related to the biological purity of the water,
(c) pH – related to the acidity or alkalinity of the water

The measurement of these parameters is not a trivial task and great care is required in the calibration and maintenance of the instruments. Frequently they are not immersed in the river itself but are placed in a small hut on the bank with the water from the river being fed to the instruments by a pump. They cannot be considered for regular widespread operational use as yet except on an experimental basis or where there is easy access.

Instrument manufacture One problem related to instrumentation that is frequently encountered in countries in the humid tropics is that a single country will have a diversity of types of instruments. This reflects the fact that different countries financing bilateral aid projects will have installed equipment from their own country. This creates long-term problems for the recipient country, as once the stocks of spare parts delivered with the different equipment have been exhausted, spares have to be bought for many types of instruments. Different types of charts also are needed and technicians need to be familiar with all the different makes of equipment.

One developing country has solved this problem by first of all standardizing on a particular type of instrument from a single manufacturer and, secondly, by making as much of the equipment as possible in their own country. For example, for flow gauging, the rods and carrying case are made in their country. Only the propeller mechanism is bought from abroad. In addition, all gauges are brought in for a complete strip-down and overhaul every four years. As a result of this policy, they have very few breakdowns and they only need to stock spares for one type of instrument. Their technicians also are fully familiar with the equipment and know which parts are most likely to need replacement. The only problem they have is that the technicians are now so adept at the repairs, it is difficult to keep them fully employed!

Staff support The above is a good example of how staff working in hydrometry have been given the financial support to carry out their task, but this is certainly not always the case. Often the staff are conscientious, but do not get this level of support. On one occasion during a site visit to a station with gauge boards, the observer was asked to show his record book. Everything appeared to be in order. However, on inspecting the site, it was noted that the board covering the level of the river on that day was missing and yet a figure had still been entered. The visitor was rather sceptical as to the accuracy of such a figure. The scepticism was sensed by the observer who proceeded to demonstrate how he was making the measurements. He stripped to his underpants and swam out to the gauge with a tape measure. There he dived under the water to hold one end of the tape to a lower level gauge board which was there, but was submerged, and then surfaced to read the level from the tape.

On another occasion on a visit to a climate station the observer, who also worked as watchman, had obviously taken great care to keep his instruments clean and to change the sunshine recorder chart regularly. However the Campbell-Stokes sunshine recorder had been badly aligned and in the afternoon the trace was off the paper and some boards in the roof of the Stevenson Screen had completely rotted away allowing sunlight to fall directly onto the instruments. These two anecdotes illustrate a common problem – good conscientious work by field personnel – but a lack of support by more senior staff which is also essential.

Conscientious work, but a lack of finance, can lead to similar results. In one country the hydrological service employed staff in the field to read and maintain a network of automatic recorders. There was a store full of new recorders to replace any that had been damaged. However, there was no hard currency with which to buy the ink and paper for the recorders and there was no fuel for vehicles to take consumables and spare parts into the field.

Telemetry

Sometimes data are required in "real time," for example, to warn of floods or to operate reservoirs or pumping systems. In such cases a system of telemetry has to be used. The four main systems are:

(a) *Telephone* The use of the standard telephone network can be a good way of transmitting data in a country with a well-developed telephone system but is rarely suitable for the humid tropics.

(b) *Radio* Systems using radio can either be automatic or use voice communication. Voice communication has many disadvantages and in general an automatic system is to be preferred. In the case of radio, care needs to be taken in the choice of frequencies to avoid interference and at the same time to have the necessary range. Radio-based systems are suitable for relatively small catchments (a few thousand of square kilometres) but for larger catchments the need to have many repeater stations makes their use prohibitively expensive.

(c) *Satellite* There are two types of satellite systems that can be used. The first is using a geo-stationary satellite, *i.e.*, one that is always above the same point on Earth. The other uses polar-orbiting satellites which only pass over the measuring stations a limited number of times per day. The use of a geo-stationary satellite allows data to be collected very frequently but is more costly than the use of a polar-orbiting satellite. However, the latter are very suitable for large catchments (hundreds of thousands of square kilometres) where the time between messages (up to six hours) is small compared to the time of concentration of floods in the catchment.

(d) *Meteor-burst* This system makes use of the fact that each time a micrometeorite burns up on entering the earth's atmosphere, a cloud of ionized gas is created which can reflect radio waves. If the cloud is positioned correctly, communication between the master station and one of the measuring stations is possible. There are several advantages of this system. For example, it can operate over a wide area, it is independent of satellite operation, and the number of micrometeorites is enough to maintain a good average transmission rate. However, it also has a number of disadvantages, including the impossibility of getting data immediately on demand (or even of actually knowing when the next message will be received), a high power requirement, and possible conflicts between transmissions if two stations are geographically close.

As well as rapid availability of data, telemetry can bring other advantages compared with non-real time methods. Included is an immediate warning if a station is non-operational, particularly if technical data on the functioning of a station are transmitted along with the hydrometric data. The use of telemetry can be a worthwhile addition to the capability of a country but if the equipment cannot be repaired in the country where it is being used, then a large stock of spare parts has to be maintained and foreign currency is needed to pay for repair charges. In some cases it has been found that when transport and handling charges have been added to the cost of repairs, the repair is not economic and replacement is the cheapest option. In one recent case, the cost of repair was four times the cost of replacement.

Data entry and processing

These days, virtually all data analysis is carried out on computers and there is, therefore, a need to transfer the data from the original medium to a computer. The most obvious method is manually, where the data from an observer's notebook or a chart are typed number-by-number into a computer. In many instances this is the only feasible method, but it does allow for additional errors to be introduced. A better method is to use a punched or magnetic tape device linked directly to the measuring system. This can then be "read" by a special device and transferred to the computer. More recently some data are stored in a solid state memory which can then be transferred to a computer. Some of these are of the type where data can only be erased by exposing the memory cartridge to intense ultraviolet light, but others can be erased by software. The better of the systems allow direct access to the data by an IBM PC-compatible computer. It is likely that this type of memory will become more common in the future, particularly if the all solid state pressure transducer type of water level measurement is used.

Associated with the entry of data is its processing for later use. This includes quality control, filing the data in an organized manner, facilities for printing it out in tables or as graphs and facilities for making it available in computer-compatible form. Recently one of the authors carried out a survey of software for performing these tasks. The three basic requirements were:

(a) the software should handle river level, rainfall, climate and water quality data and other associated data such as rating curves;

(b) it should be possible to store data at a variable time step (for the case where a small catchment might require 10-minute data during a flood but 12-hour data during the dry season); and

(c) the interface with the user should be such that only limited training would be needed.

A further feature which could have made a software package more attractive would have been the facility to store all the menu instructions in a separate file so as to allow a national language to be easily used.

Despite contacting many potential sources, no single package was found which satisfies even these three not too onerous minimum requirements. In one WMO project, new computers were recently installed and within a few hours, counterpart staff had familiarized themselves with well-known text processing and spread-sheet packages and other such software, but no one was running data processing software. This may be seen as evidence of the unattractiveness of many of the programs currently available.

The variety of storm-producing mechanisms in the humid tropics can add a complicating factor in hydrological analysis. This is most evident in flood-frequency studies where decisions must often be made as to whether to treat all floods as members of a single population or to analyse separately those of different origin.

Under the term "data processing," reference might also be made to the application of rainfall-runoff models for use in real time flood forecasting or in the derivation of various types of design data. Problems arise when the structure of the model is not appropriate to the hydrological regime found in the humid tropics or where limits imposed on certain model state variables, while necessary in temperate regions, cause distortions in the tropics. Major problems may arise during the transition from dry to wet conditions in regions subject to strong seasonality. If these problems can be avoided, then rainfall-runoff models can work well because the more consistent and high levels of catchment wetness remove some of the basic causes of error which arise when such models are applied in less humid regions.

Data availability - Present situation

To be available for later use, data must be stored in an appropriate form. Paper deteriorates rapidly in the humid tropics and can be eaten by various creatures. Computer tapes also deteriorate, but modern diskettes offer a more reliable medium of storage.

For any analysis related to water management, a frequent need is for data covering a long period of time, preferably several decades. This is the case, for example, where the long-term reliability of water resource developments have to be assessed or one wishes to study changes in runoff resulting from changes in land-use over many years. The ultimate aim of operational hydrology is to create and maintain a databank suitable for such purposes. Whereas for temperate climates it is claimed that 30 to 40 years is an adequate length of record for most purposes, the inherent variability of the climate in the tropics can demand records 50, 60 or even 100 years in length to obtain the same level of confidence in the results (WMO, 1987).

In particular, regions affected by tropical cyclones demand long records. The exception is for regions close to the equator where quite short records are usually sufficient.

In the case of countries which have become independent in the last two or three decades, the early records were collected by the former colonial governments which, in many cases but by no means all, appreciated the need for good hydrological data. In some cases, the period immediately after independence has been marked by a break in the regular collection of data. Where this did happen the reasons were complex but included:

(a) disruption resulting from civil or military conflict at or following independence;

(b) a lack of trained national engineers and technicians at independence;

(c) a lack of budgetary provision by the governments who saw their prime task as being to tackle other more urgent problems. In most cases this situation did not last for more than a few years, after which there was a resurgence of interest and hence investment in operational hydrology. National resources, together with substantial external aid or loans, funded these activities. The levels of data collection and storage rose to new heights. While these still fell short of optimum in most cases, they were adequate and provided an essential database for use in the new surge of economic growth planned by the countries.

Over the last 10 to 15 years, in the face of economic difficulties at national and international level, the trend has reversed. Even in developed countries, the investment in hydrological data collection has dropped steadily. In developing countries, the drop has been dramatic. New data are not being collected and old data are being lost. WMO has long been concerned with this and continuously monitors such developments worldwide. It is a mark of the seriousness of the situation in Africa that the World Bank recently invested a considerable sum in a study of the problem on that continent in recognition of the threat it posed for efficient planning and management of the Bank's own multi-million dollar projects. WMO has also co-operated with the Royal Belgian Meteorological Service in the DARE (DAta REscue) project whose aim is to recover as much meteorological data as possible from African countries and transfer them to stable storage media.

SOME FURTHER PROPOSALS

Instrumentation

At the most basic level, the collection of river level data is done by an observer reading a staff gauge. Even where another type of water level equipment is also used, a staff gauge is always present as the point of reference. A good staff gauge will be made of a durable material with clear lettering which will not fade or be washed off. Often the manufacture of high-quality staff gauges requires access to production technologies which are not available in many of the countries of the humid tropics. However, for many such countries, a simpler and better solution would be to make the gauge boards in their own country. These need not use sophisticated modern materials,

could even be of wood and hand painted, but at least they would enable the country to maintain a basic network entirely using their own resources.

The next stage is the automatic level station. In this case, the number of countries in the humid tropics able to manufacture all the components to the required degree of precision is probably small. Nevertheless, a lot can be done by looking at the instrument and deciding what is easy to make and what is not. The example of rods and carrying cases for current meters has already been mentioned. Another solution would be for a country to decide on a particular make of equipment and to insist that any country giving aid agree to furnish that equipment.

A further solution is to encourage the international standardization of specifications for float gauges or other instruments, down to information such as the diameter of shafts or the thread size of screws. In this, WMO and the International Organization for Standardization (ISO) would have a key role to play. This would enable spare parts to be transferable between different manufacturers' equipment and would, therefore, in the long run, reduce spare part inventories.

When it comes to some of the more sophisticated electronic technologies, then the situation becomes even more difficult. In this case, the country in which the equipment is installed should insist that circuit diagrams be delivered with the equipment and that all the components be identified by their generic names rather than by the manufacturer's order code. It should be noted that some manufacturers are reluctant to do this as a legally-enforceable system of patent protection is not always available in developing countries.

It is perhaps worth noting that, with rare exceptions, current hydrological instrumentation is still based on the same principles as were developed and used half-a-century ago. We can hope that major breakthroughs will be made in the years ahead that will lead to more accurate, robust and cheap instrumentation that will meet the needs of the humid tropics within the financial constraints that are to be faced.

A problem that many developing countries face is that salaries in the private sector are often higher than in the government sector and that someone who is trained to repair electronic equipment can find a better job in the private sector. One solution to this problem might be to move away from the idea of "in-house" maintenance to using outside firms specializing in repairing electronic equipment such as television or video recorders. Hourly rates paid to such firms may appear high but payment would be made only when the required quality of workmanship and deadlines have been met, and qualified staff need not be employed full-time when there is less than a full-time job to be done.

Data processing

These days there is a large number of sophisticated computer programs available which are relatively easy to use, at least to start with. The documentation for some fills several large volumes, thus to learn all their possibilities takes time. Their ease of use comes from the structure of the menu system (which takes a user logically through different levels of commands to get the system to do what he or she wants) the availability of "on-line context sensitive help" (which recognizes where the user is in the menu selection process and gives him relevant guidance when and if it is needed) and the use of "hot-keys" (which enables experienced users to bypass the menu system and go directly to the execution of certain commands). Unfortunately, no such easy-to-use software package yet exists for hydrometric data processing, though there is a need for it and a number of institutes are working to upgrade their software to meet this need.

One of the problems is that most of the systems available have been either developed by a particular organization for its own use or else have been developed by research institutes. In the first case, the programs do what is needed for the agency which developed them but tend not to have extra facilities for equipment not available in the country where it was produced. In the second case, the programs consist of what someone thinks the user wants – which may not be the same thing as what the user needs. Such a package could not be produced and sold commercially; the number of sales would not justify it.

A possible solution would be for the international community, working through WMO, to encourage the development of basic systems modular in concept. Such basic systems, which would include all the data storage, retrieval and presentation routines for IBM PC-compatible computers could be made available to countries at a nominal fee. Versions for other computers or extra modules to cope for non-standard digitizers, for example, would have to be paid for separately. Whatever system is developed should take into account the systems already available through HOMS, and should be compatible with other well-known systems for related data such as WMO's CLICOM system for climatic data.

HOMS, the Hydrological Operational Multipurpose Subprogramme of WMO, provides a framework for the exchange of operational technology between countries. The components of HOMS consist of guidance material, instruments and computer software. National hydrological agencies are encouraged to make use of HOMS to obtain tried and tested technology to meet their specific needs. Much of this technology has been either developed or successfully applied within the humid tropics, a fact that can be readily ascertained from the standard descriptions of the components contained in the HOMS Reference Manual (WMO, 1988).

Data quality

The above two areas of improvement are important, but they do not in themselves account for the decline in data quality in some of the humid tropic countries over the last few decades. Many possible reasons suggest themselves but on examination will be seen to be false. One possible reason might be the lack of suitable staff, but as the anecdotes quoted above illustrate,

at the field level the staff are usually conscientious and competent, even if fewer than desired. At a higher level, this is still the case. Indeed, the presence of qualified hydrologists and hydrometrists in most countries of the humid tropics is one area where the training programmes instituted by international organizations have borne fruit. This is not to say that more staff are not needed, but a shortage of staff is not the root cause.

A further reason might be the lack of repair facilities, but again this is unlikely to be that crucial. In almost every country that the authors have visited, there are television stations and video libraries. This implies that people have television sets and video machines and that these can be repaired locally. As no hydrometric equipment is a fraction as complicated as a video recorder, it is obvious that the right level of technical support should be available.

So what, then, do the authors suggest as the reason? The conclusion they have come to is that too often the effective management of water resources, and the collection of the data essential to that task, is not regarded as a priority by governments. Projects are set up with their own hydrometric networks. These last for a few years and are then closed down or fall into disrepair and contribute almost nothing to the national hydrological archives. Vast sums of money are borrowed to build dams with insufficient data to evaluate fully their effective yield. Money is paid for rehydration programmes to combat diarrhoea (a very important campaign), but inadequate funds are available to purify the dirty water supply that often causes the problem. It is relevant, in this context, to note that historians are now starting to realize that the rapid increase in health standards in Europe and the consequent increase in human productivity in the Nineteenth Century was not a product of medical science alone but was largely due to better water supplies (as discussed in Prost, this volume).

As a result of the investigations undertaken by WMO (1987), it was concluded that in comparing the humid tropics with the less humid and extra-tropical regions:

(a) there are major climatic differences;
(b) there are important differences in the hydrological characteristics;
(c) there are few fundamental differences in the instruments and techniques to be used in operational hydrology, although many modifications are to be recommended.

The root cause for the very evident differences between the humid tropics and other regions of the world is to be found in the fact that, with very minor exceptions (e.g., the north and northeast coasts of Australia), they lie within developing and not developed countries. They, therefore, suffer from a lack of funds, trained personnel and long-term economic and administrative stability. It is this, rather than any major difference in hydrological characteristics, which gives rise to the low density of station networks, poor maintenance records and low reliability of support services.

Water resource assessment and management are crucial to the development of a country. The very real possibility of a major impact on the magnitude and variability of water resources of climate change makes it vital to monitor the existing resources and the change over time. Therefore, while one could list a series of detailed technical recommendations, the essential appeal of the 1989 Townsville Colloquium and this volume, product of that meeting, is to stress the importance of water resource assessment and management to the governments of the world and call for greater priority and increased investment for this purpose.

RECOMMENDATIONS

For the humid tropics

During the study by WMO referred to earlier (WMO, 1987), it was noted that the main barriers to progress in operational hydrology in the humid tropics relate not so much to their climate as to the fact that they are composed almost entirely of developing countries. These face an uphill battle in their efforts to improve the well-being of their populations and to maintain the quality of their environment. They have the greatest need for hydrological information, but are the least able to devote the financial and personnel resources that are necessary. Beyond this important conclusion, the WMO study developed a series of recommendations for the humid tropics, which may be summarized as follows:

(a) To facilitate the solution of operational hydrology problems in the humid tropics, the climatic sub-zones must be delineated as accurately as possible. Since the limits of the sub-zones may vary with time, a relatively dense network of meteorological and hydrological stations should be installed in what are considered to be the transitional areas, and ground information should be supplemented by satellite surveys.

(b) Since rainfall-runoff models can be calibrated more easily in the humid tropics, these should be widely used to transfer information from the meteorological to the hydrological networks and, in general, to make maximum use of the data available.

(c) The relation between meteorological and hydrological characteristics, on the one hand, and physiographic characteristics on the other, should be used to improve the interpolation of observed data.

(d) Equipment used in the humid tropics should be adapted to the intensity of the meteorological and hydrological processes peculiar to those regions. Use of advanced technology for data measurement and transmission should be considered since it may be in some circumstances of great assistance in solving many of the specific operational problems of the humid tropics. Use of equipment that cannot be read and checked in the field should be avoided. Introduction of advanced equipment should, however, be made only when adequate maintenance arrangements are available.

(e) Data storage on microfilm should be encouraged to back up data storage in computer-compatible form. Catalogues of data holdings should be published and disseminated widely so that data users may have them at their disposal as early as possible. Use of microcomputers for data storage and processing should be encouraged.

(f) The analysis of cyclone meteorology and hydrology should always be carried out on an areal (regional) basis both for deterministic and statistical investigations. Satellite data should be used extensively for this purpose. Publication of consistent data on large storms by countries in the region should be strongly encouraged and co-ordinated by WMO.

(g) Probable maximum precipitation calculations should be carried out in a differentiated manner in accordance with the rain-generating mechanisms prevailing in the area (basin).

(h) The estimation of design floods and of runoff intensity in general should take into account the consequences of possible changes in land-use/land cover, which in tropical areas may have more significant consequences than elsewhere. The recorded floods should be adjusted to take into account that the inflow into a reservoir cannot be gauged at the outflow site when it is attenuated by the natural storage in the reservoir area.

(i) Caution should be exercised in estimating and using the PMF (Probable Maximum Flood) for the design of reservoirs and dams, because of the uncertainty in calculating it. Given the particular hydrological and socio-economic conditions prevailing in each country of the humid tropics, decisions on the use of PMF in design of dams and other water-resource projects should be made by the competent authorities of each country.

(j) Short-term forecasting for small basins should be carried out only in conjunction with meteorological forecasting. Short-term and medium-term (up to seven days) forecasting on larger basins should use, as much as possible, data from index basins located on small upstream tributaries.

(k) Changes in the meteorological and hydrological regime which may be due to regional water-resource projects should be carefully investigated and estimated. Research based both on theoretical considerations and the performance of existing projects should receive high priority.

(l) Water resource projects of global significance should not be initiated before carrying out the research required to make it possible to forecast their effects on the regional and global climate and meteorology.

(m) The establishment of regional institutes of tropical hydrology may be considered. However, at present, financial aid to tropical countries should aim at establishing hydrological networks (based on advanced equipment and technology); training personnel to install and operate the networks; and transferring technology. The appropriate use of satellite data should be considered in all cases.

The collection of satellite data and their free transmission to users in the humid tropics should be considered as part of international financial assistance to developing countries, provided that the appropriate reception, dissemination and operation technology is transferred to these countries. This should be one of the major objectives of WMO's HOMS.

For water resource assessment in general

In 1977, the leading experts in all aspects of hydrology and water resource development met in Mar del Plata at the United Nations Water Conference. They adopted what is known as the Mar del Plata Action Plan (MPAP) which presents a series of resolutions and recommendations for implementation at national, regional and international levels. These were designed to accelerate the development and orderly administration of water resources as a key factor in efforts to improve the economic and social conditions of mankind, especially in the developing countries.

In the preamble to its recommendation on assessment of water resources, the UN Water Conference stated that:

In most countries there are serious inadequacies in the availability of data on water resources, particularly in relation to groundwater and water quality. Hitherto, relatively little importance has been attached to its systematic measurement. The processing and compilation of data have also been seriously neglected.

To improve the management of water resources, greater knowledge about their quantity and quality is needed. Regular and systematic collection of hydrometeorological, hydrological, and hydrogeological data needs to be promoted and be accompanied by a system for processing quantitative and qualitative data

Countries should review, strengthen, and co-ordinate arrangements for the collection of basic data. Network densities should be improved; mechanisms for data collection, processing and publication should be reinforced.

Thirteen years later, in 1990, the various organizations within the UN system responsible for water matters undertook a global evaluation of progress in the implementation of the MPAP (WMO/UNESCO, 1990). With regard to water resource assessment, it was clear that "the increase in hydrological networks that occurred after the MPAP must be viewed with both encouragement and as a source of concern. The efforts of water resource assessment agencies and international agencies focused attention on the problem of scarcity of data for water resource decision-making. However, the growth that did occur took place disproportionately in the highly-developed regions of the world. More modest growth took place in developing regions, and the total number of stations remains very low. In fact, most countries fall below the minimum density guidelines established by WMO. The fact is that in many of these countries where data are needed the most, the net-

works are static or declining from already low numbers of stations. Attention must be re-focused on these regions, for without sufficient basic data networks in operation, water resource assessment activities cannot be carried out, and the concept of sustainable development will remain illusory."

The key points which were of global relevance included:

(a) The need for good legislation, coordination, and integration in water resource management and assessment.

(b) The overriding influence of the availability of financial resources as a control on the effectiveness of WRA (Water Resources Assessment) programmes, and the deteriorating situation in many countries during the 1980s.

(c) The lack of data and data collection networks for variables other than rainfall and surface water,.including water quality, groundwater,sediment; water use, and associated information such as physiography and land-use.

(d) The importance of specific purpose project data as an information resource which tends to be neglected, and which in many countries is in danger of being lost.

(e) The success of some international river basin programmes in promoting collaboration and information exchange, and the failure of others.

(f) The increasingly critical part played by modern data processing systems in ensuring delivery of data to the end-user in a suitable and timely fashion.

(g) The particular need for transfer of existing technology, supplemented by applied research and development to deal with specific circumstances and issues, such as flow measurements under difficult circumstances

(h) The severe difficulties experienced by water resource assessment agencies in recruiting and retaining staff because of unfavourable conditions of employment.

(i) The generally good progress made in education and training at university and senior technician levels, and the poorer progress at junior technician and observer levels.

(j) The need to make full use of available training opportunities by proper planning, and the effect of a lack of resources, training materials, and capable educators in some areas, particularly technician training.

(k) The beneficial impact of international and bilateral programmes in many areas, but also the large number of aspects which are undesirable or reduce their value below the possible potential.

These recommendations and key points were developed within the context of a global study, but they are equally relevant, in fact more relevant, to the humid tropics.

ACKNOWLEDGEMENTS

The authors wish to thank the Secretary-General of the World Meteorological Organization for permission to publish this paper. The views expressed in this paper are those of the authors and do not necessarily reflect those of WMO.

REFERENCES

Chow, V. T. (1964) *Handbook of Applied Hydrology.* McGraw Hill Inc., New York.

Maidment, D. R. (1990) *Handbook of Hydrology.* McGraw-Hill Inc. – in preparation.

Trewartha, G. T. (1954) *An Introduction to Climate.* McGraw-Hill Inc., New York.

World Meteorological Organization (1981) *Guide to Hydrological Practices – Volume I. Data Acquisition and Processing.* WMO Publ. No. 168. WMO. Geneva, Switzerland.

World Meteorological Organization (1983) *Guide to Hydrological Practices – Volume I, Analysis, Forecasting and Other Applications.* WMO Publ. No. 168. WMO. Geneva, Switzerland.

World Meteorological Organization (1987) *Tropical Hydrology.* Operational Hydrology Report No. 25. WMO Publ. No. 655. WMO. Geneva, Switzerland.

World Meteorological Organization (1988) *HOMS Reference Manual –* Second edition. Geneva. Switzerland.

WMO/UNESCO (1990) Water Resource Assessment WMO/UNESCO, Geneva/Paris.

4: The Problems of the Humid Tropics – Opportunities for Reassessment of Hydrological Methodology

V. Klemeš

Consultant, Victoria, B.C. Canada

(Formerly National Hydrology Research Institute, Department of the Environment, Saskatoon, Saskatchewan, Canada)

ABSTRACT

The paper presents the author's personal philosophy of hydrological methodology. It claims that the latter is biased toward mechanistic reductionism and an exaggerated orientation towards short-term technological solutions to water-related problems arising in the temperate climatic zone. It suggests that this bias stands out most clearly against the background of the specific natural conditions of the humid tropics. In this context, three aspects are emphasized as deserving major attention in the development of an adequate scientific basis for the hydrology of the region and for addressing its long-term environmental and water resource problems. These aspects are: 1) non- stationarity of the major geo- and bio-physical/chemical processes, 2) a dynamically sound macro-hydrology, and 3) quantitative eco-hydrology (i.e. the soil-vegetation-atmosphere interactions as mediated by energy and water).

A HISTORICAL PERSPECTIVE

Hydrology, as most everyone knows, is the science dealing with the dynamics of the hydrological cycle, *i.e.*, with the circulation of water on planet Earth. Not everybody realizes, however, that this dictionary definition is more a statement of a program than a statement of fact. Hydrology has never been practised in a global water context. Such an approach has only recently been contemplated as feasible, and, indeed, as necessary. In spite of this realization, it is by no means clear how such a program can be executed or how much integration is desirable and necessary among the historically-formed autonomous sciences of oceanography, climatology, meteorology and atmospheric physics, glaciology, hydrogeology, and the hydrology of the land surface, *i.e.*, hydrology in the narrow sense.

Recent interest in global hydrology has been stimulated by societal pressures reflecting the emergence of global environmental problems which refuse to remain contained within the historical boundaries of the various geo- and bio-physical disciplines. What we must realize in searching for new approaches is that the present state of scientific methodologies, the disciplinary groupings, etc., is not something *a priori* given or designed to satisfy some platonically ideal classification of the sciences, but merely a result of pragmatic rationalizations of very specific societal pressures of the past era. The development of scientific methodology is a continuing, though not necessarily continuous, process generated by an inherent dichotomy between tools shaped by the past and their inadequacy for tackling the many problems of the future (National Research Council, 1990).

While we recognize this methodological handicap in connection with hydrology in the broad or global context, we are usually less ready to acknowledge it in connection with hydrology in the narrow sense – the land surface hydrology in which we consider ourselves to be experts, in which we practise, teach, and on which we write learned volumes, guides, and textbooks. As a cursory inspection of any hydrological textbook will testify, we tend to project an image of methodological clarity, generality and power capable of "solving" almost any hydrological problem.

In my opinion, such an image is largely an illusion, or even a delusion. In fact, I consider the methodological fuzziness and lack of perspective to be much greater in the classical, land-surface hydrology, than it is in the general, or global, hydrology (Klemeš, 1986, 1988a). The problems facing the humid tropics provide an excellent background against which the deficiencies of contemporary hydrology can be seen.

In a nutshell, such is the methodological heritage of contemporary hydrology. It is a heritage from the water resource engineer solving small-scale water management problems of an industrial society that was living in a benign temperate climate and an altered ecology which was believed to represent stationary, "natural" conditions. Such conditions did not encourage the search for a better understanding. Rather, because of the stationary assumption, it favoured the description of empirical facts. Since the substance, while not well-understood, was believed to be stable over time, all that seemed to be needed was to describe the past and use the description as a model of the future.

PARADIGM SHIFT

There is one overriding difference between the social climate of the past under which the bulk of contemporary hydrological methodology developed and the social climate of the present day.

In the former case, it was a climate of humanity's self-confidence, an unshaken belief in technological progress, in the desirability of changing nature by bringing it under man's control, and a myopic naivete which did not see beyond the immediate benefit of any given action. In a way it was the culmination of the judeo-christian world-view. Such a view held that God made people the masters of nature whose sole purpose was to serve them and satisfy their material needs.

As a parenthetical but none the less hydrologically relevant observation, I may note the paradoxical fact that so far the most recent heir of this judeo–christian religious concept of humanity's position "above" nature was Stalin with his vision of building a paradise by "changing nature" by means of his grandiose "Great Constructions of Communism." Ironically, these "Constructions" were mostly water management projects: huge dams, canals and irrigation systems. For me, this is not merely a historical curiosity but a painful personal memory of a "voluntary" participation in two such "Constructions" in the early 1950's, of digging in the mud, of listening to fiery speeches about the "victory of man over nature" (sic) and of marching to the tune of a song, "We will order the rain when to fall and command the wind where to blow."

The situation today is substantially different. There has been a fundamental "paradigm shift" spreading across political boundaries, transcending religious doctrines, affecting the developed as well as the developing countries. The romantic optimism and belief in humanity's mastery over nature has been replaced by apprehension and a growing realization that the sum total of all the local "victories over nature" may add up to global defeat. The increasing affluence of the industrial nations and the struggle to provide even the basic needs for the rapidly growing population of the developing world are conspiring to produce an unsustainable stress on the world resources, the environment and eventually on the existing social and political structures.

These new concerns are perhaps most clearly apparent in the humid tropics, a geographical region which carries more than its share of the pressure of both the population explosion and the world-wide hunger for natural resources. This disproportionally high intensity of environmental stress is being manifested most visibly in the mass deforestation and the consequent land-use changes. While similar to the process that took place in the temperate regions of Europe and North America in the past, its present importance in the humid tropics is higher by an order of magnitude because of the following reasons: Firstly, unlike our forefathers, we now appreciate better the important role of forests, in particular the tropical forests, in the global environment. Secondly, the present rate of deforestation in the tropics is much faster than it was in the temperate regions. This aspect, combined with the plausible assumption that deforestation in the tropics may have a deeper impact on global environment than deforestation in the temperate regions, hints at a distinct possibility that rather drastic global changes may now result within a relatively short period of time. And thirdly, given the rapidly growing population and its expectations for a higher standard of living compared to those of a century or more ago, much more is now at stake should the steadily diminishing resources be mismanaged.

It is against this background that the hydrological problems of the humid tropics must be viewed and the adequacy of the available hydrological knowledge and methodology assessed.

A GENERAL METHODOLOGICAL FRAMEWORK

In order to advance hydrological science in the direction most useful to the contemporary needs of the humid tropics, there are three major aspects – largely absent from classical hydrology – which must be emphasized: 1) nonstationarity, 2) macro-hydrological processes and 3) bio- or eco-hydrology.

All these aspects are gaining importance in hydrology in general, but in the humid tropics their need is particularly urgent. Nonstationarity is significant because of the above average population growth and the consequently much faster rate of environmental change in comparison to other regions. Macro-hydrological processes and the eco-hydrological aspects are important because of the unique ecological importance of the region as a global source of atmospheric moisture and oxygen and as a sink for carbon dioxide.

The implications for hydrological methodology are enormous. On the one hand, the necessity to consider substantial changes in hydrological conditions renders ineffective all the standard methods for interpreting historical records in the stationary context. On the other hand, the necessity to consider the dynamics of large-scale hydrological processes and the key role of vegetation in the tropical water cycle leaves the hydrologist both empty-handed and unprepared to develop the necessary tools. The macro-hydrological and eco-hydrological methodologies are in an embryonic stage and hydrologists thus are still not receiving the education and training necessary to master these new areas. Last but not least, macro-hydrological analyses will not be feasible without a qualitatively different database than the one provided by the classical networks of sparse point measurements which hinge on local accessibility and availability of qualified personnel. The new database will have to comprise "hydrological fields", *i.e.*, time series of areal distributions of various hydrological and related variables.

Such a database is only now coming within reach through the newest remote sensing and data transmission technologies.

Because of their importance, these three major new concepts deserve to be discussed in detail.

Macro-hydrology

In the humid tropics, the overall control of hydrological processes by the global atmospheric and ocean circulation is more direct than in other regions because of the unique role of the equator. The vanishing of the Coriolis force at the equator tends, in general, to make the latter behave as a quasi-rigid boundary, *vis-a-vis* both the atmospheric and ocean circulation (the monsoon regions are the exception) and gives rise to a special class of motions in the vicinity of the equator. This feature, together with the maximum of the solar radiation input, makes the tropics the prime mover of the atmospheric and ocean dynamics and the location of the maximum water and energy fluxes. The dynamic features originating in the tropics operate on large spatial scales and on a wide spectrum of time scales which vary from diurnal (*e.g.*, convective rainstorms), to weekly (*e.g.*, hurricane generation), to seasonal (*e.g.*, monsoons) and to multi-annual (*e.g.*, the Southern Oscillation).

This stresses the need for analyzing the hydrological conditions and their changes in a broad atmosphere-ocean-land context. Such an approach also forces an explicit recognition of the fact that the key to understanding hydrological phenomena often lies outside hydrology, certainly outside the domain of the classical hydrology – in the narrow sense of the word. The major methodological consequence is a need for the development of dynamic models which capture the essential features of the atmosphere-ocean-land interactions. This is a formidable task which has been approached, though with an uneven emphasis, from all three perspectives, with the atmospheric general circulation models (GCMs) being by far the most advanced. While the need to couple the GCMs with the ocean dynamics has always been recognized, experience with their use in recent years has clearly shown that a more detailed representation of the land-surface process, *i.e.*, of land-surface hydrology in the first place, is equally important (WMO World Climate Research Programme, 1987). Because of the reasons given above, the importance of land-surface hydrology in the GCMs is accentuated in the humid tropics.

Presently, there are three major obstacles to an effective coupling of the GCMs and hydrological models, all of them more pronounced in the humid tropics than elsewhere. The first is the methodological difficulty of bridging the "scale-gap" between the GCMs which work on spatial grids to the order of hundreds of kilometres and the classical hydrological models which use river basins as spatial units and often treat the scale problem only indirectly through various empirical features like "travel times," different shapes of basin unit responses, etc. In the humid tropics, this problem is exacerbated by the fact that most hydrological models operationally available have been developed for temperate conditions and

their structure does not make them readily transferable. The second obstacle, also of a methodological nature, is the fact that, in the dynamical context, the classical hydrological model is at best a "hydraulic machine." It tends to reduce, as far as possible, all hydrological processes on the land surface to hydraulic mechanisms. Whatever is beyond this possibility is relegated to empirical, or a "systems" description. This "mechanistic bias" is a greater handicap for application in the humid tropics where the nonmechanical forcing (e.g., thermodynamics, biochemistry) exerts dominant control over the water fluxes. The third obstacle is an inappropriate database which, again, presents a much more acute problem in the tropics than in the temperate zones.

However, it is encouraging to note that important steps have been taken to overcome these problems during the past decade or so, notably in the form of international research projects. Among the most recent and relevant ones in the present context are the Global Energy and Water Cycle Experiment (GEWEX) (World Climate Research Programme, 1988) and the even more ambitious and general International Geosphere-Biosphere Programme (IGBP, 1990), also known as the Global Change Programme.

Nonstationarity

It cannot be emphasized often enough that the overriding reason for the need to develop dynamical models in general, and for the humid tropics in particular, is the condition of nonstationarity. This condition or factor, though long recognized to be important on time scales of thousands of years and beyond, has not been regarded as practically important on "human time scales" of decades and centuries until very recently. Only the recent explosive expansion of the human race and the previously unimaginable magnitude of the effects of the byproducts of its civilization brought the reality of nonstationarity right into our backyard, so to speak. It is forcing us to drastically change our ways of thinking, not only politically and economically, but scientifically as well. This is especially true in the environmental and earth sciences whose objects of study are the first potential victims of the collective aggressiveness of the human race *vis-a-vis* nature. As already mentioned, the shrinkage of the time scale on which nonstationarity is becoming significant is presently largest in the humid tropics. The changes induced there by "modern" causes (CO_2 concentrations and the long-range transport of pollutants) are happening at the same time as the drastic land-use changes. Such changes were spread over several centuries in most of the temperate zones.

This is a historically unique occurrence which has caught us ill-prepared in many areas, with hydrology among those at the top of the list. We are facing a situation for which we have neither a precedent nor a clear conceptual framework to guide us. We only have an uneasy "gut feeling" that something can go wrong, perhaps irreversibly, and can see only some of the most obvious problems that are in need of increased attention. However, recent experience has taught us that great dangers may well loom behind seemingly benign appearances or otherwise be out of our present range of sight.

The awareness of the importance of nonstationarity on relatively short time scales is eroding the very foundations of contemporary methodology of the so-called "applied hydrology," in particular its reliance on long historic records and on the modelling philosophy based on the concept of "curve fitting" and "calibration." This topic will be discussed later in a separate section. In the present context, suffice it to say that nonstationarity is giving hydrology a potent impetus to climb one rung higher on the ladder of scientific methodology: from description of observations to their casual explanation, (Bohm, 1957) *i.e.*, from empirical to dynamic modelling of physical phenomena.

Eco-hydrology

Given the importance of vegetation in the hydrology of humid tropics and the rate and volume of their present deforestation, a thorough scientific understanding of the quantitative relationships within the soil-vegetation-atmosphere segment of the hydrological cycle is indispensable for sound hydrological analysis and for any kind of medium-to-long-term hydrological assessment and prediction. This understanding also is needed for an adequate parameterization of the land-surface processes in the GCMs in order to make it possible to model the present dynamics of the water and energy fluxes in the humid tropics. Such an understanding is particularly needed to find out how much of the rainfall is generated locally, what the spatial pattern of the prevailing water fluxes is over the tropical rain forest and how the various interactions between the internal dynamics of the humid tropics and the external forcing influence the basic patterns of the boundaries between the humid and subhumid tropics. It is obvious that to address these types of questions is beyond the reach of the contemporary mechanics-and-systems-dominated hydrological methodology. Thus quantitative eco-hydrology is the key to tackling the three more general problems of immediate interest, namely (a) the local hydrological effects of the local environmental changes, (b) the external hydrological effects of the local environmental changes and (c) the local effects of external environmental changes such as global warming, the long-range transport of chemical pollutants and aerosols, etc.

SPECIFIC HYDROLOGICAL ISSUES

Apart from the general issues discussed in the preceding section, some methodological differences arise from the different aspects of hydrological processes in the humid tropics as compared to those in the temperate climate. In a nutshell, these aspects are all subsumed in the label "humid tropics" but they must be described explicitly in order to make the methodological consequences stand out clearly.

The most conspicuous feature of the humid tropics is, of course, the overabundance of both moisture ("humid") and solar energy ("tropics") which produce the exuberant vegetation. Its overall dominance of the hydrological processes was mentioned in the preceding section but it is important to point out some aspects of its function which are accentuated in the humid tropics.

One of them is the fact that evapotranspiration is seldom constrained by a limited availability of water and proceeds at a "potential" rate which, under these conditions, is then only controlled by the available energy and plant physiology. This seldom happens in the temperate zone where the reduction of the actual evapotranspiration due to a limited water supply causes the potential rate to rise and creates a gradient. This figures prominently in all temperate-zone hydrological models, though in many cases it is used improperly (Morton, 1978). Thus the problem of evapotranspiration in the humid tropics environment focuses attention on the energy and plant regimes much more than it does in temperate zones.

A related problem arises from the much more prominent role of interception in the humid tropics due to the much higher frequency (often almost regular daily occurrence) of rainfall. The ratio of the moisture-flux from the plants into the atmosphere through evaporation and transpiration is thus substantially different from the temperate regions and may necessitate a different parameterization of the evapotranspiration process in micro-hydrological studies, in GCMs and in macro-hydrological models.

A third problem involves the present relationship among evapotranspiration, the local generation of rainfall and photosynthesis – a relationship that is bound to change due to deforestation, the increasing atmospheric concentration of CO_2 and its potential climatic effects like temperature increase, changes in cloudiness, etc. Since photosynthesis "counteracts" evapotranspiration by preventing water from escaping into the atmosphere and its rate itself is controlled in part by water availability (its summary chemical equation is $6CO_2 + 6H_2O \Rightarrow C_6H_{12}O_6 + 6O_2$), the possible changes may be substantial and their direction and broader effects are by no means clear.

The role of vegetation as a mechanical dissipator of the kinetic energy of rainfall and thus as an efficient buffer protecting the soil from erosion is much more important in the humid tropics than in the temperate zones. The importance of this function can best be appreciated when we realize that rainfall intensities and volumes in the humid tropics are higher than in the temperate zone by an order of magnitude. To put it more graphically, many areas in the humid tropics are exposed many times every year to what in the temperate zone would be regarded as 100 to 1,000 year rainstorms, i.e. to conditions which would render the affected locality a "disaster area." In quantitative terms, a daily precipitation total in a humid tropics location can be larger than an annual total in a temperate location. Daily totals in humid tropics are commonly in the vicinity of 200 mm. Short-term (5 to 30 min) intensities are around 200 mm/hr. It may well be that this rainfall buffer function is the most important role of the tropical forests for ensuring a continuous long-term habitability of the region. It is not only the order-of-magnitude greater erosive power of the rainfall but also the much higher fragility of tropical soils on exposure that exacerbates the danger of their rapid degradation. The soil's fragility is caused by the rapid breakdown of organic matter, producing high CO_2 concentrations in the percolating water, thus leading to highly effective weathering. The continuing infiltration caused by the overabundant rainfall then leads to intensive leaching of chemicals, leaving behind structurally unstable and generally "poor" soils. Our quantitative understanding of these processes is still inadequate but enough is known qualitatively to pinpoint this problem as one of the central areas where new methodologies and extensive research are needed

When tackling the problems of the humid tropics, there is another aspect that calls for a revision of classical hydrological methodology. Organic matter assumes here some of the functions that the soil performs in the temperate zone, particularly by participating more directly in the cycling of nutrients and the storage of moisture.

Even from this short and sadly incomplete list, it should be abundantly evident that the problem of the humid tropics is their greater quantities of moisture and energy. Hypothetically, such quantities could be entered into the existing theoretical constructs and models as larger input values, thus producing correspondingly larger outputs. However, it can be seen that the increased quantities of moisture and energy by themselves activate processes that are dormant under lower values of these two inputs. Therefore, no methodological framework has been developed for their analysis. Most importantly, it is evident that these newly activated processes which rise to prominence in these conditions are processes of a thermodynamical, chemical and biological nature. This situation contrasts sharply with the fact that the bulk of the classical methodology developed in response to hydrological problems encountered in the temperate zone is designed for the handling of processes rooted in fluid mechanics. It may be noted that an analogous change takes place, but in the opposite direction, when the object of hydrological investigation moves from the temperate zone to the arctic or the alpine zones. There the overall decrease of energy inputs drastically reduces the intensity of chemical and biological processes, substantially slows down the fluid mechanical ones and activates those of solid mechanics, like snow drifting, avalanches, ice flow and creep in glaciers etc.

APPLICATIONS OF HYDROLOGY TO WATER MANAGEMENT

It has already been mentioned that a major component of the present-day hydrological methodology grew out of the hydraulic and water resources engineering endeavours in the temperate zone during the past 200 years. In reality, hydrology has not yet fully separated from these parent disciplines and a lot of confusion exists in regard to their respective aims and methodologies. Not even the professional community has yet fully grasped the basic differences between hydrology – an earth science – and water resource engineering, or more generally, water resource management – a technology. As a result, hydrology is often charged with solving tasks of a technological nature while technological advancements are taken for hydrological discoveries. For example, a recent Canadian occupational brochure for promoting hydrology as a profession lists almost exclusively water management tasks such as water-power generation, irrigation, water quality improvement, etc. as the "hydrologist's challenges" (see Klemeš, 1988a). Conversely, many engineering design and operational techniques, such as the stochastic generation of synthetic streamflow series for storage reservoir design or transfer-function flow forecasting models have been regarded as advances in the science of hydrology. To make this confusion seem more respectable, such activities are conveniently grouped under the term "applied hydrology" and are regarded as "bridging the gap" between theory and practice.

For the most part, hydrological concepts that are used in various water resource management applications are either empirical relationships between hydrological variables such as rainfall and runoff which are formalized through statistical or systems techniques (the unit hydrograph concept is a good example), or concepts which allow a hydrological process to be reduced to some fluid-mechanical process which is then solved with the aid of the available physical theory (e.g., flood wave propagation along river channels).

With this background information, it is relatively easy to make an assessment of the applicability of the standard hydrological methodologies to the water management problems of the humid tropics.

The least satisfactory situation is faced by the planners of long-term development of water resources. If such planning is to have any credibility, it must be based on solid estimates of regional water resources and their future variability. The available methodologies are sorely inadequate for this task since they are all based on the assumption of stationarity of the hydrological regime over a period of over 100 years – centered on the present. That is to say, they make an assessment of the water resources from the historic hydrological observations over the past 50 years or so and assume that the situation over the planning horizon (never longer than about 50 years) will not be markedly different (Klemeš, 1979a). This methodology is being increasingly attacked because of the potential for global changes within a few decades. The humid tropics probably will experience the most rapid rate of change, making the existing methodologies particularly inadequate for this region.

In this virtually unusable category belong the standard methodologies for estimating long-term mean precipitation, streamflow, groundwater levels and the associated variabilities, the estimation of the frequencies of extreme events and based on them, the design parameters of various water-related structures such as dams, levees, floodways, water supply systems, etc. (Klemeš, 1989)

Agricultural planners face a similar and perhaps even more difficult task when estimating future food production. Coping with the uncertain future of the soils, in addition to the uncertain hydrological regimes, makes such activities a guessing game.

The only hope for limiting the range of these uncertainties is rapid progress in the theoretical understanding of the dynamics of the macroscale processes and the soil-vegetation-atmosphere segment of the hydrologic cycle discussed earlier, and a reduction of the rate of the changes currently taking place globally and locally.

The situation is generally better with the applicability of the methodologies concerned with analysis of currently existing hydrological relationships. Based mostly on empirical observations described in systems or statistical frameworks, these methodologies postulate no understanding of the dynamics of the underlying processes and are readily applicable wherever the necessary observations can be made. Unfortunately, such relationships can be of help only for immediate operational applications such as flood forecasting and estimation of design parameters for local structures serving present needs. However, they have no scientific value and have to be continuously revised (recalibrated) as the natural situation changes, since all of them are based on stationarity which may be a reasonable assumption for only a very limited time span.

Methodologies concerned with issues that can be reduced to the simpler problems of applied fluid mechanics, *i.e.*, open channel flow, bank seepage, well yield, etc. have the best prospects for applicability to the humid tropics. While their empirical components such as hydraulic roughness and conductivity, and the like may be subject to changes, their theoretical bases, involving only simple mechanical processes, have a general validity. Moreover, a humid environment may sometimes be conducive to such a reduction. For example, it is well known that most of the present-generation conceptual hydrological models perform better under humid rather than arid conditions. This is most likely because the basin, for most of the time,is saturated. The antecedent moisture conditions of individual storms remain more-or-less constant, thus eliminating one variable important for the nature of the basin response. However, such an advantage may be offset by shorter lag times between precipitation and runoff with less time for effective flood forecasting, evacuation and other hazard-mitigating actions.

In closing this section, an important point should be made which is equally valid for both the humid tropics and elsewhere. Many sound water management decisions can be made with surprisingly little hydrological information. The lack of knowledge can be compensated for by a corresponding increase in the robustness and resilience of the design of the relevant facilities and by maintaining flexibility of future options. The prevailing contemporary attitude is that the highest level of scientific information is required for making every decision. Such attitudes are a comparatively recent phenomenon brought about by the development of systems theory in general, and the concept of economic optimization (usually understood as maximization of short-term economic gains), in particular. The truth is that in all but the simplest technological decisions, the concept of optimization is invalid because its underlying assumptions, including social, political, economic and other conditions, are changing rapidly and usually unpredictably. It is safe to say that most water management projects, however thoroughly their design may have been optimized, were far from optimal by the time they were put into operation. The return to conservative decisions with high safety margins and ample flexibility is particularly called for in the environmental context, and in regions such as the humid tropics which are undergoing rapid change. (Fiering & Rogers, 1989; Klemeš, 1977; 1979a,b; 1990).

CONSEQUENCES FOR HYDROLOGICAL EDUCATION

The contemporary system of hydrological education is the result of the development of hydrology over the past 200 years as a service discipline to various branches of water-related technologies in the temperate zones of Europe and North America. These technologies include transport, agriculture, forest management, water power development, flood protection, industrial and domestic water supply,and urban drainage, to name just the most important ones. As a consequence, the main aim of hydrological education has not been to prepare experts capable of advancing the hydrological science but the strengthening of these various water-related technologies with which this education has been affiliated. The minor segment provided within the framework of physical geography has been, until very recently, almost purely descriptive and paid little attention to the dynamics of the processes involved.

Because of these historical reasons, the present system of hydrological education is saddled with a fix on technology. This makes it difficult for hydrologists to address the fundamental problems of hydrology which are not of a technological nature but belong squarely within the domain of the earth sciences. The main problem of the present-day hydrological education is that it does not produce scientists but technologists motivated to applying hydrology to other ends. This has led to what has euphemistically been called "many perspectives of hydrology" but means little more than a motley collection of *ad hoc* ,empirical and often mutually inconsistent methods and techniques, whose main connection with hydrology is their use of numbers labelled with hydrological names (Klemeš, 1986; 1988a).

This situation was more-or-less tolerable as long as human interference with the environment in general, and with the hydrological cycle in particular, was on a small scale and did not appreciably alter the quasi-stationary conditions in Europe and North America. Such conditions were established after the conversion of a substantial part of their forests, prairies and other ecological entities into agricultural land. Under these conditions, there was little incentive to discover the deeper relationships in the hydrological cycle and the education system responded accordingly. It concentrated on ways of how to make use of water for the dependent technologies, and neglected to educate specialists about the science of hydrology. Hydrology is now in a situation similar to that, for example, of geology if it could be studied only via, say, geological engineering; or botany if it had to be studied only as a part of agronomy; or mathematics if it were offered only as a service discipline to actuarial studies or home economics.

The conditions in the contemporary humid tropics expose the fallacy of the technologically-based hydrological education better than anything else. The idyll of environmental stationarity is about to explode there. Suddenly, there is a need for professionals who can assess the future water resources of this vast region under changed environmental conditions and they are nowhere to be found, while the "unit hydrographs" and "flood frequency analyses" offered by the run-of-the-mill experts are of no use. The main responsibility for rectifying this situation (which will become more acute when the global changes start to affect other climatic zones more intensively) lies with the universities. Without delay, they should establish "streams" within the earth sciences which will provide appropriate backgrounds for specialists in the water-related geo- and bio-physical problems (Nash *et al.* 1990). This development should be vigorously encouraged by international as well as national environmental and water resource agencies and organizations. These entities will soon become the main targets of the societal pressures inspired by large-scale changes in water resource availability impacting the sustainable habitability of every region.

CONCLUSIONS

The geographical zone of humid tropics with its overabundant supply of solar energy and humidity, and the consequent exuberant vegetation, poses unique hydrological problems which the present methodology, developed in response to different natural and social conditions, cannot effectively tackle. The gulf between what is available and what is needed is widened by the fact that the available methodology is based on the existence of stationary environmental conditions, whilst those in the humid tropics are highly nonstationary because of the population explosion in the region and the rapid change of land-use resulting from the conversion of rain forests into agricultural land.

Three aspects have been emphasized as deserving major attention in the development of an adequate scientific basis capable of addressing hydrological and water resource problems of the region: 1) a consideration of nonstationarity in the major geo- and bio-physical processes, 2) the development of

a dynamically sound macro-hydrology and 3) the development of quantitative eco-hydrology. These aspects stand in sharp contrast to contemporary hydrological methodology which because of the underlying assumption of stationarity, favours empirical approaches based on historical observations. Because of its roots in hydraulic engineering, this methodolgy is heavily biased towards fluid-mechanistic concepts and tends not to do justice to other physical processes.

Finally, it has been pointed out that this situation has serious implications for hydrological education. In particular, it has been recommended that universities should establish "streams" in the earth science curricula that would provide adequate background in the sciences most relevant to the dynamics of the hydrological cycle. This would make it possible for the science of hydrology to be studied at a similar level to that of other geo-physical and bio-physical sciences, rather than only as a service discipline to various technological programs such as water-resource engineering, forestry, etc., as is the general practice today.

REFERENCES

Bohm, D. (1957) *Causality and chance in modern physics.* Routledge and Kegan Paul. London, England.

Fiering, M. & Rogers, P. (1989) *Climate change and water resource planning under uncertainty.* Final Report, March 1989, to the Institute of Water Resources, U.S. Army Corps of Engineers. DACW 72–88–M–D680. Environmental Systems Program, Harvard University, Cambridge, Mass. USA.

International Geosphere-Biosphere Programme (1990) *Global Change. Report No. 12.* International Council of Scientific Unions, IGBP Secretariat, Royal Swedish Academy of Sciences, Stockholm, Sweden.

Klemeš, V. (1977) Value of information in reservoir optimization,. *Water Resour. Res. 13* (5), 837–850.

Klemeš, V. (1979a) The unreliability of reliability estimates of storage reservoir performance based on short streamflow records. In: *Reliability in Water Resources Management,* 193–202. Wat. Resour. Publications, Fort Collins, Colorado, USA.

Klemeš, V. (1979b) Storage mass-curve analysis in a systems-analytic perspective. *Wat. Resour.Res. 15* (2), 359–370.

Klemeš, V. (1986) Dilettantism in hydrology: Transition or destiny? *Wat.Resour.Res. 22* (9, Supplement), 177S–188S.

Klemeš, V. (1988a) A hydrological perspective. *J. Hydrol., 100,* 3–28.

Klemeš, V. (1988b) Hydrology and water resources management: the burden of common roots. In: *Proc. VIth IWRA Congress on WaterResources,* Vol. I, 368–376, International Water Resources Assoc. Urbana, Illinois, USA.

Klemeš, V. (1989) The improbable probabilities of extreme floods and droughts. In: O. Starosolszky & O.M. Melder, (eds.), *Hydrology of Disasters,* 43–51, James and James, London, England.

Klemeš, V. (1990) Sensitivity of water resource systems to climatic variability. In: Proc. 43rd Annual Conf., Canadian Water Resour. Assoc., 233–242, Penticton, B.C. Canada.

Morton, F. (1978) Estimating evapotranspiration from potential evaporation: Practicality of an iconoclastic approach. *J. Hydrol. 38,* 1–32.

Nash, J.E., Eagleson, P.S., Philip, J.R. & van der Molen, W. H. (1990) The education of hydrologists. *Hydrol. Sci. J., 35* (6), 597–607.

National Research Council (1990) Opportunities in the Hydrological
    Sciences. National Academic Press, Washington, USA.

World Meteorological Organization (1987) *The World Climate
    Programme, 1987–1988: Second Long-Term Plan,* Part II, *2,* WMO
    No. 692, 52–73.

III. Regional Hydrology

5: Hydrology in Humid Tropical Asia

J-H. Chang

Emeritus Professor of Geography and Climatology; and Emeritus Climatologist, Water Resources Research Center, University of Hawaii, Hawaii
Present address: Chinese Culture University, Hwa Kong, Yang Min Shan, Taiwan, Republic of China

ABSTRACT

A description of rainfall, evaporation and transpiration, climatic water balance, river regimes, deltas, irrigation, groundwater, soil erosion and the hydrological effects of forest conversion are systematically presented for humid tropical Asia. Underlying themes are the implications of the monsoon circulation, e.g. typhoons and the very high population density on the hydrology and water resource management of the region.

INTRODUCTION

The humid tropics can be defined in different ways. The definition adopted by this Colloquium specifies that the humid tropics is that region where the mean temperature of the coldest month is above 18°C and the duration of the wet season exceeds 4.5 months (Chang & Lau, 1983). A wet month is defined as one that has more than 100 mm of rainfall. When the rainfall is between 60 and 100 mm, half a wet month is credited. The humid tropics has three sub-types: (1) "wet" has 9.5 to 12 wet months, (2) "subhumid" has 7 to 9.5 wet months and (3) "wet-dry" has 4.5 to 7 wet months. The total area of the humid tropics is approximately 29.4 million km^2, or 22% of the Earth's land area. It is most extensive in the Americas, accounting for 44% of the total, followed by 32% in Africa, 18% in Asia and 6% in Oceania.

The Asian sector includes nearly the entire region of South East Asia – south of the Tropic of Cancer (Fig. 1).

Over the Indian subcontinent, only a narrow strip of the west coast, a belt of the east coast of about 350 km width from southern Bangladesh to the Coromandel coast and Sri Lanka fall within the humid tropics. Much of the interior of Indian

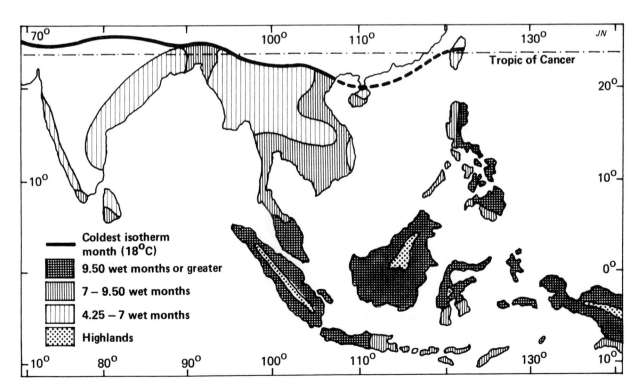

Figure 1: The humid tropics region of South and South East Asia

subcontinent is too dry during two-thirds of the year to qualify.

The southern part of South East Asia includes numerous islands. They are sometimes collectively referred to as the "maritime continent." The greater part of the maritime continent and the southern part of the Malay Peninsula belong to the wet sub-type, which accounts for 42% of the total area of the humid tropics in Asia. The eastern part of continental South East Asia, including Vietnam, Kampuchea and southern Thailand, is the subhumid sub-type. The duration of the wet season decreases towards the west, as northern Thailand, Burma (Myanmar) and much of India belong to the wet-dry sub-type. In South Asia, only southern Bangladesh, southwestern Sri Lanka, and Coimbatore in southern India have a wet season of seven months or longer.

GENERAL DESCRIPTION AND REGIONAL COMPARISON

Of the three major humid tropical regions, the Asian sector has the youngest geological formations. Africa is made up of a Pre-Cambrian shield while South America has the Guyanan and Brazilian shields in the Amazon Basin. South East Asia belongs to the more recent Alpine mountain system. In the Indian subcontinent, much of the coastal area has sedimentary deposits over the Godanava shield.

Five major rivers empty into the tropical waters between the South China Sea and the Bay of Bengal. From east to west, they are the Mekong, the Chao Phraya (Me Nam), the Salween, the Irrawaddy and the Ganges-Brahmaputra. These rivers all have broad flood plains in their lower basins, with their deltas being far more extensive than the deltas in either Africa or Latin America.

Acid, infertile Oxisols and Ultisols are the most widely distributed soils in the humid tropics. However, their distribution and characteristics vary according to geology, land forms and other factors. It has been estimated that Oxisols and Ultisols account for 82% of the total area in the American sector, 56% in Africa and only 38% in Asia (U.S. National Research Council, 1983). On the other hand, various kinds of moderately fertile, well-drained soils occupy 33% of the total area in Asia, 12% in Africa and 7% in the Americas. Fluents, Andepts, and Tropepts are especially extensive in tropical Asia, with a total area of 101 million ha, against a combined total of 55 million ha in the rest of the humid tropics.

Better soil conditions, in general, and the presence of extensive deltas, in particular, are partly responsible for the more advanced agricultural development in Asia than in the rest of the humid tropics. In 1986, arable and permanent crop fields accounted for 18% of the total land area in the Asian sector against about 7% in either Africa or Latin America [Food and Agriculture Organization (FAO), 1986]. In the Asian sector, 26% of the farmlands were irrigated, mostly in the form of paddy rice. By contrast, the percentage of agricultural lands under irrigation was only 5% in Latin America and 3% in Africa.

The original forest cover in the humid tropics has been greatly denuded. In 1986, forests accounted for 45% of the total land area in the Asian humid tropics, 55% in Latin America and 60% in Africa.

In the Asian humid tropics, the hydrological regime has been influenced greatly by changes in land-use, urbanization, soil and water conservation measures and diversion of surface and groundwater discharges. It was not coincidental that the first United Nations agencies to recognize the part played by hydrology and water resources in economic growth was the United Nations Economic Commission for Asia and the Far East (ECAFE). In 1950, the Bureau of Flood Control, later to become the Water Resources Development Division, was established.

RAINFALL

In the tropics, the four major rain-generating mechanisms are convection, convergence, orographic and cyclonic. Most thunderstorms in the tropics are caused by intense local heating in a warm, moist, unstable atmosphere. The average life span of a convective cell is about half an hour. In the entire humid tropical Asia, the number of thunderstorm days exceeds 30 in a year. In Bangladesh, southern Burma (Myanmar), southern Thailand, Malaysia and the western part of Indonesia, thunderstorm days exceed 60 in a year. In Bogor, Java, where the rugged terrain favors convection, an average of 322 thunderstorms develop in about 100 days in a year.

During the summer when the monsoon trough or northern monsoon shearline (see Manton and Bonell, this volume) is located farther away from the equator, it is referred to in India and South East Asia as the monsoon trough. The same trough in its active form is marked by towering cumulonimbus, violent turbulence and heavy rainfall. The monsoon trough in South East Asia is highly mobile and its movement determines, to some extent, the spatial distribution of rainfall.

Because of the abundant moisture content, orographic rainfall is exceedingly heavy in the tropical mountains. The Western Ghats, the Arakan Yoma and the Annamitic Cordillera are good examples. In the tropics, the altitudinal belt of maximum precipitation is well-defined and is usually much lower than mountains in the temperate zone.

Tropical cyclones have varying intensities. When the sustained wind speed exceeds 120 km hour^{-1}, they are generically known as hurricanes, or locally as typhoons in East Asia, or simply as tropical cyclones in the Bay of Bengal. With a diameter varying from 300 to 3,000 km and an average life span of five days, hurricanes can produce torrential rains over a large area. The western North Pacific Ocean and the South China Sea have an annual total of about 22 typhoons, the highest frequency in the world. They affect the northern Philippines, Taiwan, Hainan and the coast of Vietnam. Another area is the

Bay of Bengal, where intense tropical cyclones develop either before the burst of monsoon or after its retreat in the fall.

The world's record rainfall intensities for short periods of less than eight hours are observed mostly in middle latitudes; for longer durations, they occur in the tropics. Yoshino (1984) has presented a map showing the distribution of maximum 24-hour rainfall intensities in monsoon Asia. All hurricane-prone areas have values in excess of 400 mm. The lowest intensities of less than 200 mm are found in northwestern Thailand, southern Sumatra, southern Borneo and western Java. All other areas have intensities greater than 200 mm, but rarely reach 300 mm.

However, Yoshino's map is highly generalized. For the solution of many practical hydrological problems, it is necessary to have a precise estimate of probable maximum precipitation (PMP) for small areas. In hurricane-prone areas, the PMP can be estimated using global maximization as proposed by Kennedy & Hall (1981). Global values are then adjusted for regional differences as well as for local variations, including proximity to the sea and orographic effects. In areas where thunderstorms provide the most intense rainfall (e.g., Malaysia), PMP has been estimated by multiplying the 4-hour thunderstorm rainfall by a factor to obtain the values for larger durations (Kennedy, 1976). Hershfield (1965) has expressed PMP as the sum of the mean precipitation for a specific duration, plus the standard deviation, times the frequency factor K. His data suggest a mean value of 15 for K in the USA. Such an approach is useful in areas where frequency of rainfall intensity has a nearly normal distribution. However, in areas where there are several rain-generating mechanisms, and the frequency distribution is highly skewed or bimodal, the use of Hershfield's method is less accurate. In northern India, the K values vary from 9 to 14.5 (Dhar & Kulkami, 1974). It is likely that K values are distinctly different between the hurricane-prone areas and areas not affected by intense tropical cyclones.

In the temperate zone, synoptic approach of moisture maximization of storm intensity has often been used to estimate PMP. This requires a knowledge of the maximum amount of the precipitable water vapor which could be present during the storm period. In the tropics, this is difficult because aerological data are usually not available and the use of surface dew point to estimate the precipitable water vapor is inaccurate. The poor relationship between dew point and precipitable water has been demonstrated by Kennedy (1976) for Malaysia.

Stol & Sybesma (1963) have established an empirical relationship between precipitation intensity, duration and frequency occurrence by using the data of two stations in Thailand. Such a technique is useful for the estimation of rainfall intensity from one duration to another. They urged that geographical variation of the constants in the empirical formula be established.

For many agricultural and hydrological problems, a knowledge of the probability of rainfall is useful. Oldeman (1984) has presented the 75% probability rainfall as a function of the mean monthly rainfall in mm for three countries in Asia:

Indonesia (25 stations): $P_{75} = 0.82P_{mean} - 30$

Thailand (19 stations): $P_{75} = 0.86P_{mean} - 20$

Bangladesh (10 stations): $P_{75} = 0.75P_{mean} - 32$

In general, the 75% probability has the largest deviation from the mean in hurricane-prone areas, slightly smaller in areas affected by both synoptic-scale monsoon systems and numerous thunderstorms, e.g., Indonesia, and the least in the interior of the continent. In continental South East Asia and in India, the period of greatest rainfall variability is observed during the onset of monsoon (Mizukoshi, 1974). In South East Asia, the date of onset varies from 1 May to 3 June, a range of 33 days (Orgill, 1967).

Because of the large variability in rainfall, droughts have occurred even in the wet humid tropics. For instance, January 1964 was a dry month for Indonesia when the rainfall was more than 200 mm lower than in January 1963. Java also experienced a drought in 1972. A serious drought affected Malaysia in 1976 to 1977, causing heavy losses of oil palm. The seasonality of droughts in the lowlands of Sarawak has been discussed by Brunig (1969). Monthly rainfall is not necessarily a good indicator of the absence of drought in areas of high intensity. Despite a monthly total rainfall of 645 mm in January 1965 at Miri, Sarawak, agricultural drought – as indicated by soil moisture stress – was noted during the second half of the month (Baillie, 1976). Atmospheric circulations in India and Indonesia are closely related to pressure changes in the southern Pacific Ocean, in a phenomenon known as the Southern Oscillation, which in turn is related to El Niño of the equatorial Pacific dry zone. In an El Niño year, India and Indonesia usually suffer from drought and when the Pacific dry zone is drier than normal, monsoon rainfall is usually plentiful (Rasmusson & Carpenter, 1983; Tsuchiya, 1978).

EVAPOTRANSPIRATION

Measurements of potential evapotranspiration using different instruments have been carried out in the Philippines (Aglibut *et al.*, 1957; IRRI, 1965), Malaysia (Sugimoto, 1971), Laos (Kotter, 1968), Thailand (Kung, *et al.*, 1965), Bangladesh (Kung, 1961), Sri Lanka (Basnayake, 1983) and the humid tropical regions in India (Lenka, 1978; Chaudhary, 1966). Most of these studies were concerned with water needs of rice crops.

Studies by Batchelor & Roberts (1983) in Sri Lanka and by Asuncion (1971) in the Philippines have demonstrated that measurements of potential evapotranspiration agree well with the Penman formula (1948). The Penman equation consists of a net radiation energy term and an aerodynamic term. In the tropics, the former is far more important than the latter. If, indeed, the energy budget term and the aerodynamic term are proportionate to each other and the latter is relatively small, then potential evapotranspiration can be estimated from the energy term alone. By and large, this is true in the humid tropics in the absence of advection. Priestley & Taylor (1972) have demonstrated theoretically that potential evapotranspiration can be calculated by the energy term times a constant of 1.26. This constant has been confirmed by Kayane & Nakagawa (1983) in an experiment in Sri Lanka. When the observed evapotranspiration exceeds the calculated value, the difference may be attributed to the advection of sensible heat (Lang *et al.*, 1974). Advected energy is negligible during the wet season but may be appreciable during the dry season.

Net radiation records are not readily available except at a few agricultural experiment stations. Attempts have been made to estimate net radiation from global radiation. Observations at IRRI at Los Banos have shown that net radiation is about 63% of the global radiation. During the dry season when the cloud cover is reduced, the fraction of global radiation retained as net radiation may be slightly lower. Since a generalized map of global radiation is available (Chang, 1980), net radiation distribution can be estimated to within 15% on a monthly basis.

Evaporation pans, especially the U.S. Weather Bureau Class A pans, have been widely used to estimate evapotranspiration. Unlike the Penman equation, readings of evaporation pans respond to both net radiation and advection of sensible heat. Monthly data of Class A evaporation pans for several countries in South East Asia have been summarized by Oldeman & Frere (1982). Tomar & O'Toole (1979), in reviewing the ratio of potential evapotranspiration and open pan evaporation of wetland rice in South and South East Asia, concluded that, on the average, the ratio is near unity at transplanting, increasing to 1.1 at the maximum tiller stage and to 1.2 or higher during the flowering state. They have also noted that, in the absence of advection, pan evaporation rates are equivalent to net radiation. During the greater part of the year, daily evapotranspiration is about 5 to 7 mm.

In the tropics, studies of evaporation in mountain areas are scarce. As the temperature decreases with elevation, so does the psychrometric constant, which needs to be adjusted downwards for each new altitude under consideration, which means that the energy term rapidly loses its importance as the aerodynamic term assumes a dominant role in high mountains. The aerodynamic term is determined by wind speed and vapour deficit. Because the change of wind speed with elevation varies greatly in tropical mountains, evaporation rates are poorly correlated with elevation. Apparently, the decrease of temperature and net radiation with elevation is compounded by a daytime increase of global radiation, an increase in wind speed and a decrease in humidity. Observations of six stations from sea level to an elevation of 3,023 m in the western part of Java indicate that there is no simple relationship between altitude and evaporation rate (Henning & Henning, 1982). In Malaysia, evaporation rates in the northwest coastal range are higher than in the central mountains for the same altitude, primarily because of the higher humidity in the former (Scarf, 1976).

CLIMATIC WATER BALANCE

The water balance in the soil can be expressed by the following equation: Rainfall = changes in soil moisture + evapotranspiration + percolation + runoff. Climatological computation of water balance requires measurements of rainfall, an estimate of potential evapotranspiration by an appropriate method and a knowledge of the water-holding capacity of the soil. The last term is the amount of water between field capacity and permanent wilting point in a solum of soil to a depth that includes 90% of the roots. The output of the computation includes actual evapotranspiration, the water deficit required to fully satisfy the water need of the plant and surplus water, including percolation and surface runoff.

A water balance study has many applications in hydrology and in water resources management. In humid tropical Asia, it has been computed by Nieuwolt (1965) for Malaya, by Low & Goh (1972) for West Malaysia, by Maruyama (1967) for rice crops in Thailand, by Shiroma & Chuitkao (1980) for rice and sugarcane in central Thailand, by Vu (1980) for North Vietnam, by Henning & Henning (1982) for many stations in Indonesia and Malaya, by Kayane & Nakagawa (1983) for Sri Lanka and by Kayane (1971) for the entire South Asia and South East Asia region. Kayane used the annual water deficit and surplus as the criteria for delineating hydrological regions. In general, the greater part of Indonesia has no annual water deficit. Most of continental South East Asia and the coastal areas of India have an annual deficit in excess of 200 mm. Most of the Philippines and southwestern Sri Lanka have a deficit less than 200 mm. The several studies of water balance mentioned above are not exactly comparable. Some used the Penman equation; others adopted the Thornthwaite (1948) formula which expresses potential evapotranspiration as a function of temperature and day length. In general, the Thornthwaite method underestimates evapotranspiration in the tropics and is less accurate than the Penman formula. Another problem is that in a regional study, a standard soil moisture storage value is assumed for all stations. Eelaart (1973) has considered 50 mm as the average value for tropical soils, whereas followers of the Thornthwaite approach use a value of 30 mm. In reality, soil moisture holding capacity can vary from 10 mm to over 300 mm.

RIVER REGIMES

Although the monsoons govern the variability of the river flows in humid tropical Asia, there are significant regional variations. From the standpoint of river regimes, four general regions may be recognized: (1) typhoon-prone, that includes the northern part of the Philippines, Hainan, Taiwan and the coastal region of Vietnam, particularly between 16°N and 18°N; (2) equatorial, that includes the southern part of the Philippines, Indonesia, Malaysia and Sri Lanka; (3) tropical, with a distinct and prolonged dry season that includes continental South East Asia and the coasts of India and (4) Bangladesh and the neighboring areas affected by intense tropical cyclones.

In the typhoon-prone area, the maximum rainfall intensities for one or several days approach the global maximum, with the floods being very destructive. For small catchment areas, a typhoon can generate a specific discharge as high as 10 $m^3 s^{-1}$ km^{-2}. A typhoon typically produces a single flood event with a single main peak in the hydrograph lasting for several hours or days. However, the probability that a given river will be affected by such an event is much lower than the seasonal flood produced by monsoons. Annual river discharges in the typhoon-prone area are not higher than the equatorial region and the runoff coefficient can be very low. For instance, the runoff coefficient of the Pampanga River in central Luzon is only about 0.2 during the summer months (Kinosita, 1982).

In the equatorial regions of Indonesia and the southern Philippines, the dry season is very short. The rivers carry large volumes of water. Thunderstorms and disturbances developed within the monsoon atmospheric circulation are typically of short duration, producing flash-floods in small river basins. In most parts of the equatorial region, runoff exceeds 1,000 mm, reaching a value between 2,000 mm and 2,500 mm in the northeastern part of Borneo, the eastern part of Celebes and the eastern part of Mindanao.

On equatorial islands, the runoff coefficient usually exceeds 0.60, but may vary greatly depending on rainfall, land-use, soil, stream slope, roughness of the stream channel and other basin characteristics. For instance, runoff coefficients on the island of Java vary from 0.32 for the Brantas River to 0.74 for the Tjomal River. Largely because of the difference in rainfall characteristics (even in areas not affected by typhoons), models developed for temperate zones cannot be applied to the equatorial belt. In general, model parameters could be more successfully transferred between similar basins during a period of low flow than during a period of high rainfall intensity (Pickup, 1977). In general, in areas underlain by impervious granites and basalts, groundwater runoff accounts for 10 to 20% of the total, whereas in areas of highly permeable rocks, it may account for half of the total.

Rivers in Malaysia and Sri Lanka are subject to greater seasonal fluctuations than in the rest of the equatorial region. In Malaysia, major floods can occur during the northeast monsoon, such as in 1926, 1967 and 1971. In Sri Lanka, the central and the southwestern parts are known as the wet zone where the southwest monsoon brings flooding almost every year.

In continental South East Asia, the dry season and the wet season are distinct. During the onset of the summer monsoon season in May or June, much of the rainfall of the first storm is used to wet the soil to storage capacity. Oftentimes, only 10% of the rainfall runs into the river. However, the initial infiltration rate may vary greatly, depending on a number of factors, and is most difficult to simulate by models.

Central Burma (Myanmar)is located in the rain shadow of

Arakan Yoma and receives only scanty summer monsoon rainfall. For instance, the Mu River, a tributary of the Irrawaddy, has a total streamflow of only 350 mm. Central Thailand is even drier. Tributaries of the Mekong, e.g., the Mun River and the Mae Nam River, have runoff values between 200 and 250 mm. The interior of continental South East Asia is an area of minimum river runoff in humid tropical Asia. The many small river basins are responsive to intense local storms in the form of flash-floods. Hydrographs of the major rivers, however, do not show sharp fluctuations, but present themselves as a single flood occurrence of a gentle type. The Irrawaddy, the Mekong and the Chao Phraya are three major rivers in humid tropical South East Asia. The first two are, respectively, the fifth and sixth largest rivers in the humid tropics on the basis of their mean annual discharge. The Irrawaddy originates in China at an elevation of 4,500 metres and cuts a deep valley above Mandalay. In its lower reach, the river flattens and carries a large amount of silt, causing frequent meandering and expanding of its delta at the rate of 10 km^2 per year.

The Mekong also has its headwaters in the highlands of China with deep valleys above Vientiane and a relatively gentle slope below. The upper reach of the Mekong is drier; consequently, total runoff is about half that of the Irrawaddy, with the runoff coefficient being only 0.33. The Mekong reaches the highest water level in September. Excess water is stored in Pa Mong Reservoir in Laos and in Tonle Sap in Kampuchea. The former is the largest artificial reservoir and the latter the sixth largest natural lake in the tropics. Both have dams to regulate the flow of water. The development of a mathematical model to simulate runoff in the lower Mekong valley has made it possible to optimize the operation of Tonle Sap dam under conflicting requirements for fishery and sedimentation on the one hand and power generation, land reclamation, agricultural development and flood control, on the other hand.

Although the Chao Phraya is much smaller than the other two major rivers in South East Asia in terms of its length and total drainage area, it provides a productive agricultural land in the plain of Siam, which is even more extensive than the flood plains of the other two rivers. Gentle seasonal floods occur regularly which are a blessing to the paddy rice culture. For this reason, farmers do not construct protective embankments. Only rarely do severe floods occur to inflict substantial damage. Since the construction of multipurpose reservoirs, i.e., the Bhuimbol, the Chainat and the Airikit dams, catastrophic floods have been further reduced and the dry season flow augmented.

The lower Ganges and Brahmaputra join in Bangladesh to form one of the major rivers of the world with a combined discharge of 44 434 $m^3 s^{-1}$. The tributaries of the Brahmaputra have a total runoff as high as 2,000 mm in the Himalayas. In the lower basin, the orographic rainfall in the Khasi Hill area, as exemplified by the well-known station of Cherrajunji, further adds to the large flow of the river during the summer season. The mean annual rainfall at that station at an elevation of

1,313 m is 114 300 mm – of which 92% falls from April through September. As the soil is saturated and the infiltration capacity limited, almost all the rainfall ends up as surface runoff. Groundwater runoff is also very high, contributing to 17 to 28% of the total in the lower Ganges-Brahmaputra Basin.

As the soils in the mountain areas are readily susceptible to fluvial erosion, the average annual sediment yield is 1,451 10^9 kg in the Ganges and 726 x 10^9 kg in the Brahmaputra. These values rank second and third, respectively, among the rivers in the world – behind only the Yellow River in China. In response to the large amount of bed load, the course of the river shifts rapidly, often threatening to ruin human settlements.

Tropical cyclones of hurricane intensity sweep up the coast of the Bay of Bengal, either before the onset of monsoon in April and May or after the retreat of monsoon from October to December. Although the cyclones here are not as violent as those in the western Pacific and some other tropical oceans, the dense population, the expansive low-lying agricultural land and the occurrence of high tidal surges combine to make them one of the most destructive natural hazards in the tropics. For instance, on 12 November 1970, a mature tropical cyclone coincided with the occurrence of high tide which was 7 m above the normal high tide level. The surge-induced flood breached many protective works, inundated the vast coastal area, took over 200 000 lives and swept away 280 000 head of cattle. In 1985, another catastrophic flood killed more than 10 000 people when a tropical cyclone and tidal waves simultaneously hit the low islands in the Bay of Bengal – about 200 km south of Dhaka.

Some of the floods in Bangladesh can be moderated by the construction of reservoirs. However, suitable sites are not available on the main Brahmaputra and on most of its tributaries.

Floods also pose a serious problem in urban areas. Tropical deluges require extensive drainage systems to divert excess water, such as the Bukit Tinah Canal in Singapore. Where an urban drainage system is inadequate, pollution and health problems cannot be prevented or controlled. Maintenance of a good drainage system can be costly. Accelerated erosion not only may result in a change of the morphology of the channel but also may lead to siltation in downstream estuaries with a consequent impact on marine life and fisheries.

THE DELTAS

Humid tropical Asia has extensive delta areas. Almost all of Bangladesh and West Bengal has the character of a delta, even though the true, functional delta is more limited. The vast Ganges-Brahmaputra delta, with an area of 800 000 km², is the second largest in the world, about 20 000 km² smaller than that of the Amazon (Czaya, 1981). The deltas of the Mekong (55 000 km²), the Irrawaddy (31 000 km²) and in Chao Phraya 11 000 km²) also rank high in the world. In addition, there are

numerous small deltas in southwestern Luzon, Borneo, eastern Sumatra, northern Java, Malaysia and the eastern coast of India.

Deltas in humid tropical Asia are intensively used and densely populated. Their importance prompted the UNESCO Humid Tropical Research Group to organize a major international symposium (UNESCO, 1964) and the ECAFE to sponsor a series of regional conferences.

Deltas are zones of conflict between river and sea-water. Tidal waves may penetrate the lower course of a river. The extent of the penetration is dependent upon the height of the tides, the slope of the delta land and the rate of river discharge. In the Mekong delta, for instance, with an exceedingly gentle slope of about 3 x 10^{-5}, the effects of tide (about 0.3 m range) can be felt as far as 350 km from the coast during a period of minimum river flow. The major deltas in continental South East Asia generally have gentle floods in which the rise of water does not exceed 0.05 to 0.1 m a day. Gentle floods are not harmful to the rice culture, even in the absence of protective devices. On the contrary, tidal waves in the Ganges-Brahmaputra delta can build up to great heights. During the southwest monsoon season, "bores" at the mouth of the Meghna, an extension of the Ganges-Brahmaputra delta, can swell to a height of 9 m with a maximum speed between 5 to 7.5 m s^{-1}. Even in the absence of tidal waves, an onshore southwest wind can raise the sea level at the head of the Bay of Bengal by as much as 2 m.

Most deltas are plagued by sea-water intrusion, especially during the dry season when the need for fresh water is greatest. In the Ganges-Brahmaputra delta, barriers equipped with drainage sluices have been erected to partially close the estuaries and thus arrest sea-water intrusion.

Sediment yields, either expressed in total load or in weight per unit drainage area of the large rivers in humid tropical Asia, e.g., the Ganges-Brahmaputra, the Irrawaddy and the Mekong, rank among the highest in the world. Sediment deposits result in rapid evolution of the delta. The Irrawaddy delta is expanding at a rate of 10 km² per year. The old district of the city of Jakarta, which was built 300 years ago near the shore of a delta, is now 3 km inland. Typical of the process of deposition in a delta is the stratification of fluvial sediment – the separating of coarse material from the fine sands. However, the advancing edge of a delta often consists of a series of shifting islands, bordered by shoals and sandbanks, that is constantly moving. Drainage and irrigation pose very difficult problems, especially if the deltas are subject to seasonal inundation by saline water for several months.

IRRIGATION

The development of irrigation in humid tropical Asia has a long history. As early as the fifth millennium, B.C., large tanks or reservoirs were constructed in Sri Lanka for irrigation. Records of the subak irrigation system in Bali can be traced back to the first millennium A.D. However, prior to World War II, only Java, Madura, Bali and Sri Lanka had a relatively large hectarage of cultivated land under irrigation. Development in the rest of the humid tropical Asia lagged far behind East Asia.

The traditional irrigation systems were small in scale and communally financed and organized. Their primary functions were to impound river flows during the period of high rainfall and to divert the water by canals to the farmers' fields. As the water storage capacity was limited, irrigation for a second crop during the dry season was generally not possible. After World War II, governments of several countries undertook the task of developing large-scale irrigation schemes. The Upper Pamganga project in the Philippines, the Muda project in Malaysia, the Kyetmauktang dam in Burma(Myanmar), the Greater Chao Phraya project in Thailand and the multinational Mekong project are good examples. Most of the large-scale projects also generate hydroelectric power, reduce floods and facilitate river transportation.

In general, irrigation in humid tropical Asia is better developed in the island countries than on the mainland. In 1986, the amount of cultivated land under irrigation in Indonesia, Sri Lanka and the Philippines was respectively 34, 26 and 18%. On the mainland, only the Mekong delta of Vietnam and Thailand approach the percentage value of the Philippines. Because of the difficulty of controlling water in large river deltas, Bangladesh had only 15% of its cultivated land under irrigation. Burma (Myanmar)had 11%. Malaysia, which has only 15% of its cultivated land in rice production, has the lowest percentage – 8%. In humid tropical India, irrigation has been developed along the east coast in the lower valley of the Mahanadi River, in the Kistna-Godavari delta and in the Cauvery delta, but not along the west coast. The development of modern irrigation schemes has made it possible to expand cultivated areas during the dry season. In Malaysia, the area cultivated in rice production under dry season conditions grew from 1% in 1955 to 90% in 1975. In the Philippines, the area with a dry season crop increased from about 20% in 1955 to 60% by 1975. Expansion elsewhere was less dramatic but by the late 1970s, a dry season crop was grown in over one-third of the irrigated rice fields in South East Asia.

GROUNDWATER

For the study of groundwater problems, humid tropical Asia may be divided into two broad zones (Sokol, 1979). The first zone includes all small islands and the coastal areas of large islands and continents. In these areas, the groundwater is hydraulically continuous with the sea. The behaviour of groundwater follows the Ghyben-Herzherg principle. Most areas in this zone have only a short dry season. Natural groundwater recharge occurs throughout the year; consequently, water regimes do not exhibit marked seasonal variations. In deltas, the principal aquifers are composed of unconsolidated allu-

vium of variable permeability. In extensive coastal marshes, e.g., in Sumatra, Kalimantan and peninsular Malaysia, groundwater levels come to the surface.

The second zone includes the interior of the continent where the dry season is distinct and prolonged. Groundwater is subject to an annual cycle of depletion and recharge, although the variations are less marked than those of surface water. Kayane (1971) has concluded from a study of climatic water balance that even though the annual water deficit in the interior of continental South East Asia exceeds 200 mm, groundwater is a renewable resource that is rechargeable by surplus water in the wet season. However, the occurrence of groundwater is uneven and sporadic. Alluvial sediments usually form important interior aquifers but in areas of crystalline and metamorphic rocks, groundwater resources are meagre and resistant to extraction. In areas where surface water is inadequate or highly polluted, groundwater may be the major source of water supply and can even determine the pattern of rural settlement.

Groundwater tapped through springs, dug wells or drilled wells provide water for domestic, livestock and irrigation use. Groundwater is rarely used for irrigation on the continents, but its use is practiced in northern Sri Lanka, southwest Taiwan and the Philippines. The karst topography of Jaffna Peninsula is devoid of perennial rivers but groundwater utilization has made it possible to grow a crop during the dry season for centuries. Since the l960s, the enhanced withdrawal of groundwater by mechanical pumps, however, has raised the fresh-saline water interface and increased the chloride concentration (Nandakumar, 1983). Jaffna provides one of the many examples in which the withdrawal of water in excess of the annual replenishment of groundwater sources has created serious problems.

In Bangkok, which is situated 40 km from the sea and has an elevation of only 1.00 to 1.50 m above the mean sea level, extraction of groundwater from more than 8,000 wells has greatly depleted the aquifers, deteriorated water quality and caused land subsidence by more than 0.5 m (Brand, 1977; Piancharoen, 1977; Sharma, 1986). This land subsidence has so disrupted the city's drainage system that new deep stormwater drains had to be built. Although less well investigated, land subsidences have occurred in Jakarta, another highly populated urban center in humid tropical Asia, which intensively exploits groundwater as a potable water source (Sharma, 1986).

In most of humid tropical Asia, there is an urgent need to locate new groundwater resources. In the past, hydrogeological prospecting has been confined to local areas. There have been few organized regional or national surveys. These require a reasonably good coverage of geological mapping – which is still lacking in many countries. Recent development of technologies, such as isotopic and remote sensing methods, should be adopted. The identification of aquifers is only part of a comprehensive water resources management scheme which must be considered from a regional or national perspective. In

Thailand, for instance, the diversion of water in the upper tributaries of Chao Phraya during the dry season has resulted in a significant reduction of groundwater recharge in the downstream areas.

SOIL EROSION

According to a study by El-Swaify, *et al.* (1982), the overall rate of soil erosion for Asia far exceeds that of any other continent. This is particularly true in Asia's humid tropical region. For instance, the annual soil loss by water erosion has been estimated at 250 Tm ha^{-1} (Lal & Barneji, 1974), or more than 20 times higher than the "desirable" tolerance limit of 11 Tm ha^{-1} per year. The overall rate of soil erosion in South East Asia is probably comparable, as indicated by exceedingly high sediment loads in most rivers. Eroded soil particles cause aggregation of river channels, increase flood hazards, clog irrigation diversions and adversely affect water quality for industrial and domestic use.

The rate of soil erosion is primarily dependent on rainfall erosivity, ground cover, soil erodibility, slope length and erosion control measures. Rainfall initiates soil erosion by causing detachment of soil particles and transporting them overland. Wischmeier & Smith (1958) have shown that the yield of sediment is proportional to the EI_{30} index, i.e., the product of the maximum 30-minute intensity (I_{30}) and the kinetic energy (E) of the rainfall. The latter is determined by the size of raindrops and their terminal velocities. Their study was based on the experimental data obtained in the temperate climate of the USA, where thunderstorms provide most of the heavy rainfall. The average life span of a single thunderstorm cell is about 30 minutes. Typhoons and monsoon depressions in the Asian tropics can generate even heavier rainfall over a period of a day or longer. Consequently, rainfall intensity classifications devised for the USA is not applicable to tropical countries such as India (Striffler, *et al.*, 1979). Even in areas not affected by typhoons and monsoon depressions, the EI_{30} index is much higher in the tropics than in the temperate zone. For instance, it is 2,307 in Jakarta (Bols, 1978) compared to 600 for Louisiana and 780 for Hilo, Hawaii. The high value for Jakarta is indicative of its exceedingly high frequency of thunderstorms.

Oftentimes, the removal of forest as a protective cover is even more important than the destructive force of rainfall in promoting soil erosion in the humid tropics. Soil erosion under virgin forest is usually negligible. The change from forest to crop fields is accompanied by an increase of surface runoff and an acceleration of soil erosion. The exact amount of increase, depending on the nature of the new surface as well as a number of environmental factors, is difficult to assess. It is certain, however, that since the 1940s, the continued denudation of forests in the headwaters of large rivers in the Himalayas and northeastern Thailand has created serious soil erosion problems, particularly in Bangladesh and Thailand. Even well-designed cultural practices cannot obliterate erosion hazards in

tea, coffee and rubber plantations on sloped terrain. In the Philippines, the removal of natural vegetation in northern Luzon has shortened the life span of the Ambuklao hydroelectric project reservoir (Rodriguez, 1958). A similar effect has also been reported for the Karnafuli Reservoir in Kaptai, India (Chaudhury *et al.*, 1980). The problem of soil erosion following the clearing of forests in Java has been discussed by Thijsse (1977).

Soil erodibility is determined by its physical properties, such as aggregate stability, water-holding capacity, infiltration rate and the like. Soils which swell and shrink markedly on wetting and drying, and soils which have a high ratio of silica to alumina, are particularly liable to erosion. Tropical soils, for the most part, are characterized by a low silica to alumina ratio. Ultisols are generally the most abundant soils in the humid tropics. Oxisols have excellent physical properties; however, they have very limited distribution in Asia. Ultisols have less favorable physical properties than Oxisols but they still have high infiltration rates. They are the most abundant soils in humid tropical Asia, accounting for about one-third of the land including most of Malaysia, Sumatra, Kalimantan, Celebes and Mindanao. These areas include many steep slopes where orographic rainfall accelerates soil erosion and landslides.

Tropets (well-drained, or non-volcanic origin), Andepts (well-drained, volcanic origin) and Fluvents (young alluvial soils) are more extensive in Asia than in either Africa or the Americas. They are largely cultivated in rice and are not susceptible to serious erosion. Poorly drained Aquepts cover an extensive area of Bangladesh, the Irrawaddy and Chao Praya deltas and the swamps of eastern Sumatra. They are also used for rice cultivation. Psmments are deep soils with very high erodibility; they are found in parts of Sumatra and Kalimantan.

Alfisols have a higher base content than Ultisols but generally the former have a lower infiltration rate and more severe erosion hazards. Alfisols occur in isolated spots in the Philippines and Java and the Coromandel coast of India. Histosols, or organic soils, are relatively extensive in the Asian tropics, occupying 24 million ha, primarily in Sumatra, Kalimantan and peninsular Malaysia. They subside upon drainage but are not easily erodible. Spodosols, also known as groundwater Podozols, are abundant in southern Kalimantan and on Bangka and Biliton Islands. They are very susceptible to erosion.

Soils on a slope are subject to the force of gravity. Soil erosion per unit area increases at a rate of 2.5 times as the degree of slope is doubled, and increases 1.5 times as the slope length is doubled. In the humid tropics, the Asian sector has a higher percentage of rugged, sloped lands than either Africa or Latin America. However, the construction of terraces not only facilitates cropping on sloped lands, but also effectively minimizes erosion hazards. Wetland terracing originated in northern Indochina, from whence it diffused northward to China and Japan, westward to India and Ceylon, then southward to

Indonesia and eastward to the Philippines (Spencer & Hale, 1961). The Ifugao terraces in northern Luzon, which extend to an elevation of 1,700 m and provide 80 000 ha of irrigated paddy rice fields, are a remarkable example of engineering design. Terracing also offers beautiful sights in Sri Lanka, Java, Bali and northern Sumatra (Batak), but rarely appear in India and continental South East Asia.

Soil loss is greatest when the land is bare but generally decreases with the increasing completeness of a protective vegetation or crop cover. The water-covered paddy offers a surface for the effective control of erosion. On the other hand, the use of fire in slash-and-burn agriculture greatly exacerbates erosion hazards. Although shifting cultivation is practiced in many parts of tropical Asia, it is not as widespread as in Africa.

THE EFFECTS OF FOREST CONVERSION

Forest conversion initiates meteorological and hydrologic as well as edaphic changes. Forests intercept a higher fraction of rainfall than crops. Forests also help to reduce the impact of raindrops (or paradoxically the larger throughfall drops induced during the concentration of intercepted water) on the soil surface through the protective litter layer on the forest floor. Forests have higher rates of potential evapotranspiration than those of short crops because they not only retain a greater proportion of global radiation as net radiation and intercept more advected energy, but they also have a higher roughness length which facilitates upward transfer of vapor by turbulence. The forest floor has an abundance of organic matter and a relatively large proportion of retention pores in the soil profile. Therefore, forest soils have much higher infiltration rates and slightly higher moisture storage capacities than soils under crops or bare ground.

The groundwater level usually rises following deforestation because of the decreased evapotranspiration rates. However, if the infiltration rates of the new surface are altered in such a way so that the reduction of groundwater accession more than compensates for gains from lower evapotranspiration rates, the water table may manifest itself in less reliable springs and lower levels.

The conversion of forest to grassland is usually accompanied by an increase in surface runoff and/or streamflow, higher peak flows and earlier peaks in streams. This leads to greater flood damage in adjacent areas. However, if the new grasslands or agricultural lands are managed under a sound soil and water conservation program, flood damage may not be greatly aggravated. Furthermore, as the water is routed downstream to a major river channel, its effect may greatly diminish.

The possible effect of forest clearance in increasing flood frequency and severity has been mentioned in Indonesia (Wiederhold, 1957) and the Philippines (Rodriguez, 1958). However, for large drainage basins and major floods in the lower reaches of large rivers such as the Chao Phraya and the Ganges-Brahmaputra, Hamilton (1987) claims that forest conversion in the distant mountain uplands has little effect on the monsoon and other large rainfall events. In the Philippines, following the great Agusan flood of 1981, the state minister placed "30% of the blame on logging." However, it is difficult to determine the effects precisely. Hamilton with King (1983) questions that logging was responsible for a flood of this magnitude, even to the extent of 30%. On Hainan Island, flood characteristics have not been noticeably affected by deforestation – presumably because shrubs grew rapidly to replace the original forest (Qian, 1983). In the tropics, research works on watershed management are so scanty and usually cover so small an area that their results cannot be readily applied to other areas, including large drainage basins.

Another complicated and controversial subject is the possible effect of forest clearance on precipitation. Salati, et al. (1979) have measured the ^{18}O content of rainfall in the Amazon Basin. The measurements suggest that evapotranspiration from forests constitutes a large portion of moisture for rainfall in the eastern and central part of the Basin. They conclude that large-scale deforestation would lead to a decrease of rainfall in that region. However, Arulanantham (1982) has found that, based on the observations for the period from 1972 through 1977, clearing of the Sinharaja forest in Sri Lanka had no noticeable effect on rainfall in its vicinity. Meher-Homji (1980) has found that in southern India, removal of the evergreen forest is more prone to the diminishing trend than that of dry deciduous forest. Generally, rainy days are more prone to decline than is true for rainfall amounts. Exceptions are the coastal zones where high humidity, because of their proximity to the sea, seem to compensate for the loss of forest canopy. It is obvious that the precise effect is dependent on the nature of land-use following forest conversion. It is also probable that, other factors being equal, the effect of forest removal on rainfall is much smaller in the humid tropical Asia than in either the Amazon or the Congo. This is because during the summer monsoon season, South and South East Asia have much higher precipitable water vapor than the other two regions (Tuller, 1968). The supply of moisture cannot be a limiting factor for the lack of rainfall.

The original forest cover of humid tropical Asia has been greatly denuded. Only Kampuchea, Indonesia and Malaysia have more than 60% of their land areas covered by forest, followed by 56% in Laos and 47% in Burma. The forest covers in Vietnam, the Philippines and Sri Lanka all fall between 36 and 40%. Thailand has 29%, with Bangladesh having the lowest amount – 15%. The latter two countries also have the most serious soil erosion problems. Thailand (Hirsch, 1987) and the Philippines have the highest forest conversion rates in recent years. This is a result of forest conversion to rice and cassava agriculture in Thailand and shifting cultivation and grazing in the Philippines. From 1970 to 1986, the forest cover in these two countries declined by 33 and 29%, respectively. Sri Lanka and Malaysia also have relatively high rates – exceeding 15% during the same period. This situation has caused concern regarding the possibly adverse ecological and hydrological consequences, not only within their own territories but also beyond their national boundaries.

SUMMARY AND CONCLUSION

Humid tropical Asia has a very high population density. Here for many centuries water has been the life-blood of the productive agrarian economy. On the other hand, the catastrophic havoc caused by water-related natural hazards has repeatedly caused losses of human life and property. In South East Asia, 91% of the population has no reasonable access to safe water, compared to 89% for underdeveloped countries in Africa and 76% in South America (Feachem, 1975). Therefore, research initiatives in hydrological and water management problems assume an added urgency and importance in Asia, compared to other parts of the humid tropics.

The distribution of the average monthly and annual rainfall has been mapped for individual countries as well as for the region as a whole. However, methods for the estimation of probable maximum precipitation remain to be perfected. Techniques developed in the temperate zone have little application in the tropics, where rainfall mechanisms are more diverse. Within the tropics, areas affected by typhoons are basically different from those not affected by them. Rainfall variability is high, especially during the onset of the monsoon season. Even during the wet months, droughts creep in occasionally. Advances in synoptic climatology, particularly studies of telecommunication, are helpful for water resource planning – either on a short-term or long-range basis.

There have been enough experimental studies on potential evapotranspiration to offer a general understanding, particularly the consumptive use of rice crops. The Class A pan is widely used. Among the empirical methods, the Penman equation has been proven to be the most accurate. During the wet months in the humid tropics when advected energy is minimal, evapotranspiration is closely related to the net radiation. Therefore, the simplified Priestley and Taylor method is adequate. In the mountains, as the aerodynamic term assumes great importance, the estimation of evapotranspiration becomes a difficult task.

Calculation of climatic water balance has many practical applications. On a continental or regional scale, a standard value of moisture storage capacity has to be adopted. However, this may introduce a large error, especially if the purpose of the study changes. Knowledge of the depletion rates and other moisture characteristics necessary for the improvement of the budgeting procedure is insufficient.

Four major river regimes have been recognized. The high rainfall intensity and large spatial variation have not only rendered the runoff models developed in temperate zone inapplicable, but also the transfer of the model results within the tropics difficult. In general, the transfer of models between similar basins is more successful during the dry season than during a period of peak flow.

In spite of the limited plain areas, humid tropical Asia has some very large rivers. Although all of these rivers have their headwaters in the high mountains outside the tropics, they have different flood characteristics. Some have very gentle floods; others, such as the lower Ganges-Brahmaputra, have floods that are exceedingly destructive. The extensive deltas in tropical Asia are densely populated. The many and varied hydrological problems in the deltas have been thoroughly studied and discussed by the UNESCO Humid Tropics Research Group, although much engineering work and the proper methods of human adjustment have yet to be put into practice.

The development of irrigation has a long history in Sri Lanka, Java, Madura and Bali where the percentages of cultivated land under irrigation are the highest. The traditional systems were small-scale and communally financed. Since World War II, many large-scale multi-purpose projects have been constructed by governments or international groups, greatly expanding the irrigated hectarage as well as the multiple cropping index. This and other technological improvements have been the major cause for the increase of crop production in the last 40 years. In contrast, irrigation is still very limited in Africa and Latin America where the increase of food production during the same period was largely the result of an expansion of crop land.

In the humid tropics, groundwater is generally underutilized because of the abundance of rainfall and surface water. The use of groundwater is, however, more widespread in the Asian sector than in the rest of the humid tropics. This is particularly true in the islands and delta areas. The major problems associated with improper intensive groundwater development are salinization and land subsidence.

Soil erosion poses a serious problem in humid tropical Asia, as evidenced by the exceedingly high sediment loads in many rivers. Besides the very high rainfall erosivity, rapid deforestation in recent years has been a major contributing cause. Land clearing has been especially rapid in Thailand and the Philippines. Only wise government policies and well-designed soil and water conservation programs can avert the worsening of this problem.

REFERENCES

Aglibut, A. P., Catambay, A. B., Gray, H. E. & Hoff, P. R. (1957) Measurement of consumptive use of water by lowland rice. *Philipp. Agric. 41*:412–421.

Arulanantham, J. T. (1982) Effects, if any, on rainfall due to the deforestation of Sinharaja Forest. Inst. of Geosci., Tsukuba Univ., Japan. *Climatological Notes 30*:169–173.

Asuncion, M. T. (1971) Analysis of potential evapotranspiration and evaporation records at Los Banos, Laguna, Philippines. WMO/UNDP Project, Meteorological Training and Research, Manila. *Tech. Ser. 7.*

Baillie, I. C. (1976) Further studies on drought in Sarawak, East Malaysia. *J. Trop. Geogr. 43*:20–29.

Basnayake, B. K. (1983) Observing potential evapotranspiration in Sri

Lanka. *In*: M. M. Yoshino, I. Kayane & C. M. Madduma Bandara (eds.) *Climate, Water and Agriculture in Sri Lanka*. Inst. of Geosci., Univ. of Tsukuba, Japan, 139–146.

Batchelor, C. H. & Roberts, J. (1983) Evaporation from the irrigation water, foliage and panicles of paddy rice in northeast Sri Lanka. *Agric. Meterol. 29*:11–26

Bols, P. L. (1978) The iso-erodent map of Java and Madura. Belgian Tech. Assistance Proj. ATA *105*, Soil Res. Inst., Bogor, Indonesia.

Brand, E. W. (1977) Soil compressibility and land subsidence in Bangkok. In: Proc., Anaheim Symp., Int'l. Assoc. of Hydrol. Sci., Publ. 121:365–374.

Brunig, E. F. W. (1969) On the seasonality of droughts in the lowlands of Sarawak Borneo. *Erdkunde 33*:127–133.

Chang, J.-H. (1980) Some aspects of agroclimatology in South East Asia and New Guinea. *Geo J. 4*:437–446.

Chang, J.-H. & Lau, L. S. (1983) Definition of the humid tropics. Paper presented to the IAHS meeting in Hamburg, Germany.

Chaudhary, M. S. (1966) Irrigation requirement of rice crop. *In*: Central Rice Res. Inst., Int'l. and Nat'l. Training Course for Rice Breeders and Technicians. Cuttack, India.

Chaudhary, M. U., Huda, M. H. Q. & Cafur, M. A. (1980) Study of Karnafuli Reservoir and its watershed. *In: Proc., Int'l. Symp. on Remote Sensing of Environment*, Costa Rica, *3*:1409–1418.

Czaya, E. (1981) *Rivers of the World*. Cambridge Univ. Press, London.

Dhar, O. N. & Kulkarni, A. K. (1974) Estimation of probable maximum point rainfall over plain areas of north India using statistical techniques. *In: Proc., Int'l. Tropical Meteorol. Meeting*, Nairobi, 287–289.

Food and Agriculture Organization (FAO) (1986) *Production Yearbook*.

Eelaart, van der, A. L. J. (1973) *Climate and Crops in Thailand*. Report WWR–96, FAO, Bangkok.

El-Swaify, S. A., Dangler, E. W. & Armstrong, C. L. (1982) Soil erosion by water in the tropics. Coll. of Tropical Agric. and Human Resources, Univ. of Hawaii, Honolulu, USA.

Feachem, R. C. (1975) Water supplies for low-income communities in developing countries. Am. Soc. Civ. Eng., *J. Environ. Eng. Div., 101*:687–702.

Hamilton, L. S., with King, P. N. (1983) *Tropical Forest Watersheds: Hydrologic and Soil Response to Major Uses or Conversions*. Westview Press, Boulder, Colorado, USA.

Hamilton, L. S. (1987) What are the impacts of Himalayan deforestation on the Ganges-Brahmaputra lowlands and deltas? Assumptions and facts. *Mountain Res. and Development 7*(3):256–263.

Henning, I. & Henning, D. (1982) Climatological precipitation deficit in equatorial South East Asia according to a new calculation of potential evapotranspiration after Penman (modified). Inst. of Geosci., Univ. of Tsukuba, Japan. *Climatological Notes 29*:93–97.

Hershfield, D. M. (1965) Method for estimating probable maximum rainfall. *J. Am. Water Works Assoc. 75*:965–972.

International Rice Research Institute (1965) *Annual Report*.

Hirsch, P. (1987) Deforestation and development in Thailand. *Singapore J. Trop. Geogr. 8*:129–138.

Kayane, I. (1971) Hydrological regions in monsoon Asia. *In*: M. M. Yoshino (ed.) *Water Balance of Monsoon Asia*. Univ. of Tokyo Press, Tokyo, Japan, 287–300.

Kayane, I. & Nakagawa, S. (1983) Evapotranspiration and water balance in Sri Lanka. *In*: N. N. Yoshino, I. Kayane & C. M. Madduma Bandara (eds.) *Climate, Water and Agriculture in Sri Lanka*. Inst. of Geosci., Univ. of Tsukuba, Japan, 127–138.

Kennedy, M. R. (1976) Probable maximum precipitation from the northeast monsoon in South East Asia. *Proc., Symp. on Tropical Monsoons*, Poona, India, 294–303.

Kennedy, M. R. & Hall, A. J. (1981) Probable maximum precipitation in tropical Australia and WMO Region V. Paper presented to the Seminar on Hydrology of Tropical Regions, Miami, Florida, USA.

Kinosita, T. (1982) Hydrological study on floods in rivers overseas (Part 2). *Report of the National Res. Centre for Disaster Prevention, No. 29*, Ibaraki, Japan.

Kotter, E. (1968) Determination of water requirement of rice in Laos. *IRC Newsl. 17*(4):13–20.

Kung, P. (1961) Water use in rice cultivation in East Pakistan. *Agr. Pakistan 12*(1).

Kung, P., Atthayodhin, C. & Druthabandhu, S. (1965) Determining water requirement of rice by field measurement. *IRC Newsl. 14*(4).

Lal, V. B. & Barneji, S. (1974) Man's impact on erosion in the rural environment: The Indian experience. *In: Effect of Man in the Interface of the Hydrological Cycle within the Physical Environment*. Proc. of the Paris Symp. IAHS Publ. *113*:46–52.

Lang, A. R. G., Evans, G. N. & Ho, P. Y. (1974) The influence of local advection on evapotranspiration from irrigated rice in a semi-arid region. *Agr. Meteorol. 13*:5–13.

Lenka, D. (1978) Evapotranspiration and crop coefficient in rice. *Indian J. Agron. 23*:351–354.

Low, K. S. & Goh, K. C. (1972) Water balance studies and implications on water resource utilization in West Malaysia. *J. Tropic. Geogr. 35*:60–66.

Maruyama, E. (1967) Rice cultivation and water balance in Thailand. *Geophys. Mag. 33*:337–353.

Meher-Homji, V. M. (1980) Repercussions of deforestation on precipitation in western Karnataka, India. *Archiv fur Meteorologie, Geophysik und Bioklimatologie*, Ser. B, 385–400.

Mizukoshi, M. (1974) Regional distribution and seasonal change in precipitation in variability in monsoon Asia. *Geophys. Mag. 37*:175–184.

Nandakumar, V. (1983) Natural environment and groundwater in the Jaffna Peninsula, Sri Lanka. *In*: M. M. Yoshino, I. Kayane & C. M. Madduma Bandara (eds.) *Climate, Water and Agriculture in Sri Lanka*, Inst. of Geosci., Univ. of Tsukuba, Japan, 155–164.

Nieuwolt, S. (1965) Evaporation and water balance in Malaya. *J. Trop. Geogr. 20*:34–53.

Oldeman, L. R. (1984) Application of agrometeorological information in relation to rice-based cropping patterns in South East Asia. *Proc., Workshop on Need for Climatic and Hydrol. Data in Agric. in South East Asia*, Canberra, Australia, 41–52.

Oldeman, L. R. & Frere, M. (1982) A study of the agroclimatology of the humid tropics of Southeast Asia. World Meteorological Organization *Tech. Notes No. 179*.

Orgill, M. M. (1967) Some aspects of the onset of summer monsoon over Southeast Asia. Dept. Atmospheric Science, Colorado State Univ., Fort Collins, Colorado. Penman, H. L. (1948) Natural evaporation from open water, bare soil and grass. *In: Proc. Roy. Soc.*, Ser. A, *193*:120–145.

Piancharoen, C. (1977) Ground water and land subsidence in Bangkok, Thailand. *In: Proc., Anaheim Symp., Int'l. Assoc. of Hydrol. Sci.* Publ. No. *121*:355–364.

Pickup, G. (1977) Potential and limitation of rainfall-runoff models for prediction on ungaged catchments: A case study from the Papua, New Guinea Highlands. *J. Hydrol. 16*:87–102.

Priestley, C. H. B. & Taylor, R. J. (1972) On the assessment of surface heat flux and evaporation using large-scale parameters. *Mon. Wea. Rev.* *100*:81–92.

Qian, W. (1983) Effects of deforestation on flood characteristics with particular reference to Hainan Island, China. Hydrol. of Humid Tropical Regions, *IAHS Publ. 140*:249–257.

Rasmusson, E. M. & Carpenter, T. H. (1983) Relationship between eastern equatorial Pacific sea surface temperatures and rainfall over India and Sri Lanka. *Mon. Wea. Rev. 111*:517–528.

Rodriguez, F. G. (1958) Problems of the humid tropical Philippines with special reference to water resources development. Problems of Humid Tropical Regions, *UNESCO Humid Tropics Res.* 2:86–99.

Salati, E., Dall'olio, A., Matsui, E. & Gat, J. R. (1979) Recycling of water in the Amazon Basin: An isotope study. *Water Resour. Res.* *15*:1250–1258.

Scarf, F. (1976) Evaporation in peninsular Malaysia. *Water Resources Publ. No. 5.*

Sharma, M. L. (1986) *Role of Groundwater in Urban Water Supplies of Bangkok, Thailand, and Jakarta, Indonesia.* Working Paper, Environ. and Policy Inst., East-West Center, Honolulu, Hawaii, USA.

Shiroma, M. & Chunkao, K. (1980) Some aspects of agricultural water balance in the central part of Thailand. *In: Sci. Bull. of the Coll. of Agric.* Univ. of the Ryukyus, Okinawa, Japan, *27*:183–200.

Sokol, D. (1979) Groundwater development in deltas in monsoon climates. *EOS (Trans. Am. Geophys. Union) 60*(18):252.

Spencer, J. E. & Hale, G. A. (1961) The origin, nature and distribution of agricultural terracing. *Pac. Viewpoint* 2:1–40.

Stol, P. T. & Sybesma, R. P. (1963) Determination of the relationship between precipitation intensity, duration and frequency of occurrence. United Nations Econ. Commission for Asia and the Far East, *Water Resources Series 34*:60–71.

Striffler, W. D., Tejwani, K. G. & Babu, R. (1979) A note on rainfall intensity classes for tropical countries like India. *Ind. J. Soil Conserv.*

7:60–61.

Sugimoto, K. (1971) Plant-water relationship of indica rice in Malaysia. *Tech. Bull.* TARCI.

Thijsse, J. P. (1977) Soil erosion in the humid tropics. *Landbourwkundig Tidschrift* 89:408–411.

Thornthwaite, C. W. (1948) An approach toward a rational classification of climate. *Geogr. Rev.* 38:55–94.

Tomar, V. S. & O'Toole, J. C. (1979) Evapotranspiration from rice fields. *IRRI Res. Paper* Ser. No. *34.*

Tsuchiya, I. (1978) Year-to-year fluctuations of Indian southwest monsoon rainfall, cross-equatorial air flow, and low latitude atmospheric circulation from 1962–1972. *In*: K. Takahashi & M. M. Yoshino (eds.) *Climatic Change and Food Production*, Univ. of Tokyo Press, Japan, 319–328.

Tuller, S. E. (1968) World distribution of mean monthly and annual precipitable water. *Mon. Wea. Rev.* 96:785–797.

UNESCO (1964) Scientific problems of the humid tropical zone deltas and their implications. Proc. of the Dacca Symp., *Humid Tropics Res.*, No. 6.

U.S. National Research Council (1983) *Ecological Aspects of Development in the Humid Tropics.* Nat. Academy Press, Wash. D. C.

Vu, C. Q. (1980) An application of Penman's formula of potential evaporation and Thornthwaite's classification of climate to Suoi-hai Station, a station in midland region in North Vietnam. Misc. Papers, *Centrum voor Agrobiologisch Onderzoek* (Netherlands), No. *240.*

Wiederhold, Th. L. H. von (1957) Verhetering der bandjirafvoeren van het Meratus in Z. O. Kalimanten. (Flood control in the Meratus mountains of Southeast Kalimanten). *Ingenieur in Indonesie 9*(3):62–65.

Wischmeir, W. H. & Smith, D. D. (1958) Rainfall energy and its relationship to soil. *Trans. Am. Geophys. Union 39*:285–291.

Yoshino, M. M. (1984) Ecoclimatic systems and agricultural land use in monsoon Asia. *In*: M. M. Yoshino (ed.) *Climate and Agric. Land Use in Monsoon Asia*, Univ. of Tokyo Press, 81–108.

6: The Hydrology and Water Resources of Humid Northern Australia and Papua New Guinea

B. J. Stewart

Hydrology Branch, Bureau of Meteorology, Melbourne, 3001, Australia

ABSTRACT

This paper reviews the hydrology and water resources of the humid tropical regions of Papua New Guinea and Australia. Topics covered include rainfall, evaporation, runoff, water quality and water management issues.

INTRODUCTION

The UNESCO Colloquium on the Development of Hydrologic and Water Management Strategies in the Humid Tropics adopted the Chang and Lau (1983) definition of humid tropic regions. This definition identifies areas where the mean temperature of the coldest month is above 18˚C and the duration of the wet season exceeds 4.5 months. A wet month has on average more than 100mm of rainfall. When the average monthly rainfall is between 60 and 100 mm, half a wet month is allocated.

Under this definition, approximately 12% of Australia and 93% of Papua New Guinea can be classified as humid tropics. This paper provides a general description of the region and then more detailed information on the rainfall, evaporation, water balance, soil moisture, runoff, water quality, design flood and streamflow estimation procedures.

GENERAL DESCRIPTION OF REGION

Papua New Guinea

Half of the land area of Papua New Guinea is above 1,000 metres, with the highest point 4,500 metres (Fig. 1). The central range of the main island, which runs unbroken from the tip of the southeast peninsula to the border with Irian Jaya in the west, has only two passes lower than 1,500 metres. In places, the cordillera broadens into a series of parallel ridges separated by high, flat, intermontane valleys. Many aspects of the topography indicate geological youth: ungraded rivers, V-shaped

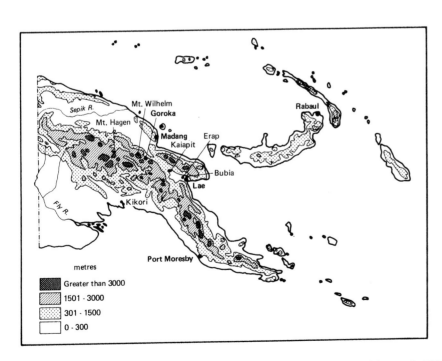

Figure 1: Physiography of Papua New Guinea (Loffler, 1977) (After McAlpine *et al.*, 1983).

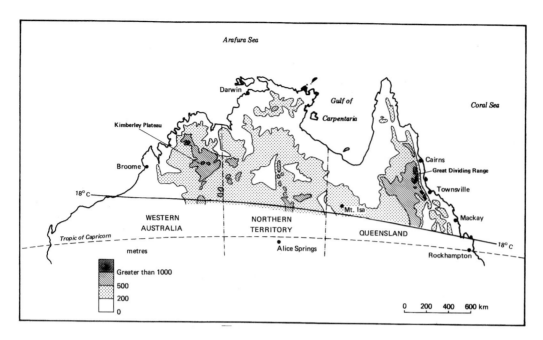

Figure 2: Physiography of Northern Australia.

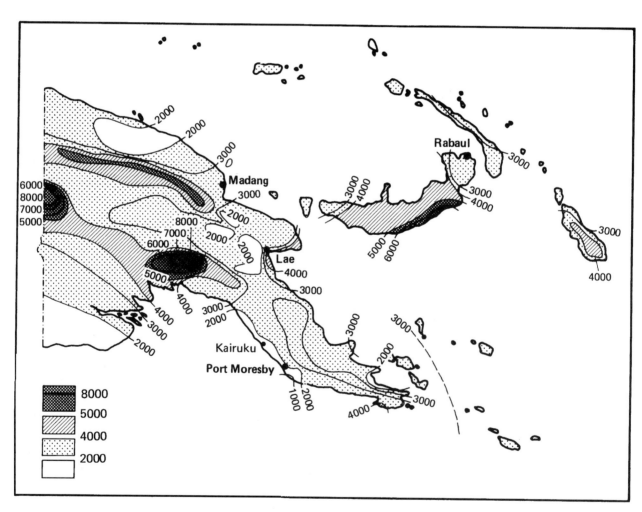

Figure 3: Mean annual rainfall over Papua New Guinea (After McAlpine *et al.*, 1983).

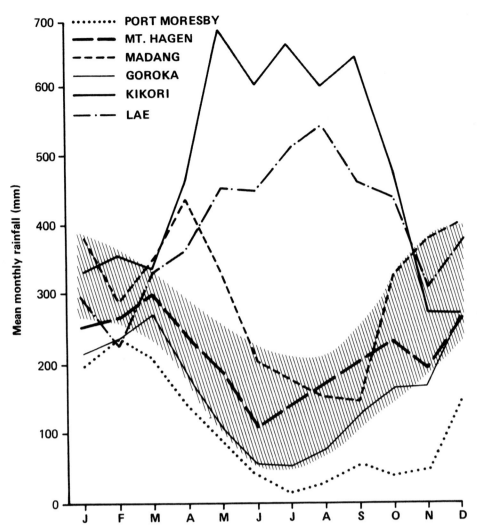

Figure 4: Mean monthly distribution of rainfall for selected stations in Papua New Guinea for a standard period (1956–1970)(After McAlpine *et al.*, 1983).

valleys, waterfalls, cliffs and frequent land slippages (Proctor, 1988).

There is a dense network of rivers and streams that dissects much of the country. Although mountains dominate, lowland (large, seasonally-flooded plains) areas are extensive. The river systems of the Fly and Sepik meander through flood plains which are largely forested. There are significant swamplands of the Fly and Sepik Rivers around the Gulf of Papua.

Northern Australia

In comparison, the relief of northern Australia is relatively low with ranges (where they exist) rising up to only 1,000 m in the Kimberley Plateau in Western Australia (Hall, 1984). There is one exception to this and that is the Great Dividing Range which rises to around 1,600 m in the Atherton Tablelands region on the east coast, near Cairns (Fig. 2).

With the exception of the most northern parts of Australia, north of 12°S, the rainfall, evaporation and runoff of Australia and Papua New Guinea often provide marked contrasts (Hall, 1984). The status of hydrology and water resource management is largely dependent upon the needs and development of

the two countries, which also differ widely in the regions being discussed.

RAINFALL

Papua New Guinea

While the four major rainfall-producing mechanisms (convection, convergence, orographic and cyclonic) in Papua New Guinea can be observed, explanations of how they interact to produce each particular rainfall regime is difficult to assess (McAlpine *et al.*, 1983). The majority of the region has a mean annual rainfall between 2,000 and 4,000 millimetres (Fig. 3) (McAlpine *et al.*, 1983), with the overall range from 900 to 9,000 millimetres. Figure 4 (McAlpine *et al.*, 1983) shows the mean monthly distribution for five stations shown on Fig. 1. Another method of indicating the different rainfall regimes is the index of seasonality. This is determined by dividing the lowest mean monthly rainfall, subtracted from the highest, by the annual mean. Figure 5 is a plot of the index of seasonality for Papua New Guinea. Overall, Papua New Guinea receives a very reliable rainfall.

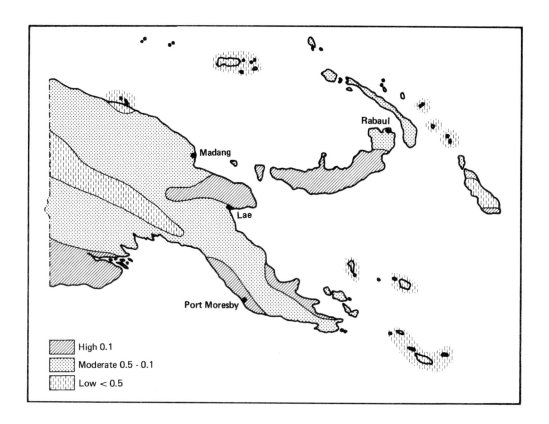

Figure 5: Distribution of the index of seasonality for Papua New Guinea (After McAlpine *et al.*, 1983).

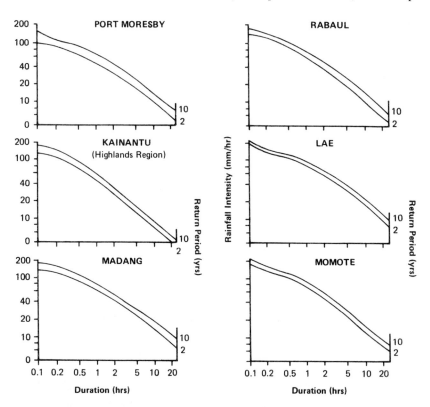

Figure 6: Rainfall intensity-duration-frequency diagrams for selected stations in Papua New Guinea (After McAlpine *et al.*, 1983).

McAlpine *et al.,* (1983) pointed out that records for Papua New Guinea are too short to permit analysis of long-term rainfall fluctuations (especially for the country as a whole). However, Magari (1980) analyzed rainfall data from Port Moresby for the period 1945 to 1976 and concluded that precipitation during the wet season could be increasing.

One-in-two-year daily rainfall intensities for Papua New Guinea range from 80 to 320 millimetres (Snowy Mountains

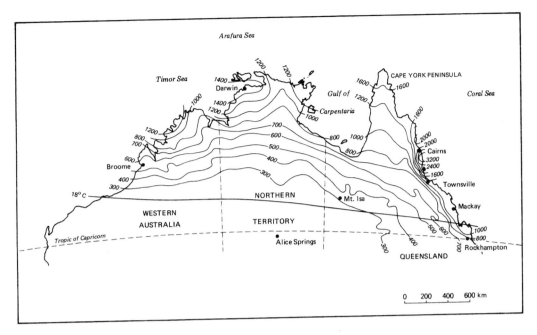

Figure 7: Median annual rainfall over Northern Australia (After Bureau of Meteorology 1973).

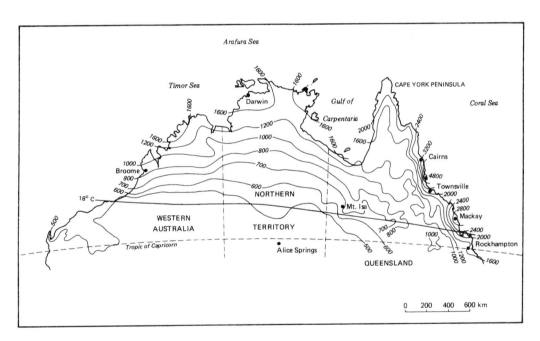

Figure 8: 90th percentile annual rainfall over Northern Australia (After Bureau of Meteorology 1973).

Engineering Corporation, 1973). Rainfall intensity frequency-duration curves have also been published by McAlpine *et al.,* (1975). Figure 6 shows curves for a number of selected stations for return periods of two and ten years.

McAlpine *et al.,* (1983) also provide the results of a range of analyses of Papua New Guinea rainfall data, including diurnal cycles and wet and dry spells.

Northern Australia

Rainfall in northern Australia is markedly seasonal, occurring during the summer monsoon. Figure 7 shows the median annual rainfall for northern Australia. In the area of interest,

the median annual rainfall ranges from 400 mm in the south to 1,600 mm in the north. The highest rainfall areas are associated with the Great Dividing Range in the east, and exceed 3,200 mm along the wet tropical coast, between Cooktown and Townsville (Bonell, 1988). Annual rainfall in this region is highly variable as indicated by the 10 and 90 percentile isolines on Figs. 8 and 9. In the highest rainfall region, the variation is from 2,800 mm to 4,800 mm. One example of the distribution of monthly rainfall is that of Darwin in the Northern Territory (Fig. 10). This displays the seasonal nature of the rainfall, but also indicates the variability through showing the maximum and minimum totals as well.

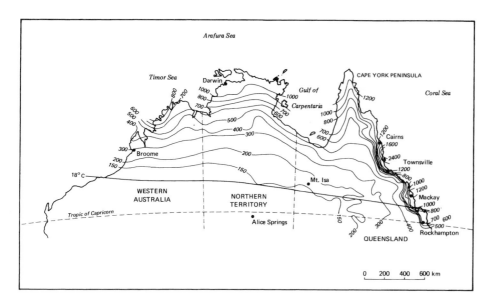

Figure 9: 10th percentile annual rainfall over Northern Australia (After Bureau of Meteorology 1973).

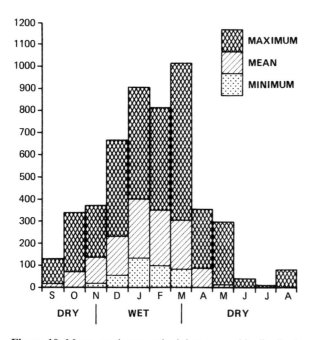

Figure 10: Mean, maximum and minimum monthly distribution of annual rainfall for Darwin, Northern Australia (After Department of Mines and Energy NT 1985).

Rainfall Intensity-Frequency-Duration (IFD) data have been derived for all of Australia (including the humid tropical region) by the Bureau of Meteorology and published in *Australia Rainfall and Runoff* (Institution of Engineers, Australia Aust, 1989). This publication contains IFD data for durations from one to 72 hours and average recurrence intervals from two to 50 years. Methods of extrapolation below one hour and up to the 200 year average recurrence interval are also available (IE Aust, 1989). In contrast to most other humid tropical areas, this is a sophisticated database (see Molion, Lal, this volume). Figures 11(a),(b),(c),(d) and 12(a), (b), (c), (d) show two examples of the data available.

Intensities for short durations are highest around the coastline and decrease towards the centre. The highest intensities tend to be caused by short duration thunderstorm activity. As duration increases, the pattern remains similar. The major exception is the eastern coastline where orographic features lead to higher intensity rainfalls (over the longer durations) around the mountainous regions near Cairns. This causes the maximum to be located farther inland from the coast. Additional factors influencing the pattern of average intensities are the decrease in atmospheric moisture levels on moving inland and the impact of cyclonic activity along the coastline.

Srikanthan & Stewart (1991) have analyzed data from between eight and 11 long-term rainfall stations in the humid tropics area for trend and jump in the mean. The data analyzed include annual, seasonal and monthly rainfall; annual, seasonal and monthly rainday data and annual maximum daily rainfall data. Figures 13, 14 and 15 show the regional mean series for the annual rainfall, annual number of raindays and annual maximum daily rainfall data. The results of statistical tests applied to these regional mean series did not support the hypothesis that the climate had changed recently. Rather, the data showed cyclic variations in mean annual precipitation.

EVAPORATION

Papua New Guinea

The measurement of evaporation in Papua New Guinea can be divided into two groups by time. Prior to 1967, the only evaporation records in Papua New Guinea were from Australian standard sunken tanks at Port Moresby and Bubia and from two lysimeters on Mt. Wilhelm (McAlpine *et al.*, 1983). SMEC (1970) and McAlpine (1970) obtained estimates of Australian standard sunken tank evaporation using Fitzpatrick's (1963) method. This method uses screen temperature and humidity data. Validation of the estimates of evapo-

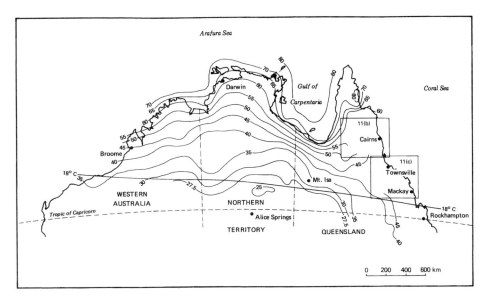

Figure 11(a): Rainfall intensity estimates (mm/hr) for 1 hour duration and 2 year average recurrence interval for Northern Australia (after IE Aust 1989).

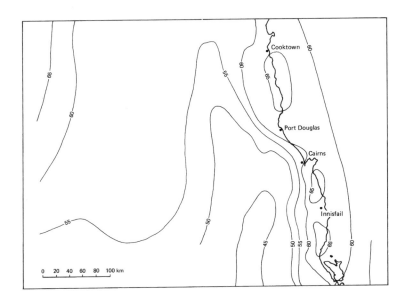

Figure 11(b): Rainfall intensity estimates (mm/hr) for 1 hour duration and 2 year average recurrence interval for Northern Australia (after IE Aust 1989). Inset to FIG. 11(a).

ration by Fitzpatrick's method was only possible at Port Moresby and Bubia, and hence its applicability in other regions is questionable.

A network of seven USA Class A pan evaporimeters, now under the control of the Papua New Guinea National Weather Service, has since been installed. Bougainville Copper Limited (BCL) has also installed three USA Class A pans. Seven of these ten instruments are at coastal locations and three are inland at elevations of 655, 750 and 1,630 metres, respectively. Keig *et al.*,(1979) made a comprehensive, broad-scale study of evaporation rates in Papua New Guinea. By applying Christiansen's (1968) method to mean monthly temperature, relative humidity, sunshine and wind data, estimates of mean monthly and annual USA Class A pan evaporation have been derived at 64 sites. The method was calibrated using the data

recorded at the seven USA Class A pan evaporimeters (Keig *et al.*, 1979).

Figure 16 shows the estimated USA Class A pan evaporation for Papua New Guinea. Rates are highest in lowland areas (>2,000 millimetres) and the inland lowland valley systems. Keig *et al.*,(1979) also demonstrated that USA Class A pan evaporation in Papua New Guinea was approximately equal to evaporation from extensive free water surfaces as estimated by the Penman equation.

Further detail on spatial and temporal variations in evaporation can be found in McAlpine *et al.*,(1983).

Northern Australia

Figure 17 shows the average annual evaporation over northern Australia, based on the standard of a USA Class A pan with

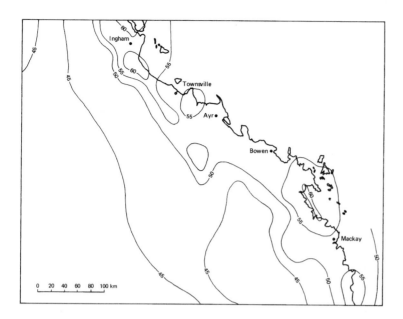

Figure 11(c): Rainfall intensity estimates (mm/hr) for 1 hour duration and 2 year average recurrence interval for Northern Australia (after IE Aust 1989). Inset to FIG. 11(a).

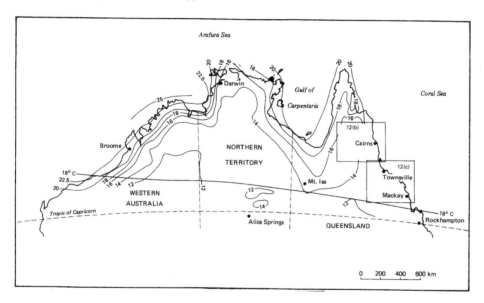

Figure 12(a): Rainfall intensity estimates (mm/hr) for 12 hour duration and 50 year average recurrence interval for Northern Australia (after IE Aust 1989).

birdguard. Mean annual evaporation varies from 3,200 mm in the south to 1,600 mm in the north and is roughly in an inverse relationship with the median annual rainfall. Figure 18 shows the monthly and inter-annual variation of evaporation for Darwin (typical of northern Australia). This indicates that evaporation is not highly variable.

WATER BALANCE

Papua New Guinea

McAlpine *et al.*,(1983) state that a number of studies, using water balance accounting techniques to estimate runoff and soil moisture, have been undertaken in Papua New Guinea for a variety of purposes. Examples include SMEC (1970, 1973)

and Ribeny and Brown (1968) for runoff and flood estimation; Pickup (1976) for daily drainage basin runoff; SMEC (1973) and Bureau of Meteorology (1972) for rainfall deficits; McAlpine (1973) and Fitzpatrick (1963) for soil moisture for drought and plant growth. Finally, Holloway (1973) used water balance accounting to assess the agricultural potential of the Markham Valley.

McAlpine *et al.*,(1983) analyzed data from a number of rainfall stations over Papua New Guinea with the aim of deriving estimates of changes in soil moisture level and estimating seasonal and annual water surpluses and runoff over a 15-year period. Figures 19 and 20 (McAlpine *et al.*,1983) show the mean monthly water balance components for Port Moresby and Madang.

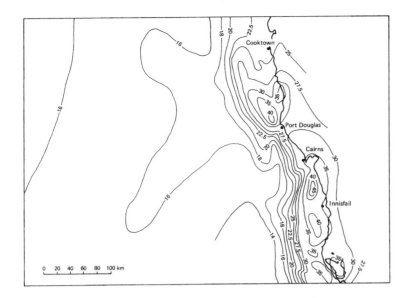

Figure 12(b): Rainfall intensity estimates (mm/hr) for 12 hour duration and 50 year average recurrence interval for Northern Australia (after IE Aust 1989). Inset to FIG. 12(a).

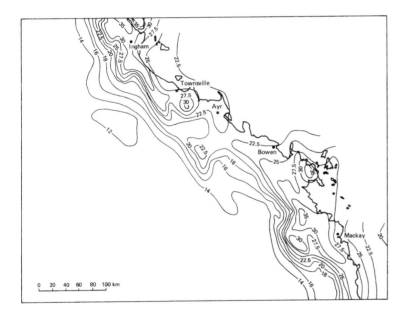

Figure 12(c): Rainfall intensity estimates (mm/hr) for 12 hour duration and 50 year average recurrence interval for Northern Australia (after IE Aust 1989). Inset to FIG. 12(a).

Northern Australia

A water balance study for the humid areas of the Northern Territory (Department of Mines and Energy, 1985) has shown that 75% of rain is lost as evaporation, 15% becomes surface runoff and the remaining 10% of rainfall becomes groundwater recharge. Elsewhere, Gilmour (1975, 1977) provided a water balance for a paired catchment study in the tropical rain forest of northeast Queensland – at Babinda, just south of Cairns. Details are mentioned in Bonell with Balek, (this volume).

SOIL MOISTURE

Papua New Guinea

Curves showing estimated mean levels of soil moisture have

been derived by McAlpine *et al.,*(1983) for a number of representative stations in Papua New Guinea (Fig. 21). This figure shows the distinct differences between the wetter and drier regions and the seasonality of soil moisture levels.

Figure 22 shows the frequency distribution of <150, <100, <50 and 0 millimetre levels of soil moisture for a number of stations in Papua New Guinea. The figures indicate that significant levels of soil moisture depletion (or drought) are both rare and brief for most of the country. The major exception is in the vicinity of Port Moresby where serious droughts occur regularly and irrigation is seen as essential to ensure agricultural production during the dry season. The limitation of this study is that a maximum soil moisture capacity of 150 millimetres is assumed. If the actual soil moisture capacity is less

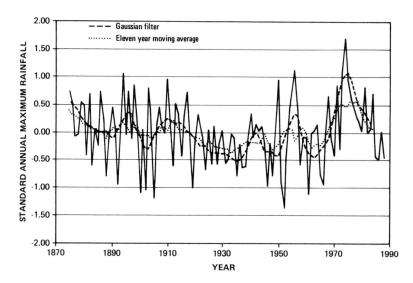

Figure 13: Standardized annual rainfall time series for the summer tropical rainfall region with eleven-year moving average (...) and gaussian filter (---) (After Srikanthan & Stewart 1991).

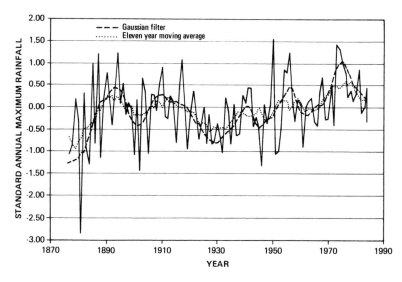

Figure 14: Standarized annual number of raindays time series for the summer tropical rainfall region with eleven-year moving average (....)and gaussian filter (---)(After Srikanthan & Stewart 1991)

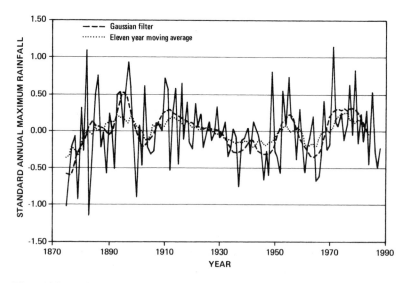

Figure 15: Standarized annual maximum daily rainfall time series for the summer tropical rainfall region with eleven-year moving average (....) and gaussian filter (---) (After Srikanthan & Stewart 1991).

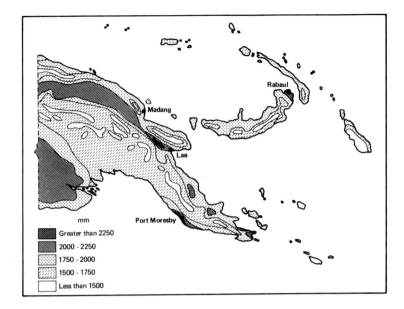

Figure 16: Estimated annual US Class A Pan evaporation (mm) for Papua New Guinea (After McAlpine *et al.*, 1983).

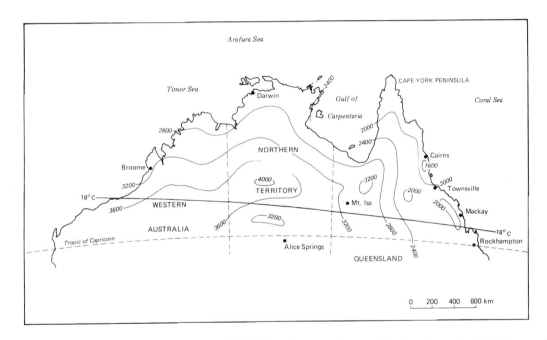

Figure 17: Mean annual US Class A Pan (with birdguard) evaporation over Northern Australia (After Bureau of Meteorology 1973).

than this upper limit (e.g., steeper slopes, poorer soils), drought occurrences would be more frequent and over longer durations (McAlpine *et al.*, 1983).

Northern Australia

Very little regional scale information is available on soil moisture levels for northern Australia. Bonell (1988) described a number of field studies undertaken in northeastern Queensland which, while principally analyzing storm runoff situations, provide some insight into the properties of the soils in that region. Bonell (1988) noted that antecedent soil moisture is high during the wet season due to the frequency of rain events and can range up to 1.2 to 1.5 metres equivalent depth of rain-

fall stored in the top three metres of soil in tropical rain forest soils. Additional information is also available in Fleming (this volume) where the runoff processes are discussed in more detail.

RUNOFF

Papua New Guinea

The Papua New Guinea Bureau of Water Resources currently operates some 93 recording streamflow discharge stations. They also hold processed data from 143 stations, totalling some 1,600 years. SMEC (1970, 1973) estimated point mean annual runoff for a large number of rainfall stations from

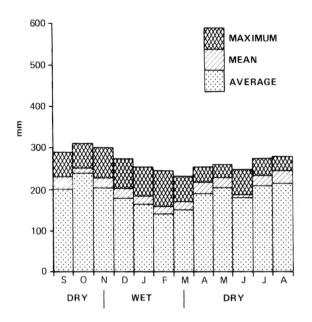

Figure 18: Mean, maximum and minimum monthly distribution of annual evaporation for Darwin, Northern Australia (After Department of Mines and Energy NT 1985).

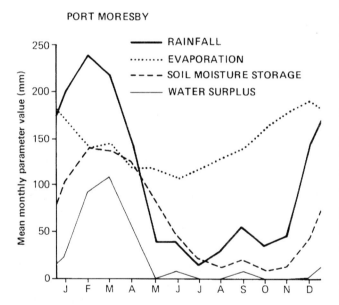

Figure 19: Mean monthly water balance component values for Port Moresby (After McAlpine *et al.*, 1983).

mean annual rainfall. These estimates were validated against recorded discharge data. A map of mean annual runoff was prepared for Papua New Guinea (Fig. 23). From this figure it can be seen that the runoff ranges from 500 to 6,000 millimetres. The mean annual runoff for Papua New Guinea has been estimated at 2,100 millimetres (Hall,1984). The runoff coefficients for Papua New Guinea range from 0.25 to 0.75.

Northern Australia

Hall (1984) states that the average annual rainfall of humid tropical Australia (above the 20°C line) is 220 millimetres. The

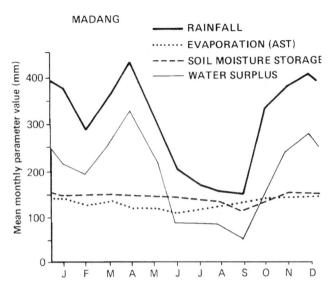

Figure 20: Mean monthly water balance component values for Madang (After McAlpine *et al.*, 1983)

variations in mean annual runoff are shown on Fig.24 (Brown, 1983). They range from over 1,000 millimetres on the east coast to under 25 millimetres inland. Comparisons of rainfall with runoff suggest runoff coefficients ranging from 0.60 on the coast to below 0.10 inland. The variability of annual runoff in northern Australia is greater than in other humid tropical regions (McMahon, 1978).

WATER QUALITY

Papua New Guinea

There is a lack of water quality data on Papua New Guinea. This situation is currently being addressed by a Bureau of Water Resources project aimed at the collection of water samples at selected sites on a regular basis.for sediment and quality analysis. Site-specific (mining, hydro) water quality analyses are being carried out.

The major water quality issue in Papua New Guinea is erosion/sedimentation. Problems of sediment/pollution caused by release and escape of waste from mining operations have been investigated. Methods of investigation included regression approaches, using channel parameters, e.g., width/depth and weighted mean silt-clay index of bed material. These proved unsatisfactory in reflecting the characteristics of the sediment load in the Fly River (Hall,1984). The regime theory-dominant discharge approach of Pickup (1985), used to predict the response of the Fly and Purari Rivers to change in sediment, has been more successful. Traveling dispersing wave models (Pickup *et al.,*1983) were found to be the most appropriate in modeling of the Kawerong Basin of Bougainville.

Northern Australia

On the east coast of northern Australia, the principal water quality issues are related to point sources of pollution. Some

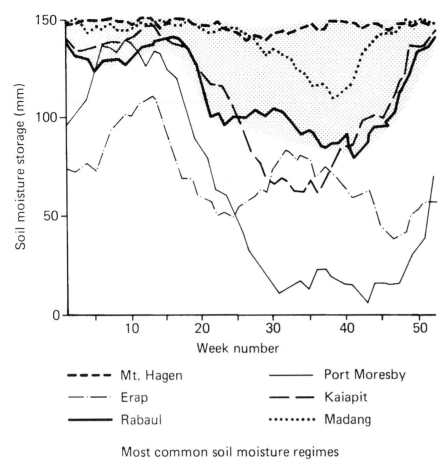

Figure 21: Mean weekly soil moisture storage for a range of stations. The hatched area indicates the range in which the most common and widespread soil moisture regimes occur (After McAlpine *et al.*, 1983).

examples include elevated concentrations of copper and zinc found in oyster flesh near Cairns, waste disposal at Innisfail, and high turbidity levels in the Herbert River, assumed to be caused by tin mining and sewage disposal at a number of sites. There has also been increasing publicity directed towards the effects of agricultural chemicals on the Great Barrier Reef – being transported to the Reef by the major rivers draining sugar-cane lands. For example, the major river basin on the northeast coast of the humid tropical region is the Burdekin River Basin. This basin contains a significant amount of beef cattle industry and irrigation for sugar-cane, rice and other intensively-raised agricultural products. Turbidity is a major water quality factor, especially from severe land degradation, resulting from over-grazing by beef cattle. The effects are evident, not only in the river and estuary, but also in the region between the Great Barrier Reef and the mainland (Garman, 1983). Coal mining has also led to some local pollution problems.

Around the Gulf of Carpentaria, no major pollution problems have been reported (Garman, 1983). The region contains a number of major mining activities including Mt. Isa (iron ore), Weipa (bauxite), Gove (bauxite) and Mary Kathleen (uranium). These have resulted in localized, but reasonably-contained instances of pollution.

Instances of pollution in river basins in the Northern Territory and northeastern Western Australia are examples of the problems associated with development in humid tropical regions. Development of irrigation in the Ord River Basin introduced significant amounts of chlorinated hydrocarbons into the environment, mainly to control pests associated with cotton growing. Investigations of concentrations of pesticide residues in water, sediment, fish and animals have been undertaken (Garman, 1983). The cessation of cotton growing virtually eliminated the continued input of pesticides into the region. However, residues were still present in some fish after a few years. Uranium mining at the Rum Jungle site on the East Finniss River resulted in heavy metal pollution of the river and its environs. The entire mine site has been rehabilitated and is monitored on a regular basis. In general, uranium mining is undertaken in tightly controlled regions in the Northern Territory (Supervising Scientist for the Alligator Rivers Region, 1983 and Department of Transport and Works, 1979). Continued environmental monitoring has been initiated by the Australian Government to ensure that environmental safeguards are met. Disposal of sewerage has been identified as a potential source of pollution at large settlements throughout the northern region of the Northern Territory.

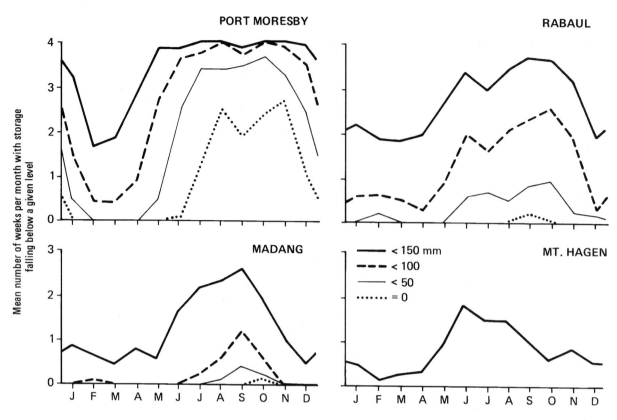

Figure 22: Frequency distribution of specified levels of soil moisture for a range of typical stations (After McAlpine *et al.*, 1983).

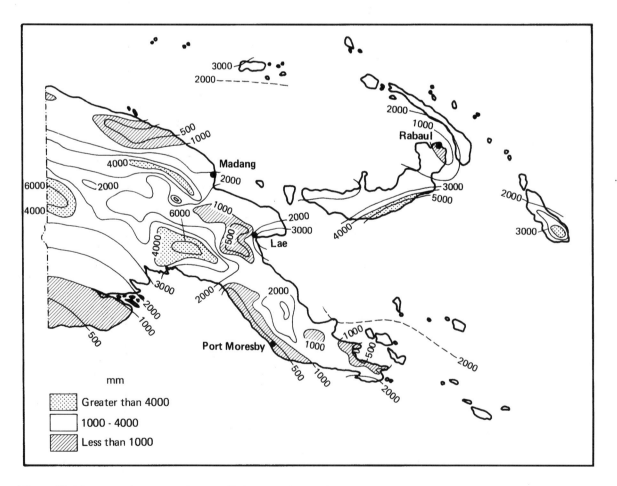

Figure 23: Mean annual water surplus (runoff) for Papua New Guinea (mm) (After McAlpine *et al.*, 1983).

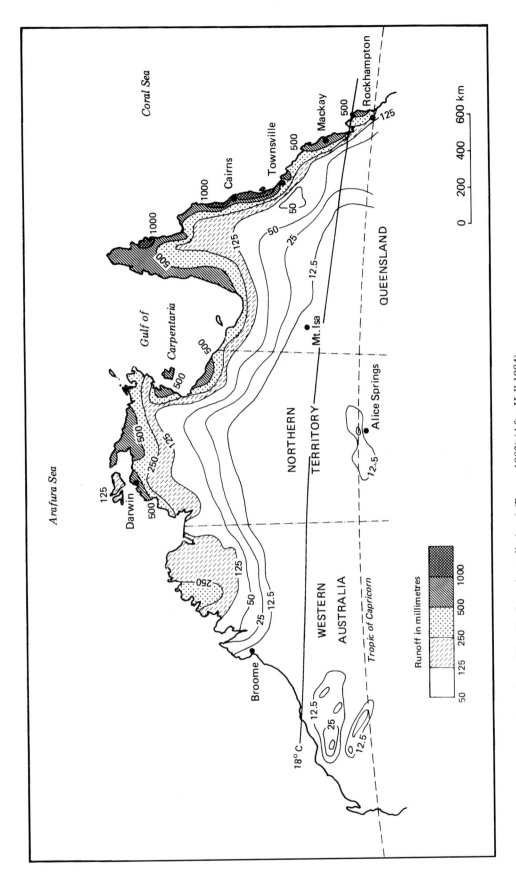

Figure 24: Observed and estimated runoff for Northern Australia (mm) (Brown 1983) (After Hall 1984).

DESIGN FLOOD AND STREAMFLOW ESTIMATION PROCEDURES

Papua New Guinea

The Snowy Mountains Engineering Corporation (1973) prepared a design flood estimation manual for Papua New Guinea. The log-normal distribution was used to determine intensity-frequency-duration data (see section on rainfall). Unit hydrographs were then used to estimate the design floods. Because of the large spatial variation in basin rainfall, the general accuracy of unit hydrograph procedures in Papua New Guinea is low. Atkins (1980) expanded on the work of Pilgrim (1972) in developing regional flood estimation techniques (peak flow *vs* catchment area) based on the results of flood frequency analysis. The log-normal distribution was selected as the best frequency distribution for Papua New Guinea streams.

Simple conceptual rainfall-runoff models and more complex models, such as the Sacramento Model (Burnash *et al.*, 1973), have been applied to catchments in Papua New Guinea, with varying degrees of success (Hall, 1984). Streamflow generation methods have also been applied in Papua New Guinea. Five different methods were used by SMEC (1977) to generate 500 years of synthetic streamflow.

Northern Australia

Australian Rainfall and Runoff (Institution of Engineers, Australia, 1989) provides methods for the estimation of design floods for all of Australia, including Northern Australia. These methods range from the simple rational method to regional parameters for runoff routing methods such as RORB (a nonlinear runoff routing method developed at Monash University in Melbourne (Mein *et al.*, 1974) and unit hydrographs. Design rainfall data are discussed in the section on rainfall.

Simple conceptual rainfall-runoff models such as the United States Department of Agriculture Model and the Boughton (1966) Model and more complex models such as the Sacramento Model have been applied to catchments in Northern Australia, but with varying degrees of success. The gamma distribution and Weibull distribution have been found suitable for flow generation models (McMahon, 1977). Gould's (gamma) procedure (Gould, 1961) of storage yield analysis has been applied with success to streamflow datasets in Northern Australia.

SUMMARY/CONCLUSION

There are distinct differences between the climatology and hydrology of Papua New Guinea and of Australia. Papua New Guinea has a comparatively uniform rainfall throughout the year which results in reliable annual streamflow by world standards (Hall, 1984). Humid tropical Australia's rainfall occurs principally in the summer months and is less reliable. Australia's streamflow is considerably more variable than those of the world's other rivers. Therefore, a shorter period of observation is necessary for an appreciation of the hydrology of Papua New Guinea, compared to Australia's.

While water resources data collection networks are not within the desirable standards set by the World Meteorological Organization, there is sufficient information available to gain an understanding of the hydrology of both regions. Design rainfall data and design flood techniques have been derived for both regions. Information is lacking in terms of the hydrological processes in the varying hydrological regimes, in different aspects of water quality, and in the links between the two systematic areas.

REFERENCES

Atkins, G. P. (1980) Regional flood frequency analysis in Papua New Guinea. Ph.D Thesis, Papua New Guinea Univ. of Technology, Lae.

Bonell, M. (1988) Hydrological processes and implications for land management in forests and agricultural areas in the wet tropical coast of northeast Queensland. In: R. F. Warner (ed.) *Fluvial Geomorphology of Australia,* Academic Press, 41–68.

Boughton, W. C. (1966) A mathematical model for relating rainfall to runoff with daily data. *Civ. Engng. Trans., Instn. Engrs., Aust., CE24 (2):* 127–134.

Brown, J. A. H. (1983) *Australia's Water Resources: Water 2000.* Consultants Report No. 1, Dept. of Resources and Energy, AGPS, Canberra.

Burnash, R. J. C., Ferral, R. L. and McGuire, R. A. (1973) *A Generalized Streamflow Simulation System: Conceptual Modeling for Digital Computers.* USA Dept of Commerce, Nat. Weather Service in cooperation with the California Dept.of Water Resources.

Bureau of Meteorology (1972) Drought in Papua New Guinea Highlands during the period June to September 1972. Working Paper No. 161, 40/169 of Dec. 1972. Bureau of Meteorology, Melbourne.

Chang, J.-h. and Lau, L. S. (1983) "Definition of the humid tropics." Paper presented to the IAHS meeting in Hamburg, Germany.

Christiansen, J. E. (1968) Pan evaporation and evapotranspiration from climatic data. *J. Irrig. Drainage* Div., *ASCE,* 94: 24–65

Department of Mines and Energy, Northern Territory (1985) *Water, Northern Territory, 1,* Water Resources Division, Dept. Mines and Energy.

Department of Transport and Works (1979) *Uranium Province Hydrology,* Vol. 1, Water Division, Dept. T & W, Northern Territory, February 1979.

Fitzpatrick, E. A. (1963) Estimates of pan evaporation from mean maximum temperature and vapour pressure. *J. Appl. Met. 2:* 280–92.

Garman, D. E. J. (1983) *Water Quality Issues: Water 2000.* Consultants Report No. 7, Dept. of Resources and Energy, AGPS, Canberra.

Gilmour, D. A. (1975) Catchment water balance studies on the wet tropical coast of north Queensland. Unpublished Ph. D. Thesis, Department of Geography, James Cook University of North Queensland, Townsville.

Gilmour, D. A. (1977) Effects of logging and clearing on water yield and quality in a high rainfall zone of Northeast Queensland. In: The Hydrology of Northern Australia (Proc. Brisbane Hydrol. Symp., June l977) Instit. Eng., Australia, Nat. Conf. Publ. 77/5, 156–160.

Gould, B. W. (1961) Statistical methods for estimating the design capacity of dams. *J. Instn. Engrs., Aust.33 (12),* 406–416.

Hall, A. J. (1984) Hydrology in tropical Australia and Papua New Guinea, *J. Hydrological Sciences, 29, 4,* 12/1984, 399–423.

Institution of Engineers, Australia (1989) *Australian Rainfall and Runoff,* A Guide to Flood Estimation, IE Aust., Canberra, 1989, (2 vols.).

Holloway, C. E. (1973) Drainage requirements in the Markham Valley *Papua New Guinea Agric., 24*: 119–30.

Keig, G., Fleming, P. M. and McAlpine, J. R. (1979) . Evaporation in Papua New Guinea. *Trop.Geog. 48*: 19–30.

Magari, K. (1980) Rainfall trend at Port Moresby from 1945 to 1976, *Weather,* 35: 110–17.

McAlpine, J. R. (1970) Climate of the Goroka-Mt. Hagan area. CSIRO Aust Land Res Ser. No. 27: 66–78.

McAlpine, J. R. (1973) A climatic classification for eastern Papua. CSIRO Aust. Land Res. Ser. No.32: 50–61.

McAlpine, J. R., Keig, G. and Short, K. (1975) Climatic tables for Papua New Guinea. CSIRO Aust. Div. Land Use Res. Tech. Paper No. 37.

McAlpine, J. R. and Keig, G. with Falls, R. (1983) *Climate of Papua New Guinea,* Australian National University Press, Canberra.

McMahon, T. A. (1977) Some characteristics of annual streamflows in northern Australia. In: The Hydrology of Northern Australia (Proc. Brisbane Hydrol. Symp., June l977) Instit. Eng., Australia, Nat. Conf. Publ. 77/5, 131–135.

McMahon, T. A. (1978) Australia's surface water resources: potential developments based on hydrologic factors. *Civ.Engng. Trans., Instn. Engrs., Aust., CE20* (2): 155–164.

Mein, R. C., Laurenson, E. M. and McMahon, T. A. (1974) Simple Nonlinear Model for Flood Estimation, *J. Hyd. Div. ASCE, 100,* HY11, 1507–1518.

Pickup, G. (1976) A self calibrating model for the simulation of daily runoff for humid tropical drainage basins. Univ. Papua New Guinea, Dept. Geog. Occas. paper No. 14.

Pickup, G. (1985) Geomorphology of tropical rivers: I, Landforms, hydrology and sedimentation in the Fly and lower Purari, Papua New Guinea, *Catena.*

Pickup, G., Higgens, R. J. and Grant, I. (1983) Modelling sediment transport as a moving wave – the transfer and deposition of mining waste. *J. Hydrol. 60*: 281–301.

Pilgrim, D. H. (1972) Flood estimation for road design in developing regions with particular reference to Papua New Guinea. *Civ. Engng. Trans., Instn. Engrs. Aust.* CE14(1): 42–48.

Proctor, P. D. (1988) "Country Report for Papua New Guinea." World Meteorological Organization, Regional Association V, Hydrology Working Group, May 1988.

Ribeny, F. M. J. and Brown, J. A. H. (1968) The application of a rainfall-runoff model to a wet tropical catchment. *Civ. Eng. Trans. Inst. Engs. Aust.,* CE10: 65–72.

Snowy Mountains Engineering Corporation (1970) *Assessment of Runoff and Hydro-Electric Potential - Territory of Papua New Guinea.* Report prepared for the Department of External Territories.

Snowy Mountains Engineering Corporation (1973) *Investigation of Flood Estimation Procedures for Papua New Guinea.* (2 vols.). Department of Public Works.

Snowy Mountains Engineering Corporation with Nippon Koei Co. Ltd and others (1977) *Purari River, Wabo Power Project, Feasibility Report.* Prepared on behalf of Dept. of Overseas Trade, Australia.

Srikanthan, R. and Stewart, B. J. (1991) Analysis of Australian Rainfall with respect to Climate Variability and Change, *Aust. Met. Mag.,* March 1991.

Supervising Scientist for the Alligator Rivers Region (1983) *Environmental Protection in the Alligator Rivers Region,* Scientific Workshop, Jabiru, May 1983, 2 vols.

7: Hydrology and Water Resources of Tropical Latin America and the Caribbean

B. Griesinger

Executive Secretary, Brazilian National Committee for the International Hydrological Programme, Brazilia, Brazil; Head of Cabinet, Brazilian Institute of Environment.

J.S. Gladwell

President, Hydro Tech International, Consultants, Vancouver, British Columbia, Canada; formerly Project Officer, Humid Tropics Programme of the International Hydrological Programme, UNESCO, Paris, France.

ABSTRACT

This chapter reviews briefly the hydrology and water resources of the humid tropic regions of Latin America and the Caribbean, following generally the definition of Chang and Lau (see the appendix to this book). For each subregion (Central America, Caribbean Islands and South America) descriptions of the climate, geomorphology and surface and groundwaters are described. It is concluded with a call for additional detailed assessments of the water resources of each country.

INTRODUCTION

The region under discussion in this section is that of the humid tropics as defined by Chang & Lau (1983). Figure 1 shows the areas of tropical Mexico, Central America, the Caribbean and tropical South America that are of concern.

The tropical regions of Latin America and the Caribbean are probably the most humid of the world. South America, for example, has some of the highest rainfall and runoff of any continent. Because most of the rivers of the region are rainfed, the seasonal distribution of the discharges is often directly related to the rainfall distribution. In the northern hemisphere, the incidence of tropical hurricanes can confuse the regionality of the rainfall amounts because of the non-uniformity with which the islands and mainland are impacted by these cyclonic events. Throughout the mainland of Latin America, the major physiographic factor affecting the water resources is the mountain chain that runs all the way from Mexico to Chile. This causes the rivers generally to be longer and larger on the Atlantic side (and contain some 84% of the land area), and shorter and with much higher gradients on the Pacific side. Some notable exceptions to this rule are to be found in Brazil and northern Bolivia and Paraguay where the Sao Francisco, Parana/Plata and Paraguay Basins have their headwaters in the central plateau of Brazil.

Because of their geographic location, the prevailing easterly and northeasterly winds are warm and humid, thus causing the typically humid conditions. The Caribbean Sea and the Gulf of Mexico, being relatively slow-moving, also become much warmer and contribute further to the humidity.

It is not possible in so short a discussion to be complete in the presentation of the hydrology and water resources of the region. But we have attempted to present a summary overview of the situation, noting the outstanding features of each of the subregions. Elsewhere in this volume, Molion provides more details, focusing on the Amazon Basin

Much remains to be done to understand more fully the hydrologic and water resources problems and issues of this region. But, for the moment, the uses of the waters are both widely diverse and highly concentrated, mainly along the coastal areas. The various water resource uses have not yet had major impacts on the larger river systems, except primarily the response to regulation of the rivers for irrigation and hydroelectric power development. Such regulation has really only become a feature of the last half century. Environmental pollution and water quality issues are emerging, however, with major cities causing heavy pollution of rivers and some of the coastal areas.

But, major land-use changes now occurring in the region suggest that in the future the impacts on the waters of the region will intensify, and become a serious concern for forestry and agriculture. The greatly increased urbanization – especially the explosion of extremely large urban areas – to one extent or another, affects the quantities of water available. But perhaps the greatest impact is, and will continue to be, the effect on the water quality. This will continue to increase in terms of ground, surface and coastal waters.

The tables contain data from selected precipitation gauges and of river discharges at selected sites.

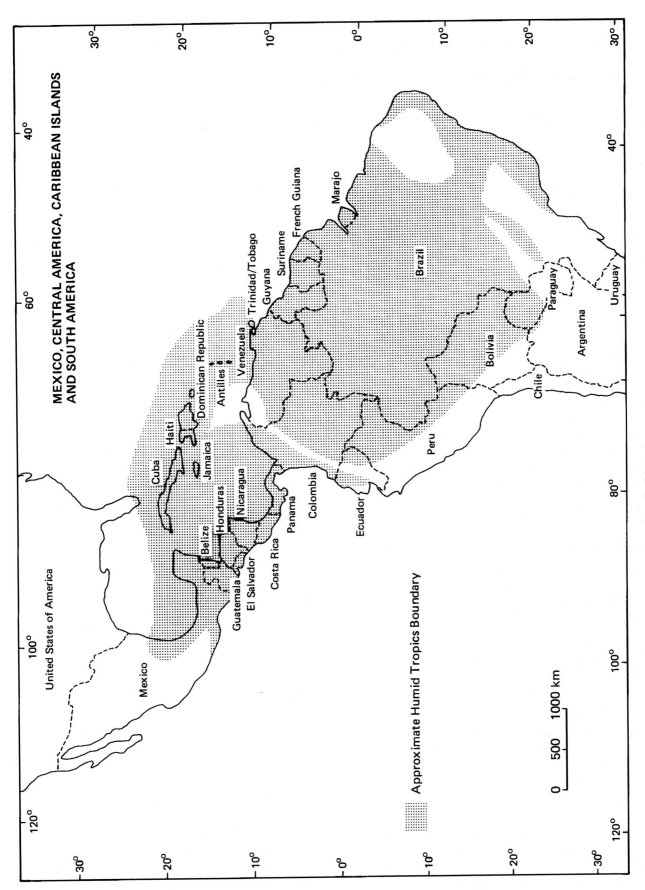

Figure 1: Approximate region of the humid tropics in Latin America and the Caribbean

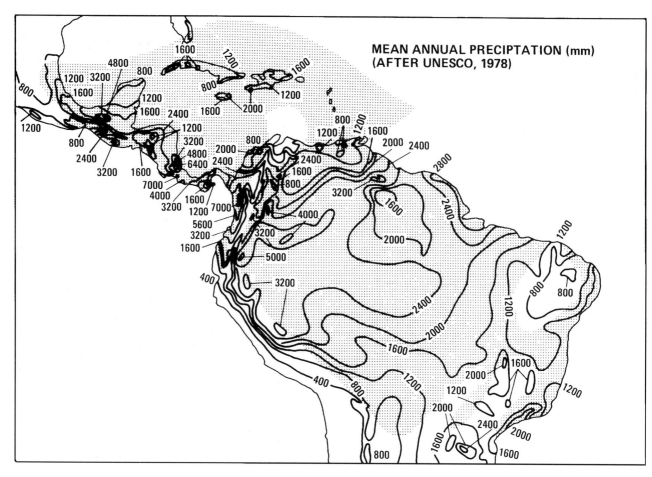

Figure 2: Mean annual precipitation, Latin America and the Caribbean (after UNESCO, l978)

THE CARIBBEAN ISLANDS, TROPICAL MEXICO AND CENTRAL AMERICA

This subregion is located between the subtropical high pressure cells of the Atlantic and the low pressures of the equatorial region for the entire year. As a result, the generally prevailing winds are the easterlies and northeasterlies, which bring moisture-laden maritime tropical air, particularly in the months of May through July when the most concentrated convective activity (see Fig. 2 of Molion, this volume) is farthest north of the equator. Figure 2 shows the regional distribution of the annual precipitation, and Fig. 3, the distribution of the annual evapotranspiration.

The Caribbean Islands

The climate Except for the larger ones, most of the islands of this area are too small to cause very much local climatic modification. The larger ones, however, can have land and sea breezes. Jamaica, Cuba and the Bahamas often have cool winds from the north. In the Caribbean island area in general, no month is rainless but the winter months are noticeably lower in precipitation than the summer months. While throughout the Caribbean, including the eastern coast of Central America, there is a tendency for there to be two maxima of precipitation, the northern islands show a pronounced

effect. Generally the summer months of May to July and the fall months of September to November show the maxima (see Table 1).

Rainfall in the islands is quite variable. For example, the western slopes of Guadeloupe can receive as much as 2,500 mm per year or more, while for the remainder of the island, the average annual rainfall is more like 1,400 mm. Similarly, sharp contrasts are also observed in the Greater Antilles. Most of the region receives between 1,000 and 2,000 mm of rainfall per year. The lowest values are found in the region of the Bahamas.

The late summer months and through October reflect the increasing frequency of disturbances in the easterlies (see Manton and Bonell, this volume) During this period, hurricanes are also likely to develop and can cause devastating consequences in the region, particularly in the island areas (see Fig. 4). The irregularity of the hurricanes can cause a great variability in the annual totals of rainfall.

The geomorphology and groundwater availability The Caribbean region contains several thousand islands which extend as an arc beginning at the western end of Cuba, between Florida (USA) and the Yucatan Peninsula (Mexico) in Central America, and terminating to the south-east along the coast of Venezuela. These islands are surrounded by deep seas

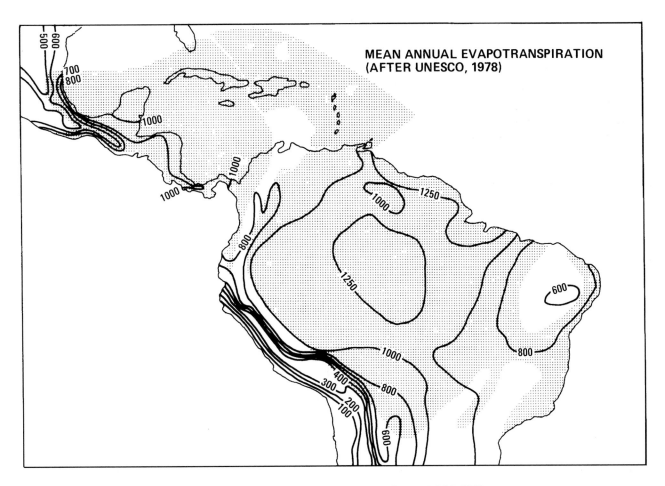

Figure 3: Mean annual evapotranspiration, Latin America and the Caribbean (after UNESCO, 1978)

– the Caribbean Sea to the south and west (4,000 m) and the Atlantic Ocean to the east (8,000 m). The islands are of various sizes. The major islands are to the northwest. The Greater Antilles include Jamaica (10 900 km²), Puerto Rico (8,800 km²) and some smaller islands, such as the Virgin Islands. The Bahama Islands are scattered to the north of Cuba.

The eastern Lesser Antilles extend north-south between Puerto Rico and the delta of the Orinoco River in Venezuela. The largest island is Trinidad (4,800 km²). The total area of the other islands is approximately 15 000 km².

North of the Venezuelan coast, the southern Lesser Antilles include to the east the Venezuelan Lesser Antilles and to the west the Netherlands Lesser Antilles.

The entire Caribbean island arc is, in fact, a partially submerged mountain chain. The highest peak is found in the Dominican Republic (3,175 m). Geological features vary widely in this vast region. The nucleus of the Greater Antilles contains igneous rocks which form the basal complex; this complex, in turn, is overlaid by sediments of marine detritic and volcanic origin (Puerto Rico and Virgin Islands). Sediments, mainly limestones, sometimes folded, cover large areas and formations of alluvium, sandstone and reef limestone are abundant in coastal areas.

Similar complex geological conditions are found in Trinidad and Tobago. Some islands are almost exclusively limestone. This is the case for Anguilla, Antigua, the Bahamas, Barbados, Barbuda, Bonaire, the Cayman Islands, Desirade, Grande Terre, Marie Galante, St. Martin and the Venezuelan Antilles. Other islands are underlaid mainly by volcanic formations. Examples are Aruba, Basse Terre, Curacao, Grenada, the Grenadines, Montserrat, Nevis, Redonda, Saba, St. Christophe, St. Eustache and St. Vincent.

In the major islands, groundwater is relatively abundant in the alluvium of river valleys and in permeable limestones and volcanic rocks. In small volcanic islands, most of the rainfall is lost through runoff to the sea. The water-yielding properties of some volcanic rocks range from mediocre to poor. Later volcanics can be good aquifers in some areas; but, if they are very pervious, rainwater infiltrates and drains rapidly to the sea.

In islands mainly underlaid by carbonate rocks, substantial amounts of rainfall infiltrate but most of it flows laterally to the sea. Sea-water intrusion can develop rapidly under the influence of pumping. Fresh water is present in the form of a relatively thin lens. Water levels in wells are practically at sea level several kilometres inland.

Many of the small Caribbean Islands are densely populated and have developed tourist facilities which consume large quantities of water. As groundwater is practically the only water resource, it has great economical and social value. Considering that it is in limited supply, it must be managed

TABLE 1. *Distribution of mean monthly precipitation at selected cities of the tropics of Latin America and the Caribbean.*
(Primary source: ECLAC, 1980) Note: peaks are underscored.

City	J	F	M	A	M	J	J	A	S	O	N	D	(MAP)
Mexico													
Campeche	17	13	13	10	61	152	206	201	221	119	54	27	(1094)
Merida	31	16	19	26	81	151	141	129	154	103	32	30	(913)
Orizaba	42	34	34	41	119	396	422	354	357	200	76	43	(2118)
San Cristobal de las Casas	8	8	18	57	116	241	166	175	242	122	35	13	(1201)
Soto La Marina	10	6	7	20	32	79	63	78	104	98	32	15	(641)
Tampico	20	14	14	18	36	153	128	104	271	127	51	36	(972)
Central America													
Balboa Hts, Pan.	23	18	18	76	201	208	185	201	206	257	269	109	(1771) **
Belice City, Bel.	136	63	38	51	105	205	163	168	235	307	209	185	(1866)
Colon, Pan.	86	38	38	104	320	353	396	386	318	394	569	295	(3297) **
Managua, Nic. Hon.	3	1	3	11	147	211	136	110	216	292	44	10	(1184)
San Jose, C.R.	11	4	13	44	221	269	213	240	325	332	149	44	(1868)
San Jose, El Sal.	—	1	6	27	107	283	227	223	274	245	38	4	(1435)
San Salvador, El Sal.	6	5	9	54	187	318	307	297	314	235	38	10	(1781)
Santiago, Pan.	33	22	20	116	344	315	255	372	377	465	269	71	(2659)
Tegucigalpa, Hond.	11	5	9	29	152	164	88	95	182	133	39	13	(920)
Caribbean Islands													
Bridgetown, Bar.	81	53	48	41	64	109	127	83	193	193	168	104	(1364) **
Camaguey, Cuba	30	37	55	86	180	230	134	170	187	170	69	37	(1385)
Havana, Cuba	62	39	50	52	104	146	105	108	150	184	84	52	(1136)
Kingston, Ja.	22	20	22	30	86	89	43	102	100	174	79	33	(800)
Nassau, Bah.	42	40	35	63	126	176	151	159	183	168	70	39	(1252)
Port au Prince, Haiti	32	53	83	162	229	99	74	146	173	168	85	36	(1340)
Port of Spain, Trin. & Tobag	69	38	46	48	89	196	224	244	188	170	180	122	(1613)**
Roseau, Dom.	130	81	75	71	101	196	275	251	228	201	205	155	(1970)
Saint Clair, TT	67	40	45	53	93	193	216	247	193	170	183	126	(1627)
San Juan, PR	106	69	69	101	159	139	150	158	155	138	162	131	(1531)
Santiago, D.R.	60	48	42	77	180	67	51	64	100	95	98	70	(950)
South America													
Aracaju, Bra.	43	97	137	182	282	211	169	100	95	61	41	52	(1470)
Asuncion, Par.	159	134	174	154	93	79	48	35	75	110	1390	116	(1316)
Barranquilla, Col.	1	0	1	11	87	103	54	102	138	202	82	65	(846)
Belem, Brazil	318	407	436	382	265	165	162	116	120	105	90	197	(2762)
Boa Vista, Bra.	40	50	100	151	304	365	346	226	110	74	66	69	(1901)
Bogota, Col.	52	55	88	124	108	58	49	50	58	151	122	74	(989)
Brasilia, Bra.	277	208	173	117	49	9	8	7	54	159	251	232	(1544)
Caetite, Bra.	134	90	99	71	27	25	14	12	12	58	182	176	(900)
Caracas, Ven.	21	11	12	36	78	103	102	109	99	108	90	47	(816)
Caucagua, Ven.	147	65	55	76	176	264	321	256	169	157	207	261	(2154)
Cayenne, FG	423	367	403	429	585	463	259	165	65	68	153	337	(3717)
Ciudad Bolivar, Venez.	30	14	13	36	103	162	205	160	99	100	65	39	(1026)
Cuiba, Brazil	249	211	211	102	53	8	5	28	51	114	150	206	(1388)
Curitiba, Bra.	199	173	124	78	85	88	81	83	119	130	105	147	(947)
Fortaleza, Bra.	89	195	325	343	188	171	43	11	15	7	12	29	(1354)
Georgetown, Guy.	222	119	138	161	297	326	269	182	87	89	160	291	(2341)
Guayaquil, Ecu.	191	229	250	173	46	10	4	7	1	2	4	19	(936)
Guiria, Ven.	36	25	13	28	62	114	123	136	124	111	105	69	(946)

Table 1 (*cont.*)

City	J	F	M	A	M	J	J	A	S	O	N	D	(MAP)
Ilheus, Bra.	149	162	254	270	170	195	191	130	97	124	189	152	(2083)
Iquitos, Peru	262	215	275	294	292	194	189	151	206	221	267	299	(2865)
Jacobina, Bra.	86	82	94	92	70	63	59	48	30	32	83	103	(842)
Magdalena, Bol.	208	192	192	99	49	15	7	24	57	111	142	202	(1298)
Manaus, Brazil	276	277	301	287	193	99	61	41	62	112	165	228	(2101)
Maracaibo, Ven.	5	1	5	28	61	54	25	50	66	122	56	23	(496)
Maracay, Ven.	6	5	6	30	98	138	120	168	127	89	47	18	(852)
Maripasoula, FG	218	214	212	249	399	278	196	136	77	63	92	234	(2368)
Medellin, Col.	41	49	75	147	178	112	105	133	129	160	114	82	(1325)
Mene Grande Ven.	36	30	61	123	185	133	146	162	185	205	171	77	(1514)
Merida, Ven.	46	38	50	160	244	164	123	127	191	258	200	88	(1689)
Paramaribo, Sur.	192	149	162	231	320	302	225	166	85	86	108	193	(2219)
Pirapora, Bra.	220	143	127	63	11	3	3	1	19	75	202	278	(1145)
Porto Alegre, Bra.	101	106	105	66	83	121	121	140	131	102	87	109	(1272)
Porto Nacional, Brazil	298	290	292	152	44	—	3	9	42	150	242	292	(1814)
Pucallpa, Peru	159	183	195	179	110	87	42	63	118	156	208	156	(1656)
Puerto Ayacucho, Venez	14	17	66	156	337	437	436	292	175	169	110	40	(2249)
Quito, Ecu.	114	127	150	171	122	47	20	23	77	125	108	101	(1184)
Recife, Bra.	47	86	174	236	289	292	248	148	63	29	25	32	(1669)
Rio de Janeiro, Brazil	137	124	135	107	78	50	43	45	59	81	100	133	(1092)
San Carlos, Ven.	207	225	216	379	399	391	338	327	249	257	313	219	(3520)
San Lorenzo, Ecu.	255	288	294	347	286	290	204	115	114	126	86	143	(2548)
Santos, Bra.	279	282	300	202	162	106	104	102	122	199	190	255	(2303)
Sipaliwini, Sur.	151	140	203	247	430	334	207	105	45	37	89	83	(2071)
Tumeremo, Ven.	71	71	34	53	92	134	115	111	54	49	60	70	(914)
Utiariti, Bra.	335	338	289	203	46	22	8	17	81	194	278	323	(2134)
Villavicencio, Col.	61	113	159	461	605	498	524	360	335	443	381	156	(4096)

**Source*: Kendrew, 1953. MAP – mean annual precipitation.

properly. The only alternative solutions left are the barging of water from the mainland or the desalination of seawater, both of which are very costly.

In the Greater Antilles, groundwater has been developed extensively for irrigation during the past few years and, as a result, the effects of overpumping are beginning to be felt.

Surface waters The smaller islands of the Caribbean region are noted for their general lack of surface waters, except where the ground level dips below the groundwater table. Where surface runoff does exist, it is usually associated with heavy precipitation and disappears quickly. The larger islands, notably Cuba, Dominican Republic/Haiti, Guadeloupe, Jamaica, Martinique, Puerto Rico and Trinidad and Tobago have some permanent and intermittent streams (see Fig. 5 and Table 2). Most discharges of rivers are variable, depending to a great extent upon the pattern of rainfall.

Tropical Mexico

As can be seen in Fig. 1, the humid tropical portion of Mexico is limited to the southern gulf coastal areas, Yucatan and the southern Pacific coastal areas.

The situation in Mexico (and, in fact, all of Central America) is somewhat more complex than that of the islands of the Caribbean. Whereas in the summer the east coast of Mexico experiences very moist tropical maritime easterly and southeasterly winds (penetrating the whole Mexican territory), in winter the influence of North America can become evident with frontal activity being felt down the eastern coast to Yucatan, and at times even further south, occasionally as far as Costa Rica. The cold can cause temperatures to fall to the freezing-point with a definite possibility of frost above the 1,300 m level. The situation in the isthmus of Tehuantepec in winter is such that the winds are generally more northerly, giving heavy rains on the north coast, but arriving relatively dry on the Pacific side.

In Mexico the east, and windward, coast has a generally more moist regime, and especially in the south, has very heavy rains. The Yucatan Peninsula rainfall decreases northward but is still liable to be impacted by the occasional hurricane. On

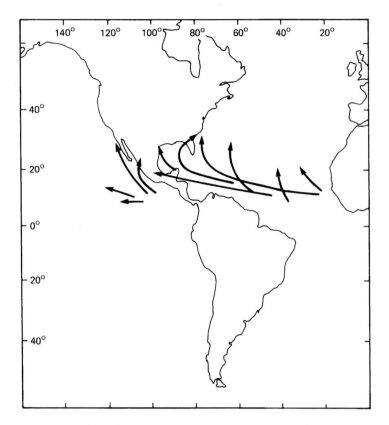

Figure 4: Principal tracks of tropical cyclones (hurricanes), Latin America and the Caribbean (modified from Palmen & Newton, 1969)

the west side of the country, the climate is, in contrast, considerably drier. The north and south sides of the isthmus of Tehuantepec show this clearly, where the rainfall on the north side (the Gulf of Mexico, and generally the windward side) is twice that of the south (the Pacific Ocean).

Geomorphology and groundwater availability Mountain ranges with west-east orientation extend to the Maya Mountains (Belize), the Sierra Madre de Chiapas (Mexico) and the central chain of Guatemala. To the east, the system extends into Honduras and northern Nicaragua.

Groundwater along the coastal plain of the Gulf of Mexico is available from marine sedimentary formations, with particularly large amounts withdrawn in the Veracruz area. Along the lower coastal areas of the Pacific, sea-water intrusion has occurred. The Yucatan Peninsula is a highly karstified region, flat and riverless. Large amounts of groundwater are pumped from the limestone aquifers.

Surface waters In general, most of the rivers that discharge into the Gulf of Mexico are short as a result of the relief of the land. In Mexico the largest surface water courses of those flowing into the Gulf of Mexico are those of the Panuco River system, the Papaloapan River and the large water system formed by the Grijalva and Usumacinta Rivers in the extreme south. The Yucatan Peninsula is a flat and riverless region with extensive karst developments. The Pacific-draining rivers are usually short. Figure 5 shows the patterns of runoff. Some of the

steep gradients shown, if correct, are probably explained by the saturated soil conditions during the rainy seasons, wherein rainfall and runoff gradients become more closely aligned.

Central America

The climate The entire area is subjected to a tropical-equatorial climate. Annual rainfalls exceed 1,500 mm practically everywhere. Many areas receive from 2,000 to 3,000 mm per year of rainfall.

With rare exceptions, Costa Rica and Panama are outside the hurricane region (see Fig. 4) but, because of the irregular hurricane rains, there can be great variability in the annual totals.

To the south, the Central American countries are a continuation of the mountainous central region that is typical of the more northern regions. These parts often have more temperate climates at heights, but similarly, are very rainy. The coasts are hot and wet. As noted above for tropical Mexico, the east coastal region (to the windward of the prevailing northeasterlies) and slopes of the mountains are the areas of higher precipitation. Annual rainfalls exceeding 2,500 mm are not unknown. The driest season is from December to April. In many areas, the dry season is still relatively wet. Most of the Central American region has two maxima of rainfall – around June and around October.

At the southernmost end of the Central American region, the Republic of Panama, because of the narrowness of the isthmus and its location between the warm waters of the Caribbean Sea

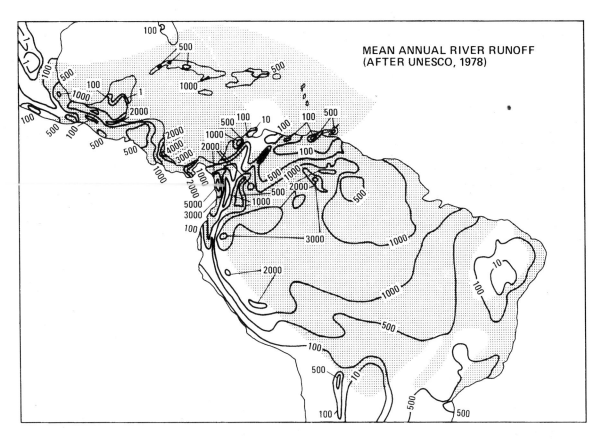

Figure 5: Mean annual surface runoff, Latin America and the Caribbean (after Unesco, 1978)

and the Pacific, has small variations in temperature and very high rainfall. On the Caribbean side, the annual rainfall often exceeds 3,000 mm, while on the Pacific side, it may be half that but still very humid. On the low mountain range, the amounts of annual precipitation on the windward side will probably exceed 5,000 mm in places, resulting in densely forested tropical forests. The drier season is usually from January through March. The entry into the dry season and out of it is abrupt. While on the Caribbean side the season is relatively dry, on the Pacific it is truly a dry season.

Geomorphology and groundwater availability Mountain ranges with a west-east orientation are in the north. They include the Maya Mountains (Belize), the Sierra Madre de Chiapas (Mexico) and the central chain of Guatemala. To the east, this system extends into Honduras and northern Nicaragua and incorporates the islands of Roatay and Guanaja (Honduras).

In the south of Central America (Costa Rica and Panama), the mountains are of Miocene age and include Cretaceous – mainly volcanic – and Cenozoic formations.

These two mountainous complexes are separated by a volcanic chain which developed during later periods. This chain has a west-northwest to east-southeast orientation in southern Guatemala, across El Salvador, southern Honduras, Nicaragua and northern Panama. Volcanoes are still active in the region.

Important fracture systems have developed in Central America. A gigantic thrust fault cuts Guatemala in an east-west direction, reaching the Gulf of Honduras. Another system, north to south, has generated the Honduras depression between the estuary of the Ulua River on the Caribbean and the Gulf of Fonseca in the Pacific.

The Nicaragua depression is occupied by large lakes which are oriented west-northwest to east-southeast. The depression is, in fact, a large graben which is still active, as evidenced by the earthquake of 1973 which destroyed part of Managua. A narrower graben, which has the same orientation, is found in El Salvador and Guatemala.

Other fracture systems cut the Isthmus of Panama into a number of compartments.

In spite of the large amount of rainfall, conditions for the development of groundwater resources are poor because of geological formations with little or no permeability. This is especially true of the metamorphic and a large fraction of the volcanic formations. However, because surface water is not available everywhere during the dry seasons, the development of groundwater resources is necessary.

Because of the complex geology and morphology, groundwater investigations and subsequent development are difficult. Most aquifers are heterogeneous and of limited extent. In fact, groundwater was not generally developed until recently. In the past few years, considerable development has taken place, especially for large metropolitan areas and for seasonal irrigation of cash crops.

TABLE 2 *Annual discharges of selected rivers of Latin America and the Caribbean*

River	Discharge to	Primary Tributary	Measurement Site	Mean Annual Discharge (m³/s)	
Bolivia					
Beni	Atlantic		Angosto del Bala	2,080.00	**
Pilcomayo	Atlantic		Villa Montes	185.00	**
Brazil					
Amazon	Atlantic		Atl. outlet	220,000.00	***
Tocantins	Atlantic		Itupiranga	9,895.00	**
Parana	Atlantic		Guaira	8,250.0.	
Sao Francisco	Atlantic		Traipu	2,943.0.	
Uruguai	Atlantic		Irai	1,190.0.	
Colombia					
Magdalena	Caribbean		Calamar	7,320.00	**
San Juan	Pacific		Cabeceras	2,500.00	
Atrato	Caribbean		Bellavista	2,420.00	**
Mira	Pacific		Caunapi	570.00	
Patia	Pacific		North of Magui	497.00	
Micay	Pacific		Zaragoza	430.00	
Naya	Pacific		at Guatala River	175.00	
Costa Rica					
Grande de Terrabe	Pacific		Palmar	306.40	
Parismina	Caribbean	Reventazon	Pascua	132.40	
Grande de Tarcoles	Pacific		Balsa	78.40	
San Juan	Caribbean	San Carlos	Jabillos	56.50	
Grande de Candelaria	Pacific		El Rey	30.30	
Cuba					
Toa	Atlantic		El Aguacate	24.30	**
Bayamo	Atlantic		La Bayamesa	14.00	**
Buey	Caribbean		San Miguel	2.68	**
Dominican Rep.					
Nizao	Caribbean		La Penita	20.00	**
Yuna	Caribbean		Los Zuemados	17.90	**
Yaque del Norte	Caribbean		Boma	17.80	**
Bani	Caribbean		El Rocodo	11.60	**
Ecuador					
Amazon	Atlantic		Yavari R. jct.	48,131.00	
Ucayali	Atlantic		Maranon R. jct.	17,685.90	
Maranon	Atlantic		Ucayali R. jct.	17,310.00	
Nanay	Atlantic		Amazon R.	8,936.00	
Esmeraldas	Pacific	Guayllabamba	Quninde	912.00	
Mira	Pacific		at Lita	111.00	
Tumbes	Pacific	Puyango	Portovelo	82.00	
El Salvador					
Lampa	Pacific		San Marco	377.50	
Goascoran	Pacific		near outlet	33.70	
Grande San Miguel	Pacific		Vado Mari	27.8	
French Guyana					
Maroni	Atlantic		Langa Tabiki	1,770.00	**
Oyapock	Atlantic		Saut Maripa	870.00	**
Guatemala					
Usumacinta	Gulf of Mexico		Boca del Cerro	1,776.00	**

Table 2 (*cont.*)

River	Discharge to	Primary Tributary	Measurement Site	Mean Annual Discharge (m³/s)	
Salinas	Gulf of Mexico		El Cedro	575.00	**
Montagua	Caribbean		Morales	181.90	**
Dulce	Caribbean	Polochic	Teleman	69.90	
Belize	Caribbean	Mopan	El Arenal	35.90	
Maria Linda	Pacific	Aguacapa	Agua Caliente	10.90	
Los Esclavos	Pacific		La Sonrisa	10.60	
Samala	Pacific		Candelaria	9.80	
Guyana					
Essequibo	Caribbean		Plantain Island	2,225.10	
Mazaruni	Caribbean		Apaikwa	746.00	**
Honduras					
Patuca	Caribbean		Cayetano	209.40	
Ulua	Caribbean		Pte. Pimienta	195.50	
Chamelecon	Caribbean		Pte. Chamalecon	43.60	
Choluteca	Pacific		Los Encuentros	33.70	
Mexico					
Papaloapan	Gulf of Mexico		Estacion Papaloapan	695.00	**
Balsas	Pacific		Presa el Infiernillo	635.00	**
Nicaragua					
San Juan	Caribbean		Los Pilares	563.60	
Coco	Caribbean		Guanas	52.20	
Grande de Matagalpa	Caribbean	Tuma	Yacica	20.80	
Tamarindo	Pacific		Tamarindo	2.90	
Panama					
Changuinola	Caribbean		Bacon Bay	203.90	
Chepo	Pacific	Bayano	Canitas	182.40	
Chiriqui	Pacific		David	123.70	
Santa Maria	Pacific		San Francisco	87.50	**
Tabasara	Pacific		Camaron	73.10	
Fonseca	Pacific		San Lorenzo	60.80	
Chiriqui Viejo	Pacific		Paso Canoa	52.90	
San Pablo	Pacific		La Mesa	52.70	
Cocle	Caribbean		El Torno	48.40	
Lake Gatun	Caribbean	Chagres	Chico	30.20	
Paraguay					
Paraguay	Atlantic		Asuncion	2,940.00	
Essequibo	Caribbean	Cuyuni	Acaribisi R. jct.	1,100.00	
Catatumbo	Caribbean		Tarra R. jct.	140.00	
Unare	Caribbean		Unare Lagoon out.	56.00	
Tocuyo	Caribbean		Carib. outlet	42.00	
San Juan	Caribbean		Carib. outlet	35.00	
Peru					
Ucayali	Atlantic	Urubamba	Tambo R. jct.	2,890.10	
Yavari	Atlantic		Amazon R.	1,565.00	
Puerto Rico					
La Plata	Atlantic		Toa Alta	8.4	**
Venezuela					
Orinoco	Atlantic	Capanaparo	Outlet	4,100.00	

Primary Source: ECLAC, 1990. ***UNESCO, 1978; ##Rodier & Roche, 1984)

The surface waters The longest river in this region is the Coco (or Segovia) River on the border between Nicaragua and Honduras. In general, most of the rivers that discharge into the Caribbean Sea are short as a result of the relief of the land. The Pacific-draining rivers are usually short. Table 2 gives information on many of the region's rivers.

The main rivers in Belize are the Hondo River (at the boundary between Belize and Mexico), the New River (the main permanent river of the country, and the Belize River (also permanent, originating in the Maya Mountains). A number of other smaller permanent and intermittent streams begin in the Maya Mountains. All rivers that originate in the Maya mountains have very rapid runoff, and are thus subject to seasonal fluctuations.

The main rivers of El Salvador, all draining to the Pacific, (Fig.5) are the Lampa, Coacoran and Paz. The annual runoff has been estimated to be 20 billion m^3 (including international waters). There are also some crater lakes and many temporary streams.

In Guatemala, some 33 large or significant drainage basins have been identified. On the Pacific side the rivers generally are shorter, with orientations perpendicular to the shore. In the interior and to the east, however, the rivers tend to be longer. Some of the large rivers are the Motagua, La Pasion and San Pedro. The larger lakes of the country are Izabal, Peten Itza and Atitlan.

Honduras is bounded on the north by an extended coastline of the Caribbean Sea, and on the south by a short coast of the Pacific. Mountain ranges extend from east to west. As a result of this mountain placement, the heavy rainfall is limited to the north slopes, where the runoff ends up in the swamps and rain forests of the coastal plain (the "mosquito coast") and the delta of the Patuca River. Most of the northern portion of the country is drained by the two main river systems of the Patuca and the Ulua. The southern drainage also has two main rivers: the Coluteca and the Nacaoma. Studies in the region of Lake Yojoa appear to indicate that large quantities of water are lost through the karstic limestones in the bottom of the lake.

In Nicaragua, about 90% of the runoff flows to the Caribbean and only 10% to the Pacific. There are three major drainage basins in the country: the Coco River, the San Juan River (which is connected to Lakes Managua and Nicaragua) and the Grande de Matagalpa River (Fig.5). All drain toward the Caribbean. On the Pacific side the 35 streams are, in general, not permanent and are all rather short. The main river on that side is the Real, discharging into the Gulf of Fonseca.

In Costa Rica the central mountain ranges form an arc that divides the country into the Caribbean and Pacific watersheds. Indirectly, the Sapoa and Drio Rivers drain to Lake Nicaragua, and the San Carlos, Sarapiqui and Chirripo Rivers into the San Juan River (between Costa Rica and Nicaragua) before draining into the Caribbean Sea. Others flow directly into the Caribbean or the Pacific. Many of these rivers are perennial, but there are other minor seasonal or temporary streams in all watersheds, especially those draining into the Pacific. Lakes are rare, with the principal ones being of volcanic origin (related to the Guanacaste Mountains and the central volcanoes). The most important lakes are Arenal, Hogote and Tortuguero.

In Panama, the rivers, in general, have courses which develop perpendicular to the coastal lines. The main water courses on the Caribbean side (which represent only about 30% of the country) are the Chinginola, Cocle del Norte and Taobre and Chagres. On the Pacific side (70% of the country) the main rivers are the Chiriqui Viejo, Chiriqui, Tabasara, San Pablo, La Villa Santa Maria, Grande, Bayano, Tuira and Chucunaque. While most of the Pacific drainage rivers are permanent, their discharges can be quite variable in response to the rainfall. The largest series of lakes in Panama are primarily man-made, having to do with the Panama Canal.

TROPICAL SOUTH AMERICA

The climate

There are basically four regions of the South American continent that have humid tropical climates: the northern Caribbean and Atlantic drainages, the northern Pacific drainages and the central drainages of the Amazon and the Chaco region of Paraguay, Bolivia and Brazil (see Fig. 1). For average annual precipitation and evapotranspiration, see Figs. 2 and 3.

Drainage systems

Tropical Pacific drainages The tropical Pacific region is a continuation of the same basic system of the western slopes of the Andes Mountain range, beginning in southern Mexico and extending to the Gulf of Guayaquil in Ecuador. The annual precipitation in the South American portion of this region, as in that of Central America, ranges from 300 to 5,000 mm, depending upon the complex orographic conditions which cause the highest amounts of rainfall to occur during the January to April period. The temperature range in this region of South America is, generally speaking, a function of the altitudes, with mean isotherms varying from 25°C along the coast to 8°C in the higher parts of the Andes range, but the annual temperature variations are small (up to 3°C).

Northeastern drainages of South America On the Caribbean side of South America in Colombia, the precipitation is highly variable and depends to a great extent on relief and low-lying air masses. The lowest rainfall is on the northernmost part of Colombia at Guajira Peninsula on the Caribbean Sea, in which part the rainy season is during May to November. The amount of annual rainfall increases generally toward the south to more than 2,000 mm in the mountains, with some areas exceeding 4,000 mm. Proceeding eastward, in the north of Venezuela, the least rainy regions are along the coast, and as with the Colombian situation, the rainiest zones are the highlands on the windward side of the mountains. The mean annual temperatures are highest in the drier portions toward the coast, and decrease with elevation. The annual temperature variation is very much determined by the amount and distribution of rainfall during the year, but in the more humid tropical regions do not amount to more than about 3.5°C.

The coastal regions of Guyana, Surinam and French Guyana have a definite two-peak rainfall as a result of the movement of the inter-tropical convergence zone. During the period May to July, the region tends to receive its highest rainfall. September through the end of the year is usually the period of lowest precipitation. The annual rainfall commonly exceeds 2,000 mm in this region. The annual temperatures tend to be high, between 25 and 29°C, with small annual temperature variations, between 1 and 2.5°C.

The Amazon region The average annual precipitation of the Amazon region is around 2,300 mm, certainly one of the highest in the world for such a large area. The Atlantic coastal region is an area with rather high annual rainfall, exceeding 2,700 mm in Belem. Toward the center of the basin, the rainfall decreases to 2,000 mm and then increases again to the north in Colombia (where mean values exceeding 8,000 mm have been claimed, and Venezuela where values in excess of 3,500 mm have been measured) and to the west toward the Andes Mountains where values in excess of 3,500 mm are found. The monthly distribution of rainfall in the Amazon Basin can be identified with four different areas (ECLAC, 1990):

The area around the equator up to 60 °W longitude, which receives the heaviest precipitation due to its nearness to the ocean (extreme north of the State of Para and of the State of Amapa);

The area to the south of the equator up to 65 °W longitude, which registers its lowest levels of precipitation in June and July;

The area west of 65 °W longitude, which has characteristics similar to the above except that its minimum levels of precipitation are higher; and

The area to the north of the equator and to the west of 60 °W longitude, which has two peak rainy seasons and no dry season.

As noted in ECLAC (1990), there are some drier zones along the western and southwestern rim of the Andes Mountains where the mean annual rainfall does not exceed 1,000 mm, and along some of the higher points along the divide where it is below 500 mm.

The Amazon Basin straddles the equator, and thus the solar radiation that is available in the outer limits of the atmosphere varies approximately from 730–885 cal cm^{-2} day^{-1}; but the amount reaching the surface is considerably less. Because of the high humidity and common cloudiness of the region, a large part of the solar radiation is absorbed and/or reflected in the troposphere. The temperature of the Amazon Basin is thus characteristically isothermic. In most of the basin mean monthly temperatures vary between 24 and 28°C, and the annual variation is only around 5°C.

The northeastern portion of Brazil is the driest part of the country, with rainfall commonly falling below the levels of potential evaporation. Absence of rain over prolonged periods has also caused the area to suffer drought. When the southerly and southeasterly winds move inland, however, they run up against orographic barriers and release their moisture on the windward side of the ranges. As a result, the eastern coastline of this region receives higher rainfall than do the inland areas. A much more tropical climate thus exists, but it is definitely limited to the coastal belt. During the winter months of the Southern hemisphere (May through July) there can be incursions of polar air masses that move up from the south. When these merge with the southeasterly trade winds, they cause heavy rainfall along the entire coast (see Molion , this volume). Much further to the south, in the area of Rio de Janeiro, the region becomes subject to the influence of the drier, more stable Atlantic tropical air mass from the south, and the lowest volume of precipitation, therefore, is recorded in the winter months.

In this coastal region the mean annual temperatures decrease from north to south and from east to west. Along the coast, the annual means range from 25.4°C (Aracaju) to 22.6°C (Rio de Janeiro).

The Chaco region The remaining portion of South America with a tropical climate is essentially the northern portion of the Plata Basin, including portions of Paraguay, Bolivia and southern Brazil – the Chaco region.

Mean annual temperatures in this region vary between 20 and 28°C. The rainy season lasts six to seven months, October through April, although rain commonly falls throughout the year. The nnual rainfall is typically in excess of 1,000 mm. At the head of the basin, the annual rainfall can exceed 1,800 mm.

Global air circulation in this portion of the continent is influenced by the interaction of the subtropical South Atlantic anticyclone and surface low pressure over the region during the summer. Such systems produce winds which, on the Brazilian side, range from northeast to northwest, according to latitude. They can deflect humid air masses from the southern Amazon Basin to the Pantanal. The lower temperatures that occur over the middle and upper Amazon Basin during the Southern hemisphere summer also provoke a deflection of humid equatorial air masses from the east of the Amazon Basin into Paraguay.

The geomorphology and groundwater availability

The structure of South America is somewhat similar to that of North America, as it has vast plateaus to the east, high mountain ranges and plateaus to the west and vast plains in between. However, due to its equatorial location, climatic features are closer to those of the African continent. In fact, as most of the lowlands – excluding those to the south – are hot and humid, a large proportion of the population is concentrated in highland areas.

The Andean mountain system (Argentina, Bolivia, Chile, Colombia, Ecuador, Peru and Venezuela) is about 7,000 km long with a width varying from 40 km (Chile) to 800 km (Bolivia). It includes several separate mountain chains in the north, which merge to the south, and also intermontane valleys and plateaus.

The Brazilian highland (Brazil-Uruguay) includes the eastern part of the Brazilian Pre-Cambrian Shield and the vast sedimentary and volcanic area which overlie its central part. (The low-rising western part of the Shield belongs to the domain of the Amazon Plain.)

The Guyana highland (Brazil, Colombia, French Guyana, Guyana, Surinam and Venezuela) is also, primarily, a Pre-Cambrian Shield area. To the north, close to the ocean, it lowers into a coastal peneplain.

The Orinoco Plain (Venezuela) stretches between the northern part of the Andes and the Guyana highland. It is covered by a thick alluvium layer.

The Amazon Plain (Bolivia, Brazil, Colombia, Ecuador, Peru and Venezuela) is the largest region of South America. It includes the alluvial flood plain itself, about 10% of the area, and the upland plains which account for 90% of the region (where Pre-Cambrian crystalline and Cenozoic formations crop out widely). To the south, the Parana and Paraguay Rivers feed many swamps, and lagoons develop from flood waters.

Groundwater plays an important role in tropical South America. This is especially true as a source of water for urban and rural water supply systems and for industry – in particular along the densely populated shores of the Guyanas and Brazil. Certain areas are experiencing salt-water intrusion in wells and rapid drops in water levels. Considering the vastness of the continent and the rapid growth of its population, it would seem that the knowledge and development of groundwater in South America are only in their early stages.

The surface waters

The primary orographic influence in South America is the Andean mountain chain. It gives rise to the Amazon River, with generally low gradients, that flows into the Atlantic, and smaller steep-gradient rivers discharging into the Pacific. Other Atlantic-oriented large watersheds not connected with the Andes Mountains include the Orinoco, and in the southern portion of the humid tropical region, the Sao Francisco, Parana/Plata and Paraguay which have their headwaters in the central plateau of Brazil. Figure 5 illustrates the regional distribution of annual river runoff.

It is interesting to note that while the drainage area of the Atlantic-draining watersheds is approximately 12 times that of the Pacific drainages, the amount of flow is closer to only ten times. This results from areas of extremely heavy precipitation on the Andean peaks. The coefficient of variation of the annual

flows, however, is much lower on the Atlantic side, approximately .05, compared with approximately .20 on the Pacific side.

The rivers of South America are primarily rainfed, with the mean annual discharges of the various rivers varying within broad limits. Large zonalities in the distribution are to be observed in the equatorial, sub-equatorial and the southern tropical regions. These include the broad Amazon lowlands, the Basin of the Orinoco and the northern part of the Paraguay/Parana Basin. Maximum flows are in the equatorial zone where in the the eastern foothills of the Andean chain many of the tributaries of the Amazon rise, and the depth of the mean annual flow over the years exceeds 3,000 mm. To the north, in the Orinoco and Guyana plateau region, the amounts of the annual runoff can vary substantially from more than 2,000 mm to less than 100 mm, depending upon how the precipitation characteristics are orographically influenced.

Northeastern Brazil The Brazilian plateau area of northeastern Brazil has a strong influence on the discharges of that region. As noted before, the coastal areas are well wetted by the incoming humid South Atlantic easterlies. Moving westward from the coast, however, the river discharges reduce from annual values of 400 to 500 mm to less than 20 mm in the lowlands of the Sao Francisco River drainage. They then gradually increase on the western part of the plateau till they reach 800 to 1,000 mm (an area in which the precipitation from the humid easterlies can exceed 2,000 mm).

The Amazon Because they are primarily rainfed, the seasonal distribution of the annual discharges is influenced by the amount and distribution of the precipitation. In the equatorial region, the highest flows occur during the three months of April to June, averaging about 50% of the total annual volume. The low water period is October to December. North and south of the equatorial zone, where the rainfall occurs mainly in torrents during their summer months, the higher river flows begin in the May-to-August period, with a maximum typically at the end of the rainy season.

The southerly tributaries of the Amazon, having their primary source in the Brazilian plateau, are characterized by very low flows during May to October. During prolonged dry periods, many of the tributaries can even approach zero flow. The maximum discharges occur in January to March. The rivers that rise in the Andes have a similar distribution of seasonal flows, but since they are also snow-fed, the December-to-March period has somewhat higher flows.

The Amazon River, at Obidos, which at that point has a contributing area of 4 680 000 km², has been estimated to have an annual discharge of 157 000 m³ s⁻¹. Nearer to the mouth of the river, and encompassing a greater area of lower precipitation, the contributing area is 6 915 000 km² (nearly half the size of Europe). The annual discharge is estimated to be 250 000 m³ s⁻¹ (Elias & Serrano, 1983); this is an estimated one-fifth of the entire world's surface fresh water. It is so massive that ocean-going vessels can traverse the river as far as Iquitos, some 3,600 km from the Atlantic. It is without doubt not only the largest river in the world in terms of volume of discharge, but in drainage area as well.

The Chaco The Chaco region, to the south of the Amazon (covering an area including parts of Bolivia, Brazil and Paraguay), received its name from its very low surface gradients. East to west it varies from 50 to 30 cm km^{-1} and north to south, from 1.5 to 3 cm km^{-1}. These low gradients account for regular flooding during the rainy season.

This marshy area (also known as the Pantanal) is largely covered by small lakes, abandoned old meanders and former river-beds, partially or completely covered by grass. The lakes and abandoned meanders are covered by a few metres of water all year long. Grass grows in the water, both at the bottom and floating on the surface. Water may flow from one lake to another during floods. During periods of low discharges, the lakes and meanders appear to be independent although sometimes overgrown with grass. Abandoned river channels, aided by permeable soil, maintain the connections. Water flows very slowly, at 1 to 5 cm s^{-1}, owing to the low gradient and the resistance of the grass. The grass which covers the area is peculiar to the region. The lakes are large but since they are almost totally covered with grass, they appear to be small. Almost all large lakes in the upper Paraguay Basin are mainly covered with grass so that they should be considered as pantanal area rather than lakes. The lakes cover an area of approximately 70 000 km^2.

An interesting feature is that as a result of the combination of factors, the Paraguay River and its tributaries have an important time lag in the filling up or emptying of the lakes at times of high and low flows, respectively. During the flood period, the rise in water level can breach the obstructions caused by the grass and the velocity of flow may then increase suddenly in the main channels.

CONCLUSION

The countries of Latin America and the Caribbean have given considerable attention to the improved use of their waters. Traditionally, the areas requiring the largest investments have been irrigation, hydro-power and drinking water supply and sanitation. It is suggested, however, that there is a need for systematic comprehensive assessments of the water resources of the countries.

As noted in ECLAC (1986), practically all of the countries of the region have put together one or more plans for the management of their water resources. But these plans have mostly been in the subsectors noted above, some of which have completely dominated the process. In those cases, mostly concerning irrigation and energy, they have come to dominate the national and regional strategies in water planning. One major

area of concern, as can be seen in the section of this book dealing with urban problems in water management (see Gladwell, this volume), involves the total water management required if the urban areas (particularly the increasingly large concentrations) are to be properly managed. The emphasis must be on total, or more properly, "integrated water management" required for both quantitative and qualitative aspects of water supply, waste management and crisis management.

But the hydrology of the region is not static. As noted by Molion (1991), most of the studies to identify possible climatic fluctuations have been inconclusive. Not only are there multiple natural variables but, in addition, human actions to change the environment are occurring at an unprecedented rate, especially in the tropics of Latin America. These may (or may not) be contributing to the variability claimed by some to be observable today. One must believe, nevertheless, that massive deforestation and consequent land-use changes must be having some effects on the hydrological cycle. There is serious concern, although certainly not yet proven, that especially because of the tropical location, such effects could have world-wide consequences on the global climate.

ACKNOWLEDGEMENT

The authors wish to give special credit for much of the information presented in this discussion to that provided in documents of the Economic Commission for Latin America and the Caribbean (ECLAC) and to the United Nations Department of Economic and Social Affairs.

REFERENCES

Chang, J-H & Lau, L. S. (1983) Definition of the humid tropics. Paper presented at the Hamburg IAHS Symp., Hydrol. of Humid Tropical Regions, Aug. 1983; and IHP Proj. Report *4.2(b)*, Jan. 1984.

ECLAC (1986) Formulation of plans for water resource management in Latin America and the Caribbean. United Nations Economic Commission for Latin America and the Caribbean – ECLAC, *LC/G.1391*, 1 April 1986, Santiago, Chile.

ECLAC (1990) *Latin America and the Caribbean: Inventory of Water Resources and Their Use.* (Vol. I: Mexico, Central America and the Caribbean; Vol. II: South America), United Nations Economic Commission for Latin America and the Caribbean - ECLAC, *LC/G.1563*, Santiago, Chile.

Elias, V., Cavalcante, S. & Jorge, A. (1983) Recent hydrological and climatological activities in the Amazon Basin, Brazil. *In: Hydrol. of Humid Tropical Regions.* IAHS Publ. *140*:365–373.

Kendrew, W. G. (1953) *The Climates of the Continents.* Oxford Univ. Press, London, England.

Molion, L. C. B. (1991) Climate variability and its effects on Amazonian hydrology. *In:* B. P. F. Braga & C. Fernandez-Jauregui (eds.) *Water Management of the Amazon Basin.* UNESCO, 261–274.

Palmen, E. & Newton, C. W. (1969) *Atmospheric Circulation Systems.* Academic Press, New York.

Rodier, J. A. & Roche, M. (1984) *World Catalogue of Maximum Observed Floods.* IAHS No. *143*, Wallingford, UK.

UN (1976) *Ground Water in the Western Hemisphere*. United Nations, Dept. of Econ. and Soc. Affairs, Natural Resources/Water Series No. *4*, New York.

UNESCO (1978) *World Water Balance and Water Resources of the Earth*. Int'l. Hydrol. Programme, United Nations Educational, Scientific and Cultural Organization (UNESCO). *Studies & Reports in Hydrol.* No. *25*.

8: Amazonia Rainfall and its Variability

L.C.B. Molion

Instituto de Pesquisas Espaciais (INPE), C. Postal 515 – Sao Jose dos Campos, 5P , Brazil

ABSTRACT

The hydrological processes in Amazonia vary widely from year to year due to natural changes in the atmospheric conditions. This paper reviews the dynamic mechanisms that produce rainfall in the region, its characteristics and fluctuations due to inter-annual variability of the large-scale atmospheric circulations associated with the El Niño-Southern Oscillation phenomenon (ENSO) and blocking patterns of the atmospheric flow. It is hypothesized that the observed trends in precipitation, as well as runoff, reported by Rocha *et al.,* (1989), may be related to a higher frequency of positive phases of ENSO and/or the presence of volcanic aerosols in the stratosphere over Amazonia. Such aerosols could influence the heating of the Andean Altiplano and, in turn, the seasonal development of the upper tropospheric anticyclone over tropical South America (Bolivian High). The effects of large-scale deforestation on local hydrology as well as the possible impacts on global climate are also discussed.

INTRODUCTION

The hydrologic cycle is an integrated product of the climate and of the biogeophysical attributes of the surface. On the other hand, it exerts an influence on climate which goes beyond the interaction between the atmospheric moisture, rainfall and runoff. It is the major single heat source for the atmosphere, in the form of latent heat which is released, mainly in the tropics, through the condensation of atmospheric moisture into clouds and rainfall.

Attempts to identify patterns in climatic fluctuations are mostly inconclusive because the climate, hence the hydrologic cycle, presents an intrinsic variability, both in space and time, which is not adequately known. At sea-level, the spatial variability of climate parameters at a global scale results mainly from the differential absorption of solar radiation by continents and oceans, unevenly distributed on the planet's surface. When the temporal and spatial heterogenity of solar radiation receipt is combined with the earth's rotational effect, this results in the general circulation of the atmosphere and oceans which redistributes heat and determines the climate over continents. There are additional external controls such as the variation of the solar energy input (solar constant) and the earth's orbital parameters, which have a longer-term impact on the general atmospheric circulation.

The local climate is further modified through the dynamic coupling between the atmosphere and the biogeophysical characteristics of the underlying surface, particularly the topography and the surface cover. Examples of these additional internal controls include the variation of surface albedo, cloudiness and the chemical composition of the atmosphere.

In addition, the anthropogenic contribution through human actions in changing the environment may inadvertently be contributing to the present variability.

This paper reviews the climatology and the inter-annual variability of rainfall in Amazonia and the associated physical mechanisms, classified by their inherent spatial scales. Rainfall was selected because it is the most significant climatological variable for the tropics and, until recently, its inter-annual variability was thought to be relatively insignificant. However, with the possibility of increasing climatic change due to human intervention in the environment, then rainfall will be a principal indicator of such change. Particular attention is also given in this paper to the impacts of a large-scale deforestation on the hydrologic cycle components.

SCALES OF RAINFALL PRODUCING MECHANISMS

The physical mechanisms that produce convection, cloud formation and rainfall in Amazonia may be classified in five broad spatial scales as described below.

Continental or large-scale

The solar and infrared radiation absorbed at the surface is primarily used for evaporating water (latent heat) and for heating the air (sensible heat) near the ground. In central Amazonia, micrometeorological studies (e.g. Molion, 1987; Shuttleworth *et al,* 1984) have shown that 80% to 90% of the available energy is used for evaporation (wet and dry canopy) and the

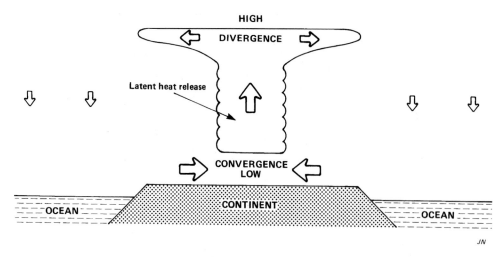

Figure 1: Schematic representation of the circulation resulting from the differential heating between continent and oceans in the summer.

rest for heating the air. Over the continent, the humid warm air is displaced upwards (convection) and during its ascent, it expands and cools. Moisture within the air condenses and releases enormous amounts of energy (latent heat) previously used in the evaporation process. Mass continuity requires the ascending air, after drying out, to sink over the adjacent oceans, therefore closing a direct or thermal circulation known as the Hadley-Walker Cell, as shown schematically in Fig. 1. The thermal circulation induces low atmospheric pressures (Low) and mass convergence at low levels, and high pressures (High) and mass divergence, with anticyclonic motion, at the upper troposphere levels.

The seasonal variation of the atmospheric circulation over Amazonia related to this main circulation cell was analyzed by Kousky (1983, pers. comm.). His mean streamline charts are reproduced in Fig. 2 for the lower (850 hPa) and the upper (250 hPa) troposphere and for two typical months: January (summer) and July (winter) . Common features in both seasons (Figs. 2a and 2c), at low tropospheric levels, are the subtropical Atlantic and Pacific anticyclones (A), also called the subtropical highs, which result from the subsiding branch of the Hadley-Walker thermal circulation cell. During the winter, the Pacific anticyclone centre is displaced slightly northward from its summer position. The Atlantic anticyclone centre, on the other hand, approaches the South American coast and extends its influence over the northeast and central parts of the continent, reducing cloudiness (Fig. 3b) and setting up the dry season for these regions. Otherwise, there seems to be no significant difference in the mean low-level tropospheric flow.

Other authors' analyses (e.g. Dean, 1971; Newell *et al.*, 1972), however, show stronger seasonal contrasts in the trade wind fields than does Kousky's analysis. Reviewing the former authors' charts, one has the impression that low-level air inflow into Amazonia is mainly associated with the Northern (Southern) Trades during the summer (winter). At high tropospheric levels (250 hPa), there is a marked seasonal variation in the mean flow. During the summer (Fig. 2b), it presents a

strong meridional component; whereas, in the winter, the motion is highly zonal (Fig. 2d). The broad and persistent diverging pattern (A) of the summer flow is a direct result of the Hadley-Walker thermal circulation cell and is the cause of the cloudiness which covers most of the tropical portion of the continent during that season (Fig. 3a). In the winter, the anticyclone centre (A) is positioned over the northwestern part of South America. The retraction of the anticyclone centre is apparent in the mean brightness maps from satellite-visible imagery (Fig. 3b) as well as in the rainfall maps (Fig. 4) published by Figueroa & Nobre (1990). Inspecting the monthly satellite brightness maps, it can be seen that only northwestern Amazonia remains cloud-covered throughout the year.

The High over Amazonia seems to be "anchored" to the Andean Altiplano, hence its denomination as the Bolivian High, since this is a heat source practically in the middle of the troposphere. The seasonal variability of the Bolivian High, both in intensity and position, appears to be responsible for the spatial and temporal distribution of the rainfall (Kousky & Kagano, 1981). During the Southern hemisphere's summer and fall (Dec.-May), the High is well-developed and covers practically the entire Amazon Basin, which, at this time, receives maximum rainfall (Fig.4a). When the High weakens and moves progressively northwestward during the winter (June-August), the southern and eastern sectors of the Basin experience their dry season (Fig. 4b), which can last between two (central) to six months (east and south). The northwest, however, is always under the influence of the High and, therefore, does not show a marked dry season (Fig. 4d). In September, the High commences edging back towards the Brazilian central plateau, moving close to the Andes Cordillera, and finally, in November, the rainy season is fully established over the Basin. Inspection of satellite daily imagery suggests that the presence of the High is not continuous but it appears to have a pulsation period of 10 to 15 days. Essentially, this system initially develops due to surface heating causing higher level divergence, which is subsequently

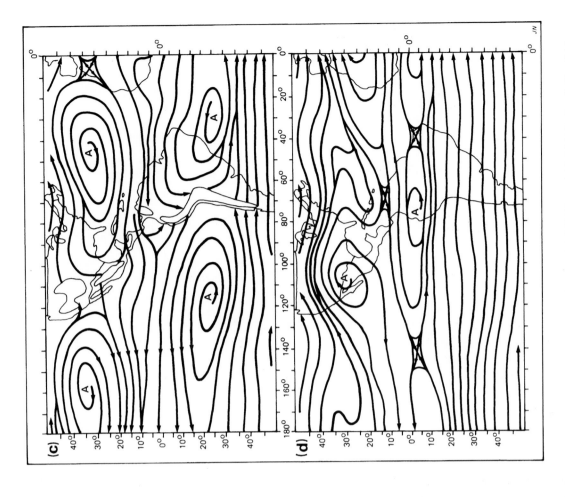

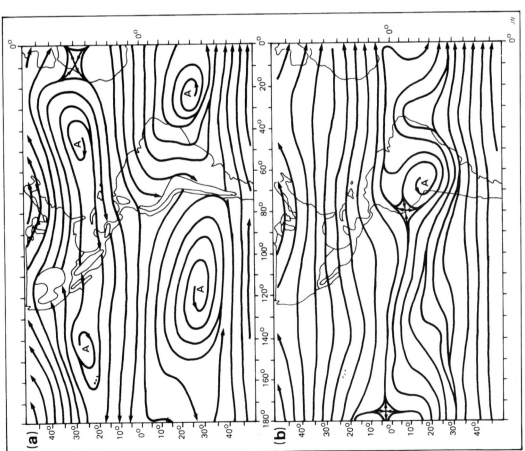

Figure 2: Streamline charts depicting the mean tropospheric flow for January (summer) at (a) 850 hPa and (b) 250 hPa level; for July (winter) at (c) 850 hPa and (d) 250 hPa level. Sign "A" refers to the location centre of anticyclonic motion (After Kousky, 1983).

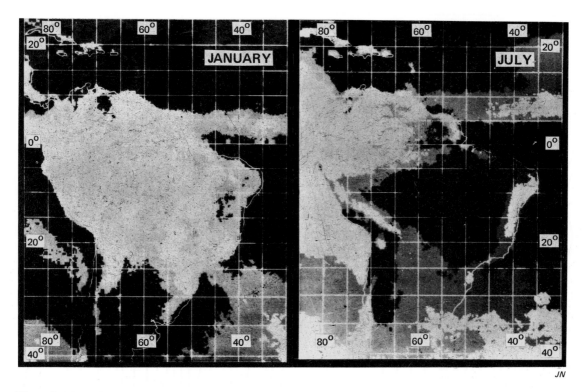

Figure 3: Mean cloudiness for January (3a) and for July (3b) over South America. Mosaic prepared using mean brightness of polar orbiting satellite imagery (14:00LT) for the period 1967–1970. (After Miller & Feddes, 1971).

maintained by latent heat release. After ten days or so, the upper High decays, probably due to persistent widespread cloudiness cutting off solar radiation and, in turn providing a possible negative feedback. The cloud-free surface heats up again and the cycle repeats itself. Therefore, superimposed on the low-frequency seasonal cycle signal, there seems to exist a high-frequency signal due to convection pulsation.

Another element of the large-scale circulation, which generates rainfall over the northeastern coast of Amazonia and northern coast of South America (Fig. 4c) is the Zonal Trough in the Easterlies (ZTE) over the Atlantic. (Sadler, 1975) (see Manton & Bonell, this volume). The ZTE is defined as the latitudinal band of confluence of Northern and Southern hemisphere trade winds, near the equator. Hastenrath & Lamb (1977) showed that the ZTE moves progressively southward from its northernmost position during the Southern Hemisphere winter (July) and reaches its southernmost position by the end of summer (March), following the apparent movement of the sun – with a time-lag of a couple of months. The ZTE is responsible for part of the annual rainfall totals during its passage over the Amazonian coast and its northern boundaries; stations under its influence present a double rainfall maxima (Fig. 4d). There is, however, no evidence of a continental ZTE during the summer, as reported by some authors, e.g. Trewartha, 1961; Ratisbona, 1976.

Synoptic scale

The characteristic length of this scale is 1,000 km and the most important dynamic mechanism of rainfall production within

Amazonia are frontal systems of temperate origin. Several authors (e.g. Trewartha, 1961; Parmenter, 1976; Ratisbona, 1976) have described the effects of the penetration of Southern hemisphere frontal systems into Amazonia during the winter of that hemisphere. Most of these descriptions emphasize the sharp 15° to 20° C decrease of temperature, which lasts for three to five days, and its consequence for the environment. Frontal systems, however, can penetrate Amazonia at any time of the year, organizing convection and rainfall. Oliveria (1986), inspecting ten years of geostationary satellite imagery, showed that many frontal systems, especially during the Southern hemisphere summer, move equatorward along eastern South America, organizing convection and intensifying rainfall over Amazonia. Figure 5, taken from her work, shows schematically the four broad classes of preferential positioning of the frontal systems. The most frequent types, 2 and 3, are generally oriented NW-SE; both influence the central Amazonia and differ only by the latitude where they cross the east coast, between latitudes 15° and 25° S. In the case of type 1, the air mass of polar origin moves equatorward close to the Andes and affects primarily the western side of Amazonia.

On the other hand, the frontal systems of type 4 travel along the east coast and influence mainly the eastern side of Amazonia. The fact that types 2 and 3 are the most common ones may explain the secondary rainfall maxima which are observed in central-southern Amazonia (Fig. 4c). In some years, the frontal systems remain stationary in this position, establishing a region of convergence with the trade wind – denominated the South American Convergence Zone (SACZ)

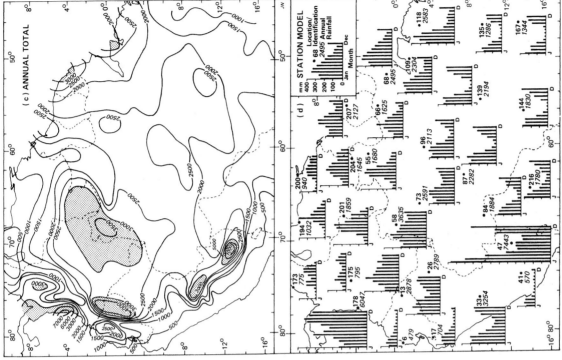

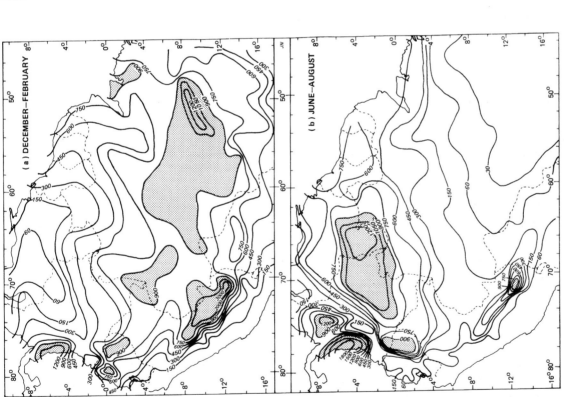

Figure 4a: Total rainfall map of Central and Western South America for the period December to February in millimetres (After Figueroa & Nobre, 1990).
Figure 4b: Total rainfall map of Central and Western South America for the period June to August in millimetres (After Figueroa & Nobre, 1990).

Figure 4c: Annual rainfall total in millimetres.
Figure 4d: Histograms of the annual distribution of monthly rainfall for selected stations. In each circle the dot shows the station location, the bold number is the station identification and the number with four digits is the annual rainfall total (After Figueroa & Nobre, 1990).

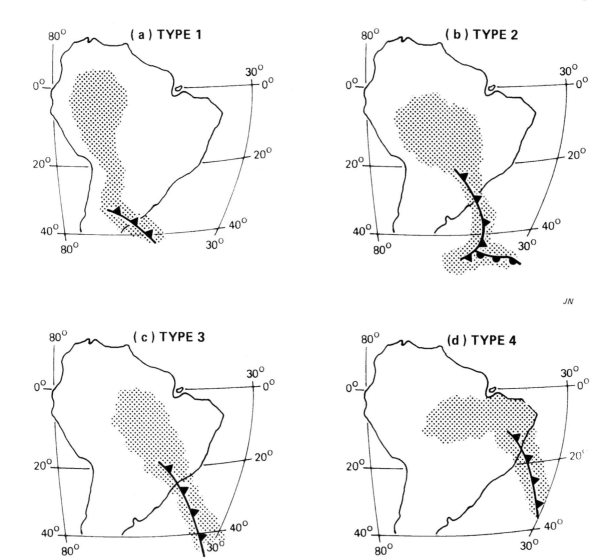

Figure 5: A schematic diagram showing the four types of preferred location of frontal systems and associated cloud bands over South America (Oliveira, 1986).

– that produces high rainfall totals. On such occasions, the Amazon right bank tributaries are subject to high flooding conditions.

Northern hemisphere frontal systems may also influence rainfall in Amazonia directly. Molion *et al.,* (1987) presented a case-study for February 1980, when the successive passage of frontal systems over the subtropical North Atlantic favoured the penetration of relatively cold and dry air of that hemisphere into Amazonia. The cold and dry air mass helped to organize an east-west convective cloud band along latitude 7°S, which was intensified by merging Southern hemisphere frontal systems. That particular event produced the highest rainfall totals in 50 years in southern Amazonia, where some stations registered over 250 mm in 24 hours and more than 600 mm in a l0-day period.

Subsynoptic scale

This scale has typical lengths of 500 km with the active atmospheric systems being the instability, or squall, lines, and the clusters of large cumulo-nimbus clouds (Cbs) associated with them. In the absence of large-scale forcing, the development of convective cells starts in the mid-morning. These cells undergo a selection process by which the larger ones grow, forming clusters or lines, whereas the smaller ones are suppressed. The formation of a line or a cluster depends on the tropospheric flow pattern. With a moderate wind field, the descending currents (downdrafts) of the original cell act as mini-frontal zones, raising the ambient humid air. The new cells, originated from this forced convection, will form in an arched line downwind (Fig. 6a). When the wind field is weak, the new cells surround the mother-cell as a ring or cluster that continues to grow at the expense of the downdrafts (Fig. 6b).

Cavalcanti (1982) demonstrated that the convergence associated with sea breeze circulation organizes convection in a linear fashion over coastal areas, near the mouth of the Amazon River. On some occasions, these lines propagate inland, reaching the Andes in about 48 hours (Molion & Kousky, 1985). Recently, Cohen *et al.,* (1989) analyzed eight

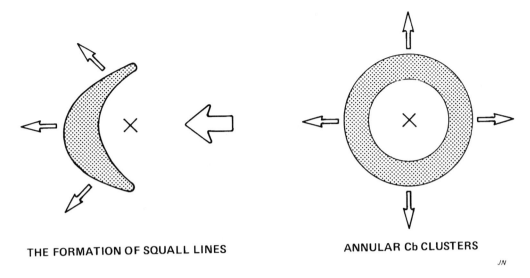

THE FORMATION OF SQUALL LINES   ANNULAR Cb CLUSTERS

JN

Figure 6: A schematic diagram showing (6a) the formation of squall lines and (6b) annular Cb clusters (Molion & Kousky, 1975).

years of geostationary satellite imagery and, with additional surface and upper-air data, were able to identify some of the physical properties of these instability lines. The highest frequency of lines occurred in July, when the ZTE is farther north. A secondary maximum was apparent in April, when the ZTE is at its southernmost position. Of the total observed number of coastal linear disturbances, only 27% propagated more than 400 km inland. Some of these reached over 2,000 km, with typical propagation speeds of about 12 to 16 ms^{-1} and remained active for some 12 to 21 hours. The characteristic length and width were 1,500 km and 170 km, respectively. Cohen *et al.,* (1989) did not monitor the contribution of these instability lines to the regional rainfall totals. They did, however, estimate that up to 45% of the eastern Amazonian rainfall may be due to such systems. Based on precipitation daily records of two stations, one in eastern and another in central Amazonia, and geostationary satellite imagery sequences every half-hour, it was estimated that these lines produce average rainfall intensities in the order of 10 to 20 mm. h^{-1} during the three to five hours they remain over the stations (Cohen *et al.,* 1989). The convergence of the sea breeze may not be the only mechanism to generate these transient disturbances, since it was observed that some of them form and develop during the night-time. Some of these lines may be associated with waves in the trade winds field, triggered by deep penetration of frontal systems over the subtropical Atlantic which sometimes reach the geographic equator. In the absence of atmospheric blocking patterns, these deep penetrations of frontal systems can occur throughout the whole year. They are particularly effective, however, during periods when the ZTE is located farthest from the equator, especially during the most northerly position in July.

Mesoscales and microscales

The mesoscale, or Global Climate Models (GCMs) grid-scale, has a typical length of 100 km. The most common weather features at this scale are clusters of Cbs, which produce intense precipitation but of relatively short duration and random distribution. Most of the time, they are immersed in frontal systems but they may develop as isolated clusters, as sketched in Fig. 6b. The larger clusters of 50 to 300 km equivalent diameter, may live typically 5 to 12 hours, with average rainfall intensities varying from 10 to 30 mm.h^{-1} and average rainfall totals reaching 50 to 150 mm in a 24 hour period. Molion & Dallarosa (1990) analyzed data sets from two paired raingauges, one, with an 11 year record, installed along longitude 56° W and another, with a ten year record, along longitude 60° W. They showed that the gauges near the Amazon River recorded an annual average rainfall 20% to 30% less than the gauge 30 to 100 km away from the large river. During the driest trimester (July-September), the time when the mid- and lower Amazon River is flooded, the difference between the near and far Amazon River location was even more striking, being 97% (at 50° W) and 35% (at 60° W), respectively. They attributed the reduction in rainfall mainly to the river breeze circulation, since it can easily develop over areas where the river is sufficiently large, i.e. wider than ten km. Additional evidence for the existence of river breeze systems is given by inspecting the high-resolution visible imagery from meteorological satellites. It is evident that clouds tend to form mainly over the river banks, leaving the main channels visible from space. The presence of the river breeze, therefore, can alter the local convection patterns. Since the raingauges in Amazonia are not randomly distributed but are mainly located near the banks of rivers, any rainfall analyses will be biased towards the lower end of the spectrum.

The microscale mechanisms have characteristic lengths of 1 km with the most common being the small convective cells, two to four km equivalent diameter, which form during the morning hours and precipitate around 1400 to 1600 hours local time. Due to the highly unstable conditions of the local air masses, small hills (100–200 metres in elevation), in com-

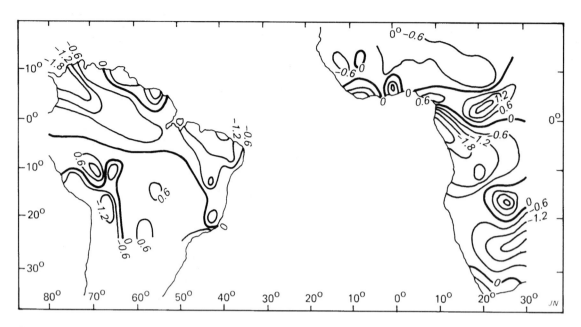

Figure 7: Isolines of rainfall deviations from the mean, normalized by the standard deviation of the series, for the period February to May 1958 (After Moura & Kayano, 1983).

bination with surface roughness, are sufficient to vertically lift the atmospheric flow to the condensation level. The small convective cells formed by this process grow selectively. They are short-lived, typically one to two hours, with average precipitation totals smaller than 5 mm. In the 11 years of record from Cachoeira Porteira, (Trombetas Basin, 56° W), for example, it was found that 57% of 365 days were rainy days, and 43% and 62% of these had rainfall daily totals smaller than 5 mm day^{-1} and 10 mm day^{-1}, respectively. In central Amazonia, the 25 year rainfall record of Reserva Ducke, 25 km northeast of Manaus (60° W), presented a similar picture, with 66% of the total being rainy days, out of which 50% and 67% rainfall daily totals were smaller than 5 mm day^{-1} and 10 mm day^{-1}, respectively.

The rainfall from microscale Cbs is more frequent but they may not be as important for the local hydrology, i.e. for generating runoff, as they may be for the existence of the tropical forest. The increasing cloudiness in mid-morning reduces the solar radiation load during the time when it is most intense and the subsequent light showers wet the canopy so that it is kept cool through evaporation of intercepted water. Consequently, the trees are not required to transpire as much as they would otherwise without clouds.

INTER-ANNUAL VARIABILITY

The physical causes of the inter-annual variability of rainfall in Amazonia are not adequately understood yet, but it is surely linked to the spatial and temporal variability of the equatorial latent heat sources. Such sources depend on various boundary conditions, that is, the sea surface temperature (SST), the available energy (net radiation) at the surface and the soil moisture. One of the phenomena that greatly modifies both the positioning and intensity of the heat sources is the El Niño-

Southern Oscillation (ENSO) event (see Manton and Bonell, this volume). During an ENSO warm phase (negative phase of SO) event, there is a reduction in rainfall because the ascending branch of the Walker Circulation (east-west component), previously located over the Amazonas Basin, is displaced westward over the abnormally warm eastern Pacific Ocean. This circulation is further intensified due to strong convection over the eastern Pacific Ocean. The descending branch covers the whole of Amazonia, the adjacent Atlantic and reaches the African continent. This corresponding subsidence suppresses convection and, thus, rainfall is dramatically reduced over these areas. Figure 7 (Moura & Kagano, 1983) shows isolines of deviations from the mean rainfall, normalized by the standard-deviation, for the period February-May 1958, a year when a strong ENSO event occurred. Over parts of Amazonia and West Africa, it is noticeable that there was a rainfall reduction in excess of 1.8 standard-deviations. For the period January-May 1983, the year of the strongest ENSO of the century, Kousky *et al.*, (1984) noted that selected stations in Amazonia recorded a reduction of rainfall in excess of 35%. Similarly, Nobre & Renno (1985), using all available station records, reported rainfall 70% below normal for January and February 1983. In fact, 1983 witnessed the driest February on record (based on 50 years of measurements). Figure 8, adapted from Kagano & Moura (1987) shows rainfall reductions in excess of 0.5 standard-deviations for the entire Amazonia during the hydrologic year October 1982 to September 1983.

Conversely, during positive phases of the SO (cold SST anomalies) – also called Anti-El Niño – or La Niña (see Falkland and Brunel, this volume) enhanced convective activity occurs, which subsequently increases rainfall over Amazonia. Typical Anti-El Niño examples are the hydrologic years 1975–1976 and 1988–1989 when the Negro River, in

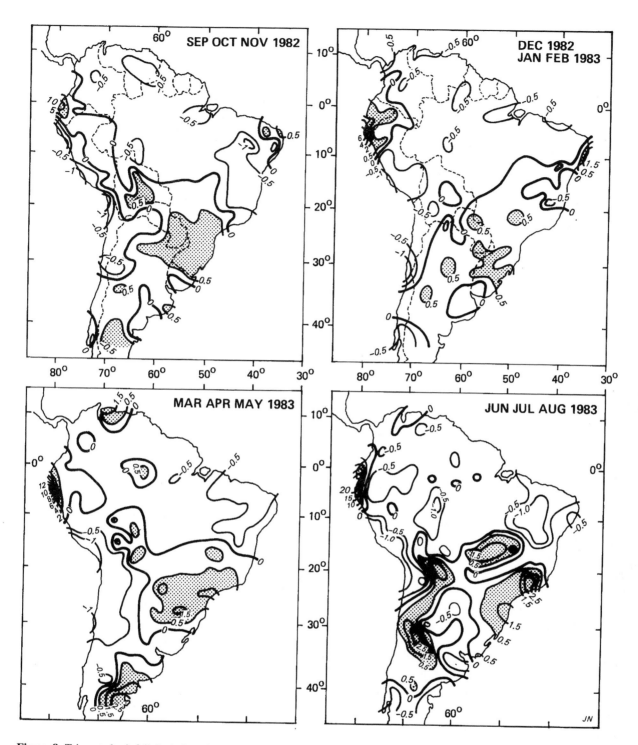

Figure 8: Trimestral rainfall deviations from the mean, normalized by the standard deviation of the series, for the period September 1982 to August 1983, year of the strongest El Nino of the century – adapted from Kayano & Moura (1987).

Manaus, recorded the second and the third highest levels of the century, respectively.

Figure 9, extracted from Rao & Hada (1990), shows isolines of correlation coefficients between the Southern Oscillation Index (SOI) and precipitation over Brazil during the trimester September-November, using 21 years of records. Positive correlation coefficients, exceeding 0.6, are found in northeast Amazonia (56° W), suggesting strong coupling between the SO and the rainfall in that part of the Basin.

Precipitation data point analysis of individual rain station records, however, do not always reflect the extent of convective activity and associated magnitude of rainfall during such SO phases. This is due to inadequate spatial distribution of gauges in the region. Molion & Moraes (1987) correlated a time series of SOI with the corresponding time series of discharges for selected rivers in South America. For the Trombetas River, a left bank tributary of the Amazon near longitude 56° W, the correlation coefficients were positive

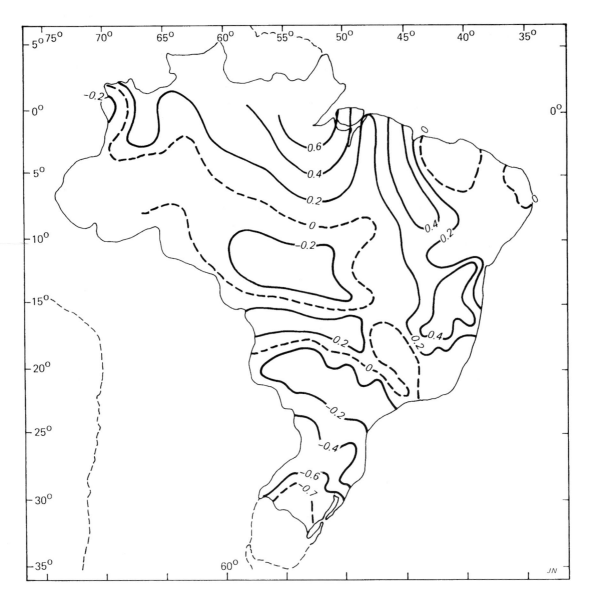

Figure 9: Isolines of correlation coefficients between SOI and rainfall for Brazil for the trimester September-November (Rao & Hada, 1990).

and higher than 0.8, implying that when the SOI is negative (El Niño years), the discharge is below the normal and vice-versa. As an example, for one particular day in February 1983, the observed Trombetas River discharge near the mouth was 47 m^3 s^{-1}, the lowest recorded so far. The long-term average discharge for this month exceeds 2,100 m^3 s^{-1}. The authors also noted that there exists a three-month time-lag between the SOI and river discharge. Molion and Morales (1987) suggested that, since the SOI is the leader, it might be used to forecast the discharge of this large river.

Other mechanisms besides ENSO may cause variations in rainfall over Amazonia. Moura & Shukla (1981) showed that when the sea surface temperatures (SST) in the North Atlantic were above normal, and simultaneously the SST in the South Atlantic were below normal, the ZTE remained northward of its normal position. At such times, the descending branch of the Hadley Cell intensifies, which causes strong subsidence

over central and eastern Amazonia, therefore reducing convection and rainfall.

As mentioned, the Southern hemisphere frontal systems are important large-scale mechanisms organizing convection in Amazonia. Their penetration may be affected by atmospheric flow blocking patterns which occur in some years over the southern part of the continent and adjacent areas of the Pacific Ocean. In years when subsiding air is more persistent, this produces a blocking mechanism which causes a reduction in the number of frontal systems penetrating Amazonia. Consequently, rainfall is reduced.

It is possible that precipitation increases are associated with the presence of volcanic aerosols in the stratosphere of low latitudes. The stratospheric aerosols reduce the solar radiation significantly in the tropics for several months during periods of strong volcanic activity (see e.g. De Luise, 1983). With reduced solar radiation absorption in low latitudes, the equator-to-pole temperature gradient is consequently reduced. This weakens the Hadley-Walker circulation, causing the subtropical anticyclones to decline in intensity and even be displaced poleward. Such ideas were also put forward by Flohn (1981). The atmospheric surface pressure becomes lower than average over the equatorial portion of the continent with convection and precipitation being enhanced. The resulting increase in soil moisture also provides a positive feedback into the atmospheric system through enhanced evaporation. Thus, above average rainfall occurs during the rainy season. Coincidentally, the Amazonian rainy seasons that followed the eruptions of Fuego (October 1974), Nyamuragira (December 1981) and Nevado del Ruiz (November 1985) were all above normal. This hypothesis, however, is difficult to verify because of the complex interactions with other mechanisms. For example, the mentioned eruptions coincided with cold phases of ENSO when convective activity would be enhanced anyway. In addition, measurements of aerosol concentrations in the stratosphere are not easily available, and the eruptions obviously have to coincide with the Amazonian rainy season for this hypothesis to have any credibility.

Other possible climatic effects related to volcanic aerosols have been described. Handler and Andsager (1990), for example, suggested that there is a statistically-significant relationship between the occurrence of warm ENSO events and the presence of volcanic aerosols in the stratosphere of low latitudes.

In addition to the observed inter-annual variations, longer-term variations also seem to be apparent. Rocha *et al.*, (1989) noticed a positive trend in rainfall at central and western Amazonian stations, which lasted between 10 and 15 years, from the beginning of the 1960s. There is no adequate explanation for these trends as yet. One possible hypothesis is that this longer-term trend may be associated with a higher frequency of negative anomalies of sea surface temperature (ENSO cold events) in the eastern Pacific. In the interval 1961–1976, in 11 out of 16 years, the SSTs of the central Pacific, between 90 °W and 160° W (Niño 3 Region), were below normal. The results of widely different GCMs provide some encouraging support for the hypothesis that tropical SST anomalies can produce significant changes in the atmospheric circulation and distribution of tropical rainfall (Shukla, 1987). Another hypothesis is that these long-term trends might be linked to the presence of volcanic aerosols in the stratosphere, as mentioned above. During the period 1963–1974, higher than normal volcanic activity and a corresponding reduction in solar radiation were observed in equatorial regions (see *e.g.*, Handler & Andsager, 1990). Such linkages possibly maintained below average atmospheric pressure over Amazonia, which in turn, resulted in wetter than normal years over the same period.

CONCLUDING REMARKS

Changes in rainfall related to human actions modifying the landscape are controversial issues, mainly because of the difficulty in isolating these changes from the natural ones at continental scales. Human effects at smaller scales, however, are better accepted. For example, extensive urbanization increases rainfall over large cities due to the enhanced surface aerodynamic roughness and the "heat island" effect. The introduction of artificial lakes can cause rainfall amounts to be increased downwind by up to 40%, because of the establishment of an additional mesoscale rain-producing mechanism, the lake breeze. Changes at a large-scale, however, are viewed more skeptically. For example, Soviet scientists claimed an increase of 1% to 2% in rainfall due to irrigation of an area of about half a million square kilometres. These percentages are well within the error limits of rainfall measurement and, therefore, cannot be statistically proven.

In Amazonia, the same controversial issues pertain to the removal of the tropical forest. Gentry & Lopez-Parodi (1980), for example, suggested that the upward trend of the Amazon River discharge at Iquitos, Peru, observed in the period 1965–1978, could be related to deforestation of the Peruvian and Ecuadorian Amazonia. As cited above, Rocha *et al.*, (1989) have shown that, in fact, there was an increase of rainfall totals over the central and western parts of Amazonia during that period, due to a natural, non-explained long-term variability. It is also worth mentioning that the highest Rio Negro stage level this century, recorded at the Manaus gauge, was in 1953 when extensive forest clearance was located only in eastern Amazonia.

Nevertheless, conversion of tropical humid forests to crop or pasture fields on a large scale over Amazonia may reduce evaporation and, therefore, local rainfall, since about half of the precipitation depends on local evaporation (Molion, 1975; Shuttleworth, 1988). Climate simulation models (GCMs) were used recently to test the effects of forest removal on the regional climate (Dickinson & Henderson-Sellers, 1988; Lean & Warrilow, 1989; Shukla *et al.*, 1990; also reviewed in Bonell with Balek, this volume). The results suggested a 20% to 30% reduction in rainfall over the Basin. Total evaporation was also reduced up to 50% in areas where high rainfall totals occurred. Another important component of the hydrologic cycle, total runoff, showed reductions between 10% to 20%. The GCMs results also implied that forest clearance may change both the spatial and temporal distribution of hydrologic variables, and increase the length of the dry season. However, most existing general circulation models cannot comprehensively parameterize the physical processes at the land surface at present, especially the role of vegetation in determining the energy and water balances. On the other hand, Shuttleworth *et al.*, (1990) recently reported that the data from the Amazon Region Micrometeorological Experiment (ARME) were satisfactorily used to calibrate land surface parameterization schemes SiB (Sellers *et al.*, 1989) and BATS (Dickinson, 1989). Much improved simulations of the surface exchanges of energy and water were possible after these calibrations. Other effects of land-use conversion include the reduction of infiltration due to soil compaction and the corresponding changes in the soil moisture regime. GCMs have suggested such changes in soil hydraulic properties to be an important climate control in the tropics. Schubart (1977), for example, compared infiltration rates of a pasture and an adjacent undisturbed forest soil and determined that infiltration rates were l0 to 20 times higher in the forest (see Bonell with Balek, this volume). Furthermore, decreased soil water availability for crops increases the frequency of plant water stress and reduces plant productivity.

Removal of the forest also deletes the canopy interception store so more rainfall is available for overland flow. In ARME (see e.g., Molion, 1987; Shuttleworth, 1988) rainfall interception by the forest was about 15% of total rainfall. So for small-scale forest clearance, overland flow increases, thus increasing erosion significantly through the exposure of the fragile soils. Jannson (1982) reviewed the literature on tropical soil erosion in Amazonia and established erosion rates up to 334 t ha^{-1} yr^{-1} may be washed away every year. Erosion causes silting in the river channels and changes the water quality and the aquatic life.

The various impacts that large-scale deforestation in Amazonia might have on the global climate are still debatable issues. These were discussed in more detail elsewhere (Molion, 1990). One possible effect is that the decline of rainfall, due to reduced evaporation, decreases the latent heat release in the Amazonian troposphere and, therefore, the power of its heat source. Then, in principle, less energy would be available for transportation to the extra-tropics, thus bringing changes to the present climate of those remote regions. Another effect, not related directly to the hydrological cycle, is the influence that the Amazonian forest may have on the "Greenhouse Effect," through the imbalance of photosynthetic absorption rates versus the release rates of carbon by forest removal and burning (Molion, 1990).

Some aspects described in this paper are from observational studies, others are speculative due to the lack of more precise information and adequate tools for studying the effect of such land-use changes. Considering the influence that the Amazonia forest and its hydrologic cycle may have on local as well as global climate, and the difficulty in isolating the anthropogenic effects from the natural climate variability, there is an urgent need to intensify the observational studies on forest-atmosphere interactions, along the lines of ARME. Such initiatives would improve various attempts at modelling the effects of large-scale forest conversion, particularly the influence of Amazonia on remote extra-tropical climates.

ACKNOWLEDGEMENTS

The author thanks M. Bonell – formerly with the Department of Geography of James Cook University, Townsville, Australia, now with the Division of Water Sciences, UNESCO – and N. Ferreira, the author's colleague at Instituto de Pesquisas Espaciais (INPE), for revising the text and their suggestions for improving it.

REFERENCES

Cavalcanti, I. F. A. (1982) Um Estudo sobre Interacoes~ entre Sistemas de Circulaçao de Escala Sinoptica e Circulacoes Locais• Tese MSc, INPE 2494–TDL/097, Sao Jose dos Campos, S.Paulo.

Cohen, J. C. P., da Silva Dias M. A. F & Nobre,C. A. (1989) Aspectos climatologicos das linhas de instabilidade na Amazonia. *Climanalise, 4* (11) 34–40, CPTEC/INPE, S.J. dos Campos, S.Paulo.

De Luise, J. J., Dutton, E. G., Coulson, K. L., DeFoor, T. E. & Mendonca, B. G. (1983) On some radiative features of the El Chichon volcanic stratospheric dust cloud and a cloud of unknown origin observed at Mauna Loa. *J Geophys. Res 88*: 67–69.

Dean, G. A. (1971) Three-Dimensional Wind Structure over South America and Associated Rainfall over Brazil. Florida State University, Department of Meteorology, Report 71–4.

Dickinson, R. E. (1989) Implications of tropical deforestation for climate: a comparison of model and observational descriptions of surface energy and hydrological balance. *Phil. Trans.Roy. Soc.Lond-. B324*:339–347.

Dickinson, R. E. & Henderson-Sellers, A. (1988) Modelling tropical deforestation: a study of GCM land-surface parameterization. *Quart J. Roy Met. Soc.,114:* 439–462.

Figueroa, S. N. & Nobra, C. A. (1990) Precipitation distribution over Central and Western Tropical South America. *Climanalise 5(6)*:–36–45, CPTEC/INPE, S.J.Campos, S.Paulo

Flohn, H. (1981) Scenarios of cold and warm periods of the past. In: *Climatic Variations and Variability: Facts and Theories*, pp.689–698, Reidel Pub. Co.

Gentry, A. H. & Lopez-Parodi, J. (1980) Deforestation and increased flooding of the upper Amazon'. *Science, 210*:1354–1356.

Handler, P. & Andsager, K. (1990) Volcanic aerosols, El Niño and the Southern Oscillation. *J. Climat* (in press).

Hastenrath, S. & Lamb, P (1977) *Climate Atlas of the Tropical Atlantic and Eastern Pacific Oceans*. University of Wisconsin Press, Madison, WI.

Jansson, M.B. (1982) Land Erosion by Water in Different Climates. UNGI Report 57, Department of Physical Geography, Uppsala University, Sweden.

Kagano, M. T. & Moura, A. D. (1987) O El Niño de 1982–83 e a precipitaçao sobre a America do Sul. *Rev. Bras. Geofisica*

Kousky, V. E. & Kagano, M. T. (1981) A climatological study of the tropospheric circulation over the Amazon Region. *Acta Amazonica, ll:*743–758.

Kousky, V. E.,Kagano, M. T. & Cavalcanti, I P. A. (1984) A review of the Southern Oscillation: oceanic, atmospheric circulation changes and related rainfall anomalies. *Tellus, 36A:* 490–504.

Lean, J. & Warrilow, A. (1989) Simulation of the regional climatic impact of Amazon deforestation. *Nature 342:* 4ll–413•

Miller, D. B. & Feddes, R. G. (1971) *Global Atlas of Relative Cloud Cover 1967–1970* US. Air Force (AWS) Department of Commerce, NOAA, Washington, DC.

Molion, L. C. B. (1975) A Climatonomic Study of the Energy and Moisture Fluxes of Amazonas Basin with Consideration of Deforestation Effects, Ph.d Thesis, University of Wisconsin, Madison, WI.

Molion, L. C .B. (1987) Micrometeorology of an Amazonian Rain Forest.

In: R.E.Dickinson (ed) *The Geophysiology of Amazonia*, 255–270, UNU, John Wiley and Sons.

Molion, L. C. B. (1990) Amazonia: burning and global climate impacts. In: *Proc. of Chapman Conference on Global Biomass Burning*: atmospheric, climatic and biospheric implications Williamsburg, VA (in press)

Molion, L. C. B. & Kousky, V. E. (1985) Climatologia da dinamica da troposfera tropical sobre a Amazonia, INPE–3560–RPE/480• Sao Jose dos Campos, S.Paulo.

Molion, L. G. B. & de Moraes, J. C. (1987) Oscilaçao Sul e descarga de rios na America do Sul Tropical, *Rev. Bras. Eng.* Caderno de Hidrologia 5 (1) 53–63.

Molion, L. C. B., Cavalcanti, I. F. A. & Perreira, M. E. (1987) Influencia da circulacao do Hemisferio Norte na precipitaçao pluviometrica da Amazonia: um estudo de caso. Anais do VII Simposio de Hidrologia e Recursos Hidricos, Salvador, BA.

Molion, L. C. B. & Dallarosa, R. L. (1990) Pluviometria da Amazonia: sao os dados confiaveis? *Climanalise 5* (3) 40–42, CPTEC/INPE, S.J. dos Campos, S.Paulo.

Moura, A. D. & Shukla, J. (1981) On the dynamics of droughts in Northeast Brazil: observations, theory and numerical experiment with a general circulation model, *.J. Atmos. Sci. 38*: 2653–2675.

Moura, A. D. & Kagano, M. T. (1983) Teleconnections between South America and Western Africa as revealed by monthly precipitation analysis. In: *Proc. of the First Int. Conf. on Southern Hemisphere Meteorology*, AMS, Sao Jose dos Campos, S.Paulo: 120–122.

Newell, R. E., Kidsen, J. W., Vincent, D. G. & Boer, G. J. (1972) *The General Circulation of the Tropical Atmosphere*, Massachusetts Institute of Technology.

Nobre, C. A. & Renno, N. O (1985) Droughts and Floods in South America due to the 1982–83 ENSO Episode. In: *Proc. of the 16th Conf. on Hurricane and Tropical Meteorology*, AMS, Houston, Texas: 131–133

Oliveira, A. S. (1986) Interacoes entre Sistemas Frontais na America do Sul e Convecçao na Amazonia, Tese MSc, INPE 4008 TDL/239. Sao Jose dos Campos, S.Paulo.

Parmenter, F. C. (1976) A Southern hemisphere cold front passage at the equator, *Bull. Amer. Met. Soc.*, 57: 1435–1440.

Rao, V. B. & Hada, K. (1990) Characteristics of rainfall over Brazil: annual variations and connections with the Southern Oscillation. *Theoretical and Applied Climatology* (in press).

Ratisbona, L. R. (1976) The climate of Brazil. In: *Survey of Climatology 12 Climates of Central and South America* W.Schwerdtfeger (ed), Elsevier, Amsterdam.

Rocha, H. R., Nobre, C. A. & Barros, M. C. (1989) Variabilidade natural de longo prazo no ciclo hidrologico da Amazonia. *Climanalise 4* (12) –36–42, CPTEC/INPE, S.J.dos Campos, S.Paulo.

Sadler, J. C. (1975) The monsoon circulation and cloudiness over the GATE area. *Mon. Weath. Rev.*, *103* (5), 369–387.

Schubart, H. O. R (1977) Criterios ecologicos para o desenvolvimento agricola das terras-firmes da Amazonia. *Acta Amazonica 7(4)* –559–567.

Sellers, P. J, Shuttleworth, W. J., Dorman, J. L., Dalcher, A. & Roberts, J. M. (1989) Calibrating the Simple Biosphere Model for Amazonian tropical forest using field and remote sensing data • *J. Appl Met.*, *28(8)*.

Shukla, J. 1987) General Circulation Modeling and the Tropics. In: R. E. Dickinson (ed) *The Geophysiology of Amazonia*, 409–458, UNU, John Wiley and Sons.

Shukla, J., Nobre, C. A. & Sellers, P. (1990) Amazon deforestation and climate change,*Science 247*:1322– 1325.

Shuttleworth, W. J. (1988) Evaporation from Amazonian rain forest. In: *Proc. Roy. Soc. Lond. B 233*: 321–346.

Shuttleworth,W. J, Gash, J. H. C., Lloyd,C. R., Moore, C. J., Roberts, J. M., Marques Filho, A. O, Fish,G., Silva Filho, V. P., Ribeiro, M. N. G., Molion, L. C. B., Nobre, C. A., Sa, L. D. A, Cabral, O. M. R., Patel, S. R. & Moraes, J. C. (1984) Eddy correlation measurements of energy partition for Amazonian Forest, *Quart. J. Roy. Meteor. Soc.110*: 1143–1162.

Shuttleworth,W. J., Gash, J. H. C., Roberts, J. M., Nobre,C. A.., Molion,L. C. B. & Ribeiro,M. N. G. (1990) Post-deforestation Amazonian climate: Anglo-Brazilian research to improve prediction. Symp. on Hydrology and Water Management of the Amazonas Basin, August 5–9, Manaus.

Trewartha, G. T. (1961) *The Earth's Problem Climate,*. University of Wisconsin Press, Madison, WI.

9: Regional Hydrology and Water Resources in the African Humid Tropics

N. B. Ayibotele

Water Resources Research Institute, Council for Scientific & Industrial Research, P. O. Box M32 Accra, Ghana

ABSTRACT

This paper reviews the current hydrological and water resources knowledge in Africa and some of the associated problems of the humid tropical region of Africa

Data availability and accuracy are identified as major constraints. Estimates of water resources vary widely. Knowledge about the lakes in the region is inadequate, apart from information on their physical characteristics. A generalized knowledge about the sediment yield for the African region has been put forward recently but is only tentative because of the data constraint. Subregional and regional syntheses are needed on water quality, droughts and low flows. Such a synthesis is available for groundwater, but needs updating with the large amount of data that has been made possible through the International Drinking Water Supply and Sanitation Decade. The impact of land degradation and climate variation on the hydrology and water resources of the region is an area of study that also needs urgent attention.

PHYSIOGRAPHY

Location

The location of the humid tropic region is the low-lying areas within the two tropics of Cancer and Capricorn – basically as defined by Chang and Lau (this volume). Figure 1 shows the African regional coverage which generally lies between latitudes 15°N and 20°S. The region is bounded on the west by the Atlantic Ocean, on the east by the Indian Ocean, on the north by the Sudano-Sahelian belt and Sahara Desert and on the south by the Kalahari Desert.

The countries covered by this area are the central African countries of Zaire, The Congo, Gabon, Central African Republic, parts or all of the West African countries of Cameroon, Nigeria, Benin, Togo, Ghana, Cote d'Ivoire, Liberia, Sierra-Leone, Guinea, Guinea Bissau and Equatorial Guinea as well as parts or all of the east African countries of Ethiopia, Uganda, Kenya, Angola, Tanzania, Zambia, Mozambique, and Burundi. It also covers the islands of Madagascar, the Comoros, Zanzibar and the islands of Equatorial Guinea.

Topography

The land surface of the humid tropical region of Africa is moderately rugged with the central part formed by a plateau and tablelands averaging 200 to 500 m in altitude. The plateau is bordered on the edges by highlands. The Atlas Mountains (4,165 m), the Futa Jalon Mountains (1,537 m) and the Cameroon Mountains (4,380 m) break the plateau terrain in the west. The highest points are found in the east – Kilimanjaro (5,895 m) and the Ethiopian highlands (4,620 m). In the east, the relief has been affected by gigantic fractures forming the Great Rift Valley, which is flanked by the Kilimanjaro, Kenyan and Ruwenzori Mountains. Figure 1 shows the relief and drainage of the continent's humid tropics, along with the rest of the continent.

Drainage

The humid tropical area is drained by the major rivers of the continent, viz. the Congo, Nile, Niger, Zambezi, Orange, Volta and the Tana (Fig. 1). The preponderance of the plateau relief has marked effects on the drainage of the continent. Rivers which rise from the bordering highlands either flow from their sources into the sea (like the rivers in East Africa) or they flow from the highlands on to the plateau and there they pass sluggishly over long distances, forming swamps and wide flood plains along their paths. Eventually, they descend as falls or rapids from the plateau into the sea. Another significant feature of the drainage of Africa is the presence of swamps which act as regulating reservoirs. The prominent swamps include the Sudd swamps of the White Nile, the Banggweulu, Lukanga, Okavango and Niver swamps and the Kafueflats.

The continent has notable lakes whose waters discharge into the sea through rivers. The most important are Lakes Victoria, Tanganyika and Rudolf. There are also a number of basins of inland drainage, like those of the Okavango, the Chad, the Gash, the Awash and the Etosha. The most notable within the humid tropics is that of Lake Malawi.

Soils

The main soil associations of Africa are shown in Fig. 2. The major soil types in the humid tropics are:

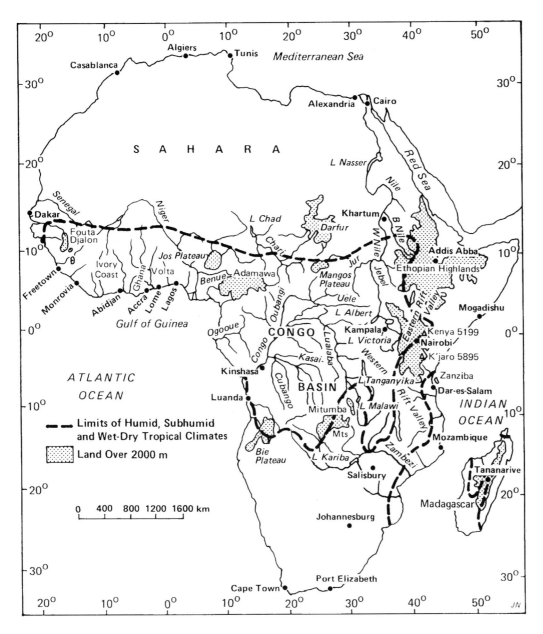

Figure l: Drainage map of Africa.

(a) Acid soils found in the lowlands of the Congo, the Niger and the Nano Rivers (Ferralsols and Acrisols). They are known to have good physical properties and are fairly resistant to erosion.

(b) Ferruginous tropical soils found in the north part of Cote d'Ivoire, in Burkina Faso, parts of Ghana, Togo and in Angola, Zimbabwe and Mozambique (ferric Luvisols and Cambisols). They are easily cultivated but they also lose their structure easily; they are able to absorb rainwater that falls on them and are thus less subject to erosion.

(c) Tropical highland soils in Kenya and Ethiopia (Nitosols and Ondosols). They occur in volcanic areas, or those sectors with basic rocks and limestones. They are erodible on steep hills.

(d) Poorly drained soils in parts of Nigeria, Cameroon, Zaire and the Congo (Greysols and Flevisols). They are of

various origins – alluvial, marine, lacustrine and fluviatile.(FAO, 1986a).

Vegetation

The vegetation type is closely related to climatic conditions which are locally modified by topography and soil. The main types are shown in Fig. 3. They are:

(a) tropical moist forests at low and medium altitudes (as found in the Sierra Leone, Liberia, Nigeria, Cameroon and Congo belt).

(b) tropical moist and dry forest on savannah mosaics (as in the transitions between the moist areas and the Sudano-Sahelian regions and also in southern Africa).

(c) mountain vegetation (as in the Futa Jalon, Cameroon, Ethiopian and East African highlands).

Almost all the forest areas have been affected by human

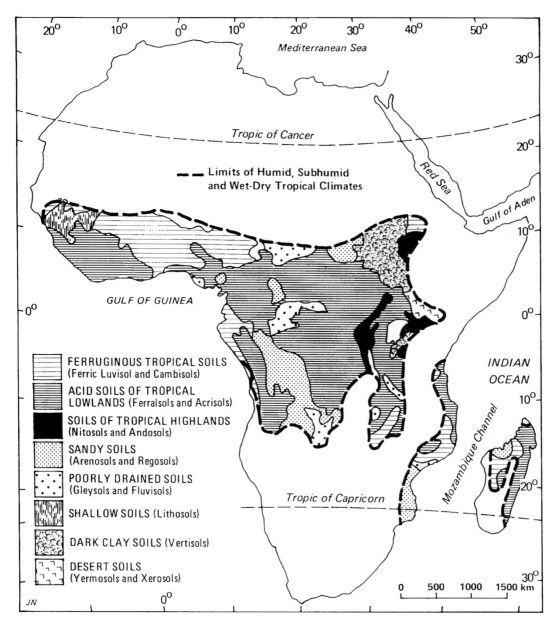

Figure 2: Main soil associations of Africa (Source: FAO, 1986).

activities. The rate of deforestation has become a matter of concern to African governments because of its impact on the climate, land and water resources (UNEP, 1985).

HYDROLOGY

The climate of the continent is characterized by its geographical position. Firstly, as the continent with the greatest land area in the tropics, it receives a large amount of solar heat (70 kcal cm^{-2} $year^{-1}$). Secondly, the low pressure belt created by the high temperatures around the equator places the continent predominantly under the influence of northeasterlies and southeasterlies blowing from the high pressure belts of the subtropics towards the equator. The subtropical parts of the continent are under the influence of the southwesterlies and northwesterlies in July and January, respectively. Thirdly, moisture is brought to the continent by the variable monsoon winds from the south Atlantic and Indian Oceans.

The simple wind system described is distorted by the north-south movement of the sun and the ensuing uneven heating of the continental land mass and the sea. In July, the sun is overhead in the Northern hemisphere. This is accompanied by the northward movement of the equatorial belt and the drawing-in of the moist monsoon winds from the Atlantic and Indian oceans. Such winds meet within the low pressure belt which has moved northwards in a zone – called the northern monsoon shearline or monsoon trough (see Manton and Bonell, this volume) – whose location from the equator varies with longitude. Rainfall is caused by the convergence of the two wind systems which forces the moist monsoon air to rise to cooler altitudes and then by condensation release its moisture as rain.

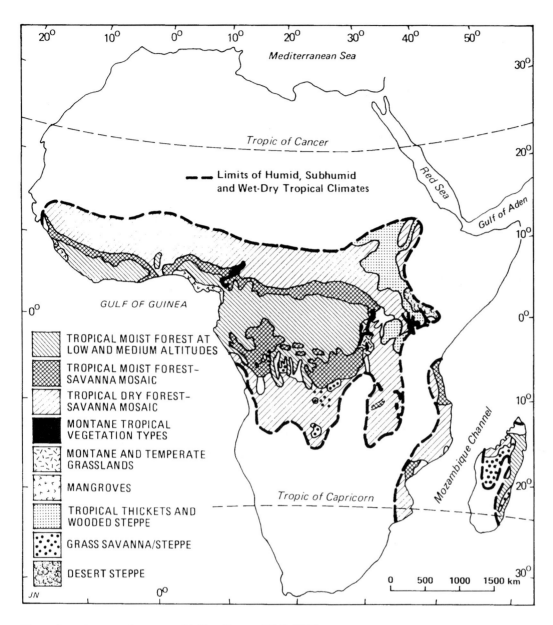

Figure 3: Main vegetation types of Africa (Source: FAO, 1986).

The Southern hemisphere of the continent is at this time in winter and due to the greater cooling of the land than the sea, a high pressure belt builds up in the continental southern section. Dry winds, therefore, move from this area towards the low pressure zone formed in the southern part of the Indian Ocean.

Elements of the water balance

In this paper only the long-term water balance at regional and basin levels is considered. As such, the discussion of the elements is restricted to rainfall, evaporation and runoff (UNESCO, 1978).

Rainfall The physical features, the north-south movement of the sun and the influence of the trade and monsoon wind system produce rainfall on the African region whose origins can be described as convectional (e.g., around the equator), convergent (as created by the movement of the monsoon

shearlines and the Maximum Cloud Zone [MCZ] of the equatorial westerlies in West Africa), orographical (found on the windward side of mountain ranges in the Cameroons), highland (as is to be found in South and East Africa and in Ethiopia) and cyclonic (as found on the eastern side of Madagascar). In some places, rainfall occurs from a combination of these types.

In general, two rainfall regimes can be identified in the humid African tropical region – humid and subhumid regions. In the subhumid region, wet and dry subregions are further distinguished. Figures 4 and 5 show the annual rainfall and annual rainfall distribution. Therefore:

(a) The Humid or Equatorial rainfall regime, located generally between latitudes 10°N and 10°S of the central and western Africa subregion is found here. Here it rains throughout the year and two rainfall maxima are observed in the year (as is obtained at Mhandanka in

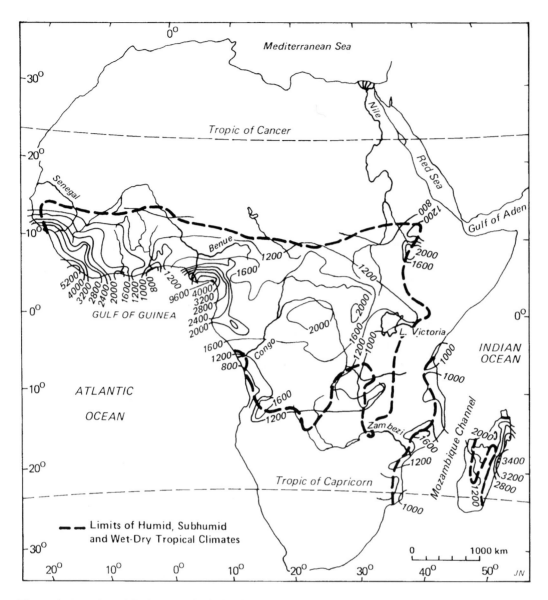

Figure 4: Annual precipitation (mm) in the humid tropics (Source: UNESCO, 1978).

Zaire). The variability of inter-annual rainfall is small in this area, which receives the heaviest amount of rainfall in the African region, with 1,600 mm to 3,200 mm as an annual mean.

(b)   The subhumid region (where it rains in the summer months and it is dry in winter) lies immediately north and south of the equatorial region and generally between latitudes 10° and 20° N & S.

  (i)   The wet subhumid part is found in West Africa, the mountain parts of East Africa and Madagascar. Here the rainfall regime has two rainfall maxima in the year (as a result of being traversed twice by the appropriate active monsoon shearline), as in Abidjan (Cote d'Ivoire). The mean annual rainfalls are between 1,200 and 1,500 mm.

  (ii)   The dry subhumid region which is further north and south of the wet subhumid region has one rainfall maximum, as exhibited in Kaduna (Nigeria).

Table 1 gives the mean monthly and annual rainfall values for some selected stations in the two sub-regions.

The results obtained by Dankwa (1974) for three selected stations in Ghana are presented in Table 2. They illustrate the rainfall intensities at various frequencies. Rainfall variability on inter-annual and intra-annual basis can be moderate in the wet subhumids to considerable in the dry subhumids, both at the same location or from place to place. The mean annual rainfall ranges from 1,200 mm to 600 mm.

Evaporation

As stated earlier, the continent, because of its position, receives a large amount of solar heat and has a positive radiation balance. The radiation balance is 70 kcal per square centimeter per year as against the global balance of 50 kcal per square centimeter per year. With this large amount of energy, evaporation is potentially high.

The potential evaporation is about 1,300 mm a year in the

TABLE 2. *Maximum rainfall intensities in mm/hour at various return periods for selected stations in Ghana.*

Kumasi (06° 43'N, 01° 36'W)

	Return Periods, Years					
Duration Hours	5	10	15	25	50	100
0.2	137.16	153.67	163.83	176.53	191.77	208.28
0.4	118.11	127.00	143.51	154.94	170 18	185.42
0.7	93.73	104.14	114.30	122.94	134.87	147.07
1.0	77.22	83.82	94.23	101.85	112.01	122.17
2.0	45.72	53.34	60.20	64.77	50.80	81.03
3.0	33.02	38.61	41.40	45.21	48.26	58.42
6.0	19.30	22.86	24.89	27.43	30.73	34.29
12.0	10.41	12.45	13.75	15.24	17.27	18.03
24.0	5.33	6.35	7.11	7.87	8.89	9.65

Axim (04° 50'N, 02° 15'W)

	Return Periods, Years				
Duration Hours	5	10	15	20	100
0.2	137.16	156.21	171.45	194.31	214.63
0.4	116.84	130.81	148.59	166.37	183.64
0.7	97.28	109.47	121.67	136.91	148.34
1.0	80.26	88.90	101.60	114.30	127.00
2.0	54.61	61.72	66.55	76.20	83.57
3.0	41.66	46.48	52.58	59.44	89.92
6.0	24.38	26.92	29.72	33.27	36.58
12.0	13.72	15.49	17.27	19.56	21.59
24.0	7.37	8.38	9.40	10.41	11.43

Tamale (09° 25'N 0° 53'W)

	Return Periods, Years					
Duration Hours	5	10	15	25	50	100
0.2	129.54	143.51	152.40	162.56	176.53	189.23
0.4	99.82	111.76	118.87	132.06	152.40	162.56
0.7	81.03	92.46	96.52	101.60	117.86	127.00
1.0	62.99	72.14	77.22	83.57	92.20	100.58
2.0	35.56	40.39	43.18	46.74	53.34	60.96
3.0	25.15	28.45	30.48	35.56	35.81	43.18
6.0	13.97	16.00	17.02	19.81	20.32	24.38
12.0	7.62	9.14	9.91	10.67	11.94	12.95
24.0	4.06	4.57	5.08	5.59	6.10	6.86

Source: Danewa, J. B. (1974)

The above observations pertain to the annual distribution of the aridity index. Greater insight into the water balance situation can be obtained if the intra-annual distribution of the index, based on either wet and dry season values or on monthly values, is computed. An example is presented in Table 4 for three selected stations (Axim, Kumasi and Tamale) in Ghana which are in the humid, moist subhumid and dry subhumid zones, respectively. The water balance at the basin scale for the Niger, Congo and Zambezi rivers at various stations were computed by Balek (1977). The results are presented in Table 5.

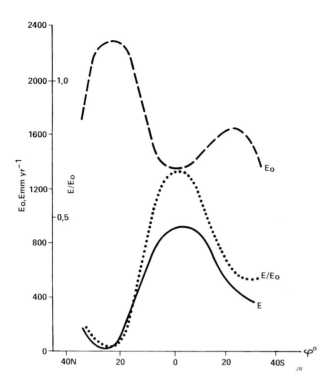

Figure 6: Mean latitudinal distribution of potential evaporation (E_o), actual evaporation (E) and their ratio ($E\,E_o^{-1}$) on the African continent (Source: UNESCO, 1978).

Problems with assessment of water balance elements

The discussion so far gives an idea of the extent of knowledge about the surface hydrology of the African humid tropical region. Balek (1977), UNESCO (1978) and ECA (1977) agree that the adequacy and reliability of data is a problem in estimating the water balance components. The longest length of record exists for rainfall, followed by evaporation and then by runoff.

There remains, however, the need for a better estimation of the water balance elements such as rainfall, evaporation and runoff and the reliability of those estimates. Also needed are estimates of their expected values, their extremes, their intra- and inter-annual variations and the distribution of these characteristics.

In addition to better assessment, there is a need to be able to predict rainfall with regard to the times of onset, the expected amounts and the areal coverage – particularly for agricultural production in the marginal or transitional regions of the continent where there is seasonal aridity. In view of the sensitivity of the aridity index to rainfall, there is a need to go back into the origins of rainfall so as to better understand the geophysical processes that bring it about. The oscillation of the monsoon shearlines and associated MCZ in Africa and their connection with the general atmospheric circulation needs to be better understood in order to have better rainfall predictions.

Considerably better understanding is needed in Africa of the effect on rainfall, evaporation, soil moisture movement and runoff processes from:

(a) deforestation arising from the removal of vegetation for timber, farming, fuelwood, food, medicine, and grazing.

(b) possible climatic variations and impending changes arising from the "greenhouse effect."

In attempting to obtain better knowledge in the areas mentioned, the role of networks for collecting data on the water cycle cannot be over-emphasized. The situation of these networks has not been satisfactory, historically, in the African continent. However, both the IHD/IHP of UNESCO and the OHP of WMO, have provided the impetus since 1965 for establishing various networks to collect water cycle data. However, since 1980 or so, the gains which were made as a result of the international programmes have started to erode for lack of data collection resources. This is a crucial problem which must be resolved if the progress that is expected in generating better knowledge of the hydrology and water resources of the continent is to be achieved.

WATER RESOURCES

Surface water

The surface water resources cover rivers, lakes and wetlands.

Rivers The rivers in the humid tropics flow over more than one hydrogeological region. For example, the Niger River has its source in the Futa Jalon highlands in the humid areas of West Africa. It flows into the dry subhumid regions of North Guinea, then through the semiarid areas of Niger and Nigeria and back into the dry and moist subhumid areas of Nigeria. Finally, the stream courses through the humid parts of Nigeria before discharging to the sea through the humid Niger Delta. Hence, for the understanding of the rivers in the humid tropics, it is important to take the whole river into account, whether or not parts originate from the semiarid areas. Almost all of the rivers have their basins located in more than one country, making it important to co-operate in water resource assessments. The rivers in the region discharge into the Mediterranean Sea in the north, the Atlantic Ocean in the west and the Indian Ocean in the east. The mean annual runoff volumes of the major rivers in the region are given in Table 6.

The inadequacy of data is a constraint in estimating the distribution parameters of runoffs. The contribution of the African countries (in terms of the lengths and gaps in data series supplied) to UNESCO publication on Discharge of Selected Rivers of the World (1969–1985) is a testimony to this constraint. For instance, out of a total number of 98 river gauging stations in Ghana, Opoku-Ankomah (1986) could only use 25 of them when determining the annual flow characteristics of major Ghanaian rivers. The results obtained for the 25 stations are presented in Table 7. The shortness of some of the data are quite evident. The inadequacy of data underlies why estimates of annual runoff of the African region taken from the literature vary widely. For instance, ECA (1977) estimated the annual runoff to be 2 481 km³. UNESCO (1978) put

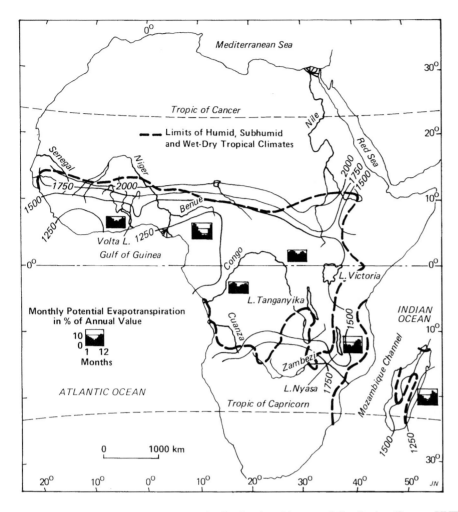

Figure 7: Annual potential evapotranspiration (mm) and its annual distribution (Source: UNESCO, 1978).

it at 4,600 km³ and quoted other works like Karisik (1970) and Zubenok (1970) who estimated it at 4,225 km³ and 7,826 km³, respectively. The assessment of mean annual runoff together with mean monthly runoff (about which even less work appears to have been done) needs to be taken up at the country levels and later synthesized at the subregional and regional levels to obtain acceptable values.

Lake water resources Information on those lakes in the humid tropics with surface areas of 500 km² and above is given in Table 8. The biggest is Lake Victoria, with a surface area of 68 800 km² and a volume of 2,850 x 10⁹ m³. Next is Lake Tanganyika, with an area of 32 900 km² and a volume of 18 940 x 10⁹ m³. The deepest lake is Lake Tanganyika with a mean depth of 700 m. The next deepest is Lake Malawi with a mean depth of 426 m.

Apart from the basic physical features that have been published on lakes in Africa, it is difficult to find data on lake level fluctuations, water quality, biological and socio-economic conditions (such as utilization), lake environment deterioration and hazards. A recent attempt to compile data on the state of the world lakes by ILEC and UNEP (1983) provided very limited information about Africa. Of the lakes in the region, Lake Victoria seems to be in crisis with regards to its quality and quantity. Studies are being planned to obtain information to plan its restoration. The same crisis affects Lake Chad, but it is located outside of the humid regions.

It is important that a first attempt be made to compile data on the characteristics referred to above. Once the gaps in knowledge have been identified, steps can be taken and information obtained for the regional management of these water bodies.

The wetlands These are the swamps, marshlands, dambos or sudds on the courses of rivers in the African region. Table 9 shows the major ones. They have a regulating effect on the flow regimes of rivers due to their storage and evapotranspiration effects. They are important water resources which can be harnessed for utilization. For instance, Egypt and the Sudan are developing the Bahr el Jebel/Bahrel Ghazal swamps on the White Nile so as to provide additional water for irrigation. Botswana is seeking to develop the Okavango swamps on the Okavango/Botletle River for agricultural purposes. Similar developments are being planned for the other wetlands. Dake(1986) has noted that the hydrological and other scientific data and information needed for planning and development are

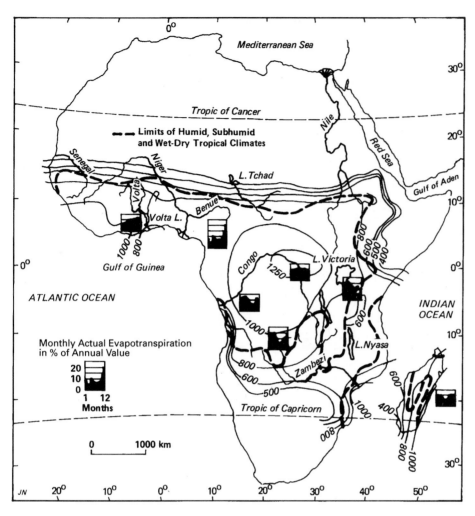

Figure 8: Annual actual evapotranspiration (mm) (Source: UNESCO, 1978).

limited. He has called for studies to be undertaken, including the use of remote sensing, of their hydrology and water resources as a basis for sound planning and development.

Extreme hydrological events

The whole of the African region, including the humid tropics, are subject to extreme hydrological events such as floods and drought. The recent droughts (1981–86) and floods which have led to deaths and socio-economic dislocation in the region are cases in point.

Floods Examples of the areas prone to flooding are low-lying areas around the banks of the Niger River in Mali; the periodic overflow of the banks of the Nyando, Nzoia and Yala Rivers in Kenya; the northern part of the Central African Republic; the flash floods in Madagascar caused by tropical cyclones and the coastal lowlands bordering the mountain ranges in Sierra Leone and Guinea.

Flood studies in the African region have been carried out either for individual projects or to obtain the regional behaviour of floods. The approaches used seem to have been dictated largely by available data. Where long-term data have been obtained, statistical methods have been used – as in the analy-

ses of the historical sequences of the Nile River maxima and minima and the flood flow frequencies of some African rivers (Andel *et al.*, 1971; Balek, 1971). Probably one of the most significant regional flood studies is that by Rodier & Auvray (1965) who studied 10-year floods of catchments with areas of 100 km² or less of some West African rivers. For the East Africa subregion, one should mention the studies of Kovacs (1971) who related the characteristics of flood discharges to catchment areas of some east African rivers. In addition to regional studies, there have been studies to analyse floods on an inter-regional basis to see whether results are comparable. Such work is reported by Balek (1977) on the analysis of Kenyan and Malawian rivers in East Africa with Nigerian rivers to establish a relation between drainage area and the 1% flood peak.

The work done by UNESCO (1976) in connection with the World Catalog of large floods is one of the most recent from which information could be obtained. However, only four countries (Congo, Gabon, Ghana and Cote d' Ivoire) from the humid tropics provided data and information. For the majority of these countries, the data provided was inadequate to obtain an updated picture of flood characteristics. Various probability distributions to which the annual floods fit were reported. The

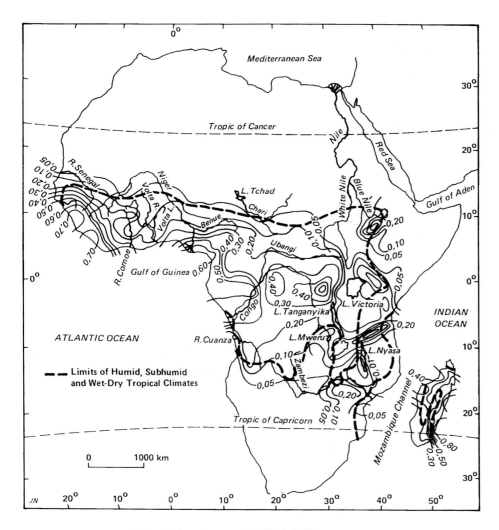

Figure 9: Annual Runoff Coefficient (Source: UNESCO, 1978).

Congo reported empirical and Gaussian distributions; Ghana, empirical and Gabon, Pearson III. Cote d'Ivoire did not indicate any type.

It would be an important contribution if an attempt were made to put together all the various approaches used to estimate floods in Africa, then analyse them to see which ones can be applied, to where and under what conditions.

With the present rate of cutting down of forests for timber, it is also important to know the impact of deforestation on flood volumes and peaks.

Drought problems It is natural that when one talks about drought in Africa one should refer to the Sahel drought which started in 1968 but which had a short respite of normal rains in 1974–1975, reaching its peak of dryness in 1981–1983. During the 1981–1983 period, it was not only the Sahel that was affected but also the rest of Africa.

There also are continental droughts of shorter duration whose impacts have been equally damaging socio-economically and ecologically. The late arrival of rains as well as insufficient rainfall have resulted in low flows of various intensities and durations.

Although rainfall has decreased in the sub-Sahara over the past 15 years, there is no overall evidence of climate trends which would indicate a higher frequency of droughts in Africa. However, certain activities such as over-grazing and deforestation (which alter the ground surface) may affect local climate and rainfall (UNEP, 1983).

Existing weather and climate information can significantly improve operational practices in agriculture, water management and energy use. National meteorological services, therefore, should be given adequate resources to ensure that reliable continuous weather data are available for both application and research in drought studies, as well as to provide support to the agricultural sector.

At present, WMO and ECA are organizing the African countries to co-operatively tackle the drought problem on a regional basis. Toward this end, the African Centre for Meteorological Applications to Development (ACMAD) is being established. It will have subregional centres in Niamey (for the Sahel countries), Nairobi (for the eastern African countries) and Harare (for the southern African countries and Madagascar).

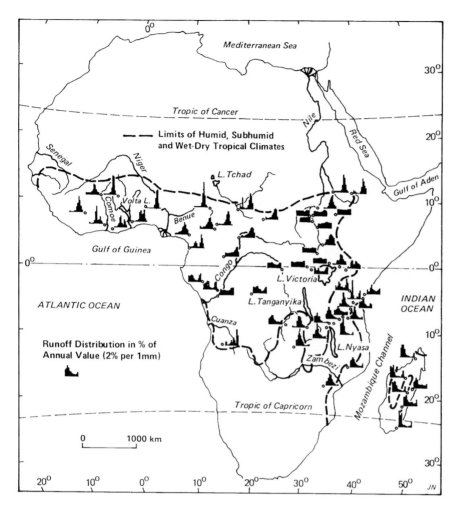

Figure 10: Annual runoff distribution (Source: UNESCO, 1987).

Erosion

L'vovich (1974) estimated the erosion products and dissolved solids carried by streams from the different continents and arrived at a result which placed the African region as the continent with the least erosion products, after Europe.

This estimate has been doubted because of the impression that ecological and climatic conditions of Africa are such that erosion should be more. The proof, however, must be provided by measurements from an adequate network of stations. Unfortunately, such a network for sediment measurement is inadequately developed in the African region.

FAO's (1986) evaluation of the erosion and land degradation over the various regions of the continent is shown in Fig. 11. It can be seen that in the humid tropics, gully erosion is generally slight to moderate in most areas, except in Kenya where it is severe. Sheet erosion is slight to moderate in all the regions.

An assessment of the sediment situation in Africa has also been made by Walling (1984). He reviewed the availability and accuracy of the data and concluded there were problems with their adequacy and reliability. He produced the results of measurements made in the Tana River at Kanbwe (Kenya) between 1965 and 1982 by nine different workers. The differences are shown in Fig. 12.

He also showed: one, that the long-term variability could be considerable, with the coefficient of variation (CV) well in excess of 1.0 and, two, that there was a need for more long-term data to arrive at meaningful estimates of the sediment yield. This variability was estimated with results from two stations on the Tana River in Kenya. (See Fig. 13.)

In the mountainous areas of East Africa, the sediment yield ranges from 100 to 1,000 t km^{-2} $year^{-1}$. A yield of 1,680 t km^{-2} year from a 150 km^2 basin in the Ethiopian highlands is reported while a yield of 200 to 400 t km^{-2} $year^{-1}$ from a 275 000 km^2 upper drainage basin of the Blue Nile and Tecazze Rivers is also indicated. In Kenya, yields of 500 t km^{-2} $year^{-1}$ and 1,000 t km^{-2} $year^{-1}$ in the Upper Tana basins, with areas of 9,520 km^2 and 31 700 km^2, respectively, are reported.

In West Africa, relatively high sediment yields were reported for northern Nigeria – in the range of 155 to 739 t km^{-2} $year^{-1}$. Also a yield of 210 t km^{-2} $year^{-1}$ for a 1 535 km^2 basin was reported in Cameroon for the Tsanage River. Again in West Africa, sediment yields of 43 and 505 t km^{-2} $year^{-2}$ from drainage basins of 57 460 km^2 and 2,120 km^2, respectively, were reported by Akrasi & Ayibotele (1984).

In Central Africa, the suspended sediment yields recorded

TABLE 3. *Distribution of mean annual aridity index for the African region.*

	20–10°W			10–0°W			0–10°E			10–20°E			20–30°E			30–40°E			40–50°E		
	P mm	E_o mm	E_o/P	P mm	E_o mm	E_o/P	P mm	E_o mm	E_o/P	P mm	E_o mm	E_o/P	P mm	E_o mm	E_o/P	P mm	E_o mm	E_o/P	P mm	E_o mm	E_o/P
40-30°N	–	–		210	1,000	8.6	374	1,700	4.7	151	1,800	9.9	84	1,750	20.8	74	1,750	23.6		–	
30-20°N	70	1,750	25.0	55	2,250	40.9	56	2,250	40.2	26	2,150	82.7	9	2,300	225.6	16	2,300	143.0		–	
20-10°N	995	2,000	2.0	595	2,100	5.0	682	2,250	3.3	412	2,100	5.0	369	2,250	6.1	472	2,200	4.7	347	1,900	5.1
10-0°N	3,380	1,350	0.4	1,620	1,400	0.9	1,920	1,400	0.7	1,730	1,280	0.7	1,670	1,620	0.97	1,065	1,700	1.6	357	1,500	4.2
0-10°S	–	–		–	–		–	–		1,460	1,380	0.9	1,575	1,500	0.95	990	1,500	1.5	792	1,500	1.9
10-20°S	–	–		–	–		–	–		772	1,640	2.1	1,000	1,650	1.70	1,090	1,500	1.4	1,920	1,400	0.7
20-30°S	–	–		–	–		–	–		163	1,640	10.1	440	1,700	3.9	875	1,600	1.8	1,210	1,350	1.1
30-40°S	–	–		–	–		–	–		505	1,500	3.0	510	1,500	2.9		–			–	

where P = Precipitation in mm

E_o = Potential evaporation in mm

$\dfrac{E_o}{P}$ = Aridity Index

Source: Ayibotele N.B. (1984)

TABLE 4. *Distribution of mean monthly aridity indices for selected stations within the humid tropics in Ghana.*

		Jan	Feb	Mar	Apr	May	Jun	Jul	Aug	Sep	Oct	Nov	Dec	Annual
AXIM														
Location: Long.02° 15'W	E_o	105.4	116.6	142.3	132.6	123.8	104.6	102.8	92.2	102.0	118.6	117.3	110.7	1,368.9
Lat. 04° 50'N	P	50.9	63.5	129.6	166.3	418.1	592.4	172.3	55.5	77.1	247.4	167.6	93.1	2,233.8
Climatic Zone: Humid	$\frac{E_o}{P}$	2.1	1.8	1.1	0.8	0.3	0.2	0.6	1.7	1.3	0.5	0.7	1.2	0.6
KUMASI														
Location: Long.01° 36'W	E_o	113.3	133.3	147.6	135.2	131.8	104.6	84.3	81.7	84.2	102.8	112.2	100.1	1,331.1
Lat. 06° 43'N	P	19.5	63.9	135.6	135.5	189.3	220.6	126.6	76.3	168.4	187.7	89.8	28.3	1,441.5
Climatic Zone: Moist Sub-Humid	$\frac{E_o}{P}$	5.8	2.1	1.1	1.0	0.7	1.5	0.7	0.4	0.4	1.0	7.6	37.1	0.9
TAMALE														
Location: Long.00° 53'W	E_o	161.2	167.4	213.9	180.0	165.1	165.8	139.7	123.8	127.5	150.2	159.8	141.1	1,895.5
Lat. 09° 25'N	P	1.8	15.9	114.7	127.0	136.8	169.4	200.8	321.6	315.3	101.2	20.9	3.8	1,529.2
Climatic Zone: Dry Sub-Humid	$\frac{E_o}{P}$	88.6	10.5	1.9	1.4	1.2	1.0	0.7	0.4	0.4	1.5	7.6	37.1	1.2

E_o = Mean annual potential evaporation in mm

P = Mean monthly precipitation in mm

$\frac{E_o}{P}$ = Aridity index

TABLE 5. *Water balance of basins within the Niger, Congo and Zambezi Rivers.*

River	Location	Drained area	Preci-pitation	Runoff	Evapo-trans-piration	Runoff coef.	Water yield	Mean annual discharge
Unit		km²	mm	mm	mm	%	1/s/km²	m³/s
Niger	Sigiri	70,000	1,640	420	1,220	0.25	13.3	931
Niger	above Benue	724,000	1,100	126	947	0.11	4.0	2,877
Niger	mouth of Gulf of Benin	1,091,000	1,250	198	1,052	0.16	6.3	6,925
Congo	confluence Lowani, Laulaba	1,085,413	1,422	249	1,179	0.17	7.7	8,358
Congo	above Ubangi	1,548,413	1,559	315	1,244	0.20	10.0	15,484
Congo	below Ubangi	2,303,243	1,569	293	1,276	0.19	9.3	21,420
Congo	above Sanga	2,312,823	1,561	293	1,268	0.19	9.3	21,528
Congo	Kwa	2,635,723	1,581	301	1,280	0.19	9.6	3,303
Congo	mouth	3,607,450	1,561	313	1,246	0.20	9.9	38,805
Zambezi	Chavuma Falls	75,967	1,288	231	1,057	0.18	7.3	555
Zambezi	Victoria Falls	1,236,580	759	30	729	0.04	1.0	1,237
Zambezi	above Kafue	1,339,960	754	34	720	0.05	1.1	1,498
Zambezi	below Kafue	1,554,816	782	38	744	0.05	1.2	1,915
Zambezi	below Laungwa	1,722,233	799	44	755	0.06	1.4	2,501

Source: Balek J. (1977)

TABLE 6. *Mean annual runoff of rivers in the African humid tropics.*

River	Drainage Area km²	Mean Annual Runoff (million m³)	Source of Info. and Remarks
North flowing rivers			
Nile (at Aswan)	2,800,000	84,000	Ministry of Industry and Energy, Algeria, 1973
East flowing rivers			
Wabi Shabelle	205,400	2,500	in Ethiopia
Tana	42,217	4,700	in Kenya
Rufiji (at Pangani Rapids)	158,000	30,000	Tanzania Hydrological Yearbook 1967
Ruvuma	155,400		
Zambezi (at Do Ana)	1,250,000	103,380	se Ataida, Service Hydraulique, 1972.
Limpopo (e.a.= 340,000 km sq)	412,000	5,330	
West flowing rivers			
Orange	650,000	11,370	South Africa Water Commission 1970.
Cuanza	121,470	26,355	Quintela-proc. Reading Symp. 1972.
Zaire	4,000,000	1,325,000	Annales Hydroloques de l'ORSTOM
Ogooue (e.a.= 203,500 km sq)	203,500	148,850	
Senaga (at Edea)	131,500	65,280	
Rivers in West Africa			
Gambia (e.a.= 42,000 km sq)	77,850	5,050	
Senegal (e.a. = 268,000 km sq)	338,000	21,800	
Niger-Bonue	1,215,000	2,215,000	Annales Hydrologiques l'ORSTOM UNESCO, 1971
Mono (e.a. = 20,500 km sq)	22,000	3,375	
Volta (at Senchi)	394,100	39,735	
Pra	2,220	7,400	
Ankobra	15,000	1,600	
Tano	8,250	8,610	

Source: ECA, 1977

are low – less than 100 t km⁻² year. The Zaire River yields 11.3 t km⁻² year while the Niver River yields 33.1 t km⁻² year⁻¹. The total drainage area of both rivers is 6.9 x 10⁶ km².

It is believed that sediment measurements have been carried out over the past few years by national hydrological services and consulting engineers in connection with many water development projects on the continent. An attempt should be made to synthesize the results so far obtained into an updated sediment yield map of the continent in general, and the humid areas in particular. The impact of various land uses on erosion and the sediment yield from deforestation, overgrazing and open cast mining also need to be carefully investigated.

Water quality and pollution

A synthesis of the quality of African surface waters, waste discharges and the subsequent pollution of these waters is difficult to come by. The Hydrological Year Books from the various services in the region hardly report on water quality parameters. Like sediment discharges, the networks for quality monitoring are not well-developed, and those that have been installed were difficult to monitor on a regular basis. In the WMO (1976) report referred to earlier, the percentage of countries without water quality networks was 41. Almost all the other countries which had such networks had densities below a desirable level. In spite of this situation, many water projects

TABLE 7. *Annual runoff characteristics of major Ghanaian rivers.*

No.	River	Station	Period of Records	Drainage Area (km²)	Mean Annual Runoff (q_n) 10⁹ cu m	Specific Yield 10³ cm/km²	Standard Deviation of 10⁹ cu m	Coefficient Variation C.V.	Standard Error Se(q)%	95% C.I. for q_n +(x10⁹cm³)	Coefficient of Skewness Cs	First Order Serial Order Coefficient.rl	95% C.I for rl
1	Volta	Yeji	1952-65	258,000	19.0	73.6	8.46	0.445	12.8	5.62	1.0602	-0.113	0.577
2	Volta	Senchi-Harcrow	1954-78	386,000	34.2	88.6	23.20	0.678	14.1	10.30	1.000	0.011	0.417
3	Volta	Kpetchu	1953-64	72,800	18.0	247.0	12.30	0.683	21.6	9.29	1.3452	-0.447	0.632
4	Black Volta	Lawra	1962-78	89,600	3.22	35.9	1.38	0.428	8.6	0.58	0.6294	0.380	0.400
5	Black Volta	Bamboi	1950-76	127,000	7.58	59.7	3.60	0.475	9.5	1.51	1.0511	-0.020	0.400
6	White Volta	Pwalungu	1952-73	56,900	3.48	61.2	1.43	0.410	9.2	0.685	0.6057	-0.315	0.447
7	White Volta	Yapei	1952-67	106,000	8.24	77.7	2.89	0.351	9.4	1.73	0.4227	-0.239	0.535
8	White Volta	Nawuni	1954-78	95,000	6.70	70.5	2.34	0.349	7.3	1.03	-0.2927	-0.051	0.417
9	White Volta	Yarugu	1962-76	41,000	2.63	64.1	1.01	0.384	10.7	0.635	-0.2051	-0.066	0.555
10	Oti	Saboba	1953-77	54,000	9.14	169.0	3.82	0.419	8.7	1.69	0.0747	-0.214	0.417
11	Bia	Dadieso	1963-76	6,040	0.860	142.0	0.349	0.406	11.7	0.232	2.1766	-0.131	0.577
12	Tana	Jomuro	1956-78	10,300	1.73	168.0	1.04	0.605	13.2	0.487	1.9073	-0.059	0.436
13	Tana	Alenda	1952-78	15,800	4.50	285.0	1.17	0.261	5.7	0.548	0.5628	-0.014	0.436
14	Ankobra	Prestea	1955-78	4,200	1.54	365.0	0.743	0.483	10.3	0.337	1.6869	-0.035	0.426
15	Pra	Twifu-Praso	1944-78	20,700	5.50	266.0	2.84	0.516	9.0	1.02	2.0500	0.140	0.348
16	Pra	Daboasi	1954-78	22,600	7.23	320.0	3.20	0.443	9.2	1.42	1.1309	0.126	0.417
17	Pra	Brenasi	1955-78	2,050	0.483	236.0	0.261	0.541	11.5	0.118	0.5063	0.387	0.426
18	Obuo/Pra	Nampong	1944-78	374	0.155	414.0	0.061	0.389	6.8	0.160	0.5871	0.227	0.348
19	Birim/Pra	Oda	1955-78	3,210	1.37	427.0	0.532	0.460	9.8	0.287	0.8548	0.576*	0.426
20	Ochi/Amissa	Dunkwa	1958-78	8,250	2.34	284.0	1.57	0.670	15.4	0.777	2.2393	-0.005	0.459
21	Ochi/Amissa	Mankessim	1956-78	1,200	0.336	280.0	0.184	0.549	12.0	0.086	1.0742	0.089	0.436
22	Ayensu	Oketsew	1960-78	696	0.190	273.0	0.121	0.637	15.5	0.064	1.5839	0.143	0.485
23	Ayensu	Okyereko	1962-78	1,640	0.347	212.0	0.254	0.732	18.9	0.146	1.5596	0.021	0.516
24	Densu	Manhia	1968-75	2,100	0.385	183.0	0.311	0.809	33.0	0.358	1.6600	0.196	0.817
25	Tordzie	Todzienu	1964-77	2,750	0.273	99.3	0.191	0.700	20.2	0.127	2.9140	-0.032	0.577

*Significant dependence

Source: Opoko-Ankomah (1986)

TABLE 8. *Lakes of Africa.*

Name	Surface Area km^2	Volume 10^9 m^3	Mean Depth m	Country
Victoria	68,800	2,750	40	Kenya, Uganda, United Republic of Tanzania
Tanzanyika	32,900	18,940	700	Burundi, United Republic of Tanzania, Zaire, Zambia
Malawi	30,800	8,400	426	Malawi, Mozambique, United Republic of Tanzania
Chad	16,317	75	4	Cameroon, Chad, the Niger, Nigeria
Bangweulu	9,850	11	4	Zambia
Turkana (Rudolf)	7,200	555[1]	73	Ethiopia, Kenya
Mobutu S.S.	5,600	140	25	Uganda, Zaire
Mweru	4,580	37	6.5	Zaire, Zambia
Tana	3,500	28	8	Ethiopia
Kyoga	2,700	20	6	Uganda
Kivu	2,699	650[1]	240	Rwanda, Zaire
Idi Amin Dada	2,300	78	34	Uganda, Zaire
Maji Mdozbe (Lepold II)	2,300	11[1]	5	Zaire
Kitangiri	1,200	6[1]	5	United Republic of Tanzania
Abaya	1,161	8.2	7	Ethiopia
Chilwa	750	1.5[1]	2	Malawi, Mozambique
Tumba	720	2.9[1]	4	Zaire
Shame	551	5.5[1]	13[2]	Ethiopia
Upemba	530	0.9	0.3	Zaire

[1] Approximate volume based on mean depth and surface area.
[2] Maximum depth.
Source: FAO, 1986

TABLE 9. *Main African swamps.*

Swamp	Country	Main Stream•	Area (km sq)
1. Babr el Jebel/Babr el Ghazal	Sudan	White Nile	64,000
2. Middle Congo Swamps	Zaire	Congo •	40,550
3. Lake Chad Swamps	Chad	Chari	32,260
4. Babr Balaaat	Chad	Chari	27,000
5. Okavango	Botswana	Okavango/Botletle	26,750
6. Upper Lualaba Swamps	Zaire	Luabala	25,750
7. Lake Kyoga Swamp	Uganda	Victoria/Nile	21,875
8. Lake Mweru Swamp	Zambia/Zaire	Luapula	17,000
9. Lake Mweru Wantipa Swamp	Zambia	Mofwe	16,750
10. Lake Bangweulu Swamp	Zambia	Zambesi	15,875
11. Kenamuke/Kabonen	Sudan	–	13,955
12 Lotagipi	Sudan/Kenya	Tarach	12,937
13. Malagarasi	Tanzania	Malagarasi	7,357
14. Nyong	Cameroon	Nyong	6,688
15 Albert Nile Swamp	Sudan	Albert Nile	5,200
16. Kafue Flats	Zambia	Kafue	2,600
17. Lukanga	Zambia	Kafue	2,600

Source: Balek J. (1977)

REGION	Country	Erosion and land degradation ▲ Slight ■ Moderate ● Severe						
		A	B	C	D	E	F	G
MEDITERANIAN AND ARID NORTH AFRICA	Algeria	■	■	▲	▲			■
	Egypt			■	▲	▲		
	Libyan Arab Jamahiriya			■	▲			
	Morocco	■	■	▲	▲	▲		■
	Tunisia	■	▲	▲	▲	▲	▲	■
SUDANO-SAHELIAN AFRICA	Burkina Faso	■	■	■	▲	■	■	■
	Cape Verde		■	■	▲	▲	▲	▲
	Chad	▲	■	●	■	■	▲	▲
	Djibouti		▲	■	▲			■
	Gambia, The	▲	▲			■	▲	▲
	Mali	▲	▲	■	▲	▲	■	■
	Mauritania	■	■	●	●	▲	■	●
	Niger, The	■	■	■	■	▲	■	●
	Senegal	▲	■	▲			■	■
	Somalia	■	■			■		■
	Sudan, The		■	■	▲	▲	▲	■
HUMID AND SUBHUMID WEST AFRICA	Benin	▲	■			■	▲	▲
	Cote d'Ivoire	▲	■			■	■	●
	Ghana	▲	■			■	▲	■
	Guinea		■			■		▲
	Guinea-Bissau		▲			▲	▲	■
	Liberia		▲			■		■
	Nigeria	■	■	▲	▲	■	▲	▲
	Sierra Leone		▲			▲		■
	Togo	▲	■			■	▲	■
HUMID CENTRAL / AFRICA	Cameroon	■	■	▲		■	▲	■
	Central African Republic	▲	■					■
	Congo		■			■	▲	▲
	Equatorial Guinea		▲			▲		▲
	Gabon		▲					▲
	Sao Tome and Principe		▲			▲	▲	▲
	Zaire	▲	■			■	▲	■
SUBHUMID AND MOUNTAIN EAST AFRICA	Burundi	■	■			▲		■
	Comoros	■	▲			■	▲	●
	Ethiopia	■	■	▲	▲	▲	▲	●
	Kenya	▲	■	▲	■	■	▲	■
	Madagascar	●	■	▲		■	▲	■
	Mauritius	▲	▲			■		■
	Rwanda	■	■			▲		■
	Seychelles		▲					■
	Uganda		■			▲		▲
SUBHUMID AND SEMI-ARID SOUTHERN AFRICA	Angola		▲	▲	▲			▲
	Botswana	▲	■	■	■	■	▲	■
	Lesotho	■	■			■		▲
	Malawi	▲	■			■	■	■
	Mozambiqaue	▲						▲
	Namibia			▲	▲			▲
	Swaziland	■	■			■	▲	▲
	Tanzania United Rep. of	▲	▲			■		▲
	Zambia	▲	■		▲	■	▲	▲
	Zimbabwe	▲	■		▲	■	▲	▲

A - Gully erosion
B - Sheet erosion
C - Wind erosion
D - Desert encroachment
E - Declining fertility
F - Soil crusting
G - Degraded vegetation

Figure 11: Erosion and land degradation in Africa (Source: FAO, 1986).

have been investigated at various levels. Water quality analyses have been part of these studies. Similarly, general investigations have been carried out in a number of countries to identify the sources of pollution – be they domestic, municipal, agricultural and industrial, including mining. ECA (1977), in assessing the water quality and pollution situation in the African region, reported there was pollution in Liberia's St. John River from the mining effluents complex there. Periodic failure was also reported of copper mining effluent treatment processes causing pollution in the Kafue River in Zambia.

Effluents from Cameroon cement and textile plants, breweries, and paint factories, etc. were being discharged as well into natural streams with little or no treatment.

The above information is *ad hoc*, not long-term, which makes it difficult to determine if trends are being established. The natural water quality status with respect to the temperatures, the surface geology, and the vegetation needs to be established for the surface waters of the continent.

In order to develop a general view of the water quality and pollution of African rivers, it would be necessary to try to

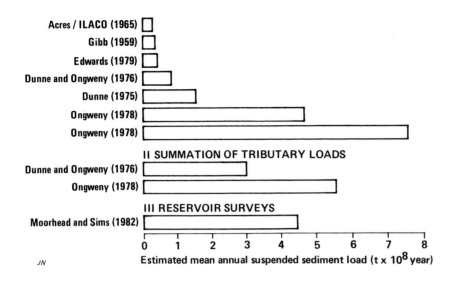

Figure 12: Estimates of mean annual sediment load, Tana River, Kenya (Source: Walling, 1984).

bring together all the scattered data and synthesize them. The work should start at the national level. Advantage could be taken of the UNEP/WMO/UNESCO/WHO Global Environmental Monitoring Project (WATER).

Groundwater

The groundwater in the humid tropical region of Africa is dominated, firstly, by the Precambrian crystalline formations in West, Central and East Africa and Madagascar; secondly, by the large sedimentary formation of Central Africa (Zaire, Congo, and Central African Republic); thirdly, by the coastal sedimentary formations in West Africa (Cote d'Ivoire, Togo, Benin, Nigeria, Cameroon and Gabon) and East Africa (Kenya and Madagascar); and, fourthly, by the intra-Cambrian and Paleozoic formation bordering between the large sedimentary formation and the Precambrian formations of the Congo and finally by the volcanic formation in the highlands of East Africa (Kenya and Ethiopia).

The Precambrian basement formations have no primary porosity. The water-bearing capability is a result of fractures, joints, fissures and weathering which have occurred over geological periods as reported in Ghana by Bannerman & Ayibotele (1984). The formations consist mainly of granites, biotite gneisses, metamorphic schists, phyllites, quartzites, sandstones, and conglomerates. Water occurs in them at various depths between 25 and 100 m. Their yields, however, are variable and generally low, but in some areas, yields as high as 45 m³ hr⁻¹ have been recorded.

The large sedimentary basin of the Congo is artesian in nature (Ministry of Geology USSR, 1982). It consists of rock formations such as gneisses, crystalline schists, conglomerate mudstone, sandstones, basalts, and sands. The groundwater occurs at depths ranging from 25 to 800 m. Many springs exist, with discharges ranging between 15 and 145 m³ h⁻¹. Highly mineralized thermal groundwater, with temperatures ranging between 53 and 100°C have been found.

The coastal sedimentary formations consist of some very productive sandstone and limestone layers (United Nations, 1988). They are found mainly along the shorelines of Guinea, Sierra Leone, Liberia, Cote d'Ivoire and Ghana; along the Benin-Togo and Ghana coastlines; in Nigeria and Cameroon, Gabon and the Central African Republic and in East Africa, along the shorelines of Kenya and Somalia. Yields averaging 18 m³ h⁻¹ are obtainable from the aquifers in Cote d'Ivoire, Benin-Togo and Togo.

Volcanic formations found in the East African rift valley areas of Kenya and Ethiopia consist of basalts, lencitites, agglomerates, tuffs, etc. Aquifers of very variable yields are found in these formations.

The current generalized picture of groundwater in the African region is based mainly on data available before 1980. Since then, due to the impetus given by the International Drinking Water Supply and Sanitation Decade (1981–1990), large-scale drillings have been undertaken in many parts to provide water for rural populations. These drillings have yielded considerable data that is being analyzed on a point basis as well as subregionally within countries. The results need to be brought together so that the generalized picture of groundwater in the African region can be updated.

Presently, information on yields and to some extent on drawdowns is available. Additional work needs to be done to determine and generalize the information on storativities and transmissivities. There also is a need to monitor groundwater levels and quality to establish whether aquifers are being recharged or contain fossil water (UNESCO, 1988). The quality of groundwater is a subject of concern. In this regard, those national institutions for groundwater exploration and research, the UNDP, UNESCO, the Inter-African Committee for Hydraulic Research, the OAU, and the Commonwealth Science Council – who are known to be active in the field – will have to pool their resources to bring the knowledge together.

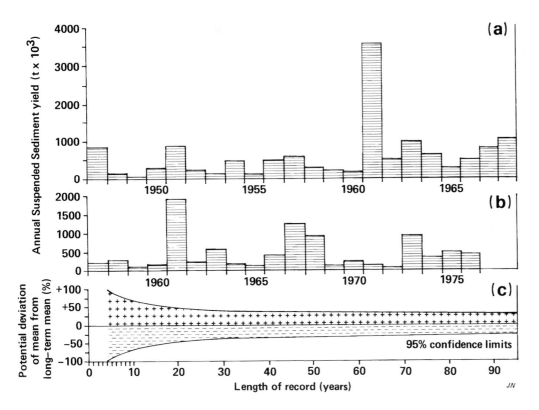

Figure 13: Long-term variability of sediment, Tana River, Kenya (Source: UNESCO, 1984).

CONCLUSIONS AND RECOMMENDATIONS

1. It is clear that the understanding of African hydrology and water resources in general suffer from inadequate or no data at all. This situation is a serious constraint to rational assessment, planning, design and operations. This shortfall in information also exists about the humid tropics of Africa. There is a need to develop networks for the collection of data, particularly on river runoff, river and lakes, sediment discharges, groundwater levels and quality. It is to be noted that the present project by UNESCO and WMO to have countries evaluate their water resources assessment activities offers the best chance of taking corrective action. The project is expected to evaluate and recommend measures for filling gaps and strengthening:

(a) Institutional frameworks for water resource assessment.

(b) Data collection, processing, storage and retrieval.

(c) Areal assessment of water balance components.

(d) Data for water resources planning.

(e) Manpower, education and training

(f) Research.

It is noted that the UNDP, the World Bank and the African Development Bank have started a similar project for that part of Africa located south of the Sahara. African countries should take advantage of this development in carrying out the UNESCO/WMO project.

2. High rates of population growth are putting pressure on the land to produce basic needs. The result is that land degradation – arising from the application of improper farming techniques, – removal of vegetation for food, timber, medicine, building materials and fuelwood – is fast increasing. The relationship between the water balance components is consequently changing and this may account for severe floods and droughts, increasing erosion and the deterioration in water quality. Such results could also be linked with the impact of other human activities outside the African region, such as the industrial emissions in the developed countries which are increasing the "greenhouse effect." To assess the impact of human activities on hydrology and water resources, it is recommended that countries within the humid tropics should:

(a) Identify those areas in their countries where land degradation is taking place and monitor their water resources so as to quantify their impacts.

(b) Take part in the project of the Committee on River and Lake Basins and the Water Resources Networks of the African Ministerial Conference on the Environment (UNEP).

(c) Take part in the Global Environmental Monitoring System for Water.

(d) Take part in the studies of the African Centre for Meteorological Applications for Development (ACMAD) with subregional headquarters in Niamey, Nairobi and Harare.

(e) Take part in the World Climate programme (Water)

3. Over 80% of the river, lake and groundwater basins of the African region are shared. Co-operation is needed at the subregional levels of hydrologists and other water scientists from different countries to obtain good knowledge about these resources at the basin level. To make this co-operation effec-

tive and sustainable, national institutions should be strengthened as a foundation for this subregional and regional co-operation.

4. National IHP committees, in collaboration with the appropriate national organizations, should undertake the updating and synthesizing the knowledge at national level, of:

(a) elements of the water balance – rainfall, evaporation/ evapotranspiration, and river runoff.

(b) extreme hydrological phenomena such as floods, drought and low streamflows.

(c) groundwater levels.

(d) water quality – both surface and groundwater.

(e) sediment discharges in rivers.

5. National IHP committees should seek the assistance of UNESCO and WMO to organize subregional and regional scientific meetings on the topics enumerated in (4) above, with the objective of updating and developing generalized knowledge at the subregional and regional levels.

REFERENCES

Akrasi, S. A., & Ayibotele, N. B. (1984) An appraisal of sediment transport measurement in Ghanaian rivers. *Proc. of the Harare Symp.: Challenges in African Hydrology and Water Resources.* IAHS Publ. 144.

Andel, J., Balek, J., & Vener, M. (1971) An analysis of historical sequences of the Nile maxima and minima. *Symp. on the role of hydrology in the economic development of Africa,* Addis Ababa. WMO 301, Geneva.

Ayibotele, N. B. (1984) Problems of African hydrology and water resources. Lecture II. River Basin Management, International Advanced Course in Water Resources Management – Water Resources Research and Documentation Centre. Perugia, Italy.

Balek, J., (1971) On extreme floods in Zambia, NCSR. Water Resources Report II., Lusaka.

Balek, J. (1977) *Hydrology and Water Resources in Tropical Africa.* Elsevier, Oxford.

Bannerman, R., & Ayibotele, N. B., (1984) Some critical issues with monitoring crystalline rock aquifers for groundwater management in rural areas. *Proc. of the Harare Symp.: Challenges in African Hydrology and Water Resources.* IAHS Publ. 144.

Dake, J. K. (1986) Swamps and dambos in African hydrologic regions. *Twentieth Intl. Symp. on Remote Sensing and the Environment.* Nairobi.

Dankwa,J. B. (1974) Maximum rainfall intensity – duration frequencies in Ghana. Dept. Note No. 32. Ghana Meteorological Services Department, Legon, Accra.

ECA (1977) Water development and management. *Proc. of the World Water Conference*, Part 2. Mar del Plata, Pergamon Press.

FAO (1986) African agriculture, the next 25 years, *Atlas of African Agriculture,Food and Agric. Organization,* Rome.

ILEC & UNEP (1983 . Survey of the state of world lakes. Interim Report (1), Otsu, Japan.

Karasik, G. (1970) *Water Balance of Africa.* Izd. VINITI

Kovacs, G. (1971) Relationship between characteristic flood discharges and the catchment areas. *Symp. Role of Hydrology in Economic Development of Africa, II*, WMO 301, Geneva.

L'vovich (1974). Water development and management, *Proc. of the World Water Conference*, Part 2, Mar del Plata, Pergamon Press, New York.

Ministry of Geology of the USSR (1982) *Hydrogeology of Africa,* "Nedra" Moscow.

Opoku-Ankomah, Y. (1986) Annual flow characteristics of major Ghanaian rivers. Water Resources Research Institute, (CSIR), Accra.

Rodier, J. & Auvray C. (1965) Estimation des debit de crues decennales pour des bass ins versants de superficie inferieure a 100 km' en Afrique occidentale, Paris ORST0M.

UNEP (1983) Report of the multidisciplinary meeting on the impact of drought on socio-economic system in Africa. Nairobi.

UNEP (1985) Report of the Executive Director of UNEP. African Environmental Conference UNEP. Cairo Dec. 1985.

UNESCO *Discharge of Selected Rivers of the World*, (1969-1985) *I–III,* Paris.

UNESCO (1976) *World Catalogue of Very Large Floods.* Paris.

UNESCO (1978) *World Water Balance and Water Resources of the Earth,* Paris.

UNESCO/WMO (1988) *Water Resources Assessment Activities.* Handbook for National Evaluation. Paris and Geneva.

United Nations (1988) *Groundwater in North & West Africa.* Natural Resources/Water Series *18.* New York.

Walling, D. E. (1984) The sediment yield of African rivers. *Proc. of the Harare Symp.: Challenges in African Hydrology and Water Resources.* IAHS Publ. 144.

WMO (1976) General Report on Hydrological Observing Networks in Africa. Meeting of Experts on Hydrological Problems in Africa. Addis Ababa, September 1976

WMO (1983) Operational Hydrology in the Humid Tropical Regions. Hydrology of Humid Tropical Regions. Proc.of SAKS Symposium held during 18th General Assembly of IUGG. Hamburg. IAHS Publ. 140.

Zubenok, L. (1970) Specified pattern of the continents water balance. Trudy. GGO.

10: Review of Hydrology and Water Resources of Humid Tropical Islands

A. C. Falkland

Hydrology and Water Resources Branch, ACT Electricity and Water, GPO Box 366, Canberra, ACT, 2601, Australia

J. P. Brunel

Institut Francais de Recherche Scientifique pour le Développement en Coopération, 2051 Ave du Val de Montferrand, 34032 Montpellier Cedex 1, France

ABSTRACT

There are many islands in the humid tropical regions of the world, most of which belong to developing nations. Islands, especially small oceanic islands, are unique in many ways. Island water resources are often very limited. Many have no surface water resources and rely on limited groundwater resources in the form of thin fresh water lenses. The exposure of islands makes them particularly vulnerable to cyclones, storm surges and droughts. This paper examines some of the major hydrological and water resources issues of small islands. The main topics discussed are fresh water occurrence, the water balance, water resource assessment, water use, and water resource development and management. Approaches to resolving some of the major problems are provided.

INTRODUCTION

Many islands in the world lie in the humid tropical region, defined as that region where the mean temperature of the coldest month is above 18°C and where the number of months with rainfall greater than 100 mm exceeds 4.5 (Chang & Lau, 1983). The land portion of many countries in the humid tropics consists entirely of islands. Examples of countries consisting of island archipelagos are Indonesia and the Philippines in South East Asia; the Cook Islands, Kiribati, the Federated States of Micronesia, Fiji, the Marshall Islands, New Guinea, Solomon Islands, Tokelau, Tonga, Tuvalu, Vanuatu and Western Samoa in the Pacific Ocean; the Maldives and Seychelles in the Indian Ocean and the Turks and Caicos and the Bahamas in the Caribbean Sea.

Some countries in the humid tropics consist of a single island or one main island; for example, the countries of Nauru and Niue in the Pacific Ocean, Mauritius in the Indian Ocean, and Barbados and Jamaica in the Caribbean Sea.

Other islands are territories of larger continental countries, for example, the Galapagos Islands (Ecuador), the Hawaiian Islands (USA) and French Polynesia (France) in the Pacific

Ocean and the Andaman and Nicobar Islands (India) in the Indian Ocean. Many of the islands in humid tropical South East Asia belong to Malaysia, Vietnam, China or Japan.

Most of the island countries within the humid tropics are developing rather than developed, with Hong Kong and Singapore being notable exceptions. Some islands have high per capita incomes such as Nauru but are still classified as developing countries.

Uniqueness of islands

Islands, especially the small oceanic islands situated far from continents or other large islands, are unique in many ways. Island resources, including land and water, are very limited. Their continual exposure to the marine environment imposes harsh conditions on materials and equipment. Islands are particularly vulnerable to major hydrological events and many islands have suffered from the effects of cyclones, storm surges and droughts. Some low-lying islands have even been overtopped by storm-driven waves.

Human influences can easily and adversely affect the natural environment of islands. In particular, water resources are particularly susceptible to human activities, including over-pumping of fragile groundwater resources causing increases in salinity of water supplies, and pollution of both surface water and groundwater resources due to urbanization, agricultural activities, mining and clearing of forests. Faecal contamination of groundwater resources used for water supply by infiltration from closely-located sanitation facilities is a common and major problem on the many crowded islands in the humid tropics. The use of pesticides, herbicides and fertilizers is a current and emerging problem for groundwater resources on many islands. Mining and forest clearing activities have caused surface water resources to become highly turbid and contaminated on a number of "high" islands which do have available surface water resources.

Because of their limited size and resources, the susceptibility of islands to natural disasters and degradation of their land and water resources is much higher than on continental land masses. Their isolation from sources of supply, the high costs of freight and often a lack of trained professional and technical staff adds to the problems experienced by island communities.

ISLAND TYPES

Islands in the humid tropics can be classified according to their geology, topography and size.

Geology

A convenient geological classification of small islands is: volcanic, limestone, coral atoll, bedrock, unconsolidated and mixed.

Volcanic islands are common in tropical regions of the Pacific Ocean (the Hawaiian Islands, many islands in Micronesia and French Polynesia) and in the Caribbean Sea. They also occur in the Atlantic Ocean and Indian Ocean (Mauritius). There are at least two sub-types of the volcanic kind: the andesitic sub-type which normally forms as island arcs on the continental sides of deep trenches, and the basaltic or oceanic sub-type which rises from the ocean floor in the middle of tectonic plates.

Limestone islands are also common in the oceans and seas within the humid tropics. Examples include old carbonate islands such as Bermuda in the Atlantic Ocean, the Bahamas in the Caribbean Sea and raised atolls such as Nauru, Niue and many of the islands of Tonga in the Pacific Ocean. Raised atolls are uplifted coral atolls that have undergone subsequent erosion and karstification.

The coral atoll type of island is common in the Pacific Ocean (the islands of Kiribati, Tuvalu and the Marshall Islands) and in the Indian Ocean (the Maldives and some of the Seychelles). There are many variants of the coral atoll type of island, as described in Scott & Rotondo (1983).

Bedrock islands are those formed by igneous or metamorphic rocks such as granite, diorite, gneiss and schist. They are mainly found on continental shelves or adjacent to large islands of similar geology. This type of island occurs in all parts of the world, including the humid tropics.

The unconsolidated type of island typically consists of sand, silt and/or mud and is generally found in the deltas of major rivers (for example, off the coast of Bangladesh).

Islands of mixed geology are common. Amongst the oceanic islands those with a mixture of volcanic and limestone rocks occur frequently (as in the Federated States of Micronesia).

Topography

Islands are often classified according to topography as either "high" or "low." This classification attempts to distinguish those with surface water resources in the form of streams and rivers from those which have no surface runoff. Volcanic islands are typically high islands and coral atolls are typically low islands.

Size

Islands have also been classified for convenience according to their size as either large or small. Although there is no exact distinction between the two sizes, a workshop on small island hydrology (Commonwealth Science Council, 1984) selected an area of 5000 km^2 as the upper limit for a small island. A smaller area of 2000 km^2 was selected as an appropriate upper limit in the course of a recent UNESCO project on the hydrology of small islands (Project 4.6 of Phase III of UNESCO'S International Hydrological Programme).

Large islands tend to have features and problems similar to those found on continents, whereas small islands often have additional or different problems. Large islands are largely addressed in the other papers dealing with regional water problems in the humid tropics in this volume (Asia: Chang; Africa: Ayibotele; Latin America and the Caribbean: Griesinger & Gladwell). The emphasis in this paper, therefore, is placed on small rather than large islands.

RECENT EXPERIENCES WITH ISLAND HYDROLOGY AND WATER RESOURCES

In recent years, there have been a number of major regional, inter-regional and international seminars, meetings and workshops on the topic of hydrology and water resources of islands, particularly small islands in the humid tropics. Included were:

*Seminar on Small Island Water Problems, organized by the United Nations Department of Technical Cooperation for Development (UNDTCD) and the Commonwealth Science Council (CSC), Bridgetown, Barbados, October 1980;

*Meeting on Water Resources Development in the South Pacific, organized by ESCAP, Suva, Fiji, March 1983;

*Workshop on Water Resources of Small Islands, organized by the CSC, Suva, Fiji, July 1984;

*Inter-regional Seminar on Development and Management of Water Resources in Small Islands, organized by UNDTCD and Government of Bermuda, Hamilton, Bermuda, December 1985;

*Southeast Asia and the Pacific Regional Workshop on Hydrology and Water Balance of Small Islands, organized by the National Committee for the IHP and UNESCO/ROSTSEA, Nanjing, China, March 1988; and

*Interregional Seminar on Water Resources Management Techniques for Small Island Countries, organized by UNDTCD and CSC, Suva, Fiji, June 1989.

Proceedings of each of the above meetings have been published and provide a good overview of the range of problems confronting islands and some of the solutions which have been implemented.

As part of Project 4.6 (Hydrology of Small Islands) of Phase III of UNESCO'S International Hydrological Programme, a review of existing knowledge regarding hydrology and the water balance of small islands was published (Diaz Arenas &

Huertas, 1986). More recently a UNESCO publication entitled *Hydrology and water resource of small islands: a practical guide* was prepared under the same project (Falkland *et al.*, 1991). While this project was not exclusively dealing with the humid tropics, it was concerned to a large degree with the humid tropical region because most small islands fall within this region. UNESCO was also involved with two research studies on the hydrological problems of small islands. The first study involved two small islands in Indonesia (Barang Lompo and Koding Areng near Ujung Pandang, Sulawesi) while the second was on Marinduque Island in the Philippines. Both studies were conducted by national organizations with the funding and review being provided by UNESCO (Daniell, 1986).

UNDTCD has been actively involved in the last two decades with the assessment and development of water resources on small islands. Projects were conducted on individual islands in the early 1970s. By the end of the 1970s, a regional project was started in the Caribbean to co-ordinate activities in 11 small island countries. In 1986, a similar regional project was launched for the small island countries in the Pacific region. As part of the project, considerable information (studies and reports) on small island hydrology and water resources has been assembled at the UNDTCD office in Fiji.

In addition to the above regional and international efforts, a number of studies and projects to assess and develop water resources on small islands have been conducted by or through national agencies such as ORSTOM (France), USAID (USA), AIDAB (Australia) and DSIR/Ministry of Works (New Zealand) and private consultants. A number of studies have also been carried out on small islands which are part of mainland countries, using their own national organizations.

FRESH WATER OCCURRENCE ON ISLANDS

Fresh water occurs naturally as either (or both) surface or groundwater on islands. The following factors influence the type and quantity of fresh water on islands: climate, size, topography, geology and vegetation.

Where suitable hydrological and geological conditions prevail on high islands, surface water occurs in the form of rivers, streams, springs, lakes and wetlands. Conditions for the occurrence of surface water are more favourable on volcanic than on limestone or coral islands. Low-lying or raised flat islands rarely have surface water, except as lakes or ponds where the permeability of the surface is sufficiently low. Volcanic islands with raised topography and low-permeability surfaces often have small streams but these are normally ephemeral in nature. Surface runoff often occurs rapidly after rainfall and may diminish to little or no flow within hours. Volcanic islands may also have perched lakes – generally found in the caldera. On many raised islands, particularly where volcanic rocks underlie limestone, perennial springs are found. These often occur around the base of the island, either slightly above or sometimes below sea-level.

Groundwater on high islands can occur in the form of elevated (high-level) or basal (low-level) aquifers. Groundwater on low islands can occur only as basal aquifers.

Elevated aquifers can either be perched aquifers or dyke-confined aquifers. Perched aquifers occur where a horizontal, low-permeability layer effectively impedes the vertical movement of percolating groundwater. Dyke-confined aquifers, found in some volcanic islands (Hawaiian Islands, French Polynesia), are formed when vertical volcanic dykes, typically in groups, trap water in the intervening compartments.

Basal aquifers consist of unconfined, partially-confined or confined fresh water bodies which form at or below sea-level. Except where permeabilities are very low, as on some volcanic and bedrock islands, most islands would have some form of basal aquifer in which the fresh water body comes into contact with sea-water. On many small islands, the basal aquifer takes the form of a fresh water lens which underlies the whole island. Fresh water lenses are a very important water supply source on many small islands and are generally more economic to develop as a primary source than other options such as rainwater catchments (unless the population is low and the rainfall is consistently high), importation or desalination. On many small islands, rainwater is an important supplementary source of high-quality potable water and, in some cases, may be the only source used except in droughts when groundwater is used.

The relative importance of surface water and groundwater on islands depends on the particular nature of the island. In general, however, groundwater tends to be the more important resource of the two. Furthermore, basal aquifers tend to be more important than perched aquifers because not all islands have the latter, and where both types occur, basal aquifers normally have greater storage volume. Basal aquifers are, however, vulnerable to saline intrusion owing to the fresh water/sea-water interaction and must be carefully managed to avoid over-exploitation with resultant sea-water intrusion. If these essential steps are not properly implemented, there is considerable risk of inducing the intrusion of saline water, as has occurred in a number of islands.

There are a number of natural features which control the formation of fresh water lenses on low-lying islands, particularly coral atolls and small limestone islands. The amount and distribution of recharge to groundwater which is dependent on rainfall and evapotranspiration patterns is most important. The size and shape of the island, particularly the island width, is a major control on the occurrence of fresh water lenses. Hydrogeological factors have a major influence on the distribution of fresh water on an island. These factors include the permeability and porosity of the sediments and reef rock, and the presence and distribution of karstic features such as small cave systems and solution cavities.

Other factors such as tidal patterns, the height of the island above sea-level, and the width of the fringing reef are impor-

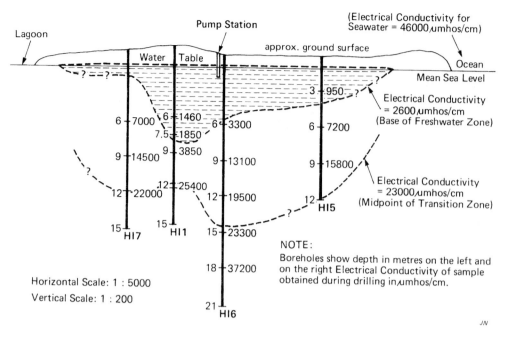

Figure l: Cross section through Home Island fresh water lens, Cocos (Keeling) Islands, Indian Ocean, November 1987 (after Falkland, 1988).

tant in that they provide some idea of the risk of overtopping by storm surges.

The term "fresh water lens" can often conjure up a misconception of the true nature of groundwater bodies underlying islands. In reality, there is no distinct fresh water body floating on sea-water but rather a gradual transition from the freshest water at the water table to sea-water at some depth below this. A zone of fresh water can, however, be defined on the basis of an objective salinity criterion such as a chloride ion concentration of 250 (or maybe 600) mg l^{-1}. Figure 1 shows the variation of salinity (expressed in terms of electrical conductivity) with depth from boreholes drilled across a coral atoll and the determination of the fresh water zone. Electrical conductivity (EC) is often used as a measure of water salinity as it can easily be measured using a portable meter. On small oceanic coral islands, EC readings of 1,500 µmhos cm^{-1} and 2,500 µmhos cm^{-1} are approximately equivalent to chloride ion concentrations of 200 mg l^{-1} and 600 mg l^{-1}, respectively.

It must be noted that drawings of fresh water lenses are normally shown with an exaggerated vertical scale which can often lead to a false impression of the depth of the fresh water zone. In practice, many fresh water zones are less than 5 m thick although the island may be 300–500 m wide. The deepest part of the fresh water zone is often displaced towards the lagoon side rather than the ocean side of coral atolls due to lower permeability sediments on the lagoon side.

The Ghyben-Herzberg theory (strictly, the Badon Ghyben-Herzberg theory after studies by Badon Ghyben, 1889 and Herzberg, 1901) has often been applied to the study of fresh water/salt-water relationships on islands. This theory assumes that the depth of fresh water below mean sea-level is approximately 40 times the height of the water table above mean sea-

level, based on the different densities of fresh and sea-water. The theory is based on the concept of immiscible fluids which is not correct in practice because the two fluids do mix as a result of both mechanical and molecular dispersion. The degree of mixing is a function of the fresh water flow from the aquifer to the sea. If the flow is large, as may occur from a continental coastal aquifer, a high degree of "flushing" (of salts) occurs and the mixing or transition zone will be thin in comparison with the fresh water zone. In this case, the Ghyben-Herzberg theory has been found to provide a reasonable approximation to reality. However, where the fresh water outflow is small, as on a small island, the degree of "flushing" is correspondingly small and the transition zone can be very thick. In many cases, the transition zone is thicker than the fresh water zone and the Ghyben-Herzberg theory does not apply. The Ghyben-Herzberg ratio of 1:40 tends to identify instead the midpoint of the transition zone. Serious errors in the estimation of the fresh water thickness can (and have) been caused by blindly applying the theory to small islands. This has often occurred because measurement of the height of the water table above mean sea-level is a lot easier than obtaining borehole salinity profiles, which provide a true indication of the fresh water/sea-water relationship.

For small coral islands, an empirical relationship has been derived (Oberdorfer & Buddemeier, 1988) between fresh water lens thickness, annual rainfall and island width:

$$H P^{-1} = 6.94 \log a - 14.38 \qquad (1)$$

where

H = lens thickness (depth from water table to sharp interface or mid-point of transition zone (m)),

P = annual rainfall (m), and,

a = island width (m).

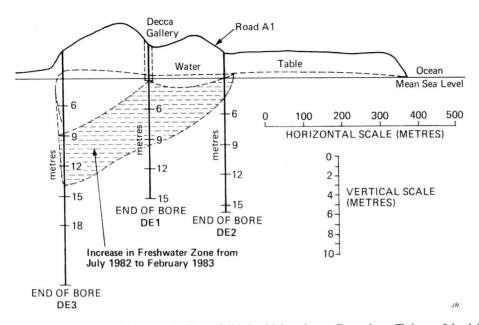

Figure 2: Increase in fresh water thickness following high recharge, Decca lens, Christmas Island, Kiribati

This equation indicates that no permanent fresh water lens can occur regardless of rainfall where the island width is less than about 120 m. Using an annual rainfall of 2000 mm (which is typical of many small coral islands in the humid tropics), the minimum island width for a small fresh water lens (say 5 m thick) to occur is a little less than 300 m. This prediction is in reasonable agreement with observed fresh water lenses.

In addition to natural controlling features, humans can induce additional influences on groundwater systems due to abstraction. The design and operation of abstraction facilities can have a major influence on the sustainability of fresh groundwater resources on small islands. Point abstraction at higher rates than the sustainable yield has caused the local destruction of fresh water lenses. This problem, however, is reversible, given sufficient recharge, as exemplified in Fig. 2. The recharge referred to in this figure resulted from about 2000 mm of rainfall which occurred on Christmas Island, Kiribati between July 1982 and February 1983 as a result of the 1982–1983 El Niño-Southern Oscillation (ENSO). By comparison, the average annual rainfall on Christmas Island is 840 mm.

WATER BALANCE

General

Islands provide an opportunity to study the full hydrological cycle over a relatively small domain. For small islands, it is common for the water balance domain to be the whole island with the boundaries being the ocean. For large islands, the water balance of individual catchments or basins is generally of greater importance than the whole island.

The water balance for an island, in many cases, can be conveniently considered in two stages: at the surface and within the groundwater system. The surface water balance has rainfall as an input and evapotranspiration and recharge to groundwater as outputs. The groundwater balance has recharge as an input and losses to the sea by outflow and dispersion, and abstraction by people, as outputs.

The water balance at the surface depends on whether the island has suitable topographical and geological characteristics to produce surface runoff. The water balance equation at the surface of an island can be expressed as:

$$P = ET_a + SR + R \pm dV \qquad (2)$$

where

P = precipitation,
ET_a = actual evapotranspiration (including interception),
SR = surface runoff,
R = recharge to groundwater, and
dV = change in soil moisture store.

Interception is sometimes treated as a separate term in the water balance, but here it has been included with ET_a since the intercepted water eventually is evaporated. On low islands, for example, coral atolls, there is no surface runoff and, hence, the SR term is deleted from the above equation. On coral atolls the main interest is recharge (R) which is the input term to the second water balance equation dealing with the groundwater system. The water balance equation at the surface ("recharge model") can be simplified and rearranged to:

$$R = P - ET_a \pm dV \qquad (3)$$

Fig. 3 is one attempt to describe the water balance processes involved for a typical coral atoll.

As shown in Fig. 3, actual evapotranspiration (ET_a) is comprised of interception losses (E_I), evaporation and transpiration from the soil zone (E_S), and transpiration of deep-rooted vegetation directly from groundwater (T_L). The first two terms can be treated as losses from storages which have upper and lower

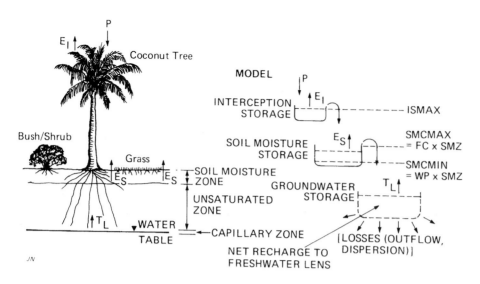

Figure 3: Recharge model for a typical coral atoll

limits. Rainfall is first used to fill up the interception store with any residual going to the soil moisture store. If this is also filled, the residual becomes gross recharge. A further loss (T_L), is incurred before net recharge to groundwater is obtained. Typically, the depth to the water table is only one to two metres and the roots of deep rooted vegetation, such as coconut trees, can penetrate to the water table. Thus, coconut trees can act as phreatophytes and draw water directly from the water table. This helps to explain why on some drier atolls, coconut trees can survive extensive droughts while shallow-rooted vegetation, which draws its moisture from the soil zone, dies off.

The water balance procedure is more complex on raised atolls, which often have large depths from the surface level to the water table (typically 30 to 100 m) and extensive karstic formations. Unlike coral (low) atolls which typically have a thin soil layer of between 0.3 to 0.5 m thick above a highly permeable unsaturated zone, typically one to two metres thick, raised atolls may have soil layers of variable thickness. In places, the soil layer is non-existent while in other places it can be metres thick. Roots of trees may penetrate through fissures and reach pockets of water at different levels. In addition, the flow paths from the surface to the water table may not be essentially vertical as with the low atoll, but rather have major horizontal components due to karstic formations (solution channels). The situation is even further complicated in mixed-type islands where the interface between limestone and underlying volcanic rock is either totally or partially above sea-level. In these cases, the groundwater flow pattern is often characterized by subterranean streams at the interface, which in some parts of the island emanate as springs, either above or below sea-level. No general model is, therefore, possible for islands of this type and each island needs to be considered on a site-specific basis. The same comment generally applies to high volcanic islands because of their complex hydrogeological structure.

The water balance at the surface should be modelled with a time step not exceeding one day because the turnover time in the soil zone is measurable on the time scale of a day (Chapman, 1985). The use of mean monthly rather than daily data will underestimate recharge. Two atoll studies (Kwajalein: Hunt & Peterson, 1980; Cocos (Keeling) Islands: Falkland, 1988) have shown that the assessed recharge is decreased by between 6% and 10% of rainfall if monthly rather than daily data is used. The water balance within the groundwater system on an island can be expressed as:

$$R = GF + D + Q \pm dS \tag{4}$$

where

R = recharge to groundwater,
GF = groundwater outflow (to the sea),
D = dispersion at the base of the groundwater body,
Q = abstraction (normally by pumping), and
dS = change in fresh water zone storage.

In Equation (4), the value of R has been estimated from the surface water balance (Equation (2)). Due to a longer turnover time, a time step of a month for the groundwater system balance is suitable. Often the turnover time of the fresh groundwater on a small island is a number of years. Even for a relatively small fresh water lens on a coral atoll, the turnover time is generally greater than 12 months.

The groundwater balance is complicated if there are perched or dyke-confined aquifers present on the island, as in the case of some volcanic islands. The equation must be modified to allow for storage in these upper aquifers in addition to storage in the lower aquifer. Obviously, the D term is only applicable in cases where a fresh water/salt-water mixing zone occurs.

The solution of the above equation has been attempted for fresh water lenses on some islands by both "sharp interface" and "dispersion" models. The relevant flow (and in the case of dispersion models, salt transport) equations are solved by

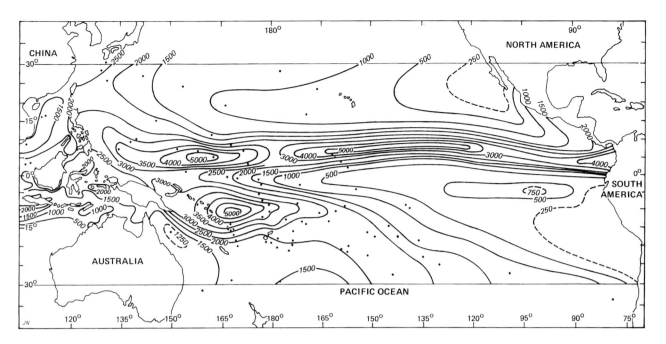

Figure 4: Distribution of mean annual rainfall (mm) over the Pacific Ocean (after Taylor, 1973).

numerical methods, expressed in either finite difference or finite element form, on a computer. Dispersion models are inherently more complex than sharp interface models as they require additional parameters to be evaluated or estimated, They are conceptually more suited to small island groundwater modelling than sharp interface models which assume that the Ghyben-Herzberg theory applies, an assumption which is not correct in small islands. However, the use of a sharp interface model can give some qualitative idea of the response of a fresh water lens which is subjected to selected (historical or simulated) natural stresses (droughts of varying lengths and severity) and artificial stresses (different pumping strategies). Therefore, they can provide in many cases an initial indication of the most important aspects to be modelled in more detail by a dispersion model. A sharp interface model has been developed for Bermuda, which is reported to work successfully (Thomson, 1985; Karranjak, 1989). To model the fresh water zone, a modified Ghyben-Herzberg ratio was selected, based on observations of the ratio of the water table above mean sea-level to the depth below mean sea-level of the fresh water limit, as defined by some objective criterion. This approach, also used in some other studies, has theoretical problems since the ratio should depend only on the densities of fresh water and sea-water (Chapman, 1985).

To properly model the fresh water and transition zone behaviour on a small island, a dispersion model should be used. This does require a much larger effort in terms of data collection and computing resources than for a sharp interface. While sharp interface models can be easily run on microcomputers, dispersion models require much larger and faster computers to achieve reasonable run times. One dispersion model, SUTRA (Voss, 1984), has been applied to the study of a number of island fresh water lenses including Oahu in the

Hawaiian Islands (Voss & Souza, 1987), Enjebi island on Enewetak atoll in the Marshall Islands (Oberdorfer & Buddemeier,1988) and Nauru (Ghassemi *et al.*, l990).

One of the major requirements needed with the groundwater balance at present is a simple yet theoretically correct method of describing the behaviour of fresh water lenses on small islands. Present methods are either theoretically inadequate and need to be "forced" to give reasonable answers, or they are too complex and beyond the resources of most organizations, particularly water resource agencies on islands.

The measurement or estimation of terms in the water balance equations are now considered.

Precipitation

On islands in the humid tropics, especially small islands, precipitation is usually synonymous with rainfall as other forms of precipitation (ice, snow) are generally negligible. Hence, the term rainfall will be used here to describe total precipitation.

The controlling mechanisms for rainfall, and climate in general, in the humid tropics have been described by Manton & Bonell (this volume). Hence, no detailed discussion is presented here. In general terms, the rainfall patterns which influence islands in the humid tropics vary considerably between locations. For instance, the annual rainfall pattern for the Pacific Ocean, as shown in Fig. 4, indicates considerable spatial variability.

The annual rainfall pattern for the Indian Ocean, as indicated in Fig. 5, also shows considerable spatial variability. Both patterns are highly interpolated owing to the limited number of data sets from island meteorological stations used in their construction. The patterns also do not take account of the considerable spatial variation of rainfall over some high

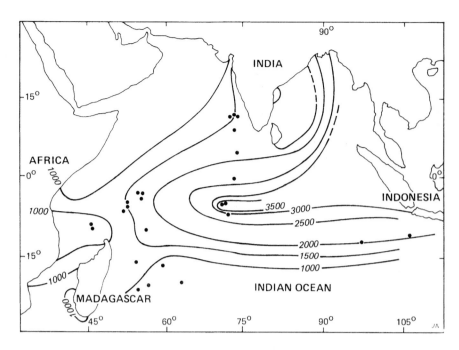

Figure 5: Distribution of mean annual rainfall (mm) over the Indian Ocean (modified from Stoddart, 1971).

islands due to orographic effects. Most raingauges are situated around the rim of high islands and thus fail to record the very high rainfalls that occur in the peaks or highlands. Where measurements have been taken in the higher altitudes, annual rainfall depths exceeding 10 m have been recorded; yet official rainfall records may show typical values of two to three m for an island. This problem is often one of logistics; it is difficult and costly to operate and maintain networks of raingauges in the high and often inaccessible parts of some islands. The advent of relatively cheap and robust electronic data-logging equipment should gradually ease this problem. If linked to satellite telemetry (for example, ARGOS), data acquisition and monitoring from remote sites can become easier. Some islands are evaluating the use of satellites for data transfer from their hydrometric networks (Solomon Islands).

Some islands within the tropics are greatly influenced by summer monsoons and have distinct wet and dry seasons and may need to contend with seasonal fluctuations in their water resources. Extensive dry seasons can and do become critical for surface water resources with streams and springs reducing to little or zero flow. Extreme variability in rainfall for particular months from year to year is another phenomenon affecting many islands, such as parts of New Caledonia (Brunel, 1981).

Islands are also influenced to different degrees by longer time-scale mechanisms such as the ENSO phenomenon (see Manton & Bonell, this volume). The influence of ENSO on island rainfall patterns in the central and eastern Pacific is very marked, as shown for one island in Fig. 6.

ENSO and anti-ENSO (also referred to as La Niña) events can produce very wet and very dry cycles to be superimposed on the normal annual cycle and, hence, exacerbate wet periods and droughts. On some islands, periods of up to six months may elapse before significant rainfall occurs. This was particu-

larly evident during the major 1982–1983 ENSO event. For instance, during the first six months of 1983, the rainfalls on Saipan and Guam in the Pacific were less than 30% of the long-term mean. Similarly, in the first five months of the same year, the rainfall on a number of islands in Micronesia was only 13% of the long-term mean (van der Brug, 1986).

In addition to the effects on precipitation, ENSO events cause a sea-level rise due to elevated sea surface temperatures and consequent thermal expansion of the ocean. This can add to the problems of small low-lying islands. For instance, a sea-level rise caused by the 1987 ENSO event resulted in damage to crops on many atolls in the Federated States of Micronesia (Maragos, 1990) adding to the effects of drought conditions also induced by that ENSO event.

In addition, the climates of many islands are influenced by random cyclonic events. Cyclones are a major problem for small island communities, often causing major wind damage, large floods, and massive hillside erosion with consequent sedimentation problems downstream. They have also been responsible for storm surges which have inundated low-lying areas and, indeed, whole islands with sea-water (for example, Funafuti in Tuvalu, some islands in the Marshall Islands and Tokelau in the Pacific Ocean, and islands off the coast of Bangladesh in the Bay of Bengal). Apart from any direct damage, the impact on fresh water lenses is extreme because they receive considerable input of sea-water. Many months may be required to naturally "flush" the salt-water from fresh water lenses and restore them to a potable condition.

On low flat islands (for example, atolls), orographic effects are negligible, but there are sometimes observed variations from one side of an atoll to another. The exact reason for this is not known but may in some cases have been due to observational errors or poor siting of gauges. It is not possible to deter-

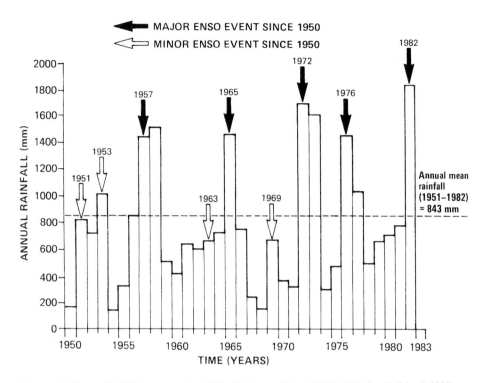

Figure 6: Effect of ENSO on annual rainfall, Christmas Island, Kiribati (after Falkland, 1983).

mine the exact reasons in some cases, due to sites having been abandoned and subsequently changed.

While the network of raingauges over the whole area of the oceans in which humid tropical islands occur is limited by the number of islands, there is a need to improve the networks on many high islands to better understand island rainfall gradients from sea-level to peaks. Further studies should take advantage of technological advances to improve the access to and reliability of the data.

Evapotranspiration

The combined process of evaporation and transpiration, often referred to as evapotranspiration, is one of the most important components of the hydrological cycle and the water balance equation. On small islands, evapotranspiration can be more than half of the rainfall on an annual basis and often exceeds the rainfall for individual months or consecutive months during dry seasons or drought periods. Despite its importance, evapotranspiration is probably the least quantified component of the water balance on small islands (Brunel, 1989).

The estimation of actual evapotranspiration (ET_a) from catchments has generally been done for small islands by a two-stage process: estimation of potential evapotranspiration (ET_p) and then estimation of ET_a from ET_p using a water balance procedure. These stages are described below.

Firstly, ET_p is estimated using a method based on climatic data, such as the Penman (or Combination) formula, or from pan evaporation data multiplied by an appropriate pan coefficient. The Penman equation has been found to be generally a good ET_p estimation method in the humid tropics (Chang, this volume). This equation has been used to estimate ET_p on a

number of small islands. For instance, Fleming (1987) used the Penman equation for estimating ET_p for Tarawa, Kiribati. In the humid tropics, the net radiation energy term dominates the aerodynamic term in the Penman equation and it has been found that the simplified Priestly-Taylor method is adequate (where ET_p is equated to 1.26 times the energy term from the Penman equation). This method was used by Nullet (1987) and Giambelluca et al.,(1988) to estimate ET_p on a number of tropical Pacific islands. Figure 7 shows the results of the earlier study for Pacific island atolls.

Secondly, actual evapotranspiration is usually determined via a water balance procedure taking into account the soil and vegetation conditions present on the island. An example for a coral atoll is indicated in Fig. 3. The components of actual evapotranspiration are direct evaporation from surfaces and transpiration by vegetation. Water which is intercepted by vegetation can be included in the model. The usual procedure is to subtract the volume of intercepted water from the evaporation demand and use the reduced demand as the effective value for calculating the transpiration (Chapman, 1985). Interception should be included as a component in the overall water balance. Failure to include it can lead to underestimation of ET_a and overestimation of recharge. For example, in the Cocos (Keeling) Islands, the difference in recharge for cases of interception stores of 3 mm and zero was 5%.

Problems with the water balance equation may occur because parameters such as depth of soil moisture store, field capacity and wilting point are not well-established for tropical island soil/vegetation associations. Further problems with spatial variation arise in complex geological conditions with variable depths and types of soil. With regard to vegetation on

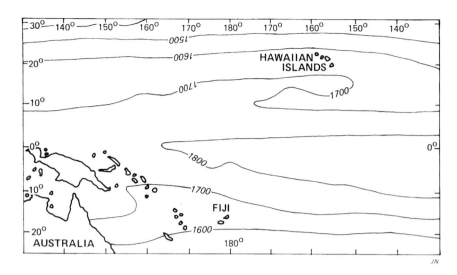

Figure 7: Distribution of mean annual potential evapotranspiration (mm) over atoll vegetated areas in the Pacific Ocean (after Nullet, 1987).

tropical islands, there is also little hard data on crop coefficients which can be applied to account for the effect of different types of vegetation on the evaporation demand. Thus, estimates based on similar vegetation from other regions have been applied. For example, crop coefficients of 1.0 and 0.8 were used to adjust ET_p for shallow-rooted vegetation and coconut trees, respectively, for some Pacific island water balance studies (AGDHC, 1982; Falkland, 1988). There is a need, therefore, for further basic research on typical soil and vegetation types on tropical islands.

An alternative approach to estimating ET_a is to carry out direct measurements using one or more techniques such as weighing lysimeters, the Penman-Monteith equation (using measured values of net radiation, wind velocity and vapour pressure deficit), and the Bowen ratio technique. These have all been used by ORSTOM at two sites in New Caledonia (Brunel, 1989). Other direct measurement methods such as the eddy correlation method also could be used.

Direct measurements of transpiration from coconut trees have added to the knowledge of the evapotranspiration process on small islands in the humid tropics. Transpiration from coconut trees, as measured in the Cocos (Keeling) Islands using a heat pulse velocity meter during a one-week study (Bartle, 1987), was 70–130 l day^{-1}. Based on this limited data, the total transpiration rate due to coconut trees is about 400–750 mm year^{-1} per tree in areas with 100% tree cover, where the tree spacing is about eight m. This has implications for water supply management and it may be prudent to selectively clear coconut trees from some fresh water lens areas to maximize the supply of water. Other forms of vegetation, particularly other large trees, need also to be assessed for their suitability in areas suitable for groundwater development.

In general, there is insufficient data available on ET_a from typical small islands. While micrometeorological techniques are available and have been used with success in a number of continental countries to study this basic process, there is yet to

be a co-ordinated study for small islands, despite their very fragile fresh water resources. It is, therefore, recommended that emphasis be placed on more detailed investigations of the actual evapotranspiration process. Existing approaches for the estimation of evapotranspiration should be re-evaluated in the light of such investigations. As outlined by Brunel (1989), the study of ET_a for islands could be attempted by a two-stage process. Firstly, a database of typical soil/vegetation/sub-climate associations, including relevant hydrological parameters, should be made for islands of the humid tropics and, secondly, measurements of ET_a should be made at selected sites to develop relationships between ET_a and typical associations. Some associations worth detailed study are coconut tree/coral sand and open grass/coral sand in the drier parts of the humid tropics where soil water deficit is prevalent in dry periods. The use of remote sensing techniques (for example, airborne sensing or satellite imagery), correlated with surface measurements of ET_a could considerably assist with the extrapolation of ET_a from point measurements to areal estimates. As the spatial scale on many small islands is not large, the problem of spatial extrapolation to an island scale is not as great a problem as on continents, except where large diversity of soils and vegetation occur.

Surface runoff

Surface runoff, as already described, only occurs on high islands with favourable topographical and geological conditions. Such islands are generally characterized by many small-sized catchments with steep slopes in which runoff occurs very rapidly. As a result, flash-floods can occur. Due to the erosive power of the streams, erosion and sedimentation problems are also prevalent. Soil and vegetation loss, high turbidity problems in streams and damage to water intakes and other on-stream structures have occurred on a number of islands.

Stream gauging stations need to be robust and the equipment needs to be well-tested in tropical environments.

Electronic data-logging devices offer some distinct advantages over the more conventional mechanical chart recorders as there are no moving parts, the data does not require digitizing, thus saving costs, and the data can be retrieved by a variety of methods, including telemetry systems (notably radio or satellite). The major disadvantage with electronic data-logging systems appears to be a reluctance on the part of field staff to deal with data acquisition systems where the recording is not immediately visible (except via a portable computer). This problem is not unique to the tropics and is only solved by proper training and education of field staff to improve their confidence and ability in the use of the equipment.

The acquisition of data from remote sites via satellite telemetry should be encouraged. While this may appear to be a "high tech" solution for small islands it has considerable advantages as data can be regularly recovered and processed at a base station, thus saving time and costs. Site visits for maintenance can be restructured to some extent to respond to known problems at stations, rather than being done solely on a periodic basis. Of course, telemetry has an additional and even more important part to play when "real time" information is required for flood warning or forecasting. It should be remembered that there is widespread acceptance of telecommunications via satellite systems, even from small islands. The use of satellites for data collection, therefore, should not be viewed as unusual. Unfortunately, this problem is fundamental in the thinking of some organizations and governments because of the low priority given to water resource assessment programmes.

Problems in obtaining surface runoff measurements on small islands in the humid tropics have been well covered in this volume by Manley & Askew and there is no need to reiterate them here.

The question of minimum networks for small tropical islands is an important one for streamflow, rainfall and evaporation stations. Hall (1983) outlines minimum densities as given by WMO (1981): streamflow (1 per 140–300 km^2), rainfall (1 per 25 km^2) and evaporation (1 per 50,000 km^2). However, these are guides only. Individual island size, topography and morphology need to be taken into account in applying them. For instance, some high islands are very much less than 140 km^2 and may require more than one streamflow station to characterize the runoff of the island. This is particularly relevant on islands which have multiple small catchments with varying geological, soil and vegetation characteristics. As with many other aspects of hydrology in small islands, generalizations are not possible. Each island, or at least, groups of similar islands, should be assessed on its or their own merits.

In dealing with hydrological data (including streamflow, rainfall and other data), it is necessary to consider data processing, archiving and reporting as well as data collection activities. The widespread use of microcomputers in recent years has resulted in the development and use of some high-quality, "user-friendly" packages for these tasks. Some examples are HYDROM (ORSTOM, France), HYDATA (Institute of Hydrology, UK), Micro-TIDEDA (New Zealand) and HYDSYS (Australia). With an amount of training matched to the computer literacy of the user(s), packages of this nature offer considerable flexibility and power to process and analyse data. There is good justification for the increased use of such packages for small island hydrology (as well as mainland hydrology) both within and outside the humid tropics. Some small islands have insufficient personnel trained in the use of computers but this problem can also be solved with appropriate training.

Recharge

The results of water balance studies on some small islands indicate the large influence of vegetation type and density on recharge to groundwater. This is confirmed by preliminary volumetric comparisons of fresh water zones beneath parts of atoll islands supporting different vegetation types. For instance, on West Island in the Cocos (Keeling) Islands, observations of salinity profiles showed that the fresh water zone below an area supporting dense coconut trees had less volume and showed greater seasonal fluctuations than below an area supporting only grass. The width of the island and the permeability of sediments at different depths at both locations were similar, thus indicating the controlling factor was the variation in ET_a between coconut trees and grass. The results of a water balance study indicated that the ET_a, and hence the recharge, varied according to the density of deeply rooted trees over a fresh water lens area. The relationship between recharge and the percentage area covered with coconut trees derived from water balance studies on Cocos (Keeling) Island is shown in Fig. 8.

Vegetation can have a direct effect on the design of abstraction facilities on small coral islands. Since deep-rooted vegetation effectively competes with abstraction by pumping for the limited fresh water resource under a small island, infiltration gallery lengths for a given pump flow rate need to be longer in areas where coconut trees are dense than where they are sparse or non-existent. This is shown in Fig. 9.

In the absence of data on ET_a, recharge to groundwater can be approximated for small low islands from results of water balance studies on other islands. Figure 10 summarizes annual rainfall versus annual recharge data derived from a selection of small islands. Once again, the effect of vegetation cover can be seen from the range of estimates of recharge for two islands with varying coverages of coconut trees. A preliminary assessment of recharge can also be made from the ratio of chloride ion in rainwater to that in shallow groundwater (Vacher & Ayers, 1980). Care must, however, be exercised with this method to avoid erroneous results (Daniell, 1983; Chapman,1985).

Groundwater

One of the main problems on many small islands is the lack of accurate knowledge of the extent and sustainable yield of fresh

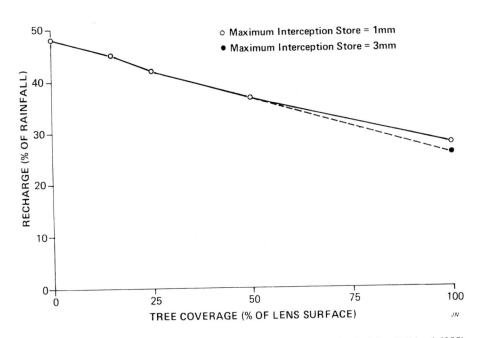

Figure 8: Recharge versus coconut tree coverage, Cocos (Keeling) Islands (after Falkland, 1988).

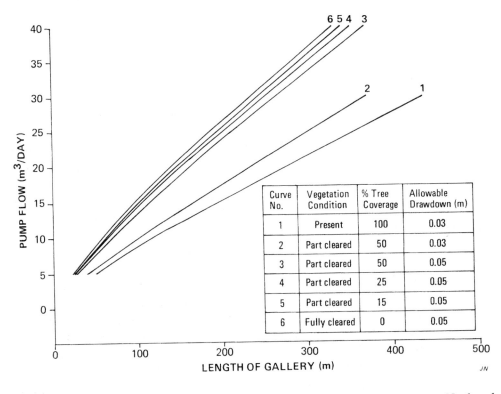

Curve No.	Vegetation Condition	% Tree Coverage	Allowable Drawdown (m)
1	Present	100	0.03
2	Part cleared	50	0.03
3	Part cleared	50	0.05
4	Part cleared	25	0.05
5	Part cleared	15	0.05
6	Fully cleared	0	0.05

Figure 9: Pump flow rate versus infiltration gallery length for different vegetation coverage, Northern lens, West Island, Cocos (Keeling) Islands (after Falkland, 1988).

water lenses. The usefulness of drilling or driving observation holes and obtaining vertical salinity profiles from the water table through to sea-water have been identified by some detailed studies (Kwajalein, Marshall Islands: Hunt & Peterson, 1980; Tarawa, Kiribati: AGDHC, 1982; Christmas Island, Kiribati: Falkland, 1983; Diego Garcia, Chagos Archipelago: Surface & Lau, 1986; Majuro, Marshall Islands: Hamlin & Anthony, 1987; South Keeling, Cocos (Keeling)

Islands: Falkland, 1988). The installation of simple monitoring systems inside these holes enable valuable post-drilling salinity profiles to be obtained. Single open boreholes, sometimes used for salinity monitoring, are not suitable as misleading results are likely to occur due to the mixing of fresh water and sea-water in the borehole. This topic has been reported from studies on mainland aquifers (Rushton, 1980; Kohout, 1980) and is just as relevant in small islands. Suitable monitoring

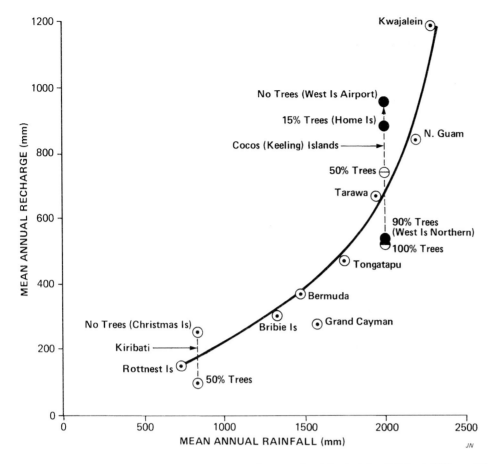

Figure 10: Annual rainfall versus annual recharge for some low islands (after Falkland, 1988, modified from Chapman, 1985).

systems are multiple open boreholes terminated at different depths (Kwajalein, Majuro), or tubes within a single bore terminated at different depths and hydraulically separated by means of bentonite plugs (Tarawa and Christmas Island in Kiribati, Diego Garcia, Cocos (Keeling) Islands). A limited number of islands now have close to a decade of salinity profile data which can be used to observe fresh water lens response to natural and artificial (pumping) stresses and to test and calibrate groundwater models. An example of this data is shown in Fig. 11.

Because of the fragile nature of fresh water lenses and the consequent acute requirement for effective monitoring systems, existing methods for obtaining salinity profiles should be critically reviewed with a view to recommending a suitable and economical method for use on small islands. This could possibly be included in the Hydrological Operational Multipurpose Subprogramme (HOMS) of the World Meteorological Organization.

Geophysical methods are well-suited to low islands with relatively small depths from the surface to the water table. Both electrical resistivity (ER) and electromagnetic (EM) methods have been successfully applied in a number of island studies in determining fresh water lens thickness. ER surveys, in particular, are well-suited to coral atolls (Dale *et al.*, 1986) and have been used on a number of atolls (Falkland, 1983; van Putten, 1988). Careful site selection is required to avoid buried objects such as pipes and cables as these can give misleading readings. It is also necessary to ensure soundings are parallel to the coastline, particularly near the edge of a lens, to minimize violation of the horizontal layering principle on which the ER method is based (Mooney, 1980). Stewart (1988) has outlined the use of EM surveys on small islands. An EM survey is known to have been used on at least one atoll (Majuro: Kauahikaua, 1987). With both ER and EM surveys, it is necessary to provide some means of independent assessment at selected locations to prevent erroneous interpretation of geophysical results. The use of geophysics is, however, recommended for preliminary reconnaissance of fresh water resources on small islands, particularly low coral islands. The application of geophysics becomes less useful in high islands, particularly where complex geological structures are present.

As previously mentioned, in the case of small islands where fresh water outflow is small, measurement of the water level fluctuations at the top of the fresh water lenses (water table) cannot be used to determine fresh water zone thickness because there is no direct relationship between the height of the water table above mean sea-level and the fresh water zone thickness. This fact has not always been realized and some incorrect assessments of fresh water storage volumes have been made in the past based on water level measurements alone. These measurements are, however, useful for monitor-

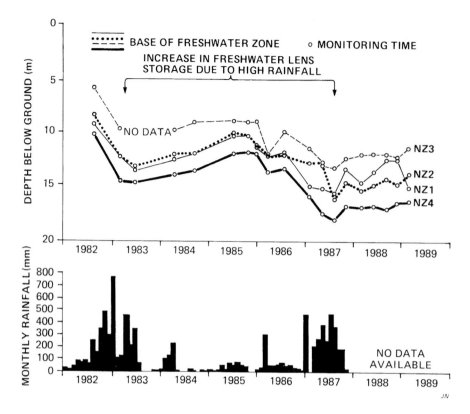

Figure 11: Groundwater salinity and rainfall data, New Zealand Airfield lens, Christmas Island, Kiribati (after Falkland, 1989).

ing the drawdown effects of pumping and setting design levels for infiltration galleries and other abstraction facilities.

Although some islands have been studied in detail, there appears to be a gap in the transfer of this knowledge to other island situations. The advantages and limitations of the various groundwater investigation and monitoring techniques should be well understood by field workers assigned to assess the groundwater resources on small islands. Where major abstraction works have been built or are being designed, resources should be provided to establish proper monitoring systems, particularly properly constructed monitoring boreholes for obtaining vertical salinity profiles. This normally requires the use of a drilling rig and an experienced drilling crew. It is essential that adequate supplies and spare parts be provided for successful drilling operations, as most small islands are located well away from supply sources in both space and time.

Moreover, there is a need for the development of simple yet accurate methods of determining sustainable yield from fresh water lenses. In the absence of detailed computer modelling for a particular island, approximate assessments of sustainable yield can be made from other studies. Based on a number of different approaches, sustainable yields of between 20% to 30% of recharge have been estimated for a number of small low-lying coral islands. It is considered that, as a first approximation, a value of 25% of recharge is considered reasonable. Using the range of percentage recharge values (approx. 25–50% of rainfall), sustainable yields of 6–12% of annual rainfall are derived. Actual vegetation conditions, however, should be allowed for in the derivation of estimates for a given atoll. Higher sustainable

yields than shown here may be appropriate in some islands, especially where the behaviour of lenses under drought and pumping conditions have been obtained and thoroughly studied.

The estimation of sustainable yield of an individual gallery system in a fresh water lens on small coral islands is another interesting and real problem. A number of approaches have been used for the design of gallery systems (Hunt & Peterson, 1980; AGDHC, 1982; Mink, 1986; Hamlin & Anthony, 1987; Falkland, 1988). However, it is apparent that further theoretical and practical research in this area is warranted. This is particularly important, given the need for simple yet effective methods for the design of gallery systems on small islands in the Pacific region as part of the United Nations water project operating from Fiji. At the 1989 UNDTCD seminar on water resources management strategies in Fiji, this matter was raised and it was recommended that further studies be initiated to develop a consistent approach for future water resource development of this nature. In the meantime, it is essential that monitoring of the behaviour of fresh water lenses using salinity profiles from properly constructed monitoring systems be continued or commenced. This will enable the response of lenses under abstraction conditions to be analyzed and calibration and testing of computer models to be undertaken.

Another potential problem with fresh water lenses is a rising sea-level associated with the "Greenhouse Effect." Some scenarios have been analyzed by Oberdorfer & Buddemeier (1988), using a dispersion type model for a typical small coral island. If the rise of the sea-level does not result in sea-water encroachment onto the land, the fresh water lens volume can actually increase because the whole lens would not only rise at the surface to the same degree but the base would tend to move into the upper and lower permeability sediments, thus slowing the fresh water outflow rate. The major issue, therefore, is whether the water level rise will lead to significant loss of available land. This is dependent on the magnitude of sea-level rise, an unanswered question at this stage. It is imperative that present sea-level monitoring programs for small islands be encouraged and expanded.

WATER USE

In general, the types of water use on small islands are similar to those experienced elsewhere in the world – water supply (for domestic purposes, tourism and industry), irrigation and hydro-electric power generation. Many small islands, however, do not have sufficient water resources or adequate land resources to allow for irrigation, or suitable topographic features to allow for the development of hydroelectric schemes. Hence, most fresh water on small islands is used for water supplies. On most large islands, irrigation is the greatest use for fresh water.

Water supply

Potable water is used for drinking, cooking, bathing, washing and cleaning. Other potable water uses include toilet flushing, cooling, heating, freezing, and drinking water for domestic animals. There are many cases, however, where non-potable water supplies, such as sea-water or brackish groundwater, are substituted for some of these purposes. In particular, non-potable water is used on some islands for such purposes as toilet flushing, fire-fighting, cooling water for power stations and ice-making for fishing industries. Brackish groundwater is sometimes used on very small islands, such as coral atolls, for all purposes except drinking and cooking. Sea-water is sometimes used for bathing while treated sewage effluent, another source of non-potable water, is sometimes used for such purposes as watering of gardens and lawns.

The per capita consumption for fresh water varies considerably between islands and within islands depending on availability, quality, type and age of water distribution systems, cultural and socio-economic factors and administrative procedures. Examples of per capita consumption expressed in term of litres per capita per day ($l\,c^{-1}\,day^{-1}$) for a number of islands are:

- Malé, Maldives: $59–88\,l\,c^{-1}\,day^{-1}$ (West & Arnell, 1975),
- Seychelles: $180\,l\,c^{-1}\,day^{-1}$ (Rooke, 1984),
- Penang, Malaysia: $265\,l\,c^{-1}\,day^{-1}$ (Harun, 1988),
- Majuro, Marshall Is: $550\,l\,c^{-1}\,day^{-1}$ (Mink, 1986),
- Tarawa, Kiribati: $20–40\,l\,c^{-1}\,day^{-1}$ (Bencke, 1980),
- Tahiti: $1000\,l\,c^{-1}\,day^{-1}$ (French Polynesia country paper, 1984),
- Barbados: $350\,l\,c^{-1}\,day^{-1}$ (Goodwin, 1984), and
- Dominica: $270\,l\,c^{-1}\,day^{-1}$ (Goodwin, 1984).

The per capita tourist water use of $500\,l\,c^{-1}\,day^{-1}$ tends to be higher in Barbados than the per capita domestic use (Goodwin, 1984). Tourist consumption of fresh water can represent the greatest use of water on many small islands.

Overall, the use of water for industrial purposes, including mining, tends to be minor on small islands. On particular islands, however, the use of water in some industries may be a major component of the overall use.

Irrigation

Much of the natural vegetation occurring on small islands in the tropical regions of the world receives adequate rainfall for growth. The natural vegetation consists of a variety of trees, particularly coconut trees, and a range of bushes and grasses. These do not require irrigation as they have adapted to local climatic conditions. Some of the natural vegetation – the coconut tree, for example – is remarkably salt-tolerant and can grow in water with relatively high salinity levels.

Irrigation schemes on small islands, where they exist, tend to be on a relatively minor scale, although there are exceptions. Many small islands, particularly coral atolls, do not have suitable soil conditions since they are both highly permeable and lacking in organic material. Relatively small-scale irrigation is possible, however, in some of the high islands.

Cultivation of root and tuber crops is practised in some islands, mainly those located in the Pacific Ocean. One important example is the cultivation of swamp taro on some coral atolls by digging a pit to the water table. This is essentially a form of irrigation as the crops have been introduced to a source of water not normally available.

At the higher water use end, cash crops such as sugar-cane are commercially grown using irrigation schemes on some islands. In the Hawaiian Islands and Fiji, for instance, the greatest use of water is for agriculture, primarily sugar-cane cultivation. On larger islands (Java, Indonesia) rice is cultivated by similar methods to those used on continents.

Hydro-power generation

There are a number of small high islands where hydroelectric power generation schemes have been implemented. These schemes tend to be on a relatively small-scale but can supply the major proportion of total power requirements (for example, Fiji). Other examples of islands where electricity is generated by hydro-power are Western Samoa, Tahiti, the Marquesas Islands, Pohnpei and Hawaii in the Pacific Ocean; Mauritius in the Indian Ocean and Dominica, Grenada and St. Vincent in the Caribbean Sea. Many other high islands have the potential for hydroelectric power generation.

WATER RESOURCE DEVELOPMENT METHODS

Many water resource development methods are used on islands. Methods which directly exploit or produce fresh water are rainwater collection, surface water collection, groundwater abstraction, desalination, and importation.

In addition, other methods assist in either the conservation of fresh water by reuse or substitution, or enhancement of the available resource. These include waste water reuse, direct substitution, non-potable water systems and potable water enhancement techniques.

Rainwater collection

This is one of the most common methods used for domestic water supply, particularly on islands with relatively high rainfall. Surfaces for rainfall collection include both roof and ground catchments with roof catchments being the most common type. Ground catchments can be either natural or artificial. Examples of artificially prepared ground catchments are airport runways (Majuro in the Marshall Islands), sealed surfaces made specifically for rainwater collection (for example, sealed rock outcrops in Bermuda and a paved hillslope in St. Thomas, USA's Virgin Islands), and synthetic liners (Coconut Island, Torres Strait, Australia).

Storages for rainwater on islands have been made of timber, steel and other metals, clay, concrete and fibreglass. Discarded containers (for example, 200 litre fuel drums) are often used.

Numerous examples of present and past methods of rainwater systems in many countries, including some islands, are provided in the proceedings of three recent international conferences on this topic (Fujimura, 1982; Smith, 1984 and Vadhanavikkit, 1987) and in Winter (1988) and IDRC *et al.,* (1989). An extensive bibliography is available from the Rainwater Harvesting Information Center of the Water and Sanitation for Health Project, which also publishes a newsletter "Raindrop." One of their publications (Edwards *et al.,*1984) considers in detail many aspects of the design and construction of roof catchment systems.

Surface water collection

Surface water development methods on islands are generally of three types: stream intake structures, dams or other storages, and spring cappings.

Stream intake structures generally consist of either in-stream weirs or buried collector pipe systems laid in, or adjacent to, the stream bed.

Water-retaining structures are constructed as dams within the stream or as "off-channel" storages. Neither are very common on small islands for a number of reasons, including unsuitable topography, geological conditions and economic considerations. Some large dams are found, however, on some larger islands (Sarawak, Malaysia).

Spring cappings typically consist of an open or covered containment structure, generally constructed from concrete or masonry. Spring flows are contained by the structure and diverted to an intake pipe.

Groundwater abstraction

Groundwater abstraction methods on islands are generally of five types: dug wells, boreholes (or drilled wells), use of natural sinkholes or cave systems, infiltration galleries and tunnels.

Dug wells, generally with diameters of between one and two metres are common on many small islands. On low islands, particularly coral atolls, such wells need only to be about two to three metres deep before the water table is reached. Where conditions are favourable, fresh water is available in moderate quantities. Dug wells also provide a source of fresh water in some areas on high islands. These areas are generally on the coastal margins in sedimentary formations. In some instances, shallow dug holes on beaches are used as a source of fresh water during low tide. These often become covered by sea-water at high tide.

Water is obtained from dug wells by means of hand bailers or pumps. Hand bailers are often buckets or discarded food or drink containers such as steel cans or plastic bottles. Pumps range from simple hand pumps, of which there are numerous designs, to more sophisticated mechanical pumps. These are driven using conventional energy sources such as diesel, or electrically powered motors, or, where conditions are favourable, renewable energy sources such as wind or solar powered pumps are sometimes installed.

Boreholes are also a common means of developing groundwater resources on islands, particularly high islands where depths to the water table are excessive or rocks are too hard for surface excavation.

On low islands, networks of boreholes have been used in the past to develop fresh water lenses. In the early 1930s, a network of boreholes was used to abstract fresh water from a lens on New Providence in the Bahamas. Due to over-abstraction of water, however, increasing salinity over the next 10 years forced the closure of the borefield (Sherman, 1980). Increases in salinity due to overpumping under similar conditions have been experienced on a number of islands. It was soon learnt that better methods of developing fragile lenses were required, and infiltration galleries became more popular.

However, where fresh water lenses are relatively thick, borehole abstraction systems have been used successfully. The main requirements are that a network of boreholes be drilled and individual pump rates are sufficiently low to prevent the occurrence of excessive localized drawdowns and consequent upconing of underlying sea-water.

On high islands, boreholes have been used to develop both high-level and basal groundwater bodies. On Oahu in the Hawaiian Islands, boreholes were drilled as early as last century to tap basal groundwater aquifers under artesian pressure (Takasaki, 1978). Often the boreholes were drilled too deep and became contaminated with sea-water. Following increased knowledge of the hydrogeological conditions, boreholes were located on or near ridges where greater yields of fresh water could be obtained from the highly-permeable basalt aquifers than in the alluvial valleys.

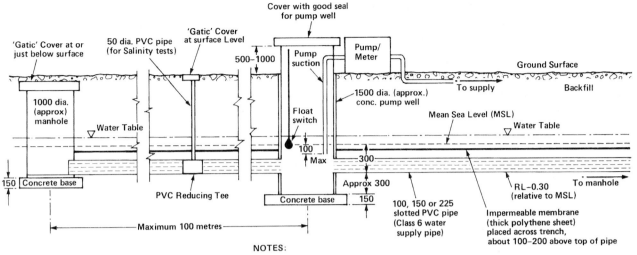

Figure 12: Cross section through an infiltration gallery (after Falkland, 1988).

Vertical and horizontal boreholes have also been used to abstract water from dyke-confined aquifers on some high islands, including the Hawaiian Islands (Peterson, 1972) and Tahiti (Guillen, 1984).

In some islands where karstic limestone is present, sinkholes or cave systems have been developed for water supply, usually by pumping with submersible pumps. Examples include Christmas Island in the Indian Ocean (Falkland, 1986), the islands of Grand Bahamas and Eleuthera in the Bahamas (Hadwen, 1988) as well as northern Guam and the Marianas Islands in the Pacific Ocean (Maragos, 1990).

Infiltration galleries, also referred to as "lens wells" or "skimming wells", have been used on low islands and in the coastal areas of high islands. This method of development is particularly effective in such conditions as relatively large volumes of water can be abstracted, compared with other methods, without causing sea-water intrusion.

On coral atolls and other low limestone islands, infiltration galleries generally consist of a horizontal conduit system permeable to water (for example, slotted PVC pipe as shown in Fig. 12). The conduit system is laid at or close to mean sea-level and allows water to be drawn towards a central pumping pit.

Infiltration galleries are generally constructed by surface excavation, using manual or mechanical methods, and subsequent laying and backfilling of a conduit system. Open trenches have been used in some islands (Bahamas; Christmas Island, Kiribati) but, while these are simple to construct, they are subject to surface pollution by crabs, birds and humans. Buried conduit systems have been installed and are successfully operating on a number of atolls including Kwajalein in the Marshall Islands (Hunt & Peterson, 1980), Diego Garcia in the Indian Ocean (Surface & Lau, 1986) and Tarawa, Kiribati

(AGDHC, 1986). On Tarawa, a yield of about 1200 m^3 day^{-1} is obtained from 17 galleries, each 300 m long, situated in two fresh water lenses.

An interesting and potentially very economical method of abstracting water from galleries has been tried in the Bahamas (Golani, 1989). Trenches have been connected to inclined pipes, allowing fresh water to flow under gravity to a deep sump towards the edge of the lens. A single pump installed there is used to pump the water to supply centres. This method avoids the need for multiple pumping systems, typically one per gallery or trench. Care must be taken with valve or weir settings at each trench to avoid excessive draining of fresh water from the upper surface of the lens, thus causing upconing of sea-water and possible local destruction of the lens.

Tunnels are probably the most technically difficult and least common method of groundwater development on islands. They have been used, however, to develop both high-level and basal groundwater bodies on high islands. In the Hawaiian Islands, tunnels or "Maui-type wells" (Chuck, 1968) have been used for many years for producing large quantities of fresh water from basal groundwater bodies in coastal areas. These tunnels were constructed by sinking a vertical or inclined shaft from ground level to a pump room just below the water table. A series of horizontal collection tunnels radiate out from the pump room, allowing water from a relatively large area to be abstracted. No new major Maui-type wells have been constructed since the early 1950s (Peterson, 1972) due to cheaper alternative boreholes.

In addition to basal groundwater bodies, perched and dyke-confined aquifers have been developed in the Hawaiian Islands using tunnels. In the case of perched aquifers, tunnelling has been done along the aquifer to collect water or divert it to perched springs. These water sources constitute a relatively small amount of the available high-level groundwater. Larger water volumes are available from the dyke-confined aquifers which are saturated to levels of several hundred metres above sea-level in some cases. Horizontal or inclined tunnels penetrate one or more dyke compartments (Peterson, 1984).

Tunnels have also been constructed at the base of large diameter (two to three m) wells or shafts at sea-level, for example at Barbados, where they have been excavated up to 60 m in length (Goodwin, 1984). Other examples occur on islands outside the humid tropics (Malta and the Canary Islands).

Desalination

Desalination plants are used on some islands for specific requirements (tourist resorts and military installations). There are few islands, however, where desalination is used as the main source of water. An exception is the USA's Virgin Islands where over 50% of the total requirement in 1980 was met with desalinated water (Coffin & Richardson, 1981). Some small islands used solely as tourist resorts also utilize desalinated water for nearly all fresh water requirements.

Desalination systems are based on a distillation or a membrane process. Distillation processes include multi-stage flash (MSF), multiple effect (ME) and vapour compression (VC) while the membrane processes include reverse osmosis (RO) and electrodialysis (ED). All types have been used on islands. For example, MSF plants with a combined capacity of a 30 000 m^3 day^{-1} operates on the island of Aruba, Netherland Antilles in the Caribbean Sea (UNDTCD, 1989). In the USA's Virgin Islands there are a number of ME plants with a combined capacity of about 25 000 m^3 day^{-1} (UNDTCD, 1989). A 1,100 m^3 day^{-1} ME plant is installed on the island of Ebeye, Marshall Islands (Toelkes, 1987). VC plants with a combined output of 2,600 m^3 day^{-1} operate in the Cayman Islands (Beswick, 1987). A number of sea-water RO plants operate in the USA's Virgin Islands, Bermuda, and on Malé in the Maldives. For instance, Bermuda has many desalination plants, including an ED plant with a capacity of 2,700 m^3 day^{-1} (Thomas, 1989).

In addition to the "high technology" desalination processes, solar stills offer a "low technology" solution in certain cases. They have been used, generally on a temporary or research basis, for the production of small quantities of fresh water from sea-water. With typical daily solar radiation values in the humid tropics of about 5 kWh m^{-2}, yields of about 3 l day^{-1} m^{-2} of solar still surface can be expected (Eibling *et al.*, 1971). Howe (1968) provides details of experiments with both transparent plastic and glass covers on a number of islands in the Pacific Ocean. He concluded that, while the stills had some major advantages such as use of readily available energy and the high-quality end product, there were some significant disadvantages and further equipment development was required. However, they can be used for emergency purposes as demonstrated by the production of 4 to 7 l day^{-1} from simple "home-made" stills in Tarawa (Harrison, 1980).

Desalination is a suitable technology in developed nations but in general is too sophisticated to be contemplated for general use in developing nations, including many islands of the humid tropics. Unless there are well-trained and skilled operators, and a ready and reliable source of spare parts and chemicals, desalination is not appropriate. Desalination has found some success on islands in the tourist industry, in military installations and as a temporary measure in the aftermath of natural disasters.

Importation

Water importation is used for a number of islands as an emergency measure during severe drought situations. In some cases, importation is used as a sole or supplementary source on a regular basis. Methods of importation include piping via fixed conduit, normally submarine pipeline, or sea transport using tankers or barges.

Hong Kong island receives about 50% of its potable water requirements via twin pipelines from the adjacent mainland (Little, 1986). The island of Penang in Malaysia also receives some of its water from the Malaysian peninsula via two submarine pipelines (Harun, 1988). Aruba in the Netherlands Antilles has received water imports by tanker from Dominica (Brewster & Buros, 1985) while approximately 30% of the total water supply for the island of New Providence in the Bahamas is presently imported by barge from a nearby and larger island, Andros (Swann & Peach, 1989). The island of Nauru in the Pacific Ocean has received most of its water as return cargo in ships used for exporting phosphate (Marjoram, 1983). Some of the smaller Fijian islands have received barged water from the larger islands since the early 1970s during drought periods. The service has become increasingly more routine in recent years.

Waste water reuse

Waste water is generally used for non-potable applications such as sanitary flushing, irrigation and industrial cooling. If waste water, which includes that discharged in either sewerage or stormwater systems, is adequately treated, it can also be used as potable water.

In Singapore, treated stormwater is used to supplement drinking water supplies. The scheme involves the collection of surface runoff from a number of urban catchments into ponds and subsequently into holding dams. Extensive treatment of the water then follows to ensure that it satisfies drinking water standards. The collection facilities are designed to collect about 70% of the runoff (Bingham, 1985).

Substitution

In extreme situations, such as severe drought conditions, substitutes for fresh drinking water have been used. The most notable is the juice from coconuts. Populations on some of the smaller and drier islands in Fiji, Kiribati and the Marshall Islands, for instance, have been known to survive on this substitute during drought periods. The coconut tree is remarkably salt-tolerant and can produce "fresh" juice from groundwater bodies which have high salinity levels – for example, chloride concentrations of at least one-third that of sea-water (Falkland, 1983).

Non-potable water systems

Non-potable water sources include sea-water, brackish groundwater and waste water. There are many examples of the use of these waters in order to conserve valuable fresh water reserves on islands. For example, sea-water is used for both toilet flushing and fire-fighting on a number of islands including St. Thomas and St. Croix in the USA's Virgin Islands (Coffin & Richardson, 1981), the major centres on Tarawa in the Republic of Kiribati (AGDHC, 1986) and the island of Hong Kong (Lerner, 1986). Many islanders, particularly on the less-developed islands, make use of sea-water or brackish well water for bathing and some washing purposes. Sea-water is also used for cooling of electric power generation plants, for ice making, in air conditioning plants, and in swimming pools.

Potable water enhancement techniques

There are a number of methods which are aimed at increasing natural or conventional supplies. These methods, primarily directed towards increasing recharge to, or storage of, groundwater supplies, include artificial recharge, sea-water intrusion barriers, groundwater dams and weather modification.

Artificial recharge aims to increase the sustainable yield from aquifers by directing surface water into pits, trenches, boreholes and infiltration basins or storages. Sources of surface water could be naturally-occurring sources (streams, springs or lakes), storm runoff from impervious surfaces, waste water and leaking pipelines. Examples of tropical islands where artificial recharge occurs are Hong Kong and Bermuda. On the island of Hong Kong, leaking water pipelines are known to contribute to groundwater recharge. This type of recharge, estimated to range from 260 to 3000 mm year^{-1} in some parts of the island, compares with an average annual rainfall of about 2000 mm (Lerner, 1986). On Bermuda, recharge from sanitation systems in unsewered urbanized areas appears to be about twice that occurring under naturally vegetated areas (Thomson & Foster, 1986).

Subsurface, artificial sea-water intrusion barriers can be constructed to impede the outflow of fresh water, or the inflow of sea-water, in basal groundwater bodies. The effect is to increase groundwater storage, at least in the short-term, thus increasing the availability of fresh water. On Miyako-jima, an island in the Ryukyu Archipelago of Japan, an experimental subsurface barrier was constructed in 1978 in a small buried valley (Sugio *et al.*, 1987). The barrier, designed to increase yields for irrigation, was constructed so as to be semi-pervious so that seepage could occur. Minimizing the concentration of agricultural chemicals in the stored groundwater was an additional goal. The barrier was found to be successful at delaying

sea-water intrusion into adjacent fresh water aquifers under pumping conditions by at least two months.

Groundwater dams have been used to store water in both Africa and India. The only known example on a small island is in the Cape Verde Archipelago, which is not actually in the humid tropics (Hanson & Nilsson, 1986). They do, however, have potential application in humid tropical islands.

Weather modification, often known as "cloud-seeding," has been the subject of research in a number of continental countries, including Australia, Israel and the USA. It has not been applied to small islands. Until further knowledge is developed in this area, it is not very relevant to this publication.

ISSUES AND PROBLEMS IN WATER RESOURCE DEVELOPMENT AND MANAGEMENT

General

The range of issues and problems related to water resources development and management on islands are in many cases similar to those found elsewhere but some are unique to islands because of their small size, limited water resources and isolation.

Broad policy issues, which will be addressed in this section, include the need for a reliable and safe source of water supply for the inhabitants, the protection of water resources from pollution and contamination (both natural and human-induced) and the rational allocation of water resources, particularly in the more developed islands.

Water resources: quantity

In some islands, all "conventional" water sources (rainwater catchments, surface water and groundwater) have been fully exploited for water supply, and other methods such as desalination or importation are required to meet demands on either a temporary or permanent basis.

Temporary problems can arise as a result of several factors, including drought, which can last for many months, or seasonally high water demands associated with the influx of large numbers of tourists.

More permanent problems have arisen, and are emerging, on some of the smaller islands, particularly atolls, with high population densities, for example Malé in the Maldives. There the problem is so severe that consumption of water exceeds recharge from rainfall (Edworthy, 1984). A number of options to increase the availability of fresh water have been implemented, including the installation of a desalination unit.

In some cases, existing sources, particularly fresh groundwater on coral atolls, are overpumped to the extent that pump wells in infiltration galleries are drained after a small number of hours. This problem can be overcome by redesigning abstraction systems by lengthening the galleries, reducing the pump rates or both. Another problem related to groundwater availability is the depletion of perched aquifers during periods of drought. Supplies from these sources are thus often unreliable in the long term. Springs emanating from perched aquifers also tend to suffer from the same seasonal problem.

Major salt-water intrusion caused by the construction of waterways and boat marinas has caused the loss of fresh water lens areas in some islands (Bahamas: Hadwen & Cant, 1980).

Rainwater systems often suffer from insufficient tank volumes or roof areas. Leaking tanks due to poor design and/or construction also cause problems in some systems.

There are many examples of failures with desalination techniques in islands, normally indicated by production rates being far below specification. Careful selection of the source water and design of appropriate pre-treatment facilities are particularly important.

Water resources: quality

Rainwater systems often suffer from water quality problems due to inferior roofing materials (for example, coconut leaf thatch can lead to taste and odour problems) or inadequate filtration (leading to bacteriological contamination). It is necessary that adequate roofing materials be used, and where seasonal patterns prevail, that simple systems to divert the "first foul flush" away from storage tank(s) be used. Rainwater bailed from tanks (or cisterns) with buckets and other containers can easily become contaminated.

The quality of surface waters is often affected by heavy sediment loads and turbidity levels, particularly in rivers and lakes following high intensity rainfall. This naturally-occurring problem is made worse by human activity, such as uncontrolled land clearing for agriculture, which causes additional sediment loads to be transported into river systems. High sediment loads cause problems such as the blockage of river or lake intakes for water supply. High turbidity levels, caused by colloidal suspensions (very fine clay particles) often associated with high sediment loads, make water treatment more costly. These problems occur on many small high islands.

Surface water quality is often affected by other pollutants such as petroleum products and general urban litter. Surface waters are particularly susceptible to pollution from toxic or other chemical spills. Concern has been expressed about the presence of timber treatment industries on some islands. These facilities use copper-chromium arsenates for preservation purposes. A spillage involving such chemicals has occurred in a small river in the Solomon Islands, leading to fish kills (Mowbray, 1984).

Many groundwater quality problems occur on islands. Detay et al., (1989) provide an overview of groundwater pollution and contamination of islands in the Federated States of Micronesia, the Marshall Islands and in Belau (Palau). Many of these problems are experienced by other islands within the humid tropical regions of the world.

The rising salinity of groundwater used for water supply and irrigation is a major problem. There are many examples of salinity increases due to sea-water intrusion as a result of over-pumping of fresh water lenses. This is generally caused by localized overpumping from wells, boreholes and infiltration galleries. In other cases, the problem is far more widespread and severe and is caused by general over-abstraction from many different systems (Malé in the Maldives: Edworthy, 1984).

Chemical, biological and microbiological pollution is associated with urban development and some rural communities, especially on heavily populated islands. Of particular concern is pollution from sewage disposal systems, including pit latrines, septic tanks, pumping stations and treatment plants located too close to groundwater sources. Nitrate levels in groundwater underlying unsewered urban areas on Bermuda were found to closely match population density (Thomson & Foster, 1986). Pollution from domestic animals such as pigs and dogs is also a problem on many islands.

Pollution from solid waste disposal areas and leakage of fuels from pipelines and storage tanks pose a major problem to fresh water lenses and other groundwater resources on many islands. Seepage from solid waste disposal areas which are located directly over, or in some cases, cut into fresh water lens areas can be a contaminant. Old quarries excavated to the water table are sometimes used as sites for solid waste disposal. On low-lying islands, large quantities of fuel are often stored near airport runways which are also often favourable locations for the development of groundwater. Oil and fuel leaks have occurred over fresh water lens areas in the Bahamas (Hadwen & Cant, 1980).

Chemical pollution can be associated with the return water from irrigation which may contain dissolved salts from fertilizers and residues of pesticides, insecticides and herbicides. Traces of potentially hazardous organic pesticides, ethylene dibromide and dibromochloropropane have been found in water supply wells on Oahu and Maui in the Hawaiian Islands (USGS, 1983; Lau & Mink, 1987). There is a potential problem with the direct spraying of insecticides onto taro which is commonly grown in pits dug down to the upper surface of fresh water lenses on Micronesian coral atolls (Brodie et al., 1984). This is particularly dangerous as no natural filter in the form of the unsaturated soil zone is present in such cases to at least partially absorb or immobilize these chemicals.

The fresh groundwater resources of coral atolls and low-lying limestone islands are, in general, particularly susceptible to pollution owing to relatively thin and highly permeable (often greater than 100 m day^{-1}) unsaturated soil zones. As a result, normally accepted minimum distances (the World Health Organization has often mentioned 30 m) from sewage disposal units (for example, pit latrines) to groundwater abstraction points are often inadequate to prevent contamination. To overcome this problem, each situation should be assessed on its own merits, giving due regard to the groundwater flow direction. On very small islands, this will generally mean that sewage disposal units should be sited close to the beach and water should be abstracted from the middle of the island.

In some cases, groundwater or surface water sources may be unpolluted, yet the water supplied to consumers may be polluted. This is due to contamination occurring in distribution systems which can be caused by the ingress of polluted groundwater through faulty joints or defective pipes when the pipeline is not pressurized. Other sources of water contamination are cross connections between potable and non-potable water distribution systems, contaminated tools and fittings during operation and maintenance activities and physical deterioration of pipes and fittings.

Water supply distribution systems

Many problems are evident with the distribution of water. Where water is collected by hand in buckets and other containers due to the lack of piped or delivered water, the water sources are sometimes a long way from village areas. The time-consuming effort of obtaining water adds to the potential problem of contamination from open water containers used to transport the water.

Piped water supply systems often suffer from major leaks due to cracks, holes, poor joints and other defects. Shortages of proper materials often lead to improvisation which may increase leakage rates. Examples include heating of PVC pipes over open fires to form bends and the laying of PVC pipes without the use of recommended jointing materials (solvent cement or rubber rings). Many water supply systems in small islands are old and in need of major rehabilitation. Losses as high as 80% of supply have been measured in some island water supply systems.

Where water deliveries are made, road tankers are often poorly maintained and consequently suffer from breakdowns resulting in disruptions to water supply services.

These problems can be resolved by efforts to install appropriate distribution systems where required and to repair and rehabilitate existing systems where leaks are excessive.

Operation and maintenance of island water supply systems are often very inadequate due to inappropriately designed and poorly constructed facilities, inadequate funding and poorly trained personnel.

By necessity, many water supply authorities on small islands operate at the level of "crisis management." Often only urgent repair jobs are done. There is generally a lack of preventative maintenance. Relatively large distances to water supply facilities such as pump stations, even on small islands, coupled with often poorly maintained tracks and vehicles, increases the problem.

A common example of the problems caused by lack of maintenance is equipment breakdown. As spare parts are often not available on the island, water supply interruptions are often experienced. There are many other examples too numerous to list here.

Allocation of water resources

A major problem, particularly on the more developed islands, is the conflict between alternative uses for the available water resources. Decisions are often required about the priorities of allocating water to various users in the domestic, industrial, tourist, agricultural and energy sectors. This problem needs to be resolved, based on local economic, social and political factors. The only generalization that can be made is the need to satisfy as a first priority, the most basic requirement, the domestic water supply.

Water legislation

Legislation dealing with water resources and supply systems on small islands varies from being almost non-existent to very complex. It is often outdated, redundant, ambiguous or difficult to enforce. Many problems have arisen due to non-recognition of the problem of having inadequate legislation or none, difficulties in enacting new legislation and/or difficulties in enforcing existing statutes. Specific problem areas are inadequate control of quantities abstracted from wells or boreholes, inadequate protection of groundwater quality, lack of procedures for the rational allocation of water to different sectors and inadequate control of the misuse or wastage of water supplied to consumers.

Even when legislation has been introduced, it is often difficult to enforce due mainly to reluctance on the part of enforcement agents to enforce legislation in a small community to which they belong.

Water administration

In many small islands, water resources and supplies are administered either by a government department concerned with much broader responsibilities or by a number of departments. There is, inevitably, intense competition for the very scarce funds and manpower. This phenomenon is not unique to small island communities and is often a major concern in large communities. The fragmentation of responsibility between a number of organizations can lead to long delays in reaching decisions and the decisions may not necessarily be based on sound technical or financial grounds.

There is often insufficient expertise to properly administer the many-faceted functions of a water supply utility, regardless of how small it may be. This problem is due to a number of basic difficulties such as insufficient training, inadequate resources, particularly funds for operation and maintenance tasks, and inappropriate technology.

Often there is little or no co-ordination between a multiplicity of agencies including water and health authorities, non-governmental organizations, bilateral and international aid agencies, and United Nations organizations.

Difficulties of transport and communications due to large distances from supply and information sources are common. This often results in long delays in obtaining necessary supplies. Large distances between islands of an archipelago add to this problem.

Reliance, in many cases, on short-term expatriate advisory and management staff often leads to lack of continuity in projects with consequent wastage of resources and inefficiency.

There is often incompatibility of materials and equipment supplied from different sources, especially for the many islands in developing countries where project assistance is obtained from different aid donors. This problem is made worse if aid donors have conditions requiring the purchase of materials and equipment from the donor country.

The largely unskilled work-forces on many small islands also means that water development and water supply projects are often not well-operated and maintained.

Natural disasters

Many small islands located in the humid tropics are susceptible to damage from the destructive forces of storms and cyclones. Water resources can be contaminated and water supply system components damaged or destroyed. In extreme cases, storm-generated waves have washed over some low islands, causing saline water to contaminate fresh water lenses. In other cases, landslides and floods have been known to damage or destroy surface water collection systems. Other destructive natural phenomena in the context of small islands are tsunamis (tidal waves) and severe volcanic activity including earthquakes and eruptions.

Following natural disasters, emergency water supplies are often required since existing supplies may be destroyed, damaged or polluted.

Climatic change and sea-level rise

Climatic change caused either by natural phenomena or by humans (for example, the "Greenhouse Effect") could dramatically influence the water resource and supply situation on small islands. Some predicted consequences of the "Greenhouse Effect" are changing rainfall patterns in some areas and a general increase in sea-level. Average rainfall on some islands is predicted to decrease while on others it may increase. A greater influence may be caused by rising sea-levels. It has been predicted that the sea-level could rise by between about 200 mm and 1.4 m (Stark, 1988) by the year 2030. Potential sea-level rises of these magnitudes are of concern to many small island nations, particularly the very low-lying coral atolls and cays (Pernetta & Hughes, 1989; McLean, 1989; Woodroffe, 1989a and 1989b; and the Malé Declaration, 1989).

In a recent review of available data, Wyrtki (1990) found that if present trends continue, the mean-sea-level of the oceans should not exceed about 50 to 100 mm in the next 50 years, which is a much lower estimate than previously presented in much of the literature. Wyrtki concludes that the problems caused by an exponentially rising population will be much more severe than those produced by sea-level rise, the "Greenhouse Effect," or any foreseeable climatic change.

APPROACHES TO WATER RESOURCE DEVELOPMENT AND MANAGEMENT

Planning

The assessment of water resources and their sustainable yields is a most important step in the planning of water resource developments. This is generally the most technically complex step in the planning phase. It must precede any substantial water resources development. Initially, available "conventional" resources, which include rainwater, surface water and groundwater resources, need to be assessed. An assessment of other options, including desalination and importation, may be required where the former are limited and where the economy can afford these generally more expensive options.

Water demand, including current use patterns and projected increases in demand due to increases in population or increasing requirements should be evaluated. Urban areas on small islands are increasing rapidly and special attention is required for these. It must be recognized that it may not be possible to economically supply water to meet all demands. Satisfaction of demands must then be made on a priority basis. Minimum requirements must first be met in the case of domestic water supplies. A minimum amount of $50 \ \mathrm{l \ c^{-1} \ d^{-1}}$ is considered appropriate in most cases, although more water is required if water-borne sewerage systems are used.

Selection of a water source is important. Where a choice between alternative sources of fresh water is available, it should be made after an assessment of engineering, economic, social, environmental, legal and administrative factors. As part of the selection process, conjunctive use of different classes of water should be considered since there is often more than one source of fresh water on small islands. Often rainwater catchments and shallow groundwater sources, either fresh or brackish, are available options even on the smallest of islands. For instance, rainwater may be used in minimum quantities for the most basic of needs, such as drinking and cooking, leaving higher salinity water for other uses such as bathing and washing. Where existing or potential water supplies are scarce, the use of dual or multi-quality supplies should be strongly considered (for example, the use of sea-water for toilet flushing).

Appropriate water quality criteria should be set. Guide-lines (WHO, 1971 and 1984) need to be adapted to suit local conditions. In particular, given the heavy dependence on water supplied from fresh water lenses by small island communities, criteria related to salinity (chloride ion concentration) need to be carefully assessed in the knowledge that island populations are often used to higher salinities in water than are specified in many guide-lines.

Co-ordination with other sectors is important. Where there is potential for alternative water uses, development of water resources must be viewed in the wider context of island development. For example, where there is potential for hydro-power generation, such development proposals have an influence on both water and energy sectors. Irrigation proposals could have strong influences on an island's economic development and land management practices, in addition to influencing the water resource. Water resource development may produce potential conflicts with other land uses. In particular, it is necessary that water supply and sanitation be seen as complementary developments. The public health of a community is strongly influenced by the level of both water supply and sanitation facilities. This aspect is particularly important on small islands where a limited land area means that there is a real threat of pollution of water supplies from sewerage systems and solid waste disposal sites since human habitation is often close to water sources. The creation of water reserves separate from village areas should be considered to minimize the pollution threat.

Evaluation of potential water resource development options should include an assessment of the technological level required. The level of technology must be appropriate to the community it is intended to serve, particularly for "rural" communities on small islands. The ability of the community to participate in implementation of water development projects and to accept responsibility for ongoing operation and maintenance are vital factors in the selection of the level of technology appropriate to the community. Community participation at all stages of the planning process is an important but often forgotten component. General principles for the selection of appropriate technology, based on a World Health Organization list (WHO, 1987), are minimal cost, participation by the village communities where possible in the selection process, capability for operation and maintenance by the local community, utilization of locally available materials wherever possible, use of local labour and compatibility with local customs.

Design criteria

When designing water resource development projects for islands, particularly small ones, certain basic criteria should be adopted.

Simple, proven designs which have been used in similar conditions should be used. Technical criteria from other regions can only be used as guides, and should be adapted to local conditions.

Locally available materials should be used where possible to minimize the cost of imported components and spare parts.

Standardization of materials and equipment is desirable to minimize the level of knowledge or experience and the variety of spare parts required for operation and maintenance. Specifying preferred and well-tested, equipment prior to receiving aid may avoid the problem of different supplies from different aid donors.

Corrosion-resistant materials should be used due to the proximity to the sea and airborne salt spray.

For rural water supply projects, operation and maintenance requirements must be minimized to enable village-level operation and maintenance (VLOM).

Use of renewable energy sources should be examined for their suitability. Such sources include solar and wind energy which can reduce the operational costs of pumping systems. Imported fuels are expensive and, where possible, designs should minimize the use of these energy sources.

Management

Water resource management can be interpreted in many ways and is a difficult term to neatly define. It can be viewed in broad terms as the rational allocation, use, control, and protection of water resources. Water resource management is a particularly important issue on islands, especially small islands, as the water resources are very scarce, easily polluted and the demands on them are often high.

In the case of independent island nations, consisting of one or more islands, the overall management of water resources should be conducted at the highest government level as it affects the livelihood of the whole community and involves interaction between a number of government agencies. The institutional framework must ensure that water resources assessment, development and management occur in the context of national planning. It must also ensure co-ordination between agencies responsible for water resources and water supplies. There needs to be a co-ordinating agency within the government for providing advice to the executive arm of government on matters of policy and planning. In the context of small island nations, this should normally be a "water resources committee" comprised of representatives from government departments with interests and responsibilities in water. For small islands which are part of large nations, policy and planning is often done by off-island government agencies in cooperation with island administrations or councils.

Typically, representatives from planning, natural resources, energy, agriculture, health, public works departments and from a water supply authority would be represented on a water resources committee. At this level, matters of mutual importance such as national water policy, allocation of water resources between various sectors (for example, water supply and agriculture) and legislative aspects need to be discussed and recommendations made. The planning department would normally take the lead role in such a committee.

There also needs to be an agency for water resource assessment, monitoring, licensing and control of water resources. In general, it is preferable that this be undertaken by a government department or ministry which is not a user agency. This ensures that there is no conflict of interest in allocating available water resources. If a single user agency such as a water supply authority is charged with this responsibility, it may favour the use of water resources in their own sector at the expense of other sectors such as agriculture. It may also lack incentive to make the most efficient use of water; such measures as leak detection and waste prevention may be given low priority as it would generally be easier to allocate more water to itself.

At the village level, the management of water issues can generally be done most effectively by a group of elected or appointed people, such as a village council, with support when needed from government agencies. Large villages and towns normally have a water utility consisting of technical and professional staff to manage their water supplies and water-related issues.

At the household level, water management often means maintenance by individuals of private wells, rainwater tanks and associated catchment areas and the allocation of water from different sources to various required uses.

Many island countries do not have a policy on development, conservation and protection of water resources. This leads to a fragmented approach to water resources. Hence, there is a need for island countries to develop national water policy plans which should outline the resources available and the policy for their development.

Water resource management issues on small islands, and especially on very small islands, tend to be dominated by the requirements for water supply for use by island inhabitants. Demands from industrial, tourist, agricultural and energy sectors are generally low or non-existent on very small islands. As island size increases and other determining factors for natural water occurrence improve, so does the demand from these other sectors increase, thus raising the potential for conflict between competing uses for the limited water resources.

While there are no set procedures which can be outlined for the development of national water policy, there is one essential requirement: the supply of an adequate and safe water supply to the island's population.

National water legislation is an essential step in ensuring the effective management of water resources. Guide-lines and case studies for islands in the Caribbean are provided in ESCAP (1983), Wilkinson (1985), United Nations (1986), Clark (1988) and Lau (1988). A summary of major points which should be covered from this and other sources is provided in Falkland *et al.,* (1991).

A reasonable approach for the drafting of water legislation for islands is to review legislation from other islands, with due consideration being given to local customs and conditions. It is essential that customary law be taken into account, especially in islands with long-standing traditions, otherwise the legislation may not be accepted. Public participation may be necessary in the legislative process. For independent small island nations, external assistance may be required, generally through international or bilateral aid agreements. The United Nations Department of Technical Cooperation for Development and the Food and Agriculture Organization have assisted a number of island countries in the Caribbean and the Pacific with a review of current legislation and the preparation of new legislation.

Although legislation should be comprehensive, it also should be framed in the simplest way possible. A good approach is for the main body of legislation not to be too detailed but to provide a general framework of powers and responsibilities and restraints (Davis, 1980). Within this framework, the detailed and exhaustive rules and regulations

can be drafted as subsidiary legislation. This approach enables repeal and amendments to be made without tedious and time-consuming parliamentary processes.

Appropriate protection and management policies for water catchments are essential for a safe water supply. It may be appropriate, especially on islands with diverse hydrological characteristics or relatively large areas, to subdivide the island into water management (or "protection") zones. This mechanism can assist in controlling extraction and minimizing pollution. Examples of islands where the water management zone concept has been adopted are Barbados in the Caribbean Sea, where 22 "groundwater units" are established (Goodwin, 1980) and Guam in the Pacific Ocean, where the Northern lens, which produces most of the island's requirements, is subdivided into 49 management zones (Goodrich & Mink, 1983). Where possible, legislation should be introduced, if not already present, to allow for land zoning and, in particular, for the establishment of reserved areas or "water reserves." Such reserves should disallow land uses which have the potential for polluting water resources, including residential, commercial and industrial development. This measure may only be partially achievable in some islands where the population is already dispersed throughout the island.

On coral atolls and other low-lying islands where the soils and geological formations are highly permeable, all development other than the necessary water supply extraction and distribution components should be sited away from known or potential fresh water lenses. In some situations, it may be possible to reserve individual islands within an atoll for water supply purposes. Another, and often more appropriate solution, is to site all residential and other development on the edge of the island or as far as practicable from the centre of the fresh water lens.

Some small islands are so developed that there is insufficient area remaining for "reserves" to be established. In these cases, the population may be living on top of its water supply with the thin permeable zone between the surface and the water table providing no barrier against pollution. This is the case in highly populated, very small islands. An example is Malé, Republic of Maldives (Mahir,1984). In these cases, the other measures listed below may be necessary.

Introducing sanitation and solid waste disposal schemes which "export" waste material from water resource areas act to improve the groundwater. Piped sewerage systems, although expensive, are effective in removing sewage. Solid waste disposal areas should generally be sited close to the edge of islands or areas where there is limited potential for fresh water to develop.

Another positive action is the introduction of measures to properly control the ownership and location of animals. Restrictions on the number and types of domestic animals may be necessary. Reserved areas for the rearing of some animals such as pigs and poultry can also assist in minimizing possible pollution of the underlying groundwater.

Introducing measures to ensure proper storage and use of potentially harmful materials and substances may also be necessary. This is important for fuel depots, mechanical workshops, hospitals, laboratories and chemical stores. Suitable measures would be to site these as far from water resource areas as possible, to minimize storage requirements and to exercise extreme care in the disposal of toxic substances.

Acceptance of the fact that the groundwater is polluted and using it only for non-potable purposes is a further option. This implies that other sources such as rainwater catchments or desalination are required for potable purposes.

The use of non-potable water for as many uses as possible may need to be encouraged, leaving basic needs to be satisfied by available potable sources. In this respect dual piped systems, one with potable and the other with non-potable water (for example, salt-water), can assist to limit demand for potable water. Other conjunctive use schemes are available when piped systems are not present.

Transmigration or resettlement of people from overcrowded islands to other locations may be necessary in extremely serious situations. It has been undertaken in the past and may be necessary in the future in the event of major natural disasters such as earthquakes, volcanic eruptions, overtopping by waves or extreme droughts. Such disasters not only affect water supplies but all the other aspects of a small island community's infrastructure.

Other water management strategies which are particularly relevant to islands are listed below.

Demand management is important for water resource management on small islands. In urban areas in particular, demand management measures should include an appropriate pricing policy and consumer education to reduce waste. Other measures may include reduction in water supply pressures to minimum levels and the use of water-conserving devices. Brewster (1989) presents further details of pricing policy, cost recovery strategies and demand management in small island environments.

As many water supply systems often have substantial leaks, an active leak detection and repair programme is essential. The savings in water can often have positive benefits in delaying the need for development of new sources.

Training at technical, professional and managerial levels is required as an ongoing requirement to improve the skills of local personnel in the assessment, development and management of their own water resources. Approaches to training was a major topic of discussion at the 1989 UNDTCD seminar on small island water resources and some solutions were offered, for example, Dale (1989).

Case studies

Case studies of approaches to water resources development and management strategies on a number of islands within the humid tropics are provided in Fujimara & Chang (1981), the proceedings of workshops and seminars listed earlier in this paper and Falkland *et al* (1991). Lau *et al.* (1989) provide an annotated bibliography of hydrology in the humid tropics, including a number of island studies.

RECOMMENDATIONS FOR FURTHER ACTION

A broad overview of the hydrology and water resources of islands, particularly small islands, within the humid tropics has been provided. Specific recommendations for further action are outlined below.

Data collection networks for water resource assessment should be expanded. In particular, there is a need to increase the coverage of rainfall recording stations (particularly in the high parts of islands), net solar radiation recording stations (due to its importance in evaporation studies), groundwater investigation boreholes for monitoring salinity profiles in fresh water lenses and sea-level monitoring stations. National agencies should be encouraged not only to install such monitoring stations, but provide sufficient resources to ensure their ongoing viability. Where local resources are unavailable, requests for external assistance to assist with this work should be made.

The use of electronic data-logging equipment should be encouraged for data collection programmes. This will normally require re-equipping and training of local staff. In addition, the acquisition of data from remote sites via satellite telemetry should be encouraged.

Detailed investigations of actual evapotranspiration using micrometeorological and/or other suitable techniques for selected vegetation/soil/sub-climate associations on small islands should be conducted. Existing approaches for the estimation of evapotranspiration should be re-evaluated in the light of such investigations. As a first stage of the investigations, a database of the typical soil/vegetation/sub-climate associations, including relevant hydrological parameters, should be assembled for islands of the humid tropics. As part of this investigation, the interception capacity of typical vegetation found on tropical islands should be studied. For small islands in the tropical zones, interception losses from dense stands of coconut trees and other vegetation are observed to be very high, but at present there is a paucity of quantitative data to confirm these observations.

Standardization of methods for the determining of salinity profiles in observation boreholes should be developed. Recognition of the problems inherent in using single open boreholes should be highlighted. Existing methods using multiple open holes or hydraulically isolated zones should be reviewed and a suitable and economical method recommended for typical island types (for example, low-lying coral atolls and elevated terrain on raised limestone islands). Recommended methods could be included in the Hydrological Operational Multipurpose Subprogramme (HOMS) of the World Meteorological Organization.

Groundwater flow models for determining the sustainable yields of fresh water lenses should be reviewed and simpler, yet theoretically correct, approaches developed for use on microcomputers. Models which allow for a transition zone, rather than sharp interface models, should be used.

Appropriate computational procedures for the design of abstraction systems in fresh water lenses should be developed based on past knowledge and further theoretical and/or practical research. Existing groundwater theory should be reviewed taking into account the special features of fresh water lenses. Existing abstraction systems on small islands should also be reviewed to determine their effectiveness and/or problems. Additional research may involve scale-model studies using sand boxes and variable density fluids and/or construction and monitoring of a prototype on a selected island.

Further research into groundwater pollution is required. Special attention has to be given to pesticide behaviour and transport, especially in terrain devoid of soil cover or with small adsorption capacity. The behaviour and transport of biological contaminants in densely populated areas, especially in thin, highly permeable soil conditions as are found on coral islands and in the coastal areas of other islands needs to be known. Guide-lines for minimum distances between sanitation and water supply facilities need to be re-evaluated for island conditions with due consideration given to groundwater flow direction, the permeability of soils and underlying geological layer(s), the rate of extraction and the type(s) of sanitation disposal.

There is scope for greater technology transfer to islands in the fields of hydrology and water resources. Options available for technology transfer are via international agencies such as FAO, UNDTCD, UNESCO, WHO and WMO, via bilateral aid agreements and through cooperative arrangements between island countries on a regional basis. The current United Nations water project operating within a number of island groups in the Pacific Ocean is a good example of an international agency approach to the problem. Seminars and workshops on the topics of hydrology and water resources of islands has assisted participants from many islands in the past and these should continue on a regular basis. Specific meetings of technical experts would help to resolve ongoing practical issues such as optimal abstraction systems and pumping rates in fresh water lenses. Sponsoring of meetings and seminars by United Nations organizations involved in water resource issues on islands in the humid tropics is particularly appropriate. Studies and reports which address the water resource issues and problems faced by islands in the humid tropics should also be encouraged. Recent examples are two UNESCO studies (Falkland et al., 1991; Falkland, 1992) and specific technical papers (for example, Underwood et al., 1992). As more information and knowledge is acquired, revised and updated publications on this topic will be required.

Training programmes for island technical and professional personnel in water sciences and engineering should be encouraged. A combination of formal training in recognized institutions often away from home, combined with specific training

in-country, or on a regional basis, is an appropriate approach. In-country training can sometimes be provided by personnel from external agencies engaged on either short- or long-term contracts. Where this type of training is unavailable, consideration should be given to regional training workshops involving the solution of practical problems along the lines of the 'REFRESHR' workshops held in the Pacific (Dale & Thorstensen, 1988a and 1988b).

Greater co-ordination is required between agencies involved with small island hydrology and water resource assessment, development and management. To assist in this regard, it is recommended that efforts to establish a library service through existing facilities be made for the collection, storage and dissemination of technical information, data and reports dealing with, or relevant to, small island hydrology. The service could be via a network of institutions on islands in the humid tropics.

REFERENCES

AGDHC, (1982) *Tarawa water resources pre-design study.* prepared by Australian Government Department of Housing and Construction for the Australian Development Assistance Bureau (unpublished report).

AGDHC, (1986) *Review of Tarawa water supply project.* prepared by Australian Government Department of Housing and Construction for the Australian Development Assistance Bureau, (unpublished report).

Badon, Ghyben W. (1889) Nota in verband met de voorgenomen put boring nabij Amsterdam. (Notes on the Probable Results of the Proposed Well Drilling near Amsterdam), *Konikl Inst. Ing. Tijdschr. 21.*

Bartle, G. A. (1987) *Report on the evaluation of the Aokautere thermo-electric heat pulse for measuring transpiration in coconut palms.* Division of Water Resources, CSIRO, Australian Government.

Bencke, W. E. (1980) *Report on visit to Tarawa and Christmas Island.* Australian Government Department of Housing and Construction (unpublished report).

Beswick, R. G. B. (1987) Cayman Islands: country situation report: non-conventional water resources development in developing countries. *United Nations Natural Resources/Water Series 22*, 338–348.

Bingham, A. (1985) Singapore traps urban run-off. *World Water.* May, 33–35.

Brewster, M. R. (1989) Water resources management in small island countries: cost recovery and demand management. *Interregional Seminar on Water Resources Management Techniques for Small Island Countries.* UNDTCD, Suva, Fiji.

Brewster, M. R. & Buros, O .K. (1985) Non-conventional water resources: economics and experiences in developing countries. *Natural Resources Forum 9* (1), 65–75.

Brodie, J. E., Prasad, R. A. and Morrison, R. J. (1984) Pollution of small island water resources. *Proc. Regional Workshop on Water Resources of Small Islands.* Suva, Fiji. Commonwealth Science Council Tech. Publ. No 154, Part 2, 379–386.

Brunel, J. P. (1981) (ed.) *Atlas de la Nouvelle Caledonie.* Carte et notice: Les elements generaux du climat. ORSTOM, Paris.

Brunel, J. P. (1989) Evapotranspiration in South Pacific tropical islands: How important?. paper prepared for *Internat. Colloquium on Development of Hydrologic and Water Management Strategies in the Humid Tropics.* UNESCO, Townsville, Australia.

Chang, J. H. & Lau, L. S. (1983) Definition of the humid tropics. Paper presented to the IAHS meeting in Hamburg, Germany.

Chapman, T. G. (1985) *The use of water balances for water resource esti-mation with special reference to small islands*. Bulletin 4, Pacific Regional Team, Australian Development Assistance Bureau.

Chuck, R. T. (1968) Groundwater resources development on tropical islands of volcanic origin. *Internat. Conf. on Water for Peace*. Washington D.C. *2:* 963–970.

Clark, S. D. (1988) *Western Samoa: a possible framework for water resources legislation*. United Nations Department of Technical Cooperation for Development, New York, INT/86/R30.

Coffin & Richardson Inc., (1981) *Water conservation under conditions of extreme scarcity: the USA's Virgin Islands*. Office of Water Research and Technology, Nat. Tech. Info. Service No. PB 82–204843.

Commonwealth Science Council, (1984) *Proc. Regional Workshop on Water Resources of Small Islands*. Suva, Fiji, Tech. Publ. No. 154, 3 parts.

Dale, W. R. (1989) Solutions for training needs. *Interregional Seminar on Water Resources Management Techniques for Small Island Countries*. UNDTCD, Suva, Fiji.

Dale, W. R. & Thorstensen, A. L. (1988a) *Review and evaluation of the subregional workshop in fresh water evaluation, Tonga*. South Pacific Commission, Noumea.

Dale, W. R. & Thorstensen, A. L. (1988b) *Review and evaluation of the subregional workshop in fresh water evaluation, Pohnpei (Micronesia)*. South Pacific Commission, Noumea.

Dale, W. R., Waterhouse, B. C., Risk, G. F. & Petty, D. R. (1986) *Coral island hydrology: a training guide for field practice*. Commonwealth Science Council Publ. Series, No. 214.

Daniell, T. M. (1983) Investigations employed for determining yield of the groundwater resources of Tarawa. *Proc. Meeting on Water Resources Development in the South Pacific*. United Nations Water Resources Series, No. 57: 108–120.

Daniell, T. M. (1986) *Report on consultancy for hydrological balance studies of small islands in Indonesia and the Philippines*. prepared for UNESCO, Australian Government Department of Territories.

Davis, C. C. (1980) Overview of water resources legislation and adminis-tration. In: Hadwen P., (ed.) (1980), *Proc. Seminar on Water Resources Assessment, Development and Management in Small Oceanic Islands of the Caribbean and West Atlantic*. Barbados, United Nations and Commonwealth Science Council, 488–503.

Detay, M., Alessandrello, E., Come, P. & Groom, I. (1989) Groundwater contamination and pollution in Micronesia. *J. Hydrol. 112*, 149–170.

Diaz Arenas A. A. & Huertas, J. B. (1986) *Hydrology and water balance of small islands: a review of existing knowledge*. Technical documents in hydrology. UNESCO, Paris.

Edwards, D., Keller, K. & Yohalem, D. (1984) *A workshop design for rainwater roof catchment systems: a training guide*. Water and Sanitation for Health, Tech. Report No. 27.

Edworthy, K. J. (1984) Groundwater development on Malé Island, Maldives. *Proc. Fifth Internat. Conf. on Water Resources Planning. 2*, Athens.

Eibling, J. A., Talbert, S. G. & Lof, G. O. G. (1971) Solar stills for com-munity use – digest of technology. *Solar Energy. 13*, 263–276.

ESCAP, (1983) Draft comprehensive programme for water resources development in the Pacific Region. Proc. *Meeting on Water Resources Development in the South Pacific. Economic and Social Commission for Asia and the Pacific*. Suva, Fiji, March 1983. Water Resources Series No 57, United Nations, 41–47.

Falkland, A. C. (1983) *Christmas Island (Kiritimati) water resources study*. prepared for Australian Development Assistance Bureau (unpub-lished report).

Falkland, A. C. (1986) *Christmas Island (Indian Ocean) water resources study in relation to proposed development at Waterfall*. prepared for Australian Government Department of Territories.

Falkland, A. C. (1988) *Cocos (Keeling) Islands: water resources and management study*. prepared for Australian Construction Services, Department of Administrative Services.

Falkland, A. C. (1989) Investigation and monitoring of freshwater lens behaviour on coral atolls. *Interregional Seminar on Water Resources Management Techniques for Small Island Countries*. UNDTCD, Suva, Fiji.

Falkland, A. C. (1992). *Small tropical islands, water resources of par-adises lost*. IHP Humid Tropics Programme Series, No. 2, UNESCO, Paris.

Falkland, A. C., Custodio, E., Diaz, Arenas A. & Simler, L. (1991) *Hydrology and water resources of small islands: a practical guide*. Studies and reports in hydrology, No. 49. UNESCO, Paris.

Fleming, P. M. (1987) The role of radiation in the areal water balance in tropical regions: a review. *Arch. Hydrobiol. Beih. 28*, 19–27.

French Polynesia country paper, (1984) *Tech. Proc. Regional Workshop on Water Resources of Small Islands*. Suva, Fiji, Commonwealth Science Council Tech. Publ. Series no. 182, 11–20.

Fujimura, F.N. & Chang, W.B., (eds.) (1981) *Groundwater in Hawaii: a century of progress*. Water Resources Research Center, Univ. Hawaii, USA

Fujimura, F. N., (ed.) (1982) *Proc. Internat. Conf. on Rain Water Cistern Systems*. Water Resources Research Center, Univ. Hawaii, Hawaii, USA

Ghassemi, F., Jakeman, A. J. & Jacobson, G. (1990) Mathematical model-ling of sea-water intrusion, Nauru Island. *Hydrological Processes. 4*, 269–281.

Giambelluca, T. W., Nullet, D. & Nullet, M. A. (1988) Agricultural drought on south-central Pacific islands. *Professional Geographer. 40* (4), 404–415.

Golani, U. (1989) Verbal presentation. *Interregional Seminar on Water Resources Management Techniques for Small Island Countries*, UNDTCD, Suva, Fiji.

Goodrich, J. E. & Mink, J. F. (1983) Groundwater management in the Guam island aquifer system. *Internat. Conf. on Groundwater and Man*. Sydney, Australia, *3*, 73–82.

Goodwin, R. S. (1980) Water assessment and development in Barbados. Country Position Paper, in Hadwen P., ed. (1980), *Proc. Seminar on Water Resources Assessment, Development and Management in Small Oceanic Islands of the Caribbean and West Atlantic*. Barbados, United Nations and Commonwealth Science Council, 255–265.

Goodwin, R. S. (1984) Water resources development in small islands: per-spectives and needs. *Natural Resources Forum. 8* (1), 63–68.

Guillen, J. A. (1984) Hydrogeological facts about dike aquifers and under-ground water circulation in Tahiti. *Tech. Proc. Regional Workshop on Water Resources of Small Islands*, Suva, Fiji, Commonwealth Science Council Tech. Publ. No. 154, 455–472.

Hadwen, P. (1988) personal communication. Former Chief Technical Adviser, United Nations Department of Tech. Cooperation for Development, Fiji.

Hadwen, P. & Cant, R. V. (1980) Sources of groundwater contamination in New Providence, a preliminary assessment. In Hadwen, P. (ed.),

Proc. Seminar on Water Resources Assessment, Development and Management in Small Oceanic Islands of the Caribbean and West Atlantic. Barbados, United Nations and Commonwealth Science Council, 598–601.

Hall, A. J. (1983) Surface water information network design for tropical islands. Proc. Meeting on Water Resources Development in the South Pacific. *United Nations Water Resources Series* No.57, 83–95.

Hamlin, S. N. & Anthony, S. S. (1987) Ground-water resources of the Laura area, Majuro Atoll, Marshall Islands. *U.S. Geological Survey Water Resources Investigation Report* 87–4047.

Hanson, G. & Nilsson, A. (1986) Ground-water dams for rural-water supplies in developing countries. *Groundwater. 24* (4), 497–506.

Harrison, G. E. (1980) *Socio-economic aspects of the proposed water supply project for South Tarawa, Kiribati.* Australian Government Department of Housing and Construction (unpublished report).

Harun, H. (1988) Water quality management in Penang Island. *Proc. Southeast Asia and the Pacific Regional Workshop on Hydrology and Water Balance of Small Islands.* Nanjing, China.

Herzberg, A. (1901) Die Wasserversorgung Einiger Nordseebader (The water supply on parts of the North Sea coast). *J. für Gasbeleuchtung und Wasserversorg*ung. *44*, 815–819 and *45*, 842–844.

Howe, E. D. (1968) Solar distillation for augmenting the water supply on low islands. *Internat. Conf. on Water for Peace.* Washington D.C., USA, *2*. 205–213.

Hunt, C. D. & Peterson, F. L. (1980) *Groundwater resources of Kwajalein Island, Marshall Islands.* Tech. Report No. 26, Water Resources Research Centre, University of Hawaii, Hawaii.

IDRC *et al.,* (1989) *Proc. Fourth Int. Conf. on Rain Water Cistern Systems.* Manila, Philippines, 2–4 August, 1989. International Development Research Centre, Canadian International Development Agency and the Philippine Water Works Association.

Karranjak, J. (1989) Verbal presentations. *Interregional Seminar on Water Resources Management Techniques for Small Island Countries.* UNDTCD, Suva, Fiji.

Kauahikaua, J. (1987) Description of a freshwater lens at Laura island, Majuro Atoll, Republic of the Marshall Islands using electromagnetic profiling. *U.S. Geol. Survey Open File Report* 87–582.

Kohout, F. A. (1980) Differing positions of saline interfaces in aquifers and observation boreholes – comments. *J.Hydrol. 48*, 191–195.

Lau, L. S. (1988) State water code – a masterpiece of compromise. *Wiliki o Hawaii* (Engineer of Hawaii). *23* (8/9).

Lau, L. S. & Mink, J. F. (1987) Organic contamination of groundwater: a learning experience. *J. Am. Water Works Assn. 79* (8), 37–42.

Lau, L. S., Moore, R. L. & Hirakawa, P. Y. (1989) *Humid tropics hydrology: an annotated bibliography.* Water Resources Research Center, Univ. Hawaii, Hawaii, USA

Lerner, D. N. (1986) Leaking pipes recharge ground water. *Groundwater. 24* (5), 654–662.

Little, M. J. (1986) New pipelines on land and across Victoria Harbour, Hong Kong. *J. Instit. of Water Engineers and Scientists. 40* (3), 271–287.

Mahir, A. M. (1984) Maldives water supply and sewage disposal. *Proc. Regional Workshop on Water Resources of Small Islands.* Suva, Fiji. Commonwealth Science Council Tech. Publ. Series No 182, Part 3, 100–103.

Malé Declaration, (1989) Malé Declaration on global warming and datalogging rise. *Small States Conference on Sea Level Rise.* Malé, Republic of Maldives.

Maragos, J. (1990) personal communication. Environment and Policy Institute, East-West Center, Honolulu, Hawaii.

Marjoram, T. (1983) Pipes and pits under the palms: water supply and sanitation in the South Pacific. *Waterlines. 2* (1), 14–17.

McLean, R. F. (1989) *Kiribati and sea-level rise. Report on a visit to the Republic of Kiribati.* University of New South Wales, Australia.

Mink, J. F. (1986) *Groundwater resources and development.* Trust Territory of the Pacific Islands, U.S. Environment Protection Agency, Region 9.

Mooney, H. M. (1980) *Handbook of engineering geophysics.* Electrical Resistivity, Bison Instruments Inc., Minneapolis, Minnesota, U.S.A., vol.2.

Mowbray, D. (1984) *A review of pesticide use in the South Pacific with a proposal for establishing a pesticide monitoring and evaluation network.* South Pacific Regional Environmental Programme, Noumea, New Caledonia.

Nullet, D. (1987) Water balance of Pacific atolls. *Water Resources Bulletin. 23* (6), 1125–1132.

Oberdorfer, J. A. & Buddemeier, R. W. (1988) Climate change: effects on reef island resources. *Sixth Internat. Proc. Sixth Coral Reef Symposium.* Townsville, Australia, *3*: 523–527.

Pernetta, J. C. & Hughes, P. J., (eds.), (1989) Studies and reviews of greenhouse related climatic change impacts on the Pacific Islands. *Meeting on Climatic Change and Sea Level Rise in the South Pacific.* Majuro, Marshall Islands. 16–20 July. South Pacific Commission, United Nations Environment Programme and Association of South Pacific Environmental Institutions.

Peterson, F. L. (1972) Water development on tropic volcanic islands – type example: Hawaii. *Groundwater. 10* (5), 18–23.

Peterson, F. L. (1984) Hydrogeology of high oceanic islands. *Tech. Proc. Regional Workshop on Water Resources of Small Islands.* Suva, Fiji, Commonwealth Science Council Tech. Publ. No. 154, 431–435.

Rooke, E. R. (1984) Seychelles country paper. *Tech. Proc. Regional Workshop on Water Resources of Small Islands.* Suva, Fiji Commonwealth Science Council Tech. Publ. Series No. 182, 185–201.

Rushton, K. R. (1980) Differing positions of saline interfaces in aquifers and observation boreholes. *J.Hydrol. 48*, 85–189.

Scott, G. A. & Rotondo, G. M. (1983) A model to explain the differences between Pacific plate island-atoll types. *Coral Reefs. 1*, 139–150.

Sherman, G. E. (1980) Water resources assessment, development and management in the Bahamas. In: Hadwen P. (ed.), *Proc. Seminar on Water Resources Assessment, Development and Management in Small Oceanic Islands of the Caribbean and West Atlantic.* Barbados, United Nations and Commonwealth Science Council, 136–144.

Smith, H. H., (ed.) (1984) *Proc. Second Internat. Conf. on Rain Water Cistern Systems.* St. Thomas, USA's Virgin Islands.

Stark, K. P. (1988) Designing for coastal structures in a greenhouse age. in *Greenhouse, Planning for Climate Change.* Pearman G.I. (ed.), CSIRO, Australia, 161–176.

Stewart, M. (1988) Electromagnetic mapping of fresh water lenses on small oceanic islands. *Groundwater. 26* (2): 187–191.

Stoddart, D. R. (1971) Rainfall on Indian Ocean coral islands. *Atoll Research Bulletin.* No 147, The Smithsonian Institute, Washington D.C. USA.

Sugio, S., Nakada, K. & Urish, D.W. (1987) Subsurface seawater intrusion barrier analysis. *J. Hydraulic Engineering. 113* (HY6), 767–779.

Surface, S. W. & Lau, E. F. (1986) Fresh water supply system developed on Diego Garcia. *The Navy Civil Engineer. XXV* (3), 2–6.

Swann, M. S. & Peach, D. W. (1989) Status of groundwater resources development for New Providence. *Interregional Seminar on Water Resources Management Techniques for Small Island Countries.* UNDTCD, Suva, Fiji.

Takasaki, K. J. (1978) Summary appraisals of the nation's ground-water resources – Hawaii region. *U.S. Geol. Survey Prof. Paper* 813–M.

Taylor, R. C. (1973) *An atlas of Pacific islands rainfall.* Hawaiian Institute of Geophysics, Data Report No. 25, HIG–73–9, University of Hawaii, Hawaii, USA.

Thomas, E. N. (1989) Water resources and supply, Bermuda. *Interregional Seminar on Water Resources Management Techniques for Small Island Countries.* UNDTCD, Suva, Fiji.

Thomson, J. A. (1985) Groundwater model of the Central Lens of Bermuda. *Interregional Seminar on Development and Management of Island Groundwater Resources.* Hamilton, Bermuda, Paper No 15.

Thomson, J. A. & Foster, S. D. (1986) Effect of urbanization on groundwater of limestone islands: an analysis of the Bermuda case. *J. Institution of Water Engineers and Scientists.* 40 (6), 527–540.

Toelkes, W. E. (1987) The Ebeye desalination project - total utilization of diesel waste heat. *Desalination.* 66, 59–68.

Underwood, M. R., Peterson, F. L. & Voss, C.I. (1992) Groundwater lens dynamics of atoll islands. *Water Resour. Res.* 28 (11), 2889–2902.

UNDTCD, (1989) Water resources of small island countries. *Interregional Seminar on Water Resources Management Techniques for Small Island Countries.* UNDTCD, Suva, Fiji.

United Nations, (1986) Water resources legislation and administration in selected Caribbean countries. Department of Technical Co-operation for Development and Food and Agriculture Organization. *United Nations Natural Resources/Water Series* No 16, New York.

USGS, (1983) National water summary 1983 – hydrologic events and issues. *U.S. Geol. Survey, Water-Supply Paper* 2250.

Vacher, H. L. & Ayers, J. F. (1980) Hydrology of small oceanic islands – utility of an estimate of recharge inferred from the chloride concentration of the freshwater lenses. *J.Hydrol.* 45, 21–37.

Vadhanavikkit, C., (ed.) (1987) *Proc. Third Internat. Conf. on Rain Water Cistern Systems.* Khon Kaen University, Thailand, 14–16 January.

van der Brug, O. (1986) The 1983 drought in the western Pacific. *U.S. Geol. Survey Open-File Report* 85–418.

van Putten, F. (1988) *A hydro-geophysical assessment of groundwater resources on the Tuvaluan islands.* Tech. Report TUV/8, UNDTCD, Suva, Fiji.

Voss, C. I. (1984) SUTRA: A finite-element simulation model for saturated-unsaturated fluid-density-dependent ground-water flow with energy transport or chemically-active single-species solute transport. *U.S. Geol. Survey. Water Resources Investigations Report* 84–4369.

Voss, C. I. & Souza, W. R. (1987) Variable density flow and solute transport simulation of regional aquifers containing a narrow freshwater-salt-water transition zone. *Water Resour. Res.* 23 (10), 1851–1866.

West, M. J. H. & Arnell, D. J. (1975) Assessment and proposals for the development of the water resources of Malé – a small tropical island. *Proc. Second World Congress, Internat. Water Resources Association.* New Delhi, vol. II: 409–427.

WHO, (1971) *International standards for drinking water.* World Health Organization, Geneva, Switzerland, 3rd. edn.

WHO, (1984) *Guide-lines for drinking water quality.* World Health Organization, Geneva, Switzerland, 3 vols.

WHO, (1987) *Technology for water supply and sanitation in developing countries.* Report of a WHO Study Group, World Health Organization, Tech. Report Series No. 742.

Wilkinson, G. K. (1985) *Final report and proposal on national water resources legislation for Tonga.* RAS/79/123, Food and Agric. Organization, United Nations, Rome.

Winter, S. (1988) *Construction manual for a ferrocement rainwater storage tank.* Appropriate Technology Enterprises for Integrated Atoll Development Project, UNDP, Suva, Fiji.

Woodroffe, C. D. (1989a) Salt water intrusion into groundwater; an assessment of effects on small island states due to rising sea level. *Small States Conference on Sea Level Rise.* 14–18 November, Republic of Maldives.

Woodroffe, C. D. (1989b) *Maldives and sea-level rise; an environmental perspective.* Report on a visit to the Republic of Maldives. University of Wollongong, Australia.

World Meteorological Organization, (1981) *Guide to hydrological practices.* 4th. edn., No. 168, vol. 1.

Wyrtki, K. (1990) Sea level rise: the facts and the future. *Pacific Science.* 44 (1), 1–16.

IV. Physical Processes

11: Recent Scientific Developments and Research Needs in Hydrological Processes of the Humid Tropics

M. Bonell

Department of Geography and Institute for Tropical Rain Forest Studies, James Cook University of North Queensland, Townsville, Q 4811, Australia
Present address: UNESCO, Division of Water Sciences, 1 rue Miollis, 75732 Paris Cedex 15, France

with

J. Balek

ENEX – Environmental Engineering Consultancy, Kopeckeho 8, Prague 6, 16900 Czech Republic

ABSTRACT

This chapter outlines some of the research developments in process hydrology within the humid tropics. Several gaps in research will be identified. Consequently, a considerable proportion of the chapter will consider technology-transfer of various methodologies and research findings from other climatic regions which may be either appropriate for application or need further testing in various humid tropical environments.

The chapter will systematically review process research connected with the water balance components: rainfall, evaporation, unsaturated zone (soil water), groundwater and runoff generation. Later, the effect of land-use impacts on some of the water balance components will be considered and an evaluation will be made of some "physically-based" modelling techniques which have potential for application in land management issues. Throughout this review, the meteorological and climatological linkages with process hydrology at different scales will be emphasized. Such linkages will particularly emerge during a consideration of recent progress and documentation of future planned "macrohydrology" projects connected with this climatic region.

INTRODUCTION

Since the 1960s, considerable progress has been made in process hydrology research in the temperate latitudes (e.g. Anderson & Burt, 1990; Calder, 1990; Ward & Robinson, 1990). In comparison, the humid tropics has received less attention, mostly for social and economic reasons (Bonell, 1991b, c). The last decade, however, has witnessed an upsurge in world-wide concern over the environmental consequences of converting tropical forests to other land uses. Part of this environmental debate has focused on the hydrological and climatic changes emanating from forest clearance, most notably related to the Amazon Basin because it is one of four major energy (latent heat) sources to the global circulation of the atmosphere (see Molion; Manton & Bonell, this volume). In response to this concern, several reviews on the hydrological effects resulting from conversion of forest to other land uses in the humid tropics have emerged (e.g. Bruijnzeel, 1989, 1990; Hamilton, 1986, 1988; Hamilton with King, 1983). It becomes evident that the limited studies undertaken in the humid tropics forces the technology-transfer of humid temperate findings in many situations. Consequently, a considerable proportion of this chapter will focus on the technology-transfer of various methodologies developed in the higher latitudes which may be either appropriate for application or need further testing in various humid tropical environments. Some of the "frontier" issues in process hydrology research will also be highlighted. In this way, it is hoped to encourage their consideration in future humid tropical research programmes. Another objective is to highlight some of the difficulties and care required in the interpretation of collected data, for example, soil hydraulic properties.

The role of process hydrology is two-fold:

(a) to assist in a more thorough understanding of the stores and fluxes of water so that simplified water balance results from input-output studies (commonly found in humid tropical literature) can be better interpreted, and

(b) to assist in the development of more physically-based models (e.g. Beven *et al.*, 1988; Beven, 1989) which are intended to have application to ungauged drainage basins.

These two objectives will be demonstrated through a systematic consideration of the water balance components. Less attention, however, will be given to groundwater because of the contribution of Foster & Chilton elsewhere in this volume. Later, the hydrological changes resulting from various land-use impacts and associated "physically-based" modelling techniques, which have potential for application in land management issues, will also be evaluated.

Where appropriate, the meteorological and climatological linkages with hydrological processes will be emphasized

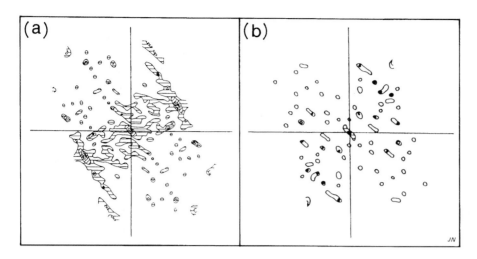

Figure 1: Examples of composite correlation diagrams for daily rainfall in north Queensland for:
A. Synoptic circulation where an active monsoon trough extends typically eastwards from the southern coast of the Gulf of Carpentaria, across the base of Cape York Peninsula and out into the Coral Sea near Cairns.
B. All synoptic weather types for the month of November.
Shaded areas have correlation coefficients greater than r = 0.0: subsequent contour interval r = 0.10 (After Sumner and Bonell, 1988).

throughout this review. Such linkages will particularly emerge in the concluding survey of recent progress and future planned projects in "macrohydrology." The challenge of parameterizing energy and water exchange across the terrestrial/atmospheric interface at different scales (for integrating with atmospheric and climatic models) is of particular interest to the humid tropics.

RAINFALL

The magnitude of rainfall is one of the driving forces in differentiating between the response of hydrological processes in the humid tropics compared with processes in higher latitudes. Berndtsson & Niemczynowicz (1988) noted, however, that globally there was a dearth of long-term rain records (> 70 years), resulting in traditional analyses concentrating on temporal patterns of rainfall at a point. Only with the advent of new techniques in the last two decades has interest emerged in the spatial domain of rainfall analysis. This has coincided with increasing attention towards mesoscale meteorology and climatology (see reviews by Browning, 1989; Houze, 1989b), and continued developments in stochastic and scale invariant rainfall models (e.g. Waymire et al., 1984; Rodriguez-Iturbe & Eagleson, 1987; Valdes et al., 1990; Olsson et al., 1992) which can input into rainfall-runoff models, taking into account the spatial and temporal distribution of precipitation over a drainage basin. The term "mesoscale" in this context refers to a regionalization of hydrological and meteorological phenomena that may range from two km^2 up to 2000–10 000 km^2 at the higher end of the spectrum (Orlanski, 1975; Atkinson, 1981).

This section will emphasize recent developments in statistical, eigentechniques, stochastic and scale invariant models for evaluating the spatial organization of rainfall, particularly at the mesoscale. The linkages with parallel work in rainfall climatology will also become apparent, including additional consideration given to measurement of rain from satellites, the dynamics of precipitation formation in mesoscale convective systems (MCSs) and differences in the preferred range of short-term rainfall amounts across the humid tropics.

The spatial organization of tropical rainfall.

The observations of Berndtsson & Niemczynowicz (1988) are even more acute in the humid tropics where, for historical reasons and a lack of resources, there is a less satisfactory network of raingauges in comparison with higher latitudes and in general, a much inferior quality of records (see Manley & Askew, this volume; Jackson, 1988a). The convective nature of rainfall and the sharp rainfall intensity gradients, in fact, demands a more dense network of raingauges cf. temperate latitudes. The economic costs are, however, prohibitive for maintaining such networks, especially in rugged terrain, e.g. Amazon Basin (Ramos et al., 1990). Problems of this nature have limited the number of actual studies at different scales of spatial and temporal rainfall across the global tropics. There is an increasing need, however, for such empirical studies as a contribution towards further advances in stochastic and scale invariant modelling (Berndtsson & Niemczynowicz, 1988). Locations that have been investigated include Tanzania (Jackson, 1969, 1972, 1974, 1978, 1985; Orchard & Sumner, 1970; Sharon, 1974; Sumner, 1981, 1983); Kenya (Baerring, 1987, 1988); Dominican Republic (Garcia et al., 1978) and Queensland, Australia (Sumner & Bonell, 1986, 1988; Lyons & Bonell, 1992a, b). Henry (1974) also analysed rainstorms (> 10 mm) at a variety of tropical locations, including South East Asia, Colombia, Panama and Costa Rica.

Spatial correlation Those studies undertaken at the mesoscale have largely been statistical in nature, using the correlation between pairwise rainfall stations as the basis. Spatial patterns have subsequently been analysed in different ways, for example, by correlating coefficient fields around a central "key" gauge (Sumner & Bonell, 1988), composite correlation maps (Baerring, 1987; Sumner & Bonell, 1988) using the method of Sharon (1974, 1978), and correlation-distance diagrams (Jackson, 1978). Other more simpler methods involving various descriptive statistics as rainfall indices have been adopted (Jackson, 1969; Sumner & Bonell, 1986). Recently, Sumner (1988) provided a comprehensive review of these methods, and with the exception of the Queensland work, the interactions between mesoscale processes and synoptic-scale weather systems have largely been disregarded. The latter approach has the advantage of identifying different rainfall characteristics across the global humid tropics. Other studies, such as the Tanzanian work, has concentrated on local controls on climate, such as topography and coastal influences.

Some interesting spatial characteristics of raincells have emerged from the preceding studies. Henry (1974), for example, found that the typical storm was about 30 km in diameter, with a spacing between two storm centres of about 60 km, except where topography interfered with the spatial distribution. Analyses in Africa (Sharon, 1974; Jackson, 1974) also suggested that storms tend to occur at preferred distances apart, based on an increase in correlation coefficients with distance beyond a certain minimum value. In addition, Sharon (1974) showed a preferred spacing for adjacent convectional cells at 40 to 45 km, using his correlation composite method. The correlation field of Sharon (1974) had a more isotropic form, however, than composites developed in a later Queensland study (Sumner & Bonell, 1988) where mesoscale banding was noted. Sumner & Bonell (1988) noted that the separation between bands varied between 200 to 250 km or 250 to 300 km, depending on either the synoptic circulation or month of the summer wet season. Figure 1a shows the WNW-ESE mesoscale banding (200–250 km apart) associated with an active monsoon trough (as defined by Sadler & Harris, 1970; see Manton & Bonell this volume). The internal rainfall macrostructure associated with the monsoon trough may be controlled at these times by the infeed of lower NW and E to SE winds either side of the trough, and the respective upper SE and NW winds representing the outflow (Sadler, 1975). The complex interaction of surface and upper winds, and mesoscale factors such as topography, in the spatial organization of rain is also demonstrated by Fig. 1b for the month of November. The NW-SE orientation of raincells was considered to be convective activity induced by orographic uplift of low-level easterly winds which are subsequently steered by upper NW winds (Sadler, 1975), that is, the rain pattern is controlled by the reversal in wind flow direction with increasing elevation. From one of the most intensive raingauge networks operated in the tropics (the Townsville area, Queensland [1

gauge per 25 km^2]), Lyons & Bonell (1992a, b) noted a similar mechanism when NW to SW upper winds prevailed. The alignment of the raincells occurred on the downwind side of relief with respect to the upper airflow, so that some of the highest rainfall occurred "downstream" from the topography. Recently, Smith (1989) proposed situations under which orographic precipitation could occur. However, the significance of wind vector reversal with height and "downstream" precipitation with respect to the upper flow was not highlighted by Smith (1989), although it was recognized earlier by Bergeron (1957). In contrast, when deep SE winds occurred, Lyons & Bonell (1992a, b) observed that the highest rain totals were more concentrated on the windward slopes.

Regionalization methods A logical continuation of spatial correlation is the establishment of precipitation areas or regions of similar precipitation characteristics. For example, if for a certain synoptic weather type, preferred areas can be defined where precipitation is likely to occur, then knowledge of such regions can be useful in both meteorological forecasting and rainfall-runoff modelling. Early attempts at regionalization involved "elementary linkage analysis" (McQuitty, 1957), using the simple correlation matrices developed previously for spatial correlation. Sumner (1988, p. 427) describes the method in some detail – which was previously adopted by Jackson (1972) for annual rainfall in Tanzania, and later for monthly rainfall (Jackson, 1985). Elementary linkage analysis, however, is rather limited because it only considers the highest correlation coefficients in the two-dimensional plane. The alternative is the increasing use of eigentechniques (e.g. Richman, 1986), such as principal components analysis, sometimes in association with cluster analysis (e.g. Willmott, 1978). Being multidimensional, these methods have the advantage of determining more than one characteristic which influences the spatial organization of rainfall. Numerous problems occur, however, not only in which broad technique to use (e.g. principal components analysis [PCA] or common factor analysis [CFA]), but also in deciding whether rotation or non-rotation to attain a "simple structure" should be used (e.g. Richman, 1986, 1987; Jolliffe, 1987). Other problems include the criteria for deciding the number of principal components (PCs) to retain (e.g. Eklundh & Pilesjo, 1990; Richman *et al.*, 1992) and where appropriate, which form of clustering strategy (see Everitt, 1980) to adopt. The valuable exchange between Richman (1987) and Jolliffe (1987) demonstrated the dichotomy of views by rainfall climatologists on the subject. Rigorous exploratory analysis on the lines described by several authors (e.g. Richman, 1986; Baerring, 1988; Eklundh & Pilesjo, 1990; Richman *et al.*, 1992) is advised before such techniques are contemplated. The major hazard still remains the degree of subjectivity required in the interpretation of different combinations of methods (D. White *et al.*, 1991; Legates, 1991; Bonell & Sumner, 1992) and the selection of the optimal solution.

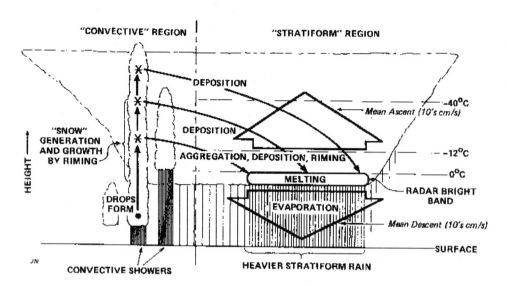

Figure 2: Schematic diagram of precipitation mechanisms in a tropical cloud system (mesoscale convective system, MCS). Solid arrows indicate particle trajectories (After Houze, 1989b).

Such comments are very appropriate to recent humid tropical work. Ogallo (1988), for example, used rotated principal components analysis (RPCA) to regionalize East African rainfall. The extracted PCs reflected the temporal and spatial variability of rain in response to the seasonal shift in the location of the monsoonal influence. In contrast, Baerring (1987, 1988) reported CFA provided a more useful technique than RPCA for explaining rainfall variability in Kenya. An 11-factor model for mapping rainfall regions was judged as being the most parsimonious for both statistical and climatological reasons, after previously comparing the outputs for 5 and 15 factors. Elsewhere, Lyons and Bonell (1993) established the importance of using near-surface atmospheric patterns within the framework of using more sophisicated regional methods as part of their intensive mesoscale study of the Townsville area. These writers also established that at that scale, the Harris-Kaiser Case II proportional ($B^T B$) oblique rotational approach (Richman, 1986, p. 319) attained the most satisfactory "simple structure" for different synoptic circulations.

Estimation of rainfall using remote sensing

The recognition that the area of cold cloud tops is strongly correlated with the amount of the area covered by rain (Houze, 1989b), is the basis for using satellites as an alternative for determining the space-time structure of rainfall fields in tropical areas with poor raingauge networks. Huygen (1989), for example, evaluated the Cold Cloud Duration (CCD) technique (Milford & Dugdale, 1987) in Zambia by comparing raingauge measurements with rainfall estimates, using METEOSAT-TIR imagery referring to 10 day periods. By means of regression analysis, various threshold temperatures of cloud tops were tested to evaluate their predictive value (using the CCD technique) in comparison with raingauge measurements. This pre-

liminary analysis indicated that it is not possible to work with a fixed algorithm to estimate rainfall. The constants in the linear regression equation varied between 10 day periods as did the optimum temperature threshold level (e.g. – 40 to – 50°C in some periods, less than – 50°C in others). Subsequent attempts to improve the CCD method by linking with the cloud dynamics of thunderstorm development did not improve the method's predictive value. Huygen (1989), therefore, concluded that several years of further calibration were required before the CCD method could be considered. Elsewhere, Dugdale *et al.* (1991) used IR METEOSAT data to estimate daily rainfall in west African areas which were just inside the humid tropics (wet-dry tropical region of Chang & Lau, this volume), over catchments and tributaries of the River Senegal. After selecting the appropriate cloud top temperature threshold for rain-producing cloud, using the methodology of Milford & Dugdale (1987) (also summarized in Dugdale *et al.*, 1991), further calibration was undertaken against all available raingauges in the catchments. Rainfall was estimated from the period of rain-producing clouds over particular sites, and such estimates were then used as inputs to selected rainfall-runoff models. Significantly, the streamflow models performed as well or even better when the satellite-derived estimates were used as inputs as against the raingauge data.

Further progress is possible from the French EPSAT-NIGER (Estimation des Précipitations SATellite- expérience NIGER) project where IR METEOSAT satellite data and a C-band weather radar system are being linked with a moderately dense raingauge network (93 gauges over a study area of 16 000 km² at different spatial and temporal scales in the Sahel, just outside the humid tropics (Guillot, 1990; Jobard, 1990; Lebel & Thauvin, 1990; Lebel *et al.*, 1991; Thauvin & Lebel, 1991). This includes the location of gauges over a regular grid with nodes

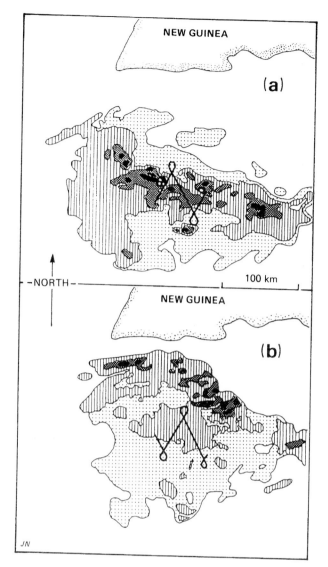

Figure 3: Radar echo patterns in a cloud cluster observed during EMEX. Data were collected with a C-band radar aboard a National Oceanic and Atmospheric Administration WP3D aircraft. Composite echo patterns were observed during (a) 2112–2142 and (b) 2314–2344 GMT 2 February 1987. Shading levels are for intensity thresholds of 1 (approximately), 20, 25, 30 and 35 dBZ. Heavy lines show flight track for the period of data collection. It is significant that the heavier rainfall has advanced northwards over the time period. An extensive area of lighter rainfall now dominates the southern part of the cloud cluster (After Houze, 1989b).

spread at 12.5 km, and a 16 gauge target area where the distances between gauges is decreased to 1 km. Lebel *et al.* (1991) present preliminary results from the joint processing of gauge and radar data related to squall lines experienced in the summer monsoon (see Manton & Bonell, this volume). The linkage between the structure (convective/stratiform) of line squalls from radar, 15 min. rainfalls, and some features of the drop size distribution have been identified for future radar calibration (Lebel *et al.*, 1991). Projects of this kind form the basis of TRMM (Tropical Rainfall Measuring Mission) planned for the

mid-1990s and will have a duration of at least three years (Simpson *et al.*, 1988; NASA, 1990). Future progress from this project may provide an economic substitution to the poor rain-gauge networks in many humid tropical countries.

Observed structure of mesoscale convective systems (MCSs)

The existing generation of rainfall stochastic models depend on the observed structure (convective and stratiform precipitation) of mesoscale convective systems (MCSs) (Waymire *et al.*, 1984). The spatial correlation and regionalization methods provide some indication of the structure of MCSs in terms of their preferred spacing and intensity. More appropriate are direct measurements based on single- and dual-Doppler precipitation radar analysis supported by rawinsondes and wind profile observations from aircraft. The economics of obtaining such measurements are expensive, but the accumulation of data from the various tropical field experiments associated with GATE (Global Atmospheric Research Programme, Atlantic Tropical Experiment) and the Global Weather Experiment, WMONEX (Hollingsworth, 1989) enabled Houze (1982) to develop an idealized conceptual model of the cloud and precipitation structure of tropical MCSs. Further detail was incorporated in the model by Houze (1989a), based on subsequent work – including results from the EMEX (Equatorial Mesoscale Experiment), as part of the Australian Monsoon Experiment (AMEX). In the mature stage (Fig. 2), the conceptual model envisages the heaviest rainfall being identified with the deep convective towers, grading to lighter rainfall associated with a stratiform region extending over a horizontal distance of 100–200 km. Embedded within the stratiform region is an area of heavier rain where the convectively generated snow particles reach the 0°C level after their passage through the stratiform cloud. After examining a wide variety of tropical MCSs, including various equatorial cloud clusters (see example in Fig. 3), Bay of Bengal depressions and hurricanes, Houze (1989b) concluded that the broad features of his conceptual model were reflected in the vertical cross-section of all such systems. The different MCSs vary, however, in their horizontal arrangement of convective and stratiform precipitation. Additional investigations on the lines described by Keenan *et al.* (1988) for the Darwin area of northern Australia will provide further opportunities for refining Houze's conceptual model and further development of the current generation of stochastic rainfall models. The spatial and temporal scales of convective and stratiform precipitation are being monitored, using a combination of radar signatures from Doppler radar and a mesoscale raingauge network. The work will also provide a ground truth station for TRMM. Other meteorological details related to MCSs were discussed elsewhere in this volume by Manton & Bonell.

Stochastic rainfall modelling, scaling and links with MCSs

There are two groups of models connected with the spatial properties of mesoscale convective systems (Olsson *et al.*,

1992). The first concerns stochastic modelling of rainfall which is based on the theory that rainfall processes are scale-dependent in space and time. In statistical terms, the parameters associated with such models are distinctly different at different scales of space and time. Consequently, the translation across time or space scale is not possible according to this theory (Orlanski, 1975; Houze, 1982; Waymire *et al.*, 1984).

Stochastic models

Recent advances in stochastic modelling of rainfall include those developed from analysis of variance procedures (e.g. Waymire *et al.*, 1984, Rodriguez-Iturbe *et al.*, 1986) which model the spatial and temporal structure of rainstorms through mathematical, multidimensional point process techniques (Berndtsson & Niemczynowicz, 1988). Such models can be considered as "conceptual models," because they incorporate the basic physical structure of rainfall expressed in terms of parameters such as raincell size and spacing, development, decay and movement, and raincell intensity, amongst others (see Table 2, Waymire *et al.*, 1984, p. 1457).

The development of the existing generation of stochastic models (e.g. Waymire *et al.*, 1984) have been based on the mesoscale rain structure of mid-latitude low pressure systems (Harrold, 1973; Harrold & Austin, 1974). Houze (1989b) demonstrated, however, that the generic picture of clusters of rainfall cells containing higher intensity rainfall embedded within clusters of lower intensity rainfall extending over larger mesoscale areas is still appropriate for the humid tropics. The principal differences, however, are in terms of the spatial organization of rainfall within temperate systems as against the various types of tropical MCSs. This has led to the testing of existing stochastic models in the tropics. Using aerial-averaged radar measurements of precipitation from GATE, Valdes *et al* (1990) developed a procedure for estimating the parameters of the Waymire *et al.*,(1984) multidimensional precipitation model. The numerical estimates of the parameters (e.g. mean number of raincells per cloud cluster) were shown to be acceptably stable and robust using different subsets of GATE information (Valdes *et al.*,1990). Similar information will become available from future space-borne sensors in the 1990s from TRMM and the Tropical Rainfall Mapping Radar (TRA-MAR). This will enable the Valdes *et al.*,(1990) estimation procedure to be tested in other tropical areas, leading possibly to the eventual use of models like Waymire *et al.*, (1984) in tropical areas with sparse raingauge networks.

Scale invariance, fractals and multifractals

The second group of models are based on the theory that the spatial and temporal properties of rainfall fields possess no characteristic spatial scale and such processes of rainfall are considered as "scale invariant" (Olsson *et al.*, 1992). The existence of "scale invariance" has received considerable recent attention (Lovejoy & Mandelbrot, 1985) and refers to the statistical properties of small and large scales which can be linked by a scale-changing operation involving only scale ratios. Over the corresponding range of scales there is no characteristic spatial size (or scale) (Lovejoy & Schertzer, 1990). Lovejoy & Schertzer (1990, p. 2021) noted, "scaling is used to indicate that certain aspects of a system (typically certain statistical exponents such as those found in energy spectra) are independent of scale." For simple scaling, the probability distributions of stochastic processes remain invariant with respect to rescaling by some mathematical function.

The term "fractals" was originally coined by Mandelbrot (1975), meaning irregular geometrical shapes (from the latin *fractus* describing a broken, irregular stone) in contrast to those of Euclid which are regular (e.g. square, cube, circle). Fractals also recognize self-similar, irregular structures (or shapes) across different scales and, therefore, can be applied to scale invariant datasets because fractal geometry provides the simplest non-trivial example of scale invariance (Lovejoy & Schertzer, 1992). The concept of fractals originates from chaos theory in the recognition that nature is controlled by a phenomena termed deterministic chaos (Mandelbrot, 1975, 1977). Whilst many environmental systems obey deterministic laws, they nevertheless behave unpredictably and their representation in modelling using Euclidean geometry proves to be quite useless (Mandelbrot, 1990). Consequently, seemingly random or chaotic processes may be governed by hidden regularities not detected by traditional analyses. The corresponding fractals are represented by a family of unsmoothed geometric shapes and patterns which can be represented by scaling behaviour (Mandelbrot, 1982). Furthermore, the dimensions of fractals are usually not whole numbers in contrast to the familiar Euclid geometry (0-point, 1-line, 2-square, 3-cube) (Mandelbrot, 1990).

There are different approaches for determining scale-invariant properties related to atmospheric phenomena. Nicolis (in press) reviews the technique of constructing a phase space "portrait" spanned by a set of variables describing the atmospheric dynamics in the form of a time series. (Nicolis & Pirogine, 1989). The method is in recognition that nonlinearity is the source of complex behaviour of atmospheric processes (including cloud and precipitation) which cause unpredictability and several outcomes or new branches of solution (known as "bifurcation" cascades) are possible when a control parameter is varied. However, many systems exhibit "dissipative" properties (as opposed to "conservative") which means that there is an irreversible evolution to a preferred set in phase space which is known as an *attractor*. The asymptoptic stability depicted by attractors ensures a certain reproducibility in behaviour. Consequently, scale-invariant properties are achieved by estimating the fractal or *correlation dimension* (v) of the obtained attractor using the expression $N_\varepsilon \approx r_t{}^v$ where the number of points N_ε on the attractor at a distance r_t from a given point varies on average with r_t as r_t goes to zero (Baker & Gollub, 1990).

Although the principal interest is in fields when concerning rainfall and not really geometrical sets, at the simplest level the temporal variability of rainfall at a point can be linked by a monofractal or unique fractal dimension in 1-D space (i.e. time) bounded by 0 and 1. This is achieved using a functional box counting method (Lovejoy *et al.*, 1987) as demonstrated, for example, by Hubert & Carbonnel (1989) and Olsson *et al.* (1992). The technique has also been used for analysing the spatial structure (in two or three dimensions) of clouds and rainfall fields (Olsson *et al.*, 1992).

The data series are viewed as sets of points and the calculation of box dimensions is achieved through counting the occurrences of rainfall over a wide range of scales. The basis of the technique is to divide the total dataset into gradually decreasing, non-overlapping segments of size r_f and for every r_f count the number of boxes, $N(r_f)$, needed to cover the set of points. Should scale invariance exist, then it can be characterized by the expression $N(r_f) = r_f^{-D_F}$, with D_F as the box (or monofractal) dimension. By taking logarithms , the plotting of log $[N (r_f)]$ as a function of log (r_f) (viz log $[N (r_f)] = -D_F$ log $(r_f) + c$ where c = constant) will produce a straight line whose slope is an estimation of $-D_F$ for a scale invariant set. For example, Hubert & Charbonnel (1989) applied this method in a comparison of two pluviometric time series of different length (45 years as against 6 months) at the station Dédougou (Burkina Faso) in the wet-dry humid tropical region of west Africa. These authors demonstrate that the occurence of rainy periods within a certain time scale (from a few days to a few months) is fractal with dimensions, $D_F \approx 0.8$. Significantly, the rain events on the time axis form a Cantor-like set (Mandelbrot, 1975, 1977) with dimensions equal to log 7/log 12 (≈ 0.8) corresponding with the wet season which is considered to last about seven months per year (Hubert & Carbonnel, 1989, p. 7).

Olsson *et al.* (1992) applied the box counting method to three different rainfall time series using respectively, minute, daily and monthly rainfall observations taken at Lund, Sweden. These authors determined the change in D_F (i.e. the slope) for different intensity thresholds as previously used by Tessier *et al.* (1988). Using a higher threshold for the minute observations removes the "stratiform-type" rain (as described by Houze, 1989b) and highlights more the "convective-type" rain in the resulting graphs. In the Olsson *et al.* (1992) study, the division of the graphs into straight sections becomes less obvious as the intensity threshold increases from 0 mm min^{-1} to 0.1 mm min^{-1}, demonstrating the sparser temporal and spatial structure of convection within the rainfall field of temperate, frontal low-pressure systems affecting that area. Some differences might be expected in parallel studies in the humid tropics where convective updrafts are more extensive in MCSs (Houze, 1989b).

Discussion so far has concentrated on 1-D space. If spatial rainfall could be shown to be scale-invariant, then there are exciting possibilities for deriving information in areas where there are no raingauges. Such steps could be achieved with the assistance of remote sensing. Rainfall parameters achieved for

large pixels would be the same for small pixels using scale invariance. However, rain fields of different intensity rarely reduce to the oversimplified binary situation of occurrence and non-occurrence. For example, Gupta & Waymire (1990) showed that departures from simple scaling can occur for both rainfall and runoff data. This included using rainfall data from GATE. They suggested that the invariance property can be better represented by the multiscaling properties associated with the cascading down of a large flux (e.g. rain) from a large scale to successively smaller scales in agreement with Schertzer & Lovejoy (1987). Multiscaling, therefore, is a new approach which includes elements from the two main theories, that is, a separation of scales but with scaling within each one. Such multiplicative cascade processes produce multifractal measures which are more suited to producing dynamical models. Multifractal measures are characterized by a scale-invariant codimension function (see Lovejoy & Schertzer, 1992, p. 11) which is an exponent function that determines how the probability distribution varies with scale. Fractal geometry is secondary to the scale fractal codimensions, in this case cf. monofractals (Lovejoy & Schertzer, 1992). Further progress on scaling properties hopefully should occur with an improved quantitative understanding of spatial rainfall variability through programmes such as TRMM.

The application of scaling models to the analysis of rain fields has shown rapid development since the 1980s. Whilst there is still a dichotomy of views on whether rainfall fields can be scaled or not both in space and time (Olsson *et al.*, 1992), it is clear from this introductory review that the testing and further development of these various types of models hold great potential for application to the humid tropics. Elsewhere, Lovejoy & Schertzer (1991 and in press) and Hubert (in press) present a more comprehensive summary of this topic.

Rainfall intensities

When considering hydrological processes, especially runoff generation, knowledge of short-term rainfall from point measurements is very significant over periods from 1 minute to 24 hours. In addition, there can be a link between different synoptic-scale rain producing systems and preferred range of short-term rainfall (Bonell & Gilmour, 1980). There is a dearth of rainfall frequency-intensity-duration information for the humid tropics (Jackson, 1988a; Lal, this volume) which can also be useful in differentiating between cyclone-prone areas, e.g. northeast Queensland, Australia and equatorial areas where convective thunderstorm clusters prevail. This is well demonstrated by available data presented for the Lower Congo by Griffiths (1972) and for Peninsular Malaysia by Lockwood (1967), compared with north-east Queensland (Babinda catchments) (Bonell, 1991a). Table 1a and 1b shows that the equatorial location (the Lower Congo) records much lower longer duration rainfall. Shorter term amounts (say, less than one hour), however, are not greatly different. It is also significant that the return period for amounts exceeding once every 10

TABLE 1a. *Lower Congo (Kinshasa) rainfall (mm) for various durations and return periods (after Griffiths, 1972)*

Return period (years)	Duration (min.)									
	10	20	30	40	50	60	70	80	90	24 h
2	23.3	37.5	46.5	56.2	62.0	66.1	67.3	69.5	69.8	
10	30.6	49.3	61.2	74.4	82.4	87.9	89.5	92.5	92.7	117
25										132
50										143

TABLE 1b. *Babinda rainfall (mm) for selected durations and return periods for a comparison with Kinshasa (after Bonell, 1991a)*

Return period (years)	Duration (min.)						
	6	12	18	30	60	3 h	24 h
2	12.7	23.1	29.3	42.2	67.5	120.9	335.8
7	20.5	28.3	40.3	58.5	90.7	209.3	497.0
14	24.7	32.0	45.5	64.5	103.9	244.0	660.9

years in Peninsular Malaysia occur at least once every 2 years in Babinda (see Lockwood, 1967). Other hydrological studies that report detailed rainfall characteristics support the preceding observations, for example, K. F. Law & Cheong (1987) (Sungai Tekam Experimental Basin, Peninsular Malaysia) and Jones (1979) (East African Rainfall Project).

The linkage between synoptic climatology – rainfall characteristics – varying storm runoff generation responses was put forward by Bonell *et al.* (1986) as a criterion for differentiating between different areas across the humid tropics. In terms of rainfall climatology, this simple distinction between equatorial areas and cyclone-affected parts of the higher tropics, however, may not be always adequate. Jackson (1986; 1988b) noted that over northern Australian stations (including the perhumid northeast Queensland) tended to record the most concentrated rainfall, that is, fewer rain days and higher mean daily intensities, in comparison with other locations across the global tropics. The fact that rain stations in central Africa have the closest rain characteristics to those found in northern Australia suggests that other factors, such as orographic uplift of moist air over high topography, may be as significant and needs further investigation (Jackson, 1988b).

Topographic interactions There is an almost complete absence of rain data from high mountain areas (Berndtsson & Niemczynowicz, 1988). Measurements have been taken at Mt Bellenden Ker Top (1561 metres above sea level), near Cairns, northeast Queensland since 1974. Mean annual rainfall there is 8065 mm (1974–89, inclusive). This station has recorded the highest daily fall (for Australia), 1140 mm in January 1979, also in the same month, the highest weekly fall (3847 mm) and the highest monthly fall (5387 mm) and the highest yearly fall

(11346 mm, 1977) (Hall, 1984). Short-term records have been collected by one of the writers since 1984, though there are intermittent gaps in the record due to data logger failure. The landfall of tropical cyclone "Ivor" to the north of Cooktown in March 1990 highlighted the typical rainfall experienced. Daily totals of 482.5 mm and 182.0 mm were recorded on the 19 and 20 March respectively. During that period, consecutive one-minute rain amounts showed sustained equivalent hourly intensities between 30–60 mm h^{-1}, with peaks up to 150 mm h^{-1} (Bonell, 1991c). It is from such mountainous areas that a major contribution is made to the recurrent disastrous flooding in the lower reaches of the principal rivers of the area, despite these areas being covered in protected undisturbed rain forest (Bonell, 1991c).

EVAPORATION

Definitions

Following Ward & Robinson (1990, p. 79), the combined processes of *evaporation*, E_i, (viz. from various water surfaces such as water intercepted on vegetation and moist, bare soil) and *transpiration*, E_t, (water vapour escaping from within plants, mostly via leaves), from a dry canopy will constitute *total evaporation*, E, in the context of this discussion. When referring to vegetation (especially forests), *wet canopy evaporation* incorporates losses from intercepted water on the wet foliage both during, and immediately after a storm (McNaughton & Jarvis, 1983, p. 4). This means that a saturation deficit is maintained over extensive areas of forest during rainfall which makes a significant contribution to the *total interception loss,* in addition to evaporation from the water-storage capacity of the canopy (McNaughton & Jarvis, 1983,

p. 10). Water losses from a dry canopy are often referred to as *dry canopy transpiration* (McNaughton & Jarvis, 1983, p. 12)

There has also been a recent re-evaluation of the concept of potential evaporation (Granger, 1989a; Morton, 1991). A stimulus was Morton's (1983) presentation of the concept of a complementary relationship between actual and potential evaporation, reviewed by Nash (1989). Granger (1989a) attempted to clarify previous ambiguities by defining systematically up to five distinct "potential evaporation parameters" which led to a further exchange of ideas in a discussion paper by Morton (1991). There are two "potential evaporation parameters" most pertinent to this discussion. The first is the "wet surface" evaporation (represented by the Penman equation (1948) which was referred to as the "Penman wet-surface evaporation" by Granger (1989a). The second is known by at least three alternative terms, namely "wet-environment" (Granger, 1989), "wet environment *areal* evaporation" (Morton, 1983) and "equilibrium evaporation" (McNaughton & Jarvis, 1983).

The difference between these two parameters is essentially one of scale (Morton, 1991). The *equilibrium evaporation* parameter represents evaporation from "... a saturated area so large that the evaporation has a controlling influence on the relatively humidity of the overpassing air ..." (Morton, 1991, p. 371) so that conceptually, advective-free conditions exist. This parameter represents the lower limit to evaporation from wet surfaces (Granger, 1989a, p. 16) and as noted by Morton (1991, p. 371) "... may eventually prove useful as causal factor or forcing function in estimating actual areal evaporation."

Technically, the equilibrium condition cannot be solved because there are four unknown parameters and only the energy balance and vapour transfer equations are available (Granger, 1989a). The Priestley-Taylor (1972) equation provides the basis of an approximation, which is identical to the first term of the Penman equation except for the inclusion of an empirical coefficient (α) to allow for some advection, viz.

$$\lambda E_{PT} = \alpha \cdot \frac{\Delta}{\Delta + \gamma} Q_n \qquad (1)$$

where λ is the latent heat of vaporization of water (J. kg^{-1}), E_{PT} is the rate of evaporation (Priestley-Taylor) (kg m^{-2} s^{-1}), Q_n is the net energy available (Wm^{-2}) for evaporation, γ is the psychrometric constant (γ = specific heat of air at constant pressure, c_p (J kg^{-1} °C^{-1}) / latent heat of vaporization of water λ (J kg^{-1}), and Δ is the slope of the saturation vapour pressure curve (kg kg^{-1} °C^{-1}). Theoretically, α becomes equal to 1 under advection free conditions and equilibrium evaporation, E_E, is attained. E_E is the asymptote, steady-state limit of water loss in a closed atmosphere, over a large expanse of vegetation with a fixed resistance. A more detailed description of the equilibrium evaporation model is provided by McNaughton & Jarvis (1983, Appendix A, p. 39).

The *Penman wet-surface evaporation*, λE_P, is the most frequently used potential evaporation parameter, and represents "the evaporation rate from a moist surface exposed to the existing available energy and atmospheric conditions" (Granger, 1989, p. 16). The Penman equation is totally independent of the surface conditions and is controlled only by energy supply and atmospheric conditions. In contrast to λE_{PT}, the Penman parameter represents "... the evaporation that would occur from a saturated surface so small that the evaporation has a negligible effect on the relative humidity of the overpassing air" (Morton, 1991, p. 371) and so conceptually, this parameter is less appropriate for application from mesoscale upwards. As noted, the Penman parameter is also often misused as a causal factor for estimating actual areal evaporation, but it is a fundamental component in the complementary relationship (Morton, 1991). Using water balance data from the Cameroons, Morton (1991, see Figs. 1 and 2 in this reference) showed that in the rainy season, actual areal evaporation $\approx \lambda E_P$, $\approx \lambda E_{PT}$, as predicted by the complementary relationship. Once the dry season was established with actual areal evaporation equal to zero, the Penman estimate was approximately twice that of the Priestley & Taylor (1972) estimate, viz $\lambda E_P = 2 \lambda E_{PT}$, in agreement with the complementary relationship. As shown in Nash (1989, p. 6), the Penman parameter is a negative index of actual evaporation under dry conditions and is "... a reflection of the energy available for evaporation, but unused because of the unavailability of water" (Nash, 1989, p. 5).

The third parameter, *potential evaporation,* follows the conditions of Van Bavel (1966) and represents the upper limit to evaporation from a moist surface. It is defined by the atmospheric conditions and the saturation vapour pressure at the actual surface temperature. The general lack of availability of surface temperate measurements has inhibited its use (Granger, 1989a).

The preceding leads into the problem of estimating actual evaporation when environmental conditions are not moist. The complementary relationship of Morton (1983) forms the basis for developing a non-empirical relationship between actual and potential evaporation as a function of soil moisture deficiency (Nash, 1989). The lack of consideration of factors such as the physiology and stomatal resistance of the plants plays a significant role in estimating actual evaporation under the condition of soil moisture deficiency. As shown by Campbell (1977) and Gates (1980) from the eco-biological point of view, a critical factor is the limited capability of plants to supply water to the leaves due to the limited amount of water in the system. Energy balance plays another significant role among the boundary conditions of the system.

Recently, several attempts have been made to describe the above processes by one or another type of formula. Granger & Gray (1989) attempted to express the mechanism by an equation similar to that of Penman by introducing the ratio of the actual evaporation to the Van Baval potential evaporating parameter, combined with the energy balance equation. In another attempt, Granger (1989b) developed a so-called complementary relationship approach for evaporation from nonsaturated surfaces, using the Penman potential evaporation parameter as the wet-surface parameter and the Van Bavel potential evaporation parameter as the potential evaporation. Morton (1991), in analysing both approaches, concluded that the relationships could not be rigorously tested. Obviously, the evaporation theory for unsaturated conditions is far from being accomplished, particularly due to the inability to describe the processes within the plants.

Measurement of evaporation

Comprehensive reviews of the various methods available for measuring evaporation at different scales of investigation were provided by Shuttleworth (1979) and Stewart (1984). These methods include the traditional water balance method, commonly used in experimental catchments, and the more sophisticated but more expensive micrometeorological, chamber and sapflow methods. Barnes & Allison (1988) also reviewed the use of environmental isotopes in estimating evaporation from bare soil. As Stewart (1984, p. 22) noted, however, "... it is very difficult, if not impossible, to measure evaporation to the accuracy often required by hydrologists ..." and "... unrealistic to hope to measure differences of 20% or less".

Water balance methods are favoured when the objective is to monitor the effect of long-term changes resulting from vegetation alteration. The problems of ungauged subterranean leakage forming a significant part of groundwater movement, and in turn, producing error in water balance calculations, is a major issue in evaluating the effects of land-use conversion in the humid tropics (Bruijnzeel, 1990). Shuttleworth (1988a, p. 322), for example, highlighted the problem of groundwater leakage in producing uncertainty in evaporation estimates from water balance work in the free-draining soils of the Manaus area of Amazonas. As he noted, underestimation of drainage easily leads to a proportionally enhanced over-estimation of evaporation.

There has been an increasing use of micrometeorological methods and they are particularly favoured when the objective is to measure evaporation over both reduced temporal and spatial scales (Shuttleworth, 1988a). Most of the remaining discussion will be devoted to evaluating these methods.

Theoretical equations

A favoured approach is the use of one-dimensional combination-type methods in which energy balance, turbulent transport and wet canopy interception considerations are combined, based on micrometeorological and interception measurements. The Rutter interception model (Rutter *et al.,* 1971, 1975) and the Penman-Monteith model (Monteith, 1965, 1973, 1981), for example, are combined (Veen & Dolman, 1989). Such a combination has been found to produce satisfactory results in the Amazon Region Micrometeorology Experiment (ARME) (Shuttleworth, 1988a, b). In contrast, Calder *et al.* (1986) used a stochastic method for modelling interception.

The stochastic method was then combined with the Penman-Monteith equation in an evaporation study of secondary lowland tropical rain forest in West Java. Other writers (e.g. Bruijnzeel & Wiersum, 1987, Shuttleworth, 1988a) have tested the appropriateness of the Gash interception model (Gash, 1979) in humid tropical forests. In terms of scale, these various theoretical equations represent a "top-down" approach, assuming that the canopy is acting as it were a "big leaf." In their extensive review, McNaughton & Jarvis (1983) demonstrate that this approximation has worked well in several humid temperate studies. Other writers, (e.g. Baldocchi, 1989; Wilson 1989), review the alternative approaches of scaling-up evaporation estimates from leaf to canopy. Whilst the "scaling-up" methods still require further development, consultation of existing work provides a useful indication of the sensitivity of certain parameters in the Penman-Monteith equation.

The following section will briefly review the relevant equations, and place particular emphasis on the reliability of estimates of selected parameters, especially under humid tropical conditions.

The Penman-Monteith model   Transpiration from dry, closed vegetation canopies is represented by the Penman-Monteith equation. Following Stewart (1989), it often is formulated as:

$$\lambda E_{PM} = \frac{[Q_n + \rho\, c_p \delta_q / r_a]}{[\Delta + c_p (1 + r_s / r_{a)/}\lambda]} \tag{2}$$

where λ is the latent heat of vaporization of water (J kg^{-1}), E_{PM} is the rate of evaporation (Penman-Monteith) (kg m^{-2} s^{-1}), ρ is the air density (kg m^{-3}), δ_q is the specific humidity deficit (kg kg^{-1}), r_a is the aerodynamic resistance to the transfer of sensible heat and water vapour from the surface to the reference level (s m^{-1}) and r_s is the surface resistance (s m^{-1}), λ, Q_n, c_p and Δ have been previously defined in equation (1). The available energy, Q_n (Wm^{-2}) can be given by:

$$Q_n = R_N - G - S_Q - P_e \tag{3}$$

where R_N is the net radiation, G is the soil heat flux, S_Q is the change in storage of energy in the canopy air and biomass, and P_e is the net photosynthetic energy.

As previously stated, the equation assumes the canopy is acting as a "big leaf." Another assumption is steady-state conditions, but the equation can be used to calculate λE_{PM} using meteorological data averaged over any period, e.g. hour, day, providing that the values of the resistances are determined over the same period. The attraction of the Penman-Monteith equation is the inclusion of the aerodynamic resistance, r_s, (or its reciprocal, aerodynamic conductance, g_a, (conveniently expressed as mm s^{-1})) and surface resistance, r_s (or its reciprocal surface conductance, g_s, (mm s^{-1})) which are lacking in other approaches such as the Priestley & Taylor (1972) equation (Veen & Dolman, 1989).

Physically, the aerodynamic resistance describes the effect of the physical roughness of the vegetation on the transfer of energy and mass from the surface to a reference level in the atmosphere. The surface resistance describes the biological control over the rate of transpiration and is particularly linked to the physiological behaviour of plants expressed through the bulk stomatal resistance (Stewart, 1989).

Physiological measurements (e.g. Roberts *et al.*, 1990; 1991a) provide stomatal conductances, g_{sto} for individual leaves which are expressed in mmol m^{-2} s^{-1} as units. On the assumption that the bulk physiological conductance is equal to the conductance of all stomata acting in parallel, i.e. the "big leaf" approximation of the Penman-Monteith equation, then a rationale emerges for estimating g_s from g_{sto} using porometer field measurements, as described in the "scaling up" procedure of Dolman *et al.* (1990). Shuttleworth (1976, 1978) earlier tested this assumption and found a fairly close agreement. As Veen & Dolman (1989) observed, however, this assumption is strongly orientated to the forest canopy being the prime source of water vapour and g_a describes the transport of water. In situations where g_s differs from bulk physiological canopy conductance, e.g. spatial and temporal variability of g_{sto} (Veen & Dolman, 1989), then observations and modelling g_{sto} based on identifying sublayers (stratified sampling procedure) in the canopy provide an alternative solution. An example is the multi-level model of Roberts *et al.* (1990).

There still remains, however, difficulties concerning the reliable estimation of r_s and r_a (Beven, 1979), of which Baldocchi (1989) and Wilson (1989) provide comprehensive reviews. It is proposed to highlight such problems and, where possible, relate them to humid tropical experience.

The surface resistance, r_s This parameter is intended to be solely a function of parallel, area-weighted sums of resistances of individual leaves. Also incorporated, however, are additional influences such as the vertical variation of available net radiation, aerodynamic influences within the canopy and soil evaporation (Baldocchi, 1989). Under certain circumstances, this can cause r_s to differ by up to a factor of two when computed from the "top-down" approach as against scaling-up from leaf to canopy or "bottom-up" approach (Finnegan & Raupach, 1987). The overall impact on the Penman-Monteith equation was succinctly stated by Baldocchi (1989, p. 32) who commented "... errors attributed to the different estimation in canopy stomatal resistance will not yield errors in (dry) canopy evaporation computed with this equation due to its non-linear dependence on canopy stomatal resistance. On the other hand, transpiration measurements should be used with caution to compute bulk canopy stomatal resistance because the associated non-linearities can result in great errors in canopy resistance estimates." These conclusions discourage the re-arranging of the Penman-Monteith equation to calculate r_s, or its reciprocal, g_s.

To accommodate this problem, Stewart (1988, 1989) proposed the incorporation of a sub-model as part of the Penman-Monteith equation to realistically reflect the effects of biological responses, and in turn, more accurately calculate r_s (or g_s). A modification of the phenomenological model of Jarvis (1976) was proposed. Because measurements of parameters in the latter model are not generally available, Stewart (1988) replaced them by microclimatological variables (viz. solar radiation, S_T in Wm^{-2}; atmospheric specific humidity deficit, δq; air temperature, T in °C), one hydrological variable (soil moisture deficit, $\delta\theta$ where θ is in cm^3 cm^{-3}) and leaf area index, *LAI* (dimensionless). These replacement driving variables were used in somewhat simplified forms of the functional relationships proposed by Jarvis (1976) to give:

$$g_s = g_{max}\, g(LAI)\, g(S_T)\, g(\delta q)\, g(T)\, g(\delta\theta) \tag{4}$$

where g_{max} is the maximum surface conductance when all the functions $g(LAI)$, $g(S_T)$, $g(\delta q)$, $g(T)$ and $g(\delta\theta)$ are equal to unity. Further details of the g functions and application of this modified sub-model for calculating, g_s are provided by Stewart (1988, 1989), Dolman & Stewart (1989) and Dolman *et al.* (1990).

Dolman *et al.* (1990) compared various models for estimating g_s, based on the micrometeorological (Shuttleworth, 1988a) and physiological (Roberts *et al.*, 1990, 1991) measurements taken within the Amazonas tropical rain forest at Reserva Ducke as part of ARME. Before this comparison is discussed further, it is worth evaluating the results of the physiological method.

Roberts *et al.* (1990) demonstrated that there was good agreement between separate estimates of transpiration based on micrometeorological measurements at the canopy top, using the eddy correlation system (Shuttleworth, 1988a), and a multi-level model incorporating profiles of temperature, radiation and leaf area combined with porometry. The stratified sampling procedure they used demonstrated that there was a g_{sto} – solar radiation relationship (S_T) which was a function of three factors, namely, attenuation of radiation in the canopy, response of stomatal conductance to radiation and a function of leaf area distribution. Furthermore, a layered response was noted (Roberts *et al.*, 1990, 1991b) whereby the measured g_{sto} decreased with decreasing height consistently across each horizontal section of canopy. The emergent trees had the highest g_{sto} – which declined rapidly during the afternoon, whilst vegetation close to the ground had lower g_{sto} with little variation during the day (see Fig. 4). This led Dolman *et al.* (1990) to argue that the close agreement between the derived transpiration estimates could be attributed to up to 80% of the variation in g_s being accounted for by variation in solar radiation (positively correlated) or humidity deficit (negatively correlated).

The results of the Roberts *et al.* (1990) study suggest that r_s (or g_s) can be calculated through re-arranging the Penman-Monteith equation, in contrast to Baldocchi's (1989) earlier cautionary remarks. Whether such findings are either unique to this particular environment or appropriate to all *continental*

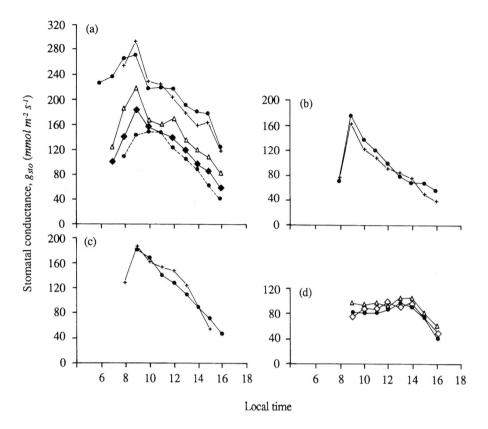

Figure 4: The diurnal variation in stomatal conductance, (g_{sto}), of the species around the sampling tower, Reserva Ducke, Manaus. Data points are mean values from all the data for the different species. (a). *Piptadenia suaveolens* at 33 m (•); *P. suaveolens*, 27.5 m (+); *Licania micrantha*, 25.6 m (Δ); *Bocoa viridiflora*, 24 m (♦); *Naucleopsis glabra*, 20 m (•). (b). *Naucleopsis glabra*, 17 m (+); *Enterolobium schumburgkii*, 13 m (•). (c). *Rinorea racemosa*, 8 m (•); *Gustavia angusta*, 8 m (+). (d). Seedling trees, 0–3 m (Δ); *Astrocaryum* sp., 0–3m (•); *Scheelea* sp., 0–3 m (◊).(after Roberts *et al.*, 1990)

rain forest climates, or are the result of improved methodologies, needs further testing through replication of similar studies at other locations. In addition, the value of canopy partitioning – multi-layer models for estimating integrated transpiration points towards the potential for more widespread use of "scaling-up" methods of this type, because of the similarities in response of g_{sto} and g_s. Nevertheless, the "scaling-up" of stomatal conductances to one single value of surface conductance still presents problems, of which knowledge of the distribution of leaf area is one of the outstanding difficulties (Dolman *et al.*, 1990).

The preceding observations led Dolman *et al.* (1990) to develop a quadratic description of g_s (computed from the Penman-Monteith equation) as a function of time to take into account diurnal variations through the external forcing of g_s by light. For the Amazonas experiment, this equation was:

$$g_s = 10.92\ (\pm 0.14) - 0.67\ (\pm 0.03)\ (\text{time -12}) - 0.21\ (\pm 0.01)\ (\text{time -12})^2 \tag{5}$$

Dolman *et al.* (1990) then compared the predictive value of g_s determined from equation (5) as against g_s determined from the more complex sub-model of Stewart (1988) in terms of E_t estimates using the Penman-Monteith equation. In addition, a third model, using a fixed estimate of g_s, formed part of the

comparison. The results showed that the simple, time-dependent model performed as well as the more complex version of Stewart (1988) in terms of comparing the average predicted, g_s and transpiration, E_t, with measured values. In contrast, the third model performed less well because it did not take into account diurnal variations in g_s, and in turn, transpiration, E_t. These results suggested that the more manageable time-dependent model works well, at least in the "constant" climatic environment of continental, equatorial rain forest which is not subjected to frequent change in air masses. Under such conditions, only solar radiation shows the main variation which, in turn, drives the diurnal variation in temperature and humidity deficits.

The aerodynamic resistance, r_a   Following Shuttleworth (1988a), the aerodynamic resistance under non-neutral atmospheric stable conditions is given by

$$r_a = \frac{b}{u_r} \tag{6}$$

and $b = \{\ln [(Z_r - d)/Z_o]\}^2 / V_K^2 \tag{7}$

where Z_r is the reference height (m), u_r is the measured windspeed at this height (m s^{-1}), d is the zero plane displacement (m), Z_o is the aerodynamic roughness length (m) and V_K is von Karman's constant (dimensionless) set to 0.41. The parameters d and Z_o are commonly expressed as functions of vegetation elevation, h (m) so that

$$d = 0.75\,h \tag{8}$$

and

$$Z_o = 0.10\,h \tag{9}$$

Thus, the aerodynamic resistance is a function of the wind speed, the aerodynamic roughness of the vegetation and the stability of the atmosphere.

There is still debate, however, whether the formulation of the aerodynamic resistance is a true representation of the physical process involved with energy transfer, especially related to wet canopy evaporation, E_i (Gash *et al.*, 1980). It is appropriate however, to highlight some of the more pertinent issues which still could affect the sensitivity of dry canopy transpiration to r_a.

The "big leaf" model assumes the validity of K-theory, otherwise known as first-order closure represented by Fick's law of diffusion (Baldocchi, 1989; Wilson, 1989). Such theory assumes that the turbulent fluxes or eddy diffusivities of heat, mass and momentum are linearly related to their vertical mean gradients. Recent work has indicated, however, that first-order closure theory often seriously fails to describe turbulent transport within plant canopies because aerodynamic resistances to mass and momentum transfer are not always identical (Wilson, 1989) because the diabatic influence on vertical transfer is not the same. In addition, an "excess" resistance at the vegetation itself affects energy transfer in comparison with momentum (Thom, 1972; Stewart & Thom, 1973). These different processes result in the transfer of heat (or mass) from a vegetation surface encountering a greater aerodynamic resistance than does the transfer of momentum (Baldocchi, 1989; Verma, 1989).

The K-theory can fail for several other reasons. These include different sources and sinks for mass and momentum transfer within plant canopies, the intermittent penetration of gusts from the Planetary Boundary Layer (PBL) affecting the length scale of vertical exchange (Wilson, 1989) and recent evidence of counter-gradient (negative) transport (Denmead & Bradley, 1985). Such issues are still areas of continuing research (see several papers in Black *et al.*, 1989), especially in temperate forests. A similar intensity of effort is required in tropical forests, although Shuttleworth's (1989a) recent review of forest micrometeorology noted that the use of expressions based on first-order closure models were, nonetheless, effective because of several fortuitous numerical relations (see pp. 306–310, Shuttleworth, 1989a). Furthermore, the within-canopy counter-gradient (negative) transport of energy flow (Denmead & Bradley, 1985) proves most favourable where the architecture of the forest is conducive to an elevated concentration of leaf foliage. Many temperate forests exhibit a leaf

area distribution in the vertical plane of this type (Shuttleworth, 1989a). In contrast, the tropical rain forest at Reserva Ducke near Manaus had a more uniform vertical structure of leaf area causing a progressive capture of incident solar radiation through the canopy. Consequently, no marked inflection in temperature with elevation was noted even where the leaf area was at a maximum – at 80% of canopy height (Shuttleworth, 1989a).

A further complication is that close to aerodynamically rough surfaces, e.g. forest canopies, the assumption of a logarithmic wind profile with associated empirical stability corrections, does not necessarily hold. Paradoxically, this can result in enhanced energy transfer just above the canopy (Raupach, 1979), which is diametrically opposite to the mentioned excess resistance to transfer of energy within the canopy. For modelling forest evaporation, Shuttleworth (1988a, 1989a) noted that this fortuitous cancellation suggests an assumed equality between the resistances for aerodynamic transfer of energy fluxes and that for momentum under neutral conditions.

A stability function, $\varnothing$ (dimensionless) (Dyer, 1974), can be incorporated within equation (7) as part of calculating for r_a to act as a correction for the effect of diabatic influences. In combination with equation (6), r_a was expressed by Stewart (1989) in the form:

$$r_a = [\,l_n\,\{\,(Z_r - d)\,/\,Z_o\,\} - \varnothing\,]^2\,/\,V_K^2\,u_r \tag{10}$$

Shuttleworth (1988a, p. 332) provides a comprehensive description of using the stability function, $\varnothing$.

Taking the preceding issues into account, it is appropriate to assess the sensitivity of r_a under dry canopy conditions. Stewart (1989) linked the sensitivity of transpiration and r_s estimates through rearranging the Penman-Monteith equation with various systematic errors in r_a. Figure 5 shows the errors in r_s and transpiration, E_t, for the quoted percentage range of r_a. For small r_a, say 10 sm^{-1} or less, typical of temperate forests (McNaughton & Jarvis, 1983, p. 10), the percentage error in both transpiration and r_s is small, being less then 10%. For larger r_a, say greater than 100 s m^{-1} at the higher end of the range for short crops in temperate areas (McNaughton & Jarvis, 1983, p. 10), the percentage errors in r_s are considerable for Bowen ratios (β_B) less than 0.3 and for λE_t when β_B exceed 1.6. The conclusions from Fig. 5 supports Shuttleworth's (1988a) observation that when using the Penman-Monteith equation for modelling forest transpiration E_t, r_a shows only limited sensitivity to the precise formulation. Such conclusions may later be confirmed from further work in higher-order closure models (Wilson, 1989).

The Priestley-Taylor Equation McNaughton & Jarvis (1983) re-wrote the Penman-Monteith equation (2) to highlight the equilibrium and advection terms which gives:

$$\lambda E_{PM} = \frac{\Delta}{(\Delta + \gamma)}Qn + \frac{\rho\,c_p[\delta_q - \delta_q E]}{(\Delta + \gamma)\,r_a + \gamma r_s} \tag{11}$$

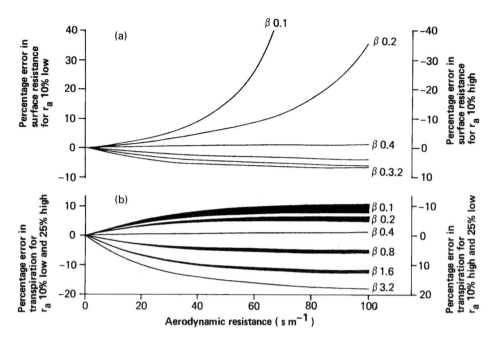

Figure 5: (a) Percentage error in surface resistance, r_s (calculated from the rearranged form of the Penman-Monteith equation – using a measurement of aerodynamic resistance (r_a) with an assumed error of 10%) for various values of r_a and β, the Bowen ratio, the ratio of sensible to the latent heat flux.

(b) Percentage error in transpiration (calculated using a surface resistance (with an error due $\pm$ 10% error in r_a) and an aerodynamic resistance with an error of $\pm$ 25%) for various values of r_a and Bowen ratio (After Stewart, 1989).

where the equilibrium saturation deficit is given by:

$$\delta_{qE} = \frac{\Delta}{(\Delta + \gamma)} \cdot \frac{\gamma r_s}{\rho c_p} Qn \qquad (12)$$

McNaughton & Jarvis (1983) noted that for short crops or grass with wet or dry surfaces, the second term of equation (11) representing advection has been determined empirically to be about 25% of the first term. According to DeBruin (1983b), a factor embodied within α is the entrainment of dry air at the top of the Planetary Boundary Layer (PBL), which is usually at the first inversion. As a result, there is a water vapour "leakage" from the PBL to the higher levels in the atmosphere, causing a lower humidity at the surface in contrast with the closed system assumed for E_E. With the use of the empirical coefficient, α to represent the advection term, equation (11) is reduced to the Priestley-Taylor (1972) equation (1). Based on mid-latitude summer-time conditions, the value of 1.26 for α to permit some advection (DeBruin, 1983b; McNaughton & Spriggs, 1989) has been found appropriate. Indirect evidence provided by Priestley (1966), and separate analysis by DeBruin (1983a), using a set of air temperature data from the humid tropics, indicate that α = 1.26 is also suitable for tropical regions. DeBruin (1983a) reasoned that the maximum air temperature shows a sharply defined upper limit of 33°C in the wet months of the humid tropics, thus indicating sensible heat flux goes to zero at this temperature. Using α = 1.26 in the Priestley-Taylor model, equation (1) predicts sensible heat flux will be zero at 32°C (Priestley & Taylor,

1972) which suggests that total evaporation, *E,* is almost equal to the water equivalent of net radiation in humid tropical regions at sea level where the temperature is close to 33°C (DeBruin, 1983a). The time of day and corresponding diurnal changes in the Bowen ratio are important in the context of these conclusions. De Bruin's (1983a) reasoning is appropriate only for a short period of the day close to net radiation and evaporation maxima (Dunin, pers. comm., 1992). During the course of presenting diurnal variations in the Bowen ratio and α, this aspect was indicated in the results of Viswanadham *et al.* (1991, Figs. 1 and 2) for the Amazon forest at the Reserva Ducke site.

The Priestley-Taylor equation was originally developed for estimating evaporation from moist surfaces on a large scale – typically using a grid network that is several hundred kilometres on a side. At that scale, Priestley & Taylor (1972) argued that radiant energy would dominate over advective effects. Later, McNaughton & Spriggs (1989) cited evidence that the Priestley-Taylor equation provides good estimates of evaporation over well-watered areas much smaller than the regional scale, that is, when environmental conditions suit the equilibrium evaporation model. Such conditions break down, however, when there are significant interactions with different air mass properties emanating from the PBL and where r_s exceeds 60 s m^{-1}, typical of forests (DeBruin, 1983b). For example, within the r_s range 0 to 60 s m^{-1}, McNaughton & Spriggs (1989) showed that α = 1.26 within $\pm$ 10% for 6 out of 9 days in a well-watered experimental area of the Netherlands. For the remaining 3 days, α significantly exceeded 1.26 because of

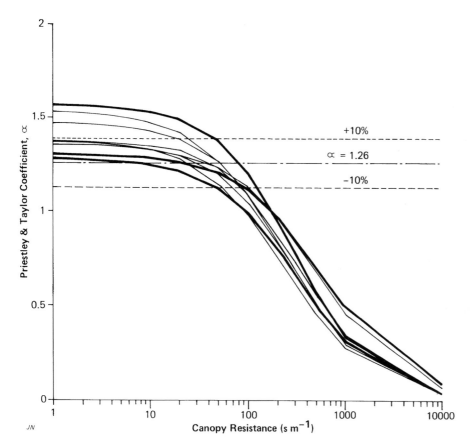

Figure 6: Priestley-Taylor coefficient α calculated for a range of r_s for each of the 9 days of the Cabauw (Netherlands) dataset (After McNaughton & Spriggs, 1989).

the influx of cooler, drier air from the PBL. For r_s values exceeding 60 s m^{-1}, the α value becomes unstable and significantly declines beyond the 10% lower limit of 1.26 (Fig. 6). The implication is that available net radiation alone is no longer an adequate parameter for measuring evaporation over forests. This led DeBruin (1989) to argue that more consideration should be given to interactions with the PBL for regionalizing evaporation. The role of atmospheric stability in causing both the Bowen ratio and α to show significant diurnal variability was indicated in the results of Viswanadham *et al.* (1991, Figs. 1f and 2f) at the Reserva Ducke, as part of ARME. During unstable and stable conditions for the daylight hours, the Bowen ratio varied from 0.10 to 0.57 and -0.71 to -0.08, respectively. The corresponding α values varied from 0.67 to 1.16 (unstable conditions) and from 1.28 to 3.12 (stable conditions). These results were based on two days of hourly measurements taken between 0700 and 1800 LST (Local Standard Time). A mean value of $\alpha = 1.16 \pm 0.56$ was calculated (1.29 ± 0.65; 1.01 ± 0.37) with the higher mean daily α value of 1.29 for the first day in indicating that the canopy surface was saturated in comparison with the second day. Under these circumstances, the canopy is freely evaporating and α is close to 1.26, the average potential value for the Priestley-Taylor equation. The indication is that α partly expresses the control which the canopy surface exerts on the water loss to the atmosphere. In contrast, α approaches unity for moderate

evaporation when the wet and dry bulb temperature differences between canopy and atmosphere are similar, as indicated in the Amazon study (Viswanadham *et al.,* 1991) and other temperate forest studies elsewhere (e.g McNaughton and Black, 1973). This point was further emphasized when Viswanadham *et al.* (1991) calculated a mean value of $\alpha = 1.03 \pm 0.13$ for 29 non-rainy days taken from three daily data sets over 1983 to 1985 of ARME.

Whilst equation (1) is more physically-based than previously considered, especially when referring to the humid tropics, it is also clear that the Priestley-Taylor model still requires further testing at different scales across a range of vegetated, humid tropical environments. Viswanadham *et al.* (1991) emphasized the sensitivity of equation (1) at "patch" scale to diurnal variations in α and recommending that it be only applied with caution to time intervals of one day or less. The sensitivity of α to advection, air intrusions from the PBL, wind speed, r_a and r_s of tropical vegetation also need further investigation. For example, based mostly on temperate experience, McNaughton & Jarvis (1983) showed that the transpiration rate over a dry forest canopy does not follow fluctuations in net radiation, and that E_t significantly departs from the equilibrium rate. When considering equation (11), the advection term becomes dominant for forests when $r_s \gg r_a$. Under wet canopy conditions, r_s and δq_E (equation 12) both go to zero, so equation (11) reduces to the Penman equation:

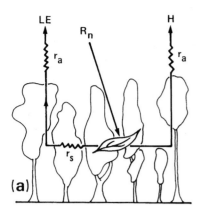 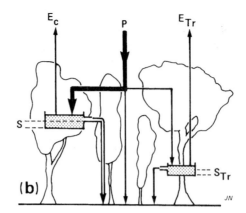

Figure 7: Elements of the micrometeorological model used in ARME to describe and interpolate the experimental data:
(a) The single source "Penman-Monteith" resistance framework. The resistance r_a is the aerodynamic transfer resistance; r_s is the surface resistance, which is equal to zero for a totally wet canopy and calibrated (by applying the model in reverse) in dry canopy conditions (see Shuttleworth, 1988a).
(b) The "Rutter" canopy water-storage model (Rutter, 1971, 1975). The precipitation is routed as direct throughfall or into stores S and S_{Tr} on the canopy and trunks respectively. Evaporation and drainage rates are related to the depth of water in these stores (After Shuttleworth, 1988a).

$$\lambda E_P = \frac{\Delta}{(\Delta + \gamma)} Qn + \frac{\rho \, c_p \, \delta q}{(\Delta + \gamma) \, r_a} \qquad (13)$$

The Rutter interception model The Penman-Monteith model (equation 2) can be linked with an interception model so that equation 2 can calculate two evaporation rates on a "stop-go" principle. The first λE_t corresponds to dry canopy conditions (transpiration) with r_s taking a value exceeding zero. The transpiration rate can be calculated using a determined r_s from one of the described sub-model options (Dolman *et al.*, 1990). Alternatively, a constant, optimized r_s value can be used, based on a combination of meteorological and soil moisture measurements (see Calder *et al.*, 1986). For example, Calder *et al.* (1986) determined an optimized r_s value of 120 s m^{-1}. The second concerns evaporation, λE_i, from a totally wet canopy when $r_s = 0$, and transpiration is assumed to be zero. The use of an interception model signals when the latter condition applies.

The Rutter interception model (Rutter, 1971, 1975) can be described as follows:

$$(1 - p - p_t) P = \lambda E_i + D_r \pm \Delta_R C \qquad (14)$$

where P is the incoming precipitation rate (mm min^{-1}), p is the proportion of rain falling directly to the ground through gaps in the canopy, p_t is that proportion diverted to the trunks, D_r is the rate of canopy drainage (mm min^{-1}) and Δ_R is the rate of change in amount of intercepted water storage, C (mm). The parameter D_r requires empirical constants modified to account for different canopies between sites. D_r is formulated as follows:

$$D_r = D_s \, \exp[b^* (C - S)] \qquad (15)$$

where S (mm) is the minimum amount of water required to wet the canopy which is determined from throughfall measure-

ments. The variable C is the amount of stored water at any particular time. When the empirical values are determined for the drainage coefficient, b^* and the drainage rate, D_s (mm min^{-1}) (when $C = S$), equation 15 returns a finite value even when C is zero. Therefore, D_s is set to zero when $C < S$ to remove this mathematical convenience (Calder, 1977; Gash & Morton, 1978). The site-specific equation used for the Amazonas work (Shuttleworth, 1988a) was:

$$D_r = 0.0014 \exp [5.25 \, (C - S)] \qquad (16)$$

Shuttleworth (1977) earlier suggested that the transition from wet to dry canopy conditions can be gradual. As an improvement over the "stop-go" principle, when C is less than S, the rate of evaporation of any rainfall intercepted by the canopy is set equal to a proportion (C/S) of the wet canopy evaporation rate, λE_i (Shuttleworth, 1988a; Veen & Dolman, 1989) so that:

$$\lambda E_i = \lambda E_i \, C/S \qquad (17)$$

likewise

$$\lambda E_t = \lambda E_t \, (1 - (C/S)) \qquad (18)$$

In operating the model, the value of C is initially set to zero, appropriate to dry canopy conditions. The change in canopy storage through time is obtained by re-writing equation (14) so that it operates as a running water balance:

$$dC/dt = P \, (1 - p - p_t) - (C/S) \, \lambda E_i - D_r \qquad (19)$$

This approach was followed in ARME (Shuttleworth, 1988a). A diagrammatic summary of the combined Penman-Monteith-Rutter model appropriate to that study is shown in Fig. 7. Lloyd *et al.* (1988) provided a description of the methodologies available for determining estimates of Rutter's parameters, including modifications to accommodate the tropical rain forest architecture.

In contrast, Calder *et al.* (1986) argued that errors in using the more simplified "stop-go" modelling principle would not be large under tropical conditions because the canopies were more likely to dry out completely between storms. Such observations might be appropriate to equatorial climates where short duration, convective storms prevail. Whether the same assumption applies to some of the outer tropical regions where longer durations of precipitation prevail remains untested. In the meantime, the approach taken in the Amazonas work would seem to be the appropriate initial strategy.

The Gash interception model The Gash interception model (Gash, 1979) – essentially a simplification of the Rutter model – focuses on rainfall occurring in a series of discrete storms, each of which comprises a period of wetting up, a period of saturation and a period of drying out to empty the canopy storage. The model requires the same Rutter variables S, p, p_t and S_{Tr} (trunk water capacity linked to D_r and D_s). In addition, a ratio of the mean evaporation rate $(\bar{E})$ to the mean rainfall rate $(\bar{R})$ for hours when the rain is falling on the canopy is introduced. It is assumed that this ratio is constant. Once this ratio has been determined from a dataset for a particular environment, the need for additional measurements of actual rates of rainfall and evaporation for other storms is removed. This is based on the assumption that the canopy is wet at such times. The derivation of $\bar{E}$ can be based on the Penman-Monteith equation for wet canopy conditions when rainfall is above a certain threshold. If one rain event per day is assumed, then the $\bar{E}/\bar{R}$ ratios can be applied to other sites where only rainfall is available. Lloyd *et al.* (1988) argue that such an assumption is reasonable in the humid tropics based on the concept of short, intense storms. This assumption may be acceptable in the equatorial regions but not in some of the outer tropical areas, where well-organized synoptic weather systems produce storms of long duration that can well exceed 24 hours, e.g. north-east Queensland (Bonell *et al.*, 1991). Work in New Zealand (Pearce & Rowe, 1981) indicated the same inadequacy of this assumption when related to the long-duration storms of the New Zealand climate.

A series of analytical forms for the various components of interception loss, E_i, were presented by Gash (1979). Information on the relevant formulae can also be obtained from later summaries elsewhere (e.g. Lloyd *et al.*, 1988; Veen & Dolman, 1989).

Experiences with the measurement and application of interception models in tropical forest environments, including linkages with the Penman-Monteith equation

Bruijnzeel & Wiersum (1987) determined reasonable predictions of interception loss, using the Gash model in an *Acacia* spp. plantation, in West Java, Indonesia. Lloyd *et al.* (1988), however, presented an interesting comparison between the performance of the Rutter & Gash models in the Reserva Ducke rain forest of Amazonas. Over nearly a two-year period

(625 days), the measured total interception loss, E_i, was 428 mm ± 173 mm standard error which amounted to 8.9 ± 3.6% of gross rainfall (4804 mm). The modelled interception loss over the same period was for Rutter, 605 ± 127 mm standard error (12.1 ± 2.6% of gross rainfall); and for Gash, 543 ± 103 mm (11.3 ± 2.2% of gross rainfall). Neither of these models were significantly different (at the 10% level) from the measured loss. So both models could be considered to have performed adequately (Lloyd *et al.*, 1988), despite the potential for various errors in the measurement of throughfall and parameters related to the forest structure.

A good review of field measurement and error assessment was provided by Lloyd & Marques-Filho (1988) and Lloyd *et al.* (1988). Earlier experience in other tropical rain forest areas (Jackson, 1971, 1975; Gilmour, 1975) established that the rotation of 20 throughfall gauges proved inadequate to monitor the spatial variability effects of canopy openings on throughfall, especially in larger storms. Lloyd & Marques-Filho (1988) increased the number of gauges to 36, randomly relocated 71 times over 494 out of 505 sampling positions in a 100 x 4 m transect. Such improvements assisted in reduction of error (see Fig. 3 in Lloyd & Marques-Filho, 1988) but the broader frequency distribution of throughfall catch (cf. temperate plantation forests) did not always allow the random fluctuations to self-cancel over shorter periods. At such times, cumulative "negative" interception was evident (e.g. days 425–453 shown in Fig. 8), thus resulting in still a relatively large error in interception measurement. The forest structure parameters, especially S, also were shown to contribute relatively large errors. Despite these considerations, it is significant that the maximum difference between modelled and measured E_i still remained less than 4% of the total measured rainfall (Lloyd *et al.*, 1988).

Nevertheless, Herwitz's (1985, 1987) work in montane rain forest of north-east Queensland highlights the need for additional studies concerning the effects of forest structure parameters, interception measurement and modelling. His work demonstrated that the assumption of constant canopy/trunk storage capacities was an oversimplification, and more account needs to be taken of the dynamic changes in storage capacities, including differences between still-air and turbulent conditions. Herwitz (1985) noted that leaf surface storage capacities averaged 1.7 mm in still-air conditions for the five canopy trees investigated, cf. 0.74 mm, Lloyd *et al.* (1988). More impressive was the high proportion of rainwater retained by bark, averaging 2.7 mm over the five species. Under turbulent conditions, bark storage could be as high as 6.6 mm (Herwitz, 1985) and storm sizes had to exceed a critical threshold (> 8.3 mm) before leaves and bark became totally saturated.

The West Java study of Calder *et al.* (1986) initially followed a Penman-Monteith-Rutter model – on similar lines to Shuttleworth (1988a). The Rutter component was subsequently discarded because the model predicted only 50% of measured interception, with optimal parameter values. These

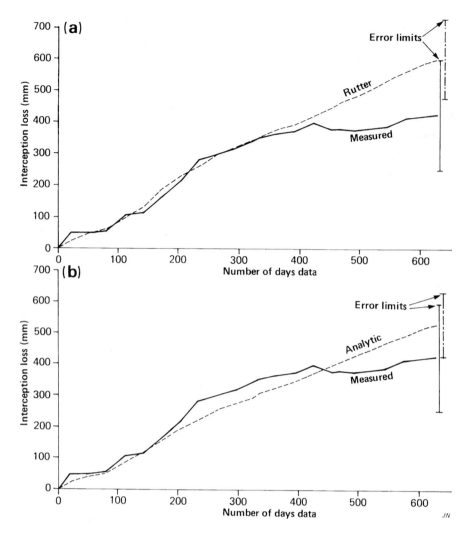

Figure 8: Cumulative measured interception loss in the Reserva Ducke, Amazonas, Brazil and cumulative interception loss by (a) the Rutter model and (b) Gash's analytical model against number of days of complete data. (After Lloyd *et al.*, 1988).

writers attributed such failure to the canopy storage parameter in which Rutter's model estimated was < 2 mm after rainfall, which was unrealistically small taking into account the results of Herwitz's study. An alternative stochastic model of interception (described by Calder 1986a) was more successful for modelling interception and integrating with the Penman-Monteith equation (Calder *et al.*, 1986). In contrast, retention of the Rutter interception model in the Amazonas experiment worked much more satisfactorily (Shuttleworth, 1988a; Lloyd *et al.*, 1988) as already indicated.

The sensitivity of r_a under wet canopy conditions within the Penman-Monteith-Rutter model was assessed by Shuttleworth (1988a) by - 50% to + 50% changes in the value of *b* in equation (5). Similar variations were also applied to the canopy storage, *S,* and the free throughfall coefficient, *p*. The *b* and *S* parameters were shown to cause considerable changes in predicted interception loss, but such changes were also shown to be somewhat self-cancelling when considering their opposite effects on transpiration (Shuttleworth, 1988a). Their overall effect on calculated total monthly evaporation, therefore, was minimal. For example, *b* affected total evaporation by only + 3% (for – 50%

change in *b*) and – 2% (for + 50% change in *b*) in monthly evaporation, respectively. Likewise, *S* showed only 3% change in monthly evaporation. Optimization of the minimum sums of squares between differences between observed and predicted throughfall, using 5 min rainfall data and the Penman-Monteith equation (r_s set to zero), in the Calder *et al.* (1986) study produced an r_a of 5 s m^{-1} for wet canopy evaporation. The same value was successfully used for dry canopy transpiration.

The contrasting experiences with interception modelling, in conjunction with the Penman-Monteith model, for West Java and Amazonas, highlights the need for replicated studies in other tropical forests. It is not clear whether such inconsistencies in canopy storage are the result of use of different methodologies of field measurement or legitimate differences (e.g. rain shedding leaf characteristics) which require new and more complex models to represent the interception process in tropical rain forests (Shuttleworth, 1989a).

Differences between coupling between canopies
Differences between evaporation rates from forests and grassland have already been suggested. In an extensive review of

temperate studies, McNaughton & Jarvis (1983, p. 10) noted that "... the rate of transpiration from forest is generally less than from grassland and arable crops, whereas the rate of evaporation of intercepted water is usually somewhat higher. The relative amounts of water transpired and evaporated over a year depend, therefore, on the proportion of time that the canopy is wet." As noted elsewhere (Calder, 1990; 1991a), average rates of wet canopy evaporation from forests can exceed those of wet, shorter vegetation by two to five times. Forests present a relatively rough surface to airflow and are more efficient in generating the forced eddy convection, which for the majority of meteorological conditions is the most dominant mechanism in vertical transport of water vapour from the wet foliage to the atmosphere. Following from this generalization, there is the suggestion that in persistently wet climates, total water losses from forests will exceed those from short crops. When soil water conditions are non-limiting, short crops are more responsive to available radiation and, in turn, show higher dry canopy losses than trees. Comparative studies between pasture and eucalypt trees in Australia provide examples. Dunin *et al.* (1985, p. 289) presented evidence that daily dry canopy losses, E, of grassland over a 5 day period exceeded that from a eucalypt forest by as much as 50%, although soil water conditions were comparable due to recent rainfall. On the other hand, there is evidence that eucalypt trees exhibit higher wet canopy evaporation, E_i, during rainfall, because of increased aerodynamically rough conditions, even though the canopy storage (0.35 mm) of these trees is less than short crops (Dunin, 1987; Dunin *et al.*, 1988). Significantly, the grassland lost more intercepted water immediately after rainfall because of the larger canopy storage (Dunin, 1987). These short-term characteristics were reflected in forest canopy loss, accounting for a greater proportion of annual rainfall (15%), than that of the grassland (11%) (Dunin, 1987). When comparing longer periods, there is a reversal in ranking in terms of total evaporation, E, between the two vegetation types (Dunin *et al.*, 1985; Dunin, 1991) indicating E_t from shallow-rooted crops is lower than for forests under soil water limiting conditions. Monthly E rates from pasture were systematically less than forest values, with the discrepancy attaining 50% when the soils were at their lowest moisture content. Thus, the greater rooting depth of the forest enabled the tapping of more available soil water. By such means the eucalypts were more buffered against heat stress and leaf loss during soil water limiting conditions, therefore, allowing sustained E_t, in contrast to the pasture. Consequently, the forest evaporated by a factor of 2, more water over longer periods (Dunin, 1991). These examples serve to demonstrate that there is a delicate balance between the duration of wet and dry canopies, and available soil moisture which controls differences in water losses between vegetation types.

Such differences in transpiration rates between vegetation types were explained in terms of the coupling between canopies, leaves and their environment in the context of scaling evaporation from leaf to canopy (McNaughton & Jarvis, 1983; Jarvis & McNaughton, 1986). They concluded that rough, tall canopies are well-coupled to their environment, which causes E_t to be more sensitive to the saturation deficit of the mixed layer and canopy surface, and less to net radiation. Short crops, in contrast, are relatively decoupled from the vapour pressure deficit of their environment so that they respond more to changes in net radiation. Consequently, both tropical and temperate forests generally have higher r_s values than shorter crops, in part, compensation for lower r_a. This is a defensive response in the cause of water conservation because of the difficulty of taller plants meeting the transpirational demand at foliage level as against water supply from the roots. Such characteristics, in turn, causes a marked compensatory response to atmospheric saturation deficit rather than to net radiation (Shuttleworth, 1989a).

The microclimate within the Amazonas tropical rain forest exhibited decoupling between the upper two-thirds and lower one-third (Shuttleworth *et al.*, 1985). Mixing in the upper canopy was evident during the day with temperature, humidity and, in turn, saturation deficit closely following the atmosphere. In the lower canopy, the lower temperatures and higher humidity produced saturation deficits only 30–50% of the canopy during daytime. Also wind speed and radiation was dramatically reduced within the lower canopy where measured solar radiation was only 1% of that measured above the canopy. The implications of this microclimate on evaporation in the Amazonas study will be considered later.

Evaporation studies undertaken in the humid tropics
Estimates of evaporation from tropical forests Bruijnzeel (1987, 1990) has provided a very comprehensive review of forest evaporation studies across the humid tropics, with particular emphasis on both lowland and highland forest environments. Consequently, only certain aspects of this work will be highlighted here, covering estimates of annual evaporation, wet canopy evaporation and dry canopy transpiration.

During his summary, Bruijnzeel (1990) noted that the lack of standardization in experimental methodology which makes such comparisons somewhat tenuous. Groundwater leakage poses a problem in many catchment water balance studies. He regarded the level of precision achievable, under the most rigorous conditions, could not be much better than within 15% of true evaporation. The alternative methodology followed by Calder *et al.* (1986), still had only a precision of 16%. In contrast, the micrometeorological/physiological approach followed by Shuttleworth and co-workers in ARME (Shuttleworth, 1988a; Dolman *et al.*, 1990; Roberts *et al.*, 1990) indicated a higher level of precision. As already noted, Shuttleworth's (1988a) sensitivity analysis showed total evaporation estimates to be stable at the 5% level of precision.

Lowland forests Figure 9 summarizes the relationships between annual total evaporation and annual total transpira-

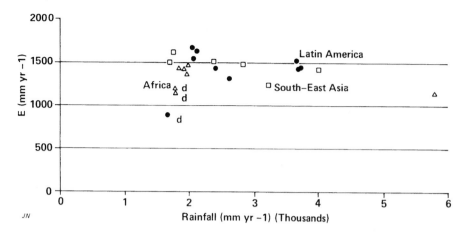

Figure 9: Annual total evaporation (*E*) versus precipitation in tropical lowland forests. Note *d* is a dry year. (After Bruijnzeel, 1990).

tion, based on 17 studies of which details are given in Bruijnzeel (1990, Table 1, pp. 23–26). Some of the scatter is accounted for by the mentioned differences in experimental methodology and precision level. Other factors affecting the scatter include the length of observation period, the occurrence of below average rainfall and the length of the dry season.

Annual evaporation, *E,* was estimated – based on two subsets of the presented data. Those studies considered the most reliable provided an annual average of 1400 mm year^{-1} (*n* = 6) whilst a larger subset gave 1430 mm year^{-1} (*n* = 11). The limited number of studies precluded any comparisons between continents.

Montane forests Whilst the dataset for montane forests is smaller, Bruijnzeel (1990) noted that some of the best long-term records are available from these areas, for example, the East African work (Blackie *et al.*, 1979). The data is summarized in Fig. 10, based on details presented in Bruijnzeel (1990, Table 2, pp. 29–30). It is evident from Fig. 10 that annual evaporation is not correlated with either elevation or annual rainfall mainly because of rather high evaporation values for some African forests.

After excluding cloud forests and some of the West Indies and African studies (Bruijnzeel, 1990), the annual total evaporation for lower montane regions (mean elevation, 1750 m a.s.l.) is around a value of 1225 mm year^{-1} (range 1155 to 1295, *n* = 5). Such estimates were surprisingly close to the annual figure of ca. 1415 mm (mean elevation 100 m a.s.l.) for lowland forests despite significant differences in elevation. As Bruijnzeel noted, there is a dearth of information on net radiation from both areas to check against these evaporation estimates.

The few details from "cloud forests" show considerably lower total evaporation (308–392 mm year^{-1}) which is attributed to increased precipitation inputs (occult precipitation) from cloud stripping (Juvik & Ekern, 1978; Zadroga, 1981) and low transpiration rates under the prevailing climatic conditions (e.g. Cavelier, 1988). The quoted figures have not been corrected for "occult" precipitation and no information was

available from the studies shown in Fig. 10. Based on adjustments from evidence elsewhere, annual total evaporation ranged between 570 and 775 mm for Colombian and Costa Rican cloud forests (Bruijnzeel, 1990). This is still considerably lower than for the figures quoted for lower montane forest.

What becomes clear from Bruijnzeel's review is the need for more studies concerning the hydrology of montane forests and especially the "occult" precipitation and evaporation aspects of "cloud forests".

Estimates of wet canopy evaporation, E_i In comparison with transpiration studies, Bruijnzeel (1987) noted over 100 "tropical" interception/throughfall studies, of which more than 70 related to natural forests. When considering the reliability of experimental methodology, this large sample is dramatically reduced to give wet canopy evaporation, E_i (interception), estimates yielding annual average values of 13% rainfall (range 4.5 to 22%) in lowland rain forests (*n* = 14), and 18% rainfall (range 10 to 24%) for montane rain forests (*n* = 6, excluding "cloud" forests). Allowance was made for stemflow (1% of annual rainfall) in these estimates (Bruijnzeel, 1990) based on studies elsewhere (e.g. Jackson, 1975; Lloyd *et al.,* 1988). Other work by Herwitz (1986) suggests this assumption may be too conservative.

Stemflow is a difficult parameter to quantify in tropical rain forests, especially when estimating the projecting contributing area amidst the profusion of vines and trees. In addition, most stemflow work has focused on equatorial rain forests where short duration, convective storms prevail. Herwitz's (1986) study suggests stemflow is not insignificant in a high rainfall (6570 mm year^{-1}) section of closed-canopy, montane rain forest of northeast Queensland, where weather systems are more organized and of longer duration. He noted that localized stemflow fluxes could be as high as 314 mm min^{-1} when the rain intensity was 2 mm min^{-1}.

Estimates of dry canopy transpiration, E_t Bruijnzeel's (1990) review indicated that there were fewer reports of transpiration

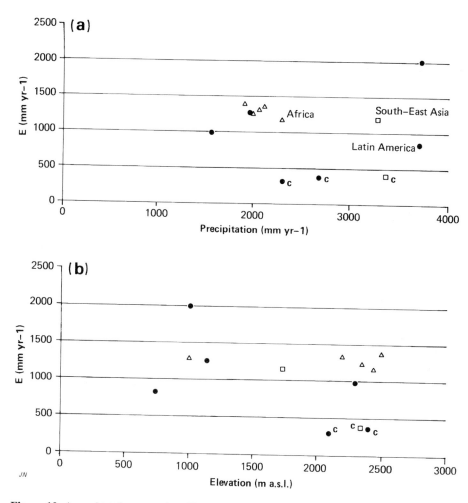

Figure 10: Annual total evaporation (*E*) versus precipitation (a) and altitude (b) for selected montane tropical forests. Note that *c* denotes "cloud forest". (After Bruijnzeel, 1990).

(dry canopy losses) for moist tropical forests cf. wet canopy evaporation. A common technique is to subtract E_i previously obtained from interception/throughfall work from total evaporation, *E,* which has been previously obtained from water balance studies for weekly or longer periods. The errors involved in interception studies (see Lloyd & Marques-Filho, 1988; Lloyd *et al.*, 1988) and drainage basin leakage place considerable doubt over many of these studies. Alternative approaches such as the zero flux plane method (Cooper, 1979), soil water balance using the neutron probe, physiological and micrometeorological methods as used in ARME (Shuttleworth *et al.*, 1991) provide better estimates of E_t.

As Bruijnzeel (1990) noted, most of the studies shown in Fig. 9 also report E_i. From this information, he calculated an average E_t of 1045 mm (*n* = 9). Individual locations indicated E_t as ranging between 885 to 1285 mm year^{-1}. In the case of studies experiencing a marked dry season (e.g. Queensland, Gilmour, 1975), E_t was below 500 mm year^{-1}. Corresponding estimates for lower montane forests varied between 560 and 830 mm year^{-1} (*n* = 4) with no relation to elevation from this small sample (Bruijnzeel, 1987).

The limited evidence available from "cloud" forests indicates E_t averages between 0.7 to 1.4 mm day^{-1}. These low val-

ues are more likely linked to the low available net radiation and vapour pressure deficits rather than to high *rs* (Bruijnzeel, 1990).

Results from selected evaporation process studies in the humid tropics with particular reference to rain forests Apart from early East African work by Callander & Woodhead (1979) in connection with tea plantations, it was not until the 1980's that process studies using micrometeorological methods, were initiated to investigate the dynamics of evaporation *per se*. The main focus has been on forest evaporation, largely undertaken by the Institute of Hydrology, UK, and reference to various aspects of that work have already been made. This section will summarize the overall research strategy of each project and the more pertinent conclusions.

Calder *et al.* (1986) initiated the first investigation over a one-year period (July 1980 to July 1981) in the Janleppa nature reserve, West Java, Indonesia containing secondary, lowland tropical rain forest. Measurements of transpiration and interception loss were reported. These were obtained using a combination of soil water balance, based on neutron probe measurements (for transpiration estimates); interception/ throughfall using plastic sheet gauges (Calder & Rosier, 1976) (which Lloyd *et al.*, 1988 found unsuitable in the Amazonas work), and above canopy automatic weather stations for input of measurements in the Penman-Monteith equation.

The most comprehensive investigation was the ARME project (Shuttleworth, 1988b) which took place over undisturbed tropical rain forest in the Reserva Ducke, 25 km northeast of the city of Manaus. The period of data collection extended for two years (September 1983 to September 1985, incl.) which provided basic hourly meteorological data, measurements of integrated rainfall interception loss and the basic soil hydraulic properties (volumetric water content, matric potential) (Shuttleworth, 1988a; Shuttleworth *et al.*, 1984, 1985; Lloyd & Marques-Filho, 1988; Lloyd *et al.*, 1988). Within this two-year period, three intensive campaigns were initiated involving the measurement of radiative components, eddy-correlation measurements of evaporation, sensible heat and momentum transfer (Shuttleworth, 1988a); temperature, humidity and windspeed profiles – and plant physiological studies (Roberts *et al.*, 1990).

It is significant that both of these studies were undertaken in equatorial climatic regimes where rainfall is more evenly distributed and moderate in annual totals (annual total, 2851 mm, Calder *et al.*, 1986; average 2636 mm, Shuttleworth, 1988a). There remains an absence of similar studies within the monsoonal outer tropics e.g. northeast Queensland, Australia, where annual rainfall can be higher (> 4000 mm) and more seasonally concentrated (Herwitz, 1986; Bonell *et al.*, 1991). Other monsoon areas, such as the coastal zone of West Africa, experience a marked dry season (Monteny *et al.*, 1985) so that tropical forests may experience water stress from the dry northeast winds blowing off the desert during the cool season. Differences between the rainfall regimes, will, in turn, affect the duration of wet and dry conditions over the year. The consequences have already been discussed in terms of coupling between different canopies (McNaughton & Jarvis, 1983). Some caution is then required in transposing the results from existing studies to the outer tropics.

Early water balance studies noted the high forest evaporation, *E*: standard Penman (1948) open water evaporation, E_p ratios determined from rain forests. Measurements of open pan losses however, are usually measured in a clearing and not from a tower over the forest. As λE_p is very sensitive to windspeed, some care is required in the interpretation of such ratios as any differences in wind regimes makes it more difficult to physically relate them to forest environments. Examples of E/E_p ratios include those from the East African experiments where Edwards & Blackie (1981) reported ratios of 0.93 for montane rain forests (average annual rainfall range, 1924 mm to 2219 mm). Gilmour (1975) noted a marginally lower ratio average of 0.87 obtained from a water balance study in the Babinda experimental catchments (average annual rainfall 4009 mm, 1970–1983, Bonell *et al.*, 1991) in northeast Queensland. Over shorter periods during this study, the E/E_p ratios were in the vicinity of 0.9 to 1.0 when soil moisture lev-

els were high following rain (typical volumetric water contents, θ, about 0.45 for the wet season). When θ declined to 0.37 (towards dry season conditions), there was a dramatic reduction in the same evaporation ratio to a value of 0.2 (Gilmour, 1975).

The high E/E_p ratios were considered by Calder *et al.* (1986) to be a feature of humid tropical forests where soil moisture limitations are rare and where the maximum daily temperature always reaches the partition temperature of 32°C. These circumstances are experienced particularly in equatorial regions where rainfall is more evenly distributed and high daily temperatures are sustained. Under such conditions, virtually all net radiation will be converted to latent heat under non-limiting soil water conditions (Priestley, 1966; De Bruin, 1983a). From the Java study, Calder *et al.* (1986) commented that the measured annual evaporation of 1481 mm required $3.58.10^9$ J ±12%, which is almost identical with the measured annual input of net radiation of $3.73.10^9$ J ±10%. In the outer tropical areas, such conditions probably only apply to the summer wet season. The ARME project also showed close agreement between measured, total evaporation (2748 mm) using the calibrated Penman-Monteith-Rutter model and the net radiation (3070 mm equivalent), giving a ratio of 0.895 over the 25-month study period. With increasing length of the dry season, and associated moisture stress, the net radiation usage will change which probably accounts for the lower mean annual E/E_p ratio in Gilmour's (1975) study.

Other significant features of ARME need mention. An average of 50% of the incoming precipitation (annual average 2636 mm from 25 months record) is re-evaporated (annual average, 1319 mm year^{-1}) and about 25% of total evaporation is from wet canopy evaporation, E_i (annual average, 328 mm year^{-1}) (Shuttleworth, 1988a). In addition, there was close agreement between the cumulative Penman (1948) λE_p estimate (2637 mm) over 25 months and measured evaporation (2748 mm), i.e. 4.1% less than the actual evaporation calculated by the calibrated model. Close agreement was also determined between the cumulative estimate, λE_{PT}, using the Priestley-Taylor equation, (1), (2868 mm over 25 months) and total evaporation from the calibrated model, i.e. 4.4% greater than actual evaporation. The close agreement between these two estimates suggests that total evaporation largely reflects available radiant energy in this environment (Shuttleworth, 1988a); that is, towards equilibrium evaporation. These findings are in close agreement with Morton (1991) applying the complementary relationship to the Cameroon data and reinforce the earlier views of De Bruin (1983a). Morton (1991) argued that the reasons why the E_p (and related E_{PT}) could break down in temperate areas are due to high wind speeds, low radiation input, low rainfall intensities of long duration and a negative sensible heat flux arising from mesoscale advection. Even if sensible heat flux above a tropical forest can occasionally be negative (see Morton, 1991, pp. 364–365), the high radiation input means that overall, net radiation is the dominant energy source which makes the Penman and related Priestley-Taylor methods applicable to tropical forests over a long period.

The close agreement between λE_P, λE_{PT} and the time-average evaporation over the 25 month period was cautiously highlighted by Shuttleworth (1988a). As mentioned earlier, evaporation rates from forests are dependent on the duration of wet and dry canopy conditions (McNaughton & Jarvis, 1983). In the Reserva Ducka study, dry canopy conditions provided evaporation losses typically 20–30% less than the Penman, λE_P or Priestley-Taylor, λE_{PT} rate. This was demonstrated over a 8 dry day period in September 1983 when actual evaporation only used $\approx$ 70% net radiation (Shuttleworth *et al.*, 1984). Significantly, the daily average of equilibrium evaporation, λE_E, 3.80 mm was a much closer approximation to measured transpiration (average 3.45 mm). Over the same days, Penman (λE_P, average 5.05 mm) and Priestley-Taylor (λE_{PT}, average 4.80 mm) exceeded actual evaporation by a factor of 1.5 (Shuttleworth *et al.*, 1984). Conversely, actual evaporation was 20–70% greater than the potential rates (λE_P, λE_{PT}) on wet days. Energy in excess of that locally available as net radiation is absorbed in the process. This led Shuttleworth (1988a, p. 342) to conclude that the "... the near-equality between average evaporation and potential evaporation in central Amazonia is, therefore, in part, fortuitous, and should only be extrapolated to other portions of the forest with caution." The fortuitous element relates partly to the combination of dry and wet canopy conditions existing over the study.

The observation that E_i can exceed net radiation, Q_n, substantially, has been noted in temperature studies, and the ratio of E_i/Q_n can vary between 0.6 to 4 for forests (see review by McNaughton & Jarvis, 1983). Substantial evaporation losses have also been observed at night-time (Pearce *et al.*, 1980). The sources of additional energy to enhance higher evaporation rates from wet canopies has still not been resolved, although several contributory elements have been discussed (e.g. Monteith, 1981; Morton, 1984; 1985; Calder, 1985; Baldocchi, 1989) such as increased turbulence above a rough forest compared with short crops, the maintenance of a saturation deficit through mesoscale, dry-air advection and the addition of heat to the air. Baldocchi (1989), for example, noted that woody biomass in forests can significantly alter the partitioning Q_n into latent and sensible heat via extra storage of heat in the woody biomass. A common source of dry-air advection are air mass changes at a synoptic scale, but this rarely occurs in the low tropics, unlike in temperate latitudes. More important, drier air can be advected at a mesoscale from surrounding areas not wetted by convective rainfall. This flux-divergence (as against vertical inputs) forms part of the self-regulating mechanism within the continental climate of Amazonia. The pattern of "pop-corn" cumulo-nimbus clouds (Paegle, 1987; Wallace, 1987) means that rainfall at one site is supported by energy entering the atmosphere from another where the canopy resistance is suppressing enhanced evapora-

tion (Shuttleworth, 1988a). Thus, there may be some compensating interaction between wet and dry canopy evaporation in terms of energy partition. Furthermore, there has also been considerable discussion concerning the role of downdrafts of "drier" air from the PBL during convective activity (McNaughton & Jarvis, 1983) in enhancing E_i but this has not been researched in the central parts of the Amazon Basin.

The experience of enhanced wet canopy evaporation over forests in temperate maritime climates led to an interesting exchange of views concerning the sources of energy (Morton, 1984, 1985; Calder, 1985). Calder (1985) presented conclusive evidence that the advection of a synoptic-scale, warmer air mass was a critical factor in the evaporation process. This mechanism is less likely to be relevant in this discussion except possibly during the cool season in the outer tropics. Another more relevant factor is the effect of mountainous areas providing "additional" evaporative power to air gained from latent heat release during orographic uplift during rain events (Thom & Oliver, 1977; Morton, 1984) or because of their penetration into dry warm upper air (Giambelluca & Nullet, 1992) above a synoptic-scale temperature inversion associated with subtropical high pressure systems (see Manton and Bonell, this volume). In the latter case, Giambelluca & Nullet (1992) measured an increase in daily evaporation with increasing elevation at three sites along a leeward slope of Haleakala crater, Maui, in the Hawaiian Islands. Using atmometers the evaporative demand during their summer observation period ranged from averages of 3.3 mm day^{-1} at 950 m, 3.6 mm day^{-1} at 1650 m and 6.4 mm day^{-1} at 2130 m. Analysis suggested that this increase was not fully explained by differences in net radiation, but instead implied strong positive heat advection from the dry air above the inversion which is warmed by the large scale subsidence. Furthermore, the night-time evaporation rate was found to increase with elevation which also confirmed a non-radiative energy source. Significantly, the four evaporation models of Penman (1948), Monteith (1965), Van Bavel (1966) and Priestley-Taylor (1972) did not account for this influence. Giambelluca & Nullet (1992) concluded that knowledge of upper air data to locate the inversion height was an important consideration in modelling of mountain evaporation, especially in the trade wind belt of the humid tropics. Such conclusions were reinforced by the close linkage between the vertical oscillation of the inversion, and the spatial and temporal variation in the enhancement effects on evaporation.

The preceding discussion highlights the need for replication of ARME-type studies in other parts of the humid tropics and also within the Amazon Basin, including the bordering mountainous regions in the west and the "maritime" coastal areas in the eastern parts of the basin (Shuttleworth 1988a, 1989b). As Shuttleworth (1988a, p. 342) acknowledged, "... it is, for instance, very probable that Amazonian regions with higher rainfall (e.g. mountainous areas of bordering the basin), where the forest spends a great proportion of the time wet, will tend to have average evaporation rates greater than the potential rate." In addition, the same potential for forests to exploit advected energy from the Atlantic Ocean in the east and enhance wet canopy losses, producing a net time-average evaporation greater than the potential rate exists. Such research needs has stimulated a new project in the Amazon Basin, which will be discussed later.

Replication of micrometeorological work, on the lines of ARME, is also required in different "maritime" tropical environments such as the equatorial climatic regime of Peninsular Malaysia and the outer tropical areas, e.g. northeast Queensland, where more organized rainfall systems occur. Existing work in the lower topography of Peninsular Malaysia has highlighted the dominance of evaporation (e.g. Abdul Rahim, 1988; Bruijnzeel, 1990, p. 23) which can account for up to 87% of annual rainfall. In view of its significance, micrometeorological work would provide more accurate estimates of evaporation. In contrast, the four-year water balance study (1970 to 1973, inclusive) of Gilmour (1975) noted evaporation accounted for about 39% (1525 mm) of a mean annual rainfall of 3900 mm. More important, wet canopy evaporation, E_i, accounted for 702 mm (46%) of total evaporation which is significantly different from the Reserva Ducke results. Very impressive visual evidence of wet canopy evaporation is common, as shown by the "stratus" cloud rising immediately above the rain forest canopy during long duration rainfall (Bonell, 1991c). Identifying the energy sources and processes of evaporation under such conditions may provide a different perspective from current work in the Amazon Basin.

The utility of the Penman-Monteith equation.
Several writers have highlighted the problems of adopting the Penman-Monteith equation for scaling-up evaporation from small plots to drainage basins, and suggested new approaches are required e.g. higher order closure models (see Morton, 1984, 1985; Baldocchi, 1989). Presentations from ARME (Shuttleworth, 1988a; Dolman *et al.*, 1990; Roberts *et al.*, 1990) have demonstrated the promising application of the Penman-Monteith-Rutter, single-layered model despite earlier objections (e.g. Morton, 1985). As Veen & Dolman (1989, p. 481) noted, "quite apart from the technical problems of validating layered models by means other than net precipitation, there is the hydrologically legitimate question of whether or not the extra effort can be proven to be worthwhile. This matter has not been settled yet, but there are indications that in many forests the single layer models perform well enough."

Stewart (1989) argued for the continued use of the Penman-Monteith equation, providing the sub-model described in equation (4) was incorporated to represent the effect of biological responses on the transpiration rate. The Stewart (1988) modification of the Jarvis (1976) sub-model still requires knowledge of the Leaf Area Index (*LAI*). Techniques of measuring *LAI* are difficult to apply in tropical rain forests and Roberts *et al.* (1990) had to rely on published *LAI* data to apply

their model. The fact that the use of the Penman-Monteith model in conjunction with the more simplified, time-dependent model (equation 5) for estimating g_s (Dolman *et al.*, 1989), performed as well as the alternate combination with the more complex sub-model (equation 4) of Stewart (1988), indicated that the latter may not be necessary in environmental conditions similar to the central Amazon climatic region. Stewart's (1988) model, however, is still recommended for use in other climatic environments (Dolman *et al.*, 1990).

Stewart's (1989) suggestion of scaling up point measurements of evaporation to larger areas probably holds the greatest potential. He proposed an area should be divided up into sub-areas of similar vegetation and soil types. Evaporation for each sub-area should be measured at least at six sites, using the Penman-Monteith equation (including the incorporation of the sub-model) based on hourly data. The summing of estimated evaporation from the individual sub-areas would provide total evaporation for the complete area. Experiences in macrohydrology projects of the 1990s (discussed later) will enable an effective evaluation of such proposals in the humid tropics.

Earlier, Dunin & Aston (1984) had tested an alternative approach (not using the Penman-Monteith equation) for extrapolating measurements from a weighing lysimeter to a 5 ha catchment, and also making parallel comparisons using a distributed (accounting for the spatial variability of soil water) and lumped (lumped soil water content) versions of a generalized water balance model, WATSIM (Aston & Dunin, 1980). The lysimeter underestimated the catchment water balance, *E,* due to spatial variability of water supply. Both model estimates of *E* performed more adequately because they were not subjected to systematic differences in soil water between the catchment and lysimeter, but the distributed version of WATSIM gave a higher level of precision in terms of variability in catchment of *E.*

Do eucalyptus species use excessive amounts of water?
In recent times, there has been considerable controversy concerning the water consumption of exotic tree species in the humid tropical countries e.g. Thailand, India, most notably *Eucalyptus* spp., in terms of depletion of groundwater resources and soil nutrients (Calder, 1986, 1991a). Fleming (this volume) makes brief reference to the problem, but it is worth further elaboration, focusing on Australian examples and later a current study in southern India.

In terms of direct measurement, as recent as 1985 Dunin *et al.* (1985, p. 272), commented that "... data on eucalypt evaporation are few, largely due to the difficulty of its measurement and hence our understanding is so far deficient for resolving these issues ..." (referring to land-use hydrology). To address this problem, a 36 tonne monolithic weighing lysimeter (3.7 m diameter, 1.5 m deep) was installed within Kioloa State Forest, located on the south coastal plain of New South Wales, Australia to provide a continuous record of water use from a regenerating natural eucalypt community, (dominated by *E. maculata* and *E. globoidea*), five years old at the time of instal-

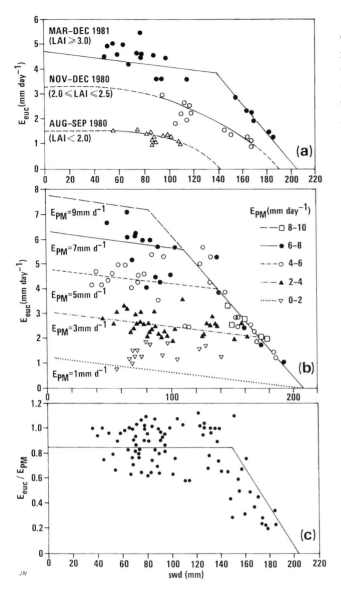

Figure 11: Relationships describing the evaporation behaviour of non-wetted canopies of the lysimeter community dominated by *Eucalyptus maculata* and *Eucalyptus globoidea* at Kiola State Forest, southern New South Wales, Australia:
(a) Eucalypt transpiration response, λE_{euc}, to combined influences of soil water supply and leaf area (*LAI*). The data points and curves shown are selected for λE_{PM} rates (defined in equation (2) of 5± 1 mm day^{-1}. The class ranges of *LAI* correspond to selected periods of canopy development during the experiment (After Dunin & Aston, 1984 and adapted in Dunin *et al.*, 1985).
(b) eucalypt transpiration, λE_{euc} response to soil water deficit (*swd*) for different ranges of λE_{PM}, with *LAI* > 3.0 (showing regression lines for each range) (After Dunin & Aston, 1984 and adapted in Dunin *et al.*, 1985).
(c) relative evaporation rate $\lambda E_{euc} / \lambda E_{PM}$ as a function of *swd* (After Dunin *et al.*, 1985).

lation. The technical and performance details of the lysimeter were provided by Reyenga *et al.* (1988), and supporting experimental methodologies (including micrometeorological studies) are reported in Dunin *et al.* (1985, 1988).

When concerning wet canopy conditions, low annual losses of interception were recorded, ranging between 10 and 15% of rainfall (Dunin *et al.*, 1988) which Calder (1991a) attributed to the smaller leaf area index (*LAI*) of *Eucalyptus* spp. compared with other tree species. Interception studies by Aston (1979, Table II) of six *Eucalyptus* spp. and one *Acacia* spp. under a rainfall simulator in Australia demonstrated low interception storages expressed as mm of water per unit ground projection area. With the exception of *E. maculata* (*LAI* 13.4), all *Eucalyptus* spp. had *LAI* less than 4.2 (cf. *LAI* 13.1, *Pinus radiata*). Interception storages ranged between 0.2 to 0.8 mm (including the *Acacia* spp. and *E. maculata*). Later, Pook *et al.* (1991), using the Gash model, determined *S* for *E. Viminalis* as being 0.25 mm in contrast to 1.2 mm for *Pinus radiata* (cf. 1.0 mm, Aston, 1979). By comparison, monthly losses were reported to be as high as 40% of rainfall in Jarrah forest (*E. marginata*) in Western Australia (Sharma, 1984), which prompted Dunin *et al.* (1988) to suggest that differences between *Eucalyptus* spp. characteristics and atmospheric influences could be important in determining interception losses. In the latter case, the typically low windspeeds (< 1 m s^{-1}) occurring above the canopy in rainy periods of the Kioloa experiment were thought to account for the low interception loss. In addition, night-time interception losses were 30% of total annual losses with an association with rain intensity. Interception loss rates could exceed 0.8 mm h^{-1} at night when rain intensity attained 6 mm h^{-1} (Dunin *et al.*, 1988).

An important biological influence, expressed in terms of changes of *LAI*, influenced dry canopy losses from the *Eucalyptus* spp. of the Kioloa experiment (Dunin & Aston, 1984; Dunin *et al.*, 1985). Figure 11a summarizes the constraints of both *LAI* and soil water deficit (*swd*) on daily transpiration rate. The outward shift of the intercept along the abscissa was interpreted as in increased root exploration of the soil water store with increased *LAI*. Therefore, there is a coupling of leaf area with that of root development in forests. More important, the data in Fig. 11a indicate that, during prolonged dry periods, the soil water store is not necessarily exhausted even by eucalypts. This point is further emphasized in Fig. 11b when comparing λE_{euc} with atmospheric demand λE_{PM}, as represented by the Penman-Monteith equation (2) and r_s equal to 60 s m^{-1}. Figure 11b shows the critical soil water deficit when λE_{euc} begins to drop markedly below λE_{PM}. The indication is that with increasing atmospheric demand the critical *swd* decreases when λE_{euc} begins to decline.

These results suggested that under Australian conditions, leaf area dynamics may be a feature of evergreen eucalypt communities with significant consequences on the transpiration rate, although Dunin *et al.* (1985) acknowledged that their work could not verify to what extent variation in *LAI* was the prime cause for water losses. A separate study on the canopy dynamics of *E. maculata* at the same Kioloa experimental site by Pook (1984a, b, 1985, 1986) provides some clues. Under non-drought conditions, Pook (1984b) noted that the *LAI* of the forest overstory (not including ground cover) was conservatively estimated to be 4. Furthermore, seasonal fluctuations of up to 0.45 in *LAI* of the eucalypt overstory were relatively small, cf. deciduous forests, pastures and crops. Thus under non-stressed conditions, the *LAI* remained relatively unchanged although changes in spatial distribution of leaf area, leaf age profile, radiative properties and interception storage capacity of canopies could have implications for *E*. Even more significant, Pook (1985, 1986) noted considerable differences in overstory canopy dynamics to two serious droughts of similar duration and overall rainfall deficiency in 1980 and 1982. Pook (1985, Fig. 8) presents evidence of unprecedented leaf-shedding during the 1980 drought which reduced the *LAI* of the overstory from about 3 to 0.8. The corresponding ratios of mean daily evaporation to Penman estimates ($\lambda E_{euc}/\lambda E_P$) progressively reduced from 0.75 to 0.08 when the water supply failed. Pook (1985), using relationships established by Aston (1979), estimated a corresponding decrease in storage capacity, *S,* for intercepted rainfall from 0.4 mm to 0.1 mm. Consequently, the benefits of shedding leaves to reduce transpiration and water conservation also extend to encouraging greater soil water recharge through *S* reduction. By comparison, the 1982 "El-Niño" drought did not show the same reduction in *LAI* (Fig. 3 in Pook, 1986). Furthermore, the $\lambda E_{euc}/\lambda E_P$ ratios were maintained, mostly within the range of 0.30 to 0.60. The contrasting responses in canopy dynamics were attributed to the sequence of occurrence of critical climatological variables. The combination of unusually high vapour-pressure deficits – coinciding with exhaustion of soil water – triggered the extensive leaf-shedding in the 1980 drought. The distribution of annual rainfall in 1982 prove fortuitously more favourable in relation to the occurrences of critical soil moisture deficits. In addition, the relatively low levels of evaporative demand permitted the eucalypts to maintain adequate water status. Consequently, there was less impact on *LAI* in the 1982 drought.

This comparison in drought response underlines the difficulty in predicting the eucalypt canopy dynamics to rainfall deficiency from limited meteorological data. It is also pertinent that the study of Dunin & Aston (1984) coincided with the 1980 drought and emphasizes the need for longer-term studies covering a wider range of rainfall, atmospheric demand and soil water conditions before the issue of changing *LAI* (especially to severe drought) becomes conclusive. Pook's work indicates, that at the Kioloa site, *LAI* otherwise remains fairly constant (Pook, 1984b). There are other complications. As Dunin & Aston (1984) observed, Figure 11a indicates that prolonged dry periods will not necessarily lead to soil water store exhaustion. Furthermore, the representation of $\lambda E_{euc} / \lambda E_{PM}$ ratios shown in Fig. 11c show that for low *swd,* dry canopy losses tend to be lower than atmospheric demand. In addition, deep tropical soils are capable of having very large soil water stores even in tropical semi-arid areas of northern Australia (e.g. Williams & Coventry, 1979, >200 mm). The coupling of

LAI and the ability of root networks to tap such stores under drought conditions has not been investigated. Finally, the representativeness of the warm temperate Kioloa experiment in relation to the marked seasonal drought of monsoonal northern Australia is uncertain. Pook (1985) noted that several eucalypt species in tropical Australia routinely shed part or all of their leaf canopy during the dry season – which can only be more exacerbated during "El-Niño" drought episodes (see Manton & Bonell, this volume).

Despite the preceding uncertainties, none of the existing experimental evidence from Australia suggests that *Eucalyptus* spp. should function much differently in the humid tropics. In a more extensive review of other studies (mostly from Australia) Calder (1986b) came to similar conclusions except in *exceptional circumstances* where selected *Eucalyptus* spp. (e.g. *E. marginata* and *E. calophylla,* Coloquhoun *et al.,* 1984) do not exhibit stomatal control (to large vapour deficits) and are able to tap freely available water in the presence of shallow water tables. Under such conditions, dry canopy losses may be much higher compared with other vegetation types. The Western Australian work by Coloquhoun *et al.* (1984) and Greenwood *et al.* (1985), in particular, were influential in highlighting different stomatal control of *Eucalyptus* spp. For example, the detailed physiological studies of Coloquhoun *et al.* (1984) led to the species they examined being classified into three types of stomatal response. Elsewhere, the high rates of annual *E* recorded in the Greenwood *et al.* (1985) study, which exceeded annual rainfall by up to a factor of four for selected species, was attributed to roots extracting water directly from groundwater (known as "water mining"). Later, Calder (1991a) provided an up-dated review, and his conclusions basically remained unchanged. Appropriately, Calder argued the need for additional research to quantify water use of eucalypts across a range of climatic and hydrological environments, and include in such work the *Eucalyptus* spp. commonly used in plantations.

Current work by the Institute of Hydrology, UK (Calder *et al.,* 1991; Hall *et al.,* 1991; Harding *et al.,* 1991; Roberts *et al.,* 1991a; Calder, 1992 a, b; Calder *et al.,* 1992a, b) in the State of Karnataka, south India so far confirm that fears of excessive water use by plantations of *E. camaldulensis* and *E. tereticornis* are unfounded. The various methodologies adopted included the same techniques as used previously in Java (Calder *et al.,* 1986) for interception (Hall *et al.,* 1991), supported by additional physiological studies (Roberts *et al.,* 1991a) and deuterium tracing for estimating transpiration (Calder *et al.,* 1991, with the method described in Calder 1991b, c) and soil water balance modelling (Harding *et al.,* 1991) to link with *E* measurements. Whilst the location of this project is marginal to the humid tropics (rainfall is less than 1000 mm year[-1] and concentrated between June and September), the results are of great interest, especially to the Wet-Dry Tropical region of Chang & Lau (this volume).

For the complete calendar years 1986 and 1990, the small

canopy capacity S (0.45 ± 0.1 mm) and annual interception losses for *E. camaldulensis* were small, being 4% (39 mm) and 12% (51 mm) of respective annual rainfall (Hall *et al.*, 1991). These results are in close agreement with the study of Dunin *et al.* (1988) and the 11.4% losses reported in later Australian work by Crockford & Richardson (1990). Even more interesting, 46% of the interception loss occurred during rainfall for the complete dataset (Hall *et al.*, 1991) which reflects similar findings of significant wet canopy losses during rainfall in Australia (Dunin *et al.*, 1988; Crockford & Richardson, 1990). High specific humidity deficits (median 2.2 g kg^{-1}) during rainfall hours were recorded although as Hall *et al.* (1991) noted, the time resolution (hourly) for such measurements was not particular fine. Modelling indicated, however, that low r_a values were also required to predict the observed interception losses. Structural features of the eucalypts such as openness of the canopy, aided by leaf flutter, were thought possible explanations for the low r_a values (Hall *et al.*, 1991). The authors also suggested large-scale gust penetration as a possibility for accounting for the high humidity deficits. If so, this supports the earlier call for greater consideration of the *PBL* in evaporation modelling (DeBruin, 1989).

In contrast, annual transpiration of ≈ 900 mm dominates water loss from the eucalyptus (Calder, 1991d). Physiological (Roberts *et al.*, 1991a) and deuterium tracing (Calder *et al.*, 1991) indicated that transpiration rates and stomatal resistances were not exceptional when compared with observations of other tree species (humid temperate or tropical). Physiological studies indicated a seasonal change in stomatal conductance, g_{sto}, with the highest values during the summer monsoon with limited diurnal change. Diurnal changes became more evident in the post-monsoon (November/December), with afternoon minima in three defined canopy levels (Roberts *et al.*, 1991a). In the pre-monsoon, the lowest g_{sto} were observed at all canopy levels and this parameter was then at least an order of magnitude smaller (e.g. < 50 mmol m^{-2} s^{-1}, or <1.3 mm s^{-1}, with significantly no diurnal variation) than either in the monsoon or post-monsoon (up to 750 mmol m^{-2} s^{-1}, or 18.8 mm s^{-1}). Roberts *et al.* (1991a) did note some differences in g_{sto} between *Eucalyptus* spp. at different canopy levels, and on a diurnal and seasonal basis. In general, however, the highest g_{sto} were in the upper canopy with a small systematic reduction towards the base.

Transpiration from the three-layer canopy model (similar to the one described in Roberts *et al.*, 1990) showed significant variations in daily loss, ranging from 6 mm d^{-1} when soil moisture was non-limiting to less than 1 mm d^{-1} in the pre-monsoon season when transpiration is severely limited by the dry soils. There was also an important linkage between *LAI* and transpiration (Roberts *et al.*, 1991a), similar to the earlier findings of Dunin and co-workers (Dunin & Aston, 1984; Dunin *et al.*, 1985).

When comparing the soil moisture observations across a variety of forest types (semi-degraded natural forest, Casuarina,

Eucalyptus), results indicated a similar water use by all types (Harding *et al.*, 1991). Furthermore, soil moisture modelling suggested that the magnitude of water use was approximately equal to the rainfall, irrespective of tree species, especially when the soils do not reach field capacity or only at this level of wetness for short periods. There were also indications that transpiration losses in excess of Penman wet-surface evaporation, λE_P occurred when soil water was non-limiting, following the monsoon.

An outcome of this project was the hypothesis that there may be a simple relationship between transpiration loss and breast height diameter (*DBH*), based on the deuterium tracing (Calder *et al.*, 1991). Simple regressions relating transpiration rate to tree basal area or total basal area per hectare were put forward by Calder *et al.* (1991) on the assumptions of limited soil moisture stress or meteorological demand, and appropriate over the age range of trees studied. As the writers noted, this hypothesis needs further testing over a range of *Eucalyptus* spp., age and spacing of trees and contrasting environmental conditions. More important, the controversy of "water mining" by eucalypts was not completely settled in this study because the species investigated did not have sustained root contact with shallow water tables. As Calder *et al.* (1991) acknowledged, further work is required to investigate and quantify any excessive water use from selected non-water stressed *Eucalyptus* spp. in other hydrological environments.

Soil evaporation

Evaporation from bare soil is particularly relevant in the context of formerly forested areas which have been cleared and subsequently degraded through erosion or areas devoted to cropping. Consideration of soil evaporation is also needed in the context of hydrological changes resulting from the rehabilitation of degraded land through reforestation. To our knowledge, apart from Cooper (1979), there has been a dearth of such studies undertaken in the humid tropics. It is relevant, however, to briefly indicate some available techniques for future application.

Assuming that the direction of moisture flow is vertical through a homogeneous soil, then two phases of soil evaporation occur (Hillel, 1975). The first is an energy limiting stage during which evaporation rate approximates one of the potential rates defined earlier. The second is a profile control stage in which evaporation declines at a rate dictated by the ability of the soil profile to transmit water to the evaporation surface, and is controlled by the hydraulic properties of the soil.

The level of sophistication in modelling soil evaporation depends on the availability of measured parameters and the computer processing time required. Available methods include a simple, one-dimensional analysis based on Darcy's Law for calculating soil water fluxes between imaginary soil compartments (van der Ploeg, 1974; Cooper, 1979; Bonell *et al.*, 1983a) and empirically-based models which would need further testing and adaptation to humid tropical conditions

(Ritchie, 1972, 1985; Ritchie & Cram, 1989). Denmead (1984), for example, described an analysis of evaporation from the floor of a pine forest – using an approach similar to that put forward by Ritchie (1972). It was determined that soil evaporation through the litter mat was reduced to about 50% of that expected if the same soil had been exposed. There are also simplified, physically-based models such as the modified Clapp model (Clapp, 1982; Clapp *et al.*, 1983; Marker & Mein, 1987) for principally homogeneous soils. The transpiration component was also accommodated in the methods of Ritchie (1972), and Marker & Mein (1987).

The physically-based models were developed partly to avoid the cumbersome computer processing time required for solving Richards equation. Marker & Mein (1987), for example, showed that their modified Clapp model produced results comparable with those from a finite difference model of an extended Richards equation. Furthermore, the writers claimed that the results could be achieved at about 5% of the total computer time required for using the finite difference model. Such claims have been challenged by Short & Dawes (1990) on the basis of Marker & Mein (1987) using an inefficient finite difference solution and subsequent improvements in available computer systems. The future trend will be to more routinely consider adopting the Richards equation with increasing sophistication in computers, e.g. personal workstations.

A useful field technique for estimating evaporation from bare soil and transpiration from vegetation is the "zero flux plane" (*ZFP*) method which was described in detail and successfully applied by Cooper (1979) in the East African experiments. From tensiometer measurements, the method relies on determining the zero flux plane (*ZFP*), that is, the inflection of the total hydraulic potential (*Φ*) versus depth (*z*) plots. Hydraulic potential is defined as:

$$\Phi = \psi - z \tag{20}$$

For practical purposes, total hydraulic (or soil water) potential (commonly expressed in dimensions of length, *l* (metres) which is the height of a liquid column corresponding to the given pressure) is the sum of matric potential, *ψ* and gravitational potential, *z* (both in units of length, *l*). Common convention is to use the ground surface as the datum level ($z = 0$). The gravitational term *z* (+ve downwards) is then subtracted and becomes negative (van der Ploeg, 1974). Such a convention is only appropriate to landscapes of low relief. Where slopes exist, say greater than 4°, then an alternative datum for *z* should be selected, such as mean sea level or lowest point in the area of interest. In that case, *z* is taken as positive upwards and summed with the matric potential so that, $\Phi = \psi + z$.

There are two hydraulic potential gradients, the first is inducing upward movement from the *ZFP* to the soil surface to maintain evaporation. The second is from the *ZFP* to deeper parts of the soil profile. With the use of neutron probe measurements for estimating volumetric water contents, a water balance for the soil profile can then be determined to calculate evaporation and deep drainage (see Cooper, 1979).

There are limitations to the *ZFP* method. It can only be used in dry periods and when the *ZFP* is below the root range of any present vegetation since water movement within roots on the average is upwards. An alternative approach which avoids such restrictions is the use of the one-dimensional unsaturated hydraulic conductivity, *K* (*θ*) – hydraulic potential (*dΦ/dz*) gradient method described for example, by Cooper (1979) and van der Ploeg (1974). Fluxes between different points in the soil profile are calculated by Darcy's Law:

$$q = -K(\theta)\frac{d\Phi}{dz} \tag{21}$$

where *q* is the flux density (l^3/l^2t^{-1} or lt^{-1}) and *K* (*θ*) is unsaturated hydraulic conductivity which varies with θ the volumetric water content (l^3l^{-3}). The minus sign indicates that water movement is in the direction of decreasing potential.

This method was used by Bonell *et al.* (1983a) in connection with a tropical rain forest experiment monitoring the movement of artificially injected tritiated water. A comparison was made between calculated soil water fluxes and upward movement of the tracer in response to evaporation during a dry period. Indications were that the tracer advanced much quicker. The discrepancy may have been due to the use of soil cores for determining *K* (*θ*) by laboratory methods rather than *in situ* techniques. Cooper (1979) also emphasized the need for good knowledge of *K* (*θ*) determined *in situ* in the field. In addition, the problems of spatial variability of hydraulic conductivity raises problems of extrapolation to other sites. Such difficulties were demonstrated by Cooper (1979) during his comparisons between the *ZFP* method and the use of Darcy's Law for one-dimensional flow.

WATER MOVEMENT IN THE UNSATURATED ZONE

Knowledge of the infiltration process and, in turn, the storm runoff generation process, requires a physical understanding – through the medium of soil physics – of soil water behaviour. In the context of hydrological modelling of field conditions, the applications of classical theory of soil water movement with confidence is fraught with difficulties. For example, the collection of accurate soil hydraulic property data at different length scales (as defined by Dagan, 1986, p. 124S) requires a very careful assessment of the limitations of available methodologies and especially the volumes of soil sampled. The choice of methodologies (theoretical and measurement) for application in field studies also depends on the scale of the problem and the questions being addressed. In addition, classical soil water theory has been mostly developed at the scale of a laboratory column under controlled conditions, "... through the continuum approach to soil water movement as embodied in Darcy's Law and Richards' equation in which the macroscopic soil-water movement is considered and not the microscopic flow pattern in the complex pore network. However, soils are not generally the uniform inert porous materials which this approach assumes" (Youngs, 1988, p. 425). Under field conditions, soil hydraulic properties exhibit heterogeneity, both in time and space, from natural and man-induced disturbances. This produces non-uniformity of soil water flow which violates the stable flow conditions assumed by the classical Richards' flow theory. Swelling and shrinking of soils induces the formation of fissures within the soil matrix. Similarly, large channels are created by roots and soil fauna activity (Beven & Germann, 1982), which become especially important when soils are close to saturation in the infiltration process (I. White *et al.*. 1992 – *Note that all subsequent referencing to White refers to I. White*). Both situations may lead to preferential flow through the noncapillary macropores which are not amenable to one-dimensionless flow models. Instability of flow also occurs where there are abrupt and gradual increases in hydraulic conductivity with depth – commonly resulting from either surface compaction or layering of superficial deposits. Water repellancy of the solid phase (Wallis *et al.*, 1991) and compression of air ahead of an wetting front during infiltration (Raats, 1973; Philip, 1975; Collis-George & Bond, 1981), are also causes of unstable flow. In areas where the compressed air cannot escape, such as flatlands, further entry of water into pores is impeded and, therefore, this mechanism becomes an overriding control over infiltration. Under these circumstances, water entry is greatly reduced or ceases (Youngs, 1988).

The preceding conditions of heterogenity lead to unstable infiltration flows in the field which can be explained by the theory of viscous fingering (Philip, 1975; White, 1988). For example, a surface crusted soil underlain by preferential flow paths means that not all the soil will be wetted uniformly and "fingers" of infiltration water may occur below the crust.

Summary of physical theory of soil water movement

Good reviews of physical theory of soil water movement have been presented elsewhere in several sources (e.g. Philip, 1970; Feddes *et al.*, 1988; Hillel, 1980; Marshall & Holmes, 1979; Youngs, 1988). The physical theory of soil water movement had its origins with the development of Darcy's equation describing water flowing down saturated columns of sand. Darcy's equation for saturated flow can be written in general form as:

$$v = -K_s \cdot \nabla \Phi \tag{22}$$

where v is the macroscopic velocity of water (lt^{-1}), K_s is the saturated hydraulic conductivity (lt^{-1}) and $\nabla \Phi$ is the gradient of the total hydraulic potential (Φ) previously described in equation (20). Philip (1970) provides a more detailed account elsewhere of the fluid mechanics of Darcy's equation, however, it is pertinent to note that equation (22) is an empirical equation describing the average macroscopic flow velocity in

porous material only and K_s is mathematically a proportionality constant (Youngs, 1988). Throughout this text K^* (lt^{-1}) appears more frequently and denotes field-saturated, or satiated, hydraulic conductivity. Based on field (*in situ*) determinations, K^* is usually lower than K_s determined under sorption conditions in the laboratory. Bouwer (1966) showed that K^* may be as low as $0.5 K_s$.

Subsequently, Richards (1931) extended the application of Darcy's Law to unsaturated soils, so that in generalized form

$$v = -K(\theta) \nabla \Phi \tag{23}$$

where $K(\theta)$ denotes hydraulic conductivity which now varies with (θ) the volumetric water content. An expanded version of this equation was presented earlier when concerning soil evaporation (equation 21). Richards then combined equation (23) with the conservation of mass or continuity equation. Thus,

$$\text{dir } \bar{v} = -\frac{\partial \theta}{\partial t} - S_w \tag{24}$$

where:

t = time
S_w = the sink (or negative source) of soil water (e.g. water extraction by roots).

For one-dimensional vertical flow in the z direction with a matric potential gradient, equation (24) reads as:

$$\frac{\partial v}{\partial z} = -\frac{\partial \theta}{\partial t} - S_w \tag{25}$$

Combining equation (23) with equation (25) yields the non-linear Richards equation.

$$\frac{\partial}{\partial z} [K(\theta) (\frac{\partial \varphi}{\partial z} - 1)] = \frac{\partial \theta}{\partial t} + S_w \tag{26}$$

Soil water flow is, therefore, highly non-linear because both hydraulic conductivity and the matric potential (or soil water pressure head) depend on soil water content. Equation (26) forms the basis of modelling the water dynamics of the unsaturated zone, using analytical and numerical approaches (viz. finite differencing, finite element) (see review of Feddes *et al.*, 1988). Exact analytical solutions are only possible when simplified flow cases are considered under a number of restrictive assumptions and imposed initial and boundary conditions (Feddes *et al.*, 1988). Such conditions, however, have provided insight into the physics of water movement and when faced with the problems of heterogeneity give useful engineering approximations. It is most important that such assumptions and conditions are closely evaluated before field application is contemplated, and some appreciation of the underlying mathematical theory is essential as part of that evaluation. The physically-based infiltration equations considered later will fall under this category.

Spatial and temporal heterogeneity of soil hydraulic properties

One of the major problems in hydrological modelling is accounting for spatial and temporal heterogeneity of soil hydraulic properties. Since the 1970s, there has been an upsurge in studies connected with the spatial heterogeneity of hydraulic properties connected with agricultural and forest lands (e.g. Rogowski, 1972; Nielsen *et al.*, 1973; Baker, 1978; Sharma *et al.*, 1980; Talsma & Hallam, 1980; Bonell *et al.*, 1987). In contrast, temporal variability of soil hydraulic properties such as infiltration parameters have received only limited attention under field conditions (e.g. Tricker, 1981; Gish & Starr, 1983; Smettem, 1987; White & Perroux, 1989), although the general nature of the problem is well known (Skaggs & Khaleel, 1982; ASAE, 1983).

Disruptions to the soil fabric by such factors as biological activity, various management practices and intense rainfall compacting the surface (producing crusts over bare soil), combined with the change in hydraulic status influence temporal heterogeneity of hydraulic properties at a point (McIntyre, 1958a, b; Beven & Germann, 1982; Skaggs & Khaleel, 1982; Tarchitzky *et al.*, 1984). Bonell & Williams (1986a) monitored changes in the "S" (sorptivity) and "A" (transmission) parameters (see Lal, this volume) related to the Philip infiltration equation (Philip, 1957a, b; 1969) over three wet seasons in a tropical, semi-arid environment of Queensland, Australia. They showed statistically that temporal heterogeneity in these parameters is at least significant as spatial variability. Temporal variability at a sampling point was attributed to both biological activity (both fauna and flora), raindrop impact and the subsequent desiccation and cracking of thin (c. 1 mm) surface crusts between storms. This highlights the problem of using measurements at one point in time as representative estimates of soil hydraulic parameters as inputs to infiltration models.

Whilst the work of Bonell & Williams (1986a) was undertaken in a tropical semi-arid environment, such conclusions should also be relevant to cleared areas within the humid tropics where bare soil is exposed. The exceedingly high surface permeabilities associated with tropical rain forests (Nortcliff & Thornes, 1981; Bonell *et al.*, 1983b), however, means that any temporal heterogeneity should have little impact on the infiltration and drainage process unless disturbed by management practices.

For a comprehensive description of surface hydraulic properties, knowledge, therefore, is required of both their temporal and spatial variability, especially in agricultural and degraded areas. Similar comments apply to the hydraulic properties of subsoils with high clay content where there are drastic changes in structure resulting from swelling and shrinkage. This issue will be addressed in more detail when discussing macropores.

Scaling and kriging of soil hydraulic properties

Various techniques have been developed to address the problems of characterizing heterogeneity over space. One approach is to apply scaling theory that is based on the similar media concept (Miller & Miller, 1956), which assumes that for different soils their internal geometry differs only by the characteristic size. Such material is assumed to have identical porosities, and the same relative particle and pore size distributions. Scaling theory is then concerned with the microscopic dimension, viz. particle, pore or aggregate size or some average of one of these parameters for a particular soil sample. If λ is the microscopic dimension and $\overline{\lambda}_M$ be of some reference value, such as the microscopic dimension characterizing a reference soil, then a dimensionless scaling factor a is defined as:

$$a = \lambda_M \ / \ \overline{\lambda}_M \tag{27}$$

The objective is then to express spatial variability in terms of a single, physically-based parameter. The scaling factor a can also be used as a single parameter to approximate the $\theta\,(\psi)$, K^* and infiltration curve of a soil from these characteristics of the reference soil. Various ways of using, the scaling factor, a are adopted e.g. $1/a$, a^2, a, $1/a^3$, depending on the soil hydraulic variable or time variable considered. Peck *et al.* (1977) provide such details.

Initially, scaling theory was applied to soils, notably clean sands, under controlled laboratory conditions (Peck *et al.*, 1977). Application and testing of such theory to field soils commenced in the late 1970s by several writers, with reasonable success (Warrick *et al.*, 1977; Peck *et al.*, 1977; Sharma & Luxmoore, 1979; Sharma *et al.*, 1980). On the basis of "S" and "A" parameters of the Philip infiltration equation, Sharma *et al.* (1980), for example, evaluated scaling factors calculated by different methods in producing a scaled, cumulative infiltration versus time, $I(t)$, curve for an entire drainage basin. Field soils, however, do not fully satisfy all the conditions required by similar-media theory with respect to porosity, due to the consequences of shrinkage and swelling and preferential flow. For example, Sharma *et al.* (1980) highlighted the problem in the context of scaling the sorptivity, "S" ($lt^{-1/2}$), and transmission, "A" (lt^{-1}), parameters of the Philip infiltration equation. In addition, Warrick *et al.* (1977) noted that the scaling factors calculated from $\psi\,(\theta)$ were not the same as those calculated to scale $K\,(\theta)$ data referring to the same set of soils. Furthermore, scaling factors calculated from $\psi\,(\theta)$ were more effective in scaling the $K\,(\theta)$ data, than vice versa. Therefore, soil water properties based on scaling theory are only an approximation. Nevertheless, they provide a framework for studying the effects of soil heterogeneity on the hydrological response of land surfaces. Sharma and Luxmoore (1979) provided an example in the context of water balance simulation.

The difficulty of transferring scale factors from one set of hydraulic properties to another in field soils led Youngs & Price (1981) to provide an alternative method for scaling soil-water behaviour, without the restrictions of similar geometry. Furthermore, Miller (1980) pointed out that scaling of macroscopic flow equations requires a macroscopic, as well as microscopic length. White (1988) noted that an inherent soils-based scaling length, which could fulfill both macroscopic and microscopic requirements, was the macroscopic capillary (or sorptive) length, λ_c (*l*) introduced by Philip (1985). Philip's work was based on the quasi-linear analysis of steady unsaturated soil water flow, represented as

$$K(\psi) = K_s \exp(\alpha\psi) \qquad (28)$$

$$\alpha_s > 0 \; \psi \leq 0$$

with α called the "sorptive number" with dimensions (l^{-1}), $2\,\alpha_s^{-1}$ as the "sorptive length" and $\lambda_c = \alpha_s^{-1}$ (White & Sully, 1987). The λ_c parameter is essentially a flow-weighted mean soil-water potential and has a scaling length (*l*) for both distance and soil water potential, and is not dependent on equation (28) (White *et al.*, 1992). The relationship between λ_c and a characteristic microscopic, mean pore radius λ_m (*l*) was provided by Philip (1987). White & Sully (1987) summarized the relationship between λ_c and λ_m, and also showed that λ_c can be directly related to field measurable sorptivity and hydraulic conductivity using the disc permeameter (Perroux & White, 1988). The λ_m parameter was suggested as an *in situ* measure of soil structure (White & Sully, 1987) which offers a hydraulic approach to studying the impact of environmental and land-use changes. For example, with the aid of the disc permeameter method, White & Perroux (1989) used λ_m to quantify changes in soil structure caused by intense drought-breaking rains falling on a drought-exposed, unprotected soil surface. The results showed that the soil matrix remained essentially unchanged, but the larger pores had been infilled.

Geostatistical techniques such as the several estimation procedures embodied in the general term "kriging" (see reviews by Hohn, 1988; Krige *et al.*, 1989; Oliver *et al.*, 1989a, b), provide a method of interpolating and mapping of soil hydraulic properties from a limited number of sampling points. Such an approach is essentially the spatial equivalent of time series analysis (White, 1988) so that structure functions become semi-variograms. In the context of determining effective parameter values at a grid scale (e.g. 250 x 250 m) from point measurements, Beven (1989, p. 163) signalled a note of caution in the use of this technique. He noted that "... the theory of block kriging is a linear process. It may not lead to the same results as the hydraulic averaging of nonlinear processes that occurs in real catchments." The writer continues that "... this type of spatial interpolation and lumping assumes that we are dealing with a stationary random function. This may not be true for field soil properties and non-stationarity of the mean may affect parameter values at the grid scale ..." Temporal variability of infiltration properties provides an example of the problems of non-stationarity. Subsoil K^* values, on the other hand, are more stable over time and may be more appropriate for applying kriging. White (1988), however, noted that kriging is not an "optional" interpolator as claimed, but pointed out it is one of many methods of interpolating spatially (or serially) correlated data.

Preferential flow

In the introduction to this section, attention was given to factors causing non-uniformity of flow within the unsaturated zone. Forest soils, in particular, are noted for the proliferation of macropores at the surface in response to the high density of roots and soil fauna activity associated with the decomposition of leaf and branch litter. The physical and hydraulic properties of the surface soils, therefore, are extensively modified which makes the field application of most hydrological models for the unsaturated zone tenuous, in the context of their assumptions of soil being isotropic and homogeneous.

As White *et al.* (1992) observed, the importance of preferential flow paths in the rapid redistribution of surface water and solute transport was recognized as long ago as 1889 (Moore, 1889) but it was not until the 1970s – coinciding with increasing field experimentation – that the importance of this mechanism was widely recognized (see reviews of Beven & Germann, 1982 and White *et al.*, 1992). Although there is a diversity of quantitative definitions of macroporosity (see for example Table 1, Beven & German, 1982), several authors have alerted to the importance of such soil structural features in terms of preferential, downward movement of free water being at least partially independent of hydraulic conditions in the smaller pores (Beven & Germann, 1982; Bouma & Dekker, 1978; Germann & Beven, 1981). Macropores also dominate infiltration under ponded conditions and even have a significant role after matrix ponding has occurred during flux infiltration (Clothier & Helier, 1983; Smetten & Collis-George, 1985). Other writers have noted preferential flow occurring through large pores in an unsaturated soil matrix, a process described as bypass flow (Smettem & Trudgill, 1983) or short-circuiting (Bouma & Dekker, 1978; Bouma *et al.*, 1981; Hoogmoed & Bouma, 1980). In sloping forest soils, this flow-type has been observed in humid temperate areas when the soil matrix is unsaturated (Whipkey, 1969; Blake *et al.*, 1973; Cheng *et al.*, 1975). Gilmour *et al.* (1980, pp. 45–50) described the hillslope runoff response to an intense pre-monsoon rainstorm in northeast Queensland tropical rain forest. A protracted dry period had reduced volumetric moisture contents (θ) to as low as 0.14 in the surface soils prior to this event (under wet season conditions, such values are usually between 0.45 and 0.60). Normally, saturation overland flow (Kirkby, 1978, p. 373) prevails in this environment during the wet season, but none was recorded in the pre-monsoon event. The bulk of the response was subsurface stormflow (Kirkby, 1978, p. 373), associated with bypass flow in the macropores of the top 0.25 m.

Philip and co-workers (Philip, 1986a, 1987; Philip *et al.*, 1989) have recently introduced a theory to support the preceding empirical evidence of by-pass flow in macropores under unsaturated conditions. They analysed steady downward unsaturated seepage in a uniform soil, interrupted at some depth by a hole. Initially, the dry hole behaves as an obstacle to flow. Calculations showed that if the downward seepage is fast enough, there is a subsequent build-up of water pressure until at some point on the cavity walls, the seepage pressure equals that of the local soil atmosphere. Water then enters the hole. More important, Philip *et al.* (1989) established that the *larger* the hole, the more vulnerable it is to water entry from unsaturated seepage inflow. Such findings are contrary to conventional understanding drawn from capillary statics that larger cavities are least likely to hold water in a matrix field of unsaturated flow. These writers' theory also showed that cavity shape influences seepage inflow.

Preferential flow *becomes more prevalent* under conditions of intense rainfalls and low matric potentials, and can occur simultaneously with interstitial piston flow in heterogeneous soils (Foster & Smith-Carrington, 1980, Bonell *et al.*, 1984a). Bonell *et al.* (1982, 1983a, 1984a) describe such mechanisms in detail in tropical rain forest of northeast Queensland through the monitoring of subsoil water movement artificially tagged with tritiated water. However, unlike interstitial piston-flow, recharge attributed to this mechanism could not be quantified in that work due to the nature of the experimental design.

Feddes *et al.* (1988) reviewed recent attempts at incorporating preferential flow into unsaturated zone models. Most models are essentially of the two-domain concept whereby soil water is partitioned between soil matrix and macropore flow, and the fate of water flowing downward through macropores is approached differently by various methods (e.g. Hoogmoed & Bouma, 1980; Bronswijk, 1988). Water movement in the matrix domain may be modelled with a good approximation of Richards' equation based on Darcy's Law (Beven & Germann, 1982; Youngs, 1988). Water movement through the sides of cracks into the soil matrix can be considered as one-dimensional absorption as described for example, by Philip (1957a, p. 348; 1969, p. 235). This reduces the nonlinear diffusivity version of Richards' equation for specific boundary conditions to give:

$$x\,(\theta,\,t)\ =\ \varphi\,(\theta)\,t^{-1/2} \qquad\qquad (29)$$

where x is the horizontal Cartesean coordinate (l) and φ is a similarity variable for one-dimensional absorption ($lt^{-1/2}$), which is the solution of equation (34) in Philip (1969, p. 235) (or equation 12 in Philip, 1957a) – using the Boltzmann transformation. Flow in macropores can be modelled as laminar Poiseuille flow (e.g. Hillel, 1980), although not all available data supports the predictive value of this relationship (Beven & Germann, 1982). Part of the problem is that little is known of the hydraulics of flow in macropores. It is thought that the roughness of the pores may induce quasi-turbulent or turbulent flow conditions, accentuating non-Darcian flow conditions.

Beven & Germann (1982) questioned the adoption of the two-domain concept when there is a continuum of pore size ranges, as under these conditions macropore is a relative term. They noted that the domain concept of modelling would be most suitable where there is a distinct bimodal pore size distribution with a high degree of continuity in the large pores (such

as the 0.2 m depth cores of cracked soils modelled reasonably accurately by Hoogmoed & Bouma, 1980). Consequently, these writers (Beven & Germann, 1984; Germann & Beven, 1985, 1986) opted for the use of kinematic wave theory to model water flow through soils with different sized macropores, including the effect of water sorbance into the soil matrix. As noted by Feddes *et al.* (1988), this approach predicts outflow rates of unsaturated soil cores, but does not yield profiles of soil moisture.

Interest in applying such models under field conditions has particularly concentrated on cracking clay soils (e.g. Jarvis & Leeds-Harrison, 1987; Bronswijk, 1988). Bronswijk (1988), for example, calculated swelling and shrinkage, and corresponding cracking and subsidence in relation to changes in water content. Factors taken into account in Bronswijk's model include the area of cracks at the surface for preferential flow, and maximum infiltration rate of the soil matrix between the cracks. An additional consideration in such modelling is the linkage between matrix and preferential flow through horizontal infiltration through the walls of macropores or alternatively a sink due to evaporation. Kabal *et al.* (1988, described by Feddes *et al.*, 1988) are currently developing such a model through the addition of one extra term in the one-dimensional Richards equation for vertical flow. However, the quantification of such a term poses considerable difficulties (Feddes *et al.*, 1988). In contrast, Smettem (1986) showed that the approximate analysis of Philip (1985) satisfactorily described flow out of individual water-filled cylindrical macropores under both artificial and natural conditions where the channels were embedded in a soil matrix and did not form part of an extensive network. Smettem (1987) used the term "passive" flow to describe this situation as distinct from "dynamic" flow in macropore networks which are not controlled by conditions in the surrounding soil.

It is clear that taking macropores into account in the modelling of unsaturated flow is still in the infancy of development. The significance of such flow in modelling of field situations is also dependent on the scale of interest and the connectivity of the macropore system. Over small areas, such as those represented by 0.7 m^2 *in situ* infiltration rings (at a scale similar to the laboratory scale for deriving Darcy's equation, Dagan [1986]), the effectiveness of macropores on infiltration can be highly significant, causing physically-based infiltration equations to break down in their applications because assumptions and boundary conditions are violated (Bonell & Williams, 1986a). This highlights the inappropriateness of small-scale measurements if the interest is the runoff generation process along entire hillslopes. It is possible that the concept of the "repetitive unit" (Bear, 1979) may reduce the problem of macropores generating variability at the scale of a hillslope. The *repetitive unit* aims at regarding the behaviour of an inhomogeneous material as an equivalent homogeneous material if the length scale of observation is larger than the characteristic length of the repetitive unit (see Fig. 12). As was noted by

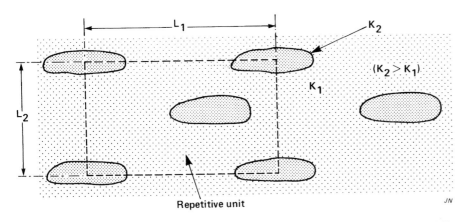

Figure 12: Diagram illustrating the concept of the "repetitive unit" which is aimed at regarding the behaviour of an inhomogeneous material as an equivalent homogeneous material if the length scale of observation exceeds the characteristic length (L_1 and L_2) of the repetitive unit (After Bear, 1979 and reproduced in Williams & Bonell, 1988).

Bevan & Germann in the early 1980s (1982, p. 1322), "... there has been no significant theoretical advances in modelling hillslope flows that involve macropores ..." They added "... even using a two-domain concept similar to the infiltration case, the multiplication of flowpaths that water can take downhill would make specific interactions between macropores and matrix very complex". They suggested that as a first approach, the problem could be decoupled into one of vertical unsaturated flow components (e.g. Hoogmoed & Bouma, 1980) and one of lateral saturated flow components. A relatively simple model of lateral flow, based on Darcian principles, could be used on the assumption that both macropores and matrix must be saturated before significant lateral flow can take place. O'Loughlin (1986) followed this procedure in the development of a topographic-wetness digital terrain model. Furthermore, use of a catchment average, K^* in O'Loughlin's model as against taking into account the spatial variability of K^*, showed little difference in the output for headwater catchments (Short, 1990, pers. comm.). Increasing the scale of interest beyond the "repetitive unit" might have contributed to this finding. Other writers (Binley *et al.*, 1989a, b) caution against using simple mean hydraulic conductivities based on random variations in K^* because they disregard the development of preferential flow networks.

Kirkby (1988) provided a comprehensive review of more recent developments in modelling of hillslope runoff. He concluded that subsurface properties, particularly the existence of lines of preferential flow – including macropores and pipes – influence the total response of a hillslope and call into question whether it can be treated as a Darcian flow system. Pipe systems pose a significant challenge to forecasters since their density, or indeed their presence, can only be sampled sparsely for most modelling applications (Kirkby, 1988, p. 336). Pipes can vary from less than 10 mm to more than 2 m diameter, especially in semi-arid areas (Kirkby, 1988) and their role in runoff generation has been particularly examined in humid temperate climate areas, notably in mid-Wales (Gilman & Newson, 1980; Jones, 1971, 1986). *Bypass pipes* (which carry ponded water at the surface through macropores) (Gilman & Newson,

1980; Kirkby, 1988) can transfer turbulent water flows (non-Darcy type) at velocities on parity with those of open channels, such as the perennial pipes described in Wales. The role of *seepage pipes* (originating within the saturated zone, so that inflows occur under appreciable pressure) is less clear (Kirkby, 1988), although a recent study using environmental isotopes by McDonnell (1990, p. 283) indicated the role of such pipes in quickly dissipating transient water tables downslope. This mechanism produces a rapid throughflow response of pre-storm event water. In terms of modelling, the connectivity of pipes is unknown in most catchment studies (the study of Jones, 1986 is an exception), which impedes any practical consideration of their role in hillslope runoff modelling.

There is a dearth of detailed studies on the role and extent of pipeflow within the humid tropics. During the course of reviewing runoff processes, Walsh (1980) noted that pipeflow had been reported from a variety of tropical environments, ranging from highly seasonal to perennially very wet, e.g. West Bengal (Banajee, 1972), Colombia (Feininger, 1969), Sarawak (Baillie, 1975) and in his own studies within Sarawak. Baillie (1975) describes pipes in some detail, and the geomorphic and pedological requirements for their occurrence, based on an upland Sarawak study. Walsh (1980) also noted a variable occurrence of pipes, depending on the soil type, in his study of runoff processes in different climatic zones of Dominica. Walsh still concluded, however, that when considering runoff process patterns "... in particular, more information on pipes in the tropics and their role in runoff generation is needed" (Walsh, 1980, p. 198). More recently, the study of Elsenbeer & Cassel (1990) in rain forest of western Amazonia, (Peru) is one of the few which have mapped the position of pipeflow outlets and discussed their role as a source of exfiltration and concentrated overland flow in the storm runoff generation process. The subterranean connectivity, however, was not surveyed.

In the Babinda catchments of northeast Queensland (Gilmour *et al.*, 1980), water can be observed emanating from seepage pipes at the head of some first-order streams which maintain almost perennial flow. Hydraulic conductivity measurements taken above one such head, for example, showed a two-order increase in K^* (Bonell, 1991a, p. 51) from similar measurements on a nearby experimental site (upper slope tracing site, Bonell et al., 1982, 1984a). The seepage pipes were associated schistosed lenses of higher K^* resulting from differential weathering of basic metamorphic rock (Bonell, 1991a). This led to a detailed K^* survey of the top of the impeding soil layer to prevailing rainfalls (0.2–0.5 m depth), in both disturbed and undisturbed catchments, to determine the nature and extent of these lenses (Bonell et al., 1987 and unpublished data).. If they proved extensive, then their presence would have a significant influence on the runoff generation process. Supplementary measurements of K^* at specific locations were also taken at greater depths. The results of this survey indicated that such schistosed lenses may not be as extensive as was first thought. Nevertheless, their connectivity still remains unknown. In addition, exposures of partially decomposed rock in some sections adjoining the main stream precluded determination of K^*. Consequently, their hydraulic significance could not be evaluated. In contrast, only a few pipes along the lines described by Elsenbeer & Cassel (1990) have been observed.

More important, the hydrological literature has placed little emphasis on the effects of diurnal and seasonal change in root water content causing shrinkage and swelling, and, therefore, providing annular space for macropore flow. Discussion on this subject, as presented for example in Kozlowski (ed.) (1981, p. 111), emphasize the few data available for tree roots in comparison with herbaceous plants. During the course of taking runoff measurements in the tropical rain forest (Bonell & Gilmour, 1978), it was observed that most subsurface stormflow emanated from the annular space surrounding functioning (not decayed) large root systems.

INFILTRATION

As Chapman (1990, p. 20) recently commented "... the process of infiltration has probably received more research effort than any hydrological process, particularly since it has been seen as the key to the hydrological changes resulting from manipulation of the soil surface by agricultural practices." Extensive reviews of both physically-based or approximate, time-dependent models and "empirical" infiltration models (including details of equations) have been provided, most of which describe infiltration under continuous ponding (e.g. Philip, 1969; Hillel, 1980; Skaggs & Khaleel, 1982; ASAE, 1983; see also Lal, this volume). Despite such concentrated effort, the successful field application of physically-based, vertical infiltration models, however, is still questionable and considerable skill in their interpretation of infiltration parameters is required (Bristow & Savage, 1987). The purpose of this section will be to evaluate the application of some of these models in the context of the humid tropics, placing particular attention on the physically-based models.

Physically-based models, such as those of Philip (1957a, b, 1958, 1969) and Green-Ampt (1911), are attractive in that most input parameters can be measured in the field (e.g. Perroux & White, 1988; Wierda et al., 1989; Smettem & C. Kirkby, 1990). The main problem is ensuring that the initial boundary and soil profile conditions assumed in the analytic solutions are closely met under field conditions. Such comments particularly apply to the Philip equation, because there has been recent adaptations of the Green-Ampt model to incorporate variable intensity rainfall and to account for layered soils (Mein, 1980; Brakensiek & Rawls, 1983; Chu et al., 1986). In the case of the Philip infiltration equation, for example, the restrictive boundary conditions are embodied in the "constant concentration" (Philip, 1973) assumption and concerns "instantaneous" surface ponding of a uniform soil having a uniform moisture content. This makes the Philip equation theoretically sound for one-dimensional, isotropic systems only and limits its application in the field to areas where Horton-type (infiltration-excess) overland flow (Horton, 1945; Kirkby, 1978) regularly occurs or to irrigation schemes.

As a later section will outline, there is a dichotomy of evidence whether such flow type occurs in tropical rain forests. The majority of studies note that the highly transmissive surface soil layers (Walsh, 1980; Bonell et al., 1981; Nortcliff & Thornes, 1981, Elsenbeer & Cassel, 1990) prohibit the existence of infiltration-excess overland flow, thus precluding the application of ponded infiltration theory. On the other hand, Casenove et al. (1984) and Wierda et al. (1989) present evidence of Horton-type overland flow in the Tai Forest National Park in Cote d'Ivoire. Dubreil's (1985) review also indicated similar occurrences in West African work.

When considering cleared, or degraded, former rain forest land, the combination of, one, compaction reducing the infiltration rates and, two, high rainfall intensities, should make Hortonian overland flow more possible over these areas. There is, however, a dearth of detailed rainfall records and associated soil hydraulic properties to comprehensively support such notions (Spaans et al., 1989; Bonell, 1991a). Lal (1981, 1983) measured an increase in overland flow from various agricultural treatments in Nigeria, but no soil hydraulic properties were presented to verify that it was infiltration-excess overland flow. On the other hand, Prove (1991) in northeast Queensland provided comprehensive evidence of Hortonian overland flow occurring within the interrows of planted sugar-cane on krasnozems (Stace et al., 1968) in a no-tillage treatment. Compaction of soil in the interrows by agricultural machinery was the cause. Also in northeast Queensland, as a later section will outline, measurements of K^* in formerly logged and cleared rain forest of the Babinda catchments showed that the surface K^* of an intergrade of Inceptisols-Ultisols (Red Podzolic, Stace et al., 1968) had been significantly reduced to enable Hortonian-type overland flow to occur, but only under the highest rainfall intensities, (exceeding 180 mm h^{-1}) (Bonell, 1991a, c). Elsewhere in Costa Rica, Spaans et al.

(1989) measured a reduction in K_s from 416 mm h^{-1} under forest, to 21 mm h^{-1} under pasture only 3 years after conversion by hand and chainsaw – in a relatively old Hunoxic Tropohumult soil. In contrast, K_s was unchanged at 29 mm h^{-1} under both types of land-use on a Oxic Humitropept soil, despite the forest conversion to pasture taking place 35 years previously. No rainfall intensities are mentioned but Hortonian overland flow, (and, therefore, the application of ponded infiltration theory) would seem possible in all types of land except for the forest related to the relatively old Hunoxic Tropohumult soil. Contrasting conclusions were reached from work by Dias & Nortcliff (1985) in the Amazon Basin. They described the comparison of physical properties of Oxisols at a virgin forest site, a site cleared by traditional methods and a site cleared mechanically by bulldozer. Surface bulk densities (0–20 cm) did not change between the virgin forest and slash-and-burn site, but a marked increase was measured due to compaction by the bulldozer. Also, at this bulldozer site there was a significant increase in penetration resistance to a depth of 15 cm below the surface (Nortcliff & Dias, 1988). In spite of a corresponding five-fold reduction in infiltration rates, however, these were still not sufficient to produce overland flow and erosion (i.e. virgin forest >2000 mm h^{-1} and bulldozed site, 390 mm h^{-1} (Dias & Nortcliff, 1985). In contrast, disturbed areas, such as access tracks in the Reserva Ducke, near Manaus show considerable visible evidence that the surface soil fabric collapses and seals on disturbance, resulting in semi-permanent depression storage. Elsewhere in Sabah (Malaysia), Malmer (1990) reported a marked reduction in steady-state infiltration capacities from 154 to 0.28 mm h^{-1} (related to Orthic Acrisol soils with high clay content) arising from tractor disturbance following logging. Furthermore, such low infiltration rates persisted 5–9 years after logging. Stadtmueller (1990) also reported similar results during his review of the effects of forest disturbance in nearby East Kalimantan, Indonesia. The implications on the hillslope hydrology were not discussed, however, by either Malmer (1990) or Stadtmueller (1990).

The idea of immediate surface ponding is more suited to border irrigation. Even under high rainfall intensities experienced in the tropics, the time to ponding can be sufficiently appreciable (Bonell & Williams, 1986a). The resultant delay causes a change in surface hydrological conditions which are not taken into account, for example, in the case of the Philip infiltration equation. Mein & Larson (1973), however, extended the Green & Ampt model to handle a constant rainfall input at a rate less than the infiltration capacity of the soil. Thus, the two-stage model incorporates pre-ponding as well as post-ponding infiltration. Mein (1980) reviewed the field applications of this extended model and also tested the input of a variable rainfall with reasonable success. Skaggs & Khaleel (1982) also discuss the use of the Green & Ampt model for unsteady rain.

More problematical is the assumption of uniformity in water content and hydraulic properties in connection with physi-cally-based equations. Crusting, layered materials, macropores and the mechanism of shrink/swell all contribute to heterogeneity which requires considerable caution in the interpretation of infiltration experimental data (Bristow & Savage, 1987). For example, Bonell & Williams (1986a) found that despite excellent agreement between measured and predicted cumulative infiltration-time plots using the Philip equation, the determined constants were empirical and not the physically-based "*S*" and "*A*" parameters of Philip (1969). They found that disruption to the soil surface fabric by soil fauna activities and raindrop compaction associated with high rainfall intensities were sufficient to destabilize conditions, and thus obviate any physical interpretation of the parameters "*S*" and "*A*". Bonell & Williams (1986a) determined, however, that cumulative infiltration (*I*), expressed in units of length (*l*), could be best described by Philip's (1969) "profile at infinity" condition, viz. $I = K*t$ (the product of field-saturated hydraulic conductivity and the time from ponding) in the circumstances of prevailing high rainfall intensities and hydraulic properties of that particular soil (Red Earth, Stace *et al.*, 1968).

Smiles & Knight (1976) described a simple graphical test to confirm the appropriateness of the Philip infiltration equation. They suggested that data routinely should be plotted with $I\,t^{-1/2}$ versus $t^{1/2}$ as shown in Fig. 13. If the plots are similar to either the "short" or "long" time situations, then the "*S*" and "*A*" parameters can be determined using simple regression analysis. Bristow & Savage (1987) presented an alternative, least squares method for determining the same parameters to avoid the problem of self-correlation between $t^{1/2}$ on both axes. The latter procedure was followed by Bonell & Williams (1986a), after it was found that their graphical plots closely resembled the long-time solution in Fig. 13.

As Mein (1980) noted, a "piston-flow" model is assumed in the case of the Green-Ampt model, with water displacing air in the soil voids. Crucial to this assumption is a sharp wetting front which separates an almost saturated surface layer from a lower unwetted zone (Skaggs & Khaleel, 1982; Wierda *et al.*, 1989). Such conditions are more suited to coarser rather than finer textured soils (Mein, 1980) and become less appropriate where extensive macropores exist. Davidson (1984), however, noted that an irregular wetting front does not invalidate the Green-Ampt approach, as it has been shown to work in a cracked soil.

Wierda *et al.* (1989) applied the Mein & Larson's (1973) extension of the Green-Ampt model in Côte d'Ivoire rain forest with reasonable success, but they did encounter problems. Corrections had to be made to $K*$ to accommodate an irregular wetting front. They also included an extra surface storage parameter to the Mein-Larson equation to accommodate a marked change in porosity with depth – resulting from biological activity. Methods of determining the $K*$ correction and the surface storage parameter were obtained from the infiltration rate-cumulative infiltration rate curves. Wierda *et al.* (1989) did, however, encounter problems with the Mein-Larson equa-

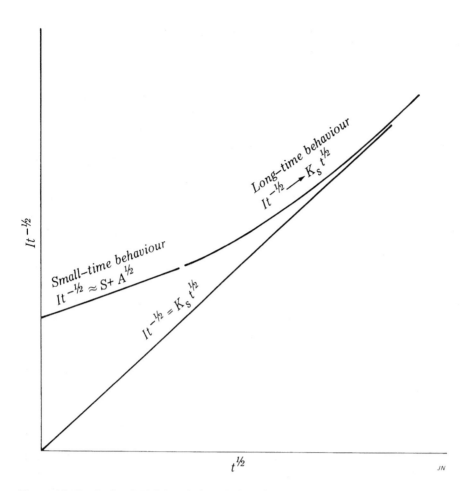

Figure 13: The Smiles & Knight (1976) analysis to determine sorptivity (S) and transmission (A) parameters from data measuring cumulative infiltration (I) into a uniform soil and time since ponding (t), where $I = St^{1/2} + At$ becomes $It^{-1/2} = S + At^{1/2}$. The intercept and slope of the initial linear portion are respectively S and A (After Smiles & Knight, 1976). The self-correlation of $t^{1/2}$ on both axes makes the procedure followed by Bristow & Savage (1987) more suitable but this figure provides a simple graphical test to confirm the appropriateness of the Philip equation.

tion where there was a gradual change in soil porosity with depth. This led them to highlight the need for an equivalent multilayered model. In fact, such a model had already been proposed elsewhere (Brakensiek & Rawls, 1983; Chu et al., 1986) which was later carefully evaluated by Silburn et al. (1990). A three-layer soil system was tested. An upper soil layer was subject to surface sealing in the top few mm's, which effectively created two layers. These horizons, in turn, overlayed an infinitely deep, homogeneous sub-layer. Inclusion of surface sealing is appropriate to cultivated, or degraded lands consisting of extensive bare soil areas.

Infiltration into the upper layer was described by the Green-Ampt equation in the form presented by Mein (1980). The surface seal was represented by the equations of Brakensiek & Rawls (1983), including provisions for a cumulative, rainfall-crusting energy term. The lower layer infiltration capacity was described by applying the constant hydraulic conductivity Green-Ampt equation for a second time. The energy-based, transient Green-Ampt model was found to adequately represent infiltration for the limited number of soils tested by Silburn et al. (1990). However, twelve parameters are required

for input to the model, of which only six can be directly measured. An additional four are obtained using a rainfall simulator and the remaining two from published values based on soil texture. The latter concerned the respective, average matric potentials of the wetting front for the upper and lower layers,. These values were taken from Mein (1980). Sensitivity analysis of parameters used in the model showed that in rank order, the initial and final surface seal hydraulic conductivities were the most affected, followed by upper layer volumetric moisture deficit and average matric potential of the wetting front . The sensitivity of the cumulative, rainfall-crusting energy parameter (and its method of determination) appeared less crucial. The model was applied in a distributed parameter runoff model ANSWERS, described by Beasley et al. (1980) (see Rose, this volume) and good predictions of runoff hydrographs were obtained for small catchments. Silburn et al. (1990) pertinently observed, however, that a major impediment to the widespread application of this modified Green-Ampt model was the dearth of in situ hydraulic conductivity data.

The recent improvements to the Green-Ampt make it a more practical proposition for use in the field compared to Philip's approach. The described modifications are a step forward in removing the dependence of a physically-based equation on "ideal soil" properties, a factor which has invalidated previous use of equations of this type in modelling field hydrology. However, caution should be exercised in the application of the modified Green-Ampt model in soils where significant subsurface stormflow occurs on steep slopes, thus violating the one-dimensional vertical flow assumption. Wierda *et al.* (1989, p. 379) were careful to check first that the possible occurence of subsurface stormflow did not explain the overestimation of cumulative infiltration in their study before they settled on the need for a K^* correction.

Empirical equations (e.g. Horton, 1940; Holton & Lopez, 1971), or use of the Philip equation treated empirically (because assumed boundary conditions are severely violated), provide the alternative approach when field conditions prevent the correct use of physically-based models. Such equations are of little predictive value and the determined parameters have no physical interpretation. They can, however, be useful as indices in comparing various treatment effects (e.g. conversion of forests, various agricultural tillage treatments) on soil physical conditions by performing ponded infiltration measurements in time and space. After the procedure of Smiles & Knight (1976) had shown the invalidity of interpreting the Philip parameters physically, Prove (1991) adopted the empirical equations of Kostiakov (1932) and Ostiachev (1936), for evaluating various sugar caneland tillage treatments. The determined coefficients proved useful indices for such comparisons. Elsewhere, Dixon (1976, p. 116) had highlighted several advantages of Kostiakov's equation in making it free from the usual restrictions identified with the physically-based, approximate models.

The future trend will be the increasing adoption of numerically based models (Skaggs & Khaleel, 1982, p. 155; Ross, 1990a; Ross & Bristow, 1990; Grigorjev & Iritz, 1991) for estimating infiltration based on the Richard's equation. Rapid developments in computer capabilities are making these models a more practical proposition. Such models, however, are still faced with the same problems of instability of soil water flow, especially near saturation; heterogeneity in both space and time and the demand for adequate *in situ* soil hydraulic property information. A recent exchange (Smith & Parlange, 1989; White & Broadbridge, 1989) suggests that the Smith and Parlange (1978) infiltration equation presents a more rigorous alternative (cf. Green & Ampt, Philip) for pre-ponding and post-ponding conditions, and should now be seriously considered for testing in the humid tropics. With the inclusion of a soil-averaged λ_c (macroscopic capillary length), White & Broadbridge (1989, in their equations 14 and 15) considered that the modified Smith & Parlange infiltration functions to be adequate for many field situations (e.g. infiltration under conditions which are ponded from commencement of infiltration, post-ponding variable rate rainfall infiltration), provided large depths of ponded water did not occur.

FIELD MEASUREMENT OF SOIL HYDRAULIC PROPERTIES

The preceding discussion has emphasized that one of the limitations to applying soil water movement theory under field conditions is limited knowledge of field soil hydraulic properties and limited knowledge of the spatial and temporal distribution of these properties. Such information is also fundamental to the understanding of the impact of land-use change. Changes in magnitude of the critical soil hydraulic parameters can indicate, for example, possible changes in preferred pathways of storm runoff, which, in turn, have ramifications on the erosion process and associated land management. Unfortunately, *in situ* hydraulic properties are poorly documented across the humid tropics (UNESCO, 1978, Lal, 1980) despite the urgency in understanding the impact of land-use change. There is, however, some detailed information related to small-scale catchment studies, for example, in East Africa (Cooper, 1979), northeast Queensland (Bonell *et al.*, 1983a, b; 1987), the Reserva Ducke, Amazonas (Nortcliff & Thornes, 1981) and in the western Amazonas (Elsenbeer & Cassel, 1990).

Recent improvements in field techniques now make it more possible to carry out replicated measurements. The field measurement of sorptivity and field-saturated hydraulic conductivity from infiltrometer experiments for use in the Philip infiltration equation owes much to the early contribution of Talsma (1969). A simple linear plot of I versus $t^{1/2}$ gives S, based on the assumption that the At term can be neglected in the initial 1–2 min of infiltration. This assumption emerged after the individual time limits for a range of soils had been calculated (Talsma, 1969, pp. 274–275). The method, however, can lead to markedly enhanced values of S in certain soils unless A is small (Collis-George, 1977 and confirmed in the study of Bonell & Williams, 1986a). Consequently, other analytical techniques have to be considered on the lines previously described (Smiles & Knight, 1976; Bristow & Savage, 1987). Dunin (1976) described in more detail the constant head permeameter method for measuring K^* used by Talsma (1969).This involves the careful extraction of an undisturbed core and testing it above ground level on a porous, horizontal surface. The method is particularly suitable for testing cohesive soils with high clay contents, and also is suitable for use on steep slopes.

The constant head permeameter method was widely used in the tropical rain forest studies in northeast Queensland (Bonell *et al.*, 1983b) where slopes are commonly steep and the soils cohesive. Even so, many samples could not be incorporated in the dataset because of suspected disruption to the soil fabric during insertion of the ring through the surface rootmat. An alternative is to insert infiltrometers in the ground, allow them to consolidate, followed by intermittent measurements over time. This method is only suited to areas of low relief, and where surface permeabilities are lower than in forests, e.g. disturbed areas. Under these circumstances, one-dimensional flow can be ensured before the wetting front emerges at the ring base (Bonell & Williams, 1986a). Prove (1991), however, extended the technique to sugar caneland with an overall slope of 12%, using an improvement of Talsma's method (Ross *et al.*, 1984). He found no significant differences in K^* determined from ring placement, either vertically or parallel to the soil surface. Therefore, subsequent measurements were made with the ring placed vertically, for ease of operation.

The more recent development of the disc permeameter (Perroux & White, 1988) makes possible the measurement of sorptivity, K^* and macroscopic capillary length, λ_c (White & Sully, 1987) with minimal soil disturbance. The method is also useful for measuring the hydraulic properties of field soils containing macropores and preferential flow paths appropriate to soil management studies, and should now be considered for humid tropical research. White *et al.* (1992) provided a comprehensive review of the various applications of the disc permeameter. They noted that the estimation of surface area of soil exposed by macropores can be achieved by taking sorptivity measurements at two different supply potentials (ψ), one ponded and the other unsaturated (summarized in equation 24 of White *et al.*, 1992). The method assumes that macropores fill quicker compared with the rate of sorption of water into the soil matrix and that the respective sorptivities are measured within the range of tension saturation. Elsewhere, Smettem & Ross (1992) used a combination of disc permeameters and ponded rings of dissimilar radii to determine the flow properties of unsaturated and saturated soil under two tillage treatments. The macropore-matrix dichotomy in both tillage treatments was characterized by an order of magnitude change in K^*, λ_m and λ_c as ψ moved over the narrow range, -30 mm to zero (Smettem & Ross, 1992, Table I and II). On the other hand, "S" only increased by 25% over the same range which was attributed to the increase in surface area for absorption provided by a few large cylindrical macropores. These writers also noted that hydraulic conductivity measured under a slight negative potential also provides a more appropriate matchpoint than K_s when using ψ (θ) to determine K (θ), or K (ψ) of field soils. Smettem & Ross (1992) also demonstrated the use of these field hydraulic properties in comparing "time to incipient ponding," using the models of Broadbridge & White (1988) and Ross (1990b) SWIM (Soil Water Infiltration and Movement).

Whilst the disc permeameter is a significant improvement over other methods which risk greater disturbance to the soil environment, such as the ring infiltrometer (Talsma, 1969), the new technique still has many limitations common to other approaches. These include the ramifications arising from simplifying assumptions of the analysis so as to obtain the required soil hydraulic properties, for example, the soil be uniform (structure, texture, moisture content) and non-swelling. Nonuniformity can give negative values of K^*, especially under conditions of early-time infiltration (White *et al.*, 1992).

There are also problems with steep slopes, freshly cultivated soils and the time taken to reach steady-state in heavier-textured soils. White *et al.* (1992) provide a constructive review of such limitations, but also itemize the positive features of the disc permeameter.

Good reviews of other recent developments in measuring *in situ* hydraulic properties (including $K(\theta)$, soil-water diffusivity ($l^2 \, t^{-1}$), effective mean pore values, indexing hydrophobicity) are provided in several sources (e.g. Clothier, 1988; Feddes *et al.*, 1988; Hendrickx, 1990; Smettem & C. Kirkby, 1990; White, 1988; Wallis *et al.*, 1991; White *et al.*, 1992). However, soil cores may still, be acceptable in the context of $K(\theta)$ linked with measuring evaporation. Under these conditions, Richard's equation behaves quite well. The paper by White (1988) is also a very useful up-dated review of analytical techniques especially pertaining to heterogeneity, (e.g. scaling, kriging) and highlighting the problem areas in the field measurement of soil-water properties.

An alternative method of determining soil hydraulic properties is the use of simple, empirical relationships proposed by Campbell (1974). He developed relationships for calculating unsaturated hydraulic conductivity, $K(\theta)$ directly from a moisture retention function based on the ψ (θ) curve and K_s. Such relationships were used for example, by Bristow & Williams (1987) in a sensitivity analysis of simulated infiltration to changes in hydraulic properties. Clapp & Hornberger (1978) used Campbell's relationship to calculate sorptivity and wetting front potential for use in Green & Ampt's infiltration model. Beven (1982a, b) also used Campbell's (1974) relationships to solve subsurface flow problems in hillslope hydrology where both K_s and separately θ decrease with depth. Talsma (1985) compared experimental $K(\theta)$ relationships with those calculated from modified Campbell equations for five laboratory soils and nine field soils. A total of eight out of the 14 soils showed good agreement between observed $K(\theta)$ relationships and those predicted from their water retention curves. The remaining comparisons were either inadequate (partly due to preferential flow paths) or only suitable for the wetter end of the range.

The application of field-measured soil hydraulic properties in calculating soil water flux in the unsaturated zone, using simple soil physics principles, was well demonstrated by Talsma & Gardner (1986a, b) in the context of a hillslope hydrology project. An important hydraulic property in the runoff generation process, however, is subsoil K^* which highlights the presence of any impeding layers to vertical percolation (Bonell *et al.*, 1983b; Elsenbeer & Cassel, 1990). Talsma & Hallam (1980) provided an improved field method based on the earlier "shallow-well pump-in" technique (Bouwer & Jackson, 1974, Boersma, 1965) for determining subsoil K^* of the unsaturated zone in the absence of a shallow water table. Known as the CHWP method (Constant Head Well Permeameter), it was originally developed in Australia by Talsma & Hallam (1980) and later refined by Canadian researchers (see review of Hendrickx, 1990). This method is particularly suited to the deeply weathered tropical soils where water tables are commonly at a considerable depth below the surface and the H/r ratio is ideally at least 10 (H is the wetted length of the auger hole, r is the hole radius). Later, Reynolds *et al.* (1983) proposed a numerical and analytical solution of the Laplace equation to correct the calculated K^* from the existing Glover solution. For example, when the H/r ratio is 10 the correction factor is 1.65 which results from the C value ratio in Table 1 of Reynolds *et al.* (1983) of 3.3 (Numerical) / 1.998 (Glover). This correction factor is not particularly sensitive to changes in the H/r ratio, and certainly not between 5 and 10 (Reynolds *et al.*, 1983).

Subsequently, there was exchange of views on the method concerning its apparent neglect of absorption effects on K^* beyond the bulb-shaped region of saturated soil (Philip, 1985, 1986b; Reynolds *et al.*, 1985). Such concerns were particularly influenced by low α_s values (10 to 0.1 m^{-1}) (in equation 28) being computed from laboratory experiments on repacked soils. Subsequent work connected with *in situ* soils in Australia determined much higher α_s values (Talsma, 1987; White & Sully, 1987). Comparisons of other field studies, both by Talsma (1987) and White & Sully (1987), noted no α_s values <1 m^{-1}. In fact, White & Sully (1987) went further by suggesting an α_s value in the order of 10 m^{-1} would not be unreasonable for a wide range of undisturbed soils. Consequently, overestimation in calculating K^* should not then be seriously in error as earlier suggested (Reynolds *et al.*, 1985; Philip, 1985). More significant is the problem of smearing or pore closure during augering of holes, which more than offsets any overestimates from capillary effects (Talsma, 1987). A useful comparison was made by Talsma (1987) between K^* determined by the CHWP and the standard pumping test known as the *auger hole method* (Bouwer & Jackson, 1974) for the same soil in the presence of a seasonal water table. He noted that the ratio of hydraulic conductivities (CHWP/auger hole method ratio) ranged between 0.40 and 0.75, so that K^* calculated from the CHWP underestimates the true K^* by a factor ranging between 1.3 to 2.5. This led Talsma (1987) to suggest an additional correction factor of 2 should be applied to accommodate auger hole smearing (after the Reynolds *et al.*, [1983] numerical correction has previously been made). Much wider testing in a variety of soil types is necessary, however, before a correction value for smearing can be applied with confidence in a particular environment.

Alternative techniques to counter the smearing problem have been put forward. Reynolds *et al.* (1983), for example, suggested the use of a reversed bevel cutting edge to dig the well and to ream the well with a large test tube brush. Elsewhere, Koppi & Geering (1986) suggested the use of a quick-setting epoxy resin to prepare an unsmeared soil surface. These writers had measured 2.5 to 6 times more rapid entry of water with an unsmeared soil surface. Talsma (1987) noted, however, that it would be difficult to apply the Koppi &

Geering (1986) method to deep subsoil.

In situations where soils, and their associated unsaturated zones, are deep (typical of some upper slopes in tropical rain forests) then the use of the CHWP is limited to application in the top 1 m. Furthermore, its small reservoir capacity means that determinations for wetted depths (H) exceeding 0.5 m is restricted to low K^*. Vertessy *et al.* (1991, pp. 23–24) describe the modification of an instrument originally designed by Bell & Schofield (1990) (known as the "Schofield" permeameter) to permit measurements to be made in deep soils of high permeability.

Bouwer & Jackson (1974), and earlier Boersma (1965), provided comprehensive reviews of both laboratory and field methods for measuring saturated hydraulic conductivity in the presence of a shallow water table. Recently, both Hendrickx (1990) and Jenssen (1990) provided succinct up-dated summaries of available techniques. Field methods are preferred and are particularly suitable for humid tropical areas with seasonally shallow water tables. The determined K^* can then be later compared with measurements made using the CHWP along the lines of Talsma (1987). Some care is required in using such techniques, however, when saturated zones are only transitory in the period immediately after storms. Bonell & Gilmour (1978) tried to use *the piezometer method* when a perched water table developed in the top 1 m from a short, intense storm in tropical rain forest. They found, however, that the hydraulic potential head was too unsteady (on the basis of adjacent piezometer responses) during the pumping test to confidently apply the method.

The combination of high rainfall, and fully wetted soil profiles for a large proportion of the year makes K^* a meaningful parameter in humid tropical environments. Bonell *et al.* (1981, 1983a, b) linked seasonal changes in short-term rainfall intensity to vertical changes in K^*, using the methods of Talsma (1969) and Talsma & Hallam (1980), to determine the runoff generation process across a spectrum of soils in northeast Queensland. The auger hole method was also used in the Babinda experimental catchments when the water table could be located at greater depths (Bonell & Gilmour, 1978). Ternan *et al.* (1987) followed a similar procedure in an agroforestry management study in Grenada, West Indies.

A representative sample size

Spatial heterogeneity in soil hydraulic properties can also be enhanced by the size of the sample tested. For example, Sisson & Wieringa (1981) found a substantial reduction in spatial variability upon increasing the diameter of infiltration rings. Similarly, Anderson & Bouma (1973) reported a similar reduction in the variability of K^* upon increasing the length equivalent of soil core. The concept of the *representative elementary volume* (REV) (Bear, 1979) has been discussed by several authors (e.g. Youngs, 1983; Bouma, 1983; Baveye & Sposito, 1984; Williams & Bonell, 1988; Jenssen, 1990). A summary of the concept is presented in Fig. 14.

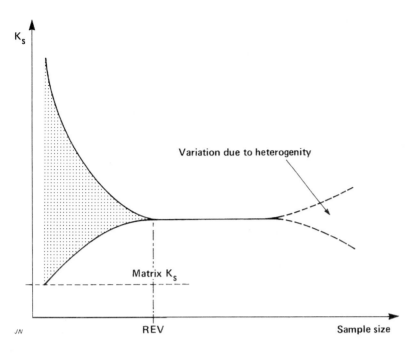

Figure 14: Suggested relationship between sample size and the saturated hydraulic conductivity (K_s) (expressed in mm h^{-1}) (After Jenssen, 1990).

Considerable variability occurs with small sample sizes, and the magnitude of variability also depends on whether they are measuring soil matrix or macropore flow (Smettem & Collis-George, 1985). As the sample size increases, variability decreases until the REV is attained. From thereon, K^* or K_s should not vary with increasing sample size unless the material is inherently heterogeneous.

Bouma (1983) suggested hypothetical REV's for different soils. In superficial deposits with extensive fissures, large-scale field tests (characteristic length >50 cm, sampling volume $\geq 10^6$ cm^3) may be required (Jenssen, 1990). When using various borehole techniques above and below the water table, Jenssen (1990) suggested a minimum characteristic length of 25 to 50 cm ($10^4 - 10^5$ cm^3).

The repetitive unit of Bear (1979) is also linked with the REV in terms of the length scale of observation. Williams & Bonell (1988) compared the "A" and K^* parameters of Philip (1969) between 0.07 m^2 (0.3 m dia) infiltrometer rings permanently located in the ground with the same parameters determined from adjacent to 250 m^2 unbounded, runoff plots (discussed in a later section). They found that the spatial and temporal variability of ring estimates for A and K^* parameters were approximately 4 to 10 times greater than for the large plots. The rings would need to be 1–2 m in diameter to reduce spatial variability to the estimated repetitive unit. Nevertheless, they found that the rings located in bare soil between grass tussocks gave estimates of "A" and K^* which were approximately twice that of the plots as against some six or seven times for the grass tussocks. This suggested that the bare soil was dominating the overland flow hydrograph during the experiment. The fact that the bare ring and plot estimates

were in the same order of magnitude was encouraging and also indicated the need for hydraulic property measurements to be determined from large areas following the repetitive unit concept.

Other writers (e.g. Dagen, 1986; Gelhar, 1986) have inferred doubts about the generality of REV. Dagen (1986, p. 125S), for example, put forward some limitations of the concept. These included the ill-defining of REV near boundaries of contrasting media (e.g. pervious, impervious) and quantifying the "averaging volume," V_{REV} when the "space average" of the variable of interest, e.g. K^*, loses its dependence on V_{REV}. An alternative statistical view of sample size (that is, a stochastic model) was presented by Dagan (1986). In this approach, the sample volume becomes representative when the measured quantity and the ensemble average are interchangeable. White (1988, pp. 75–76), however, outlined the difficulties of field testing stochastic models generally. One problem is the impractical number of field measurements demanded, for example, by the approach of Dagen (1986).

Providing a reliable framework for field measurement is thus a difficult one. Essentially, the selection of the appropriate methodology depends on the scale and type (e.g. unsaturated or saturated flow) of the questions being addressed, and the conceptual framework used to tackle such questions. A common assumption is that the selected measurement scale is much larger than heterogeneities within the sample, but for practical reasons much smaller than the scale of interest. An example is the reliance on "small, bounded" runoff plots to sample a hillslope in runoff generation. Such an approach is required in stochastic models (e.g. Dagan, 1986). Less commonly, the sample size is directed towards the scale of interest such as catchment (e.g. Burch *et al.*, 1987). When considering unsaturated or saturated flow, there is evidence (e.g. Perroux & White, 1988) to subscribe to the notion that the scale of measurement has to increase when moving from the unsaturated to the saturated phase. For example, at ponded conditions the large spatial variability of surface soil hydraulic properties can be attributed to preferential flow in macropores. Such variability (and preferential flow participation) should decrease as the matric potential (ψ) becomes more negative. Consequently, under humid temperate conditions of rainfall infiltration (where ponded conditions occur less frequently), measurement techniques have been devised for determining soil hydraulic properties at $\psi < 0$ (negative phase) (White *et al.*, 1992). In contrast, the higher prevailing rainfall intensities associated with the humid tropics encourage ψ to approach zero more frequently, especially arising from various management practices. Techniques associated with ponded conditions and associated larger sample sizes should be considered. *The comments of Youngs (1983) are very pertinent for the humid tropics when he suggested that more attention should be given to measurements of "bulk properties of the whole system." Analysis of large unbounded plots or small catchments is one method of doing this.*

GROUNDWATER

Groundwater is an important component of the hydrological cycle because it integrates over relatively large spatial and long temporal scales, and, therefore, can buffer the hydrological system against rapid change. For example, consistent isotopic compositions of various streams at low flows noted by Pearce (1990) indicates extremely well-mixed groundwater outflow which is buffered against large fluctuations in rainfall isotopic composition. There are fundamental differences in groundwater behaviour, specifically in the location and timing of recharge and discharge under humid and sub-humid conditions. Groundwater simulation models theoretically are well-developed. However, existing groundwater models are often complex and have substantial data requirements. There is a need to develop more generic groundwater models capable of simulating:

(1) changes in groundwater recharge caused by climatic or vegetation changes in the recharge zone;
(2) changes in long-term groundwater storages;
(3) groundwater transfers along preferred pathways; and
(4) changes in human use of groundwater relative to scenarios of future climatic or social changes.

At the continental scale, groundwater movement is controlled by lithological and structural factors (e.g. rock porosity and permeability, the presence of fissures and faults), and by subsurface "topography" (which may bear little relation to the surface topography). Groundwater recharge and discharge are controlled primarily by climatic factors and the nature of the vegetation cover, although topography and the geomorphic structure of the land surface influence the size and location of the recharge/discharge areas.

Natural recharge of groundwater may occur by precipitation, or from rivers, canals and lakes. Practically all the water from these sources is of meteoric origin. What is known as juvenile water (of volcanic, magmatic and cosmic origin) only exceptionally contributes to the groundwater recharge.

The process of recharge is very complicated. Perhaps one of the most important factors is the time delay between the time when the meteoric water enters the soil profile, and the time when it is manifested as an effectively exploitable groundwater source.

In principle, the following types of natural groundwater recharge can be recognized:

(1) Short-term recharge, which may occur occasionally after a heavy rainfall, mainly in regions without marked wet and dry seasons.
(2) Seasonal recharge, which usually occurs regularly, e.g. during the wet period. Occasionally, when there has been an unfavourable development of soil moisture, the recharge may be poor or may not occur at all.
(3) Perennial recharge, which may occur in tropical humid regions with an almost permanent downward flow of water.
(4) Historical recharge, which occurred a long time ago and contributed to the formation of the present groundwater resource. This phenomenon is closely linked with what is known as groundwater residence time. The residence time is defined as the time which has elapsed between the time at which a given volume of water was recharged and the time when it reaches the groundwater table. Sometimes the time taken by a given volume of water to be transformed into baseflow is decisive for the assessment of the residence time.

A knowledge of how the groundwater regime behaves under given environmental conditions is essential for further groundwater resource development. Any groundwater surveys, data collection, analysis, experimental measurements, simulation methods and other means of the systems approach should always be performed with respect to the practical objectives in the given aquatic ecosystem. Thus an assessment of groundwater availability extends far beyond the stage of data collection. It also involves an analysis of groundwater flow and its interaction with the stream network, an assessment of the hydrological balance, and of the residence time. However, it also includes forecasting the future use of water resources based on various exploitation alternatives. The economical and social aspects must be taken into account as well as the technical and hydrological factors (Balek, 1989). The evaluation of future demands should be based upon the results of a comprehensive survey among the users. Such an approach will vary from region to region, and, thus, the criteria involved cannot be blueprinted for a great number of countries. Also, approaches can be expected to differ in humid and arid regions. In a humid region, flood control may have priority in water planning, and this may lead to the conclusion that there is always abundant groundwater – or because surface waters are available, the groundwater is considered to be a less significant source. However, in these regions, groundwater is often the only source of supply during a long or short dry period, or when surface water becomes heavily polluted and its treatment expensive. Groundwater is always a reasonably source of safe water supply for rural areas and for domestic use (Wurzel, this volume). Groundwater becomes a more significant source in the wet-dry tropical region of Chang & Lau (this volume) where it is often the only reliable water resource available throughout the year. In such areas, even a costly, large-scale survey which includes aerial photography, a field geological survey, groundwater data collection and evaluation, a geophysical investigation and isotope studies is usually profitable.

When trying to establish the critical amount which can be extracted from aquifers as part of renewable resources, the use of existing observations is essential. Only after they have been evaluated can newly established observational programmes be planned together with other survey methods to fill the gaps in existing data and identify the results.

The immediate challenges within the field of groundwater management and development (Balek, 1989) can be identified as follows:

(a) to provide a progressive build-up of knowledge of groundwater conditions in the area,

(b) to provide a reliable background for water resource master planning and the selection of the most promising areas for groundwater resource development,

(c) to optimize the number of wells, drilling depth, and well construction,

(d) to protect groundwater from pollution,

(e) to control groundwater quality,and

(f) to improve the management of groundwater resources.

RUNOFF GENERATION PROCESSES

As Kirkby (1988) noted, greater than 95% of the water in streamflow has passed over or through a hillside and its regolith before reaching the channel network. There are various flow mechanisms whereby net precipitation can be transferred downslope during or immediately after storms. These mechanisms have already been introduced and include *infiltration-excess overland flow* (or Hortonian overland flow), *saturation overland flow* (or saturation-excess overland flow) and *subsurface stormflow*. An additional mechanism is *return flow* (defined by Kirkby, 1978, p. 371) when subsurface stormflow is constrained to flow out of the soil as exfiltration, contributing to overland flow.

This section will initially review the variable source area concept of runoff generation and assess the various delivery mechanisms from both hydrometric and environmental isotope studies that were based on humid temperate experience. Later,

focus will be on the limited number of studies that have been undertaken in the humid tropics. Emphasis will be given to any shift in the delicate balance between rainfall intensity-soil hydraulic properties-topography in causing more wide-ranging hillslope runoff responses across the humid tropics compared with humid temperate areas. This will include a detailed comparison between two hillslope hydrology studies in northeast Queensland and western Amazonia. A similar delivery mechanism occurs at both sites, but the dominant controls are different, i.e. high prevailing rainfall intensities at one site as against soil hydraulic properties expressed in the form of a very shallow impeding layer at the other.

The variable source area concept

Prior to 1961, the simplistic infiltration-excess model of Horton (1933) prevailed in the literature. Excess rain water which was unable to infiltrate into the soil was considered the sole source of quickflow (defined by Kirkby, 1978, p. 371) and such overland flow emanated from widespread areas of a drainage basin. Water which infiltrated during storms and subsequently recharged the water table was the sole source of delayed flow (defined by Kirkby, 1978, p. 366). Horton's model reinforced the unit hydrograph theory of Sherman (1932), which delayed the introduction of other ideas until the variable source area model was introduced by Hewlett (1961) and later expanded by Hewlett & Hibbert (1967). The controversial history of this conceptual model's development was discussed by Hewlett (1974) and Ward (1975). The model is based on the assumption that the area of a drainage basin contributing to storm runoff is not fixed but rather is dynamic. This contributing area can vary in size between storms and during the course of an individual storm.

The initiative for this concept was in recognition that the high surface infiltration capacities associated with undisturbed humid temperate areas are rarely exceeded by the prevailing rainfall intensities, which results in the absence of Hortonian overland flow. Such developments initiated a considerable number of field experiments from the 1960s onward with particular emphasis on humid temperate environments. Dunne (1978, 1983) provided a comprehensive review of results from the early field studies, with the early ramifications for modelling being summarized by Freeze (1972) and later Freeze (1978), amongst others, in the benchmark publication of Kirkby, (ed.) (1978). By the late 1970s, acceptance of the variable source area concept was universal, but the momentum of research in hillslope hydrology has continued as before because many details concerning process mechanisms (or delivery mechanisms) of storm runoff generation are still not fully understood in headwaters of drainage basins (Dunne, 1983). Such work is a requirement for improving physically-based models incorporating runoff generation (see papers in the edited publications of Anderson & Burt, 1985a, 1990).

At this stage, it would be appropriate to summarize some of the experiences from humid temperate areas. By the early

1970s, it became apparent that the variable source area model could not be generalized on the basis of a single delivery mechanism into streams. Freeze (1974) summarized three versions of Hewlett's general model viz.:

(a) *variable source area-saturation overland flow*, whereby the source areas are located adjacent to stream channels which expand and contract according to hydrometeorological factors. These source areas or riparian areas are represented by a shallow water table co-axial with a stream channel. During storms, the water table rises to the surface and delivers storm runoff from a combination of exfiltrating subsurface stormflow (Hewlett, 1974) or return flow (Dunne & Black, 1970) from upslope (Hewlett & Troendle, 1975), and direct precipitation onto the saturated area.

(b) *partial area-overland flow*, proposed by Betson (1964) and Betson & Marius (1969), whereby runoff is generated from fixed source areas, e.g. shallow soils, as opposed to variable ones.

(c) *variable source area-subsurface stormflow*, whereby the delivery mechanism to the stream relates to an expanding and contracting subsurface, saturated wedge located usually above a relatively impermeable subsoil horizon (e.g. Weyman, 1973). Essentially, this was the original model proposed by Hewlett and co-workers (Hewlett, 1961; Hewlett & Hibbert, 1967), which also took into account the effect of channel interception on the storm discharge hydrograph.

When considering certainly models (a) and (c), Burt's (1989, p. 12) view that "... subsurface stormflow is now viewed as *the* major runoff-generating mechanism, both because of its influence on saturation-excess overland flow (e.g. Dunne & Black, 1970), and as an important contributor to stormflow in its own right (Anderson & Burt, 1978) ..." is appropriate to relatively undisturbed humid temperate environments.

Delivery mechanisms

Work has continued to evaluate the delivery mechanisms at different scales of investigation,with recent reviews being presented by Kirkby (1988) and Burt (1989).

Infiltration and water transfer through the unsaturated zone
Previous consideration has been given to infiltration and the role of macropore/matrix flows. Within the unsaturated zone, flow vectors are normally vertical (Kirkby, 1988) and for any lateral flux, soils must exhibit considerable anisotropy. The major role of macropores, e.g. root holes, is to allow vertical bypassing of the unsaturated matrix and reach the saturated zone more quickly than through the unsaturated soil matrix.

Saturated lateral subsurface stormflow Zaslavsky & Rogowski (1969), and Zaslavsky & Sinai (1981), considered the effects of soil anisotropy and soil layering in generating lateral subsurface stormflow. Downslope flow predominates, especially in the

regolith, with the hydraulic gradient assumed to be parallel, and, therefore, sensitive to the topographic gradient.

The means by which subsurface stormflow is delivered to streams can be viewed as routing through the soil matrix (matrix flow), macropore flow and pipeflow. The size of voids and the scale of porous media controls whether the lateral flow can be described by Darcian flow or not. A previous section (water movement in the unsaturated zone) has already introduced their role in hillslope hydrology. It should be added, however, that macropores are particularly effective when rainfall intensities exceed the matrix or pedal hydraulic conductivity. Studies by Germann (1986), and others reviewed by Burt (1989), demonstrated that at lower rainfall intensities, no macropore flow occurs and is appropriate when depth of wetting is less than the depth to an impermeable horizon. The exception is when the matric potentials are close to zero when no more water can be stored in the solum peds so that by-pass flow is produced. Another threshold identified was antecedent soil moisture. If the soil is too dry, then flow in the macropores is absorbed into the soil matrix along the lines described by Kirkby (1988) and Feddes *et al.* (1988), and no by-pass flow occurs. A volumetric water content (θ) of at least 0.30 was put forward as the threshold by Germann (1986) for by-pass flow to occur, but this is dependent on the prevailing size of pores.

As earlier highlighted, the connectivity of macropores in the downslope direction still remains unresolved (Kirkby, 1988), to apportion the significance of such voids to rapidly conduct the water down the slope. Hewlett & Hibbert (1967) originally envisaged that the sideslope delivery of moisture to a water table at the foot of the slope was by displacement of pre-existing soil water through the soil matrix (viz. translatory flow), based on the work of J. H. Horton & Hawkins (1965). Other writers have termed similar recharge mechanisms as "piston-type flow" (Goel *et al.*, 1977) and "interstitial piston flow" (Foster & Smith-Carrington, 1980). However, the occurrence of lateral subsurface stormflow via matrix flow was considered too slow to produce significant volumes of quickflow. More recently, a "groundwater ridging" mechanism (Ragan, 1968; Sklash & Farvolden, 1979; Abdul & Gillham, 1984; Gillham, 1984) has been described in areas adjacent to streams where the capillary fringe or tension-saturated zone above the water table is close to the surface. Such conditions should be more prevalent in finer-textured soils. Under these circumstances, only small amounts of rain are required to convert small negative matric potentials into positive pressures. This causes the water table to rapidly rise and, in turn, steepen the hydraulic gradient. If the K^* of the soil is sufficiently large, then significant volumes of subsurface stormflow will be discharged through the banks into the stream – most of which will be "pre-existing" soil water. Despite some field evidence (Abdul & Gillham, 1989) for this water table ridging mechanism, further experimentation is necessary to verify the generality of this process. Field retention data, for example, have failed to observe significant zones of tension-saturation even in clays and loams (Perroux *et al.*, 1982, Figs. 3 and

4). Clothier & Wooding (1983) even question its existence in laboratory media, despite the idea of soil having a tension-saturated zone emanating from earlier laboratory experiments (Clothier, 1988 pers. comm. to McDonnell). In contrast, Germann (1990) noted the role of the capillary fringe mechanism in the use of kinematic wave models for both vertical and lateral components of macropore flow in hillslopes. Flow along preferred paths connected with water stored in highly saturated capillary fringes was indicated, but Germann (1990, pp. 357–358) appropriately observed that "... it is not clear yet to what degree water is indeed flowing through macropore systems at volume flux densities suggested by the (kinematic) model or whether energy waves simply push water out from the vicinity of the drainage face and water flow trails way behind the energy wave." These issues are currently being examined as part of a stable isotope and hydrogeochemical study in the Babinda catchments in northeast Queensland.

The continuum of infiltration-excess overland flow, saturation overland flow, and subsurface stormflow Modelling work by Smith & Hebbert (1983) supported the idea that infiltration-excess (Hortonian) overland flow, saturation overland flow, along with subsurface stormflow, can be three parts of a continuum in hydrological response within the same drainage basin when layered soils exist. For example, the two surface runoff generating mechanisms are not mutually exclusive in their spatial distribution so that both mechanisms can occur at different times on certain sections of a slope. These depend on changes in storm rainfall intensities in relation to $K*$ of a layered soil (Smith & Hebbert, 1983, p. 994). Smith & Hebbert (1983) also demonstrated the important role of vertical movement towards the surface of the perched water table in terms of its influence on quickflow as against the draining of lateral, perched aquifer flow via subsurface stormflow. Their modelling results established that the contribution to the flood hydrograph from saturation overland flow emanating from a small saturated area (primarily in the near stream region) was far more important than the draining of the perched aquifer within an entire hillslope. Kirkby (1988) was also of the view that subsurface flows were too delayed to contribute to peak flows.

The role of topography

Hewlett's conceptual model acknowledged the role of topography in determining the location of variable source areas through downslope movement of moisture. Other work has emphasized both the two-dimensional (Freeze, 1972) and three-dimensional (Anderson & Burt, 1978) role of topography such as hillslope hollows or convergent headwater areas in encouraging the convergence of lateral soil-water movement and resulting subsurface stormflow/saturation overland flow. Sideslopes may, however, cover a larger area of a catchment. Thus they can still dominate the discharge hydrograph by producing more runoff per unit area, especially if the slopes are steep (Beven *et al.*, 1988).

Recent developments based on digital terrain models (O'Loughlin, 1986; Beven *et al.,* 1988) have demonstrated that spatial variations in soil moisture, and, therefore, in turn, runoff producing areas are driven by topographic gradients. It is clear that morphological controls on runoff generation are more significant than previously suggested by Dunne (1978) who emphasized climate and soils as being the dominant controls. Topography was previously considered an important secondary control at a subcatchment scale (Dunne, 1978).

Summarizing the humid temperate experience from hydrometric studies

The humid temperate experience in *undisturbed* environments is then one of slowly expanding and contracting source areas, with subsurface stormflow dominating hillslope responses unless the soil becomes saturated to the surface, especially at the slope base. Such a description relates to environments where surface infiltration capacities are high compared with precipitation rates which are small in magnitude by tropical standards. Studies in western Europe (e.g. Weyman, 1973; Bonell *et al.*, 1984b; Bevan, 1986a; Quinn *et al.*, 1989), New Zealand (Pearce *et al.*, 1984, 1986; Taylor & Pearce, 1982) and in the Coweeta catchments of the United States (Swift *et al.*, 1988) quote typical short-term, rainfall intensities below 10 mm h^{-1}. This caused Hewlett *et al.* (1977; 1984) to strongly question the value of rainfall intensity as a predictor of stormflow based on statistical analyses.

In disturbed environments where the surface $K*$ has been much reduced in magnitude by compaction, such as by the use of agricultural machinery or intensive grazing by farm animals, then infiltration excess (Hortonian) overland flow can occur even though prevailing rain rates may be small by temperate (and tropical) standards.

Experience using environmental isotopes in humid temperate areas

The preceding discussion has concentrated on findings from hydrometric studies. The underlying assumption is that quickflow and delayed flow or more recently known as slowflow (as defined by Kirkby, 1978; Jakeman *et al.*, 1990; Robson & Neal, 1991) is, respectively, "new (or event)" and "old" (or pre-event) (defined by Rodhe, 1987) water emanating from saturation/Hortonian overland flow and deeper interflow/groundwater discharge, respectively. In addition, macropores and pipes were considered to be the main arteries for delivering predominantly "new" water via shallow, subsurface stormflow. Recent use of environmental isotopes in humid temperate areas have presented results contrary to the above, with "old" water surprisingly dominating storm hydrographs (see detailed reviews by Rodhe, 1987, pp. 33–41; Sklash, 1990). Even in the very wet areas, such as the west coast of New Zealand (Miamai catchments), where storm hydrographs are very responsive with high quickflow : gross precipitation ratios; between 75–97% of storm hydrographs consisted of "old" water (Pearce *et al.*, 1986; Sklash *et al.*, 1986). Bonell *et al.* (1990) reported some of the highest storm volumes of "new" water in another humid temperate area of New Zealand (Otago) where annual rainfall was only half of that in the west coast study. Topographic controls were isolated as the main cause for such differences, with the Otago study having more extensive "wetlands" arising from concave slopes – which were capable of producing greater volumes of saturation overland flow during the larger storms. Nevertheless, the stream hydrographs of smaller storms consisted of "old" water only. Other recent work in the UK using environmental isotopes (Ogunkoya & Jenkins, 1991: Cairngorm, Scotland; M. Robinson, 1991, pers. comm.: Holderness, eastern England) and more conservative chemical tracers, as well as Oxygen-18 (Robson & Neal, 1990, 1991; Robson *et al.*, 1992: central Wales) have continued to indicate the domination of storm hydrographs by pre-event water. Robson and Neal (1991) noted that for *moderate* or *small* rain events the "new" water component of storm hydrographs could be neglected because the rainfall signature has no relationship with the stream chemistry on an event basis. The "quickflow" component was considered to have a mixed composition of acidic "soil water" as one endmember, and chemically well-buffered "deep water" (greater than 1 m depth) as the other endmember. The latter was associated with the slow flow component through some displacement mechanism. The chemical hydrograph separation between soil water and deep water was undertaken using the conservative tracer, stream Acid Neutralisaiton Capacity, ANC (defined in Robson & Neal, 1990; Robson *et al.*, 1992). The corresponding predictions of stream ANC using TOPMODEL (discussed later) were good, except during the protracted hydrograph recession when a significant divergence between actual and predicted ANC emerged (Robson *et al.*, 1992). At such times no allowance had been made for mixing of the well-buffered, deep water stores with the upper acidic soil waters towards the end of prolonged rainfall. More pertinent was the need to reconcile the water chemistry of the stream with the substantial contributions of flow prediced by TOPMODEL that emerge from the saturated areas. The chemical composition of this runoff source was unknown, but the flow dynamics indicated that it should be a combination of rainwater and well-mixed subsurface waters. Reference to the preceding stream chemistry suggested that the direct contribution of rain, i.e. the "new" water component, is small. Therefore flow from the saturated contributing areas probably had a well-mixed composition of soil and deep water.

There are, however, legitimate concerns about the validity of the assumptions implicit in the mass balance method used for hydrograph separation (for detailed discussions concerning this important issue, see DeWalle *et al.*, 1988, 1990; Genereux & Hemond, 1990; Kennedy *et al.*, 1986; Littlewood & Jakeman, 1991; Ogunkoya & Jenkins, 1991). Amongst the more recent contributions, Bonell *et al.* (1990) list and

appraise the assumptions. The most critical is that storm runoff can be partitioned into only two-end members or reservoirs, viz. event (rainfall) and pre-event (groundwater). Depending on the environment, a major weakness can be the neglect of the soil water (unsaturated zone) isotopic composition which can be significantly different from groundwater (i.e. not similar as assumed), or the contribution of the unsaturated zone to pre-event water can be more significant than previously envisaged (DeWalle *et al.,* 1988, 1990; Kennedy *et al.*, 1986). McDonnell *et al.* (1991) evaluated both the traditional two-end member and three-component model of DeWalle *et al.* (1988). Whilst the differences in groundwater only and combined groundwater plus soil water contributions to storm hydrographs did not differ considerably, they concluded that if sufficient water samples are taken, both in space and time, then an alternative time series approach for flow component identification may be preferred (e.g. Turner & MacPherson, 1990; Stewart & McDonnell, 1991). Also, the concentration in rainfall varies through storms rather than a random, temporal variability around the weighted mean (obtained from a bulk sample) as was shown by Bonell *et al.* (1990). An alternative is the use of the "incrementally adjusted weighted mean concentration" method of McDonnell *et al.* (1990) which allows for the continuous adjustment of rainfall isotopic concentration, and was used in the humid temperate studies of Bonell *et al.* (1990) and Ogunkoya & Jenkins (1991). The main advantage of the technique is that later rain in a storm event does not influence the analysis for the period up to the time of streamwater sampling.

A favoured explanation for the domination of storm hydrographs by "old" water is the "groundwater ridging hypothesis" (Ragan, 1968) adopted by Sklash & Farvolden (1979), but as previously indicated, this mechanism still requires more extensive validation under field conditions. Even more important, there is a need for additional studies on the lines followed by McDonnell (1989) to reconcile hydrometric hillslope studies with results obtained from using environment tracers. McDonnell (1989, 1990), working in the same catchments (Maimai) as Pearce *et al.* (1986), established that the input water (rainwater) isotopic signatures attained an "old" water status very quickly, due to the large soil water store relative to rainfall input for individual storms (McDonnell, 1990; p. 2828; Ogunkoya & Jenkins, 1991, p. 280) and the mixing with a near-saturated soil matrix along crack and pipe walls. Consequently, water emanating from macropores and pipes was dominated by "old" water, although the process for isotopic signature exchange between the pipe/matrix water could not be fully explained. McDonnell (1990, pp. 2829–2830) put forward a conceptual, crack-pipe model appropriate to various upslope zones which could explain the contribution of substantial volumes of "old" water to the stream hydrograph. A typical upslope zone source of "old" water was from pipes in hollow zones (small zero-order basins), in addition to near-stream zones of first-order channels. The mechanisms for connecting these sources were also described (McDonnell, 1990, p. 2829). Significantly, McDonnell (1989, 1990) failed to establish that the soils in the Maimai catchments exhibited capillary fringe characteristics cited elsewhere as an integral part of groundwater ridging (Abdul & Gilham, 1984, 1989). The prevailing low matric potentials (ψ), however, proved highly sensitive to rainfall inputs because of a break in the capillary continuity by macropores, thus limiting matrix drainage (Clothier, pers. comm. 1988 to McDonnell) (known as "the limited storage effect"). When rainfall inputs exceed the soil matrix hydraulic conductivity, microscale ponding occurs, leading to macropore flow filling up "the limited storage effect". Soil water overlying the near-stream groundwater is, therefore, rapidly converted to positive ψ which causes a rapid, vertical water table response (similar to the description of Smith & Hebbert, 1983, p. 998). McDonnell (1990, p. 2829) then considered that the resident groundwater begins to discharge into the channel, "assisted by groundwater ridging (or mounds) along the channel margins," thus displacing soil water into the valley bottom channel. The latter subsurface near-stream response is in addition to valley floor saturation overland flow and on-channel precipitation. The combination of the described upslope and valley bottom zone mechanisms resulted in hillslope flow being dominated by "old" water contributions, ranging between 70 and 100%, with the proportion of "old" water grading to 100% as soil moisture moved downslope (McDonnell *et al.*, 1991; Stewart & McDonnell, 1991). The smaller "new" water contributions emanated from the narrow flood plain area and headwater subsurface flow. These results explained why "old" water dominated the storm hydrographs in the Miamai catchments (McDonnell, 1989).

Elsewhere, work in the UK (Ogunkoya & Jenkins, 1991; M. Robinson, 1991, pers. comm.) has observed that as rain intensities increase under dry antecedent conditions, the proportion of event water in storm hydrographs increases, probably via preferential pathways. The difficulties of ascribing the appropriate mechanism for the continued release of substantial volumes of "old" water (54% in Ogunkoya & Jenkins, 1991) still remains. The common reasoning that the equivalent depth of rain stored in wet soils, relative to the much smaller input from individual rain events, accounts for the observed "old" water tracer signal in streams becomes open to question under dry conditions. The role of surviving, more localized "wetter" sources might, however, contribute to such patterns.

The dilemma of hydrograph separation

The preceding section has highlighted the arbitrary nature of various hydrograph separation procedures into "quickflow" (or "storm runoff") and "delayed flow" (or "baseflow" or "slow flow") (described, for example, in Linsley *et al.*, 1988; Ward & Robinson, 1990). Furthermore, the use of environmental isotopes immediately dismisses the simplistic assumption (emanating from Horton's infiltration-excess model) that "storm runoff" is event or "new" water. Therefore, attaching

any physical interpretation to the separated hydrograph components is totally meaningless.

As noted, however, the use of conservative chemical tracers in hydrograph separation are not devoid of their own problems in terms of the weaknesses in the widely-used two-component mass balance model. Littlewood & Jakeman (1991), for example, noted disparities in the proportions of slow flow and "deep water" at peak streamflows within the same catchment between the use of their time series technique (Simple Refined Instrumental Variable, SRIV, described in Jakeman *et al.*, 1990) and earlier work of Robson & Neal (1990) based on the mass balance model. The main contribution from environmental isotopic studies is that substantial proportions of storm hydrographs consist of pre-event water, at least within the framework of antecedent catchment moisture conditions and prevailing rain intensities experienced in humid temperate latitudes. At the time of completing this review (December 1991), there are no corresponding publications pertaining to the humid tropics to indicate similar conclusions. Early results from northeast Queensland however, indicate that under certain hydrological conditions substantial volumes of event or "new" water can contribute to the storm hydrograph.

The challenge to the hydrograph separation problem is an understanding of environmental tracer data in terms of linkages with the physics of hillslope hydrology. Recently, Beven (1991) provided an illuminating critique of the current methods of hydrograph separation (e.g. traditional methods, time series analysis techniques, using environmental tracers). It is worth citing his conclusions because they are well-stated. In connection with the domination of hydrographs by pre-event water, Beven (1991, p. 3.6) noted that this component "... must be the result of subsurface displacement mechanisms caused by rapid pressure wave propagation (at least within a Darcian framework). What is this old water? Stormflow, or baseflow or both? ..." and went on "... I would suggest that no physical interpretation or naming of processes should be entertained." Beven further asserts that "... the temptation to equate the slow component (of hydrograph separation) with groundwater and the fast component with surface runoff, and go on to do chemical mixing for water quality calculations must be resisted as revisiting the interpretational sins of the past."

Hillslope hydrology studies in the humid tropics

In comparison with the humid temperate areas, only a limited number of field experiments have been undertaken in the humid tropics (Dunne, 1983; Walsh, 1980). Also the intensity of investigation and experimental methodology in deducing the runoff process shows considerable variation. The studies in the Reserva Ducke, Amazonas (Nortcliff & Thornes, 1981), in the Babinda catchments of northeast Queensland (Gilmour *et al.*, 1980; Bonell *et al.*, 1981) and in western Amazonas, Peru (Elsenbeer & Cassel, 1990) represent some of the more intensively monitored experiments. Other investigations have evaluated runoff sources and processes from stream hydrographs

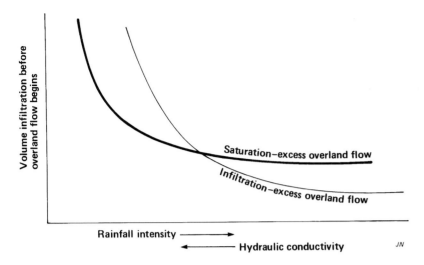

Figure15: The relationship between the volume of rainfall which infiltrates before overland flow begins and rainfall intensity. Soils with low hydraulic conductivity will be dominated by infiltration-excess overland flow; those with high hydraulic conductivities by saturation-excess overland flow (After Kirkby, 1978).

and hydrochemical studies (e.g. Bruijnzeel, 1983) and deductions of the runoff process from the determination of K^* for different soil horizons (e.g. Ternan *et al.*, 1987).

Most field experiments have been connected with native tropical forests (see reviews by Walsh, 1980; Bruijnzeel, 1990), but some work has also been undertaken in plantations (Bruijnzeel, 1983). Agricultural areas have also received some attention, concentrating on the impact of forest conversion and the effects of various agricultural treatments (e.g. Lundgren, 1980; Lal, 1981; 1983; Othieno, 1979; Prove *et al.*, 1986; Prove, 1991). Such investigations have commonly monitored small basins (<4 ha) or slope sections as part of runoff-erosion studies. Recently, Lal (1990) reviewed the limited knowledge on runoff-erosion connected with agroforestry.

When dealing with the humid tropics, some differences in runoff generation might be expected, based on the higher rainfall intensities in comparison with the humid temperate areas. The *rainfall* section of this paper also alerted to differences in rainfall intensity-frequency-duration between the equatorial areas where convection and orographic uplift prevail, in contrast to the more well-organized rain-producing systems of the outer tropics. Other spatial differences in rainfall characteristics were highlighted (Jackson, 1986; 1988b). On the basis of rainfall properties alone, differences in hillslope hydrology across the humid tropics should be expected (Bonell *et al.*, 1991).

Kirkby (1978) identified the critical role of hydraulic conductivity in relation to rainfall intensity through the recognition of domains dominated by the combination of saturation-excess (saturation) overland flow and subsurface storm flow as against infiltration-excess, Hortonian overland flow (Fig. 15). Bonell & Gilmour (1978) and Bonell *et al.* (1981) strongly emphasize the interaction between prevailing rainfalls and vertical changes in K^* down the profile in explaining their observations of hillslope stormflow response. The same principles

were used in the empirical extrapolation of their catchment study results through detailed K^* measurements taken across other soil types and across the rainfall gradient of the wet tropics of northeast Queensland (Bonell *et al.*, 1983b). Later, Ternan *et al.* (1987) in Grenada and Elsenbeer & Cassel (1990) in western Amazonia followed a similar methodology. These studies emphasize that measurements of soil hydraulic properties are an integral part of hillslope hydrology. Unfortunately, such knowledge is lacking in most hillslope studies in the humid tropics.

Tropical forests Walsh (1980) provided the first comprehensive review of work in the humid tropics, most of which was undertaken in tropical forest. He proposed that four runoff models in terms of delivery mechanisms occurred, viz.: (i) widespread saturation overland flow model, (ii) localized saturation overland flow dominant/subsurface stormflow model (the variable source area-saturation overland flow model of Freeze, 1974), (iii) dominant subsurface storm flow model, (iv) rapid, subsurface stormflow model via pipes. The latter two models compare with the Freeze's (1974) variable source area-subsurface stormflow model.

Later, a brief, updated review was provided by Bruijnzeel (1990). He envisaged a continuous spectrum of hillslope responses broadly controlled by nature of the soil substrate, in terms of permeability and topography. At one end of the spectrum, he considered the subsoil to be sufficiently permeable to prevent any type of overland flow occurring on hillsides, except under extreme rainfalls. Amongst the examples cited were those studies in Amazonas (Nortcliff & Thornes, 1981), Tanzania (Lundgren, 1980) and Colombia (Vis, 1989) where normally overland flow is usually less than 1% of rainfall. The delivery mechanisms to streamflow during storms were envisaged to follow similar descriptions for humid temperate studies. In the case of environments with steep slopes adjoining narrow valley bottoms (e.g. Walsh, 1980, Dominica; Bruijnzeel, 1983, Indonesia; Vis, 1989, Colombia), the quick-flow component was dominated by subsurface stormflow resulting from both translatory (flow and by-pass flow in macropores. In valleys with wider "riparian" areas, locally generated saturation overland flow was considered to be the principal delivery mechanism (e.g. Nortcliff & Thornes, 1981, 1984, Reserva Ducke, Amazonas) in response to the more concave slopes.

At the other extreme of the spectrum, impeding soils which show a sudden decrease in permeability with depth are capable of frequently producing widespread overland flow on the hillslopes, (not just the valley bottoms) in conjunction with subsurface stormflow. Examples cited (Bruijnzeel, 1990) included from northeast Queensland (Bonell et al., 1981; Herwitz, 1986), western Amazonia (Elsenbeer & Cassel, 1990) and Panama (Dietrich et al., 1982). More recently Ross and co-workers (S.M. Ross et al., 1990; Nortcliff et al., 1990) also inferred a similar runoff mechanism in terra firme forest of Maracá Island, northern Roraima, Brazil. As mentioned, Herwitz (1986) demonstrated that funnelling of stemflow at the base of trees in northeast Queensland, could also produce "local" infiltration-excess overland flow to add to the prevailing saturation overland flow.

Not all soils with an impeding layer, however, necessarily generate frequent saturation overland flow. Vertical changes in K^* (using identical field methods to those studies in northeast Queensland) in a Grenada catchment (Ternan et al., 1987), favoured a subsurface stormflow dominant model or less frequently, a localized saturation overland flow/subsurface stormflow model in line with Walsh's (1980) earlier observations. Three factors possibly contribute to the lack of widespread saturation overland flow. The top 0.2 m of soil is more permeable than elsewhere (cf. Bonell et al., 1983b; Elsenbeer & Cassel, 1990) with higher available soil water storage capacities. In addition, the synoptic weather systems, and related rainfall characteristics, are different from some of the other studies, e.g. northeast Queensland and, of course, differences in topography.

Not highlighted in Bruijnzeel's (1990) review are the more recent reports of infiltration-excess (Hortonian) overland flow occurring in some West African tropical rain forests (e.g. Dubreil, 1985; Wierda et al., 1989) where the impeding layer is at the surface. In undisturbed rain forest, Wierda et al. (1989) noted that low saturated infiltration rates (approximately K^*) ranged from 7 to 12 mm h^{-1} over a catenary slope in Côte d'Ivoire. These infiltration rates declined from the upper to the lower slope. When the infiltration rates were compared to rainfall data, it was concluded that Hortonian overland flow frequently occurred, especially on the middle and lower slope sections. Wierda et al. (1989) study indicates that infiltration-excess overland flow is not confined to arid-to-sub-humid climates as previously suggested in schematic representations of factors affecting storm runoff processes (e.g. Dunne,

1978, p. 289; Walsh, 1980, p. 197), but can occur in undisturbed tropical rain forest as well.

It is significant that the influence of different rainfall characteristics across the humid tropics was neglected by Bruijnzeel (1990) as one of the underlying criteria responsible for the spectrum of hillslope responses. Perhaps the reason is that the bulk of the studies reviewed reflect similar runoff processes and a variable source area response closely aligned to the description for the humid temperate areas. Also, most of these studies were located in the equatorial regions where annual rainfalls are moderate (by humid tropical standards), short-duration convective storms prevail and, in many examples, there are no shallow impeding layers. Nortcliff & Thornes (1981) presented probably some of the more reliable K^* data from the group, which is summarized in Fig. 16a. If we assume that typical, maximum 1 minute rainfalls for the convective storms are c. 60 mm h^{-1} (no detailed short-term rainfall intensity information is given) then these Oxisols have a deep, transmissive layer down to 0.90 m depth (K^* range 61.3 to 156.7 mm h^{-1}) before K^* lowered to 21.7 mm h^{-1} (0.90 – 1.15 m depth). Below 0.90 m would be the "impeding" layer and probably would encourage saturated, lateral subsurface flow. The *short duration rainfalls,* however, would be insufficient to cause the perched water table to emerge to the surface. Consequently, saturation-excess overland flow would not occur on the hillslopes. It is also significant that the K^* of the top 0.15 m of these Amazonas soils (K^*, 921.3 mm h^{-1}) differed little from those reported in northeast Queensland. Under these conditions, it is clear that infiltration-excess overland flow cannot occur.

Disruptions to the soil fabric from surface disturbance of the Amazonas Oxisols, however, might have greater ramifications on the runoff process compared with other environments where shallow, impeding soils naturally occur. Compared with Fig. 16b, the predisturbance "throttle" (in Fig. 16a) is more remote from the surface and the upper transmissive layer is much deeper which acts as a buffer to either form of overland flow. Despite the earlier observations of Dias & Nortcliff (1985), Bonell (1991a,b) suggested that disturbance arising from forest conversion could transfer the impeding layer to the surface. Such a change in soil hydraulic properties of the surface might cause the delivery mechanism to transform from deep, lateral subsurface flow to infiltration-excess overland flow and, therefore, dramatically increase the volume of hillslope runoff (Bonell, 1991a,b). As highlighted, visible observations suggest that the surface fabric of the Reserva Ducke soils easily collapses on disturbance.

In contrast, Fig. 16b conceptually shows the runoff generation process that commonly occurs in the undisturbed (South Creek, 25.7 ha) and disturbed (North Creek, 18.3 ha), paired catchment study near Babinda, northeast Queensland. The K^* presented emanated from the survey across both catchments (details in Bonell et al., 1987; Bonell, 1991a, c). Despite the dramatic reduction in surface K^* (0 – 0.1 m) in North Creek,

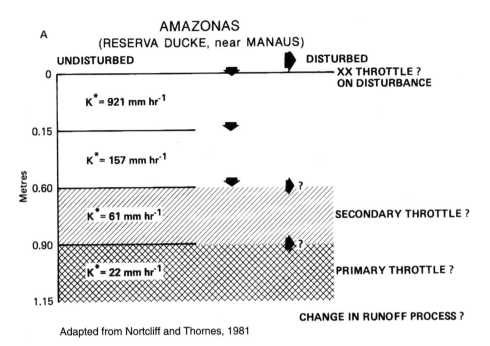

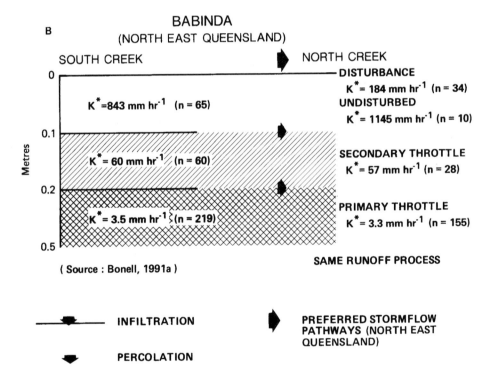

Figure 16a: A schematic diagram showing the changes in field-saturated hydraulic conductivity, K^* with depth in Oxisols of the Reserva Ducke, Amazonas (After Nortcliff & Thornes, 1981). In addition, the suggested position of the throttle layer pre-disturbance to prevailing rainfall and corresponding delivery mechanism of storm runoff is also shown. *A possible scenario* of the post-disturbance situation (no available data) is also highlighted (Bonell, 1991a,b) based on visual observations of persistent depression storage on walking and vehicular tracks in the Reserva Ducke.

Figure 16b: A schematic diagram for the Babinda catchments (South Creek, undisturbed tropical rainforest; North Creek, disturbed) showing the delivery mechanisms of storm runoff in relation to field-saturated hydraulic conductivity, K^* (expressed in mm h^{-1}), which are logarithmic means based on the sample sizes shown, n, for the 0–0.1 m, 0.1–0.2 m and 0.2–0.5 m layers. In the case of North Creek, the K^* were determined between 1984 and 1986 when the former cleared section contained natural rainforest regeneration. Note that there is no basic change to the runoff generation process on disturbance, except occasional very high short-term rainfall intensities which can produce infiltration-excess (Horton-type) overland flow (After Bonell, 1991c).

even after more than 10 years of natural forest regeneration, the runoff process, however, remains essentially unchanged. The shallow impeding layer below 0.2 m continues to encourage saturation-excess (saturation) overland flow, and subsurface stormflow in the top 0.2 m. Extreme short-term rainfalls may also produce infiltration-excess overland flow over North Creek, especially as temporal variability has not been taken into account in the surface K^* estimates so that they could be lower resulting from raindrop compaction during periods of persistent rainfall (Bonell & Williams, 1986a). The prevalence of overland flow probably explained why Gilmour (1975, 1977) earlier had found no detectable changes in quickflow volume, quickflow duration or time to peak following logging and clearing of this catchment (Gilmour *et al.*, 1982). By contrast in a different rainfall regime (Maracá Island, northern Roraima), Nortcliff *et al.* (1991) noted little difference in runoff characteristics between an undisturbed and partially cleared forest slope, but an average increase from 6 to 16% of rain in overland flow between these treatments and a third completely-cleared forest slope. No detail is provided on surface changes in K^*, so that it is unclear whether this increase reflects a transformation from saturation-excess to infiltration-excess overland flow.

Comparison of hillslope hydrology studies in Babinda, Queensland and western Amazonia where shallow impeding layers exist

The long-term Babinda study (Bonell, 1991a, c; Bonell *et al.*, 1981; Gilmour *et al.*, 1980, 1982) and the recent investigation by Elsenbeer (Elsenbeer & Cassel, 1990) in western Amazonia provide an interesting comparison of runoff generation at locations where the soil has shallow impeding layers. The frequent occurrence of widespread saturation overland flow occurs in both studies but the prime mechanism is different. At the Babinda site, the overriding control is the prevailing high rainfall intensities, whereas in the western Amazonia work it is K^*.

The Babinda study was stimulated by Gilmour's (1975) earlier observation that average runoff from South Creek was 2616 mm from a mean annual rainfall of 4175 mm between 1970–1975. About 47% of annual runoff was quickflow. But for individual monsoon storms, the quickflow to gross rainfall response ratio could be as high as 74%, with 45% commonly exceeded. In addition, peak instantaneous discharges up to 70 mm h^{-1} were occasionally recorded. Time series analysis also demonstrated that the lag response between rain and streamflow was 0.4 h, irrespective of storm amount or duration under wet conditions (Bonell *et al.*, 1979; 1981). The combination of high rainfalls and the soil hydraulic properties was put forward as an explanation for the observed runoff process. About 63.5% of the annual rainfall occurs between December and March. Even more important, the nature of the synoptic meteorology, by way of well-organized circular disturbances on the monsoon trough results in long duration rainfalls, with a large proportion of the annual rainfall occurring only on a few days. The average maximum daily rainfall for raindays

exceeding 100 mm was 270.5 mm from 7.1 days per year over the 1970–1984 period (Bonell, 1988). Furthermore, it is common for a significant percentage of the yearly rain total to occur on consecutive days because of the near-stationarity of vortices on the monsoon trough. The most extreme example was 2602 mm from 14 days of continuous rain in 1981 (3–17 January) which amounted to 48.9% of the 1981 total of 5325 mm (see Manton & Bonell, this volume).

The deeply weathered parent material (basic metamorphics) is characterized by kaolin-dominated silty clay loam to clay loam soils up to 6 m in depth. These heavy-textured soils can store between 1.2 to 1.5 m equivalent depth of rain in the top 3 metres during the wet season (Gilmour, 1975). The prevailing matric potentials are very close to the saturation level for long periods so that widespread saturation can redevelop almost instantaneously with the onset of intense storms (Cassells *et al.*, 1985). Figure 17 shows a continuous record of soil water pressure changes for selected depths during tropical cyclone "Ivor." This diagram demonstrates the rapid responses in matric potential at the onset of rain and that during the two large rain pulses, all pressure-transducers show positive matric potentials. The steep catchment slopes (0.3 m m^{-1}) and efficient drainage density (0.23 m m^{-2}, which does not include the numerous shallow swales activated during a storm) are able to tap a large proportion of the saturation overland flow. Furthermore, considerable volumes of saturation overland flow can occur over steep, convex slopes bordering a very narrow flood plain which is contrary to humid temperate experience (Freeze, 1972; Dunne, 1978) where subsurface stormflow is considered to dominate the storm hydrograph (Dunne, 1978, p. 289). The extreme climatic conditions of this study also demonstrated that forests do not act as infinite "sponges" but can generate flood-producing runoff which is contrary to many of the widely-held beliefs in many developing countries (e.g. Nautiyal & Babor, 1985). The total storm discharge was 2096 mm and 1880 mm from North Creek (disturbed) and South Creek (undisturbed) respectively over the period 3–17 January 1981 associated with the sustained monsoon rains (Bonell *et al.*, 1991).

Preliminary statistical analysis showed that storm runoff response was highly sensitive to rainfall intensity in the undisturbed forested catchment (Bonell & Gilmour, 1978; Bonell *et al.*, 1981) in contrast to Hewlett *et al.* (1977, 1984). Work has just been completed analysing all storms on the lines of the Hewlett *et al.* model for both North and South Creek (unpublished results). The results confirm the significance of rain intensity for quickflow volumes exceeding 10 mm for both catchments. For the largest storms (>100 mm quickflow), total rain is the most significant because the upper soils' storage capacity is full so that most rain immediately is discharged as saturation-excess overland flow. This sensitivity to rainfall intensity led to the conclusion that a major reason why runoff generation differed from other reports from elsewhere (Walsh, 1980) was due to the synoptic and rainfall climatology of the area (Gilmour *et al.*, 1980; Bonell *et al.*, 1991). With the aid of examples,

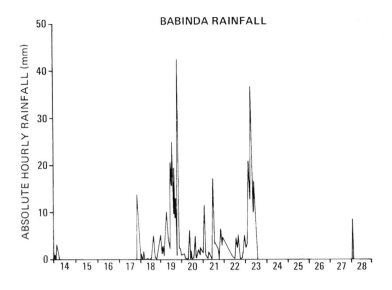

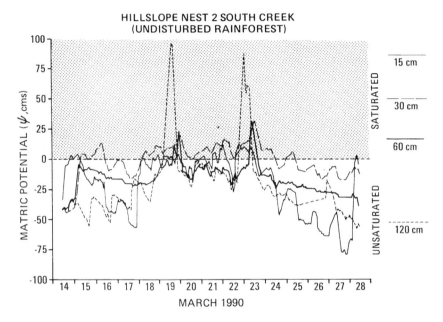

Figure 17: A continuous record of soil water pressure changes during tropical cyclone "Ivor" in undisturbed rainforest (South Creek, Babinda catchments). Note that the depths shown refer to depths of the soil water pressure-transducers (After Bonell, 1991c).

Gilmour *et al.* (1980) demonstrated the importance of rainfall intensity by showing that saturation overland flow declined in frequency of occurrence at selected, unbounded runoff plots within South Creek, from the monsoon season (December-March), through the post-monsoon season (April-June) to the winter/spring period from July onwards. The preferred range of maximum short-term (0.1 h) rainfalls successively declined across these three seasons so that from July onwards when SE trade wind "stream" showers prevail, saturation overland flow becomes much less significant because 0.1 h rainfalls are usually below 20 mm h⁻¹. The soils become transmissive, and matric potentials also become increasingly more negative from September onwards as the frequency of rainfall reduces (Bonell, 1988). Because this study is located in the outer tropics, the seasonal rhythm of monsoonal and upper atmospheric disturbances (of temperate origin) means that in the one location, the variable

source area concept ranges from the extreme wet end of the spectrum (dominated by widespread saturation overland flow in large summer monsoon/post-monsoon storms) to the more localized saturation overland flow or subsurface stormflow model later in the year, as found elsewhere in humid tropical and temperate areas (Dunne, 1978; Walsh, 1980). The latter is caused by the much weaker rainfall intensities associated with 'stream' showers embedded within the SE trade winds as well as the more frequent occurrence of disturbances of temperate origin. At such times, statistical analysis suggests that antecedent stream discharge is the most significant parameter for quickflow volumes less than 2 mm. Rain intensity surprisingly still takes over as the most significant variable for the occasional quickflow volumes in excess of 2 mm.

The view of Walsh (1980, p. 181) that of all the tropical rain forest areas then investigated, the Babinda catchments had "...

the only *distinctively tropical* runoff process pattern ..." due to the anomalously high cyclonic rainfalls was originally supported by Bonell *et al.* (1986). More recently, however, Elsenbeer & Cassel (1990) noted a similar hillslope response in a 1 ha headwater catchment in western Amazonia, Peru where the annual rainfall is about 3300 mm. This is not a climatic environment where well-organized, long duration rainfall associated with cyclonic vortices occurs. Maximum storm rainfall is much lower than for Babinda over a range of duration from 0.1 to 1 h, with equivalent hourly intensities typically occurring between 2 to 27 mm h^{-1} (see Elsenbeer & Cassel, 1990, Table 4). Total storm rainfall in Elsenbeer & Cassel's study was also very low, with the upper 75% quartile showing only 13.2 mm. The relationship between rainfall intensity and K^* showed, however, why saturation overland flow occured. The much lower rainfall intensities are compensated in this environment by the impeding soil layer being both more shallow than in the Babinda study (i.e. at 0.1 – 0.2 m depth, log mean $K^* = 6.8$ mm h^{-1}) and also the K^*, which was an order of magnitude lower below 0.3 m depth (i.e. 0.22 mm h^{-1}) (Elsenbeer & Cassel, 1990, Table 5). In addition, the frequent rainfall caused almost year-round near-saturation as was shown by matric potentials being close to zero (Elsenbeer & Cassel, 1990). Elsenbeer & Cassel (1990) also noted that pipeflow (the by-pass pipes of Kirkby, 1988), was a significant source of return flow.

It becomes clear that any shift in the delicate balance between rainfall intensity – soil hydraulic properties – topography can provide widely different runoff responses – both within and between various tropical forests. Recent studies have highlighted that Walsh's earlier schematic summary of runoff processes does not completely hold (Table 6, Walsh, 1980). Hortonian overland flow can occur in the ever-wet tropics (Wierda *et al.*, 1989), widespread saturation overland flow is not just confined to environments such as "extensive catchment flats," seasonal climatic areas with prevailing high rainfall intensities of long duration (compare Dietrich *et al.*, 1982 and Elsenbeer & Cassel, 1990 with Bonell *et al.*, 1981) concave slopes (cf. convex slopes in Bonell & Gilmour, 1978; Bonell *et al.*, 1981), or to disturbed environments as previously suggested by Walsh (1980). It is evident, however, that further field research initiatives across a wider range of humid tropical environments are still required.

The need for studies using environmental isotopes and other conservative chemical tracers
There is a need for environmental isotope and other conservative tracer studies in the humid tropics along the lines of Pearce *et al.* (1986), Sklash *et al.* (1986) and McDonnell (1989, 1990) in New Zealand, and Robson & Neal (1990) in the UK to provide further understanding of the runoff process, and for assisting studies in water-borne nutrient cycling.

As mentioned, such work has commenced in northeast Queensland in the Babinda catchments. Preliminary results (Bonell and Barnes, unpublished data) indicate that the quick-

flow component emanates from at least two distinct reservoirs, viz., surface and shallow, subsurface water and a deep permanent water table. There is a marked non-linearity in both catchment responses as expressed by the varying contributions of "new" and "old" water to streamflow from storm to storm. Such responses would seem partly to depend on antecedent wetness and maximum rain intensities. However, the hypothesis that under wet antecedent conditions, the stream hydrograph is dominated by "new" or event water is proving correct during monsoon storms – in contrast to previous findings from humid temperate work. There is almost a convergence in the isotopic concentration of the stream at the flood peak with the corresponding signature of the most intense rainfalls. Even more surprising, the undisturbed South Creek releases much larger quantities of "new" water in comparison with the disturbed North Creek under certain antecedent conditions.

These early results in the Babinda study are also highlighting that recharge to the deeper groundwater body during large monsoon events is also considerably greater than previously thought. The vertical response of the water table beneath a convex slope increases away from the stream edge. This causes a steepening of the hydraulic gradient and, therefore, allows "pre-existing" water to be pushed into the stream, especially during the period of stream hydrograph recession. Under these conditions, the widely quoted "groundwater – capillary fringe" ridging mechanism (Sklash & Farvolden, 1979, Abdul & Gillham, 1989) is not confirmed. The role of groundwater also possibly contributes to the non-linearity of the storm hydrograph response. In complex storms, where there is more than one peak in the hyetograph, the corresponding stream hydrograph progressively shows larger volumes of water for each successive peak. In addition to larger volumes of saturation and subsurface stormflow emanating from more extensive source areas, there are indications from well hydrographs that the deeper groundwater body makes increasingly significant contributions soon after the final stream hydrograph peak. At such times, preliminary analysis of the stream water isotopic contents indicate that substantial quantities of pre-existing water, of possible different residence times, supplements contributions to the stream hydrograph. In addition, the isotope results show clear evidence for mixing of vertical inputs with the water table. There is, however, considerable spatial variability in groundwater recharge with the soil water isotopic concentrations down to 1.2 m depth indicating three different types of behaviour. The first has nearly uniform stable isotopic concentrations with depth, with the concentration changing after each major rainfall event. The second shows that concentrations change between events, down to 0.45 to 0.60 m, but are invariant below that. The third shows that concentration profiles do not change with time. This is interpreted respectively as evidence of rapid percolation (possibly to the water table) along structural cracks associated with the medium to fine blocky soil structure; of partial penetration with subsequent lateral subsurface stormflow/saturation overland flow (both within and over the upper soil layer); and of mostly saturation overland flow. In the latter case, subsurface stormflow is likely to occur in the most transmissive top 0.1 m where no isotopic samples were taken. Furthermore, under very dry antecedent conditions, weak storm hydrograph responses show no evidence of change in isotopic concentration from pre-storm background levels (Barnes & Bonell, unpublished data). In the latter case, similar characteristics have already been noted in the review of humid temperate work (cf. Bonell *et al.*, 1990). Analysis is proceeding to explain these contrasting responses.

Agricultural areas

Lal (1981, 1983) presented data from a Nigerian study showing that forest clearance, the method of land clearing and development and the type of tillage system significantly affect the magnitude of runoff and erosion. The forested basin showed no significant runoff and erosion due to thick undergrowth and surface litter. Traditional shifting cultivation produced only a minimal runoff response – which significantly increased when permanent clearing occurred, using manual or mechanical methods. However, the most effective water and soil conservation system was manual clearing followed by no-tillage. Conventional tillage operations produced the largest volume of runoff (and erosion). Lal (1983) does not discuss the hillslope runoff processes in detail, but he noted that both the overland flow (type not indicated) and subsurface stormflow components increased following a deterioration in the infiltration properties after forest clearance.

Recent results from a sugar caneland study in the perhumid, northeast Queensland environment by Prove (1991) provides further insight into the runoff processes for different management treatments. He compared the runoff hydrology and erosion (see Rose, this volume) between mechanized, conventional tillage (tillage extended to 0.15 m depth) and no-tillage management, along a 100 m section of krasnozem soil, having an overall slope of 12%. In the inter-sugar-cane row area of the conventional tillage, and within the rows of sugar-cane for both management types, all develop perched water tables during storms, which subsequently rise to the surface to produce saturation overland flow in a similar manner to that described for the nearby Babinda study. Similar subsoil K^* values were also determined for the impeding layer (2.5 to 4 mm h^{-1}).

The reduced infiltration associated with the inter-sugar-cane row area of the no-tillage treatment (due to surface compaction from agricultural machinery), generates infiltration-excess (Hortonian) overland flow and saturation overland flow at different times, depending on the prevailing matric potentials and rainfall intensities. Such findings are in agreement with the concept illustrated in Fig. 15 after Kirkby (1978) and the work of Smith & Hebbert (1983) and Burt (1989, p.13). This more complex runoff process explained why greater runoff volumes occurred from this management system, especially when matric potentials were considerably negative in the opening stages of the wet season (Prove, 1991).

An improved methodology for measuring hillslope hydrology

There seems to be a considerable amount of humid tropics information (a large proportion of which is not published in journals) which describe runoff and erosion losses from *bounded* plots (e.g. Othieno, 1979). The use of artificially bounded plots, however, has major disadvantages, and results from this experimental design are doubtful in value. For example, the boundaries interfere with the natural processes by diverting overland flow and transported sediment. The results collected only refer to the area enclosed and do not facilitate confident extrapolation. In addition, the trough lengths are often short of the "repetitive unit" and the area monitored below the "representative elementary volume" (REV).

As a starting point, runoff trough lengths should exceed the repetitive unit, and the plot remain unbounded to accept all overland flow, subsurface stormflow and sediment transport from upslope. A new cascade system of troughs, 10 m long and offset at 25 m intervals, was described for a tropical semi-arid area (Bonell & Williams, 1986b, 1987; Williams & Bonell, 1987, 1988). Similar designs should be considered in future humid tropical studies (Fig. 18), especially where overland flow is frequent. The design is based on the assumption that the overland flow observed on the upslope trough is an estimate of the runon to a 250 m^2 plot, and that the run-off from the plot is monitored by the downslope trough. The runoff-runon (mm) is calculated by a simple continuity model, as the difference between the troughs over a given time period, viz:

$$X = Q_d - Q_u \qquad (30)$$

where Q_u is the upslope trough and Q_d is the downslope overland flow. (A similar model for estimating erosion and deposition can be used, see Bonell & Williams, 1987.) It follows that X is positive for runoff and negative for runon. Cumulative infiltration, I (mm) can be calculated by:

$$I = R - X - d_S - d_V \qquad (31)$$

where R is the rainfall in the given period, and change in surface detention store and depth in surface water is given by d_S while d_V represents the change in water interception by vegetation. For short time intervals, say 1 min., d_S and d_V can be assumed to be zero. Bonell & Williams (1987) demonstrated that despite considerable volumes of overland flow measured in their study, considerable redistribution occurred over a 100 m transect, so that net discharge downslope was small. Had the plots been bounded, the high overland flow volumes recorded would have been misleading.

Williams & Bonell (1988) treated the Philip equation empirically to determine the "S" and "A" constants from the calculated I in equation (31), following the least squares procedure of Bristow & Savage (1987). Also, K^* was determined from the linear portion of the $I(t)$ plot (again using regression analysis) because the "profile at infinity" conditions of Philip (1969)

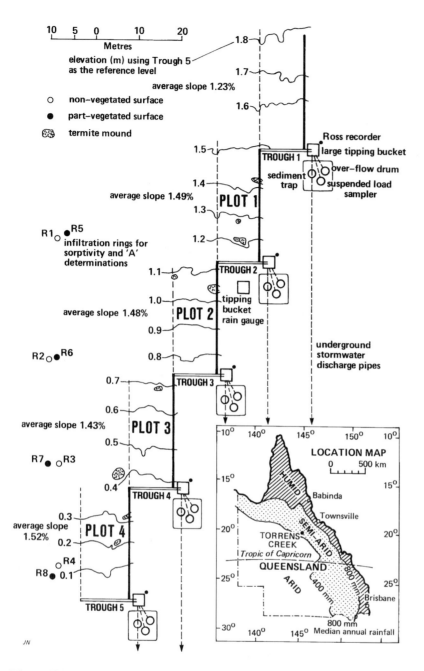

Figure 18: A cascade system of troughs for estimating runoff-runon and erosion-deposition following the design originally used at Torrens Creek, central north Queensland (Source: Bonell & Williams, 1986b, 1987; Williams & Bonell, 1987, 1988). Note that the very low relief and nature of overland flow movement did not require the bounding of the left side of each plot by an aluminium strip (Bonell & Williams, 1989). It is necessary however, to insert this additional bounding strip on steeper slopes and where significant cross-slope flow occurs.

were rapidly attained on similar lines to the infiltration rings (Bonell & Williams, 1986a). Later, Prove (1991) successfully applied this simple model to determine the infiltration properties (K^*, coefficients for equations of Kostiakov (1932) and Ostiachev (1936)) for different management treatments of sugar-cane along 100 m hillslope transects.

RUNOFF REGIMES

The river regimes in the humid tropics can be considered in two broad categories, viz:

(a) equatorial rivers of the humid tropics
   (1) with one peak
   (2) with two peaks.
(b) rivers of wet and dry regions.

Regimes of equatorial rivers

Regimes of the equatorial rivers with one peak such as the Brantas River in Indonesia (Fig. 19) are produced by heavy annual precipitation of over 1750 mm, without a pronounced dry season. The precipitation distribution may have two periods of increased rainfall, but only one peak is produced.

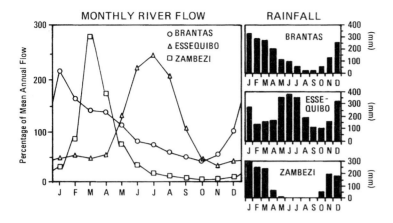

Figure 19: Three types of rainfall and runoff regimes in the humid tropics (After Oyebande & Balek, 1989).

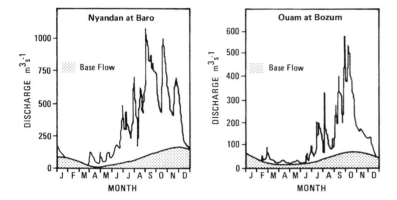

Figure 20: Streamflow and base flow for two rivers in wet/dry regions not affected by monsoons or trade winds (After Oyebande & Balek, 1989).

Equatorial rivers with two peaks are produced by precipitation regimes with monthly totals over 100 mm and an annual total over 1750 mm. Tropical forest dominates in such basins. The River Essequibo in Guyana is an example (Fig. 19) of such a stream with one main and one secondary peak.

The described equatorial regimes are typical of the continuously wet climate. However, the seasonal fluctuation of the river regime usually is more pronounced than of the rainfall regime because the drier months experience higher total evaporation, E. Another consideration is the effect of the dense stream network. The largest tropical rivers, such as the Amazon and the Congo, collect a significant part of the runoff from the equatorial zone. However, their regimes are affected by other factors including the effects of different rain producing weather systems (see Molion, this volume) and as such they cannot be considered as rivers with a distinct equatorial pattern.

Regimes of tropical wet/dry regions

Typical river regimes of the tropical wet and dry regions, such as the Zambezi (Fig. 19) are associated with basins with a prolonged dry season. The seasonal effect of the rainfall is also reflected in the runoff regime, with the length of the dry season being a decisive factor. Usually such basins are covered by a savanna-woodland type of vegetation. The runoff pattern is

delayed by several weeks to several months beyond the rainfall pattern, depending on the replenishment of the soil water deficit and formation of storage in the river channels and in the aquifers (Fig. 20).

With the increase in length of the dry season, streams with ephemeral river regimes are found at the rims of the humid tropics, especially in the smaller basins. In such areas, the various processes of erosion increase (described in Rose, this volume) owing to frequent wetting and drying of soils and the less dense vegetational cover (Falkenmark & Chapman, 1989).

Regimes of monsoonal regions

Large variations in the runoff regimes are experienced in the monsoonal regions of southeast Asia. The variability depends on the origin and extent of the monsoonal effects and/or trade winds. For example, the regime of the Huai Bang Sai in Thailand and the Mekong at Phnom Penh (Fig. 21) (Volker, 1983) reflect the difference in rainfall occurrence – both spatially and temporally over each basin as well as differences in their geomorphic characteristics. Tropical cyclones play a significant role in modifying these river regimes. The flood hydrographs resulting from tropical cyclones occur as random pulses superimposed on more or less regular hydrographs during the cyclone season.

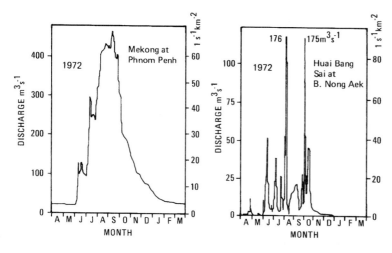

Figure 21: Contrasting hydrographs of two rivers in southeast Asia (After Volker, 1983).

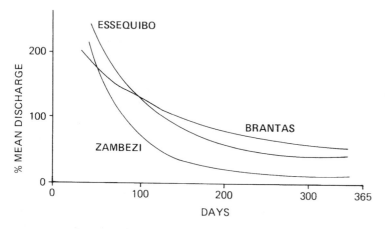

Figure 22: Flow duration curves for three rivers in the humid tropics (After Oyebande & Balek, 1989).

Differences between the tropical river regimes can also be traced in the shape of the flow duration curves (Fig. 22), which show that the river regimes of the equatorial rivers are more stable than those of the wet and dry regions (Balek, 1983).

The role of swamps and wetlands

Swamps and wetlands also play a significant role in the surface hydrology. Intermittent headwater swamps, known in Africa as dambos (Balek & Perry, 1973; see Bullock, this volume) can be considered as significant contributors to catchment response. As they remain wet during the first part of the dry season, they have a stabilizing "sponge" effect on the outflowing streams. Although the headwater swamps are rather small in size, they are great in total number. For example, some $10^4 -$ 10^5 of dambos exist in Africa. A typical hydrograph composition of dambo outflow is shown in Fig. 23.

Wetlands and swamps are also found elsewhere in the humid tropics such as the "flooded" forest adjoining some of the Amazon tributaries e.g. the Negro near Manaus. Their hydrological regime is affected by the morphological structure and prevailing type of the vegetation. Essentially, each swamp/wetland is a unique hydrological structure.

FLOODS

Many attempts have been made to define extremes, or at least the boundaries between extremes and a situation which can be considered normal. In simple schemes, a difference is made between so-called high water, the low water and the normal water years. From such simple verbal descriptions, the development of the analysis of extremes has become more and more sophisticated, with regional formulae being drawn up. However, only long-term observations of extremes can form a solid basis for such a work.

Knowledge of exceptionally large floods is essential to the solution of many problems in water management and also to assess the susceptibility to flooding of any structure to be constructed in the vicinity of a water course. The choice of a design flood value is generally difficult and delicate. It necessitates the use of the appropriate hydrological method of estimation (if possible, several methods should be used and the results compared) and good knowledge of the condition of operation of the structure to be studied in order to choose correctly the risk. The latter is closely related to frequency – for those cases where it is possible to speak about frequencies.

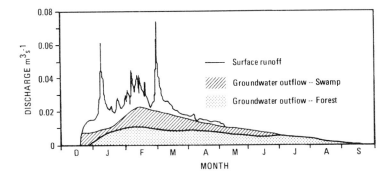

Figure 23: Typical hydrograph components of a headwater swamp (After Oyebande & Balek, 1989).

Flood analysis is closely related to consideration of risk, and reliability in hydrological design can be considered as most fundamental by answering the following questions:

(a) What are the loadings on the structure?

(b) What degree of safety should be incorporated in the design?

From the hydrological point of view, the loading means the flood which can pass through the structure. If the flood estimate in the design is incorrect, then the benefits of subsequent hydraulic and structural designs will be largely nullified. It is typical that increased safety can be obtained by increasing the capacity of the structure – at a considerably increased cost. This holds good for small as well as large structures.

In the tropics, major studies are usually undertaken for estimating the spillway design floods for important dams, but very little effort is made to estimate properly design floods for small structures. However, as has been found in Australia, more than 60% of all the expenditure is on water works, which taken individually are of minor importance. Therefore, the design frequency analysis is a most important step in the design of flood-passing structures. For economic reasons, only very large and important structures can be designed on the basis of the probable maximum flood.

The selection of design frequency depends upon many factors. Design based on a flood of selected probability implies that some damage will occur at every occurrence of a flood of lower probability, thus it is an economic question as to what sort of damage can be permitted for the construction and how often.

The formulae for the calculation of flood peaks of certain probability can be traced back to the middle of the nineteenth century. The main difficulty with the application elsewhere of various types of empirical formulae is the proper selection of the parameters involved. The extension of the validity of the formulae to tropical conditions is even more difficult because of their derivation in hydrological and meteorological conditions very different from those in the tropics. However, many types of formulae and methods for the estimation of flood peaks of various probability developed in regions of "moderate" rainfall have been applied in tropical countries.

It would be impossible to present all formulae available with an explanation of them. These can be found in the basic hydrological literature but examples of the more limited applications in the humid tropics, especially related to Africa, were reviewed by Balek (1977, 1983). The selection of the proper formula should be verified on a selected river in the region to be investigated for which a longer reliable record is available.

Obviously, the results of flood studies performed directly in tropical regions are more valuable. Particularly useful are studies with conclusions on the regional validity of the results. Significant flood studies in west Africa have been carried out by French teams, particularly by ORSTOM. Several experimental basins in west Africa had already been established for this purpose before the International Hydrological Decade started and many results related to each particular basin have since been published.

Attempts were made from the beginning to establish the relationship between effective rainfall (after taking into account the soil water balance) and the formation of hydrographs. For example, the flood hydrograph response to different effective rainfalls were studied in the Sahel region. A difficulty in carrying out such work lies in the non-availability of autographic records; normally, only daily totals are available. Therefore, in accordance with the estimation of permeability, the limits of infiltration conditions were established which when combined with the daily totals, give basic information on the rainfall-runoff relationships. For impermeable soils, an infiltration value of 10 mm h^{-1} and for permeable soils a value of 40 mm h^{-1} were estimated. In such regions with an annual rainfall of 300–1000 mm, it was calculated that effective rainfall, up to 85% of the daily total, can fall in 90 minutes on soils accepting 20 mm h^{-1} ; 75% of the daily total can fall in 55 minutes on soils accepting 40 mm h^{-1}. Providing the rainfall record is long enough, such a study can serve as a basis for the calculation of effective rainfall with a certain probability of occurrence.

As a simple approach, the empirical probability curve is often used for the calculation of flood frequency. The application of more sophisticated methods can be considered, however, few if any take into the account land-use and climate changes anticipated in the future. Therefore, all methods, which have been applied in the humid tropics need to be re-examined under various scenarios of expected land-use and climate changes.

LAND-USE IMPACTS – WATER YIELD CHANGES AFTER FOREST CONVERSION

The escalating intrusion into forests because of increasing socio-economic pressures (some are outlined in Pereira, 1991), combined with a dichotomy of scientific beliefs arising from the hydrological impacts of forest conversion, provides a clear need for an increased number of controlled experiments in the humid tropics (Hamilton with King, 1983; Pereira, 1989).

As early as the 1950s, consolidation of existing scientific evidence (e.g. F. Law, 1956) suggested that forests consume more water than shorter vegetation. The "classical" method of controlled drainage basin experiments or the "paired catch-ment method" (Hewlett *et al.*, 1969; Hewlett & Fortson, 1983) was the principal source of this information. Later, Cassells *et al.* (1987) reviewed this methodology. The method was initi-ated in the Wagon Wheel Gap experiments, USA (Bates & Henry, 1928) and later more extensively in the Coweeta Hydrologic Laboratory, North Carolina, USA (see reviews in Swank & Crossley, 1988, eds.). Hibbert (1967) provided the first review of 39 such experiments. Bosch & Hewlett (1982) updated Hibbert's review with the incorporation of an extra 55 studies to make the total 94. These additions did not alter Hibbert's earlier generalizations on vegetal effects on water yield and total evaporation, viz:

– reduction of forest cover increases water yield
– establishment of forest cover on sparsely vegetated land decreases water yield.

Bosch & Hewlett (1982) did, however, challenge Hibbert's (1967) earlier third generalization that the magnitude of water yield response to reforestation (or deforestation) was unpre-dictable. Despite the variability of data precluding the setting of any meaningful error limits (due to variations in experimen-tal conditions and results presentation), Bosch & Hewlett (1982) suggested that the following changes would occur (with the + or – signs showing the *respective* trends in increase or decrease):

– coniferous and eucalypt cover cause approximately ± 40 mm change in annual water yield per $\mp 10\%$ change in forest cover
– deciduous hardwoods cause approximately ± 25 mm change in annual water yield per $\mp 10\%$ change in forest cover
– low brush forests cause approximately ± 10 mm in annual water yield per $\mp 10\%$ change in forest cover.

Furthermore, none of the experiments evaluated in this review reported any change in annual streamflow when the change in forest cover affected less than 20% of the total catchment area (Bosch & Hewlett, 1982). Whilst Pereira (1989, 1991) rightly considered the results of this synthesis had no predictive value, the work did indicate trends in water yield changes and was of significant empirical value to land-use planners.

More important, Bosch & Hewlett's (1982) summary high-lighted the deficiency of work in the humid tropics, with the inclusion only of four experiments from such areas. It is the lack of such studies which has led to unsubstantiated claims and the generation of myths (mentioned but not concurred with in Hamilton with King, 1983) concerning the impact of tree planting on the hydrological cycle in developing countries of the humid tropics. For example, it is commonly stated that forests act as "sponges," so that subsequent conversion increases flooding. Some writers have addressed on more scientific grounds myths of this type (e.g. Hamilton, 1986, 1988; Hamilton with King, 1983; Bruijnzeel with Bremmer, 1989). An additional claim is that forests attract rainfall, when the explanation is simply that there is more surviving natural forest in areas of higher topography. As Pereira (1989, p. 57) noted, foresters conveniently generated this myth in defence of their trees. However, in areas where there is a significant input of "occult" precipitation or cloud interception (e.g. Hawaii, Juvik & Ekern, 1978; Costa Rica, Zadroga, 1981) associated with montane forested areas, then conversion of these forests would lose access to these additional inputs (Hamilton, 1986; 1988).

In response to the world-wide concern of the 1980s over the environmental consequences of conversion of tropical forests to other land-uses, several rigorous reviews related to the hydrological and sediment-transfer effects have emerged (Bruijnzeel, 1989, 1990; Cassells *et al.,* 1987; Hamilton, 1986, 1988; Hamilton with King, 1983; Veen & Dolman, 1989). Evident from these reviews was the need for technology-trans-fer of humid temperate findings because of the limited number of *controlled* experiments undertaken in the humid tropics, even though environmental conditions are different and the certainty of application somewhat tenuous (Hamilton with King, 1983). Even more limited is the number and scale of inquiry of hydrological process work to assist in the interpreta-tion of water balance studies (Bonell, 1991b).

The first controlled experiments, and most extensive in terms of sampling a range of environments, were the East African Catchment Experiments established by the UK in 1957 and 1958. Different parts of the experiment were termi-nated between 1969 and 1974 due to socio-economic difficul-ties. Edwards & Blackie (1981) summarized the results of this programme for the period 1958 to 1974. More comprehensive details were presented by Blackie *et al.* (1979). Three experi-mental locations were selected that covered a range of land-use conversions. Included were montane rain forest to tea plantations (Kericho Experiments, Kenya), with a sub-project assessing the water balance of bamboo forest; bamboo forest to coniferous plantations and Kikuyu grass (Kimakia Experiments, Kenya) and montane rain forest to cultivated crops (Mbeya Experiment, Tanzania).

All the catchments at Kericho, Kimakia and Mbeya were on volcanic soils with high infiltration rates so that quickflow was only 2–3% of annual stream discharge in indigenous forest. Process hydrology work, therefore, concentrated on measuring evaporation by the eddy correlation technique from a mature tea crop (Callander & Woodhead, 1979), and by the zero flux plane method and hydraulic conductivity-potential gradient method (Cooper, 1979).

In 1969, the paired Babinda catchments (viz South Creek, 25.7 ha, Control; North Creek, 18.3 ha, 70% cleared) were established in lowland tropical rain forest of northeast Queensland, Australia, following the controlled experimental methodology (Gilmour, 1975, 1977). Later, a series of more intensive phases of process hydrology work were initiated – focusing on the runoff generation process. The first review of how this work assisted the interpretation of the previous water balance work was presented by Gilmour *et al.* (1982). More recently, Bonell (1991c) provided an updated overview of the Babinda research programme.

Process hydrology work has not yet been linked with the few remaining controlled experiments, although there are plans to commence such studies in Malaysia (Abdul Rahim, pers. comm., 1989). A programme of controlled experiments was initiated in Malaysia from the 1970s, from which water balance results have already been presented for the Sungai Tekam Experimental Basin study, Peninsular Malaysia (Abdul Rahim, 1987, 1988; K. F. Law & Cheong, 1987). The objective of this work was to determine the hydrological changes arising from the conversion from logged-over forest (following extraction of commercial timbers) to agricultural plantations, namely oil palms and cocoa (Abdul Rahim, 1988).

With the possible exception of the ORSTOM-operated catchments in the tropical moist forests of French Guyana (Roche, l982a, b; Fritsch, 1987; Fritsch *et al.*, l987), there is a dearth of long-term, controlled experiments integrated with process hydrology work within Central and South America. Fritsch (1992) reported the effect on the total water yield of several different land-use treatments following forest conversion – using both heavy equipment and the traditional trash and burn cultivation. Process studies undertaken elsewhere were taken into consideration during the interpretation of the results. Salati (1987) provided a review of the results of all short-term water balance studies in the Amazon. The unknown magnitude of groundwater leakage in particular, (Shuttleworth, 1988a) adds to the uncertainty in their interpretation and may contribute to the broad range of quoted annual runoff coefficients (19 to 39%, Salati, 1987).

There are no reports of parallel studies being undertaken on tropical islands.

Forests and total water yield

Bruijnzeel (1987, 1990) reviewed the existing literature on water yield changes from controlled, humid tropical experiments. Some of the conclusions which emerged were as follows:

– removal of forest cover may result in considerable increases in water yield (up to 800 mm yr^{-1}), supporting the general findings of Bosch & Hewlett (1982).

– regardless of type of conversion, the highest increases are in the first year after treatment, followed by a more or less regular decline, depending on the time associated with the establishment of the new cover and the rainfall patterns.

– the response of the post-treatment streamflow levels varies, depending on the replacement vegetation cover. For example, in the East African experiments, replacement of rain forest by a tea estate at Kericho (Blackie, 1979a) resulted in an overall reduction in water use; whereas, replacement of bamboo forest by pine softwood plantations at Kimakia (Blackie, 1979b) showed no significant changes in water yield, once the pine canopy had closed. In contrast, at Mbeya the replacement of evergreen forest by smallholder cultivation on very steep slopes resulted in a large increase in water yield (Blackie *et al.*, 1979; Edwards & Blackie, 1981).

Forests and delayed flow

Evidence from controlled experiments indicates that the largest relative increases in streamflow after forest clearance occurs with the delayed flow component and is most marked in the dry season (Bruijnzeel, 1990). Such characteristics are evident even in the contrasting equatorial and cyclone-prone environments where there is c. 2000 mm difference in mean annual precipitation but where both still have a marked dry season in common. Edwards (1979a), for example, presented evidence for higher depletion of available soil moisture in the forested section of the control catchment of the Mbeya experiment, which corresponded with the dry season (Fig. 24). The deepest layer at 3 m did not return to a fully-wetted state until the end of January. This time lag, in turn, affected the delayed flow response in the stream (Edwards, 1979a). Observed delayed flow levels were, on average, twice as high in the cultivated catchment. Similarly, Gilmour (1975; Gilmour *et al.*, 1982) noted an increase in soil moisture levels in the 70% cleared section of North Creek catchment because of reduced "evaporation" demand. The corresponding significant ($p = 0.05$) increase in monthly streamflows in North Creek, amounting to 10% annually (293 mm) over two water years, and a highly significant ($p = 0.001$) rise in minimum weekly discharge after clearing also supported the increase in soil moisture levels.

Some caution, however, should be exercised in extrapolating the level of increase recorded in both studies until the hydrological processes accounting for the changes in yield have been investigated. Based on results related to the clearing of temperate evergreen forests in New Zealand and Oregon Coastal Range, for example, (Pearce & Rowe, 1981), an increase of c. 700 mm in water yield might have been expected following the 70% clearing of the Babinda catchment. As groundwater leakage in the experimental catchments seemed unlikely (Gilmour, 1975), the lower than "expected" measured yield increase in Babinda could be due to:

– higher net radiation available for evaporation than in the temperate latitudes.

– persistent SE trade winds which produce light "stream" showers even during the "dry" season. Such activity

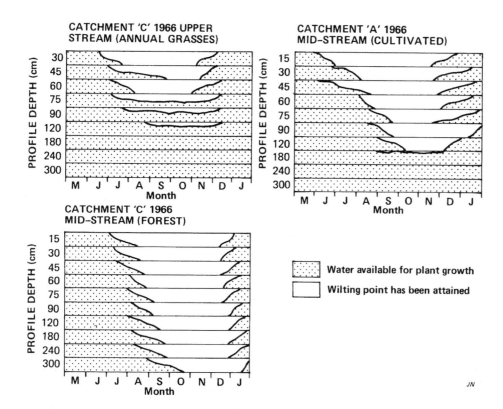

Figure 24: The contrasting patterns of water extraction and associated depletion of available water storage capacities as measured by gypsum resistance blocks for specific soil layers in the Mbeya catchments (Tanzania). Catchment "C" is the "Control," consisting of evergreen forest with annual grasses near the headwaters. Catchment "A" is the cultivated catchment (After Edwards, 1979a p. 239).

maintains "wet" surface soils for long periods, thus encouraging high soil evaporation rates. The 293 mm coincides with the last 3 months or so of the dry season when surface soils dry out, corresponding with increases in net radiation, reduced rainfall and continued consumption of effluent groundwater seepage by the riparian rain forest within the buffer strips.

The need for additional process hydrology work was also highlighted by Edwards & Blackie (1981) in relation to the Kericho experiment where annual rainfall is more evenly distributed. They noted that process and conceptual modelling studies implied some changes in groundwater recharge and, therefore, seasonal delayed flow distribution. In contrast, the observed differences in delayed flow between catchments (Table V, Blackie, 1979a) were small in practical terms.

The controversial issue of the removal of forests and effects on dry season streamflow   Contrary to the preceding experimental evidence, Pereira (1991) reported that there were widespread qualitative observations from the tropical developing world where the conversion of forests on steeplands, and resulting exposure to severe overgrazing, caused peak flows to increase during the rainy season whilst dry season flows decline. Examples are cited from Kenya and the depletion of water supply for the reservoirs of the Panama Canal. Similarly, Bruijnzeel (1989) noted that the existing controlled experiments did not involve immediate soil degradation after conver-

sion, whereas the long occupancy of landscapes in the humid tropics means that degraded agricultural landscapes are widespread and more typical. In such areas, both Bruijnzeel (1989) and Pereira (1991) suggested that surface infiltration rates are more likely to be reduced through dispersion and sealing by heavy rain, and accentuated by trampling and compaction by livestock, thus increasing overland flow during rainy periods. Water is thus lost for deep percolation, and, therefore, reduces streamflow. Increased infiltration opportunities arising from reforestation may increase percolation and, therefore, groundwater recharge, which could cause dry weather streamflow to be more sustained.

These opposite trends in streamflow changes to those described from "paired-catchment" studies emanate especially from experimental observations in Indonesia (Bruijnzeel, 1989). However, the existing evidence from a comparison of non-controlled catchments is obscured by such factors as differences in catchment area, leading to the potential for catchment leakage. Nevertheless, Bruijnzeel's (1989) review did lead to hypotheses which need further investigation. It was suggested that the conflicting water yield changes hinged on the shift in the delicate balance between surface infiltration capacity (and its relation to prevailing rainfall intensities) and total evaporation associated with forested and non-forested environments. Thus:

– if infiltration opportunities after forest removal decrease so that the amount of water leaving an area immediately as stormflow exceeds the gain in delayed flow associated with savings in evaporation, then diminished dry season flow could result. Improved soil infiltration through reforestation, for example, would not necessarily lead to increased streamflow, however, especially in severely eroded landscapes where available soil moisture capacities have been reduced.

– if surface infiltration characteristics are reasonably maintained, because of stable soil aggregates, immediate soil conservation measures or low rainfall erosivity, then the effect of reduced evaporation after clearing will result in increased delayed flow.

The determination of *in situ* soil hydraulic properties in a variety of forested and non-forested environments, linked with short term rainfall intensities on the lines described by Gilmour *et al.* (1987), would provide a preliminary "empirical" test of these hypotheses. Subsequently, more expensive micrometeorological methods, could be contemplated in conjunction with a "paired-catchment" approach examining the long-term changes resulting from reforestation of degraded lands (Bonell, 1989, 1991b).

Forest and stormflow

There is considerable publicity in the humid tropics that devastating floods have been caused by forest clearance in the headwaters of drainage basins, for example, in China (Sharp & Sharp, 1982), in the Philippines (Corvera, 1981; quoted by Hamilton, 1990) and more recently in Thailand in November 1988 (Nation, 1989). Such statements are simplistic when considering available scientific evidence related to the runoff generation process.

The East African catchments at Kericho, Kimakia and Mbeya were located on volcanic soils with extremely high infiltration rates. Surface runoff only constituted a small percentage of streamflow (2 to 3%) from the indigenous cover, and this did not significantly change after conversion (Blackie & Edwards, 1979, Edwards & Blackie, 1981). There was, however, a temporary dramatic change (still small in absolute terms), at the time of conversion from forest to tea plantations at Kericho (Dagg & Pratt, 1962). At Kimakia, the fact that the pine plantation rapidly developed a protective canopy and deep litter layer helped maintain high surface infiltration rates, and so no significant rise in stormflow occurred (Blackie, 1979b). At Kericho, the inclusion of soil and water conservation measures prevented significant storm runoff changes (Blackie, 1979a).

In the Mbeya experiment, Edwards (1979a, Table II, p. 236) noted a slightly higher percentage of rainfall leaving the cultivated catchment as storm runoff compared with the forest. This difference was most marked for lower intensity storms. All percentage values were low, however. Ironically, the maximum storm runoff recorded emanated from the forested catchment when 7.8% of a 53 mm rainfall event was counted as quickflow.

Abdul Rahim (1987; 1988) reported that no significant changes in quickflow volumes occurred in the Sungai Tekam experiment. There were, however, apparent increases in peak specific discharges (up to 37%) and decreased time to peaks (up to 50%) (not statistically proven), especially after some realignment of the stream. Also, there was an apparent decrease in peak discharge following the establishment of cover crops. No change in time to peak was associated with the increase in peak discharge following clear felling (Abdul Rahim, 1988). Whilst no process hydrology studies have yet been undertaken, qualitative soil descriptions suggest, however, that the soils are very permeable (Law & Cheong, 1987). Consequently, the annual runoff coefficient of the undisturbed forest remains low, ranging from 6 to 17% across the three basins (Abdul Rahim, 1988). Quickflow only accounted 21% of annual runoff (K. F. Law & Cheong, 1987). Bruijnzeel (1989) suggested that the relative absence of increased stormflows – even following forest clearance – indicated that significant overland flow along the hillslopes still did not occur in that environment.

Elsewhere, Fritsch (1992) presented contrasting evidence for significant increases in storm flow volumes and peak flows from clear-cut, exposed soil catchments during the first year of conversion. Volume increases ranged between +66 and +200%, depending upon the basin. With the exception of the pasture treatment, storm runoffs were not significantly different from pre-disturbed levels after three to six years, following either reforestation (eucalyptus and pine), natural regrowth or planting of a pomelo orchard. Similarly, peak flow increases ranged from +17 to +166% immediately following conversion, but subsequently returned to pre-disturbed levels with the exception of the pasture treatment.

In the per-humid environment of Babinda, (Gilmour, 1977; Gilmour *et al.,* 1982) noted that whilst peak discharges increased slightly following logging and clearing, the statistical evidence, however, still remained weak. In addition, there were no significant changes in duration of quickflow, quickflow volumes or time to peaks. The runoff generation studies of the 1970s and the subsequent K^* survey across both catchments greatly assisted the interpretation of these results. It becomes evident that the position of the throttle (or impeding) layer in relation to the prevailing rain is very important, however, in determining the magnitude of quickflow response (cf. Fig. 16b).

The existing "paired catchment" studies do not signal large absolute increases in streamflow, but the spatial coverage is limited and more replications are required. The Babinda experiment does highlight, however, that floods can emanate from forests and that they do not act as infinite "sponges" (see Bonell *et al.,* 1991, for example of January, 1981). Further west of the Babinda catchments on the Bellenden Ker range, more impressive rain amounts are recorded (Mt Bellenden Ker Top, mean annual rainfall 8065 mm, 1974–89 incl., Bonell 1991c). Herwitz (1986) presented K^* results for a sample plot on the Mt Bellenden Ker range which showed the same trend (down the profile) in magnitude to that described for Babinda.

Consequently, widespread saturation overland flow can occur during the summer monsoon season, in addition to localized Hortonian overland flow from stemflow (Herwitz, 1986; Bonell, 1991c). It is from such mountain areas covered in "pristine" forests that a major contribution to the recurrent disastrous flooding is made by affecting the lower reaches of the major rivers of the wet tropical coast (Bonell, 1991a). Annual runoff coefficients also range between 0.58 and 0.90 (Bonell, 1988).

Available scientific evidence then suggests that floods can occur from undisturbed forests when storm events increase in intensity and duration – a point that was also highlighted by Pereira (1991, p. 141). For example, the rainfall of 450 mm to 750 mm on two days, and even over 1000 mm, reported in the southern Thailand flood of November 1988 (Rao, 1988) are more towards the typical rainfall regime experienced in northeast Queensland. As Hamilton (1990) noted, damage was more likely caused by the combination of heavy rainfall, compounded by landslips temporarily blocking stream channels, rather than forest removal. It is worth including the following quotation: "... while, indeed, there maybe increased stormflow volumes and peaks in streams emanating from a logged area in smaller, frequent storm events, as storm events increase in intensity and longevity, and as one moves further down the watershed into increasingly larger basins, the effects of logging dwindle to insignificance for major storms and on large river systems..." He adds, "... it is a question of scale, ... microscale-yes, macroscale-no ..." (Hamilton, 1990, p. 5). Hamilton did concede however, that increased local flooding aggravated by forest clearance can have adverse effects, viz increased erosion, if good management practices are not adopted.

Future research priorities and the role of agroforestry

Bruijnzeel (1987) pertinently observed that there was a dearth of studies dealing with the impacts on streamflow of shifting cultivation or converting natural forest to annual cropping, as these two activities account for a major portion of tropical forest destruction. The Mbeya experiment (Edwards, 1979a, b) is one of the most appropriate, although the environment (long dry season; high infiltration capacity of surface soils) may be atypical for the humid tropics. Nevertheless, the study established a significant decrease in water use by the cultivated catchment. As the dry season progressed, the differences between the forest and cultivated catchment increased. During the wet season, however, water use was similar.

The major challenge is the restoration and rehabilitation (as defined by Schreckenberg *et al.,* 1990, p. 18) of degraded, former forested land to take the pressure off the remaining relatively undisturbed forest in both the developing world (Schreckenberg & Hadley, 1989; Schreckenberg *et al.,* 1990) and the "developed" humid tropics (e.g. northeast Queensland, Bonell, 1991c). Bonell (1991b, c) noted a dearth of controlled catchment experiments in the humid tropics to evaluate the

impact of reforestation and the corresponding time lag in recovery of the hydrology/erosion (including total water yield) to predisturbed levels. It was suggested that this should be one of the research priorities related to the World Heritage Area of northeast Queensland. In addition, there is a major gap in knowledge concerning the hydrological changes resulting from the introduction of agroforestry (Hamilton with King, 1983), despite such practices being well-established in, for example, South and South East Asia and through the Pacific Islands (Rao, 1989). Recently, Lal (1990) reviewed the limited knowledge concerning the role of agroforestry systems in soil and water conservation within the humid tropics. Whilst not following the usual controlled experiment catchment conditions, data presented (Lal, 1990, p. 340) provided an example that properly maintained hedgerows of *Leucaena* in conjunction with the alley cropping of maize can effectively reduce total runoff and erosion. Lal (1990) concluded that such systems are effective by:

–   reducing runoff velocity by creating a barrier formed by a closely-spaced hedge of shrubs
–   decreased runoff amount by allowing more time for water to infiltrate into the soil
–   minimizing raindrop impact and sheet erosion due to increasing protection of ground cover, by e.g. leaf fall mulch, shrub canopy
–   curtailing sediment transport in overland flow.

SELECTED SIMULATION AND MODELLING TECHNIQUES FOR TROPICAL HYDROLOGY

Input-output or "black box" models

Black box models solve the relationship between a given input and output when the physical background of the process is not fully understood. That is, they contain no physically-based transfer function to relate input to output such as rainfall and streamflow. Consequently, there is no attempt to describe the physical processes involved with the transfer of water through the catchment system e.g. regression analysis, unit hydrograph or real-time forecasting models (Anderson & Burt, 1985b; Wood & O'Connell, 1985).

Conceptual models

Conceptual models describe the physical processes in the catchment by mostly empirical formulae supported by field and laboratory experiments. Each important physical process which takes place within the system is mathematically defined; the description usually reflecting recent knowledge of the existing processes and their interaction. In conceptual simulation, the hydrological cycle or parts of it are described by mathematical and/or empirical formulae. Because most of the formulae applied contain parameters characterizing the physical features and properties of a modelled region, the conceptual models are sometimes called *parametric models*. The lack of detailed knowledge of the drainage basin, such as the physical and biological processes governing water movement, forces several simplifications to be made. The most common simplification made in catchment modelling is *lumping* or spatial averaging. As Blackie & Eeles (1985, p. 313) commented "... the implication is that the catchment system, its inputs and response can be represented mathematically, using only the dimensions of depth and time ..." They continued, "... no account is taken of variations within the catchment of precipitation, vegetation, soils, geology or topography." Blackie (1979c) described the application of the Institute of Hydrology conceptual model in connection with the East African catchment at Kericho (Blackie, 1979a). Blackie's (1979c, p. 42) concluding points highlight the reasons for exercising caution in using lumped models "... to be of real value ... the functions and parameters in question must have values firmly established from process studies."

Physically-based models

Since the 1980s, there has been increasing attention devoted to the development and testing of physically-based distributed models, such as the Institute of Hydrology Distributed Model (IHDM) (Beven, 1985) and the Systéme Hydrologique Européen (SHE) model (Abbott *et al.*, 1986a, b; Bathurst, 1986a, b). A good review of distributed models was provided by Beven (1985). The attraction of these models is that they intend to use parameters which have physical interpretation and have apparent theoretical rigour when related to hydrological processes of the environment. Beven (1985) outlined the potential applications of such physically-based models, including the prediction of the effects of land-use change by changing selected parameter values and forecasting the hydrological response of ungauged catchments. By contrast, lumped parameter models (Blackie & Eales, 1985) require an acceptable length of meteorological and hydrological records for their calibration – which may not always be available, especially in the humid tropics. Consequently, this aspect alone makes physically-based models more attractive. Furthermore, calibration of lumped models involves curve fitting which makes physical interpretation of fitted parameter values very difficult (Beven, 1989).

From the previous section, it is evident that providing that a long period of hydrological data is available for parameter calibration, modelling the rainfall-runoff process can be achieved satisfactorily through the use of input-output or blackbox models, e.g. unit hydrograph or lumped conceptual models which are quasi-physical in nature (Anderson & Burt, 1985b). For example, the use of unit hydrograph theory in drainage basins of high infiltration capacity e.g. forests, may successfully model the storm hydrograph for engineering hydrological applications, but this theory has no physical basis when related to flow pathways on hillslopes for application in geochemical or erosion studies (Beven *et al.*, 1989). Attention has been redirected towards more complex, deterministic models or distributed, physically-based models (Wood & O'Connell, 1985)

along the lines of the proposed Multiple Independent Pathways (MIPS) model (Beven *et al.*, 1989) for routing subsurface flows on a hillslope, using probability density functions of the component processes.

There has been a dearth of experimental catchment studies using physically-based models in the humid tropics, apart from isolated testing such as the work of Storm *et al.*, (1987) and Sonsin (1989) in different parts of Thailand. Sonsin (1989) reported preliminary trials of the SHE model to quantify the hydrological behaviour and effects of land-use change on the hydrological regime of three experimental catchments, northeast of Bangkok. Only moderate success was obtained in simulating the hydrological regime. Discrepancies between simulated and measured stream discharge records were attributed mainly to an inadequate description of the spatial variations of catchment rainfall. Earlier, Storm *et al.* (1987) established that they could adequately simulate the streamflow from an upland catchment of forest and agricultural land-use, despite the difficulties of spatial representation of meteorological inputs and soil data. Experiences from humid temperate studies, however, suggest that the confident application of the current generation of physically-based models still poses problems, and that further theoretical developments and associated field testing are required (Beven, 1989). The problem of using small-scale field measurements and associated equations depicting small-scale physics of homogeneous systems, e.g. infiltration, and scaling-up such sub-grid processes to heterogeneous, model grid scale (250 x 250 m blocks in the case of the SHE model) still presents a major challenge. In effect, there is no present theory for the lumping of subgrid processes in hydrology. The use of small-scale physical equations, or even say a single point infiltration equation with "effective parameters" for predicting surface runoff production over heterogeneous soils, may not be successful when applied at the model grid scale (Beven *et al.*, 1988; Beven, 1989). There has been several studies searching for alternative, consistently "effective" parameters which are able to reproduce the spatially variability of subgrid processes (e.g. Sharma *et al.*, 1987; Binley *et al.*, 1989a, b) with only limited success (Beven *et al.*, 1988) when dealing with catchment runoff response. Such effective parameters are difficult to relate to field-measured parameter values because of scale and limited number of points of measurement. In addition as the scale increases, different interactions between different hydrological processes influence the "effective" parameter values which makes their calibration difficult (Beven *et al.*, 1988).

An alternative is to increase the scale of measurement which led Wood *et al.* (1988) to propose the concept of *Representative Elementary Area* (REA) (analogous with the concepts of REV and "repetitive unit" (Bear, 1979) presented earlier) for predicting runoff production in the context of modelling at the catchment scale. For their particular USA study, Wood *et al.* (1988) determined c.1 km^2 as the scale of minimum variance, based on simplistic assumptions of the runoff

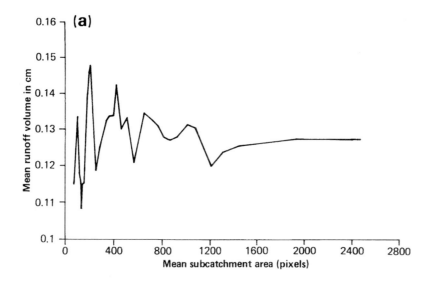

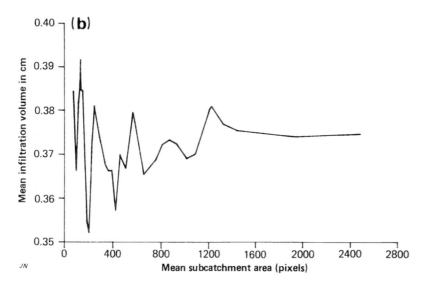

JN

Figure 25: Runoff production and infiltration volume from subcatchments of different size (1 pixel = 30 x 30 m), based on simulation results using uniform soil properties and rainfall so that variability arises only from topography (from Wood *et al.*, 1988). (a) Cumulative runoff volume; (b) cumulative infiltration volume. In this example, the "Representative Elementary Area" (REA) (or scale of minimum variance) is of the order of 1 km^2 (After Wood *et al.*, 1988).

generation process (Fig. 25). Beyond the scale of REA, increases in variances between different areas are more legitimately due to nonstationarity in either soil, rainfall or morphology (Beven *et al.*, 1988). Consequently, the suggestion is that measurements taken at the REA scale may be able to simplify the representation of catchment responses whilst still retaining the important effects of heterogeneity on hydrological processes (Beven *et al.*, 1988). When consideration is given to extension of results from gauged to ungauged catchments, the concept of hydrological similarity is now being tested. This is based, for example, on dimensionless catchment similarity parameters (Sivapalen *et al.*, 1987) which are intended for use in physically-based models representing runoff production.

The potential for useful application of the current generation of physically-based models to real catchments would seem still limited, based on the benchmark review of Beven (1989).

Problems associated with errors in model structure, estimation of parameter values, specification of initial and boundary conditions, the need for a theory for lumping of subgrid scale processes and rigorous assessment of uncertainty in model predictions means that such physically-based models do not overcome the disadvantages of lumped conceptual models. The relevance on "effective" grid scale values of parameters and variables means that so-called physically-based models are themselves lumped conceptual models because they do not account for the effects of spatial heterogeneity (Beven, 1989). Consequently, existing progress maybe appropriately described as an illusion (O'Loughlin, 1990a).

Earlier, Dooge (1986) had remarked on the futility of developing scaling laws for hydrological behaviour ranging from microscale to macroscale. He called for a change, with the need to develop new hydrological laws for the mesoscale of interest. With the possible exception of scaling theory applied to soil physics (Youngs, 1988), these remarks are appropriate to all aspects of hydrological behaviour. In that respect, recent developments based on digital terrain models (DTMs) for hillslope hydrology, and the associated different scales (and methods) of parameterization, are going some way towards addressing Dooge's (1986) ideas (O'Loughlin, 1990a). O'Loughlin (1990a), however, emphasized some of the existing difficulties in merging hydrological behaviour across a variety of scales even with these alternative developments. Nevertheless, DTMs hold great potential for application in the humid tropics and require further detailed consideration.

Digital terrain models for runoff production

When considering runoff production, topographically- and physically-based hydrologic models, based on digital terrain models (Moore *et al.*, 1991), offer an alternative to the problems associated with the present generation of other "physically-based" models (Quinn *et al.*, 1989). As Moore *et al.* (1991, p. 21) commented, "... the last five years has seen an increasing emphasis on the need to predict spatially variable hydrologic processes at quite fine resolutions. We are now in the era of *spatial modelling*." Such models recognize the morphology controls on water flow pathways which are supported both theoretically (e.g. Zaslavsky & Sinai, 1981) and experimentally (e.g. Anderson & Burt, 1978; Beven, 1978) so that lateral flow of water widely occurs. This even applies in uniform soils without a subsoil layer of lower permeability, providing that a topographic gradient exists (O'Loughlin *et al.*, 1989). The topographic controls allow the lateral flow of water to redistribute soil moisture, which, in turn, dictates the landscape's response to storm runoff. The digital terrain models then aim to predict patterns of soil saturation and their corresponding relationship with the production of saturation overland flow, Hortonian overland flow (where appropriate) and subsurface storm flow. They can also accommodate spatial differences in soil, vegetation and land-use as expressed in terms of estimated or measured transmissivity (O'Loughlin 1986).

Moore *et al.* (1991) provided a detailed review of the current status of digital terrain modelling, including a description of the various digital elevation models (DEM) for structuring a network of elevation data using either square-grid, triangular irregular- or contour-based networks. Topographic attributes can be divided into primary and secondary (or compound) attributes, with primary attributes such as elevation and slope being calculated from elevation data. As Moore *et al.* (1991, p. 11) noted, "... compound attributes involve combinations of the primary attributes and are indices that describe or characterize the *spatial variability of specific processes occurring in the landscape* such as soil water content distribution or the potential for sheet erosion." Compound attributes are derived empirically within the framework of a simplification of physical processes. A summary of the appropriate primary topographic attributes and derivation of analytically derived compound topographic indices were presented by Moore *et al.*

(1991, pp. 11–17). A framework for understanding this procedure is shown in Fig. 26 which summarizes the application of a DTM, using TOPOG as an example. The DTM component is represented in the TOPOG Kernel by Sections 1B (Digital Elevation Model) and 1C (Terrain Analysis Routines). These provide the primary attributes which are required in the subsequent physically-based models associated with application modules 2 to 5.

Following the introduction of the Geomorphic Instantaneous Unit Hydrograph (GIUH) by Rodriguez-Iturbe and Valdez (1979), other topographically-based hydrologic models have emerged, all of which have continued to be adapted (including the GIUH) for various geomorphological and biological applications (see Moore *et al.*, 1991, pp. 13–24; Rose, this volume). When considering the spatial modelling of the runoff process, the TOPMODEL of Beven *et al.*, (1984), which was derived from earlier ideas of Beven and Kirkby (1979), and TOPOG (O'Loughlin *et al.*, 1989), which was the culmination of parallel developments by O'Loughlin (1981; 1986), both have great potential for application in land management planning. These models also have several features in common. Under hypothetical, steady-state conditions, they utilize a hydrologically-based compound topographic index to determine soil moisture distributions and, in turn, to predict transient surface or shallow subsurface runoff responses. This index is achieved for specific points across a drainage basin by dividing the topographic contours up into elemental strips and calculating the lateral drainage flux from upslope per element of contour length based on the variables: drained specific catchment area, transmissivity when the soil profile is saturated, and the local slope angle.

TOPMODEL or TOPOG can cope with the limited environmental databases associated with many developing countries of the humid tropics, and at the same time, address some of the more urgent land management issues in these areas. Therefore, it is worth expanding in more detail on some of the theoretical aspects of these models.

TOPMODEL Quinn *et al.* (1989) provided an up-dated, thorough review of the theoretical basis and applications of TOPMODEL. Essentially, TOPMODEL was developed for flood forecasting purposes, and is topographically driven – commonly using a grid-based method of DEM analysis in combination with a distributed hydrological model. Recent versions of TOPMODEL also consider Hortonian overland flow (Beven, 1986a, b). The topographic/soil index is expressed as $A_s / (T_o \tan \beta)$, where A_s (l^2) is the area drained per unit contour length at a point, T_o is the saturated transmissivity ($l^2\,t^{-1}$) of the soil at the point and $\tan \beta$ is the local slope angle. This index is used to predict a local soil moisture deficit, S_i^* (in units l, usually metres of deficit) at any point i from a catchment average moisture deficit $\bar{S}^*$, as:

$$S_i^* = \bar{S}^* - m \ln (A_s / T_o \tan \beta)_i + m\gamma^* \qquad (32)$$

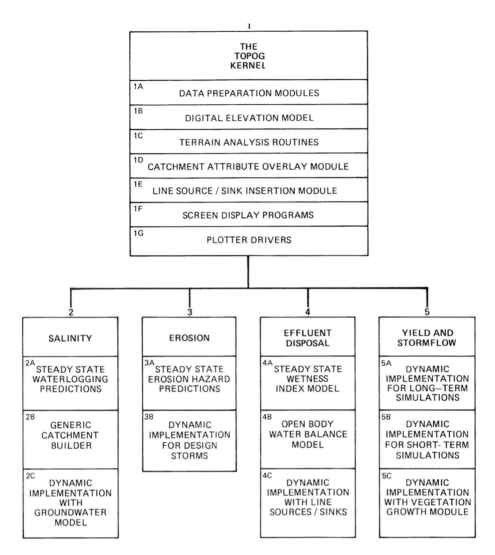

Figure 26: Components of the TOPOG modelling package. Four application modules are shown, each connected to a central "computing engine" known as The TOPOG Kernal. Some of these modules were in the process of development as at November 1991 (O'Loughlin, 1991, pers. comm., Australian Centre for Catchment Hydrology / CSIRO Division of Water Resources, Canberra, Australia).

where m is a parameter of the exponential decline in transmissivity with soil depth (expressed in metres) and γ^* is the expected value of $\ln (A_s / T_o \tan \beta)$ in the catchment or subcatchment. The higher the value of $\ln (A_s / T_o \tan \beta)$, the lower the soil moisture deficit, and, therefore, the more likely the soil will saturate to the surface. A major assumption is that similar $\ln (A_s / T_o \tan \beta)$ values respond hydrologically in a similar manner for uniform inputs.

Equation (32) may be rewritten in the form:

$$\frac{S_i^* - \bar{S}^*}{m} = [\ln (T_o)_i - \ln (T_o)] - [\ln (A_s / \tan \beta)_i - \Lambda] \qquad (33)$$

where Λ is the expected value of $\ln (A_s / \tan \beta)$ in the catchment (formula defined in Beven & Kirkby, 1979; Quinn *et al.*,

1989). In the context of uniform inputs of rainfall, equation (33) may be described as the scaled deviation in storage deficit being equal to a soil transmissivity deviation from the expected value in $\ln (T_o)$ minus a topographic deviation from the expected value in $\ln (A_s / \tan \beta)$. Two important aspects pertain to this equation. The greater the negative deviation from the mean storage deficit, the more likely the soil will reach saturation. Even more significant, Beven *et al.* (1988) highlighted that this equation demonstrated the overriding influence of topography over transmissivity in shaping the distribution of soil moisture deficits, unless there was a strong correlation between T_o and $\ln (A_s / \tan \beta)$. As Beven *et al.* (1988, p. 363) noted, "given the variability of several log units in $\ln (A_s / \tan \beta)$ index (Fig. 3 Beven *et al.*, 1988), in general, the right hand side of equation (33) will be dominated by the topographic deviations, even assuming large variances (say 1 log unit) for the soil transmissivity." Between storms, however, the basin mean transmissivity will control lateral drainage and the mean storage deficit.

Equation (33) is rearranged for S_i^* and used as part of the calculation of subsurface lateral fluxes to discharge through the stream bank.

$$Q_b = \sum_{l_i}^{N} l_i \ (T_o \tan \beta) \exp \ (-S_i * / m) \qquad (34)$$

where the summation is occurring over n individual river reaches of length l_i, and for both banks. This stream discharge is simulated over time.

In the UK example given by Quinn *et al.* (1989) where subsurface flows dominate because of the high surface transmissivities and weak rainfall intensities, the infiltration excess option (Beven, 1986a,b) in TOPMODEL was not used. In the humid tropics, however, this option would be necessary in certain environments (e.g. Wierda *et al.*, 1989).

Subsequent testing of TOPMODEL by Quinn *et al.* (1991) using information from a small catchment (Booro-Borotou, 1.36 km²) in Côte d'Ivoire is of particular interest here. This catchment is located in the seasonally-wet humid tropics where the runoff process is a mixture of saturation (saturation-excess) overland flow and subsurface stormflows. In contrast to humid temperate work, Quinn *et al.* (1991) noted that the assumption of subsurface stormflow rates and pathways being proportional to a hydraulic gradient approximately equal to the surface slope is not always correct. The Booro-Borotou catchment consisted of deeply weathered soils which encouraged a deep saturated zone on the upper slopes, a feature typical of many humid tropical headwater areas. The *concept of a reference level* was introduced into the model which catered for the deviation in water table levels from the soil surface, and the manipulation of field experience concerning the water table gradient within the upper slope areas under wet conditions. Where the water table intersects the surface, the surface elevation values resume their definition of the reference level as before. Once the appropriate reference level has been established for the whole catchment, it is used within the TOPMODEL structure to obtain the ln $(A_s / T_o \tan \beta)$ distribution and flow directions as for a surface DTM. As noted by Quinn *et al.* (1991, p. 74) the introduction of the reference level concept "... was vital in this example since it was known that a deep unsaturated zone was acting as a large long-term reservoir for water in the soil for much of the rainy season's volume of precipitation." Furthermore, the difference in land surface elevation and reference level elevation can be used to estimate storage and travel times in the unsaturated zone. The writers acknowledged that with little field information available, the definition of an appropriate reference level, however, will be difficult and subjective. The difficulties of determining the depth-transmissivity relationship were also addressed (Quinn *et al.*, 1991, p. 74).

During the same study, the sensitivity of a multiple flow direction algorithm (rather than the commonly used undirec-

tional algorithm) and changes in DTM grid scale (reduced from 50 m to 12.5 m) in TOPMODEL were also evaluated. Both aspects had an important and improved effect on model predictions of moisture status. Such improvements are essential for simulating the hydrological response of small catchments (Quinn *et al.*, 1991).

TOPOG   In a similar manner to the early development of TOPMODEL, the TOPOG model simulates the re-distribution of moisture by subsurface storm flow dictated by flux-gradient equations. The assumption of the analysis is once again a steady-state behaviour of the perched water tables, which continues in terms of efflux into the stream as the "delayed flow" component of the storm hydrograph.

The basic criterion in O'Loughlin's (1981, 1986) work is that for the development of local surface saturation on a hillslope, the accumulated drainage flux, (the product of upslope partial catchment area A and the areal drainage flux $q^*(x, y)$ passing across a contour of length B exceeds the product of soil transmissivity T and the local slope M (Fig. 27). This is expressed as follows:

$$Aq^* (x, y) / B \geq TM \qquad (35)$$

The transmissivity, $T \ (l^2 \ t^{-1})$, is the depth integral of the field-saturated hydraulic conductivity, K^*, through the soil profile (which can be adjusted, depending on the thickness of the saturated layer), and the "net lateral drainage flux," $q^*(x, y)$ is a slowly varying residual of rainfall less transpiration and deep drainage losses.

This criterion can be rewritten in terms of a normalized wetness function, $w(x, y)$, that can be calculated at any point in the catchment:

$$w \ (x, y) = \frac{1}{MBL} \ (\frac{\overline{T}}{T(x, y)}) \int \frac{q^*(x, y)}{\overline{q}^*} \ dA \geq$$
$$\frac{\overline{T} A_t}{Q_o L} = W \qquad (36)$$

Here, L is a reference length, conveniently the mean hillslope length, and A_t is the total hillslope or catchment area generating the outflow $Q_o \ (lt^{-1})$. The relationship between the residual drainage flux from a hillslope and the hillslope outflow, Q_o, is

$$\int_{At} q^*(x, y) \ dA = Q_o \qquad (37)$$

Assuming steady-state conditions, where $q^*(x,y)$ everywhere was uniform with a value of $\overline{q}^*$, then $\overline{q}^* = Q_o/A_t$. Consequently, for an entire drainage basin when information on K^* and the delayed flow component determined from storm hydrographs are available, the right-hand side of equation (36) can be used, represented by the dimensionless quantity, W, as part of the prediction of the saturated zones. The attraction of O'Loughlin's (1986) approach is that at the base level, it does not require the input of any other information beyond detailed topographic data. The topography is divided up into elemental

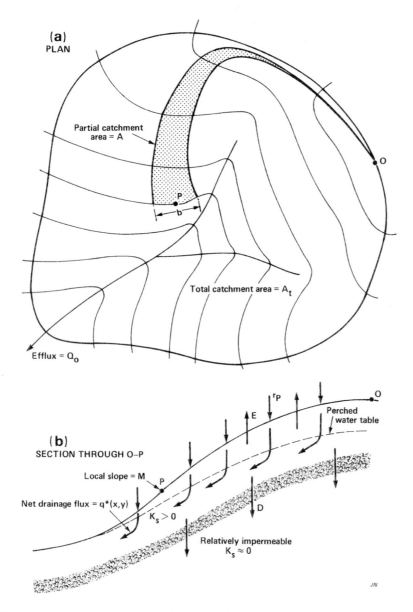

Figure 27: Definition sketches for (a) plan of draining terrain and (b) section along transect of partial catchment area. The "net lateral drainage flux," $q^*(x, y)$, is a spatially varying quantity. This residual flux is calculated by subtracting the vertical components of flow (deep drainage D and evaporation, E) from a steady-state percolation, r_p. All quantities q^*, D, E and r_p have dimensions $[lt^{-1}]$ (After O'Loughlin, 1986, modified to include symbols in O'Loughlin *et al.*, 1989).

strips, following a routine described by Dawes & Short (1991, in preparation). Mathematical manipulation and assumptions (such as a constant transmissivity across the drainage basin, $\bar{T}/T = 1$, so that there is no spatial heterogeneity and uniform drainage flux $q^*(x, y)$) avoids the need for hydrological information such as K^* to determine the wetness function. Absolute values of $q^*(x,y)$ are not required, instead the ratio $q^*(x, y) / \bar{q}^*$ acts as a weighting factor that expresses the local drainage flux as a fraction of the average flux over the catchment.

TOPOG can also cater for spatially varying properties such as soil transmissivity, infiltration properties or vegetation cover by digitizing a series of polygons (described in O'Loughlin, 1986; O'Loughlin *et al.*, 1989) and assigning the appropriate values automatically to each terrain element that falls within the polygon – using a "point-in-polygon" tech-

nique (O'Loughlin, 1990b). The method also estimates evaporation as part of the determination of the net lateral flux (O'Loughlin *et al.*, 1989).

Later, equation (36) was re-written (O'Loughlin *et al.*, 1989; O'Loughlin, 1990b) to generalize the criterion for soil saturation as follows:

$$W(x, y) \; = \; \frac{1}{MT}\int q^*(x, y) \; dA_L \geq 1 \qquad (38)$$

where $W(x, y)$ is a non-dimensional "wetness index" (w/W in equation (36)) and dA_L has dimensions $[l]$. Essentially, the wetness index is a ratio between subsurface lateral flow within the perched water table of any point and the capacity of the soil to conduct the flow. Where the wetness index exceeds

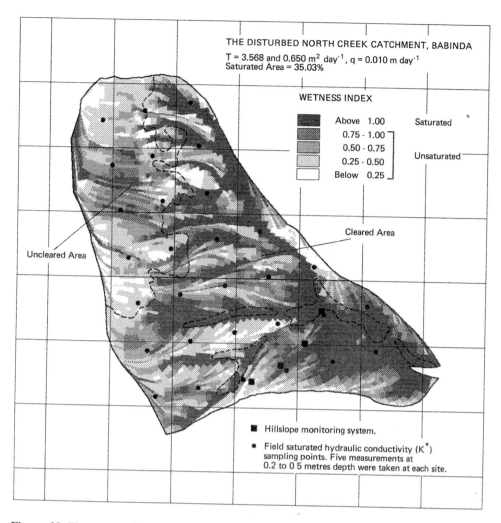

THE DISTURBED NORTH CREEK CATCHMENT, BABINDA
T = 3.568 and 0.650 m^2 day^{-1}, q = 0.010 m day^{-1}
Saturated Area = 35.03%

WETNESS INDEX

	Above 1.00	Saturated
	0.75 - 1.00	
	0.50 - 0.75	
	0.25 - 0.50	Unsaturated
	Below 0.25	

Cleared Area

Uncleared Area

■ Hillslope monitoring system.

● Field saturated hydraulic conductivity (K*) sampling points. Five measurements at 0.2 to 0.5 metres depth were taken at each site.

Figure. 28: The pattern of the wetness index using TOPOG over the disturbed North Creek catchment for a steady-state drainage over the catchment of 10 mm day^{-1} (after Bonell, 1991c). Notes: The hillslope monitoring system provides hydrometric data on changes in soil water (matric potential) and groundwater pressure. Such information could also form part of the "ground truth" of TOPOG. Soil water and groundwater samples are also collected for environmental isotopic analysis. Transmissivity ($T = K^*$. *depth*) is the depth-integral of K^* for the top one metre. For the tropical rainforest $T = 3.568$ m^2 day^{-1}. For the remaining cleared area, $T = 0.650$ m^2 day^{-1}.

unity, water-logging and saturation overland flow occurs. At locations where the index is less than unity, it is a measure of the degree of saturation within the profile (O'Loughlin *et al.,* 1989). This means that the spatially and temporally varying projected sources of storm runoff and, in turn, inferences concerning the spatial origin of hillslope eroded material, can be determined from the delayed flow flux (i.e. average flux throughout a storm hydrograph) between different storms (e.g. Moore and Burch, 1986; Moore *et al.,* 1988a, b; see Rose, this volume).

A software package and corresponding manual (ACCH, 1990) have been developed for the practical application of TOPOG. An example of its use is presented in Fig. 28. The distribution of the wetness index for daily delayed flow (Q_o) of 10 mm per day from the disturbed, North Creek catchment of the Babinda experiment (Bonell, 1991a, c) is shown. Included in this analysis is the use of the polygon technique to differentiate between the transmissivities of the surviving rain forest and the disturbed area. The higher surface transmissivity asso-

ciated with the remaining undisturbed rain forest in the headwaters, and within the rain forest buffer strips of the stream channels, is reducing the wetness index for that particular daily discharge. When $Q_o = 20$ mm per day (frequent in monsoon storms), however, TOPOG predicts that a large part of North Creek (68%) is showing an index greater than 1.

If the user wishes to calculate saturation overland flow from saturated areas using the steady-state version, and compare the model predictions with observed hydrographs, then TOPOG still depends on an external arbitrary method (whose problems were previously discussed) for partitioning each storm hydrograph into quickflow and delayed flow. The procedure commonly used prior to the application of TOPOG relates to a low-pass digital filter method (Lyne & Hollick, 1979; restated in Moore *et al.,* 1986), emanating from signal processing, which, in the absence of any other physical information, is intuitively the most appropriate method compared with other arbitrary techniques (Ward & Robinson, 1990, pp. 234–237). The digital filtering method is particularly suitable to highly responsive drainage basins, where deeper, permanent ground-water contributes only a very small proportion of storm hydrographs and shallow perched water tables respond relatively quickly to rainfall (Moore *et al.*, 1986).

Such conditions conform with the Babinda experiment, especially under monsoonal rainfalls. Whilst a deep ground-water body sustains perennial, dry weather flow, dramatic increases in storm discharge (up to two orders of magnitude from pre-storm discharge) emanating from the transient, shallow perched water table and associated saturation overland flow makes this environment more suitable for using filtering techniques (Bonell & Gilmour, 1978; Bonell *et al.* 1981). Furthermore, the widespread occurrence of the shallow perched water table, especially in rainfall events of long duration with wet antecedent conditions, suggests that the continued use of the soil surface as the reference level is perhaps more valid in this environment, compared with the experience of Quinn *et al.* (1991) in Côte d'Ivoire. These assumptions, however, require some caution based on preliminary results from the current hydrogeochemistry study. Previous discussion has indicated that the deeper groundwater body is influencing the storm hydrograph under selected conditions of antecedent wetness and rainfalls more significantly than previously thought (Bonell & Gilmour, 1978, Bonell *et al.,* 1981). In addition, this deeper, permanent water table may occupy a more transmissive layer than assumed in TOPOG. Furthermore, this groundwater does not intersect the soil surface (as shown in Fig. 27 and Eqn. 35) except through the stream bank. These aspects are currently not taken into account in TOPOG and require further consideration. In addition, subjectivity still prevails throughout the hydrograph separation procedure, including the choice of filter factor and time increments by which the unfiltered runoff rates are interpolated (Bonell, 1991c). Further progress in the use of environmental isotopes, when linked with improved hydrograph separation procedures, may provide a more physically-based procedure in the future.

Comparison of TOPMODEL and TOPOG In their present form, TOPMODEL has the advantage of simulating stream hydrographs whilst TOPOG has the ability to predict a range of wetness functions, in the absence of a hydrological database. O'Loughlin (1990b) provided a comparison between the approaches. He noted that the applications of TOPMODEL have been directed more towards flood runoff studies and total runoff area, rather than the spatial distribution of soil moisture (and runoff production areas) *per se*. Nevertheless, TOPMODEL has these capabilities during the course of predicting local storage deficits across hillsides. Even more significant to the humid tropics are the recent adjustments to TOPMODEL so that changes in the reference level from the soil surface to the water table can be included in circumstances where the unsaturated zone is substantial in depth, especially at upslope locations associated with deeply weathered soils. The inclusion of the multiple flow direction algorithm and smaller grid scale also improves TOPMODEL (Quinn *et al.*, 1991). On the other hand, O'Loughlin (1990b) detailed advantages in using the TOPOG procedure for subdivision of the landscape into topographic units. TOPOG, for example, showed considerable success in simulating storm hydrographs (Moore *et al.*, 1986) through the use of a lumped parameter model to calculate rapid storm runoff. This was based on the predicted saturated areas by TOPOG being effectively impermeable to rainfall and supplied by rapid subsurface stormflow from upslope. In addition, both Moore *et al.* (1986) and Burch *et al.* (1987) showed how the *effective transmissivity* of both small forested and grassland catchments could be estimated by using the wetness index versus percent saturated source area relationship in combination with observed stormflow data. They used an analogy with the Theis (1935) graphical method for determining the transmissivity of a groundwater aquifer under conditions of unsteady flow.

TOPMODEL has always been dynamic. For example, the simplified theory presented in Beven (1986a) showed how the subsurface moisture storage could be represented as a sequence of steady states, being good approximations to storage relationships under transient conditions, by the use of equation 32 (expressed in favour of $\bar{S}*$) in a continuous accounting model. Such an assumption depends on the selection of the optimum time step which is of sufficient duration for upslope inputs to be transmitted to the channel bank, and, paradoxically, short enough so that the dynamics of the variable contributing area within a storm period can be simulated. Hourly accounting periods were used (Beven, 1986a). Furthermore, the dynamics of the model catered for the range of delivery mechanisms of runoff production described earlier (Freeze, 1974). By contrast, O'Loughlin and co-workers have only so far published versions of TOPOG appropriate to steady-state conditions. Nevertheless, prolonged dry periods commonly associated with the dry season in certain humid tropical areas or where prevailing soil matric potentials are close to zero in conjunction with long duration storms events, e.g. northeast Queensland (Bonell, 1988), makes TOPOG in particular an attractive proposition for indicating soil moisture and saturated zones encouraging saturation overland flow. Under stormflow conditions where Horton (infiltration-excess) overland flow develops, then the steady-state version is clearly inadequate. TOPOG has now been extended within application module 5 (shown in Fig. 26) to incorporate an unsteady-state component, and is associated with the transient-wetting up of a catchment in the early part of storms, which subsequently grades into the steady-state version. In addition, this fully dynamic version now computes the *total* hydrograph and its components (subsurface stormflow, saturation-excess overland flow from saturated areas, Hortonian overland flow) within the Yield and Stormflow module 5 (O'Loughlin, 1991, pers. comm.) which makes the model more versatile and practical.

Another aspect of TOPOG is that it also takes into account the evaporation process, driven by net radiation, for predicting antecedent moisture (O'Loughlin *et al.*, 1989, O'Loughlin, 1990b).

Application of DTMs to the humid tropics It can be concluded that of all the so-called physically-based models, the described DTMs probably hold the greatest potential for application in the humid tropics and are most suitable for application to small or medium (<1000 ha) headwater catchments. As Moore *et al.* (1991) noted, in many developing countries, there is a general lack of environmental data. The first information that usually comes to hand is a topographic map which enables the preliminary application of DTM's through the calculation of topographic attributes from the elevation data. For example,, the wetness function of O'Loughlin's model, which avoids the need for basic soil hydraulic information, is ideal for use in the humid tropics where hydrological data bases are deficient and in ungauged catchments. As noted by O'Loughlin (1988), this function effectively determines the areas which are most liable to soil water-logging controlled by topographic convergence/ divergence areas and is an important consideration in decision-making on alternatives in land management where agencies do not have access to supporting scientific infrastructure. The output from such models are also fundamental to the future planning and location of data collection networks as the economics for establishing more scientific infrastructure because more feasible. Further improvements to the existing terrain-based indices and the incorporation of more physically-based erosion models, when landscape management is the focus, can be considered later, once hydrological and pedological information become available.

In the meantime, further testing of the performance of these topographically-driven process models, amongst others, in a broad spectrum of environments remains mandatory. The same note of caution was given by Beven *et al.* (1988, p. 365) who used the work of Burt & Butcher (1985) as an example for demonstrating that observed saturated zones exhibit a more dynamic response than were predicted by the generation of topographically-driven process models up to that time. In this regard, the role of field process hydrology studies is critical through the "ground-truthing" of hillslope responses against model predictions. Figure 20 shows the locations of hillslope monitoring systems in North Creek for ground-truthing TOPOG by way of continuous measurement of hydraulic potential in both the unsaturated/saturated zones and extraction of unsaturated/saturated zone water for environmental isotopic measurement.

Macrohydrology – the Interface between Hydrological Processes, Atmospheric General Circulation Models (AGCMs) and Scale

Progress up to the mid-1980s The recognition from modelling that the earth's climate could be sensitive to large changes in surface properties, notably albedo, roughness and soil moisture (Shuttleworth, 1988b) has led to the linkage between the escalating rate of tropical forest conversion and the results from AGCMs. Henderson-Sellers (1987) provided a very honest, critical appraisal (including of her own work) of existing model simulations based on the studies throughout the humid tropics (Potter *et al*., 1975, Wilson, 1984) and for the Amazon Basin only (Lettau *et al.*, 1979; Henderson-Sellers & Gornitz, 1984). Widely varying conclusions were obtained in terms of the impact of tropical forest conversion on precipitation and surface temperature. Apart from the model of Lettau *et al.* (1979) the remainder indicated decreases in precipitation – ranging from 100 to 800 mm year^{-1} for Amazonia, whilst Wilson (1984) suggested a 200–600 mm year^{-1} decrease for the Congo basin and an overall decrease of 230 mm year for the 5°N to 5°S zone. There was also a dichotomy of results concerning temperature, ranging from no systematic change in two of the models (Wilson 1984, Henderson-Sellers & Gornitz, 1988) to a 0.55°C increase for Amazonia (Lettau *et al.*, 1979) and a decrease of 0.4°C in 5°N – 5°S zone (Potter *et al.*, 1975).

Henderson-Sellers (1987) came to several conclusions which are worth mentioning in the context of more recent progress and future planned activities, viz.

- changes in the surface hydrology were at least as important as changes in surface albedo.

- results are sensitive to land parameterizations inherent in such models and were highly simplistic, especially when considering the hydrology. For example, rainfall was generally assumed to "fill the soil" with water, with any excess effectively disappearing from the computational scheme in a "runoff" term. This simplistic scheme, termed a *bucket hydrological model*, was used, for example, in the work of Wilson (1984) through the use of a single soil layer.

- AGCMs which have spatial resolutions between 2° and 10° are designed to address issues concerning large-scale atmospheric circulation, not mesoscale climatology revolving around the environmental impact of change in land-use at local to regional scales. There is, then, the need for mesoscale climatological submodels to be "embedded" within AGCMs.

In summary, Henderson-Sellers (1987) presented a pessimistic future of the capabilities of such models. In that regard, results from the Amazon Region Micrometeorology Experiment (ARME) proved to be a major step forward in addressing the deficiency of surface hydrology measurements directly relevant to climate models for tropical rain forest. This project also marked the opening phase of *Macrohydrology* experimentation (Shuttleworth, 1988b). Significantly, the spatial scale of interest in *process macrohydrology* is much larger than considered so far in this paper, with scales typically being 400 x 400 km. Paradoxically, the time scales of interest are much shorter (commonly 12 minutes) than associated with many hydrological studies (Shuttleworth, 1988b).

During a review of existing macrohydrological experiments, including HAPEX (The Hydrological Atmospheric Pilot Experiment) and FIFE (the First ISLSCP Field Experiment) undertaken in humid temperate areas, Shuttleworth (1988b) evaluated the level of complexity required to provide a one-dimensional description of the surface hydrology relevant at a large space scale but conversely at a short time-scale. He concluded that with this combination of scales, the criterion of minimum complexity holds in process macrohydrology and continued "... the parameters controlling the model must be allowed to lose some of the physical and physiological relevance they possess at a point scale, and may well require redefinition" (Shuttleworth, 1988b, p. 40). Similarly, Dyck & Baumert (1991) observed that modelling of the hydrological cycle at global and mesoscales can be only feasible if spatial and temporal averages are considered. The challenge is to develop algorithms which include empirical allowance for subgrid scale processes in conventional, one-dimensional process description which incorporate the interactive water and energy transfer processes between soil, vegetation and atmosphere as embodied in SVATS (Soil / Vegetation / Atmosphere Transfer Schemes) models (Dyck & Baumert, 1991). The concept of "effective" infiltration, for example, was linked to the local rainfall rate within the rain-covered portion of a grid by a probability density function, although actual values are currently poorly defined (Shuttleworth, 1989b). Similarly, Entekhabi & Eagleson (1989) have developed land-surface hydrology parameterizations for representing subgrid hydrologic processes in AGCMs, using probability density functions of the spatial variability of soil moisture and precipitation. Such functions were incorporated, along with deterministic equations, for a description of the basic physics of soil moisture so that expressions for the hydrological processes could be derived. In recognition of the important role of vegetation, Dyck & Baumert (1991, Fig. 1, p. 35) presented a hierarchy of one-dimensional evaporation models across local (1 km^2), catchment (100 km^2), mesoscale (10 000 km^2) and continental scales, with each model of the hierarchy being capable of being derived from the higher (larger scale) by defined physical idealizations. At the top of the hierarchy are the radiation-dominated models of Priestley and Taylor (1972) and Turc (1961). Earlier, Shuttleworth (1988b) discussed the prospect that at the AGCM scale, simple "energy balance" models such as Priestley & Taylor (1972) might eventually prove to be adequate for a one-dimensional description of areal average evaporation. At this scale, the evaporation process is primarily driven by the imposed net radiation input with some allowance in Priestley & Taylor equation for the diurnal growth of the PBL (De Bruin, 1983b, 1989) and advection. The randomness of convection rainfall, for example, in the Amazon Basin may well have simultaneous compensating effects at a grid scale between wet and dry canopy evaporation to "dampen" concerns of spatial variability normally associated with smaller scales.

When concerning the earlier "bucket models," Shuttleworth (1988b) recommended the incorporation of an additional vegetation canopy store and at least two soil moisture stores to improve AGCMs. The function of the vegetation canopy is important as part of simulating the short-term response necessary in AGCMs in terms of inhibiting evaporation under dry canopy conditions through the canopy resistance and yet providing easy accessible water in the upper soil moisture store. Even if the vegetation canopy and soil moisture stores can be considered uniform at a grid scale, the spatial variability of mesoscale meteorological processes as expressed by convective rainfall have still to be included in modelling considerations.

The problems of parameterizing hydrological behaviour across different scales for AGCMs The previous problems of "scaling-up" that were highlighted in earlier sections (e.g. evaporation, unsaturated zone, runoff generation, physically-based models) become prominent in macrohydrology, especially from the catchment scale upwards for parameterizing at the sub-grid scale (mesoscale) nested within the AGCM grids. Various strategies have been proposed (Dyck & Baumert, 1991; Kavvas *et al.*, 1991; Shuttleworth, 1991; Veen *et al.*, 1991), but clearly more than one approach is required, especially at the catchment scale, in the scaling-up procedure to cross-check closure of the water balance by one-dimensional SVAT models. For example, Dyck & Baumert (1991) cited three approaches for "scaling-up," involving a statistical treatment of heterogeneities for formulating probability density functions, identification of sub-areas with dominant hydrological processes and characteristic parameter configurations (similar to the proposal of Stewart (1989) concerning evaporation) and the integrated pixel information from remote sensing. In this context, the problem of regionalizing evaporation from heterogeneous surfaces at different scales was discussed by both Veen *et al.* (1991) and Kruijt *et al.* (1991). A field experiment has been set up in the Netherlands to focus on the adjustment of the surface boundary layer (the lowest 10% of the PBL) and the length scale adjustment of water vapour and energy fluxes across sharp discontinuities between forest and non-forest. As Kruijt *et al.* (1991) observed, there is a lack of understanding in both the areal integration of processes and the local edge effects between different surface types. A strategy is currently underway using one-dimensional modelling (the Penman-Monteith equation) for each land unit, and then allowing for horizontal interactions between land units by correcting for advection (Veen *et al.*, 1991). A tentative conclusion is that evaporation from small patches of forest may exceed evaporation from a equivalent area of forest in larger patches, due to higher interception losses near the forest edges. This "conclusion" needs consideration in the "scaling-up" procedure. Whilst this experiment is presently confined to the "local-scale" of Dyck & Baumert (1991), and located outside the humid tropics, such studies are fundamental contributions

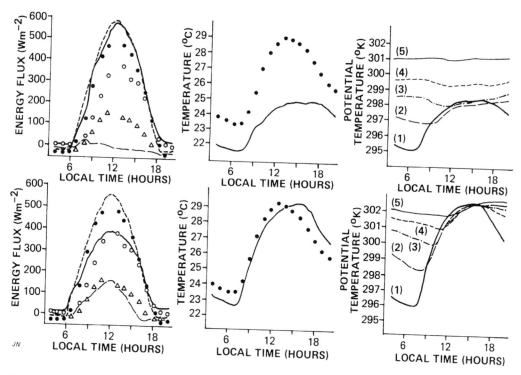

Figure 29: Comparison between model simulations made with a simple bucket model of land/atmosphere interactions (top row) and those made with SiB (bottom row). Comparisons are against average field data measured for fine days and are made for surface-energy fluxes (left hand side), near-surface air temperature (centre), and air temperature at several levels (right hand side). Data are given as points, model simulations as lines. In the surface-energy diagram, the full circles and hashed line are net radiation, the open circles and full line are evaporation and crosses and broken line are sensible heat. (After Sato *et al.*, 1989 and also presented in Shuttleworth *et al.*, 1991).

towards understanding the scale problem of heterogeneous surfaces. These measures form part of the "ground truthing" which will be necessary in the current ABRACOS and future HAPEX-type programmes (e.g. HAPEX-Amazonia, Shuttleworth, 1991), associated with areas of cleared and uncleared forest. As noted by Milly (1991, p. 6) the Netherlands programme "... is a prime example of an elemental physical interaction that needs to be understood before the scaling problem can be solved properly."

Recent progress linking the macrohydrology experiment, ARME, with AGCMs The impact of improved land surface parameterization arising out of ARME in connection with AGCMs was demonstrated by Shuttleworth *et al.* (1991). They showed a comparison, between simulations using a simple bucket model and the Simple Biosphere model (SiB) (described in Sellers, 1987) calibrated against ARME data (Fig. 29). The previous poor simulations of surface exchange fluxes are evident which have clear consequences on climatic simulations when projecting environmental impacts of deforestation. The revised calibrated SiB model also provides a more realistic diurnal growth in the PBL (Shuttleworth *et al.*, 1991).

Shuttleworth (1989b) appropriately emphasized the limitations of using only the single site ARME data in AGCM predictions for the effect of land-use change, and also the absence of similar high-quality data for cleared or replacement forest.

Improvements in predictive accuracy are, nevertheless, possible by at least calibrating model schemes for uncleared forest. In doing this, there is acceptance that the single point representation of mid-continental forest will not address the spatial variability of convective rainfall on interception (Shuttleworth, 1989b). For example,, following improved land surface parameterization schemes arising out of calibrating SiB with ARME data, much improved simulation of the surface exchange was then possible, as shown in Fig. 30 (Sellers *et al.*, 1989). This led to more credible predictions concerning the impact of forest conversion in the Amazon Basin (Shukla *et al.*, 1990) (Fig. 31). Indications are that evaporation will fall by 20% ($\pm$ 10%), that air temperature will rise by 2°C ($\pm$ 1°C) and precipitation will fall by 30% ($\pm$ 20%).

Future programmes The 1990s will witness more ambitious projects linking process macrohydrology with AGCMs within the Amazon Basin. The *Anglo-BR*azillian Amazonian Climate Observation Study (ABRACOS [abracos means embrace in Portuguese]) study is currently underway (details provided in Shuttleworth *et al.,* 1991; BHS Circulation, 1989, p. 11) which will address three of five experimental components in Shuttleworth's (1989b) future research agenda, viz:

– the atmospheric interaction of cleared forest areas
– the difference in near-surface climate at adjacent cleared and uncleared sites

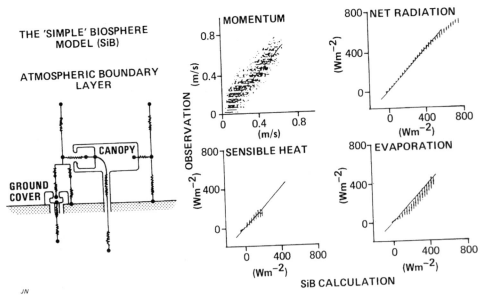

Figure 30: Diagram illustrating the satisfactory simulation of measurements given by SiB following its calibration against ARME data. (After Sellers *et al.*, 1989).

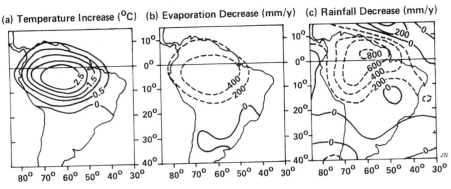

Figure 31: Predictions of the changes in post-deforestation Amazonian climate given by Shukla *et al.* (1990) with a GCM which includes the SiB after calibration against ARME data. (After Shuttleworth *et al.*, 1991).

—   the effect of continental position on the atmospheric interaction of existing forest. This includes consideration given to possible exploitation of advected energy from the Atlantic Ocean in the evaporation process pertaining to near coastal areas (Shuttleworth, 1989b).
ABRACOS also will address the need for incorporation of mesoscale processes in AGCMs.

Other major hydrology initiatives co-ordinated through the World Climate Research Programme as part of the Global Energy and Watercycle Experiment (GEWEX) and in conjunction with the International Geosphere-Biosphere Project (IGBP) through the Core Project "Biospherical Aspects of the Hydrological Cycle" (IGBP-BAHC), are in planning for the 1990s (Dyck & Baumert, 1991; Shuttleworth, 1991). These experiments will incorporate detailed mesoscale field experiments on similar lines of HAPEX (covering a 100 x 100 km area) and FIFE (15–20 km squares). The methodology of such programmes were originally summarized in Shuttleworth (1988b) and later up-dated in Shuttleworth (1991). The pro-

posed densely instrumented networks will encompass mixed regions of cleared and uncleared forest, investigating thereby the effects of tropical forests conversion (HAPEX-Amazonia) to follow-up the research findings from ABRACOS (Shuttleworth *et al.*, 1991). Hopefully, all these new initiatives will address Henderson-Sellers (1987) earlier legitimate criticisms of the first generation of AGCMs, and considerably improve the prediction of post-clearance Amazonian climate.

CONCLUSIONS

This review has highlighted that there are major gaps in several areas of knowledge concerning hydrological processes in humid tropics. Existing work, nevertheless, supports the earlier comments of Hamilton with King (1983, p. 2) who noted that "... there is no intrinsic reason why tropical systems should respond to different laws of nature than do temperate ones. Yet it must be recognized that both the nature of the driving variables (e.g. precipitation intensity or temperature) and the response of a system (e.g. sediment yield or evapotranspiration loss) may be quite different in the tropics." The higher prevailing rainfall intensities, for example, increase the potential for more frequent occurrence of widespread surface ponding and overland flow in disturbed areas, and even in undisturbed areas (Bonell, 1991a, c; Elsenbeer & Cassel, 1990; Wierda *et al.,* 1989). In equatorial regions under high radiation inputs and non-limiting soil water conditions, virtually all of the high net radiation is converted to latent heat (Calder *et al.,* 1986; Shuttleworth, 1988a)

When considering individual water balance components in more detail, there have been significant advances in the understanding of the evaporation process from tropical rain forests in Amazonas and Java, and *Eucalyptus* spp. in south India, largely through the Institute of Hydrology, UK collaborative programmes with various national agencies. Replication of this micrometeorological and physiological work on the lines of ARME is expensive. This review has indicated, however, the need for additional studies in other tropical rain forest environments, such as equatorial rain forest environments in the "maritime continent" (Ramage, 1968) and the outer tropical areas where more organized rainfall systems occur. Other areas that need closer attention include the contrasting experiences with interception modelling (Shuttleworth, 1989a), the conflicting role of stemflow (Herwitz, 1986; Lloyd & Marques-Filho, 1988) between equatorial and cyclone-prone areas and the need for more detailed studies concerning "occult" precipitation and evaporation aspects of montane "cloud forests" (Bruijnzeel, 1990). The current expansion of the Amazon Basin work through the ABRACOS programme, taking into account cleared forest areas and other parts of the climatic spectrum in that basin, will ensure continuing progress in evaporation processes. Similarly, the "water mining" by *Eucalyptus* spp. controversy should be closer to being settled if Calder and co-workers extend their study in south India to include species which have sustained root contact with shallow water tables. Calder's (1986b, 1991a) reviews highlight, however, the strong role of previous Australian experience in eucalypt evaporation research. It is both surprising and unfortunate that Australia to date has not participated in technology-transfer of this research interest into the humid tropics.

Several studies have been undertaken in hillslope hydrology which have advanced our appreciation of a broader spectrum of runoff responses compared with the humid temperate experience. With the exception of the Australian work, most of these studies, however, have been short-term and not integrated with a longer term water balance programme addressing land management problems. More important, this review has emphasized the dearth of soil hydraulic property measurements and related use of soil physics principles to aid interpretation and extrapolation in several reports. The use of simple field techniques for measuring soil hydraulic properties in degraded areas would also go some way to addressing the con-

troversial issue of the removal of forests and the effects on dry season streamflow and stormflow.

Despite recent advances in our understanding of the spatial organization of tropical rainfall, a continuing drawback remains the poor raingauge networks across most of the humid tropics which has hampered the use of rainfall-runoff models, especially in the more remote locations. Continued developments using space-borne sensors (on the lines of Dugdale *et al.,* 1991) and planned projects for the mid-1990s such as the Tropical Rainfall Measuring Mission (Simpson *et al.,* 1988) should go some way to providing a substitution for the poor ground networks. Such initiatives still need an effective "ground-truthing" strategy to develop the appropriate algorithms to give grid-average estimates from point observations as required in AGCMs. Work over three decades ago (e.g. Bleasdale, 1959), demonstrated the methodology and accuracy level of existing rainfall measurements are still inadequate because ground-level gauges (Rodda, 1967, 1970) have not been widely adopted. The incorporation of a dense network of such gauges within both TRMM and HAPEX-type field experiments in the humid tropics would be an ideal strategy. Calibrated information from space-borne sensors hopefully would also extend to the testing of stochastic and scale invariant (include multiscaling) models applied to rainfall fields in humid tropical areas, with resulting hydrological applications in rainfall-runoff models. A continuing deficiency, however, will be rainfall frequency-intensity-duration information which is useful for evaluating the runoff generation process in conjunction with measured soil hydraulic properties.

During considerations given to technology-transfer, it becomes evident that future research directions in the humid tropics are very much integrated with current challenges of scientific inquiry into hydrological processes identified from work in other climatic regions. For example, the need to address the problems of scale, both in time and space, emerges throughout this account in terms of appropriate measurement methodologies (e.g. soil hydraulic properties) and theoretically in the development of new algorithms (e.g. macrohydrology). Furthermore, the continued development in topographically- and physically-based hydrologic models, based on digital terrain models, are probably even more relevant to the humid tropics where hydrological databases are more commonly deficient.

The applied focus in experimental process hydrology will continue to be in support of improving our understanding of the hydrological impacts of land-use conversion at different scales. One aspect that emerges through technology-transfer considerations is the need for reconciling hydrometric and hydrogeochemistry studies in terms of the origin of storm runoff linked with hydrograph separation. Significantly, no contributions devoted to this subject emanated from the humid tropics in recent special issues of *Journal of Hydrology* (vol. 116, 1990) and *Water Resources Research* (vol. 26/12, 1990). New initiatives in the humid tropics are required which would also have considerable applications in the water quality and erosion problems identified in other contributions to this volume (e.g. Hufschmidt, Roche, Rose). Such work would also contribute to the global research effort on this topic. Pearce (1990) for example, provided a succinct review of the challenges and directions for future research, using the application of environmental isotopes as an illustration. Pearce's (1990, p. 3046) concluding statement is particularly relevant to the humid tropics "... we need to concentrate our collective efforts on a modest number of field locations which are already well documented. The objective should be to produce physically and chemically based models which work for those well-documented conditions. Then we can branch out and modify tested models or derive new models for other sets of conditions." Evident in the present review is the limited number of experimental drainage basins in the humid tropics which are adequately documented and still operational to follow Pearce's recommendations. The Babinda experimental catchments in north-east Queensland is one example. Hydrogeochemistry studies are currently in progress there. The experimental drainage basins operated by the Forest Research Institute and the Department of Irrigation and Drainage in Malaysia also offer good prospects (e.g. Abdul Rahim, 1988; K. F. Law & Cheong, 1987). Should the ORSTOM-operated catchments in French Guyana (mentioned in Fritsch, 1987; Fritsch *et al.*, 1987; Fritsch, 1992) continue to be monitored, they also would be suitable sites for such experimentation. It is clear, however, that new initiatives at other locations to cover the spectrum of humid tropical environments (including selected tropical islands) need to be considered. Furthermore, the socio-economic climate has to be sufficiently stable to conduct long-term experimental catchment programmes of this kind. In this context, it is unfortunate that the UK East African work had to be truncated.

A persistent theme throughout this account has been the need to consider technology-transfer from the higher latitudes because process hydrology research in the humid tropics has either been lagging behind the "cutting-edge" of the science (the few exceptions have been acknowledged) or it has been non-existent. This trend will begin to change in the 1990s, should projects such as the macrohydrology initiative HAPEX-Amazonia proceed as planned. The work will make a major contribution towards the global problem of parameterizing hydrological behaviour across different spatial and time scales, and also appropriately interface with meteorology. New initiatives of this kind will also address some of Philip's (1991, pp. 94–96) criticisms related to natural science in general becoming increasingly dependent on computer-based modelling at the expense of supporting the field (and laboratory) experimentation required for model validation. However, further improvements in the experimental design of such large-scale field experiments could be considered. Existing macrohydrology campaigns in the Amazon Basin, for example, has been biased towards the requirements of the AGCM scale and remote sensing. Hydrological parameterization has depended mostly on point measurements (usually evaporation and energy balance), even though the methodology of measurement has been comprehensive. In previous basin work, there has been a neglect of nesting at different scales, scientific outputs emanating from process hydrology undertaken in conjuction with the more traditional catchment studies. Such initiatives would provide another important avenue for field validation of the appropriate algorithms for defining the surface hydrology. There are also still questions whether small-scale heterogeneity dominates the averaged values of surface parameters used in AGCMs or that runoff, infiltration, soil moisture, and deep drainage can be accurately calculated from AGCMs which have a relatively coarse vertical and horizontal resolution (Rind *et al.*, 1992). Future HAPEX-type studies planned for the humid tropics, therefore, would greatly benefit from another "layer" of more intensive investigation of the surface hydrology, through the inclusion of catchment studies at different scales.

When considering land management applications for process hydrology research, there are additional land management issues emerging which need urgent attention to diversify from the persistent focus concerning land-use impacts of forest conversion. These concern the concept of "sustainable development" and the issue of the rehabilitation of degraded lands through reforestation mentioned earlier in the chapter (e.g. Goudberg & Bonell with Benzaken, 1991; Schreckenberg *et al.*, 1990; Poore & Sayer, 1991; WCED, 1987). The word "sustainable," however, has become a catchword, an article of faith commonly mentioned in the agenda and policies of various international agencies (e.g. the International Tropical Timber Organization, ITTO Action Plan, 1990) as well as in other disciplines as the environmental sciences (Bonell, 1991d). Dogse (1989), for example, listed about 60 definitions concerning the concept of sustainability in the literature and noted "... the scientific community has a long way to go before a consensus is reached on how sustainability should be defined" (Dogse, 1989, p. 2). The problem is that we have no rigorous scientific criteria for providing tolerable thresholds in terms of changes, for example, in hillslope runoff and related soil hydraulic properties which have obvious ramifications for erosion and water quality.

When considering rehabilitation, there is an urgent need for improved understanding and further expansion of research (on the lines of post-ABRACOS) on the hydrological processes connected with human-impacted landscapes (e.g. secondary forests, degraded lands) (Bruijnzeel, 1989; Gradwohl & Greenberg, 1988; Pereira, 1991; Schreckenberg *et al.*, 1990). Such work would provide baseline information in assessing the time lag in recovery to pre-disturbed forest levels of water transfer through various reforestation practices, under controlled experimental conditions e.g. agroforestry, plantations (Bonell, 1991c). Successful reforestation with native hardwood species, however, involves a complex array of environmental variables, e.g. solar radiation, slope, aspect, soil nutrient status and hydrology, as well as the optimum tree plantings (including protective cover crop) (Applegate & Bragg, 1991; Vanclay, 1991). The use of spatial modelling (Moore *et al.*, 1991) such as TOPOG (O'Loughlin, 1990a, b) would be a very useful tool in developing the appropriate strategy for tree species selection, required density for planting and need for underplanting with a cover crop. It is through the application of such methods that more effective management techniques for these human-impacted systems can be devised so that the pressure can be taken off the remaining "pristine forests" (Bonell, 1991d). Placing greater emphasis on biophysical research strategies (including process hydrology), which are directed more towards sustainable management and rehabilitation of human-impacted zones, may well hold the key to land management problems in the humid tropics.

GLOSSARY OF MAJOR SYMBOLS

Roman Symbol	Description	Defining Equation No. (if applicable)
A	partial catchment area	35
"A"	transmission parameter in Philip infiltration equation	
A_s	drained area per unit contour length at a point	32
A_t	total hillslope or catchment area generating the outflow Q_o	36
a	scaling factor	27
B	contour length	35
b	aerodynamic roughness parameter defined in equation 7	7
$b*$	drainage coefficient	15
C	amount of stored water at any particular time	15
c_p	specific heat of air at constant pressure	1 and 11
D	deep drainage	
D_F	monofractal dimension related to scaling of rainfall	
D_r	rate of canopy drainage	14
D_s	drainage rate when C is equal to S	15
d	zero plane displacement	7
d_s	change in surface detention store and depth in overland flow	31
d_v	change in water interception by vegetation	31
E	total evaporation, that is the sum of $E_i + E_t$	
$\overline{E}$	mean evaporation rate used in the Gash interception model	
E_E	equilibrium evaporation, which is the steady state limit of water loss in a closed atmosphere over a large expanse of vegetation with a fixed r_s	1
E_{euc}	rate of evaporation from *Eucalyptus* spp	
E_i	evaporation from various water surfaces, water intercepted on vegetation and moist bare soil	

E_P	Penman wet surface evaporation	13
E_{PM}	rate of evaporation (Penman-Monteith)	2
E_{PT}	rate of evaporation (Priestley-Taylor)	1
E_t	transpiration, that is water escaping from within plants mostly via leaves	
G	soil heat flux	3
g_a	aerodynamic conductance which is the reciprocal of r_a	
$g(LAI)$	bulk surface conductance dependence on Leaf Area Index (LAI)	4
g_{max}	maximum surface conductance	4
g_s	bulk surface conductance which is the reciprocal of r_s	4
$g(S_T)$	bulk surface conductance dependence on solar radiation	4
g_{sto}	stomatal conductance	
$g(T)$	bulk surface conductance dependence on air temperature	4
$g(\delta\theta)$	bulk surface conductance dependence on soil moisture deficit	4
$g(\delta q)$	bulk surface conductance dependence on specific humidity deficit	4
H	wetted length of an auger hole connected with the CHWP method	
h	vegetation elevation	8
I	cumulative infiltration	
i	i as a subscript for any spatial point in a catchment used in equation 32	32
$K*$	field saturated, or satiated, hydraulic conductivity	
K_s	saturated hydraulic conductivity	22
$K(\theta)$	unsaturated hydraulic conductivity which varies with volumetric water content of soil	21
L	reference length, usually the mean hillslope length	36
LAI	Leaf Area Index	4
l	length, referring to unit of measurement	
l_i	length increment of river reach	34
M	local slope (expressed as a decimal of ratio)	35
m	parameter of the exponential decline in transmissivity with soil depth	32
N	N individual river reaches of length l_i	34
$N(r_f)$	number of boxes of non-overlapping segments of size τf to determine D_F of rainfall using the functional box counting method	
N_ε	number of points on the attractor in phase space connected with atmospheric scaling	
n	statistical sample size	
P	incoming precipitation rate	14
P_e	net photosynthetic energy	3
p	proportion of rain falling directly to the ground through canopy gaps	14
P_t	proportion of rain diverted to tree trunks	14
Q_b	subsurface lateral flux contribution to discharge	

	through a stream bank	34
Q_d	overland flow measured in downslope runoff trough	30
Q_n	net radiation energy available for evaporation	1
Q_o	areal lateral drainage outflow generated from a total hillslope or catchment area	36
Q_u	overland flow measured in upslope runoff trough	30
q	flux density of soil water movement in Richards' (1931) extension of Darcy's Law to unsaturate soils	21
$\bar{q}^*$	average areal lateral drainage flux for an area A	36
$q^*(x, y)$	net areal lateral drainage flux after taking into account transpiration and deep drainage losses	35
R	total rainfall in given period	31
$\bar{R}$	mean rainfall rate used in the Gash interception model	
R_N	net radiation	3
r	auger hole radius connected with CHWP method	
r_a	aerodynamic resistance to the transfer of sensible heat and water vapour	2
r_p	steady state percolation	
r_s	surface resistance	2
r_t	distance from a given point to the number of points N_ε on an attractor in atmospheric scaling (phase space portrait)	
S	minimum amount of water required to wet the canopy	15
"S"	sorptivity parameter in Philip infiltration equation	
S_Q	change in storage of energy in the canopy air and biomass	3
S_W	the sink of soil water	24
S_{Tr}	trunk water capacity linked to D_r and D_s	
$\bar{S}^*$	catchment average moisture deficit	32
S_i^*	local soil moisture deficit at any point i from a catchment average moisture deficit, $\bar{S}^*$	32
swd	soil water deficit	
T	transmissivity which is the depth integral of K^* through the soil profile adjusted depending on the thickness of the saturated layer	35
$\bar{T}$	average of transmissivity (T) for an area A	36
T_o	saturated transmissivity of the soil at a point	32
t	time. Also forms part of a unit of measurement throughout the text	24
$\tan\beta$	local slope angle	32
u_r	measured wind speed at a reference height, Z_r	6
V_K	von Karman's constant	7
V_{REV}	"averaging volume" when the "space average" of soil hydraulic property variable of interest loses its dependence on volume of soil sampled (Dagan, 1986)	

v	macroscopic velocity of water in Darcy's Law	22
$W(x, y)$	areal non-dimensional wetness index (W)	36 and 38
$w(x, y)$	areal normalised wetness function	36
X	runoff or runon	30
x	horizontal Cartesian coordinate	29 and 35
y	horizontal Cartesian coordinate	
Z_o	aerodynamic roughness length	7
Z_r	reference height	7
z	Vertical Cartesian coordinate or depth reference level for calculating hydraulic potential (Φ) of soil or groundwater	20 and 25

GREEK SYMBOLS

α	empirical coefficient to allow for advection in equation 1	1
α_s	sorptive number	28
β	slope angle	32
β_B	Bowen ratio	
γ	psychrometric constant	1
γ^*	expected value of $\ln(A_s/T_o \tan\beta)$ in equation 32	32
Δ	slope of the saturation vapour pressure curve	1
Δ_R	rate of change in amount of intercepted water storage	14
δ_q	specific humidity deficit	2
δ_{qE}	equilibrium saturation deficit	12
Λ	expected value of $\ln(A_s/\tan\beta)$	33
λ	latent heat of vaporisation of water	1
λ_c	macroscopic capillary (or sorptive) length which is essentially a flow-weighted mean soil water potential	
$\bar{\lambda}_M$	reference value such as the microscopic dimension characterizing a reference soil in scaling	27
λ_m	characteristic, microscopic mean pore radius	
v	fractal v correlation dimension related to atmospheric scaling (phase space portrait)	
ρ	air density	2
Σ	summation sign	
Φ	total hydraulic potential	20
φ	similarity variable for one-dimensional absorption	29
$\varnothing$	stability function to act as a correction for the effect of diabatic influences in equation 7	10
ψ	matric potential	20
$\psi(\theta)$	matric potential – volumetric soil water content relationship	

OTHER ABBREVIATIONS

ABRACOS	The Anglo/BRazillian Amazonian Climate Observation Study
AGCM	Atmospheric General Circulation Model

AMEX	Australian Monsoon Experiment
ARME	Amazon Region Micrometeorology Experiment
CCD	Cold Cloud Technique
CFA	Common Factor Analysis
CHWP	Constant Head Well Permeameter
EMEX	Equatorial Mesoscale Experiment
FIFE	The First ISLSCP Field Experiment
GATE	Global Atmospheric Research Programme, Atlantic Tropical Experiment
GEWEX	Global Energy and Watercycle Experiment
HAPEX	The Hydrological Atmospheric Pilot Experiment
IGBP	The International Geosphere-Biosphere Project
IGBP-BAHC	Core project of IGBP "Biospherical Aspects of the Hydrological Cycle"
ISLSCP	International Satellite Land Surface Climatology Project
MCS	Mesoscale Convective System
ORSTOM	Office de la Recheche Scientifique et Technique Outre-Mer but now known as "L'Institut Français de Recherche Scientifique Pour Le Développement En Coopération"
PBL	Planetary Boundary Layer
PC	Principal Component
PCA	Principal Components Analysis
REV	Representative Elementary Volume
RPCA	Rotated Principal Components Analysis
SVATS	Soil/Vegetation/Atmospheric Transfer Schemes
TRAMAR	Tropical Rainfall Mapping Radar
TRMM	Tropical Rainfall Measuring Mission
WMONEX	Winter Monsoon Experiment
ZFP	Zero Flux Plane method

ACKNOWLEDGEMENTS

Some sections of this paper greatly benefited from the constructive comments made by Frank Dunin, CSIRO Division of Plant Industry, Canberra, Australia (evaporation); Emmett O'Loughlin, CSIRO Division of Water Resources, Canberra, Australia (digital terrain models for runoff production); Keith Smettem, CSIRO, Division of Soils, Townsville, Australia and Ian White, CSIRO, Division of Environmental Mechanics, Canberra, Australia ("water movement in the unsaturated zone" through to the "runoff generation process" section). In addition, Pierre Hubert, Ecole des Mines de Paris, and John Roberts, Institute of Hydrology, UK supplied to the senior author, respectively, rainfall (multifractals) and evaporation literature at the pre-publication stage or only recently published.

Compiling this manuscript was a major exercise and the infinite patience of Rosie Aziz, Secretary, Department of Geography, James Cook University, in word processing is greatly appreciated. Also, thanks to Moya Tomlinson, Assistant to the Director, Institute for Tropical Rainforest Studies, for providing a second check of the bibliography listing and cross referencing with the manuscript.

Finally, Professor Yoshinori Tsukamoto, Department of Environmental Science and Natural Resources, Tokyo University of Agriculture and Technology, is thanked for providing the facilities for M. Bonell to complete this manuscript in January 1992 whilst he was in receipt of a Japan Society for the Promotion of Science (JSPS) Fellowship.

REFERENCES

Abbott, M. B., Bathurst, J. C., Cunge, J. A., O'Connell, P. E. & Rasmussen, J. (1986a) An introduction to the European Hydrological System – Système Hydrologique Européen, 'SHE', 1. History and philosophy of a physically-based, distributed modelling system. *J. Hydrol.* 87:45–59.

Abbott, M. B., Bathurst, J. C., Cunge, J. A, O'Connell, P. E. & Rasmussen, J. (1986b) An introduction to the European Hydrological System – Système Hydrologique Européen, 'SHE', 2. Structure of a physically-based, distributed modelling system. *J. Hydrol.* 87:61–77.

Abdul, Rahim, N. (1987) Impact of forest conversion on water yield in peninsular Malaysia. Paper presented at the Workshop on *Impact of Operations in Natural and Plantation Forests on Conservation of Soil and Water Resources*, Universiti Pertanian, Malaysia, Serdan, 23–26 June 1987.

Abdul, Rahim, N. (1988) Water yield changes after forest conversion to agricultural land use in peninsular Malaysia. *J. Trop. Forest Sci.* 1:67–84.

Abdul, A. S. & Gillham, R. W. (1984) Laboratory studies of the effects of the capillary fringe on streamflow generation. *Wat. Resour. Res.* 10(6):691–698.

Abdul, A. S. & Gillham, R. W. (1989) Field studies of the effects of the capillary fringe on streamflow generation. *J. Hydrol.* 112:1–18.

ACCH (Australian Centre for Catchment Hydrology (1990) *TOPOG: Terrain Analysis and Steady-State Modelling*, Version 1.4a and Version 2.0i, User Manual, CSIRO Div. of Water Resources, Canberra, Australia.

Anderson, J. L. & Bouma, J. (1973) Relationships between saturated hydraulic conductivity and morphometric data of an agrillic horizon. *Proc. Soil Sci. Soc. Am.* 37:408–413.

Anderson, M. G. & Burt, T. P. (1978) The role of topography in controlling throughflow generation. *Earth Surf. Processes* 3:331–344.

Anderson, M. G. & Burt, T. P. (eds.) (1985a) *Hydrological Forecasting*. Wiley, Chichester.

Anderson, M. G. & Burt, T. P. (1985b) Modelling strategies. *In*: M. G. Anderson & T. P. Burt (eds.) *Hydrological Forecasting*. Wiley, Chichester, 1–13.

Anderson, M. G. & Burt, T. P. (eds.) (1990) *Process Studies in Hillslope Hydrology*. Wiley, Chichester.

Applegate, G. & Bragg, A. (1991) Agroforestry and land rehabilitation research for rational land use management. *In*: N. Goudberg & M. Bonell, with D. Benzaken (eds.) *Tropical Rainforest Research in Australia: Present Status and Future Directions for the Institute for Tropical Rainforest Studies*. (Proc. Townsville Workshop, May 1990), Inst. for Tropical Rainforest Studies, James Cook Univ. of North Queensland, Townsville, Australia, 127–132.

ASAE (1983) Advances in infiltration. In: *Proc. Nat. Conf. on Advances in Infiltration*, Dec. 12–13, 1983, Chicago, Ill. Am. Soc. Agric. Engrs., St. Josephs, Mich.

Aston, A. R. (1979) Rainfall interception by eight small trees. *J. Hydrol.* 42:383–396.

Aston, A. R. & Dunin, F. X. (1980) The prediction of water yield from a 5 ha experimental catchment. *Aust. J. Soil. Res.* 18:149–162.

Atkinson, B. W. (1981) *Mesoscale Atmospheric Circulation.* Academic Press, London.

Baerring, L. (1987) Spatial patterns of daily rainfall in central Kenya: Application of principal component analysis, common factor analysis and spatial correlation. *J. Climatology* 7:267–289.

Baerring, L. (1988) Regionalization of daily rainfall in Kenya by means of common factor analysis. *J. Climatology* 8:371–389.

Baillie, I. C. (1975) Piping as an erosion process in the uplands of Sarawak. *J. Trop. Geogr.* 41:9–15.

Baker, F. G. (1978) Variability of hydraulic conductivity within and between nine Wisconsin soil series. *Wat. Resour. Res.* 14:103–108.

Baker, G. & Gollub, J. (1990) *Chaotic Dynamics.* Cambridge University Press, Cambridge.

Baldocchi, D. (1989) Canopy-atmosphere water vapour exchange: Can we scale from a leaf to a canopy? *In*: T. A. Black, D. L. Splittlehouse, M. D. Novak & D. T. Price (eds.) *Estimation of Areal Evapotranspiration.* Proc. Vancouver Symp., Aug. 1987, Int'l. Assoc. of Hydrol. Sci. Publ. 177:21–41.

Balek, J. (1977) *Hydrology and Water Resources in Tropical Africa.* Elsevier, Amsterdam.

Balek, J. (1983) *Hydrology and Water Resources in Tropical Regions.* Elsevier, Amsterdam.

Balek, J. (1989) *Groundwater Resources Assessment.* Elsevier, Amsterdam.

Balek, J. & Perry, J. (1973) Hydrology of African headwater swamps. *J. Hydrol.* 19:227–249.

Banejee, A. K. (1972) Morphology and genesis of pipe structure in ferralitic soils in Midnapore. *J. Ind. Soc. Soil Sci.* 20:399–402.

Barnes, C. J. & Allison, G. B. (1988) Tracing of water movement in the unsaturated zone using stable isotopes of hydrogen and oxygen. *J. Hydrol.* 100:143–176.

Bates, C. G. & Henry, A. J. (1928) *Forest and Streamflow Experiments at Wagon Wheel Gap, Colorado.* U.S. Weather Bureau Monthly Weather Review Suppl. *30.*

Bathurst, J. C. (1986a) Physically-based distributed modelling of an upland catchment using the Système Hydrologique Européen. *J. Hydrol.* 87:79–102.

Bathurst, J. C. (1986b) Sensitivity analysis of the Système Hydrologique Européen for an upland catchment. *J. Hydrol.* 87:103–123.

Baveye, P. & Sposito, G. (1984) The operational significance of the continuum hypothesis in the theory of water movement through soils and aquifers. *Wat. Resour. Res.* 2:521–530.

Bear, J. (1979) *Hydraulics of Groundwater.* McGraw-Hill, New York, N.Y.

Beasley, D. B., Huggins, L. F. & Monke, E. J. (1980) Answers: A model for watershed planning. *Trans. Am. Soc. Agric. Engrs.* 23:938–944.

Bell, R. W. & Schofield, N. J. (1990) Design and application of a constant head well permeameter for shallow high saturated hydraulic conductivity soils. *Hydrol. Processes* 4:327–342.

Bergeron, T. (1957) Periodic and irregular disturbances of the tropospheric models. *In*: C. L. Godske, T. Bergeron, J. Bjerkness & R. C. Bundgaard (eds.) *Dynamic Meteorology and Weather Forecasting.* Published jointly by the Am. Met. Soc. and the Carnegie Inst. of Wash., Wash. D.C.

Berndtsson, R. & Niemczynowicz, J. (1988) Spatial and temporal scales in rainfall analysis – some aspects and future perspectives. *J Hydrol.* 100:293–313.

Betson, R. P. (1964) What is watershed runoff? *J. Geophys. Res.* 69:1541–1552.

Betson, R. P. & Marius, J. B. (1969) Source areas of storm runoff. *Wat. Resour. Res.* 5:574–582.

Beven, K. J. (1978) The hydrological response of headwater and sideslope areas. *Hydrol. Sci. Bull.* 23:419–437.

Beven, K. J. (1979) A sensitivity analysis of the Penman-Monteith actual evapotranspiration estimates. *J. Hydrol.* 44:169–190.

Beven, K. J. (1982a) On subsurface stormflow: An analysis of response times. *Hydrol. Sci. Bull.* 4:505–521.

Beven, K. J. (1982b) On subsurface stormflow: Predictions with simple kinematic theory for saturated and unsaturated flows. *Wat. Resour. Res.* 18(6):1627–1633.

Beven, K. J. (1985) Distributed models. *In*: M. G. Anderson & T. P. Burt, (eds.) *Hydrological Forecasting.* Wiley, Chichester, 405–435.

Beven, K. J. (1986a) Runoff production and flood frequency in catchments of order *n*: An alternative approach. *In*: V. K. Gupta, I. Rodriguez-Iturbe & E. F. Wood (eds.) *Scale Problems in Hydrology.* Reidel, Dordrecht, 107–131.

Beven, K. J. (1986b) Hillslope runoff processes and flood frequency characteristics. *In*: A. D. Abrahams (ed.) *Hillslope Processes.* Allen and Unwin, Boston, Mass., 187–202.

Beven, K. J. (1987) Towards the use of catchment geomorphology in flood frequency predictions. *Earth Surf. Processes Landf.* 12:69–82.

Beven, K. J. 1989) Changing ideas in hydrology – the case of physically-based models. *J. Hydrol.* 105:157–172.

Beven, K. J. (1991) Hydrograph separation? *In*: *Third National Hydrol. Symp.* Proc. Univ. Southampton, Sept. 1991, British Hydrol. Soc., Instit. Hydrol., UK, 3.1–3.7.

Beven, K. J. & Germann, P. F. (1982) Macropores and water flow in soils. *Wat. Resour. Res.* 18:1311–1325.

Beven, K. J. & Germann, P. F. (1984) A distribution function model of channelling flow in soils based on kinematic wave theory. *In*: J. Bouma & P. A. C. Raats (eds.) *ISSS Symp. on Water and Solute Movement in Heavy Clay Soils.* ILRI Publ. 37:89–100.

Beven, K. J. & Kirkby, M. J. (1979) A physically-based variable contribution area model of basin hydrology. *Hydrol. Sci. Bull.* 24:43–69.

Beven, K. J., Hornberger, G. & Germann, P. F. (1989) Hillslope hydrology: A multiple interacting pathways model. *In*: *Second Nat. Hydrol. Symp.*, (Proc. Univ. Sheffield, Sept. 1989), British Hydrol. Soc., Inst. Hydrol., UK. 1.1–1.8.

Beven, K. J., Kirkby, M. J., Schofield, N. & Tagg, A. F. (1984) Testing a physically-based flood forecasting model (TOPMODEL) for three UK catchments. *J. Hydrol.* 69:119–143.

Beven, K. J., Wood, E. F. & Murugesu S. (1988) On hydrological heterogeneity – catchment morphology and catchment response. *J. Hydrol* 100:353–375.

Binley, A., Beven, K. & Elgy, J. (1989b) A physically-based model of heterogeneous hillslopes. 2. Effective hydraulic conductivities. *Wat. Resour. Res.* 25(6):1227–1233.

Binley, A., Elgy, J. & Beven, K. (1989a) A physically-based model of heterogeneous hillslopes. 1. Runoff production. *Wat. Resour. Res.* 25(6):1219–1226.

Black, T. A., Spittlehouse, D. L., Novak, M. D. & Price, D. T. (1989) *Estimation of Areal Evapotranspiration.* (Proc. Vancouver Symp., Aug.

1987). Int'l. Assoc. of Hydrol. Sci., Publ. *177*.

Blackie, J. R. (1979a) The water balance of the Kericho catchments. *In*: J. R. Blackie, K. A. Edwards, & R. T. Clarke (Compiled) *Hydrological Research in East Africa. E. Afr. Agric. For. J. Special Issue 43*:55–84.

Blackie, J. R. (1979b) The water balance of the Kimakia catchments. *In*: J. R. Blackie, K. A. Edwards & R. T. Clarke (Compiled) *Hydrological Research in East Africa. E. Afr. Agric. For. J. Special Issue 43*:155–174.

Blackie, J. R. (1979c) The use of conceptual models in catchment research. *In*: J. R. Blackie, K. A. Edwards & R. T. Clarke (Compiled) *Hydrological Research in East Africa. E. Afr. Agric. For. J. Special Issue 43*:36–42

Blackie, J. R. & Eales, C. W. O. (1985) Lumped catchment models. *In*: M. G. Anderson & T. P. Burt (eds.) *Hydrological Forecasting*. Wiley, Chichester, 311–345.

Blackie, J. R. & Edwards, K. A. (1979) General conclusions from the land-use experiments in East Africa. *In*: J. R. Blackie, K. A. Edwards & R. T. Clarke (Compiled) *Hydrological Research in East Africa. E. Afr. Agric. For. J. Special Issue 43*:273–277.

Blackie, J. R., Edwards, K. A. & Clarke, R. T. (Compiled) (1979) *Hydrological Research in East Africa. E. Afr. Agric. For. J. Special Issue 43*.

Blake, G., Schlichting, E. & Zimmermann, U. (1973) Water recharge in a soil with shrinkage cracks. *Proc. Soil Sci. Soc. Am. 37*:669–772.

Bleasdale, A. (1959) The measurement of rainfall *Weather 14*: 12–18.

Boersma, L. (1965) Field measurement of hydraulic conductivity above a water table. *In*: C. A. Black (ed.) *Methods of Soil Analysis*. Am. Soc. Agron., Madison, Wisconsin. *Agron. 9*:234–252.

Bonell, M. (1988) Hydrological processes and implications for land management in forests and agricultural areas of the wet tropical coast of north-east Queensland. *In*: R. F. Warner (ed.) *Fluvial Geomorphology of Australia*. Academic Press, 41–68.

Bonell, M. (1989) *Keynote address presented at the FRIM-IHP-UNESCO Regional Seminar on Tropical Forest Hydrology*, Kuala Lumpur, 4–9 Sept. 1989.

Bonell, M. (1991a) The applications of hillslope hydrology in forest land-management issues: The tropical north-east Australian experience. Previously presented at the *UNESCO Regional Seminar on Tropical Forest Hydrology* (Kuala Lumpur Symp., Malaysia 4–9, Sept. 1989). *In*: B. P. F. Braga & C. A. Fernandez-Jauregui (eds.) *Water Management of the Amazon Basin*. Proc. Manaus Symp., Aug. 1990, UNESCO (ROSTLAC), Montevideo, Uruguay, 45–82.

Bonell, M. (1991b) Progress in runoff and erosion research in forests. *In*: J. Parde & G. Blanchard (eds.) *Forests, A Heritage for the Future*. Proc. 10th World Forestry Congress, Paris, Sept. 1991, Revue Forestière Française, Hors Série No. 2 (Proc. 2), ENGREF, F-54042 Nancy Cedex, 101–113 (English), 114–126 (French), 127–138 (Spanish). (Bibliography not included, available from author.)

Bonell, M. (1991c) Progress and future research needs in water catchment conservation within the wet tropical coast of north-east Queensland. *In*: N. Goudberg & M. Bonell, with D. Benzaken (eds.) *Tropical Rainforest Research in Australia – Present Status and Future Directions for the Institute for Tropical Rainforest Studies*. Proc. Townsville Workshop, Inst. for Tropical Rainforest Studies, James Cook Univ., Townsville, Australia, May 1990, 59–86.

Bonell, M. (1991d) Opening Address. *In*: N. Goudberg & M. Bonell, with D. Benzaken (eds.) *Tropical Rainforest Research in Australia: Present Status and Future Directions for the Institute for Tropical Rainforest*

Studies. Proc. Townsville Workshop, May 1990, Inst. for Tropical Rainforest Studies, James Cook Univ. of North Queensland, Townsville, Australia, 199–203.

Bonell, M. & Gilmour, D. A. (1978) The development of overland flow in a tropical rainforest catchment. *J. Hydrol. 39*:365–382.

Bonell, M. & Gilmour, D. A. (1980) Variations in short-term rainfall intensity in relation to synoptic climatological aspects of the humid tropical north-east Queensland coast. *Singapore J. Trop. Geog. 1*:16–30.

Bonell, M. & Sumner, G. (1992) Autumn and winter daily precipitation areas in Wales, 1982–1983 to 1986–1987. *Int. J. Climatol. 12*:77–102

Bonell, M. & Williams, J. (1986a) The two parameters of the Philip infiltration equation: Their properties and the spatial and temporal heterogeneity in a red earth of tropical semiarid Queensland. *J. Hydrol. 87*:9–31.

Bonell, M. & Williams, J. (1986b) The generation and redistribution of overland flow in a massive oxic soil in a eucalypt woodland within the semiarid tropics in north Australia. *Hydrol. Processes 1*:31–46.

Bonell, M. & Williams, J. (1987) Infiltration and redistribution of overland flow and sediment on a low relief landscape of semiarid tropical Queensland. *In*: R. H. Swanson, P. Y. Bernier & P. D. Woodard (eds.) *Forest Hydrology and Watershed Management*. Proc. Vancouver Symp., Aug. 1987, Int'l. Assoc. of Hydrol. Sci. Publ. *167*:199–211.

Bonell, M. & Williams, J. (1989) Reply to P.I.A. Kinnell's comments on "The Generation and Redistribution of Overland Flow on a Massive Oxic Soil in a Eucalypt Woodland within the Semiarid Tropics of North Australia," Mike Bonell & John Williams. *Hydrol. Processes 3*:97–100.

Bonell, M., Cassells, D. S. & Gilmour, D. A. (1982) Vertical and lateral soil water movement in a tropical rainforest catchment. *In*: E. M. O'Loughlin & L. J. Bren (eds.) *The First National Symposium on Forest Hydrology*, 11–13 May, Melbourne, Vic., Inst. Engrs. Aust., Canberra, A.C.T., 30–38.

Bonell, M., Cassells, D. S. & Gilmour, D. A. (1983a) Vertical soil water movement in a tropical rainforest catchment in north-east Queensland. *Earth Surf. Processes Landf. 8*:253–272.

Bonell, M., Gilmour, D. A. & Cassells, D. S. (1983b) A preliminary survey of the hydraulic properties of rainforest soils in tropical north-east Queensland and their implication for the runoff process. *In*: J. de Ploey (ed.) *Rainfall Simulation, Runoff and Soil Erosion, Catena Suppl. 4*:57–78.

Bonell, M., Cassells, D. S. & Gilmour, D. A. (1984a) Tritiated water movement in clay soils of a small catchment under tropical rainforest in North-East Queensland. *In*: J. Bouma & P. A. C. Raats (eds.) *ISSS Symp. on Water and Solute Movement in Heavy Clay Soils*. ILRI Publ. *37*: 197–201.

Bonell, M., Cassells, D. S. & Gilmour, D. A. (1987) Spatial variations in soil hydraulic properties under tropical rainforest in north-eastern Australia. *In*: Yu-Si Fok (ed.) *Proc. Int. Conf. on Infiltration Development and Application*. Wat. Resour. Res. Center, Univ. of Hawaii at Manoa, Jan. 1987, 153–165.

Bonell, M., Gilmour, D. A. & Cassells, D. S. (1986) The storm runoff response to various rainfall systems on the wet tropical coast of north-east Queensland. *East-West Center, Environ. and Policy Inst. Working Paper*, Honolulu, Hawaii.

Bonell, M., Gilmour, D. A. & Cassells, D. S. (1991) The links between synoptic climatology and the runoff response of rainforest catchments on the wet tropical coast of north-eastern Queensland. *In*: P. A. Kershaw & G. Werran (eds.) *Australian National Rainforests-Study Report* Vol. 2, Australian Heritage Commission, Canberra, Australia, 27–62.

Bonell, M., Gilmour, D. A. & Sinclair, D. F. (1979), A statistical method for modelling the fate of rainfall in a tropical rainforest catchment. *J. Hydrol.* 42:241–257.

Bonell, M., Gilmour, D. A. & Sinclair, D. F. (1981) Soil hydraulic properties and their effect on surface and subsurface water transfer in a tropical rainforest catchment. *Hydrol. Sci. Bull.* 26:1–18.

Bonell, M., Hendriks, M. R., Imeson, A. C. & Hazelhoff, L. (1984b) The generation of storm runoff in a forested clayey drainage basin in Luxembourg. *J. Hydrol.* 71:53–77.

Bonell, M., Pearce, A. J. & Stewart, M. K. (1990) The identification of runoff-production mechanisms using environmental isotopes in a tussock grassland catchment, eastern Otago, New Zealand. *Hydrol. Processes* 4:15–34.

Bosch, J. M. & Hewlett, J. D. (1982) A review of catchment experiments to determine the effect of vegetation changes on water yield and evapotranspiration. *J. Hydrol.* 55:3–23.

Bouma, J. (1983) Use of soil survey data to select measurement techniques for hydraulic conductivity. *Agric. Wat. Management* 6:177–190.

Bouma, J. & Dekker, L. W. (1978) A case study on infiltration into dry clay soil, I. Morphological observations. *Geoderma* 20:27–40.

Bouma, J., Dekker, L. W. & Muilwijk, C. J. (1981) A field method for measuring short-circuiting in clay soils. *J. Hydrol.* 52:551–557.

Bouwer, H. (1966) Rapid field measurements of air entry value and hydraulic conductivity of soils as significant parameters in flow system analysis. *Water Resour. Res.* 2:729–738.

Bouwer, H. & Jackson, R. D. (1974) Determining soil properties. *In*: J. van Schilfgaarde (ed.) *Drainage for Agriculture*. Am. Soc. Agron., Madison, Wisconsin. *Agron.* 17:611–666.

Brakensiek, D. L. & Rawls, W. J. (1983) Agricultural management effects on soil water process, Part II: Green and Ampt parameters for crusting soils. *Trans. Am. Soc. Agric. Engrs.* 26:1753–1757.

Bristow, K. L. & Savage, M. J. (1987) Estimation of parameters for the Philip two-term infiltration equation applied to field soil experiments. *Aust. J. Soil Res.* 25:369–375.

Bristow, K. L. & Williams, J. (1987) Sensitivity of simulated infiltration to changes in soil hydraulic properties. *In*: Yu-Si Fok (ed.) *Proc. Int. Conf. on Infiltration Development and Application*, Wat. Resour. Res. Center, Univ. of Hawaii at Manoa, Jan. 1987, 555–564.

British Hydrological Society (BHS) (1989) *Circulation Newsletter* 24:11–12.

Broadbridge, P. & White, I. (1988) Constant rate rainfall infiltration: A versatile nonlinear model. 1. Analytic solution. *Wat. Resour. Res.* 24:145–154.

Bronswijk, J. J. B. (1988) Modeling of water balance, cracking and subsidence of clay soils. *J. Hydrol.* 97:199–212.

Browning, K. A. (1989) The mesoscale database and its use in mesoscale forecasting. *Quart. J. Roy. Met. Soc.* 115:717–762.

Bruijnzeel, L. A. (1983) Evaluation of runoff sources in a forested basin in a wet monsoonal environment: A combined hydrological and hydrochemical approach. *In*: *Hydrology of Humid Tropical Regions with Particular Reference to the Hydrological Effects of Agriculture and Forestry Practice*. Proc. Hamburg Symp., Aug. 1983. IAHS Publ. 140:165–174.

Bruijnzeel, L. A. (1987) On the hydrology of moist tropical forests: With special reference to the study of nutrient cycling. *Paper presented at the British Ecol. Soc. Symp. on Mineral Nutrients in Tropical Forest and Savannah Ecosystems*, Stirling, 9–11 Sept. 1987.

Bruijnzeel, L. A. (1989) Review Paper. (De)forestation and dry season flow in the tropics: A closer look. *J. Trop. Forest Sci.* 1:229–243.

Bruijnzeel, L. A. (1990) *Hydrology of Moist Tropical Forests and Effects of Conversion: A State of Knowledge Review*. UNESCO IHP, Humid Tropics Programme, Paris.

Bruijnzeel, L. A. with Bremmer, C. N. (1989) *Highland-Lowland Interactions in the Ganges Brahmaputra River Basins*. ICIMOD Occasional Paper No. 11, Kathmandu, Nepal.

Bruijnzeel, L. A. & Wiersum, K. F. (1987) Rainfall interception by a young *Acacia auriculiformis* A. Cunn. plantation forest in West Java, Indonesia: Application of Gash's analytical model. *Hydrol. Processes* 1:309–319.

Burch, G. J., Bath, R. K., Moore, I. D. & O'Loughlin, E. M. (1987) Comparative hydrologic behaviour of forested and cleared catchments in south-eastern Australia. *J. Hydrol.* 90:19–42.

Burt, T. P. (1989) Storm runoff generation in small catchments in relation to the flood response of large basins. *In*: K. J. Beven & P. Carling (eds.) *Floods – Hydrological, Sedimentological and Gemorphological Implications*. Wiley, Chichester, 11–35.

Burt, T. P. & Butcher, D. (1985) Topographic controls of soil moisture distribution. *J. Soil Sci.* 36:469–476.

Calder, I. R. (1977) A model of transpiration and interception loss from a spruce forest in Plynlinon, central Wales. *J. Hydrol.* 33:247–265.

Calder, I. R. (1985) What are the limits on forest evaporation? – Comment. *J. Hydrol.* 82:179–184.

Calder, I. R. (1986a) A stochastic model of rainfall interception. *J. Hydrol.* 89:65–71.

Calder, I. R. (1986b) Water use of eucalypts – A review with special reference to South India. *Agric. Wat. Management* 11:333–342.

Calder, I. R. (1990) *Evaporation in the Uplands*. Wiley, Chichester.

Calder, I. R. (1991a) Water use of eucalypts – A review. *In*: *Growth and Water Use of Forest Plantations*. Proc. Int. Symp., Bangalore, India, 7–11 Feb. 1991. (Available from Inst. of Hydrol., UK).

Calder, I. R. (1991b) Development of the deuterium tracing method for the estimation of transpiration rates and transpiration parameters of tracers. *In*: *Growth and Water Use of Forest Plantations*. Proc. Int. Symp., Bangalore, India, 7–11 Feb. 1991. (Available from Inst. of Hydrol., UK).

Calder, I. R. (1991c) Implications and assumptions in using the "Total Counts" and convection-dispersion equation for tracer flow measurements – with particular reference to transpiration measurements in trees. *J. Hydrol.* 125:149–158.

Calder, I. R. (1991d) A water use and growth model for *Eucalyptus* plantation in water limited conditions. *In*: *Growth and Water Use of Forest Plantations*. Proc. Int. Symp., Bangalore, India, 7–11 Feb. 1991. (Available from Inst. of Hydrol., UK).

Calder, I. R. (1992a) A model of transpiration and growth of *Eucalyptus* plantation in water-limited conditions. *J. Hydrol.* 130:1–15.

Calder, I. R. (1992b) Deuterium tracing for the estimation of transpiration from trees, Part 2. Estimation of transpiration rates and transpiration parameters using a time-averaged deuterium tracing method. *J. Hydrol.* 130:27–35.

Calder, I. R. & Rosier, P. T. W. (1976) The design of a large plastic-sheet net-rainfall gauges. *J. Hydrol.* 30:403–405.

Calder, I. R., Swaminath, M. H., Kariyappa, G. S., Srinivasalu, N. V., Srinivasa Murthy, K. V. & Mumtaz, J. (1991) Measurements of transpiration from *Eucalyptus* plantation, India, using deuterium tracing. *In*: *Growth and Water Use of Forest Plantations*. Proc. Int. Symp., Bangalore, India, 7–11 Feb. 1991. (Available from Inst. of Hydrol., UK).

Calder, I. R., Kariyappa, G. S., Srinivasalu, N. V. & Srinivasa Murty, K. V. (1992a) Deuterium tracing for the estimation of transpiration from trees, Part I. Field calibration. *J. Hydrol. 130*:12–17.

Calder, I. R., Swaminath, M. H., Kariyappa, G. S., Srinivasalu, N. V., Srinivasa Murty, K. V. & Mumtaz, J. (1992b) Deuterium tracing for the estimation of transpiration from trees, Part 3. Measurements of transpiration from *Eucalyptus* plantation, India. *J. Hydrol. 130*:37–48.

Calder, I. R., Wright, I. R. & Murdiyarso, D. (1986) A study of evaporation from tropical rainforest – West Java. *J. Hydrol. 89*:13–31.

Callender, B. A. & Woodhead, T. (1979) Eddy correlation measurements of convective heat flux and estimation of evaporative heat flux over growing tea. *In*: J. R. Blackie, K. A. Edwards & R. T. Clarke (Compiled) *Hydrological Research in East Africa. E. Afr. Agric. For. J. Special Issue 43*:85–101.

Campbell, G. S. (1974) A simple method for determining unsaturated conductivity from moisture retention data. *Soil Sci. 117*:311–314.

Campbell, G. S. (1977) *An introduction to environmental biophysics.* Springer Verlag, New York.

Casenave, A., Flory, J., Mahieux, A. & Simon, J. M. (1984) *Etude hydrologique des bassins de Tai, campaign 1981.* ORSTOM, Centre d'Adiopodoumé, Côte d'Ivoire.

Cassells, D. S., Bonell, M., Hamilton, L. S. & Gilmour, D. A. (1987) The protective role of tropical forests: A state of knowledge review. *In*: N. T. Vergara & N. C. Briones (eds.) *Agroforestry in the Humid Tropics – Its Protective and Ameliorative Roles to Enhance Productivity and Sustainability.* Environ. and Policy Inst., East-West Center, Honolulu, Hawaii, South-East Asian Regional Center for Grad. Study and Res. in Agric. College, Laguna, Philippines, 31–58.

Cassells, D. S., Gilmour, D. A. & Bonell, M. (1985) Catchment response and watershed management in tropical rainforests in north-east Australia. *Forest Ecol. and Management 10*:155–175.

Cavelier, J. (1988) *The Ecology of Elfin Cloud Forests in Northern South America.* Unpublished Report, Trinity College, Univ. of Cambridge, Cambridge.

Chapman, T. G. (1990) Construction of hydrological models for natural systems management. *J. Maths. & Comp. in Simulation 32*:13–37.

Cheng, J. D., Black, T. A., de Vries, J., Willington, R. P. & Goodell, B. C. (1975) The evaluation of initial changes in peak streamflow following logging of a watershed on the west coast of Canada. *Int'l. Assoc. of Hydrol. Sci. Publ. 117*:475–486.

Chu, S. T., Onstad, C. A. & Rawls, W. J. (1986) Field evaluation of layered Green-Ampt model for transient crust conditions. *Trans. Am. Soc. Agric. Engrs. 29*:1268–1272, 1277.

Clapp, R. B. (1982) *A Wetting Front Model of Soil Water Dynamics.* PhD dissertation, Univ. of Virginia, Charlottesville.

Clapp, R. B. & Hornberger, G. M. (1978) Empirical equations for some soil hydraulic properties. *Wat. Resour. Res. 14*(4):601–604.

Clapp, R. B., Hornberger, G. M. & Cosby, B. J. (1983) Estimating spatial variability in soil moisture with a simplified dynamic model. *Water Resour. Res. 19*:739–745.

Clothier, B. E. (1988) Measurement of soil physical properties in the field – Commentry. *In*: W. L. Steffen & O. T. Denmead (eds.) *Flow and Transport in the Natural Environment: Advances and Applications.* Springer-Verlag, Heidelberg, Germany, 86–94.

Clothier, B. E. & Heiler, T. (1983) Infiltration in slot-mulch tillage: Simulation and field verification. *In*: *Advances in Irrigation*, ASAE, Michigan, USA. Publ. *11–83*:275–283.

Clothier, B. E. & Wooding, R. A. (1983) The soil diffusivity near satura-

tion. *Soil Sci. Soc. Am. J. 47*:636–640.

Collis-George, N. (1977) Infiltration equations for simple soil systems. *Wat. Resour. Res. 13*:395–403.

Collis-George, N. & Bond. W. J. (1981) Ponded infiltration into simple soil systems: 2. Pore air pressures ahead of and behind the wetting front. *Soil Sci. 131*:263–270.

Colquhoun, I. J., Ridge, R. W., Bell, D. T., Lonergan, W. A. & Kuo, J. (1984) Comparative studies in selected species of *Eucalyptus* used in rehabilitation of the northern jarrah forest, western Australia. I. Patterns of xylem pressure potential and diffusive resistance of leaves. *Aust. J. Bot. 32*:367–373.

Cooper, J. D. (1979) Water use of a tea estate from soil moisture measurements. *In*: J. R. Blackie, K. A. Edwards & R. T. Clarke (Compiled) *Hydrological Research in East Africa. E. Afr. Agric. For. J. Special Issue 43*:102–121.

Crockford, R. H. & Richardson, D. P. (1990) Partitioning of rainfall in a eucalypt forest and pine plantation in south-eastern Australia, IV. The relationship of interception and canopy storage capacity, the interception of these forests, and the effect on interception on thinning the pine plantation. *Hydrol. Processes 4*:169–188.

Dagan, G. (1986) Statistical theory of groundwater flow and transport: Pore to laboratory, laboratory to formation, and formation to regional scale. *Water Resour. Res. 22*(9):120S–134S.

Dagg, M. & Pratt, M. A. C. (1962) Relation of stormflow to incident rainfall. *E. Afr. Agric. For. J. 27*:31–35.

Davidson, M. R. (1984) A Green-Ampt model of infiltration in a cracked soil. *Wat. Resour. Res. 20*:1685–1690.

Dawes, W. & Short, D. (1991) Contour-based topographic analyses for modelling catchment hydrology. (in prep.)

De Bruin, H. A. R. (1983a) Evapotranspiration in humid tropical regions. *In*: *Hydrology of Humid Tropical Regions with Particular Reference to the Hydrological Effects of Agriculture and Forestry Practice.* Proc. Hamburg Symp., Aug. 1983. IAHS. Publ. *140*:299–311.

De Bruin, H. A. R. (1983b) A model for the Priestley-Taylor parameter a. *J. Clim. Appl. Meteorol. 22*:572–578.

De Bruin, H. A. R. (1989) Physical aspects of the planetary boundary layer with special reference to regional evapotranspiration. *In*: *Estimation of Areal Evapotranspiration.* T. A. Black, D. L. Splittlehouse, M. D. Novak & D. T. Price (eds.) Proc. Vancouver Symp., Aug. 1987, Int'l. Assoc. of Hydrol. Sci. Publ. *177*:117–132.

Denmead, O. T. (1984) Plant physiological methods for studying evapotranspiration: Problems of telling the forest from the trees. *Agric. Wat. Management 8*:167–189.

Denmead, O. T. & Bradley, E. F. (1985) Flux-gradient relationships in a forest canopy. *In*: B. A. Hutchinson & B. B. Hicks (eds.) *The Forest-Atmosphere Interaction.* D. Reidel Publ. Co., Dordrecht, 421–442.

DeWalle, D. R., Swistock, B. R. & Sharpe, W. E. (1988) Three-component tracer model for stormflow on a small Appalachian forested catchment. *J. Hydrol. 104*:301–310.

DeWalle, D. R., Swistock, B. R. & Sharpe, W. E. (1990) Tracer model for stormflow on a small Appalachian forested catchment – Reply. *J. Hydrol. 117*:381–384.

Dias, A. C. D. C. P. & Nortcliff, S. (1985) Effects of two land-clearing methods on the physical properties of an Oxisol in the Brazilian Amazon. *Tropic. Agric. 62*:207–212.

Dietrich, W. E., Windsor, D. M. & Dunne, T. (1982) Geology, climate and hydrology of Barro Colorado Island. *In*: E. Leigh, A. S. Rand & D. M. Windsor (eds.) *The Ecology of Tropical Forest: Seasonal Rhythms and*

Long-Term Changes. Smithsonian Inst., Wash., D.C, 21–46.

Dixon, R. M. (1976) Comment on "Derivation of an equation of infiltration," by H. J. Morel-Seytoux and J. Khanji. *Wat. Resour. Res.* *12*:116–118.

Dogse, P. (1989) *Sustainable Tropical Rainforest Management: Some Economic Considerations.* Dept. Econ., Univ. Stockholm, Sweden.

Dolman, A. J. & Stewart, J. B. (1989) Modelling forest transpiration from climatological data. *In*: R. H. Swanson, P. Y. Bernier & P. D. Woodards (eds.) *Forest Hydrology and Watershed Management*, Proc. Vancouver Symp., Aug. 1987. Internt'l Assoc. of Hydrol. Sci. Publ. *167*:319–327.

Dolman, A. J., Gash, J. H. C., Roberts, J. & Shuttleworth, W. J. (1990) Stomatal and surface conductance of tropical rainforest. *Agric. Forest Meteorol. Special Issue* *54*:303–318.

Dooge, J. C. (1986) Looking for hydrologic laws. *Wat. Resour. Res.* *22*(9):46S–58S.

Dubreuil, P. L. (1985) Review of field observations of run-off generation in the tropics. *J. Hydrol.* *80*:237–264.

Dugdale, G., Hardy, S. & Milford, J. R. (1991) V: Daily catchment rainfall estimated from METEOSAT. *Hydrol. Processes* 5:261–270.

Dunin, F. X. (1976) Infiltration: Its simulation for field conditions. *In*: J. C. Rodda (ed.) *Facets of Hydrology*. Wiley-Interscience, New York, N.Y. 199–227.

Dunin, F. X. (1987) Run-off and drainage from grassland catchments. *In*: R. W. Snaydon (ed.) *Managed Grasslands, B. Analytical Studies.* Elsevier Sci. Publ., Amsterdam, 205–213.

Dunin, F. X. (1991) Extrapolation of "point" measurements of evaporation: Some issues of scale. *Vegetatio 91*:39–47 and *In*: A. Henderson-Sellers & A. J. Pitman (eds.) *Vegetation and Climate Interactions in Semiarid Regions*. Kluwer Academic Publ., Belgium, 39–47.

Dunin, F. X. & Aston, A. R. (1984) The development and proving of models of large-scale evapotranspiration: An Australian study. *Agric. Wat. Management* 8:305–323.

Dunin, F. X., McIlroy, I. C. & O'Loughlin, E. M. (1985) A lysimeter characterization of evaporation by eucalypt forest and its representativeness for the local environment. *In*: *The Forest-Atmosphere Interaction*. B. A. Hutchinson & B. B. Hicks, (eds.) Reidel, Dordrecht, 271–291.

Dunin, F. X., O'Loughlin, E. M. & Reyenga, W. (1988) Interception loss from eucalypt forest: Lysimeter determination of hourly rates for long-term evaluation. *Hydrol. Processes* 2:315–329.

Dunne, T. (1978) Field studies of hillslope flow processes. *In*: M. J. Kirkby (ed.) *Hillslope Hydrology*. Wiley, Chichester, 227–293.

Dunne, T. (1983) Relation of field studies and modeling in the prediction of storm runoff. *J. Hydrol.* 65:25–48.

Dunne, T. & Black, R. D. (1970) An experimental investigation of runoff production in permeable soils. *Wat. Resour. Res.* 6:478–490.

Dyck, S. & Baumert, H. (1991) A concept for hydrological process studies from local to global scales. *In*: G. Kienitz, P. C. D. Milly, M. Th. van Genuchten, D. Rosbjerg & W. J. Shuttleworth (eds.) *Hydrological Interactions Between Atmosphere, Soil and Vegetation*, Proc. Vienna Symp., Aug. 1991. Int'l. Assoc. of Hydrol. Sci. Publ. *204*:31–42.

Dyer, A. J. (1974) A review of flux-profile relationships. *Boundary Layer Meteorol.* 7:363–372.

Edwards, K. A. (1979a) The water balance of the Mbeya experimental catchments. *In*: J. R. Blackie, K. A. Edwards & R. T. Clarke (Compiled) *Hydrological Research in East Africa. E. Afr. Agric. For. J. Special Issue* 43:232–247.

Edwards, K. A. (1979b) Sediment yields at Mbeya. *In*: J. R. Blackie, K. A. Edwards & R. T. Clarke (Compiled) *Hydrological Research in East*

Africa. *E. Afr. Agric. For. J. Special Issue 43*:248–253.

Edwards, K. A. & Blackie, J. R. (1981) Results of the East African Catchment Experiments, 1958–1974. *In*: R. Lal & E. W. Russell (eds.) *Tropical Agricultural Hydrology – Watershed Management and Land Use*. Wiley, Chichester, 163–188.

Eklundh, L. & Pilesjo, P. (1990) Regionalization and spatial estimation of Ethiopian mean annual rainfall. *Int. J. Climatology 10*:473–494.

Elsenbeer, H. & Cassel, D. K. (1990) Surficial processes in the rainforest of western Amazonia. *In*: R. R. Zimmer, C. L. O'Loughlin & L. S. Hamilton (eds.) *Research Needs and Applications to Reduce Erosion and Sedimentation in Tropical Steeplands*. Proc. Fiji Symp., June 1990, Int'l. Assoc. of Hydrol. Sci. Publ. *192*:289–297.

Entekhabi, D. & Eagleson, P. (1989) Land surface hydrology parameterization for atmospheric general circulation models including subgrid scale spatial variability. *J. Climate. Am. Met. Soc.* 2:816–831.

Everitt, B. (1980) *Cluster Analysis*, 2nd Edition. Halsted Heinemann, London.

Feddes, R. A., Kabat, P., Van Bakel, P. J. T., Bronswijk, J. J. B. & Halbertsma, J. (1988) Modelling soil water dynamics in the unsaturated zone – State of the art. *J. Hydrol. 100*:69–111.

Feininger, T. (1969) Pseudokarst on quartz diorite, Colombia. *Z. Geomorphol. N. F. 13*:287–296.

Finnagan, J. J. & Raupach, M. R. (1987) Modern theory of transfer in plant canopies in relation to stomatal characteristics. *In*: E. Zeiger, G. Farquhar & I. Cowan (eds.) *Stomatal Function*. Stanford Univ. Press, 385–429.

Foster, S. S. D. & Smith-Carrington, A. (1980) The interpretation of tritium in the chalk unsaturated zone. *J. Hydrol. 46*:343–364.

Freeze, R. A. (1972) The role of subsurface flow in generating surface runoff. 2. Upstream source areas. *Wat. Resour. Res.* 8(5):1272–1283.

Freeze, R. A. (1974) Streamflow generation. *Rev. Geophys. Space Phys.* 12:627–647.

Freeze, R. A. (1978) Mathematical models of hillslope hydrology. *In*: M. J. Kirkby (ed.) *Hillslope Hydrology*. Wiley, Chichester, 177–225.

Fritsch, J. M. (1987) Ecoulements et érosion sous prairies artificielles après défrichement de la forêt tropicale humide. *In*: R. H. Swanson, P. Y. Bernier & P. D. Woodard (eds.) *Forest Hydrology and Watershed Management*. Proc. Vancouver Symp., Aug. 1987. Int'l. Assoc. of Hydrol. Sci. Publ. *167*:123–129.

Fritsch, J-M. (1992) *Les Effets Du Defrichment De La Foret Amazonienne Et De La Mise En Culture Sur L'Bydrologie De Petits Bassins Versants – Operation ECEREX En Guyane Francaise*, Editions de l'ORSTOM, Institut Francais De Recherche Scientifique Pour Le Developpement En Cooperation, Collection Etudes et Theses, Paris, 1992.

Fritsch, J. M., Dubreuil, P. L. & Sarrailh, J. M. (1987) De la parcelle au petit bassin-versant: Effet d'échelle dans l'écosysteme forestier amazonien. *In*: R. H. Swanson, P. Y. Bernier & P. D. Woodard (eds.) *Forest Hydrology and Watershed Management*. Proc. Vancouver Symp., Aug. 1987. Int'l. Assoc. of Hydrol. Sci. Publ. *167*:131–142.

Garcia, O., Bosart, L. & DiMego, G. (1978) On the nature of the winter season rainfall in the Dominican Republic. *Mon. Weather Rev.* 106:961–982.

Gash, J. H. C. (1979) An analytical model of rainfall interception in forests. *Quart. J. Roy. Met. Soc. 105*:43–55.

Gash, J. H. C. & Morton, A. J. (1978) An application of the Rutter model to the estimation of the interception loss from Thetford Forest. *J. Hydrol.* 38:49–58.

Gash, J. H. C., Wright, I. R. & Lloyd, C. R. (1980) Comparative estimates of interception loss from three coniferous forests in Great Britain. *J. Hydrol.* 48:89–105.

Gates, D. M. (1980) *Biophysical Ecology*. Springer Verlag, New York, USA.

Gelhar, L. W. (1986) Stochastic subsurface hydrology from theory to applications. *Water Resour. Res.* 22(9):135S–145S.

Genereux, D. P. & Hemond, H. F. (1990) Three-component tracer model for stormflow on a small Appalachian forested catchment – Comment. *J. Hydrol.* 117:377–380.

Germann, P. F. (1986) Rapid drainage response to precipitation. *Hydrol. Processes 1*:3–14.

Germann, P. F. (1990) Macropores and hydrologic hillslope processes. *In*: M. G. Anderson & T. P. Burt (eds.) *Process Studies in Hillslope Hydrology*. Wiley, Chichester, 327–363.

Germann, P. F. & Beven, K. J. (1985) Kinematic wave approximation to infiltration into soils with sorbing macropores. *Wat. Resour. Res.* 21:990–996.

Germann, P. F. & Beven, K. J. (1986) A distribution function approach to water flow in soil macropores based on kinematic wave theory. *J. Hydrol.* 83:173–183.

Germann, P. F., & Bevan, K. J. (1981) Water flow in soil macropores I. An experimental approach. *J. Soil Sci.* 32:1–13.

Giambelluca, T. W. & Nullet, D. (1992) Evaporation at high elevations in Hawaii. *J. Hydrol. 136*: 219–235.

Gilham, R. W. (1984) The effect of the capillary fringe on water-table response. *J. Hydrol.* 67:307–324.

Gilman, K. & Newson, M. D. (1980) *Soil Pipes and Pipeflow. A Hydrological Study in Upland Wales*. Geobooks, Norwich.

Gilmour, D. A. (1975) *Catchment water balance studies on the wet tropical coast of North Queensland*. Unpublished PhD Thesis, Dept. of Geography, James Cook Univ. of North Queensland, Townsville, Australia.

Gilmour, D. A. (1977) Effects of logging and clearing on water yield and quality in a high rainfall zone of north-east Queensland. *In: The Hydrology of Northern Australia*. Proc. Brisbane Hydrol. Symp., June 1977, Inst. Engrs., Canberra, Australia. *Nat. Conf. Publ. 77/5*:156–160.

Gilmour, D. A., Bonell, M. & Sinclair, D. F. (1980) An investigation of storm drainage processes in a tropical rainforest catchment. Australian Government Publ. Service, Canberra, A.C.T. *Australian Water Resources Council Tech. Paper 56*.

Gilmour, D. A., Cassells, D. S. & Bonell, M. (1982) Hydrological research in the tropical rainforests of north Queensland: Some implications for land use management. *In*: E. M. O'Loughlin & L. J. Bren (eds.) *First National Symp. on Forest Hydrol.*, Melbourne, May 1982, Instit. Engrs., Canberra, Australia. Nat. Conf. Publ. *82*–6:145–152.

Gilmour, D. A., Bonell, M. & Cassells, D. S. (1987) The effects of forestation on soil hydraulic properties in the Middle Hills of Nepal: A preliminary assessment. *Mountain Research and Development, 7*(3): 239–249.

Gish, T. J. & Starr, J. L. (1983) Temporal variability of infiltration under field conditions. *In: Proc. Nat. Conf. on Advances in Infiltration*, 12–13 Dec. 1983, Chicago, Ill. Amer. Soc. Agric. Engrs., St. Joseph, Mich., 122–131.

Goel, P. S., Datta, P. S. & Tanwar, B. S. (1977) Measurement of vertical recharge to groundwater in Haryena State (India) using tritium tracer. *Nordic Hydrol.* 8:211–224

Goudberg, N. & Bonell, M. with Benzaken, D. (eds.) (1991) *Tropical Rainforest Research in Australia: Present Status and Future Directions for the Institute for Tropical Rainforest Studies*. Proc. Townsville

Workshop, May 1990, Inst. for Tropical Rainforest Studies, James Cook Univ. of North Queensland, Townsville, Australia.

Gradwohl, J. & Greenberg, R. (1988) *Saving the Tropical Forests*. Earthscan Publications.

Granger, R. J. (1989a) An examination of the concept of potential evaporation. *J. Hydrol. 111*:9–19.

Granger, R. J. (1989b) A complementary relationship approach for evaporation from nonsaturated surfaces. *J. Hydrol. 111*:31–38.

Granger, R. J. & Gray, D. M. (1989) Evaporation from natural nonsaturated surfaces. *J. Hydrol. 111*:21–29.

Green, W. H. & Ampt, G. (1911) Studies of soil physics, Part I. – The flow of air and water through soils. *J. Agric. Sci.* 4:1–24.

Greenwood, E. A. N., Klein, L., Beresford, J. D. & Watson, G. D. (1985) Differences in annual evaporation between grazed pasture and *Eucalyptus* species in plantations on a saline farm catchment. *J. Hydrol.* 78:261–278.

Griffiths, J. F. (1972) *Climates of Africa*. Elsevier, Amsterdam.

Grigorjev, V. Y. & Iritz, L. (1991) Dynamic simulation model of vertical infiltration of water in soil. *Hydrol. Sci. J. 36*:171–179.

Guillot, B. (1990) Rainfall estimation using satellite imagery: The EPSAT Program (Abstract). *In: Measurement Modelling and Forecasting of Rainfall in Space and Time*. Proc. XV General Assembly, Copenhagen, 23–27 April 1990. European Geophys. Soc., *135*.

Gupta, V. K. & Waymire, E. (1990) Multiscaling properties of spatial rainfall and river flow distributions. *J. Geophys. Res. 95*:1999–2009.

Hadley, M. & Schreckenberg, K. (1989) Contributing to sustained resource use in the humid and sub-humid tropics: Some research approaches and insights. *MAB Digest 3*. UNESCO, Paris.

Hall, A. J. (1984) Hydrology in tropical Australia and Papua New Guinea. *Hydrol. Sci. J. 29*:399–423.

Hall, R. L., Calder, I. R., Rosier, P. T. W., Swaminath, M. H. & Mumtaz, J. (1991) Measurements and modelling of interception loss from eucalypt plantations in southern India. *In: Growth and Water Use of Forest Plantations*. Proc. Int. Symp., Bangalore, India, 7–11 Feb. 1991. (Available from Inst. of Hydrol., UK).

Hamilton, L. S. (1986) Overcoming myths about soil and water impacts of tropical forest land uses. *In*: S. A. El-Swaify, W. C. Moldenhauer & A. Lo Akeny (eds.) *Soil Erosion and Conservation*. Soil Conservation Soc. of Am., 680–690.

Hamilton, L. S. (1988) The environmental influences of forests and forestry in enhancing food production and food security. Presented at *Expert Consultation on Forestry and Food Production/Security*, Bangalore, India, 14–20 Feb. 1988. FAO, Rome (1987).

Hamilton, L. S. (1990) Tropical Forests: Identifying and clarifying issues. *An Overview Paper on Issues for the Tropical Forests Task Force of the Pacific Economic Cooperation Council*, Kuala Lumpur, 25–29 Sept. 1990. (Available from Environ. and Policy Inst., East-West Center, Honolulu, Hawaii.)

Hamilton, L. S. with King, P. N. (1983) *Tropical Forested Watersheds: Hydrologic and Soils Response to Major Uses or Conversions*. Boulder, Westview Press.

Harding, R. J., Hall, R. L., Swaminath, M. H. & Srinivasa Murthy, K. V. (1991) The soil moisture regimes beneath forest and an agricultural crop in southern India – Measurements and modelling. *In: Growth and Water Use of Forest Plantations*. Proc. Int. Symp., Bangalore, India, 7–11 Feb. 1991. (Available from Inst. of Hydrol., UK).

Harrold, T. W. (1973) The structure and mechanism of widespread precipitation. *Quart. J. Roy. Met. Soc. 99*:232–251.

Harrold, T. W. & Austin, P. M. (1974) The structure of precipitation systems – A review. *J. Rech. Atmos.* 8:41–57.

Henderson-Sellers, A. (1987) Effects of change in land use on climate in the humid tropics. *In*: *The Geophysiology of Amazonia – Vegetation and Climate Interactions.* Wiley/UNU, 463–493.

Henderson-Sellers, A. & Gornitz, V. (1984) Possible climatic impacts of land cover transformations, with particular emphasis on tropical deforestation. *Climatic Change* 6:231–258.

Hendrickx, J. M. H. (1990) Determination of soil hydraulic properties. *In*: M. G. Anderson & T. P. Burt (eds.) *Process Studies in Hillslope Hydrology.* Wiley, Chichester, 43–92.

Henry, W. K. (1974) The tropical rainstorm. *Mon. Weather Rev.* 102:717–725.

Herwitz, S. R. (1985) Interception storage capacities of tropical rainforest trees. *J. Hydrol.* 77:237–252.

Herwitz, S. R. (1986) Infiltration-excess caused by stemflow in a cyclone-prone tropical rainforest. *Earth Surf. Processes Landf.* 11:401–412.

Herwitz, S. R. (1987) Raindrop impact and water flow on the vegetative surfaces of trees and the effects of stemflow and throughfall generation. *Earth Surf. Processes Landf.* 12:425–432.

Hewlett, J. D. (1961) Watershed management. *In*: *Report for 1961 Southeastern Forest Experiment Station.* U.S. For. Serv., Asheville, N.C., USA.

Hewlett, J. D. (1974) Comments on letters relating to "Role of subsurface flow in generating surface runoff, 2. Upstream source areas," by R.A. Freeze. *Wat. Resour. Res.* 10:605–607.

Hewlett, J. D. & Fortson, J. C. (1983) The paired catchment experiment. *In*: J. D. Hewlett (ed.) *Forest Water Quality.* School of Forest Resources, Univ. of Georgia, Athens, GA 11–14.

Hewlett, J. D. & Hibbert, A. R. (1967) Factors affecting the response of small watersheds to precipitation in humid areas. *In*: W. E. Sopper & H. W. Lull (eds.) *International Symp. on Forest Hydrol.* Pergamon, Oxford, 275–290.

Hewlett, J. D. & Troendle, C. A. (1975) Non-point and diffused water sources: A variable source area problem. *In*: *Watershed Management Symp.* Comm. Watershed Manage., Irrig. Drain. Div., Am. Soc. Civ. Engrs., Logan, Utah, 21–46.

Hewlett, J. D., Fortson, J. C. & Cunningham, G. B. (1977) The effect of rainfall intensity on stormflow and peak discharge form forest land. *Wat. Resour. Res.* 13:259–266.

Hewlett, J. D., Fortson, J. C. & Cunningham, G. B. (1984) Additional tests on the effect of rainfall intensity on stormflow and peak flow from wildland basins. *Wat. Resour. Res.* 20:985–989.

Hewlett, J. D., Lull, H. W. & Reinhart, K. G. (1969) In defense of experimental watersheds. *Wat. Resour. Res.* 5:306–316.

Hibbert, A. R. (1967) Forest treatment effects on water yield. *In*: W. E. Sopper & H. W. Lull (eds.) *Int'l. Symp. on Forest Hydrol.* Pergamon, Oxford, 527–543.

Hillel, D. (1975) Simulation of evaporation from bare soil under steady and diurnally fluctuating evaporativity. *Soil Sci.* 120:230–237.

Hillel, D. (1980) *Fundamentals of Soil Physics.* Academic Press, New York.

Hollingsworth, A. (1989) The global weather experiment – 10 years on. *Weather* 44:278–285.

Holm, M. E. (1988) *Geostatistics and Petroleum Geology.* Van Nostrand Reinhold, New York, USA.

Holtan, H. N. & Lopez, N. C. (1971) *USDAHL-70 Model of Watershed Hydrology.* USDA-ARS Tech. Bull. *1435.*

Hoogmoed, W. B. & Bouma, J. (1980) A simulation model for predicting infiltration into cracked clay soil. *Soil Sci. Soc. Am. J.* 44:458–461.

Horton, R. E. (1933) The role of infiltration in the hydrological cycle. *Trans. Am. Geophys. Union* 14:446–460.

Horton, R. E. (1940) An approach toward a physical interpretation of infiltration capacity. *Soil Sci. Soc. Am. Proc.* 5:399–417.

Horton, R. E. (1945) Erosional development of streams and their drainage basins: Hydrological approach to quantitative morphology. *Bull. Geol. Soc. Am.* 56:275–370.

Horton, J. H. & Hawkins, R. H. (1965) Flow path of rain from the soil surface to the water table. *Soil Sci.* 100:377–383.

Houze, R. A. (1982) Cloud clusters and large scale vertical motions in the tropics. *J. Meteorol. Soc. Jap.* 60:396–410.

Houze, R. A. (1989a) Convective and stratiform precipitation in the tropics. *In*: *Tropical Precipitation Measurements.* Proc. Tokyo Symp., Sci. and Tech. Corp., Hampton, VA (in press).

Houze, R. A. (1989b) Observed structure of mesoscale convective systems and implications for large-scale heating. *Quart. J. Roy. Met. Soc.* 115:425–461.

Hubert, P. (in press) Fractal and multifractal studies of rainfall time variability. *In*: *Space and Time Scale Variability and Interdependencies in Various Hydrological Processes* (Feddes, R. A., ed), Proc. UNESCO-IAHS G. Kovacs 1992 Colloquium, UNESCO – Cambridge University Press.

Hubert, P. & Carbonnel, J-P. (1989) Dimensions fractales de l'occurrence de pluie en climat soudano-sahélien, *Hydrol. Continent (ORSTOM)* 4: 3–10.

Huygen, J. (1989) *Estimation of Rainfall in Zambia Using METEOSAT-TIR Data.* The Wind Staring Centre, Wageningen, The Netherlands. Report *12.*

International Tropical Timber Organization (ITTO) (1990) *ITTO Action Plan/ Criteria and Priority Areas for Programme Development and Project Work.* Int'l. Tropical Timber Council, Ninth Session, Yokohama, 16–23 Nov. 1990.

Jackson, I. J. (1969) Tropical rainfall variations over a small area. *J. Hydrol.* 8:99–110.

Jackson, I. J. (1971) Problems of throughfall and interception assessment under tropical rainforest. *J. Hydrol.* 12:234–254.

Jackson, I. J. (1972) The spatial correlation of fluctuations in rainfall over Tanzania: A preliminary analysis. *Arch. Met. Geoph. Biokl. Ser. B,* 20:167–178.

Jackson, I. J. (1974) Inter-station rainfall correlation under tropical conditions. *Catena* 1:235–256.

Jackson, I. J. (1975) Relationships between rainfall parameters and interception by tropical rainforest. *J. Hydrol.* 24:215–238.

Jackson, I. J. (1978) Local differences in the patterns of variability of tropical rainfall: Some characteristics and implications. *J. Hydrol.* 38:273–287.

Jackson, I. J. (1985) Tropical rainfall variability as an environmental factor: Some considerations. *Sing. J. Trop. Geog.* 6:23–34.

Jackson, I. J. (1986) Relationships between raindays, mean daily intensity and monthly rainfall in the tropics. *J. Climatology* 6:117–134.

Jackson, I. J. (1988a) *Climate, Water and Agriculture in the Tropics.* Longman, UK.

Jackson, I. J. (1988b) Daily rainfall over northern Australia: Deviations from the world pattern. *J. Climatology* 8:463–476.

Jakeman, A. J., Littlewood, I. G. & Whitehead, P. G. (1990) Computation of the instantaneous unit hydrograph and identifiable component flows with application to two small upland catchments. *J. Hydrol.* *117*:275–300.

Jarvis, P. G. (1976) The interpretation of the variations in leaf water potential and stomatal conductance found in canopies in the field. *Phil. Trans. Roy. Soc., London, Ser. B*, *273*:593–610.

Jarvis, N. J. & Leeds-Harrison, P. B. (1987) Modelling water movement in drained clay soils. *J. Soil Sci.* *38*:487–509.

Jarvis, P. G. & McNaughton, K. G. (1986) Stomatal control of transpiration: Scaling up from leaf to region. *Adv. Ecol. Res. 15*:1–49.

Jenssen, P. D. (1990) Methods for measuring the saturated hydraulic conductivity of tills. *Nordic Hydrol. 21*:95–106.

Jobard, I. (1990) Satellite infrared radiances related to the precipitations in the tropical West Africa (Abstract). *In*: *Measurement, Modelling and Forecasting of Rainfall in Space and Time*. Proc. XV General Assembly, Copenhagen, 23–27 April 1990, European Geophys. Soc., p. 138.

Jolliffe, I. T. (1987) Rotation of principal components: Some comments. *J. Climatology 7*:507–510.

Jones, J. A. A. (1971) Soil piping and stream channel initiation. *Wat. Resour. Res. 7*:602–610.

Jones, J. A. A. (1986) Some limitations of the a/s index for predicting basin-wide patterns of soil water drainage. *Z. Geomorphol., Suppl. Band 60*:7–20.

Jones, M. E. (1979) Rainfall intensities of East African storms. *In*: J. R. Blackie, K. A. Edwards & R. T. Clarke (Compiled) Hydrological Research in East Africa. *E. Afr. Agric. For. J. Special Issue 43*:261–264.

Juvik, J. O. & Ekern, P. C. (1978) *A Climatology of montane fog on Mauna Loa Hawaii Island*. Univ. Hawaii, Water Resources Res. Center. Tech. Report *118*.

Kavvas, M. L., Soong, S. T., Saquib, M. N. & Chen, Z. (1991) Integrated modeling of hydrologic and atmospheric processes at regional scale. *In*: G. Kienitz, P. C. D. Milly, M. Th. van Genuchten, D. Rosbjerg & W. J. Shuttleworth (eds.) *Hydrological Interactions Between Atmosphere, Soil and Vegetation*. Proc. Vienna Symp., Aug. 1991. Int'l. Assoc. of Hydrol. Sci., Publ. *204*:21–29.

Keenan, T. D., Holland, G. J., Manton, M. J. & Simpson, J. (1988) TRMM ground truth in a monsoon environment: Darwin, Australia. *Aust. Met. Mag. 36*:81–90.

Kennedy, V. C., Kendall, C., Zellweger, G. W., Wyeman, T. A. & Avenzino, R. J. (1986) Determination of the components of stormflow using water chemistry and environmental isotopes, Mattole River Basin, California. *J. Hydrol. 84*:107–104.

Kirkby, M. J. (1988) Hillslope runoff processes and models. *J. Hydrol. 100*:315–339.

Kirkby, M. J. (ed.) (1978) *Hillslope Hydrology*. Wiley, Chichester.

Koppi, A. J. & Geering, H. R. (1986) The preparation of unsmeared soil surfaces and an improved apparatus for infiltration measurements. *J. Soil Sci. 37*:177–181.

Kostiakov, A. N. (1932) On the dynamics of the coefficient of water percolation in soils and the necessity of studying it from the dynamic point of view for the purposes of amelioration. *Trans. Sixth Comm. Int. Soc. Soil Sci.*, 17–21.

Kozlowski, T. T. (ed.) (1981) *Water Deficits and Plant Growth*, Vol. VI (Woody Plant Communities). Academic Press, New York, USA.

Krige, D. G., Guarascio, M. & Camisani-Calzolari, F. A. (1989) Early South African geostatistical techniques in today's perspective. *In*: M. Armstrong (ed.) *Geostatistics, Vol. I*. Kluwer, Dordrecht, 1–19.

Kruijt, B., Klaassen, W., Hutjes, R. W. A. & Veen, A. W. L. (1991) Heat and Momentum fluxes near a forest edge. *In*: G. Kienitz, P. C. D. Milly, M. Th. van Genuchten, D. Rosbjerg & W. J. Shuttleworth (eds.) *Hydrological Interactions Between Atmosphere, Soil and Vegetation*. Proc. Vienna Symp., Aug. 1991, Int'l. Assoc. of Hydrol. Sci. Publ. *204*:107–115.

Label, T. & Thauvin, V. (1990) Spatial and temporal sampling of Sahelian squell line rainfall: Some experimental features. *In*: *Measurement, Modelling and Forecasting of Rainfall in Space and Time*. Proc. XV General Assembly, Copenhagen, 23–27 April 1990, European Geophys. Soc., 139.

Lal, R. (1980) Physical and mechanical composition of Alfisols and Ultisols with particular reference to soils in the tropics. *In*: B. K. G. Theng (ed.) *Soils with Variable Charge*. New Zealand Soc. of Soil Sci., Soil Bureau, Dept. of Sci. and Indust. Res., Lower Hut, New Zealand, 253–279.

Lal, R. (1981) Deforestation of tropical rainforest and hydrological problems. *In*: R. Lal & E. W. Russell (eds.) *Tropical Agricultural Hydrology – Watershed Management and Land Use*. Wiley, Chichester, 131–140.

Lal, R. (1983) Soil erosion in the humid tropics with particular reference to agricultural land development and soil management. *In*: *Hydrology of Humid Tropical Regions with Particular Reference to the Hydrological Effects of Agriculture and Forestry Practice*. Proc. Hamburg Symp., Aug. 1983, Int'l. Assoc. of Hydrol. Sci. Publ. *140*:221–239.

Lal, R. (1990) Agroforestry systems to control erosion on arable tropical steeplands. *In*: R. R. Ziemer, C. L. O'Loughlin & L. S. Hamilton (eds.) *Research Needs and Applications to Reduce Erosion and Sedimentation in Tropical Steeplands*. Proc. Fiji Symp., June 1990, Int'l. Assoc. of Hydrol. Sci. Publ. *192*:338–346.

Law, F. (1956) The effect of afforestation upon the water yield of water catchment areas. *J. British Waterworks Assoc. 38*:489–494.

Law, K. F. & Cheong, C. W. (1987) Effects of land use changes on the hydrological characteristics of Sungai Tekam experimental basin. Paper presented at UNESCO/MRP Workshop on *Impact on Operations in Natural and Plantation Forest on Conservation of Soil and Water Resources*, 23–26 June 1987, Universiti Pertanian, Malaysia.

Lebel, T., Sauvageot, 0., Hoepffner, M., Desbois, M., Guillot, B. and Hubert, P. (1991) Estimation des precipitations au Sahel: l'expérience EPSAT-NIGER. *Hydrol. Continent (ORSTOM) 6*, 133–143. (An English version of this article is published in *Hydrological Sciences Journal* 37 (3) June 1992.)

Legates, D. R. (1991) The effect of domain shape on principal components analyses. *Int. J. Climatol. 11*:135–146.

Lettau, H., Lettau, K. & Molion, L. C. B. (1979) Amazonia's hydrologic cycle and the role of atmospheric recycling in assessing deforestation effects. *Mon. Weath. Rev. 107*:227–238.

Linsey, R. K., Kohler, M. A. & Paulhus, J. L. H. (1988) *Hydrology for Engineers* (SI Metric edn.). McGraw-Hill.

Littlewood, I. G. & Jakeman, A. J. (1991) Hydrograph separation into dominant quick and slow flow component. *In*: *Third National Hydrol. Symp.* Proc. Univ. Southampton, Sept. 1991, British Hydrol. Soc., Inst. Hydrol., UK, 3.9–3.16.

Lloyd, C. R. & Marques Filho, A. de O. (1988) Spatial variability of throughfall and stemflow measurements in Amazonian rainforest. *Agric. For. Meteorol. 42*:63–73.

Lloyd, C. R., Gash, J. H. C., Shuttleworth, W. J. & Marques Filho, A. de

O. (1988) The measurement and modelling of rainfall interception by Amazonian rainforest. *J. Hydrol.* *43*:277–294.

Lockwood, J. G. (1967) Probable maximum 24-hour precipitation over Malaya by statistical methods. *Met. Mag. 96*:11–19.

Lovejoy, S. & Mendelbrot, B. B. (1985) Fractal properties of rain and a fractal model. *Tellus 37A*:209–285.

Lovejoy, S., Schertzer, D. & Tsonis, A. A. (1987) Functional box-counting and multiple elliptical dimensions of rain. *Science 235*: 1036–1038.

Lovejoy, S. & Schertzer, D. (1990) Multifractals, universality classes and satellite and radar measurements of cloud and rainfields. *J. Geophys. Res. 95*:2021–2034.

Lovejoy, S. and Schertzer, D. (1991) Multifractal analysis and rain and cloud fields from 10^{-3} to 10^{-6}. In: D. Schertzer and S. Lovejoy, eds. *Scaling, fractals and non—linear variability in geophysics,* Kluwer, 111—144.

Lovejoy, S. and Schertzer, D. (1992) *Multifractals in Geophysics,* AGU-CGU-MSA Spring Meeting, May , 1992.

Lovejoy, S. and Schertzer, D. (in press) Multifractals and rain. In: Rundeczick, ed. *Scaling in Hydrology,* Birkhauser.

Lundgren, L. (1980) Comparison of surface runoff and soil loss from runoff plots in forests and small scale agriculture in the Usambera Mts., Tanzania. *Geografiska Annaler 62*:113–148.

Lyne, V. D. & Hollick, M. (1979) Stochastic time-varying rainfall-runoff modelling. *In: Hydrol. and Water Resour. Symp.*, Perth, Inst. Engrs., Canberra Australia, 89–92.

Lyons, W. F. & Bonell, M. (1992a) Tropical mesoscale patterns: The opening phase of the monsoon in the Townsville area of north-east Queensland, Dec. 1988. *Aust. Geog. Studies 30*:185–205.

Lyons, W. F. & Bonell, M. (1992b) Daily mesoscale rainfall in the tropical wet/dry climate of the Townsville area, north-eastern Queensland during the 1988/89 wet season: Synoptic scale airflow considerations. *Int. J. Climatol. 12*:655–684.

Lyons, W. F. & Bonell, M. (1993) Regionalization of daily mesoscale rainfall in a tropical wet/dry climate of north-east-Queensland during 1988/89. *Int. J. Climatol.* (in press).

Malmer, A. (1990) Stream suspended sediment load after clear-felling and different forestry treatments in tropical rainforest, Sabah, Malaysia. *In: Research Needs and Applications to Reduce Erosion and Sedimentation in Tropical Steeplands.* Proc. Fiji Symp., June 1990, Int'l. Assoc. of Hydrol. Sci. Publ. *192*:62–71.

Mandelbrot, B. B. (1982) *The Fractal Geometry of Nature,* Freeman, Cooper, San Francisco, California., USA.

Mandelbrot, B. B. (1975) *Les Objets Fractals, Forme, Hasard et Dimension.* Flammarion, Paris.

Mandelbrot, B. B. (1977) *Fractals: Form, Chance and Dimension.* Freeman, San Francisco.

Mandelbrot, B. B. (1990) Fractals – a geometry of nature. *New Scientist* (15 September 1990), 37–43.

Markar, M. S. & Mein, R. G. (1987) Modelling of evapotranspiration from homogeneous soils. *Wat. Resour. Res. 23*:2001–2007.

Marshall, T. J. & Holmes, J. W. (1979) *Soil Physics.* Cambridge, Univ. Press, Cambridge.

McDonnell, J. J. (1989) *The age, origin and pathway of subsurface storm-flow in a steep humid headwater catchment.* PhD Thesis, Dept. of Geography, Univ. of Canterbury, Christchurch, New Zealand.

McDonnell, J. J. (1990) A rationale for old water discharge through macropores in a steep, humid catchment. *Wat. Resour. Res. 26*:2821–2832.

McDonnell, J. J., Bonell, M., Stewart, M. K. & Pearce, A. J. (1990) Deuterium variations in storm rainfall: Implications for stream hydrograph separations. *Wat. Resour. Res. 26*:455–458.

McDonnell, J. J., Stewart, M. K. & Owens, I. F. (1991) Effect of catchment-scale subsurface mixing on stream isotopic response. *Wat. Resour. Res. 27*:3065–3073.

McIntyre, D. S. (1958a) Permeability measurements of soil crusts formed by raindrop impact. *Soil Sci. 85*:185–189.

McIntyre, D. S. (1958b) Soil splash and the formation of surface crusts by raindrop impact. *Soil Sci. 85*:261–266.

McNaughton, K. G. & Black, T. A. (1973) A study of evapotranspiration from a Douglas fir forest using the energy balance approach. *Water Resour. Res. 9*: 1579–1590.

McNaughton, K. G. & Jarvis, P. G. (1983) Predicting effects of vegetation changes on transpiration and evaporation. *In*: T. T. Kozlowski (ed.) *Water Deficits and Plant Growth.* Academic Press, New York 7:1–47.

McNaughton, K. G. & Spriggs, T. W. (1989) An evaluation of the Priestley and Taylor equation and the complementary relationship using results from a mixed-layer model of the convective boundary layer. *In*: T. A. Black, D. L. Splittlehouse, M. D. Novak & D. T. Price (eds.) *Estimation of Areal Evapotranspiration.* Proc. Vancouver Symp., Aug. 1987, Int'l. Assoc. of Hydrol. Sci. Publ. *177*:89–104.

McQuitty, L. L. (1957) Elementary linkage analysis for isolating orthogonal and oblique types and typal relevances. *Educ. and Psych. Measurement 17*:207–217.

Mein, R. G. (1980) Recent developments in modelling infiltration: Extension of the Green-Ampt model. *Hydrol. and Water Resources Symp.*, Adelaide, 4–6 Nov. 1980, Instit. Eng., Canberra, Aust. Nat. Conf. Publ. *88*(9):23–28.

Mein, R. G. & Larson, C. L. (1973) Modelling infiltration during a steady rain. *Wat. Resour. Res. 9*:384–394.

Milford, J. R. & Dugdale, G. (1987) *Rainfall Mapping over Sudan in 1986.* Report to the UK Overseas Development Administration on Research Scheme R3636. NASA (1990) *Tropical Rainfall Measuring Mission (TRMM) Science: Research Opportunities,* NRA-90-OSSA-15, 10 July 1990, Wash. D.C..

Miller, E. E. (1980) Similitude and scaling of soil-water phenomena. *In: Applications of Soil Physics.* Academic Press, New York, 300–318.

Miller, E. E. & Miller, R. D. (1956) Physical theory for capillary flow phenomena. *J. Appl. Phys. 27*:324–332.

Milly, P. C. D. (1991) Some current themes in physical hydrology of the land-atmosphere interface. *In*: G. Kienitz, P. C. D. Milly, M. Th. van Genuchten, D. Rosbjerg & W. J. Shuttleworth (eds.) *Hydrological Interactions Between Atmosphere, Soil and Vegetation.* Proc. Vienna Symp., Aug. 1991, Int'l. Assoc. of Hydrol. Sci., Publ. *204*:3–10.

Monteith, J. L. (1965) Evaporation and environment. *Symp. Soc. Exp. Biol. 19*:205–234.

Monteith, J. L. (1973) *Principles of Environmental Physics.* Arnold, London.

Monteith, J. L. (1981) Evaporation and surface temperature. *Quart. J. Roy. Met. Soc. 107*:1–27.

Monteny, B. A., Barbier, J. M. & Bernos, C. M. (1985) Determination of the energy exchanges of a forest-type culture: *Hevea Brasiliniensis. In*: B. A. Hutchinson & B. B. Hicks (eds.) *The Forest-Atmosphere Interaction.* D. Reidel, Dordrecht, 211–233.

Moore, E. C. S. (1889) *Sanitary Engineering: A Pactical Treatise on the Collection, Removal and Final Disposal of Sewage and the Design and Construction of Works of Drainage and Sewerage.* Batsford, London.

Moore, I. D. & Burch, G. J. (1986) Modelling erosion and deposition: Topographic effects. *Trans. Am. Soc. Agr. Engrs.* 29:1624–1630, 1640.

Moore, I. D., Burch, G. J. & Mackenzie, D. H. (1988b) Topographic effects on the distribution of surface soil water and the location of ephemeral gullies. *Trans. Am. Soc. Agr. Engrs.* 31:1098–1107.

Moore, I. D., Grayson, R. B. & Ladson, A. R. (1991) Digital terrain modelling: A review of hydrological, geomorphological and biological applications. *Hydrol. Processes 5*:3–30.

Moore, I. D., Mackay, S. M., Wallbrink, P. J., Burch, G. J. & O'Loughlin, E. M. (1986) Hydrologic characteristics and modelling of a small forested catchment in south-eastern New South Wales, prelogging condition. *J. Hydrol. 83*:307–335.

Moore, I. D., O'Loughlin, E. M. & Burch, G. J. (1988a) A contour-based topographic model for hydrological and ecological applications. *Earth Surf. Processes Landf. 13*:305–320.

Morton, F. I. (1983) Operational estimates of areal evapotranspiration and their significance to the science and practice of hydrology. *J. Hydrol. 66*:1–76.

Morton, F. I. (1984) What are the limits on forest evaporation? *J. Hydrol. 74*:373–398.

Morton, F. I. (1985) What are the limits on forest evaporation? – Reply. *J. Hydrol. 82*:184–192.

Morton, F. I. (1991) A discussion of four papers on evaporation in Volume 111. *J. Hydrol. 124*:363–374.

Nash, J. E. (1989) Potential evaporation and the complementary relationship. *J. Hydrol. 111*:1–7.

Nation (1989) Government Revokes Logging Concessions Nationwide. Bangkok, *Headlines*, Vol. *14*, No. 4241:1.

Nautiyal, J. C. & Babor, P. S. (1985) Forestry in the Himalayas – How to avert an environment disaster. *Interdisciplinary Sci. Reviews 10*:27–41.

Nicolis, C. (in press) Predictability of the atmosphere and climate: towards a dynamical view. *In: Space and Time Scale Variability and Interdependencies in Various Hydrological Processes* (Feddes, R. A., ed.), (Proc. UNESCO-IAHS G. Kovacs 1992 Colloquium), UNESCO – Cambridge University Press.

Nicolis, G. & Prigogine, I. (1989) *Exploring complexity*. Freeman, New York.

Nielsen, D. R., Biggar, J. W. & Erh, K. T. (1973) Spatial variability of field measured soil water properties. *Hilgardia 42*:215–259.

Nortcliff, S. & Dias, A. C. D. C. P. (1988) The change in soil physical conditions resulting from forest clearance in the humid tropics. *J. Biogeography 15*:61–66.

Nortcliff, S. & Thornes, J. B. (1981) Seasonal variations in the hydrology of a small forested catchment near Manaus, Amazonas, and the implications for its management. *In*: R. Lal & E. W. Russell, (eds.) *Tropical Agricultural Hydrology – Watershed Management and Land Use*. Wiley, Chichester, 37–57.

Nortcliff, S. & Thornes, J. B. (1984) Floodplain response of a small tropical stream. *In*: T. P. Burt & D. E. Walling (eds.) *Catchment Experiments in Fluvial Hydrology*. Geo-Books, Norwich, 73–85.

Nortcliff, S., Ross, S. M. & Thornes, J. B. (1990) Soil moisture, runoff and sediment yield from differentially cleared tropical rainforest plots. *In*: J. B. Thornes (ed.) *Vegetation and Erosion*. Wiley, Chichester, 419–436.

O'Loughlin, E. M. (1981) Saturation regions in catchments and their relations to soil and topographic properties. *J. Hydrol. 53*:229–246.

O'Loughlin, E. M. (1986) Prediction of surface saturation zones in natural catchments by topographic analysis. *Water Resour. Res. 22*:794–804.

O'Loughlin, E. M. (1988) Hydrology of changing landscapes. *Aust. Civil Engrs. Trans. CE30*(4):163–173.

O'Loughlin, E. M. (1990a) Perspectives on hillslope research. *In*: M. G. Anderson & T. P. Burt (eds.) *Process Studies in Hillslope Hydrology*. Wiley, Chichester, 501–516.

O'Loughlin, E. M. (1990b) Modelling soil water status in complex terrain. *Agric. Forest Meteorol. 50*:23–38.

O'Loughlin, E. M., Short, D. L. & Dawes, W. R. (1989) Modelling the hydrological response of catchment to land use change. *In: Hydrology and Water Resources Symposium, Comparisons in Austral Hydrology*. Inst. Engrs., Canberra, Australia, 28–30 Nov. 1989, Christchurch New Zealand, 335–340.

Olsson, J., Niemczynowicz, J., Berndtsson, R. and Larson, M. (1992) An analysis of the rainfall time structure by box counting – some practical implications. *J. Hydrol. 137*: 261–277.

Ogallo, L. J. (1988) The spatial and temporal patterns of the East African seasonal rainfall derived from principal components analysis. *J. Climatology 9*:145–167.

Ogunkoya, O. O. & Jenkins, A. (1991) Analysis of runoff pathways and flow contributions using deuterium and stream chemistry. *Hydrol. Processes 5*:271–282.

Oliver, M., Webster, R. & Gerrard, J. (1989a) Geostatistics in physical geography. Part I: Theory. *Trans. Inst. Br. Geogr. N.S. 14*:259–269.

Oliver, M., Webster, R. & Gerrard, J. (1989b) Geostatistics in physical geography. Part II: Applications. *Trans. Inst. Br. Geogr. N.S. 14*:270–286.

Orchard, A. Q. & Sumner, G. N. (1970) East African rainfall project. *Network Report No. 4.*

Orlanski, I. (1975) A rational subdivision of scales for atmospheric processes. *Bull. Am. Met. Soc. 56*:527–530.

Ostiachev, N. A. (1936) The law of distribution of moisture in soils and methods for study of same. *Int. Congr. Soil Mech. Found. Engrs. Res. 20*:203–211.

Othieno, C. O. (1979) An assessment of soil erosion on a field of tea under different soil management procedures. *In*: J. R. Blackie, K. A. Edwards & R. T. Clarke (Compiled) *Hydrological Research in East Africa. E. Afr. Agric. For. J. Special Issue 43*:122–127.

Oyebande, L. & Balek, J. (1989) Humid warm sloping land. *In*: M. Falkenmark & T. Chapman (eds.) *Comparative Hydrology*. UNESCO, Paris.

Paegle, J. (1987) Interactions between convective and large-scale motions over Amazonia. *In*: R. T. Dickenson (ed.) *The Geophysiology of Amazonia – Vegetation and Climate Interactions*. Wiley, New York/UNU, 347–387.

Pearce, A.J. (1990) Streamflow generation processes: An Austral view. *Water Resources Res. 26*, 3037–3047.

Pearce, A. J. & Rowe, L. K. (1981) Rainfall interception in a multistoried, evergreen mixed forest: Estimates using Gash's analytical model. *J. Hydrol. 49*:341–353.

Pearce, A. J., Rowe, L. K. & O'Loughlin, C. L. (1984) Hydrology of mid-altitude tussock grasslands, upper Waipori catchment: II – Water balance, flow duration and storm runoff. *J. Hydrol. (New Zealand) 23*:60–72.

Pearce, A. J., Rowe, L. K. & Stewart, J. B. (1980) Night time, wet canopy evaporation rates and the water balance of an evergreen mixed forest. *Wat. Resour. Res. 16*:955–959.

Pearce, A. J., Stewart, M. K. & Sklash, M. G. (1986) Storm runoff generation in humid headwater catchments 1. Where does the water come from? *Wat. Resour. Res. 22*:1263–1272.

Peck, A. J., Luxmoore, R. J. & Stolz, J. L. (1977) Effects of spatial variability of soil hydraulic properties in water budget modeling. *Wat. Resour. Res. 13*:348–354.

Penman, H. L. (1948) Natural evaporation from open water, bare soil and grass. *Proc. Roy. Soc. London, Ser. A, 193*:120–145.

Pereira, H. C. (1989) *Policy and Practice in the Management of Tropical Watersheds.* Westview, Boulder.

Pereira, H. C. (1991) The role of forestry in the management of tropical watersheds. *In*: J. Parde & G. Blanchard (eds.) *Forests, A Heritage for the Future.* Proc. 10th World Forestry Congress, Paris, Sept. 1991. Revue Forestière Française, Hors Série No. 3 (Proc. 3), ENGREF, F-54042 Nancy Cedex, 139–150 (English), 151–160 (French), 161–170 (Spanish). (Bibliography not included, available from author.)

Perroux, K. M. & White, I. (1988) Designs for disc permeameters. *Soil Sci. Soc. Am. J. 52*:1205–1215.

Perroux, K. M., Raats, P. A. C. & Smiles, D. E. (1982) Wetting moisture characteristic curves derived from constant-rate infiltration into thin soil samples. *Soil Sci. Soc. Am. J. 46*:231–234.

Philip, J. R. (1957a) The theory of infiltration: 1. The infiltration equation and its solution. *Soil Sci. 83*:345–357.

Philip, J. R. (1957b) The theory of infiltration: 4. Sorptivity and algebraic infiltration equations. *Soil Sci. 84*:257–264.

Philip, J. R. (1958) The theory of infiltration: 7. *Soil Sci. 85*:333–337.

Philip, J. R. (1969) Theory of infiltration. *Adv. Hydrosci. 5*:215–296.

Philip, J. R. (1970) Flow in porous media. *Amer. Rev. Fluid Mech. 2*:177–204.

Philip, J. R. (1973) On solving the unsaturated flow equation: 1. The flux-concentration relation. *Soil Sci. 116*:328–335.

Philip, J. R. (1975) Stability analysis of infiltration. *Soil Sci. Soc. Amer. Proc. 39*:1042–1049.

Philip, J. R. (1985) Approximate analysis of the borehole permeameter in unsaturated soil. *Wat. Resour. Res. 21*:1025–1033.

Philip, J. R. (1986a) Quasilinear unsaturated soil-water movement: Scattering analogue, infiltration, and watertightness of cavities and tunnels. *In: Proc. Ninth Australasian Fluid Mech. Conf.*, Univ. of Auckland, Auckland, New Zealand, 140–143.

Philip, J. R. (1986b) Correction to Approximate analysis of the borehole permeameter in unsaturated soil by J.R. Philip. *Wat. Resour. Res. 22*:1162.

Philip, J. R. (1987) The quasilinear analysis, the scattering analog and other aspects of infiltration and seepage. *In*: Y.-S. Fok (ed.) *Proc. Int. Conf. on Infiltration Development and Application.* Water Resources Res. Center, Honolulu, Hawaii, 1–27.

Philip, J. R. (1991) Soils, natural science and models. *Soil Sci. 151*:91–98.

Philip, J. R., Knight, J. H. & Waechter, R. T. (1989) Unsaturated seepage and subterranean holes: Conspectus, and exclusion problem for circular cylindrical cavities. *Wat. Resour. Res. 25*:16–28.

Pook, E. W. (1984a) Canopy Dynamics of *Eucalyptus maculata* Hook. I. Distribution and Dynamics of Leaf Populations. *Aust. J. Bot. 32*:387–403.

Pook, E. W. (1984b) Canopy Dynamics of *Eucalyptus maculata* Hook. II. Canopy Leaf Area Balance. *Aust. J. Bot. 32*:405–13.

Pook, E. W. (1985) Canopy Dynamics of *Eucalyptus maculata* Hook. III. Effects of Drought. *Aust. J. Bot. 33*:65–79.

Pook, E. W. (1986) Canopy Dynamics of *Eucalyptus maculata* Hook. IV. Contrasting Responses to Two Severe Droughts. *Aust. J. Bot. 34*:1–14.

Pook, E. W., Moore, P. H. R. & Hall, T. (1991) Rainfall interception by trees of *Pinus radiata* and *Eucalyptus viminalis* in 1,300 mm rainfall

area of south-eastern New South Wales: II. Influence of wind-borne precipitation. *Hydrol. Processes 5*:143–155.

Poore, D. & Sayer, J. (1991) *The Management of Tropical Moist Forest Lands: Ecological Guidelines*, 2nd Edition. The IUCN Forest Conservation Programme, Gland, Switzerland and Cambridge, UK.

Potter, G. L., Ellsaesser, H. W., MacCracken, M. C. & Luther, F. M. (1975) Possible climatic impact of tropical deforestation. *Nature 258*:697–698.

Priestley, C. H. B. (1966) The limitation of temperature by evaporation in hot climates. *Agric. Meteorol. 3*:241–246.

Priestley, C. H. B. & Taylor, R. J. (1972) On the assessment of surface heat flux and evaporation using large-scale parameters. *Mon. Weath. Rev. 100*:81–92.

Prove, B. G. (1991) *A Study of the Hydrological and Erosional Processes under Sugar Cane Culture on the Wet Tropical Coast of North Eastern Australia.* Unpublished PhD Thesis, Dept. of Geography, James Cook Univ. of North Queensland.

Prove, B. G., Truong, P. N. & Evans, D. S. (1986) Strategies for controlling caneland erosion in the wet tropical coast of Queensland. *Proc. Aust. Soc. Sugar Cane Tech. Conf.*, 77–84.

Quinn, P. F., Beven, K. J. Chevallier, P. & Planchon, O. (1991) The prediction of hillslope flow paths for distributed hydrological modelling using digital terrain models. *Hydrol. Processes 5*:59–79.

Quinn, P. F., Beven, K. J., Morris, D. G. & Moore, R. V. (1989) The use of digital detain data in the modelling of the response of hillslopes and headwaters. *In: Second National Hydrological Symposium.* Proc. Univ. Sheffield, Sept. 1989, British Hydrol. Soc., Instit. Hydrol., UK, 1.37–1.47.

Raats, P. A. C. (1973) Unstable wetting fronts in uniform and non-uniform soils. *Soil Sci. Soc. Am. Proc. 39*:1049–1053.

Ragan, R. M. (1968) An experimental investigation of partial area contribution. *Int. Assoc. Sci. Hydrol. Publ., 76*: 241–249.

Ramage, C. S. (1968) Role of a tropical "maritime continent" in the atmospheric circulation. *Mon. Weath. Rev. 96*:365–370.

Ramos, F., Coimbra, R. M. & de Oliveira, E. (1990) A bacia do Rio Amazonas e sua rede hidrologica cenario atual e futorea. *In: Hydrology and Water Management of the Amazon Basin Symp.*, 5–9 Aug. 1990, Manaus, Brazil, ABRH/IWRA/UNEP/UNESCO, 10 pp.

Rao, Y. S. (1988) Flash floods in southern Thailand. FAO, Bangkok. *Tiger Paper 15*:1–2.

Rao, Y. S. (1989) Why, what, how and where of agroforestry in the Asia Pacific region. FAO, Bangkok. *Tiger Paper 16*(2):1–10.

Raupach, M. (1979) Anomalies in flux-gradient relationships over forests. *Boundary Layer Met. 16*:467–486.

Reyenga, W., Dunin, F. X., Bautovich, B. C., Rath, C. R. & Hulse, L. B. (1988) A weighing lysimeter in a regenerating eucalypt forest: Design, construction, and performance. *Hydrol. Processes 2*:301–314.

Reynolds, W. D., Elrick, D. E. & Clothier, B. E. (1985) The constant head well permeameter: Effect of unsaturated flow. *Soil Sci. 139*:172–180.

Reynolds, W. D., Elrick, D. E. & Topp, G. C. (1983) A re-examination of the constant head well permeameter method for measuring saturated hydraulic conductivity above the water table. *Soil Sci. 136*:250–268.

Richards, L. A. (1931) Capillary conduction of liquids through porous mediums. *Physics 1*:318–333.

Richman, M. B. (1986) Rotation of principal components. *J. Climatology 6*:293–335.

Richman, M. B. (1987) Rotation of principal components: A reply. *J. Climatology 7*:511–520.

Richman, M. B., Angel, J. R. and Gong, X. (1992) Determination of dimensionality in eigenalysis. In: *5th Internatiohnal Meeting of Statistical Climatology*, 22–26 June 1992, Toronto, Canada, Environment Canada, 229–235.

Rind, D., Rosenzweig, C. and Goldberg, R. (1992) Modelling the hydrological cycle in assessments of climate change *Nature 358* (9 July 1992), 119–122.

Ritchie, J. T. (1972) Model for predicting evaporation from a row crop with incomplete cover. *Wat. Resour. Res. 8*:1204–1213.

Ritchie, J. T. (1985) A user-orientated model of the soil water balance in wheat. *In*: W. Day & R. K. Atkin (eds.) *Wheat Growth and Modelling*. Proc. NATO Advance Research Workshop, Bristol, April 1984. Plenum, New York, 293–305.

Ritchie, J. T. & Cram, J. (1989) Converting soil survey characterization data into IBSNAT crop model. *In*: J. Bouma and A. K. Berg (eds.) *Land Qualities in Space and Time*. Proc. Wageningen Symp., Aug. 1988. Pudoc, Wageningen, The Netherlands, 155–167.

Roberts, J., Cabral, O. M. R. & De Aguiar, L. F. (1990) Stomatal and boundary-layer conductances in an Amazonian terra firme rain forest. *J. Appl. Ecol. 27*:336–353.

Roberts, J., Cabral, O. M. R., Molion, L. C. B., Moore, C. J. & Shuttleworth, W. J. (1991b) Transpiration from an Amazonian rainforest calculated from stomatal conductance measurements (in preparation).

Roberts, J. M., Rosier, P. T. W. & Srinivasa Murthy, K. V. (1991a) Physiological studies in young *Eucalyptus* stands in southern India and their use in estimating forest transpiration. *In*: *Growth and Water Use of Forest Plantations*. Proc. Int. Symp., Bangalore, India, 7–11 Feb. 1991. (Available from Inst. of Hydrol., UK).

Robson, A. & Neal, C. (1990) Hydrograph separation using chemical techniques: An application to catchments in mid-Wales. *J. Hydrol. 116*:345–363.

Robson, A. & Neal, C. (1991) Chemical signals in an upland catchment in mid-Wales – some implications for water movement. *In*: *Third National Hydrol. Symp*. Proc. Univ. Southampton, Sept. 1991, British Hydrol. Soc., Inst. Hydrol., UK, 3.17–3.24.

Robson, A., Beven, K. and Neal, C. (1992) Towards identifying sources of subsurface flow: A comparison of components identified by a physically based runoff model and those determined by chemical mixing techniques. *Hydrol. Processes 6*:199–214

Roche, M. A. (1982a) Evapotranspiration réelle de la forêt amazonienne en Guyane. Cahiers ORSTOM, *Série Hydroilogie 19*:37–44.

Roche, M. A. (1982b) Comportements hydrologique comparé et érosion de l'écosystème forestier amazonien à Exérex, en Guyane. Cahiers ORSTOM, *Série Hydrologie 19*:81–114.

Rodda, J. C. (1967) The rainfall measurement problem. *International Association Scientific Hydrology* (Proc. Berne Assembly) 78: 215–231.

Rodda, J. C. (1970) On the questions of rainfall measurement and representativeness. *Symposium on the World Water Balance*, IAHS Publ. 92: 173–186.

Rodhe, A. (1987) *The origin of streamwater traced by Oxygen-18*. Uppsala Univ., Dept. of Physical Geography, Div. of Hydrol. (Doctoral Thesis), Report Series A, No. 41.

Rodriguez-Iturbe, I. & Valdes, J. B. (1979) The geomorphic structure of hydrologic response. *Wat. Resour. Res. 15*:1409–1420.

Rodriguez-Iturbe, I. & Eagleson, P. S. (1987) Mathematical models of rainstorm events in space and time. *Water Resour. Res. 23*:181–190.

Rodriguez-Iturbe, I., Cox, D. R. & Eagleson, P. S. (1986) Spatial modelling of total storm rainfall. *Proc. R. Soc., London, Ser. A, 410*:269–288.

Rogowski, A. S. (1972) Watershed physics: Soil variability criteria. *Wat. Resour. Res. 8*:1015–1023.

Ross, P. J. (1990a) Efficient numerical methods for infiltration using Richards' equation. *Wat. Resour. Res. 26*:279–290.

Ross, P. J. (1990b) SWIM – A simulation model for soil water infiltration and movement. *Reference Manual*, CSIRO Div. Soils, Australia.

Ross, P. J. & Bristow, K. L. (1990) Simulating water movement in layered and gradational soils using the Kirchhoff transform. *Soil Sci. Soc. Am. J. 54*:1519–1524.

Ross, P. J., Bridges, B. J., Fergus, I. F., Prebble, R. E. & Reeve, R. (1984) Studies in landscape dynamics in the Cooloola-Noosa River area, Queensland. 2 Field measurement techniques. *CSIRO Div. of Soils Report 74*:27–29.

Ross, S. M., Thornes, J. B. & Nortcliff, S. (1990) II. Soil hydrology, nutrient and erosional response to the clearance of terra firme forest, Maracá Island, Roraima, northern Brazil. *Geog. J. 156*:267–282.

Rutter, A. J., Kershaw, K. A., Robins, P. C. & Morton, A. J. (1971) A predictive model of rainfall interception in forests. I. Derivation of the model from observations in a stand of Corsican pine. *Agric. Meteorol. 9*:367–384.

Rutter, A. J., Morton, A. J. & Robins, P. C. (1975) A predictive model of rainfall interception in forests. II. Generalization of the model and comparison with observations in some coniferous and hardwood stands. *J. Appl. Ecol. 12*:367–380.

Sadler, J. S. (1975) *The Upper Tropospheric Circulation over the Global Tropics*. Dept. of Meteorol., Univ. of Hawaii, Honolulu.

Sadler, J. S. & Harris, B. E. (1970) The Main Tropospheric Circulation and Cloudiness over South East Asia and Surrounding Areas. Hawaii Inst. of Geophys., Univ. of Hawaii, Honolulu.

Salati, E. (1987) The forest and the hydrological cycle. *In*: R. E. Dickinson (ed.) *The Geophysiology of Amazonia – Vegetation and Climate Interactions*. Wiley/UNU, 273–296.

Sato, N., Sellers, P. J., Randal, D. A., Schneider, E. K., Shukla, J., Hoo, Y.-T., Kinter, J. L. & Albertazzi, E. (1989) The effects of implementing the Simple Biosphere Model (SiB) into a GCM. *J. Atmos. Sci. 46*:2757–2782

Schertzer, D. and Lovejoy, S. (1987) Physically-based rain and cloud modelling by anisotropic, multiplicative turbulent cascades *J. Geophys. Res. 92*: 9693–9714.

Schreckenberg, K., Hadley, M. & Dyer, M. I. (eds.) (1990) *Management and restoration of human-impacted resources: Approaches to ecosystem rehabilitation*. MAB Digest 5. UNESCO, Paris.

Sellers, P. J. (1987) Modeling effects of vegetation on climate. *In*: *The Geophysiology of Amazonia – Vegetation and Climate Interactions*. Wiley/UNU, 297–344.

Sellers, P. J., Shuttleworth, W. J., Dorman, J. L., Dalcher, A. & Roberts, J. M. (1989) Calibrating the Simple Biosphere Model for Amazonian tropical forest using field and remote sensing data. Part I: Average calibration with field data. *J. Applied Met. 28*:727–759.

Sharma, M. L. (1984) Evaporation from a eucalyptus community. *Agric. Wat. Management 8*:41–45.

Sharma, M. L. & Luxmoore, R. J. (1979) Soil spatial variability and its consequences on simulated water balance. *Wat. Resour. Res. 15*:1567–1573.

Sharma, M. L., Gander, G. A. & Hunt, C. G. (1980) Spatial variability of infiltration in a watershed. *J. Hydrol. 45*:101–122.

Sharma, M. L., Luxmoore, R. J., DeAngelis, R., Ward, R. C. & Yeh, G. T.

(1987) Subsurface water flow simulated for hillslopes with spatially dependent soil hydraulic characteristics. *Water Resour. Res.* 23:1523–1530.

Sharon, D. (1974) The spatial pattern of convective rainfall in Sukumaland, Tanzania – A statistical analysis. *Arch. Met. Geoph. Biokl. Ser. B*, 22:201–218.

Sharon, D. (1978) Rainfall fields in Israel and Jordan and the effect of cloud-seeding on them. *J. Appl. Meteor. 17*:40–48.

Sharp, D. & Sharp, T. (1982) The desertification of Asia. *Asia 2000 1*(4):40–42.

Sherman, L. K. (1932) Steamflow from rainfall by unit-graph method. *Engng New Record 108*:501–505.

Short, D. L. & Dawes, W. R. (1990) Checking the Clapp wetting front model. (Manuscript in preparation, CSIRO Div. Water Resources, Canberra, Australia).

Shukla, J., Nobre, C. & Sellers, P. (1990) Amazonian deforestation and climate change. *Sci. 247*:1322–1325.

Shuttleworth, W. J. (1976) A one-dimensional theoretical description of the vegetation-atmosphere interaction. *Boundary-Layer Meteorol. 10*:273–302.

Shuttleworth, W. J. (1977) Comments on Resistance of a partially wet canopy: Whose equation fails? *Boundary-Layer Meteorol. 12*:385–386.

Shuttleworth, W. J. (1978) A simplified one-dimensional theoretical description of the vegetation-atmosphere interaction. *Boundary-Layer Meteorol. 14*:3–27.

Shuttleworth, W. J. (1979) *Evaporation Report No. 56*, Inst. of Hydrol., Wallingford, UK.

Shuttleworth, W. J. (1984) Observations of radiation exchange above and below Amazonian forest. *Quart. J. Roy. Met. Soc. 110*:1163–1169.

Shuttleworth, W. J. (1988a) Evaporation from Amazonian rainforest. *Proc. Roy. Soc. London, Ser. B*, 233:321–346.

Shuttleworth, W. J. (1988b) Macrohydrology – The new challenge for process hydrology. *J. Hydrol. 100*:31–56.

Shuttleworth, W. J. (1989a) Micrometeorology of temperate and tropical forest. *Proc. Roy. Soc. London, Ser. B*, 324:299–334.

Shuttleworth, W. J. (1989b) *Priorities in Climate-Related, Hydrological Research in Amazonia.* Instituté of Hydrology, Wallingford, UK, Unpubl. Report.

Shuttleworth, W. J. (1991) The role of hydrology in global science. *In*: *Hydrological Interactions Between Atmosphere, Soil and Vegetation.* G. Kienitz, P. C. D. Milly, M. Th. van Genuchten, D. Rosbjerg & W. J. Shuttleworth (eds.) Proc. Vienna Symp., Aug. 1991, Int'l. Assoc. of Hydrol. Sci., Publ. *204*:361–375.

Shuttleworth, W. J., Gash, J. H. C., Roberts, J. M., Nobre, C. A., Molion, L. C. B. & De Nazare Goes Ribeiro, M. (1991) Post-deforestation Amazonian climate: Anglo-Brazilian research to improve prediction. *In*: B. P. F. Braga & C. A. Fernandez-Jauregui (eds.) *Water Management of the Amazon Basin Symp.* Proc. Manaus Symp., Aug. 1990. UNESCO (ROSTLAC), Montevideo, Uruguay, 275–288.

Shuttleworth, W. J., Gash, J. H. C., Lloyd, C. R., Moore, C. J., Roberts, J., Marques Filho, A. de O., Fisch, G., De Paula Silva Filho, V., De Nazare Goes Ribeiro, M., Molion, L. C. B., De Abreu SA, L. D., Nobré, J. C. A., Cabral, O. M. R., Patel, S. R. & Carvalho De Moraes, J. (1984) Eddy correlation measurements of energy partition for Amazonian forest. *Quart. J. Roy. Met. Soc. 110*:1143–1162.

Shuttleworth, W. J., Gash, J. H. C., Lloyd, C. R., Moore, C. J., Roberts, J., Marques Filho, A., Fisch, G., De Paula Silva Filho, V., De Nazare Goes Ribeiro, M., Molion, L. C. B., De Abreu SA, L. D., Nobré, J. C. A.,

Cabral, O. M. R., Patel, S. R. & Carvalho De Moraes, J. (1985) Daily variations of temperature and humidity within and above Amazonian rainforest. *Weather 40*:102–108.

Silburn, D. M., Connolly, R. D. & Glanville, S. (1990) Processes of infiltration under rain on structurally unstable cultivated soils and prediction using a Green and Ampt model. *In*: *Conference on Agric. Engng.* Proc. Toowoomba Symp., Nov. 1990, Inst. Engrs. Aust., Canberra, Australia, 182–186.

Simpson, J., Adler, R. F. & North, G. R. (1988) A proposed Tropical Rainfall Measuring Mission satellite. *Bull. Amer. Met. Soc.* 69:278–295.

Sisson, J. B. & Wierenga, P. J. (1981) Spatial variability of steady-state infiltration rates as a stochastic process. *Soil Sci. Soc. Am. J. 46*:20–26.

Sivapalan, M., Beven, K. J. & Wood, E. F. (1987) On hydrologic similarity 2. A scaled model of storm runoff production. *Water Resour. Res.* 23:2266–2278.

Skaggs, R. W. & Khaleel, R. (1982) Infiltration. *In*: C. T. Haan, H. P. Johnson & D. L. Brakensiek (eds.) *Hydraulic Modelling of Small Watersheds.* Am. Soc. Agric. Engrs., *ASAE Monograph 5*:121–166.

Sklash, M. G. (1990) Environmental isotope studies of storm and snowmelt runoff generation. *In*: M. G. Anderson & T. P. Burt (eds.) *Process Studies in Hillslope Hydrology.* Wiley, Chichester, 401–435.

Sklash, M. G. & Farvolden, R. N. (1979) The role of groundwater in storm runoff. *J. Hydrol. 43*:45–65.

Sklash, M. G., Stewart, M. K. & Pearce, A. J. (1986) Storm runoff generation in humid headwater catchments. 2. A case study of hillslope and low-order stream response. *Wat. Resour. Res. 22*:1273–1282.

Smettem, K. R. J. (1986) Analysis of water flow from cylindrical macropores. *Soil Sci. Soc. Am. J. 50*:1139–1142.

Smettem, K. R. J. (1987) Characterization of water entry into a soil with a contrasting textural class: Spatial variability of infiltration parameters and influence of macroporosity. *Soil Sci. 144*:167–174.

Smettem, K. R. J. & Collis-George, N. (1985) The influence of cylindrical macropores on steady-state infiltration in a soil under pasture. *J. Hydrol. 79*:107–114.

Smettem, K. R. J. & Kirkby, C. (1990) Measuring the hydraulic properties of a stable aggregated soil. *J. Hydrol. 117*:1–13.

Smettem, K. R. J. & Ross, P. J. (1992) Measurement and prediction of water movement in a field soil: The matrix-macropore dichotomy. *Hydrol. Processes 6*:1–10.

Smettem, K. R. J. & Trudgill, S. T. (1983) An evaluation of fluorescent and non-fluorescent dyes in the identification of water transmission routes in soils. *J. Soil Sci. 34*:45–56.

Smiles, D. E. & Knight, J. H. (1976) A note on the use of the Philip infiltration equation. *Aust. J. Soil Res. 14*:103–108.

Smith, R. B. (1989) Mechanisms of orographic precipitation. *Met. Mag. 111*:85–88.

Smith, R. E. & Hebbert, R. H. B. (1983) Mathematical simulation of interdependent surface and subsurface hydrologic processes. *Wat. Resour. Res. 19*:987–1001.

Smith, R. E. & Parlange, J.-Y. (1978) A parameter-efficient hydrologic infiltration model. *Water Resour. Res. 14*:533–538.

Smith, R. E. & Parlange, J.-Y. (1989) Comment on "Constant rate rainfall infiltration: A versatile nonlinear model, 2, Applications of solutions." Presented in *Water Resour. Res. 23*:155–162 by I. White and P. Broadbridge. *Water Resour. Res. 25*:1051–1053.

Sonsin, A. (1989) Effects of land use changes on the hydrology of catchments in Thailand. Paper presented at the *Reg. Seminar on Tropical*

For. Hydrol., FRIM-UNESCO, Kuala Lumpur Symp., Malaysia, Sept. 1989.

Spaans, E. J. A., Baltissen, G. A. M., Bouma, J., Miedeme, R., Lansu, A. L. E., Schoonderbeek, D. & Wielemaker, W. G. (1989) Changes in physical properties of young and old volcanic surface soils in Costa Rica after clearing of tropical rain forest. *Hydrol Proc. 3*:383–392.

Stace, H. C. T., Hubble, G. D., Brewer, R., Northcote, K. H., Sleeman, J. R., Mulcahy, M. J. & Hallsworth, E. G. (1968) *A Handbook of Australian Soils.* Rellim Tech. Publ., Glenside, S.A.

Stadtmueller, T. (1990) Soil erosion in East Kalimantan, Indonesia. *In*: *Research Needs and Applications to Reduce Erosion and Sedimentation in Tropical Steeplands.* Proc. Fiji Symp., June 1990. Int'l. Assoc. of Hydrol. Sci. Publ. *192*:221–230.

Stewart, J. B. (1984) Measurement and prediction of evaporation from forested and agricultural catchments. *Agric. Wat. Management 8*:1–28.

Stewart, J. B. (1988) Modelling surface conductance of pine forest. *Agric. For. Meteorol. 43*:19–35.

Stewart, J. B. (1989) On the use of the Penman-Monteith equation for determining areal evapotranspiration. *In*: T. A. Black, D. L. Splittlehouse, M. D. Novak & D. T. Price (eds.) *Estimation of Areal Evapotranspiration.* Proc. Vancouver Symp., Aug. 1987, Int'l. Assoc. of Hydrol. Sci. Publ. *177*:3–12.

Stewart, J. B. & Thom, A. S. (1973) Energy budgets in pine forest. *Quart. J. Roy. Met. Soc. 99*:154–170.

Stewart, M. K. & McDonnell, J. J. (1991) Modeling baseflow soil water residence times from deuterium concentrations. *Water Resour. Res. 27*:2682–2693.

Storm, B., Jørgensen, G. H. & Styczen, M. (1987) Simulation of water flow and soil erosion processes with a distributed physically-based modelling system. *In*: R. H. Swanson, P. Y. Bernier & P. D. Woodard (eds.) *Forest Hydrology and Watershed Management.* Proc. Vancouver Symp., Aug. 1987. Int'l. Assoc. of Hydrol. Sci. Publ. *167*:595–608.

Sumner, G. N. (1981) The nature and development of rainstorms in coastal East Africa. *J. Climatology 1*:131–152.

Sumner, G. N. (1983) The use of correlation linkages in the assessment of daily rainfall patterns. *J. Hydrol. 66*:169–182.

Sumner, G. N. (1988) *Precipitation – Process and Analyses.* Wiley, Chichester.

Sumner, G. N. & Bonell, M. (1986) Circulation and daily rainfall in the north Queensland wet seasons, 1979–1982. *J. Climatology 6*:531–549.

Sumner, G. N. & Bonell, M. (1988) Variation in the spatial organization of daily rainfall during the north Queensland wet seasons, 1979–1982. *Theoretical and Applied Climatology 39*:59–74.

Swank, W. T. & Crossley, D. A. (eds.) (1988) *Forest Hydrology and Ecology and Coweeta.* Springer-Verlag, New York.

Swift, L. W., Cunningham, G. B. & Douglas, J. E. (1988) Climatology and Hydrology. *In*: W. T. Swank & D. A. Crossley (eds.) *Forest Hydrol. and Ecology at Coweeta.* Springer-Verlag, New York,USA, 35–55.

Talsma, T. (1969) *In situ* measurement of sorptivity. *Aust. J. Soil Res. 7*:269–276.

Talsma, T. (1985) Prediction of hydraulic conductivity from soil water retention data. *Soil Sci. 140*(2):184–188.

Talsma, T. (1987) Re-evaluation of the well permeameter as a field method for measuring hydraulic conductivity. *Aust. J. Soil Res. 25*:361–368.

Talsma, T. & Gardner, E. A. (1986a) Soil water extraction by a mixed eucalypt forest during a drought period. *Aust. J. Soil Res. 24*:25–32.

Talsma, T. & Gardner, E. A. (1986b) Groundwater recharge and discharge response to rainfall on a hillslope. *Aust. J. Soil Res. 24*:343–356.

Talsma, T. & Hallam, P. M. (1980) Hydraulic conductivity measurement of forest catchments. *Aust. J. Soil Res. 18*:139–148.

Tarchitzky, J., Banin, A., Morin, J. & Chen, Y. (1984) Nature, formation and effects of soil crusts formed by water drop impact. *Geoderma 33*:135–155.

Taylor, C. H. & Pearce, A. J. (1982) Storm runoff processes and subcatchment characteristics in a New Zealand hill country catchment. *Earth Surf. Processes Landf. 7*:439–447.

Ternan, J. L., Williams, A. G. & Solman, K. (1987) A preliminary assessment of soil hydraulic properties, and their implications for agroforestry management in Granada, West Indies. *In*: R. H. Swanson, P. Y. Bernier & P. D. Woodard (eds.) *Forest Hydrology and Watershed Management.* Proc. Vancouver Symp., Aug. 1987. Int'l. Assoc. of Hydrol. Sci. Publ. *167*:409–421.

Tessier, Y., Lovejoy, S., & Schertzer, D. (1988) Multifractal analysis of global rainfall from 1 day to 1 year. *Nonlinear Variability in Geophysics 2*, abstract volume, Paris.

Thauvin, V. & Lebel, T. (1991) IV: EPSAT-Niger study of rainfall over the Sahel at small time steps using a dense network of recording rain-gauges. *Hydrol. Processes 5*:251–260.

Theis, C. V. (1935) The relationship between the lowering of the piezometric surface and the rate and duration of discharge of a well using groundwater storage. *Trans. Am. Geophys. Union 16*:519–524.

Thom, A. S. (1972) Momentum, mass and heat exchange of vegetation. *Quart. J. Roy. Met. Soc. 98*:124–134.

Thom, A. S. & Oliver, H. R. (1977) On Penman's equation for estimating regional evaporation. *Quart. J. Roy. Met. Soc. 103*:345–357.

Tricker, A. S. (1981) Spatial and temporal patterns of infiltration. *J. Hydrol. 49*:261–276.

Turc, L. (1961) Evaluation des besoins en eau d'irrigation evapotranspiration potentielle. *Ann. Agron. 12*:13–50.

Turner, J. V. & MacPherson, D. K. (1990) Mechanisms affecting streamflow and stream water quality: An approach via stable isotope, hydrogeochemical and time series analysis. *Wat. Resour. Res. 26*:3005–3019.

UNESCO (1978) *Tropical Forest Ecosystems: A state of knowledge report.* UNESCO, Paris, 258–261.

Valdes, J. B., Nakansto, S., Shen, S. S. P. & North, G. R. (1990) Estimation of multidimensional precipitation parameters by areal estimates of oceanic rainfall. *J. Geophys. Res. 95*:2101–2111.

Van Bavel, C. H. M. (1966) Potential evaporation: The combination concept and its experimental verification. *Wat. Resour. Res. 2*:455–467.

Van der Ploeg, R. R. (1974) Simulation of moisture transfer in soils one-dimensional infiltration. *Soil Sci. 118*:349–356.

Vanclay, J. K. (1991) Research needs for sustained rainforest resources. *In*: N. Goudberg & M. Bonell, with D. Benzaken (eds.) *Tropical Rainforest Research in Australia: Present Status and Future Directions for the Institute for Tropical Rainforest Studies.* Proc. Townsville Workshop, May 1990. Inst. for Tropical Rainforest Studies, James Cook Univ. of North Queensland, Townsville, Australia, 133–143.

Veen, A. W. L. & Dolman, A. J. (1989) Water dynamics of forests: One-dimensional modelling. *Prog. Phys. Geog. 13*:471–506.

Veen, A. W. L., Hutjes, R. W. A., Klaassen, W., Kruijt, B. & Lankreijer, J. M. (1991) Evaporative conditions across a grass-forest boundary: A comment on the strategy for regionalizing evaporation. *In*: G. Kienitz, P. C. D. Milly, M. Th. van Genuchten, D. Rosbjerg & W. J. Shuttleworth (eds.) *Hydrological Interactions Between Atmosphere, Soil and Vegetation.* Proc. Vienna Symp., Aug. 1991. Intt'l. Assoc. of

Hydrol. Sci. Publ. *204*:43–52.

Verma, S. B. (1989) Aerodynamic resistances to transfers of heat, mass and momentum. *In*: *Estimation of Areal Evapotranspiration*. T. A. Black, D. L. Splittlehouse, M. D. Novak & D. T. Price (eds.) Proc. Vancouver Symp., Aug. 1987. Int'l. Assoc. of Hydrol. Sci. Publ. *177*:13–20.

Vertessy, R. A., Hatton, T. J., O'Loughlin, E. M. & Brophy, J. H. (1991) *Water Balance Simulation in the Melbourne Water Supply area using a Terrain Analysis-Based Catchment Model*. Australian Centre for Catchment Hydrol., CSIRO Div. of Water Resour., Canberra, Australia. *Consultancy Report 91/32*.

Vis, M. (1989) *Processes and patterns of erosion in natural and disturbed Andean forest ecosystems*. PhD Thesis, Univ. of Amsterdam.

Viswanadham, Y., Silva Filho, V. P. & André, R. G. B. (1991) The Priestley-Taylor parameter α for the Amazon forest. *Forest Ecology and Management 38*: 211–225.

Volker, A. (1983) Rivers of South East Asia: Their regime, utilization and regulation. *In*: R. Keller (ed.) Proc. Hamburg Symp., Aug. 1983. *Internt'l Assoc. of Hydrol. Sci.*, Publ. *140*:127–138.

Wallace, J. M. (1987) Comments on "Interactions between convective and large-scale motions over Amazonia." *In*: R. T. Dickenson (ed.) *The Geophysiology of Amazonia Vegetation and Climate Interactions*. Wiley, New York/UNU, 387–390.

Wallis, M. G., Scotter, D. R. & Horne, D. J. (1991) An evaluation of the intrinsic sorptivity water repellancy index on a range of New Zealand soils. *Aust. J. Soil Res. 29*:353–362.

Walsh, R. P. D. (1980) Runoff processes and models in the humid tropics. *Z. Geomorphol. N. F. Suppl.-Bd. 36*:176–202.

Ward, R. C. (1975) *Principles of Hydrology*. McGraw-Hill, London, 2nd Ed., 244–256.

Ward, R. C. & Robinson, M. (1990) *Principles of Hydrology*. McGraw-Hill, Maidenhead, UK.

Warwick, A. W., Mullen, G. J. & Nielsen, D. R. (1977) Scaling field-measured soil hydraulic properties using similar-media concept. *Wat. Resour. Res. 13*:355–362.

Waymire, E., Gupta, V. K. & Rodriguez-Uturbe, I. (1984) A spectral theory of rainfall intensity at the Meso-ß Scale. *Wat. Resour. Res. 20*:1453–1465.

Weyman, D. R. (1973) Measurements of the downslope flow of water in a soil. *J. Hydrol. 20*:267–288.

Whipkey, R. Z. (1969) Storm runoff from forested catchments by subsurface routes. *In*: *Floods and Their Computation, Studies and Reports in Hydrology*. Leningrad, UNESCO-IASH-WMO *3*:773–779.

White, D., Richman, M. & Yarnai, B. (1991) Climate regionalization and rotation of principal components. *Int. J. Climatol. 11*:1–26.

White, I. (1988) Measurement of soil physical properties in the field. *In*: W. L. Steffen & O. T. Denmead (eds.) *Flow and Transport in the Natural Environment: Advances and Applications*. Springer-Verlag, Heidelberg, 59–85.

White, I. & Broadbridge, P. (1989) Reply to comment by R. E. Smith & J.-Y. Parlange. *Water Resour. Res. 25*:1054–1059.

White, I. & Perroux, K. M. (1989) Estimation of unsaturated hydraulic conductivity from field sorptivity measurements. *Soil Sci. Soc. Am. J. 53*:324–329.

White, I. & Sully, M. J. (1987) Macroscopic and microscopic capillary length and time scales from field infiltration. *Wat. Resour. Res.*

23:1514–1522.

White, I., Sully, M. J. & Perroux, K. M. (1992) Measurement of surface-soil hydraulic properties: Disk permeameters, tension infiltrometers and other techniques. (In) Advances in Measurement of Soil Physical Properties: Bringing Theory into Practice. *SSSA Special Publication No. 30*, SSSA, Madison, Wisc. 69–103.

Wierda, A., Veen, A. W. L. & Hutjes, R. W. A. (1989) Infiltration at the Tai rain forest (Ivory Coast): Measurements and modelling. *Hydrol. Processes 3*:371–382.

Williams, J. & Bonell, M. (1987) Computation of soil infiltration properties from the surface hydrology of large field plots. *In*: Yu-Si Fok (ed.) *Int'l. Conf. on Infiltration, Development and Application*. Univ. of Hawaii, Water Resour. Center, 272–281.

Williams, J. & Bonell, M. (1988) The influence of the scale of measurement on the spatial and temporal variability of the Philip infiltration parameters – An experimental study in an Australian savannah woodland. *J. Hydrol. 104*:33–51.

Williams, J. & Coventry, R. J. (1979) The contrasting soil hydrology of red and yellow earths in a landscape of low relief. *In*: *The Hydrology of Areas of Low Precipitation*. Proc. Canberra Symp., Dec. 1979. Int'l. Assoc. of Hydrol. Sci. Publ. *128*:385–395.

Willmott, C. J. (1978) P-mode principal components analysis, grouping and precipitation regions in California. *Arch. Met. Geoph. and Biokl. Ser. B*, 26:277–295.

Wilson, J. D. (1989) Turbulent transport within the canopy. *In*: T. A. Black, D. L. Splittlehouse, M. D. Novak & D. T. Price (eds.) *Estimation of Areal Evapotranspiration*. Proc. Vancouver Symp., Aug. 1987. Int'l. Assoc. of Hydrol. Sci., Publ. *177*:43–80.

Wilson, M. F. (1984) *Construction and use of land surface information in a general circulation climate model*. PhD Thesis, Univ. of Liverpool, UK.

Wood, E. F. & O'Connell, P. E. (1985) Real-time forecasting. *In*: *Hydrological Forecasting*. M. G. Anderson & T. P. Burt (eds.) Wiley, Chichester, 505–558.

Wood, E. F., Sivapalan, M., Beven, K. J. & Band, L. (1988) Effects of spatial variability and scale with implications to hydrologic modeling. *J. Hydrol. 102*:29–47.

World Commission on Environment and Development (WCED) (1987) *Our Common Future*. (Brundtland Report).

Youngs, E. G. (1983) Soil physical theory and heterogeneity. *Agric. Wat. Management 6*:145–159.

Youngs, E. G. (1988) Soil physics and hydrology. *J. Hydrol. 100*:411–431.

Youngs, E. G. & Price, R. I. (1981) Scaling of infiltration behaviour in dissimilar porous materials. *Wat. Resour. Res. 17*:1065–1070.

Zadroga, F. (1981) The hydrological importance of a montane cloud forest area of Costa Rica. *In*: *Tropical Agricultural Hydrology – Watershed management and Land Use*. R. Lal & E. W. Russell (eds.) Wiley, Chichester, 59–73.

Zaslavsky, D. & Rogowski, A. S. (1969) Hydrologic and morphologic implications of anisotropy and infiltration in soil profile development. *Soil Sci. Am. Proc. 33*:594–599.

Zaslavsky, D. & Sinai, G. (1981) Surface hydrology: I. Explanation of phenomena, II. Distribution of raindrops, III. Causes of lateral flow, IV. Flow in sloping layered soil, V. In-surface transient flow. *Proc. Am. Soc. Civil Engrs. J. Hydraulics Div*. HY *1*:1–93.

12: Groundwater Systems in the Humid Tropics

S.S.D. Foster and P. J. Chilton

British Geological Survey (Groundwater & Geotechnical Surveys Division), Keyworth (Nottingham) NG12 S66 and Wallingford (Oxford) OX1O 8BB, United Kingdom

ABSTRACT

Aquifers underlie large areas of the humid tropics. Because of their widespread distribution, they can often be developed close to the location of demand, with modest pumping and reticulation costs, to produce groundwater of excellent chemical and microbiological quality requiring little treatment. These factors are leading to increased interest and rapid development of groundwater resources. Groundwater systems characteristic of the humid tropics fall into five main types, major alluvial formations, basement regoliths, intermontane valley-fill, active volcanic arcs and karstic limestones. The principal features of each of these are described and the main problems which arise in each case in the exploitation and management of groundwater resources are indicated. In the humid tropics, the close association between groundwater and surface water and the generally shallow water table make groundwater a valuable resource which, in some cases, is highly vulnerable to pollution from a range of human activities.

INTRODUCTION

Significance of groundwater

Aquifers underlie geographically large areas of the humid tropics. Because of this widespread distribution, their groundwater resources can often be tapped close to the location of water demand, thereby minimizing reticulation costs, at least for supplies up to moderate volume. Groundwater has other advantages for water supply development. Given adequate protection, it has excellent microbiological and organic quality, and requires minimal treatment. Although certain naturally-occurring dissolved chemical constituents can impart an unacceptable quality or a health risk for domestic supply, groundwater with these characteristics are of limited geographical distribution. The capital cost of groundwater development is thus relatively modest and the land requirements are minimal. Finally, the resource lends itself to flexible development, capable of being phased with rising demand.

Despite the historical tendency to favour the exploitation of surface water in both urban and rural areas, these factors are leading to increasing interest and rapid development of groundwater resources in the humid tropics, even given the relative abundance of surface water. The treatment costs to obtain a supply of equally-high quality from a surface water source are often excessive because of colouration and other problems associated with high natural organic contents or intermittently heavy suspended sediment loads. As a result, in some countries a major proportion of the potable water supplies are obtained from aquifers. In nations such as Costa Rica, the rate of growth of groundwater exploitation has been rapid (Fig. lA).

Many humid tropical areas have a significant dry season of three to seven months duration, and occasionally experience failure of wet season rains. Given that the volume of drainable storage of many aquifer systems is very large, there is growing interest in groundwater development for supplementary irrigation to act as drought insurance, especially for more valuable crops. This tendency is clearly illustrated by the statistics on water well drilling for the Upper Cauca Valley of Columbia (Fig. lB).

In addition to their importance in water supply, groundwater systems are an integral element of the humid tropical ecosystem (animal-plant-soil-water) because of the intimate relationship between surface and groundwater and the frequently shallow groundwater table with abundant phreatophytic vegetation in such environments.

Occurrence of groundwater

The land area of the humid tropics includes a wide range of geological build which interacts with the prevailing climate to produce distinctive geomorphological features and hydrogeological regimes. The principal objective of this chapter is to provide a preliminary review of these hydrogeological regimes in comparison with those of more temperate areas (which have been far more intensively researched), and to discuss their special groundwater exploitation and management problems.

Although it is perhaps dangerous to over-simplify, groundwater systems developed in, and characteristic of, the humid tropics tend to fall into one of five main types (Table 1). The

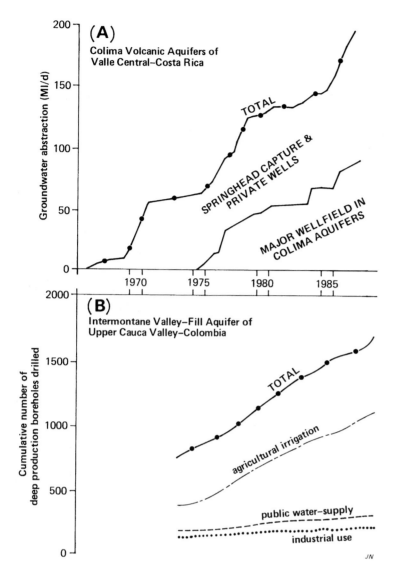

Figure 1: Growth of groundwater exploitation in two areas of the humid tropics: (A) the Valle Central of Costa Rica and (B) the Upper Cauca Valley, Colombia.

major alluvial formations and basement regoliths both occupy very extensive land areas. Groundwater systems developed in intermontane valley-fill, active volcanic arcs and karstic limestones are of more limited geographical distribution, but they are of major importance in some areas because they are capable of supporting large wellfields of high-yielding production boreholes. This broad subdivision of groundwater systems is useful for a global discussion of the hydrogeological characteristics, groundwater development and management problems of the humid tropics, but it must be recognized that it is more difficult to generalize about the groundwater systems of these regions than about their climate, vegetation and hydrology.

Moreover, it should be noted that other aquifer types, such as geologically older sedimentary basins, also occur in parts of the humid tropics. These have been omitted from the classification since they are similar to the (well-described) aquifer systems developed in more temperate regions, with regional aquifers strongly confined between argillaceous formations acting as near-aquicludes.

Small tropical islands also do not fit readily into the classification. They are normally formed as atolls of karstic or sub-karstic limestones (Fig. 2a), and thus have affinities with that group, or of volcanic material of relatively recent origin, with similar hydrogeological characteristics to the eruptive cones of active volcanic arcs (Fig. 2b). In some islands, fringing coral reefs or limestone terraces adjoin a core of volcanic rock or deep ocean sediments.

Characteristic hydrogeological processes
The spatial variation of precipitation in the humid tropics can be very high, with variation in altitude and in distance from the ocean, for example. Rainfall and excess rainfall (groundwater recharge in areas of highly-permeable soil) can also exhibit marked temporal variation. The results of a detailed analysis of agro-meteorological data for the 10-year period 1974-1983, from a small permeable catchment south of Mombasa, Kenya, with an average precipitation of 1,110 mm year^{-1}, showed that the rate of diffuse rainfall recharge varied from 0–905 mm

TABLE 1. *Summary of principal groundwater systems characteristic of the humid tropics.*

Geological Build	Major alluvial formations	Intermontane valley-fill	Karstic limestones	Active volcanic arcs	Basement regoliths
- distribution	numerous very large river basins and many important coastal regions	elongated tectonic valleys of globally-limited distribution	mainly coastal regions of globally-limited distribution	elongated areas often bordering fertile valleys of globally-limited distribution	very extensive inland areas
- examples	Amazon & Orinoco basins, Guayana coast, lower Niger & Volta valleys; West African coast, Congo basin, Ganges and Mekong deltas, parts of Thailand, and Malaysian & Indonesian coasts	Cauca and Magdalena valleys of Colombia, Guayas valley of Ecuador; Brahmaputra, Irriwaddy and other valleys of SE Asia	Yucatan-Mexico; Cuba & Puerto Rico north coasts; Jaffna-Sri Lanka	Central America; Colombia-Cordillera Central; Indonesia-Java Highlands; Phillippines-parts of Luzon	areas of NE Brasil, India (Orissa & Andhra Pradesh), and west, central and east Africa (Ghana, Nigeria, Uganda, Kenya, Malawi, etc)
Hydrogeological Features					
- infiltration capacity	variable with much potential recharge rejected on lower ground where aquifers discharge to rivers and swamps	variable becoming high along lateral margins of valley	extremely high, no surface water other than phreatic ponds and lagoons	very variable, surface watercourses exhibit complex influent-effluent relations with groundwater	relatively low, much of excess rainfall generates intermittent surface runoff
- depth to water table	widely 0-5 m except distant from watercourses, phreatophytic vegetation over extensive areas	varies considerably from shallow (<5 m) along present alluvial tract to deep (sometimes >50 m) along lateral margins	shallow (<5 m) along coastal plains but can increase inland to >50 m towards margins of formation	variable but can be deep (>50 m) on higher ground	generally shallow and rarely exceeding 10 m in dry season
- hydraulic gradients	generally low (<0.1%) but steepening towards margins of systems	moderate-to-steep with flow often directed perpendicular to valley axis	universally very low (often <0.01%)	always steep and can be very steep (>1%)	low-to-moderate and generally sub-parallel to land surface
- aquifer type (T = tranwmissivity)	thick complex multi-aquifer systems with variable T (usually in range 100-1000 m²/d) and large groundwater storage	comparable to 'major alluvial formations' but higher T often developed in fan deposits along valley margins	highly heterogenous, overall very high T (sometimes >10,000 m²/d) but limited storage	very variable with frequent perched aquifers, locally high T (>1000 m²/d) and can have significant storage if pyroclastic deposits interbedded with lavas	relatively thin aquifers of low T (normally <10 m²/d) and limited storage
- natural groundwater chemistry	generally good with low-to-moderate TDS but tendency for DO absent and high Fe and Mn at depth	comparable to 'major alluvial formations'	good, but relatively high Ca-Mg hardness	often excellent, low TDS but high SiO₂; locally natural toxic constituents (As, F, B, Se) may be in solution	generally good but variable, with type of basement rock, locally can have high F, Mg, SO₄, Fe and/or Mn

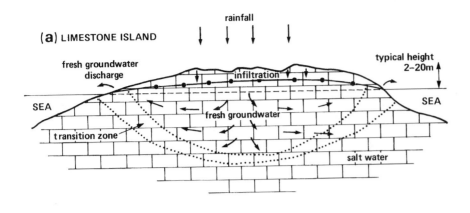

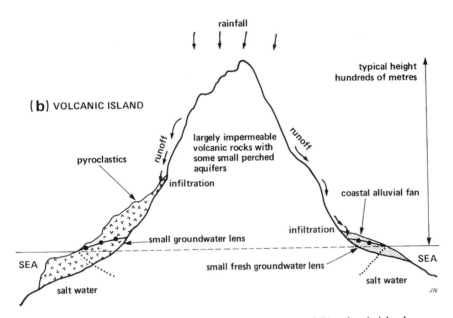

Figure 2: Generalized section through a typical (a)limestone and (b) volcanic island.

year^{-1}. During the wettest months (May and June) there was a 45% probability of significant recharge and a 20% probability of the recharge rate exceeding 100 mm month^{-1}. The longest period without significant recharge was as long as 33 months.

In the humid tropics, the mechanisms of groundwater recharge and discharge are often closely interrelated (Fig. 3) and exert a strong influence on the overall surface water behaviour of a catchment in terms of rainfall-runoff response. The subject has been reviewed in some detail for natural forest vegetation by Bruijnzeel (1990). This is a consequence of the generally shallow groundwater table developed over large areas of most, although not all, geological builds in the humid tropics.

Groundwater recharge mechanisms

Throughout most of the non-alluvial areas of the humid tropics, thick lateritic soil profiles have developed as a result of deep weathering and strong leaching by infiltrating meteoric water. Although the characteristics of the residual mantle of alteration vary significantly with underlying geology, this

layer normally includes some horizons of low vertical permeability, for example, those rich in kaolinitic clays or hardened by iron oxides.

Given the frequent occurrence of high-intensity precipitation, excess rainfall often exceeds the soil-profile infiltration capacity and results in temporary soil-water perching. In consequence, a variable (and often high) proportion of the excess rainfall generates shallow soil interflow (Weyman, 1973) or overland flow (Dunne, 1978) to land surface depressions, from where it either evaporates or runs off in surface water courses. This phenomenon further complicates the estimation of diffuse aquifer recharge in many hydrogeological environments of the humid tropics. An exception to this general scheme is the karstic limestone areas, which have thinner residual soils and normally have sufficiently high infiltration capacity to accept all the excess rainfall, either directly or via swallow-hole solution features.

In the humid tropics, the lower-lying areas of alluvial basins and coastal plains invariably have a shallow groundwater table. Aquifers tend to fill up rapidly in the wet season with the

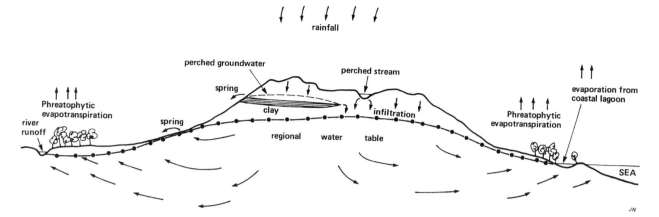

Figure 3: Mechanisms of groundwater recharge and discharge in the humid tropics.

water table virtually reaching the land surface, producing a characteristic truncated form to groundwater hydrographs. Further excess rainfall will then be rejected because of the absence of storage space and, once again, will result in extensive overland flow (Dunne, 1978; Ward, 1984).

In the upper parts of some catchments and towards the lateral margins of groundwater systems, the tributary streams themselves will often be perched above the regional groundwater table (Fig. 3), and streambed recharge to underlying aquifers will be a frequent and significant process.

Aquifer discharge mechanisms

The natural vegetation of the humid tropics is equatorial or tropical rain forest, or the more richly-vegetated type of savannah grassland. All these vegetation groups include phreatophytic species which, at least in part, draw their moisture from the water table and, as a result, evaporate large quantities of groundwater. Phreatic evapotranspiration is a very common process throughout areas with groundwater table depths of less than 5 m, and can continue when it is at a substantially greater depth. It is this process which keeps these areas green during the dry season, when the available shallow soil moisture has been exhausted. The rate of aquifer discharge via this route can be difficult to calculate, however, because of uncertainties and inaccuracies in the overall hydrological balance of such areas.

Aquifers also discharge in large volume by seepage to riparian areas and other surface depressions such as swamps, ponds and coastal lagoons. Because of the always shallow groundwater table in such areas, this discharge may increase rapidly in volume following incidents of groundwater recharge from excess rainfall (Pearce et al.,1986) and contribute to the peak runoff response of tropical catchments. In the case of areas of significant relief underlain by recent volcanic lavas and karstic limestones, groundwater discharge also occurs by spring flow, sometimes of prodigious volume.

Effect of vegetation change

If the natural forest vegetation is cleared for agricultural cultivation, excess rainfall will, in general, increase as a result of reductions in evapotranspiration and perhaps also by excess supplementary irrigation in the dry season. Whether this, in turn, will result in an increased rate of groundwater recharge will depend upon the overall soil-profile infiltration capacity and on the depth to the groundwater table.

A recent review suggests that diverse responses may occur (Bruijnzeel, 1990). In some cases there is evidence of an increase in groundwater recharge with a rising water table and potential soil water-logging problems. More commonly, compaction of the superficial soil layers during deforestation and land preparation substantially decreases infiltration capacity, resulting in increased peak surface runoff and lowered groundwater levels in shallow aquifers – with associated reductions in some well yields, spring discharges and dry weather streamflows.

Controls over groundwater quality

The natural groundwater chemistry of aquifers in the humid tropics is distinguished by two prominent processes:

(a) Relatively rapid dissolution of minerals associated with the high rates of circulation of infiltrating meteoric water, leading quite commonly to high dissolved SiO_2 concentrations, for example.

(b) Very large dilution as a result of the high groundwater recharge rates, and thus low concentrations of salts (such as NaCl and $CaSO_4$), which have become concentrated in the soil through plant evapotranspiration.

Important aspects of the natural groundwater chemistry are inadequately understood, notably the controls on pH in non-carbonate systems and on Eh (redox potential of the groundwater). In relation to the latter, the consumption of DO (dissolved oxygen) in tropical soil profiles appears generally to be rapid, as a result of the oxidation of organic material and/or inorganic minerals. Thus, anaerobic groundwaters may be relatively widespread.

Elevated DOC (dissolved organic carbon) concentrations and TC (total coliform) counts have been reported from aquifers in the humid tropics in areas that appear to be free from surface contamination (see also Prost, this volume). This

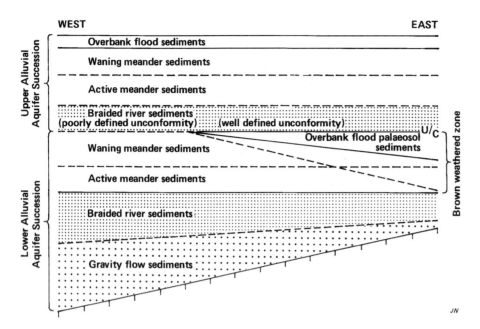

Figure 4: Schematic cross section of the Lower Ganges plain (northwest of Dhaka) to show sedimentary sequence (after Davis, 1989).

suggests that these materials may arise naturally and be derived from unusually deep (biologically- active) soil profiles.

FEATURES OF PRINCIPAL GROUNDWATER SYSTEMS

Major alluvial formations

The humid tropics include some of the world's largest rivers and associated alluvial deposits. The river basin alluviums, together with the extensive alluvial deposits developed along some tropical coastlines, normally form complex, thick, multi-aquifer systems. They exhibit considerable vertical and lateral variations in lithology from horizons of coarse sand to thick deposits of silts and clays, which act as aquitards and often become predominant downstream in estuarine, deltaic and coastal areas.

The associated aquifer systems are characterized by shallow water tables, low hydraulic gradients, slow groundwater flow and very large groundwater storage. The frequent presence of horizons containing a relatively high content of organic material often leads to the elimination of dissolved oxygen from the very slowly circulating groundwater, at least over certain depth intervals, with a consequent tendency for increased solubility of Fe, Mn and other minerals subject to redox processes.

The alluvial groundwater system of the Ganges Delta in Bangladesh is a good example of this category. The multi-aquifer system is sedimentologically complex (Davies, 1989) with a shallow water table aquifer developed over wide areas, underlain by a deeper semi-confined and more productive formation (Fig. 4).

Intermontane valley-fill

These groundwater systems differ from those of the classical alluvial systems of the humid tropics in that they are of much more limited extent, being confined within the active fold mountain chains of South America and South East Asia.

Valley development and sedimentary deposition, for the most part, are tectonically-controlled. The margins of the valleys are thus often formed by highly-heterogenous alluvial and co-alluvial fan deposits, forming complex high-transmissivity aquifers. These normally give way to somewhat finer deposits beneath the lower ground along the valley axis, where there may be evidence of drainage having been periodically interrupted, leading to periods of lacustrine deposition with the formation of aquitard horizons.

The Upper Cauca Valley of Columbia, in the vicinity of Cali, is an example of such a groundwater system (Fig. 5). Very high recharge rates from excess rainfall and from influent streams originating in the adjacent mountains lead to steep hydraulic gradients and large groundwater flow towards the centre of the valley with discharge to the Cauca River.

Karstic limestones

A high rate of biological productivity and calcareous deposition is characteristic of some tropical seas. This has resulted in the formation of major Tertiary-Quaternary limestone formations which form important aquifers in some parts of the humid tropics.

The aquifer properties are controlled by the degree of diagenesis and cementation of the calcareous deposits and also by secondary solution features of the limestone rock formed. When exposed to the heavy rainfall of humid tropical areas, which also often have organic (slightly-acid) soils, large-scale solution can result with extensive underground drainage networks and intense karstification of the landscape.

The extensive karstic limestone aquifer of the Yucatan Peninsula in Mexico is a classic example of such a groundwater system (Fig. 6). Of low elevation and in hydraulic continu-

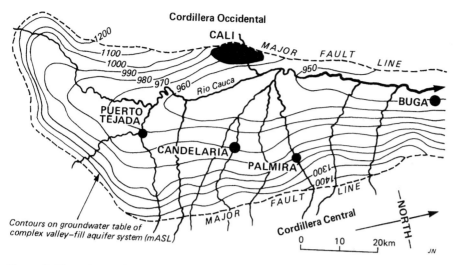

Figure 5: Groundwater flow regime in the Upper Cauca Valley, Colombia.

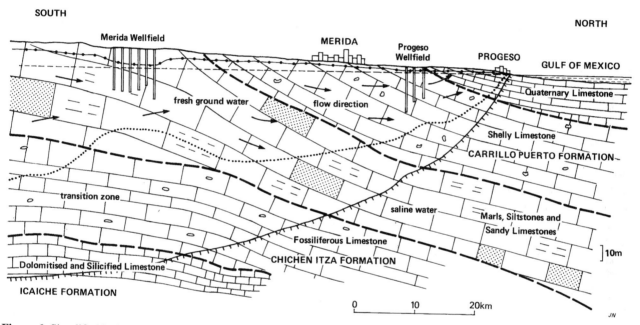

Figure 6: Simplified hydrogeological section through part of the Yucatan Peninsula.

ity with the Caribbean Sea, the formation is invaded by saline water over very large distances from the coast. However, the high rates of seasonal excess rainfall lead to the development of a relatively thick lens of fresh water in which horizontal groundwater flow is controlled by major solution features along joints, but groundwater storage is significantly influenced by the relatively high primary porosity preserved in some parts of the limestone formation.

Active volcanic arcs

Important active volcanic arcs, such as those of Central America and Indonesia, lie within the humid tropics. Eruptive episodes there have led to the deposition of large quantities of relatively viscous lavas, interbedded with pyroclastic deposits known geologically as tuffs and ignimbrites, of andestic-to-rhyolitic composition.

The complex interbedding of brecciated and fractured lava, with thin, more welded, volcanic deposits of low permeability leads to the development of frequent perched aquifers. The Valle Central of Costa Rica illustrates this situation (Fig. 7) (Foster et al.,1985; Parker, et al.,1988). Here and elsewhere, lava aquifers are overlain by porous pyroclastic deposits of large groundwater storage, which combine to form a groundwater system of high yield and drought reliability.

The volcanic cones are subject to rapid erosion by heavy tropical rainfall, with the main surface watercourses often becoming deeply incised in canyons. The steep immature relief and sloping base of aquifers can result in remarkably steep hydraulic gradients for high transmissivity formations. These gradients are sustained by continuous vertical leakage from pyroclastic aquitards and high rates of excess rainfall on the upper slopes of the volcanic cones.

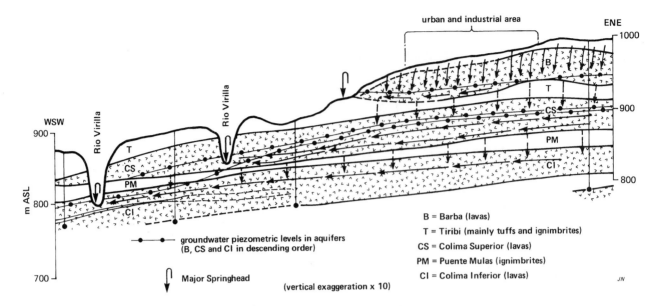

Figure 7: Simplified hydrogeological section of the volcanic aquifers along the lower part of the Valle Central of Costa Rica (after Parker *et al.*, 1988).

While the natural quality of groundwater in volcanic terrain is normally excellent, the presence of potentially toxic constituents such as arsenic, fluoride, boron and selenium, associated with the volcanicity itself should be carefully checked.

Basement regoliths

Extensive regions of Africa, and to lesser extent, South America and Asia, are directly underlain by crystalline basement rocks of the continental shield. In the humid tropics, these rocks have a very deep weathered mantle or regolith. Minor but extensive aquifers are developed in this regolith and the fractured basement rocks themselves (Fig. 8). The regolith generally appears to be the more consistent and reliable of the two interconnected aquifers (Chilton & Smith-Carington, 1984; Omorinbola, 1984). Regolith aquifer development will be a function of geomorphological setting, bedrock type and present/past climatic regimes, since these factors will exert an interactive control on the depth of weathering, the present groundwater level and the extent of any regolith removal by erosion. In particular, as shown in Fig. 8, deep weathering is encouraged by structural features such as faults and fracture zones, and the concentration of runoff and infiltration at the foot of inselbergs (Wright, 1990). Groundwater flow is often controlled by these structural features and by dykes and major quartz veins. These may provide the restricted zones of enhanced permeability in the fresh bedrock which occasionally make it locally a more important aquifer.

Natural groundwater quality problems may be encountered rather frequently in some areas. These include those related to low pH and Eh, such as Fe, Mn and Al mobility, as well as the presence of elevated concentrations of Mg and SO_4 associated with the weathering and oxidation of sulphide minerals (Chilton & Smith-Carington, 1984).

GROUNDWATER EXPLOITATION AND MANAGEMENT PROBLEMS

A comprehensive discussion of groundwater exploitation and management is outside the scope of this review. Nevertheless, it is of relevance to give an indication of the type of aquifer development problems that may be encountered and of resource management issues that are likely to arise.

In all hydrogeological environments, the average rate of active groundwater recharge is difficult to quantify with precision. The aquifers of the humid tropics are no exception in this respect. However, this is not generally critical to development, provided that the initial stages of exploitation are adequately staged and properly monitored. Closely monitored operational experience can generate key data with which to refine estimates of the recharge element of the groundwater resources, but all too often this valuable opportunity is lost.

More caution in development is urged in the case of groundwater systems in which even modest lowering of the groundwater table may lead to coastal saline intrusion, the interference of flow to captured springheads, or the extensive lowering of the groundwater table in areas with phreatophytic vegetation.

The high rates of precipitation in the humid tropics cause rapid leaching of pollutants from urban and industrial wastes disposed in (or on) the land surface – resulting in serious pollution risk for vulnerable aquifers. Increasing NO_3 and/or DOC concentrations within and downstream of urban areas normally indicate the incipient signs of such contamination (Fig. 9), (Foster, 1985; Thomson & Foster, 1986; Foster, 1990).

Major alluvial formations

If relatively high yields are required, production borehole construction is costly and requires sound investigation and design (Stoner *et al.*, 1979). Even in the case of lower-yielding bore-

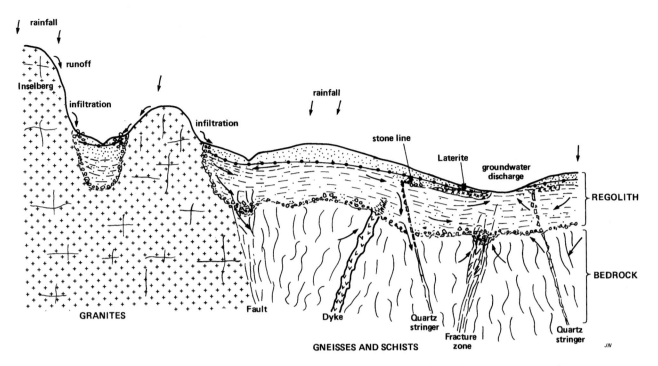

Figure 8: Schematic hydrogeological section through a typical basement rock sequence in humid tropical areas.

holes, attention to detail during construction is important to avoid subsequent operational problems and reduction in useful well life.

Adequate logging of alluvial sequences to identify the most productive horizons and the possible presence of either shallow perched aquifers, or deeper saline aquifers, is the first requirement. Subsequently, careful selection of well screen size and associated gravel pack (Driscoll, 1986) implies adequate sampling and testing, and is necessary to avoid sand pumping and borehole infilling by silt.

The occurrence of groundwaters of low Eh, at least in some parts of alluvial sequences, is likely to lead to the presence of soluble Fe, which can result in rapid biofouling and encrustation of wells with associated deterioration in hydraulic performance and useful life. Appropriate construction materials and regular maintenance will be necessary to avoid much more costly rehabilitation problems later, or even loss of production boreholes.

The shallowest aquifer unit in the multi-aquifer sequence, commonly developed in thick alluvial formations, is rather vulnerable to pollution from human activities at the land surface, given its shallow water table. Diffuse pollution from agricultural practices and unsewered sanitation in urban and more densely populated rural areas is fairly widespread. In some cases, surface inundation of flood plain areas can also result in direct wellhead pollution.

Deeper, semi-confined, aquifers within the alluvial system are more protected against surface pollution. However, where they underlie highly polluted water table aquifers, they may suffer the same fate as a result of inadequate production borehole construction leading to direct vertical leakage or of heavy pumping inducing local vertical leakage with the penetration of more mobile and persistent contaminant species.

The uncontrolled exploitation of these aquifers can result in even more serious problems, including the dewatering of overlying water table aquifers and major interference with shallower production boreholes. In addition, under certain conditions, dewatering and compaction of aquitards with serious subsidence of the land surface, such as is occurring in the vicinity of Bangkok and Djakarta, also can take place.

In coastal areas, saline intrusion frequently occurs where uncontrolled exploitation is allowed to cause local reversal of aquifer hydraulic gradients. Coastal sections of the Guyanas, Cuba, Thailand, Indonesia, and southern India are among the many examples.

Intermontane valley-fill

Much of the preceding discussion also applies to some degree to intermontane valley-fill aquifers. Production borehole drilling, particularly towards the margins of intermontane valleys, often presents major problems as a result of the presence of coarse granular deposits, including boulders, which can be difficult to penetrate or can result in lost drilling fluid circulation.

The marginal areas may also be highly vulnerable to pollution, especially if they are subjected to rapid and uncontrolled urbanization, which sometimes occurs as a result of the limited land area of such valleys that are otherwise suitable for settlement. Minor surface watercourses are often strongly influent and if these become polluted as a result of indiscriminate effluent discharges, groundwater quality deterioration in underlying aquifers is likely to result.

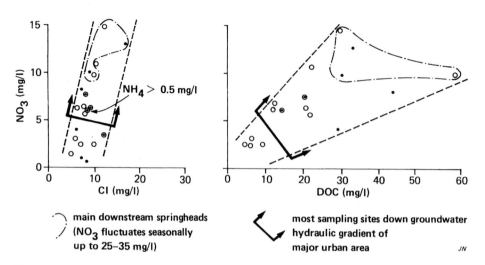

Figure 9: Evidence of incipient contamination of the volcanic aquifers of the Valle Central of Costa Rica, probably as a result of urban sanitation and waste disposal (after Foster, 1990).

Karstic limestones

While production borehole siting, design and construction present few problems in such formations, borehole yields, which can be very high, may be somewhat unreliable in drought. Moreover, this type of aquifer system is the most vulnerable to both coastal saline intrusion and pollution from the land surface (Chilton et al.,1990). Groundwater resources therefore, require very careful management if they are not to suffer serious or irreversible deterioration.

Preferential flow in the unsaturated zone and very rapid fissure flow in the saturated aquifer must always be expected, leading to high vulnerability to pollution of all types. In many formations, the characteristic fissure size is such that colloidal transport of otherwise immobile pollutants is also a distinct possibility. The relatively low hydraulic gradients associated with many formations mean that the groundwater flow direction can be easily modified by pumping. Thus, the definition of catchment areas to be protected against pollution can prove difficult.

The effects of pollution may not only be felt by those exploiting groundwater resources for water supply. Such aquifers also are commonly characterized by major discharges of fresh water along the coast into fragile coral reef environments. If these discharges become polluted by nutrients from agricultural cultivation or ground disposal of human or animal wastes, significant coastal pollution also could arise. In this context, it is the total flux of pollutant discharge which is important, not simply its concentration in the groundwater.

Active volcanic arcs

The exploitation of groundwater in volcanic terrains is greatly complicated by the highly heterogenous character of such groundwater systems. The most productive aquifer horizons,

normally associated with brecciated and fractured lava flows, are often discontinuous, difficult to locate and to correlate. Considerable effort in hydrogeological exploration is needed for production borehole siting and this tends to increase development costs. Moreover, in the often steeply-sloping volcanic terrain, both groundwater levels and the more productive aquifer horizons may be very deep, giving rise to significant production borehole drilling problems.

The exploitation of shallower perched volcanic aquifers presents less technical difficulty and requires reduced investment, but production boreholes and springhead captures in such aquifers may suffer substantial reductions in yield during drought. Groundwater recharge mechanisms, and groundwater transfers between perched and deeper aquifers, are complex processes in volcanic areas. Thus, groundwater resources are difficult to quantify.

The vulnerability of volcanic aquifers to pollution from the land surface is extremely variable. Where brecciated and fractured lavas outcrop at the land surface or along the beds of influent streams, the risk of groundwater pollution will be high if strict measures to avoid soil and surface water pollution are not taken. Where a surface cover of porous pyroclastic deposits or a well-developed lateritic soil mantle are present, the vulnerability of the groundwater to pollution will be more restricted and associated with leaching or discharge of highly mobile and persistent contaminants (Foster *et al.* 1985).

Basement regoliths

Although not very productive, these aquifers are of increasing importance for rural water supplies, especially given improved hand pump technology. The available drawdown to the most productive (permeable) regolith horizon in drought is a critical factor for groundwater development (Fig. 10) (Foster, 1984).

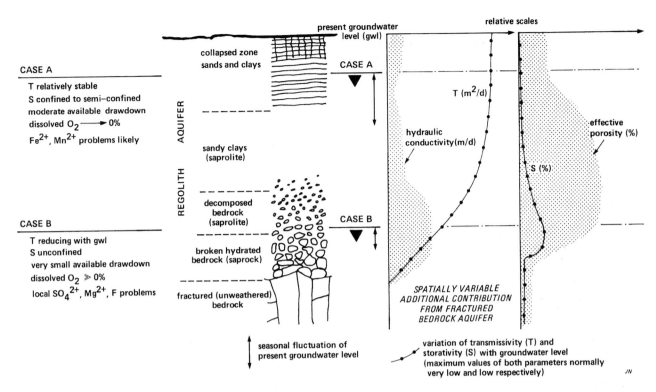

Figure 10: Schematic section of regolith showing vertical variation in aquifer properties and importance of water level in groundwater development (after Foster, 1984).

Case A in Fig. 10 is the more favourable and more typical situation in humid basement areas of western and central Africa, whereas Case B is perhaps more typical of the drier parts of western Africa and large areas of southern Africa.

In some areas, the weathered mantle and fractured bedrock may possess sufficient transmissivity to allow borehole yields adequate for motorized pumping plants. The prospect then arises of groundwater utilization to supply small urban areas and/or small-scale supplementary irrigation. These borehole yields, however, can only be sustained if both aquifer storage and recharge rates are adequate.

The location of boreholes to give sufficient yields to be exploited by hand pumps to supply rural domestic and livestock demands (more than $0.2 \ \mathrm{l \ s^{-1}}$) presents few problems in the humid tropics. However, if higher yields for small-scale irrigation or minor urban water demands are required, the investment in exploration rises steeply and the likelihood of success still may not be high.

The more detailed attention that is paid to borehole design and construction, the greater is the likelihood of success in all cases because of the shallow nature of the aquifer system and the limited available drawdown. Three points are particularly important: first, not to seal off permeable horizons by solid lining tubes; second, to make reliable predictions of minimum drought groundwater levels and, third, to avoid, as far as possible, any horizons containing groundwater of unacceptable quality.

In addition to the natural water quality problems already referred to, the regolith aquifers are also significantly vulnerable to pollution because of their shallow water tables and the possibility of shrinkage cracks developing in the dry season, thus presenting pathways for preferential penetration of pollutants from the land surface. The most commonly encountered pollution problems relate to the ground disposal of human and animal wastes.

CONCLUSIONS

1. Groundwater resources are widely and favourably distributed in the humid tropics and are likely to be subjected to increasing exploitation for water supply.

2. Groundwater systems of the humid tropics fall into a number of distinct groups according to geological build, which also exerts an important control on soil type, surface water regime, and thus has considerable influence on the vegetation and ecosystem.

3. The groundwater systems of the humid tropics have not been studied in much detail, but in view of the above factors, detailed investigation of selected type areas is strongly recommended.

4. Because of the close association between groundwater and surface water, development of groundwater resources in the humid tropics can have potentially major environmental implications, which need to be more carefully and consistently considered when development projects are under consideration.

5. Groundwater should be regarded as a valuable, but potentially fragile, resource in the humid tropics which, in

some cases, is highly vulnerable to pollution from the uncontrolled disposal of urban and industrial liquid effluents and solid wastes, as well as from intensive agricultural cultivation, and saline intrusion in coastal areas due to local over-exploitation.

ACKNOWLEDGEMENTS

This contribution is published by permission of the Director of the British Geological Survey, a component institute of the Natural Environment Research Council. The authors wish to acknowledge that their knowledge of the hydrogeology of the humid tropics, which is still far from complete, has been considerably enhanced from the experience of various colleagues in the British Geological Survey-Hydrogeology Group, notably Brian Adams, Judy Parker, Jeff Davies and Nick Robins. The groundwater projects with which they themselves have been associated in various nations of the humid tropics have been mainly funded by the British Overseas Development Administration or the WHO Pan American Health Organization.

REFERENCES

Bruijnzeel, L. A. (1990) Hydrology of moist tropical forests and conversion – a state of knowledge review. UNESCO-IHP Humid Tropics Programme (Paris).

Chilton, P. J. & Smith-Carington, A. K. (1984) Characteristics of the weathered basement aquifer in Malawi in relation to rural water supplies. IAHS Publn 144: 57–74

Chilton, P. J., Vlugman, A. A. & Foster, S. S. D. (1990) A groundwater pollution risk assessment for public water-supply sources in Barbados. Proc. AWRA Congress "Tropical Hydrology & Caribbean Water Resources" (San Juan-July 1990): 279–289.

Davies, J. (1989) The geology of the alluvial aquifers of Central Bangladesh. British Geological Survey Technical Report WD/89/9.

Driscoll,. F. G. (1986) Groundwater and wells, 2nd ed., Johnson , St. Paul, Minnesota, USA.

Dunne, T. 1978. Field studies of hillslope flow processes. Hillslope Hydrology (Wiley-New York): 227–293.

Foster, S. S. D. (1984) African groundwater development – the challenges for hydrogeological science. IAHS Publn 144: 3–14.

Foster, S. S. D. (1985) Potable groundwater supplies and low-cost sanitary engineering – how compatible? UN Nat Res Forum 9: 125–132.

Foster, S. S. D. (1990) Impact of urbanization on groundwater. IAHS Publn 198: 187–207.

Foster, S. S. D., Ellis, A. T., Losilla-Penon, M. & Rodriguez-Estrada, H. (1985) Role of volcanic tuffs in the groundwater regime of the Valle Central, Costa Rica (CA). Ground Water 23: 795–802.

Omorinbola, E. O. (1984) Groundwater resources in tropical African regoliths. IAHS 144:15–24.

Parker, J. M., Foster, S. . D. & Gomez-Cruz, A. (1988) Key hydrogeological features of a recent andestic volcanic complex in Central America. Geolis 2:13–23.

Pearce, A. J., Stewart, M. K. & Sklash, M. G. (1986) Storm runoff generation in humid headwater catchments – where does the water come from? Water Res Research 22:1263–1272.

Stoner, R. F., Milne, D. M. & Lund, P. J. (1979) Economic design of wells. Quat. J. Eng. Geol.12 (2) 63–78.

Thomson, J. A. M. & Foster, S. S. D. (1986). Groundwater pollution risks on limestone islands: an analysis of the Bermuda case. J Inst Water Engrs. Scientists 4O: 527–540.

Ward, R. C. (1984) On the response to precipitation of headwater streams. J. Hydrol. 74: 171–189.

Weyman, D. R. (1973) Measurements of the downslope flow of water in a soil.J. Hydrol.: 20 267–288

Wright, E. P. (1990). Groundwater occurrence and groundwater flow systems in basement aquifers. Proc. Intl. Symp. Groundwater Exploration and Development in Crystalline Basement Aquifers. (Harare), Commonwealth Science Council, CSC (89) WMP-13 TP 273(2): 251–256.

13: Perspectives on the Hydrology and Water Resource Management of Natural Freshwater Wetlands and Lakes in the Humid Tropics

A. Bullock

Institute of Hydrology, Wallingford, Oxfordshire OX10 8BB United Kingdom

ABSTRACT

A review is presented of the hydrology and water resource values of humid tropical wetlands and lakes. Issues discussed in relation to natural lakes are long-term changes in water levels, aquatic weeds and modelling lake/river interactions. The status of tropical wetland research is presented with reviews of different modelling strategies, empirical water balance studies, and wetland influences upon downstream flows. Current issues involving wetlands in the water resource arena are discussed, and strategies for optimal wetland management and research strategies are identified.

INTRODUCTION

Wetlands occupy approximately 6% of the earth's land surface, and can be found on each of the tropical continents. Table 1 illustrates that tropical wetlands are most extensive in South America, with similar areal extents between Asia and Africa. Balek (1977) reports that the 340 000 km^2 of wetlands within Africa are composed of some 10^4 to 10^5 individual wetland systems alone. Table 2 illustrates that wetlands can cover up to 10% of individual countries on the African continent and up to 30% on the Asian continent. Early estimates of wetland proportions in the tropics are being revised upwards as improved surveys are carried out in more isolated regions of the world (Maltby, 1988). Furthermore, consideration of wetland extent in the past has tended to under-represent the number of small seasonal wetlands, such as the dambo, pan and black clay types (Howard-Williams & Thompson, 1985), which can occupy over 25% of the land surface areas in many regions of southern Africa. With over 3,000 individual wetlands having been mapped recently in central Togo alone (Runge, 1991), and with an estimated 10 000 individual wetlands in Zimbabwe, estimates of the number of African wetlands will continue to be revised upwards as more national inventories are completed.

The importance of wetlands lies in their many functions which can be summarized by a set of wetland values (e.g.,

Dugan, 1990). These are presented in Table 3, with examples from the humid tropics. Wetland values do not derive from their hydrological functions alone as many wetlands also represent significant forest, wildlife, fisheries, forage and energy resources. Conversely, wetlands can contribute detrimentally to the environment through the promotion of disease by acting as impediments to construction and by the release of periodic nutrient flushes.

Yet, for a component of the hydrological cycle which is so extensive, and a natural resource of such considerable potential, the hydrology and water resource aspects of wetlands in the humid tropics have not received the research attention commensurate with their values. Most attention has been directed towards the larger wetlands and the RAMSAR Convention global list of internationally important wetlands, which comprises 464 sites. Many of these sites owe their status to the conservation of wildlife. Thousands of smaller wetlands have no formal Convention status, and have little value to wildlife. Rather, it is the recognition that wetlands are of considerable significance in meeting the water and agronomic demands of local communities that has attracted research workers more to the smaller wetlands. Most of the recent hydrological research effort has been directed towards these smaller wetlands, especially within southern Africa.

This paper presents the state of knowledge of the hydrology and water resource values of humid tropical wetlands and lakes, albeit with a stronger emphasis on the former. Amongst the issues most pertinent to natural lakes are long-term changes, ecological stress, management problems and modelling. Examples from the humid tropics of each of these aspects are presented. The status of tropical wetland research as presented at international and national conferences is described to trace the development of our knowledge of wetland systems. Reviews are presented of the current status of research in relation to different modelling strategies, empirical water balance studies and wetland influences on downstream flows. Current issues involving wetlands in the water resource arena are discussed, and strategies for optimal wetland management and research strategies are identified.

TABLE 1. *Extent of swamps in tropical parts of the continents.*

Continent	Swamp area (km²)
South America	1,200,000
Asia	350,000
Africa	340,000
Australia	2,000

After Balek (1989)

TABLE 2. *National wetland coverage in example countries and regions.*

Country/region	Feature	Proportion of Surface Area	Source
Amazon	Varzea, Igapo	3%	Pires (1973)
Tanzania	Wetland	10%	Mwanyika (1982)
Kenya	Wetland	3%	Mavuti (1982)
Uganda	Wetland	6%	Lind (1956)
Uganda	Wetland	10%	IUCN (1990b)
Zambia	Dambo	4.6%	Kalapula (1986)
Zimbabwe	Dambo	3.6%	Whitlow (1985)
Indonesia	Wetland	>25%	Maltby (1986)
Sumatra	Wetland	30%	Maltby (1986)

PERSPECTIVES ON TROPICAL LAKES

The Symposium on the Hydrology of Natural and Manmade Lakes in Vienna (Schiller *et al.,* 1991) identified four key issues: long-term changes, lakes under stress, management problems and modelling (including background processes, water quantity and quality). These are common problems from around the world and examples of these issues can be found in humid tropical regions.

Lake level variations

Explaining long-term lake level variations, whether rises or falls, has not always been found to be easy, as evidenced by the cases of Lakes Malawi, Victoria and Toba. The level of Lake Malawi has varied widely throughout this century, and increased dramatically in the period 1976–1980 (Drayton, 1984). The causes of the change in lake level were investigated, using simple numerical models based on the lake's water balance. Drayton shows that the recorded changes in rainfall were sufficient to have caused the observed changes in level, and that human-made changes in runoff and outflow have been comparatively unimportant. Eccles (1984) states that the greater than normal annual rises were probably due to increased direct rainfall on the lake, possibly resulting from changes in either the mean position or the intensity of the monsoon shearlines and the Maximum Cloud Zone (see Manton & Bonell, this volume). Increased runoff resulting from extensive deforestation does not explain the appearance of a bimodal distribution of increases in level. Applying four methods of analysis comprising water balance methods, time series analysis, probability analysis and multivariate statistical techniques, Neuland (1984) concludes that rainfall in the catchment, catchment runoff and rainfall on the lake rank in descending order as causes of the abnormal high lake levels.

The sharp rise in Lake Victoria between 1961 and 1964 had been found difficult to explain in terms of the components of the water balance. However, after reviewing lake inflows and the method of calculating lake rainfall from lakeside gauges, Piper *et al.,* (1986) were able to reproduce the historic lake water balance. The rise in lake level can be explained through rainfall and resulting tributary inflows, allowing projections of future levels to be made by analysis of rainfall series.

Lake Toba in the northern mountains of Sumatra covers an area of 1,120 km² and is situated within a volcanic caldera, whose inner slopes provide a drainage area of 3,565 km². Toba is drained by the River Asahan which falls steeply to the coastal plain, making an ideal site for hydroelectric power generation. The construction of two generating stations in 1984 was followed by a drop in the lake level by 2.5 metres between 1984 and 1987, as illustrated in Fig. 1.

Several explanations were proposed to account for the sudden drop, including deforestation of the caldera slopes, increases in the area of paddy fields, leakage from the lake along a geological fault and tectonic deformation of the lake bed. A water balance of the lake was developed (Meigh *et al.,* 1990) in which the change in the volume of water stored in the

TABLE 3. *Examples of hydrological functions and water resource values of wetlands in the humid tropics.*

WETLAND VALUE	EXAMPLE
Groundwater discharge	Dugan (1990) reports the example of the Amboseli National Park in Kenya where the most important source of water for wildlife is a series of groundwater springs which have their source on Mount Kilimanjaro and which, after percolating through porous lava soils, re-emerge in the Amboseli Basin in a series of small swamps.
Flood control	Balek (1989) illustrates the reduction of river flows in Central Africa with increasing proportions of wetlands within catchments under different annual rainfall regimes.
Sediment/toxicant retention	Kalk *et al.,* (1979) contrast the sediment load of two rivers draining the inselberg massif of Mount Mulanje into Lake Chilwa in Malawi, with similar geological features. The Phalombe River, which meanders through a cultivated plain, has an annual silt load of 1,693 tons km^{-2} whereas the Sombani River which passes through dense swamps along its course has an annual yield into Lake Chilwa of 2 tons km^{-2}.
Nutrient retention	Howard-Williams & Thompson (1985) identify the value of a Phragmites swamp in the removal of faecal bacteria, increasing water clarity and a summer removal of nitrate, citing the example of the removal of nutrients by a papyrus swamp from the treated effluent from the sewage works at Kampala, Uganda.
Biomass export	Shepherd (1976) estimated that 500 kg $year^{-1}$ of animal dung was dropped on every 100 m stretch around the Bangula Lagoon in Malawi, believed to contribute to the exceptional production in the lagoon.
Water supply	Howard-Williams & Thompson (1985) collate data which show that three major wetland systems, the Barotse, Kafue (both in Zambia) and Shire (Malawi) flood plains together support over 700 000 cattle. Russell (1971) identifies domestic water supply as the dominant use of dambo wetlands in Malawi.
Water transport	Dugan (1990) cites the example that in the Bangweulu Basin, water transport by public and private canoes provides a cheap and readily available means for moving both people and freight in what is otherwise an isolated region.
Recreation/Tourism	Heyman (1988) estimates the annual cash income from Coroni Swamp in Trinidad to be US$2 million, as wetlands attract interest from persons engaging in sport hunting, fishing, ornithology and nature photography, amongst other activities.

lake, dV, must be equal to the total input minus the total output:

$$dV = (Q_{cat} + P_{lake}) - (E + Q_{out} + L) \qquad (1)$$

where Q_{cat} is the volume of runoff from the catchment, P_{lake} is the volume of precipitation falling on the lake, E is the evaporation from the lake, Q_{out} is the volume of outflow from the lake and L is the volume of losses through the lake bed.

Direct measurement of the lake evaporation, the precipitation and lake outflow, and estimation of the inflows from analysis of gauged data in the vicinity, allowed calibration of the water balance model of the lake. The series of net inflow, outflow and storage are illustrated in Fig. 2.

A series of observed lake levels were available, and the close agreement between observed and simulated lake levels for the period 1973 to 1989 (Fig. 3) demonstrates the satisfactory performance of the model. Between 1980 and 1984, the bed of the River Asahan was dredged and regulating dams were constructed along the water course, but these operations did not initiate changes in lake levels beyond their normal range of variability. The period from mid-1984, when power generation began taking water at a steady rate of approximately 100 m^3 s^{-1}, coincided with a period when rainfall, and consequently lake inflows, were well below average. After this period, a return of net inflows to near normal and a reduction in the water offtake rate resulted in a considerable recovery in the lake level. No evidence was found to indicate that water had been lost through other causes, such as tectonic deformation or lake bed fractures. Changes in land-use, particularly deforestation, do not appear to have been on a sufficiently large scale to have adversely affected the water balance.

The significance of lake level variations are well illustrated by the impacts upon the chemical and biological limnology of Lake Murray in Papua New Guinea (Osborne *et al.,* 1987). Lake Murray, with a surface area of 647 km^2, and a highly convoluted shoreline, is the largest lake in Papua New Guinea and exhibits marked seasonal fluctuations in water level. The fall in water level of four metres between April and December 1982 was accompanied by a marked rise in pH, conductivity, total hardness and filterable residue. Clearly, impacts caused by seasonal variations in lake levels are more severe when associated with long-term variations in levels.

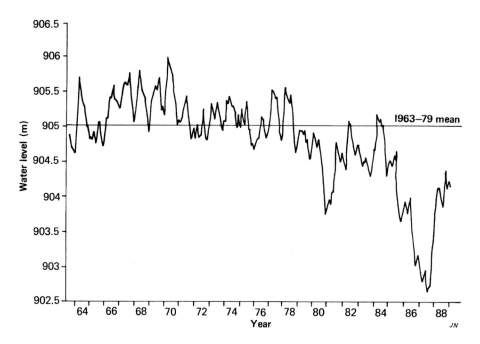

Figure 1: Drop in water levels in Lake Toba, Indonesia. After Meigh *et al.,* (1990).

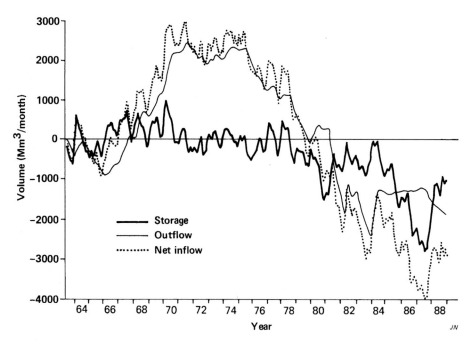

Figure 2: Net inflow, outflow and storage – cumulative deviations from the 1963–79 mean. After Meigh *et al.,* (1990).

Aquatic weeds

One key issue that places lakes under stress, and which has become one key element of lake management is the problem of aquatic weeds. The control of aquatic weeds is vital to the provision of safe and adequate water supplies and for the preservation of wetland and lake characteristics. Three aquatic weeds are notable in introducing detrimental impacts into wetlands and lakes. *Salvinia Molesta* (Kariba Weed), indigenous to South America, has become one of the most widely-distributed aquatic weeds in tropical countries, existing in India, Malaysia and Australia. Kariba Weed was first reported in the Zambezi in 1948 before becoming widespread in many

African countries. *Salvinia* forms thick, impenetrable mats of small, free-floating fern with trailing roots and a narrow rhizome. The mats have the capacity to cause severe ecological alterations which impact the quantity and quality of water and other environmental resources, including stifling plant competition, removal of nutrients, decreasing oxygen content of the water, increasing evaporation, increasing sedimentation and destroying fish life. Ase (1987) reports that the evaporation and transpiration from samples of *Salvinia* approximately equal the potential evaporation from a free water surface under similar conditions at Lake Naivasha, Kenya. Evapotranspiration represents the only loss from that lake, which

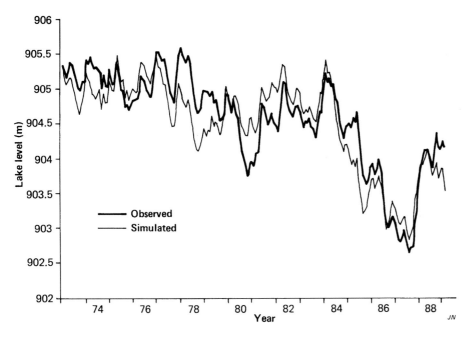

Figure 3: Observed and simulated levels of Lake Toba. After Meigh *et al.,* (1990).

possesses no natural outlet, and therefore the magnitude of losses has increased significance for lake levels in this case. *Salvinia* can be successfully controlled by the introduction of the weevil *Crytobagous salviniae*, by blocking water inflows, pumping water from infested lagoons, diversion of animal and human movement, and by manual raking and burning.

Another indigenous weed of South America which has become widespread throughout the tropics, and which is persistent in Zimbabwe, for example, is *Eichhornia Crassipes* (Water Hyacinth), a species commonly regarded as the world's worst water weed. This reputation is partly founded on the plant's capability to double its mass and the area which it occupies every five days, and also on the impact which the plant introduces by reducing oxygen concentrations, pH content and temperature, and increasing evaporation rates. Van der Wert & Kamerling (1974) report values of evapotranspiration to potential evaporation in the range of 3.0 to 4.0 for standing water series covered by water hyacinth.

Less widespread in tropical areas, but significant in tropical Africa at least, is *Pistia Stratiotes* (Water Lettuce), which can exert similar impacts on the physical composition of wetlands and lakes as *Salvinia* and the Water Hyacinth.

Modelling lake/river interactions on the Mekong
The Great Lake, Thailand introduces a complication into the simulation of Mekong River flows because of its dual function of storage and regulation. During the flood season, there is a net inflow to the Great Lake from the Mekong through the Tonie Sap. The flow in the Tonie Sap changes direction as the Mekong recedes. Under these circumstances flows from the lake augment the dry season flows into the Mekong delta. The dry season contribution of the lake to the delta inflows is then less dependent on Mekong flows. This dual function of storage

and regulation needed to be overcome in order to simulate time series of flows into the Mekong delta (Institute of Hydrology, 1988). To estimate a long-term time series of daily flows in the Tonie Sap, multiple regression analysis was used, based on continuous records of lake storage and Mekong flows at Kratie, upstream of the Tonie Sap confluence (Fig. 4).

Daily water levels for the Great Lake, recorded at Kompong Chnnang, were converted to storage volumes using an elevation-storage relationship. Continuous records of daily flow at Kratie (Q_k) and the Tonie Sap (Q_{ts}) and the storage in the Great Lake (S_{gl}) were available for the period 1960 to 1969. Initially, a single multiple regression relationship for the whole year with Q_{ts} as the dependent variable and Q_k and S_{gl} as the independent variables was carried out. Reasonable results were obtained. Further analyses were aimed at dividing the dataset into two parts. The start of the first period was set at the time when the flow at Kratie, Q_k, fell during the recession to a given threshold. This period continued until Kratie flows began to rise again, thus reducing the outflow at the Tonie Sap. The end of this period was set at the time when Q_{ts} fell below a given threshold as the lake drains into the Mekong. The threshold discharges were set to be $Q_k < 15\ 000\ \mathrm{m^3\,s^{-1}}$ and $Q_{ts} < 750\ \mathrm{m^3\,s^{-1}}$.

The correlation matrices for the three variables for each of these periods are as follows:

PERIOD 1 (Commences at the first day after October 1st when $Q_k < 15\ 000\ \mathrm{m^3\,s^{-1}}$ and continues until the start of period 2)

	Q_{ts}	Q_k
Q_k	-0.581	
S_{gl}	0.094	0.701

PERIOD 2 (Commences at the first day after January 1st when $Q_{ts} < 750\ \mathrm{m^3\,s^{-1}}$ and continues until the start of period 1)

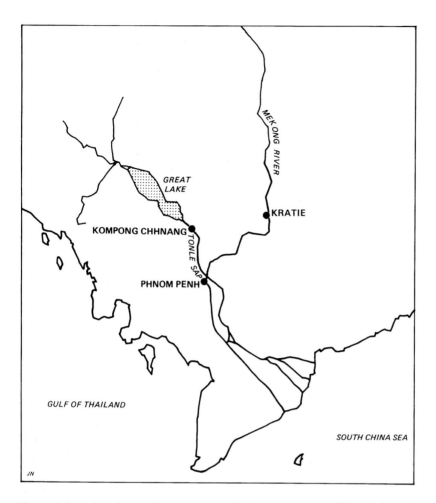

Figure 4: Location of measuring stations used in the investigation of Great Lake outflows. After Institute of Hydrology (1988).

	Q_{ts}	Q_k
Q_k	0.800	
S_{gl}	0.924	0.890

During the first period when the flow in the Mekong is high, the storage in the lake has little effect on the Tonie Sap flows. However, later in the dry season when the water level in the Mekong is low, and there is little backwater effect on the draining of the lake through the Tonie Sap, the storage in the lake becomes more significant.

Regression analyses, using both Q_k and S_{gl} as independent variables, were used to predict Q_{ts}, as shown below:

PERIOD I $Q_{ts} = 114 - (0.369\ Q_k) + (0.0200\ S_{gl})\ R^2 = 78\%$ (2)

PERIOD II $Q_{ts} = 157 - (0.146\ Q_k) + (0.0231\ S_{gl})\ R^2 = 86\%$ (3)

Using the regression equations, a time series of daily flows was generated for the period 1924 to 1960. A comparison of the observed and predicted flows in the Tonie Sap, downstream of the Great Lake is presented in Fig. 5 for the period 1960 to 1969. The predicted data were combined with the more recent observed data to create a composite long-term record at a site just downstream of Phnom Penh, and before the river divides into the Bassac and the Mekong.

STATUS OF RESEARCH OF THE HYDROLOGY OF TROPICAL WETLANDS

The First International Wetlands Conference in Delhi in 1982 (Gopal *et al.*, 1982, p. xii) recognized that at that time "very little is yet known of the wetland hydrology in even temperate areas, and almost nothing in tropical and sub-tropical regions." Although the Delhi Conference and subsequent Minsk Conference on Ecosystem Dynamics (UNEP/SCOPE, 1982) included several papers that recognize the importance of wetland hydrology, the absence of hydrological data and experiments in the published proceedings is striking. Wetland research a decade ago placed more emphasis upon nutrient and energy cycling and primary production than hydrological and water resource problems, as illustrated by the major conference themes summarized in Table 4. Recent wetland conferences in Finland (Academy of Finland, 1988) and Seville (Consejo Superior de Investiagaciones Cientificas, 1988) have focused more on hydrological themes, but the geographical scope of the conferences was limited to temperate and cold regions and to semi-arid and arid regions, respectively. Models and approaches adopted in other regions of the world are, of course, pertinent to the humid tropics. In this respect, Dugan (1988) identifies that there are global lessons to be learnt from

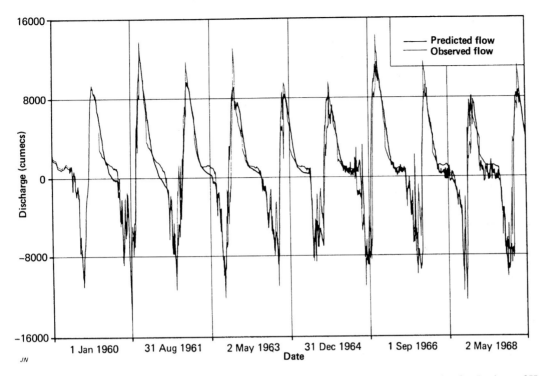

Figure 5: Tonie Sap model results for the simulation of flow from the Great Lake, Thailand. After Institute of Hydrology (1988).

temperate wetland studies through the application of scientific methodologies to conservation and sustainable development through a north-south transfer of expertise and through the possible extension of temperate wetland adjustment to global impacts on tropical environments. It is indicative of the lack of research activity that there has been no single international forum which has addressed the hydrology of tropical wetlands. There have, however, been national conferences on wetlands with discussion on wetland hydrology. Table 5 cites examples from Zambia, Zimbabwe and Amazonia. The convention of national meetings has arisen more through the necessity to solve local problems than to organize a seminal forum.

Dugan (1990) assesses that "there is a broad consensus amongst scientists that wetland loss occurs most often because of failure to apply effectively the information that is already available," and that "not only is much available information left unused, but a lot of the work that is carried out is of greater scientific interest than applied value." This assessment may be appropriate to wetland hydrology in the United States, for example, which, amongst nations, has devoted the most resources to wetland hydrology. However, the advice is somewhat inappropriate in the context of tropical regions, where there is fundamentally little or no understanding (Gopal *et al.*, 1982) and, worse still, misconceptions of wetland values (Bullock,1992a,1992b) in tropical regions. A mere three papers containing research investigations specific to the hydrology of tropical wetlands have been published in the Journal of Hydrology within the past 20 years (Balek & Perry, 1973; Sellars, 1981; and Faulkner & Lambert, 1991). Hydrological Sciences Publications have focused on mountainous, karst, forested and urban regions of the world in

recent years, but it is not since the publication of the "Hydrology of Marsh-ridden Areas" in 1975 that a publication was devoted solely to the hydrology of wetlands. If keyword searches of international library databases are considered a trustworthy source, then there is virtually nothing published on the hydrology of South American and Asian wetlands despite tropical wetlands being most extensive on these two continents.

This geographical imbalance may be due to a bias towards the retrieval of anglophone research reports and to the background of past reviewers, but nonetheless there does appear to have been a much more significant research effort within the African continent. It is also true that a significant amount of research work remains unpublished, as recognized by the convenors of the National Workshop on Dambos in Zambia (Ministry of Agriculture and Water Development/FAO, 1986)) who sought to collate existing information and make it available to scientists and officers involved in wetland planning. However, despite these two qualifying statements, it is certainly true that, despite the widely-stated importance of wetlands, the topic of tropical wetland hydrology has not attracted the research effort commensurate with its extensive distribution and contributions of functions and values to dependent communities. Many disciplines are reliant upon hydrological information, including agriculture, aquaculture, irrigation engineering, biologists, ecologists, and social scientists in studies of wetlands. The development and application of future conservation policy and water resource strategies on a poor understanding of hydrologic process and function could have damaging consequences to the wetlands themselves and to the dependent communities.

TABLE 4. *Themes discussed at International Wetland Conferences.*

Wetlands Ecology and Management, First International Wetlands Conference, New Delhi, India (1980)	1. Evaluation of the present state of knowledge of wetlands 2. Status of existing wetlands and guidelines for conservation and management 3. Establishment of an international program of comparative studies of wetlands
Ecosystem Dynamics in Freshwater Wetlands and Shallow Water Bodies, Minsk, USSR (1981)	1. Ecological, botanical, zoological and hydrological studies 2. Methodology, theory and practice of constructing mathematical models of ecosystem dynamics 3. Use of wetlands in the national economy
Symposium on the Hydrology of Wetlands in Temperate and Cold Regions, Joensuu, Finland (1988)	1. Classification, geomorphology and properties of wetlands 2. Hydrology of wetlands and human influence upon it 3. Water quality of wetlands and human influence upon it
Symposium on Hydrology of Wetlands in Semi-Arid and Arid Regions, Seville, Spain (1988)	1. Geomorphology and pedology of wetlands 2. Ecological classification and inventory of wetlands 3. Hydrology and hydrogeology 4. Hydrogeochemistry 5. Environmental impacts of development on wetlands

TABLE 5. *Themes discussed at National Wetland Conferences.*

National Workshop on Dambos, Nanga, Zambia (1986)	1. Dambos – a national inventory 2. Detailed characteristics of dambos 3. International studies and experiences on dambos 4. Suitability of different types of dambo for agricultural development 5. Engineering and hydrological considerations 6. Ecological considerations for dambo development 7. Strategy for the development and management of dambos for food protection
The Use of Dambos in Zimbabwe's Communal Areas, Harare, Zimbabwe (1986)	1. Review of dambos in Zimbabwe 2. Wetlands soils and maize cultivation 3. Socio-economic aspects of dambo cultivation in Zimbabwe 4. Environmental aspects of dambo cultivation 5. History of dambo cultivation
Wetlands in Amazonia, Belem, Brazil (1990)	1. Use and conservation of Amazonian varzeas 2. Small-scale fisheries in lakes Igarapes and Igapos 3. Utilization of wetlands by Indian communities 4. Impact of large development projects on Amazonian wetlands

One issue that may have contributed to the past failure to integrate wetland studies is the lack of a single universally-adopted classification scheme to provide a unifying framework. Because wetlands can be composed of complex assemblages of different soil and vegetation types, have very different geomorphological origins and occur in different locations within a drainage basin, they are subject to definition by disparate physical criteria. This has led to several classification

Table 6. *Ramsar Convention Classification of Freshwater Wetlands.*

2.1	*Riverine*		
	Perennial	i)	Permanent rivers and streams, including waterfalls
		ii)	Inland deltas
	Temporary	i)	Seasonal and irregular rivers and streams
		ii)	Riverine flood plains, including river flats, flooded river basins, seasonally flooded grassland
2.2	*Lacustrine*		
	Permanent	i)	Permanent freshwater lakes (>9 ha), including shores subject to seasonal or irregular inundation
		ii)	Permanent freshwater ponds
	Seasonal	i)	Seasonal freshwater lakes (>9 ha), including floodplain lakes
2.3	*Palustrine*		
	Emergent	i)	Permanent freshwater marshes and swamps on inorganic soils, with emergent vegetation whose bases lie below the water table for at least most of the growing season
		ii)	Permanent peat-forming freshwater swamps, including tropical upland valley swamps dominated by Papyrus or Typha
		iii)	Seasonal freshwater marshes on inorganic soil, including sloughs, potholes, seasonally flooded meadows, sedge marshes and dambos
		iv)	Peatlands, including acidophilous ombrogenous or soligenous mires covered by moss, herbs or dwarf shrub vegetation, and fens of all types
		v)	Alpine and polar wetlands, including seasonally flooded meadows moistened by temporary waters from snowmelt vegetation
		vii)	Volcanic fumaroles continually moistened by emerging and condensing water vapour
	Forested	i)	Shrub swamps, including shrub-dominated freshwater marsh, shrub carr and thickets, on inorganic soils
		ii)	Freshwater forest, wooded swamps on inorganic soils
		iii)	Forested peatlands, including peat swamp forest

schemes which have emphasized merely one element of these characteristics. One was based on soil properties, using the FAO soil classification (FAO/UNESCO, 1974), another on vegetation types (e.g., Thompson, 1985), another on geomorphological criteria (Raunet, 1985) and still another on topographic location (e.g., Turner, 1985). Those classification schemes which have been applied to wetlands often suffer from deficiencies of non-unique classes and a failure to cover the full range of wetland types. The classification scheme adopted by the Ramsar Convention (Table 6) does combine hydrologic, soil and vegetation criteria. Although the scheme is largely descriptive and resorts in part to simple geographic location, it represents the most internationally accepted classification of freshwater wetlands. However, unless there are sharper definitions of the physical properties it is unlikely that the scheme can fulfil adequately the role for the transfer and extrapolation of data in comparative wetland studies.

MODELLING APPROACHES IN TROPICAL WETLAND HYDROLOGY

Observed hydrological data for wetland systems are generally sparse, and tend to comprise river flow gaugings at inflow and outflow points on the major flood plain wetlands and water surface elevations within wetlands. Wetland hydrologists are faced with the complexities of a vast number of wetlands lacking any observed data, the internal hydrological complexities of major wetlands, and the difficulties of measurement of inflows into headwater wetlands lacking well-defined inflow channels. Consequently, the hydrologist must rely more frequently upon the modelling of wetland systems than the analysis of empirical data. Yet there have been relatively few efforts to model wetlands (Carter et al., 1978) in temperate areas, let alone in tropical regions. Carter et al., (1978) propose three strategies which are appropriate to wetland modelling; firstly, the consideration of wetlands as part of a larger system, in particular the catchment unit; secondly, the modelling of an individual wetland, using either conceptual models or more physically-based deterministic models to understand the function of the different components of a single wetland; and thirdly, modelling of an individual component or process of wetland hydrology, for example, seasonal water level.

Each of these approaches has been applied in tropical regions with varying degrees of success. The International Hydrological Decade gave impetus to the establishment of experimental and representative catchment studies, and this led to the initiation of specific studies of catchments with significant wetland components, for example, at Luano in

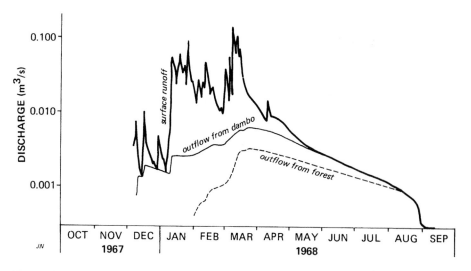

Figure 6: Hydrograph of outflow from different components of one of the Luano catchments (1967–1968). After Balek (1977).

Zambia. The Luano catchments continue today to yield data series in excess of 20 years long and have contributed much to the understanding of the hydrology of headwater wetlands in Africa (Balek & Perry, 1973). Numerous other wetland studies have assembled data series of much shorter duration than the Luano study, as identified later in this paper. Although there are observations of various hydrological parameters in each case, the data to calibrate deterministic models are sparse and may not always be appropriate for transfer to other geographical domains.

Because wetlands constitute a naturally integrated component of a catchment, wetland studies cannot adopt the classic pre- and post-data calibration phases commonly applied to land-use replacement studies. No hydrological study has yet adopted the strategy of completely excavating or artificially constructing a wetland system within a catchment for comparative scientific investigation in the way that forests have been planted and cleared. A second experimental strategy popular in identifying differences in catchment response is the paired catchment approach. Again, this is not appropriate since it is not possible to attribute differences in catchment response to wetland processes alone in a paired catchment study comprising a wetland and a non-wetland basin without first establishing the specific differences in catchment processes which led to the evolution of a wetland within one basin and not the other. There have been several regional multivariate studies based on catchments containing different proportions of wetland which also consider other catchment factors as significant determinants of catchment response (Drayton *et al.*, 1980, Bullock 1992a). However, this type of model, frequently regression-based, suffers difficulties in terms of, one, considering wetlands as "black-box" elements within the hydrological system; two, in correlation of catchment variables; three, in ascribing physical meaning to model parameters; and, four, in the instability of model parameters.

There is a significant array of models which have been used to simulate wetland hydrology in North America (Mitsch *et*

al., 1982) and in the former USSR (Koryavov, 1982). Mitsch *et al.*, classify the different types of models as (a) ecosystem models which describe a water budget for a homogeneous individual wetland and do not consider uplands as part of the model, (b) regional models which present overall gains and losses for a large-scale watershed that includes wetlands, and (c) hydrodynamic transport models which deal with short time periods over large areas.

Balek (1977) describes a distributed conceptual model of a water balance in a tropical basin, developed to describe the water storage and its fluctuation in the main components of the hydrological cycle. The model is based upon a general tropical catchment, consisting in its simplest form of a shallow region, representing the alluvial or wetland zone of rapid groundwater level response, and a second deeper region. Three vertical zones are modelled within each region: a groundwater zone, a zone of capillary rise and the upper moisture zone, with a surface vegetation zone. The model simulates processes of interception, infiltration, evapotranspiration, vertical recharge and groundwater flow, and has been calibrated and applied in the investigation of the Luano catchments to simulate daily river flow and water level data and monthly evaporation data (Balek & Perry, 1973). Figure 6 illustrates the contribution of different components of one of the Luano experimental catchments predicted by this model. However, the dense observational network required for model calibration and for the simulation of daily data means that the model has not been applied elsewhere.

Faulkner & Lambert (1991) describe a conceptual model of a dambo wetland which was developed to explain observed rainfall, evapotranspiration, river flow and piezometric data at Chizengeni dambo in Zimbabwe. The model is used to describe in a semi-deterministic manner the groundwater occurrence and groundwater movement within the experimental catchment, and is applied to an investigation of the effect of dambo cultivation upon water levels and streamflow. The proposed model is based on water balance data and is not the type

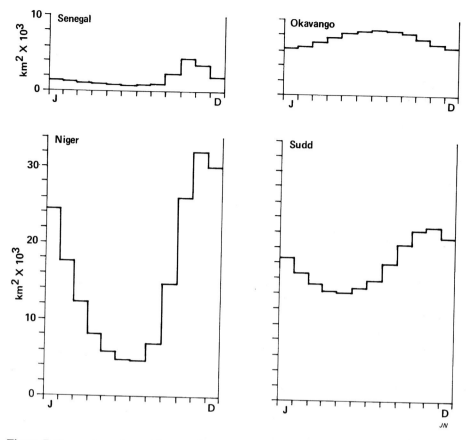

Figure 7: Comparison of monthly mean flooded areas in four major African wetlands. After Sutcliffe & Parks (1989).

of numerical model developed by Balek which defines the component boundaries and the algorithms of water fluxes. The model developed at Chizengeni is recommended as appropriate for assessing the impact of irrigation on water table levels on catchments displaying broadly similar features to the experimental catchment.

Verry *et al.,* (1988) recognize that one way to extend the understanding of streamflow response from wetlands in cold and temperate regions is to combine long-term climatic data, which tend to be widely available, with a functional model of basin streamflow response in a variety of landscape configurations. Verry *et al.,* recommend that the Peatland Hydrologic Impact (PHIM) Model (Guertin *et al.,* 1987) is a suitable vehicle for achieving that understanding. PHIM is a functional model of basin streamflow response which can handle a spatial and size mix of mire and upland mineral soil, and produces an hourly or daily water budget coupled with an iterative reservoir routing algorithm. The strategy of model application with different wetland configurations under different climatic regimes will be a valuable contribution to the understanding of streamflow response of tropical wetlands. However, the model of Balek (1977) requires the calibration of too many parameters for easy replication elsewhere.

Not surprisingly, most success in the development of transferable techniques amongst tropical wetland systems has been achieved when emphasis has been placed upon the modelling of an individual hydrological parameter, rather than upon the

whole catchment or combinations of hydrological processes. Examples are cited below of the applications in the tropics of models and techniques to investigate the water level and prediction of extent of flooded areas, evaporation, hydraulic conductivity, and vegetation resistance.

Water level and prediction of extent of flooded areas

A flood plain storage model was developed to simulate changes in the flooded area with time in the Upper Yobe River of northern Nigeria (Sellars, 1981). The relationship between the flood plain storage and flooded area was investigated using an idealized cross-section comprising both surface and groundwater storage to produce a theoretical relationship between storage and flooded area. The model, with monthly time steps, was calibrated using observed data of storage, inflows, losses and flooded area to predict annual open water and swamp evaporation losses. The model underestimated the extent of the flooded area to a maximum of 6% and overestimated to a maximum of 2%.

A simple relation between flooded area and volume is included in a water balance model applied to four major African wetlands – Senegal, Niger, Sudd and Okavango (Sutcliffe & Parks, 1987). In the water balance model, the swamp is treated as a reservoir whose storage volume is calculated from cumulative inflow less cumulative outflow, with allowance made for soil moisture recharge on newly inundated areas. Linear relationships between area and volume are used

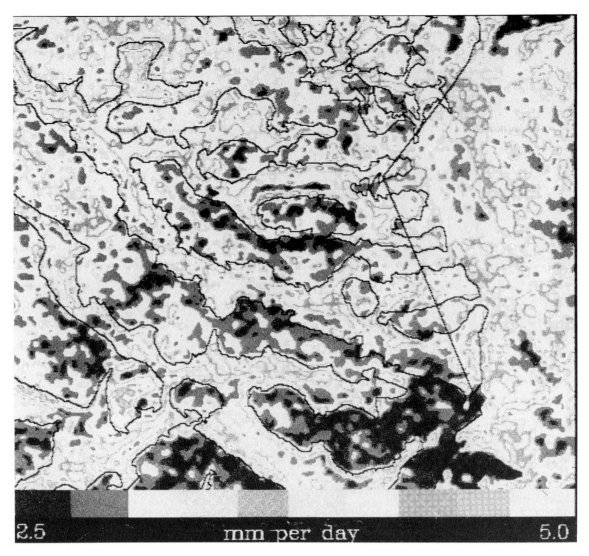

Figure 8: Areal estimation of evaporation using remotely sensed data over wetland and non-wetland zones of the Nyatsime River, Central Zimbabwe. After British Geological Survey & others (1989).

in order to estimate monthly series of flooded areas over the period of record of inflows and outflows. The monthly mean flooded areas (Fig. 7) show great differences in the proportions of permanent and seasonal swamp; the Niger and Senegal exhibit largely seasonal flooding, with a ratio of seasonal to permanent flooding of 6:1, whilst the Sudd and Okavango display a large proportion of perennial flooding, with a ratio of 0.6:1 for the Sudd and less for the Okavango.

A water budget model was developed for the tropical Magela flood plain (Vardevas, 1989), which is located downstream from the Ranger uranium mine in Northern Territory, Australia. Using input data on rainfall and water discharge from Magela Creek, the model provides daily estimates of the volume of surface water on the flood plain, and the rate of discharge at the outlet of the flood plain. In addition to utilizing daily rather than monthly time steps, the components of the storage of the flood water are modelled in a more complex manner than by Sutcliffe & Parks, by means of surface depression storage and wedge and prism stores. The hydrological model involves nine parameters which determine the volume

of surface water on the flood plain. Parameter values, which represent surface storage capacity, interception storage capacity, constants and flood plain albedo to solar radiation, were set by optimizing the agreement between four years of measured and simulated water depths. The model predicts daily estimates of the volume and surface area of inundation and, in addition, the discharge from the flood plain. The model was validated by testing its ability to predict the measured daily water depth at the outlet channel. Comparison of prediction with 16 years of observed data produced an average error of 16%.

Dincer *et al.,* (1987) developed a simple mathematical model of the Okavango Swamp, Botswana, which despite its development beyond the humid tropics, possesses a methodology which is readily transferable to other regions. The model consists of a number of cells along a flow line for each distributary system. The water balance of each cell is calculated, starting with the upstream cell, and the outflow is distributed when there is more than one cell downstream. The model is calibrated using measured discharges, water levels in the

TABLE 7. *Published permeability values for dambo and non-dambo zones.*

	Non-dambo	Dambo
1. Luano, Zambia	$1.5 - 3.0$ m d^{-1}	
2. Chimimbe, Malawi	$0.7 - 10^{-2}$ m d^{-1}	$10^{-2} - 10^{-4}$ m d^{-1}
3. Chizengeni, Zimbabwe	100 m d^{-1}	$10^{-2} - 10$ m d^{-1}

Source
1. Mumeka & Mwasile (1986)
2. British Geological Survey & others (1989)
3. Faulkner & Lambert (1991)

swamp and satellite imagery. The model shows changes in the flow distribution in the Okavango Swamp, and can predict the effect of human-made changes on the inflow and on the swamp.

Evaporation

Stewart (British Geological Survey & others, 1989) describes a technique using satellite data for the estimation of areal evaporation – as applied to a dambo wetland in Zimbabwe. Until recently, the only alternative to assuming that there was no spatial variation in evaporation was replication of the measurement systems. Stewart describes how the combination of point measurement and satellite imagery can be utilized to determine evaporation over larger areas. The experimental area was within the Chizengeni dambo, in the Nyatsime catchment, south-east of Harare, Zimbabwe. Measurements of surface temperature were made with an infra-red thermometer with two sensors, each measuring an area of 0.1 m^2 within 100 m of the central channel of the dambo. Measurements of the components of the energy budget

$$R_n = lE + H + G \tag{4}$$

where R_n is the net all-wave radiation (W m^{-2})
l is the latent heat of vaporization (J kg^{-1})
E is the evaporation (mm s^{-1})
H is the sensible heat flux
and G is the heat flux into the soil (W m^{-2})

were made using the Bowen ratio method. The sensible heat flux can be expressed as

$$H = rc_p g_a(T_s - T_a) \tag{5}$$

where r is the density of air (kg m^{-3})
c_p is the specific heat of air (J kg^{-1} $^{\circ}$C^{-1})
g_a is the transfer coefficient (m s^{-1}) between the surface and the height at which T_a is measured
T_s is the temperature of the surface ($^{\circ}$C)
T_a is the temperature of the air ($^{\circ}$C).

On 8 August 1989, observations were synchronized with the overpass of the Landsat TM sensor. The satellite data were used to determine T_s over an area 200 km^2, based on the cali-

bration of the pixel values using the observed T_s from the infra-red thermometers. Calculation of H was based on measurement of g_a and the difference between T_s and T_a. The former was observed by the meteorological instrumentation. The temperature difference was based on the calculation of air temperature from the surface temperature of a forest stand in the vicinity of the wetland on the imagery – on the basis that the surface temperatures of wooded areas differ by less than two degrees from air temperatures. Once H was determined, the evaporation could be calculated, using the energy balance if measurements or estimates of R_n and G could be attained. Since there were no direct measurements of R_n at the time of the satellite overpass, previously observed values of 450 W m^{-2} for net radiation and 50 W m^{-2} for G were assumed to apply. Therefore, the energy available for the sum of the sensible and latent heat fluxes was 400 W m^{-2}. Over the Chizengeni dambo, the estimated latent heat flux varied between 178 and 284 W m^{-2} with a mean value of 229 W m^{-2}, equivalent to evaporation rates of 3.0, 4.8 and 3.9 mm day^{-1} respectively. The mean value of evaporation over all dambo regions within the 200 km^2 region was estimated to be 3.5 mm day^{-1} and 3.2 mm day^{-1} for the area outside of the dambos. The transformation of Landsat data into images portraying daily evaporation rates (Fig. 8) identifies the dambo edges as the zones of highest evaporation, being approximately 80% of the potential rate at the margins and 64% on the drier dambo slopes. The application of the methodology developed by Stewart for the areal estimation of evaporation has errors in the absolute values of estimated evaporation, largely due to inadequate ground control of both surface temperatures and evaporation rates. However, the capability to identify relative differences in this case between dambo zones and dambo and non-dambo areas, and the potential for greater precision in absolute estimates with improved calibration data, clearly establishes the merit of the methodology for advancing investigations of evaporation losses from wetlands.

Sharma (1988) assessed the relative suitability of the Penman, Thornthwaite and Morton methods (see Bonell with Balek, this volume) in estimating evaporation from the Lukanga swamps of Zambia, as adjudged against evaporation calculated by the water balance method. Although the Morton

TABLE 8. *Different sources of inputs into wetland systems.*

Source of input	Factors controlling relative dominance of input
a. Direct precipitation	i. Density of wetland within catchment
b. Surface runoff from upslope	i. Density of non-wetland portion within catchment ii. Evaporation losses iii. Infiltration properties of hillslope
c. Subsurface runoff from upslope	i. As per surface runoff ii. Groundwater pathways bypassing wetland
d. Channel contributions from upper catchment	i. Transmission losses during confined channel flow ii. Frequency and magnitude of overbank flooding

method was found to be most appropriate for estimating losses from woodland, the Penman potential method can be successfully used to estimate open-water evaporation from central African swamps, with the Morton lake method ranking second.

Seyhan *et al.,* (1983) identified diurnal fluctuations in streamflow which were superimposed on the general recession trend of streamflow from a small research catchment in Zululand, Republic of South Africa. The daily additional reduction was assumed to be the result of greater evaporation losses from marshland adjacent to the river channel. The reduction in flow was calculated for 76 days and related to hydrometeorological variables which could be expected to be correlated with evapotranspirational losses. Factor analysis, correlation analysis and regression analyses were used to develop equations to estimate the daily additional reduction in streamflow. These equations were found to satisfactorily explain the additional losses, with hydrograph stage being the most important variable, explained in physical terms by the water which was available for evaporation in the riparian marshland.

Hydraulic conductivity (permeability)

Table 7 summarizes reported permeability rates for dambo and non-dambo portions of catchments in Malawi, Zimbabwe and Zambia. The presented measurements identify a significant distinction in permeability between wetland and non-wetland portions of the catchments, with implications for the buffering of hydrological pathways and transmission losses. Comparison of the observed permeabilities within Chimimbe and Chizengeni dambos illustrate that there can be orders of magnitude of difference between permeability amongst individual wetlands. Indeed,the Chizengeni dambo soils are equivalent in permeability to the non-dambo portion at Chimimbe and Luano. There can also be considerable local variations in permeability within a single wetland, with variations reported from 0.9 10^{-6} m s^{-1} up to 5.8 10^{-6} m s^{-1} within a distance of 10 metres within the Chizengeni dambo.

Overland flow and vegetation resistance

Kadlec (1990) provides a development in modelling the resistance contributed by emergent wetland vegetation to the flow of surface water, which requires a knowledge of statistical distributions of wetland ground elevation, depth and velocity. Parameters in the model can be estimated with sufficient accuracy from vegetation and soil surveys, when combined with relatively little hydrologic data. Because wetland flows are often in the transition zone between laminar and turbulent, the Mannings equation to flow resistance is deemed inappropriate. Since stems are typically spaced many diameters apart, fluid friction should be computed from the effect of single objects rather than from channel or packed bed equations, such as Mannings, which apply to situations where the bottom drag is controlling. Complications arise from the vertical variation of vegetation density and nonoriented spatial variation of soil elevations. Data from subtropical wetlands are included in the examination of vegetation resistance amongst different vegetation types in this study.

WATER BALANCE OF TROPICAL WETLANDS

The water balance of tropical wetlands is considered here in relation to hydrological inputs into wetland systems, water table regimes and changes in storage; evaporation losses from wetlands compared with non-wetland areas; and impacts of wetlands upon downstream flow regimes. A comparison is made between humid tropical wetland and temperate wetland influences on downstream river flows.

Inputs into wetland systems

The quantification of system inputs is the key to understanding how individual wetlands function and how wetland systems differ. In this respect, the evolution of wetland hydrology in the tropics is handicapped by the rudimentary numerical foundation of hydrological process, fluxes and water balances of the tropical landscape in general. The two principal inputs into

TABLE 9. *Groundwater fluctuations, Chizengeni, Zimbabwe May-Sept 1986. After Dambo Research Unit (1987).*

Zone	Drop in groundwater level (metres)
DAMBO	
Margin	0.35
Upper	0.25
Lower	0.46
INTERFLUVE	
Upper regions	1.64
Mean	1.29

a wetland system are direct precipitation onto the wetland and water from the surrounding or upstream land area. Maltby (1988) uses the terms "meteoric" water and "telluric" water to establish the distinction. Because wetlands are frequently the valley bottom component of the catena, and are rarely located in watershed situations, the telluric inputs are controlled by hillslope catchment processes. Broadly speaking, hillslope water can arrive at the wetland interface as surface runoff or subsurface throughflow at a range of depths in the soil profile. Table 8 summarizes the factors which are likely to control the relative dominance of sources of water into wetlands.

The relative importance of direct precipitation is clearly a function of the density of the wetland within a catchment as a whole. While 100% of direct precipitation over the wetland area, termed W_{dp}, can be considered as an input, precipitation onto non-wetland portions, termed NW_{dp}, is clearly subject to evaporation losses from vegetation and losses from depression and soil moisture stores along the pathway to the wetland. In the most simple case of a small headwater wetland, the sum of the inputs into the wetland, I_w, can be considered as

$$I_w = W_{dp} + (NW_{dp} * c) \qquad (6)$$

where c is a runoff coefficient dependent on the properties of the non-wetland portion of the catchment. For example, the Luano catchments of northern Zambia, which are discussed further in the water balance section of this paper, receive an average annual rainfall of 1,390 mm, and evaporation losses from the non-wetland portions of the catchment are 1,340 from woodland and 1,070 mm from the transition zone. The residual 50 mm and 320 mm are runoff and contribute to streamflow through the wetland system. Even though the wetlands in this area are a relatively small portion of the catchment, being around 5% in areal extent, the high evaporation losses from the wooded slopes means that direct precipitation onto the wetlands is the dominant contributory source of water to the wetland.

The separation of non-wetland contributions into overland flow and subsurface flow is clearly a function of the hydrological properties of the hillslope (see Bonell with Balek, this volume). For example, Walsh (1980) proposes a model in which the relative importance of rapid throughflow and overland flow,

with intermediate processes, is controlled by soil factors, including topsoil permeability, topsoil depth and subsoil permeability, and other factors including topography, geology, rainfall intensity and disturbance by people. In general, however, water balance components from the main vegetation zones throughout the world (Falkenmark, 1989) suggest that the ratio of groundwater to surface runoff is approximately 1:3 in drier subtropical zones comprising savanna vegetation, increasing to 2:3 in wet tropical savanna and wet monsoon forests, rising to 1:1 under perennially wet evergreen forests. Clearly, the understanding of runoff processes on hillslopes is a topic of wider concern than to wetlands alone, and is a topic covered elsewhere in this volume. However, it is clear that the quantification of tropical hillslope processes is a prerequisite to understanding the sources of contributions into wetland systems.

However, not all runoff generated from hillslopes above a wetland will necessarily constitute an input to a wetland. Groundwater pathways may cause water to bypass the wetland, or else to be routed beneath the wetland system and contribute to streamflow downstream of the wetland outflow. For example, on Basement Aquifer terrain in Malawi and Zimbabwe, the existence of quartz stringers and fracture zones can constitute important hydrological pathways which are not necessarily in a perpendicular downslope direction, and a proportion of subsurface flow is believed to pass beneath the dambo clays (British Geological Survey & others, 1989).

The importance of different sources of water for a wetland will have a strong seasonal perspective. Whilst wet season direct precipitation onto the wetlands may be the dominant bulk input by volume, upslope subsurface contributions are likely to be the only significant input during dry seasons. On an event basis, overland flow from significant intense rainfalls may be greater than subsurface contributions, although they may represent only a minor source on an annual basis. Nortcliff & Thornes (1981) identify that the dominant and very rapid response in the stream hydrograph of the Barro Branco catchment near Manaus, Brazil, derives essentially from saturation overland flow on the flood plain. The direct contribution made from the hillslope to the stream hydrograph is mainly in terms of base flow, because of the low significance of the lateral flux components and because the gradient across the flood plain is very small. Only under very dry conditions in the Barro Branco catchment is the channel hydrograph dominated by the hillslope contributions.

Direct precipitation onto wetlands and upslope contributions are likely to be the more important bulk sources in headwater wetlands. In the case of flood plain wetlands, direct precipitation will decrease in significance with the size of contributing catchment above the wetlands. The dominant source of water into flood plain wetlands is likely to be exogenous river flows. River flows can contribute as transmission losses through the channel bank where the bank material is porous, depending on the coincidence of river level and soil moisture deficits in the flood plain soil. It is more likely that overbank flow is the more common source of water into flood plain wetlands, with the volumes depending upon the frequency and magnitude of flows exceeding channel capacity within the river regime.

Water table regimes and changes in storage

Using a network of piezometers at selected sites on the dambo and hand-dug wells on the interfluve, the Dambo Research Unit (1987) compared dry season groundwater fluctuations in the dambo compared with fluctuations in the upper catchment. The results, presented in Table 9, illustrate that groundwater levels on the upper regions of the catchment drop by 1.3 metres on the average, compared with drops of less than 0.5 metres throughout the dambo. Fluctuations within the dambo are more significant in the central portion of the dambo than towards the dambo margins. Using estimates of aquifer storativity, the Dambo Research Unit estimated a change of aquifer volume of $0.36 \text{ m}^3 \times 10^6$ in the dambo, compared with $2.43 \text{ m}^3 \times 10^6$ in the non-dambo zone during the dry season period.

Water level variations in the Chimimbe catchment of Malawi were monitored with a network of piezometers (British Geological Survey & others, 1989) from August 1986 to June 1987. Fluctuations in the dambo bottom are up to 1.5 metres, and the water table is only within 0.5 metres of the surface in the period from November to March, rising to a peak in early December. The water table at the interfluve crest drops by over 2 metres between August and the middle of February, before rising to a peak in the middle of March, three months after the peak in the dambo bottom, and a subsequent drop. Piezometers in intermediate locations upslope exhibit annual fluctuations of approximately 1.5 metres, and a tendency for the maximum water table level to be achieved at a later date further up the slope profile.

Filius (1986) describes water table fluctuations within Ichiumfwa dambo, Zambia within a single wet to dry season sequence in 1985. In March, water tables are at the surface at the dambo margin, at around 30 cm (below surface) in the main dambo grassland zone and at 20 cm in the central zone. From 1st March until 1st September, the water table remains at the surface at the dambo margin and only drops by 10 cm within the subsequent two months. Water table levels drop consistently in the other dambo zones until the end of July and then remain constant at approximately 40 to 50 cm in the other dambo zones.

Balek & Perry (1973) and Mumeka & Mwasile (1986) identify a clear pattern in groundwater fluctuations within the Luano dambos with rainfall. The general pattern is that the water table rises to the surface during the first month after the onset of the rainy season and remains constant throughout the rainy season, keeping the dambo waterlogged. The groundwater level starts to fall more than one month after the end of the rainy season. Water tables within the Luano catchments, therefore, are more responsive and more persistent than the regime described for Chimimbe dambo. Whilst rainfall

depths are likely to be a factor in determining water table regimes, Acres *et al.,* (1985) report variations attributable to dambo soil type, drawing a distinction between predominantly sandy soils and predominantly clay soils. Sandy soils are saturated earlier in the wet season and are more liable to rapid fluctuations, compared to clay soils which rise and fall more slowly.

Evaporation losses from wetland systems

Losses from the wetland system can either be as evaporation or as contributions to river flow. Gopal *et al.,* (1982) identifies a key question in wetland hydrology to be whether the evaporation losses of a catchment are increased or decreased by the presence of wetland vegetation, with implications for flow regimes. Table 10 presents data from recently published reports in different countries which enable Gopal's question to be addressed.

In the three examples dealing with dambo wetlands, annual evaporation losses from dambos are lower than losses from other catchment vegetation types. The data cited by Balek & Perry (1973) indicate that woodland evaporates nearly three times more water than the dambo grasslands. The magnitude of this distinction is not duplicated in the Bua catchment, where the dambo losses are more similar to the losses from the Zambian dambos (640 mm compared with 500 mm) than the woodland losses, which are notably lower in Malawi (700–750 mm) than in Zambia (1,000–1,350). The lower losses from the dambos than from the woodland are despite the higher bulk availability of water in the dambos, which receive not only direct precipitation but also contributions from the upslope vegetation communities.

Conversely, the actual evaporation losses from the flood plain swamps, which lie in the middle and lower reaches of rivers, are higher than from the non-wetland portions. This is because of the supply of water for evaporation from exogenous river flooding. Unlike the headwater dambos where there is a water availability control upon the evaporation process, the example of the Lukanga Swamp in Zambia illustrates that swamps can evaporate at rates equivalent to potential rates when water availability is less of a limiting factor. Balek (1977, 1989) cites two further examples of wetlands in middle reaches of rivers where evaporation from the wetlands is higher than from non-wetlands, resulting in river outflows being lower than river inflows into wetland systems. In the Luapula catchment of the Upper Congo, evaporation from non-wetland areas is 890 mm compared with evaporation of up to 2,180 mm from the Bangweulu swamp area. This represents a water loss to the outflowing Luapula River of 60% of the inflow. In the Kafue River, a tributary of the Zambezi River in Zambia, evaporation from non-wetland areas is 785 mm and 1,005 mm from the Kafue Flats, representing a water loss to the river caused by the passage of the river through the flats of 4% of inflows.

Evaporation from the dambo grassland exhibits strong seasonality, with actual rates as high as 75% of potential during

TABLE 10. *Comparison of published average annual evaporation losses from wetland and non-wetland communities.*

Balek & Perry (1973) at Luano catchments, Zambia			
Mean annual rainfall (1967–1969)	1 387 mm		
Potential evaporation E_o	1 722 mm		
Actual evaporation E_t from			
Dambo	491 mm	E_t/E_o	0.29
Brachystegia woodland	1 336 mm	E_t/E_o	0.78
Transitive zone	1 067 mm	E_t/E_o	0.62
Oyebande & Balek (1989) at dambo (Zambia)			
Mean annual rainfall	1 330 mm		
Potential evaporation E_o	1 710 mm		
Actual evaporation E_t from			
Non-wetland area	1 320 mm	E_t/E_o	0.77
Dambo	1 075 mm	E_t/E_o	0.63
Smith-Carrington (1983) at dambos in Bua catchment, Malawi			
Mean annual rainfall	900 mm		
Potential evaporation E_o	1 900 mm		
Actual evaporation E_t from			
Wooded interfluve	760 mm	E_t/E_o	0.40
Fallow interfluve	692 mm	E_t/E_o	0.36
Dambo	640 mm	E_t/E_o	0.34
Sharma (1988) at Lukanga Swamp, Zambia			
Mean annual rainfall	1 250 mm		
Potential evaporation E_o	1 790 mm		
Actual evaporation E_t from			
Swamps	1 800 mm	E_t/E_o	1.01
Non-swamp	950 mm	E_t/E_o	0.53
Faulkner & Lambert (1991) at Chizengeni dambo, Zimbabwe			
Mean annual rainfall	922 mm		
Potential evaporation E_o	1 550 mm		
Dry season evaporation from			
Dambo	$0.66 – 2.25$ mm d^{-1} *		
Interfluve	0.45 mm d^{-1}		

* 0.66 mm d^{-1} from dry dambo bottom

1.12 mm d^{-1} from lower dambo

2.25 mm d^{-1} from upper dambo

the wet season, but well below 10% in the dry season (Balek, 1977). It is during the dry season that Stewart has illustrated the wetland losses to be higher than from the non-wetland zone in Zimbabwe. Those remotely sensed data are supported by the observed dry season evaporation data at Chizengeni of Faulkner & Lambert, which further identifies the upper dambo as the zone of highest evaporation losses.

The seasonal distribution of evaporation in the Bangweulu Swamp exhibits considerably less variability throughout the year than the dambo type of wetland (Balek, 1977) because of the more constant inputs. Whereas 7% of evaporation occurs in the dambo wetland in the dry season (May to September), 25% of evaporation occurs in the equivalent period from the Bangweulu Swamps.

Impacts of wetlands upon downstream river flows

An important distinction exists between the impact of headwater wetlands and that of wetlands in the middle and lower reaches of rivers. In the former, the influence of wetlands upon catchment response is upon the generation of runoff and the buffering of hillslope processes. These factors assume a much lesser importance in the middle and lower reaches where it is the effect of

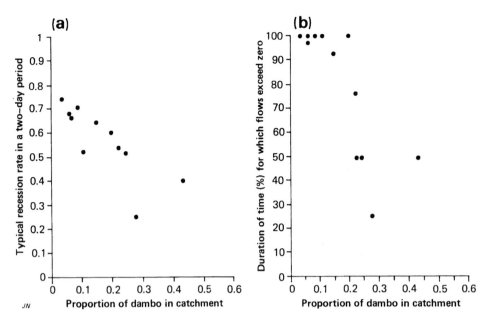

Figure 9: Increase of recession rates with higher proportion of dambo headwater wetland (a) and corresponding reduction of low flows (b) in Zimbabwe. Recession rates (a) are indexed by the 80th percentile from a cumulative frequency curve of 2-day recessions. Low flows (b) are indexed by the frequency of flows greater than of zero. After Bullock (1992a).

routing exogenous flows through a wetland which influences the downstream flow regime. The impact of the two wetland types is discussed further in terms of the impact exerted upon total runoff, storm runoff and floods and low flows.

Headwater wetlands The data presented in Table 10 have identified lower evaporation losses from headwater wetlands compared to non-wetland areas, which must be to the benefit of streamflow volumes in catchments with higher proportions of wetlands. However, the consistent suggestion from smaller catchment studies of higher streamflows with higher proportions of headwater wetlands has not been identified in regional river flow data. Bullock (1992a) analyses runoff and evaporation data from 109 catchments in Zimbabwe containing different proportions of headwater dambos and concludes that dambos are an indiscriminatory factor in determining volumes of catchment runoff. Balek & Perry (1973) conclude that the annual volume of runoff from the four study catchments at Luano, Zambia are independent of their proportional extent of dambos.

In relation to storm runoff, Dubreuil (1985) identifies different runoff processes operating between wetland and non-wetland soils on granites in humid tropical environments. The ferruginous soils of the tablelands and slopes are dominated more by subsurface processes while the hydromorphic soils of the valley bottoms are associated more with the generation of saturation overland flow. Similarly, in his scheme of the factors affecting storm runoff processes in humid tropical areas, Walsh (1980) identifies increasing extent of catchment flats within a catchment being associated with the domination of runoff generation less by rapid throughflow and more by localized and widespread saturation overland flow. The models of

Dubreuil and Walsh are both consistent with the model of variable source areas of runoff generation (Hewlett & Hibbert, 1967; Dunne, 1983).

Dubreuil (1985) reports that the minimum rainfall to produce runoff on tropical hydromorphic soils is between 9–12 mm, exceeding the minimum rainfall required to produce runoff on other soil types on schist geologies. The same minimum rainfall of 9–12 mm for poorly drained soils on granites is, conversely, lower than the minimum rainfall for other granitic soil types, for example, 15–20 mm on ferrallitic soils (Dubreuil, 1985). Headwater wetlands, therefore, can be identified as introducing heterogeneity into the runoff generation process for storm events. The data presented by Dubreuil suggest that wetlands will respond less frequently to rainfall events than non-wetland soils on schist geologies, and will respond more frequently on granitic geologies.

The dynamic aspect of runoff generation of wetlands is also illustrated by runoff generation data from different studies collated by Dubreuil (1985). The data identify that on hydromorphic vertisols, the minimum rainfall required to generate runoff at the middle and end of the rainy season is less than that required at the beginning of the wet season, as saturation conditions are achieved. Because the minimum rainfall remains more or less constant on the permeable granites, the wetlands are likely to exhibit a more seasonally-varying response to rainfall events than the granite regolith.

In the case of headwater wetlands in West Africa, Albergel (1989) identifies that while the interfluves soils are very permeable (5×10^{-5} to 11×10^{-7} m s^{-1}), permanent flooding of the valley bottoms reduces infiltration to a minimum. The flooded wetlands are calculated to have an annual coefficient of runoff of 80%, and 100% in the case of 10-year flood events (Olivry

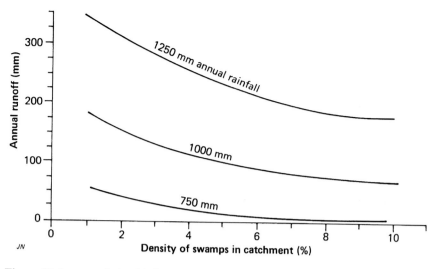

Figure 10: Inverse relationship between annual runoff and the proportion of catchments covered by swamps. After Balek (1989).

& Dacosta, 1984). However, the downstream impact of preferential runoff generation on the wetlands is diminished in the large alluvial channels.

Balek & Perry (1973) identify that a higher proportion of headwater wetlands generate a greater proportion of stormflow. On the other hand, Bullock (1992a) concludes that dambo extent is an insignificant factor in determining specific mean annual flood discharges, the coefficient of variability of annual maximum floods and the shape of standardized flood frequency curves amongst 49 catchments with different dambo densities in Zimbabwe.

Dambos are reported to have a stabilizing effect on the regime of the river headwaters, delaying the surface runoff process (Oyebande & Balek, 1989). Surface runoff in the Luano catchments does not commence until the dambos have become saturated in mid-January. Wetland grasses have been attributed a role in the attenuation of surface runoff because of the high resistance which they offer to overland flow (Balek & Perry, 1973; Mumeka & Mwasile, 1986). Furthermore, the retention of surface runoff in depression stores associated with cracks in montmorillonitic wetland soils is another factor in the retardation of runoff. However, although dambos are reported to exert a stabilizing effect on the regime, there is no significant relationship between the coefficient of variation of either the mean annual flood or total annual runoff with dambo density amongst catchments in Zimbabwe (Bullock, 1992a).

In a review of 12 conceptual reports of the impact of headwater wetlands on streamflow response from 6 different countries in southern Africa, Bullock (1992b) identifies that the one impact upon which there is apparent consensus is the increase in the duration of base flow from groundwater storage. Other studies based on observed hydrological data portray a wider range of conclusions (Bullock, 1992b). However, in reassessing the empirical data from past studies, Bullock establishes that no foundation exists to support the concept of dambo wetlands storing and releasing water to the benefit of dry season

flows. Instead, by integrating data from different methodologies, scales and countries, a case can be established for dambos to be considered as reducing low flows under certain circumstances (Fig. 9). These circumstances depend upon the associated geology and regolith characteristics, with those permeable regoliths which yield dry season base flow contributing to evaporation at the dambo margins.

Flood plain wetlands Balek (1989) demonstrates an inverse relationship between the annual runoff from central African catchments and the proportion of the catchment covered by swamps (Fig. 10). Evaporative losses from swamps are estimated in the case of the Sudd to be 50% of the total Nile input (Hurst, 1933), to be 60% of the total inflow to Lake Bangweulu (Balek, 1977), 85% from the Okavango swamps (Wilson & Dincer, 1976), and 4% from the Kafue flats and 8% from the Lukanga swamps in Zambia (Balek, 1977). Flows from the Lukanga swamp at Mswebi are estimated to be 77% of the flows which would occur in the absence of the swamps (Sharma, 1988). Sellars (1981) calculates open-water and swamp evaporation losses to be approximately 45% of the total runoff in the Upper Yobe River of northern Nigeria. Hill & Kidd (1980) developed a regional rainfall-runoff relationship from 38 catchments in Malawi and concluded that the impact of the flood plain dambos is to reduce total annual runoff by 6.4 mm for every 1% of the catchment that is dambo, an impact different from the headwater dambos of Zambia and Zimbabwe.

Drayton *et al.* (1980) propose – from a regional analysis based on 30 catchments in Malawi – that the effect of higher dambo densities is to reduce the magnitude of the mean annual flood peak. This reduction of flood magnitudes by the flood plain dambos is different from the indiscriminatory effect of headwater dambos in Zambia and Zimbabwe, and is attributed to the greater capacity for overbank flood plain storage introducing a more sluggish wet season response.

Ledger (1969) identifies differences in dry season flow regimes between rivers in West Africa possessing extensive flood plains (Niger, Logone and Black Volta) and those rivers which do not (Bia, Milo, Alibori, Oueme, Senegal and Benue). The shallow slopes of many rivers in the northern part of West Africa mean that they are unable to transport contributions from their headwater areas, and consequently large shallow flood plain lakes are formed. This storage process has several important effects on the dry season flow regimes of the rivers whose downstream reaches are in humid tropical areas. Firstly, it flattens and elongates the flood wave so that the length of the period of low base flow diminishes with distance downstream. Secondly, it delays the rate at which the flood peak moves downstream so that the early part of the dry season is a period of rising, rather than falling, discharge. Thirdly, it increases the amount of base flow so that the minimum flow in these rivers exceeds the minimum flow into them from the headwater areas. As a result, over 50% of total flows in these basins occurs during the dry season, although there are very high evaporation losses from the flood plains within the arid and semi-arid inland reaches.

Comparison of humid tropical wetland and temperate wetland influences on downstream river flows The status of knowledge concerning the hydrological significance of wetlands in temperate regions is similar to that in humid tropical regions. There are few quantitative studies, a cluster of research activity in one region – the glaciated northern states of the USA – and again a widely-held belief that wetlands maintain low flows despite contrary evidence (Carter *et al.*, 1978). In temperate regions, as in humid tropical regions, wetlands have been assigned the dual roles of flood attenuation and production. In general, it is reported that temperate wetlands attenuate flood peaks and storm flows by temporarily storing surface water. For example, Novitski (1978) illustrates that the net effect of wetlands or lakes is to smooth (or attenuate) flood peaks by as much as 60% to 80% depending on the proportion which they represent in a particular catchment. The location of the wetlands within the basin is recognized to be important in influencing flow distribution and the time/flow response, especially the role of tributary wetlands in desynchronizing tributary and main channel flood peaks. The importance of temperate wetlands in flood runoff generation has been identified by Verry & Boelter (1978).

The arguments linking base flows and wetlands is also equivalent in temperate and humid tropical regions. Carter *et al.*, (1978) state that although it is often suggested that wetlands store water and release it during dry periods to maintain low flows, wetlands do in fact reduce low flows, and cite several examples to support their case. However, Newson (1981) does present contrary data from Wales which illustrates higher low flows from upland catchments containing greater expanses of peat. Littlejohn (1977) shows that swamps help to stabilize aquifer storage and discharge into Naples Bay, Florida.

The behaviour of humid tropical wetlands does not, therefore, appear to be fundamentally different from wetlands in other regions of the world in the manner in which they influence floods and low flows in rivers. In both temperate and tropical regions, there is conflicting evidence about positive and negative impacts on both aspects of the flow regime.

CURRENT ISSUES IN TROPICAL WETLAND HYDROLOGY AND WATER RESOURCES

Resource assessment

A prerequisite to the consistent assessment of wetland resources at an international scale is the adoption of a standardized classification for wetland systems. The standardized classification should address the issues of overlap, exclusion and the adoption of more multi-disciplinary criteria. Dugan (1990) reports that regional wetland inventories have been completed for the Western Paleartic (Carp, 1980), Latin America and the Caribbean (Scott & Carbonell, 1986), Asia (Scott, 1989) and Africa (Mepham & Mepham, in press). IUCN (1990b) reports that an inventory of the wetlands of Oceania and Australasia has been initiated.

Human impacts on wetlands

Many wetlands throughout the world are threatened by human impacts. Dugan (1990) gives several examples from around the world of wetland losses and deterioration in wetland functions. Table 11 illustrates examples from the humid tropics of loss of hydrological function, which may have consequences for other wetland values which depend upon the hydrological regime. In some cases, alteration of the wetland function may have benefits to other water resource demands within the catchment.

Often it is the problem of wetland loss that gives an impetus to research, or to the collation of information. The Kafue Flats in Zambia are an excellent example of threatened wetland loss attracting considerable research attention, not only in assessing hydrological impacts but also the implications for wildlife (Obrdlik *et al.*, 1989). The Kafue Hydroelectric Scheme, closed in 1972 and flooding an area of 800 km^2 permanently, has changed the natural flow pattern of the river. Despite temporal changes in flow and the high evaporation on the Kafue Flats, some 7,000 km^2 of wetland has remained after the regulation. It is generally agreed that the amount of evaporation from the flooded area on the Flats is considerably higher than that from the non-flooded areas. Sharma (1984) estimates 3.8 mm d^{-1} from December through March in Zambian headwaters, compared to 5.1 mm d^{-1} on the Kafue Flats (Ellenbroek, 1987). The flooded area of the Flats is estimated to oscillate between 622 and 4,745 km^2 during June with operation of the Itezhi-tezhi Reservoir (Turner, 1984). Flows from the scheme are in excess of the natural flow regime, in terms of both the total flow and of each monthly total, resulting in the formation of permanent lagoons where temporary habi-

TABLE 11. *Loss of wetland hydrological values to human impacts.*

Threat	Example
Groundwater abstraction	Groundwater abstraction for irrigation of crops on Chizengeni dambo in Zimbabwe lowers the level of catchment aquifer by up to 20mm. Extension of the irrigated area to 10% of the wetland area could lower the table by up to 270mm, with implications for early wet season base flow (Dambo Research Unit, 1987). This implication is adjudged insignificant against natural water fluctuations and against the social benefits which would accrue.
Dams	Construction of a hydroelectric dam on the Turkwel River in Kenya is likely to have significant adverse impacts on *Acacia tortilis* riparian forest as flood attenuation will reduce recharge of the flood plain aquifer (Adams, 1989).
	Construction of the Itezhi-tezhi and Kafue Gorge Reservoirs on the Kafue river, in Zambia has reduced conductivity and concentrations of sulphates and chlorides, threatening fish populations and dependent wildlife on the Kafue flats (Obrdlik *et al.*, 1989).
Drainage	Russell (1981) identifies the drainage of headwater swamps as a key area of concern, with little foundation for understanding the deterioration in the river flow regime likely to be associated with wetland drainage for arable cropping.
Gullying	McFarlane & Whitlow (1991) identify that wetland gullying is extensive in some parts of Zimbabwe and Malawi. However, because gullying is attributed more to subsurface water breaching the wetland clay than to overland flow, management policy based upon the wetland instead of the whole catchment is shown to be inappropriate.
Urbanization	The construction of urban areas in the Avondale Stream basin in Harare, Zimbabwe has led to the replacement of dambo channels of density of between 0.35 and 0.80 km km^{-1} with a channel network including urban storm water drains of 3.15 km km^{-1}, with wetland loss caused by average channel widening of 1.7 times and bank erosion rates of 0.33 m year^{-1} (Whitlow & Gregory, 1989)

tats existed before. Total evaporation in a typical dry year is identified as increasing from $1 \times 850 \times 10^6$ m^3 prior to construction (1963/64 water year) to $7,872 \times 10^6$ m^3 (in 1972/73). In response to possible adverse environmental impacts to the unique flood plain ecosystem, ecologists founded the Kafue Basin Research Committee in 1967 at the University of Zambia. It was proved that the spatial and temporal variations of the flood regime were to the detriment of the abundance of wild herbivores, especially the endemic Kafue lechwe whose numbers were reduced by 50%, and some water birds which were affected by the inhibited growth of water-lilies upon whose seeds they feed. Additionally, fish production was reduced, possibly because of inaccessibility of fish camps or changes in the physiochemical status of the regulated water.

Wetlands and food security

Irrigation of wetlands has a crucial role to play in the attainment of food security. The relative importance of sustainable small-scale irrigation in wetlands over conventional irrigation practice has recently been emphasized in Zimbabwe (Faulkner & Lambert, 1991). Conventional irrigation in Zimbabwe's commercial farming sector is commonly based on artificial farm dams within wetlands. Within the informal sector of the Communal Lands, areas of cultivation successfully using

microscale irrigation on wetlands exceed areas conventionally irrigated by a ratio of three to one. On conventional irrigation it is frequently necessary to grow cash crops rather than food crops in order to pay irrigation fees, with the consequence that malnutrition can be higher than in micro-scale irrigated areas. The amount and variety of crops allows for an improved nutritional status of families with dambo gardens. Men are usually the direct beneficiaries of conventional irrigation schemes, but women benefit more from small-scale schemes through the sale and use of the crops.

With proper environmental safeguards, the present level of micro-scale irrigation on dambos in Zimbabwe could be increased three-fold. The safeguards are based on the six recommendations that: (a) dambos should not be permanently drained, (b) cultivation should not take place on the dambo bottom or around the edges, but on the permanently wet upper dambo areas where organic topsoils are not less than 10cm thick, (c) a grassed strip, at least 60 metres wide, running along the bottom of the dambo, should be left uncultivated on intensively used dambos, (d) broad, grass-covered alleys connecting the bottom grass strip to the upper catchment must be maintained at regular intervals to act as protected drainage lines for storm flow during heavy rains, (e) the total cultivated area of the dambo should not exceed 30% of the total dambo area or 10% of the catchment area, whichever is the smaller in

TABLE 12. *Protecting America's wetlands: an action agenda. After National Wetlands Policy Forum (1988).*

I.   Creation of a Coherent Framework
     * Establishment of a National Goal
     * Planning for Wetland Protection and Management
     * Implementation of National Recommendations
II. Promoting Private Stewardship
     * Public Education
     * Increasing Incentives for Private Protection – Tax Incentives, Subsidies, Relocating Development
III. Improving Regulatory Programs
     * Defining Wetlands Consistently
     * Refining Existing Regulatory Programs
     * Improving Mitigation Policies
     * Filling Regulatory Gaps
IV. Establishing Government Leadership
     * Reducing Government Alterations
     * Reducing Government Inducements
     * Acquiring and Managing Wetlands
V.   Providing Better Information
VI. Restoring and Creating Wetlands
VII. Financing Wetlands Protection and Management

order to protect groundwater resources, and (f) motorized pumps can be used, provided the area cultivated and the number of wells is controlled, but the use of watering cans or hand pumps is very unlikely to significantly affect dambo water resources.

Boundary problems for the modeller

Boundary definition is recognized by Newson (1988) to be one particular problem facing the modeller seeking to simulate wetland hydrology. Problems are associated with selection of the physical parameter to be used for boundary definition; hydrologic parameters, vegetation, or pedologic properties will each provide a unique delimitation. Even when a single or suite of quantifiable descriptors for the wetland boundary are accepted, the modeller must account for the temporal variability of this boundary.

Boundary problems for the wetland manager

Successful management of wetlands that is based on jurisdictional procedures awaits the defensible delineation of wetland boundaries. Equally, boundaries associated with wetland values pose dilemmas to those with management aspirations (Mitchell, 1990). Boundary problems arise in situations in which two or more management interests overlap. For example, a situation may occur when one agency of government wishes to preserve or expand wetlands in order to enhance wildlife habitat whereas another agency wishes to drain them in order to increase agricultural production. If these interests are to be integrated into a systematic management strategy, it is essential to develop mechanisms through which different values and interests can be identified and articulated, and then decisions made which reflect these different and often conflicting aspirations.

Wetlands and climate change

The possible impacts of climate change upon wetlands has become one element of the Wetlands Programme in the Global Change initiative, which will focus upon the problems of flood plain management, impact upon river systems, and the effects on selected biotopes (IUCN, 1990a).

Political hydrology

Hollis (1988) argues that it is not enough for hydrologists to be concerned with the refinement of the various methods of assessing the hydrological impacts of developments on wetlands. Rather, hydrologists should become more involved in wetlands and their sustainable utilization and management. The traditional exclusion of hydrologists from this sphere has tended to result in the advancement of single function priorities because of the dominant stance of irrigation engineers, agriculturalists and aquaculturalists. The case for an elevated involvement by hydrologists is founded, Hollis argues, on a systems approach to the hydrological cycle within whole catchments. To achieve a more substantial involvement in research, management, legislation, training, EIA (Environmental Impact Assessment) and economic analysis, Hollis advocates the development and implementation of "political hydrology" as the necessary associate of "scientific hydrology".

TABLE 13. *Examples of applications of remote sensing to tropical wetland studies.*

Cook (1981)	Application of photographic interpretation to an integrated resource survey in the Lower Rufiji flood plain, Tanzania.
Whitlow (1984)	National survey of dambo density from panchromatic aerial photography in Zimbabwe.
Rollet (1986)	The wetlands of Guadeloupe in the Lesser Antilles are typified through vegetation interpretation on panchromatic 1:20 000 aerial photography.
Estrin (1986)	The delineation and classification of inland wetlands utilizing false colour infra-red imagery
Miller (1986)	The quantification of flood plain inundation by the use of LANDSAT and Metric Camera information in Belize, Central America.

STRATEGIES FOR THE ADVANCEMENT OF WETLAND HYDROLOGY AND WATER RESOURCES IN THE HUMID TROPICS

Getting hydrology onto the wetland management agenda

The technical content of this review has illustrated a strong imbalance in the state of knowledge of the hydrology of wetlands in Asia and South America compared to a relative abundance, albeit recent, of data from the African continent. Results of the African studies have challenged the perception of the nature and significance of wetland functions, and the considerable diversity that can exist between different wetland systems. Such results must have implications for future wetland management, and must surely initiate a reassessment of current water resource management thinking in many African countries with significant wetlands. The need for strategic hydrological research to precede sustainable wetland management policies is clear. There exists a potential danger that inappropriate management schemes will be implemented if hydrological research is subordinated to unproven assumptions.

Once hydrological and other physical functions are established, then the political dimension must come to the fore, with the action agenda (National Wetlands Policy Forum, 1988) for protecting the United States of America's (USA) wetlands (Table 12) representing a potential blueprint for the adoption by other nations.

Remote sensing

Monitoring of hydrological processes is complicated by the fact that individual wetlands can represent very large areas, smaller headwater wetlands can be distributed over wide areas, and hydrological processes can be very dynamic. These characteristics place demands of spatial resolution, extensive coverage and relatively short duration time-scales upon data collection, which means that the major advances in wetland hydrology will be derived from the wider application of remote sensing techniques. Panchromatic, colour, infra-red and thermal infra-red photography have all been used successfully in wetland mapping (Engman & Gurney, 1991). The improved resolution of sensors broadens the applications of satellite data, especially for smaller wetlands. Table 13 pre-

sents examples of the application of remote sensing to wetland studies in the humid tropics.

Many inventories have been carried out by using vegetation characteristics to delineate wetland boundaries from aerial photography. One current research impetus is towards delineating wetlands from satellite data. An example of the application of satellite imagery for wetland delineation for regional water resource assessment is described by Drayton (1986). Landsat MSS imagery was used to define the aerial extent of dambos in undisturbed and cultivated catchments in the Dwanga and Bua catchments of Malawi. Using false colour composites based on Landsat Bands 4, 5 and 7, dambo boundaries could be interpreted by tonal differences due to the grass cover and moist soils of the dambos contrasting with the forested interfluves of the Dwanga catchment and the dry bare soil of the Bua catchment. Drayton estimates that the dambo areas calculated from the remotely sensed imagery were 110% of those areas calculated from dambos plotted on 1:50 000 maps. Proportional areal extent of dambo (*DAMBO*) from satellite imagery was combined with estimates of stream frequency (*STMFRQ*) and catchment area (*AREA*) from equivalent satellite sources within multiple regression equations for the estimation of the mean flow (*ADF*) and mean annual flood (*MAF*) at ungauged sites throughout Malawi. The best regression equations, using only those variables obtainable from satellite data, were:

$$ADF = 0.035 \, AREA^{0.75} \, STMFRQ^{0.41} \, DAMBO^{1.000} \quad R^2 = 81\% \quad (7)$$
$$\text{s.e.e.} = 0.24$$

$$MAF = 7.52 \, AREA^{0.37} \, STMFRQ^{0.72} \, DAMBO^{6.75} \quad R^2 = 40\% \quad (8)$$
$$\text{s.e.e.} = 0.39$$

Units: *ADF* in $m^3 \, s^{-1}$, *MAF* in $m^3 \, s^{-1}$, *STMFRQ* = network junctions per km^2, *DAMBO* is dimensionless. *DAMBO* is adjusted by the addition of 1.000 to all catchment values prior to the application of the logarithmic transformation.

Current research emphasis is directed towards the automatic delineation of wetlands in Malawi, using computer-based supervised classifications. Beyond the definition of wetland boundaries, Engman & Gurney (1991) describe the application of Landsat data to estimate wetland water volumes and to monitoring human-induced changes in wetland systems. Further advances are to be made in the extrapolation of key hydrological variables, as described in the case of the extrapolation of point measurements of evaporation, especially in the measurement of soil moisture data and vegetal biomass.

In the case of surface water bodies, Engman & Gurney describe the application of remote sensing to the measurement of morphometric indices of lakes, for example, length and width characteristics, orientation, shoreline properties and the area of water surface for different elevations. Estimation of water surface area can be applied in conjunction with an area/volume relationship to estimate water storage and its variations over time.

Integrated wetland studies

Significant advances in recent years in the hydrology of tropical wetlands have been made by two multidisciplinary studies. The Dambo Research Unit at Loughborough University combined staff from the Water Engineering and Development Centre, the Department of Geography and the Department of Civil Engineering with the University of Zimbabwe Civil Engineering Department to assess the use of dambos in rural development in Zimbabwe. By combining soil science and hydrology with socio-economic studies, the group was able to conclude that the adverse environmental impacts of increasing dambo irrigation were minor but yielded significant improvements to rural livelihoods. However, the group concluded that to achieve the identified potential of a five-fold increase in dambo cultivation, there is a need for clarification of national policy regarding restrictive legislation and for assistance with the provision of agricultural advice and credit.

The Basement Aquifer project (1984–1989) of the British Geological Survey & others (1989) combined expertise in the fields of hydrogeology, geomorphology, geochemistry, structural geology, geophysics, hydrology and remote sensing to study the system of an aquifer which outcrops widely in Africa, Brazil and South East Asia. A report on this project contributes good quality data and conclusions about wetland and regolith behaviour with bases founded in diverse disciplines. For example, groundwater recharge estimates in Malawi and Zimbabwe by base flow separation, chloride balances of well waters and base flow, seepage loss, flow net analysis and water level changes each have errors associated with the technique, but together they provide consistent estimates of between 8% and 16% of discharge. Furthermore, the need to relate hydrological data to the explanation of data series from other disciplines forces a more deterministic framework to evolve than might be achieved within a single discipline.

Reflecting the diversity of interests in wetland resources, a interministerial committee steering the Wetlands Conservation and Management Programme in Uganda brings together 15 different ministries. In the hydrological sphere, the project is concentrating upon rice agriculture, urban water supply and waste water treatment, with a strong emphasis on sustainable development.

The ongoing Wetland Utilization Research Project in West Africa combines international expertise from different disciplines; the French organization ORSTOM in the sphere of hydrology, the Dutch University of Wageningen in agronomy and the Belgian University of Gembloux in the application of remote sensing techniques, coordinated through the InterAfrican Committee for Hydrological Studies.

International cooperation

The new three-year plan of the Asian Wetland Bureau, which commenced in 1989, contains four main thrusts: biological diversity, institutional strengthening and awareness, water resources and environmental management and policy. The impetus in the sphere of water resources is aimed at the evaluation of the hydrological properties and physical functions of wetland systems and the impact of human intervention, but initial studies are restricted to Malaysia. However, studies of biological diversity under this programme have included coordination of such work in Malaysia, Philippines, Indonesia, Brunei, east China, South Korea and Vietnam. In time, it is hoped that the hydrological research work will gain such extensive regional coverage.

IUCN (1990b) reported that in July 1990, the 10 countries of the Southern African Development Coordination Conference (SADCC) commenced the initial phase of a regional wetland conservation project which will extend from Namibia to Tanzania. Funded by the Norwegian NORAD and the Finnish agency FINNIDA, the initial phase of the programme aims primarily to focus upon a survey of wetlands in the SADCC region, building national, regional and international awareness of the importance of wetlands and their appropriate use, and the development of a regional policy and an action programme for the conservation and wise use of the region's wetlands.

Elsewhere, SI-A-PAZ is an international programme to undertake a series of conservation activities in the watershed of the San Juan River along the border between Costa Rica and Nicaragua. The SI-A-PAZ collaborative project will address the issue that in Costa Rica, some of the wetlands, which are protected within national parks, are under serious pressures from encroachment, as well as assessing wetland resources in Nicaragua. One resolution made at the RAMSAR Convention on Wetlands at Montreux, Switzerland in 1990 was the establishment of a wetland conservation fund "to provide assistance to developing countries, upon official request from a competent national authority for activities in furtherance of the purposes of the (RAMSAR) Convention." It is envisaged that this fund will be fully operational in 1991.

CONCLUSIONS

There has been too little research into the hydrology of tropical wetlands for the development of a truly international perspective on the issue. However, much has been learnt about the hydrology of southern African wetlands as a result of mostly recent research initiatives. In combination, these studies have ensured that this region has become the fulcrum of tropical wetland research. The data that have become available have allowed perspectives to be formed on the water balance, water table regimes, and losses from wetland systems. However, it is clear from comparison of the results that there is much diversity amongst the hydrology of these wetland systems, as illustrated by examples of studies of evaporation, permeability, water table fluctuations and influences upon flow regimes. Current initiatives in wetland research in West Africa and the South East Asia regions will broaden the data base in the near future.

Advances have been made in the techniques available to the wetland hydrologist, including models for the prediction of flooded areas, evaporation modelling, permeability measurement and modelling vegetation resistance. Replication of such techniques in different regions will assist in the transfer of conclusions, and should replace the current situation of comparing results from different experimental designs.

If there is to be a hydrologic foundation upon which to construct future wetland policy, then there must be an extension of strategic research into tropical wetlands. This is especially applicable to the South American and Asian continents where tropical wetlands are most extensive, but results are sparse. Research needs for identifying and quantifying hydrologic functions of wetlands (Carter *et al.*, 1978) include: (a) improving existing measurement techniques, (b) accurate measurements of all hydrologic inputs and outputs to representative wetland types, (c) quantifying the soil-water-vegetation relationships in wetlands, (d) long-term, in-depth wetland studies, and (e) developing wetland models. Application of remote sensing techniques and the adoption of integrated studies will further the development of wetland hydrological science.

Modelling of wetlands themselves poses no insurmountable hydrologic problems, although boundary problems introduce major difficulties. Significant steps can be advanced if existing hydrological expertise in data collection, analysis and model development gathered in other tropical biomes were to be redirected. In this regard, however, the relative importance of the instrumentation and modelling of wetland sites must be compared against forest or agricultural sites, since tropical wetlands are not unique in requiring expanded research and basic data collection.

Tropical wetlands have been shown to function differently, according to their location within a river catchment, and do not differ in this regard from temperate wetlands. Headwater wetlands differ from main river wetlands in their sources of water, and in their influence upon downstream flows. Hydrological processes amongst headwater wetlands alone have been shown to vary considerably. Significant advances in the extrapolation of process results and knowledge will be best achieved by the development of a classification which incorporates key physical properties to replace existing schemes that are based upon qualitative description. With detailed registers of vegetation species for the major wetland types, the tropical wetland biolo-

gist is far ahead of the hydrologist who lacks an adequate framework for classification. A major step forward would be through the linking of hydrological properties with soil properties, and thence the extrapolation of hydrological function through internationally adopted soil classifications.

The stimulus to further understand the hydrological science of tropical wetlands will emerge as increasing pressures from wetland destruction, food security and climate change threaten wetland values. The role of the hydrologist in future wetland issues is sign-posted and encouraged by Hollis (1988), as follows:

It is time for the hydrologists and hydrogeologists to become more involved in wetlands, and their sustainable utilization and management. Irrigation engineers, agriculturalists and aquaculturalists, for example, have tended to dominate the consideration of water in wetlands and they have advanced the view that a single function is the proper use of wetland resources. There have been so many problems with the environmental impact of such developments that, today, it is important that all of the functions and values of the wetland are appreciated, conserved and utilized. Hydrologists and hydrogeologists have a specially relevant view of wetlands, based on a systems approach to the hydrological cycle within whole catchments, which ought to be more in evidence in wetland management.

REFERENCES

Academy of Finland (1988) Proceedings of the Int. Symp. on the Hydrology of Wetlands in Temperate and Cold Regions. Joensuu, Finland, 6–8 June 1988. 2 vols., 320 & 105, Academy of Finland.

Acres, B. D., Blair-Rains, A., King, R. B., Lawton, R.M., Mitchell, A. J. B. & Rackham, L. J. (1985) "African dambos: Their distribution, characteristics and use." *Z. Geomorph. N.F.*, Suppl.-Bd. 52, 63–86.

Adams, W. M. (1989) "Dam construction and the degradation of flood plain forest on the Turkwel River, Kenya." *Land Degradation and Rehabilitation, 1*, 189–198.

Albergel, J. (1989) "Fonctionnement hydrolique des Bas-fonds -synthese preliminaire." Annex to 'Mise en valeur des Bas-fonds en Afrique de l'Ouest', Zeppenfeldt, T. and Vlaar, J. C. J. (eds.). Comite Interafricain d'Etudes Hydrauliques, Ouagadougou, Burkina Faso, June 1990, 28.

Ase, L. E. (1987) "A note on the water budget of Lake Naivasha, Kenya: especially the role of *Salvinia molesta* Mitch. and *Cyperus papyrus.*" *Geografiska Annaler, Series A*, 69A(3–4), 415–429.

Balek, J. & Perry, J. E. (1973) "Hydrology of seasonally inundated African headwater swamps." *J. of Hydrol. 19*, 227–249.

Balek, J. (1977) *Hydrology and Water Resources in Tropical Africa.* Elsevier.

Balek, J. (1989) Chapter, "Humid warm flatlands" In: K. Falkenmark & T. Chapman (eds.)"*Comparative Hydrology: an Ecological Approach to Land and Water Resources,*" , UNESCO, Paris, 353–369.

British Geological Survey & others (1989) The Basement Aquifer Research Report 1984–1989: final report to the Overseas Development Administration. Technical Report WD/89/15, Wallingford, U.K. 158.

Bullock, A. (1992a) "The role of dambos in determining river flow regimes in Zimbabwe." *J. Hydrol.134*: 349–372

Bullock, A. (1992b) "Dambo hydrology in southern Africa: Review and reassessment." *J. Hydrol.134*: 373–396.

Carp, E. (1980) *Directory of wetlands in the Western Paleartic.* IUCN, Gland, Switzerland.

Carter, V., Bedinger, M. S., Novitski, R. P. & Wilen, W. O. (1978) "Water resources and wetlands." In: P. E. Greeson, J. R. Clark & J. E. Clark (eds.) *Wetlands Functions and Values: the State of Our Understanding,*" American Water Resources Association, Minneapolis, 344–376.

Consejo Superior de Investiagaciones Cientificas/International Association of Hydrogeologists (1988) Proc. Int. Symp. on *Hydrology of Wetlands in Semi-arid and Arid Regions.* Seville, Spain, May 1988.

Cook, A. (1981) "Integrated resource survey as an aid to soil survey in a tropical flood plain environment." In: J. R. G. Townshend (ed.) *"Terrain Analysis and Remote Sensing",*.Geo. Allen and Unwin, London, 184–203.

Dambo Research Unit (1987) The use of dambos in rural development with reference to Zimbabwe. Final Report to the Overseas Development Administration. Loughborough University, University of Zimbabwe.

Dincer, T., Child, S. & Khupe, B. (1987) "A simple mathematical model of a complex hydrological system – Okavango Swamp, Botswana." *J. of Hydrol. 93*, 41–65.

Drayton, R. S., Kidd, C. H. R., Mandeville, A. N.. & Miller, J. B (1980) A regional analysis of river floods and low flows in Malawi. Instit. of Hydrology Report No. 72. Wallingford, U.K.

Drayton, R. S. (1984) "Variations in the level of Lake Malawi." *Hydrol. Sci. J. 29(1)*, 1–12.

Drayton, R. S. (1986) "Dambo hydrology – an application of satellite remote sensing to water resource studies in the Third World". Proc. Symp. on *Mapping from Modern Imagery,* Inter. Society for Photogrammetry and Remote Sensing Commission IV and the Remote Sensing Society, Edinburgh, Scotland, 8–12 Sept., 125–134.

Dubreuil, P. (1985) "Review of relationships between geophysical factors and hydrological characteristics in the tropics." *J. of Hydrol. 87*, 201–222.

Dugan, P. J. (1988) "Conservation of wetlands: A global concern." Proc. Symp. on the *Hydrology of Wetlands in Temperate and Cold Regions,* 2, 1–4. Joensuu, Finland 6–8 June l988, Academy of Finland.

Dugan, P. J. (ed.) (1990) Wetland conservation: A Review of Current Issues and Required Action. IUCN, Gland, Switzerland.

Dunne, T. (1983) "Relation of field studies and modelling in the prediction of storm runoff." *J. of Hydrol. 65*, 25–48.

Eccles, D. H. (1984) "On the recent high levels of Lake Malawi." *South African J. of Science 80(10)*, 461–468.

Ellenbroek, G. A. (1987) Ecology and productivity of an African wetland. The Kafue Flats, Zambia. W.Jung, Den Haag.

Engman, E. T. & Gurney, R. J. (1991) *Remote sensing in hydrology.* Chapman and Hall, London.

Estrin, S. A. (1986) "The delineation and classification of inland wetlands using fcir (false colour infra-red) stereo imagery" In: M. C. J. Damen, *et al.*,(eds.) *Remote sensing for resources development and environmental management.*" Proc 7th ISPRS Commission VII symposium, Enschede, 1986, Vol. 2, (Balkema), 713–716.

Falkenmark, M. (1989) Chapter, "Comparative hydrology – a new concept" In: M. Falkenmark & T. Chapman (eds.) *Comparative*

Hydrology: an Ecological Approach to Land and Water Resources. , UNESCO, Paris, 10–42.

FAO/UNESCO (1974). *Soil Map of the World.* UNESCO, Paris.

Faulkner, R. D. & Lambert, R. A. (1991) "The effect of irrigation on dambo hydrology: A case study". *J. of Hydrol. 123*, 147–161.

Filius, P. C. (1986) "A one-year experience of cultivation in an upland dambo in Luapula Province." Proc. National Workshop on Dambos, Nanga, Zambia. Ministry of Agriculture and Water Development (Zambia) and Food and Agriculture Organization, Rome, 65–80.

Gopal, B., Turner, R. E., Wetzel, R. G. & Whigham, D. F. (eds.) (1982) Wetlands: Ecology and management. Proc. of the First Int. Wetlands Conference, New Delhi, 10–17 Sept. 1980. National Institute of Ecology and International Scientific Publications.

Guertin, D. P., Barten, P. K. & Brooks, K. N. (1987) "The peatland hydrologic impact model: development and testing." *Nordic Hydrology 18*, 79–100.

Hewlett, J. D. & Hibbert, A. R. (1967) "Factors affecting the response of small watersheds to precipitation in humid areas." In: W. Sopper & H.Lull, (eds.) *Forest Hydrology.*. Pergamon, Oxford, 275–290.

Heyman, A. M. (1988) "Self-financed resource management: A direct approach to maintaining marine biological diversity." Proc. *Workshop on Economics,* IUCN General Assembly, 4–5 Feb. 1988, Costa Rica.

Hill, J. L. & Kidd, C. H. R. (1980) Rainfall-runoff relationships for 47 Malawi catchments. Report No. TP7, Water Resources Branch, Malawi.

Hollis, G. E. (1988) "Environmental Impacts of Development on Wetlands", International Symp. on *Hydrology of Wetlands in Semi-arid and Arid Regions,* Seville, Spain, May 1988.

Howard-Williams, C. & Thompson, K. (1985) "The conservation and management of African wetlands," chapter in: P. Denny (ed.) *"The Ecology and Management of African Wetland Vegetation",* . W. Junk, Dordrecht,.203–230.

Hurst, H. E. (1933) "The Sudd Region of the Nile." *Journal of the Royal Society of Arts, 81,* 721–736.

Institute of Hydrology (1988) Water balance study Phase 3: Investigation of dry season flows. Report to the Interim Committee for Coordination of Investigations of the Lower Mekong, 2 vols., p.67.

IUCN (1990a) Wetlands Programme Newsletter No.1, June 1990. IUCN, Gland.

IUCN (1990b) Wetlands Programme Newsletter No.2, Nov. 1990. IUCN, Gland.

Kalapula, E. S. (1986) "Dambos and their possible development for agriculture." *Area 18 (3)*.

Kalk, M., McLachlan, A. J., & Howard-Williams, C. (1979) (eds.) *Lake Chilwa: Studies of Change in a Tropical Ecosystem.* W. Junk, Monographiae Biologicae 35, The Hague.

Kadlec, R. H. (1990) "Overland flow in wetlands – vegetation resistance." *Journal of Hydraulic Engineering, 116 (5)*, 691–706.

Koryavov, P. P. (1982) "Mathematical modelling of the hydrology of wetlands and shallow water bodies." Proc. *Ecosystem Dynamics in Freshwater Wetlands and Shallow Water Bodies,* Minsk, July 12–26 1981, Vol. 2. 297–310. SCOPE, UNEP, Moscow.

Ledger, D. C. (1969) "The dry season characteristics of West African rivers." chapter in: *Environment and Land Use in Africa,* M. F. Thomas & G. W. Whittington (eds.). Methuen, London, 83–102.

Lind, E. M. (1956) "Studies in Uganda swamps." *Uganda Journal 20*, 166–176.

Littlejohn, C. B. (1977) "An analysis of the role of natural wetlands in regional water management." In: C. A. S. Hall & J. W. Day (eds.)."*Ecosystem Modelling: Theory and Practice,*" John Wiley and Sons, New York, 451–476.

Maltby, E. (1986) *Waterlogged Wealth.* Earthscan, London.

Maltby, E. (1988) "Properties, geomorphology and classification of peatlands." Proc. Symp. on the *Hydrology of Wetlands in Temperate and Cold Regions.* Joensuu, Finland, 6–8 June, *2,* 5–13. Academy of Finland.

Mavuti, K.M. (1982) "Wetlands and shallow water bodies in Kenya." In: D. O. Logofet & N. K. Lukyanov (eds.) *Ecosystem Dynamics in Freshwater Wetlands and Shallow Water Bodies.* SCOPE/UNEP, Moscow 1982, 188–204.

McFarlane, M. J. & Whitlow, J. R. (1991) "Key factors affecting the initiation and progress of gullying in dambos in parts of Zimbabwe and Malawi." *Land Degradation and Rehabilitation 2 (3),* 215–236.

Meigh, J. R., Acreman, M. C., Sene, K. J. & Purba, J. B. (1990) "The water balance of Lake Toba." Int.Conf.on Lake Toba, Jakarta, Indonesia, May 1990. BPP Teknologi, Jakarta.

Mepham, R. H. & Mepham, S. (In press) *A Directory of African Wetlands of International Importance.* IUCN, Gland, Switzerland.

Miller, T. S. (1986) "The quantification of flood plain inundation by the use of Landsat and Metric Camera information, Belize, Central America." In: M. C. J. Damen *et al.,*(eds.)*, "Remote Sensing for Resources Development and Environmental Management"*, Proc 7th ISPRS Commission VII symposium, *2,* Enschede, 1986, (Balkema), 733–738.

Ministry of Agriculture and Water Development/Food and Agric. Organization (1986) Proc. of the National Workshop on Dambos. Nanga. 22–24 April, 1986.

Mitchell, B. (ed.) (1990) *Integrated Water Management: International Experiences and Perspectives.* Belhaven Press, London.

Mitsch, W. J., Day, J. W., Taylor, J. R. & Madden, C. (1982) "Models of North American Freshwater Wetlands – a review." Proc. *Ecosystem Dynamics in Freshwater Wetlands and Shallow Water Bodies*, Minsk, July 12–26,1981, *2,* SCOPE, UNEP, Moscow,5–32.

Mumeka & Mwasile (1986) "Some aspects of the hydrology of dambos in Zambia." Proc. National Workshop on Dambos, Nanga, April 1986. Ministry of Agriculture and Water Development (Zambia) and Food and Agric. Organization, 111–127.

Mwanyika, N. A. (1982) "Human use of wetlands in Tanzania: The case of the Lower Rufiji flood plain." Proc. *Ecosystem Dynamics in Freshwater Wetlands and Shallow Water Bodies,* Minsk, July 12–26, 1981, *2.* SCOPE, UNEP, Moscow 357–379.

National Wetlands Policy Forum (1988) Protecting America's Wetlands: An Action Agenda. The Conservation Foundation, Washington, D. C. USA.

Neuland, H. (1984) "Abnormal high water levels of Lake Malawi – an attempt to assess the future behaviour of the lake water levels." *GeoJournal 9(4),* 323–334.

Newson, M. D. (1981) "Mountain Streams." Chapter in *British Rivers*, J. Lewin (ed.). Geo. Allen and Unwin, London, 59–89.

Newson, M. D. (1988) "From Description to Prescription: Measurements to Management." Chapter in *Horizons in Physical Geography.* M. J. Clark, K. J. Gregory, & A. M. Gurnell (eds.).MacMillan Education, Basingstoke U.K., 353–366.

Nortcliff, S. & Thornes, J. B. (1981) "Seasonal variations in the hydrology of a small forested catchment near Manaus, Amazonas, and the implications for its management." In: R. Lal & E. W. Russell (eds.)*Tropical Agricultural Hydrology,.* John Wiley & Sons, 37–58

Novitski, R. P. (1978) "The hydrological characteristics of Wisconsin's wetlands and their influences on floods, streamflow and sediment." In:

P. E. Greeson, J. R. Clark & J. E. Clark (eds.).*Wetlands Functions and Values: the State of our Understanding.* American Water Resources Association, Minneapolis, 377–388.

Obrdlik, P., Mumeka, A. & Kasonde, J. M. (1989) "Regulated rivers in Zambia – The case study of the Kafue River." *Regulated Rivers: Research and Management, 3 (1),* 371–380.

Olivry, J. C. & Dacosta, H. (1984) Le marigot de BAILA, Bilan des apports hydriques et evolution de la salinite. ORSTOM DAKAR.

Osborne, P. L.., Kyle, J. H., & Abramski, M. S. (1987) "Effects of seasonal water level changes on the chemical and biological limnology of Lake Murray, Papua New Guinea." *Australian J. of Marine and Freshwater Research 38(3),* 397–408.

Oyebande, L. & Balek, J. (1989) "Humid warm sloping land," Chapter in: *Comparative hydrology: an ecological approach to land and water resources*, M. Falkenmark & T. Chapman (eds.). UNESCO, Paris, 224–273.

Piper, B. S., Plinston, D. T. & Sutcliffe, J. V. (1986) "The water balance of Lake Victoria." *Hydrol. Sci. J.31(1),* 25–37.

Pires, J. M. (1973) Tipos de Vegetacao de Amazonia. Publ. avulsas, Museu Goeldo, Belem 20, 179–202.

Raunet, M. (1985) "Les bas-fonds en Afrique et a Madagascar." *Z. fur Geomorphologie.* N.F., Suppl.-Bd. 52, 25–62.

Rollet, B. (1986) "Photo-interpretation of wetland vegetation in the Lesser Antilles (Guadeloupe)" In: M. C. J. Damen,*et al.,* (eds.) Remote Sensing for Resources Development and Environmental Management, Proc 7th ISPRS Commission VII symposium, Enschede, 1986, *1,*(Balkema), 499–504.

Runge, J. (1991) "Geomorphological depressions (bas-fonds) and present-day erosion processes on the planation surface of Central Togo, West Africa". *Erdkunde, Band 45,* 52–65.

Russell, R. G. (1971) Dambo Utilization Survey. Unpubl. Report of Bunda College of Agriculture, Lilongwe, Malawi.

Russell, E. W. (1981) "Role of watershed management for arable land use in the tropics." In: R. Lal & E. W. Russell (eds.) *Tropical Agricultural Hydrology*, John Wiley & Sons Ltd. 11–16.

Schiller, G., Lemmela, R. & Spreafico, M. (eds.) (1991) *Hydrology of Natural and Man-made Lakes.* IAHS Publication No. 206.

Scott, D. A. (ed..) (1989) *A Directory of Asian Wetlands.* IUCN, Gland, Switzerland and Cambridge, U.K.

Scott, D. A. & Carbonell, M. (eds.) (1986) *A Directory of Neotropical Wetlands.* IWRB, Slimbridge and IUCN, Gland, Switzerland and Cambridge, U.K.

Sellars, C. D. (1981) "A flood plain storage model used to determine evaporation losses in the Upper Yobe River, Northern Nigeria". *J. of Hydrol.52,* 257–268.

Seyhan, E., Hope, A. S. & Schulze, R. E. (1983) "Estimation of streamflow loss by evapotranspiration from a riparian zone." *South African Journal of Scienc*e Vol.79, 88–90.

Sharma, T. C. (1984) "Some hydrologic characteristics of the Zambian headwaters." *Zambian J. of Sci. Technol. 7,* 12–21.

Sharma, T. C. (1988) "An evaluation of evapotranspiration in tropical central Africa." *Hydrol. Sci. J. 33* (1).31–40.

Shepherd, C. J. (ed.) (1976) Investigation into fish productivity in a shallow freshwater lagoon in Malawi. Report prepared by the Ministry of Overseas Development, London for the Malawi Government.

Smith-Carrington, A. K. (1983) Hydrological Bulletin for the Bua Catchment: Water Resource Unit Number Five. Report of the

Groundwater Section, Department of Lands, Valuation and Water, Lilongwe, Malawi.

Sutcliffe, J. V. & Parks, Y. P. (1987) "Comparative water balances of selected African wetlands". *Hydrol. Sci. J. 34 (1)*, 49–62.

Thompson, K. (1985) "Emergent plants of permanent and seasonally-flooded wetlands".Chapter in: *Ecology and Management of African Wetland Vegetation*. P. Denny (ed.) W. Junk, Dordrecht, 43–108.

Turner, B. (1984) "The effect of dam construction on flooding of the Kafue Flats." In: D*evelopment on the Kafue Flats – The Last Five Years*. W. L. Handlos & G. J. Williams (eds.) KRBC, University of Zambia, Lusaka, 1–9.

Turner, B. (1985) "The classification of fadamas in central northern Nigeria." *Z. Geomorph.* N.F. Suppl.-Bd. 52, 87–113.

UNEP/SCOPE (1982) *Ecosystem Dynamics in Freshwater Wetlands and Shallow Water Bodies*. D. O. Logofet & N. K. Lukyanov, (eds.). Minsk, July 12–26, 1981.SCOPE/UNEP, Moscow 1982, 2 vols.

Van der Wert, R. & Kamerling, R. G. (1974) "Evapotranspiration of water hyacinth (*Elchhornia-Grassipes*)." *J. Hydrol. 22*, 201–212.

Vardevas, I. M. (1989) "A water budget model for the tropical Magela flood plain." *Ecological* M*odelling 46*, 165–194.

Verry, E. S. & Boelter, D. H. (1978) "Peatland Hydrology." In: P. E. Greeson, J. R. Clark & J. E. Clark, (eds.) *Wetlands Functions and Values: The State of Our Understanding*, . American Water Resources Association, Minneapolis, 389–402.

Verry, E. S., Brooks, K. N. & Barten, P. K. (1988) "Streamflow response from an ombotrophic mire." Proc. Symp. on the *Hydrology of Wetlands in Temperate and Cold Regions*, Joensuu, Finland 6–8 June, *1*, . Academy of Finland, 52–59.

Walsh R. P. D. (1980) "Runoff processes and models in the humid tropics." *Z. Geomorph*. N.F. , Suppl. 36, 176–202.

Whitlow, J. R. (1984) "A survey of dambos in Zimbabwe". *Zimbabwe Agricultural Journal 81 (4)*, 129–138.

Whitlow, J. R. (1985) "Dambos in Zimbabwe – a review." *Z. Geomorph*. N.F., Suppl.-Bd. 52, 115–146.

Whitlow, J. R.. & Gregory, K. J. (1989) "Changes in urban stream channels in Zimbabwe." *Regulated Rivers: Research and Management, 4*, 27–42.

Wilson, B. H. & Dincer, T. (1976) *An Introduction to the Hydrology and Hydrography of the Okavango Delta*. Okavango Delta Symposium, Botswana Society, Gaborone, 33–47.

14: Erosion and Sedimentation

C.W. Rose

Faculty of Environmental Sciences, Griffith University, Nathan, Brisbane, Queensland, Australia 4111

ABSTRACT

This chapter outlines some of the issues involved in erosion and sedimentation and the major approaches adopted by research investigators in dealing with them. Particular emphasis is given to recent developments in quantitatively describing soil erosion and deposition processes, and to elucidating the consequences of such description in the form of comprehensive mathematical models. How such models are used in practice to assess soil erodibility and depositability is illustrated, using data collected at the scale of runoff plots.

Approaches to investigating and representing erosion and deposition at the scale of catchments are also reviewed, although somewhat more briefly.

INTRODUCTION

Scientific study of soil erosion by water has a long history in the geographic and geomorphic sciences, where much of the emphasis is on erosion as one of the natural landscape-forming processes. Human activity, and especially the widespread and expanding activities of agriculture, has led to an acceleration of soil erosion commonly associated with agricultural practices. It is not surprising, therefore, to find that early agriculturally-focused research on soil erosion depended on successful agronomic research methodologies. These methodologies were typified by planned experimentation followed by quantitative (often statistically guided) analysis of the results obtained.

Following a significant period in which this fruitful agronomic type of approach has proved useful, research personnel now appear to be commonly using alternative methodologies in which more call is being made on physical theory to provide a framework in which experimental data are analysed. In this approach, parameters, which still have to be experimentally determined, are sought which are more closely related to the processes believed to be importantly involved in erosion and deposition.

Whilst this new approach to soil erosion and deposition is beginning to prove useful at the scale of runoff plots and the units of agricultural operations, the ability of this approach to assist in interpreting erosion and deposition on larger scales is now being questioned with increasing urgency. This urgency in part arises from increasing recognition that it is sediment delivered from non-point sources to streams and rivers which is a substantial cause of widespread decline in water quality.

All these issues are given a new urgency by the rapid conversion of tropical forests to meet the need for new land uses. The expansion of more intensive and permanent forms of agriculture than the slash and burn traditional systems of forest use and increasing areas of plantations for export crops have combined to produce extensive forest clearing in the tropics (IBSRAM, 1987).

Studies of erosion in the humid tropics have recently received quite extensive reviews by Bruijnzeel (1990) and Ziemer *et al.* (1990). Thus, the outcome of such studies, especially at the scale of small erosion plots, will receive limited attention in this review. New methodologies for soil erosion plot studies in which rates of runoff are measured will be described in this review since these methodologies have been used recently in the tropics. Also towards the end of this review, there will be some discussion of catchment-scale studies in the humid tropics and factors affecting the delivery of sediment to streams.

There is growing recognition of the desirability of interacting erosion experimentation with the development of models of the processes involved. Experimentation is essential to provide the database which models must be able to comprehend. However, models of erosion processes are not simply based on data, but on basic physical theory applied to the interpretation of such processes. Nevertheless, the experience of interacting models based on theory with experimental data during their development has been shown to be most helpful, and progress in such interaction is taking place most rapidly in simply slope erosion studies.

One purpose of process-based models is to obtain the value of physically-defined erosion parameters by analyzing data collected at a particular site and time using the model. However, despite this need for data in order to obtain location-specific erodibility parameters, an adequate description of

TABLE 1. *Examples of on-site and off-site damage associated with water erosion and sedimentation.*

On-Site	Off-Site
Loss of Plant Nutrients	Siltation of Streams, Rivers, Estuaries
Loss of Organic Matter	Siltation of Dams
Damage to Soil Structures	Damage to Crops, Roads, Culverts, etc.
Subsoil Exposure	Deposition of Soil Pollutants

processes is of general application. Hence a physically-based model of erosion processes has a capacity to extrapolate and generalize beyond the database used to check its utility.

A major focus of this review is to provide sufficient information on these developments in erosion process modelling and their applications so that the potential of these approaches to assist in the particular context of the humid tropics can be more effectively assessed. The implications of these newer approaches for field data collection methodologies are also indicated, as technology transfer of these methodologies to the humid tropics is already underway.

ISSUES IN EROSION AND SEDIMENTATION

Problems associated with soil erosion and sedimentation may be divided into those which are on-site and off-site in character. Typically, on-site problems are associated with a net loss of sediment, and off-site problems with a net gain of sediment (at least temporarily). Table 1 illustrates these types of problems, and indicates that off-site problems include those associated with water quality, which may be adversely affected by sediment, nutrients and other chemicals associated with the sediment, as well as soluble chemicals.

On-site problems of net erosion are of most widespread concern in rural areas where the sustainability of rural land uses can be threatened. The focus of off-site problems often concern damage to infrastructure such as roads, streams, rivers and dams.

Erosion and sedimentation are also of concern in urban areas, particularly during building construction and road-making activities. Also, erosion and sedimentation are an issue during open-cut mining and in the restoration and stabilization of the landscape following such mining. Sometimes it is the mining activity itself which leads to the problems but the disposal of wastes following processing (as in copper mining in Bougainville) can also cause problems for rivers and near-shore areas. Field use of toxic processing chemicals can present particular hazards, as can occur in gold mining.

Much more research has been reported for on-site and in-stream studies than for the succession of temporary storage and subsequent movement of sediment which must generally occur between net on-site loss and delivery to streams. This emphasis (or lack of it) will be reflected in this review.

Off-site damage is sometimes more obvious than on-site damage and the social and economic costs associated with off-site damage are commonly easier to identify and quantify.

Nation-wide surveys of land degradation associated with soil erosion are not commonly available but have been carried out by some countries such as Australia (DEHCD, 1978) which has some part of its land mass in the humid tropics. The global extent of soil degradation by various processes is considered in a major study by FAO/UNEP/UNESCO (Riquier, 1982). Other land degradation problems such as secondary (or human-induced) salinity can contribute indirectly to soil erosion. Salt-affected soils lead to poor vegetation growth and a deterioration in soil structure and infiltration, all of which tends to lead to severe erosion. Human activity can also exacerbate the problem of naturally occurring acid soils which are common in some soil orders of the humid tropics (Craswell & Pushparajah, 1989). Plant production and thus erosion protection can be poor on such soils.

Whilst research related to on-site erosion problems is substantial, much less attention has been given to conveyance from the upland catchment to permanent streams. The transport of sediment in streams by bed load and sediment load has again received considerable attention in research (Hadley *et al.*, 1985). This concentration of research at the source and sink ends of the problems is in evidence even in recent major symposia where encouragement was given to bridge this gap (DeCoursey, 1990).

While the use of isotope tracer studies holds promise in following the fate of eroded sediments (Sklash, Moore & Burch, 1990), the complex geometry and the wider spatial and longer time scales involved in such studies have been daunting. Nevertheless, there is widespread recognition that dealing with erosion and sedimentation concerns at a catchment scale does require methodology development and a co-ordination of land management practices at that scale.

Considering practices which conserve soil and reduce sedimentation problems in lowlands, rivers and reservoirs, we shift from a biophysical level of abstraction to a level where economic and culture-dependent issues become of great significance. In most countries, there is widespread concern at the rate of adoption and implementation of land management practices promoted by government agencies as desirable by stabilizing land against erosion.

Most governments accept a responsibility for developing policy designed to achieve sustained production and minimize the off-site consequences of erosion. Some policies introduced to encourage good land management have produced distortions, such as benefiting least those less able to implement desirable changes in land management.

Thus, the management of environmental issues of erosion and sedimentation are typically complex and multi-disciplinary. The type of skills required include an understanding of

TABLE 2. *Effects of methods of deforestation on runoff and erosion.*

Clearing treatment	Runoff (mm y^{-1})	Soil Erosion (t ha^{-1} y^{-1})
Traditional farming (incomplete clearing)	3	0.01
Manual clearing	35	2.5
Shear blade	86	3.8
Tree pusher/root rake	202	17.5

Source: Lal, 1987a

the processes involved, the ability to apply this understanding in the design of effective methods of management which need also to be economic and socially acceptable, to gain the acceptance and adoption of more land-conserving management practices and to provide an institutional and policy framework which encourages such social changes, involving attitudes as well as action.

This review focuses on erosion and sedimentation processes and the use of this process knowledge in the design and evaluation of methods of mitigating the problems involved. Before embarking on this major focus, some of the on-site and off-site issues of particular concern in the humid tropics will be briefly considered.

On-site issues in the humid tropics

The climax vegetation of the humid tropics is typically tropical rain forest. The soils supporting tropical rain forests are highly leached and weathered, such as the oxisols and ultisols, with most minerals (except quartz) weathered to kaolinite and non-soluble products such as iron and aluminium oxides (Lal, l987a). Since kaolinite has a low cation-exchange capacity, the capacity to retain nutrients depends a great deal on the humus content. Indeed, in rain forests nutrients are concentrated in the plant biomass, the litter layer and the associated organic matter. Since this organic fraction declines rapidly following clearing, due to erosion as well as other reasons, good plant growth is not sustained without substantial fertilizer application (Bruijnzeel, 1990). Thus, the potential for a substantial decline in productivity is one of the most serious on-site consequences of erosion in the humid tropics.

Lal (1987a) has reviewed the consequences typical of deforestation in the tropics which can lead to substantial acceleration of soil erosion. However, the soil loss per unit land area is greatly affected by the clearance methods employed. The incomplete clearance associated with many traditional farming practices may cause only limited acceleration in soil loss. However, more complete clearance, especially if carried out with heavy equipment involving substantial soil disturbance, can lead to a substantial increase in rates of soil loss, as is illustrated in Table 2.

Bruijnzeel (1990) recently reviewed the effects of deforestation on sediment production, indicating surface erosion can vary from year to year by two orders of magnitude in some situations involving tree crop systems where the climax vegetation was tropical forest. This review also showed that soil loss is often minimal in cultivation systems where the soil surface was protected by a well-developed layer of litter and herbs.

There is increasing evidence that clearance of the rain forest canopy alone, if the understory and litter layer can be left intact, does not lead to any substantial increase in soil loss (Wiersum, 1985). This is not to deny that a dense tree canopy is commonly associated with little surface erosion but does indicate that it may well be the cover in close contact with the soil surface, often dominated by leaf and litter fall which is essential to protection against surface erosion (Bruijnzeel, 1990; Besler, 1987). Herbs, moss and other low-growing understory vegetation commonly add to this protection in forests of the humid tropics.

The use of heavy equipment to extract timber from tropical forests is commonly associated with considerable disturbance of the surface soil and litter layer. Associated works, such as roughly cleared access roads, logging tracks, log landings and "snig tracks" are a major source of soil loss. Tree roots provide considerable mechanical reinforcement to soil in the root layer, and can provide significant protection against shallow forms of mass soil movement on sloping lands.

However, as reviewed by Bruijnzeel (1990) for the humid tropics, deep-seated mass wasting and landslides appear to be little influenced by the presence even of a well-developed root network.

Despite the importance of the various forms of mass movement in the humid tropics, often associated with extreme rainfall events, these forms of soil movement will receive little attention in this review, which focuses on surface erosion.

Off-site issues in the humid tropics

Further comment about off-site issues will be given in a later section concerned with erosion and deposition on a catchment scale.

In the humid tropics, the commonly high rates and large total quantities of rainfall and runoff can lead to high sediment concentrations and sedimentation in streams following land clearing (IBSRAM, 1987). In some tropical countries, such as Malaysia, forest clearing is very extensive (Zakaria *et al.*, 1987) and the desired rate of clearing has required the use of heavy machinery.

Much of the soil eroded from a hillslope moves into at least temporary storage in footslopes where slopes are reduced, or may be trapped in uncleared areas. The ratio between the sediment carried by a stream at the outlet from a basin and on-site erosion within the basin is called the "sediment delivery ratio." Sediment may be much reduced in magnitude and long delayed in time before soil eroded on-site reaches an off-site stream. These attenuation effects increase markedly, in general, with the size of the basin considered (Walling, 1983, 1988).

Measuring sediment transport in rivers is made more difficult because whilst finer sediment is transported as a suspended load carried throughout the entire volume of the river, the coarser fraction of sediment is transported as a highly concentrated layer close to the base of the river, and referred to as bed load (Walling, 1988). Measuring the bed load component presents substantial practical difficulties, and for this reason, data on river sediment transport are often for suspended sediment only. Sediment carried in streams or rivers may not all come from surface erosion within the land surfaces of the basin, but may include significant contributions from the stream bank or a degrading stream bed. Studies of sediment transport in the Amazon River (Meade *et al.*, 1985) indicated substantial sediment storage on the river bed could occur, followed by resuspension from storage under flood conditions. Substantial fluctuations in sediment transport through time and lack of a unique relationship between even the suspended load component of sediment transport and river discharge also make for difficulty in longer-term estimates of sediment discharge (Walling, 1988).

Lal (1987b) has reviewed measurements of the effect on runoff and soil erosion of the conversion of tropical rain forests to a variety of land uses. Logging using typical practices can lead to substantial increases in the sediment load of streams draining logged catchments, though recovery can occur with reforestation. This recovery and reduction in rate of soil loss appear to be strongly associated with recovery in cover in close contact with the soil surface ("surface contact cover"), which can be accelerated by seeding a tropical cover crop, a common practice in establishing plantation and perennial crops (Lal, 1987b).

The effect on basin sediment yield of deforestation of various forms in the moist tropics has been recently reviewed by Bruijnzeel (1990). He indicates that many of the general features and concerns associated with off-site erosion damage in temperate regions also apply to the humid tropics. The importance to sediment transport of less frequent very high rainfall events may be even more pronounced in tropical as compared to temperate environments.

Because of the long delays in sediment delivery to streams due to the substantial storage of sediment, especially in larger basins, evaluating the effect of different land clearing or management practices on sediment yield at exit from the basin is by no means a simple or short-term task (Bruijnzeel, 1990). Nevertheless, there is accumulating evidence from both on-site and off-site studies of the substantial control of soil erosion made possible through practices which ensure maintenance of good surface contact cover aided by some density of trees which increase surface stability through the mechanical soil support of their deeper root systems. However, in situations of geologically unstable terrain and extreme rainfall events the effects of human activities may be completely overridden by natural erosive processes.

APPROACHES TO SOIL EROSION BY WATER

Forms of soil erosion

A basic distinction can be made between erosion as a natural landscape-forming process or series of processes and erosion where the rate has been significantly accelerated by human action. It is acceleration of the natural erosion processes which is of concern.

Considering soil erosion as a natural geomorphic process, there are a number of erosion mechanisms of importance. There are types of erosion in which flowing water and water as raindrops are directly involved. Whilst these types of erosion will receive almost exclusive attention in this review, there is another broad class of erosion in which water plays only the indirect role of weakening the soil, with the direct cause of erosion being the weight of the soil itself. This weight, due to gravity, can give rise to stresses in the soil fabric which are too high to be resisted so that failure occurs. The resulting erosion is referred to as "mass movement" (Statham 1979).

Examples of mass movement are landslides and landflows, where the soil material can move over substantial distances and involve vast quantities of material. Mass movement is more common and extensive in geologically younger landscapes (O'Loughlin, 1984).

However, human action can trigger or accelerate mass movement of soil, examples being after clear felling of forests (Tsukamoto & Minematou, 1987) and instability of a hillside induced by excavation in road construction (Megahan & Kidd, 1972; Megahan, 1977). In some circumstances, cultivation of steep land can lead to the loss of most soil in the cultivated layer during a major rainfall event (Bruijnzeel, 1990).

Understanding the various forms of mass movement has been sought more by geomorphologists and engineers with a background in soil mechanics than by agricultural, environmental and soil scientists and hydrologists, who have tended to give more attention to other forms of soil erosion.

When soil erosion is not due to gravity directly, as in mass movement, but is carried in water moving over the soil surface, there can be more than one process at work. Sometimes, an approximately uniform depth of soil is removed in erosion. More commonly, water finds some preferential paths on which to flow over the land surface. Then a common, though not universal, characteristic of erosion is for these flow concentrations to deepen the paths in which they flow and then form an easily recognizable small channel, referred to as a rill. Rills can be formed in a variety of shapes and sizes, and can substantially concentrate overland flow, in the way that rivers do on a much larger scale. Water erosion in which rills are formed is often referred to as rill erosion. Rills are ephemeral in the sense that their visual form can be readily removed by cultivation. In some literature, relatively large rills are referred to as ephemeral gullies.

The term gully erosion is generally used to describe a feature arising from concentrated flow which is too deep to be removed by mechanical cultivation equipment and may be of such a depth as to be impassable by such equipment.

Following an erosion event, it is clear whether or not rills have developed. If rilling has occurred, then the surface can be described as consisting of rills and interrrill regions. It is common to refer to rill erosion and interrill erosion. It is also common to assume that distinct erosion processes operate, or at least predominate, in rills and interrills. Rill erosion is then associated with overland flow processes, and interrill erosion with raindrop impact.

There is little doubt that rill erosion is dominated by flow processes. Furthermore, if rills are close together, so that overland flow across interrill regions is soon captured by a rill, then it is probably correct that erosion in this region is dominated by rainfall impact. However, typical distances between rills vary considerably, with the soil type appearing to be one factor associated with such variation. Distance between rills can be sufficiently great so that the possibility should be kept open that not only rainfall impact but also flow processes may play a role in interrill erosion. A particular case of this is where significant erosion occurs, with overland flow evidently being an erosive agent, but there is no clear evidence of rill formation.

It follows from such field observations in different parts of the world, including the humid tropics, that use of the terms rill erosion and interrill erosion to imply two quite distinct sets of erosion processes may not always be correct. Situations exist where this clear distinction is valid but this clear distinction is not universal. This chapter gives a description of erosion processes in which it is not necessary to assume that erosion processes in interrill regions are quite distinct from those which operate in rills.

The description of erosion processes and their consequences given later will omit all forms of mass movement and also gully erosion in which mass movement may play a part.

The role of models in soil erosion research

When scientists seek to understand something about a rather complex phenomenon (such as soil erosion in an agricultural context) they tend to set up a series of realistic experiments which cover the range of inputs and variables. This methodology was, and still is, in common use in agronomy. It was also adopted in the first major scientific investigation of soil erosion by water which was carried out in the agriculturally important regions of the USA. Recognizing that loss of soil from a cultivated field could vary from year to year, the objective of this research was to experimentally find how such loss – averaged over a number of years – depended on factors believed to be important at the time these investigations were planned.

Though this statement requires further historical justification, it seems that the research of Ellison (1952) was not only important in itself, but influenced the mental model held by the scientists responsible for designing this extensive research program centred in the mid-west of the USA. This mental

model can be described as follows: "Raindrops detach soil, and overland flow simply transports this previously removed sediment over the soil surface." Since raindrops rather than overland flow were assumed to be the active erosion agent, then rainfall rate was measured, but not the runoff rate. It is true that at the time of this period of extensive experimentation, technology was better developed for measuring rainfall rate than runoff rate. However, it seems likely that the choice of instrumentation was influenced more by the conceptual model held about the rate-limiting erosion process than the technical difficulty of measuring rate of runoff.

The factors investigated (Wischmeier & Smith, 1978) were the largely uncontrollable factors of rainfall rate, soil type and land slope. Reflecting agricultural practice at the time (post-second World War), soils were cultivated, rather more frequently than today's common standards. The experiments were subject to a variety of more controllable land management factors such as the amount of crop residue left on the soil surface to provide protection to the soil against erosion, and the effective length of slope over which water could flow relatively unimpeded by soil conservation structures such as graded banks.

These experiments were carried out on a large number of experimental runoff plots of standard length located over the agricultural lands of the USA. The experimenters were presented with a very large body of data which needed to be summarized in some way in order to be generally useful. Since soil erosion was seen as related to rainfall, and since there would be no erosion on absolutely flat land, then a zero value for either of these two factors should result in no soil loss. Whether for this or other reasons, a factor-product type of model was tried and found to be successful as a way of summarizing the data.

Sub-experiments were also carried out in order to evaluate as far as possible the separate effects of slope (S) and slope length (L). The effect of cropping practices on soil loss was also investigated and represented by a factor C related roughly to the degree of exposure or lack of protection provided by any cropping practice.

Proceeding in this way, it was found possible to provide a good and efficient summary of the great body of data collected, using the factor product equation or mathematical model of the following form:

$$A = RKLSCP \tag{1}$$

where A is average annual soil loss per unit area; R is a rainfall erosivity factor related to rainfall rate; K is a factor believed to depend chiefly on soil type; LS is a combined slope length and slope factor; C is a cropping practice factor; and P is a factor depending on management practices adopted which affect soil loss.

A full description of equation (1), referred to as the Universal Soil Loss Equation or USLE, is given by Wischmeier and Smith (1978). Since sub-experiments provide guidance to the values of all terms in equation (1) except K, this soil erodibility factor can be evaluated, once a value of A has been experimentally determined.

This situation where one (or more) parameters have to be experimentally determined using runoff plots is typical of most models of soil erosion. Since the database from which the USLE was derived was extensive in respect to agriculture in the USA at that time, this was not a major limitation. However, it implied that the values of the parameters in the USLE refer to the locations where the data was collected, and raised the question as to how transferable these parameters may be to other regions in the USA, or other countries. Dunne (1984), for example, refers to these problems with respect to the tropics. A nomogram was developed, based on the original USLE database, in which a value for K in equation (1) could be estimated from soil characteristics measurable without requiring runoff plots (Wischmeier and Smith, 1978). It is now recognized that such aids, whilst very useful in the region yielding the database, are not universal in nature.

Several things can be said about the type of model represented by the USLE or equation (1).

Firstly, it has proved to be a useful way in which data collected over a number of years at one place can be summarized.

Secondly, in environments where annual soil loss is very variable (e.g., especially in semiarid subregions), measurement may need to continue for such a long period of time that the usefulness of the methodology in such environments is questionable.

Thirdly, because the general importance of runoff rate to erosion is now recognized, the value of parameters in the USLE must include the effect of whatever the correlation may be at the experimental site between rainfall characteristics and runoff rate.

Probably the most extensive study of the utility of the USLE outside the USA has been carried out in the state of New South Wales (NSW), Australia. Data on runoff and soil loss were recorded in excess of 30 years in some cases at six research centres of the Soil Conservation Service of NSW. These six centres covered a 700 km transect of the state's agriculturally important regions, where average annual rainfall varied from approximately 550 mm to 750 mm and wheat was a commonly grown crop. The most thorough and complete analysis of this data was carried out by Edwards (1987).

Conclusions from this extensive study included the following: Soil losses were highly variable in space and time, some 10% of the runoff events contributing 90% of the total soil loss, mostly occurring when the ground cover was low. A lack of suitable parameter values with which to operate the USLE, doubts as to the applicability of some of its factors, and the long time-scale of experimentation needed in these environments were factors seen as limiting the application of the USLE to Australia.

Despite these shortcomings or limitations, the USLE has continued to find useful application, sometimes enhanced by a computer-interactive format, as illustrated by Rosewell & Edwards (1988).

The objective of the USLE was to provide a useful summary of an existing substantial database, and the methodology has proven itself useful in this regard. The investment required for the long-term measurements needed to yield term A in equation (1) can be a deterrent in using this methodology. But are there any other reasons for developing an alternative approach to the USLE for interpreting soil loss data, and in moving toward a more predictive phase in erosion research?

One such reason was a growing recognition that it was not an objective of the USLE to develop a model of soil erosion which incorporated the processes involved, nor were physical principles used to develop a model of the combined processes believed to be involved. An approach in which processes are described would still be expected to require experimental evaluation of defined parameters in the model, but it would also be expected that such parameters would have a more definite physical meaning than the K factor in the USLE. Attempts to elucidate the general physical meaning of this term have been somewhat frustrating, though not without some success.

Thus, it may be argued that one reason for the development of more physically-based models is the hope that the parameters defined in such models may have a more definite physical meaning than the K factor in the USLE. If so, there is an expectation that such physically-defined parameters may be able to be obtained independently of runoff plot experimentation. At least there is hope that such parameters could be inferred from more readily measured soil characteristics with sufficient accuracy for the management decisions needed to reduce soil loss to acceptable levels. Such a development holds out the hope (as yet not fully realized) that the assessment of soil erodibility may be made in a survey mode, or at least not requiring the installation of instrumented runoff plots to measure soil and water loss, except as a check where this is feasible.

Controlled experimental studies on erosion and deposition processes

One of the difficulties associated with field studies of erosion and sedimentation is that the driving variables are uncontrolled and it is not always possible for the erosion event to be visually observed. If the objective is to better understand component processes, then there is an important role for controlled studies. Typical of such studies is the use of controlled simulated rainfall falling on soil placed in a tilted flume.

Another motivation for such controlled studies has been the assumption that standardizing test procedures might reduce some of the confusion surrounding the term "soil erodibility" to which Bryan *et al* (1989) has drawn attention. Invalid assumptions commonly made about soil erodibility have been given as one reason for such confusion (Bryan *et al.*, 1989), and lack of general agreement on specific vocabulary in describing the erosion process does not help. Govers *et al.*

(1990) have also shown how important the initial water content of soil can be in erosion experiments, so that this should be fully reported in flume studies.

Adequate resolution of the confusion in the literature on what is meant by soil erodibility may not be possible without relation to theory of the processes at work, and a recognition that sediment concentration depends not only on erosion processes (and thus on "soil erodibility") but also on deposition processes (and thus on "soil depositability"), a term introduced by Rose & Hairsine (1988).

Using simulated rainfall studies in a 3 m long flume with a constant water flux, Moss (1979) showed that the sediment concentration of poorly sorted sand at exit from the flume was insensitive to slope (S) up to $S \approx 0.03$ or 3%. In this regime, it was concluded that rainfall was the only agent of sediment detachment, with the surface flow acting simply to transport and not generate sediment. This regime has been called "rainflow transportation" by Moss *et al.* (1979). Rain-flow transportation rates were found to be greatest at flow depths corresponding to 2 to 3 drop diameters, decreasing rapidly with further increase in water depth, indicating protection of the bed from raindrop impact by the increased depth of water (Moss & Green, 1983).

In the experiments of Moss (1979), the ability of overland flow alone to entrain and transport sediment rose rapidly as the bed slope increased beyond 0.01, equalling that of rainflow transport at S of approximately 0.07. In these experiments, entrainment by surface flow became the dominant mechanism contributing to sediment transport for $S > 0.07$ approximately, though these results are strongly affected by the nature of the soil material (Singer & Walker, 1983).

As slope and discharge increase, there is a common though not universal tendency for some form of surface irregularity or channels to develop in initially uniform and flat beds (Moss *et al.*, 1982). Some soils, such as loess, tend to form incised rills. Moss & Walker (1978) suggested this tendency in loess is related to the general absence of particles greater in size than that of silt.

Moss *et al.* (1980) investigated the transport of loose sandy detritus in shallow water flows in a flume and found a linear relationship between sediment flux and stream power (Bagnold, 1977), with sediment concentration being proportional to slope. Also working with non-cohesive quartz materials, so that the rate of sediment transport may be assumed to be at the maximum net erosion rate called the "transport capacity", Govers (1987) investigated how this transport capacity could be related to a range of physically-based hydraulic parameters for five materials which covered a wide range of discharges and flume slopes. Useful power-type relationships were found to exist between sediment transport capacity and shear stress, effective stream power and unit stream power. Coefficients in these relationships were found to vary with the grain size of the particles, which would indicate differences in deposition or settling velocity characteristics. Such studies as the one of Govers (1987), indicate that at the transport capacity, both hydraulic and sediment size-related characteristics are important. However, it appears that continued exploration of empirical relationships, while helpful, may require the assistance of theoretical development in order to overcome the types of confusion commented on by Bryan *et al.* (1989).

Despite the attractions of working with net sediment transport quantities as outlined by Foster (1982), Rose (1985) argues the case for the separate explicit representation of the deposition process. This is partly because deposition modifies the increase in sediment concentration due to erosion processes. The rate of deposition depends on the size distribution of the sediment involved, or more exactly on the settling velocity distribution, which is illustrated in the next section.

Lovell & Rose (1991) have shown that there is interaction between settling particles during sedimentation, especially when – as is characteristic of soil – there is a wide range of size and settling velocities involved. Thus, it appears preferable to determine settling velocities for soils by using a technique where this interaction can occur. Such interaction does occur in bottom withdrawal tube techniques (Lovell & Rose, 1988a). Though terminal velocities are rapidly achieved, it is uncertain whether this is so for larger aggregates, and whether settling velocities are affected by the turbulence induced particularly by raindrop impact on shallow water layers.

Controlled experiments of the type mentioned above in this subsection are simplifications of the field situation. Despite flume lengths being often shorter than distances of uninterrupted overland flow in the field, such experiments are not "scale models" of the field situation. Hence, the proper concerns associated with scale modelling are not normally involved. Despite such reassurance, the major use made of controlled experimental work in the following sections of this review is to aid the conceptualization of processes. Theory is then built on this conceptualization. It is field experimentation which provides the test of how useful this theory is when applied to field data.

The following section concentrates on the development of theory for soil erosion and deposition processes acting together.

SOIL EROSION AND DEPOSITION PROCESSES

Flux relationships

Consider sediment-laden water flowing down a sloping planar land surface as shown in Fig. 1. The flow does not have to be as spatially uniform as implied by the figure, but averaged across the width of the plane. Such flows are indicated per unit width of the plane, and thus are referred to as fluxes.

Let q = volumetric flux of water in units of $m^3\,m^{-1}\,s^{-1}$.

Let c = sediment concentration, the mass of sediment per unit volume of flowing water ($kg\,m^{-3}$).

Then it follows from the definition of q and c, that the mass of sediment crossing unit width per second (the sediment flux, q_s) is given by:

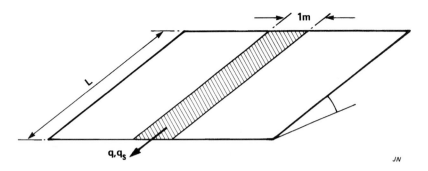

Figure 1: Representing rate of flow for water (q) and sediment (q_s) from unit strip width of a sloping planar land element of length L. (From Rose, 1988)

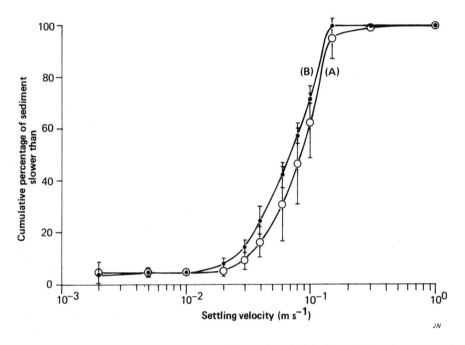

Figure 2: Settling velocity distributions for sediment deposited during a field environment and analysed whilst still moist ($\theta_g = 0.476$ kg kg^{-1}, curve A); for the same sediment allowed to air dry $\theta_g = 0.133$ kg kg^{-1}) and re-wet by immersion (curve B). Each distribution is the mean of three replicates; 95% confidence limits are shown. Soil is a Udic Pellustert in the Vertisol soil order.

$$q_s = q\, c\ (\text{kg m}^{-1}\,\text{s}^{-1}) \tag{2}$$

From equation (2) it follows that the total soil loss per unit area is given by

$$\int \frac{q_s}{L} dt\quad (\text{kg m}^{-2}) \tag{3}$$

where the integral or summation in equation (3) is over the duration of the erosion event.

Equation (2) indicates that in interpreting or predicting soil loss, the hydrology (term q) is of equal importance to factors which affect the sediment concentration c, which is the main focus of this review.

Please note that all major symbols used in this review are given in a glossary at the end of the chapter.

Deposition

Whilst some sediment can temporarily float in water, most sediment sinks due to its positive immersed weight in water. Thus, the majority of eroded sediment will return again to the soil surface after being carried some distance downslope in overland flow. This process whereby sediment falls through a water layer is called "sedimentation", and, on reaching the surface, "deposition."

The characteristic of soil which governs the rate of deposition is called the "settling velocity characteristic." This characteristic of soil can be readily measured experimentally by determining the proportion of soil mass with any particular settling velocity. Soil eroded by water consists largely of soil aggregates, so that this characteristic is dominated by the degree of aggregation of the soil and its stability. The characteristic can be measured with rather simple equipment called a modified bottom withdrawal tube – described by Lovell & Rose (1988a), who also gave the theory of operation of the equipment. A computer program is helpful in analysing results as the calculations are tedious (Lovell and Rose, 1986).

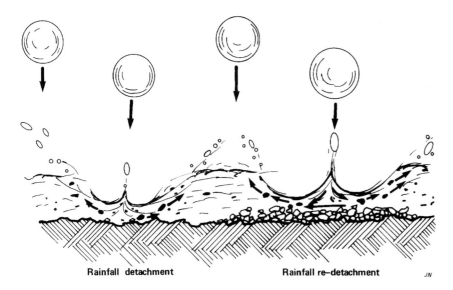

Rainfall detachment   Rainfall re-detachment   JN

Figure 3: Schematically illustrating the distinction between rainfall detachment of the original soil matrix, and re-detachment of previously eroded and deposited sediment.

Figure 2 illustrates settling velocity characteristics determined using the modified bottom withdrawal tube. In Figure 2, data is presented as a cumulated percentage of sediment which settles with a velocity slower than any settling velocity. The figure shows that no sediment in this soil settled with a velocity faster than somewhere between 0.1 and 0.2 m s⁻¹. Any clods must be excluded from the withdrawal tube, the upper limit size or settling velocity effectively being set by the ability of the equipment to handle it. However, this is not an important limitation since clods play no role in deposition. The upper limit to the size of soil aggregates which can take part in deposition would be somewhat less than the depth of water on the soil surface (which can vary spatially).

Figure 2 refers to sediment taken from a deposit in the field and presents results both if the sediment was still moist, or if it had dried and was re-wet. A settling velocity curve such as Fig. 2 for original soil can be regarded as a soil characteristic in which the size distribution of stable soil aggregates plays a dominant role.

There is also a lower limit to the size of sediment which will deposit, at least in water with any degree of turbulence. In water, settling sedimentary units (which can be ultimate soil particles or aggregates of soil particles) quite quickly achieve a characteristic "settling velocity," or "terminal velocity." At this velocity, the downward acting immersed weight and upward drag force resisting the motion are equal and oppositely directed. Let us denote the settling velocity of any size fraction i of particles by v_i. What is required for deposition to occur is for v_i to be significantly greater than the typical velocity (u_*) of whirling eddies in turbulent flow. This velocity u_* is called the "friction velocity." If $v_i < u_*$ then before such a particle has settled very far, it will experience the velocity u_* of the small eddies and be pushed in somewhat random directions. Thus, this fine fraction of sediment is unlikely to deposit until a situation of much quieter and less turbulent water is reached.

This fine fraction of sediment is similar in behaviour to the "suspended load" in streams or rivers, where it is distinguished from the "bed load" which remains close to the bed of the stream even when it is moving.

In such erosion theory for equilibrium or near-equilibrium conditions (e.g., Rose & Hairsine, 1988), it is shown that a single quantity derived from the settling velocity characteristic represents the role played by deposition in controlling sediment concentration. This single quantity, called the "depositability" of soil, is calculated by reading off the settling velocity at regular steps of the ordinate in Fig. 2, and then finding the mean of the sum of such settling velocities. The depositability is defined as $\sum v_i\, I^{-1}$, where I is the total number of equal mass intervals into which the ordinate in Fig. 2 is arbitrarily divided.

The limits used in calculating the summation $\sum v_i\, I^{-1}$ from data (such as in Fig. 2) are of significance if the summation is to accurately represent the role of deposition in the entire erosion/deposition process. The upper and lower limits discussed in earlier paragraphs should be used in this summation. In practice, recognition of the upper limit in this summation is more important than the lower limit because of the proportionately far greater contribution made to $\sum v_i\, I^{-1}$, by larger aggregates.

Let c_i denote the sediment concentration of the class of particles (or aggregates) whose settling velocity is v_i. Then it follows from the definition of these terms that the rate of deposition (d_i, defined as the mass of sediment reaching the soil surface per unit area per second) is given by:

$$d_i = v_i\, c_i\ (\text{kg m}^{-2}\,\text{s}^{-1}) \tag{4}$$

Equation (4) holds for all sediment for which $v_i > u_*$.

Sediment concentration depends on the difference between the rate of sediment addition to overland flow by the erosion processes active and the rate of deposition, given by the sum-

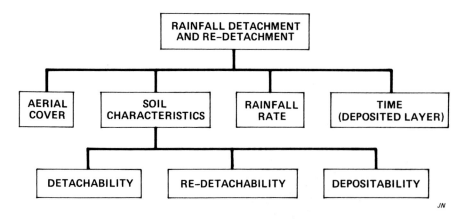

Figure 4: Schematically categorizing factors affecting rainfall detachment and redetachment.

mation of d_i in equation (4) over the total velocity range of depositing sediment.

Erosion processes, and factors on which their rates depend, will be considered in the sections which follow. In general, more than one type of erosion process is active during any erosion event.

Rainfall detachment and re-detachment

The removal of soil for the first time in an erosion event by the impact of raindrops is referred to as rainfall detachment. Such detachment is caused by the locally intense shear stresses generated at the soil surface unless it is protected by a significant depth of water or by other cover. The process is represented diagrammatically in Fig. 3. Some of the sediment removed in rainfall detachment is splashed into the air; but if there is a covering water layer, this layer captures much of the detached sediment.

Figure 3 illustrates the fact that in time some of the detached sediment settles back to the soil surface (or deposits), covering at any time some fraction (*H*) of the original soil surface. This layer of deposited material is also subject to rainfall impact, and its removal by rainfall impact is referred to as "re-detachment" (Rose & Hairsine, 1988).

A major reason for distinguishing between detachment of original soil and re-detachment of the deposited layer lies in the strength of the material being detached, which can have a major effect on its rate of removal. The original soil will typically possess some cohesive strength. Previously detached sediment deposited on the soil surface typically has little time to re-build cohesive links with neighbouring particles, and thus the deposited layer is very weak (Rose *et al.*, 1990). Thus, the energy needed to remove unit mass of sediment is much greater for the original soil (if cohesive) than for the deposited layer.

Another reason for drawing attention to the deposited layer is that, at least for a significant period of time, the deposited layer is coarser than the original soil, though this difference is reduced as processes move towards a dynamic equilibrium. During this period of time, eroded sediment produced by rain-

fall detachment and re-detachment is substantially finer in the runoff than the original soil. This fineness can have important consequences in enhancing the loss of sorbed nutrients or other chemicals over what might be expected on the basis only of soil loss.

Rainfall detachment and re-detachment can be the only erosive processes on relatively flat regions of small extent. For such small low-slope regions, the conceptual model mentioned earlier embodied within the USLE is indeed adequate, with any slow-moving overland flow playing only a passive role. This passive role is in simply transporting sediment whose concentration is determined by the net outcome of rainfall detachment and re-detachment adding sediment to the water layer, and deposition removing sediment.

Figure 4 illustrates factors on which the rate of detachment and re-detachment depend. Any cover which intercepts raindrops, breaking them up into very fine drops, or which collects rainfall and allows the water to reach the ground in a non-erosive manner is directly effective in reducing the rate of rainfall detachment or re-detachment. The role of soil and rainfall characteristics listed in Fig. 4 will be discussed below. Time (Fig. 4) plays a role, even within erosion events, because of changes which take place in the deposited layer, and perhaps because of the development of other surface changes which can lead to the development of surface crusts on drying. The strength, degree of aggregation and other soil characteristics can also change with time.

The theory given in Hairsine & Rose (1991a) shows why, at equilibrium, the sediment concentration (**c**) due to the interaction of rainfall detachment, re-detachment and deposition, is given approximately by:

$$c = \frac{H \, a_d \, P}{\Sigma \, v_i \, I^{-1}} \quad (\text{kg m}^{-3}) \qquad (5)$$

where *H*, the fractional coverage of the soil surface by the deposited layer, appears to have a value of approximately 0.9 at equilibrium (Proffitt *et al.*, 1991), a_d is a re-detachability coefficient and *P* is rainfall rate. Thus, at equilibrium (which

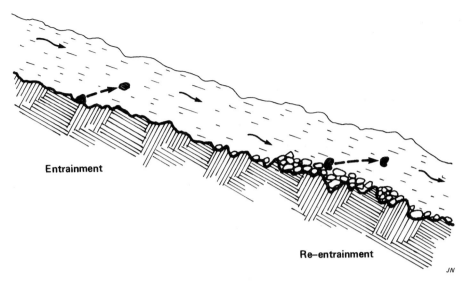

Entrainment

Re-entrainment

JN

Figure 5: Schematically illustrating the distinction between runoff entrainment of the original soil matrix, and re-entrainment of previously eroded and deposited sediment.

may take some 10–15 minutes to achieve even under steady rainfall), c is proportional to the rainfall rate and inversely proportional to the depositability $\sum v_i I^{-1}$.

At equilibrium, the detachability of the original soil plays a relatively minor role in affecting the sediment concentration, though it affects the time required to achieve an equilibrium situation in which – under rainfall of constant rate – the sediment concentration and settling velocity characteristic are constant.

Runoff entrainment and re-entrainment

If water runs over soil with sufficient velocity, even if rainfall has ceased, it can be a potent agent of erosion. If the overland flow of water is concentrated into a channel or rill, it can be especially erosive. The turbid sediment-laden waters common in floods are testimony to the erosive power of flowing water, as is coastal erosion.

It is, therefore, clear that the conceptual model behind the USLE in general must be expanded to acknowledge the role of overland flow itself as an erosion agent. The expanded word model is then that: "Raindrops detach, and overland flow both erodes and transports sediment." As will be shown below, the greater the rates of erosion, the less important the processes of rainfall detachment and re-detachment become relative to the erosive action of overland flow.

When water flows over a sloping land surface, a shear stress develops between the land surface and the flowing water. At equilibrium or non-accelerating flow, the stress acting upslope on the flow must be equal and opposite to the component of the weight of water acting down the plane. Also, the stress acting upslope on the water has an equal and opposite reactive stress acting downslope on the soil. Removal of original soil from the surface due to this stress can be called "entrainment." (Terminology is not uniform, and the term "runoff detachment" is also in use to describe this process).

In considering rainfall detachment and re-detachment, we saw that erosion altered the nature of the soil surface, in particular by partly covering the original soil surface by a deposited layer of sediment, generally weaker and coarser than the original soil. This layer, which partly blankets the original soil surface, forms even more quickly when runoff entrainment is the erosion process. Sediment is removed from this deposited layer due to exactly the same mutual shear stresses which, acting between the soil and overland flow, cause entrainment. Such removal from the deposited layer is referred to as "re-entrainment."

Distinction between entrainment and re-entrainment is important because of the weaker nature of deposited sediment. A significant amount of work has to be done in entraining the original soil and tearing it loose from the cohesive forces holding it to its neighbouring soil. In contrast, the deposited sediment does not have time to re-establish cohesive bonds before it is removed again. The two processes are illustrated schematically in Fig. 5.

Bagnold (1977) introduced the concept of "stream power" (Ω), defined as the rate of work of the shear stress (τ). Thus, if V is the velocity of flow, then

$$\Omega = \tau V \, (\mathrm{W\ m^{-2}}) \tag{6}$$

By equating the shear stress (τ) acting up the plane on the water and the component of the weight of water down the plane, it follows that:

$$\tau = \rho g S D \, (\mathrm{W\ m^{-2}}) \tag{7}$$

where ρ is the density of water, g the acceleration due to gravity, S the slope of the soil surface (the sine of the slope angle), and D is the depth of water.

Substituting for τ from equation (7) into (6) gives:

$$\Omega = \rho g S D V \, (\mathrm{W\ m^{-2}}) \tag{8}$$

A process model of re-entrainment

The theory given here is similar to that in Rose & Hairsine (1988). When water flows down a non-erodible surface such as concrete, all the loss of potential energy must reappear as heat, because energy in all forms is conserved (the principle of conservation of energy). Thus, when water flows over erodible soil, it would be expected that not all the potential energy lost by the water in doing work against the opposing shear stress (τ) would be used in entraining or re-entraining sediment. Suppose the fraction of the stream power effective in such erosive processes is F (where the value of F must be somewhere between zero and one, and be determined experimentally).

Work carried out in the Griffith University Tilting Flume Simulated Rainfall facility has found that $F = 0.1$ approximately for turbulent flow. This value has been shown (Proffitt, 1988) not to depend on soil type and to increase to about 0.2 for laminar flow. Thus, fraction F appears to be a characteristic of fluid behaviour and does not depend, for example, on whether either entrainment or re-entrainment is the dominant erosion process or on the type of soil being eroded. The rate of working of the shear stress effective in eroding soil is thus $F\Omega$.

Since only some 10% stream power goes into eroding soil, the remaining 90% of stream power (or $(1 - F)\Omega$) ends up as heat.

It follows that the effective stream power in entrainment and re-entrainment is $F\Omega$, and that, from equation (8):

$$F\Omega = F\rho gSDV \text{ (W m}^{-2}) \tag{9}$$

It is assumed that if the soil surface is completely covered by a layer of freshly deposited sediment (i.e. $H = 1$), then entrainment will not occur and re-entrainment is the only erosion process. It is also assumed that freshly deposited sediment is of negligible strength, since it has had no opportunity to develop cohesive bonds with neighbouring recently deposited sediment. Thus, the work required to re-entrain sediment is only that needed to lift the sediment against its immersed weight in water (whereas, in entrainment also work would be done against cohesive forces in removing sediment from neighbouring sediment).

The immersed weight of a volume V_s of sediment in water is less than its weight (mg, where m is its mass and g the acceleration due to gravity). This reduction is because of the upward buoyancy force.

Following Archimedes, this buoyancy force is equal to the weight of fluid displaced by the volume V_s of sediment, this fluid weight being $V_s\rho g$, where ρ is the density of water.

The volume of sediment $V_s = m\,\sigma^{-1}$, where σ is the sediment density. Thus, the immersed weight of sediment (W_s) is given by:

$$W_s = mg - (m\,\sigma^{-1})\rho g \tag{10}$$

$$= mg\left(\frac{\sigma - \rho}{\sigma}\right) \quad \text{(N)},$$

where N is the unit of force (Newton).

In flow of depth D, the mass of sediment per unit area is Dc, where c is the sediment concentration. Substituting Dc for m in equation (10), the weight of sediment per unit area is:

$$Ws = Dc\ g\left(\frac{\sigma - \rho}{\sigma}\right) \quad \text{(N)}. \tag{11}$$

If we temporarily assume all the sediment has a settling velocity v (m s^{-1}), then the rate at which mechanical work has to be done per unit area against its immersed weight W_s

$$= force \times \frac{\text{distance}}{\text{time}}$$

$$= W_s \times v, \text{ which from equation (11) is}$$

$$Dc\ g\left(\frac{\sigma - \rho}{\sigma}\right) v \tag{12}$$

The source of energy to do this work is the effective stream power $F\Omega$.

Equating the expression for $F\Omega$ given in equation (9) to the expression for the rate work expenditure per unit area in raising re-entrained sediment against its immersed weight (equation (12)), then finally

$$c = \frac{F\rho}{v}\left(\frac{\sigma}{\sigma - \rho}\right) SV \quad \text{(kg m}^{-3}). \tag{13}$$

Equation (13) has been derived assuming a complete coverage of deposited sediment so that all sediment is eroded by re-entrainment. Thus, equation (13) is an expression for the *maximum* sediment concentration that can result due to overland flow. If some of the energy $F\Omega$ is used in entraining original cohesive soil, then the sediment concentration will be less than that given by equation (13).

Recognizing that sediment consists of aggregates or particles having a wide range of distribution of settling velocities in water, and not the single velocity v assumed in equation (13), sediment can be considered to consist of a number (I) of sediment size classes of equal mass, each sediment class having a characteristic settling velocity v_i. Then it can be shown that equation (13) still holds if is replaced by the depositability, or

$$\sum_{i=1}^{I} v_i\ I^{-1}.$$

Equation (13) also shows that the maximum sediment concentration increases with slope (S) and velocity of flow (v), which is also affected by slope, but particularly by the formation of rills, which will lead to an increase in , compared with uniform overland flow without rills.

The form of equation (13) is in agreement with the summary of a large number of experiments carried out with cohesionless (sand) material in flumes, those by Yang (1972) being a notable example. The equation given by Yang to summarize

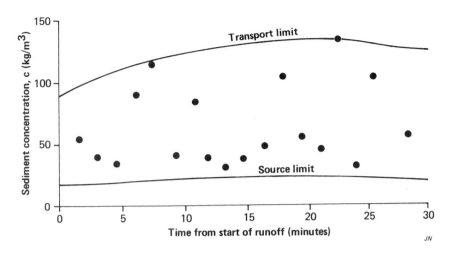

Figure 6: Total concentration for sediment eroded with active rilling from a soil, initially air dry with a plane surface, wet with simulated rainfall, and then subject to 100 mm h^{-1} rainfall and clear-water overland inflow at the top of a 6 m long flume. The flume slope was 6%, and streampower 0.5 W m^{-2} at the flume outlet. Soil was a Udic Pellustert in the Vertisol soil order. Continuous curves are theoretically structural relations for the (upper) transport limit and (lower) source limit. (From Rose *et al*, (1990)).

his experiments on non-cohesive sediments of limited size range, whilst not identical to equation (13), is closely proportional to the product SV in that equation. It may also be shown that there is quite good agreement between equation (13) and the equation of Yang (1972). The product SV is the rate of decrease of the potential energy of unit weight of water flowing on a slope S with velocity V. Yang (1972) termed this "unit stream power".

Foster (1982) drew particular attention to the experimental finding that there is an upper limit to the sediment concentration even in situations where entrainment and re-entrainment are dominant, and introduced the term "transport limit" to describe this phenomenon. This term is widely used in the soil erosion literature, and denoting it by c_t, then:

$$c_t = \frac{F\rho}{\Sigma v_i \, I^{-1}} \left(\frac{\sigma}{\sigma - \rho}\right) SV \quad (\text{kg m}^{-3}). \qquad (14)$$

The derivation given above provides a theoretical basis for such a concentration limit. This derivation indicates that if any higher value of c_t was artificially produced, for example, by externally adding sediment, then the deposition process would quickly ensure that the concentration fell to c_t, since all the available stream power is used in sustaining sediment at that concentration against the process of deposition which continuously acts to decrease it. In terms of this theoretical derivation, the term "deposition limit" would provide a more appropriate description than "transport limit."

The derivation of equation (14) assumed the soil surface to be completely covered by previously entrained and deposited material, also assumed to be effectively cohesionless. Complete coverage by a deposited layer (i.e. $H = 1$) is unlikely to be sustained for any substantial time period, since with $H = 1$ the only source of sediment is the deposited layer itself,

which is likely to become at least partially depleted in a short period of time unless it is being continuously fed sediment by processes such as active headcutting or rill wall collapse. Thus, the upper or transport limit to sediment concentration would be expected to be an unstable upper limit. Figure 6 illustrates that, indeed, this can be the case, where there is some indication of experimental support for an upper limit, even though the concentration of the majority of samples collected at the end of this flume experiment was less than this upper limit.

Once the deposited layer becomes less than complete in its coverage, then the original cohesive soil is exposed, and some of the stream power is expended in entraining such cohesive sediment.

Theory for entrainment and re-entrainment acting together

Since all soils except pure sands usually exhibit some cohesive strength, they offer some resistance to the shear stresses attempting to entrain them. Thus, there must be some energy expended in entraining any mass of cohesive soil.

Rose & Hairsine (1988) introduced as a fundamental erodibility parameter (J) the energy per unit mass of soil entrained, or the specific energy of entrainment. The energy expended in entraining unit mass of soil must be greater than the energy used to simply lift sediment against its immersed weight. Thus, the sediment concentration (c) measured when entrainment as well as re-entrainment is taking place must be less than the upper limiting value c_t given by equation (14). Thus, we can write, in general terms, that:

$$c = c_t - f(J) \, (\text{kg m}^{-3}) \qquad (15)$$

where $f(J)$ is a function of J, which is not known *a priori* and the function would be expected to depend on other variables in addition to J.

Figure 6 illustrates the likelihood that there is a lower as well as an upper limit to sediment concentration. Theory can be developed (Rose & Hairsine, 1988; Hairsine & Rose, 1991b) which indicates a basis for such a lower limit, set by the value of the erodibility parameter J defined above (and other known parameters). This lower limit depends on the value of J for the source of sediment, which is the original soil matrix, and this limit has therefore been called the "source limit."

In less erosive situations than that for Fig. 6, the sediment concentration may not only never reach the upper or transport limit, but may not show any of the oscillatory or limit cycle behaviour hinted at in Fig. 6 at all. In fact, it is quite common for sediment concentration to stay close to the source limit. Theory shows that the source limit concentration does not occur with the deposited layer fractional coverage (H) = 0, but rather H takes an equilibrium value for any particular erosive situation, soil, and value of J.

Using equation (15), if c is measured, c_t calculated from equation (14), and the form of $f(J)$ is known from theory, then the specific energy of entrainment (J) can be calculated. This is how this fundamental erodibility parameter is determined (Rose *et al.*, 1990).

Theory for the function $f(J)$ in equation (15) can be derived by considering the equation for mass conservation of sediment. For situations close to equilibrium, the time rate of increase of storage of sediment can be neglected relative to the spatial rate of increase of sediment transport ($d\,q_s\,dx^{-1}$). Then the mass conservation equation simplifies to equating $d\,q_s\,dx^{-1}$ to the sum of expressions for the rate of entrainment and re-entrainment, from which the rate of deposition (given in equation (4)) must be subtracted.

Solution of the ordinary differential equation for J requires numerical solution of a differential equation. A program called GUEST (Griffith University Erosion System Template) (Misra & Rose, 1990) is available to aid such evaluation.

However, a relatively simple analytic solution in algebraic form of the differential equation involved can be obtained, assuming the value of 2 for the parameter m in the kinematic flow approximation to describe the volumetric flux (q) on a plane:

$$q = KD^m \ (\text{m}^3 \ \text{m}^{-1} \ \text{s}^{-1}) \tag{16}$$

where K is dependent on slope and roughness of the plane surface over which the flow is occurring, and D is the depth of flow. Taking $m = 2$ indicates a state of flow somewhat closer to turbulent conditions (for which $m = 1.67$), than laminar conditions (when $m = 3$). The other assumption is that overland flow occurs on a planar land surface without rills.

For $m = 2$, the function $f(J)$ in equation (15) is given by Rose & Hairsine (1988) as:

$$f(J) = \frac{3\phi\psi}{2} \ (1 - \frac{\psi}{x^{1/2}} + \frac{\psi^2}{2x}), \tag{17}$$

where $\phi = F\rho S(KQ)^{1/2} \sigma / [(\sigma - \rho) \sum v_i \ I^{-1}]$, $\quad$ (18)

and $\psi = J\phi \ (F\rho g S)^{-1}$, $\qquad\qquad\qquad\qquad$ (19)

and x is the downslope distance from the commencement of overland flow, and Q is the runoff rate per unit land area. In these equations, threshold effects, which can be of minor importance, are ignored.

It may be shown that, using the term ϕ from equation (18):

$$c_t = \phi x^{1/2}. \tag{20}$$

It is expected that J will have a strong dependence on soil shear strength. It follows from equation (15) and the form of equations (17) and (19) that sediment concentration c will decrease as J (and so soil strength) increases. From equation (18), an inverse dependence on the soil depositability ($\sum v_i \ I^{-1}$) is still present, as for c_t in (20).

A less obvious consequence of this analysis provides a solution to the problem that different experimenters have found different dependencies of soil loss (and thus c) on slope (S) and slope length (x). It is a consequence of equations (15) and (17)-(18) that as J increases, the dependence of c on slope and slope length decreases. Thus, there is no completely general form of dependence on these geometrical plot parameters as has been assumed in the USLE; for example, the form of dependence is seen to depend on J, and thus on soil strength.

Equations (15) and (17)-(18) are illustrated in Fig. 7 for the assumed values of J shown in the figure. The value of other parameters are taken from Proffitt (1988) for a vertisol (or black earth): $\sum v_i \ I^{-1} = 0.06$ m s^{-1}, $\sigma/(\sigma-\rho) = 2$. Assuming K in equation (16) is given by $K = S^{1/2} \ n^{-1}$, then Manning's $n = 0.03$ and $S = 0.05$ is assumed. Also the runoff rate per unit area, Q, is taken as 50 mm h^{-1} (or 1.39 x 10^{-5} m s^{-1}).

Figure 7 shows c plotted as a function of stream power (Ω), firstly for the transport limit, when $c = c_t$ and J is zero, but also for increasing values of J. Since entrainment and re-entrainment are the only erosion processes represented, c is shown as falling to zero for $\Omega = 0$, whereas there could be a significant contribution to c from rainfall for all Ω, including $\Omega = 0$.

The same concepts which lead to the above equations (17)-(19) for uniform flow also hold when rilling occurs, with the increased complexity in equations arising from the more complex geometry involved with rilling. Because of the availability of computer programs (Misra & Rose, 1989), this complexity is hidden in their use. This analysis provides a theoretical interpretation of the well-known fact that rilling increases sediment concentration when it occurs, other factors being equal or similar (e.g., Proffitt, 1988). The basic reason for this increase is that channelling flow into rills considerably increases the (local) stream power of flow in rills, compared to the same overall flow rate per unit area in uniform flow. Though rilling is not a universal feature, it is quite common.

Following cultivation, soil can gradually regain strength, leading to a decline in sediment concentration, as well as a decline in dependence on land slope and slope length.

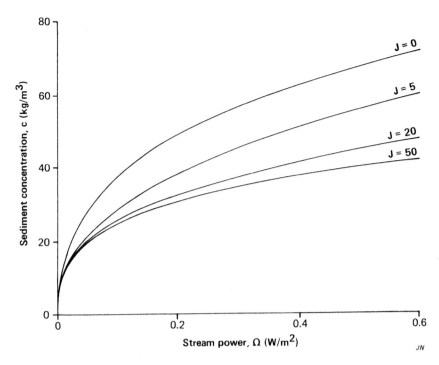

Figure 7: Illustrating the form of variation with streampower in sediment concentration at the transport limit (denoted by $J = 0$) for particular assumptions regarding soil, slope and runoff characteristics. Also illustrated in the lower family of curves is the variation in sediment concentration at the source limit defined in equation (15), calculated assuming the values shown for the parameter J ($J \, kg^{-1}$) which is defined in the text, and is related to soil strength.

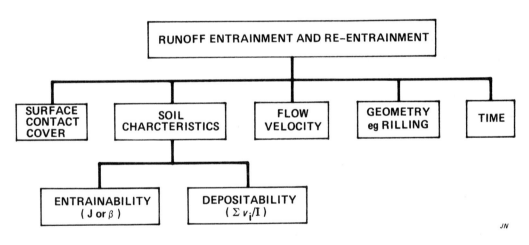

Figure 8: Schematically categorizing factors affecting entrainment and re-entrainment by overland flow.

Figure 8 indicates the factors on which runoff entrainment and re-entrainment depend. Again, cover is important, but for cover to be effective in reducing these processes it has to hug the soil as it needs to be sufficiently close to the soil surface to be immersed in the overland flow. Such cover can be described as "surface contact cover." The aerial cover provided by the upper parts of erect plants may well interrupt raindrops, but their height above the ground surface may be too much to share any of the stresses which lead to entrainment and re-entrainment, and so impede overland flow. This destruction between surface contact cover and aerial cover is of considerable practical importance in soil conservation.

As with the erodibility factor K in the USLE, the specific energy of entrainment (or entrainability, J, Fig. 8) requires runoff plots for its evaluation. One major reason for moving toward erosion models in which processes are represented is that the parameters defined in them should have some physical meaning. The nature of the relationship between J and soil strength is being explored for that reason. There is hope that some readily portable equipment can provide an appropriate measurement of soil strength, leading to the possibility of evaluating soil erodibility in a survey mode. Depositability ($\Sigma v_i I^{-1}$, Fig. 8) can be determined without the need for runoff plots, as will be described below.

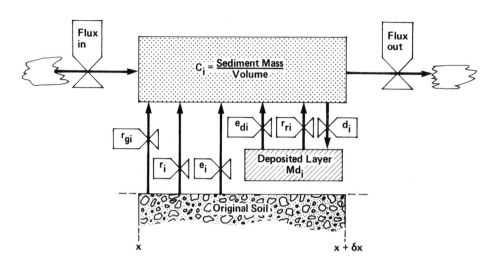

Figure 9: Flow diagram (after the style of Forrester, 1970) describing the interaction of erosion processes between the sediment flux, the original soil and the deposited layer for the range of erosion processes and deposition. Rates of processes (for size class i) exchanging sediment are shown by valve symbols. The water layer and deposited layer are shown artificially elevated for clarity. Rates shown are for deposition (d_i), rainfall detachment (e_i), rainfall re-detachment (e_{di}), entrainment (r_i) re-entrainment (r_{ri}), and gravity-driven processes (r_{gi}).

There is a close and very important interconnection between the factors of flow velocity and rill geometry in Fig. 8. The preferred pathways with rilling substantially increase flow velocity and its ability to entrain soil. The formation of rills of whatever geometry, as with streams and rivers on a larger scale, is testimony to the substantial increase in erosivity associated with a concentration of flow.

Experience in erosion analysis has indicated the great importance of including observational data on rill frequency in particular, but also rill geometry or shape in erosional analysis. The value of J is quite directly affected, as it should be, by such information. Even though the ability to predict the rill behaviour of soil is limited, such behaviour appears to be reasonably repeatable and even typical. Since rill development is visually obvious after an erosion event, it is readily recordable, and post hoc observation appears to be adequate to provide a useful interpretation of erosion phenomena and the evaluation of J.

Relation to unit stream power theory

It was noted earlier in commenting on equation (13) that the product SV, termed "unit stream power" by Yang (1972), is the rate of decrease of the potential energy of unit weight of water. This follows since the decrease in potential energy of a mass in falling through a distance h is given by mgh, where g is the acceleration due to gravity. Unit weight is given by taking $mg = 1$, and so the rate of decrease of unit weight of water is the height through which it falls in one second, which is given by SV for water travelling at velocity V over a surface of slope S.

Yang (1973) showed that for non-cohesive alluvial sediment the sediment concentration was closely proportional to unit stream power, or to SV. Moore & Burch (1986a) noted that there was considerable variability in some of the pub-

lished database for the USLE, but that the form of the LS factor in the USLE (equation 1) was in reasonable agreement with what would be expected if sediment concentration was proportional to SV or unit stream power. This would seem to indicate that the experimental base selected to develop the LS factor must have been for situations where the soil had a low cohesive strength, probably reflecting the recent cultivation characteristic of this experimentation.

Moore & Burch (1986b) developed expressions for unit stream power in situations where flow occurs on rills as well as sheet flow, and showed that sediment transport predicted using Yang's unit stream power equations provided on upper limit to observed transport on a range of soil types, all of which were in an effectively non-cohesive condition. Results also suggested that rill initiation in non-cohesive soils may occur when a critical unit stream power is exceeded.

Since, as noted earlier, there is good agreement between equation (13) and that of Yang (1972), then the data reviewed by Moore & Burch (1986a,b) also provide support for equation (13). These data for soils with a range of settling velocities should be reviewed to see if equation (14) provides even better agreement.

Erosion and deposition processes all acting together

Whilst each process has so far been considered almost in isolation, all processes, of course, commonly act together at the same time. Thus, both the rainfall and runoff driven erosion processes add to sediment concentration, with deposition reducing it. The ability to imagine all the processes acting simultaneously may be helped by likening it to a juggler throwing balls into the air with one hand, catching them with another and throwing them into the air again. When this act continues in a steady way, there will be a constant number of balls in the air. The analogy is a steady rate erosion situation, where the rate of addition of sediment by erosion processes is constant, as is the deposition rate, so that the sediment concentration is also constant. This is illustrated schematically in Fig. 9, where the rates are shown as valve symbols and the sediment is shown artificially elevated above the soil surface to allow the rate processes to be shown more clearly.

It is the outcome of all these processes acting together which is of practical concern. However, since the individual processes depend on quite different soil characteristics or external factors, penalties are paid if the processes are not separately described. Discussion of gravity-aided processes will be given later.

Relative importance of rainfall and runoff erosion processes

The relative importance of rainfall detachment and re-detachment in comparison to runoff and entrainment has been shown by Proffitt and Rose (1991) to depend on the magnitude of the stream power of overland flow. Stream power is defined in equation (8) and is shown to increase both with land slope and the product DV which is the flux of water per unit plane width (q) for uniform flow, for which $q = QL$, the product of runoff per unit area and slope length. As stream power increases, the importance of rainfall impact as an erosive process diminishes relative to those associated with overland flow.

When rills form, and if much of the overland flow occurs in rills, then the stream power is substantially higher than it would be on a nonrilled surface. This was taken into account in the generalization by Proffitt et al. (1991) given above. Due to the irregular micro-topography of natural or cultivated land surfaces, some fraction of the soil surface will normally be exposed to raindrop impact and make some contribution to sediment loss. The relative importance of rainfall and runoff-driven erosion processes is also related to the contribution to total soil loss which comes from rill and interrill regions. This was the subject of a field study in Belgium by Govers and Poesen (1988).

Gravity-driven processes and mass movement of soil

A process model of re-entrainment was derived earlier in which a fraction F of the stream power was assumed to be used in re-entraining sediment which had already been made available by some process which could produce a complete layer of effectively non-cohesive sediment. Especially if sediment at the transport limit is to be sustained, it is clear that entrainment cannot be the source of this previously eroded sediment now forming a deposited layer completely blanketing the original soil matrix. What must be the type of process or processes which can give rise to the sediment forming this complete deposited layer?

In situations of substantial erosion, it has been argued that rainfall detachment is unlikely to be more than a very minor contributor to such sediment. Thus, the only possible answer

to this question of the source of a layer of deposited sediment which is complete – at least in the bottom of rills – must lie in movement of the soil mass by gravity, even though other processes may play a part. The common observation of rill wall collapse, sometimes stimulated by bank undercutting or resulting from rill wall strength failure, is one such gravity-driven process. Rill headcutting is another.

Thus, if sediment concentration is to achieve the transport limiting value, and be sustained at this limit, then there must be a rate of addition of sediment due to processes deriving their energy directly from gravity. Such a rate process is shown schematically on the top of Fig. 9. It is likely that such processes as rill wall collapse and headcutting will be stochastic in nature but may occur with sufficient frequency for the deposited layer to completely blanket the original soil matrix, and so for the transport limit concentration to be sustained.

Such gravity-aided processes are expected to play an even greater role in more extreme forms of erosion which lead to gullies. From this point of view, gully erosion may be considered to be an extreme form of rill erosion when the scale of rilling has so increased that gravity-aided processes dominate entrainment and re-entrainment, and certainly rainfall detachment and re-detachment. Sometimes rather vertical erosion features not unlike rills can form on gully walls, so it may not necessarily be the case that entrainment is completely negligible, but simply quite overshadowed by other erosion processes.

Apart from gullies, other forms of erosion involving mass movement and the shear strength of soils can occur. These can take quite a wide variety of forms and have been well described in Hadley et al. (1985). Mass movement involves the downslope movement of earth material under the direct influence of the weight of this material, the motion not involving fluid as the transporting agent. Mass movement can be slow (creep), but can take a number of qualitatively different fast forms, including earthflows, debris, landslides or soil avalanches, debris flows and even more catastrophic and rapid mass movements in avalanches or the volcanically created "lahar."

In the humid tropics, very high soil losses of cultivated soil on sloping land during rainfall events of high rate and substantial quantity have been reported. Prove (1991) has carried through a thorough study of such soil loss in the wet tropical coast of north-eastern Australia where sugar cane is grown on land of such slope that it is mechanically cultivated in rows up and down slope. Soil losses of 400 $t\ ha^{-1}$ in a single intense rainstorm were recorded. Prove (1991) brought together a number of strands of evidence and theory strongly indicating that mass movement is involved for soil weakened by cultivation, and saturated with water forming a perched water table in the cultivated inter-row zone. In these same intense rainfall events, the higher strength of uncultivated soil and associated differences in hydrology lead to at least an order of magnitude reduction in soil loss, compared to that from cultivated slopes in the same soil.

In these circumstances where soil strength and hydrology dominate, the role of surface cover, so important in reducing soil loss in situations where mass movement of soil is of limited importance, is very secondary in reducing soil loss.

In such saturated situations, Prove (1991) showed that it was largely the cohesive strength of soil which was important, the angle of internal friction not being affected by cultivation and the small effective stress on surface soils making the frictional strength component of less importance.

The work of Prove (l991) and its conclusions are of significant importance to soil conservation in the humid tropics. Theory appears to be adequately developed to indicate conditions under which mass movement of a cultivated layer is likely to occur (Skempton & De Lory, 1957). If this occurs, then loss of the entire cultivated depth of soil will occur.

THE DETERMINATION OF ERODIBILITY AND DEPOSITABILITY PARAMETERS DEFINED IN THE PREVIOUS SECTION (GUEST MODEL)

Depositability

The depositability of soil or sediment was defined earlier as $\Sigma v_i\, I^{-1}$, which can be calculated, once the settling velocity characteristic of the soil has been experimentally determined (e.g., Fig. 2). At least two different types of equipment have been used to measure this characteristic: the top entry tube and the bottom withdrawal tube.

In the top entry tube, illustrated by the equipment described by Hairsine and McTainsh (1986), sediment is added to the top of a column of water at a known time. The mass of soil arriving at the bottom of the tube at measured time intervals is measured, yielding the settling velocity characteristic. Since the faster-settling component of sediment always outstrips the slower component, there is no interaction involved whereby slower settling sediment is overtaken by faster. This may be approximately the case when sediment is depositing in reservoirs and lakes, and perhaps in flood plains and deltas.

In shallow overland flow involving active erosion, deposition occurs in a situation where sediment is being continually added to the water layer. In this situation, sediment is depositing as a polydisperse system in which the faster settling sediment moves through slower settling components. Significant interactions occur in this more complex system, with the wake of faster settling sediment capturing slower sediment which results in a faster rate of settling than would be measured in a top entry tube (Lovell & Rose, 1991). Settling velocity characteristics measured in this more complex or polydisperse mode are more relevant to deposition in shallow overland flow. Such characteristics can be measured in a modified bottom withdrawal tube (Lovell & Rose, 1988a). This tube has one end open and one fitted with a device allowing settling sediment samples to be periodically withdrawn.

The settling velocity characteristic obtained can be affected to some extent by the moisture content of the sample prior to immersion in the tube, if drier than soil with a pore water suction of about 1.5 MPa (Lovell & Rose, l988b). Whilst such effects as slaking can be controlled by gentle wetting procedures, the wetting of dry soil by rainfall in the field may be almost as severe as wetting by sudden immersion in water. Rainfall impact itself can have some effect in breaking down larger structural aggregates. Thus, possible sample pre-treatment prior to the determination of settling velocity characteristics of samples taken dry from the field is a question requiring consideration, and perhaps some investigation. Despite these comments, the settling velocity characteristic of soil is as stable a soil characteristic as many other soil characteristics, though reflecting structural stability.

The modified bottom withdrawal tube technique involves the addition of a sediment sample to the tube. Clods of such a large size that they would clog the withdrawal tube should not be added; in any case, it is unlikely that stable clods of such size will take part in deposition. For deposition to occur, aggregates need to be of a size less than that which results simply in rolling along the bed, and the aggregate size would need to be less than the depth of water in overland flow. An average depth of water for any runoff event, where this average pays attention to the degree and shape of rills (if they occur) is given as an output of GUEST (Misra & Rose, 1990a). This depth then can be used to set an upper limit to the calculation of the effective depositability ($\Sigma v_i\, I^{-1}$) appropriate for that particular runoff event.

Following introduction of the sample into the withdrawal tube, a technique involving manually tipping and rotating the tube full of water and sediment is required in order to distribute the sediment sample uniformly along the length of the tube. This technique is described in Lovell & Rose (1986). When this initial distribution is visually judged to be as uniform as possible the tube is quickly placed (or rotated in a cradle) to an upright position, which gives the time for the commencement of the experiment. At a series of subsequent times, samples are withdrawn into separate containers from the controlled outlet at the bottom of the tube. These withdrawal times can be arbitrary, but are best taken more rapidly in the beginning. Subsequently, the withdrawal times are increased so that an approximately logarithmic form of sampling is produced. This form of sampling times provides a better definition of the settling velocity distribution than if withdrawals were regularly spaced in time.

Voice announcement to a tape recorder when a sample is withdrawn enables the experiment to be conveniently carried out by one person. Provided a stand to hold the tube vertical is provided, the operation can be done in the field, though transport of samples to a laboratory situation is usually more convenient.

Sample withdrawal from the bottom of the vertical tube removes any sediment deposited there since the last withdrawal, together with sediment not yet deposited but taken with the withdrawn sample. The volume of sample withdrawn and the oven-dry mass of sediment it contains are recorded. An alternative to measuring sample volume is to record the height of water remaining in the tube following each sample withdrawal.

Analysis of this raw data to yield the settling velocity curve illustrated in Fig. 2 is tedious to do manually, but can be conveniently assisted by the computer program GUDPRO (Misra & Rose, 1990b) which is a development of the program described and given in Lovell & Rose (1986). GUDPRO also calculates the "effective depositability" ($\Sigma v_i I^{-1}$) if given an effective upper limit to v_i appropriate to a particular runoff event, or calculates the depositability as the same sum but with the upper limit to v_i set by the maximum size of water-stable aggregates or by the size of exit from the tube.

Rainfall detachability and re-detachability

The determination of depositability was described first since the theory outlined earlier indicates that the type of erodibility parameters defined in this theory require for their evaluation a knowledge of depositability (or rather the effective depositability appropriate to a particular recorded erosion event).

The theory of sediment concentration and transport at equilibrium or near-equilibrium conditions in the absence of any flow-driven processes is given by Hairsine & Rose (1991a). This theory defines as one erodibility parameter the detachability of the original soil matrix to rainfall impact. This detachability (a) is defined by:

$$e = aP \quad (\text{kg m}^{-2}\,\text{s}^{-1}) \tag{21}$$

where e is the rate of detachment per unit area of soil by rainfall of rate P. The value of a is a maximum for shallow water depths, but decreases for depths greater than a breakpoint depth of the order of a couple of raindrop diameters.

However, as described in the previous section, after a time much of the soil surface becomes covered by a deposited layer. At equilibrium, this coverage is so relatively complete (>90%) that more important than the soil detachability given in (21) is the re-detachability (a_d), given in relation to equilibrium sediment concentration in equation (5).

Whilst the detachability (a) would be affected by soil strength, the re-detachability (a_d) would not be so affected and there is reason to believe it may bear some inverse type of relationship to the depositability. The value of a_d can be calculated from the measured sediment concentrate at equilibrium (c) under a rainfall rate (P) using equation (5). In that equation, the best current advice is that $H = 0.9$ approximately, and the effective depositability ($\Sigma v_i I^{-1}$) is calculated for the appropriate water depth, as described above.

From here on in this section, the methodology of erodibility parameter determination is that recommended for and used in Project 8551 of the Australian Centre for International Agricultural Research (ACIAR). This project is entitled "The Management of Soil Erosion for Sustainable Crop Production"

and involves two collaborating groups in the Philippines, and one each in Malaysia and Thailand. The project is collaboratively directed by Griffith University, Brisbane, and the Queensland Department of Primary Industries. The advice given below has been greatly aided by the experience of all collaborators in that project.

The detachability parameters a and a_d given in equations (21) and (5) respectively are defined assuming rainfall impact is the erosion mechanism, not overland flow. Thus, to determine these parameters, sediment concentration needs to be determined in a situation where the erosive effect of overland flow is negligible, as expressed, for example, by stream power (Ω, equation 8), . Thus the experimental situation needs to be such that the product SVD (equation 8) is small. This could be ensured by measuring sediment concentration for bare soil exposed to rainfall and with a small surface slope (S), and small slope length L (since $DV = QL$, where Q is the runoff rate per unit area for flow over a uniform surface).

Figure 10 illustrates the type of equipment used in the ACIAR project to measure the sediment concentration under the restrictions outlined in the previous paragraph. The equipment, called a "detachment tray", provides a convenient way of catching all sediment shed in runoff from a small area of soil. Soil in the tray should be in a condition as similar as possible to the soil in the runoff plot to which the detachability parameter values will be applied.

The modest scale of equipment sketched in Fig. 10 requires justification. On two sides of the detachment tray a splash guard is provided so that soil is not only lost by splash from the central area of the tray from which sediment is collected, but it can also be returned. As the area of soil impacted by drops gets smaller, the net loss by splash increases. However, once there is an effective depth of overland flow, most of the detached material is added to the surface water and is not splashed out of the area (Walker et al., 1978). Whilst the case has not been established that the detachment tray in Fig. 10 is too small, it should probably be regarded as the minimum recommended size. If constructed any smaller, net splash loss may become sufficiently significant to affect the measured sediment concentration, and thus the detachability parameter values.

The tray is supported at a small angle of a few per cent to the horizontal, enabling sediment to flow to the sediment collection tray. Exactly what this angle is should be of no consequence, provided the threshold stream power for entrainment is not exceeded.

The value of a_d can be calculated using equation (5) from measured c and P (assuming $v_i I^{-1}$ known from the settling velocity characteristics, and $H = 0.9$). Theory (Hairsine & Rose, 1991a) then shows that:

$$a = Q\left[P\,c^{-1} - \{(\Sigma v_i I^{-1})\,a_d^{-1}\}\right]^{-1} \;(\text{kg m}^{-3}) \tag{22}$$

Whilst the value of a does not have as dominant an effect as a_d on the concentration achieved at equilibrium, it has a more

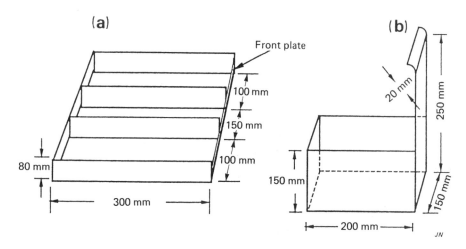

Figure 10: Illustrating construction of the detachment tray used to measure the soil detachabilty characteristics under rainfall. The detachment tray (a) is filled with soil in its centre section, and the two outer guard areas are designed to mimize net loss of soil by splash. Sediment which runs off the centre section of the detachment tray (which is placed at a small angle to the horizontal) is caught in the collection tray shown in (b).

substantial effect on the rate at which that equilibrium is achieved.

Erodibility parameters for entrainment and re-entrainment

Determination of the erodibility parameters involved in entrainment and re-entrainment typically involves what is called a "runoff plot." A runoff plot is a segment of land so established that the runoff measured leaving the plot can be said to have originated from rainfall received by the plot, at least with a considerable degree of certainty. Thus the upper end of the plot, and the sides of the plot which run up and down slope, are defined in such a way as to prevent surface flow across them from occurring. A sketch of such a plot is given in Fig. 11. The upper end of the runoff plot may be defined by a ditch and side walls by a mound of soil, as illustrated in Fig. 11. In some experiments, these boundaries have been defined by a strip of galvanized iron driven into the soil.

Such runoff plots were used in the USLE experimentation discussed earlier, with the total amount of sediment and water leaving the plot being measured as described, for example, in Hudson (1971) or Morgan (1986).

The theory of entrainment and re-entrainment outlined in the earlier section involves not just the total volume of runoff, but the rate of runoff, which varies with time. The theory also describes how sediment concentration will vary when rates of runoff and rainfall vary throughout an erosion event. The rate of runoff has been typically measured by passing water through some standard form of flume and recording the height of water at a suitable section in the flume (Turner, 1984).

In the ACIAR experiments previously referred to, this method of measuring runoff rate has been used, but more commonly the "tipping-bucket" type of flow rate measuring technology, similar to that described by Bonell & Williams (1986) and Williams & Bonell (1987) has been used. This technology

is a large-scale version of the tipping bucket commonly employed in the measurement of rainfall rate. Water directed to this measuring device proceeds to fill one of two buckets into which the collecting device is divided. When filled to a certain level, this over-centre device tips, shedding the volume of water required to cause tipping, and exposing the second collecting bucket to the directed flow. Tipping thus indicates that a known volume of water has been collected. The information automatically collected on some form of digital electronic recorder is the time at which the bucket device tipped. The recorder is fed an electrical signal as the bucket tips, generated, for example, by a magnetic switch.

From this basic recorded information on the time of tipping, and from calibration of the volume of water required to cause tipping (which varies with tip rate), the volumetric flow rate can be calculated.

Measurement of sediment concentration as a function of time will be discussed below. More commonly, only the total quantity of sediment lost from the runoff plot is measured. In the ACIAR experiment referred to, sediment leaving the plot is collected in either a concrete apron or metal channel constructed to have a low slope of the order of 1%. This low slope (compared to that of the plot) leads to net deposition of a substantial fraction of the sediment leaving the plot, which is collected and weighed, and subsampled for the determination of sediment water content (used to convert from a wet to over-dry basis).

Sediment which did not deposit in this low slope collecting area or channel was passed through the flume or tipping bucket device and a small subsample withdrawn by gravity into a container. The sampling device used is simply a copper pipe in which a slot was cut with a hack-saw. This slotted pipe is located near to the exit from the collecting apron and near to the entry to the tipping bucket flow rate measuring equipment. The rate of this suspended sediment collection would increase

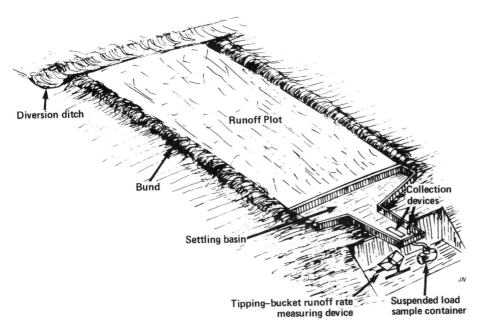

Figure 11: Illustrating a hydrologically defined runoff plot. Water entry at its upper end is prevented by the diversion ditch. Runon or runoff can be prevented from occurring laterally to or from the plot by a low bund (illustrated) or galvanized iron sheeting, for example. In the AClAR experiments referred to in the text, the settling basin was concrete and of lower slope than the plot, allowing coarser sediment (or "bed load") to settle out on it for later collection. A (flow-weighted) sample of "suspended sediment", defined as that which is lost from the settling basin, can be collected from a split pipe shown located at the end of the settling basin, and stored in a container shown located beneath the settling basin. The other collection device shown is a slot in the end of the settling basin through which runoff falls into a tipping-bucket flowrate measuring device. The time of bucket tipping is recorded on an electronic data-logger. (The end of the settling basin need not be cantilevered as shown; beneath it is shown cut away for clarity).

with flow rate so that the average concentration of sediment collected would be weighted for flow rate in some way. The total of this suspended sediment lost during the erosion event was then calculated from the mean event sediment concentration in the collected sample and the total volume of water shed by the plot during the event.

Total soil lost by the plot is then the sum of this loss of "suspended" sediment, and the coarser fraction left on the collecting apron (in some ways analogous to the "bed load" of streams), following a similar procedure to Bonell & Williams (1987).

The fraction of stream power used in the processes of entrainment or re-entrainment was given the symbol F in the theory outlined in the previous section. The magnitude of fraction F can be evaluated using equation (14), or an equivalent appropriate expression when rilling occurs, provided all other terms than F are measured, including the sediment concentration at the transport limit.

Ensuring that the measured sediment concentration is at the transport limit is vital to the evaluation of F. Since this concentration is an unstable upper limit, ensuring that it has been achieved commonly requires sediment samples to be taken as a function of time, as was illustrated in Fig. 6 – where the sampling was carried out manually. Manual sampling in the field requires the presence of observers during the erosion event, which is not easy to achieve. Equipment which is automatically triggered by an erosion event, and which automatically

takes samples from an erosive flow according to some pre-programmed instructions, does exist, though development in terms of reliability and suitability for this purpose can be said to be continuing. Earlier available equipment was designed more for taking water quality samples where the sediment load is small. Some modification of such equipment has proved necessary.

An advantage of manual sampling is that the timing of sample collection can be guided by the sampling person to include samples when the sediment concentration is likely to be high. Times of active rill development, or times when sediment is delivered following rill wall collapse, would be sampling times when the transport limit is likely to be achieved. Thus, the presence of an observer able to take manual samples can increase the likelihood of samples at the transport limit being collected. However, plotting the sediment concentration of automatically collected samples can also provide information on the maxima achieved though there is then no corroborative visual evidence of very active erosion, and so less certainty that these maxima do represent a transport limit concentration.

Whilst data from the field in ACIAR experiments are accumulating, data obtained using the Griffith University Tilting Flume Simulated Rainfall (GUTSR) facility still probably provide the soundest database for the value of F. This facility is shown in Fig. 12. The experiments of Proffitt (1988), indicating the same pattern of behaviour and values of F for two contrasting soil types, suggest that the value may be, as its defini-

Figure 12: The Griffith University Tilting Flume Simulated Rainfall Facility (GUTSR): soil is placed in the 6.0 m long by 1.0 m wide flumee which can be tilted over a wide range of angles. Simulated rainfall falls through 9 m and is protected from wind effects by side curtains. Water can be added at the top of the flume.

tion suggests, more to do with fluid behaviour than with soil properties. For both soils, F was found to have a value of close to 0.1 at values of stream power higher than about 0.5 W m^{-2}, when the flow would be expected to be fully turbulent (Fig. 13). At lower stream powers, the value of F increased to a value of approximately 0.2 (Fig. 13).

This evidence suggests that the value of F may be related to the degree of turbulence in the flow, with the increase in F at lower stream powers from about 0.1 to 0.2 (Fig. 13) possibly corresponding to the decrease in turbulence toward non-steady laminar flow at low stream powers.

Whilst some support from field studies for the values of F shown in this figure has been obtained, it may take some time

before field studies can provide any improvement on Fig. 13. The reason for this is as follows: To evaluate F, it must be ensured that transport limiting conditions are achieved. This requires good observation during an erosion event in which such conditions are achieved, which is not an easy task. Thus, since Fig. 13 provides the most reliable data on F under equilibrium conditions, these data are currently used in analysis.

It is not uncommon for the maximum sediment concentration reached in an erosion event to be less than the transport limit. If so, the question arises, how can the value of the specific energy of entrainment, J, be evaluated for such an event when, as shown in equation (15), the value of f (J) is given by

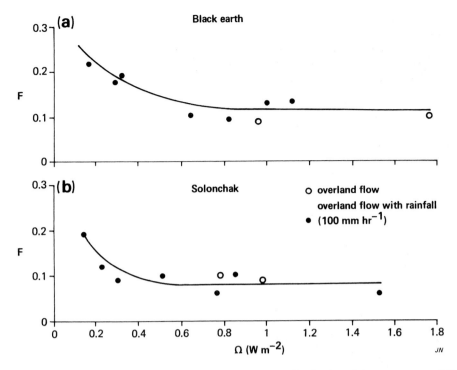

Figure 13: Illustrating values of F as a function of the localized value of the streampower (Ω) for two different soil types ((a) and (b)), determined with rainfall of rate 100 mm h^{-1} or for overland flow without rainfall. Relations were very similar for either soil type. (After Proffitt, 1988).

$(c_t - c)$, the difference between c_t and the corresponding value of c at the source limit (of Fig. 6). Once the value of $f(J)$ at any time during the erosion event is known, then it follows from equations (17) – (19) (for uniform overland flow) that J can be determined. Indeed, the value of J can be determined at all times through an erosion event for which the transport and source limiting sediment concentrations are known. The only caveat to this statement is that the number, size and shape of rills can change during an erosion event, and information on such rill features is not commonly available, except in such controlled experimentation as is possible in the GUTSR shown in Fig. 12.

However, the number, size and shape of rills found following an erosion event would be expected to closely reflect the situation during the most erosive period of the event when the flow rate would also be close to its maximum recorded value. In the ACIAR experimentation, it is this post-event evidence on rill characteristics which are assumed to have been typical for the event as a whole.

Whilst it is possible to measure sediment concentration as a function of time in the field, either by hand or using automatic equipment, ACIAR project experience has been that such data are far less common than data where soil loss measurement is restricted to total soil loss for the event (obtained as described above). Thus, an approximate method of analysis is required in which the data requirements are restricted to total soil loss. This method for determining an approximate erodibility parameter for the processes of entrainment and re-entrainment will now be described.

Empirical Erodibility Parameter for the Entrainment and Re-entrainment Processes

At the present stage of technology in soil erosion research, the measurement of sediment concentration as a function of time during an erosion event is not common. This currently limits evaluation of the fundamental erosion parameter, J, the specific energy of entrainment. However, the insight gained through the development of entrainment theory involving J, and the role of J in affecting sediment concentration as is illustrated in Fig. 7, has provided a basis for the definition of an empirical erodibility parameter, β, which is related to J if sediment concentration stays close to the source limit. The advantage of this empirical erodibility parameter β is that its evaluation does not require the measurement of sediment concentration as a function of time, rather, only total soil loss is required.

With a knowledge of total soil loss, the average sediment concentration, $\bar{c}$, can be calculated by dividing the total soil loss by the total volume of water lost. Strictly, the total volume of runoff used in this calculation should be the runoff for which the rate of runoff exceeds the threshold for entrainment, though this may be a very minor difference in major erosion events. In the ACIAR experiments, a computer program (DUFLO) is available which determines the time of commencement and completion of effective runoff when the threshold stream power for entrainment to take place is exceeded. Whilst the magnitude of this threshold stream power does not seem to vary too greatly for cultivated soils, it is preferable to determine it experimentally. This same program

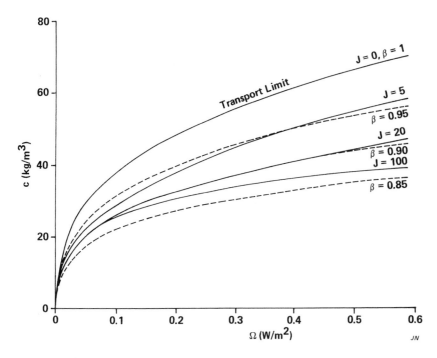

Figure 14: Sediment concentration (c) calculated as a function of streampower (Ω) for particular values of parameters J and ß defined in the text. The curves for $J > 0$ and ß < 1 are calculated assuming that sediment concentration remains at the source limit corresponding to the chosen values of J.

DUFLO for duration of effective flow) thus allows determination of the effective volume of runoff during the effective time period (or periods) during the erosion event.

The empirical erodibility parameter β is defined from the following equation:

$$\bar{c} = c_t^{\beta} \text{ (kg m}^{-3}\text{)}, \tag{23}$$

where the value of the parameter β should be less than one. If a value of β is calculated using equation (23) to be greater than unity, then this would indicate that some other process, such as mass movement, would be taking place in addition to re-entrainment. Thus, the empirical erodibility parameter β indicates the extent to which $\bar{c}$ is less than the sediment concentration at the transport limit.

Equation (14) shows that c_t is proportional to the velocity of flow, V. For uniform overland flow the volumetric water flux per unit strip width, q, is given by:

$$q = KD^m \qquad \text{from equation (16)}$$
$$= DV \qquad \text{by definition,}$$

from which it follows that

$$V = KD^{m-1}$$
$$\alpha q^{(m-1)/m}$$

If turbulent flow is assumed, $m = 5/3$, so that:

$V \alpha q^{0.4}$, so that from (14)

$c_t \alpha q^{0.4}$, and since $q = QL$, then

$$c_t \alpha Q^{0.4}, \tag{24}$$

where Q is the rate for unit area, which is known as a function of time in all ACIAR experiments. It follows from equation (24) that in calculating c_t, the appropriate average value of Q to use is that which is obtained by averaging values of $Q^{0.4}$, and not by averaging Q. This appropriate average value of Q is calculated in the ACIAR program DUFLO. A sister program (XDUFLO) carries out the second type of calculation if rilling or concentrated flows are observed to occur.

That there can be an approximate relationship between the empirical erodibility parameter β, and the specific energy of entrainment, J is illustrated in Fig. 14 for the same set of conditions as applied for Fig. 7. Figure 14 illustrates that the relationship between c and Ω for any particular chosen value of J can be closely fitted by selecting a particular value of β. Note that in Fig. 7 no contribution to c from rainfall was assumed.

The reason why it was said above only that there *can* be a relation between β and J is nothing to do with the degree of approximation in fitting the curves as illustrated in Fig. 14, but for the following reason. As was discussed in relation to Fig. 6, the value of J is evaluated using the source and transport limiting values at any time during the event. The source limit is not known in the data situation when only mean sediment concentration is measured. From equation (23) it follows that the value of β reflects that mean sediment concentration, which in Fig. 6 will depend on the whole history of concentration excursion between the transport and source limits. Thus, in a highly erosive situation such as that illustrated in Fig. 6, the value of β will not correspond to the value of J evaluated using the source limit. In a less erosive context, where the sediment concentration is mostly close to the source limit, then the value of

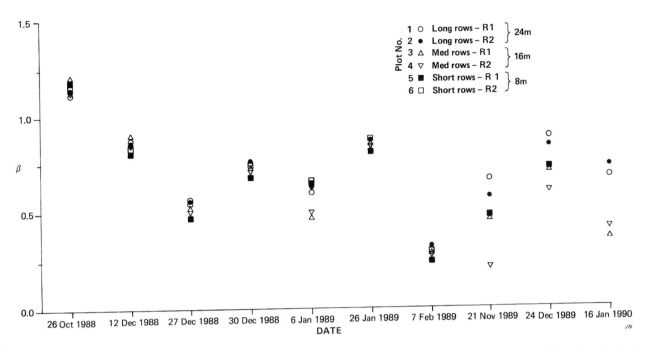

Figure 15: Variation over a 15 month period in values of the erodibility parameter ß defined in the text, calculated from data obtained in the ACIAR experiments referred to in the text. Soil is a gravelly lithosol under pineapple cultivation on steep (~30%) slopes. Two replicates are shown for each row length of up and downslope cultivation (8m (□,■); 16m (Δ,∇); and 24m (o,•) with a deeply incised rill developing on the 24m plots after 12/12/88. Site: Walker's Farm, W3. Data from C. Ciesiolka (pers. comm.).

β obtained will reflect the value of J, and the apparent inverse type of relationship between the two parameters (illustrated in Fig. 14) will be physically meaningful.

Thus, even in a situation where c as a function of time is known, so that J can be evaluated (e.g., Fig. 6), determination of the empirical parameter β is still very useful for predictive purposes. The value of β is determined using equation (23), the measured value of $\bar{c}$, and c_t calculated, using measured values of runoff rate. With the value of β thus determined, the average sediment concentration can be predicted for other situations in which the value of β is appropriate, using information on runoff rate. If infiltration or runoff characteristics can be defined, then the value of β can be used for a longer term prediction (Ward & Rose, 1989). However, the value of β will change with time, for example, tending to fall following cultivation, as is illustrated in Fig. 15 from the ACIAR program. The value of β for the first measurement date (26/10/88) is greater than unity. Some mass movement on this highly-sloping pineapple land (average slope of 30%) is not impossible, especially when freshly cultivated, and this may be why $\beta > 1$ was obtained.

The tendency for β to decline with time (following the cultivation which occurred prior to the first observations) probably reflects the strengthening of the soil associated with consolidation, and with the loss of fines from the soil surface illustrated by the development of some armouring by sand-sized particles. Growth of pineapple plants and the associated development in aerial cover probably played a minor role in this reduction as well. With pineapples being planted on the top of ridges, they provide little cover close enough to the soil sur-

face to affect entrainment or re-entrainment, which were expected here to be the dominant erosion processes.

Whether or not it is likely that the value of β obtained will give a usefully accurate estimate of J may be judged by the severity of the erosion event, as illustrated by the degree of rilling and by the value of β itself. In general, the lower the value of β, the more likely it is that the value of β gives a physically meaningful indication of the value of J. If so, then the relationship between β and J as is illustrated in Fig. 14 can be used to infer J from β. This can be done whether or not rilling takes place.

The previous section has outlined the theory and understanding of erosion and deposition processes developed in the Division of Environmental Sciences at Griffith University, Australia. This section has illustrated how this theory has been applied using the computer program GUEST to evaluate erodibility parameters, and GUDPO to evaluate the depositability parameter. (These programs are available from the author).

Once erodibility parameters and depositability have been determined, it is a simple matter to use the same equations in a predictive sense, using the values of these now available parameters. The program converted into this predictive format is called GEMS (Griffith University Erosion Management System). This program includes the important effects of cover in reducing soil loss.

The following section proceeds to outline an alternative approach to erosion modelling developed in the USA, though the reader will find similarities and differences between the two approaches.

WEPP PROFILE MODEL

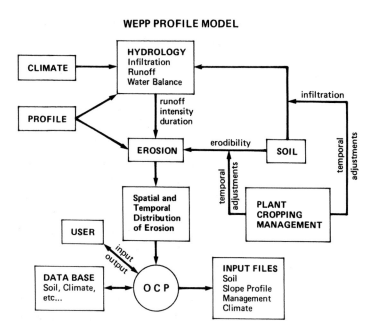

Figure 16: Flow chart illustrating the components of the hillslope profile version of the USDA/WEPP computer model (After Lane and Nearing (eds) 1989).

REPRESENTATION OF SOIL EROSION PROCESSES USED IN THE USA WATER EROSION PREDICTION PROGRAM (WEPP)

The USDA Water Erosion Production Project (WEPP) arose from the decision to develop a new generation of water erosion prediction technology for use by the USDA's Soil Conservation Service and Forest Service, and other organizations involved in soil and water conservation and environmental planning.

It is because the USLE has become a legal instrument in the USA that the need has arisen with some urgency to update the erosion technology. The responsibility for this replacement has been placed on a substantial team of scientists supporting the WEPP program. The aim is to have in place three versions of the technology appropriate for use at different scales and for different objectives: a hillslope profile version, a watershed version and a grid version.

This section will be restricted to the hillslope profile model documented by Lane & Nearing (1989). Comment on the grid version will be given in a later section.

This hillslope profile model deals with net erosion or net deposition on a two-dimensional hillslope of arbitrary shape. The WEPP model addresses the need for evaluation of long-term effects caused by the variation in climatically-driven erosion events by using a continuous simulation modelling approach, though it can also be run on a single-storm basis. Use of continuous simulation, driven by a stochastic climate generator, allows erosion prediction to be made as a function of time. Representation of hillslope geometry also allows prediction of regions in the hillslope where a net loss of soil is expected and regions where net deposition is expected to occur.

Runoff is generated from rainfall input using a Green-Ampt type infiltration equation. The daily water balance is calculated, recognizing the changes in total evaporation accompanying plant growth (see Bonell with Balek, this volume). The decay of plant residue is also simulated. The erosion and deposition component of the WEPP model depends directly on the extensive previous research of Foster (1982). A flow chart for the WEPP computer model is given in Fig. 16.

The rill behaviour of soil, though not universal, is a very common feature of erosion, especially on the agriculturally important soils of the USA. The observational experience of this common behaviour, and recognition of its importance if it occurs, is basic to the history of the approach developed by Foster and used in WEPP (Lane and Nearing, 1989). Thus, soil erosion processes are defined by the morphological description of the eroding soil surface as "rill" or "interrill."

The WEPP model considers sediment eroded from interrill areas to be fed to a rill, the distance between rills being taken as constant for the hillslope. The rate of interrill erosion per unit area (D_i) is taken to be the product of an interrill erodibility parameter and the square of an effective constant rainfall rate calculated for the rainfall event. In quantitative terms:

$$D_i = K_i I_e^2 C_e G_e (R_s w^{-1}) \ (\text{kg m}^{-2} \text{ s}^{-1}) \qquad (25)$$

where K_i is an interrill erodibility parameter, I_e is an effective rainfall intensity, C_e is the effect of canopy on interrill erosion, G_e represents the effect of ground cover on interrill erosion, and R_s is the spacing of rills of rill width w. Empirical relations have been developed for C_e, G_w and w.

The rill accepts eroded sediment from its supplying interrill area. Any rill is assumed to have an upper limit to the rate at which it can transport sediment, called the transport capacity

(T_c). If T_c is greater than the sediment flux in the rill (G), then a net rill erosion rate (D_r) is assumed to be proportional to $(T_c - G)/T_c$. Furthermore, the net erosion rate in the rill is assumed proportional to the excess of hydraulic shear stress (τ) over a critical shear stress (τ_c). Thus:

$$D_r = K_r(\tau - \tau_c)\left[\frac{T_c - G}{T_c}\right] \quad (\text{kg m}^{-2}\,\text{s}^{-1}) \qquad (26)$$

where K_r(s m^{-1}) is a rill erodibility parameter. The *f* transport capacity (T_c) in equation (26) is evaluated using a simplified sediment transport equation which is:

$$T_c = k_t\,\tau^{3/2}\,(\text{kg m}^{-1}\,\text{s}^{-1}) \qquad (27)$$

where k_t is a transport coefficient and τ the hydraulic shear stress acting on the soil.

Net deposition in a rill is calculated to occur when G is greater than T_c, making D_r negative (equation 26).

However, when $G > T_c$, an alternative equation to (26) is used to calculate D_r, namely:

$$D_r = \left(\frac{\beta v_f}{q}\right)(T_c - G) \quad (\text{kg m}^{-2}\,\text{s}^{-1})\ (G > T_c) \qquad (28)$$

where the experimentally determined parameter β is taken as 0.5, v_f is an effective fall velocity for the sediment, and q is volumetric flow per unit slope width. It may be noted that $v_f q^{-1}$ in equation (28) is the reciprocal of the downstream distance moved by a sedimentary particle between its removal from and return to the soil surface.

Thus, in the situation where net deposition occurs, D_r is negative, and equation (28) is used which acknowledges the role of settling velocity. However, v_f is not present in equation (26). This illustrates one major conceptual difference between the WEPP model and that described previously and used in GUEST. In GUEST, deposition is represented as a continuous ongoing process which occurs whether or not there is net deposition. Since gravity is always acting, this would seem to be justifiable. In WEPP the deposition process is explicitly recognized only when it occurs at a rate greater than the erosion processes. If this commentary is accepted, then it implies that v_f would be a factor affecting K_r in equation (26), indicating that it is a compound parameter containing both erodibility and depositability elements within it.

Steady-state mass conservation requires that:

$$\frac{dG}{dx} = D_r + D_i \quad (\text{kg m}^{-2}\,\text{s}^{-1}) \qquad (29)$$

where G is sediment flux (kg m^{-1} s^{-1}) and x is distance downslope (m). Thus if D_r is negative and greater than D_i, then $dq_s dx^{-1}$ will be negative, indicating net deposition.

Runoff rate in the WEPP program is assumed to be constant at the peak rate predicted from rainfall rate – using a Green-Ampt type infiltration equation. The effective duration of runoff is calculated by dividing the total volume of runoff for the erosion event by the peak rate of runoff.

The slope profile is represented, taking slope in any section as a linear function of downslope distance. Equation (29) is solved for each slope section, with the output from each section becoming the input to the next section downslope. This procedure allows the WEPP model to deal, not only with change in slope with downslope distance, but also with downslope variation (within homogeneous elements) in factors such as surface roughness, cover and erodibility parameters.

When flow is routed through a slope section in which net deposition is calculated to have occurred (so that equation (28) applies), the effect of selective deposition of coarser sediment is represented. The particle size distribution of eroded sediment is calculated as described in Foster *et al.* (1985). A rather complex numerical procedure is then used to compute a new particle size distribution for sediment emerging from a region of net deposition. This procedure is described by Foster *et al.* (1989). Also the effective duration of runoff is calculated and a typical value of runoff rate for the event used in the analysis.

Mass conservation of sediment then requires that, for each hillslope element of constant slope, the rate of increase of sediment flux with distance downslope is given by the sum of the interrill (D_i) and rill (D_r) net erosion rates.

The approach to deposition in WEPP presents a contrast to that given in earlier sections where deposition was explicitly represented as a continuous process. In the theory of Foster (1982), deposition is conceived in net terms, rather than as a continuous process in its own right. This approach is also followed in WEPP. Thus, if in equation (25) $G > T_c$, net deposition is indicated rather than net erosion, and whilst net deposition rate is taken as proportional to $(G - T_c)$, other terms in equation (25) are replaced by a term which is the reciprocal of the downstream distance moved by a sedimentary particle between its removal from and return to the soil surface.

An extensive field research program, using a mobile rainfall simulator across most major soil types in the USA, was used to gather data from which parameter values such as K_r in equation (25) can be estimated. The WEPP program also has the capacity to evaluate conservation systems where change in conditions throughout the year and from year to year are simulated. This is achieved in part by separate components of the program which generate simulated climate change and the growth of plants and residue management (Fig. 16).

COMMENTS ON THE WEPP HILLSLOPE PROFILE MODEL AND THE DESCRIPTION OF EROSION PROCESSES GIVEN EARLIER

Past members of the WEPP erosion team have themselves provided a critique of soil erosion research and its use in the WEPP model (Nearing *et al.*, 1980). The self-critique of WEPP provided by these authors will be used as the basis for comments in this section.

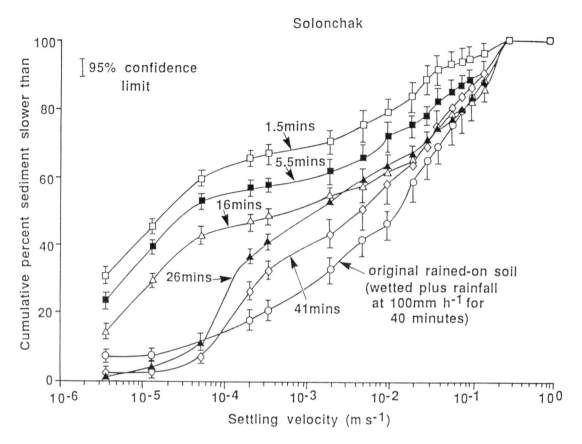

Figure 17: Mean settling velocity distributions of the original rained-on aridisol (or solonchak) soil, and of eroded sediment at the selected sampling times shown following the commencement of 100 mm h^{-1} rainfall. Mean water depth = 5 mm, surface slope 0.4%. (After Proffitt, Rose and Hairsine (1991)).

Detachment processes between rills

A major deficiency these authors see in representing detachment processes between rills is the inability of equation (25) to indicate that the size distribution of particles coming from interrill areas indicates finer sediment than that coming from rills. These authors note the importance of this lack in estimating chemical transport associated with sediment.

Is this an equally valid criticism of the description of erosion and deposition processes given earlier? There is some experimental support for the assumption that rainfall detachment of the original soil matrix is non-selective with respect to the size of particles and aggregates (Proffitt, 1988). Suppose sediment is divided into an arbitrary number (I) of size or settling velocity classes, with an equal mass of some in each class. Also, noting that a fraction (H) of the original soil is covered or shielded from raindrop impact by the deposited layer which forms through time, then equation (21) can be written more generally (Hairsine & Rose, 1991a) as:

$$e_i = (1 - H)\, a\, P\, I^{-1} \;(\text{kg m}^{-2}\,\text{s}^{-1}) \tag{30}$$

where e_i is the rate of rainfall detachment of particles of size class i.

Re-detachment of sediment from the deposited layer is essentially the same process as detachment from the original soil matrix. The difference lies in the weaker nature and size

distribution of sediment forming the deposited layer. For sediment of size class i in the deposited layer, denote the mass per unit area by M_{di}, the total mass per unit area being M_{dt} where:

$$M_{dt} = \sum_{i=1}^{I} M_{di} \quad (\text{kg m}^{-2}).$$

Just as for rainfall detachment, it may be assumed that re-detachment will not be selective with respect to size distribution of sediment in the deposited layer. Thus the rate of re-detachment would be expected to be proportional to the mass per unit area of material present of any size range, and thus proportional to ($M_{di} M_{dt}^{-1}$) for sediment class i. Hence, as in Hairsine & Rose (1991a), the rate of re-detachment by rainfall (e_{di}) may be written as:

$$e_{di} = H a_d P (M_{di} M_{dt}^{-1}) \;(\text{kg m}^{-2}\,\text{s}^{-1}) \tag{31}$$

where a_d is a re-detachability co-efficient (the detachability of the deposited layer).

Mass conservation of sediment of size class i on a plane surface with water depth D then requires that:

$$\frac{\partial q_{si}}{\partial x} + \frac{\partial (c_i D)}{\partial t} = e_i + e_{di} - d_i \quad (\text{kg m}^{-2}\,\text{s}^{-1}) \tag{32}$$

where q_{si} is sediment flux per unit width of plane in size class i, t is time, c_i is sediment concentration for size class i, and d_i is given by equation (4).

Equation (32) can be solved numerically, and yields results which are similar in form to the experimental data given in Fig. 16. This figure illustrates the common observation that, on average, sediment generated by rainfall detachment and re-detachment (e.g., interrill sediment) is finer than sediment generated in rills, the latter being similar in settling velocity characteristic to that of the original soil matrix.

Figure 17 shows further that under the constant rainfall conditions of these experiments, sediment in the runoff becomes progressively coarser with time until at long time periods, the settling velocity characteristics become similar to those of the original rained-on soil (for which any structural breakdown produced by rainfall impact is acknowledged).

Nearing *et al.* (1990) sees the ability to be able to describe the fineness of sediment generated by interrill processes (and illustrated in Fig. 17) as an important goal for erosion-prediction technology, which is not yet in place in the WEPP model. It would seem that the separate but interrelated description of rainfall detachment (in equation (30) and of re-detachment (in equation (31), together with recognition of the strongly size-selective process of deposition (equation (4)), may be necessary to fully describe such fineness.

Since solution of equation (32) is not trivial (and, indeed, such an equation must be solved for each size class), a full interpretation of the dynamics of the change in size or settling velocity characteristics of sediment eroded by detachment and re-detachment is complex. The simple solution given in equation (5) applies only when an equilibrium is reached, which may take 10 minutes or more under steady rainfall, depending on soil and rainfall characteristics.

Rill processes

Nearing *et al.* (1990) comment that perhaps the greatest limitation to the representation of rill erosion processes in the WEPP model is that it assumes the shear stress generated by overland flow to be the only erosion mechanism at work. They recognize that other erosion processes can be active, such as rill sidewall sloughing and headcutting, and suggest that explicit representation of such processes could improve erosion models.

This lack of explicit representation of such processes is a general feature of most current erosion models, but this omission may not be as great a limitation as Nearing *et al.* (1990) suggest, for the following reasons: Suppose the transport limiting sediment concentration is achieved when the original soil matrix is completely shielded by a layer of deposited sediment, and the dominant process is re-entrainment of this deposited material, as described earlier. If so, then the role of the possible range of gravity-aided processes not explicitly described could simply be to ensure the completeness of this blanket of deposited sediment. Thus, provided such gravity-aided processes are effective in providing this complete coverage by weak and effectively non-cohesive sediment, on which the effective stream power can work, lack of explicit description of such processes may not imply error or limitation in the theory.

Using this conceptual model, the transport limiting situation is one in which the rate of re-entrainment of deposited sediment and the rate of deposition are equal, but oppositely directed processes. Furthermore, theory based on this assumption was developed, leading to equation (14). In this derivation, no recourse was made to the literature describing the similar transporting mechanisms of non-cohesive sediment in streams. Nevertheless equation (14) was found to be in quite good agreement with data obtained in such situations.

Nearing *et al.* (1990) also comment on the concept of transport capacity, used in the WEPP model as the maximum flux of sediment a flow can carry without net deposition occurring. Their critique is that estimation of the transport capacity may not be entirely appropriate as it is based on simplification of relationships developed for flow in streams. The relationship used in WEPP is given by equation (27). Nearing *et al.* (1990) note that the use of this method for estimating transport capacity (requiring the evaluation of the k_t in equation (27) produces results accurate to only an order of magnitude.

As noted above, the derivation of equation (14) describing the sediment concentration at the transport limit was made using physical principles and assumptions. Equation (14) also provides a basis for interpreting the observed effect of aggregate density on sediment concentration, and thus the soil loss. The component in this equation given as $\sigma (\sigma - \rho)^{-1}$ can alternatively be written as $1 (1 - \rho \sigma^{-1})^{-1}$, showing more readily that the magnitude of this term will increase as aggregate density σ decreases towards ρ, the density of water. Thus, equation (14) predicts that, other things being equal, sediment concentration at the transport limit will be higher for sediment of lower density.

Equation (14) also indicates that the only other soil characteristic required to be known is its depositability ($\Sigma v_i \, I^{-1}$). However, the characteristics of rilling, which will affect velocity V in equation (14), are also affected by soil type and soil strength, amongst other factors (Moore & Burch, 1986b).

Nearing *et al.* (1990) have also noted that the ability of the WEPP hillslope program to predict changes with slope in the particle size distribution requires the development of theoretical underpinning and experimental testing. This remains a challenge for the future.

WHAT IS KNOWN AND CAN BE TRANSFERRED FROM TEMPERATE EXPERIENCE TO THE HUMID TROPICS

The participants of the related Erosion and Sedimentation Workshop (Appendix B) recognized that on-site effects of erosion threaten biological sustainability in a number of ways

(Table 1). Loss of topsoil can reduce the depth of soil available for effective exploration by plant roots with effects that include a reduction in the amount of stored water available for plant growth. Erosion also is commonly associated with a decline in soil structural stability, and a deterioration in infiltration characteristics.

However, it also is most significant that, together with the loss of soil, is a loss of plant nutrients required for plant growth. The loss of plant nutrients is often greater than would be expected from multiplying the soil loss by the concentration of nutrients in the soil. This excess is due to enrichment effects. The loss of sediment enriched in nutrients (and any other chemicals sorbed to or closely associated with the soil) can be of particular significance when rainfall detachment and re-detachment are important erosion processes when compared with entrainment and re-entrainment (Rose & Dalal, 1988).

When soil is lost from the soil surface by rainfall impact or overland flow, it is known that this can be reduced to low levels by maintaining cover so close to the soil surface that it impedes the rate of overland flow of water, as well as provides protection against rainfall impact.

The discussion of soil erosion and deposition processes, and the incorporation of knowledge gained concerning such processes into predictive models described earlier in this review, gave scant attention to soil conservation practices. However, such knowledge can make an important contribution to the evaluation of soil conservation methodologies and to the design of soil conserving systems. Indeed, the capacity to design or aid the conceptualization of practical and flexible soil conserving systems is necessary for productive and sustainable uses of land to be achieved.

This review has described how both rainfall impact and the shear stresses imposed on the soil surface can result in the erosion of soil. In forested areas, rainfall is intercepted both by the foliage and at ground level by the mat of leaves, twigs and the low-growing vegetation typically present. Thus, having vegetation above the soil surface and on the soil surface are both commonly present in natural forest systems. However, this protective combination is not always present in agricultural systems where it is particularly helpful to distinguish between the above ground or "aerial cover" and the cover very close to the soil surface or "surface contact cover."

This distinction between the two types of cover is useful because aerial cover is effective only in reducing raindrop detachment and re-detachment. The type of aerial cover provided by trees may not even be complete in this protection. The reason for this reservation is that quite large water drops can be formed by tree leaves. If these drops fall through a height of several metres, then they can be just as erosive as raindrops.

Also, high levels of aerial cover in agricultural systems, where the raindrops are virtually all intercepted by vegetation, can still be accompanied by very serious erosion caused by runoff. Especially in non-tilled crops, such as sunflower, if no

residue is available from a previous crop, then bare soil under the crop is unprotected from runoff entrainment and re-entrainment

To be effective in reducing erosion due to entrainment and re-entrainment processes, surface contact cover is required so that the cover is so close to the soil surface that overland flow is impeded. Thus, surface contact cover is that cover which is so close to the soil surface that this distance is comparable to the depth of water flowing over the surface. In agricultural systems, this contact cover can be provided by residue or non-harvested components of a previously grown crop.

In alley cropping or hedgerow systems, trimming the hedgerow is necessary to prevent its excessive growth. These trimmings can be returned as a surface mulch to the cropped segments between hedgerows. A common practice is to incorporate such trimmings into the soil. Doing so probably speeds up the processes which lead to the release of plant-available forms of nitrogen from the leguminous hedgerow trimmings. However, at least in the humid tropics, this nutrient cycling will still be quite rapid even if the trimmings are not incorporated but left on the soil surface between the hedgerows. In the ACIAR experiments – referred to earlier in this review – carried out in the Philippines with hedgerows, the soil loss was substantially reduced if hedgerow trimmings were left on the surface, rather than being incorporated into the soil.

The reasons for the effectiveness of soil surface cover are several. Firstly, such cover inhibits both rainfall and runoff-driven erosion processes, whereas aerial cover is effective only against detachment by rainfall. Surface contact cover bears some, perhaps a large part, of the shear stresses exerted on the surface by overland flow. This relates to a second reason for the protection provided by surface contact cover, namely that the velocity of overland flow is substantially reduced by the cover. The degree depends on the amount and type of such cover. It also appears that surface contact cover reduces the likelihood of rill formation, perhaps by inhibiting the tendency of water to develop preferred channels of flow. If surface contact cover is provided by material with a root system (alive or dead) this can also reduce the likelihood of rill formation and erosion. Surface contact cover is also associated with higher rates of infiltration and thus the reduced runoff.

Figure 18 summarizes the general form of relationship resulting from many investigations into the effectiveness of cover in reducing sediment concentration, and thus soil loss. The ordinate in Fig. 18 shows mean sediment concentration with some measured level of cover – normalized by dividing by the sediment concentration measured on the same soil in a bare condition. Such measurements typically are made using runoff plots. The typical non-linearity in this relationship shown in Fig. 18 is good news in the sense that only limited fractional cover helps reduce sediment concentration much more than if the relationship was linear.

Based on the understanding of erosion processes given in this review, one would expect the non-linearity in the form of

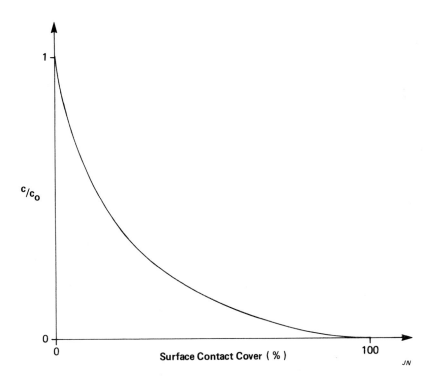

Figure 18: Generalized form of relationship illustrating the reduction in sediment concentration in runoff from plots with varied surface contact cover. c represents sediment concentration with any percentage surface contact cover, and c_0 the concentration with no cover (i.e. bare soil).

Fig. 18 to be highest if entrainment and re-entrainment were the dominant processes, and that the relationship would tend to be more linear if rainfall detachment and re-detachment dominated. There is certainly some variety in the form of the relationship reported in the literature as is illustrated in Fig. 18. The above expectation may be a significant cause of such variation, though this is worthy of a fuller and more focused examination.

If some form of mass movement is the dominant cause of soil loss, then, as mentioned earlier in the review, the surface cover plays only a minor role in affecting soil loss.

Since the velocity of flow (V in equation 14) increases with slope length, then limiting the length of slope over which water can flow could be expected to reduce the sediment concentration achieved, and so to reduce soil loss. The same conclusion is reached for sediment concentration at the source limit (defined in equations 15 and 17–19) rather than the transport limit (equation 14), even though the dependence on slope length is more subdued in the case of the source limit.

In soil conservation, this restriction in slope length can be achieved in various ways. The dense row of shrubs in alley cropping systems is one method. In larger-scale mechanized agriculture, this is commonly achieved by the construction of a bank approximately across the slope to interrupt the runoff. The collected water is usually channelled at a low slope to safe disposal on a grass-covered waterway that leads to a natural stream. Such structures are sometimes called graded terraces or contour banks.

Such terraces or banks perhaps may best be regarded as a type of insurance policy should an erosive event occur when the soil surface has inadequate protection. However, some forms of terrace are a common ingredient in soil conservation systems, such as the "fanya-juu" systems used extensively in east Africa and elsewhere (Hurni, 1986).

The theory given in this review can assist in the interpretation of experience gained in the desirable spacing of such terraces or banks, and could improve their design.

There is considerable experience in the design of soil-conserving systems in some climatic, economic and cultural systems. However, this experience is not all readily transferable to quite different systems. A knowledge of the basic processes is useful in that it can aid the development of effective, acceptable and economically feasible systems, if the wider issues involved are also recognized.

Thus, for example, whilst the dominant importance of maintaining an effective level of surface contact cover is recognized, the development of management systems in which this cover is maintained must involve active participation by farmers and other land users. This is essential to achieve acceptability. One vital aspect of acceptability is that changes in management practices to ensure land protection must be tied to direct economic benefit of the land user. Securing this active participation in devising effective, practical and economic management systems appears to be the major bottleneck in extensive implementation of soil-conserving and biologically-sustainable practices.

There is more uncertainty and less well-established knowledge that can be transferred in dealing with off-site problems caused by the transport and deposition of sediment. This may also be the case for erosion and sedimentation in urban situations, which have received less study than in a rural context.

EROSION AND DEPOSITION ON A CATCHMENT SCALE

Introduction

As the scale of concern is increased from the hillslope to the catchment, it is not clear to what extent it is feasible to retain in models anything like the degree of process representation sought on the hillslope scale. Also, as the scale is increased, the difficulty of direct measurement of sediment fluxes is greatly increased. There is a considerable gap in measurement techniques between runoff plots and sediment transport in rivers draining catchments, where indirect methods such as the use of environmental isotopes (Sklash *et al.*, 1990) are providing useful information.

The escalating increase in interest in surface and groundwater quality, and the effect of non-point sources arising from land uses such as agriculture, have added to the pressure for expansion in scale of erosion and deposition interpretation. Because of the difficulty and expense involved in monitoring processes at a catchment scale, a great deal of emphasis and responsibility has been placed on non-point source water quality modelling, as is illustrated by the proceedings of the international symposium on that topic (De Coursey, 1990).

Approaches to interpreting the delivery of sediment to streams at a catchment scale covers a very wide range of disciplines and literatures. The review given below is, therefore, quite selective, has a physical bias, and is chosen to have some links with the foregoing parts of this review.

Modellers' perspectives on non-point source runoff and sediment yield models have been given by Rose *et al.* (1990) and Leavesley *et al.* (1990), and a users' perspective on such models by Oliver *et al.* (1990) and Seip & Botterweg (1990).

Larger-scale studies in catchment or watershed erosion and sediment yield to rivers have been reviewed by Hadley *et al.* (1985) and this aspect has received little attention in this review. Rather, some prominence has been given to methodologies based on the digital terrain analyses developed in Australia, initially by O'Loughlin (1986).

Topographic effects on erosion and deposition

The methodology used in the WEPP hillslope model to describe the effect of two-dimensional topography on erosion and deposition was described earlier. In this subsection, models are reviewed which investigate erosion hazards in two-and three-dimensional topography, which assume sediment concentration is always proportional to unit stream power or the product of slope (*S*) and flow velocity (*V*), (i.e. to *SV*). Notable examples of this approach are given by Moore & Burch (1986b), and Burch *et al.* (1986). Before illustrating some of the interesting outcomes of this work, the assumptions made in this approach will be considered.

Firstly, the derivation of equation (14) describing sediment concentration at the transport limit assumes an equilibrium condition in which the rates of re-entrainment and deposition are equal and oppositely directed. At the catchment scale, there are spatial variations in *S* at least, and time variations in rainfall and runoff rates. The significance of dynamic effects introduced by such spatial and temporal variations is currently not well understood, but such uncertainty probably does not seriously invalidate the general outcome of this type of landscape analysis where a major objective (Moore & Burch, 1986b) is to locate regions of erosion hazard.

The second assumption made in using term *SV* alone in studying topographic effects from equation (14) is that it neglects changes in the depositability ($\Sigma v_i \, I^{-1}$) which will occur for at least two different kinds of reasons in erosion on natural landscapes. Firstly, as mentioned in the section concerned with application of the GUEST model, the upper limit to v_i in the summation calculation is determined by the depth of water, since sediment larger than the depth of water will not take place in deposition. The depth of water will clearly vary with space and time. The second reason for variability in $\Sigma v_i I^{-1}$ is that the size or settling velocity characteristics of sediment will be at least somewhat different in regions of net erosion and net deposition in the landscape. Larger or faster settling aggregates tend to be removed from the flow in regions of net deposition. The sediment emerging from such regions then possess a lower value of depositability than on entry.

These comments point to a general lack of adequate, physically-based theory to predict not only the sediment concentration, but also the size or settling velocity distribution of sediment in situations where the sediment flux experiences changes in slope.

The work of Moore & Burch (1986b) also assumes steady state runoff is occurring on a bare soil surface which does not develop rills or other preferred pathways of flow. Storage-attenuation effects of topography on volumetric water flux are also neglected.

Using Manning's equation for uniform turbulent kinematic sheet flow:

$$V = (S^{1/2} \, n^{-1}) \, D^{2/3} \; (\text{m s}^{-1}), \tag{33}$$

where *n* is Manning's roughness coefficient and recognizing that by definition the volumetric water flux per unit width (*q*) is given by:

$$q = DV \, (\text{m}^3 \, \text{m}^{-1} \, \text{s}^{-1}), \tag{34}$$

it follows that unit stream power (*VS*) is given by:

$$VS = q^{0.4} \, S^{1.3} \, n^{-0.6} \; (\text{m s}^{-1}). \tag{35}$$

Since sediment concentration (*c*) is taken to be proportional to *VS*, and since sediment flux $q_s = qc$ (equation 2), then:

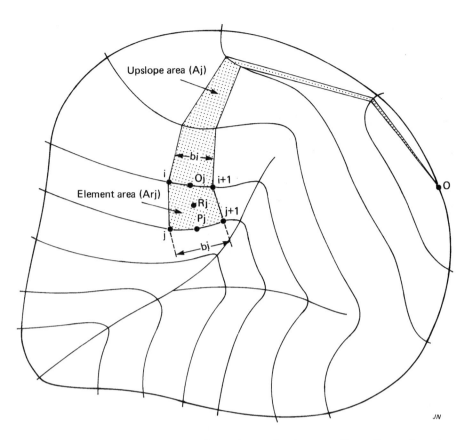

Figure 19: Idealized topographic map of a catchment showing two adjacent flow trajectories encompassing a typical element area (A_{rj}). The element is bounded by vertices i and $i + 1$ on one contour, and vertices j and $j + 1$ on the next lower contour. (After Moore, O'Loughlin and Burch (1988)).

$$q_s \propto q^{1.4} S^{1.3} n^{-0.6} \text{ (kg m}^{-1} \text{ s}^{-1}). \qquad (36)$$

If $\partial q_s/\partial x$ (where x is distance downslope) is positive on a given element of hillslope, then net erosion is indicated; conversely if $\partial q_s/\partial x$ is negative, then deposition is predicted to occur.

Using the assumptions listed above, equation (35) can then be widely used on catchments where the spatial variation in q and S are known. Moore & Burch (1986c) use this approach to investigate the effect of idealized two-and three-dimensional topographic variations on relative erosion or deposition rates.

Considering two-dimensional hillslope profiles while using this analysis, Moore & Burch (1986c) investigated the spatial variation in erosion or deposition on profiles with sequences of convex or concave curvature. Sequences of hillslope segments with net erosion followed by net deposition were predicted for such curved profiles, a feature found in field studies using radionuclide methods (e.g., McHenry & Bubenzer, 1985). Of the simple hypothetical profiles investigated, those which terminated downslope in a convex form gave a higher relative sediment flux than the equivalent plane, and those which terminated in a concave form yielded a lower relative sediment flux.

Applying the same methodology to idealized three-dimensional (conical) hillslopes, (Moore & Burch (1986c) demonstrated the powerful effect of convergent or water-gathering surface types in increasing relative erosion when compared with a non-convergent (i.e. plane) surface of the same slope, particularly for large slope lengths. As expected, the divergent or water-spreading surface of the same slope reduced relative erosion rate in comparison with a plane surface.

Extension of this and related methods to actual catchments is considered in the following section.

Modelling erosion hazard in natural catchments
Topographic analysis developed by O'Loughlin (1986), Hutchinson (1988) & Moore *et al.* (1988) provide information on local slope and the upslope area which can contribute to runoff across any contour element in three-dimensional landscapes (Fig. 19) (as is also reviewed in Bonell with Balek, this volume). The basis of topographic analysis is a topographic map, from which digital information on location and height are read into a computer. Computer programs take this information and generate the information needed for hydrologic analysis, generally making similar assumptions to those listed in the previous section. Equations such as (35) and (36) can then be applied to estimate the sediment entering and leaving any of the element areas such as that denoted A_{rj} in Fig. 19, yielding the net erosion or deposition rate for that element. Because of the computer implementation of this analysis, effects of areally

varying soil or land use characteristics, if known, can be incorporated.

Moore & Burch (1986c) & Burch *et al.* (1986) applied this methodology to a sub-catchment and found that the areas where relatively severe erosion was predicted (even though interspersed between regions of predicted net deposition, leading to a banded representation) were similarly located to areas with gullies or severe sheet erosion.

Using such digital terrain analysis, Vertessy *et al.* (1990) have illustrated how greater realism can be introduced into predicting erosion hazard areas in catchments by utilizing spatially distributed data on soil type, soil transmissivity, and pasture cover. They have also shown that the methodology is not restricted to the assumed dependence of sediment concentration on unit stream power (*VS*, equation (35)) and that other erosion models can be used in the application of digital terrain methodology. This type of application is a significant step forward, since evaluating erosion hazard purely on topographic attributes, whilst useful, would be expected to be susceptible to significant error when applied to catchments where variability in cover, soil type and condition will substantially modulate spatial patterns of runoff generation and erosion.

This type of methodology has the potential to aid land management planning by identifying areas in catchments where the erosion hazard is of concern. It seems that areas of high hazard often contribute large amounts of sediment which are out of proportion to their restricted area. Thus, the prior identification of such hazard areas could lead to the possibility of avoidance behaviour, though this raises wider extension and educational issues than those covered in this review.

Similar opportunities exist in comparing the possible improvement achievable by alternative remedial measures designed to recover from a situation of unacceptable erosion, deposition or sediment delivery. Given the costs involved in such remediation activity, improving its efficiency could be justified economically.

Grid or grid and channel approaches to erosion and sediment delivery in catchments or watersheds

There is a wide variety of models which have been developed, most commonly in the USA, to assist in assessing the expected delivery of sediment to streams. These methodologies have been as much driven by the concern for water quality as for on-site damage by soil erosion or deposition. Hence, such models commonly are as much or more concerned with the transport of nutrients or other water pollutants as with sediment itself. This reflects the fact that many pollutants are strongly sorbed to or associated with soil or its organic matter.

A partial list of such agricultural nonpoint source pollution models is given in Table 3 (Rose *et al.*, 1990). Most models include a hydrologic model to give overland flow, from which in piggyback fashion is added models of sediment erosion and transport and chemical delivery. The real possibility of delivery of dissolved or colloidally-bound forms of pollutants to streams through groundwater is addressed by other classes of models.

Nonpoint source pollution models can vary substantially in their objectives. Some models are not intended to make use of all available knowledge, but are designed for use as screening tools to aid planning by indicating possible effects of land use or management on pollution delivery to streams. Models described by Dickinson *et al.* (1984) and Haith & Tubbs (1981) are examples of this type. Other more detailed and hydrologically-based models attempt to give more recognition to knowledge of processes, hoping to relate catchment and environmental characteristics to system behaviour. Knowledge concerning processes is most often gained at the scale of a field or runoff plot. Extrapolation of such process information or knowledge to a watershed or catchment scale is fraught with difficulty, requiring judgement concerning the way processes can be represented at scales larger than that at which they have been observed.

At the watershed scale, models are usually of a distributed parameter rather than a lumped parameter type. An example is now briefly described of a distributed-parameter model in which the watershed is subdivided into cells or elements which are assumed uniform in characteristics such as soil type, management and slope. This is the ANSWERS model described by Beasley *et al.* (1980), with a user's manual by Beasley & Huggins (1981).

ANSWERS is a deterministic and primarily event-oriented model intended to simulate the behaviour of agricultural watersheds. For application of ANSWERS, a watershed is divided into square "elements" with average parameter values ascribed to that element. Those elements which contain channels are identified, otherwise elements are considered to be plane surfaces possessing some interception store. The hydrologic response of each element to water input is calculated with the overland flow, leaving any element becoming an input to downslope adjacent elements.

Alternative erosion or transport models can be selected for use in ANSWERS (Beasley & Huggins, 1981) but erosion and deposition is dealt with in a manner similar that described previously for the WEPP model, which depends on the work of Foster (1982) and colleagues. Sediment transport is taken to depend on volumetric flux and landslope.

ANSWERS can cope with spatial variability in relevant characteristics, but obtaining such information appropriate to each element is often beyond the available resources, requiring judgements to be made from a more limited database. Depending on the scale of the watershed and choice of element size, operation of the model can be computationally demanding, though this is bound to become less of a problem as large computer capability increases. ANSWERS has been found to give hydrologic results of useful accuracy, but it appears more remains to be done in testing its ability to predict sediment delivery. This testing will continue since ANSWERS, or some

TABLE 3. *Partial list of agricultural nonpoint source pollution models.*

Model Acronym and Name	Source/Basics	Reference
PTR-Pesticide, Transport and Run-off	Stanford/Hydrocomp	Crawford &Donigian 1973
ARM-Agricultural Runoff Model	" " "	Donigian *et al.*, 1977
NPS-Nonpoint Simulation Model	" " "	Donigian & Crawford 1977
HSPF-Hydrologic Simulation Program – FORTRAN	" " "	Johanson *et al.*, 1984
ACTMO-Agricultural Chemical Transport Model	USDA/ARS	Frere *et al.*, 1975
CREAMS – Chemicals, Runoff, and Erosion from Agricultural Management Systems	" "	Knisel (ed.) (1980)
AGNPS-Agricultural Nonpoint Pollution Model	" "	Young *et al.*, 1987
ANSWERS-Areal Nonpoint Source Watershed Environment Response Simulation	Purdue	Beasley *et al.*, 1980
UTM-TOX-Unified Transport Model for Toxics	Oak Ridge	Patterson *et al.*, 1983
LANDRUN-Overland Flow and Pollution Generation Model	Wisconsin	Novotny *et al.*, 1979
GAMES/GAMESP-Guelph Model for Evaluating the Effects of Agricultural Management Systems in Erosion and Sedimentation/Phosphorus	Guelph	Cook *et al.*, 1985
Land Directorate-Method for Targeting of Agricultural Soil Erosion and Sediment Loading to Streams	Environment Canada/GAMES	Snell, 1984
EPA Screening Procedure –	EPA	McElroy *et al.*, 1976

Source: Rose *et al.*, (1990)

modification or further development of it, is being actively considered for larger-scale use as part of the catena of models in the WEPP program.

Development, testing and application of such models will continue. However, it has already been shown that the availability of data required by such models can quickly become a major limitation to their application. Predictive use of such models depends on the generation of runoff from rainfall, and this depends on the ability to represent spatial and temporal variation in infiltration characteristics, or in situations where runoff is limited to areas of saturation, and the prediction of such areas (O'Loughlin, 1986).

It is a common finding that substantial quantities of sediment and associated chemicals are lost or delivered to streams in a single major erosion event. Thus, to test the predictive ability to such models requires the collection of data on hydrology and sediment delivery at a time scale well within the duration of such a major storm event. This is very demanding, and suitable reliable equipment to obtain such data automatically is in need of further development.

Isotopic tracer methods for evaluating soil erosion and deposition

One method of testing models of erosion and deposition on a catchment or watershed scale, or of providing independent evidence of the integrated result of such processes over time, is the use of isotopic tracers of natural or human origin. Obtaining measurement of net erosion or deposition on a sizeable catchment over a significant time period using direct traditional measurement techniques can be inaccurate or not feasible. Isotopic tracer techniques have their own limitations but do offer a feasible solution to this challenging problem (Sklash *et al.*, 1988).

Caesium-137, a product of nuclear explosions, is the longest and most widely used isotope in studies of net erosion and deposition. This isotope, released in above-ground explosions, is distributed around the world's hemispheres, and brought to ground by rainfall where it is strongly absorbed to soil particles. It also is redistributed by erosion and deposition processes, becoming depleted in sites of net erosion, and concentrated in the sediments of accumulation sites.

The usefulness of ^{137}Cs in studying erosion and deposition was recognized by Rogowski & Tamura (1970), McHenry *et al.* (1973), Ritchie *et al.* (1974), & McCallan & Rose (1977).

The input of ^{137}Cs appears to spatially uniform at the scale of modest catchments, though it is desirable to measure input by sampling at sites believed to be undisturbed and uneroded (e.g flat hilltops) as close as possible to the catchment under study (McCallan *et al.* 1980). The isotope has a half-life of 30 years, and input of ^{137}Cs to the atmosphere in the Southern hemisphere fell to about 0.1 mCi km^{-2} in the late 1970s (Longmore *et al.*, 1983). Concentrations of ^{137}Cs are substantially higher in the Northern than the Southern hemisphere where the measurement counting time allows only one or two measurements per day per detector without extracting ^{137}Cs from the soil. Thus, at least in the Southern hemisphere, the sample counting time coupled with detector cost are the main factors limiting the scale or number of studies that are feasible.

Figure 20a shows the topography and soil core sampling sites for a ^{137}Cs study of erosion and accumulation of a cultivated paddock in the south Darling Downs, some 100 km SW of Brisbane, Australia. The paddock was bounded on its upslope by a road, and on its downslope by a stream (Swan Creek), and contained a fence line. Based on the ^{137}Cs survey at locations indicated, a map was drawn (Fig. 20b) in which lines of equal caesium concentration, termed isocaes, were drawn. Regions with less than the input value of 23.5 mCi km^{-2} were areas of net erosion, and areas with a concentration greater than this figure experienced net accumulation.

The caesiographic evidence of net erosion on the steeper parts of the paddock, and accumulation on the southern flatter part of the paddock were expected. Less predictable, although understandable, was minor accumulation above and straddling the fence line. Quite unexpected was the highly eroded tongue of land adjacent to creek, with less than 10mCi km^{-2}. The farmer who lived on the property indicated that this was where the creek sometimes broke its bank in flood, thus providing an explanation of the data.

The purpose of giving this detail is not its intrinsic importance, but to indicate that, provided an adequate density of sampling and measurement is achieved, a fairly detailed interpretation of net erosion and accumulation is possible with this technique, covering a time period which was some 35 years in this study since the commencement of ^{137}Cs fallout.

Soileau *et al.* (1990) illustrate more recent studies using ^{137}Cs. The challenge in calibrating ^{137}Cs measurements for quantifying erosion rates were presented by Walling & Quine (1990). Sklash *et al.* (1990) show how conjointly measuring ^{137}Cs and ^{7}Be can provide independent evidence on the form of erosion (e.g., rill, interrill or ephemeral gully erosion) which may or may not be visually observed and recorded.

Catchment scale sediment behaviour and management in the humid tropics

Although rates of rainfall and runoff are typically much higher in the humid tropics than in temperate regions, with sediment transport rates consequently being higher also, there is no inherent reason why methodologies outlined earlier in this section cannot be more widely applied in the tropics. There are typical differences in hydrologic behaviour at the catchment scale between these temperate and tropical regions as reviewed by Bonell with Balek (this volume). These differences have direct implications for erosion and sedimentation. On cultivated land in the tropics, Horton-type overland flow can be more common and extensive than in temperate regions; under tropical rain forest, natural erosion by stemflow and saturation overland flow can be greater than in temperate forests (Bonell with Balek, this volume).

The type of models which are based on three-dimensional topographic information are reaching the stage of development where their use in the management of activities such as selective logging, or other forms of possible forest conversion, is feasible. For example, digital terrain models based on the work of O'Loughlin (1986), despite assumptions outlined earlier in this section of the review, have proved capable of interpreting topographic and subsurface controls of water movement and concentration of a quality quite adequate for forest management purposes. Whilst testing of this capacity has been more extensive in temperate regions, the physical process basis of such approaches indicates that the initial successful indications for tropical application is likely to continue. Recommendations based on such topographic modelling approaches appear likely to provide much better guidance on areas which should not be logged due to erosion danger or water-logging than arbitrary width recommendations for buffer strips surrounding streams.

Since subsurface wetness is a major factor in failures resulting in mass movement, topographic modelling should prove of value in avoiding not only excessive surface erosion but mass

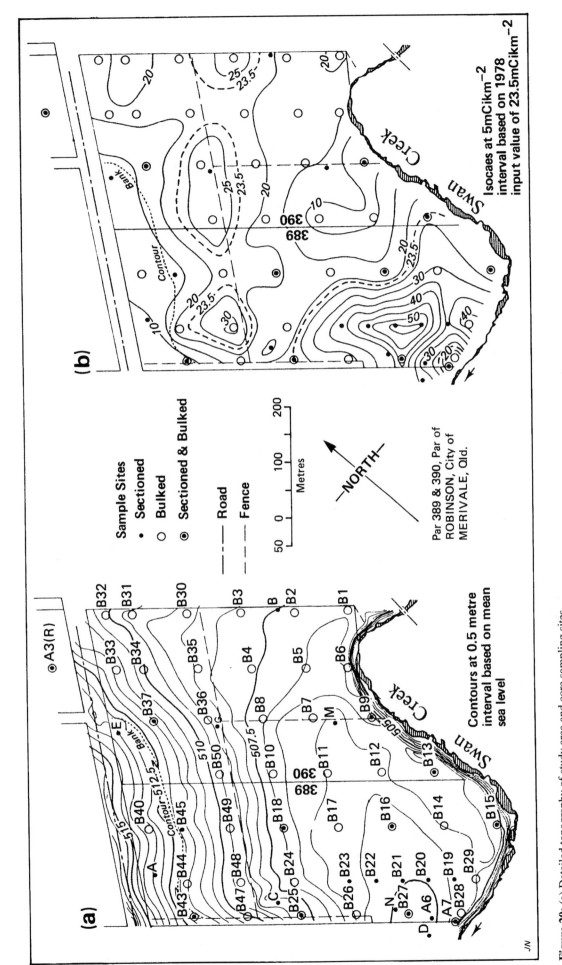

Figure 20: (a) Detailed topography of study area and core sampling sites.

(b) Caesiographic map of same study area as in (a). (After Longmore *et al* (1983)).

movement from deeper-seated failures in soil strength. Quantitative representation of subsurface wetness requires information on the depth to a layer impeding water movement, and the hydraulic transmission characteristics of the layer above it (O'Loughlin, 1986). Talsma & Hallam (1980) have described a practical methodology – called the Constant Head Well Permeameter method – to derive such data, which is both simple and rapid, given an accessible field site. Substantial variation in hydraulic conductivity, both spatially and within a given profile, characteristically nearly log-normally distributed, implies the desirability of significant replication in measurement to obtain mean data for a catchment (e.g., tropical rain forest, Bonell et al.,1987). However, even if such information is not available but has to be guessed at, the methodology still provides useful guidance, based on quantitative topographic data alone.

Whilst the application of such methodologies hold out hope for the future in providing information useful to guide the management of humid tropic catchments, what are some of the types of results obtained from studies at the catchment scale on sediment yield from such catchments?

Two recent reviews by Bruijnzeel (1990) & Ziemer et al. (1990) which provide good coverage of this topic were previously discussed. These reviews illustrate that most data reported on sediment load in streams are restricted to suspended loads. Malmer (1990), for example, gives data in forested catchments in Sabah, Malaysia. From the discussion of erosion and deposition processes which forms the centre of this review, it will be clear that the relative importance of bed load and suspended load delivered to streams will depend at least on the settling velocity characteristics of the catchment soil, the geometry of the mix of overland flow paths experienced by eroded sediment enroute to the catchment stream, and the degree and kind of litter layer traversed during such pathways. Whilst the expected significance of these factors can be stated, ability to predict the mean settling velocity characteristics of sediment reaching the stream (or streams) draining the catchment is currently very restricted by both theoretical and data limitations.

Even if such characteristics of sediment delivered to the stream system were known, there are significant within-stream processes which will modify the relationship between bed load and suspended load. Under equilibrium conditions there may be some rather simple relationships between the size (or settling velocity) distributions of a bed load and suspended load in streams (Rose, 1985; Rose et al., 1990). However, the relative magnitude of the bed load and the suspended load fluxes, and how this may vary in aggrading and degrading stream conditions presents considerable challenge to current knowledge.

It is suggested that it is the kind of issues discussed in the above two paragraphs which require more attention and progress to be made in them before we will be in a position to provide major advances in interpreting and predicting sediment delivery ratios, a concept introduced early in the review.

Until such progress is made, reviews such as that by Walling (1988) provide such information and guidance as is currently available.

There have been considerably fewer long-term than short-term studies of the effects of humid tropical forest conversion on the suspended load in streams draining the modified catchment (Malmer, 1990). The new sources of sediment mobilized by road construction, logging or other soil exposing activities in a catchment generally go through a sequence of exposure, mobilization, storage and evacuation in streamflow from the catchment (Douglas et al., 1990). Care in the location, construction methods and timing of disturbance activities in tropical forest catchments has been shown by many studies to reduce the yield of suspended sediment in catchment streams (Gilmour et al., 1982; Cassells et al., 1984).

It is now recognized (Spencer et al., 1990) that natural disturbances in rain forests, such as treefall, has significant implications for sediment sources and sinks, as well as for understanding the structure, diversity and dynamics of rain forests. Unless such natural disturbance is close to a stream bank, the sediment released by intense local disturbances in the forest is likely to be largely captured and at least temporarily stored in litter and other organic obstacles (Spencer et al., 1990). The transport of nutrients and carbon associated with sediment loss from catchments subject to disturbance is also of great importance (Anderson & Spencer, 1991), but is not covered in this review.

CONCLUSION

There are important "on-site" and "off-site" consequences of soil erosion and sedimentation. The location of effective management action to minimize damage due to these processes may be described as dominantly "on-site." This is one reason for the dominant hillslope scale focus of this review. If processes at this scale are understood, then management action at this scale may often be the most effective and economic way of minimizing or avoiding the suite of problems associated with accelerated rates of erosion or deposition, whether these problems are on- or off-site.

Reasons are given for a significant and widely-adopted change in soil erosion modelling methodology. This is illustrated by developments beyond the Universal Soil Loss Equation towards a representation of the physical processes involved, as they are currently understood. A substantial focus in this chapter is to review the current state of the art in slope erosion modelling and its field data requirements. In this review there is a bias towards methodologies with which the author is familiar, and which are being applied in the humid tropics.

An outcome of the review is a case for the recognition and representation of deposition as a process in its own right in a more explicit manner than has been common in process-oriented models. Reasons for this are that the soil characteristics which determine rate of deposition are different to and not necessarily correlated with the characteristics which affect soil erodibility.

The chapter outlines analytical development which provides a theoretical basis for understanding concepts such as the "transport limit" which have been used in the erosion literature for some time.

How such theoretical developments relate to alternative methodologies is explored to some extent, with particular attention given to the substantial Water Erosion Prediction Program (WEPP) in the USA.

How the theoretical developments are being applied in the humid tropics and elsewhere to determine the erodibility and depositability parameters on which net erosion depends is illustrated by a case study project supported by ACIAR (the Australian Centre for International Agricultural Research). Attention is also given to the question of how what has been learnt about soil conservation management from temperate experience can be applied or transferred to the humid tropics.

Finally, erosion and deposition at the catchment scale has been considered, giving particular attention to topographic effects. Implications of the substantial recent developments in hydrologic models based on terrain and subsurface information are considered for erosion control management. The role of isotopic tracer methods in evaluating the spatial distribution of net erosion and deposition over longer time scales is also reviewed.

Generalization on sediment behaviour at the catchment scale in the humid tropics available from recent reviews is summarized. Some of the challenges for the future in better understanding the delivery of sediment in streams which drain catchments are outlined.

ACKNOWLEDGMENTS

The author would like to thank UNESCO for the opportunity to participate in the Colloquium and, in particular, the participants in the Workshop on Erosion and Sedimentation who provided much stimulus and sound advice on the varied aspects of the Workshop Topic.

Since a section of this review draws particularly on humid tropical experience gained in jointly directing a project funded by the Australian Centre for International Agricultural Research, I would like to acknowledge that support and thank the project participants for the opportunity to learn by working together.

GLOSSARY OF MAJOR SYMBOLS

Roman Symbol	Description	Defining Eq. No. (If applicable)
a	Detachability of the soil by rainfall	21
a_d	Re-detachability coefficient	5
c	Sediment concentration	2
c_t	Sediment concentration at the transport limit	14
$\bar{c}$	Average sediment concentration for erosion event	23
d_i	Sediment deposition rate of sedimentary particles of size range i	4
D	Depth of overland flow	
D_i	Rate of interrill erosion per unit area	25
D_r	Net rill erosion rate	26
e	Rainfall detachment rate	21
e_i	Rainfall detachment rate for particles of size class i	
e_d	Rainfall re-detachment rate	
e_{di}	Rate of re-detachment by rainfall for particles of size class i	31
F	Fraction of streampower effective in erosive processes	
g	Acceleration due to gravity	
G	Sediment flux in rill	
H	Fractional coverage of soil surface by by deposited layer	
i	as a subscript: Refers to a particular sediment size range	
I	Number of sediment size ranges	
J	Specific energy of entrainment	
K	Hydraulic coefficient ($= S^{1/2} n^{-1}$)	
L	Length of plane	
m	Mass (Eqn. 10); hydraulic parameter (Eqn. 16).	
M_{di}	Mass per unit area of size class i sediment in the deposited layer	
M_{dt}	$= \sum M_{di}$, summed over all I size class intervals	
n	Manning's roughness coefficient	
P	Rainfall rate	
q	Volumetric water flux per unit width of plane	2
q_s	Sediment flux per unit width of plane	2
q_{si}	Value of q_s in sediment size class i	
Q	Runoff rate per unit plane area	
S	Slope of the plane (the sine of the angle of land surface inclination)	
t	Time	
T_c	Transport capacity for sediment flow	
u_*	Friction velocity	
V	Velocity of overland flow	
V_s	Volume of sediment	
v_i	Settling velocity of sedimentary particles of size range i	4
W_s	Immersed weight of sediment	10
x	Distance downslope from the top of the plane	

GREEK SYMBOLS

β	An empirical erodibility parameter	23
ρ	Density of water	
σ	Density of sedimentary material	

Σ	Summation sign	
τ	Shear stress between soil surface and overland flow	
Ω	Stream power	6,8

REFERENCES

Anderson, J. M. & Spencer, T. (1991) *Carbon, Nutrient and Water Balances of Tropical Rain Forest Ecosystems Subject to Disturbance: Management Implications and Research Proposals.* MAB Digest 7. UNESCO, Paris.

Bagnold, R. A. (1977) Bedload transport by natural rivers. *Water Resour. Res. 13:* 303–311.

Beasley, D. B., Monke, E. J. & Huggins, L. F. (1980) ANSWERS: A model for watershed planning. *Trans. Am. Soc. Agric. Engrs. 23* (4): 938–944.

Beasley, D. B. & Huggins, L. F. (1981) ANSWERS – User's Manual *U.S. EPA Report No. EPA 905/9–82–001* U.S. Environmental Protection Agency. 230 South Dearborn St., Chicago, IL 60604. 54.

Besler, H. (1987) Slope properties, slope processes and soil erosion risk in the tropical rain forest of Kalimantan Timur (Indonesian Borneo). *Earth Surf. Processes and Landforms 12:* 195–204.

Bonell, M. & Williams, J. (1986) The two parameters of the Philip infiltration equation: Their properties and spatial and temporal heterogeneity in a red earth of tropical semi-arid Queensland. *J. Hydrol., 87:* 9–31.

Bonell, M. & Williams, J. (1987) Infiltration and redistribution of overland flow and sediment on a low relief landscape in semi-arid tropical Queensland. *In:* R. H. Swanson, P.Y. Bernier, & P.D. Woodward (eds.). *Forest Hydrology and Watershed Management,* IAHS Publ No. 167:199–211.

Bonell, M., Cassells, D. S. & Gilmour, D. A. (1987) Spatial variations in soil hydraulic properties under tropical rain forest in north-eastern Australia. *In:* Yu-ji Fok, (ed.) Int. Conference on Infiltration Development and Application, University of Hawaii, Water Resources Center, January 1987. 155–165.

Bruijnzeel, L. A. (1990) *Hydrology of Moist Tropical Forests and Effects of Conversion: A State of Knowledge Review.* UNESCO IHP, Humid Tropics Programme, Paris

Bryan, R. B., Govers, G. & Poesen, J. (1989) The concept of soil erodibility and some problems of assessment and application. *Catena 16:* 393–412.

Burch, G. J., Moore I. D., Barnes, C. J., Aveyard, J. M. & Barker, P. J. (1986) Modelling Erosion Hazard: A total catchment approach. *In:* Proc. Hydrology and Water Resources Symp., (at Griffith University, Brisbane, 25–27 November, 1986). The Institution of Engineers, Australia, National Conference Publ. No. 86/13, 345–349.

Cassells, D. S. , Gilmour, D. A. & Bonell, M. (1984) Watershed forest management practices in the tropical rain forests of north-eastern Australia. *In:* C. L. O'Loughlin & A. J. Pearce (eds.) Proc. of IUFRO Symp. on Effects of Forest Land-Use on Erosion and Slope Stability IUFRO, Vienna/East-West Center, Honolulu, Hawaii. 289–298.

Cook, D. J., Dickinson, W. T., & Rudra, R. P. (1985) GAMES – The Guelph Model for Evaluating the Effects of Agricultural Management Systems in Erosion and Sedimentation. User's Manual Technical Report No. 126–71, School of Engineering, University of Guelph, Guelph, Ontario, Canada.

Craswell, E. T. & Pushparajah, E. (1989) *Management of Acid Soils in the Humid Tropics of Asia.* ACIAR Monograph No. 13 (IBSRAM Monograph No. 1), GPO Box 1571, Canberra, ACT, 2601, Australia.

Crawford, N. H. & Donigian, A. S., Jr. 1973. Pesticide Transport Runoff Model for Agricultural Lands. EPA 600/274–013. Office of Research and Development, U.S. Environ. Prot. Agency, Washington, D.C., USA.

De Coursey, D. G (Ed.) (1990) *Proc. of the Int. Symposium on Water Quality Modelling of Agricultural Non-Point Sources,* Part I and Part 2. U.S. Department of Agriculture, Agricultural Research Service, ARS-81.

DEHCD (Department of Environment, Heritage and Community Development) (1978) Commonwealth and State Governments Collaborative Soil Conservation Study 1975–77. Report 1. A basis for soil conservation policy in Australia. *Aust. Govt Printer,* Canberra, Australia

Dickinson, W. T., Pall, R. & Wall, G. J. (1984) GAMES – a method of identifying sources and amounts of soil erosion and fluvial sediment. *Proc. Canadian Hydrology Symposium on Water Quality Evolution within the Hydrological Cycle of Watersheds.* Quebec, PQ. 805–824.

Donigian, A. S., Jr., Beyerlein, D. C., Davis, H. H. Jr. & Crawford, N. H. (1977) Agricultural Runoff Management (ARM) Model version II: Refinement and Testing. U.S. Environ. Prot. Agency. EPA 600/3–77–098. Environmental Research Laboratory, Athens, GA.

Douglas, I., Greer, T., & Mong, W. W. (1990) The impact of commercial logging on a small rain forest catchment in Ulu Segama, Sabah, Malaysia. *In:* R. R. Ziemer, C. L. O'Loughlin & L. S. Hamilton (eds.). 165–173. *Proc. of Fiji Symposium, June 1990.* IAHS Publ. No. 192.

Dunne, T. (1984) The prediction of erosion in forests. *In:* C. L. O'Loughlin & A. J. Pearce (eds.) Proc. of IUFRO Symp. on Effects of Forest Land-Use on Erosion and Slope Stability, May 1984 IUFRO, Vienna/East-West Center Honolulu, Hawaii.

Edwards, K. (1987) *Runoff and Soil Loss Studies in New South Wales.* Soil Conservation Service of NSW. and Macquarie University Technical Handbook.No.10, Sydney, NSW, Australia.

Ellison, W. D. (1952) Raindrop energy and soil erosion. *Empire J. Exp. Agr. 20,* 81–97.

Forrester, J. E. 1970. *Industrial Dynamics.* MIT Press, Cambridge, MA, USA.

Foster, G. R. (1982) Modelling the erosion process. *In:* C. T. Hann (ed.). *Hydrologic Modelling of Small Watersheds. Am. Soc. Agr. Eng. Monogr.* No. 5, 297–379, St. Joseph. MI.

Foster, G. R., Lane, L. J., Nearing, M. A., Finkner, S.C. & Flanagan, D. C. (1989) Erosion component. Chapter 10. in L. J.Lane & M. A. Nearing (eds.).*USDA – Water Erosion Prediction Project: Hillslope Profile Model Documentation. NSERL Report No. 2,* National Soil Erosion Laboratory. USDA – ARS, W. Lafayette, IN.

Frere, M. H., Onstad, C. A. & Holton, H. N. (1975) ßß – An Agricultural Chemical Transport Model. US Dept. of Agric., Agric. Res. Ser. ARS-H-3. Hyattsville, MD.

Gilmour, D. A., Cassells, D. S. & Bonell, M. (1982) Hydrological research in the tropical rain forests of north Queensland: Some implications for land-use management. *In:* E.M. O'Loughlin & L. J. Bren (eds.) First National Sym. on Forest Hydrology, Melbourne, May, 1982), Instit. Eng., Australia, Canberra, Nat. Conf. Publ. No. 82/6: 145–152.

Govers, G. (1990) Empirical relationships for the transport capacity of overland flow. *In: Erosion, Transport and Deposition Processes* (Proc. Jerusalem Workshop, March-April 1987).IAHS Publ. No. 189: 45–63.

Govers, G. & Poesen, J (1988) Assessment of the interrill and rill contributions to total soil loss from an upland field plot. *Geomorphology. 1:* 343–354.

Govers, G., Everaert, W., Poesen, J., Rauws, G., De Ploey, J. & Lautridou, J. (1990) A long flume study of the dynamic factors affecting the resistance of a loamy soil to concentrated flow erosion. *Earth Surf. Proc. and Landforms.* 15: 313–328.

Hadley, R. F., Lal, R., Onstad, C. A., Walling, D. E. & Yair, A. (1985) *Recent Developments in Erosion and Sediment Yield Studies.* Technical Documents in Hydrology, IHP, UNESCO, Paris.

Hairsine, P. B., & McTainsh, G. (1986) The Griffith Tube: A simple settling tube for the measurement of settling velocity of soil aggregates. AES Working Paper 3/86 (Griffith University, Nathan, Qld, 4111).

Hairsine, P. B. & Rose C. W. (1991a) Rainfall detachment and deposition: Sediment transport in the absence of flow-driven processes. *Soil Sci Soc. Am. J* 55: 320–324.

Hairsine, P. B. & Rose, C. W. (1991b) Modelling water erosion due to overland flow using physical principles I. Uniform Flow. *Water Resources Res.*(submitted)

Haith, D. A., & Tubbs, L. J. (1981) Watershed loading functions for non-point sources. *J. Environ. Eng.* 107 (EE1): 121–137.

Hudson, N. (1973) *Soil Conservation.* B.T. Batsford Ltd. London.

Hurni, H. (1986) *Soil Conservation in Ethiopia.* Community Forests and Soil Conservation Development Department, Ministry of Agriculture, Ethiopia.

Hutchinson, M. F. (1988) A new procedure for gridding elevation and stream line data with automatic removal of spurious pits. *J. Hydrol.* 106: 211–232.

IBSRAM (International Board for Soil Research and Management) (1987) *Tropical Land Clearing for Sustainable Agriculture: Proc. of an IBSRAM Inaugural Workshop*, IBSRAM Proc. No. 3, Bangkok, Thailand.

Johanson, R. C., Imhoff, J. C., Davis, H. H. & others. (1984) User's manual for Hydrological Simulation Program – FORTRAN (HSP): Release 7.0 U.S. Environ. Prot. Agency. Athens, Georgia, USA.

Knisel, W., Jr. (1980). CREAMS: A Field-Scale Model for Chemicals, Runoff, and Erosion from Agricultural Management Systems. US Dep. Agric., Conser. Res. Rep. No. 26.

Lal, R. (1987a) Need for, approaches to, and consequences of land clearing and development in the tropics. *In: Tropical Land Clearing for Sustainable Agriculture: Proc. of an IBSRAM Inaugural Workshop.* 15–27. IBSRAM Proc. No. 3, Bangkok, Thailand.

Lal, R. (1987b) Chapter 13.in *Tropical Ecology and Physical Edaphology.* John Wiley and Sons, New York.

Lane, L. J. & Nearing, M. A. (eds.) (1989) *USDA Water Erosion Prediction Project: Hillslope Profile Model Documentation. NSERL Report No.2*, National Soil Erosion Laboratory, USDA-ARS, W. Lafayette, Indiana, USA.

Leavesley, G. H., Beasley, D. B., Pionke, H. B. & Leonard, R. A. (1988) Modelling of agricultural nonpoint – source surface runoff and sediment yield – A review from the modelers' perspective, *In*: G. De Coursey (ed.) *Proc.International Symp. on Water Quality Modeling of Agricultural Non-Point Sources*, Part I, USDA Agricultural Research Service, ARS–81: 171–194.

Longmore, M. E. (McCallan), O'Leary, B. M., Rose, C. W. & Chandica, A. L. (1983) Mapping soil erosion and accumulation with the fallout isotope Caesium –137. *Aust. J. Soil Res.* 21: 373–385.

Lovell, C. J. & Rose, C. W. (1986) Measurement of the settling velocities of soil aggregates, using a modified bottom withdrawal tube. AES Working Paper 4/86, Griffith University, Nathan, Brisbane.

Lovell, C. J. & Rose, C. W. (1988a). Measurement of soil aggregate set-tling velocities I. A modified bottom withdrawal tube method. *Aust. J. Soil Res.* 26: 55–71.

Lovell, C. J. & Rose, C. W. (l988b) Measurement of soil aggregate settling velocities. II. Sensitivity to sample moisture content and implications for studies of structural stability. *Aust. J. Soil Res.* 26: 73–85.

Lovell, C. J. & Rose, C. W. (1991) Wake-capture effects observed in a comparison of methods to measure particle settling velocity beyond Stoke's range. *J. Sedimentary Petrology.* 61: 575–582.

McCallan, M. & Rose, C. W. (1977). The construction of a geochronology for alluvial deposits on the Condamine Plain and the estimation of the aerial variation in erosion intensity in the Upper Condamine drainage basin. *At. Energy Serv. Tech Rep.* 2/77.

McCallan, M. E., O'Leary, B. M. & Rose C. W. (1980) Redistribution of Caesium-137 by erosion and deposition on an Australian soil. *Aust. J. Soil Res.* 18: 119–128.

McElroy, A. D., Chiu, S. Y., Nebgen, J. . & others (1976) Loading Functions for Assessment of Water Pollution from Nonpoint Sources. US. Environ. Prot. Agency, EPS-600/2–76–151.

McHenry, J .R., & Bubenzer, G. D. (1985) Field erosion estimated from ^{137}Cs activity measurements. *Trans. Am. Soc. Agric. Engrs.* 28: 480–483.

McHenry, J. R., Ritchie, J. C. & Gill, A. C. (1973) Accumulation of fallout caesium-137 in soils and sediments in selected watersheds. *Wat. Resour. Res.* 9: 679–686.

Malmer, A. (1990) Stream suspended sediment load after clear-felling and different forestry treatments in tropical rain forest, Sabah, Malaysia. In: R. R. Ziemer, C. L. O'Loughlin & L. S. Hamilton (eds.) *Proc. of Fiji Symp., June 1990.* IAHS Publ. No. 192: 62–71.

Meade, R. H., Dunne, T., Ritchie, J. E., Santos, U de M. & Salati, E. (1985) Storage and remobilization of suspended sediment in the Lower Amazon River of Brazil. *Science.* 228: 488–490.

Megahan, W. F. (1977) Reducing erosional impacts of roads. In: S. H. Kakle (ed.) *Guidelines for Watershed Management.* FAO Conservation Guide 1, Rome. 237–261.

Megahan, W. F. & Kidd, W. J. (1972) Effects of logging and logging roads on erosion and sediment deposition from steep terrain. *Forestry* 70: 136–141.

Misra, R. K. & Rose, C. W. (1990a) Manual for use of program GUEST, *Division of Australian Environmental Studies Report*, Griffith University, Brisbane, Australia, 4111.

Misra, R K., & Rose, C. W. (1990b) GUDPRO (Version 1.2) Manual for IBM-PC users. *Division of Australian Environmental Studies Report*, Griffith University, Brisbane, Australia, 4111.

Moore, I. D. & Burch, G. J. (l986a) Physical basis of the length-slope factor in the Universal Soil Loss Equation. *Soil Sci. Soc. Am. J.* 50: 1294–1298.

Moore, I. D. & Burch, G. J. (1986b) Sediment transport capacity of sheet and rill flow: Application of unit stream power theory. *Wat. Resour. Res.* 22: 1350–1360.

Moore, I. D. & Burch, G. J. (1986c) Modelling erosion and deposition: Topographic effects. *Trans. Am. Soc. Agric. Engrs.* 29: 1624–1630.

Moore, I. D., O'Loughlin, E. M. & Burch, G. J. (1988) A contour-based topographic model for hydrological and ecological applications. *Earth Surf. Processes and Landforms.* 13: 305–320.

Morgan, R. P. C. (1986) *Soil Erosion and Conservation.* Longman Scientific and Technical, Harlow, England.

Moss, A. (1979) Thin-flow transportation of solids in arid and non-arid areas: A comparison. *In: Symp. on the Hydrology of Areas of Low Precipitation.* IAHS Publ. No. 128: 435–445.

Moss, A. J. & Green, P. (1983) Movement of solids in air and water by raindrop impact: Effects of drop-size and water-depth variations. *Aust. J. Soil Res. 21*: 257–269.

Moss, A. J. & Walker, P. H. (1978) Particle transport by continental water flows in relation to erosion, deposition, soils and human activities. *Sedimentary Geology. 20*: 81–139.

Moss, A. J., Walker, P. H. & Hutka, J. (1979) Raindrop-simulated transportation in shallow water flows: An experimental study. *Sedimentary Geology. 20*: 81–139.

Moss, A. J., Walker, P. H. & Hutka, J. (1980) Movement of loose, sandy detritus by shallow water flows: An experimental study. *Sedimentary Geology. 25*: 43–66.

Moss, A. J., Green, P. & Hutka, J (1982) Small channels: Their experimental formation, nature and significance. *Earth Surface Processes and Landforms 1*: 401–415.

Nearing, M. A., Lane, L. J., Alberts, E. E. & Laflen, J. M. (1990). Prediction technology for soil erosion by water: Status and research needs. *Soil Sci. Soc. Am. J. 54*: 1702–1711.

Novotny, V., Chin, M. & Tran, H. V. (1979) LANDRUN – An Overland Flow Mathematical Model: User's Manual, Calibration, and Use. International Joint Commission, Windsor, Ontario, Canada.

O'Loughlin, C. L. (1984) Effectiveness of introduced forest vegetation for protection against landslides and erosion in New Zealand's steeplands. *In*: C. L. O'Loughlin & A. J. Pearce (eds.) Proc. of IUFRO Symp. on Effects of Forest Land-Use on Erosion and Slope Stability, IUFRO Vienna/East-West Center, Honolulu, Hawaii.275–280.

O'Loughlin, E. M. (1986) Prediction of surface saturation zones in natural catchments by topographic analysis. *Wat. Resour. Res. 22*: 794–804.

Oliver, G., Burt, J., & Solomon, R. (1990) The use of surface runoff models for water quality decisions: A user's perspective. *In:* D. G. De Coursey (ed.) *Proc. of the Int. Symp. on Water Quality Modelling of Agricultural Non-Point Sources,* US Department of Agriculture, Agricultural Research Service, ARS 81. *Part 1*:197–204.

Patterson, M. R., Sworski, T. J., Sjoreen, A. L. & others. (1983) A User's Manual for UTM-TOX, A Unified Transport Model. Prepared by Oak Ridge National Laboratory, Oak Ridge, TN, for US EPA Office of Toxic Substances, Washington, D.C.

Proffitt, A. P. B. (1988). The influence of rainfall detachment, runoff entrainment and sediment deposition processes on sediment concentration and settling velocity characteristics of eroded soil. A thesis submitted for the Degree of Doctor of Philosophy, Griffith University, Brisbane, Australia (unpublished).

Proffitt, A. P. B. & Rose, C. W. (1991) Soil erosion processes l. The relative importance of rainfall detachment and runoff entrainment. *Aust. J. Soil Res. 29*: 671–683.

Proffitt, A. P. B., Rose, C. W. & Hairsine, P. B. (1991) Rainfall detachment and deposition: Experiments with low slopes and significant water depths. *Soil Sci. Soc. Am. J. 55*: 325–332.

Prove, B. G. (1991) A study of the Hydrological and Erosional Processes Under Sugar Cane Culture on the Wet Tropical Coast of North Eastern Australia. Thesis submitted for the degree of Doctor of Philosophy, James Cook University of North Queensland, Townsville, Australia (unpublished).

Riquier, J. (1982) A world assessment of soil degradation. *Nature and Resources, 18*: 18–21.

Richie, J. C., Sparberry, J. A., & McHenry, J. R. (1974) Estimating soil erosion from the redistribution of fallout [137]Cs. *Soil Sci. Am. Proc., 38*: 137–139.

Rogowski, A. S. & Tamura, T. (1970) Erosional behaviour of caesium-137. *Health Phys. 18*: 467–477.

Rose, C. W. (1985) Developments in soil erosion and deposition models. *Advances in Soil Science 2*: 1–63. (Springer-Verlag, New York, USA).

Rose, C. W. (1988) Research progress on soil erosion processes and a basis for soil conservation practices.*In: Soil Erosion Research Methods.* (Chapter 6) Soil and Water Conservation Society, Ankeny, Iowa, USA.

Rose, C. W. & Dalal, R. C. (1988) Erosion & runoff of nitrogen. In: J. R. Wilson,(ed.) *Advances in Nitrogen Cycling in Agricultural Ecosystems* (C.A.B. International: Wallingford, UK). 212–233.

Rose, C. W., Dickinson, W. T., Ghadiri, H. & Jorgensen, S. E (1990) Agricultural non-point source runoff and sediment yield water quality (NPSWQ) models: Modeller's perspective. *In*: D. G. DeCoursey (ed.) *Proc. of the Int. Symp. on Water Quality Modelling of Agricultural Non-Point Sources,* USDA Agricultural Research Service, ARS-81. *Part 1*:145–169.

Rose, C. W., & Hairsine, P. B. (1988) Process of water erosion. *In*: W. L. Steffen & O. T. Denmead (eds.)*Flow and Transport in the Natural Environment: Advances and Applications,* Springer Verlag, Berlin. 312–326.

Rose, C. W., Hairsine, P. B., Proffitt, A. P. B. & Misra, R. K. (1990) Interpreting the role of soil strength in erosion processes. *Catena Supplement 17*: 153–165.

Rosewell, C. J. & Edwards, K. (1988) SOILOSS: A program to assist in the selection of management practices to reduce erosion. Soil Conservation Service of NSW, Technical Handbook No. 11, PO Box 198, Chatswood, NSW, Australia.

Seip, K. L. & Botterweg, P. (1990) User's experiences and the predictive power of sediment yield and runoff models. *In*: D. G. de Coursey (ed.) *Proc. of the Int. Symp. on Water Quality Modelling of Agricultural Non-Point Sources,* USDA Agricultural Research Service, ARS-81.*Part 1*: 205–224.

Singer, M. J. & Walker, P. H. (1983) Rainfall-runoff in soil erosion with simulated rainfall, overland flow and cover. *Aust. J. Soil. Res. 21*: 109–122.

Skempton, A. W. & De Lory, F. A. (1957) Stability of natural slopes in London clay. *Proc. 4th Int. Conf. Soil Mech. 2*: 378–381.

Sklash, M. G., Moore, I. D., & Burch, G. J. (1990) Environmental isotope tracer studies of catchment processes: Tools for testing integrated water quality models. *In*: D. G.de Coursey (ed.) *Proc. of the Int. Symp. on Water Quality Modelling of Agricultural Non-Point Sources,* USDA Agricultural Research Service, ARS-81. *Part 1*: 459–478.

Snell, E. A. (1984) A Manual for Regional Targeting of Agricultural Soil Erosion and Sediment Loading to Streams. Lands Directorate Working Paper No. 36, Environment Canada. Ottawa, Ontario, Canada

Soileau, J. M., Hajek, B. F. & Touchton, J. T. (1990) Soil erosion and deposition evidence in a small watershed using fallout caesium-137 *Soil Sci. Soc. Am J. 54*: 1712–1719.

Spencer, T., Douglas, I., Green, T. & Sinun, W. (1990) Vegetation and fluvial geomorphic processes in South-East Asian tropical rain forests. Chapter 27 in J. B. Thornes.(ed.) *Vegetation and Erosion* John Wiley and Sons Ltd, New York.

Statham, I. (1979) *Earth Surface Sediment Transport.* Clarendon Press, Oxford, England.

Talsma, T. & Hallam, P. M. (1980) Hydraulic conductivity measurement of forest catchments. *Aust. J. Soil Res. 18*: 139–148.

Tsukamoto, Y. & Minematou, H. (1987) Evaluation of the effect of deforestation on slope stability and its application to watershed management

In: R. H. Swanson, P. Y. Bernier & P. D. Woodward, (eds.) *Forest Hydrology and Watershed Management.* (Proc. Vancouver Symp., August 1987). IAHS Publ. No. 167: 181–189.

Turner, A. K. (Ed.) (1984) *Soil-Water Management.* Int. Development Program of Australian Universities and Colleges Ltd (IDP). Canberra, ACT, 2601, Australia.

Vertessy, R. A., Wilson, C. J., Silburn, D. M., Connoly, R. D. & Ciesiolka, C. A. (1990) Predicting erosion hazard areas using digital terrain analysis. IASH. *Proc. Int. Symposium on Research Needs and Applications to Reduce Erosion and Sedimentation in Tropical Steeplands*, Suva, Fiji, 11–15 June (In press).

Walker, P. H., Kinnell, P. I. A., & Green, P. (1978) Transport of a noncohesive sandy mixture in rainfall and runoff experiments. *Soil Sci. Soc. Am. J. 42*: 973–801.

Walling, D. E. (1983) The sediment delivery problem. *J. Hydrol. 65*: 209–237.

Walling, D. E. (1988) Measuring sediment yield from river basins. Chapter 3 in *Soil Erosion Research Methods.* Soil and Water Conservation Society, Ankeny, Iowa, USA.

Walling, D. E. (1988) Erosion and sediment yield research – some recent perspectives. *J. Hydrol. 100*: 113–141.

Walling, D. E. & Quine, T. A. (1990) Calibration of caesium-[137] measurements to provide quantitative erosion rate data. *Land Degradation and Rehabilitation 2*: 161–175.

Ward, D. P. & Rose, C. W. (1989) Planning for soil erosion protection in an uncertain hydrologic environment. *In: Proc. National Environmental Engineering Conference.* The Institute of Engineers, Australia, (Sydney, March 20–22, 1989).180–181.

Wiersum, K. F. (1985) Effects of various vegetation layers in an *Acacia auriculiformis* forest plantation on surface erosion in Java, Indonesia. *In*: S. El-Swaify, W. C. Moldenhauer & A. Lo (eds.) *Soil Erosion and Conservation.* . Soil Conservation Society of America,Ankeny, Iowa, USA 79–89.

Williams, J. & Bonell, M. (1987) Computation of infiltration properties from the surface hydrology of large field plots. *In*: Yu-Si Fok (ed) *Proc. Int. Conf. Infiltration Dev. Appl.* , Water Res. Centre, University of Hawaii, Honululu, Hawaii. 272–281.

Wischmeier, W. H. & Smith, D. D. (1978) Predicting rainfall erosion losses – a guide to conservation planning. Agriculture Handbook No. 537 (U.S. Department of Agriculture, Washington, DC).

Yang, C. T (1972) Unit stream power and sediment transport. *J. Hydraul. Div Am. Soc. Civ. Eng. 78*, (HY10): 1805–1826.

Yang, C. T. (1973) Incipient motion and sediment transport. *J. Hydraul. Div. Am. Soc. Civ. Eng. 99* (HY 10): 1679–1704.

Young, R. A., Onstad, C. A., Bosh, D. B., & Anderson, W. P. (1987) AGNPS, Agricultural Non-Point-Source Pollution Model. USDA Agricultural Research Service Conservation Research Report 35.

Zakaria, M. N., Yew, F. K., Pushparajah, E. & Karim, B. A. (1987). Current programs, problems, and strategies for land clearing and development in Malaysia. In: *Tropical Land Clearing for Sustainable Agriculture: Proc. from an IBSRAM Inaugural Workshop.* (IBSRAM Proceedings No. 3, Bangkok, Thailand) 141–152.

Ziemer, R. R., O'Loughlin, C. L. & Hamilton, L. S. (eds.) (1990) *Proc. of Fiji Symp.*, June IAHS Publ. No. 192.

15: Water Quality Issues in the Humid Tropics

M-A. Roche

Institut Français de la Recherche Scientifique pour le Développement en Coopération (ORSTOM), 911 Av. Agropolis, 34032 Montpellier, France

ABSTRACT

This chapter covers the broad scientific issues associated with water quality control in response to the various socio-economic activities within the humid tropics. The account will emphasize the dearth of rigorous scientific field studies and the associated problem of validation of water quality models in this climatic type. Consequently, a substantial part of the work will highlight research gaps and needs. Where possible, however, examples of water quality research will be highlighted, many of which relate to the experiences from ORSTOM projects. An underlying theme will be the linkages between water quality research and control and the institutional, social and economic aspects of water management. In addition, the various types of urban and rural pollution emanating from domestic and industrial sources will be described. Particular attention will be paid to the microbiological aspects because of their effects on human health.

INTRODUCTION

This chapter will review the broad scientific issues associated with water quality control in response to the various socio-economic activities within the humid tropics. In common with other themes of Section IV, the dearth of scientific field studies and the associated problem of validation of water quality models will be given strong emphasis. Consequently, the research gaps will be highlighted and some attention devoted to various methodologies that require further testing under humid tropical conditions.

Any consideration of water quality problems cannot be separated from the topic of water management. Therefore, the linkages between water quality control and institutional, social and economic aspects of water management will be an underlying theme throughout this chapter and will receive particular attention in the early stages of the work. The type of issues that need consideration include the lack of standardization between countries concerning "acceptable" threshold concentrations for various pollutants, which are further aggravated where national boundaries are not coincident with the topographic boundaries of drainage basins. The enforcement of national or internationally agreed upon standards is even more difficult to apply in many humid tropical countries where commonly there is insufficient laboratory infrastructure and the appropriately trained human resources. Similarly, water quality monitoring networks are often lacking in equipment and personnel. Various suggestions are put forward for the improvement of such networks so that adequate baseline data is available for testing various water quality models and to alert the responsible authorities of pollution sources. The objective here is also to make the reader aware that such deficiencies in infrastructure (laboratories, monitoring networks) and the necessary skills (human resources, education and training) can be a major impediment to the future progress in scientific research on various water quality problems.

Managing the microbiological aspects of water quality control is perhaps *the* major issue in the humid tropics in terms of the direct effects on human health (see also Prost, Wurzel, this volume). The later stages of this chapter will describe the various sources of microbiological pollution. Consideration also will be given to the vexing issue of the most appropriate indicators, especially when related to faecal contamination.

Another aspect is the various types of pollution emanating from industry. In an effort to broaden their economic bases, many developing countries have embarked on a policy of rapid industrialization. Such steps have commonly been at the expense of the enforcement of anti-pollution controls on industries in order to maximize profits. Consequently, the concluding stages of this work will review the different types of industrial pollution in both urban and rural areas, and suggest possible technical measures that can be taken within the existing economic constraints.

ROLE OF SCIENCE AND TECHNOLOGY IN WATER QUALITY CONTROL

The fundamental basis of sustainable development is to minimize any adverse impacts on the quality of the atmosphere, water and terrestrial environment in general. The environmental degradation of water resources (Biswas, 1978) has been aggravated by rapid international economic patterns, along with social and political changes which are causing an exponentially expanding urban population. In most developing countries, in urban as well as rural areas, control of water pollution is still either lacking or grossly inadequate.

There is an urgency to the question of whether the fields of hydrochemistry, hydrological science and technology, and water resource management have the appropriate methods to meet the rising water demands of the humid tropics.

The best contribution scientists can offer is to make available to the various parties involved (agencies, governments, social groups, etc.) the maximum amount of relevant information in a well-digested and structured manner. The authorities then can base their decision-making on more scientific and technical advice rather than postponing critical decisions until later.

Gaps in information and methodology must be first detected in order to establish which research need has priority (Stewart & Verhoeven, 1987), as well as make effective use of existing scientific and technical knowledge (transfer, cooperation, related social and economical aspects), and to set-up recommendations for action in research, education and training, and institutional strengthening. Institutional topics, as well as human resource needs must also be analyzed as important basic issues. These problems can be solved only by a multidisciplinary approach, especially the inclusion of hydrobiologists, in addition to emphasizing the hydrological and chemical aspects.

Water quality problems cannot be solved by isolated means, or for only a particular factor or region. Water quality is a global problem that needs examining from all points of view. Although some problems are more acute in some countries than others, there are some essential common problems such as drinking water supplies and sanitation treatment. Therefore, at the beginning of such an international analysis, the baselines of the most important topics must be presented in order to, one, classify them by priority, and, two, propose strategies for the whole situation. It also must be considered as to how the humid tropics are really different from the temperate developed zones where considerable experience on water quality control already exists.

An additional question is whether "high" and "expensive" technology is necessary or can the objectives be better reached by less expensive and "appropriate" technology. Or are both needed?

WATER QUALITY PROBLEMS IN THE HUMID TROPICS

The most urgent problems of water quality in the humid tropics are associated with:

(a) Aquatic vectors and larvae responsible for water-borne endemics and epizootics, and the pesticides used in vector control programmes and agriculture.

(b) Organic urban (faecal) and industrial waste water, and sanitation systems.

(c) Drinking and recreational water supplies in rural and urban areas.

(d) Industrial microtoxics (heavy metals).

(e) Change in salt and nutrient cycles and contents, and the spreading of water-related diseases because of land and water resource management. This includes new methods in agriculture (deforestation/forestation, new crop strains and fertilizers, reservoir irrigation and drainage).

(f) Other causes of water pollution that may occur locally are notably related to injections of hot water or specific compounds into the receiving medium.

During the next few decades, the consequences – already very serious in some rural and urban areas – will lead to an even more acute water quality crisis in many tropical countries if efficient and co-ordinated actions are not taken now (National Research Council, 1982).

NATIONAL AND INTERNATIONAL INSTITUTIONAL ASPECTS AND COOPERATION

Between the years of 1960–70, most tropical countries became aware of the pollution in their environment and the significance of water quality for a growing economy. As pointed out by Prost (this volume), sustainable development, the promotion of health, and the rational use of water resources are absolutely inseparable. However, in many of these countries, water quality control has been divided among numerous national ministries and agencies. Linkages between these national planning, investigatory and coordinating services seem difficult to establish and maintain.

Such pluralism may allow for more specialization (Ngunya, 1975) but the lack of efficient river basin authorities, coupled with local and national committees, leads to incoherent actions. Such authorities can make use of a water law in which water quality is given priority.

Many such national services try to adapt parts of their water law that relate to their responsibilities over water quality. However, parts of the law may conflict with their mandate and policy. For example, some services fear that measures against water pollution may hurt economic growth and increase what is perceived as desirable "low" public taxes. Such situations hinders rational decision-making.

Even when a "Department of the Environment" exists, it is usually given little funding. The national laboratories do not have enough analytical capacity to systematically assess their country's water quality. Such infrastructure is an indispensable precursor to any progress on the overall problem.

Water quality control, in spite of numerous international meetings, has traditionally been considered to be an internal problem. Today, however, the population and industrial growth, as well as the technological specifications, require us to take a more comprehensive international outlook (Soon,

1982). Tropical countries may attempt to solve their water quality problems by themselves, but this will result in useless and unnecessary duplication of work and expenditure of funds. The experience which has been gained in developed countries for more than 25 years (Lamb, 1988), provides an important technological tool for solving most of the problems in the humid tropics. There also is a need to transfer specific tropical scientific knowledge and technology among the developing tropical countries, as their problems and technical adaptations are similar.

Many hydrosystems cross or extend over various humid tropical countries. Therefore, they are affected by upstream, downstream, or common conditions dealing with water quality control. Such countries generally have understood their reciprocal responsibilities and rights, and have organized multinational commissions to coordinate their actions. However, the different aspects of managing water resources do not always include the problems of water quality. Few, or none, of the multinational structures are able to study and control pollution efficiently. There are two main problems. First, there is the need for internationally appropriate legislation along with an inventory of the existing pollution. Second, homogeneous standards for maximum pollutant loads are needed. The African Convention, established by OUA (Organisation de l'Unité Africaine) for Nature Conservation, could be the framework for such a survey and enactment of legislation in the humid tropical countries of Africa. Various nations, however, have not yet signed this convention.

Some countries already have agencies and laboratories that are able to conduct inventories and administer a water quality control monitoring network (Tan, 1978; Quano, *et al.* 1987) but many do not. International programmes can offer them a realistic way of participating as well as profiting from international networks (Global Environmental Monitoring system GEMS/WATER, 1974; International Geosphere-Biosphere Programme, IGBP, 1988). Databases, interpretative software, mathematical computer models and inter-laboratory calibrations can be exchanged in this manner. Research on eutrophication in warm water bodies, groundwater pollution by nutrients or toxic materials and others also could be carried out in the framework of international programmes. In addition, related education and training requirements must rely on international cooperation. The developed countries, which have a great deal of experience in water quality control, can play a large role in this area.

Laboratory analyses

Many countries have facilities for making isolated or routine analyses of water samples, including physical, chemical, bacteriological and biological tests. Sometimes, all these facilities exist in one country; but are dispersed, making co-ordination of a complex project very difficult. Advanced technological instruments also may be available in one laboratory but they are not used due to the lack of a specialized operator or organization of research projects (Stephenson, 1988).

It is imperative that every humid tropics country have at least one functional laboratory able to do complete analyses, including all the main water quality standards. The basic needs are appropriate infrastructure, equipment, specialized staff and financial support. Part of the latter could come from the commercialization of the analyses.

Such laboratories must regularly receive certificates of their capabilities in analytical measurements. This means a national and international system of checking samples, methods and instruments. These laboratories also must be recognized as official entities of the country's water quality control system and thoroughly connected with all the authorities in its field. They may be directly integrated in the service of a ministry of health, for instance, or indirectly through the municipal/departmental authorities responsible for local sanitation, or as an adjunct of a university.

There has been a recent surge of interest in increasing the role of those university research institutes that have analytical laboratories. These institutes are generally able to rigorously use and adapt specific techniques and methodologies dealing with environmental quality measurements. Involvement of these university institutions can provide the necessary global view of problems, multidisciplinary approaches and the appropriate technology. Moreover, a larger participation of such institutions in water quality control work would reflect directly on educational actions, which are of major importance in the fight against pollution.

Standards

Control of the quality of water must be based on standards for incoming and effluent water (Anon. 1971, 1975, 1976, 1977).

What is allowable in a water supply has been deduced from knowledge of the effects of chemical constituents on animals and humans, and also from empirical methods (Evison & James, 1977). However, the toxic effects of specific pollutants have not been studied extensively in the humid tropics but their proliferation is increasing. Therefore, as in other areas, investigations are needed to obtain precise and useful threshold standards. With the present state of knowledge, it is possible to think that carcinogenic or mutagenic substances and heavy metal components could take specific forms in the humid tropics area. Furthermore, local conditions – such as higher temperatures and solar radiation inputs, lower oxygen contents, the frequency of toxic discharges, their type and amount of accumulation at various levels in the sediments and the tropical biotope – may well call for a modulation of standards.

The most important standards for human health deal with drinking water and recreational water uses. Drinking water standards must guarantee a harmless consumption. The World Health Organization (WHO) has published guidelines which are the basis for drinking water quality control in many countries (EFP/82.39, EFP/83.58, Guide for Drinking Water Quality, 1986), although others often have adapted their own standards.

There have been few studies concerning the appropriate salt levels in natural water which suit the human physiological response to the thermal and humidity conditions experienced in the humid tropics (Pescod, 1977). Such work may introduce some modification to existing standards. Meanwhile, an inventory of the standards used in all countries of the humid tropics, and the justification of their values, would be highly beneficial from a water quality management standpoint (Moore & Christy, 1978).

From the point of view of the developing nations, the most significant shortcoming is the lack of flexibility in the standards (Kilani, pers. comm. 1989). Many governments find themselves investing large sums of money in projects designed to attain high standards in communities where much lower quality standards may be acceptable. Thus, unattainability of such standards could be partially responsible for the "failure" of most water supply projects.

EDUCATION AND TRAINING

Education

Education, in the framework of integrated water quality control, is one of the most important factors for development in the humid tropics (Scotney, 1980). Only the full awareness by the individual, from the earliest age, of the importance of a safe environment, can provide a better basis for a long-term education. Consequently, aquatic environmental control should be a prime ingredient in natural science programmes in elementary and high schools. Such an educational programme has to be achieved by television (cartoons), newspapers (comic books), conferences, exhibitions, posters, stamps, and contests/prizes for children and adults. Likewise, pedagogical information should be made available to teachers and professors.

Informing the media as well should be the responsibility of the water quality control services. It also could be UNESCO's role to organize contests for aquatic protection, with prizes for pupils, investigators, laboratories, industries, farmers and towns.

Training

Training should be the concern of universities, specialized research schools and water-related agencies. A census should be made of the existing training programmes in every country, not just those situated in the humid tropics. International training prospects thus could be considered.

Each tropical country also must foresee its needs for scientific, technical and administrative staffs in order to insure essential education and training for hydrologists, specialists in hydrodynamics, water quality and modelling, biologists, hydrobiologists, statisticians, engineers, technicians, documentalists and writers.

Information

Often national works and, *a fortiori*, foreign works are not available at the right time, or not known by all persons in the water quality field. Thus, there is a need for most tropical countries to hold and to disseminate a national and international bibliographic index on ecology, hydro-ecology, hydrochemistry and water quality control. This indispensable tool must be in microcomputer format.

WATER QUALITY MODELLING AND HYDRODYNAMICS

Development of modelling

The evolution of suspended and dissolved components in space and time depends mainly on the water dynamics involved with the transport of such components. Despite recent developments, there is still the need for additional integration of studies in dissolved and suspended load transport with process hydrology for incorporation in water quality modelling (James, 1984; Whitehead, 1984; Orlob, 1986; Trudgill, 1986). Furthermore, there is an urgent need for such models to be field tested in the humid tropics.

Structure of water quality models

These models apply to unitary or complex hydrosystems, such as streams, lakes, lagoons, reservoirs, urban watersheds, groundwater and estuaries. They couple a chemical model to a hydrological model of transport. The following physico-chemical phenomena should be taken into account:

(a) substance production functions (e.g. sediment production by erosion and nitrogen production in the root horizon);

(b) physical or geochemical sedimentation (e.g. soil, streams, reservoirs);

(c) physico-chemical, microbiological or biochemical trapping and transformation, interactions with the bottom flora, fauna; and

(d) concentration by evaporation and dilution by rainfall.

Role and limits of water quality models in decision-making

With simulation, it is becoming possible to predict the dilution of a product, the self-purification capacities of the stream, the sedimentation of dissolved and suspended substances and the risks of eutrophication or concentration of salts, nutrients, pesticides, etc. downstream from a disposal point (Tchobanoglous & Schroeder, 1985). However, the adaptation of models to humid tropical conditions must be realized through the calibration of the various algorithms and the reliability of the results. Technology to evaluate the confidence levels of predictions also must be developed in order to account for uncertainties in the decision-making process. In addition, various specific chemical and biochemical processes must be modelled to address the results in the best way (Trudgill, 1986).

The effects of farm irrigation and the massive use of fertilizers constitute a significant area for applying such models, particularly for their impact on groundwater. Likewise, in urban and industrial situations, the models could be applied to how the drainage networks transport organic and inorganic pollutants in storm runoff and waste water. Another application could be the evaluation of water purification systems and optimizing the functioning of treatment stations.

Some examples

Temperature is one of the parameters that differs the most between the temperate and tropical zones. It has a direct influence on the physical, chemical and biological properties of an aquatic medium. A water temperature model has been incorporated within a water quality model for application to both natural and disturbed environments (Girard, *et al.*, 1972). A further advantage is that both have been coupled to a conceptual model by Girard *et al.*(1972).

Later, Girard (1988) presented a joint modelling exercise of water and nitrogen in a groundwater system where the different yields of nitrogen are developed in the form of a production module (rainfall, fertilizers, animal deposition). The author mentions that the main thrust of the modelling is to determine the nitrogen flow from the non-saturated root zone to the aquifer. The model used agricultural parameters – such as the type of crops, the fertilizer dosages, and the chemistry of the products – so that it is possible to simulate preventive decision-making for decreasing the pollutant, or determining curative measures such as the dilution of a contaminated aquifer by the injection of clean water. Nevertheless, some biochemical and surface transport processes still need to be included in such modelling in order to improve the results. Likewise, Girard (1988) insists on the need for better agricultural data (annual crop maps).

The Vollenweider-OECD model (Vollenweider & Kerekes, 1980) has been used to predict the phosphorus concentrations in lake water from known inputs, and also the eutrophication possibilities. In the Project on Climatology and Hydrology of Bolivia (PHICAB) (as part of the UNESCO IHP), Benavidez (1988) applied a conceptual model (HYMO 10) to predict the effect of soil management on soil erosion and sediment transport in a watershed of the Andean Amazon of Bolivia.

Estuarine water quality modelling merits special attention because of the very marked hydrodynamics and changes of dissolved and suspended contents that occur in such systems which can lead to unsatisfactory predictions . The modelling of waste water sewage and disposal, urban floods, and the economic aspects of water quality control must also be examined (Servat, 1985, 1987).

These references are not exhaustive, as it is an area of intense activity at present. However, it would be extremely useful from the standpoint of decision-making in the humid tropics to have a complete and detailed listing of all operational models. The most adequate models could then be tested and adapted to local problems. Specific modelling, of course, must be developed for unique cases.

Direct use of hydrodynamic data

As previously acknowledged, water quality modelling ideally requires information on the hydrological processes of water transfer, especially the runoff generation process and groundwater movement (see Bonell with Balek; Chilton and Foster, this volume). In most humid tropical environments, such information is not available for the successful application of various models. The alternative is to make management decisions concerning various water quality issues based on direct measurements. Such data collected should include the quantity and intensity of rainfall, the trajectories and magnitudes (quantity and quality) of the horizontal, vertical and transversal movements, stratification, mixing, etc. (Roche, 1972, 1976). Measurements may also include water transit velocities through the use of natural tracers (dissolved constituents, salinity, isotopes) or artificial tracers (radioelements, colorants, float-balls, chemicals, etc.) or by other appropriate hydrometrical methods. Such information is generally related to the evaluation of discharges, volumes and transfer times (Roche, 1975, 1977). It is also necessary to know turn-over times and the mean age of the water as well as infiltration and evapotranspiration information that represents other terms of the water balance equation.

INVENTORY AND ANALYSIS OF THE PRESENT NATIONAL STATUS OF WATER QUALITY, AND FUTURE TRENDS

It is a matter of urgency for most countries in the humid tropics to establish a scientific and technically-detailed inventory and analysis of their water resources.

Some syntheses are available that deal with the hydrosystems of regions that include the humid tropics. These publications include the ecology and utilization of African inland waters (Leveque & Burton, 1981); a draft review of the state of aquatic pollution of East African inland waters (Alabaster, 1981); la pollution des eaux continentales africaines (Dejoux, 1988);and the Philippines environmental report (Anon., 1985).

In some underdeveloped countries, valuable studies on the water chemistry and quality of one or several hydrosystems already exists (George & El Moghraby, 1978; Brockman, 1986). However, the projects have generally not been exhaustive enough to draw a complete picture of each nation's aquatic environment, including the fauna and flora. Often only the major ions were discussed and if the contamination was not highly visible, the remaining pollution was ignored. Thus, only the most serious problems have been taken into consideration.

This incomplete knowledge of the water quality is mostly due to inadequate organizational infrastructure at the national levels for conducting detailed inventories and scientific analyses of existing resources (Chia, 1987). It is necessary to scientifically study the present level of pollution across all types of hydrosystems (Gower, 1980; Oluwande *et al.*, 1983) that are impacted by human activities to provide a comprehensive understanding of the problem. Such information will provide a reference base that will permit the detection of future trends

(Kaoma & Salter, 1979; Robarts, 1975; Panswad *et al.*, l988). A second study later would show the trends versus the original state (Secretaria de desarrollo Urbano y Ecologia, 1985). These surveys of water quality in the humid tropics are urgently required as the basis for developing appropriate water quality policies.

In some humid tropical countries, unpolluted or almost unimpacted hydrosystems can still be found (Keller, 1983; Roche, 1973). A study of these areas is urgent. They can furnish baseline information for numerous similar scientific areas elsewhere whose aquatic ecosystems are threatened and which will not exist within a few years (Roche, 1982; Roche *et al.*; 1986; Guyot *et al.*, 1990). It is important to study these undisturbed hydrosystems (Anon, 1978) to learn how they function and to ensure them the highest level of protection so that they can provide baseline information over a long period (Roche & Canedo, 1984; Roche & Fernandez, 1988).

A global database of the precise national water quality situations in the humid tropical countries would help detect common or similar problems. Such issues could then be properly addressed by international or bilateral co-operative actions (Anon. 1979).

To avoid inconsistent methodologies in water quality monitoring between different countries, a working plan could be developed as a guideline for encouraging projects to be done in a more homogeneous manner between each country. Such measures would permit a more useful global synthesis of the subject across the humid tropics. Where necessary, however, national conditions should permit some appropriate modifications to these methodologies. In common with recommendations from water management policy (see Hufschmidt, this volume), large drainage basins should be the geographical unit used whenever water quality surveys are undertaken.

WATER QUALITY MONITORING NETWORK

Purposes of monitoring water quality

Parallel to doing an inventory every five or ten years, daily to monthly data also should be collected from a water quality network (Steele, 1987). Such observations provide continuous monitoring which ensures that, one, no sudden deterioration takes place in the water's quality from bad management practices and, two, that appropriate quality levels are maintained for the various uses (Aziz, 1986).

Although a little different in design, the inclusion of an additional warning system is also based on monitoring, specifically in real time. Nevertheless, both systems (inventory and warning) permit the identification and source of contaminants which are essential inputs for rapic decision-making

Monitoring detects the long-term trends of water quality, from local to global levels, by acquisition and interpretation of long temporal series of analytical data, including extreme values and the short-term shock loads of pollutants (Hadley, 1986; Jones, 1982; Lau & Mink, 1984). Such information

improves the calculation of hydrogeochemical balances (Roche, 1982; Guyot *et al.*, 1988). The observations also improve the mathematical water quality models, and their validation.

International experiences of water quality monitoring

Various long-term networks have been proposed or set up to monitor water quality locally. The Global Environmental Monitoring System (GEMS) was organized in 1974 by a group of international organizations: UNESCO, WHO, WMO and UNEP. The GEMS/WATER component was established, including a water quality monitoring system. Physico-chemical and biological analytical data are stored in a reference center. Analyses and periodic interpretations are made. If desired, national participation can be limited to sampling and analysis. Calibration of methods and results of analysis can be helped by the reference laboratory – in particular by sending reference solutions. Not all countries participate in GEMS/WATER, though contacts are being made for a broader international participation. This program, however, will only become more meaningful when considered over a long-term period and with a sufficient density of stations contributing to the network. However, some participating countries already are detecting fluctuations in their water quality not previously known (Chantraine & Dufour, 1983).

Among other world programmes that include a water quality network, the recently established International Geosphere-Biosphere Programme (IGBP) should be mentioned. Its goal is, one, to develop a coordinated action to describe and understand the interactions and integrative role of the physical, chemical and biological processes which regulate the environmental conditions on Earth, and, two, to predict how human activities could disrupt this system.

At a regional level (Dejoux, 1988), the following programmes can be mentioned as examples: Commission of Lake Chad Basin (CBLT); Committee for Continental African Fisheries, 1972 (CPCA); East Environmental Pollution Research Committee (EEPRC); East African Federal Fisheries Organization (EAFFRO); and in South America, the PHICAB Project (Climatological and Hydrological Project of Bolivia ORSTOM – IHH – SENAMHI – CONAPHI).

The latter project will assure for some years the monitoring of water quantity and quality (salinity, major ions) in the Amazon Basin of Bolivia. A reconnaissance of water quality in the entire highlands of the Andes and in Lake Titicaca and Poopo Basins (Roche *et al.*, 1986; Guyot & Herail, 1989; Guyot *et al.*, 1988) has also been carried out.

National and international compliances with monitoring of water quality

To be realistic, the successful operation of a long-term water quality network appears to be very difficult, especially in developing countries. Many factors, essentially the same ones that have been previously mentioned for a water quality inventory and analysis, are even more serious for the proper management of a monitoring network. The shortage of funds needed for instrumentation, their installation and continued maintenance is particularly critical, especially in many humid tropical countries of Africa. The economic crisis of the 1980's has weakened the quality and density of the networks, while the delays in the collection and processing of the data are getting longer and longer. National hydrological organizations are having more and more difficulty in properly managing their networks.

The possibilities of rehabilitating the density and the efficient functioning of these water quality networks needs to be urgently explored. But this can only be achieved with a continuous supply of ample funds. One way to achieve this goal is for countries to integrate their own actions with international programmes. The ability of developing nations to fund the monitoring of water quality and quantity networks – as well as the ability and reasons for the developed nations to share the costs – is an important issue. As an example, one can mention the large participation of the French ORSTOM (Institut Francais de Recherche Scientifique pour le Développement en Coopération) in the monitoring of national hydrometrical networks, often accompanied by hydrochemical studies, in West and Central Africa from the 1940's. Similar programmes were carried out in the 80's in the Bolivian and Brazilian Amazon. In past years, remote sensing hydrometric stations also have been installed in West and Central Africa and Brazil. Recently, the European Economic Community (EEC) partly funded some telemetry equipment in these projects, supplemented by assistance from Canada. The World Meteorological Organization (WMO) and the World Health Organization (WHO) also support this line of endeavor with their own hydrometric monitoring projects.

A major difficulty in devising an effective water quality monitoring network is taking into account the proliferation of more complex chemical compounds which are contributing to pollution. Can observation networks and the usual analytical control systems truly detect the existence of all these harmful substances in hydrosystems? Consequently, difficult decisions have to be made on what pollutants need to be monitored, why and how. Specification of objectives helps determine the precision and accuracy to be attained, as well as the number and frequency of the sampling, whilst keeping in mind the cost-effectiveness of the operation.

Still a major concern, however, are the large areas in the humid tropics without observational networks – particularly in Africa, the Amazon, and Borneo. Ideally, water quality monitoring systems in real time would be most appropriate, assisted by telemetry using satellites.

In the meantime, sensors that respond to requested water quality standards must continue to be designed, improved, tested, built, and widely distributed. The experience acquired by the Hydrological Watch Organization (HYWA) with the participation of ORSTOM in West and Central Africa, under international co-ordination, permits data collection and dissemination from a selected number of hydrological stations (Le Barbé *et al.*, 1987). This is just the beginning of what must be done in various areas for water quality control, including the pairing of hydrometric and quality measurements (Bader *et al.*, 1987; Gautier *et al.*, 1987; Le Barbe *et al.*, 1987). Satellite telemetry is already widespread in the region as demonstrated by the following :

* Linkage with the ARGOS System. The WMO-Hydroniger Program on the Niger Basin is operating 65 stations. WHO is operating nearly 100 stations in West Africa for the purpose of the onchocerciasis control program. The Office de Mise en Valeur du Fleuve Senegal (OMVS) has installed a six-station network, and was to have 10 stations in operation in 1989 on the Senegal River and its main tributaries. National hydrologic organizations are also using satellite telemetry facilities, such as Benin (10 stations) and Guinea (5 stations). Twenty-three stations have been installed in Brazil (Amazon) by ORSTOM-DNAEE independently of HYWA.

* Linkage with the METEOSAT System: The Société Nationale d'Electricite du Cameroun (SONEL) is operating a ten-station network on the Sanaga River. A Meteosat network (12 stations) also was to be installed in 1989 in the Zaire-Congo-Oubangui Basin, co-ordinated by ORSTOM.

The HYWA structure should be first of all, a co-ordination organization to ensure proper collection, processing and dissemination of the data by electronic mailbox – and executed under international control to ensure effective data-sharing.

Taking the humid tropics as a whole, the water quality data that is available is not exploitable because the primitive means of storage do not go beyond the stage of tables, accompanied by a few comments in internal reports. They are of little or no use for anyone but a specialist. Cullen (1987) reports that considerable funds are spent in monitoring water quality, yet there appears to be a poor return for this investment. Stewart & Verhoeven (1987) justly mention that too little emphasis is placed on the ultimate use of data. The main reason for this situation is the lack of computerized data-processing. Data collection starts in the field and laboratory. Encouraging the use of personal computers (PC's) at this level would help generate a more workable data storage base, possibly from the analytical instrument to the computer. The interfacing of such databases with appropriate software needs to be encouraged in order to produce basic statistical analyses and outputs quickly from expert systems.

There is considerable interest in coupling chemical-physical monitoring with biological monitoring. Determining the impact on life is the main purpose of water quality studies (Environmental Health Division, 1986). Biological changes induced by pollution seldom or never are directly indicated by chemical and physical measurements. Once defined, sensitive biological communities may be more useful indicators of trends in water quality than the classical pollution measurements (Ngoile *et al.*, 1987; Noble *et al.*, 1971). In the last few decades, taxonomy has lost its importance in the biological sciences. However, the pollutional impact on aquatic biota is based on the identification of plants and animals, and their appropriate groups. Therefore, the significance of taxonomy must once again be recognized. Further research in this discipline in conjunction with water pollution studies should be encouraged.

AQUATIC TROPICAL DISEASE VECTORS, PARASITES AND PESTICIDES

Control in health and agricultural sectors

General aspects Two types of interrelated pollution affect tropical inland waters, creating serious hazards for human health and the aquatic environment. Large-scale human or animal endemics and the epizootics are widespread in the humid tropics: onchocerciasis (Prost, 1986), malaria, yellow fever, dengue, dracunculiasis, trypanosomiasis, leishmaniasis, bilharziazis, etc. The vectors or their larvae live in water, as well as the parasite itself sometimes (bilharziazis). This is a form of biological water pollution (Prost & Vaugelade, 1981), whose extensive control by insecticides, molluscicides and herbicides leads to another type of contamination of the hydrosphere.

In agriculture, disease and predation control treatments against various parasites or animal (micro-parasites, insects, birds, rodents, etc. and land or water weeding are also accomplished by insecticide, fungicide or herbicide spraying so that the yields of rice, cotton, sugar-cane, citrus, palm, soya, etc. can be improved. In addition, the same substances are often used in vectorial and agriculture control.

Thus, pesticides are among the most prevalent pollutants in many hydrosystems of the humid tropics. Considerable literature on the toxicity of pesticides has been published since the 1960's, but they have been scarce for the humid tropics. The World Health Organization has published information on spraying conditions for pesticides, their degradability in the environment, exposure type for humans (air, water, food, contact) and their metabolism and effects on health. Dejoux (1985, 1988) performed an exhaustive study on the impact of vectors and agriculture control programmes on the aquatic environment in Africa.

Chemistry, permanence and relative toxicity of pesticides The products used are classified into different groups:

Organochlorides: They were the first synthesized, and are the most ubiquitous. They are not very soluble in water, but are soluble in oil, and accumulate in greases. They have generally long residence times and ,consequently, are a major hazard. The main chemical species are: DDT; aldrin; dieldrin; chlordane; hexachlorobenzene; heptachlor; lindane; metoxychlore and 2, 4-D. The HCH isomers (i.e. lindane) are among the most soluble in water.

Organophosphorus: These are less permanent. Examples are parathion, malathion, TEPP and phosdrine.

Pyrethroids: There are natural and synthetic forms that are resident in water for some period of time. They attach quickly to organic matter, and their toxicity is generally high. They account for a high percentage of pesticides sold in the world market. Pyrethrum is a tropical chrysanthemum and the use of pyrethroids is sometimes mentioned as a biological control.

Organic or organo-metallic compounds: Used as herbicides, they are derived from urea, thioural, tiazenes, etc. Carbamates and dithio-carbamates are fungicides. Some are natural.

Mineral substances: Although "old pollutants," they are still in use : Compounds of mercury, copper (sulphates), tin and sulphur are examples. Mercury salts are used the most, despite their high residual toxicity.

Bacterial products: These are "biological insecticides," such as *Bacillus thuringiensis israelensis*. The latter is a bacterium whose spores produce a toxic crystallized protein. H14 serotype destroys the larvae of some insects (mosquitoes, simulies). It is not very toxic (Dejoux, 1979).

The works of Pimentel (1971), Albaster (1969), Anon. (1981), Dejoux (1988), have established a relative scale of toxicity for dozens of pesticides. Thus, some products may be mildly toxic for the environment (Temephos/Abate/Bacillus t.i.) while others (endrin, dieldrin, endosulfan, pyrethrinoides, etc.) can strongly affect the aquatic medium.

The toxicity of a pesticide depends on its formulation; nevertheless, organo-chlorides are more toxic than the organophosphorus pesticides. Herbicides are generally less toxic, at least towards fish. However, the proliferation of new chemical compounds and the increase in the quantities being used is so large (Smith & Lossey, 1981) that the consequences of many pesticides and their behaviour in the hydrosystems (Benson, *et al.*, 1971) are not sufficiently known. The residual products may be more or less toxic than the original substances.

It is known that some products, whose application is prohibited in industrialized countries (DDT, HCH, etc.), are still being used in developing countries because of their large scope of action and their low price.

The amounts and species of pesticides used are often not accounted for at a national level in the humid tropics (Whittemore, 1977). They would, however, be in the order of thousands of tons for the more advanced countries.

Behaviour and effects of pesticides in humid tropical hydrosystems In the last two or three decades, important control programmes for the eradication of vectors (breaking of the cycle) have been worked out, most often with the sponsorship of international organizations, and sometimes over a twenty-year period (Prost & Prescott, 1984). Similarly, pesticide treatments in agriculture are becoming more systematic. Public opinion (sometimes in the form of groups or organizations) has exerted some pressure against the intensive use of pesticides, thus causing governments to face a dichotomy of policy choices: Protection of public health on the one hand, economic productivity on the other. This is why the impact of large-scale vector eradication programmes or agricultural treatments are considered with all the attention they deserve. Nevertheless, the parallel scientific operations associated with them are still insufficient. The chemical substances used often have a very high toxicity which contaminates directly or indirectly the aquatic medium and, therefore, lead to a high mortality of non-targeted organisms.

The chemical by-products of vector eradication schemes and agricultural treatments accumulate in sediments, weeds and all along the food chain at higher concentrations than in the water – sometimes in the order of 1000 times or more. The accumulation begins in microorganisms, then in the phytoplankton, the zooplankton and the invertebrates, ending up in fish and humans (Dejoux, 1975, 1985, 1988). These contaminations can irreversibly damage the ecological balance on which depends human health or sources of proteins (Eichelberger & Lichtenberg, 1971).

The effects of pesticides on inland hydrosystems can also have indirect effects such as the diminution of plankton, the dissolved oxygen level, and a modification of the pH and CO_2 (Pesson, 1976). Such problems were only examined in humid tropical countries long after the use of various chemical treatments. Scientific documentation is minimal. Dejoux (1988) details the investigations carried out in West Africa, especially as a part of the onchocerciasis control programmes. The impact on the environment also has been studied by Dejoux (1988) as part of a programme for selecting the appropriate insecticides. Elsewhere, studies on the impact of other pesticides have been carried out in other countries (the Victoria Nile in Uganda, Sudan, etc.). Generally, it appears that most of the products tested have strong toxicity.

Factors of transfer of pesticides through the hydrosystems
Many factors affecting the transfer of pesticides through the hydrosystems are closely linked. These can be itemized as follows:

The chemical characteristics of the pesticide, such as solubility and resistance to physical and biochemical degradation, are a major factor. Sub-products from the chemical breakdown can be more or less toxic and even more resistant than the original compound.

The circumstances of injection determine the dissemination (concentration of use, non-point or point injection, duration and frequencies).

The air and water dynamics determine the transit of pollutants – the meteorological conditions during the treatment (wind, evaporation, humidity), elevation of application if aerially applied and intensity of rainfall, run-off and solid transport flows (Bader *et al.,* 1987; Le Barbe &

Gioda, 1987). Agricultural treatments using pesticides at variable distances from water bodies can result in less toxicity because of both adsorption and absorption within the soil system during the transit of the pollutant to various aquatic systems. However, in vector control, running or stagnating waters are directly treated, therefore the pollutant risk is increased. As indicated, the nature of the soil, the chemistry of the water, the air temperature and humidity, all have an influence on the transfer of chemical substances. Interactions with the mineral and biological medium govern largely the gradual transformation of the chemical compounds (Mathur & Saga, 1975). For example, organochlorides are easily adsorbed on suspended matters and sediments, and then can progressively be washed-out or removed by erosion.

Ways of controlling the agents of tropical diseases and pollution by pesticides
Integrated control. Only the search for a set of diverse, yet complementary methods (integrated control) adapted to countries of the humid tropics, will allow the control of the vectors and disease-causing parasites of humans, animals or plants. The treatment with pesticides must be considered only as one of these methods. Thus within the framework of integrated control, if a method allows a reduction in the expansion of pesticides, this should help in decreasing the contamination of the hydrosystems. For example, control of schistosomiasis in the tropical zone illustrates, among other things, the methodology for an integrated control. The World Bank sponsored more than thirty operations for the control of this disease (Lake Volta, Ghana, etc.) led by the WHO.

The choice of pesticides, technologies and conditions of their use is an area where considerable work remains to be done. We need to have better knowledge of the toxicity of many products, beyond their absorption in rats or fish. Investigations are required to determine the minimum amounts of pesticides to be effective. The residence times of the products, and their impact on the fauna and flora in tropical media also must be better assessed. New application methods, including the use of screen traps, aimed at limiting the propagation of pesticides in the environment, also must be investigated.

Greater use of biological controls is a possibility that needs more investigation. Predators (fishes, gastropods, insects, etc.) and competitors (closely related species of insects, molluscs, radiosterilized males, etc.) need to be considered. Weakening the targeted species before and after treatment allows the amount of pesticide to be cut. Such methods have yet to be widely adapted to the conditions of each humid tropical country. The example of the rice-eating gastropods illustrates that the introduction of allochtonous species can be very tricky.

Knowledge of the hydrological regimes (rainfall, height of rivers, discharges, flooded areas and depths, etc.) is indispensable for integrated control to work. Consequently, an inventory of the whole aquatic situation is necessary to avoid rein-fection of the treated zones from the non-treated ones. All this information must be acquired before and during the treatment programmes. Where sufficient parameters have been measured, premodelling the effects of treatment operations may be attempted (Pouyaud, 1987; Pouyaud and Le Barbe, 1987a) but their reliability is still open to question because their theoretical basis does not take into account all the factors associated with chemical transfers, e.g. selective bioaccumulations, auto-purification, etc.

Cleaning the aquatic medium through public works is a basic method in taking preventive measures or as a form of rehabilitation. The following mechanical actions represent a fundamental contribution toward such objectives (Highton, 1970; Hervouet, 1983):

* Deviation, sweeping, alternated drying of natural channels or opened sewers, depending on the life span of the targeted and non-targeted organisms.
* Improvement of drainage and water transit speeds through the draining of small dams and swamps.
* Building of vertical wall spillways; removal of rocky thresholds, collapsed works, natural pits upstream and downstream.
* Removal of over-abundant plants.

Treatment of drinking water can eliminate or diminish the contents of pesticides through the use of flocculation, decantation and ultra-violet light. Activated carbon (powder or beads) can contribute to pollutant trapping; but high doses are sometimes necessary. Nevertheless, even apparently used-up carbon can still contribute to the elimination of pesticides. To conclude, integrated control of water quality, related to the control of water-related or water-borne diseases and chemical pollution (vectors, parasites, pesticides) is a wide field for multidisciplinary research in the humid tropics in which hydrology must take an important place.

MICROBIOLOGICAL ASPECTS OF WATER QUALITY CONTROL IN THE HUMID TROPICS – DRINKING AND RECREATIONAL WATER

General aspects

Despite the momentum created by the International Drinking Water Supply and Sanitation Decade, progress has hardly matched the increase in population size (Prost, this volume). Drinking water supply of acceptable quality and quantity for every person is far from being reached. However, it must remain a high priority for the year 2000, as one of the most important factors of human health.

In many countries of the humid tropics water kills, whereas it should be a source of life and health. The first objective of any program to manage water quality therefore should be to identify how many people die – and why – from water-related problems and seek the most effective approaches to reduce the number of deaths.

Microbiological aspects that deal with water quality control are numerous (Smith, 1969). Bacteria, viruses, fungi, insects, larvae, worms, etc. are the causes of the water-associated diseases which are very common in the humid tropics, more so than chemical and micro-toxic pollution causes.

Vectorial diseases, whose parasitical cycles include an aquatic phase, together with faecal pollution, are the main causes of peoples' poor health in the humid tropics. Hundreds of millions of people are contaminated. The aspects of vectorial disease control, and its consequences on water pollution, are the object of chapters elsewhere in this volume (Prost, Wurzel).

Faecal bacterial contamination patterns in the humid tropics

Faecal water-borne diseases include cholera, dysentery, gastroenteritis, other diarrhoeas, typhoid, infectious hepatitis, and some vectorial diseases such as bilharziasis.

Knowledge of faecal bacterial contamination patterns is necessary for decision-making when dealing with water quality control. Consequently, the characteristics of contamination patterns will be examined in both rural and urban areas.

Rural areas Faecal pollution, though often localized, presents some serious problems in rural areas. Bad drainage and no protection are observed around drinking water wells. Consequently, some wells become contaminated with *Escherichia coli* – emanating originally from human and animal excrement. Even the ropes and buckets used at these wells can be contaminated at each withdrawal of water.

There is a lack of individual and community latrines. Excrement is dropped everywhere, particularly on the river banks by people and animals, or directly into the water (boat people) at the same place where water is used for drinking and personal bathing. The resulting pollution can then be transported downstream by river flow, especially during floods. At such times, dilution is more effective. However, during the dry season, the low dilution increases the content of the faecal material.

Watering of vegetables with faecally (or chemically) contaminated water brings diseases to the town where these products are sold. Agricultural re-use of municipal effluents commonly is a source of contamination. However, a WHO scientific group recommended that the bacterial guidelines for waste water use in agriculture be relaxed (WHO,1989). They recommended, on the basis of epidemiological evidence and of available water treatment technologies, that water used for the irrigation of edible crops which contains less than 10,000 faecal coliforms per liter be allowed. Furthermore, the irrigation of such crops, and trees was to be permitted after the retention of waste water for 10 days in stabilization ponds.

In urban areas Waste water open channels are used in the same way in urban areas as the river banks in the countryside.

Moreover, they are used for disposal of wastes which remain *in situ* between major floods (Feachem, 1977; see Gladwell, this volume).

Unsatisfactory waste water treatments lead the non-treated urban effluents to natural lagoons or streams where faecal contamination is high. This is one of the major sources of pollution. Even the stream sediments can be contaminated with pathogenic germs from urban effluents. If subsequently dredged for construction use, they become a hazard for the building workers.

Septic tanks are rare. The commonly used latrines also can contaminate neighboring water wells through groundwater movement. Even treated water coming from a waste water treatment station can remain contaminated by pathogenic germs.

Appropriate microbiological indicators of faecal and other pathogenic pollution

The difficulty of finding significant faecal indicators is a major problem. If faecal microorganisms appear in drinking water, an alert normally must be given, even though in some circumstances, the organisms may not be pathogenic (Bremond & Perrodon, 1979). One harmful group are generally faecal coliforms and faecal streptococci; they may also contain enteroviruses (Bockemuhl, 1985). Other bacteria such as coliforms or sulphite-reducing *Clostridium* are ubiquitous, and are certainly not indicators of faecal pollution. Moreover, other microorganisms, for instance fungi, actinomicetes, etc., may be more responsible for human health troubles than coliforms, for example. Oxidizing or reducing bacteria can be damaging to water supply and waste water networks.

Indicators are used for public health purposes, or as tests to estimate the efficiency of a treatment. For instance, the resistance of coliforms to disinfection is similar to that of *Salmonella* (Jimenez *et al.*, 1989). Thus, they are basic tools to guarantee the public health of the people living in the tropics (Bialkowska-Hobranska, 1987; Evison, 1988). However, the problem is not simple. Research is continuing to find the most suitable microbiological indicators of faecal and other pathogenic pollution in the most advanced humid tropical countries. However, such work needs to be expanded and the results considered on a global scale. The research challenge is to identify the appropriate humid tropical species which can be used as indicators such as total and faecal coliforms, identification and isolation of *Salmonella*, and identification of enteroviruses, fungi, actinomicetes. The complexity of this issue is highlighted in the research of Fujioka and co-workers (Fujioka & Schizumura, 1985; Fujioka *et al.*, 1980, 1981, 1988, 1989; Hardina & Fujioka, 1991), Hazen *et al.* (1980, 1987) and Hazen (1988, 1989).

Hazen (1989), in mentioning Bonde's (1977) eight criteria of an ideal faecal-pathogen indicator, reviews various studies related to the following bacteria as candidates for this purpose, viz. *E. coli*, faecal streptococci, *Bifidobacterium* spp., *Clostridium perfringens*, viruses, and bacteriophages. He concluded that the standard faecal indicator, *E. coli*, is unacceptable.

With few studies having reported the use of faecal indicators (other than *E. coli* in tropical source water), objective evaluations of the efficacy of these alternate indicators is difficult. At present, the obligate anaerobes, or their phages seem the best candidates, primarily due to their inability to survive extraenterically.

However, all of these indicators have the inherent difficulty that they or their host may survive under some conditions and that the media used for bacterial indicator enumeration may allow the growth of false-positive background flora. The viable but non-culturable phenomena reported for many pathogens in both temperate and tropical waters suggests that indicators may only rarely be correlated with disease risks in source waters (Colwell *et al.*, 1985; Hazen *et al.*, 1987; Baker *et al.*, 1983). Thus the best indicator may be no indicator, i.e. direct enumeration of selected resistant pathogens may be required. This would allow a more realistic estimation of the health risk.

Immunofluorescent staining can detect densities of pathogenic bacteria as low as 10 cell per ml, a density which may give no cultural counts (Colwell *et al.*, 1985). The use of monoclonal antibodies makes this technique specific. However, as a result of cross-reactivity when using immunofluorescence (even with mono-clonal antibodies) the most specific and sensitive methods for detecting pathogens may be the nucleic acid probes (DNA or RNA). DNA probes have already been developed and tested for enterotoxigenic *E. coli* (Bialkowska-Habrzanska, 1987; Moseley *et al.*, 1982) and *Salmonella* spp. (Fitts *et al.*, 1983). Thus, direct detection of pathogens currently is possible.

Common enteric pathogens which could be enumerated are poliovirus and *Salmonella typhimurium*. Detection of either one of these in tropical source water would indicate a risk of human disease. Instead of enumeration, maximum contaminant levels could be based on detection only. One potential problem with this approach is that the presence or absence of one pathogen may have little bearing on other pathogens. A multi-species test for two or more of the more resistant and common pathogens found in tropical waters may be necessary.

In Hawaii, many of the national regulations such as the use of bacterial indicators to assess water quality are not applicable because fecal coliform (*Escherichia coli*) and fecal streptococci bacteria are naturally present in the streams of the island of Oahu . The concentrations of these two sanitary indicator bacteria commonly exceed the new U.S. Environment Protection Agency (USEPA) recreational water standard in freshwater of 126 *E. coli* per 100 ml or 33 enterococci per 100 ml. *Escherichia coli* was shown to multiply in stream water samples (but not enterococci). The soils are considered to be the most likely source for the high concentrations of these indicator bacteria. Further, the higher temperatures in stagnant pools of stream water exposed to sunlight are thought to encourage rapid multiplication of the *Escherichia coli* (Fujioka *et al.*, 1985, 1988, 1989; Hardina & Fujioka, 1991). It is significant that indicator bacteria were recovered from the surface soil as well as from samples of soil down to a depth of 36 cm at different locations adjoining a stream near Honolulu (Hardina & Fujioka, 1991). Another feature is that aquatic organisms selected for toxicity assessment by the USEPA are not present in Hawaii.

APPROPRIATE MICROBIOLOGICAL WATER QUALITY STANDARDS AND LEGISLATION FOR HUMID TROPICAL DRINKING AND RECREATIONAL WATER

Several sources (Anon. 1975, 1976, 1986) propose water quality standards that are often adapted by each country for their specific conditions (National Research Council, 1982). Most humid tropical countries do not have legislation that deals with the bacteriological quality of drinking water from treatment stations. In this case, the recommended standards are only guidelines. Other microorganisms can be responsible for poor human health (for instance, cutaneous diseases in recreational water), without inevitably being of faecal origin. As highlighted, the use of faecal indicators does not contribute a universal test, even though the standards commonly adopted by sparse legislation allow for maximum threshold values.

In temperate countries, water is generally acceptable for drinking only if it contains neither faecal nor pathogenic germs and only a very small number of cultivatable germs. The adoption of such standards by the EEC countries has recently improved the quality of drinking water. However,the studies of Fujioka and Hazen have emphasized that the temperate standards have no significance in tropical waters. Fujioka (1989, pers. comm.) mentions that conditions in tropical countries will never be fully understood and that these countries will continue to be forced to operate under assumptions which often are not applicable to their own conditions.

Also, these standards often are not respected in rural areas, where treatment stations do not exist. Drinking water is taken from the common well or, more often, from a stream or lake at the same place that is utilized for personal washing, swimming or animal watering. Thus, it becomes increasingly difficult to meet WHO standards, especially in respect to faecal coliforms. Even in big cities, the performance of the existing treatment stations is sometimes unsatisfactory because of technical or human faults.

According to Kilani (1989, pers. comm.), there is a need to relax the existing water quality standards in Kenya. Experience has shown that the use of FC (indicator organisms) as an indication of faecal contamination and, therefore, the presence of pathogens, is a luxury that is no longer economic in rural water supply operations.

Considerable additional research is required to identify the bacterial flora found in water sources and in the tropical distribution systems. Epidemiological work is needed on the hazards to health from different standards of water, including the relationship between cyanobacteria (and phytoplankton) in water storages and their effects on human health. Epidemiological statistics "in real time" could be also considered as efficient warning indicators.

THE ROLE OF MICROORGANISMS IN THE BIODEGRADABILITY OF ORGANIC MATTER, CHEMICAL SUBSTANCES AND INDUSTRIAL WASTE

This aspect of microbiology is more significant in the humid tropics because the temperatures found there are more favourable for biodegradability – in spite of the decrease in the dissolved oxygen content. Biodegradability in an anaerobic medium is also a research theme of great interest elsewhere because it is able to provide power alternatives (Environmental Health Division, 1986).

The main items to consider in the microbiological and biochemical area are the microbiology of degradation in humid tropics water and soil: the aerobic and anaerobic processes, the metanogenic and non-metanogenic bacteria, the methods of analysis and tests, peculiar kinetic processes, reaction velocity characteristics, kinetics and control of parameters, mixture regimes, and operational problems.

METHODS OF BACTERIAL WATER POLLUTION CONTROL IN THE HUMID TROPICS

The methods of controlling bacterial pollution in the humid tropics are quite diverse and complementary. Some are cheap, others are expensive, but they must be applied together at each corresponding level. The cheapest ways are capable of efficiently alleviating the most urgent water quality problems. Nevertheless,the appropriate agencies need educational help and grants from their national governments and international organizations for their successful application.

Methods for controlling bacterial contamination of waste water

Sewage disposal systems (collection and conveyance networks), must be built, as well as facilities for treatment and disposal of domestic waste water (Anon., 1975; Feechem *et al.*, 1977; McKendrick, 1982; Shuval *et al.*, 1981). However, until a satisfactory method of removing faecal pollution is reached, the comprehensive treatment of collected wastes remains difficult, and the resulting effluent water will contribute to the pollution of hydrosystems (Aziz, 1981). In this case, the practice of discharging such effluent into the sea is preferable, because most of the bacteria cannot survive in salt water. In other instances, the effluents of the waste water treatment stations should be disinfected downstream with oxidants. Controlled "lagoonage" is another solution to be considered as often as possible. Strong solar radiation favours the destruction of germs.

Faecal pollution can be reduced by the construction of small and cheap individual or municipal systems (unsewered sanitation). In rural areas, communal facilities can be planned, such as compost systems, but the social and cultural aspects of some societies may hinder these measures.

Methods for preventing bacterial contamination of drinking and recreational water in rural areas

The protection of water used for drinking and recreational use incorporate many of the issues highlighted elsewhere in connection with watershed management (see Hufschmidt, this volume). The escalating social and economic pressures in humid tropic countries means that the conflict between wanting to protect watersheds for water supply and recreation as against economic development is even more acute than in many developed countries. Similar comments apply to many of the global water quality management issues.

A diverse range of actions need to be taken. For example, analyses should be carried out to determine the main contamination problems and to locate where they are most severe (Munoz, 1981) in order to ensure a safer water supply. Individual and community swimming pools need operational methods for treatment (flocculation, filtration, disinfection), quality control and legislation.

The municipal treatment stations and other supply systems should use the classic methods, including sedimentation, flocculation and filtration (Anon. 1983; Richard, 1987; Masschelin, 1988). Chlorination and cleaning of water tanks can be improved by other oxidants such as ozonization. Other factors that must be kept in mind include the local production of oxidant, the place and time of injection, equipment, continuous and emergency disinfection, super chlorination, and contact times. Nevertheless, bacteriological standards do not guarantee protection against such viruses as cause typhoid and infectious hepatitis. Sand filters and chlorination would not eliminate hepatitis viruses (Smith, 1969). Classic analyses cannot detect some toxic substances (emanating from agriculture) or viruses, in the same way as classic treatments cannot remove them.

To administer domestic water supplies more safely (Elmendorf, 1987) municipal administrations should be encouraged to install piped and tap water supplies in households, as they are better than the public standpipes (Jordan *et al.*, 1982; Merrick, 1985; Mendoza, 1989). Where such improvements are not economically possible, alternate simple and more inexpensive systems (Elmendorf, 1987) need to be considered with the assistance of the national government (e.g. Thailand, Government of Thailand, 1989) and international organizations. These include the building, setting-up and maintenance of hand-pumps/small boreholes with various protective methods (described in Junkin, 1979) to avoid pollution of the aquifer, either by direct entry of polluted surface water

into the well or from percolation through the unsaturated zone. In the latter case, sandy unconfined aquifers are particularly vulnerable (Sharma, 1986).

One of the commonly used domestic supply systems is the collection and storage of rain from roof run-off (Nongluk,1989). Studies, however, need to be carried out to determine the amount of roof area needed to furnish a reliable water supply – based on the water demand. It is then necessary to devise the piping arrangement and the type of tank: concrete (with bamboo), metal or plastic.

The withdrawal system from these tanks and their hygiene are also important issues. Prost (this volume) points out that small amounts of water usually are drawn during the day from the main tank or jar, using small cups kept nearby. It is obvious that each collection contaminates the reserve. In time, bacteria proliferate, increasing the risk to health. Periodic disinfection of the water tanks (ground or roof cistern) should be part of the routine maintenance. In addition, the use and local production of kitchen gravity filters – made of porous ceramic – also improves the quality of the water supply. This is a significant issue because in many areas, piped and tap water is not supplied. Even if it is, the quality is generally dubious, needing a final treatment just before consumption. The small amount of water drunk per family per day could be safely provided by such simple filters and they could be manufactured locally at a low cost.

Better methods of constructing and maintaining small flocculation tanks and sand tanks with carbon filters can improve the water storage situation for villagers (Schulz & Okun, 1984; Van dijk & Oomen, 1981).

DOMESTIC AND INDUSTRIAL POLLUTION CONTROL: ORGANIC MATTER, NITROGEN AND PHOSPHORUS EUTROPHICATION; SANITATION, SEWAGE AND WATER DISPOSAL

Types of urban and industrial pollution

The rapid growth of cities in the humid tropics has produced a significant deterioration in the water quality of surface and subsurface water resources. The principal sources are untreated domestic waste water and solid wastes which are producing a high rate of faecal contamination, and hydrocarbon and inorganic toxic materials from the accompanying development of industries (Broche & Peschet, 1983).

Causes and aspects of organic pollution

Biodegradable organic pollution originates from two sources, namely, domestic waste from the high population density of many urban locations (DuFour & Maurer, 1979) and a diverse range of industrial operations. The latter include slaughterhouses, dairies (Van der Berg, 1988), canneries, pulp and cellulose mills (Lointier & Roche, 1988, Roche, 1976,1977), wood processors, breweries, fruit juice factories, oil mills and soap factories, sugar-cane, coffee, rubber, cocoa and rice processors, refineries, and garages. (Debatisse, 1988; El Hinnawi, 1980).

The frequent urban runoff from tropical downpours "sweep" the organic pollutants from the various point sources into drainage networks and so further concentrates the level of pollution in various surface water bodies. This causes eutrophication due to the oxygen depletion and the high production of nitrogen and phosphorus (Thorton & Nduku, 1982; Toerien, 1975). Phosphorus (phosphate and phosphoric acid) and nitrogen, which are heavily used in the agricultural food industry are major contributors to eutrophication (Vermaak *et al.*, 1981). Self-purification is not attainable, particularly where the turn-over is slow of further inputs of water (Marshall & Falconer, 1973; Unesco, 1980). Eutrophication has several effects on the aquatic ecology, including the demise of many fish species and the encouragement of aquatic weeds (Steyn *et al.*, 1976). The anaerobic environment encourages the concentration of undesirable gases from the chemical process of reduction, and the deposition of organic suspended matter.

There are numerous instances of pollution by hydrocarbons in the humid tropics (Orekoya, 1978; Dejoux, 1988) either directly from the oilfields and refineries or train accidents during transportation. Intended disposals of waste oil, often directly in the aquatic medium, are in many environments, the most important form of water pollution by hydrocarbons. Lagoons with a low turn-over time of new water inputs are particularly affected by this type of pollution.

The effect of hydrocarbons is to form an asphyxiating film over the water, thus limiting oxygen exchange through the surface. However, some water is still able to evaporate, leaving a compact residue behind. The soluble phases of this residue are rich in aromatic compounds that have a toxic potential. Degradation of this material by bacteria can produce substances that are even more toxic than the original crude oil, especially when they are dissolved in the upper part of the water.

Refined products (kerosene) also are generally more toxic than the crude oil, and, moreover, can have an anti-bacterial action that inhibits biodegradation.

Waste and waste water disposal

The problem of waste and waste water collection, disposal and treatment by appropriate processes for humid tropical cities, towns and villages is one of the most important issues for human health and the environmental quality of life (Tay & Ong (1986), see also Gladwell, this volume). Several studies concerning the city of Abidjan by Colcanap & Dufour (1982), Broche and Pejchet (1983), Chantraine and Dufour (1983), and summarized by Dejoux (1988), are examples of research that needs to be undertaken. The focus of such work was concerned with the collection and treatment of the waste waters from this large urban centre which are concentrated in a tropical lagoon. Modelling (in terms of BOD) of pollution coming from the various urban and industrial sectors, and its transport

through different systems (sewers, surface drainage, runoff) to the outlet lagoon was undertaken. This included the development of software by several specialized services for simulating the effect of various rainfalls (including actual storms). Flood routing of runoff from various urban watersheds are estimated via waste water networks and their associated hydraulic structures, e.g. spillways.

Ways of controlling nitrogen, phosphorus and organic pollution

The methods of controlling organic pollution in villages and cities include those systems previously mentioned when bacterial pollution control was discussed.

Individual or community waste collection and waste water treatment systems. These remain the best methods to avoid organic pollution of the hydrosphere with high contents of nitrogen and phosphorus. However, design and management of such systems require an adequate understanding of many biochemical phenomena involved (Dangerfield, 1983; Valiron, 1983). There has been less experience in the humid tropics in developing the appropriate treatment systems (Edeline, 1988) for different types of organic pollution. For example, further research is needed to study the various approaches to flocculation of sludge and its subsequent deposition and decomposition (Vedry, 1987). Likewise, the conditions which foster the biodegradability of specific substances and the metabolization of hydrocarbons and phenolic products should be investigated.

An alternative is the deposition of waste water in both natural (like the Abidjan example) and artificial lagoons, in artificial and natural wetlands and in septic tanks. Such practices can be improved by research focused on the mixing of polluted water (urban and industrial), the selection of microflora for optimal degradability and the effects of the addition of nitrogen and phosphorus (for the elimination of phenols).

Anaerobic digesters of waste and sludge is another avenue of possible disposal, but operational problems and potential harmful effects need investigating. For example, the subsequent incineration of the dessicated waste products demands a good knowledge of their composition (toxics, heavy metals, and microbiology) to avoid release of toxic fumes.

Whichever treatment is adopted, the subsequent use of the final products in agriculture – by spreading them in a liquid or paste form – needs close monitoring to assess their impacts on crops, forest and soil (D'Itri *et al.*, 1981). Particular attention should be given to the bacteriology and biochemistry processes that are associated with the decomposition of sludge over different soil types. The risks of concentrating enteric diseases needs to be monitored as well.

Chemical and biological pathways In order to predict eutrophication and to assess management strategies, a better understanding is required of the main pathways of products resulting from the biochemical and physical chemical processes occurring in natural or artificial media. Consequently, there is a close linkage between this issue and studies in hydrological processes, referring to the use of hydrogeochemistry for tracing flow paths in storm runoff generation (see Bonell with Balek, this volume). However, further research is required on the various roles of microbiological populations (pathogenic protozoa, rotifers, nematode threadworms, algae, fungi), the dissolved oxygen balance (including the effects of thermal pollution) and photosynthesis in their influence on rational purification of polluted waters. A related issue is the chemistry and role of the various nitrogen, phosphorous, carbon and sulphur cycles and metallic ions of iron and manganese in the exchange of pollutants between water and sediments in natural and contaminated media.

The Vollenweider-OECD model attempts to incorporate the interactions between organic matter, nutrients and oxygen. However, its effectiveness in the humid tropics needs checking. Cullen (1987) highlights some of the major problems in water quality research, including issues related to modelling.

CONTROL OF INDUSTRIAL MICROTOXICS AND CHEMICALS

General causes of pollution

Microtoxics and chemical pollution come mainly from industrial operations, but in some cases they also can originate from natural weathering processes and erosion. The toxic sources, especially those of natural origin, are not always easily identified. In Kenya, for instance (Kilani, pers. comm.; Ongwency, 1973), high background levels in fluoride content are found in the drinking water in several areas. The source of this natural pollutant is closely related to the characteristics of the water supply aquifers. Ironically, this is one of the major water quality problems in Kenya.

Industrial pollution in the humid tropics, as in the rest of the world, is characterized by the diversity of the effluents and their toxicity, as well as the variability in amounts of solid wastes. This is because each type of industry has its own technical processes that lead to different types of waste water and solid wastes (Ong, 1987). Furthermore, the degree of environmental contamination varies with the degree of development, the type of primary material resources and the exploitative technology used (Martin, 1977).

The race for development in the humid tropics is not always favourable for the installation and control of purification methods for waste water. The economics of foreign competition do not encourage the use of costly, sophisticated purification systems. Therefore, waste water treatments, when available, are often inadequate. Moreover, the desire within humid tropical countries to quickly set up factories which provide the most economic yields usually comes at the expense of the environment in the form of less rigid application of waste water standards.

Chemical characteristics

Heavy metals such as mercury, lead, zinc, arsenic, manganese, copper, tin, nickel, cobalt, iron, uranium, and other constituents such as cyanide, fluorine, acid, and soda, chloride compounds are very toxic. Thus, priority must be given to them in water quality control. They originate in mining activities, metallurgy, tanneries, surface treatments, pulp mills, and cellulose and wood processing. Heavy metals pollution occurs only at selected places in the humid tropics, mostly from point sources such as mines in the headwaters of drainage basins, e.g. the Amazon (Vicente & Beckett, 1986).

Waste oils, by their often uncontrolled spreading, also contribute to raising the level of heavy metals in aquatic systems.

Pesticides (organo-halogens, organo-phosphorated), used in cellulose, wood processing and construction, are often considered as inorganic toxics because of their weak degradability and behaviour patterns in the aquatic environment (Dappen, 1976). In addition to the point-source pesticides originating from industry, there are the nonpoint sources associated with their widespread applications in agriculture, and in termite and vectorial control.

Behaviour in the aquatic environment

Microtoxic and chemical contamination is harmful even in small degrees above the toxicity threshold standards. Heavy metals and pesticides accumulate in sediments, flora and fauna, leading to barely reversible situations (Waldron, 1975). The effects of these materials on the aquatic environment and human health are not well known, outside of the consequences of accidents and laboratory experiences.

The kinetics of microtoxic and chemical accumulation depend on the nature and form (e.g. dissolved) of the constituents and the characteristics of the aquatic environment, notably those of the sediment, physico-chemistry (pH, temperature, salinity) and flow. The reactions, interactions, absorption, adsorption (chelation with organic matter, humic acids and detergents), all control the patterns of deposition or dissolution of the constituents in the water. For example, salinity and pH levels have a major influence on the kinetic evaluation of various components with alkalinity favouring precipitation of hydroxides, carbonates and sulfates whereas an acid pH or low dissolved oxygen encourage dissolution.

Environmental detection of toxic substances

The substances to be detected are numerous (Mills *et al.*, 1982). The following is an incomplete listing: aldrin, dieldrin, endrin, isodrine, endosulfan, chloronitrobenzene, trichlorobenzene, hexachlorobenzene, chlorosaniline, parathion, benzene, 1.1.1.- trichloroetane, 1.2.- dichloride, chloroform, PCB, phosphorus, ammonium and adsorbable organo-halogen compounds.

The accumulation of these materials and their impact on tropical organisms is not sufficiently known. Their detection can be achieved by toxicity assays on aquatic mussels and algae, bacteria, zooplankton, fish (Lesel *et al.*, 1979), and sometimes on humans. Tests for the inhibition of cholinesterase and mutagenic assays (Simmon *et al.*, 1977) also need to be carried out.

Ways of controlling industrial microtoxic and chemical pollution

The difficulty in eliminating harmful microtoxic pollution in aquatic systems stresses the need for more effective measures of controlling pollution at the sources. Prevention is better than cure. Therefore, integration of efficient anti-pollution systems in the conception and building of industrial systems is greatly needed.

In-plant control New production processes are being considered to reduce the quantity and quality of the wastes and the characteristics of contamination. An example is the new technology for paper pulp processing and associated bleaching. to cleaner technology, according to the local conditions. These newer methods separate water as much as possible at each manufacturing phase. Minimal amounts of water also are used in order to minimize any subsequent treatment.

Waste water treatment Further development is still needed in the design and operation of infrastructures for various waste water treatments, taking into account the following: physical, chemical and biological processes, activated sludge, activated carbon, solvent extraction and oxidation, etc. (Eckenfelder, 1970). Recovery of wastes and elimination of contaminant-leakage during production obviously improves the product yield and profits in the long-term, thus paying for the cost of the specific control technology (Eckenfelder, 1980; 1982).

The methodology of setting up the treatment infrastructure (pre-treatment, primary and secondary treatments) also is an important aspect. Ponds and stabilization basins can be directly integrated with the aquatic systems (Desjardin, 1988). However, sludge by-products from mining operations are still a serious challenge.

Self-control The times of the "ashamed polluter," or the "cautious polluter," are gone. If it is natural to produce, then it should be natural to depollute. The manufacturer is the one who knows best what are the effluents being emitted, especially their peaks and fluctuations. It is thus necessary to promote the notion of self-control of these discharges. They may be monitored with instantaneous and integrated counting stations on site and in the receiving medium. Instantaneous transmission of data can facilitate the modulation of the effluents or raise an alert of excessive emissions taking place.

Administrative pressure through the principle of "the polluter must pay" must be applied at every level as the beginnings of water quality control. Consequently, the self-control

systems must be the subject of continous monitoring and visits from the responsible environmental services.

Special treatment and security system Few measures are actually adopted in the humid tropics to reduce toxic waste water effluents by specific treatments that destroy harmful wastes. There is an urgent need for specialized treatment centers to be set up to deal with different types of wastes and problems such as oily emulsions, acid neutralization, solidification of liquid waste, and treatment of sludge, etc.

Preventing accidental leakage of toxic materials also must be a priority. Efforts must be directed towards improving the reliability of installations which for social and economic reasons receive little preventive maintenance due to a shortage of spare parts and specialists in that field. Security systems can be complex, but arrangements must be planned for the detection of leakages through observation basins, retention and drainage areas, basins for solid waste, pumping wells encircling the factory to prevent the risk of contaminant propagation into the water table, upstream control and indicator wells, watertight gutters, waterproof areas and separative sewers.

Ways of controlling natural chemicals in drinking water

Previous mention has been made of the high fluoride levels occurring naturally in drinking water supplies of Kenya (WHO,1970; Kilani, pers, comm.). Because of the consequent adverse effects on human health, a major effort is underway in that country to develop a technically feasible, cheap and simple method of reducing these fluoride levels. Such simple technologies include the use of clay pots, activated alumina, bone charcoal and alum-flocculation. All of these show potential for practical application. In the meantime, the WHO standard of a fluoride content of not more than one mg l^{-1}, has been found to be too difficult to attain in rural water supplies.

INDUSTRIAL THERMAL POLLUTION

The sources of thermal pollution are largely power generation, industrial cooling, and geothermal plants. The effects of temperature increases on the aquatic ecology of various surface water systems has received little attention within the humid tropics (Catalan & Alfonso, 1973).

WATER QUALITY CONTROL RELATED TO LAND MANAGEMENT

The conversion of forests to other land uses, mining exploitation and various management practices connected with agriculture (Guyot & Herail, 1989),all increase erosion and the turbidity and sedimentation in streams and lakes (Roche, 1982; Wilson & Henderson, 1983; Giambelluca, 1986). The weaker penetration of light diminishes the photosynthesis of aquatic plants and the functions of phytoplankton. Subsequent sedimentation of the suspended load also modifies the physico-chemistry of the bottoms, disturbs the gas exchanges with plants and animals (fish) and limits growth and reproduction at different levels of the food chain. Suspended matter also absorbs microtoxics whose effects can be more harmful, depending on the mineralogy of the particles (Garman, 1983). Such phenomena require more detailed investigation .

Land alterations also modify the chemical and nutrient cycles and balances. These important aspects should be studied within the framework of an experimental drainage basin project involving process hydrology studies. The same is true for the management of irrigated perimeters, where natural chemical and nutrient cycles are very much modified, with the added risk of salinization (Kovda, 1973; Watts & Hanks, 1978; Laraque, 1988).

WATER QUALITY CONTROL AS RELATED TO WATER MANAGEMENT

Various kinds of water management have been previously mentioned as a means of pollution control. Close linkages with land management also exist. For example, the recyling of waste water in agriculture within the appropriate standards and control remain crucial water and land management issues (WHO, 1989). Water quality is also generally changed by such other water infrastructures as reservoirs, irrigated areas, channels for transport or drainage, intensive pumping and the exploitation of quarries (Australian Water Resources Council, 1988). The effects of such progressive river regulation imposes a new regime on the hydrochemistry.

The strongest physical, chemical and biological impacts result from the creation of human-made lakes (Kinawi & El Chamr, 1973; Ita & Petr, 1983), and the drainage of flatlands – in the area itself and downstream (UNESCO, 1964; Deon, 1982). To minimize destruction of these aquatic ecosystems (Leveque & Burton, 1981) and degradation of water quality, multidisciplinary studies are required at the design stage and at each implementation step of such projects. By the nature of the problem, this is a long-term process extending over tens of years because of the time lag in reaching new steady states.

Biocenotic changes also appear with human-made structures, including the development of vectorial endemics (insects, gastropods, larva), favoured by human migration, and the accumulation of organic matter (Philippon & Mouchet, 1976). Ecological guidelines need to be developed for the design of water storage and drainage projects in order to minimize the danger of disease vectors (Deschiens, 1970; Entz, 1980).

The abnormal development of macrophytes, plankton and toxic algae is also a consequence of water impounds that needs controlling (Gaudet, 1979). For example, water management can modify the ecology of plankton and macrophytes, thus favouring the occurrence of nuisance species such as toxic algae (*Microcystis* and *Anabaena*). Factors related to succession and growth at both the cellular and ecosystem lev-

els (the N/P ratio, iron and climate) also must be assessed. Toxins should be detected and methods developed to remove or inactivate them. The role of riparian vegetation and buffer strips adjoining streams, lakes and flatlands in protecting surface water quality should be further investigated, as well as their implications on the organic vegetation cycle. Modification of the hydrochemistry of soils, ponds, riverbanks and wetlands by floods must also be assessed (Talling, 1980; Rangeley, 1985).

Intensive aquaculture has shown considerable development in tropical areas since the beginning of the 1980s (Saad, 1980). Two types of water quality problems are frequently encountered with this activity. The first concerns the "self-pollution" of ponds through the aquacultural operations. As a result of constantly high temperatures, it has become common practice to encourage one breeding cycle after another. Large quantities of food, often in excessive amounts, are thrown into the ponds. This accumulation of organic matter causes the aquacultural ponds to become polluted (Martin *et al.*, 1992). The characteristic of this pollution is oxygen depletion, which prevents the accumulated organic matter from being adequately mineralized. This leads, in turn, to the formation of compounds (H_2S, NH_3) that have a toxic effect on living organisms. They often have a irreversible deterioration in the water quality unless very costly remedial measures are taken such as draining the ponds and leaving them to dry for long periods, or mechanical dredging.

The second problem encountered concerns the discharge of effluents from the aquacultural ponds. These effluents are discharged into adjacent waters, and consequently often have an adverse impact on the quality of the receiving water body (Martin *et al.*, 1992). This impact is even greater if the recipient water body is highly confined. In most cases, the aquacultural farms are located in aquatic sites where water movement is low (lakes, bays, lagoons, marshes). Consequently, there is little scope for the mechanisms of waste assimilation and dilution and the processes of oxygenation to operate in this low discharge environment. There also is a negative feedback from the supply of poor quality water to the aquacultural ponds.

Therefore, it is clear that the proper development of aquacultural activities in the humid tropics is dependent on two main factors: (1) appropriate management of ponds and of aquacultural practices, with particular regard to production levels that must not be exceeded; and (2) suitable selection of sites for the installation of ponds with a high potential for assimilation of effluents, little or no confinement, and good oxygenation. Research on water quality in aquaculture should be carried out prior to and concomitantly with any development in this field.

CONCLUSIONS

Countries in the humid tropics have become very aware of the value of their natural heritage as an essential factor for sustainable development. Water is a natural resource of prime importance that cannot be considered fully renewable in the sense that its quality, and that of contiguous land areas, can be dangerously impaired for long periods of time from various human activities.

An appropriate institutional infrastructure needs to be set up in all countries to study, monitor and control water quality. Areas of responsibility at all levels must be clearly defined, and international co-operation should be broadened.

The media should be used to raise awareness and influence opinions on water quality issues. The education of children in this regard is seen as an important strategy for the future. Institutions such as universities, which can both conduct research and dispense education, have a large role to play. The appropriate faculties must be encouraged to keep their course contents in line with the most recent issues and new practices. International agencies should continue to provide traditional and new computer information systems on the subject and also support the development of water quality models for the humid tropics.

Countries in the humid tropics must continue to conduct or instigate national assessments of water quality and quantity, using appropriate guidelines such as those produced by ESCAP. Long-term monitoring networks should be implemented for data collection and to maintain the accuracy of the data required. Sampling, field test kits and advanced technology should be used according to the areas and issues involved.

Among the major water pollutants are fertilizers and pesticides. Complementary techniques adapted for use in humid tropical countries must be researched to diminish the use of these products. Guidelines for water quality, such as those of WHO, can be misleading. They need to be adjusted for local conditions and to the proposed use of the water resource.

Release of untreated sewage into hydrological systems causes major biological and chemical pollution. Development of water supply, sanitation and waste disposal systems must be urgently pursued in a coordinated fashion. Treatment of industrial wastes should be undertaken within the factories producing them and chemical standards established legally in terms of the concentrations and amounts of effluents subsequently released into the environment.

The re-use or recycling of water must be strongly supported with the qualification that monitoring be continued to ensure that no concentration of diseases or viruses are being encouraged. Municipal wastes should be treated and where possible reused. If it is not possible to reuse these materials, landfills should be constructed to maximize their containment through effective sealing so that the groundwater is not polluted.

Preventive techniques which control the collection and disposal of toxic wastes must be researched and strongly supported. Legislation must be introduced in all countries to ensure that the polluters are held financially responsible for all pollution they cause. International agencies should also prepare guidelines for the control and handling of microtoxins.

It is also clear from this review that a better understanding of water quality is intimately linked with the need for process hydrology studies in both urban and rural areas (Bonell with Balek, Gladwell, this volume). Furthermore, the interrelationships with urban and rural water management has also become apparent. In the latter case, the development and encouragement of low-cost technologies for rural areas needs much more attention. An inventory of farming technology and practices that reduce erosion should be prepared as part of an international program.

Urban drainage systems also are being subjected to over-use by increasing populations. Countries should strive to improve their drainage systems and reconstruct sewage and other waste disposal facilities in order to reduce pollution.

This account has highlighted the various kinds of water and land management practices that can be used for improving pollution control. On the other hand, water management projects can significantly change water quality in a negative way. To minimize the destruction of hydrosystems and the degradation of their water quality, multidisciplinary studies are required at the design stage of the project, and at each step of implementation. Such work should continue over several decades to make certain that sustainable management can be assured.

REFERENCES

Alabaster, J. S. (1969) Survival of fish in 164 herbicides, insecticides, fungicides, wetting agents and miscellaneous substances. *Int. Pest Control 11(2)*:29–35.

Alabaster, J. S. (1981) Review of the state of aquatic pollution of East African inland waters. *CIFA Occas. Paper 9.*

Anonyme (1971) Standard methods for examination of water and waste-water. Amer. Public Health Assoc., Inc. A.P.H.A. AWWA-WPCF, 13th Ed.

Anonyme (1975) Directives du conseil des communautés européennes concernant la qualité des eaux de baignade. 8 Dec. 1975.

Anonyme (1976) Directives du conseil des communautés européennes concernant la qualité des eaux superficielles destinées à la production d'eau alimentaire. 16 Juin 1975.

Anonyme (1977) Drinking water and health. Nat. Res. Council, Wash. D.C., Nat. Academy of Sci.

Anonyme (1978) Water quality surveys. A guide for the collection and interpretation of water quality data. Prepared by IHD/ WHO working group on the quality of water. UNESCO.

Anonyme (1979) Paramètres de la qualité des eaux. Ed. Ministère de l'Environnement et du Cadre de Vie, Paris.

Anonyme -SPVAG- (1981) Répertoire des produits sanitaires disponibles en Martinique, Guadaloupe et Guyane. Publ. Serv. Prot. Vég. Antillas et Guyane.

Anonyme (1983) Alimentation en eau de petites collectivités. Technologies appropriees pour les petites installations d'alimentation en eau dans les pays en voie de développement. Centre de Formation Internationale à la Gestion des Ressources en Eau-France, Lavoisier, Paris. *Doc. Tech. 18.*

Anonyme (1985) National Environmental Protection Council. The Philippines Environment Report 1984–1985.

Anonyme (1986) Guo, P. H. (1986) Water Quality Management in the Philippines. Assignment Report as Adviser, WHO Western Pacific Regional Center for the Promotion of Environmental Planning and Applied Studies (PEPAS).

Australian Water Resources Council (1988) Proc. of the Nat. Workshop on Integrated Catchment Management. Univ. of Melbourne, May. *AWRC Conf. Series 16*, VGPO.

Aziz, M. A. (1981) Some public health aspects of water supply and waste-water disposal in hot climate countries. Proc. 9th Federal Convention of Australian Water and Wastewater Assoc., Instit. of Engineers. Australia, April 6–10.

Aziz, M. A. (1986) Certain aspects of water quality monitoring. Proc. Int. Conf. Wat. & Waste. Mgt. Asia, Singapore. 124–137.

Bader, J. C., Delfieu, G., Koudjou, A. & Wome, K. A. (1987) Etudes hydrologiques menées dans le cadre du programme de lutte contre l'onchocercose. Rapport Final de la Campagne 1986. Vol.2, ORSTOM, Lomé.

Baker, R. M., Singleton, F. L. & Hood, M. A. (1983) Effects of nutrient deprivation on Vibrio cholerae. *Appl. Environ. Microbiol. 46*:930–940.

Benavidez, C. F. (1988) Influencia de los cambios en el uso del suelo sobre el escurrimiento y la erosion en la cuenca del Rio Pirai – Amazonia andina, Bolivia. *Publ. PHICAB*, tesis UMSA, La Paz.

Benson, W. R. (1971) Photolysis of solid and dissolved dieldrin. *J. of Agric.& Food Chem. 19*:66–72.

Bialkowska-Hobrzanska, H. (1987) Detection of enterotoxigenic Escherichia coli by dot blot hybridization with biotinylated DNA probes. *J. Clin. Microbiol. 25*:338–343.

Biswas, A. K. (1978) Water Development, Supply and Management. United Nations Water Conf. Pergamon Press.

Bockemuhl, J. (1985) Epidemiology, etiology and laboratory diagnosis of infectious diarrhea diseases in the tropics. *Immun. Infekt. 13*:269–275.

Bonde, G. J. (1977) Bacterial indication of water pollution. *Adv. Aquatic Microbiol. 1*:273–364.

Bourrier, R. (1985) Les réseaux d'assainissement. Lavoisier.

Bremond, R. & Perrodon, C. (1979) Paramétres de la qualité des eaux. Ministére de l'environnement et du cadre de vie, France.

Broche, J. & Peschet, J. L. (1983) Enquête sur les pollutions actuelles et potentielles en Cote d'Ivoire. ORSTOM Multig.

Brockmann, C. E. (1986) Perfil ambiental de Bolivia. La Paz, Bolivia.

Chantraine, J. M. & Dufour, P. (1983) Réseau national d'observation de la qualité des eaux marines et lagunaires en Côte d'Ivoire. Doc. Min. Environ. Francais et Ivoirien. Multig.

Chia, L. S. (ed.) (1987) Environmental Management in Southeast Asia: Directions and Current Status. Faculty of Science, Nat. Univ. of Singapore.

Colcanap, M., Dufour, P. (1982) L'assainissement de la ville d'Abidjan. Evaluation, recommandations et propositions d'alternatives. Ministére de l'Environnement-ORSTOM, Paris.

Colwell, R. R., Brayton, P. R., Grimes, D. J., Roszak, D. B., Huq, S. A. & Palmer, L. M. (1985) Viable but non-culturable Vibrio cholerae and related pathogens in the environment: Implications for release of genetically engineered microorganisms. *Biotech. 3*:817–820.

Cullen, P. (1987) A Review of Research Opportunities in Water Quality. Nat. Water Res. Seminar. Discussion Papers, AWRAC, DRE, Sept. 81–99.

Dangerfield, B. J. (1983) Water supply and sanitation in developing countries. Inst. Water Engineers & Scientists, London.

Dappen, G. (1976) Pesticide analysis from urban storm run-off.

Springfield, Virgine. Dept. of the Interior. *Report No.PB*. 238–593.

Debatisse, M. L. (1988) Industries du sucre et amidons. Lavoisier.

Dejoux, C. (1975) Action du molluscicide Frescon (R) sur certains éléments de la faune non cible des lacs tropicaux. Cah. ORSTOM, *Ser. Ent. Med. et Parasitol 8(2)*:81–83.

Dejoux, C. (1979) Recherches preliminaires concernant l'action de Bacillus thuringiiesis israelensisde Barjac, sur la faune d'invertébrés d'un cours d'eau tropical. WHO/VBC/79, 721.

Dejoux, C. (1985) Incidence des pesticides dans la pollution des eaux continentales africaines. *Verh. Internat. Verein. Limnol. 22*:2452–2456.

Dejoux, C. (1988) La pollution des eaux continentales africaines. ORSTOM, Paris.

Deon, J. (1982) Mise en valeur des ressources hydriques et santé. Bibliogr. selective. WHO/PDP/ 82.2, Multig.

Deschiens, R. (1970) Les lacs de retenue des grands barrages dans les régions chaudes et tropicales, leur incidence sur les endémies parisitaires. *Bull. Soc. Path. Exot. 63(1)*:35–51.

Desjardins, R. (1988) Le traitement des eaux. Ed. de l'Ecole Polytechnique de Montreal, Lavoisier, Paris.

D'Itri, F. M., Aguirre, M. J. & Athie, L. M. (1981) Municipal Wastewater in Agriculture. Academic Press, New York.

Dufour, P. & Maurer, D. (1979) Pollution organique et eutrophisation eu milieu tropical saumâtre. *Biologie-Ecologie Mediterranéenne 6(3–4)*:252.

Eckenfelder, W. W. (1970) L'eau dans l'industrie. Pollution, traitement, recherche de la qualité. Techn. et Doc., Lavoisier.

Eckenfelder, W. W. (1982) Gestion des eaux usées urbaines et industrielles. Caractérisation, Techniques d' épuration, Aspect économique. (Traduit de l'americain), Lavoisier, Paris.

Eckenfelder, W. W., Jr. (1980) Principles of Water Quality Management. CBI Publ. Co. Inc., Boston, Massachusetts.

Edeline, F. (1988) L'épuration biologique des eaux résiduaires. Théorie et Technologie. CEBEDOC, Lavoisier, Paris.

Edwards, C. A. (ed.) (1975) Persistent pesticides in the Environment. C.R.C. Press.

Eichelberger, J. W. & Lichtenberg, J. J. (1971) Persistence of pesticides in river water. *Environ. Sci. & Technol. 5*:541.

El Hinnawi, E. E. (1980) The state of the Nile environment: An overview. *Water Supply Management 4*:1–11.

Elmendorf, M. (1987) Water quality and women in small systems. Proc. Int. Symp. Small Sys. Wat. Sup. & Waste. Disposal, Singapore. 131–141.

Entz, B. A. G. (1980) Ecological aspects of Lake Nasser – Nubia. The first decade of its existence, with special reference to the development of insect populations and the land and water vegetation. *Water Supply Management 4*:67–72.

Environmental Health Division (1986) Review of Water Quality Monitoring in Thailand. Dept. of Health.

Environmental Health Division (1986) Manual on design, construction, operation and maintenance of upflow anaerobic filter treatment system for hospital wastewater. Dept. of Health. IDRC *Report 5*, December.

Evison, L. M. (1988) Comparative studies on the survival of indicator organisms and pathogens in fresh and sea water. Proc. Internat. Conf. on Water and Wastewater Microbiol. *2*:50 (1–7), Newport Beach, CA.

Evison, L. M. & James, A. (1977) Microbiological criteria for tropical water quality. *In*: R. Feachem, M. McGarry, & D. Mara (eds.) Water Wastes and Health in Hot Climates. John Wiley, New York. 30–51.

Feachem, R. G. (1977) Infectious disease related to water supply and excreta disposal facilities. *Ambio. 6*:55–58.

Feechem, R., Mc Garry, M. & Mara, D. (eds.) (1977) Water, Wastes and Health in Hot Climates. John Wiley & Sons, London.

Fitts, R., Diamond, H. C. & Neri, M. (1983) DNA-DNA hybridization assay for detection of Salmonella spp. in foods. *Appl. Environ. Microbiol.* 46:1146–1151.

Fujioka, R. S., Loh, P. C. & Lau, L. S. (1980) Survival of human enteroviruses in the Hawaiian ocean environment: Evidence for virus inactivating microorganisms. *Appl. Environ. Microbiol.* 39:1105–1110.

Fujioka, R. S., Ueno, A. A. & Narikawa, O. T. (1989) Unreliability of the KF Agar technique to recover fecal streptococcus from marine waters. Accepted for publication in *J. Water Poll. Control Fed.*

Fujioka, R. S. & Shizumura, L. K. (1985) Clostridium perfringens, a reliable indicator of stream water quality. *J. Water Poll. Control Fed.* 986–992. Abstracts of the 83rd Annual Meeting, Amer. Soc. for Microbiol. 6–11 March 1983, New Orleans, Louisiana

Fujioka, R. S., Tenno, K. & Kansako, S. (1988) Naturally occuring fecal coliforms and fecal streptococci in Hawaii's freshwater streams. *Toxicity Assessment* 3:613–630.

Fujioka, R. S., Hashimoto, H. H., Siwak, E. B. & Young, R. H. F. (1981) Effect of sunlight on survival of indicator bacteria in seawater. *Appl. Environ. Microbiol.* 41:690–696.

Garman, E. E. J. (1983) Part I: Water Quality Issues in Australia, Water 2000; Consultants Report No. 7, Water Quality Issues, DRE, AGPS, Canberra, 1983.

Gaudet, J. J. (1979) Aquatic weeds in African man made lakes. *PANS* 25(3):279–286.

Gautier, M., Pépin, Y., Etienne, J., Lapetite, J. M. (1987) Installation de balises Argos/Chloé de télétransmission des données hydrologiques en Guinée et Côte d'Ivoire. ORSTOM – OMS – OCP, Montpellier.

George, T. T. & El Moghraby, A. I. (1978) Status of aquatic pollution in the Sudan, its control and protection of the living resources. Sixth FAO/SIDA workshop in aquatic pollution in relation to protection of living resources. *PIR;TPLR/78/Inf. 20*, Multig.

Giambelluca, T. W. (1986) Land-use effects on the water balance of a tropical island. *Nat. Geogr. Res.*2:125–151.

Girard, G. (1988) Modélisation conjointe du cycle de l'eau et du transfert des nitrates sur un systéme hydrologique. Journées Hydrologiques ORSTOM, Montpellier.

Girard, G., Morin, G. & Charbonneau, R. (1972) Modèle précipitations-débits à descrétisation spatiale. Cah. ORSTOM, *Ser. Hydrol.* IX(4):35–52.

Government of Thailand (1989) Thailand Country Profile on Drinking Water Supply and Sanitation. March.

Gower, A. M. (ed.) (1980) Water Quality in Catchment Ecosystems. John Wiley, New York.

Guyot, J. L. & Hérail, G. (1989) Mining operations and modification of the physical chemical nature of the waters of the Rio Kaka drainage basin (Andes, Bolivia). Sediment and the enviroment, IAHS Third Scientific Assembly, Baltimore, May. *IAHS Publ. 184*:115–121.

Guyot, J. L., Roche, M. A. & Bourges, J. (1988) Etude de la physico-chimie et des suspensions des cours d'eau de l'Amazonie bolivienne: l'exemple du Rio Beni. *Journées Hydrologiques de l'ORSTOM*, Montpellier, Septembre.

Guyot, J. L., Roche, M. A., Noriega, L., Calle, H. & Quintanilla, J. (1990) Salinities and sediment loads on the Bolivian Highlands. *J. Hydrol. 113*: 147–162.

Hadley, R. F. (1986) Long-term monitoring of natural and man-made changes in the hydrological regime and related ecological environmental Project IH8 – II. UNESCO *A.3.1.*

Hardina, C. M. & Fujioka, R. S. (1991) Soil: The environmental source of *Escheria coli* and enterococci in Hawaii's streams. *Environmental Toxicology and Water Quality: An International Journal 6*; 185–195.

Hazen, T. C. (1988) Fecalcoliforms as indicators in tropical waters: A review. *Toxicity Assessment 3*:461–477.

Hazen, T. C. (1989) Tropical Source Water. Savannah River Laboratoy, Aiken, South Carolina.

Hazen, T. C., Santiago-Mercado, J., Toranzos, G. A. & Bermudes, M. (1987) What does the presence of fecal coliforms indicate in the waters of Puerto Rico? A Review. *Bol. Puerto Rico Med. Assoc. 79*:189–193.

Hazen, T. C., Fuentes, F. A. & Santo Domingo, J. W. (1980) In situ survival and activity of pathogens and their indicators. *ISME Proc. IV*:406–411.

Hervouet, J. P. (1983) Aménagement hydro-agricole et onchocercose: Loumana (Haute-Volta). *In*: De l'Epidémiologie à la Géographie Humaine. Editions ACCT/CNRS, Paris. *Travaux et Documents de Géographie Tropicale 48*:271–276.

Highton, R. B. (1970) The influence of water conservation schemes on the spread of tropical diseases in Kenya. E. Afr. Med. Res. Council, Sci. Conf., Nairobi, Jaun.

Ita, E. O. & Petr, T. (1983) Selected bibliography on major African reservoirs. *CIFA Occ. Paper 10.*

James, A. (1984) An introduction to Water Quality Modelling. John Wiley.

Jiménez, L., Muñiz, I., Toranzos, G. A. & Hazen, T. C. (1989) The survival and activity of Salmonella typhimurium and Escherichia coli in tropical freshwater. *J. Appl. Bacteriol.*

Jones, G. P. (ed.) (1982) Improvement of methods of long-term prediction of variations in groundwater resources and regimes due to human activity. *IAGS Publ. 136.*

Jordan, P., Unrau, G. O., Bartholomew, R. K., Cook, J. A. & Grist, E. (1982) Value of individual household water supplies in the maintenance phase of a schistosomiasis control programme in Saint Lucia, after chemotherapy. *Bull. WHO 60*:583–588.

Junkin, M. C. (1979) Pompes à mains destinées à l'approvisionnement en eau potable dans les pays en voie de développement. Lavoisier, Paris. *Doc. Tech. 10.*

Kaoma, C. & Salter, E. F. (1979) Environmental pollution in Zambia. *In*: D. S. Johnson and W. Roder (eds.) Proc. of the Nat. Seminar on Environ. and Development, Lusaka. *Zambia Geogr. Assoc. Occas. Stud. 10*:181–216.

Keller, R. (ed.) (1983) Hydrology of humid tropical regions. Hamburg Symposium. *IAHS Publ. 140.*

Kinawy, I. Z., El Chamr, O. A. (1973) Some effects of the high dam on the environment. ll ème, Congrés des Grands Barrages, Madrid, 1973. *Q 40, R 59*:959–974.

Kovda, V. A. (1973) Irrigations, Drainage and Salinity. An international source book. Hutchinson Publ. Group, London, FAO.

Lamb, J. C. (1988) Water Quality and Control. John Wiley, New York.

Laraque, A. (1988) L'évolution hydrochimique de retenues collinaires du Nordeste brésilien. *Journées Hydrologiques ORSTOM de Montpellier*, Septembre.

Lau, L. S. & Mink, J. F. (1986) Indentifying and correcting organic chemical contamination of groundwater sources: A learning experience. Proc. Amer. Water Works Assoc., 1986 Annual Conf., Denver, Colorado.

Le Barbe, L. & Gioda, A. (1987) Epandage d'insecticides dans les rivières. Portées et dosages optimae. ORSTOM, Montpellier.

Le Barbé, L., Gioda, A., Delfieu, G. & Rome, K. A. (1987) Etudes hydrologiques menées dans le cadre du programme de lutte contre l'onchocercose. Etude expérimentale de la propagation des insecticides dans les rivières. Rapport Final. ORSTOM, OMS, OCP, Montpellier.

Lesel, R., Landragin, G. & Kolz, H. (1979) La biodetection. Agence Financiere de Bassin Rhin-Mense.

Lévêque, C. & Burton, M. (1981) Fishes. *In*: Symoens, J.J. Burgis, M. J. & Gaudet, J. (eds.) The Ecology and Utilization of African Inland Waters. UNEP, Nairobi. 69–79.

Lointier, M. & Roche, M. A. (1988) Salinités et suspensions des estuaires de Guyane. Méthodes et résultats. *Journées Hydrologiques ORSTOM*, Montpellier, Septembre.

Marshall, B. E. & Falconer, A. C. (1973) Eutrophication of a tropical African inpoundment (Lake Mc Ilwaine, Rhodesia). *Hydrobiologic 43(1–2)*:109–123.

Martin, J. L. Hussenot, T. & Guelorget, O. (1992) La matière organique et son impact dans un système d' élevage de crevettes en milieu inter-tropical. Agriculture Europe 16, 51.

Martin, M. (1977) Devenir des polluants organiques daus le milieu naturel en fonction des traitements et par rapport aux cycles de vie de ces produits. (PCB, pesticides, agents de surface). Univ. Rennes. Coll. Recherche Environment. *Journées de Montpellier 8*.

Masschelein, W. J. (1988) L'ozonation de l'eau, manuel pratique. Lavoisier, Paris.

Mathur, S. P. & Saha, J. G. (1975) Microbiologic degradation of lindane C-14 in a flooded sandy loam soil. *Soil Sci. 120*:301.

McKendrick, J. (1982) Water supply and sewage treatment in relation to water quality. *Monogr. Biol. 49*:201–217.

Mendoza, G. G. (1989) Abastecimiento de agua potable a los centros urbanos e industriales. Presented in Querétaro, Qro. Febrero.

Merrick, T. W. (1985) The effect of pipes water on early childhood mortality in urban Brazil, 1970 to 1976. *Demography 22*:1–24.

Mills, W. B., and others. (1982) Water quality assessment: A screening procedure for toxic and conventional pollutants. *Doc. EPA–600/6–82–004a.*

Moore, G. & Christy, L. (1978) Legislation for control of aquatic pollution. Part I. International aspects. Sixth FAO/SIDA Workshop on aquatic pollution in relation to protection of living resources. *FIR:TPLR/78 6*:1–12.

Moseley, S. L., Echeverria, P., Seriwatana, J., Tirapat, C., Chaicumpa, C., Sakudaipeara, T. & Falkow, S. (1982) Identification of enterotoxigenic Escherichia coli by colony hibridization using three enterotoxin gene probes. *J. Infect. Dis. 145*:863–869.

Muñoz, C. (1981) UNEP'S related activities on the international drinking water supply and sanitation decade. UNEP Report, first joint FAO/UNEP/WHO Panel of Experts on Environmental Managment for Vector Control, Géneve, *Multig*:9–11.

National Research Council (1982) Ecological aspects of development in the humid tropics. Nat. Academy Press, IABS. Washington, D.C.

Ngoile, M. A. K., Challe, A. E. & Mapunda, R. R. (1987) Aquatic pollution in Tanzania. Sixth FAO/SIDA Workshop on aquatic pollution in relation to protection of living resources. *FIR:TPLR 78/19*, Multig.

Ngunya, E. A. (1975) Some administrative aspects of establishing a pollution control organization in a developing country. *Progress in Water Tech. 7(2)*:83–91.

Noble, R. G., Pretorius, W. A., Chutter, F. M. (1971) Biological aspects of water pollution. *Suid Afrikaanse Tydskrift vir Wetenskap*. 132–136.

Nongluk, T. (1989) Water supply and water user behavior: The use of

cement rainwater jars in northeastern Thailand. Thailand-Australian Northeast Village Water Resource Project. *Report 134*, June.

Oluwande, P. A., Sridhar, K. C., Bammeke, A. O. & Okubadejo, A. O. (1983) Pollution level in some Nigerian rivers. *Water Res. 17*:957–963.

Ong, S. E. (1987) Water pollution control in Singapore – Historical development and future prospects. Proc. Int. Symp. Small Sys. Wat. & Waste. Disposal, Singapore. 15–25.

Ongwency, G. S. O. (1973) The significance of the geographic and geologic factors in the variation of groundwater chemistry in Kenya. M.Sc., Univ. Nairobi.

Orekoya, T. (1978) Oil pollution abatment. Getrie FAO/SIDA Workshop on aquatic pollution in relation to living resources. Nairobi and Mombasa, Multig. 102–112.

Orlob, G. T. (ed.) (1986) Mathematical Modelling of Water Quality (Streams, Lakes and Reservoirs). John Wiley.

Panswad, T., Polprasert, C. & Yamamoto, K. (eds.) (1988) Water pollution control in Asia. Proc. Second IAWPRC Asian Conf. on Water Pollution Control. Bangkok, Thailand,.9–11, November.

Pescod, M. B. (1977) Surface water quality criteria for tropical developing countries. *In*: R. Feachem, M. McGarry & D. Mara (eds.) *Water, Wastes and Health in Hot Climates*. John Wiley, New York. 52–72.

Pesson, P. (1976) La pollution des eaux continentales. *In*: Gauthier-Villars (ed.) Incidences sur les Biocénoses Aquatiques. Paris.

Philippon, B. & Mouchet, J. (1976) Repercussions des aménagements hydrauliques à usage agricolesur l'épidémiologie des maladies à vecteurs en Afrique intertropicale. *In*: *Cahiers da CENECA, Doc. 3–2, 13*, Multig.

Pimentel, D. (1971) Ecological effects of pesticides on non-target species. Publ. Exec. Off. of the Pres., Off. of Sci. & Tech., Cornell Univ.

Pouyaud, B. (1987) Télétransmission satellitaire au service du programme de lutte contre l'onchocercose en Afrique de l'Ouest. Colloque sur la télémesure et la transmission des données en hydrologie, 23–27 mars, Toulouse.

Pouyaud, B. & Le Barbé, L. (1987) Onchocercose, hydrologiques et télétransmission. Water for the future: Hydrology in perspective. Proc. of the Symp., April, Rome. *IAHS* 239–244.

Prost, A. (1986) The burden of blindness in adult males in the Savanna villages of West Africa exposed to onchocerciasis. *Trans. of the Royal Soc. of Trop. Med. & Hygiene 80*:525–527.

Prost, A. & Prescott, N. (1984) Cost-effectiveness of blindness prevention by the Onchocerciasis Control Programme in Upper Volta. *Bull. WHO 62*:795–802.

Prost, A. & Vaugelade, J. (1981) La surmortalité des aveugles en zone de savane ouest-africaine. *Bull. WHO 59*:773–776.

Quano, E. A. R., Lohani, B. N. & Thanh, N. C. (eds.) (1987) Pollution control in developing countries. Proc. Internat. Conf., Bangkok, Thailand, 21–25 Feb.

Rangeley, R. (1985) Irrigation and drainage in the world. Proc. Internat. Conf. on Food and Water. May 26–30, College Station, Texas.

Richard, Y. (1987) Vade-mecum du chef d'usine de traitement d'eau destinée á la consommation. AGHTM, Lavoisier.

Robarts, R. D. (1975) Water pollution in Rhodesia. *Rhod. Sci. News 9*:328–332.

Roche, M. A. (1972) Traçage hydrochimique naturel du mouvement des eaux dans le lac Tchad. Sect. Hydrol. ORSTOM, Paris, Octobre. C. R. Symp. sur l'Hydrolgie des lacs, A.I.H.S., Helsinki, Septembre 1973. 18–27.

Roche, M. A. (1973) Tracage naturel salin et isotopique des eaux du système hydrologique du lac Tchad. Thèse de Doctorat ès-Sciences, Octobre 1973, Paris VI. Trav. et Doc. de l'ORSTOM, 1980.

Roche, M. A. (1975) Geochemistry and natural ionic and isotopic tracing: Two complementary ways to study the salinity regime of the hydrological system of Lake Chad. *J. of Hydrol.* 26:153–171. Présenté Coll. Hydrochimie des eaux Naturelles, Burlington (Canada), août.

Roche, M. A. (1976) Méthodologie de mesure de la dynamique des eaux, des sels et des suspensions en estuaire. C. R. Quatorzièmes Journées de l'Hydraulique, Soc. Hydrotechnique de France, Question III, *Rapport 1*:1–8.

Roche, M. A. (1977) Hydrodynamique et évaluation du risque de pollution dans un estuaire à marées. Cah. ORSTOM, *Sér. Hydrol. XIV(4)*:345–382.

Roche, M. A (1982) Comportements hydrologiques comparés et érosion de l'écosystème tropical humide à Ecérex, en Guyane. Cah. ORSTOM, *Ser. Hydrol.*:81–114.

Roche, M. A. & Canedo, M. (1984) Programa Hidrológico y Climatológico de la Cuenca Amazónica de Bolivia. Folleto de presentación del PHICAB. Publ. PHICAB, La Paz.

Roche, M. A., Abasto, N., Tolède, M., Cordier, J. P. & Pointillart, C. (1986) Mapas de las salinidades de los ríos de la Cuenca Amazónica de Bolivia. PHICAB, ORSTOM.

Roche, M. A. & Fernandez, J. C. (1988) Water resources, salinity and salt exportations of the rivers of the Bolivian Amazon. *J. of Hydrol. 101*:305–331, Elsevier, Amsterdam.

Saad, M. A. H. (1980) Eutrophication of Lake Mariut, a heavily polluted lake in Egypt. Int. Atomic. Energy Agency Vienne. 153–163.

Schulz, C. R. & Okun, D. A. (1984) Surface water treatment for communities in developing countries. John Wiley & Sons, New York.

Scotney, N. (1980) Developing a health education component for the UNICEF water and sanitation program in Sudan. UNICEF Report, Nairobi, Multig.

Secretaria de Desarrollo Urbano y Ecologia (1985) Evaluación de cuencas hidrológicas de acuerdo a su grado de contaminacion. Dirección General de Prevención y control de la contaminación. Ambiental, Mexico.

Servat, E. (1985) Etude de la DB05 et de la DCO du ruissellement pluvial urbain. Essai de modelisation. L.H.M., Universite des Sciencies, Montpellier. *Note 11/85*.

Servat, E. (1987) Contribution à l'étude de la pollution du ruissellement fluvial urbain. Thése Montpellier, ORSTOM.

Sharma, M. L. (1986) Role of groundwater in urban water supplies of Bangkok, Thailand and Jakarta, Indonesia. Working Paper, Environ. and Pol. Inst., East-West Center, Honolulu.

Shuval, H. I., Tilden, R. L., Perry, B. H. & Grosse, R. N. (1981) Effect of investments in water supply and sanitation on health status: A threshold-saturation theory. *Bull. WHO 59*:243–248.

Simmon, V. F.,Kauhanen, K. & Tardiff, R. G. (1977) Mutagenic activity of chemicals identified in drinking water. *Develop. in Toxicol. & Environ. Sci. 2*:249–258.

Smith, L. S. (1969) Public health aspects of water pollution control. *Water Pollut. Control 68*:544–549.

Smith, A. & Lossey, O. (1981) Pesticides and equipment requirements for national vector control program in developing countries. *WHO/VBC/81.4*.

Soon, C. H. (1982) Pollution Control in Singapore. *Malayan Law Review 24(2)*:213–219.

Steele, T. D. (1987) Water quality monitoring strategies. *IAHS 32*:2.

Stephenson, D. (1988) Water and Wastewater Systems Analysis. Development in Water Science, Elsevier.

Stewart, B. J. & Verhoeven, T. J. (1987) Surface Water Research Needs in Australia. SWCC, AWRC, PAWA, Sept.

Steyn, D. J., Toerrien, D. F. & Visser, J. H. (1976) Eutrophication levels of some South African inpoundements. Part 3 and 4, *Roodeplaat Dom. Water S.A. 2*:1–6.

Talling, J. A. (1980) Some problems of aquatic environments in Egypt from a general viewpoint of Nile ecology. *Water Supply Manage. 4*:13–20.

Tan, T. H. (1978) Water pollution control. Proc. Sem. on Pollution Prevention and Environ. Conservation in Singapore. 11–13.

Tay, J. H. & Ong, S. L. (eds.) (1986) Water and wastewater management in Asia. Proc. Internat. Conf. on Water & Wastewater Management in Asia. 26–28 February, Singapore.

Tchobanoglous, G. & Schroeder, E. E. (1985) Water Quality Characteritics, Modelling and Modification. Benjamin Cummings Publ. Co. Inc., New York.

Thorton, J. A., & Nduku, W. K. (1982) The aqueous phase: Nutrients in run-off from small cachetments. *Monogr. Biol. 49*:71–77.

Toerien, D. F. (1975) South African eutrophication problem: A perspective. *Wat. Pollut. Control 76*:136–162.

Trudgill, S. T. (1986) Solute Processes. John Wiley.

UNESCO (1964) Scientific problems of the humid tropical deltas and their implication. *Nat. Resources Res. 6.*

UNESCO (1974) Humid tropical Asia. *Nat. Resources Res. 12.*

UNESCO (1980) Dispersion and self-purification of pollutants in surface water systems. A report by UNESCO IHP Working Group 6.1. P.G. Whitehead & T. Lack, Chief Editors, UNESCO, Paris.

Valiron, F. (1986) Mémento de l'exploitant de l'eau et de l'assainissement. Lyonnaise des eaux, Lavoisier, Paris.

Van Der Berg, T. (1988) Dairy technology in the tropics and subtropics. Lavoisier.

Van Dijk, J. C. & Oomen, J. H. C. M. (1981) La filtration lente sur sable pour l'approvisionement en eau collective dans les pays en développement. Centre International de Reférence pour l'approvisionmement en eau collective et l'assainissement, Lavoisier Paris. *Doc. Tech. 11.*

Vedry, B. (1987) L'analyse écologique des boues activées. SEGETEC, Lavoisier, Paris.

Vermaak, J. F., Swanepoel, J. H., Schoonbee, H. J. (1981) The phosphorus cycle in Germiston Lake. I. Investigational objectives and aspects of the limnology of the lake. *Water S.A. 7(3)*:52–57.

Vicente, V. A. & Beckett, R. (1986) Trace Metal Speciation in Laguna de Bay. Nat. Sci. Res. Inst. (Project), Univ. of the Philippines, Diliman.

Vollenweider, R. A. & Kerekes, J. (1980) The loading concept as a basis for controlling eutrophication: Philosophy and preliminary results of the OECD programme on eutrophication. *Trog. Wat. Technolog. 12*: 5–38.

Waldron, H. A. (1975) Health standards for heavy metals. *Chem. Br. 11*:354–357.

Watts, D. G. & Hanks, R. J. (1978) A soil water nitrogen model for irrigated corn on sandy soils. *Soil Sci. Soc. Amer. J. 42*:492–499.

Whitehead, P. G. (1984) The application of mathematical models of water quality and pollutant transport: An international survey. UNESCO, *SC.86/WS/10.*

Whittemore, F. W. (1977) Technical, economic and legislative factors determining choice of pesticides for use in developing countries. Rapp. FAO, *WS/D 8662, Multig.*

WHO (1970) Fluoride and human health. *WHO Monogr. Series 59*, Geneva.

WHO (1989) Health guidelines for the use of wastewater in agriculture and aquaculture. Report of a WHO scientific group. *WHO Publ. Tech. Report Series 778*, Geneva.

Wilson, M. F. & Henderson-Sellersa. (1983) Deforestation impact assessment: The problems involved. *In*: R. Keller (ed.) Hydrology of Humid Tropical Regions. Hamburg Symposium. *IAHS Publ. 140*:273–283.

16: Ecological Characteristics of Tropical Fresh Waters: An Outline

H. L. Golterman
Station Biologique de la Tour du Valat, 13200 Arles, France
M. J. Burgis
London Guildhall University, Old Castle Street, London, E1 7NT, U. K.
J. Lemoalle
ORSTOM, B. P. 5045, 34032 Montpellier Cedex, France
J. F. Talling
Freshwater Biological Association, The Ferry House, Ambleside, LA 22 0LP, U. K.

ABSTRACT

The characteristics and general ecology of tropical fresh waters are outlined, with reference to the physical and chemical regulation of aquatic environments and the differentiation and dynamics of their communities.

Solar radiation income is the most distinctive and driving variable which leads to relatively high and seasonally-maintained temperatures, with relatively small differentials with depth. This thermally controlled density stratification is sensitive to changes in the energy balance and wind stress, which, in turn, controls chemical and biological stratification. Indefinitely prolonged stratification is common in very deep lakes, and well developed seasonal cycles are frequent, even in equatorial water bodies.

The primary chemical inputs are dominated by rock weathering and atmospheric gaseous exchange, although numerous secondary pathways and controls exist. Dissolved constituents are separable into major, minor and trace elements; gases; and organic compounds. The total ionic content varies widely, but is generally low in the humid tropics. Important interactions between gaseous (CO_2) and ionic constituents occur via the $CO_2/HCO_3^-/CO_3^{2-}$ system, with implications for pH buffering, Ca^{2+} removal as $CaCO_3$, P removal as apatite, and the biological supply of CO_2. Inorganic forms of the nutrient elements N, P and Si are subject to biological depletion; in tropical regions low concentrations of nitrate and high ones of dissolved Si are prevalent. Anoxia is associated with accumulations of the chemically reduced gases CH_4, H_2S and NH_3 with their derivatives, both in very deep and very shallow (swamp) waters. "Black water" rivers carry high concentrations of organic material. The relationship of concentration with river flow rate varies with different chemical constituents, as do upstream-downstream relationships.

Primary (plant) production in tropical fresh waters is sustained by generally high levels of solar energy input and temperature. Biomass development can be large. It is often affected by seasonal variations in water supply, although other factors – such as vertical mixing and nutrient supply – are influential. Annual cycles of phytoplankton biomass are common and often involve a succession of species. Light penetration underwater is varied, and influences the depth-distribution of biomass and photosynthetic activity. Rates per unit biomass tend to be high. Biomass also induces physical and chemical reactions upon the medium, including turbidity, gases, organic matter and nutrients. Recycling of nutrients is critical for further production; vertically separated zones and episodes of increased mixing are often involved.

Consumer (animal) communities are functionally ordered by feeding relationships, with successive levels represented by decreasing biomass. The resulting food-web involves transfers of varying length with some shortened forms being prominent in tropical fresh waters. Both "bottom-up" and "top-down" controls are possible. Constituent species are typically numerous, especially of invertebrates which occur predominantly either as bottom-living *benthos* or open water *plankton*. In flowing waters benthos is dominant. Tropical fish exhibit a wide variety of feeding niches. Seasonal changes are rarely limited by temperature, but not infrequently by water input and river flow. Extensive flood plains with seasonal overspill are ecologically important.

Interactions involving man are illustrated by invertebrate carriers of human diseases (e.g. onchocerciasis, schistosomiasis), biological consequences of reservoir creation, human introductions and cultivations of fish, and naturally-based fisheries.

INTRODUCTION

The aim of this chapter is to provide an outline of the biological and ecological features related to the management of fresh waters in the humid tropics. These features are among the subject matter of fresh water science (limnology), for which a number of text and reference books exist. Attention may be drawn to the comprehensive multi-volume work of Hutchinson (1957, 1967, 1975, and in press) on lakes; to the more introductory text of Payne (1986) on tropical lakes and rivers; and the similarly introductory text of Moss (1988) that includes both tropical and extra-tropical examples. Basic features of tropical limnology are reviewed by Lewis (1987). Serruya & Pollinger (1983) provide a richly documented account of natural and man-made lakes of the tropics and subtropics; Beadle (1981) refers to African lakes and rivers, and Talling (1992) to the environments of shallow African lakes. The "Ecology and Utilization of African Inland Waters" is summarized by Symoens, Burgis & Gaudet (1981). The greatest river of the humid tropics, the Amazon, is the subject of a recent monograph by Sioli (1984). Greenwood & Lund (1973) summarized an extensive study on a shallow, productive tropical water body, Lake George. For other tropical African lakes, monographs exist for the deep Lake Tanganyika (Coulter, 1991) and the shallower Lake McIlwaine (Thornton, 1982), Lake Chilwa (Kalk *et al.*, 1979) and Lake Chad (Carmouze *et al.*, 1983). The River Nile has also received monographic treatment (Rzóska, 1976), as have West African rivers (Grove 1985), the Sri Lankan lake of Parakrama Samudra (Schiemer, 1983) and the Malayan swamp Tasek Bera (Furtado & Mori, 1982). A directory and a bibliography of African wetlands and shallow water bodies have been published by Burgis & Symoens (1987) and Davies & Gasse (1988), respectively.

The broad scale of the subject means that the following account cannot be regarded as comprehensive. This chapter, however, attempts to introduce some major aspects of fresh waters in terms of their physical control and chemical make-up and of various levels of their plant and animal communities. Although much material applies to fresh waters in general, tropical examples and features distinctive of the tropics are emphasized.

ENERGY TRANSFER AND THE PHYSICAL ENVIRONMENT

The regime of energy transfer is the distinctive and controlling feature of tropical environments, mainly conditioned by latitudinal and directional correlates of short-wave solar radiation. The relationship between the latter and seasonal temperature cycles in tropical fresh waters (see examples in Talling, 1992) is affected by other components of the energy budget. These include long-wave net back-radiation, evaporative loss and its controlling factors (vapour pressure deficit, wind) and the exchange of sensible heat by conduction and convection between water and atmosphere. All components are sensitive to the differences between the humid and dry tropics. There is also a fairly regular relationship between deep-water temperature and altitude in tropical waters, with values around 30°C at sea level, around 20°C at 1800 metres, and around 10°C at 3,500 metres.

Much of the limnological information and classification has evolved from temperate regions where the lacustrine environment, and particularly the vertical structure of a water body, depends on the seasonal cycle. In these regions, the seasonal energy input in spring and summer will heat the superficial layers of lakes, which – unless very shallow – are then vertically divided into two main parts: the upper part, or *epilimnion,* which is separated from the deeper part, or *hypolimnion,* by a relatively steep temperature-dependent density gradient, the *thermocline.* The epilimnion remains in contact – if not equilibrium – with the air, receives more illumination and has thus a larger photosynthesis, while gases like CO_2 and O_2 may exchange between water and air. In the hypolimnion, anoxia and low redox potential may occur if bacterial activity has extensively decomposed organic matter. In the autumn, cooling of the upper layers in contact with the atmosphere results in a weakening and disappearance of the thermocline; the now cool surface water mixes with the deeper water mass of the hypolimnion and the lake becomes homogeneous.

In deep tropical lakes, the situation is different. First, deep remnants of cold winter water are absent and so vertical differences in temperature are relatively small. Second, within the range of temperatures most commonly found in the tropics, a given small temperature difference results in a much larger difference in water density than in cooler temperate regions. Less energy input (or loss) is thus needed to generate (or destroy) a strong thermal density gradient. Most lakes or reservoirs in the humid tropics, therefore, are either intermittently or permanently stratified; given sufficient depth or wind-shelter, vertical mixing is seldom complete. Examples are Lakes Tanganyika and Kivu (Beauchamp, 1964). The chemical consequences of this permanent stratification are given later in this chapter. If the lake is less deep, complete mixing can be periodic (e.g. annual) or the lower limit of the thermocline may just reach the bottom (e.g. Lake Victoria, Beauchamp, 1964). Talling (1969) compares several other African examples.

Although there is often a strong seasonal rhythm in the rainfall in the humid tropics, the intra-annual temperature variation is strongly reduced by latitudinal effects. In shallow tropical lakes, another pattern of stratification is often pronounced, linked to a daily temperature cycle where the vertical structure of these lakes is often determined by the daily wind regime and the diurnal difference in irradiation. Heating during the daytime may cause a stratification, the extent and duration of which depend on the wind. For example, daily stratification in a shallow lake has been well-documented for Lake George, Uganda (Viner & Smith, 1973; Ganf & Horne, 1975). Evening

TABLE 1. *Chemical composition of important rocks arranged in order of decreasing solubility.*

Non-silicates

Rock salt (Halite)	$NaCl$
Gypsum	$CaSO_4$
Calcite	$CaCO_3$
Dolomite	$MgCO_3$, $CaCO_3$ pearl-spar

Silicates

Feldspar	$KAlSi_3O_8$ orthoclase
Soda-lime feldspar	a series from $NaA1Si_3O_8$ (plagioclase), albite to
	$CaAl_2Si_2O_8$ anorthite
Basalt	$NaAlSi_3O_8$ plagioclase
	$RSiO_3$ pyroxene (R = Ca^{2+}, Mg^{2+}, or Fe)
	$NaAl(SiO_3)_2$ alkali-pyroxene
	$(Fe,Mg)_2SiO_4$ olivine
Granite	SiO_2 quartz
	$NaAlSi_3O_8$
	$KAlSi_3O_8$
	$Al_2Si_2O_5(OH)_4$ kaolin
	$K(Fe,Mg)_3Si_3AlO_{10}(OH)_2$ biotite
	$KAl_2AlSi_3O_{10}(OH,F)_2$ K mica or K muscovite
	and more complex structures

winds and nocturnal cooling destroy the stratification which developed previously during the day. Many of the flood plain lakes of the Parana River (South America) have a very similar behaviour.

CHEMISTRY

Introduction

The chemical composition of fresh waters contributed by the dissolved and the suspended matter is rarely constant, but changes in time and space. Regional differences in the chemical composition are caused by the geology of drainage basins, together with the nature of the erosion processes and the hydrological characteristics of rivers. All suspended inorganic matter in rivers originates from weathering and erosion of rocks. The dissolved compounds are formed from these erosion products by ionic dissolution and an ongoing hydrolysis of the original rock material. Most rocks are composed of a number of minerals, although some important types consist of only one (Table 1). By weathering of igneous rocks ("the silicates"), their components are set free: the alkaline elements (Na^+ and K^+) are easily dissolved and ultimately appear in rivers and lakes, although K^+ is also adsorbed onto clay. Whereas, in the *temperate regions,* the main solid product of weathering is some form of hydrous aluminium silicate, i.e. the clay minerals, under *tropical conditions* the aluminium and iron may also remain in solid form, mostly as the hydroxides $Fe(OOH)$ and $Al(OH)_3$ – the

so-called lateritic weathering process. Quartz is mainly physically disintegrated, especially in temperate regions, and may give rise to layers of sand. Sedimentary rocks, such as evaporites, carbonates (marine limestones) and hydrolysates, are more readily soluble than the igneous rocks, depending on their degree of recrystallisation. The spatial and temporal variability of temperature and rainfall, together with the relief influence the nature of the weathering and erosion processes, while vegetation and soil type again may modify the chemical dissolution. Vegetation may, for example, modify the time that the rainwater is in contact with the soil, and hence the mechanism to attain to an equilibrium composition. Rainwater itself, with dry deposition of aerial solids locally increased by burning of vegetation, is a not insignificant source of chemical input (e.g. Lewis, 1981 for Venezuela).

The compounds that occur dissolved in natural waters can be subdivided for convenience into the following categories: major, minor and trace elements; gases; and organic compounds (Table 2). The terms major and minor refer to the quantities usually found, and not to their biological meaning as nutrients. Generally speaking, the major elements control which organisms may occur in fresh water, whereas the minor elements (some of which are major nutrients) largely determine the quantity. Several compounds also occur in fresh water in a particulate form – as suspended matter. Of these, the "silts" and "clays" are usually responsible for the adsorbing capacity of the suspended matter.

TABLE 2. *Major, minor, trace elements, organic compounds and gases normally found as solutes in natural fresh waters (the percentages refer to relative quantities calculated from "mean global" fresh water composition).*

Major elements				Minor elements	Trace elements	Gases
Ca^{2+}	64%	HCO_3^-	73%	N (as NO_3^- or NH_4^+)	Fe, Cu, Co, Mo,	O_2, CO_2, N_2, H_2, CH_4
Mg^{2+}	17%	SO_4^{2-}	16%	P (as o–P)	Mn, Zn, B, V	
Na+	16%	Cl–	10%	Si (as SiO_2 or $HSiO_3^-$)		
K^+	3%					
H^+						
(Fe^{2+})						
(NH_4^+)		F^-				

Concentrations usually found:
0.1 – 10 mmol^{-1} .001 – < 1 mg l^{-1} up to a few µg l^{-1}
(Mean 2.4 mmol^{-1})   (Si sometimes higher)

organic compounds:
e.g. as humic compounds,
excretion products
vitamins, other metabolites

Chemical components

Major elements The presence and the concentration of the major elements, as ions (listed in Table 2), are reflected in the "electrical conductivity" of the water. It varies widely, ranging in values between 10 and > 100 000 µS cm^{-1} and is roughly proportional to the total ionic concentration. An approximate estimate of the quantity of dissolved ionic matter expressed in mg l^{-1} in a water sample may be made by multiplying the conductivity (at 25°C) by an empirical factor varying from 0.55 to 0.9, depending on the nature of the dissolved salts. A similar estimate in meq l^{-1} (of cations or anions) (*We use the old unit of milliequivalent, meq, which is equal to the "SI" mmol/electric charge of the ion*) may be made by multiplying the conductivity in µS cm^{-1} by 0.01. Talling & Talling (1965) gave examples of many African waters with exceptionally low or high ionic concentrations. They divided the lakes in Africa into three classes, according to their conductivity. Class I had a conductivity (at 20°C) < 600 µS cm^{-1}, Class II 600–6000 µS cm^{-1}, and Class III > 6,000 µS cm^{-1}. Class II lakes often contain concentrations of Ca^{2+} plus HCO_3^- in concentrations depending on the equilibrium with air of the CO_3^{2-}/HCO_3^- system. Class II includes Lake Turkana (formerly Lake Rudolf) – one of the most saline in this group – that occupies a closed basin. Formerly, the same basin contained a larger lake with an outflow to the Nile. The salinity in Class II lakes is often mainly due to Na+, Cl$^-$ and HCO_3^- with Ca^{2+} largely lost by precipitation as carbonate. Class III includes lakes in closed basins which have permanently no outflow. Under these conditions, salts may accumulate. The highest conductivity that has been measured is 160 000 µS cm^{-1} with titration alkalinity (viz. acid neutralizing capacity) up to 1,500 – 2,000 meq l^{-1}.

In humid tropical regions, however, normally the Class I prevails. Some of the lowest conductivities, indeed, are found either in the Congo Basin (Zaire) or in lakes whose inflow drained through swampy regions. These lakes are frequently dark-coloured, organically rich and are called "black waters." The entire range, and its major ionic composition, are illustrated in Fig. 1.

Tropical fresh waters are notably diverse in total ionic concentration for both geological and hydrological reasons. The land masses involved are generally very old, being part of the ancient supercontinent of Gondwanaland and having a long history of chemical denudation undisturbed by recent glaciation. All these features favour a low denudation rate per unit area. When combined with high rainfall and rapid transit times (e.g. Amazonia), waters of low ionic concentration result, as found in the basins of the Rio Negro and Zaire Rivers. However, high evaporation rates in basins of low (or even zero) water throughput are also common, especially in the more arid tropics, where a higher salinity series can develop. Central and East Africa, including Ethiopia, provide many examples (in Fig.1) that include lakes in the east and west limbs of the Rift Valley. In addition to ongoing evaporative concentration, geologically localized sources of solutes may also be influential, as in the western Rift Valley. Here the predominant anion is generally bicarbonate (HCO_3^-), and a salinity series results in alkaline "soda lakes" with Na+ as the dominant cation. This situation contrasts with the "mean" global fresh water composition (Table 2; Rodhe, 1949) showing that, on a global basis, Ca^{2+} and HCO_3^- predominate as the most abundant ions. Later calculations (Golterman & Kouwe, 1980) have shown that if the larger African lakes like Tanganyika are

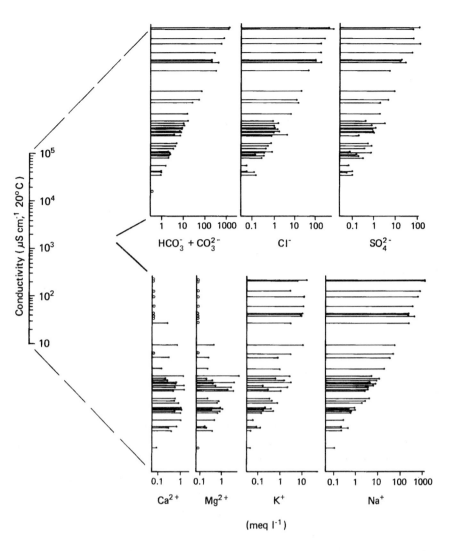

Figure 1: Analyses of African lake waters arranged on a salinity-related scale of ascending conductivity to show concentrations of major anions and cations (from Fryer & Talling, 1986).

included, the global fresh water composition is less Ca^{2+} dominated.

The sum of the major cations in Table 2, except those placed between brackets, should equal the sum of the major anions if both concentrations are expressed in mmol l^{-1}/ ionic charge (the old "meq"). Fe^{2+} and NH_4^+ normally occur only in anaerobic water, and in alkaline water they are not present in the ionic state. The impact of the presence of these reduced compounds on the biota will be discussed later. H^+ is always present, of course, but it contributes significantly to the ionic balance only in water at pH < 4. OH^- will be increased in strongly alkaline waters but only together with much larger concentrations of HCO_3^- and CO_3^{2-}. It will, therefore, never contribute significantly to the ionic balance.

The presence of large quantities of HCO_3^-, with balancing ions such as Ca^{2+} or Na^+, is important in determining and buffering the pH of the water. Furthermore, the CO_2/HCO_3^- system can provide CO_2 for photosynthesis, drawing upon the total CO_2, which is present in equilibriated quantities of the CO_2, HCO_3^- and CO_3^{2-} (Fig. 2). A comprehension of the

$CO_2/HCO_3^-/CO_3^{2-}$ system is, therefore, essential for understanding pH-related processes in fresh waters, such as photosynthesis and the effects of acid rain.

When based on Ca^{2+}, the system is fundamentally limited and constrained by the solubility of CO_2 in water and the solubility product of $CaCO_3$. For the buffering capacity, furthermore, the equilibria:

$$CO_3^{2-} + H^+ \leftrightarrow HCO_3^- \text{ and } HCO_3^- + H^+ \leftrightarrow CO_2$$

are important.

The content of CO_2 in equilibrium with air shows a linear relationship with the partial CO_2 pressure (~ 0.035% in air) and depends on the temperature (Examples in mg l^{-1}: 0.4 at 30°C, 0.6 at 20°C and 1.2 mg at 0°C). The following equilibria occur:

$$(CO_2)_w + H_2O \leftrightarrow H_2CO_3 \tag{I}$$
$$H_2CO_3 \leftrightarrow H^+ + HCO_3^- \tag{II}$$
$$HCO_3^- \leftrightarrow H^+ + CO_3^{2-} \tag{III}$$
$$Ca^{2+} + CO_3^{2-} \leftrightarrow CaCO_3 \tag{IV}$$
$$H_2O \leftrightarrow H^+ + OH^- \tag{V}$$

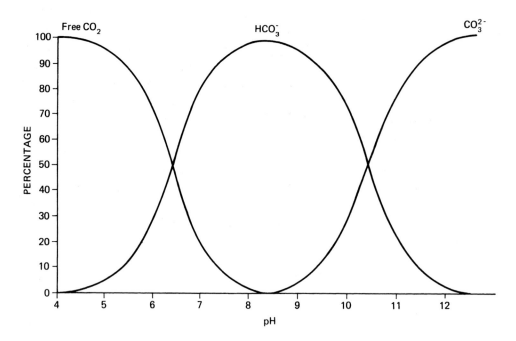

Figure 2: Relation (at 25°C) between pH and per cent of total "CO_2" as free CO_2, HCO_3^- and CO_3^{2-} for waters of low total ionic concentration.

All equilibria are controlled by dissociation constants (which are dependent on temperature). As there are as many chemical species as reactions, there is only one solution for all equilibria at a given moment to fulfill quantitatively the equations: the system has no degree of freedom.

If only the solubility of CO_2 is considered, the concentration of HCO_3^- will remain very low. However, the following reaction brings more HCO_3^- into solution from solid $(CaCO_3)_s$:

$$(CaCO_3)_s + 2H^+ \rightarrow Ca^{2+} + 2\,HCO_3^-$$

with the H^+ derived from reaction (II). It can be shown that, in equilibrium with air, the solubility of $Ca(HCO_3)_2$ is about 2 mmol l^{-1}; at higher CO_2 pressures, it is higher (and the pH lower) and at lower pressures (which may occur in water by photosynthetic consumption of CO_2) it is lower. High CO_2 pressures may occur in nature by the intrusion of volcanic gases, or by rainwater percolating through soils with bacterial activity, or in deep water of productive and stratified lakes.

If some CO_2 is removed from an air-equilibrated solution, reaction (I) will shift leftwards, causing the pH to decrease. Therefore, reactions (II) and (III) will shift to the right, causing the concentrations of OH- and CO_3^{2-} to increase. If the water was saturated with respect to $CaCO_3$, more of this compound will be formed (reaction IV) and will start to crystallize. If water that is supersaturated with CO_2 comes into equilibrium with air again, the excess CO_2 will escape and $CaCO_3$ will be formed and sediment. This may also happen when CO_2 escapes due to increasing temperature, or when it is taken up by plants (phytoplankton or macrophytes) during photosynthesis. Often the quantity of sedimenting $CaCO_3$ may be so large that it covers plants in the littoral region.

The amount of CO_3^{2-} formed as the results of a shift in pH can be calculated, but it is easier to use Fig. 2 for an approximate graphical solution.

In most natural waters, where the Ca^{2+} is the predominating cation, the solubility product of $CaCO_3$ determines the amount of Ca^{2+} and CO_3^{2-} (and thus of HCO_3^-) which can coexist in solution. (The solubility product is a constant at a given temperature and is obtained as $[Ca^{2+}]x[CO_3^{2-}]$ for a saturated solution. At 15°C the value is 0.99 x 10^{-8}, at 25°C it is 0.87 x 10^{-8}). Combined with the influence of the temperature on the CO_2 solubility, the $Ca^{2+}/HCO_3^-/CO_3^{2-}$ system depends strongly on the temperature. In tropical waters, the saturation concentrations and thus the pH-buffering capacity will be lower. In many lakes, conditions of supersaturation may persist for a long time, with Ca^{2+} concentrations greatly exceeding theoretical values. The degree of supersaturation may depend on the pH. However, in many tropical lakes, Na^+ is the preponderant cation, and the solubility of Na_2CO_3 is much higher. Consequently, alkaline soda lakes can reach high concentrations of HCO_3^- and CO_3^{2-} (approximately 2 mol l^{-1}) and in the sequence of evaporative concentration, the Ca^{2+} is largely precipitated as $CaCO_3$ at an early stage. Mg^{2+} follows later, as illustrated by Talling & Talling (1965) and Wood & Talling (1988) for African lakes.

Chloride in river and lake water may come from natural erosion processes, but usually in low concentrations. Higher concentrations may be due to leaching of marine evaporites, to aerial transport, volcanic activity or human pollution. Sea spray ("ocean spray") has been shown to be a source of chloride in coastal areas. In deltaic regions, chloride may be enhanced in fresh water due to the movement of some seawater still remaining underground or percolating upward.

Sulphate may have its source in the erosion of natural gypsum, but nowadays larger quantities enter many (largely temperate) fresh waters from the industrial disposal of H_2SO_4 or $CaSO_4$, or from acid rain.

In standing waters, under natural situations, the concentrations of the major elements do not usually show considerable temporal variability, except for an increase of pH and the decrease of the Ca^{2+} and HCO_3^- concentrations during periods of photosynthesis in near-surface waters and the reverse in deeper stratified waters ("hypolimnia"). Important increases may occur in newly-made reservoirs (e.g. Lake Kariba) due to leaching from freshly flooded areas. In flowing water, seasonal changes of major elements are common, typically with some dilution associated with periods of higher flow rates.

Minor elements Three minor elements in fresh water, namely N, P and Si, are of main concern, as they may control plant primary production. (*In this chapter, in general, simple chemical formulae are used without reference to the exact ionic state, e.g. NH_3 = ammonia plus its ionic derivates, notably NH_4^+; SiO_2 = silicic acid plus its ionic derivates, notably the silicates; o-P = dissolved orthophosphate, i.e. $H_3PO_4 + H_2PO_4^- + HPO_4^{2-} + PO_4^{3-}$; but if concentrations are mentioned, they are the concentrations of the element.*)

Nitrogen may occur in the form of $NH_3(+NH_4^+)$, NO_2^{1-}, NO_3^- and as organic N, in which form it has the same oxidation state as in NH_3. NO_3^- is thermodynamically the most stable form. In the latter stage, many plants can use it after reduction as a source of N. In unpolluted waters of the tropics and the subtropics, the concentrations are usually very low, of the order of $0.01 - 0.1$ mg l^{-1}. Elsewhere, due to sewage disposal and agricultural runoff, concentrations may be very high (> 5 mg l^{-1}).

Besides occurring as an organic fraction, phosphorus also occurs in an inorganic form, mainly as $H_2PO_4^-$ and HPO_4^{2-}. Traces of PO_4^{3-} occur, increasing with pH. The presence of this ion is important, because it is this ion which is involved in the solubility product of apatite $\{Ca_5(PO_4)_3OH\}$, and may control phosphate solubility in many alkaline waters. As described for the $CaCO_3$ system, an increase of pH will induce a shift towards PO_4^{3-} and may cause apatite to precipitate. Often the apatite will co-precipitate and co-sediment with $CaCO_3$. Concentrations of phosphate in some natural waters may be as low as < 0.001 mg l^{-1} of P, but values > 1 mg l^{-1} have been found in some soda lakes and polluted waters.

The ultimate origin of silicate is rock erosion. It occurs in solution by hydrolysis of amorphous SiO_2 and as its ionization products (silicates), depending on the pH (*A discussion on the chemical composition of SiO_2 is beyond the limits of this chapter. For details see Stumm and Morgan, 1981*). It also may occur as colloidal SiO_2 or multimeric species. The prevailing concentrations tend to be higher in tropical regions, where mineral weathering is distinctive and temperature has an important influence. Golterman (1975, quoting Livingstone,

1963) calculated a mean value for African waters of 10.5 mg l^{-1} of Si and a relatively high mean value, i.e. around $10 - 20$ mg l^{-1} of Si, for South America. These high concentrations of Si probably reflect the different erosion patterns, but volcanic influences are included as well. The latter, for example, are probably responsible for high average Si concentrations in the natural waters of Japan and Iceland.

The presence of silicate in water is important for growth of diatoms, which need considerable quantities of silica. Often diatom populations will die after silicon depletion.

In lakes, all three of these nutrients typically show considerable seasonal changes of concentration caused by phytoplankton growth. In some new reservoirs (e.g. Lake Kariba) considerable problems of excessive plant growth have been observed during the first $5 - 10$ years due to leaching of these nutrients from flooded soils and the decomposition of drowned vegetation.

Gases Diffusion at the air-water interface leads to a dissolved content of atmospheric gases, notably CO_2, O_2 and N_2. The content of (physically) dissolved gases depends on partial pressure and temperature. Thus the solubility of these gases is lower in warmer tropical waters than in temperate waters. O_2 and CO_2 are actively involved in photosynthesis, which alters their concentrations considerably in productive waters. CO_2 is also involved in regulating the concentration of other ions, notably HCO_3^-, CO_3^{2-}, Ca^{2+} and Mg^{2+} (see above). N_2 is relatively inert although it can be used as a source of N by some bacteria and algae. Under certain conditions it may be produced by other bacteria during a nitrate-reducing process called denitrification. Other gases which may occur under anoxic conditions are H_2S, NH_3 and CH_4. The first two are always in equilibrium with other chemical forms, changing their solubility. CH_4 (methane) may strongly accumulate at depth in those tropical lakes where no complete annual vertical mixing occurs (e.g. Lake Kivu) and under dense vegetation cover in some (e.g. papyrus) swamps.

Variability in time and space

Horizontal variability: aspects of river chemistry

Most of the processes, as described above, also operate in flowing water. Thus, the pH of rivers is as much buffered by the $CO_2/HCO_3^-/CO_3^{2-}$ system as in lakes, but many tropical rivers are poorly buffered and are slightly acid (pH ~ 5). A feature different from lakes, of course, is the flow or current of a river, which makes the temporal variability of concentrations much larger. Also the chemical composition may differ in space with chemical concentrations often increasing towards the river mouth, but not always.

In principle, three different situations may control the concentration of a particular compound as a function of flow rate (m^3 s^{-1}):

(a) The concentration remains more or less constant. This will happen if the concentration of the compound is controlled by the solubility. $Ca(HCO_3)_2$ is a typical case. If the water percolating the marine limestone is saturated with $CaCO_3$, a change of the flow rate cannot alter the concentration. The constancy may often be obscured by a difference in temperature, concomitant with a change in flow rate. Both temperate and tropical rivers often have a higher flow rate in relatively cooler periods. For the tropics, data on the relation between flow rate and chemical concentration are scanty.

(b) The concentration is decreasing with increasing flow rate. This mechanism is found when the input of the compound is constant, as when the input comes from an anthropogenic release. Phosphate and nitrate often showed this pattern in the past, but nowadays, with a relatively larger input by agriculture, this is often no longer true. In principle, the product of flow rate (Q) and concentration (C) should remain constant: $QC = K$, but due to secondary effects and imprecision of data, the relationship between Q and C often becomes linear.

(c) The concentration is increasing with increasing flow rate. This is usually the case with suspended matter whose transport is enhanced by the increased erosional force of a large flooding. It has often been found that the relation between sediment concentration S_c and flow rate Q can be described by:

$$S_c = A\,Q^B, \qquad\qquad (VI)$$

where A and B are constants for a given river and watershed geology.

It has been shown that when the flow rate is rapidly increasing, the sediment load is larger than at decreasing flow rates. This hysteresis effect will cause the relation between sediment load and flow rate to be more an envelope than a regression line.

Silt concentrations in tropical and subtropical rivers can be relatively large. Keulder (1974) found concentrations as high as 1300 mg l^{-1} during the wet season in the Orange River, where as much as 90% of all components was delivered during the wet season only. Values of up to 2.5 g l^{-1} have been recorded in the Nile (Omer Ali Bedri, cited in Symoens *et al.*, 1981). Furthermore, most of the suspended matter in tropical rivers may be transported during a flood period. For example, Viner *et al.* (in Symoens *et al.*, 1981) recorded that a river in Morocco transported as much as 98% of the P and 74% of the N (both elements being mainly transported as particulate matter) annual load in only four days.

In many rivers this increase in concentration with flow rate also applies to nitrate. One suggested mechanism is that heavy rainfall will displace considerable volumes of groundwater into the river, which contain large amounts of nitrate (often due to agricultural use). Bonell & Balek (this volume) discuss this mechanism in connection with the chemistry of "old" water in stream hydrographs. More important to the tropics are the first rains of the rainy season which cause a nitrate "flush"

from leaching from soils of products accumulated over the preceding dry season.

A special feature of river water in the humid tropics is probably the large difference that can be found in concentrations of organic matter. Thus for the Amazon, Sioli (1984) distinguished three classes, the white waters, the black waters and the clear waters. White waters are relatively rich in nutrients and electrolytes, with a neutral pH. They originate from the Andean and pre-Andean areas and transport large quantities of fine inorganic particles. In general, the ions HCO_3^- and $Ca^{2+} + Mg^{2+}$ dominate in white waters. Black-water rivers are extremely poor in nutrients and electrolytes, with a low pH (< 5). Humic substances from flooded forests, and probably associated dissolved iron, cause a brown colour of the transparent (particle-poor) water. Clear water rivers are transparent and greenish, with a more varied composition, depending on the catchment area. Most of them originate from Central Amazonia, or from the Precambrian shield of the Guyanas, with gentle slopes. In black and clear waters, non-carbonate anions dominate, together with Na^+ and K^+. The nutrient status of clear waters is intermediate between black and white waters.

Some prominent examples of rivers with low concentrations of major elements come from South America. A summary of the water chemistry of the Amazon and the Paraguay Rivers is given by Golterman (1975), from which review the following is taken. The Upper Parana River has a conductivity of 44 (range 32 – 72) µS cm^{-1}, with total cation and total anion concentrations around 1 mmol l^{-1}. The Paraguay River, containing about 1 mM of NaCl extra, has otherwise a not very different composition. Both the Upper Parana and the Paraguay Rivers contain more Mg^{2+} and Na^+ than Ca^{2+}, and a solid content at least twice the dissolved content. Gibbs (1972) calculated the chemical concentration of the Amazon, also a river with low conductivity. He indicated a value of 50 – 60 mg l^{-1} for the dissolved load, with a solid load about 1.5 times higher. In temperate hard waters, the dissolved/solid load ratio is usually around 0.2. Gibbs (1967a, b) demonstrated that salinity of the Amazon decreases from 120 – 140 mg l^{-1} in the Andes to very low values at the mouth (36 mg l^{-1}), with the dilution being caused by rivers coming from the tropical savanna which is underlain by Precambrian igneous rock. Garner (1968), recalculating Gibbs' data, showed that the climate and alluvial valley sediments of the eastern Andes have a larger influence on the chemical composition of the Amazon than Gibbs believed.

In Africa, the Congo or Zaire River is the major river within the humid tropics; it also is characterized by low ionic concentrations and black waters. Probst *et al.* (1992) reported on the dissolved major elements in the catchment of this river. They measured a mean dissolved load of about 34 mg l^{-1} in the lower Congo River and 42 mg l^{-1} in the Ubangi River, its main tributary. In both rivers Ca^{2+} and Mg^{2+} were the dominant cations, whereas HCO_3^- represented the dominant anion, reflecting the geological substratum, which is mainly composed of crystalline and metamorphic rocks. Silicate represented a considerable part, about 30%, of the dissolved load. For both rivers, the total dissolved solids and its main constituents – except silicate – varied inversely with flow rate, often showing nearly theoretical dilution. Another major African river, the Nile, has higher concentrations that are differentiated upstream-downstream for geological, hydrological and climatic reasons (Talling, 1976).

Vertical variability: the chemical consequences of thermal stratification Prolonged stratification may lead to strong chemical differences between the hypolimnion and the epilimnion. Thus, Beauchamp (1964), for example, noted that Lake Tanganyika has a pronounced vertical stratification with 90% of the water permanently deoxygenated. Another example he described is Lake Kivu, where in the hypolimnion so much CO_2 and CH_4 has accumulated that any attempt to draw this water results in a continued gush like a geyser or self-driven air-lift pump. Another gas likely to accumulate in the deeper hypolimnia is H_2S, the occurrence of which depends on the ratio of SO_4^{2-} to Fe^{3+} originally present in the lake. A final example is Lake Nyos, where so much CO_2 accumulated in the hypolimnion that when the lake mixed vertically the large volumes of escaping gas caused the death of c. 1,700 people. Other more widespread important accumulations in deep layers of lakes are those of nutrients, including ortho-phosphate and ammonia.

PRIMARY PRODUCTION

Introduction

In the tropics, as elsewhere, bodies of fresh water generally provide a range of environments for plant growth of varied complexity. This growth is *primary production,* sustained by the photosynthetic fixation of solar energy and with a demand upon inorganic nutrients. The complexity depends upon recruitment from either lower or higher forms of vegetation, and upon relationships with the water mass and its upper (air/water) and lower (e.g. sediment/water) boundaries. For example, free suspensions of algal cells, filaments or colonies form the *phytoplankton* and bottom-related aggregates of these constitute the *benthic algae* or *periphyton.* Larger plants *(macrophytes)* may be entirely submerged, floating as rafts (e.g. Nile cabbage, *Pistia stratiotes)* or dense mats (e.g. papyrus, *Cyperus papyrus),* or attached below but emerging above the water surface as in many reedswamp species.

All these plant communities are potentially sensitive to hydrological manipulation and management. Water depth determines, together with the light penetration, the proportion of the water column in which light-dependent production by phytoplankton is possible and whether bottom-related vegetation can survive and develop. Indirectly, with outflow and draw-off, it governs the retention time of the water-mass and hence the liability of phytoplankton to wash-out. Also indi-

rectly, with wind stress, it determines the likelihood of thermal stratification whose persistence or interruption can modify phytoplankton abundance and qualitative composition. Various forms of water use can affect the supply and concentration of plant nutrients and hence the character of both macro- and micro-vegetation. In the tropics, especially the humid tropics, a winter-limitation of growth for physical reasons of low incident light and temperature is absent, and other controls acquire a greater significance for seasonal plant growth.

In the following outline, emphasis is placed upon the biomass and activity of the dispersed phytoplankton throughout the water mass and their implications for water quality. The lesser reference to larger aquatic vegetation can be supplemented from recent surveys of the subject for Amazonia (Sioli, 1984) and Africa (Denny, 1985).

Biomass development

Biomass can be assessed and compared from various general indices – fresh weight, dry weight, the content of carbon, nitrogen and chlorophyll, and cell volume. The last two are the most widely used for phytoplankton, and usually expressed as a quantity per unit water volume. Figure 3 illustrates the variation in a number of African waters, which range from unproductive to highly productive and visibly discoloured. It also shows, by suitably aligned logarithmic scales, the associated contents of cellular C, N and P that might be expected, given two suggested forms of mean interrelation between the various indices.

For more general comparison of plant stands, measures of biomass per unit area are required. Selected examples from tropical fresh waters appear in Fig. 4 that comprise larger plants, a submerged macrophyte, and a dense phytoplankton. The units shown include both chemical indices as well as dry weight. Floating mats of the sedge *Cyperus papyrus* and the grass *Paspalum* spp. can exceed 1 kg dry weight m^{-2}, as can the reedswamp plant *Typha domingensis,* but even an unusually dense phytoplankton is an order of magnitude lower.

With larger plants, the horizontal distribution of biomass cover and community structure is largely determined by topography and water depth (Fig. 5), current velocity and tributary drainage. Many authors (see Denny, 1985) have published cross-sections of tropical swamp vegetation on a sloping shoreline. The obstructive qualities of floating vegetation have often attracted attention; examples include the Sudd swamp with papyrus of the southern Sudan, the invasive water fern *Salvinia molesta* in Lake Kariba, and the now pan-tropical occurrences of the similarly invasive *Eichhornia crassipes* (water hyacinth).

With phytoplankton, typically more mobile and passively distributed, major and sustained patterns of spatial distribution may develop from several causes. Large elongate reservoirs (e.g. Volta, Kariba) and lakes with strong throughput often show longitudinal gradients of phytoplankton abundance (Fig.

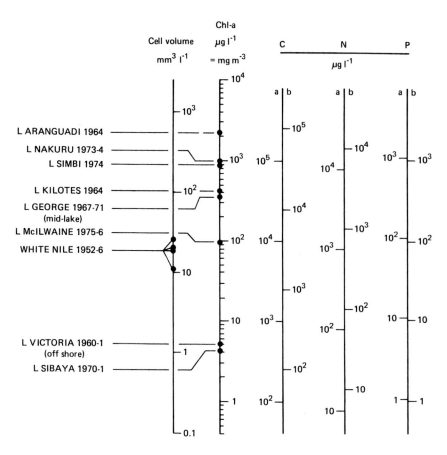

Figure 3: Approximate interrelations, read horizontally across vertical logarithmic scales over four orders of magnitude, of concentrations of two indices of phytoplankton biomass (cell volume, chlorophyll *a*) as maxima observed in nine African waters, and estimates of associated quantities of cellular C, N and P. The interrelations are based on a chl-a content per unit cell volume of 4 µg mm^{-3}, C/chl-a and C/N and C/P ratios that are (a) the mean values for Lake George (Uganda), recorded by Viner (1977), (b) a C/chl-a mass ratio of 40, and the generalized Redfield values for C : N : P. Adapted from Talling (1981).

9a), related to its growth with time of water retention but modified by local tributary sources of nutrient renewal. A differentiation between offshore and inshore phytoplankton is also common, especially where – as in Lake Victoria (Talling, 1987) – the shoreline is of convoluted outline with opportunities for localized water masses, or – as in Lake George, Uganda – there is evidence of some concentric water circulation (Viner & Smith, 1973).

The biomass of vegetation stands, macro or micro, is generally changing in time. The variations may be responses to irregular events, or be expressed as regular seasonal cycles of increase and decrease, or form part of longer-term changes. The last can constitute a process of colonization, as of a newly-created habitat such as a reservoir, or of a river-lake-system into which a previously alien species has been introduced (e.g. the floating macrophyte *Eichhiornia crassipes*). Climatic and hydrological trends may also induce long-term changes, but more rarely in the humid than in the arid tropics where the African lakes Chad and Chilwa are examples. Regular seasonal cycles of aquatic plant cover or abundance are common in the tropics, but – unlike in the temperate situation – are less often induced by direct influences of temperature and radiation income. Seasonal rainfall and hydrologic regime are generally

more influential, often with major effects on the aquatic environment as regards nutrient concentrations, turbidity and wash-out.

In deeper waters, the stratified structure of the water column often undergoes an annual cycle that is susceptible to seasonal wind regime and components of the energy budget. As a result, cycles of phytoplankton abundance can appear that are of wide amplitude. African examples, reviewed by Talling (1986), are dominated by external causes that are hydrological with control by water input-output (e.g. Nile reservoirs), or internal causes with control by water-column structure and composition (e.g. Lake Victoria), or some seasonal combination of these (e.g. Lake Volta). The qualitative composition of the phytoplankton is another variable, a seasonal succession of species being common. Thus, a phase of deep mixing from the surface often favours a predominance of diatoms, whereas a shallow stratification promotes that of blue-green algae. Examples appear in Fig. 6. Blue-green algae include many species that are buoyant, at least periodically. They then may form dense accumulations near the water surface during calm weather, with adverse effects upon the water quality. This process may be rapid, within a few hours, and then owes more to biomass redistribution than to growth itself.

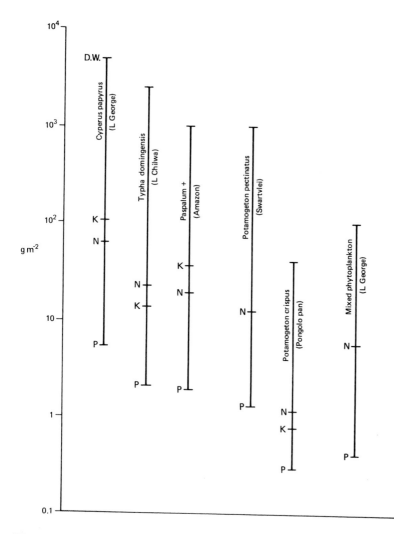

Figure 4: Stocks per unit area of the elements K, N and P in dense stands of reedswamp (*Typha*), floating macrophytes (*Cyperus* and *Paspalum*), submerged macrophytes (*Potamogeton* spp.) and phytoplankton, of varied dry weight (D. W.) (from Talling, 1992).

However, the primary development of biomass by growth has requirements for light-energy and nutrient utilization. Adequate supplies of both are thereby implied. At least above the water surface the tropical light energy input is considerable at all seasons, although below it may be soon extinguished in waters of high colour (e.g. parts of Amazonia) or a heavy silt load. Such extinction can preclude plant growth, and in the humid tropics may be a seasonal component related to rainfall, as in Lake Kainji, Nigeria, where a "white flood" may inhibit primary production for some months after rainfall (Grove, 1985).

Nutrient inputs and resulting concentrations have already been discussed. Low concentrations are not necessarily an indication of poor conditions for growth, as they may coexist with active regeneration by mineralization and recycling. In north-temperate regions, there is much correlative and experimental evidence that phosphorus is often the most limiting nutrient for phytoplankton growth, although situations of limitation by N and Fe are locally documented and Si has been shown to limit diatom growth very often. No such generalization can yet be proposed for tropical fresh waters, where the nitrogen balance is often distinctive (e.g. prevalence of low nitrate concentrations) and the volcanic sources of P are more widespread. Although there is a trend towards higher concentrations of Si in tropical waters, examples of pronounced depletion by diatom growth are also known.

For water management, a possible capacity of the system to develop dense suspensions of phytoplankton is clearly an important consideration. Possible triggering effects include: (i) a new nutrient input, such as a major sewage outfall, (ii) a redistribution by vertical mixing of nutrients previously accumulated and unavailable below a seasonal thermocline and (iii) the combination of nutrient-rich with silt-rich water, followed by sedimentation with clarification that removes the obstacle to light-dependent growth.

Photosynthetic activity

For phytoplankton, a further assessment of the central process of photosynthesis is possible by rate measurement on experimental samples of water exposed at their original depths. Either carbon dioxide uptake or oxygen evolution can be used; for the former, labelling with a radioactive isotope, ^{14}C, is gen-

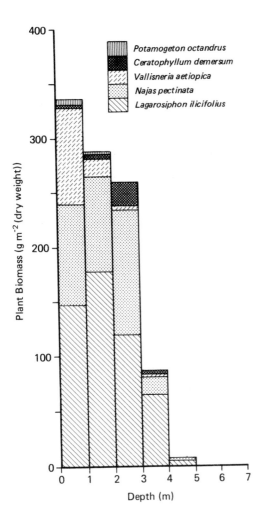

Figure 5: Variation with depth in standing stock of submerged macrophytes in Lake Kariba (from Machena & Kautsky, 1988).

erally employed. With increasing depth, the attenuation of the radiant energy or photon flux ultimately induces a similar near-exponential decline of rates of photosynthetic production. This decline can be predicted closely from suitable measurements of light penetration; rates are usually small or insignificant below a depth at which the flux of available energy (spectral region approx. 400 – 700 nm) is less than 1% of the surface daytime values. The water column above this depth is the physically defined *euphotic zone,* the main location of photosynthetic production. A rough assessment of this depth can also be obtained by multiplying by 2.5 the transparency measured by the depth of disappearance of a white disc (Secchi disc). Sometimes, in physically stratified waters, deep thin layers of photosynthetic organisms develop (e.g. around a thermocline) and produce there a localized peak of photosynthetic activity. There are as yet few descriptions of this feature in the tropics.

Towards the water surface, rates of photosynthetic production usually reach a maximum, a level at which the process is light-saturated. Higher still, at least under conditions of strong sunlight, reduced rates are typical in a zone of "surface inhibition." Examples of entire depth-profiles of photosynthesis, that

incorporate the three regions of light-inhibition, light saturation and light limitation of rate, are shown in Fig. 7. Their depth-extension is variable, according to rates of light attenuation. The other (horizontal) dimension, expressing the rates of activity per unit volume of water, also varies but in an inverse manner. This is because the deeper profiles of clear water are generally associated with low concentrations of photosynthetic biomass. Waters with high biomass concentrations are typically highly active around the optimum depth. Activity per unit biomass is typically higher in warm tropical than in cold temperate waters (Lemoalle, 1981). However, such dense biomass quantities inevitably have a self-shading influence and, in addition, they are often developed in waters that already have a strong background absorption and appear yellowish in bulk.

For making comparisons, primary production is best expressed per unit area of water surface and the rates over a depth-profile can be integrated as the area enclosed by the profile. Examples expressed per m^2 and h are included in Fig. 8. This areal measure, however, is ultimately constrained by the self-shading behaviour of dense populations, in which high but localized rates per unit water volume are confined to a shallow euphotic zone. There is a parallel constraint to the maximum possible biomass in the euphotic zone. Expressed in terms of chlorophyll *a,* the most common index, this is of the order of 200 – 300 mg m^{-2} – a magnitude approached in many tropical fresh waters (e.g. Lemoalle, 1981), including the example of Lake George as shown in Fig. 7.

Reactions upon the medium

The presence of plant (including algal) biomass, its synthesis and its decomposition all affect water characteristics and so provide examples (Fig. 9) of reactions upon the medium. The light absorption (and hence turbidity) introduced by phytoplankton has already been mentioned, as has the possibly adverse effect of surface accumulations of blue-green algae. Mechanical obstacles are also posed by the many distinctively tropical species of floating macro-vegetation.

Photosynthetic activity enriches water in dissolved oxygen and depletes it in carbon dioxide. In consequence, productive waters are often supersaturated with respect to atmospheric oxygen by day, at least in the more illuminated surface layers. In such layers, production exceeds losses by respiratory uptake and transfer to the atmosphere. By night, losses prevail and by dawn a minimum oxygen concentration is often found within the day-night cycle. This may have undesirable biological effects, e.g. upon fish. This phenomenon is more severe in the tropics than in temperate waters, as the O_2 solubility in warm waters is much lower than in colder temperate waters (e.g. at sea level about 7.6 mg l^{-1} at 30° C against 10.1 mg l^{-1} at 15°C). The accompanying but inverse changes of carbon dioxide concentration appear less influential, but lead to higher pH by day that may be 1 or even 2 units above that expected at the CO_2 air-equilibrium. Effects on CO_2 concentration are again more severe in tropical waters as CO_2 solubility is lower and

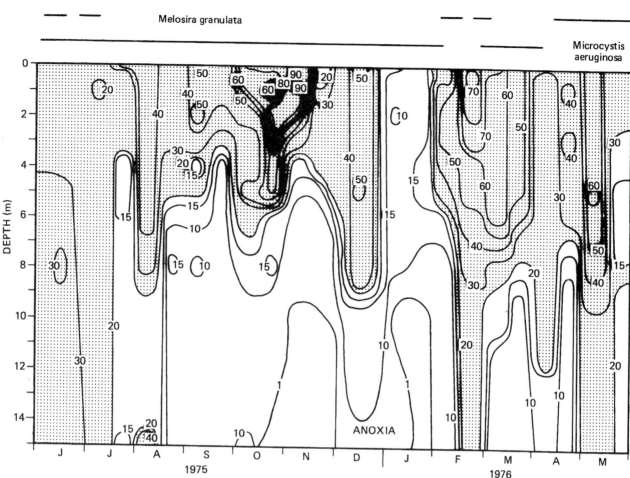

Figure 6: Lake McIlwaine, Zimbabwe: the variation with depth and season of phytoplankton abundance, expressed by chlorophyll *a* concentration (μg l^{-1}) (areas above 20 μg l^{-1} stippled). Also, above, periods of abundance of the blue-green alga *Microcystis aeruginosa* and the diatom *Melosira granulata,* the latter correlated with periods of greater mixing that enhance the deep occurrence of chlorophyll *a*. Adapted from Talling (1986), based on Robarts (1970).

the different constants for the chemical equilibria of the HCO$_3$/CO$_3$ system are also influenced by temperature, and in the same direction.

The growth of aquatic plants, including phytoplankton (cf. Fig. 3), incorporates nutrients – notably the elements N, P and Si – that may be correspondingly depleted in the water medium. Nutrient depletion can have growth-limiting implications. It may be enhanced or dominated by other pathways of chemical loss, such as bacterial denitrification for N and adsorption on sediments for P.

The converse process of decomposition allows a net regeneration of nutrients. This "recycling" of nutrients probably often takes place mainly in the upper water layers rather than in or near the bottom sediments – although it will continue there. Few data on tropical lakes are available but it may be expected that this recycling is more intense in warmer than in cooler lakes (Golterman & Kouwe, 1980). A part of the dying biomass, however, is not recycled in the upper layers and will deposit as sediment. Because of this sedimentation, decomposition will continue in deep water and near the water-sediment

boundary. These situations favour the development of anoxia, especially if exchange with the atmosphere is further impeded by thermal stratification or, in swamps, by floating mats of such plants as papyrus. Then products of nutrient regeneration will include considerable amounts of NH$_4^+$ rather than the oxidized form NO$_3^-$, which are often accompanied by reduced forms of iron (Fe^{2+}), manganese (Mn^{2+}) and sulphur (H$_2$S). Such products accumulate at depths where their re-utilization in primary production is obstructed by insufficient light. This combination of reduced compounds can be seen, at least seasonally, in most deep and at least moderately productive tropical lakes and reservoirs. If the thermal stratification is ended by surface cooling, wind stress or both, deep accumulations of nutrients can be dispersed to include surface layers where primary production is stimulated and biomass concentrations increased. Some form of seasonal mixing and biomass response by phytoplankton is widespread in tropical lakes and reservoirs. Lake Victoria (Talling, 1966) and Lake Kariba provide examples. Besides its effect via nutrient redistribution, a seasonal or otherwise periodic mixing may also create light

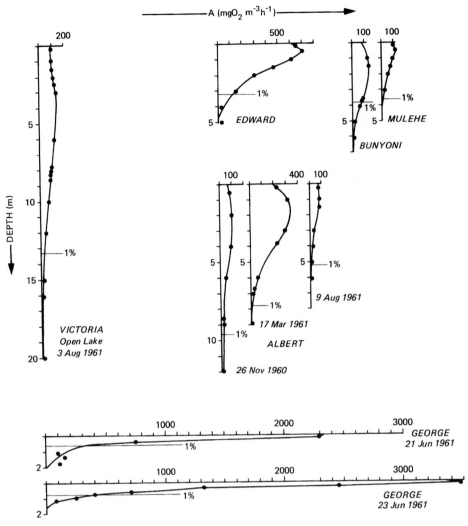

Figure 7: The depth-distribution of photosynthetic rates per unit volume of water (A), measured in six African lakes. Also indicated is the depth at which 1% of the surface radiation available for photosynthesis was found (from Talling, 1965).

conditions and turbulent suspension possibilities that are selectively favourable for diatom growth among phytoplankton (cf. Fig. 6).

THE CONSUMER COMMUNITY IN TROPICAL WATERS

Introduction

The consumer community of animals can be described as a "pyramid of numbers" (or of biomass) in which the quantity of organic material at each trophic (feeding) level is represented by a block (Fig. 10), the size of which decreases from primary producers to primary consumers to secondary consumers. It is not a simple, one-species, one-level structure; different stages of animals may occur at different levels and each level may comprise many different species. The pyramidal concept, however, does illustrate how, during each transfer of biomass between levels, energy is lost through respiration and thus the amount of energy (biomass) stored in each higher level is less ($< 10\%$ in some cases) than in the preceding level. Aquatic

habitats, which frequently have long food chains, therefore often have a relatively low efficiency of energy transfer from the primary producers to the highest trophic levels. The exceptions are those cases where some fish species eat primary producers directly (see below).

The detailed structure of a consumer community can be described as a food web, i.e. the interrelationships of what eats what and how much. In tropical lakes and rivers, the food webs are not fundamentally different from those at higher latitudes. Seasonal changes in these relationships may be controlled by different factors and the diversity of species involved is likely to be much greater. Two major differences from temperate waters are the occurrence in warm waters of fish species which feed directly on plant material (both macro and micro) and the absence of a period of quiescence (winter). The latter situation is presumably due to the fact that the solar radiation income and the water temperature never fall to levels at which primary production is critically low and consumers must either migrate, diapause (i.e. enter a resting phase) in some form or subsist on stored body fat. In tropical waters

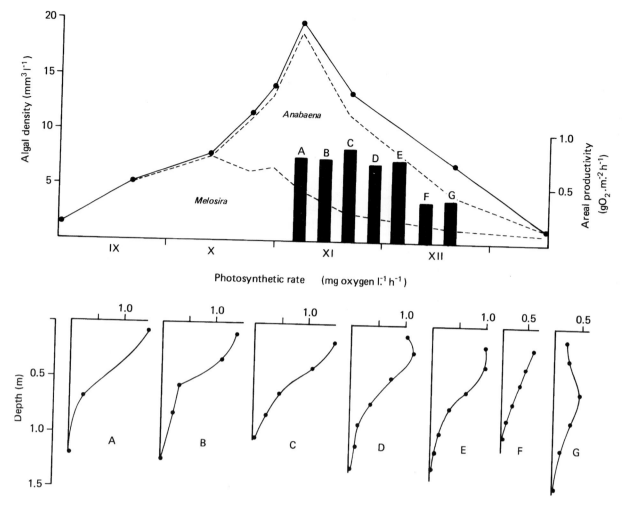

Figure 8: The growth and decline during 1953–54 of phytoplankton in the White Nile near Khartoum, with estimates (as histograms) of photosynthetic productivity per unit area derived from (below) depth-profiles of photosynthetic activity measured by oxygen production. (Adapted from Prowse & Talling, 1958).

food is available for consumers all the year round, although not necessarily the same producers or consumers at all seasons. This should not be taken to imply that seasonality is lacking since there are marked seasonal changes in many tropical waters, such as those caused by rainfall. Here, especially in shallow waters, diapause may be induced by drought.

Components of the consumer food web

Consumers in aquatic environments can be divided into two major groups: 1) in terms of their taxonomic group, i.e. invertebrates and fish (other vertebrates are significant in some places but not universally), or 2) in terms of where they live and feed (on the bottom or in the open water), which is the approach taken here. However, it is important to realize that different species will be found in the equivalent habitats of different geographic regions (Fig. 11).

Benthic animals live on the bottom and either burrow into the substratum or live on the surfaces it provides. This mode of life predominates among river invertebrates since open water invertebrates would be swept downstream and out to sea. Nevertheless, the open water of large rivers, particularly those

with lakes upstream, may contain large quantities of drifting plankton.

Functionally, the benthic invertebrates may be grouped into four major feeding types (Cummins, 1974): a) grazers – scraping food, either algal or bacterial, off surfaces, b) shredders – feeding on large particles or organic matter such as leaves, c) collectors – feeding on fine particulate organic matter either sieved from the flowing water or from the surface of the sediment, and d) predators – feeding on other animals. The relative abundance of these groups changes from the headwaters downstream (Fig. 12). In the headwaters, shredders and predators tend to predominate and the majority of organic matter comes from outside the system (allochthonous). Collectors and grazers increase as the organic matter is reduced in size and the quantity of autochthonous (internally produced) organic matter increases. In the lower reaches of the river, where the water is deep, the channel wide and the water laden with both organic and inorganic particles, collectors predominate (Vannote *et al.*, 1980). There is no reason to suppose that this pattern, as described for temperate rivers, is different in the tropics.

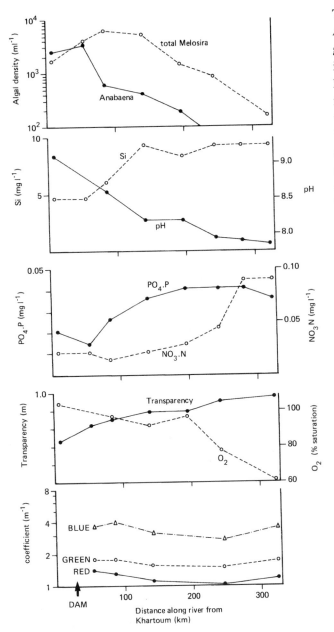

Figure 9: Longitudinal sections along a reservoir on the White Nile above Khartoum, showing (a) the increase of two major components of phytoplankton towards the dam, with correlated depletions of nutrients (b,c), elevation of pH (b) oxygen concentration (d) and light extinction (e), and reduction of transparency (d). (Adapted from Prowse & Talling, 1958).

The main difference is the huge variety of species which makes up the invertebrate communities of the tropics and of the species of fish which prey upon them. The macro-invertebrate community includes worms, molluscs, crustaceans, insects and many other taxa (Table 3). Most spend their entire lives in the aquatic environment but many aquatic insects are juveniles which, when adult, emerge from the water before mating and laying eggs back into the water. Altogether there are more than 7,000 species of fresh water fish. On the Indian sub-continent, for example, there are about 400 indigenous species of riverine fish and about 700 in Malaysia and

Thailand (Fernando, 1980, cited by Moreau & de Silva, 1991). Among this great variety of tropical fresh water fish there is a greater variety of specialized feeding niches than among those in temperate waters. Examples of every conceivable feeding habit can be found, from the Amazonian species which eat fruits that fall from forest trees into the river, to the cichlids which scrape scales from the flanks of other species in Lake Victoria. Nevertheless, of the approximately 2,500 species that occur in Africa, only about 160 are characteristically primary consumers, i.e. feeding on primary producers or detritus (Bowen, 1988). The majority of species depend on invertebrates as their link to the photosynthetic food base.

In lakes, the greatest diversity of benthic macro-invertebrates is found in the littoral zone, where the surface area available for attachment and the variety and extent of refuges from predators may be greatly increased by the presence of macrophytes. In many systems, the structural role of these large plants is more important than as a direct source of food for animals, although in the tropics, unlike in the temperate zones, there are fish such as the grass carp *Ctenopharyngodon idella*, which do feed directly on large aquatic plants. The macrophytes also provide an important source of food when they die, disintegrate and are eaten by detritivores.

Beyond the water depth in which macrophytes can grow, the benthic environment and its invertebrate community are less varied. How much of the lake floor provides suitable habitat for animal life depends on the extent to which oxygen reaches the bottom of the lake. Some invertebrates can survive quite long periods of deoxygenation but usually do not survive permanent anoxia – as in the depths of some deep tropical lakes such as Tanganyika.

Away from the shore and the bottom of the lake is the seemingly uniform environment of the open water habitat, or pelagic zone. Here the primary source of food is the phytoplankton and a complex array of planktonic animals comprise the food web for which the algae form the base. Among the smallest are the Rotifera (Table 3). Some are predatory although the majority are herbivorous, extracting the smallest algae from the water. There are three main groups of planktonic Crustacea: the filter-feeding Cladocera, and the calanoid and cyclopoid Copepoda. The latter grasp and chew their food and are thus able to take a wide variety of particles. Some are strictly carnivorous, others omnivorous. The planktonic Cladocera are mostly herbivores, as are many of the calanoid copepods. Some species of the latter are predatory.

The Rotifera and Cladocera are parthenogenetic (females can reproduce without the need of males) and are thus able to increase their populations very rapidly when conditions are favourable. For the copepods, sexual reproduction is obligatory and this, plus their more complex life cycle, makes for lower rates of population increase.

Although zooplankton is unable to resist major water movements, many populations migrate vertically within the water column, rising to the surface at night to feed on algae in

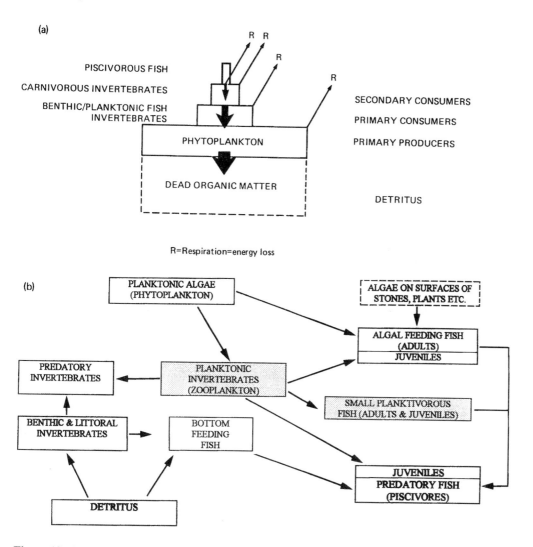

Figure 10: a) A hypothetical pyramid of biomass for the main trophic levels in an aquatic ecosystem. b) A simplified diagram to illustrate the main elements of food webs in fresh water ecosystems. All elements would be present in lakes; the shaded elements will usually be less significant in rivers.

the euphotic zone, but sinking down to the darkness during the day to avoid visual predators in the upper layers. One exceptional insect, *Chaoborus*, is even a member of both planktonic and benthic communities in many lakes. Its larvae spend the day buried in the sediment, then rise into the plankton at night to feed on the planktonic Crustacea. Like the chironomid (midge) larvae which are the dominant components of many benthic communities, *Chaoborus* larvae eventually pupate and then emerge en masse from the water as flying insects which, in East and Central Africa (and probably elsewhere), form clouds over the water and adjacent land, providing food for birds, bats and other terrestrial predators (including humans, around the shores of Lake Malawi: Beadle, 1981). All these invertebrates, both benthic and planktonic, provide food for fish.

Invertebrate vectors of human disease

In some circumstances it is not enough to know and study the invertebrate community only in terms of feeding types. The most studied invertebrate groups in tropical lakes and rivers

are the vectors of human diseases such as *Simulium damnosum*, the blackfly which transmits the filarial worm, *Onchocerca volvulus* that causes onchocerciasis or river blindness. Its larvae are confined to running water because they are collectors, straining fine organic particles from the flowing water. The control of this disease involves the dosing of rivers with insecticides to kill the insect larvae and this may have significant effects on the rest of the ecosystem. As there are hundreds of species of *Simulium,* only a few of which carry onchocerciasis, it is important to know which species are present in a particular system. Similarly, the schistosomes which cause bilharzia (schistosomiasis) are species-specific with respect to the aquatic snails they infect so it is essential to know which species occurs where and details of their ecology.

Controlling factors and seasonal changes

In tropical fresh waters, temperature is not normally a limiting factor in the growth and production of aquatic invertebrates or fish, except at high altitude. Of the abiotic factors, rain and wind are more significant, particularly in relation to mixing of

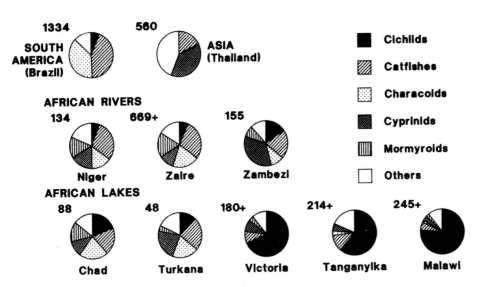

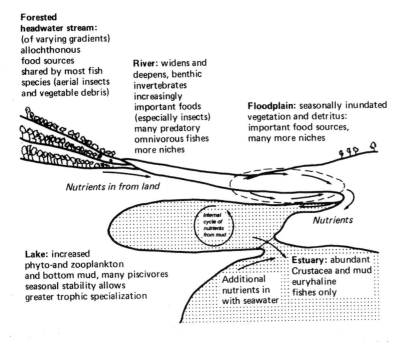

Figure 11: The proportional composition of the fresh water fish faunas of South America (Brazil), South East Asia (Thailand), African rivers and African lakes, based on the number of species indicated (from Lowe-McConnell, 1987). Note the marked differences between the individual lakes, many of whose species are endemic to that lake.

Figure 12: The linear succession of dominant food sources for fishes in a tropical river system (from Lowe-McConnell, 1987).

the water column and rates of flow and flood. These may affect the consumer community directly or indirectly (via the primary producers).

Despite much recent building of dams, many tropical rivers are still free to inundate their flood plains and turn lakes, which were isolated basins of still and shallow water during the low-water season, into deeper, more extensive, flowing water systems for the duration of the flood. In Lake Jacaretinga, one of thousands of "varzea" lakes on the Amazon flood plain, only one zooplankton species *(Moina reticulata)* of the five or six present in the open water throughout the low water period flourishes during the high water phase (Fig. 13). Experimental

evidence suggests that this species is better able than any of the others to grow and reproduce when food is short (diluted by the increased volume of the lake) and the inorganic turbidity is high (Hardy, 1989). Fish, too, are profoundly affected by the seasonal expansion of rivers and lakes to a far greater extent than in non-tropical systems. As the waters spread out over terrestrial vegetation, new sources of food develop as both plants and animals are drowned. Refuges from predators also become temporarily available among land vegetation usually inaccessible to aquatic organisms. Many fish species breed at this time in the shallow flood water where there is a ready supply of food for their young and some of the larger predators

TABLE 3. *A simple summary to show the taxonomic position of the groups and species of animals mentioned in the text (species names are underlined and only the relevant subdivisions of each phylum are included.*

INVERTEBRATES

Phylum	*Class*	*Examples used in the text with English names and an indication of size*
ROTIFERA		rotifers (<1 mm)
NEMATODA		round worms
PLATYHELMINTHES		flatworms and flukes e.g. *Schistosoma* (bilharzia)
ANNELIDA		worms and leeches (5–100 mm)
MOLLUSCA		snails, bivalves (1–100 mm)
ARTHROPODA	CRUSTACEA	Cladocera (c. 1 mm) e.g. *Moina reticulata* Copepoda – Cyclopoida – Calanoida
	INSECTA	Insects = e.g. caddis flies mayflies dragonflies beetles true flies e.g. *Chaoborus* (3-10 mm) *Simulium* (ca 5 mm)
VERTEBRATES	PISCES Fishes	CICHLIDAE (the cichlids including the tilapias and haplochromines) CLUPEIDAE (the herring family – - mostly marine) e.g. *Limnothrissa* and *Stolothrissa* "Tanganyika sardines" CYPRINIDAE (the carp family) e.g. *Ctenopharygodon idella*, the grass carp CHARACIDAE e.g. *Hydrocynus brevis* CENTROPOMIDAE e.g. *Lates* spp
	AMPHIBIA REPTILA AVES (birds) MAMMALIA	

cannot reach them. Failure of the flood may result in poor recruitment in that year of fish species which breed on the flood plain. This has been demonstrated for *Hydrocynus brevis* in the central delta of the Niger and for a number of species in the Senegal River (Welcomme, 1979). Elimination of the flood (by dam building or flood control measures) may endanger the continued survival of such species, most of which are of great commercial importance.

As is clear from the food pyramid model, the abundance and community composition of herbivores and detritivores is con-

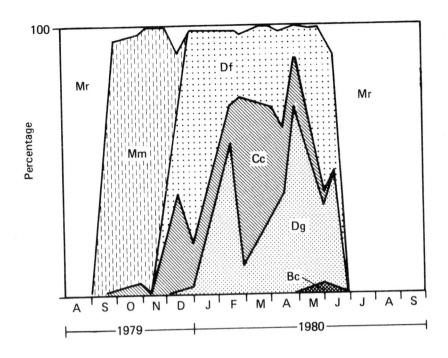

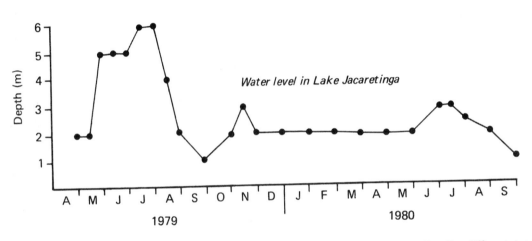

Figure 13: The species composition of the zooplankton in Lake Jacaretinga, Amazonas, Brazil at different stages of the flood cycle. Mr = *Moina reticulata* which occurs throughout the year but is dominant during the high water period; Mm = *Moina minuta*, which dominates in the transitional phase; Df = *Diaphanosoma* spp., Cc = *Ceriodaphnia cornuta;* Dg = *Daphnia gessneri;* and Bc = *Bosmina chilensis* which all occur only in the low water phase (from Hardy, 1989).

trolled not only from the "bottom-up" (by physical and chemical factors and food supply) but also from the "top-down" by predation. This is illustrated in a generalized form for Lake Lanao in the Philippines (Fig. 14) where the highest percentage of mortality per day due to predation falls on the largest prey species (Lewis, 1979). The primary carnivores in this lake are larvae of *Chaoborus* whose feeding accounts for 95% of total losses from the planktonic herbivore populations. The herbivores are squeezed between these two sets of controls from "above" and "below."

Specialized plankton-feeding fish occur in the pelagic zone of many lakes. For example, the "sardines" of Lake Tanganyika are endemic to that lake, living their entire lives in the open water. There are two species: *Stolothrissa tanganicae* occurs

more offshore than *Limnothrissa miodon* which has a rather less specialized diet and is more frequent in inshore waters (Lowe-McConnell, 1987). These two species, plus the juveniles of much larger predators of the genus *Lates*, migrate towards the surface of the lake at night, following the zooplankton and are themselves prey to adult *Lates* and other piscivores as well as forming the basis of a pelagic fishery. The equivalent feeding niche is filled in Lake Malawi by the "utaka" group of cichlids which are a quite different taxonomic group from the Tanganyika sardines (which are in the herring family) but lead a similar lifestyle. It has been suggested that "Tanganyika sardines" should be introduced to Lake Malawi but such a move could be disastrous for the Lake Malawi community. One reason why the introduction of "sardines" to the newly-formed (by

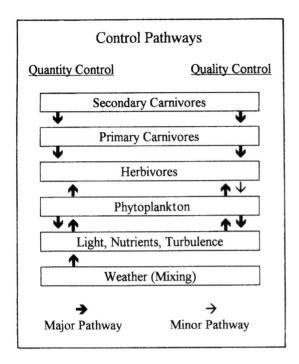

Figure 14: The control pathways for the plankton community of Lake Lanao, Philippines as deduced from a detailed analysis of all the major components throughout the year (from Lewis, 1979).

damming of the Zambezi) Lake Kariba was successful was because they filled a vacant niche there. There were no plankton nor planktivores in the river which could expand into the previously non-existent pelagic zone.

The disastrous introduction of *Lates niloticus* (the Nile perch) to Lake Victoria around 1958 provides proof that predation can influence the structure of an aquatic community. The piscivorous *L. niloticus* grows up to 2 m long. Before 1976, this species comprised about 5% of the fish taken from the lake and its introduction seemed not to have caused a problem. By 1981, it had increased to 50–60% of the catches. Numbers of other predatory fish caught had declined and the huge numbers and variety of much smaller species of haplochromine cichlids have all but disappeared (Lowe-McConnell, 1987). This has had profound repercussions for the local people and is a great zoological loss since almost all the 300+ species of Lake Victoria haplochromine cichlids are endemic to that lake. It should, however, be said that, as pointed out by Ogutu-Ohwayo (1992), not everyone views introductions as disastrous since they may lead to increased catches of fish as food for local people.

Flood plains

Because of the periodic changes in the land-water boundary caused by the water level fluctuations, flood plains are highly dynamic transition zones (ecotones) through which aquatic and terrestrial systems exchange nutrients and organic material through the production-decomposition cycle. Our knowledge of flood plain ecology, however, is still very limited and the

regulation of a large proportion of major rivers means that the natural ecology of flood plains has been lost. The impounding of rivers and conversion of flood plains and other land uses are in many cases the outcome of economic mistakes. The costs and benefits of such environmental changes still need to be evaluated.

Although very large flood plains are most often cited as case studies, attention should also be given to medium and small-sized inundatable areas which are essential to the ecology and productivity of many river systems in the tropics. A general presentation of flood plain ecology has been given by Junk (1982), based mostly on central Amazonia, while the general fisheries ecology of these systems has been dealt with in detail by Welcomme (1979). The paragraphs below are based on their reviews.

The main characteristic of a flood plain is its annual hydrological cycle. During peak flood, the area is inundated and becomes an aquatic system. When the water level decreases, the flood plain becomes a complex mosaic of terrestrial and aquatic habitats. The dry areas may be covered by grass, as found in many African flood plains (Rivers Chari and Niger), but also with the River Mamore in the upper Amazon catchment. In the central Amazon, a forest covers the higher parts of the flood plain. Distributed over the flood plain, lakes of different origins act as refugia – or traps – for aquatic organisms during the drought. These organisms have often developed survival strategies for unfavourable periods, including high reproductive rates and short life cycles. A schematic graph of the nutrient cycle and food webs is given in Fig. 15 (from Junk, 1982), where, in fact, a double cycle, one aquatic and one terrestrial, successively develop in a single year. The herbaceous plants, which are directly consumed only to a limited extent, contribute to the detritus food web and, as such, support much of the fish community. The intensity of the nutrient or energy food web is largely dependent upon the water quality of the river. With nutrient-rich white waters (as the Amazon), much of the imported clay particles (illite and montmorillonite) which settle in the flood plain have a high exchange capacity; these fertile flood plains are called "varzea." Clear and black waters are associated with lower productivity.

The direct relationship between inundated area, fish production and fisheries yield, for instance, has been clearly demonstrated for the North Cameroon Yaere, a medium-sized flood plain (10 000 km²) of the Rivers Chari and Logone, south of Lake Chad. It has also become clear that modifying a flood plain has close and/or distant impacts on both the related terrestrial and aquatic ecosystems. At present, the main uses of flood plains are fisheries, animal husbandry, agriculture and tree exploitation. Other controversial land use conversions or modifications which are occurring relate to aquaculture and intensive agriculture. The polluting effects of these activities are usually underestimated and in the long-term can have negative effects.

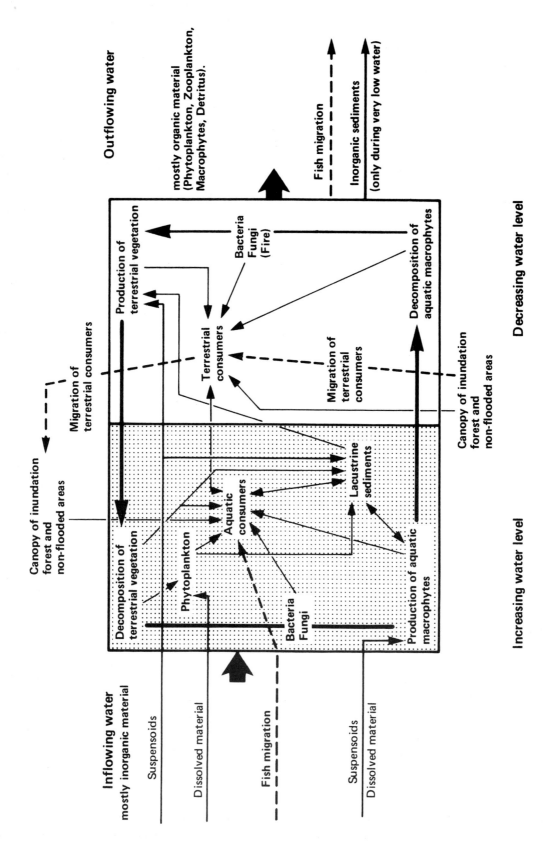

Figure 15: Schematic figure of the nutrient cycle and food-web in an Amazonian flood plain. (From Junk, 1982).

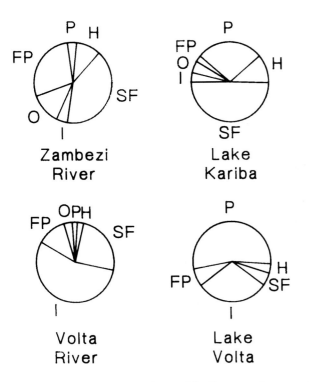

Figure 16: The relative abundance of feeding types, in terms of percentage biomass, among the fish assemblages of two rivers and two man-made lakes resulting from their impoundment. P = plankton feeders, H = herbivores, SF = substrate feeders, I = predators on invertebrates, FP = predators on fish, O = omnivores (From Payne,1986; reprinted by permission of John Wiley and Sons, Ltd).

Fisheries

Fresh water fisheries are a very significant source of food in many tropical countries. In Africa, the vast areas of the lakes have supported fisheries of various kinds for many years, but increasing human populations and the introduction of more intensive modern fishing methods are putting pressure on the fish populations. Many examples could be cited where nets with progressively smaller mesh sizes have been used and the average size of the fish caught has declined and the catch per unit of effort has plummeted. Both are sure signs of over- fishing. In Asia and tropical America, where there are fewer lakes, fisheries depended more on riverine species (Welcomme, 1979), although riverine fisheries have generally been poorly developed in most of Asia (Moreau & De Silva, 1991).

With a natural paucity of lacustrine species and the increasing construction of reservoirs, it is not surprising that many exotic species, such as tilapias from Africa, have been successfully introduced to many Asian waters. There are already 20 introduced fish species in Philippines, 19 in Sri Lanka and 14 in Thailand (Moreau & De Silva, 1991). These fish feed mainly on algae and/or organic detritus and, although they may compete for food or other resources (e.g. breeding sites) with native species, thus altering the balance of the community composition, their introduction is less likely to do irreparable damage than the introduction of piscivorous feeders as

described above. They are also more efficient as a basis for exploitation since they are converting primary production and/or detritus – situated near the bottom of the food chain – directly into fish flesh. This contrasts with the alternative of several transfers of energy, with loss at each stage, through a multi-stage chain such as algae – invertebrate – small fish – piscivores – humans. When rivers are dammed to form reservoirs, the nature of the aquatic environment is profoundly changed and species adapted to a flowing water environment may be unable to survive in the new lake. Some of the more flexible species will adapt and both the structure and functioning of the community as a whole will change (Fig. 16). When tropical reservoirs first begin to fill, there is an overall increase in productivity, based on the nutrients released from the flooded land, which is reflected in increased fish yields. However, the highest levels are maintained for only a few years before productivity declines as the new lake equilibrates. Subsequent fish yields may well be greater than those from the original river but there is plenty of evidence to show that the initial high yields should not be used for long-term planning. Indeed, any gain in fish production may well be offset by the loss of seasonally-inundated wetland areas, resulting in an overall loss of production (taking fish, pasture and agriculture together) (Dugan, 1990).

CONCLUSIONS

The ecology of tropical waters is quantitatively less well understood than that of temperate waters, but it is clear that these ecosystems are at least as sensitive to perturbations as are temperate systems. Better information is particularly needed concerning:

1. The influence of a catchment's geochemistry on the chemical composition of its waters.

   The chemical composition of waters also depends very much on the climate and altitude of the catchment. Many of the largest rivers of the world have the whole or part of their catchment in the humid tropics, but some of the observed characteristics of their waters may originate from outside this climatic zone.

2. The waters of true "humid tropics" (i.e. forested) areas which are mostly black waters.

   In these black waters, much of the food web relies on allochthonous organic detritus since primary (autochthonous) production of organic matter is low. These waters are also very poor in ionic content and nutrients, and are thus poorly buffered and easily susceptible to alteration (pollution).

3. The seasonal patters of phytoplankton succession and the relations between primary production and its limiting nutrients.

   For very many tropical waters, it is even not known whether primary production is predominantly limited by P or N.

4. The communities of aquatic animals, which are still very poorly understood even for fish.

    The details of many tropical aquatic food webs still remain to be clarified, although it is evident that they are extremely sensitive to human-made changes such as the introduction of exotic species, especially predators.

5. Management of flood plains needs careful planning, including consideration of their importance for both indigenous fisheries and wildlife.

6. How and where strategies for the conservation of potential resources should be applied.

    Since it is widely accepted that much of the natural genetic resources of the world lies within the tropics, there is great interest in the study of both species diversity and within-species (genetic) diversity in these areas.

REFERENCES

Beauchamp, R. S. A. (1964) The Rift Valley Lakes of Africa. *Verh. int. Ver. Limnol. 15*:91–100.

Beadle, L. C. (1981) *Inland Waters of Tropical Africa: An Introduction to Tropical Limnology*. (2nd Edition). Longman, London.

Bowen, S. H. (1988) Detritivory and herbivory. *In*: C. Lévêque, M. N. Bruton and G. W. Ssentongo (eds.) *Biology and Ecology of African Fresh water Fishes*. ORSTOM, Paris. *Travaux et Documents 216*:243–247.

Burgis M. J. & Symoens, J. J. (eds.) (1987) *African Wetlands and Shallow Water Bodies. Directory*. Editions de l'ORSTOM, Paris. Collection *Travaux et Documents 211.*

Carmouze, J.-P., Durand, J.-R. and Lévêque, C. (1983) Lake Chad. Ecology and productivity of a shallow tropical ecosystem. *Monographiae Biologicae 53*. Junk, The Hague.

Coulter, G. W. (ed.) (1991) *Lake Tanganyika and Its Life*. Oxford Univ. Press.

Cummins, K. W. (1974) Structure and function of stream ecosystems. *Bioscience 24*:631–641.

Davies, B. & Gasse, F. (1988) *African Wetlands and Shallow Water Bodies. Bibliography*. Editions de l'ORSTOM, Paris. Collection *Travaux et Documents 211.*

Denny, P. (ed.) (1985) *The Ecology and Management of African Wetland Vegetation*. Junk, Dordrecht.

Dugan, P. J. (ed.) (1990) *Wetland Conservation. A Review of Current Issues and Required Action*. IUCN, Gland, Switzerland.

Furtado, J. I. & Mori, S. (1982) Tasek Bera, the ecology of a fresh water swamp. *Monographiae Biologiae 47*. Junk, The Hague.

Ganf, G. G. & Horne, A. J. (1975) Diurnal stratification, primary production and nitrogen fixation in a shallow equatorial lake (Lake George, Uganda). *Freshwater Biology 5*:13–39.

Garner, H. F. (1968) Geochemistry of the Amazon River system: Discussion. Bull. *Geol. Soc. Am. 79*:1081–1086.

Gibbs, R. J. (1967a) The geochemistry of the Amazon River system: Part I. The factors that control the salinity and the composition and concentration of the suspended solids. *Bull. Geol. Soc. Am. 78*:1203–1232.

Gibbs, R. J. (1967b) Amazon River: Environmental factors that control its dissolved and suspended load. *Science N. Y. 156*:1734–1737.

Gibbs, R.J. (1972) Water chemistry of the Amazon River. *Geochim. Cosmochim. Acta 36*:1061–1066.

Golterman, H. L. (1975) Chemistry (of rivers). *In*: B. A. Whitton (ed.) *River Ecology*. Blackwell Scientific Publications, Oxford.

Golterman, H. L. & Kouwe, F. A. (1980) Chemical budgets and nutrient pathways. *In*: E. D. Le Cren & R. H. Lowe-McConnell (eds) *The Functioning of Freshwater Ecosystems*. IBP Nr 22. Cambridge Univ. Press, Cambridge.

Greenwood, P. H. & Lund, J. W. G. (1973) A discussion on the biology of an equatorial lake: Lake George, Uganda. *Proc. R. Soc. Lond.(B) 184*:227–346.

Grove, A. T. (ed.) (1985) *The Niger and Its Neighbours*. Balkema Publ., The Netherlands.

Hardy, E. R. (1989) *Effect of temperature, food concentration and turbidity on the life cycle characteristics of planktonic cladocerans in a tropical lake, Central Amazon: Field and experimental work*. PhD Thesis, Univ. of London.

Hutchinson, G. E. (1957, 1967, 1975 and in press) *A Treatise on Limnology*, Vols 1–4. Wiley. N.Y.

Junk, W. J. (1982) Amazonian flood plains: Their ecology, present and potential use. *Rev. Hydrobiol. Trop. 15*:285–301.

Kalk, M., Howard-Williams, C. & McLachlan, A.J. (1979) Lake Chilwa. Studies of change in a tropical ecosystem. *Monographiae biologicae 35*. Junk, Dordrecht.

Keulder, P. C. (1974) The influence of silt on primary production. *In*: E. M. van Zinderen Bakker (ed.), *The Orange River* (cited in Symoens, Burgis & Gaudet, 1981).

Lemoalle, J. (1981) Photosynthetic production and phytoplankton in the euphotic zone of some African and temperate lakes. *Rev. Hydrobiol. Trop. 11*:31–37.

Lewis, W. M. (1979) *Zooplankton Community Analysis*. Springer-Verlag, N.Y.

Lewis, W. M. (1981) Precipitation chemistry and nutrient loading by precipitation in a tropical watershed. *Water Resour. Res., 17*: 169–181.

Lewis, W. M. (1987) Tropical limnology. *Ann Rev. Ecol. Syst., 18*: 159–184.

Livingstone, D. A. (1963) Chemical composition of rivers and lakes. *In*: Data of Geochemistry. *Geol. Surv. Prof. Pap. 440*:Gl–G64.

Lowe-McConnell, R. H. (1987) *Ecological Studies in Tropical Fish Communities*. Cambridge Univ. Press.

Machena, C. & Kautsky, N. (1988) A quantitative diving survey of benthic vegetation and fauna in Lake Kariba, a tropical man-made lake. *Freshwater Biol. 19*:1–14.

Moreau, J. & De Silva, S. S. (1991) Predictive fish yield models for lakes and reservoirs of the Philippines, Sri Lanka and Thailand. *FAO Fisheries Tech. Paper 319*. FAO, Rome.

Moss, B. (1988) *Ecology of Fresh Waters. Man and Medium*. 2nd edition. Blackwell.

Ogutu-Ohwayo, R. (1992) The purpose, costs and benefits of fish introductions: With specific reference to the Great Lakes of Africa. *Mitt. Intern. Verein. Limnol. 23*:37–44.

Payne, A. I. (1986) *The Ecology of Tropical Lakes and Rivers*. Wiley, Chichester, UK.

Probst, J. L., NKounkou, R. R., Krempp, G., Bricquet, P. P., Thiébaud J. P. & Olivry, J. C. (1992) Dissolved major elements exported by the Congo and the Ubangi Rivers during the period 1987 – 1989. *J. Hydrol., 135*: 237–257.

Prowse G. A. & Talling, J. F. (1958) The seasonal growth and succession of plankton algae in the White Nile. *Limnol. Oceanogr. 3*:222–238.

Robarts, R. D. (1970) Underwater light penetration, chlorophyll *a* and primary production in a tropical African lake (Lake McIlwaine, Rhodesia).

Arch. Hydrobiol. 26:423–444.

Rodhe, W. (1949) The ionic composition of lake waters. *Verh. int. Ver. Limnol.* 10:377–386.

Rzóska, J. (ed.) (1976) The Nile, biology of an ancient river. *Monographiae Biologicae 29.* Junk, The Hague.

Schiemer, F. (ed.) (1983) *Limnology of Parakrama Samudra – Sri Lanka.* Junk, The Hague.

Serruya, C. & Pollinger, U. (1983) *Lakes of the Warm Belt.* Cambridge Univ. Press.

Sioli, H. (ed). (1984) The Amazon. Limnology and landscape ecology of a mighty tropical river and its basin. *Monographiae Biologicae 56.* Junk.

Stumm, W. & Morgan, J. J. (1981) *Aquatic Chemistry.* Wiley, New York.

Symoens, J. J., Burgis, M. & Gaudet, J. J. (eds) (1981) *The Ecology and Utilization of African Inland Waters.* United Nations Environment Programme, Reports and Proc. Series *No. 1,* Nairobi.

Talling, J. F. (1965) The photosynthetic activity of phytoplankton in East African lakes. *Int. Rev. Hydrobiol.* 50:1–32.

Talling, J. F. (1966) The annual cycle of stratification and phytoplankton growth in Lake Victoria (East Africa). *Int. Rev. Hydrobiol.* 51:545–621.

Talling, J. F. (1969) The incidence of vertical mixing, and some biological and chemical consequences, in tropical African Lakes. *Verh. int. Ver. Limnol.* 17:998–1012.

Talling, J. F. (1976) *Water Characteristics. In:* J. Rzóska, 1976, pp. 357–384.

Talling, J. F. (1981) The conditions for high concentrations of biomass. *In:* J. J. Symoens, M. Burgis & J. J. Gaudet (eds.) *The Ecology and Utilization of African Inland Waters.* United Nations Environmental Programme, Nairobi.

Talling, J. F. (1986) The seasonality of phytoplankton in African lakes. *Hydrobiologia* 138:139–160.

Talling, J. F. (1987) The phytoplankton of Lake Victoria (East Africa). *Arch. Hydrobiol. Beih., Ergebn. Limnol.* 25:229–256.

Talling, J. F. (1992) Environmental regulation in African shallow lakes and wetlands. *Rev. Hydrobiol. Trop.* 25 :87–144.

Talling, J. F. & Talling, I. B. (1965) The chemical composition of African lake waters. *Int. Rev. Hydrobiol.* 50:421–463.

Thornton, J. A. (1982) Lake McIlwaine. *Monographiae Biologicae 49.* Junk, The Hague.

Vannote, R. L., Minshall, G. W., Cummins, K. W., Sedell, J. R. & Cushing, C. E. (1980) The river continuum concept. *Can. J. Fisheries Aquat. Sci.* 37:130–137.

Viner, A. B. & Smith, I. R. (1973) Geographical, historical and physical aspects of Lake George. *Proc. R. Soc. Lond. (B) 184*:235–270.

Welcomme, R. L. (1979) *Fisheries Ecology of Flood-plain Rivers.* Longman, London.

Wood, R. B. & Talling, J. F. (1988) Chemical and algal relationships in a salinity series of Ethiopian inland waters. *Hydrobiologia 158*:29–67.

V. Physical Processes – Human Uses: The Interface

17: Challenges in Agriculture and Forest Hydrology in the Humid Tropics

R. Lal

Department of Agronomy, Ohio State University, Columbus, Ohio, USA

ABSTRACT

This paper focuses mainly on the humid tropical and parts of the subhumid tropical subregions of Chang & Lau (1986), associated with the narrow geographic band of about 5° to 7° north and south of the equator. In the humid tropical regions, where precipitation exceeds evaporation for at least 9.5 months per year, Tropical Rain Forest (TRF) is the dominant vegetation in undisturbed conditions and tropical cyclonic influences are absent (Lal, 1987). Occasional reference, however, will be made to the hydrological characteristics of the wet-dry tropical region. The gaps in research knowledge related to the water balance, erosion and sedimentation and change in land-use are addressed. Brief reference is also given to the applications of remote sensing and geographic information systems, and to the problems of scale.

INTRODUCTION

The climax vegetation of the humid tropics regions is tropical rain forest (TRF), which is a diverse and complex system that occupies approximately 10% of the world's area (Golley *et al.,* 1975). The vegetation of TRF is characterized by high plant biomass, concentration of nutrients within the plant biomass and efficient and rapid rates of nutrient recycling which enable this eco-system to sustain itself. The vegetation biomass ranges from 200 to 400 t ha^{-1} and the number, size and height of tree species within a TRF are highly variable and diverse. In addition, mature TRF has a multi-storey canopy of different strata (Lal, 1986) with a leaf area index (LAI) commonly reported as ranging from 7 to 28.

This paper focuses mainly on the humid tropical and parts of the subhumid tropical regions of Chang and Lau (1986), associated with the narrow band of about 5° to 7° north and south of the equator. In such areas, tropical cyclonic influences are absent (Lal, l987). Occasional reference, however, will be made to the hydrological characteristics of the wet-dry tropical region. Both Stewart and Fleming (this volume) provide more detail on the latter region, using northern Australia as an example. The gaps in research knowledge related to the water balance, erosion and sedimentation, and land-use impacts will be briefly surveyed.

HYDROLOGICAL CHARACTERISTICS OF THE HUMID TROPICS

Hydrological characteristics of humid tropical climates are determined by rainfall, runoff, soil water storage and evapotranspiration, as shown in Equation 1:

$$P = R + E + \Delta\phi + \Delta D + T + W \tag{1}$$

where P is rainfall, R is runoff, E is evapotranspiration, $\Delta\phi$ is change in soil water storage, ΔD is change in surface detention, T is interflow, and W is groundwater recharge. The interrelationship among these variables is shown in Fig. 1. Alterations in any one factor influences the magnitude and rate of the others. The following is a description of the hydrological parameters of regions not usually subjected to tropical cyclones.

Rainfall

Rainfall characteristics of TRF are described by Richards (1952), Pinker (1980), Lawson *et al.,* (1981), and Ghuman & Lal (1987). These regions are characterized by a large amount of annual rainfall (> 2,000 mm per year), falling in rain bursts of high intensity (F). Rainfall intensities (F) exceeding 150 mm h^{-1} over short periods of five to ten minutes are not uncommon. Such F values are partly due to the large drop sizes observed for tropical rains (Lal, 1986). In an undisturbed primary rain forest, the proportion of rain reaching the ground as throughfall ranges from 60–90%, depending on the canopy characteristics and leaf area index of the predominant species (Lawson *et al.,* 1981). Nortcliff & Thornes (1977) observed that 83% of the free falling rain reached the ground under mature tropical rain forest. The mean value of stemflow commonly observed is about 10%, although a range of 5–25% is reported in the literature.

There are few reliable data based on long-term observations for rainfall intensity, drop size distribution and the energy load of rainfall in humid tropical regions. These rainfall-energy

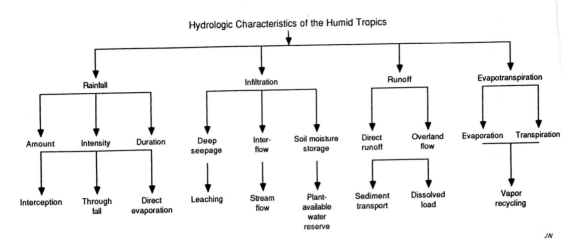

Figure l: Hydrologic parameters and their interaction with principal processes.

(RE)intensity relations are generally expressed in the form shown in Equation 2:

$$RE = a + b \log_{10}F \qquad (2)$$

where a and b are regression coefficients. However, these relations are not known for most eco-regions of the humid tropics. Furthermore, little is known about the intensity-duration relationship for these regions. Research information on stemflow, throughfall, canopy drip, and size distribution of primary *vis-a-vis* secondary drops and their terminal velocity is also scanty. In addition, few attempts have been made to assess the throughfall distribution at different levels within the forest canopy.

Scanty as the observations on rainfall characteristics are at local and small-drainage basin scales, even less is known about the vapour movement in humid tropical climates at regional or global scales. It is believed that more than 50% of the rainfall is returned to the atmosphere through evapotranspiration (Salati *et al.,* 1979). There has been little verifiable data indicating the exact proportion of rainfall recycled within a given region. In fact, it is very difficult to evaluate the contribution of recycled water vapour over a reasonably large area, such as a river basin (see Bonell with Balek, this volume). A severe constraint is the heterogeneous hydrometeorological conditions at that scale. Studies of this nature, by necessity, have to be conducted on relatively small and supposedly homogeneous areas over a range of seasons, and then extrapolated to the entire basin. A major challenge to hydrometeorologists lies in evaluating the amount of recycled rainfall and its fluctuations due to change in vegetation, terrain, land-use, and other anthropogenic perturbations.

Another important but difficult challenge to be addressed in evaluating rainfall characteristics is that of instrumentation and station density. Do we know the optimum station density for measurement of rainfall, throughfall, stemflow, drop size and intensity for a region where complexity and heterogeneity is the rule rather than the exception? Is one station sufficient for 1, 10, 100 or 500 km²? Furthermore, what is the ideal fre-quency of measurement – daily, weekly, or monthly?

Recommended minimum station density varies with the objective and the resources available. Accessibility to remote areas covered by undisturbed vegetation poses another major logistical problem.

Infiltration

Infiltration, the volume flux of water entering soil through the soil-air interphase, is an important hydrological process that governs runoff and soil-water storage (Fig. 1). Infiltration capacity (or the steady state or equilibrium infiltration rate) decreases exponentially with time. The profile-controlled infiltration process can be expressed mathematically by Kostiakov's (1932) or Philip's (1957) models which are given below in Equations 3 and 4, and 5 and 6, respectively,

$$i = at^b \qquad (3)$$

$$di \, dt^{-1} = ct^d \qquad (4)$$

$$i = st^{1/2} + At \qquad (5)$$

$$di \, dt^{-1} = 1/2 \, St^{-1/2} + A \qquad (6)$$

where i is cumulative infiltration, $di \, dt^{-1}$ is instantaneous infiltration, S is sorptivity, A is transmissivity, t is time and a, b, c and d are empirically determined constants. The above equations, however, do not apply to flux-controlled or rainfall-controlled infiltration in situations where soil surface attains saturation very rapidly. For these situations, Bonell & Williams (1986) observed that the cumulative infiltration can be adequately described by the product of field saturated hydraulic conductivity, K^* and time (Equation 7):

$$i = K^*t \qquad (7)$$

This implies that field-saturated hydraulic conductivity under ponded conditions is the dominant parameter. The latter is, however, highly variable over time and space (Williams & Bonell, 1988).

Horton (1933) described the process as occurring in two parts: (a) a period in which the soil-water reserve is being filled, and (b) a period in which a limiting value or the equilibrium rate is being reached. The latter is equal to the rate of transmission of water through the soil, or the soil water transmissivity described by parameter "A" in Equations 5 and 6 above. The equilibrium infiltration rate "$di\,dt^{-1}$" may be profile-controlled or flux-controlled. With high-intensity rains, when rainfall intensity (F) exceeds the infiltration capacity ($di\,dt^{-1}$), the infiltration is profile-controlled. It is limited by soil characteristics for the transmission of water through the profile. "Infiltration envelope" refers to the time taken by a given soil to reach ponding or the profile-controlled condition (Smith, 1973).

There are many issues pertaining to infiltration that are relevant to soils of the humid tropics which need to be addressed by hydrologists and soil physicists. An important one is the assessment of infiltration capacity over a watershed. Assessment of infiltration capacity for a soil series can be made by flooding or a sprinkling infiltrometer. However, it is difficult to extrapolate these results to a watershed with diverse soil types, a high soil heterogenity and a distinct profile stratification. The infiltration capacity and infiltration envelopes depend upon these factors as well as upon the vegetation types, their rooting characteristics, and the ground cover. Consequently, infiltration characteristics of a watershed are highly variable in time and space, and its parameters are difficult to generalize. Most infiltration models are not applicable for layered soils. Above all, there are severe sampling problems that make it virtually impossible to describe infiltration characteristics of a watershed with a reliable set of values (see Bonell with Balek, this volume).

Surface Runoff

The excess of rainfall intensity over infiltration capacity is the surface runoff. Horton (1945) proposed Equation 8:

$$q = (F - (di\,dt^{-1}))A \qquad (8)$$

to describe the initiation of runoff, where q is the runoff rate per unit contour length, F is rainfall intensity, ($di\,dt^{-1}$) is infiltration capacity and A is the area drained. In addition to the problems involved in reliable assessment of infiltration, application of the model described in Equation 8 to soils of the humid tropics is beset with many other obstacles. For example, surface runoff can be generated even if rainfall intensity did not exceed the infiltration capacity. This occurs in stratified soils where the soil-water transmissivity of the subsoil horizons is drastically lower than that of the surface layer. This is a common phenomenon in most soils of the humid tropics because of the illuviation of clay from the surface to the subsoil layer (see Fleming, this volume).

Overland flow can also be initiated by saturation of the superficial horizon due to a rapid occurrence of "infiltration envelope", development of a perched water table, or lateral movement of subsurface water called interflow. The effect of saturation of the surface layer, on initiation of runoff, is exemplified by the experiments conducted in the cyclone-prone, humid tropical coast of Queensland (Gilmour, 1977a; Bonell *et al.*, 1979; Gilmour & Bonell, 1977, 1979; Bonell & Gilmour, 1978) where a perched water table rapidly developed soon after the onset of storms.

A third confounding factor is the change in infiltration characteristics due to slaking of the surface layer. Raindrop impact, both by primary drops in a cleared, unprotected soil or secondary drops from canopy drip in a forest/crop cover, leads to detachment of aggregates and blockage of transmission pores by primary particles and micro-aggregates. Rapid decline in infiltration capacity during a rainstorm leads to initiation of runoff even at low levels of rainfall intensity. Interaction between raindrop impact and shallow overland flow enhances the slaking process and accelerates the runoff.

Estimating runoff over a watershed is also confounded by the heterogeneity of soil properties and their spatial and temporal variability. While some soils may not generate runoff even at high intensities, others may do so even at low intensities. Even over a small watershed, there are regions of variable runoff rates and amounts. For the same rainfall intensity, some areas may have a runoff coefficient of 0.0 while others have a value of 100.

Runoff prediction for humid tropical regions is also constrained by insufficient information on vegetation characteristics. There is a rapid rate of conversion of TRF to plantations, seasonal crops and pastures (Lal, 1986). Not only is TRF complex and heterogeneous but little is known regarding the effects of tropical crops and plantations on the hydrological properties of soil and on runoff rates and amounts (see the "CHANGE IN LAND-USE AND HYDROLOGICAL PROBLEMS" part of this chapter). There is scanty information about canopy and root characteristics, water use, and other physiological properties for banana, plantain, teak, gmelina, oil palm, rubber, cassava, yam, or other tropical crops and pastures. These crops have a varying influence on throughfall, stemflow, surface detention capacity, infiltration capacity and overland flow, and these effects are not known.

Interflow

In a stratified soil, where saturated hydraulic conductivity of the surface horizon(s) exceeds that of the layers beneath, lateral flow occurs beneath the surface at the junction of the two horizons. Interflow is that part of the runoff that travels beneath the soil surface to reach a stream channel or to the point along the hillslope where the slowly-permeable subsoil horizon is exposed to the surface. The term interflow may be used synonymously with throughflow. Some hydrologists describe lateral flow above the water table under saturated conditions as "saturated interflow."

Interflow is a common phenomenon throughout the humid tropics. The rate and amount of interflow depends on other parameters shown in Equation 1. Other factors remaining the same, the rate, amount and duration of interflow depends on evapotranspiration, especially on the amount of water uptake by deep-rooted perennials. Change of land-use from TRF to arable land-use or pasture is likely to change the rate and amount of interflow. However, the trends in these changes, increase or decrease, depend on external factors. Hibbert (1967) and Pereira (1973) observed an increase in both surface runoff and interflow by deforestation. However, in Kenya, Hursh (1953) reported that forested watersheds favored maximum sustained water yield. In southwestern Nigeria, Lawson *et al.,* (1981) observed a drastic increase in interflow by conversion of a forested watershed to arable land-use. Watershed management experiments conducted at IITA, Ibadan, Nigeria showed that an intermittent stream became a perennial stream by deforestation. Furthermore, interflow increased with time after deforestation as the regrowth of trees and shrubs also eventually ended.

It is now established that interflow can occur even when there is no overland flow. In fact, interflow is an important mode of water flow on hillsides in the humid tropics. Many geomorphologists now visualize the drainage basin or watershed as an extended plane. This is the basis of the "variable area source concept" or "partial area concept", which recognizes the regions close to the drainage channel that contribute to water runoff (see Bonell with Balek, this volume). As the rainy season progresses and the soil becomes increasingly wet, the saturated zone contributing to interflow progressively extends upwards. There are some measurements of interflow rate for watersheds under different land-uses in the tropics (Roose, 1977). However, there are not many long-term records from watersheds of different characteristics (geologic, hydrologic and land-use). Until empirical data have been developed for a large number of watersheds, it is difficult to validate some of the models being proposed.

Evaporation and soil moisture storage

Evaluating plant-available water reserves over a watershed remains as a major challenge to soil scientists and hydrometeorologists. In general, soils of the humid tropics have a low level of plant-available water reserves (Lal, 1979). Consequently, short-duration and shallow-rooted crops suffer from periodic drought stress even during the rainy season. While soil surface management and agronomic practices can increase water storage in the root zone, it is important to know the water requirements of different crops, especially at critical stages of growth. Research data on evapotranspiration for some important seasonal and perennial crops are not available. Some empirical models have been developed and tested for tropical watersheds. However, research information on actual evapotranspiration, E, in relation to pan evaporation, E_o ($E{:}E_o$) for different crops, soils and agro-ecological regions is not readily available.

Reliable assessment of soil moisture reserves in the root zone for large areas has always been a major problem. Some of the non-destructive techniques (neutron moisture metering and time domain refractometry) are expensive, labour-intensive, and may pose health hazards. In view of the limitations of these methods of *in situ* determinations of soil moisture content, remote sensing is a potentially useful technique to assess soil moisture for a surface layer about five cm thick (see later in this paper for further details). The data on soil moisture for a shallow layer can be useful in evaluating moisture content for seed germination and seedling establishment. Attempts also have been made to relate moisture content in the surface layer to that of the root zone – about one m deep (Jackson *et al.,* 1982; Blanchard *et al.,* 1981). It is important to realize, however, that remote sensing does not provide an accurate nor as deep a measurement of soil moisture as can be obtained by conventional *in situ* methods. Nonetheless, remote sensing offers a technique to measure the average moisture content over a large watershed. This type of data also can improve runoff forecasting for large basins in addition to developing information on crop water use.

SEDIMENT TRANSPORT AND SOIL EROSION

A considerable amount of research data is available on soil erosion measured on field plots established on different soils, slope lengths and gradients, and management variables. However, long-term records of erosion and sediment transport are not available for medium- and large-sized watersheds. Sediment transport from a well-defined hydrologic unit is influenced by change in land-use. Deforestation and conversion of TRF to arable land-use increases the rate and total amount of sediment transport (Lal, 1986).

Modelling Sediment Load

Many empirical models have been developed to relate sediment yield to the characteristics of rainfall, runoff, drainage density and geomorphology (Anderson & Burt 1985). Fournier (1960) related sediment yield to the climatic coefficient ($p^2 P^{-1}$), altitude and slope (Equation 9):

$$\log SY \,(\mathrm{t\,km^{-2}\,yr^{-1}}) = 2.65 \log p^2 P^{-1}(\mathrm{mm})$$
$$+ 0.46 \log H \tan f - 1.56 \qquad (9)$$

where SY is suspended sediment yield, H is the mean height, f is the mean slope in a drainage basin, p is the rainfall (mm) in the month of the maximum rainfall and P is the mean annual rainfall. Douglas (1967) developed a similar relationship relating suspended load to p^2P^{-1} for watersheds in eastern Australia (Equation 10) and to runoff (Equation 11):

$$\log SY = 0.737 \log p^2 P^{-1} + 0.380, \quad r = 0.69 \qquad (10)$$

$$\log SY = 0.527 \log Q + 0.067 \quad r = 0.498 \qquad (11)$$

where SY is sediment yield in *$m^3\ km^{-2}\ year^{-1}$* and Q is the runoff in mm and r is the correlation coefficient. Because runoff decreases the proportion of rainfall retained on the watershed, sediment yield can also be related to the effective rainfall. One such model for some drainage basins in Queensland, Australia, was developed by Douglas (1968). In addition, Jansen & Painter (1974) proposed empirical models relating sediment yield from tropical watersheds to runoff discharge, area, relief and mean annual temperatures. Rose (this volume) discusses in detail recent developments in erosion process modelling.

Although accelerated soil erosion is widely recognized to be a serious problem in the humid tropics, reliable assessment of the magnitude of erosion and its impact remains a major challenge. Information readily available in the literature often is based on reconnaissance surveys and extrapolations based on sketchy data. Most techniques are expensive, labour-intensive and time-consuming. Therefore, methods that utilize remote sensing data and deterministic models can potentially supply new information to help estimate sediment load (Lyon *et al.*, 1988a,b; Lyon, 1987). In comparison with conventional methods of measuring sediment transport and erosion, remote sensing and aerial photographs offer an alternative. There are some limitations of this technique for use in the humid tropics. Firstly, perpetual cloud cover is a problem that can possibly be solved by selecting an appropriate spectrum for obtaining the photograph. Secondly, satellite water-resource studies are often limited to one date of coverage. However, to develop potential operational uses of satellite data for measurement and modelling of water resources requires multiple-day coverage. High frequency data of water characteristics also is necessary to assess sediment transport.

The rate of deforestation of TRF and percentage of ground cover also can be assessed by remote sensing (Booth, 1989). It is a useful tool, provided that photographs for the same site can be obtained for consecutive years. The images from LANDSAT data, however, may be expensive. Each frame, with a resolution of 1 kilometre, costs about USA $3,600. It would, for example, require more than 200 frames just to survey the Amazon Basin (Booth, 1989). Therefore, it would be more appropriate to combine LANDSAT images with ground-truthing, or other remote sensors that are more economical.

Delivery ratio

The term "delivery ratio" is defined as the ratio of sediment delivered at the watershed outlet to gross erosion within the watershed. Information on delivery ratio is a necessary prerequisite to convert the data on sediment load to average soil erosion over the watershed. The delivery ratio depends on many factors, including watershed area, predominant soil types, slope, parent material, geomorphological and environmental factors, vegetation cover and land-use. Some empirical relationships have been developed between "delivery ratio" and watershed characteristics. However, these empirical relationships are locally specific and vary widely from one watershed to another (Walling, 1983). The sediment delivery ratio may range from as low as 5% for large watersheds to about 70% for small plots. Erroneous results can be obtained by assuming a wrong value of the delivery ratio.

Research information on delivery ratio for representative watersheds in the tropics is not known. Furthermore, there is a need to develop a reliable model to estimate delivery ratio for a wide range of watersheds representative of important geomorphological, environmental and land-use variables.

CHANGE IN LAND-USE AND HYDROLOGICAL PROBLEMS

Population of the TRF ecosystem is expected to make up almost 33 per cent of the total world population of about 6.5 billion by the year 2000 (Gladwell & Bonell, 1990). Consequently, the rain forest ecosystem has and is rapidly being changed to meet the ever-increasing demands to other land-uses, *e.g.* human-made forests, perennial crops and plantations, arable and pasture lands. Alterations in land-use have drastic effects on different components of the hydrologic cycle. Possible changes are conceptualized in Figure 2. The magnitude of these changes, however, depends on several interacting factors, *e.g.* rainfall amount and its distribution, soil properties and horizonation, the land-use adopted and its intensity, terrain characteristics, the proportion of source area and the riparian zone, and the management system(s) adopted.

The concepts outlined in Fig. 2 have not been widely evaluated within the TRF. Although not very well established, it is now recognized by some that deforestation may affect local and regional circulation patterns of the air and the amount of rainfall received (Salati & Vose, 1984; Meher-Homji, 1988; Shukla *et al.*, 1990). If the land-use is human-made forest or plantation crops, the long-term effect may be negligible. The effects on rainfall may, however, be measurable if the land-use is pasture, food crop seasonals or annuals. Experiments to evaluate those effects have been conducted for relatively short periods and at few locations. Therefore, the results should be interpreted with caution.

The effects of deforestation on water yield have been the subject of interest for several studies (Lal, 1983, 1987; Lal & Russell, 1981; Bruijnzeel, 1990). Results of some relevant studies are outlined in Tables 1 and 2. Major findings of these studies are summarized below:

(a) Reduction of forest cover and conversion to arable land-use:

    (i) increases the frequency and rate of peak flow, duration and amount of interflow, and total annual discharge or accumulative water yield;

    (ii) increases source area contribution;

    (iii) raises ground water, and may increase the soil-water storage in the subsoil; and

    (iv) decreases the soil water storage in the surface layers.

(b) Low-intensity disturbances (selective logging, harvesting of minor products or establishing perennial crops with full canopy cover and deep root system may:

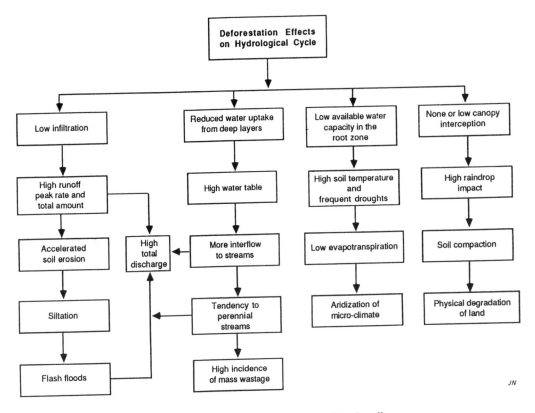

Figure 2: Deforestation-induced changes in hydrological cycles and land attributes

(i)   have little long-term effect on runoff rate and amount, and total annual discharge;

(ii) cause drastic initial increase in water yield including high-overland flow and interflow; and

(iii) lead to transient changes in landscape stability, causing mass wastage and high sediment load in overland flow.

(c) Afforestation of scrublands and non-forested catchments may:

(i)   decrease water yield;

(ii) decrease peak flow rates and total annual discharge;

(iii) decrease interflow volume and rate; and

(iv) decrease rill-interrill erosion, stabilize the landscape and reduce mass wastage.

(d) Surface erosion in TRF ecosystems is generally low (Lal, 1983, 1987; Roose, 1977; Wiersum, 1984). Conversion of forest to crop land and grazed pastures or grasslands may:

(i)   increase sediment density;

(ii) increase dissolved load in surface runoff and interflow;

(iii) accentuate soil compaction and structural degradation; and

(iv) decrease soil organic matter, species diversity and activity of soil fauna.

Results of these finds have not been widely adopted in the development of practical technology for sustainable management of soil and water resources.

REMOTE SENSING, GEOGRAPHIC INFORMATION SYSTEMS AND HYDROLOGIC PROCESSES

Remote sensing involves deriving information from images about objects at a certain distance without making contact with the object. It relies on measurements of electromagnetic energy reflected or emitted by the object. Furthermore, objects can be distinguished from one another due to their spectral differentiation. Remote sensing, therefore,can be used to evaluate soil colour and wetness, inundated areas, colour of the flow and of the large bodies of water. Soil colour evaluated by remote sensing can be related to organic matter content, iron oxide content and other soil properties relevant to hydrological processes (Latz *et al.*, 1984). Remote sensing is also a useful tool to evaluate the *C*-factor of the Universal Soil Loss Equation, spatial and temporal changes in landscape due to mass wastage and ephemeral gully erosion, evaluation of gully formation patterns and estimating the areas of eroded and deposited soil (Thomas & Welch, 1988). Lyon *et al.* (1988a) also used remote sensing techniques to assess suspended sediment in streams and lakes. Monitoring changes in terrain profile is another useful application of remote sensing (Ritchie & Jackson, 1989).

Computer-coded land information systems can be used to reduce the difficulties involved in manual overlay techniques. Computer-aided modern techniques are rapid and accurate (Tomlinson, 1976; Marble, 1984a). A computer-coded land

TABLE 1. *Annual water yield as influenced by deforestation and change in land-use.*

Land-use	Country	Year of Observation After Land-use Change	Change in Annual Water Yield (%, mm/yr)	References
1. Selection logging	Malaysia	3rd	+ 44 – 72%	Baharuddin (1988), Gilmour (1977a,b), Luvall (1984)
2. Plantation crops				
(i) Eucalyptus	Guyana	3rd	+ 40 mm	Fritsch (1983), Fritsch & Sarrailh (1986)
(ii) Pinus	Guyana	3rd	+ 210 mm	Fritsch (1987)
(iii) Cocoa	Malaysia	4th	+ 158%	Abdul Rahim (1987, 1988)
(iv) Oil Palm	Malaysia	4th	+ 470%	Abdul Rahim (1987, 1988)
(v) Tea	Kenya	6th (mean	+ 150 mm	Blackie (1979a)
(vi) Bamboo	Kenya	7th (mean)	+ 5 mm	Blackie (1979b)
3. Arable landuse				
(i) Seasonal crops	Nigeria	3rd	+ 140 mm	Lal (1983)
(ii) Cropping	Zambia	5th	+ 56 – 74%	Mumeka (1986)
4. Grassland				
(i) Paspalum	Congo	?	+ 14%	Focan and Fripiat (1953)
(ii) Imperata	Philippines	4 yr(mean)	+ 95%	Dano (1990)
(iii) Burned grassland	Madagascar	11 yr (mean)	+ 100 mm	Bailly et al. (1974)

Modified from Bruijnzeel (1990)

information system that is designed to input, store, manipulate and display this information is called a "Geographic Information System" (GIS) (Marble, 1984b; Burrough, 1986). Several GIS systems are available for use in hydrological studies (Marble, 1984b).

THE PROBLEM OF SCALE

Hydrological measurements of runoff, interflow, water and sediment yields, evapotranspiration, etc., are beset with the problems of scale. Field plots of 10 to 500 m² are used to measure runoff and erosion along a hillslope on a "microscale." Effects of agricultural practices of soil and crop management on hydrological parameters are usually assessed on small watersheds of 0.5 to 5 ha on a "mesoscale." Effects of climate, parent rocks, lithology, and land-use on runoff and sediment transport are measured on river basins of thousands of km² on a "macroscale." Methodology, units of measurement, and instrumentation involved are different at different scales with the choice of appropriate scale of measurements apparently depending on the objectives. However, budgetary considerations play a major role in the decision-making process.

The results of hydrological measurements are scale-depen-

dent, and the extrapolation of results from one to another scale poses a major research challenge (see Bonell with Balek; Rose, this volume). Hydrological parameters and processes at one scale are usually not important and/or predictive at another scale. Hydrological problems often require the extrapolation of plot-scale measurements for evaluation of watershed-scale phenomena. Furthermore, understanding hydrological systems and the effects of anthropogenic perturbations on spatial and temporal heterogeneity requires evaluations and integrations across units that differ in their size, shape and arrangement. Despite the widely acknowledged importance of scale-dependence of hydrologic processes, scaling rules have not been developed and limits to extrapolation have been difficult to identify. It is, therefore, important to develop research techniques that will represent information across scales or quantify the loss of relevant information with changing scales from micro- to macro-levels. Such methods are necessary for hydrological insights of the processes involved between spatial and temporal scales. Gross errors can be made if the results of erosion measurements of field plots are applied to a watershed or a river basin. Similarly, using information on sediment yield without consideration of the delivery ratio can lead to erroneous interpretation of the erosion hazard.

TABLE 2. *Peak flow rates and storm flow volume as influenced by change in land-use.*

Land-use	Country	Storm Flow Volume (mm/yr or %)	Peak Flow (%)	Reference
1. Clearcut logging	Taiwan	Not significant	+ 48%	Hsia (1987)
2. Plantation				
(i) Tea	Kenya	Small absolute increase	Small absolute increase	Dagg & Pratt (1962)
(ii) Eucalyptus*	India	−28%	−73%	Mathur et al. (1976)
(iii) Clearing for Eucalyptus	Guyana	− 330 to 410 mm	–	Fritsch (1983, 1987)
(iv) Clearing for Pine	Guyana	+ 385 to 495 mm	–	Fritsch & Sarrailh (1983)
(v) Cocoa	Malaysia	− 26 to 21%	+ 280%	DID (1986, 1989)
(vi) Oil Palm	Malaysia	+ 19 to 37%	+ 17 – 65%	DID (1986, 1989)
3. Grassland				
(i) Pasture	Guyana	+ 235 mm	–	Fritsch (1983, 1987)
(ii) Burned grassland	Madagascar	+ 100 mm	+ 66%	Bailly et al. (1974)
4. Arable				
(i) Improved cropping	Madagascar	− 75 mm	+ 53%	Bailly et al. (1974)
(ii) Mechanized farming	Nigeria	Increase	Increase	Lal (1983)

* Eucalyptus plantation was established on degraded scrub land.
  Modified from Bruijnzeel (1990)

RESEARCH PRIORITIES

Land degradation is a serious problem in the humid tropics. It undermines capacity of tropical ecosystems to feed their growing human and cattle populations, and to preserve the environment. FAO/UNEP (1983) estimated that at present rates, five to seven million ha of cultivated area (0.3–0.5%) are being lost every year through degradation, of which a large proportion takes place in the humid tropics. The different processes of land degradation prevalent in the humid tropics are vegetation degradation due to clearance of TRF and conversion to other uses, degradation and pollution of natural waters including those of wetlands and swamps, and soil degradation due to erosion, high cohesion through hardsetting or drying, thus reducing infiltration and percolation and fertility depletion. Clearance of TRF may also have a significant effect on the emission of radiatively-active gases into the atmosphere. Important gaseous emissions include CO_2, CH_4, N_2O and NO_x. These gases are partly released by burning and decomposition of biomass, but mostly by decomposition and mineralization of soil organic matter and humified fractions of the biomass.

Land degradation and hydrological processes are interlinked through cause-effect relationships. For example, alterations in different components of hydrological balance (runoff, interflow, evaporation, or soil-water storage) change the vegetation, cause the erosion or leaching of nutrients, and disrupt the cycles of carbon, nitrogen, sulfur, and other major elements. Although research into the atmospheric and environmental impacts of anthropogenically-induced perturbations in hydrologic parameters is at an incipient stage, it has far-reaching global consequences in terms of the "greenhouse" effects through emission of CO_2 from biota and CH_4 from marshes and wetlands.

There is a need to develop research networks to study hydrological processes in tropical drainage to begin to address some of these issues. In addition to basic research needs and knowledge gaps discussed in each section, there is a need to organize *multidisciplinary* teams to study hydrological processes, especially those related to agriculture, forestry, and wetland eco-systems. Investigations of hydrological parameters of eco-systems that are being restored/rehabilitated also should be given a priority. It is relevant to establish quantitative hydrological evidence of the rate and extent by which "stability" of tropical drainage basins and other eco-systems may be improved under increasing demographic pressure. Above all, there is a need to develop methodology for extrapolating hydrological processes across scales, and to improve our predictive capabilities through conceptual or process-driven models. Remote sensing is a tool that should be used extensively in studying hydrological processes, their temporal and spatial variation over a watershed, and their effects on land degradation and the environment.

REFERENCES

Abdul Rahim, N. (1987) The impact of forest conversion on water yield in Peninsular Malaysia. Paper presented at the workshop on Impact of Operations in Natural and Plantation Forests on Conservation of Soil and Water Resources. Serdang, 23–26 June 1987.

Abdul Rahim, N. (1988) Water yield changes after forest conversion to agricultural land-use in Peninsular Malaysia. *J. of Tropical Forest Science 1*: 67–84.

Anderson, M.G. & Burt, T.P. (1985) *Hydrologic Forecasting*, J. Wiley & Sons, New York.

Baharuddin, K. (1988) Effect of logging on sediment yield in a hill dipterocarp forest in Peninsular Malaysia. *J. of Tropical Forest Science 1*:56–66.

Bailly, C., Benoit De Cognac, G., Malvos, C., Ningre, J. M. & Sarrailh, J. M. (1974) Etude de l'influence du couvert naturel et de ses modifications à Madagascar; experimentations en bassins versants élémentaires. Cahiers Scientifiques du Centre Technique Forestier Tropical No. 4:1–114.

Blackie, J. R. (1979a) The water balance of the Kimakia catchments. *East African Agricultural and Forestry J. 43*:155–174.

Blackie, J. R. (1979b) The water balance of the Kericho catchments. *East African Agricultural and Forestry J. 43*:55–84.

Blanchard, B. J., McFarland, M. J., Schmugge, T. J. & Rhoades, E. (1981) Estimation of soil moisture with API algorithms and microwave emission. *Water Resour. Bull. 17*:767–774.

Bonell, M. & Gilmour, D. A. D. A. (1978) The development of overland flow in a tropical rain forest catchment. *J. Hydrol. 39*:365–382.

Bonell, M. & Williams, J. (1986) The two parameters of the Philip infiltration equation: Their properties and spatial and temporal heterogeneity in a red earth of tropical semi-arid Queensland. *J. Hydrol. 87*:9–31.

Bonell, M., Gilmour, D. A. & Sinclair, D. F. (1979) A statistical method of modelling the fate of rainfall in a tropical rain forest catchment. *J. Hydrol. 42*:251–267.

Bruijnzeel, L. A. (1990) Hydrology of moist tropical forests and effects of conversion: a state of knowledge review. IHP, UNESCO, Paris.

Burrough, P. A. (1986) Principles of Geographic Information Systems, Methods and Equipment for Land-use Planning. USGS, Reston, VA.

Booth, W. (1989) Monitoring the fate of the forests from space. *Science 243*:1428–1429.

Chang, J-h. & Lau, L. J. (1986) *Definition of the Humid Tropics.* UNESCO, Div. Water Sciences, IHP 4.2, Paris, Annex 2.

Dagg, M. & Pratt, M. A. C. (1962) Relation of stormflow to incident rainfall. *East African Agricultural and Forestry J. 27*:31–35.

Dano, A. M. 1990) Effect of burning and reforestation on grassland watersheds in the Philippines. IAHS Publ. 192:53–61.

DID (1986) Sungai Tekam Experimental Basin. Transition Report July 1980 to June 1983. Water Resources Publ. No. 16. Drainage and Irrigation Department, Ministry of Agriculture, Kuala Lumpur, Malaysia.

DID (1989) Sungai Tekam Experimental Basin. Final Report July 1977 to June 1986. Water Resources Publ. No. 20. Drainage and Irrigation Department, Ministry of Agriculture, Kuala Lumpur, Malaysia.

Douglas, I. (1967) Natural and man-made erosion in the humid tropics of Australia, Malaysia, and Singapore. *Int. Assoc. Sci. Hydrol. 75*:17–30.

Douglas, I. (1968) Erosion in the Gungei Gombak Catchment, Selangor, Malaysia. *J. Trop. Geogr. 26*:1–16.

FAO/UNEP (1983) Guidelines for the control of soil degradation. Food and Agric. Organization (Rome) and United Nations Env. Program. (Nairobi).

Focan, A. & Fripiat, J. J. (1953) Une année d'observation de l'humidité du sol á Yangambi. *Bulletin des Seances de l'Institut Royal Colonial Belge 24*:971–84.

Fournier, F. (1960) *Cimat et Erosion: La relation entre l'erosion du sol par l'eau et les precipitations atmospheriques*, PUF, Paris.

Fritsch, J. M. (1983) Evolution des écoulements, des transport solides a l'exutoire et de l'érosion sur les versants d'un petit bassin après défrichement mécanisé de la forêt tropical humide. *IAHS 140*:197–214.

Fritsch, J. M. (1987). Ecoulements et érosion sous prairies artifielles après défrichement de la forêt tropicale humide. *IAHS 167*:123–130.

Fritsch, J. M. & Sarrailh, J. M. (1986) Les transports solides dans l'écosystème forestier tropical humide en Guyane: les effets di défrichement et de l'installation de paturages. Cah. ORSTOM, *Série Pédologie 22*:93–106.

Gilmour, D. A. (1977b) Logging and the environment, with particular reference to soil and stream protection in tropical rain forest situations, Conservation Guide No. 1, Food and Agric. Organization, Rome. 223–235.

Gilmour, D. A. (1977a) Effect of rain forest logging and clearing on water yield and quality in a high rainfall zone of northeast Queensland. Institute of Engineers of Australia: Symp. on the Hydrology of Northern Australia, Brisbane, Qld. Natl. Conf. Publ. No. 77/5:155–160.

Gilmour, D. A. & Bonell, M. (1977) Streamflow generation processes in a tropical rain forest catchment – Preliminary assessment. Institute of Engineers of Australia: Symp. on the Hydrology of Northern Australia, Brisbane, Qld. Natl. Conf. Publ. No. 77/5:178–179.

Gilmour, D. A. & Bonell, M. (1979) Runoff processes in tropical rain forests with special reference to study in northeast Australia. In: A. F. Pitty (ed.)"*A Geographical Approach to Fluvial Processes*," Geobooks, Norwich.

Gladwell, J. S. & Bonell, M. (1990) An international programme for environmentally sound hydrologic and water management strategies in the humid tropics. Paper presented at the Symp. on Tropical Hydrology, San Juan, Puerto Rico, 23–27 July, 1990.

Golley, F. B., McGinnis, J. T., Climents, R. G., Child, G. I. & Duever, M. J. (1975) *Mineral cycling in tropical moist forest ecosystems.* Univ. Georgia Press, Athens, Georgia, USA.

Hibbert, R. A. (1967) Forest treatment effects on water yield. Proc. Intl. Symp. on Forest Hydrology. Penn. State Univ., Pergamon Press: 537–543.

Horton, R. E. (1933) The role of infiltration in hydrological cycles. *Trans. AGU* IU:446–460.

Horton, R. E. (1945) Erosional development of streams and their drainage basins: hydrological approach to quantitative morphology. *Bull. Geo. Soc. Am. 56*:275–330.

Hsia, Y. J. (1987) Changes in storm hydrographs after clearcutting a small hardwood forested watershed in central Taiwan. *Forest, Ecology & Management 20*:117–134.

Hursh, C. R. (1953) Land-use and stream flow. *East African Agricultural and Forestry J. 53*:139–145.

Jackson, T. J., Schmugge, T. J. & Wang, J. R. (1982) Passive microwave sensing of soil moisture from surface measurement. *J. Irrig. and Drain. Div.*, IR–2, ASCE: 81–92.

Jansen, J. M. L. & Painter, R. B. (1974) Predicting sediment yield from climate and topography. *J. Hydrol. 21*:371–380.

Kostiakov, A. N. (1932) On the dynamics of the coefficient of water percolation in soils and on the necessity for studying it from a dynamic point of view for purposes of amelioration. Trans. VIIth Int. Cong. Soil Sci. Part A: 17–21.

Lal, R. (1979) Physical characteristics of soils in the tropics: Determination and management. In: R. Lal & D. J. Greenland (eds.) *"Soil Physical Properties and Crop Production in the Tropics:*, J. Wiley & Sons, U.K. 7–46.

Lal, R. (1983) Soil erosion in the humid tropics with particular reference to agricultural land development and soil management. IAHS *140* 221–239.

Lal, R. (1986) Conversion of tropical rain forest: Agronomic potential and ecological consequences. *Adv. Agron. 39*:173–264.

Lal, R. (1987) *Tropical Ecology and Physical Edaphology*. J. Wiley & Sons, U.K.

Lal, R. & Russell, E. W. (eds.). (1981) *Tropical Agricultural Hydrology*. J. Wiley & Sons, U.K.

Latz, K., Weismiller, R. A., Van Scoyoc, G. E. & Baumgardner, M. F. (1984) Characteristic variations in spectral reflectance of selected eroded Alfisols. *Soil Sci. Soc. Am. J. 48*:1130–1134.

Lawson, T. L., Lal, R., & Oduro-Afriyie, K. (1981) Rainfall redistribution and microclimatic changes over a cleared watershed. In: R. Lal and E. W. Russell (eds.) *"Tropical Agricultural Hydrology,"*, J. Wiley & Sons, U.K.: 141–151.

Luvall, J. C. (1984) Tropical Deforestation and Recovery: The Effects on the Evapotranspiration Process. Ph.D. thesis, University of Georgia, Athens, GA, USA.

Lyon, J. G. (1987) Use of maps, aerial photographs, and other remote sensor data for practical evaluations of hazardous waste sites. Photogrammetric Eng. & Remote Sensing *53*(5):515–519.

Lyon, J. G., Bedford, K. W. & Yen, J. C. C. (1988) Determination of suspended sediment concentrations from multiple day Landsat and AVHRR data. *Remote Sensing of Environment 25*:107–115.

Lyon, J. G., Mitchell, C. A. & Zobeck, T. M. (1988b) Impulse radar for identification of features in soils. *J. Aerospace Eng.* 1(1):19=8–27.

Marble, D. F. (1984a) Geographic Information Systems. An Overview. Proc. Spatial Information Technologies for Remote Sensing Today and Tomorrow, 2–4 Oct. 1984, Sioux Falls, SD: 18–24.

Marble, D. F. (1984b) Introduction In *Basic Readings in GIS* (D. F. Marble, H. W. Calkins, and D. J. Peuquet, (eds.), SPAD, Williamsville, NY.

Mathur, H. N., Ram Babu, Joshi P. & Singh, B. (1976) Effect of clear felling and reforestation on runoff and peak rates in small watersheds. *Indian Forester 102*:219–226.

Meher-Homji, V. M. (1988) Effects of forests on precipitation in India. In:

E. R. C. Reynolds & F. B. Thompson (eds.). *Forests, Climate and Hydrology: Regional Impacts*. 51–77. The United Nations University, Tokyo.

Mumeka, A. (1986) Effect of deforestation and subsistence agriculture on runoff of the Kafue river headwaters, Zambia. *Hydrol. Sci. J. 31*:543–554.

Nortcliff, S. & Thornes, J. B. (1977) Water and cation movement in a tropical rain forest environment. I. Objectives, experimental design, and preliminary results. London School of Economics, Graduate School of Geography, Discussion Paper 62.

Pereira, H. C. (1973) *Land-use and water resources*. Cambridge Univ. Press.

Philip, J. R. (1957) The theory of infiltration. *Soil Sci. 83*:345–357, 435–448, *84*:163–177, 257–264, 329–339; *85*:278–286, 333–336.

Pinker, R. (1980) The microclimate of a dry tropical forest. *Agricultural Meteorology 22*: 249–265.

Richards, P. W. (1952) *The Tropical Rain Forest*. Cambridge Univ. Press, London.

Ritchie, J. C. & Jackson, Th. L. (1989) Airborne laser measurements of the surface topography of the simulated concentrated flow gullies. *Trans. ASAE 32*:645–649.

Roose, E. J. (1977) Application of the USLE in West Africa. In: D. J. Greenland and R. Lal (eds.) *"Soil Conservation and Management in the Humid Tropics,"* J. Wiley & Sons, U.K.: 177–188.

Salati, E., Dall 'Olio, A., Matsui, E. & Gat, J. R. (1979) Recycling of water in the Amazon Basin: an isotopic study. *Wat. Resour. Res. 15*:1250–1258.

Salati, E. & Vose, P. B. (1984) Amazon Basin: a system in equilibrium. *Science 225* (4658):129–138.

Shukla, J., Nobre, C. & Sellers, P. J. (1990) Amazon deforestation and climatic change. *Science 247*:1322–1325.

Thomas, A. W. & Welch, R. (1988) Measurement of ephemeral gully erosion. *Trans ASAE 31*:1723–1728.

Tomlinson, R. E., Calkins, H. W. & Marble, D. F. (1976) Computer handling of geographical data. National Resources Research Report No. 13, UNESCO (Paris).

Walling, D. E. (1983) The sediment delivery problem. *J. Hydrol. 65*:209–237.

Wiersum, K. F. (1984a) Surface erosion under various tropical agroforestry stems. In: C. L. O'Loughlin & A. J. Pearce (eds.) Proc. Symp. on Effects of Forest Land Use on Erosion and Slope Stability. IUFRO, Vienna, and East-West Centre, Honolulu, Hawaii.

Williams, J. & Bonell, M. (1988) The influence of scale of measurement on the spatial and temporal variability of the Philip infiltration parameters in an experimental study in the Australia savannah woodland. *J. Hydrol. 104*, 33–51.

18: The Impact of Land-use Change on Water Resources in the Tropics: An Australian View of the Scientific Issues

P. M. Fleming

Division of Water Resources, CSIRO, Australia

ABSTRACT

Northern Australia only includes the wet-dry tropical subregion of Chang and Lau's humid tropics; the remainder is within the dry tropics. However, the continent experiences some of the highest short-term rainfalls in the humid tropics which have significant ramifications for the hydrology and requirements for land conservation. This paper describes the limited Australian experience in change in land-use hydrology within the wet-dry tropics, and also devotes some attention to temperate Australian work, e.g., hydrology of eucalypt trees, to evaluate the potential for technology transfer. The account will commence with a brief statement of the four basic balance equations in hydrology and the characteristics which are relevant in the humid tropics and the general nature of Australia's research in these fields.

INTRODUCTION

The colloquium was held in Townsville in Northern Queensland at a latitude of 19°S and so was at the poleward limit of the humid tropics. However, the combination of geomorphology and seasonal wind patterns means that true humid tropic climate reaches within 100 km of Townsville. Significant hydrologic research on both basic hydrology and landscape management in the humid tropics also is based in Townsville.

This paper complements the basic paper on this topic by Lal (this volume) and presents an Australian viewpoint. It was prepared originally at short notice because ill-health seemed likely to preclude any contribution to the colloquium by Rattan Lal. The paper has now been modified to emphasize the nature of Australian research into land-use impacts generally. The limited nature of tropical rain forest and humid tropical hydrology in Australia means that this paper will focus in particular on the possibilities of technology transfer. Because of the nature of the continent, most Australian hydrology studies have concentrated on subhumid and semiarid climates and landscapes. Australia has the advantage, however, that virtually all its landscapes have been managed for agriculture and the pasturage of domestic animals for less than 150 years. Cultural transformations are all very recent. Subhumid and semiarid landscapes are inherently sensitive to land-use changes and because of the tectonic stability of the continent, the soils and land surfaces are old, and fragile when disturbed. Additionally, there is clear evidence that climatic change over the last 250 000 years has meant that landscapes and vegetation over much of northern Australia have changed from those typical of the humid tropics to seasonally dry tropics. Thus, landscapes currently with true humid tropical rain forest were, less than 10 000 years ago, dry sclerophyll *Eucalyptus* woodland (Kershaw, 1978). However, these transitions may well be the total humid tropic experience (Colinvaux, 1989).

Hydrological research in Australia has always had a strong theoretical base because it has been our experience that many of the empirical and conceptual assumptions of northern hemispheric humid temperate hydrology are not valid in Australia (Finlayson & McMahon, 1988). This is also true of tropical hydrology.

Therefore, this paper will commence with a brief statement of the four basic balance equations in hydrology and the characteristics which are relevant in the humid tropics and the general nature of Australian research in these fields. We will then address a series of topics closely related to those dealt with by Lal but concentrating on Australian examples as far as possible.

THE THEORETICAL BASIS OF TROPICAL HYDROLOGY

There are basically four balance equations which have to be understood in the light of any hydrological response or research. The essence of the scientific method is, of course, to conduct actual experiments which attempt to constrain as many elements as possible and allow variation in the remainder to be closely controlled or observed.

Many, indeed most, of the processes involved in these equations are highly non-linear, with complex feedback paths to other elements. However, the simple linear expressions of the balance equations are valid and sufficient to set the broad framework of understanding.

The balance equations

(i)   The water balance
(ii) The energy balance
(iii) The salt balance
(iv) The sediment balance

These all interact and modulate the whole hydrological cycle, and its expression in any situation. I will now state the balances as simple linear equations.

The water balance equations

$$P = E + SRO + IF + GWF + \Delta SS + \Delta MS + \Delta GWS \qquad (1)$$

$$= E + RO + \Delta S \qquad (1a)$$

where P = precipitation
E = evaporation
SRO = surface runoff or rapid response flow
IF = interflow runoff or intermediate response flow
GWF = groundwater flow or baseflow
ΔSS = change in surface storage
            which includes IS interception store
            SDS surface depression store
            FWS free water surfaces
ΔMS = change in mantle storage
            which includes
            RZ root zone store
            TZ transitional zone store
            PGW perched groundwater store
ΔGWS = change in the regional groundwater.

In equation (1a) SRO, IF and GWF are combined in RO and ΔSS, ΔMS and ΔGWS in ΔS for simplicity.

The energy balance equation

$$LE = RN - A - \Delta HS \qquad (2)$$

where LE = latent heat energy of evaporation
RN = net radiation energy
A = sensible heat loss to the airstream
ΔHS = change in stored energy
             (Principally applies to water in ponds, lakes and reservoirs.)

The net radiation energy can be broken down into three components.

$$RN = RS(1\text{-}a) + RLD - RLU \qquad (3)$$

where RS = incoming solar radiation
a = the albedo or short wave reflection
RLD = long wave (heat) radiation from the atmosphere
RLU = long wave radiation emitted from the surface.

The salt balance equation

$$P*PC = SRO*SRC + IF*IFC + GWF*GWC + \Delta MSS \qquad (4)$$

where P, SRO, IF, GWF are the elements of the water balance equation
PC, SRC, IFC and GWC are the respective concentrations of salts
ΔMSS is the change in mantle salt storage.

The most important point to note is that evaporation appears in equations (1) and (2) but not in (3) because the evaporation flux can carry no salt. It is, however, the key element in salinization because of this very fact.

The sediment or material balance equation

This is not so simply expressed because the fluxes do not easily convert to losses per unit area or unit time. However, it can be expressed as follows for simple water transport or material through a gauging station.

$$\Delta SL = SRO*SSC + BL \qquad (5)$$

where ΔSL = change in soil loss on the whole catchment
SSC = the suspended sediment concentration in the surface runoff
BL = the bed load which is also transported by high flow surface runoff.

In general, interflow and baseflow do not transport material. Even more than water or salt there is redistribution in the landscape and the concept of a sediment delivery ratio is widely used which is the ratio of ΔSL to the actual volume stripped from eroding areas. Sediment delivery ratios in the short-term are often quite small with eroded material being redeposited in lower parts of the landscape or in streams and often only moved out of the catchment in major episodes (see Rose, this volume).

There is, however, one subsurface process which does give rise to significant sediment transport, and that is pipeflow. The classical extreme event is piping failure of earth dams but in subhumid and semiarid regions, with texture contrast soils, gully head movement is commonly by successive piping failures. See Eyles (1977) for a southern Australian example. The northern hemisphere humid experience is not dissimilar (McCaig, 1984), and there is anecdotal evidence in tropical forests (Herwitz, 1986) and Bonell (pers. comm.).

Changes in land-use often initiate major erosion events as do extreme natural events such as very heavy rainfall, very strong winds, earthquake movements and wildfires. Although wildfires do not occur often in the humid tropics, they are much more common in the wet-dry tropics and the dry tropics. In Australia fire management is an essential part of land management (Walker *et al.*, 1986; Gill *et al.*, 1981).

Discussion of the water balance equations

Let us quickly look at equations (1) and (1a) in the humid tropical situation, both in the natural and in the man-made landscapes. We will also look at wetting-up events and drying-out events separately because precipitation and evaporation do not seriously interfere with each other, particularly in the humid tropics.

Wetting-up It is apparent, as E is ignored, in equation (1a) that RO is closely linked to P if P is large and ΔS is small. Conversely, if ΔS is large, then short-term changes in RO can be mainly controlled by the storage characteristics and there may be no apparent relationship between individual storms and runoff events. This is not uncommon in the tropics, particularly during the drier seasons of the year.

To look at equation (1) in greater detail, in tropical regions with tropical rain forest ΔSS can be quite large with storage in the canopy, the litter layer and the upper soil zone. If the rain forest is converted to rice paddy in terraces, then ΔSS is again very large but of different form. If it is converted to row crop or pasture, then ΔSS is greatly reduced and so the hydrograph peaks are enhanced.

We can link *IF* interflow with changes in ΔMS, although changes in mantle storage also contribute to ΔGWF and so GWF. Thus, good baseflow behaviour requires large values of MS and GWS with high infiltration capacity and good permeability at depth.

Drying conditions Here P is zero and so E and RO draw on the various stores. Evaporation first draws on the surface stores SS and through vegetation and capillary suction to the soil surface, also on MS. If E is high all the year around, as it is in the humid tropics, then during dry intervals, SS and MS may be depleted even if quite large, and sustained streamflow depends on drainage from a large groundwater store.

Evaporation can only draw on GWS if it is close to the surface, which it must be where GWS contributes to GWF. If GWF is highly concentrated as in a spring, then the influence of E may be quite small. Frequently, however, GWF emerges from diffuse sources in swamps and seepage zones, and deep-rooted vegetation bordering such regions can draw on GWS and thus reduce GWF. This influence often appears as a diurnal variation in streamflow.

Discussion of the energy balance equation

Evaporation requires energy and in the humid tropics water is so freely available that the simple one-dimensional energy balance equations such as equation (2) are nearly always valid. If we only consider the land surface, then on a daily or weekly basis there is little storage of energy. Therefore, net radiation is partitioned into latent energy of evaporation and sensible heat involved in heating or cooling the air. These latter energy fluxes are determined by gradients in vapour pressure or wet-bulb depression and temperature, respectively, and appropriate transfer equations.

A useful equation can be derived by combining the transfer equations and the energy balance to give the Combination or Penman Equation (Fleming, 1987; Monteith, 1985).

$$LE = \frac{S}{S+g}\ RN + \frac{g}{S+g}\ f(u)\ (\text{s.d.}) \qquad (6)$$

$$\text{or}\ \frac{S}{S+g}\ RN + h\ D_z \qquad (6a)$$

which is often expressed also in the Priestley-Taylor form

$$LE = \alpha \frac{S}{S+g}\ RN \qquad (6b)$$

where $\dfrac{S}{S+g}$ and $\dfrac{S}{S+g}$ are temperature weighting functions; $f(u)$ and h are transfer functions; (s.d.) is the saturation deficit of the atmosphere at a reference height or the difference between actual vapour pressure, e_a, and saturation vapour pressure, e_{sat}; D_z is the wet bulb depression at the reference height. In the humid tropics $S/(S+g)$ is effectively constant at 0.75.

The coefficient α, sometimes called the Priestley-Taylor coefficient, is usually considered to lie between 1.2 and 1.3. This means that in the humid tropics about 90% of the radiant energy absorbed becomes latent heat, and only 10% is used in heating the air. The small diurnal range of air temperatures confirms this.

If we look at equation (3), albedo a and surface temperature are constant as is the term RLU. Therefore, net radiation RN depends on both solar radiation and downward long wave radiation which are principally modulated by cloudiness. Clouds reduce incoming solar radiation but increase RLD.

Therefore, the key to estimation of evaporation in the humid tropics is the measurement of net radiation, but failing that, solar radiation, which in turn can be estimated from sunshine hours or cloudiness and astronomical data (Fleming, 1987; Fleming & Wells, 1989).

Discussion of the salt balance equation

As indicated earlier, this equation cannot be considered in isolation since the role of evaporation which removes water but on the other hand not salt is the key to its use to explain the appearance of salt at the surface of bare soil or in the root zone. In the humid tropics, this does not often occur because on a weekly-to-monthly-basis there is usually a water surplus which ensures leaching of soils and low values of MSS. However, as the humid tropics grade into the seasonally dry tropics, this equation becomes important in explaining such diverse water-use problems as the water requirements of salt production mud-flats, and local salinity outbreaks (Fleming & Wells, 1989).

Discussion of the sediment balance equation

Equation (5) is the basis of the classical method of determining sediment outflow from a catchment. The problem lies in the simultaneous measurements of discharge rates and integrated suspended sediment concentration. The normal procedure is to

measure stage height and use a rating curve to determine discharge rate and measure some index quantity like turbidity to
infer the sediment concentration. This requires some experimental data to establish good correlations.

Bed load is almost impossible to measure and is usually
calculated from formulae based on flume experiments.
Different formulae provide answers which may differ by a
factor of 5 commonly and up to 20 in cases where bed armouring can occur. Equation (5) should then be treated with great
caution. Isotope methods for partitioning contributions at
stream junctions are now becoming established and when
worked upstream from a reservoir where there is an absolute
collection of bed load may provide real bed load data at last
(Wasson & Murray, CSIRO Division of Water Resources,
pers. comm.).

SOME HYDROLOGICAL CHARACTERISTICS AND PROCESSES IN THE HUMID TROPICS AND IN AUSTRALIA GENERALLY

The introduction has expressed the idea that tropical hydrology is different and must be interpreted in the light of the four
balance equations. I have already established that evaporation
is almost a constant and high proportion of net radiation. This
means that correlations with the temperature, day length and
saturation deficit used commonly in temperate regions are
likely to be misleading.

The classical humid tropics are in the zone 7°N to 7°S.
Consequently, this zone is only rarely involved with
tropical storms linked with the monsoon trough such as
tropical cyclones. In contrast, these systems dominate the
rainfall of northern Australia. However, the definition
adopted for this symposium is a development by Chang
and Lau (Chang, this volume) and Stewart, (this volume).
The tropics are defined as the region having a minimum
mean monthly temperature in excess of 18°C. There are four
wetness classifications. Tropical Australia only has the two
lowest:
(i)   Dry tropics – total wet months less than 4–1/2.
(ii) Wet-dry tropics – wet months total 4–1/2 to 7.

Whilst the rainfall amounts fall below evaporation demand
in the drier months, rainfall events during the wet season can
be of much greater sustained intensity than the humid tropi-
cal/subhumid tropical subregions. In the latter areas, single
daily events – of a few hours duration at most – occur from
local convection cells.

In contrast, the convergence associated with monsoon
troughs, tropical lows and tropical cyclones, as well as just
persistent onshore winds in the higher latitudes, give rise to
successive rain cells which make daily totals in excess of 500
mm not uncommon (Stewart, this volume; Bonell, 1989). In
fact, 500 mm in 24 hours has a return period of about seven
years in Babinda near Cairns, northeast Queensland (see
Stewart, this volume).

Rainfall characteristics

The largest single coherent area of the humid tropical subregion lies in the Amazon Basin where rainfall from locally
organized convection events prevail. Some extraordinary
claims have been made with respect to proportion of rainfall
which is recycled in this Basin (Salati *et al.,* 1979). However,
an annual inflow from the North and South Atlantic equal to
the discharge of the Amazon is indicated and much of this
runoff originates in the deep interior. Molion (this volume)
presents a more detailed discussion.

Because tropospheric depth is greatest within the equatorial
zone and there is maximum conversion of latent to sensible
heat, very deep convection cells exist, and thus very high
intensity rainfall takes place. It is, however, also a characteristic of these rainfalls that they are very localized in time and
space because the weak Coriolis force precludes well-organized meteorological systems. Thus event and even daily and
weekly rainfall totals are poorly correlated over very short distances. This makes for very difficult hydrologic research and
modelling, and in respect of larger catchments, makes flood
warning and catchment behaviour studies even more difficult
than in temperate humid regions (Jackson, 1969).

Conversely, the very deep convection and localized nature
of the events means that equatorial geostationary satellites can
provide useful information about storm location and intensity
through monitoring the rate of change and actual cloud top
temperatures.

As indicated above, the Australian tropics, because of their
location between 10° and 20°S latitude are subject to monsoonal
influences and tropical storms and cyclones. The statistical characteristics have been very usefully set out by Pilgrim &
Canterford (1987) and the broader characteristics extensively
discussed by Sumner & Bonell (1986) and Bonell *et al.,* (1991).

Annual raindays in northern Australia are much less than in
the humid tropical subregion, but 24-hour totals greatly exceed
those of the near-equatorial areas so that the opportunity for
runoff generation is greatly increased. The heaviest rainfalls
are associated with tropical vortices, for example, tropical
cyclones; these are often quite distant from the track of the
storm centre. Extreme events usually add orographic effects to
storm circulation effects.

Interception and rainfall transformation

The primary vegetation of the humid tropics is Tropical Rain
Forest, *TRF,* which has a very complex layered structure with
a great diversity of species. It is typically subject to constant
change over all time scales from many impacts (see
Colinvaux, 1989 for a recent popular account). As Lal (this
volume) has indicated, leaf area ratios – *LAI,* or the ratio of
surface area of plant surfaces to ground cover, for *TRF* is in the
range of 7 to 25. This is three to six times that of a typical agricultural crop or even lush tropical pastures.

Rainfall entering a natural *TRF* will have less chance of
direct contact with the ground so interception volumes can be

expected to be quite large. Recent work by Herwitz has quantified the value of interception volumes in the north Queensland rain forest (Herwitz, 1985, 1986, 1987). Earlier work by Jackson (1975) in Tanzania showed that high intensity and large droplets make interception a function of rain intensity but as a proportion of volume it is quite small. Interception should always be examined as a volume per storm or per day and not as a percentage.

Herwitz has shown that the action of *TRF* interception is to redirect water into regular and characteristic paths with quite high volumes appearing as stemflow and concentrated driplines which often produce local regions of overland flow from modest rainfalls. Because most of the energy transfer in the humid tropics is as latent heat, there is little sensible heat remaining to heat the air and so be available to enhance evaporation. Therefore, we would expect little change in total runoff following clearing of *TRF*.

The data from Cassells (1984) on the Babinda Forest indicate an increment of annual runoff after clearing of nearly 300 mm. This probably reflects the fact that in the non-monsoon period significant drying of the mantle stores can occur. However, Bonell & Gilmour (1978) confirm that clearing made little difference to the storm volumes from high intensity events where changes in interception volumes had almost no effect.

Infiltration, deep percolation and runoff generation

The plant canopy protects the surface from direct rainfall energy effects, but generates large droplets, drip points and stemflow streams which localize water at the surface. The litter layer and surface humus layers have almost infinite infiltration capacity and most mineral soils in the tropics have large hydraulic conductivities. Therefore infiltration equations are largely meaningless (Bonell with Balek, this volume).

Besides high instantaneous rainfall rates, tropical storm events have very large total volumes. Thus, during significant events, the total available mantle and surface stores are filled, i.e. the profile is saturated. Gilmour (1975), and later, Bonell & Gilmour (1978), have shown that profile saturation is widespread and uniform in the Babinda catchments which are capable of generating widespread overland flow . This is believed to be the general surface runoff generation process, especially under monsoon rainfalls, in the humid tropics of northeast Queensland, both in forests and cleared cropland and grassland. I believe it should be termed Saturation Overland Flow, *SOF*. For example, Bonell *et al.,* (1983) looked at a wide range of tropical soil profiles in north Queensland and, on a basis of the hydraulic properties, showed it must be the normal runoff generation mechanism in this region.

I believe that this mechanism should be distinguished from the common mechanism of the humid temperate zone. In the latter zone, rainfall intensities and storm volumes are significantly less but so is evaporation. Most rainfall enters the mantle storage zones and moves downslope to eventually emerge as exfiltration in saturated zones of the lower landscape, contributing interflow and baseflow. These saturated, exfiltrating zones usually are close to stream courses and generate storm runoff because infiltration cannot occur.

This is the *Partial Area Concept* of Betson, which is better described now in the terminology of Hewlett and Dunne as the *Variable Source Area Concept* for the generation of stormflow runoff (See Bonell with Balek, this volume). I believe this should be called exfiltration zone overland flow, *EXOF*. O'Loughlin (1986) has shown how some simple assumptions with respect to slope characteristics can quantify the area of exfiltration zones and how this area can be correlated with pre-storm baseflow discharge rates.

This process of *EXOF* is probably relevant to small volume or low intensity storms in the humid tropics and in tropical grasslands and crop lands. The O'Loughlin analysis requires very detailed contour information which at present is unavailable in *TRF*. It is becoming possible to develop effective relief models from images created by side-looking airborne radar, *SLAR*. It is certainly possible to define stream nets in *TRF* from *SLAR*, which is a significant advance. The discussion in Bonell (1989) suggests that *EXOF* may well be a minor component in streamflow generation, especially in the headwater catchments of forested steeplands associated with the cyclone-prone humid tropics of northeast Queensland, and that the Bonell process of profile saturation is the more general mechanism.

Because of the fact that in *TRF* most of the nutrient pool is in the vegetation mass and high infiltration rates ensure rapid leaching, conversion to pasture or crop can often rapidly reach a stage of low fertility and changed hydraulic condition. This can mean that conventional *Hortonian Overland Flow, HOF,* can now occur where the rainfall intensity exceeds the infiltration rate and so generate rainfall excess long before the profile is saturated. Overland flow tens of centimetres in depth can be generated under these conditions from humid tropics rainfall.

The common crop of the humid tropics is, of course, rice, which is usually grown in bunded fields with cascading overflows. Here the great bulk of the landscape has a runoff capacity of 100%; however, the hydrograph is shaped by the overflow characteristics of the control weir on each bund and the storage capacity of the individual bunds. Such terrain, again, is a very different hydrologic regime from most natural temperate landscapes. It can be modelled today with cellular hydraulic models.

Models also exist for the behaviour of bunded landscapes during non-overflow periods (Walker & Rushton, 1984). It is believed, in part because of the widespread practice of paddy puddling, that most seepage flow occurs through and under the bunds and not through the floors of the paddy fields.

Rice was developed from natural grains occurring in humid tropical wetlands, with much of the lowland paddy being a conversion of natural wetlands. Water management in the humid tropics, therefore, involves a great deal of drainage control in the periods of excessive rainfall as well as water storage and supply during the drier periods to achieve maximum and preferably year-round production. The international institutes – IRRI in Los Banos, Philippines and IIME in Kandy, Sri Lanka – are addressing these problems.

The crop lands of tropical Australia grow some rice under irrigation but the most common crop is sugar-cane, most of which is harvested mechanically. The limitations of such harvesting impose restrictions on row gradients, regularity and also drainage requirements. However, the restrictions on duration of water-logging are the principal drainage criterion (Smith & Rixon, 1982). Bonell (1988) summarizes a great deal of experimentation and the impacts of various cultural practices on runoff characteristics and erosion amounts.

The other area of cultivation in the Australian tropical zone is in the Northern Territory. Some research is in progress and has been reported briefly in Perrens (1989).

Empirical local solutions have been developed slowly at most locations in the humid tropics. The challenge today is to provide viable solutions in locations without a previous local tradition or often without any previous development at all. There is an additional problem that at many of these locations there is minimal hydrologic and climatic data. A workshop under the auspices of the United Nations University was held in Canberra in 1983 on the need for climatic and hydrologic data in agriculture in South East Asia. The proceedings (Fitzpatrick & Kalma, 1989) offer some case studies and suggestions of relevance.

LAND-USE CHANGE AND THE AUSTRALIAN EXPERIENCE

The first land-use change in the humid tropics was swidden cultivation for primitive horticulture, a land-use still active in some areas but one never practised in Australia. This practice produces a limited spatial change in land cover and the secondary forest to which it reverts probably behaves hydrologically like the primary forest.

Very early – perhaps 5,000 or more years ago – wetland modification for horticulture began. It appears to have been discovered and rediscovered in all parts of the humid tropics, with priority perhaps in New Guinea and Thailand. Again, a limited change occurred (Thorne & Raymond, 1989).

The most recent population pressures have resulted in much more extensive land-use changes, with both clearing and reforestation of cleared land following reduced productivity. The current knowledge of all aspects of this has been summarized by a series of publications (many written, edited or sponsored by one of the organizers of this colloquium, L.S. Hamilton). Some of the important publications are Williams & Hamilton (1982), Hamilton and King (1983), Vergara & Briones (1985) and the RAPA report 1986/3.

Australian experience, except for Bonell & Gilmour's work at Babinda, is not in the humid tropic environment. However, it is instructive to briefly review some of the implications of land-use changes in the Australian environment generally, and so avoid blind transference of results and/or concepts to the humid tropics.

Clearing of forest and woodland for arable farming

Clearing will be dealt with in respect to four ecological/hydrological zones:

(a) *The semiarid woodlands of the Australian tropics and subtropics.* Here clearing or killing of trees in a region where they can only exist by drawing on deep mantle stores during the winter dry season often raises the regional groundwater and generates perennial streams where there are none. These are often mildly brackish, more so after a heavy wet season.

If there is no regional groundwater, the water content below one metre is increased and a vertical flux to the root zone of grasses and shrubs increases the salinity in the upper soil profile.

(b) *Semiarid and subhumid woodlands and forests of temperate Australia.* These are the regions cleared most for Australian agriculture. Basically, the favourable moisture conditions of autumn, winter and spring are exploited for grain crops and annual pastures. The trees explored a significant depth of mantle store and ensured recharge rates to regional groundwater as low as 0.5 mm per annum, which under agriculture can rise to 50 to 100 mm.

Particularly polewards of 30°, widespread salinity is occurring from raised saline regional groundwater. The low flow salinity in streams is considerable. Higher rainfall areas tend to have lower salinity and, of course, proportionately less increment in groundwater. Summer rainfalls of high intensity on landscapes of minimal vegetative cover can give rise to extensive erosion.

(c) *The temperate wet eucalypt forests of Victoria.* These forests have high winter rainfalls, no overland flow of almost any form and are the tallest hardwoods in the world, with a natural lifetime of hundreds of years. They are subject to wildfires and, indeed, require them for even-aged stand regeneration.

Following clearing or wildfire of a mature or overmature forest, the annual runoff increases for a few years by 150 to 200 mm and then declines so that at age 50 to 100 years there is a yield decrement of the order of 600 mm (Kuczera, 1987). The full biological explanation of this extraordinary change is incomplete but requires certainly a capacity to extract sensible heat energy for evaporation. The simple one-dimensional energy balance indicates a downward flux of energy from the air.

(d) *The Mediterranean climate forests of Western Australia.* The mechanism here is the same as the subhumid and semiarid temperate forests of eastern Australia. However, the winter seasonality of rainfall is much stronger and very strong rainfall gradients are associated with the Darling Escarpment.

Clearing of forests and woodlands in the areas of rainfall less than 700 mm has caused almost universal salinization of streams and drainage lines within tens of years of clearing.

Widespread replanting of forestry strips is being advocated to reduce saline inflows to reservoirs (Loh, 1989).

In the rainfall zone of 700–1000 mm per annum, salinization is not universal. Some experimental thinning to enhance yield is occurring. Much of this region has an average annual runoff less than 10 mm. In the zone of the Jarrah Forest greater than 1,000 mm there is controlled forestry but not clear felling. However, there also is extensive mining of bauxite with a requirement for regeneration of exotic eucalypt forest in the mine pits. Extensive research and monitoring of the changes associated with mining is occurring with experiments also being carried out in the lower rainfall zones, forbidden at present to mining.

The karri forests, which usually require greater than 1,200 mm annual rainfall, are being clear felled for wood chipping. Indications are for behaviour similar to the wet eucalypt forest of Victoria.

Afforestation and plantation forestry

Australia is currently carrying out a number of major projects to increase forest and woodland cover. Indeed, there is a national objective to plant one billion trees by the year 2000. Major investigations and possible methodologies are under study to locate the most favourable sites for amelioration of water resource changes.

The most extensive work has been carried out in Western Australia and there is some evidence to suggest that strip afforestation immediately up-gradient of groundwater emergence achieves some reduction in discharge (Schofield, 1989).

Development of the techniques outlined in O'Loughlin (1986) has been used to undertake theoretical impacts of planting and this is currently being used to locate experimental plantings throughout the Murray-Darling Basin in southeast Australia. The same technology is being used to study water movement in the mantle and development of surface saturation in both cleared and forested tropical catchments (O'Loughlin, 1990) and is also outlined in Bonell with Balek (this volume).

Plantation forestry has been examined in a number of locations. In humid temperate regions, differences in runoff seem principally related to differences in interception storage with secondary effects related to changes in leaf area index and the potential/actual evaporation behaviour of different species. Australian native vegetation has relatively low *LAI* and reduces *LAI* in response to drought stress, even in regions supporting forest and dense woodland (see Smith *et al.,* 1974; Aston & Dunin, 1980). The results may not be relevant in the humid tropics.

Most plantation forestry in Australia has been carried out with exotic species, a common experience world-wide. This even extends to eucalypt species where eastern Australian species have been shown to be most effective in Western Australia (Sharma, 1984). Australian eucalypts are far more extensively established in plantations outside Australia and, indeed, more intensively studied.

The water-logging tolerance of *E. camaldulensis* or River Red Gum has led to its widespread planting as a reclamation species in marshes and other areas with a high water table. This has led to a reputation for using vast amounts of water (not entirely true) which has been extended to all eucalypts.

Determination of water use by natural stands of trees is very difficult but extensive experimentation in forests now has produced fairly unequivocal results which indicate that eucalypts, by and large, use less water on an annual basis than most trees because their interception capacity is low and they respond to stress by reducing canopy. See Sharma (1984) for a major group of papers on the subject, and Dunin *et al.* (1985) for a special lysimeter study.

Calder (1986a,b) has discussed the possible mechanisms with respect to the difference in evaporation rates between forests and pasture and the characteristics of eucalyptus. Studies comparing exotic pine plantations and eucalypt forests in Australia show significantly higher annual water use by pine plantations. Most is believed to be related to different interception characteristics and lower albedo (Smith *et al.*, 1974).

Tillage and cropping

This mostly applies to impacts at the smaller catchment level. Indeed, most of the quantitative work has been done at the plot and contour bay scale. Australian experience has been that our soils behave in a quantitatively different manner from those described in the world literature. This is because of their age and the widespread incidence of what we call duplex and gradational soils. These soils are characterized by light, textured soils lying over heavy clay soils, often at shallow depths or increasing clay contents with depth. However, the processes and procedures applied are within the normal range of the world literature. See Lal (1989) and Perrens (1989) for a broad review of the Australian scene. Bonell *et al.* (1983) and Bonell & Williams (1986) have reviewed the behaviour of tropical soils in northeast Australia.

It is stated in Perrens (1989) that over 1 million ha of cropped land in Queensland has been treated with soil conservation measures and that on over 40% of land needing treatment, contour banks have been established. Contour tillage and crop residue management is now common. This can be placed in world context by reference to Lal (1989).

Pasture and agroforestry forestry

Much of the land clearing in tropical Australia has been undertaken to enhance the growth of pasture and, indeed, over much of Australia. The response has already been discussed in an earlier section. However, the long-term response usually requires some manipulation of the pasture, particularly the introduction of exotic grass and legume species.

The most extensive experiments with respect to tropical savannah have been carried out in the Upper Burdekin catchment west of Townsville and reported by Bonell & Williams (1989). In common with many other experiments in southern

Australia and world-wide, it is apparent that stocking rate and grazing pressure are the principal determinants of the rate of runoff generation and erosion.

In subtropical Queensland, which is subject to rainfall intensities comparable with tropical Australia, there has been very significant experimentation but the effects can be summarized as follows: Increased soil cover reduces both runoff and sedimentation concentration in runoff. See Freebairn & Boughton (1985) and the reports in Perrens (1989).

CONCLUSIONS

Land-use change has, in general, a radical impact on the hydrological properties of the soil mantle and nowhere more so than in the tropics. In addition, it alters the key elements in the water, salt and energy balances. Cultural practices also can enhance natural runoff paths or radically restrict or otherwise manage them.

Lal, in the companion paper to this one, explicitly discusses some of the processes involved in the balance equations set out here. It is believed that experimental data and design methods should in the first place be reviewed for their relevance to problems in tropical hydrology, using the simple balance equations for water, energy, salt and sediment.

Secondly, the use of *appropriate* process modelling should be a priority in assessing methods of analysis. Also, once again, the assumptions and simplifications inherent in the development of simplified and empirical expressions should be examined. A simple example is the close phase-coupling of the annual cycles in temperature, humidity and radiation to establish correlations in temperate regions which can be very misleading in tropical regions where the annual amplitude is small and phase-coupling often very different.

Therefore, we should attempt to develop predictive models which are parsimonious with respect to the detailed specification of environmental boundary conditions but are realistic with respect to relevant processes. This should allow the extension of data from experimental sites concerned with manipulation of a limited suite of external variables.

REFERENCES

Aston, A. R. & Dunin, F. X. (1980) Land-use hydrology: Shoalhaven, New South Wales. *J. Hydrol. 48*, 71–87.

Bonell, M. (1988) Hydrological processes and implications for land management in forests and agricultural areas of the wet tropical coast of northeast Queensland. In: R. F. Warner(ed.) *Fluvial Geomorphology of Australia*, 41–68. Academic Press: Sydney.

Bonell, M. (1989) Applications of hillslope process hydrology in forest land management issues: The tropical northeast Australian experience. UNESCO Regional Seminar on Tropical Forest Hydrology, Sept. 1989, Kuala Lumpur, Malaysia.

Bonell, M. & Gilmour, D. A. (1978) The development of overland flow in a tropical rain forest catchment. *J. Hydrol. 39,*. 365–82.

Bonell, M. & Williams, J. (1986) The generation and redistribution of

overland flow on a massive exic soil in a eucalypt woodland within the semiarid tropics of north Australia. *Hydrol. Proc. 1*, 31–46.

Bonell, M. & Williams, J. (1989) A comparative analysis of runoff-producing mechanisms in three contrasting tropical climates and under different land-use treatments. In: *Comparisons in Austral Hydrology*, Hydrology and Water Resources Symposium, Inst. Engineers, Australia/Instit. Professional Engineers, NZ/NZ Hydrological Society, Christchurch, New Zealand, November 1989, 386–391.

Bonell, M., Gilmour, D. A.& Cassells, D. S. (1991) The links between synoptic climatology and the runoff response of rain forest catchments on the wet tropical coast of north-eastern Queensland. In: P. A. Kershaw and G. Werran (eds.) *The Rain Forest Legacy. 2.* Conservation Status of the Rain Forest of North Queensland, in press (AGPS: Canberra for Australian Heritage Commission).

Bonell, M., Gilmour, D. A. & Cassells, D. S. (1983) A preliminary survey of the hydraulic properties of rain forest soils in tropical northeast Queensland and their implications for the runoff process, Rainfall Simulation, Runoff and Erosion: *Catena* Special Supplement 4, 57–78.

Calder, I. R. (1986a) What are the limits on forest evaporation? – A further comment. *J. Hydrol. 89*, 33–36.

Calder, I. R. (1986b) Water use of eucalypts – a review with special reference to south India. *Agric. Water Manage. 11*, 333–42.

Cassells, D. S. (1984) Forest and water: just what is known? *Queensland Agric. J. 110*, 49–53.

Colinvaux, P. A. (1989) The past and future Amazon. *Sci. Am. 260*, 68–74.

Dunin, F. X., McIlroy, I. C. & O'Loughlin, E. M. (1985) A lysimeter characterization of evaporation by eucalypt forest and its representativeness for the local environment. In: B. A. Hutchinson and B. B. Hicks (eds.) *The Forest-Atmosphere Interaction*, 271–91. (D. Reidel)

Eyles, R. J. (1977) Birchams Creek: The transition from a chain of ponds to a gully. *Aust. Geogr. Stud. 15*, 146–57.

Fitzpatrick, E. A. & Kalma, J. D. (compilers) (1989) Need for climatic and hydrologic data in agriculture in South East Asia: Proc. of United Nations University Workshop. CSIRO Division of Water Resources Technical Memorandum 89/5.

Finlayson, B. L. & McMahon, T. A. (1988) Australia vs. the World: a comparative analysis of streamflow characteristics. In: (ed.) R. F. Warner, *Fluvial Geomorphology of Australia,,* 17–40. Academic Press: Sydney.

Fleming, P. M. (1987) The role of radiation in the areal water balance in tropical regions: a review. *Arch. Hydrobiol. Beih. 28*, 19–27.

Fleming, P. M. & Wells, A. T. (1991) A review of the effect of land-use on the water resource: a report. Victorian Department of Water Resources and AWRAC.

Freebairn, D. M. & Boughton, W. C. (1985) Hydrologic effects of crop residue management practices. *Aust. J. Soil Res. 23*, 23–35.

Gill, A. M., Groves, R. H. & Noble, I. R. (eds.) (1981) *Fire and the Australian Biota.* Aust. Acad. Sci.: Canberra.

Gilmour, D. A. (1975) Catchment water balance studies on the wet tropical coast of north Queensland. Unpublished Ph.D thesis, Department of Geography, James Cook University of North Queensland.

Hamilton, L. S. & King, P. N. (eds.) (1983) *Tropical Forested Watersheds: Hydrologic and Soils Responses to Major Uses or Conversions.* (West View Press: Boulder, Colorado.)

Herwitz, S. R. (1985) Interception storage capacities of tropical rain forest canopy trees. *J. Hydrol. 77*, 237–52.

Herwitz, S. R. (1986) Infiltration-excess caused by streamflow in a cyclone-prone tropical rain forest. *Earth Surface Proc. Landforms 11*, 401–12.

Herwitz, S. R. (1987) Raindrop impact and water flow on the vegetative surfaces of trees and the effects on streamflow and throughfall generation. *Earth Surface Proc. Landforms 12,* 425–32.

Jackson, I. J. (1969) Tropical rainfall variations over a small area. *J. Hydrol. 8,* 99–110.

Jackson, I. J. (1975) Relationships between rainfall parameters and interception by tropical forest. *J. Hydrol. 24,* 215–38.

Kershaw, A. P. (1978) Record of the last interglacial-glacial cycle from north-eastern Queensland. *Nature 272,* 159–61.

Kuczera, G. (1987) Prediction of water yield reductions following a bushfire in ash-mixed species eucalypt forest. *J. Hydrol. 94,* 215–36.

Lal, R. (1989) Conservation tillage for sustainable agriculture: tropics versus temperate environments. *Adv. Agron. 42,* 85–197.

Loh, I. C. (1989) The history of catchments and reservoir management on Wellington Reservoir catchment W. A. In: P. Laut and B.Taplin (eds.).*Catchment Management in Australia in the 1980s,* CSIRO Division of Water Resources Divisional Report 89/3.

McCaig, M. (1984) Soil properties and subsurface hydrology. In: K. S. Richards, R. R. Arnett and S. Ellis, (eds.) *Geomorphology and Soils,* 121–40. (George Allen and Unwin: Boston/Sydney.)

Monteith, J. L. (1985) Evaporation from land surfaces: progress in analysis and prediction since 1948. In: *Advances in Evapotranspiration,* 4–12. ASAE Publication 14–85. ASAE: St Joseph Michigan.

O'Loughlin, E. M. (1986) Prediction of surface saturation zones in natural catchments by topographic analysis. *Water Resour. Res. 22,* 794–804.

O'Loughlin, E. M. (1990) Modelling soil water status in complex terrain. *Agric. For. Meteorol. 50,* 23–38.

Perrens, S. J. (ed.) (1989) Needs and priorities for research into the hydrology of agricultural catchments. Proc. Workshop, Sydney, 1988. University of New England and NSCP of Dept. of Primary Industry and Energy, Armidale.

Pilgrim, D. H. & Canterford, R. P. (eds.) (1987) *Australian Rainfall and Runoff.* 2 vols. (Inst. of Engineers Aust.: Canberra.)

RAPA (1986) Land-use, watersheds and planning in the Asia-Pacific region. RAPA Report 1986/3. Environment and Policy Inst., East-West Centre, Honolulu, Hawaii and Regional Office for Asia and the Pacific (RAPA) of Food and Agric. Organization, Bangkok.

Salati, E., Dall'olio, A., Matsut, E. & Gat, J. R. (1979) Recycling of water in the Amazon Basin: an isotope study. *Water Resour. Res. 15,* 1250–58.

Schofield, N. J. (1989) Stream Salinity and its Reclamation in Southwestern Western Australia. Report of the Steering Committee for Research on Land Use and Water Supply. Water Authority of Western Australia Report WS52. WAWA, Leederville, W.A.

Sharma, M. L. (ed.) (1984) *Evapotranspiration from Plant Communities. Developments in Agriculture and Managed-forest Ecology. 13.* (Elsevier: Amsterdam.)

Smith, M. K., Watson, K. K. & Pilgrim, D. H. (1974) A comparative study of the hydrology of radiata pine and eucalypt forests at Lidsdale, NSW. Civ. Eng. Trans., I.E. Aust., *CE16,* 82–86.

Smith, R. J. & Rixon, A. J. (eds.) (1982) Proc. Symp. on Rural Drainage in Northern Australia. (Darling Downs Soil and Water Studies Centre DDIAE: Toowoomba.)

Sumner, G. N. & Bonell, M. (1986) Circulation and daily rainfall in the north Queensland wet season. *J. Climatol. 6,* 531–49.

Thorne, A. & Raymond, A. (1989) *Man on the Rim.* (Angus and Robertson: Sydney.)

Vergara, N. T. & Briones, N. D. (eds.) (1987) *Agroforestry in the Humid Tropics: Its Protective and Ameliorative Roles to Enhance Productivity and Sustainability.* Environment and Policy Inst., East-West Centre, Honolulu, Hawaii and Southeast Asian Regional Centre for Graduate Study and Research in Agriculture, Los Banos, Philippines.

Walker, F. H. & Rushton, K. R. (1984) Verification of lateral losses from irrigated rice fields by a numerical model. *J. Hydrol. 71,* 335–61.

Walker, J., Raison, R. J. & Khanna, P. K. (1986) Fire. In: J. S. Russell and R. F. Isbell (eds.) *Australian Soils: The Human Impact,* , pp. 185–216. (University of Queensland Press: St. Lucia, Queensland.)

Williams, J. & Hamilton, L. S. (1982) *Watershed Forest Influences in the Tropics and Subtropics:* A Selected, Annotated Bibliography. Environment and Policy Inst., East-West Centre, Honolulu, Hawaii.

19: Urban Water Management Problems in the Humid Tropics: Some Technical and Non-technical Considerations

J. S. Gladwell

President, Hydro Tech International, Consultants, Vancouver, Canada; formerly Project Officer, Humid Tropics Programme, International Hydrological Programme, UNESCO, Paris, France.

ABSTRACT

This chapter reviews the present situation within the humid tropics with respect to urban sanitation, drainage, disaster preparedness, health, and water supply. It stresses the point that urban water management requires a broad perspective of the impacts. It discusses the technical and non-technical aspects that are both causing the problems and are necessary for their solution. It reviews briefly some of the approaches available to the urban water experts who will be presenting solutions.

INTRODUCTION TO URBAN WATER MANAGEMENT

The theory

The rapid growth of population centers in the tropics continues, and the various resources required by the populations put ever-greater demands upon the environment. It seems clear that the growth must have its limits. Yet it is difficult to see at the moment, at least, any concerted effort to stop the deterioration of the landscape that is already obvious. The problem also is by no means limited to those urbanizing areas of the developing countries, with the concept of "bankruptcy" now being extended, in theory at least for the moment, to several cities in well-developed countries. The problems of maintaining, let alone improving, the infrastructures of these cities for the management of their water resources is of ever-increasing concern.

But, while the pressures of increased population create increased problems, the cities do offer some of the solutions. Because, as difficult (some would say impossible) as the situation appears, cities provide answers. As noted by Lugo (1991), " . . . when human activity is concentrated, human needs are easier to satisfy, human environmental impacts are easier to manage, and the standard of living and quality of life are easier to improve. Properly designed and located cities may be the best tool available for conserving large areas of wilderness."

Lugo points out, however, that if the cities grow too much, they can have significant negative impacts on the natural resources. He also notes that cities have become synonymous with social and environmental problems that can require intensive management if the damages are to be minimized. Several factors that are needed by a growing city are:

(a) the maintenance of a healthy and productive internal environment,

(b) the proper handling of the natural resources required and waste products produced by the urban activities in an environmentally acceptable manner,

(c) the capacity of the institutions within and without the cities to manage the effects of natural hazards,

(d) the tolerance of the people to the situations in the cities, and

(e) the maintenance of the regional and national infrastructures so that they are able to supply the cities' needs from the outside.

Noting the above, it should be evident that water management is but one aspect of the total management of an urban center, albeit perhaps the single most important one. It means, though, that successful water management must consider all of the related aspects. Integrated water management is required within the total perspective. The development of a water supply must consider total waste disposal, surface drainage and other issues of a social or environmental nature. It is also critical that the right sort of information be obtained when planning begins for water projects. In all too many urban and urbanizing areas of the developing world, the water, sanitation and natural hazard management is now too often being carried out with little and/or poor (sometimes even "no") information (Gladwell & Low, 1991). As noted by Geiger, *et al.,* (1987) planning, design and operation should be done within the total urban water system. But to achieve such a goal, the designer has to be aware of the total economic, financial, institutional and management (and one should add, social and cultural) implications of the technical considerations. Engineers as well as the policy-makers and planners need to avoid the tendency to view water-related issues in isolation from total development.

The fact

Although there has been a great deal of effort expended to reverse the situations, most urban centers in the developing world still lack the facilities that are adequate for the proper collection and disposal of domestic and industrial wastes. Urban runoff is typically highly polluted with pathogenic and organic substances that can have potentially serious health consequences. Industrial pollution also is looming ever larger.

Perhaps less than half of the urban population in these countries has adequate access to sewage treatment and disposal systems. Most of those are poorly operated, with the wastes being discharged directly to receiving waters with little or no treatment whatsoever. Many urban areas use natural water bodies as the direct means of getting rid of wastes, without even the semblance of a sewage collection system. And in too many other cities, septic systems are subject to flooding (during which the contents are discharged to natural water bodies), or else they are potential hazards to groundwater. Garbage, domestic and otherwise, is often either dumped directly into water bodies or disposed of on roadsides or other equally unsuitable areas from where they can often later be washed into streams and lakes.

As noted by the World Health Organization (WHO) (1988), the mere existence of a sewerage system in a city should not be deemed to imply that there is proper treatment of the wastes. Bartone (1990) emphasizes the poor experience in South America with O&M (operation and maintenance) of existing sewage treatment plants. He reports that a survey in one country revealed that 80 per cent of over 230 such plants were not functioning because of operational problems. In another country, 33 of 42 mechanical treatment plants had been found in a survey to be out of service. Furthermore, the direct outflow of sewage into rivers is common. He notes that, given the tendency to use river water as sources of drinking-water, the problem of infection is increased.

Why is this situation being tolerated?

A large portion of many urban areas in the humid tropics is required to rely upon itself for the supply of basic water supply and sanitation facilities. This continues to cause related problems: over-exploitation of groundwater sources, disposal of untreated wastes in open canals or in open areas, land subsidence, pollution of the groundwater, and flooding of low-lying areas. With the rapid and presently uncontrolled increase in urban populations and the resulting creation of extremely large metropolitan areas, the solutions to these problems are becoming very expensive. Delay in their implementation, however, merely puts off to another day a worse situation which then will be less economically achievable.

In many areas, sewerage systems are simply non-existent, with human faeces and other wastes being deposited directly into surface drainages. In some cases, the very same drainage is one used by a good portion of the population for bathing and washing their clothes. Large volumes of solid wastes in rivers

are a common sight in the tropics. Why should a large proportion of the population choose to live this way?

The answer is that these people have no other alternative. Those who live in the predominantly slum and squatter areas of the urbanized regions are the poorest classes. They stay there because they are typically the least sought-after lands, and are thus more easily available to them. The rivers and drainage canals provide convenient, if not generally acceptable, means of sewage and garbage disposal. Like it or not, they also provide cheap recreation and all too often convenient water for drinking and bathing. The essence of the problem is financial.

The people in these cities usually have no choice in the matter of pollution control. Nor, in fact, too commonly, do the water engineers and the public works authorities that have to deal with the problems. They work under the severe constraint of tight budgets imposed upon them by the politicians. This further restrains the possibility of cooperation between agencies that is so necessary if a single agency is not to have the total responsibility for the overall management of water. The politicians, in turn, are constrained more often than not by the economic conditions of their countries.

In developing countries, water problems can be affected by two sometimes very contradictory policy categories. The first deals with the underdevelopment of the resources; the second with those caused by activities involved in economic development. The coexistence of these water issues are sometimes difficult to resolve. The politicians are, therefore, often caught in the difficult situation of trying to increase the industrial development of the country in order to improve its economic standing, but the industries they encourage are often the very same ones that are relatively unconstrained in the polluting of the waters and impairing the health of their people.

As a result, many developing countries do not have well-developed infrastructures, and, therefore, the environmental problems are increasing. Basic needs are often unmet with the economy often not permitting rapid improvements. In this situation, global approaches to ecological and environmental problem solutions are too often ranked low in priority (Niemczynowicz, 1991).

As pointed out by Nickum & Easter (1989), the perceived cost of enforcing environmental regulations may also often exceed the apparent benefits to be received. Thus, in the case of industrial facilities, at least, there may be little incentive to enforce the rules. They note that different "tools" can provide different levels of compliance. Whereas, an outright ban, with severe penalties, may act as a forceful deterrent to the pollution situation, it may also be a deterrent to the sharing of information by those who might violate the rules. On the other hand, standards without significant penalties will encourage the sharing of information but will not be particularly effective in ensuring compliance. The different policy options have varying strengths and weaknesses, and the option chosen may very well depend upon the transaction costs involved in enforcement, obtaining the desired information and monitoring the situations.

But many of the countries of the humid tropics lack any criteria and standards at all for water quality for beneficial uses. As long as these are not established, control over the water bodies is impossible.

Although it has not yet come generally to the stage where the shortage of water could impose limits on population growth and economic expansion, the signs are ominous. While in most countries water and environmental legislation exists, monitoring and enforcement seem to be the weakest links. They depend on a whole different set of factors such as trained manpower and capital which many developing countries of the humid tropics can ill afford.

The social and environmental problems are made even more difficult to control as a result of generally low literacy levels, a lack of media coverage and a general lack of community participation in discussions of possible solutions. It is an unfortunate conclusion, therefore, that given the rapid pace of urbanization in these countries – with all too rare exceptions – urban water problems will probably escalate rather than attenuate in the foreseeable future.

Whether or not the urban water problems will eventually be solved depends as much upon the existing social and economic situation (the willingness to do something about the conditions) as they do upon the availability of affordable technical means. Lack of skilled or dedicated experts is not always the problem; but the solutions do depend upon technical personnel (engineers, geologists, hydrologists, etc.) who are able to look at the problems within a total social and environmental context.

Structural or non-structural approaches?

In the now nearly classic publication, "Water and the City," Lindh (1983) discusses the issues of structural versus non-structural approaches to water management in the urban areas. He notes that most of the existing cities of the world (and certainly of the humid tropics) are near water bodies. And, while the location of future urban areas is probably not so nearly dependent upon such a location, the fact remains that the present cities need good water management that must take the existence of the water bodies into serious consideration.

In the case of flood protection, one might consider the building of dams in the upper reaches of the river – accompanied by levees along the river banks in the vicinity of the urban area – an approach followed throughout most of history. But these structural approaches can be very expensive, and have proven to be more and more out of the economic reach of all but the most wealthy nations.

Furthermore, the structural approach can never guarantee the complete elimination of the threat of devastating floods, as many communities have learned to their great sorrow when dams and levees have failed. The design of these structures, after all, is based upon hydrologic data that in almost no case is

to the total satisfaction of the design engineer, and in any case is dominated by economics. What level of protection can the city or region afford to buy?

Because of the economic implications, different approaches (non-structural) have come into being. These include flood plain zoning, flood insurance, and better advance warning systems. These approach the problem of flood protection from the prevention and loss minimization point-of-view, rather than the remedial. Such approaches emphasize that through land-use planning, regulations, and educational efforts, the highly capital-intensive costs of the structural option can be avoided while reducing the flood damages.

Another option, which falls somewhere between the structural and non-structural options calls for keeping the flood-prone area free of all buildings. Such "green belts" can form excellent recreational areas and when they are connected, a network of open space is maintained through which the flood waters can pass with a minimum of damage to the urbanized area.

Deciding which option to choose is not altogether straightforward from a "cost" point of view. From an economic perspective, one must be able to put a value on such aspects as reduced inconvenience, improved aesthetics and an increased sense of security. Tangible benefits include a reduction in damages to public facilities and fewer lives lost.

However, the traditional governmental approach to water use conflicts has been to build another project to satisfy the perceived demands. Besides easing the controversy among users, this supply-oriented structural approach is usually easier for the governments to implement. But, as Nickum & Easter (1989) point out, the costs, while giving the appearance of being low, are in fact often borne by non-beneficiaries.

In the case of water supply aspects of urban water management, as more and more cities begin to face water shortages, they will have to turn to demand management tools such as water charges and promotion of low water-using technologies. These should in turn generate more revenue, providing the urban areas with more funds to develop new water supplies or to reduce the water losses in the existing systems (Nickum & Easter, 1989).

THE HUMID TROPICS SETTING

Some general considerations about tropical hydrology and urbanization

A number of misconceptions concerning the characteristics of the hydrology and water balances of the humid tropics have been generated because it is perceived that this region receives a greater proportion of rainfall and energy than the temperate and arid zones. The fact is that annual rainfall volumes are large, and the events can be quite intense, but they can also be highly variable, with periods during which little or no rain occurs. Because of this variability, streams and canals can have high ratios of high to low flows. The low flows can occur

Table 1. *Rainfall intensities (mm per hr.) for two-year return period.*

City	5 min.	15 min.	30 min.
Niamey (Africa)	160	110	79
Kuala Lumpur (Asia)	250 (approx)	148	110
Paramaribo (S. America)	–	124	88
Paris (Europe)	82	41	27
Montpellier (Europe)	126	69	48

Source: Modified from Desbordes & Servat, 1987)

over extended periods, resulting in low capacities for natural purification of biodegradable wastes to occur. The predominating high temperatures can further reduce the capability of self-purification. In general, the capacities of countries of the humid tropics to respond to the challenges are also hampered by a number of inadequacies in knowledge of the natural systems existing in the region and in the attitudes to the preparation and implementation of management policies.

Table 1 shows some typical rainfall intensities in selected cities of the humid tropics. Note the comparison with Paris and Montpellier which are in the temperate zone.

Even at the two-year return period, such rainfall intensities would be expected to exceed the natural infiltration rates of soils. Thus, even unpaved areas of urbanized areas will contribute to runoff (Bouvier, 1990). These rainstorms will also typically exceed the infiltration rates of soils that have been disturbed in the development of the urbanized areas. They will contribute to flooding, and thus to the incidence of erosion.

However, the rainfall-intensity regimes for which tropical urban stormwater control systems should be designed for generally are poorly known. There are few automatic raingauges in most cities. Furthermore, what records do exist are usually short and discontinuous because of the inevitable equipment failures with lack of follow-up service. In addition, many of those recording instruments allow for time intervals of only down to 15 minutes. For the analysis of runoff from small urban drainage areas, 1- to 5-minute rainfall intensities are usually required (Dunne, 1986).

While the humid tropics is known to be a region of heavy precipitation, there are typically also periods of "dry seasons" of variable lengths. There are also the normal fluctuations of annual precipitation. As noted by Ojo (1987), in the case of Lagos – as in many other tropical cities of Nigeria over the past two decades – there have been persistently dry conditions. In many of these cities, severe hydrometeorological phenomena related to precipitation and the water balance frequently occur, causing great damage, leading to the loss of lives and property. He cites the following average trends for rainfall in tropical Nigeria: fairly normal conditions during the decade of 1951—1960; generally wetter conditions during 1961–1970; and generally persistently dry conditions during the period 1971–1985.

Because of their violent nature, duration, and the extensive area they can affect, tropical "cyclones" also can be among the most damaging natural phenomena of the humid tropics (Bassan & Luscombe, 1991). The disasters result from the violent winds, excessive rainfall, and rising seas, and can continue to cause problems for two weeks or more.

Coastal cities are, of course, more susceptible to the damaging effects of these cyclones, especially where land reclamation projects have effectively encouraged settlements in low-lying areas, or where the land-use change has also removed protective vegetation, reefs or sand dunes (Clark, 1991).

If a tropical cyclone moves inland or along a coastline, the area affected can be extensive, often involving many countries. It is not possible to prevent tropical cyclones, but radar and satellite and radio permits them to be tracked, with warnings being issued in advance of their approach.

When the urbanization and industrialization process reaches a certain degree of development, changes in the overall environment become inevitable. These changes will no doubt affect the hydrologic cycle within the boundaries of the cities, but may also have an effect well beyond those boundaries, including incidences of flooding, reduction of base flows and groundwater recharge, and the inevitable downstream pollution (Geiger, *et al.*, 1987).

Growth of cities and their problems in the humid tropics
General problems of growth In 1800, only one per cent of the world population lived in cities. But, as can be seen in Table 2, in 1970, 37 per cent of the entire world's population was urban. In 1990, it was about 46 per cent. By the year 2000, the urban population is expected to reach 51 per cent of the total. The urban population of the world will have increased almost two and a half times during the period from 1970 to 2000.

It is important, however, to be aware that the urban population is growing differently in various parts of the world. Looking again at the projected trends in more developed countries, we can see that the urban population is expected to have increased by about 64 per cent between 1970 and the year 2000. In contrast, the corresponding urban population growth in the less developed areas has been estimated to have increased by 239 per cent in the same period. In real terms, between 1990 and the year 2000 the developed countries' urban population will probably increase by about 150 million, whereas those of the less-developed regions will probably jump by almost 700 million.

When we consider urban areas with populations greater than 1 million, the trend toward urbanization becomes even more obvious. In 1820, London, for example, was the only city in the world with one million people. But by 1910 there were 11 cities with more than one million inhabitants, six of which were in Europe.

Only 50 years later, this number had increased to 75 cities; 24 of which were located in the less-developed countries. By 1985, it was estimated that there were 270 urban areas of such

Table 2. *Urban and rural populations, 1965 to 2000.*

	Year				
	1965	1970	1980	1990	2000
Urban Population (millions)					
World Total	1158	1352	1854	2517	3329
More developed regions	651	717	864	1021	1174
Less developed regions	507	635	990	1496	2155
Rural Population (millions)					
World Total	2131	2284	2614	2939	3186
More developed regions	386	374	347	316	280
Less developed regions	1745	1910	2267	2623	2906

Source: UNESCO, 1979

size; 140 of which were located in less-developed countries. Table 3 shows a number of the tropical metropolitan areas with populations of one million or more in 1980.

Table 4 shows that even after the end of the International Drinking Water Supply and Sanitation Decade (1980–1990) effort to reduce the number of people in developing countries who had no access to good water or sanitation facilities, the number remaining is staggering. It should be particularly noted that only in the rural areas has the number of the unserved decreased.

Recent trends in the management of urban water has indicated that there is a clear transition from supply management, characterized by increasing supplies to meet the growing demands of the users, to demand management, which limits water use growth through demand management or conservation. This trend is obvious not only in the countries with limited water resources, but even under conditions of relative water abundance of the humid tropics. The concept reduces problems at the source, the urban system itself. At the "source," the benefits can include the reduction of land subsidence due to cuts in groundwater pumping, and the lessening of the need to invest large sums of funds to provide water supply systems. Conservation can also help alleviate the problems caused by the pollution of water, the disposal of sewage and sludge (because of the direct connection between water supply and sewage), and the need for funding for sewage treatment works (Marsalek, 1990a).

It is evident that the urban areas of the humid tropics need to apply the concept of systems analysis toward the solution of the problems of water supply and sanitation.

In the Asia-Pacific region, the growth of cities has intensified the urban/rural interactions. In most cases, these interactions are evidently mutually beneficial. The urban areas, by drawing more and more heavily on the food, labor and natural resources of the rural sector, are able to provide them with higher incomes, employment and cultural influences. However, the demands of the urban sector on the water of the rural sector, and the latter's creation of domestic and industrial

wastes is providing a major conflict. It is particularly critical in the Asia-Pacific region because most of the larger cities there are surrounded by irrigated agriculture, and thus are competing with them for low-cost water instead of facing the alternative of expensive (or unavailable) water sources such as groundwater or inter-basin transfers (Nickum & Easter, 1989).

One of the common features of the growth of urban areas of the humid tropics is the spontaneous settlements that have developed on land ill-suited to housing, e.g., on hillsides prone to landslides as in Caracas, Guatemala City, La Paz and Rio de Janeiro; and on land prone to flooding or to inundation, e.g., Bangkok, Bombay, Guayaquil, Lagos, Monrovia, Port Moresby, and Recife (WHO, 1988).

Connections to water and electricity supplies are typically illegal and precarious. Many inhabitants rely on vendors who sell water of often questionable quality, and at prices that can exceed by ten or more times those that are paid by the middle or upper income groups in the same cities for piped water. In Sao Paulo, two-thirds of the slum population lives in areas prone to flooding and landslides. Sixty-six per cent of the houses have no lighting, 98 per cent have no access to sewers or septic tanks, and about 80 per cent have no drinking water (WHO, 1988).

In Bogota, a study made in 1973 estimated that 59 per cent of the population were living in houses or shacks built on illegal subdivisions (so-called "pirate developments"), but that only one per cent were living in squatter areas. In the survey of 135 of the "pirate subdivisions" made four years later, it was found that more than half of them lacked sewers, more than a third of them lacked water and electricity, and a fifth of them were without water, sewers, electricity, streets, or pavements (WHO, 1988).

In Nouakchott, Mauritania, an estimated 64 per cent of the inhabitants (in 1982) lived in largely self-built communities, with more than two-thirds of them having no direct access to water (WHO, 1988).

In 1978, there were almost two million people living in 415 squatter sites throughout the urban area of Manila. Most of the

Table 3. *Tropical metropolitan areas with > 1 million inhabitants, 1980.*

Central City#		Country	Population x 10^6	Annual Rainfall (mm, rounded)
C	Abidjan	Cote d'Ivoir	1.6	1980
C	Accra	Ghana	1.0	790 +
	Addis Ababa	Ethiopia	1.2	1070
C	Ahmadabad	India	1.7	820 +
C	Bandung	Indonesia	1.7	1900
C	Bangkok	Thailand	5.2	1440 +
	Belo Horizonte	Brazil	2.4	1060 +
	Bogota	Colombia	4.0	1470 +
C	Bombay	India	8.2	2080 +
	Brazilia	Brazil	1.1	1560
C	Calcutta	India	9.1	1600 +
C	Caracas	Venezuela	3.0	840 +
C	Chittagong	Bangladesh	1.1	2730 +
	Dacca	Bangladesh	3.0	2010 +
	Delhi/New Delhi	India	5.2	670 +
C	Fortaleza	Brazil	1.3	1370 +
	Guadalajara	Mexico	2.3	900 +
*	Guangzhu	China	2.9	1720
	Guatemala City	Guatemala	1.6	1310 +
C	Guayaquil	Ecuador	1.1	1100
C	Hanoi	Vietnam	2.6	1800 +
C	Havana	Cuba	2.0	1230
C	Ho-Chi Min City	Vietnam	3.4	1980 +
C*	Hong Kong		5.1	2270
C	Jakarta	Indonesia	6.5	1800 +
	Kinshasa	Zaire	2.2	1390 +
C	Kuala Lumpur	Malaysia	1.0	2410
C	Lagos	Nigeria	1.7	1830 +
C	Madras	India	4.3	1210
	Manila	Philippines	5.5	2070
	Mazatlan	Mexico	2.1	810
*	Nanking	China	2.4	1320 +
	Nova Iguacu	Brazil	1.1	n.a.
C*	Porto Alegre	Brazil	2.3	1330
	Quezon City	Philippines	1.0	n.a.
C	Rangoon	Myanmar	2.7	2620 +
C	Recife	Brazil	2.3	1610 +
	Rio de Janeiro	Brazil	5.1	1080 +
C	Salvador	Brazil	1.8	1900
C	San Juan	Puerto Rico (USA)	1.1	1630
C	Santo Domingo	Dominican Republic	1.0	1420
	Sao Paulo	Brazil	8.3	1430
C	Singapore	Singapore	2.4	2420
C	Surabaya	Indonesia	2.4	1780
C*	Taipei	Taiwan (R. of China)	3.5	2180 +

(Modified from Landsberg, 1986)

C – indicates coastal city; * – outside the Tropics but has a tropical climate;
+ – rainfall shows a very pronounced annual (monsoonal) variation.
# – note that the populations given are for the central cities; the total urbanized area is always greater. See also Table 2, Low, this volume.

Table 4. *Water supply and sanitation coverage, 1980-1990, for urban and rural areas of developing countries (population in millions).*

Global/ Sector	1980				1990			
	Popula-tion	Per Cent coverage	Number served	Number unserved	Popula-tion	Per Cent coverage	Number served	Number unserved
Global totals								
Urban water	933.47	77	720.77	212.70	1332.22	82	1088.52	243.70
Rural water	2302.99	30	690.25	1612.74	2658.51	63	1669.79	988.72
Urban San.	933.47	69	641.39	292.08	1332.23	72	955.22	377.00
Rural San.	2302.99	37	860.64	1442.35	2658.51	49	1294.72	1363.79

After: Christmas & de Rooy (1991)
(Note that the populations for 1980 and 1990 differ slightly from those of Table 2)

households in these settlements had an extremely inadequate supply of water (often of poor quality) and little or no provision for sanitation, disposal of household refuse, or community services and facilities (WHO, 1988).

In Latin America, the primary objective of water resources management has always been to provide a supply of drinking water and to protect the population against the effects of flooding. Environmental issues are only considered as far as they may influence health conditions. However, the results of the rapid spontaneous developments have brought about a major change in the social structure that is explosive.

While air pollution problems that are caused by automobiles, trucks and motorcycles might be assumed to be relatively less serious in the developing countries' cities, they may in fact be greater. The combination of traffic congestion, narrow streets, lack of wind to disperse pollutants, old and poorly maintained vehicles, plus high levels of lead additives in gasoline, often create higher concentrations of lead and carbon monoxide than are found in the developed countries (WHO, 1988).

The operation of motor vehicles in urban areas produces such materials as leaked fuels, lubricants and coolants; particulate exhaust emissions; dirt, rust and decomposed paint falling off the vehicles; and vehicle components broken by vibration or impact. Such material contributes substantially to the presence of hydrocarbons, including toxic metals in run-off (Marsalek, 1990b).

While the present situation appears to be sufficiently formidable, the future doesn't give any feeling of consolation. Every forecast indicates that there will continue to be migration of people from the rural areas to the cities, with fewer people involved in farming. But what is worse is the general malaise of the situation. As noted by Landsberg (1986), "Anyone who has visited the cities in the tropics learns quickly that planning entered only marginally into lay-out and growth patterns of the urban areas, and climatic planning seems to have been virtu-

ally non-existent." In terms of urban water management in the humid tropics, it seems to be primarily a "catch-up" game.

Water supply and sanitation The sanitary conditions in many humid tropic countries may be visualized by considering the situation in Nigeria, a very urbanized country with a population of about 70 million. It is to be noted that in 1975 there was not a single Nigerian town or city with a central sewage treatment system, although there were many serving institutions, housing estates and army barracks. In the major cities and towns, the majority of the inhabitants had no water-serviced toilet systems. In the smaller towns and the rural areas, where more than 80 per cent of the population lived, the flush toilet was almost totally absent (Geiger, *et al.*, 1987).

As an example of drainage problems aggravated by flooding, with consequent sanitary consequences, one might look at Nudu'alofa, a town of 22 000 inhabitants (in 1978) located on the north coast of Tongatapu Island (in the Kingdom of Tonga) in the South Pacific. During heavy rainfalls, heavy runoff is generated throughout the town and the runoff drains into a valley in the center of the town. Consequently, during and following rainy periods, parts of the town are flooded. The accumulated water often remains on the surface for days or weeks. There is no sewage collection network and properties rely on individual waste water disposal facilities. Sanitation consists only of septic tanks and pit latrines, with many of the properties having no latrine at all. Thus, when storm runoff passes through the town, it becomes polluted with overflows from the septic tanks and pit latrines (Geiger, *et al.*, 1987).

The city of Dar es Salaam in Africa is also experiencing problems in the disposal of waste waters that are typical of the humid tropics. During the rainy season, parts of that city are also flooded to the point of impairing the efficiency of certain production activities, thereby causing a substantial loss of revenue to the country. It has been reported that all but two of the 17 sewage pumping stations in the city had stopped operating before 1979, due to the lack of routine maintenance of the sewerage systems (Gondwe, 1990).

It is also stated that there is a high wastage of water in the city's distribution system due to faulty valves, broken pipes, etc. It is thought to be in the range of 30–40 per cent (Gondwe, 1990). It seems evident that the lack of metering and low pricing are encouraging some people to consume excessive amounts of the city's water.

But, as high as those unaccounted for "losses" may seem, they are dwarfed by other cities. In 1981 48 per cent of the distributed water from the Bangkok system was "lost." In 1982 it was 45 per cent. In Jakarta, "losses" were estimated to be 53 per cent of the total water produced in 1983 (Sharma, 1986).

Foster (1990) reports that in 18 Latin American cities such leakage was put in the range of 15–25 per cent, with the average being 17 per cent. Of course, such "losses" are not all truly the result of pipe leakages. Much, in fact, may be due to illegal taps. Even in recently installed and well-maintained systems, the unaccounted for "losses" are unlikely to be less than 10 per cent.

While the unaccounted-for-water represents major losses of revenue to the water supply organizations, the water itself may be a significant input to groundwater recharge. Of course, the use of purified water for groundwater recharge is an expensive procedure. Even if the water is later recovered by pumping from local wells, the cost of such pumping can be considerable (Foster, 1990).

Groundwater typically has a good natural quality and is less vulnerable to pollution than are surface waters. But once polluted, the rehabilitation of an aquifer can be a long-term process. In addition, the consequences of groundwater pollution can be adverse economic, social and environmental effects involving the expenditure of enormous funds.

In earlier times, scant attention was paid to the protection of groundwater. The public was mostly unaware that there was a threat to this resource, as it was invisible. Groundwater protection measures in the humid tropics are still at an early stage and will require considerable refinement.

What protection efforts do exist are mainly concerned with preventing the pollution of existing public water supplies. They largely ignore soil quality, the shallow, vulnerable aquifers, and the currently unused groundwater sources.

While unsewered or *in situ* sanitation does under many circumstances provide adequate protection levels for the disposal of excreta in urban areas (under favorable site conditions), those kinds of sanitation measures can significantly influence the quality of groundwater. Table 5 summarizes the impacts of some urbanization processes on groundwater (Foster, 1990).

Typical of the world-wide situation in the developing countries, and particularly in the humid tropics, is the fact that only a small percentage of the population has access to sewerage systems, with many being served by on-site sanitation facilities. These cause many water quality problems.

In Jakarta, the surface water pollution has become so great that difficulty is often found in bringing the domestic supply to an acceptable quality, even with the treatment available (Sharma, 1986).

In many countries of the tropics, extensive peri-urban areas remain without sewerage, or even water supplies (as noted earlier). Increasing numbers of industries, such as textiles, metal processing, vehicle maintenance, laundries, printing, tanneries, photoprocessing, etc. tend to be located in these areas. Many of these industries typically generate liquid effluents that are discharged directly to the soil. With the variety and complexity of chemicals used in these industries, they can offer serious threats to the groundwater's quality. They are often quite mobile and persistent (Foster, 1990).

While the shortage of easily available and uncontaminated water is a common problem among developing countries, the lack of planning and resultant rapid urbanization has led to the deterioration of the existing water systems, natural and man-made. Especially in the densely populated areas where facilities for the disposal of household and industrial wastes are inadequate (or non-existent), bacterial contamination of water by these wastes is becoming a major problem. Table 6 shows diseases related to water supply and sanitation, and their control measures.

The importance of water to the transmission of diseases can be seen also in many areas of developing countries where what were previously basically rural diseases are now becoming endemic in the cities. Malaria, yellow fever, dengue, and schistosomiasis are among these disease. Sao Paulo and other Brazilian cities have also shown a relationship between the transmission of schistosomiasis and the migration of individuals from rural areas where it is endemic.

The close relationship between water supply and waste water disposal means that improvement in one of these areas without a corresponding improvement in the other is unlikely to have a major impact on increased health. The use of sewers to dispose of human wastes not only depends on the availability of pipes, but also on having sufficient water. Likewise, the inability to dispose of human wastes has a strong possibility of endangering that water that is available. There is little doubt but that strong action by governments can affect health (See Table 7). With the growth of the food-processing, metallurgical, chemical and petrochemical, pulp and paper and other industries, wastes from these facilities are being increasingly concentrated. Thus, the saturation levels of the environment in many areas of the humid tropics (the ability to assimilate and dilute those wastes) has often been reached. These wastes may very well include toxic materials, which can find their way into the municipal waste systems.

Furthermore, the large industrial complexes often use large volumes of water in their processes. They also use lagoons for handling and concentrating their effluents. The use of shallow oxidation ponds for the treatment of urban waste waters is also being used more and more prior to their discharge to natural water systems (see the discussion, later, on the technical means of domestic waste treatment). These lagoons have the

Table 5. *Summary of impact of urbanization processes on groundwater.*

Urban Process	Recharge Modification			Implications for Quality	Principal Contaminants
	Rates	Area	Time Base		
Surface Imper- meabilization	reduction	extensive	permanent	minimal	none
Stormwater * Soakaways	increase	extensive	intermittent	marginally negative	Cl, HC, (ClHC spills)
Mains * Drainage	reduction	extensive	intermittent to continuous	none	none
Modification of Surface Water Flow	marginal reduction	lineal	variable	none	none
Local Groundwater Supply Abstraction	minimal	extensive	continuous	minimal	none
Imported Mains Water-Supply	increase	extensive	continuous	positive	none
Unsewered * Sanitation	increase	extensive	continuous	negative	N–(NO$_3$),FP DOC, ClHC
Mains * Sewerage	marginal increase	extensive	continuous	marginally negative	FP,DOC N–(NH$_4$)
Modification Surface Water Quality *	none	lineal	variable	negative	variable
Land Storage- Disposal * Effluents & Residues	marginal increase	restricted	continuous	negative	N–(NH$_4$), DOC ClHC, HM
Irrigation of increase Amenity Areas	restricted	seasonal	variable	variable	N–(NO$_3$)N, Cl

*	important industrial component	DOC	dissolved organic carbon
Cl	chloride and other major ions	HC	hydrocarbon fuels
N	nitrogen compounds (nitrate or ammonium)	Cl HC	chlorinated hydrocarbons
HM	heavy metals	FP	faecal pathogens

Source: Foster, 1990

possibility of having considerable impact on local groundwa- ter quality, especially in relation to nitrogen and organic com- pounds (Foster, 1990).

Pollution of drinking water sources and the infection of the population can result from numerous causes. Among them are: pollution of the water source with sewage; contamination of the groundwater source by garbage dumps, septic tanks and cesspools; inadequate network design of the supply system; poor maintenance of the system; low pressure in the supply system causing the intrusion of polluted groundwater into the pipes; organic growth in the systems; and general public igno- rance by the public about the importance of water and personal hygiene (Bjorth, 1991).

In the Caribbean, less that l0 per cent of the total domestic waste water receives treatment before disposal. In most cases, the untreated domestic waste water, mixed with industrial wastes, is released into streams and coastal waters and estuar- ies, contributing thereby to the eutrophication and to reduction of the oxygen content of those water bodies. With the common seasonal physical processes, the sedimented particulate organic material is typically resuspended, regularly causing oxygen depletion. This phenomena commonly occurs in the bays of Havana and Cartagena (Linden, 1990a).

While the percentage of cities in the humid tropics with sewage treatment facilities is growing (although still low), a more important factor is the percentage of those facilities that

Table 6. *Diseases related to water supply and sanitation: control measures.*

Disease	Type and importance of control measures[a]					
	improvement in water quality	improvement in water supply quantity/convenience	personal and domestic hygiene	wastewater disposal/ drainage	excreta disposal	food hygiene
Diarrhoea:						
viral diarrhoea	••	•••	•••	–	••	••
bacterial diarrhoea	•••	•••	•••	–	••	•••
protozoaldiarrhoea	•	•••	•••	–	••	••
Poliomyelitis and hepatitis A	•	•••	•••	–	••	••
Worm infections:						
ascaris, trichuris	•	•	•	•	•••	••
hookworm	•	•	•	–	•••	–
pinworm, dwarf tapeworm	–	•••	•••	–	••	•
other tapeworms	–	•	•	–	•••	•••
schistosomiasis	•	•	–	•	•••	–
guinea-worm	•••	–	–	–	–	–
other worms with aquatic hosts	–	–	–	–	••	•••
Skin infections	–	•••	•••	–	–	–
Eye infections	•	•••	•••	•	•	–
Insect-transmitted diseases:						
malaria	–	–	–	•	–	–
urban yellow fever, dengue	–	–	•[b]	••	–	–
bancroftian filariasis	–	–	–	•••	•••	–
onchocerciasis	–	–	–	–	–	–

[a] Importance of control measures ••• high •• medium • low to negligible.

[b] Vectors breed in water storage containers.

Source: WHO, 1988.

are actually operating satisfactorily. According to Prost (1991) the maximum is in Asia and the Pacific, where it is estimated that only half operate satisfactorily. In Sub-Saharan Africa, the number is put at 30 per cent. This is obviously a serious matter.

If efficient management of water supply and sanitation systems is to be achieved, there is a need for reconsideration of the institutional structures. Too commonly the responsibilities are widely distributed among competing agencies. In Bangkok, for example, at least six separate agencies are involved (Sharma, 1986).

In Nigeria, Tokun (1983) reports that town planning authorities and the ministries of housing and environment bear the responsibility of controlling physical development in the cities. However in spite of these boards, cities have largely been left to grow on their own, with very little physical planning, inadequate surface water drainage, hardly any central sewerage and poorly coordinated system layouts. The efficiency of these agencies is hampered mostly by the lack of adequate manpower and by socio-political factors.

Urban drainage Certainly one of the most significant water management problems in many tropical urban areas is the control and safe disposal of storm runoff.

An increase in storm runoff from urban areas is usually accompanied by a decrease in the amount of water infiltrating into the soil. While some of the water would have evaporated back to the atmosphere under natural conditions, and a portion of the infiltrated water still does so – even in a city – the net effect is usually to lower the water table and decrease dry-

Table 7. *Links between health and government action at different levels to improve housing conditions in urban residential areas.*

Health Risk Level	Action at Individual and Household or Community Level	Public Action at Neighbourhood Level	Action at City or District	Action at National Level
Contaminated water-typhoid, hepatitis, dysenteries, diarrhoea, cholera, etc.	Protection of water supply to house, promotion of knowledge of hygienic water storage	Provision of water supply infrastructure; promotion of knowledge and motivation in community	Plans to undertake action described and resources to do so	Ensuring that local and city authorities have the power, funding base, and trained personnel to implement action at household
Inadequate disposal of human wastes-pathogens from excreta contaminating food, water, or fingers, leading to faecal-oral transmission of diseases or intestinal worms (e.g. hookworm, tapeworm, roundworm, schistosomiasis)	Support for construction of easily maintained latrine/WC matching physical conditions, social preferences, and economic resources washing facilities, promotion of hand-washing	Mix of technical advice and installation, servicing and maintenance of equipment (mix dependent on technology used)	Plans to undertake action described plus resources, ensuring availability of trained personnel and finance for servicing and maintenance	neighbourhood, city, and district levels, reviewing and where appropriate, changing legislative framework and norms and codes to allow and encourage action at lower levels and ensure that infrastructure standards are appropriate to needs and available resources, support for training courses and seminars for architects, planners, engineers, etc. on the health aspects of their work
Waste water and garbage -water-logged soil ideal for transmitting diseases like hookworm, pools of standing water becoming contaminated, conveying enteric diseases and providing breeding-ground for mosquitos spreading filariasis, malaria, and other diseases, garbage-attracting disease vectors.	Provision of storm/surface water drains and spaces for storing garbage that are ratproof, catproof, dogproof, and childproof	Design and provision of storm and surface-water drains, advice to households on materials and construction techniques to make houses less damp	Regular removal, or provision for safe disposal of household wastes, organization of framework and resources for drains	
Insufficient water, facilities for washing and for personal hygiene-ear and eye infections (including trachoma), skin diseases, scabies, lice, fleas	Adequate water supply for washing and bathing, provision for laundry at household or community level	Health and personal hygiene education for children and adults, facilities for laundry at this level, if not within individual houses	Support for health education and public facilities for laundry	Technical and financial support for educational campaigns, coordination of housing, health, and education ministries
House sites subject to landslides or floods as result of no other land being affordable to lower-income groups	Regularization of each household's tenure if dangers can be lessened, relocation through offer of alternative sites as last resort	Action to reduce dangers and encourage upgrading or offer of alternative sites	Ensuring availability of safe housing sites that lower income groups can afford	National legislation and financial and technical support for interventions by local and city governments in land markets to support lower level action getting training institutions to provide needed personnel at each level

Many of the above measures to improve the quality of the services and facilities a house contains will often have to be modified in the case of houses, flats, rooms, or house sites that are rented. A high proportion of the lowest income individuals and households in most Third World cities rent their accommodation, and improving its quality and reducing the health risks it presents will need governmental programmes and actions that are not summarized in this table.

(Modified from WHO, 1988)

weather flows. As Dunne (1986) remarks, in many cities in the tropics, storm drainage problems are extreme for a number of reasons. First, of course, rainstorms are often intense. Second, the urban growth has frequently been rapid and uncontrolled, so that neither time nor resources have been available for planning to mitigate the runoff problems. These become particularly acute when a city spreads from a lowland into the surrounding hills. The resulting uncontrolled storm runoff surcharges the already typically inadequate storm-sewer system of the flatter downtown area, which thus is frequently flooded. And, third, the impact of sediment deposited in drains and natural channels can be particularly severe because of the typically highly erosive rains of the tropical regions and the unpaved and heavily-used roads.

The hydrology of urban areas, dealing with relatively small catchments compared with rural areas, requires data with very fine time and space resolutions. Such data are usually not available from the national meteorological services. For example, rainfall data necessary for meaningful runoff calculation must have time resolutions of the order of single minutes. Dense networks of gauges are necessary to achieve required spatial representativity of data. Such data are usually not available without special measurements. This is especially the case in the countries situated in the humid tropics (Gladwell, 1991).

Rainstorms of high intensity and long duration in the humid tropics can cause serious erosion problems on soils disturbed in the process of urbanization and in land uses external to the urbanized area, such as the conversion of tropical forests. When sediments are deposited on flat stretches of rivers near a city, they can result in increases in flooding because of the raised river bed levels. In such instances, the channels become useless for protection.

As Ruiter (1990) reported, flooding causes economic injury and inconvenience to a wide range of interests, including property, business, transportation, and social life. As a result, the normal political influence is subject to numerous conflicting interests. He suggests that for Asia, at least, the traditional pragmatic approaches to the solution of flooding problems are no longer acceptable. He notes that Asian cities have three main problems: One is the lack of overall project organization and clear allocation of responsibilities. Another is that adequate detailed urban land-use planning and enforcement also is missing, often resulting in bad routing of piping systems. The last is that a planning capability able to embrace all phases and aspects, both technical and non-structural, is lacking.

But that less-than-adequate situation is certainly not limited to Asia. As noted by Desbordes & Servat (1988), urban storm drainage needs to be more seriously considered in urban planning in Africa. Sewer systems are frequently not well adapted to the fast urban growth. New drainage facilities which are now used in developed countries can be taken as references, but there is need for study to decide which are in agreement with the socio-economic conditions in African cities.

As Tokun (1983) points out, owing to inadequate and ineffi-

cient drainage networks in Nigeria, and the fact that they are used as garbage collectors, the natural carrying-capacities of these channels are drastically reduced. It is not surprising, therefore, that the cities flood annually, accompanied by heavy damage to property and the loss of human lives. At the city of Lagos, located practically at sea level, no matter how little rain falls, the runoff creates ponding which stagnates traffic on the already congested roads. Buildings regularly are flooded and made non-inhabitable.

Desbordes & Servat (1988) confirm that most of the cities of western and central Africa have undeveloped drainage systems which are made worse by the high rates of population growth. As they point out, however, the classical designs of sewer systems could lead to expensive solutions with difficult maintenance problems, much of which would be caused by erosion/sedimentation processes and sanitary disorders.

Yen & Yen (1991) point out that, on tropical islands, the local storage of stormwater is not only desirable but also practical. Cisterns can augment the often precarious water supply and also reduce the requirements for drainage facilities. But they note also that in tropical areas, ditches which are apparently perfect places for the collection of street wastes also serve as breeding grounds for bacteria and for the biological digestion of organic materials. They are the primary source of the region's undesirable odors.

No matter whether sewerage systems are designed as separate or combined systems, they are commonly used for disposal of all kinds of urban wastes, particularly in the case of open channels. With the climatic conditions of the tropics, with their periods of dry weather, this fact can lead to consolidation of deposits which will prevent self-cleaning of the sewers, even during the following wet weather periods. As a result, under such conditions the problem of pollution, even from separate systems, must be considered (Geiger, 1988).

Because the drainage of urban areas is so closely related to waste water disposal, in particular in those areas served by combined sewers, new concepts in urban drainage management include conservation. The idea is to use control measures designed to reduce both the discharges and pollutant loads of drainage effluents – in addition to physical solutions. The problem is, of course, that the physical options are often difficult to implement in existing areas because they can be structurally extensive, involve expensive acquisitions of properties, cause disruption of economic activities, and will almost always be required to deal with the problem of aging infrastructures.

Disasters As cities grow, their land prices tend to increase, thus putting pressure on incentives to develop marginal lands – e.g., those in valley bottoms, and those too close to seashores. With the cyclonic dangers evident in the humid tropics, these tendencies are a matter for strong concern. Considering the possibility of sea-level rise that could accompany a global warming, there is certainly a need to reconsider land-use poli-

Table 8. *Measures to remove or reduce hazards from various climatic hazards.*

Climatic Hazards (Descending order of mortality)	% of Total Natural Disasters reported in a given year	Cause of Death	Vulnerability of Houses and Settlements	Removal of risk		Reduction of risk	
				Possible Solution	Major Difficulties	Possible Mitigation Measures	Major Difficulties
Drought	15%	Malnutrition resulting from lack of food/water/resistance to simple infections	Overall location relative to food/water supply	Maintain supply of staple foodstuffs/water	Logistics; market prices	Advance warning of crop failures (i.e. market price fluctuations in staple foods)	Resources to monitor drought indicators
Tropical Cyclones	20%	Drowning/Exposure; injuries sustained when parts of buildings/trees/walls fall on people	Location of settlement adjacent to coastlines	Relocation of communities	Impossible in any general sense (may be possible at local level); location related to occupations and income levels	Cyclone shelters; evacuation policies; shelter breaks; stronger buildings	Economics, logistics, social consequences
Floods	40%	Drowning/Exposure	Location of settlements on low-lying ground in river basins/estuaries	Relocation of communities	Impossible in any general sense (may be possible at local level); location related to occupations and income levels	Flood control measures; evacuation policies; stronger buildings	Economics, logistics, social consequences
Other	10%	Various					

Note: The overall total percentage in the above list excludes a 15% figure for esarthquakes which has been omitted since this is obviously not a climatic hazard.

Source: Davis, 1986

Table 9. *Alleviation measures for climatic hazards.*

Hazard	Warning System	Environmental Measures to Alleviate Hazard Impact (Excluding specific building measures
River Flooding	Monitoring rainfall in valley, advising on anticipated flood levels for downstream communities.	Two measures are normally adopted: *Storage*: By means of flood plain emergency storage lakes to attenuate peak discharges. *Conveyance*: Building up levels and by-pass channels.
Flash River Flooding	The breaking of cables stretched across river valleys by water flow offering very short warnings to downstream communities.	The above are unlikely to offer significant protection against the sudden onslaught of flash flooding
Coastal Surges	Monitoring by satellite and aircraft surveillance linked to a detailed preparedness plan for community evacuation and protection of property.	Raising up the level of coastal villages at risk. Building sea walls or levees. Excavating emergency conveyance channels to absorb the influx of the surge. Constructing raised earthen mounds for people and their cattle to occupy for the duration of a surge.
Cyclones	As above.	As above, with the addition of the planting of trees and shrubs as shelter breaks around settlements, or in a continuous coastal band in areas that are cyclone-prone.

Source: Davis, 1986.

cies in coastal cities. Enormous costs would have to be incurred, either by damages or by the construction of protective structures (Bjorth, 1991).

In Bangladesh, the problem of coastal zone habitation is perhaps carried to the ultimate, with the resulting major disasters periodically recurring. Despite the rapid progress in satellite warning systems for detecting cyclones, the linkages between the various groups involved in disseminating a warnings to those at risk remains poor. Even with advanced warn-

ing, however, there is the question of whether or not to evacuate. Severe logistical problems can be posed by mass evacuation even if the problem has been thought out.

Franceschi & Rodrigues (1983) report that in Venezuela, towns have grown rapidly, occupying both land and major drains in very narrow mountain valleys. The slopes are very steep, and the streets – acting as major drains – concentrate the floods rapidly in the drains. Many of these drains have been occupied by highways, buildings and other types of

Table 10. *Water supply service levels and options for excreta and sullage disposal in urban areas.*

Water Supply Service Level	Typical Water Consumption L/cap./day	Options for Excreta Disposal	Options for Sullage Disposal
Standpipes	20 – 40[1]	Pit latrines Pour-flush toilets[2] Vault toilets	Soakage pits
Yard taps	50 – 100	Pit latrines Pour-flush toilets[2] Vault toilets Sew. pour-flush toil. Septic tanks	Soakage pits Stormwater drains Sew. pour-flush toil. Septic tanks
Multiple tap in-house connections	> 100	Sew. pour-flush toil. Septic tanks Conventional sew.	Sew. pour-flush toil. Septic tanks Conventional sew.

[1]Consumption depends on standpipe density.
[2]Feasible only if sufficient water carried home for flushing.

Source: Mara, 1982; Geiger, 1988.

development. Such occupancy of the stream courses and the rights-of-way of the major natural drains have resulted in deaths and major property damage.

There is also the lingering problem of the cost of maintaining the readiness of flood protection measures such as protective dikes, overflow canals and temporary storages to absorb the excess flow. Furthermore, as a result of upstream land-use conversions (e.g., the problems of deforestation followed by agricultural use of the lands), the canalization of streams and the encroachment of buildings into the flood plain, the flood levels have increased in many instances. As a result, there is the problem of keeping up with the increasing flood levels (Davis, 1986). Table 8 shows measures that can remove or reduce hazards from various climatic hazards; Table 9 gives some alleviation measures.

Rivers and coastal areas Linden (1990b) reports an estimate has been made that by the year 2000, some 75 per cent of the world's population will live on a narrow strip (up to 60 km wide) along the shores of the continents. In South East Asia, 65 per cent of all major cities (over 2.5 million) are now located along the coasts. See also Table 2.

With so many of the major cities of the tropics located along the coasts, or along rivers that will eventually lead thereto, many of the problems of the urban areas are transported out of their boundaries. The wastes that are flushed – mostly untreated in any manner whatsoever before discharge to the river, estuary or coast – have major impacts on those environments.

There is no doubt but that the discharge of stormwater and waste water into nearby water courses, if approached ade-

quately, may represent an acceptable method of disposal in those urban areas with lower population densities. But it must be recalled that due to the high variability of tropical rainfalls, streams usually have very high ratios of maximum to minimum discharges and that the low flows can occur for extended periods. The low flows will produce little turbulence, with a reduction in the natural re-aeration rates that are required for the biological breakdown of organic matter. Furthermore, high temperatures not only reduce the amount of oxygen that can be dissolved in the water, but also influence the rate at which oxygen is used by the microorganisms (Geiger, 1988).

In many tropical countries, the river flow may actually be near zero during some of the dry periods. In that case, no water is available for dilution of the waste effluents. Under this condition, what was once a river becomes an open sewer, and may become anaerobic. Even with some natural flow, the quality of the water can be severely affected, and its use for domestic water supply or even for irrigation seriously impaired.

The impacts of stormwater and combined sewer overflows are generally associated with biodegradable organic matter, nutrients, settleable solids, bacteria, and toxic substances. The time scale of the impacts vary, from generally short-term in the case of bacteria, to long-term in the case of phosphorus or toxics. The scouring and resuspension of bottom deposits near outfalls can frequently create two to three orders of magnitude greater impacts on oxygen demand than that of undisturbed sediment (Marsalek, 1990b).

The impact of the urban effluents can be extremely hard on standing waters as well. The latter may be even more susceptible to the introduction of organic wastes or dissolved phosphorus. Where the mixing of the water body by wind or by nocturnal overturn is insignificant, the use of the water for domestic supply may be impossible due to the accumulation of hydrogen sulphide, ammonia, dissolved manganese and iron in the lower layers. In such cases, waste treatment is certainly called for (Uhlmann, 1990).

Along coastal areas, the discharge of untreated or inadequately treated wastes has caused eutrophication with consequent oxygen depletion, algal blooms, "red tides" and outbreaks of jellyfish. Linden (1990a) reports that microbial pollution, related to the release of sewage, constitutes a public health threat along most tropical coasts. He also notes that significant upwelling is rather uncommon in tropical areas of the world's oceans. This situation magnifies the importance of the productivity of the shallow coastal regions where habitats such as mangrove forests, seagrass beds, coral reefs, estuaries and coastal lagoons encourage marine productivity, including fish production. It is to be noted that with the high human population levels along the coasts, urban dependence upon the protein provided through fisheries is usually high.

But shallow coastal bays often attract development because they offer a protected anchorage. The adjacent land is often flat and thus is easier to build upon. As a result, many such bays are heavily polluted by industrial and domestic wastes. In addition, coastal lagoons shoreward from barrier beaches are often sites of cities, with their sewage often being discharged to the lagoons through outfalls that are more easily constructed in the relatively calm and shallow waters. The environmental balance of these lagoons is more sensitive than those of the less-constricted embayments (Gunnerson, 1988).

One result of coastal area pollution is a threat to human health through contact with water and the consumption of seafood. Areas such as the lagoon at Lagos (where oysters are cultured) have high faecal coliform bacteria, both in the oysters' flesh and in the water. In the coastal waters off Dakar and the Volta estuary, high coliform numbers have also been reported. Diseases such as infectious hepatitis, dysentery, polio, typhoid and cholera are common in these areas (Linden, 1990b).

In the Caribbean Sea, the bays of Havana and Cienfuegos off Cuba (both partially enclosed water bodies) receive huge quantities of mainly untreated industrial effluents from all types of industries. These cause anoxic conditions, particularly during the summer. Columbia's Bay of Cartagena receives effluents from a refinery, and petrochemical, fertilizer, food processing, and chlor-alkali industries, among others. The bay is particularly bothered by heavy metal pollution, such as mercury. Linden (1990b) reports that its contamination with petroleum hydrocarbons is typical of many other areas in the Caribbean. Fish kills are common there. In Havana Bay, most of the area appears to be permanently without any fish.

While there are certain areas in West and East Africa, that suffer from industrial pollution, it is not a dominant phenomenon. The degree of sewage treatment in East Africa being relatively small, the local incidences of pollution by domestic sewage and industrial pollution are frequent, however. In West Africa there is, in general, no sewage treatment, with raw sewage often being discharged directly into estuaries and lagoons very close to the cities. As a result the discharge of municipal sewage constitutes the main pollution load in all countries of the region. As a consequence, human health is threatened (Linden, 1990a).

Pollution from sewage is no longer a local phenomenon in the coastal waters of the Indo-Pakistani subcontinent, becoming more or less obvious over large areas of the coastal waters. The main pollution in southern Asia is from three metropolitan areas in India: Bombay, Madras and Calcutta. Linden (1990b) reports that it has been estimated that about 1 million tons per year of untreated sewage and industrial effluents are released into the Indian Ocean each year from these three cities alone. As a result, the oxygen concentrations in the nearshore areas off Bombay are around zero, with similar conditions prevailing in the lower Ganges estuary as a result of the release of the waste waters from Calcutta.

Areas such as Manila Bay and the coastal waters of Djarkata are more or less permanently anoxic. Large parts of the bottom waters of the Straits of Malacca and the Gulf of Thailand show the same pattern. It has been estimated that the Gulf of Thailand receives over 300 tons of BOD per day from rivers, most of it from the sewage flowing from the greater Bangkok area. The South China Sea coastal areas now have much more frequent algal blooms, and the incidence of red tides and blooms of blue-green algae have also become much more common recently. Many people have become ill due to paralytic shellfish poisoning, and while there is as yet no direct proof of a relationship, the problems are evidently of recent origin. In Malaysia, for example, such cases had not been reported before 1970. There also may be as many as 10 000 to 50 000 individuals being poisoned in the islands of the Pacific. Relatively frequent outbreaks of viral hepatitis and typhoid in Malaysia, Indonesia and Vietnam have been related to the sewage discharge problem (Linden, 1990a).

A large percentage of all of the fish sold in the markets in most developing countries are species directly dependent on shallow coastal habitats. The drop in fish catches experienced in many shallow areas in South and South East Asia, in East and West Africa and in Central America, points to the destruction of these productive coastal habitats and the indiscriminate dumping of wastes as a major factor.

Considering that it has been estimated that almost two-thirds of the people in developing countries obtain between 40 and 100 per cent of their animal protein from fish, this means that in Asia alone over one billion people depend totally on fish for their protein. It is, of course, generally the poorest people who are the most dependent. The consequences of this high a level of loss of marine productivity in tropical and subtropical oceans could mean the difference between life and death for many of these individuals (Linden, 1990b).

SOME TECHNICAL CONSIDERATIONS

Waste treatment and management

While developing countries of the humid tropics might wish to aim for the standards they have seen in the highly industrialized countries, those expectations are more often than not counteracted by the realities of their local economies. As a result, many have taken no steps at all toward improving sanitation. It remains, nevertheless, that the health objective of excreta disposal programmes should be their number one priority. Depending upon the situation, that objective can be achieved by sanitation technologies that can be considerably less costly than waterborne sewage. Table 10 offers a variety of approaches, the selection of which is governed by, among other factors, the water supply service and water consumption (Geiger, 1988).

The use of unsewered sanitation technology greatly increases the rate of groundwater recharge. In those areas where most of the water is supplied from local wells, large-scale recycling of groundwater can occur. Foster (1990) notes that in the more densely populated parts of the high water use districts of central Bermuda that overlie a lens of fresh groundwater, the use of unsewered sanitation techniques to dispose of waste water, together with stormwater soakaways, doubles the net rate of local groundwater recharge.

But while the recharge of the groundwater is a positive factor, some types of unsewered sanitation technology risk is the penetration of pathogenic bacteria and viruses to the aquifers under certain hydrogeologic conditions. This most often occurs in densely populated settlements, but can also occur where individual homes have both private wells and septic tanks. Such penetrations have been a proven vector of pathogen transmission in numerous disease outbreaks.

It is evident, then, that protection of groundwater requires much more serious consideration, particularly in the developing countries where the problem is not always seen as being serious "at this time." This will require the development of institutional arrangements with the necessary powers and resources for the creation, coordination and implementation of a comprehensive groundwater strategy and policy. In turn, legislation is needed to regulate the management and control of groundwater protection and quality conservation programmes (UNESCO, 1992).

But groundwaters are also in danger of contamination by the land disposal of urban solid wastes. The most serious risk, according to Foster (1990), can occur where the dumping of wastes is uncontrolled, as opposed to controlled sanitary landfills. Groundwater is also endangered where hazardous industrial residues are inappropriately disposed – often, unfortunately, with no records being kept of the amount and character of the wastes.

If an urban water supply is heavily based on groundwater, the associated pumping can cause over-development of the resource, with lowering of the groundwater levels or of the pressure of confined aquifers. This can cause reductions in well production and/or land subsidence. Along coastal areas, it could be the cause of saline water intrusion. This situation is typical of the Bangkok region where about 50 per cent of the total water used is being pumped from deep aquifers. No doubt as a direct consequence, that part of the city with subsidence greater than 10 cm per year – and from which at least 50 per cent of the groundwater was withdrawn – is considered a critical area. Because the ground surface in this area was only 0.1 to 1.0 m above mean sea level to begin with, it is subject to frequent flooding which will increasingly cause problems of sanitation (Sharma, 1986). Restrictions on pumping in certain areas has recently reduced the rate of subsidence (UNESCO, 1992).

In the Jakarta region, land subsidence has been accompanied by salt water intrusion. It has been observed as far as 8 km from the coast in northern Jakarta, and while only the first unconfined aquifer has so far been affected, it is only a matter of time before the others are also contaminated.

In the tropics, the climatic conditions are usually favorable for a full use of biological processes such as photosynthetic oxygenation, biochemical flocculation of polyaromatic compounds and biofiltration by zooplankton or zoobenthos. A low-waste technology or an extensive treatment which makes full reuse of both the waste and the nutrients may be the only way of reducing the pollution of water bodies and bridging any gap that might exist between supply and demand. As Uhlman (1990) notes, "A proper integration of process engineering and ecological engineering may give the best solution to the treatment of organic wastes."

In this respect, if land is the primary limiting factor, although they can have very high construction and O&M costs, conventional aerobic processes such as activated sludge can be used. High-rate biological filtration with plastic media is also suitable for coping with high BOD and with the shock loads of industrial wastes. Both systems have the drawback that they will become unstable due to power failures or to personnel problems. In those instances, it may require a period of days or even weeks until the desired effluent quality again can be established (Uhlman, 1990).

For the above reasons, as well as for its many advantages, stabilization ponds are normally considered to be the most economical method for the purification of waste waters in the tropics. In the case of concentrated industrial wastes, a combination of anaerobic reactors or high-rate biofiltration with stabilization ponds as intermediate or final stages may be considered. But for normal urban waste systems, stabilization ponds have been found to be particularly robust, flexible and fail safe as compared to other techniques of biological treatment. High efficiency is combined with comparatively low costs of construction and relatively low costs for O&M. And whereas the aerobic technologies require electrical energy input, the stabilization pond is based on photosynthetic aeration combined with solar energy, which in the tropics is usually available all-year-round. The effluent products from multiple stabilization ponds, provided the water temperature does not go below 18 or 19 C°, is excellent, with the removal efficiency of fecal coliforms and *Salmonella* reaching a level of 99.99 per cent. If the ponds are to be used to eliminate dissolved phosphorus or nitrogen, they will probably work better under tropical conditions than those in temperate zones (Uhlman, 1990).

The reuse of treated waste waters should also be considered as a part of the treatment system. Some uses include irrigation, cooling water, fish rearing and direct reuse in industrial processes.

But, under most circumstances there will be a need to "get rid of" the treatment plant effluents. Depending upon the degree of treatment (or non-treatment), the receiving waters will be variously affected. In this respect, it is important to consider the sewerage system by which the waste waters reach the treatment plant. If storm and waste water systems are separate, the problem is reduced. But, if the systems are combined, that is both storm and waste waters are conveyed by the same pipes, the problem can be magnified during storm periods when most of the flow will be diverted directly to receiving waters – including, of course, the untreated urban sewage.

The bacterial contamination of the receiving waters from the combined sewer overflows is important from a public health perspective. These discharges can lead to high bacteria levels that can persist for some time after the discharge has ceased. Also, since the bacteria tend to attach themselves to suspended particles, some of the disappearance of the bacteria may be due only to sedimentation. Thus, while the levels of bacteria in the receiving waters certainly diminish as a result of dilution and die-off, those that are removed to the bottom sediment may have extended periods of survival. The concern with toxic substances entering water bodies is that, even though they may enter at levels well below acute toxicity levels, they can accumulate and consequently reach chronic toxicity levels after long-term accumulations (Marsalek, 1990b).

An option for waste water disposal that is available to coastal cities is the use of ocean outfalls. The technology has been used world-wide, and can be applied successfully, provided that it is properly done. Outfalls also can be used in bays. However, this choice must be very carefully considered because of the latter's limited assimilative capacities. They should be limited to small discharges or used in combination with a high degree of treatment.

The concept of the outfall option is that the waste will be diluted sufficiently such that, in combination with the natural die-off of bacteria and viruses, there will be no adverse ecological or health effects. It is thus obviously necessary to avoid any situations that will lead to accumulations of non-biodegradable material. In that respect, then, the motion of the coastal currents is important. Whatever the source of the current's motion, the average current's flow is generally parallel to the coastline and to the depth contours. Nevertheless, onshore/offshore components (which must be of concern) will arise periodically due to waves, tides and wind action. However, effluent plumes are seldom carried out to sea. (Gunnerson, 1988)

Urban drainage

Until recently, urban storm drainage design served only to devise measures to protect urban development from stormwater. It usually consisted of evaluating the peak runoff rate and designing a network of sewers and ditches to collect and convey the stormwater downstream, just away from the urbanized area. Flood plains and sewered drainage systems were considered separately, with urban runoff being regarded as merely an adjunct consideration in land-use planning. The urban development and industrialization of recent years, however, has tended to influence new public demands and policies which, if they are to be satisfied, require the use of true comprehensive planning, including integrated land and water management. Greater emphasis will have to be placed upon the total impact of projects and the overall optimal solution as opposed to optimizing individual devices for presumed sets of criteria (Geiger, *et al.*, 1987).

While planning and design of urban drainage systems, one must consider many aspects beyond those of a purely technical matter. Those have been discussed earlier in this chapter. However, the practice of engineering design of the systems eventually comes into play. This requires a correct appreciation of the physical processes involved.

Increases in knowledge, in conjunction with modern advances in computational methods, can allow the designer to take into account many of the factors that could only have been partially considered in earlier days. It is now possible to consider such factors as surface storage and attenuation, flood wave movement in pipes, and flows in surcharged pipes, as well as the selection of the eventual designs based on assessments of costs and benefits.

The selection of the actual methods to be used in urban hydrologic studies should consider three factors: the size and importance of the urban area, the type of drainage system to be used, and the availability of appropriate data. The size and importance of the urban area should consider the expansion of the city limits, zoning, rates of growth of the population and densities, the existing drainage network, and damages due to flooding. With respect to the types of drains, reference is made to surface runoff and the secondary drains or major drains, each of which should be expected to perform its functions properly. No less important, and possibly too often underemphasized, the methodology should always consider the availability (in kind and quality) of the data. As Franceschi & Rodrigues (1983) emphasize, too much effort is wasted on rainfall-runoff modelling by failing to realize the importance of the basic data sources.

In fact, all but a small fraction of the storm sewers in the world have been sized by fully empirical methods. Given the lack of evidence of superior methods, these simplistic procedures proved adequate when the primary purpose of storm sewers was to drain the land and quickly transport the surface runoff to receiving waters. Out of sight, out of mind. But once restraint or containment of flows and their pollution loads became additional primary objectives, the traditional procedures of analysis were no longer adequate. The use of mathematical simulation is undoubtedly an effective means for analyzing alternative urban runoff control strategies; its most important use, however, is in the assessment of expected system performance. The models can provide qualified quantitative information to serve as one input into the decision-making process. To ensure the validity of planning models, their verification against local field data or data from catchments with similar geographical, climatological and hydrological characteristics is essential (Geiger, *et al.*, 1987).

But models based on the detailed reproduction of observed phenomena would normally be much too complex for simulation in the typical humid tropic setting. The problem is, of course, complicated in the urban drainage design when one realizes that we are considering imaginary events on a future catchment. It is complicated further because the physical laws are not simple and require unlikely knowledge of initial conditions. Furthermore, the individual microtopography of sub-catchments draining to inlets will probably be beyond description (Department of Environment, 1981).

Without appropriate data, it is not possible to make accurate calculations, or to model the runoff – without which reasonable water management and design of drainage systems in cities are very difficult. Without experimental data, it also is not possible to obtain the basic knowledge about the hydrological processes governing the hydrological cycle of urban areas. Such knowledge is necessary in order to develop the methods of calculation and design procedures relevant to the local climatic, geological and socio-economic conditions of the humid tropics. Most of the data and knowledge, however, have been acquired in developed countries with much different climatic conditions.

The development of modern urban water management in developed countries (situated mainly in the temperate zone) was preceded by a period of learning about basic hydrological and hydraulic processes going on in urbanized areas. Gathering this basic knowledge was facilitated by the establishment of several urban experimental basins where different hydrological elements were accurately measured on the appropriate time and space scales. There are no such experimental basins, fully equipped with instrumentation, situated in the humid tropics. The establishment of such research basins was strongly encouraged at the Humid Tropics Colloquium, Australia, 1989 (Gladwell, 1991).

There are many technical options to alleviate urban flood and drainage problems, among which are:
– upstream storage reservoirs,
– flood plain clearance,
– river modifications and levees along urban streams,

– retention basins/reservoirs in or near cities,
– diversion channels around affected areas,
– conversion of low-lying urban areas into polders,
– flood pumping stations,
– storm drains/sewers,
– construction of streets to serve as drains.

In some cases, there can be no lowering of technical standards as the result of failure could be disastrous. But in many instances, failure does not mean disaster, and thus the use of approaches appropriate to the local situation is more than adequate.

Desbordes & Servat (1988) suggest that good solutions in developing countries should have the following characteristics:
– they must not induce strong constraints for the users,
– they must be easily and quickly constructed. In that way they may be quickly adapted to a changing socio-economic environment,
– they must not induce new problems and/or nuisances, especially in the sanitary domain,
– they must not require the importation of costly materials or equipment. Rather, they should lead to local economic developments, if possible,
– they should not be regarded as poor or cheap techniques, but rather as solutions better adapted than those that have been designed for very different climatic and socio-economic conditions.

As an example, these authors note that streets and roads can be used in a design scheme by considering them as major components of a storm drainage system. Under this concept, one is led to "minor" systems which would be the classical drainage system designed for frequent rainfalls (a two-year return period, for example), and "major" systems consisting of the streets themselves being designed for larger storm events (perhaps a 50-year or more return period). The city of Tahoua, Niger, has proposed that paved streets be used for storm drainage, which if compared to a classical sewer system, has a number of advantages, according to Desbordes & Servat(1988):
– no specialized staff nor unusually heavy earthworking equipment would be required,
– maintenance would be limited to pavement repair,
– it would be unnecessary to import construction material and workers would be found locally,
– the construction of the system would be less expensive than classical asphalt streets with classical drainage systems using open channels (which generally leads to serious sanitary problems due to misunderstandings by the populace of their function) or covered drains.

In the planning and designing of storm runoff control systems, it is usually necessary to be able to predict the volume, peak discharge and the timing of urban runoff under a variety of circumstances. Most of the techniques used have been extended from procedures used earlier in rural areas. Some of these techniques are discussed below.

Time-Area Method The total time that it takes an elemental volume of water, originating from a point within an urban watershed, to reach a specific point of interest within the catchment includes the sum of the times required for the excess precipitation (runoff) to flow over the slopes, through the gutters and into and through the drainage networks. The watershed can be divided into incremental areas a_j with boundaries defined by lines connecting points in the drainage area space which have equal travel times (isochrones). The time required for a particle of water to move from the most hydraulically remote point to the point of interest in the basin (the lowest part of the drainage, as defined) is equal to $n\Delta t = t_c$, where t_c is the time of concentration, n is the number of isochrones and Δt is the time of travel between adjacent isochrones. The area a_j ($j=1,n$) plotted against the time is called the time-area histogram, the base of which is t_c (Jovanovic, 1986).

If the excess precipitation is uniformly distributed over the watershed area, the runoff at the point of interest after time $k\Delta t$ is:

$$Q_k = \sum_{j=1}^{k} a_j i_{k-j+1}$$

(where $i(1)$, $i(2)$,...$i(n)$ are the rainfall intensities within time increments Δt)

If the duration of the excess precipitation is $t_r = t_c = n\Delta t$, all of the drainage area will be contributing to the runoff, and as a result the peak discharge will be:

$$Q_n = Q_{max} = \sum_{j=i}^{n} a_j i_{n-j+1}$$

and the falling limb of the hydrograph will be represented by:

$$Q_{k=n+c} = \sum_{j=k+1-n}^{n} a_j i_{k-j+1}$$

for $c = 1,2, \ldots ,n$

This method of calculating surface runoff is referred to as the isochronal method (or genetic method), and is the basis of many storm drainage design methods.

If the intensities of excess rain are uniform in time, $i(1)=i(2)=...= i$ then it can be shown that this transforms into the well-known Rational Method.

It should be noted that this genetic approach does not take into account storage effects. Because of that fact, it tends to over-predict the peak rate of direct runoff.

The Rational Method Although this method is widely criticized, it remains widely and successfully used, even by those who criticize it. It is to be found hidden within many complex computer models.

The Rational Method is used to predict peak discharge rates, using data on the contributing area, rainfall intensity, and the characteristics of the land surface, from:

$$Q_p = 0.28 \, CIA$$

where,

Q_p is the peak flow at point of interest in $m^3 \, s^{-1}$,
C is the coefficient of runoff,
I is the rainfall intensity in $mm \, hr^{-1}$,
A is the contributing area of the basin in km^2

The equation is best used for basins smaller than about 100 hectares. These basins could be expected to approach equilibrium between the input rainfall and output runoff in a short and intense storm. With the added assumptions about the average shapes of hydrographs in the region (which is easily done), the method could be extended for the design of small detention reservoirs. Its extension for use in tropical urban areas requires that the contributing drainage areas be determined. It should be noted that this is not necessarily an easy task in urbanized areas because the existence of surface channels or buried drains may transfer runoff across the obvious boundaries. It also requires that short-period rainfall intensities be available or measured, and that the C (perhaps obtained from other regions and urban areas) be checked and modified for the tropical conditions. The latter can be done by the simultaneous measurement of rainfall intensities and peak runoff rates in small basins (Dunne, 1986).

In general, it can be said that there is a wide variation in losses due to infiltration and depression storage. The assumption of 100 per cent runoff from impervious surfaces and zero runoff from previous surfaces is a simplification that leads, in general, to over-estimation of runoff for a given storm input. This is true even for those large events comparable with the magnitude of a design storm. There is more storage available above ground than is usually considered (Department of Environment, 1981).

To reflect the effects of reservoir-type storage in the system as a whole, runoff hydrographs predicted by the Rational Method generally require attenuation. However, Bouvier (1990) reports that, in Africa, runoff coefficients cannot be considered as the ratio of impervious areas. He highlights this by noting the cases of Niamey and Cotonou where, for the same imperviousness of 30 per cent, the runoff coefficients are equal to 60 per cent in Niamey and 20 per cent in Cotonou for a one-year return period rainfall. Such results emphasize the need for careful consideration before transferring information. As he notes, considering rainfall intensities, it is clear that the contribution of natural grounds must be taken into account for a proper evaluation of both runoff volumes and peaks.

Unit hydrograph Another way of deriving a discharge hydrograph from rainfall data is through the derivation of a unit hydrograph. The unit hydrograph for a drainage basin defines the time distribution of a unit of excess precipitation (after all losses are removed) generated by a rainstorm (in theory) uniformly over the drainage area at a uniform rate during a specified unit of time. The theory is based principally on the following criteria:

(1) For a given watershed, storms producing excess precipitation over equal durations will produce discharge hydrographs with approximately the same time bases, irrespective of the intensity of the rain.

(2) For a given watershed, the magnitude of the ordinates of the produced hydrograph (the discharge) will be proportional to the volume of the surface runoff produced.

(3) For a given watershed, the time of the distribution of the discharge is independent of the effects of any previous or subsequent rainfall incidents.

One then speaks of a T-hour unit hydrograph, where "unit" refers to the volume of excess precipitation, e.g., one centimeter, or one millimeter, etc.

The concept, though clearly only approximate, has been found to be useful because many of the drainage basin characteristics, such as area, shape, and channel gradient, which obviously affect the timing of the runoff are constant from storm to storm. This is particularly true of urban areas. When unit hydrographs have been derived for a number of drainages, parameters of the shape (e.g., peak discharge, time to peak discharge, duration of storm runoff, etc.) can be correlated with physical characteristics of the drainage. The resulting relationships can then be used to generate synthetic unit hydrographs for ungaged basins where the requisite information has been measured. It is also possible to derive from one unit hydrograph, unit hydrographs of other durations.

The approximation permits the ordinates of the unit hydrograph, which is derived by trial-and-error, to be multiplied by the ratio of the depth of runoff generated in the design storm and the unit depth of the runoff under the unit hydrograph. The design hydrograph is thus derived.

Curve numbers Another method of predicting runoff as well as the hydrograph is the U.S. Soil Conservation Service Method that uses curve numbers. These are empirical indices of the hydrologic response of soil and vegetation covers which were developed from runoff records of small basins in the United States of America (USA). Clearly, then, the values will need to be modified on the basis of tropical measurements, but they represent a good starting point on which to base preliminary calculations as well as analyses of databases of runoff. The method has been extended to urban areas where the curve number, and, therefore, the volume of runoff per unit of rainfall increases with the extent of impervious cover. It is, nevertheless, evident from the discussion above (see "The Rational Method") that field measurements will have to be made to determine the relationship of the pervious vs impervious area contributing to runoff.

Flood routing Of interest in urban hydrology is the value of delaying the runoff. This can be accomplished by diverting it

through natural lakes, ponds, or wetlands or by constructing small dams. There may also be a need to study the effects of some alteration of the land surface or of storage on the flood hydrograph downstream. This requires the consideration of how the various storages affect the timing and shape of the flood wave. These calculations are necessary to establish heights of the flood peaks at specific locations, estimating the protection that might result from construction of a reservoir, and for determining the height of a levee for flood protection, for example. Procedures developed to make these studies are known collectively as flood routing. There are basically two such procedures: hydraulic routing and hydrologic routing.

Hydrologic routing is the simpler of the two, and can be usually computed on the basis of a few specifications or field measurements. The principles underlying it are as follows: the continuity of mass (i.e., inflow to a reservoir or channel reach minus outflow being equal to change in storage during some time interval), and that there is a specifiable relationship between the storage volume and the outflow. The latter relationship is defined on the basis of engineering design of some structure of known hydraulic characteristics in the case of a dam and spillway, and by field measurements of water elevation and discharge (Dunne, 1986).

Hydrologic simulation What is a model? All of the above techniques are attempts to model different aspects of the hydrologic system. Without the aid of computers, the early urban catchment runoff models were, however, inevitably forced to be simple in computational requirements. With the advent of the digital computer, the modelling philosophy changed considerably. More and more detail and computational complexity has been made possible. With that has come the realization that the limitation is shifting to the lack of understanding of the physical, chemical and biological phenomena and insufficient data for model calibration and verification. In recent years, there has been a trend toward the development and use of general-use models rather than special-purpose models (Yen, 1986).

A model consists of mathematical representations of processes so as to transform an input to produce an output. One philosophy emphasizes the physical process-based modelling, following as closely as possible the spatial and temporal sequences of the physical system process. Another philosophy considers the transformations to be analogous to something else, not to the true physical process, but adequately simulating the transformation to produce "satisfactorily approximate" answers.

There are many urban rainfall-runoff models in existence. The majority of them have been developed with the intention of being used in real-world situations. A summary (far from exhaustive) of the important features on the urban overland surface runoff portion of selected models, mostly non-proprietary, is given by Yen (1986) for physically based models and for conceptual and unit-hydrograph based models. It is obvious that modern hydrologic simulation techniques will become more and more important in urban water management in the humid tropics, as elsewhere.

CONCLUSION

While water management is but one aspect of urban development, it is quite evident from the many examples of consequences of improper management throughout the humid tropics that a much more integrated approach is required. Decision-makers, planners, engineers and the public need to be aware that the results of narrowly focused and limited approaches to the solution of the various water-related problems can have consequences far beyond the obvious limits of the urban area. It is also evident that even apparently non-water-related urban activities can have adverse effects on the water resources and subsequently the health of the people of the cities and of the region in general.

It is important that those who will be making the decisions on water management issues take into account the need for an all-encompassing perspective and not continue to "solve" the problems as has been done in the past. That approach has not worked. And unless there is a change of perspective, the outlook for the urban areas of the humid tropics is not good.

REFERENCES

Bartone, C. R. (1990) Water quality and urbanization in Latin America. *In: Water Int., J. of the Int. Water Resources Assoc.* IS (1990) 3—14.

Bouvier, C. (1990) Concerning experimental measurements of infiltration for runoff modelling of urban watersheds in Western Africa. *In: Hydrol. Processes and Water Management in Urban Areas.* (Invited lectures and selected papers, Duisburg Conf., Urban 88, April 1988). *IAHS Publ. 198*:43–49.

Christmas, J. & de Rooy, C. (1991) The decade and beyond: At a glance. *In: Water Int., 16 Sept. 1991; J. of the Int. Water Resources Assoc.*

Clark, J. R. (1991) Coastal zone management. *In: Managing Natural Disasters and the Environment.* (Selected papers from the Washington D.C. Colloquium, June 1990) Environ. Dept., The World Bank, 115–144.

Davis, I. R. (1986) The planning and maintenance of urban settlements to resist extreme climatic forces. *In: Urban Climatology and Its Applications With Special Regard to Tropical Areas.* Proc. of the Mexico Tech. Conf., Nov. 1984, World Climate Programme. World Meteorol. Organization, *WMO-No. 652*:277–312.

Department of Environment, National Water Council, UK (1981) *Design and Analysis of Urban Storm Drainage, the Wallingford Procedure, Vol. 1: Principles, Methods and Practice.* Nat'l Water Council, London, England.

Desbordes, M. & Servat, E. (1988) Towards a specific approach of urban hydrology in Africa. *In: Hydrological Processes and Water Management in Urban Areas.* Proc. Duisburg Conf., Urban Water 88, April 1988). Int'l. Hydrol. Programme, 231–237.

Dunne, T. (1986) Urban hydrology in the tropics: Problems, solutions, data collection and analysis. In: *Urban Climatology and Its Applications With Special Regard to Tropical Areas.* Proc. of the

Mexico Tech. Conf., Nov. 1984, World Climate Programme. World Meteorol. Organization, *WMO-No. 652*:405–434.

Foster, S. S. D. (1990) Impacts of urbanization on groundwater. *In: Hydrological Processes and Water Management in Urban Areas.* (Invited lectures and selected papers, Duisburg Conf., Urban 88, April 1988.) *IAHS* Publ. *198*:187–207.

Franceschi, A. & Rodriguez, I. (1983) Current status of urban hydrology in Venezuela. *In: Urban Hydrology.* Proc. Baltimore, May/June 1983. Am. Soc. of Civil Engrs., New York, NY, 184–192.

Geiger, W. F., Marsalek, J., Rawis, W. J. & Zuidema, F. C. (1987) *Manual on drainage in urbanized areas, Vol. I. Studies and Reports in Hydrol.* No. *43*, UNESCO, Paris, France.

Geiger, W. F. (1988) Appropriate technology for urban drainage in developing countries. *In: Hydrological Processes and Water Management in Urban Areas.* Proc. Duisburg Conf., Urban Water 88, April 1988. Int'l Hydrol. Programme, UNESCO, Paris, France, 257–268.

Gladwell, J. S. (1991) A programme of study and research on urban drainage in the humid tropics. *In: New Technologies In Urban Drainage, UDT `9l.* Proc. of the Dubrovnik, Int'l Conf. on Urban Drainage and New Tech., IRTCUD. Elsevier Sci. Publ. Ltd., Essex, England, 337–346.

Gladwell, J. S. & Low Kwai Sim (1991) Urban water issues/Strategies in the humid tropics. *In: Water Resources: Planning Management and Urban Water Resources.* Proc. New Orleans, 18th Annual Conf. and Symp., May 1991. Am. Soc. of Civil Engrs., New York, NY, 358–364.

Gondwe, E. (1990) Water management in urban areas of a developing country. *In: Hydrological Processes and Water Management in Urban Areas.* Invited lectures and selected papers, Duisburg Conf., Urban 88, April 1988. *IAHS* Publ. *198*:315–322.

Gunnerson, C. G. (ed.) (1988) *Wastewater Management for Coastal Cities, The Ocean Disposal Option.* UNDP Proj. Management Report No. 8, The Int'l. Bank for Reconstruction and Development/The World Bank, Washington, D.C., USA.

Hassan, E. M. & Luscombe, W. (1991) Remote sensing and technology transfer in developing countries. *In: Managing Natural Disasters and the Environment.* (Selected papers from the Wash. D.C. Colloquium, June 1990). Environ Dept., The World Bank, 141–146.

Hjorth, P. (1991) Urban water policy – a conceptual framework. *In: Water awareness in societal planning and decision-making.* Proc. Skokloster Int'l. Workshop, June/July 1988. Swedish Council for Building Res., D4:l99l, 143–154.

Jovanovic, S. (1986) Hydrologic approaches in urban drainage system modelling. *In: C. Maksimovic & M. Radojkovic (eds.) Urban Drainage Modelling.* Proc. of the Dubrovnic Symp., April 1986. Pergamon Press, 185–208.

Landsberg, B. E. (1986) Problems of design for cities in the tropics. *In: Urban Climatology and Its Applications With Special Regard to Tropical Areas.* Proc. of the Mexico Tech. Conf., Nov. 1984, World Climate Programme. World Meteorol. Organization, *WMO-No. 652*:461–472.

Linden, O. (1990a) Environmental threats against fish-producing tropical coastal waters. *In: Water Resources Management and Protection in Tropical Climates.* (Selected papers from Havana First Int'l. Symp., Feb. 1988.) Res. Centre for Hydraulic Resources (Havana) and Swedish Environ. Res. Group, Stockholm, 389–402.

Linden, O. (l990b) Human impact on tropical coast zones. *In: Nature & Resources,* UNESCO, 26(4):3–11.

Lindh, G. (1983) *Water and the City.* UNESCO, Int'l. Hydrol. Programme, Paris, France.

Lugo, A. E. (1991) Cities in the sustainable development of tropical land-scapes. *In*: *Nature and Resources 27*(2), UNESCO, Paris, France.

Mara, D. (1982) Appropriate technology for water supply and sanitation. Sanitation alternatives for low-income communities – a brief introduction. Wash., D.C., The World Bank.

Marsalek, J. (1990a) Integrated water management in urban areas. *In*: *Hydrological Processes and Water Management in Urban Areas*. (Invited lectures and selected papers, Duisburg Conf., Urban 88, April 1988.) *IAHN* Publ. *198*:315–322.

Marsalek, J. (1990b) *Sediment in urban areas: Concerns, sources and control*. Nat'l. Water Res. Inst., *NWRI* Contribution *90–94*, Burlington, Ontario, Canada.

Niemczynowicz, J. (1991) Environmental impact of urban areas – the need for paradigm change. *In*: *Water Int., J. of the Int. Water Resources Assoc.*, *16*:83–95.

Nickum, J. E. & Easter, W. R. (1989) Water use conflicts in Asian-Pacific metropolises: Institutional concepts. *In*: *Integrated water management and conservation in urban areas*. Proc. Nagoya Int'l. Symp.-cum-Seminar, Aug./Sept. 1989. IHP Japanese Nat'l. Committee, Int'l. Hydrol. Programme, UNESCO, Paris, France, 200–213.

Ojo, O. (1987) Recent trends in precipitation and the water balance of tropical cities: The example of Lagos, Nigeria. *In*: *Hydrological Processes and Water Management in Urban Areas*. Proc. Duisburg Conf., Urban Water 88, April 1988. Int'l. Hydrol. Programme, UNESCO, Paris, France, 95–106.

Prost, A. (1991) Personal communication. World Health Organization, based on internal memo RUD/EG/es, Geneva.

Ramaseshan, S. (1983) Progress since 1979 in India. *In*: *Urban Hydrology*. Proc. Baltimore, May/June 1983. Am. Soc. of Civil Engrs., New York, NY, 143–149.

Ruiter, W. (1990) *Watershed: Flood protection and drainage in Asian Cities. Land & Water Int'l. 68*:17–19.

Sartor, J. (1987) The interaction of flood water flows in sewer networks and small river systems. *In*: *Hydrological Processes and Water Management in Urban Areas*. Proc. Duisburg Conf., Urban Water 88, April 1988. Int'l. Hydrol. Programme, UNESCO, Paris, France, 113–120.

Sharma, M. L. (1986) Role of groundwater in urban water supplies of Bangkok, Thailand, and Jakarta, Indonesia. (Paper prepared for the Environ. and Policy Inst., East-West Center, Honolulu, Hawaii.)

Tokun, A. (1983) Current status of urban hydrology in Nigeria. *In*: *Urban Hydrology*. Proc. Baltimore, May/June 1983. Am. Soc. of Civil Engrs., New York, NY, 193–207.

Uhlmann, D. (1990) Alternative methods for the treatment of organic wastes in tropical climates. *In*: *Water Resources Management and Protection in Tropical Climates*. (Selected papers from Havana First Int'l. Symp., Feb. 1988.) Res. Centre for Hydraulic Resources (Havana) and Swedish Environ. Res. Group, Stockholm, 325–343.

UNESCO (1979) *Impact of Urbanization and Industrialization on Water Planning and Management*. Report of the Zanvoort Workshop, Paris. UNESCO Press. (Studies and Reports in *Hydrol. No. 26*).

UNESCO (1992) Ground water: Managing the invisible resource. *Environmental and Development Briefs*, No. 2, Paris, France.

World Health Organization (1988) *Urbanization and Its Implications for Child Health: Potential for Action*. WHO, Geneva, Switzerland.

Yen, B. C. (1986) Rainfall-runoff process on urban catchments and its modeling. *In*: C. Maksimovic & M. Radojkovic (ed) *Urban Drainage Modelling*. Proc. of the Dubrovnic Symp., April 1986. Pergamon Press, 3–26.

Yen, B. C. & Yen, C. L. (1991) Urban drainage in tropical islands. *In*: *New Technologies In Urban Drainage, UDT `91*. Proc. of the Dubrovnik, Int'l. Conf. on Urban Drainage and New Tech., IRTCUD. Elsevier Sci. Publ. Ltd., Essex, England, 325–335.

20: The Management of Water Resources, Development and Human Health in the Humid Tropics

A. Prost

World Health Organization, Avenue Appia, 1211 Geneva 27, Switzerland

ABSTRACT

Over the past twenty years, economists and planners have questioned the benefits of water supply, while other health interventions have competed with the water sector for resource allocation. This has created a credibility gap which the medical profession and sanitary engineers have had difficulty in bridging. Evaluation of the health impact of water management in tropical developing countries has likewise been excessively oriented towards adverse effects, further impeding the development of the necessary resources. Despite the momentum created by the International Drinking Water Supply and Sanitation Decade, coverage of the world population with supply services is far from satisfactory. Progress can hardly keep pace with the increase in population. The majority of rural people in poor tropical countries are not served and there is little prospect of achieving universal coverage in the foreseeable future. Moreover, the generalization of cost mechanisms may shift resources from other vital sectors (nutrition, health) and so offset the benefits of a safe water supply.

Innovative approaches can overcome the professional barriers between health specialists and decision-makers. The main health benefits accrue from the availability of unlimited quantities of water, whatever its quality. Unnecessarily stringent quality standards may be counter-productive since they may reduce the quantities available, delay the supply or increase its cost. Recycling waste water is a necessity to increase the resources, and appropriate health safeguards can be developed with respect to each usage category, depending on selected application techniques.

Finally, engineering techniques and nonmedical interventions in water management can be shown as the most cost-effective measures for controlling diseases of economic importance. Examples of dracunculiasis and onchocerciasis illustrate the point.

THE CONCEPTUAL APPROACH: RECENT DEVELOPMENTS

Historical background: the traditional approach

Since the Fifth Century B.C., when the treatise on Airs, Waters and Places became part of the Hippocratic corpus, and until the Nineteenth Century, swamps and humid places had been considered insalubrious. It had been established that there was a relationship between such places and malaria – and, to a certain extent, diarrhoea – although water had not emerged as being one of the most important determinants of health and disease.

It did so the day John Snow removed the handle of the Broad Street pump in London during the 1854 cholera epidemics, thereby inducing a striking decrease in the number of new cases in the families who used this public supply (Fig. 1). It really emerged as an essential determinant after Pasteur's discoveries and the establishment of microbiology as a scientific discipline. By the end of the Nineteenth Century, a corps of hygienists had been created in almost all countries. They were not usually medically trained, but focused their activities primarily on water quality and sewerage, setting up stringent standards and broadly advocating a "germ free" approach.

One hundred years later, it is difficult to recognize the sectoral impact of new approaches, and even more difficult to evaluate the benefits associated specifically with water quality. A study on the decline in mortality in three French cities (Paris, Lyons, Marseilles) in the Nineteenth Century provides some insight (Preston & Van de Walle, 1978). Data indicate that mortality first declined in Lyons around 1850, followed by Paris ten years later, and finally by Marseilles in the 1890s. Figure 2 illustrates the time sequence of the gains in female life expectancy in the three cities. The authors argue that the observations are consistent with the multiplier effect of the completion of both a generalized water supply system and a

Figure l: "Death" at the Broad Street Pump during the cholera outbreak, London, 1854. (Nineteenth Century etching, courtesy A. Dodin, Institut Pasteur, Paris.)

sewage system in Lyons by 1855 and with slow but continuing progress in service delivery in Paris throughout the period 1855–1900. A satisfactory water supply was not available in Marseilles before 1898. However, similar trends could not be identified in other historical series, especially in England, and Preston's presentation is merely an assumption.

More recently, a study in urban Brazil has estimated that access to piped water accounted for about one-fifth of the reduction in child mortality between 1970 and 1976 (Merrick, 1985).

The documented impact on cholera and typhoid, the decline in mortality rates and the long-lasting campaign for a hygienic environment have led to the firm belief that safe water is one of the basic needs for human life, that its health benefits "go without saying", and that looking for further evidence would be a waste of time and resources.

The economic reassessment of the l970-l980s: consequences and shortfalls

With the onset of the economic crisis, investment in the so-called "soft" social sectors has been increasingly scrutinized. Priority was to be given to sectors which were likely to produce economic and financial returns.

It was argued that resources were not sufficient to implement the complete package of primary health care interventions, and that a selective approach was necessary (Walsh & Warren, 1979). In particular, it was argued that the cost per

infant death averted through water supply and sanitation programmes was l0 to 15 times higher than the cost per infant death averted through a selective package which includes oral rehydration therapy, basic immunization, malaria treatment and promotion of breast-feeding (Table 1).

Similar results were derived from a comparison of life expectancy in countries with low and those with high water and sanitation coverage. The "threshold-saturation theory" (Shuval *et al.,* 1981) suggests that improvements in water supply and sanitation would have little effect on health in populations at both low and high levels of socio-economic development. In poor countries, other pathologies offset the benefits which can accrue from the provision of safe water (malnutrition, infectious diseases, etc.). In rich countries, minor improvements in health necessitate high expenditures (e.g., complex sewage systems, treatment plants, etc.). A direct impact is measurable only in middle-level countries (Fig. 3). Thus, many development agencies have considered that investments in water supply and sanitation were not cost-effective below a certain level of development, and they subsequently shifted resources towards other interventions such as those proposed in the previously mentioned "selective package."

At the same time, concerns were expressed about the adverse health effects of water management in the tropics. The gross irrigated area of the world doubled between 1950 and 1970 (Rangeley, 1985). Almost 40% of all reservoirs and dams over 30 metres high were completed between 1960 and 1980, and the proportion would exceed 50% if smaller dams were included (Table 2).

The spread of water-related diseases raised questions about the actual benefits of such development schemes for the populations. For example, the first Aswan dam in Egypt was responsible for an increase in the prevalence of schistosomiasis from l–ll% in 1934 to 44–75% in l937 in nearby population groups. In 1968, one year after the completion of the Akosombo dam in Ghana, 90% of children aged 10–14 years in the waterside villages carried *Schistosoma* parasites, whereas the carrier rate did not exceed 10% before construction of the dam.

Irrigation schemes can be even more harmful. In Burkina Faso, the irrigation of 1,200 hectares of rice fields in the Loumana Plain created conditions that induced an outbreak of onchocerciasis, resulting in the desertion of the settlement and in the dilapidation of the system five years later (1957–1962), after 15% of the women and 20% of the men became blind (Hervouet, 1983). Earlier irrigation in Upper Egypt had created breeding sites suitable for the mosquito *Anopheles gambiae*. The pullulation of this previously uncommon vector of malaria, which was possibly unknown in this area, was the direct cause of about 130 000 malaria deaths registered in 1942–1943.

Numerous reports of this nature were used by financiers to challenge the rationale of major investments in water manage-

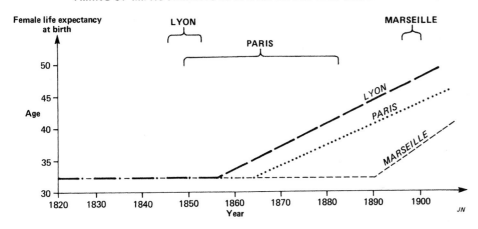

Figure 2: Mortality decline in three cities of France during the Nineteenth Century (Data from Preston & Van der Walle, 1978).

Table 1. *Estimated annual costs of different systems of health intervention.*

Intervention	Cost *per capita* (US$)	Cost per infant (I) and/or child death (C) averted		
Basic Primary Health Care	2.00		700	(I)
Mosquito Control for Malaria	2.00		600	(I)
Water Supply and Sanitation	30-54	3600 –	4300	(I,C)
Nutrition Supplementation	1.75	213 –	3000	(I,C)
Selective Primary Health Care	0.25	200 –	250	(I,C)

Source: Walsh & Warren, 1979

ment. It increased the separation between developers and health professionals, the latter being considered rather as brakes on the pace of development and sometimes even as obstacles to the development process.

In the mid-1980s, we were facing an apparent divorce between engineers and health professionals. On the one hand, the reluctance to reevaluate the health impact of water supply programmes has been detrimental to the entire sector. There is a lack of good-quality data to challenge the conclusions of those who advocate a shift in resource allocation. There is even some reluctance to argue that methodological biases distort the conclusion, or that it is not legitimate to evaluate the impact of water supply programmes on a single outcome, e.g., infant deaths averted, which is not the primary objective of those programmes and which is one among many health benefits that are derived from it. Moreover, on these grounds it is irrelevant to compare water supply with programmes such as oral rehydration therapy, which are designed explicitly to reduce infant mortality. Cost-effectiveness analyses should encompass all demonstrated outcomes and should, as discussed by Briscoe (1984), validate cost figures more accurately.

On the other hand, adverse health effects of certain types of water management have been used excessively in a negative way. This has been detrimental to the credibility of the health profession, since negative recommendations, which are perceived as obstacles, are usually overcome and often discarded. The mistake of health advisers is twofold: first, they have not been skillful enough in determining which minor and often low-cost adjustments to project designs could have the greatest effect in minimizing the health risks. In this respect, experiments on the profiles of overflow channels to prevent blackflies (vectors of onchocerciasis) from establishing breeding sites near reservoirs remain an exception (Quelennec *et al.*, 1968) (Fig. 4).

Second, negative opinions are not properly balanced with the positive outcome of development projects. Risk assessment is often carried out as an evaluation of the risk factors added to the present situation. It does not consider the benefits derived from the project, e.g., the health effects of higher income, improved nutrition, housing, etc. This topic will be dealt with later in this paper.

The epidemiological reassessment of the health risks associated with water

Significant progress in the conceptual approach of health risks associated with water has followed the innovative analysis

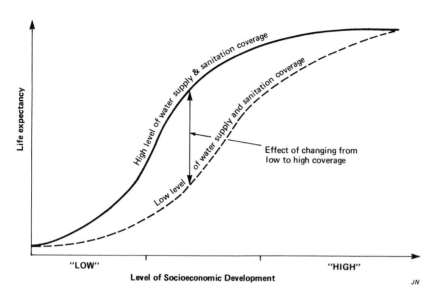

Figure 3: The threshold-saturation theory (Adapted from Shuval *et al.*, l981).

Table 2. *Large dams over 15 metres high.*

	Number of dams completed		
	before 1961	1961–1980	After 1981 and under construction
Africa	251	404	96
America	2673	1361	196
Asia	2235	2027	584
Europe	2033	1562	377
Oceania	216	202	63
TOTAL	7408	5556	1316

Source: World Resources Institute, New York, 1987.

Note: About 20 000 dams between 15 and 30 metres high in China, India, Japan and the USA, and about 3000 in the USSR are excluded from these statistics.

developed by White *et al.,* (1972). It departs from the century-old exclusive concern for the quality of water. It introduces the notion of the importance to health of water quantity, whatever its quality. It presents a comprehensive picture of the relationship between health and water as a necessary component of the human environment: uses for drinking and for hygiene; contact during work and recreation; distant effect through the consumption of water products (fish, etc.) and through the development in water of arthropods and snails harmful to man. It places the human-water relationship in an ecological perspective.

Table 3 summarizes the four categories of disease: water-borne, water-washed, water-based, and water-related.

Water-borne diseases, which are related to the quality of water, are no longer the most important category, but they rep-

resent a limited proportion of all pathological conditions related to water. Moreover, they are not exclusively dependent upon water for their transmission. Viruses are rapidly and more often transmitted through interhuman contacts. Their presence in water, in which they do not multiply, is merely an indicator that they have already spread in the community. Bacteria can use many routes of entry, with contaminated food as a much more common source of infection than contaminated water. Vegetables are involved in the transmission of typhoid fever and cholera, shifting attention from drinking water to water used in irrigation.

The category of *water-washed diseases* is a new and important concept. It comprises all diseases which are sensitive to hygiene practices, mainly skin diseases, external otitis and eye diseases. Pyodermitis (infectious skin disease) is the first dis-

Figure 4: Experimental dam at M'Para, Burkina Faso. The different shape of each overflow channel is designed to determine which profile can prevent blackfies from breeding. (See Quelennec, 1968.)

ease to be strikingly reduced when water is made available in unlimited quantities. It might be an early indicator of the beneficial health impact of water supply projects. In eye diseases, there is no evidence that water plays a significant role in conjunctivitis, but lack of water is associated with an increased, perhaps doubled, risk for trachoma, a blinding condition (Prost & Negrel, 1989). More surprising is the observation that water quantity also influences diarrhoea and other conditions which were previously considered to be exclusively dependent upon the quality.

Water-related and water-based diseases are the broad categories of interest in irrigation projects and reservoir construction. Two diseases are on the front line: malaria and schistosomiasis. Abundant scientific literature has been produced about them in relation to water management projects in the tropics.

The conclusions can easily be summarized: any project which increases the area of surface water results in the development of the *Anopheles* mosquito vectors of malaria and of one of the fresh water snail vectors of schistosomiasis. Such increases in surface water area occur in connection with irrigation schemes, canals, dams,and reservoirs, and there is no satisfactory engineering technique to prevent this situation. The dominance of malaria and schistosomiasis in risk-forecasting, based on sound epidemiological evidence and historical episodes, has relegated other diseases to low priority concerns. This is not entirely correct. Leptospirosis, for example, is an occupational risk in any place where man and animals live together in the vicinity of water collections which can be contaminated with rodent faeces. The expansion of irrigated rice cultivation is held responsible for the spread of Japanese encephalitis over southern Asia and westwards to India. There are examples of onchocerciasis outbreaks in relation to irrigation (Hervouet, 1983), although water management usually reduces the speed of the flow of water to values which do not permit the survival of blackfly larvae. Rift Valley fever outbreaks have developed suddenly after the completion of large reservoirs, *e.g.*, one million people infected and 18 000 clinical cases in 1976–77 in Upper Egypt near the Aswan lake, and an unexpected outbreak in Mauritania in 1987 which followed the completion of the Diama Dam on the Senegal River (Provost, 1989).

The case of schistosomiasis is indisputable. It has been documented in almost all water development projects, especially in the smaller ones (Fig. 5) (Hunter, 1981). Malaria, however, deserves a more cautious approach. In general, irrigated agriculture and reservoir construction will result in a more intense transmission that is more regularly spread over a longer period of the year. Its consequences for human disease are not unequivocally accepted, especially the assumption that increased transmission results in a worsening of the clinical symptoms and in greater severity. It is argued, on the basis of convincing evidence, that the severity of clinical malaria is not related to the number of infective bites, but to the immune status of the people. Repeated infections, at short intervals, stimulate the immune response and thus contribute to the mainte-

Table 3. *Diseases associated with water.*

* WATER-BORNE	BACTERIAL	Salmonella (typhoid), Enterobacteria (*E. coli*, Campylobacter), Cholera, Leptospirosis, etc.
water acts as a passive vehicle for infective agent	VIRAL	Hepatitis A, Poliomyelitis, Rotaviruses, Enteroviruses
	PARASITIC	Amoebiasis, Giardiasis, Intestinal protozoa, *Balantidium coli*
	ENTERIC	E.g., a proportion of diarrhoeas and gastroenteritis
* WATER-WASHED	SKIN	Scabies, Ringworm, Ulcers, Pyodermitis
Infections that decrease as a result of in- creasing the volume of available water	LOUSE-BORNE	Typhus and related fevers
	TREPONEMATOSES	Yaws, Bejel, Pinta
	EYE & EAR	Otitis, Conjunctivitis, Trachoma
* WATER-BASED	CRUSTACEANS	Guinea Worm, Paragonimiasis
A necessary part of the life cycle of the infective agent takes place in an aquatic organism	FISH	Diphyllobothriasis, Anisakiasis, Flukes
	SHELLFISH	Flukes, Schistosomiasis
* WATER-RELATED	MOSQUITOS	Malaria, Filariasis, Yellow Fever, Dengue, Haemorrhagic Fever
Infections spread by insects that breed in water or bite near it	TSETSE FLIES	Trypanosomiasis (Sleeping Sickness)
	BLACKFLIES	Onchocerciasis

Definitions by White *et al.* 1972

nance of a high immunity which counters the onset of severe clinical symptoms.

Changes in the water environment also induce changes in the mosquito populations which, as far as we know, are more diverse than the snail populations. New strains adjust to chang-ing ecological conditions and may be more efficient or less efficient vectors of the parasite. In the irrigated rice area of the Vallee du Kou, Burkina Faso, the density of *Anopheles* mos-quitoes is two times higher than in the surrounding dry savanna zone. However, the transmission of malaria is signifi-cantly lower (about three times). Tentative explanations assume a greater zoophilic preference: a reduction of the life span of the vectors in the irrigated ecosystem or a change in the *Anopheles* strain, or both (Robert *et al.*, 1985). On the other hand, irrigation on the Kano Plateau in Kenya has resulted in a reduction of the number of human-biting mos-quito species, but has induced the replacement of outdoor bit-ing species with indoor biting mosquitoes, thus quadrupling exposure to malaria transmission.

Major waterworks do not exclusively increase the area of surface waters. Often many have a component, or an objective, of flood control. Floods are more harmful to health than reser-

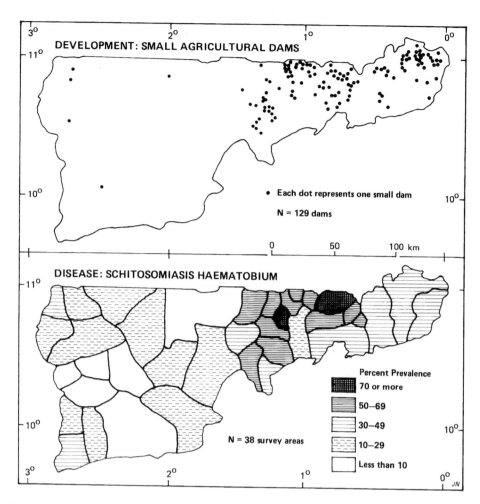

Figure 5: Distribution of small agricultural dams and of schistosomiasis disease in the upper region of Ghana (Source: Hunter, 1981).

voirs and canals, even if their effect on agriculture and housing and the disruption to normal life are disregarded. The containment of the Yang-Tsé and Huang-Hé Rivers in China has not only brought to an end a recurrent series of catastrophes, but has also significantly contributed to the reduction of the incidence of malaria to very low levels. In Shandong Province, located at the mouth of the Huang-Hé, an average of only 75 000 cases of malaria has been reported annually since the late 1950s, except for 1960 (over two million cases) and 1971 (3.2 million cases) when major floods overflowed the dikes and sometimes ruptured them (personal data). Drainage of the Rhine Valley was one the major factors which led to the disappearance of malaria from northeastern France, Germany and the Netherlands.

Finally, concerns about water storage in relation to health are not limited to reservoirs and large dams. Domestic storage has emerged as a key factor in the chain of determinants. Several studies, many of them carried out in Bangladesh by the International Centre for Diarrhoeal Disease Research, have shown that the impact of safe water supply on diarrhoea is maximal if tap water is supplied in the household itself. It decreases with the distance between the source and the household if water is supplied through public standpipes and is negligible when supply sources are further than 200 metres. In the

case of household tap supply, water is readily and permanently available. There is no storage. The farther the source, the less frequent the water collections and the longer the storage. Small amounts of water are drawn all day long from the main jar, using small cups deposited nearby. Each collection contaminates the reserve, and the multiplication of bacteria after a certain period of time (depending upon temperature) is sufficient to create a significant risk.

Longer storage (or permanent domestic systems) also influences diseases such as yellow fever. The vector, an *Aedes* mosquito, breeds in small water collections in the domestic environment, as well as in tins, broken pots, jars, etc. The pattern of the urban yellow fever outbreak in Oyo State, Nigeria, in 1987 has been clearly determined by domestic water storage practices (Cordellier, 1990).

The environmental reassessment

By the middle of the Twentieth Century, it was no longer possible to consider that water was a natural commodity available in unlimited quantities. Most Middle Eastern and sub-Saharan African countries suffer chronic water shortages, exacerbated by recurrent drought periods. So do parts of the United States of America, Australia, China, and the Soviet Union. Per capita water availability varies from about 122 000 cubic metres per

Table 4. *Faecal coliforms in rivers included in the Global Environment Monitoring System (Six-year period 1979-1984).*

number of faecal coliforms per litre	Mean North America	South and Central America	Europe	Asia and Pacific
Less than 100	8	0	1	1
100 – 1000	4	1	3	2
1000 – 10,000	8	10	9	14
10,000 – 100,000	3	9	11	1O
100,000 – 1 million	0	2	7	2
over 1 million	0	2	0	3
Total number of rivers	23	24	31	32

Source: Adapted from Global Pollution and Health, WHO/UNEP publication, Yale Press Ltd, UK, 1987.

year in Canada to 70 cubic metres per year in Malta (Forkasiewicz & Margat, 1980). On average, it is estimated that per capita water use rose from 300 cubic metres in 1950 to 800 cubic metres in 1980. Consumption per person per year averages 650 cubic metres in European OECD countries and exceeds 2,200 cubic metres in the United States of America. In many countries – and in many more each year – the demand for water exceeds available resources. For them, a 100% coverage of the population with satisfactory water supplies is unlikely.

Nowadays, we face the evidence that water is a scarce commodity which cannot be wasted. It is unlikely that a decrease in demand will occur. On a global scale, needs for irrigation account for about two-thirds of all human uses, and many irrigation activities have associated water losses, estimated to be between 50% and 80%. Water for irrigation does not require the same degree of purification as domestic water. Thus, a conscious water policy should consist of measures to reduce the exploitation of fresh water from natural sources (a critical level is apparently reached if more than one-third of the annual available water is utilized), to preserve natural waters from human-made pollution, and to recycle waste water for agricultural and industrial use when appropriate. The latter measure has a compounded effect on the two previous ones.

The subject of chemical contaminants in water and the associated risks has again been raised. Chronic lead intoxication, once rather common, has disappeared with the replacement of lead pipes and devices in supply systems. Water sources with naturally-occurring arsenic contents have been identified and disposed of. Fluorosis, a bone and joint deformation associated with high fluoride content in water, is being eliminated from natural foci, although it still occurs in India, China and central Asia, in the Arabian peninsula, and in selected places of eastern and southern Africa. New issues of concern are, *inter alia*, inorganic mercury, which accumulates in the waters and sediments of lakes and reservoirs; nitrates, which contaminate

groundwater as a result of the use of fertilizers in agriculture and from animal and human organic wastes; acidic deposition resulting from the chemical photo-conversion of sulphur oxides and nitrogen oxides emitted as air pollutants; pesticides and herbicides which are washed out in rivers. Many other chemicals and heavy metals may occur, especially in industrialized areas; all of them create new hazards for human health. In addition, the lack of sanitation, or inadequate facilities, results in widespread microbiological contamination of fresh and sea waters. Of the 110 major rivers included in the WHO/UNEP Global Environmental Monitoring System, almost one-half contain more than 10 000 faecal coliforms per litre, which is the upper limit tolerable for bathing waters (Table 4).

Therefore, at a time when greater attention should be devoted to the management of water resources, a large proportion of these resources are spoiled and unsafe. Groundwater itself is not spared. The most important source of pollution is the release of raw waste water into streams without any preliminary treatment.

It could be considered that recycling water, if possible, would both increase the available supply and improve the quality of environmental water. In addition, domestic waste water contains organic matters that have a fertilizing potential. The policy of agricultural reuse of municipal effluents is already advocated in countries such as Israel, Cyprus, Spain and in some North African states. Limited schemes have existed in the United Kingdom since 1840, in Germany and France since the end of the Nineteenth Century, and more recently in India and the United States of America. In 1987, a World Health Organization (WHO) scientific group recommended that the bacterial guidelines for waste water use in agriculture be relaxed (WHO, 1989). On the basis of epidemiological evidence and of available water treatment technologies, it recommended that water used for the irrigation of edible crops

Table 5. *Access to water supply services: World population coverage.*

per capita income categories	Total population (millions)			Population served (millions)			Percentage of total population		
	urban	rural	total	urban	rural	total	urban	rural	total
High income	807			806			100		
(over $ 5000)		283			280			99	
37 countries			1090			1086			100
Upper middle income	347			312			90		
($1800 to 5000)		196			125			64	
24 countries			543			437			80
Lower middle income	282			218			77		
($500 to 1800)		444			200			45	
46 countries			725			418			58
China and India	427			295			69		
($300)		1408			858			61	
			1835			1153			63
Low income	144			93			65		
(less than $500)		518			168			32	
42 countries			662			261			39
TOTAL	2007	2849	4855	1724	1631	3355	86	57	69
151 countries									

Note: The table does not include countries and territories with a population of less than 0.1 million. Also excluded are South Africa, Namibia, Comoros, Equatorial Guinea, Cambodia, Macau and Taiwan for which no data are available (total population excluded is about 55 million).

Reference year and sources: Mid-1986 population and 1986 GNP are from the 1988 World Development report (World Bank, 1988) as well as the proportion of urban population. Countries which are not members of the World Bank have been listed within the relevant category according to their own estimates. Water supply coverage figures are taken from reports to WHO received in 1988 as part of the monitoring of the Health for All Strategy. They are relevant to the 1985–87 period. For non-reporting countries, end-l985 data provided for the monitoring of the International Drinking Water Supply and Sanitation Decade have been used.

China: There is no aggregated coverage figure for China. A 60% coverage has been assumed for the urban population and 43% coverage for the rural population on the following grounds: In 1983, 221 of the 242 major cities in the country had piped water serving an estimated 50% of the urban population, whereas adequate supply was available to 350 million rural people, i.e. 44% (World Bank, 1984) At the same time, information collected in two of the most populated provinces of China provided the following population coverage figures: in Shandong Province, 85% in urban and 27% in rural; in Sichuan Province, 75% in urban and 32% in rural. In the small Ningxia autonomous region, 63% of rural population had still to rely on surface water (Prost, unpublished World Bank reports). Thus, we have retained a slightly increased figure of 360 million rural people served in 1986, and a conservative 60% coverage figure for urban populations.

should contain less than 10 000 faecal coliforms per litre, and that irrigation of fodder crops, industrial crops, and trees be permitted after the retention of waste water for l0 days in stabilization ponds. The consensus reached by the group of experts indicates that health is not threatened by the proposed practices; the stringent purity standards imposed for irrigation in many areas (e.g., the state of California in the United States of America) seem excessive. Recycling, a key element of the environmental management of water resources, is feasible at no risk with minimal precautionary measures.

Reuse of waste water is only one of the recommended practices for a comprehensive management of water resources. Others are to be developed, in close collaboration between engineers and public health specialists, with the same objec-

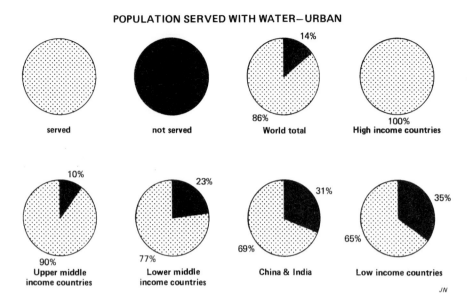

POPULATION SERVED WITH WATER–URBAN

served

not served

World total

High income countries

Upper middle income countries

Lower middle income countries

China & India

Low income countries

Figure 6: Urban population served with water

tives: savings in water extracted from the environment, and protection of environmental water from pollution, both measures that contribute to the increase in available resources.

POPULATION COVERAGE WITH WATER AND SANITATION SERVICES: GEOGRAPHICAL AND ECONOMICAL DISPARITIES.

Water supply

In 1986, which is the average year of reference for the most recent reports, two-thirds of the planet's five billion population had access to adequate quantities of water of reasonably good quality. The remainder, i.e., 1.5 billion people, had to rely upon surface water and other unsafe sources. For this one-third of the earth's people, one of the basic needs for a healthy life was not met. The inequity line divides the planet between continental entities that are adequately served (North America, Europe) and those that are not, between affluent countries and poor ones, between urban and rural areas, and even between regions which have achieved different population coverages within the same country and between cities and their slum suburbs. Table 5 and Figs. 6 and 7 summarize the situation for urban and rural populations in 151 countries – representing about 99% of the world's population. Since regional breakdowns result in merging together countries with very different levels of service, countries have been distributed in four per capita income categories, with China and India separated from the relevant group because of the size of their population. Table 5 provides valuable information about the unserved, who they are, and where they live.

They live in the poorest countries Table 5 shows the declining water coverage with decreasing average per capita income. In the poorest one-third of the world's nations, only one out of

every three persons has access to some kind of safe water supply. At the other extreme, the billion people who live in the affluent industrialized and the major oil exporting countries enjoy universal coverage.

They live in rural areas About one half of the world's rural population lacks coverage compared with only 15% of urban dwellers. The relationship between lower income and lower service coverage is stronger in the rural as opposed to the urban population, with a rapid and greater decrease in the proportion of people served. The noticeable discrepancy in the China-India group results from efforts specifically directed towards the rural sector, for political and economic reasons. India and Thailand consistently report higher coverages in rural areas. The only other example is Côte d'Ivoire which, with strong political backing, has since 1975 implemented an infrastructure programme in villages and small townships.

Urban dwellers enjoy a better coverage because they are richer, because they belong to the modern segment of the society, and because decision-makers are more sensitive to the needs of a category to which they belong. In addition, one should bear in mind that decisions based on economic analyses favour the urban sector. As the criterion for effectiveness is usually the additional number of persons served, cost-effectiveness analyses always arbitrate in favour of investments in the most densely populated areas.

They live in tropical areas All countries in the lowest income group, and with the lowest level of service coverage, are situated between the Tropics of Cancer and Capricorn, with the exception of Afghanistan and China. On the other hand, tropical countries which have achieved 90% coverage all have some particularity, being either small countries (Costa Rica)

POPULATION SERVED WITH WATER– RURAL

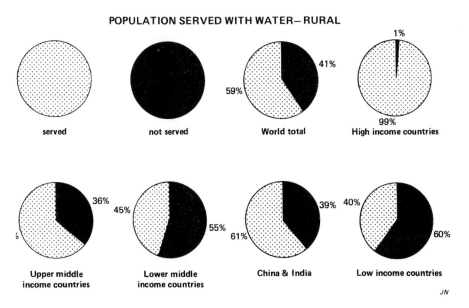

Figure 7: Rural population served with water

and even small islands (Fiji, Samoa, Jamaica, Trinidad, Tobago), oil exporters (Brunei, Trinidad) or industrialized city states (Hong Kong, Singapore). Thus access to safe water is often not available in the part of the world where it is most needed, i.e., in the tropics, where human pathogens develop rapidly in warm and humid conditions.

The vast majority live in a limited number of countries Eight countries account for 75% of the total number of unserved population: China, India, Indonesia, Nigeria, Pakistan, Bangladesh, Ethiopia, and Vietnam, each with over 30 million people unserved. If attention is focused on the urban segment of the unserved population, eight countries also account for 75%, with Brazil and Argentina replacing Pakistan and Ethiopia in the previously listed countries. Considering coverage as the target and cost-effectiveness as the rationale, the implication is that programmes should preferably be developed in this limited number of countries.

Sanitation

It is more difficult to estimate the coverage with sanitation facilities. Definitions vary from country to country; private facilities outnumber the public sewerage systems, thus enlarging the confidence limits of the reports. In most communities, sanitation is not valued as much as water supply, nor is it considered an essential element of life, thus decreasing the demand for services.

Overall, the coverage of the world population with adequate sanitary facilities is lower than the coverage with water services. Available WHO estimates for 1986 indicate that, on average, 75% of urban residents and 30% of rural people have access to such facilities. Urban coverage is over 90% in Europe, North America and western Pacific countries, close to 80% in the eastern Mediterranean region, and within the range

of 40–60% elsewhere. Rural coverage is excellent in Europe and North America (90%), satisfactory in the western Pacific countries (75%), and below 25% in the rest of the world, with figures of 10% and even lower in many countries.

Trends of the 1980s and beyond: issues and challenges

The year 1980 was a benchmark. It was the starting date of the International Drinking Water Supply and Sanitation Decade launched by the United Nations – acting upon the recommendation of the 1977 Mar del Plata Conference. The Decade has been instrumental in sensitizing nations to the needs of the sector and in mobilizing resources for a leap forward.

A comparison between figures for 1980 and 1988 indicates that in developing countries other than China, about 535 million people have gained access to water services during the period (310 million rural population and 225 million urban). Meanwhile, 325 million people have gained access to appropriate sanitation (75% in urban areas).

By combining national targets to obtain a global view, it is clear that the 1990s will provide governments with formidable tasks to achieve their goals. These call for more than doubling the implementation rates of urban water achieved since the beginning of the Decade, and raising the rate for rural sanitation by a factor of five. Rural sanitation will be the most difficult challenge. Unfortunately, the achievement of these ambitious objectives is unlikely in the present economic situation where countries in greatest need are facing major financial constraints.

The first challenge is the difficulty in coping with population growth. Between 1980 and 1985, the annual average population growth rate has been 1.9% in the low income group of countries, and 2.3% in the middle income group. An annual increase of 1.9% and 2.1%, respectively, is projected between 1985 and 2000. It means that about 55 million additional people are to be provided with services every year in developing countries, together with another 15 million in China. Thus, an increase of a few percentage points in coverage implies that progress in the provision of services would have to exceed the population growth. This is the reason why many countries have trimmed their plans. Between 1980 and 1985, the number of developing countries looking for 100% urban water supply coverage fell from 50% to 40%. Nowadays, no more than 15% of them have hopes of completing rural coverage before the year 2000. Population growth offsets the progress made in terms of additional coverage, thus conveying the impression that the International Decade has been a partial failure.

Another feature is the weighted effect of a few large countries in the world-wide statistics. As previously indicated, an effective programme aiming at the improvement of world coverage figures should concentrate on fewer than ten countries which account for 80% of the unserved people. On the other hand, if one considers that half the states of the planet are below decent coverage figures, corrective action may consist of implementing programmes in countries with a small population, thus maximizing the effect on the indicator while limiting the size of investments. The consideration of equity issues would tend to reorient programmes towards rural areas, whereas for the sake of cost-effectiveness, they focus on urban and densely populated areas. There is no clear-cut determination of what should be the thrust of future action. It should simply be noticed that the selection of criteria is based on implicit value judgements that may distort insidiously the apparent rationality of choices.

Considering the high cost of investments in the water supply and sanitation sectors, economists tend to foster community financing, which consists of contributions to the investment (free labour, raw materials), community responsibility for operation and maintenance and, more frequently, charges to users. This paper is not the right place to discuss the legitimacy of the approach, but two consequences should be pointed out. First, the introduction of cost-recovery schemes orients new investments towards places where there are people with the cash to cope with the requirements. This reinforces the preference given to urban areas or, more generally, to those parts of countries being developed and modernized, further neglecting the traditional and the poor segments of the population which are rural and peri-urban. Second, it can be considered that the elasticity of the demand for water services is limited, perhaps negligible. It means that the quantity of water used per capita changes very little when the cost of water increases or decreases.

Beyond an incompressible minimum threshold, water needs are determined by usage patterns which depend on behaviour, education and other cultural determinants, and less on family income. Adrianza & Graham (1974) have clearly demonstrated that, in the suburbs of Lima, the poorer the residents, the higher the share of water expenses in the total expenditure of the household. Therefore, the result of charging users for

water can be to divert towards this sector resources which would otherwise have been spent on lower priority items. There is a high risk that expenditure for health and nutrition might be severely restricted, thus offsetting the benefits of the provision of safe water.

THE ECONOMIC CONSEQUENCES OF WATER-ASSOCIATED DISEASES

It is almost impossible to evaluate the magnitude of the pathology associated with unsafe water, with the insufficient availability of water, and with human pathogens which depend on water during the whole or part of their life. In almost all diseases, it is difficult, and sometimes impossible, to obtain reliable incidence figures, i.e., the number of new cases that occur within a period of one year. At best, we can use global estimates within broad confidence limits. Another uncertainty results from the assumption of the proportion of cases attributable to water. Whereas leptospirosis is constantly associated with water contaminated with rodent faeces, poliomyelitis and viral diseases are more often transmitted through inter-human contacts. WHO estimates that, in the developing world (excluding China), over one billion episodes of diarrhoea occur annually in children under five years of age, with a death toll of five to seven million. About one-third of these episodes might be prevented by improvements in water supply (27% according to the review of the studies considered most satisfactory by Esrey *et al.*, 1985).

The above discussion illustrates the difficulty in carrying out sound cost-effectiveness and cost-benefit analyses in the water sector. There is no consensus on the benefits gained from interventions, nor is there a consensus on the nature and importance of indirect benefits. There is little or no data available on the target diseases, and there is no information on the proportion of cases attributable to water in the specific situation under consideration. Thus, effectiveness indicators are usually of poor quality. It is difficult to get an accurate evaluation of unit costs, and even more so to identify whether the relevant unit for evaluation is the individual ("per capita" figures), the household, or the village. Finally, data on water use are frequently nonexistent, although it is well-known that prerequisites for an impact are that a system should be operating and used by its potential beneficiaries.

The assessment of individual benefits is even more difficult. Yet, these may be the most important. The time saved by housewives on water collection seems to be reallocated mainly to child care and to agriculture. Thus, child nutrition improves and morbidity decreases, as does infant mortality. Water availability enhances vegetable gardening, thus improving the quality of the diet.

In the following section, two diseases will be discussed in relation to their economic impact: guinea-worm disease, which is both a water-based and a water-borne disease, and which is the primary target disease to be eliminated in the foreseeable future through adequate water supply; and onchocerciasis, an example of a water-related disease which has attracted a substantial investment from the international community on the basis of its economic justification.

Guinea-worm disease (or dracunculiasis)

Dracunculiasis is a parasitic disease caused by a filarial worm about one metre long. Gravid females migrate to the skin and produce a cutaneous blister that bursts in contact with fresh water into which embryos are released. These embryos are ingested by *Cyclops* (small crustaceans of the copepod family) in which they develop to the infective stage. Human infection results from swallowing infected *Cyclops* with drinking water collected in ponds, unprotected wells, or similar sites. Following the lesion of the skin made by adult worms, secondary infection seriously aggravates the condition, producing abscesses of the lower limbs. The disease usually results in total disablement of the patient for 4 to 28 weeks, with an average duration of symptoms of 12–13 weeks (Smith *et al.*, 1989). Complications include arthritis, ankylosis, and sometimes a fatal tetanus infection.

About 10 million people are infected each year, out of a population at risk of about 150 million. Eighteen countries in Africa north of the equator, from Senegal to Ethiopia, account for the bulk of reported cases. Large areas of the Indian subcontinent are also infested, and small foci may still exist in Saudi Arabia and Yemen. Elimination of the disease has been achieved in southern USSR, the Islamic Republic of Iran, and India's Tamil Nadu State. The disease is on the verge of being eliminated from Pakistan. Because the disease occurs seasonally at the worst possible time of the year in terms of peak demand for agricultural labour, its economic impact is enormous. A study of the opportunity cost of time in Kenya has shown that the value of rural labour time in the lowest income category is eight times higher during the wet season than in the dry season (Mwabu, 1988). The ratio in the highest income category – those who are not usually affected by guinea-worm disease – is only 2:1.

In a study carried out in three Burkina Faso villages with 3,376 inhabitants, 433 patients infected with guinea-worm experienced a total of 7,596 days of total disablement resulting in agricultural losses of about US$ 21 000 during the three-month agricultural season of 1983 (Guiguemde, 1984). The cost to the villages in economic terms (adjusted productivity losses and treatment cost) was estimated at US$ 39 000 and the cost of the disease to the whole country was tentatively extrapolated at US$ 5 million. For comparison, the 1983 per capita GNP in Burkina Faso was US$ 180, although this figure does not provide any information on the real income in rural areas.

Another study of 87 households, in a rice-growing area of southeastern Nigeria in 1987, estimated that the annual losses in profit from rice production alone amounted to US$ 22 500, resulting from a loss of 11% of productive man-days in a population in which one out of every five persons suffered from guinea-worm disease at the time of the survey (Edungbola *et al.*, 1987). A tentative extrapolation of these results to the 1.6 million rice-growers in the guinea-worm stricken area of the southeastern states of Nigeria indicates that losses may amount to over US$ 20 million annually.

In India, each episode of dracunculiasis among economically active people results in about 70 days of wages lost, with a total loss for the country estimated at about 12 million work days annually (Rao, 1985).

The elimination of dracunculiasis rests mainly with engineers, not health professionals. There is no medical cure for the disease, nor is there any preventive drug. Nevertheless, it is the primary target for disease eradication in the foreseeable future. *Cyclops* are relatively large animals which can be removed through filters made of simple cotton, nylon, or polyester cloth. Simple engineering and construction designs to prevent people from dipping their lower limbs into drinking water reservoirs would render impossible the release of the embryos and thus the infection of the intermediary host, *Cyclops*. Vector control, using small amounts of temephos (Abate) can be used as a back-up intervention. The cost of control measures will be less than the economic losses due to the disease.

Rapid eradication is feasible if efforts are somehow coordinated at the national level. In the Dimbokro district of Côte d'Ivoire, the prevalence of dracunculiasis fell from 50% to 5% on average between 1977 and 1979 as a result of the implementation in that area of a nation-wide water supply programme to villages (personal communication of the district medical officer, 1979). India, which began its eradication programme in 1980, reduced the number of reported cases from 45 000 in 1983 to 17 031 cases in 1987, despite the improvement of the reporting system and the active search for cases. Several West African countries are embarking on eradication programmes, raising hopes that this scourge will be eliminated in the 1990s.

Onchocerciasis

Onchocerciasis is also a filarial disease. Contrary to dracunculiasis, it is not transmitted through drinking water but through the bites of blackflies of the *Simulium* species that breed in fast-flowing water courses. Thus, it is a water-related disease, not a water-borne one. The severity of onchocerciasis results from the blinding complications which can occur in heavily infected patients after several years of exposure to repeated bites. An estimated 18 million people are actually infected with onchocerciasis, resulting in 340 000 blind world-wide, the vast majority of them in West Africa (WHO, 1987). In that part of the world, blindness reduces life expectancy by at least 13 years at age 30 (Prost & Vaugelade, 1981). Thus, prevalence surveys misrepresent the actual burden which blindness imposes on communities. Data collected in Burkina Faso indicate that, in hyperendemic villages, 46% of males and 35% of females aged 15 were likely to become blind before they died (Prost, 1986).

By the end of the 1960s, it was clear that onchocerciasis was an obstacle to the agricultural development of relatively fertile river valleys which were not populated – or were only sparsely so – because of onchocerciasis. An international control programme was set up in 1974, covering seven West African countries and 15 million people at risk, with WHO as the executing agency. It is one of the very few health programmes that have been established so far on the basis of an economic finality. Since its inception (1971), the programme used vector control techniques in the larvae breeding sites, mostly by weekly spraying of insecticides on selected portions of the rivers.

The Onchocerciasis Control Programme has attracted much interest from bilateral and multilateral development agencies. Annual contributions rose from US$ 10 million in 1976 (the first year of full operation) to about US$ 34 million in 1987, making it one of the most important investment programmes aimed at the control of a water-related disease. After 12 years of operation, the programme is highly successful: the number of new cases of blindness has declined faster than expected, being negligible since 1981 in the core of the programme area; children born since the beginning of operations have been totally spared from infection but for a few isolated cases; and the spontaneous death of adult filarial parasites in patients infected before the beginning of the programme has resulted in the absence of infection (zero prevalence) in previously hyperendemic villages.

A cost-effectiveness analysis was made tentatively (Prost & Prescott, 1984), using data from Burkina Faso, because this was the only country completely covered by the Onchocerciasis Control Programme operations, and because better quality data allow for control of confounding factors.

It was estimated that about 60 000 years of healthy life had been added annually to the community by preventing the onset of onchocerciasis blindness, with most of these healthy years accruing in the productive period of life. It is important to note that this number of years of healthy life added do not actually occur in the year to which they are attributed. Rather, the measure represents an estimate of the future number of healthy years that are added by preventing the onset of blindness during the year of reference.

The cost-effectiveness analysis shows that onchocerciasis control may be compared favourably with other prevention programmes, such as measles immunization. The cost of one productive year of life added was estimated at US$ 20 with the Onchocerciasis Control Programme, and US$ 15-17 with measles immunization in the same region. The application of discounting factors to account for time preference even switched the cost-effectiveness ranking in favour of onchocerciasis control.

The two examples above, guinea-worm disease and onchocerciasis, illustrate the fact that two water-associated diseases which are out of reach of current therapeutic methods can be effectively controlled through interventions in the water sector which combine various proportions of engineering management and chemical application. Moreover, the proposed strategies are highly successful with good prospects for the elimination of a disease (dracunculiasis) or with a cost-effectiveness ratio which compares favourably with popular medical interventions (both diseases).

WATER RESOURCES DEVELOPMENT AND HEALTH

The achievement of sustained development, the promotion of health, and the rational use of water resources are inseparable. Disregard one and, sooner or later, the other two will collapse. If water is scarce, or if resources are abused, both health and development suffer. If water management is not conducive to health promotion, if it ignores elementary health safeguards, development fails. If standards for health protection are unduly strict, and if they are not set up in a dynamic perspective, development efforts are hampered.

Health protection measures should be planned at the design stage of water development schemes. A first reason is that small changes at this stage may be the most cost-effective way of alleviating adverse effects, or at least some of them. A second is that health protection is necessary from the very early stages of the settlement of populations, since such social and behavioural changes are always associated with the weakening of individual defence and host resistance to environmental aggressions.

The key factor for success is the accuracy of risk forecasting and the comprehensive evaluation of all interacting factors in the new setting. Too often, risk assessment is based upon a situation analysis of existing water-associated diseases, with a magnification factor based on the size of the project. Thus, it is not surprising that the outcome is rather pessimistic.

Table 6 tentatively lists some of the components of a comprehensive evaluation, each having potential adverse and beneficial effects. For example, increased human-water contacts may favour schistosomiasis, but will reduce the incidence of water-washed diseases; migrants may bring new diseases to the area, but increased population density will create a barrier to several vector-borne diseases; development schemes usually include provision of domestic water through public standpipes, thus reducing man-water contacts; proper design of new housing may prevent mosquitoes from entering the houses, thus reducing the malaria transmission, although an increase would have been expected from the expansion of surface water areas.

The primary outcome is the identification of the elements which could be built into the project design and which could maximize the health benefits: for example, the slope of overflow channels and periodicity of water discharges in dam construction, procedures to facilitate the maintenance of canals, housing design and materials, sanitation and waste disposal in

Table 6. *Examples of considerations in risk assessment for water development schemes.*

Amplification of existing risk factors

Additional risks owing to the increase of the area of surface water

Modifications resulting from changing ecological conditions (vectors)

Additional risk factors introduced by migrant settlers (new diseases)

Greater sensitivity of immigrants to local risk factors (lack of immunity)

Effects of increased population density (adverse and beneficial effects)

Changes in human-water contacts (increases and decreases)

Impact of higher income and cash availability

Impact of changes in housing

Impact of nutritional changes and modification of agricultural patterns

Impact of water supply and provision of sanitation facilities

Behavioural and occupational changes

Creation of new occupational hazards

the settlements and water supply. The impact of such measures is not trivial. In Saint Lucia, a Caribbean island, the installation of tap water in each household and the installation of public baths and community wash places have decreased the incidence of schistosomiasis by 60% in the absence of any sanitation measure (Jordan *et al.*, 1982). It has been the direct effect of the reduction of human contact with the water.

A second outcome of this comprehensive approach is the consideration of the elements which influence the living standards, and thus indirectly derive health benefits. By definition, a development project aims at improving the income of the resident population. Rises in income level are consistently the most important contributing factor to the improvement of the health status, even though no medical intervention specifically takes place. Higher income increases food consumption and improves its quality and variety; it always results in an improvement of housing conditions; it gives access to drugs sold in drugstores, but also in market-places and through traditional commercial channels, thus facilitating self-medication practices outside the control of the medical profession.

An example of an assessment of the risks of malaria integrating the various considerations could be as follows: the irrigation scheme will certainly result in a striking increase in the number of mosquitoes, and in an appreciable increase in malaria transmission. However, the maintenance of a relatively good immunity in the population over five years of age renders unlikely the occurrence of a greater number of complications or a worsening of the clinical picture. Since higher income will generate a circulation of chloroquine tablets in commercial channels, and will also make the heads of house-

holds more receptive to messages concerning the need to rest under mosquito nets, the proposed project will probably not result in any health impairment because of malaria.

Unfortunately, it is not possible to foresee all changes and their possible impact. The unexpected replacement of mosquito species by more dangerous ones (Kano, Kenya) or by less dangerous strains (Kou Valley, Burkina Faso) has been discussed above. Such changes cannot be quantified in advance, even though they might be suspected. Moreover, social and economic changes necessitate behavioural adjustments which may not take place – or will be slowly introduced – with resulting dramatic consequences on health.

The example of an irrigated rice scheme in an African country illustrates the point. At the time of project design, malaria and schistosomiasis had been targeted as the major risk factors. A local health network had been set up with relevant priorities for action. After a certain time of operation, an evaluation showed that priority health problems were malnutrition and alcoholism. The sequence of causal factors proved to be the following. In this region, men were in charge of cereal crops (millet, maize) and women responsible for vegetables and condiment gardening which bring most of the essential micronutrients to the diet. Rice cultivation caused a brutal shift from a subsistence economy to a cash market economy. Males were earning salaries instead of collecting cereal crops for household consumption; they spent a large amount on nonessential items (beverages, appliances) at the expense of food security for the family. At the same time, women could no longer grow fruits and vegetables, all the land in the project area being used for rice. Thus, major nutritional deficiencies occurred, especially in children. Who could have foreseen that a successful agricultural project would result in qualitative and quantitative nutritional deficiencies?

The complexity of the interactions makes forecasting a difficult exercise. Assumptions on possible health impact should never be presented as definitive. There are behavioural adjustments, or maladjustments, which make a situation tolerable, or intolerable, in apparently similar conditions.

In all water management projects, the demand for health services among residents is usually not related to water-induced diseases. The demand is for perceived priority needs which do not differ from those expressed in other communities of the region: it may be a demand for surgical treatment of hernias, for improved delivery facilities, for treatment of acute respiratory infections. The concept of risk-forecasting is misleading in the sense that it seems to be limited to risks derived from the project and does not take into account the overall health situation of the population. In fact, people become sensitive to health issues during social adjustment periods, and they complain when their demands are not adequately met. Health care provisions based upon risk assessment studies usually fail to meet the demand because they address health problems which are not perceived as priority ones.

The development of tropical areas requires innovative approaches. One such approach is the shift from risk assessment studies focusing narrowly on the impact of environmental changes to an evaluation of the trade-off between risks and benefits, including social, demographic, behavioural, and economic changes in the community. Interdisciplinary collaboration is therefore essential to provide decision-makers with an appropriate background for "informed choices."

REFERENCES

Adrianza B. T. & Graham, G. G. (1974) The high cost of being poor: water. *Archives of Environmental Health* 28 312–315.

Briscoe, J. (1984) Water supply and health in developing countries: Selective primary health care revisited. *American J. of Public Health* 74, 1009–1013.

Cordellier, R. (1990) La fièvre jaune en Afrique occidentale, 1973–1987. Faits observés – Etudes réalisées, lutte, prévention et prévision. *Rapport trimestriel de Statistiques sanitaires mondiales 43* (2), 52–67.

Edungbola, L. D., Braide, E. I., Nwosu, A. B., Arikpo, B., Genmade, E I. I. & Adeyeme, K. S. (1987) Guinea-worm control as a major contributor to self-sufficiency in rice production in Nigeria. Unpublished report UNICEF/WATSAN/NIG/GW/2/87.

Esrey, S. A., Feachem, R. G., & Hughes, J. M. (1985) Interventions for the control of diarrhoeal diseases among young children: improving water supplies and excreta disposal facilities. *Wld Hlth Org. 63*, 757–772.

Forkasiewicz, J. & Margat, J. (1980) Tableau mondial de données nationales d'économie de l'eau. Ressources et utilisations. *Document du Bureau de Recherches Géologiques et Minières,* Département d'Hydrologie, Orléans, France.

Guiguemde, T. R., Kagone, M., Compaore, T., Meda, P., Lozachmeur, P. Sokal, C. D., Roux, J., Orivel, F. & Millot, B. (1984) Les conséquences

socio-économiques de la dracunculose: esquisse d'une méthode d'évaluation du coût économique de cette maladie. OCCGE Informations (Bobo Dioulasso, Burkina Faso) *89* 73–90.

Hervouet, J. P. (1983) Aménagement hydro-agricole et onchocercose: Loumana (Haute-Volta). In: *De l'Epidémiologie à la Géographie Humaine* (Travaux et Documents de Géographie Tropicale) No *48,* 271–276. Editions ACCT/CNRS, Paris.

Hunter, J. M. (1981) Past explosion and future threat: Exacerbation of Red Water Disease (*Schistosomiasis haematobium*) in the Upper Region of Ghana. *Geo. J. 5.4*,305–313.

Jordan, P., Unrau, G. 0., Bartholomew, R. K., Cook, J. A., & Grist, E. (1982) Value of individual household water supplies in the maintenance phase of a schistosomiasis control programme in Saint Lucia, after chemotherapy. *Bull. Wld Hlth Org. 60*, 583–588.

Merrick, T. W. (1985) The effect of piped water on early childhood mortality in urban Brazil, 1970 to 1976. *Demography 22*, 1–24.

Mwabu, G. M. (1988) Seasonality, the shadow price of time and effectiveness of tropical disease control programmes. In: N. Herrin & P. L. Rosenfield (eds.) *Economics. health and tropical diseases,* University of the Philippines Publ.

Preston, S. H. & Van de Walle, E. (1978) Urban French mortality in the Nineteenth Century. *Population Studies 32*, 275–297.

Prost, A. (1986) The burden of blindness in adult males in the Savanna villages of West Africa exposed to onchocerciasis.*Trans. Roy. Soc. Trop.Med. Hyg. 80*, 525–527

Prost, A. & Négrel, A.-D. (1989) Water, trachoma and conjunctivitis. *Bull. Wld Hlth Org. 67*, 9–18

Prost, A. & Prescott, N. (1984) Cost-effectiveness of blindness prevention by the Onchocerciasis Control Programme in Upper Volta. *Bull. Wld Hlth Org 62*, 795–802.

Prost, A. & Vaugelade, J. (1981) La surmortalité des aveugles en zone de savane ouest-africaine. *Bull. Wld Hlth Org. 59*, 773–776.

Provost, A. (1989) La fièvre de 1a Vallée du Rift. *La Recherche 20* 254–256.

Quélennec, G. Simonkovich, E. & Ovazza, M. (1968) Recherche d'un type de déversoir de barrage défavorable à 1' implantation de *Simulium damnosum (Diptera, Simuliidae). Bull. Wld Hlth Org. 38*, 943–956.

Rangeley, W. R. (1985) Irrigation and drainage in the world. *Proc. Int. Conference on Food and Water* (College Station, Texas, May 1985) p. 8.

Rao, C. K. (1985) Epidemiology of dracunculiasis in India. In: *Workshop on opportunities for the control of dracunculiasis* (Washington D.C., June 1982) 25–32. Contributed papers, National Academy Press, Washington D.C.

Robert, V., Gazin, P., Boudin, C., Molez, J. F., Ouedraogo, V., & Carnevale, P. (1985) La transmission du paludisme en zone de savane arborée et en zone rizicole des environs de Bobo Dioulasso (Burkina Faso). *Ann. Soc. belge Med. trop.* (suppl. 2), 201–214.

Shuval, H. I., Tilden, R. L., Perry, B. H. & Grosse, R. N. (1981) Effect of investments in water supply and sanitation on health status: a threshold-saturation theory. *Bull. Wld Hlth Org 59*,243–248.

Smith, G. S., Blum, D., Huttly, S. R. A., Okeke, N., Kirkwood, B. R. & Feachem, R. G. (1989) Disability from dracunculiasis: effect on mobility. *Ann. trop. Med. Parasit. 83*,151–158.

Walsh, J. A. and Warren, K. S. (1979) Selective primary health care. An interim strategy for disease control in developing countries. *New England J. of Medicine 301*, 967–974.

White, G. F., Bradley, D. J. & White, A. U. (1972) *Drawers of water.* Chicago, Chicago University Press.

World Bank. (1984) China, the health sector. World Bank Publication Washington, D.C.

World Bank. (1988) *World Development Report*. Oxford University Press, New-York.

World Health Organization. (1987) Expert committee on onchocerciasis: Third report. Geneva: World Health Organization technical report series No. 752.

World Health Organization. (1989) Health guidelines for the use of waste water in agriculture and aquaculture. Report of a WHO scientific group. Geneva: World Health Organization, technical report series No. 778.

21: Water Supply and Health in the Humid Tropics with Particular Reference to Rural Areas

P. Wurzel

UNICEF, Maputo, Mozambique

ABSTRACT

This paper addresses the issues related to water supply and health in rural areas. Consideration will be first given to various means of interrupting the transmission cycle of water-borne and water-washed diseases. Such interventions can contribute significantly to improved health in the developing world. Finally, the trends, strategies and philosophy for the future management of rural water supplies are considered, with the emphasis on low-cost and appropriate technologies. Particular emphasis will be directed towards groundwater as the primary water resource.

INTRODUCTION

This paper concentrates largely on rural water supplies and sanitation as they relate to health in the humid tropics. By interrupting the transmission cycle of water-borne and water-washed diseases, these interventions can contribute significantly to improved health in the developing world. The issues, trends, strategies and philosophy for the future are explored, with the accent on low-cost and appropriate technologies. Particular emphasis is given to groundwater as the primary water resource.

The history of the human race is dominated by its quest for a safe water supply. Wars have been fought and massive migrations have taken place to gain and to guarantee access to safe water. Archaeology has revealed the monumental aqueducts, sewerage disposal systems, dams and other costly irrigation projects of the ancient civilizations which attest to the fact that the availability of water and the disposal of waste have always been among the principal societal concerns. It was taken for granted that a safe and ample water supply and adequate sanitation facilities were a condition *sine qua non* for healthy and happy living. Indeed, the statement "Health for All" does not originate from a recent United Nations (UN) primary health-care publication nor indeed from a recent forum on "Poverty, Water and Health" but rather from the wise lips of Hippocrates, over 2,000 years ago.

If the firm relationship between water and health was seen to be true in the world 2000 years ago with a relatively minute population, this relationship can be ignored only at our peril in today's world with a population of five billion people, of whom two billion do not have access to a safe, ample, potable water supply.

When we consider the startling figures issued by the World Health Organization, viz. 70–80% of hospital beds in developing countries (the majority of which lie in the humid tropics) are occupied by patients with water-related diseases, the link between water and health appears indisputable. Despite such a strong subjective correlation, however, we are passing through a period where, in some agencies, the cost-efficacy *vs.* health benefits of water supply and sanitation programmes are being questioned.

The Alma Ata Declaration of 1978, which included the provision of "an adequate supply of safe water and basic sanitation" in its declared goal of comprehensive primary health care for all by the year 2000, met with criticism from the Rockefeller Foundation. In the face of limited financial and human resources, the Foundation viewed the goal set at Alma Ata as irreproachable but unattainable because of its very scope.

Walsh & Warren (1979) examined and compared the costs of relevant interventions in the developing world, namely, comprehensive primary health care (which includes general development as well as all systems of disease control), basic primary health care, multiple disease-control measures (e.g., insecticides, water supplies), selective primary health care, and research and the health benefits which accrued from each. Since children under five years of age were recognized as the group most at risk, costs were evaluated on the basis of infant and child deaths averted. No other benefits for which interventions may have been responsible were measured because they were considered much more difficult to quantify.

Walsh & Warren (1979) concluded that a selective primary health care package was the most cost-effective method at $200 to $250 per child death averted. The package included measles and diphtheria – pertussis – tetanus vaccination for children over six months old, tetanus toxoid to all women of child-bearing age, encouragement of long-term breast feeding, provision of chloroquine for episodes of fever in children

Table 1. *Reduction in diarrhoeal morbidity rates due to improvements in water supply and sanitation.*

Improvement in		% reduction
Water quality	. . . .	16
Water availability	. . . .	25
Both quality and availability	. . . .	25
Excreta disposal	. . . .	22

(After S.A. Esrey, RG Feachem and JM Hughes, 1985)

under three years old in areas where malaria is prevalent and, finally, oral rehydration packets and instruction. Water supply and sanitation programmes which were calculated to cost about $4,000 per child death averted were considered prohibitively expensive.

Another factor to be considered is the influential policy-oriented analysis of Shuval *et al.* (1981), who made national-level comparisons of life expectancy among countries with low and high water supply and sanitation coverage. This study suggested that at both low and high levels of socio-economic development, improvements in water supply and sanitation conditions would have relatively little effect on health, and that it was in the "middle level" countries that the effect would be greatest. Briscoe (1987), for example, stated that "as a result of these analyses, the *de facto* policy of several international agencies has been that water supply and sanitation interventions may occasionally be appropriate at relatively advanced stages of the development process, but that they are not cost-effective at the earlier stages where other interventions such as immunizations, oral rehydration and family planning are believed more sensible. Thus, for instance, in the Asia region, the United States Agency for International Development (USAID) may give consideration to a water supply programme in Thailand "a middle-level country" but not in Bangladesh "a poor country."

At the start of the International Drinking Water and Sanitation Decade, it was implicitly claimed that diseases among children in developing countries would be reduced by 80% if water supply and sanitation conditions improved. The exaggerated nature of such claims led to a backlash and it was subsequently suggested in some quarters that water supply and sanitation programmes had little effect on health. Walsh & Warren (1979), noted that the approximate cost of US$4,000 per child death averted related largely to the provision of public standpipes. They concluded that, given only outside house connections, *Shigella*-caused diarrhoea decreased by just 5%. When sanitation and washing facilities were available within the home, these diseases decreased by 50%. Briscoe (1983) challenged this conclusion, citing that sound, available studies show reductions in diarrhoeal diseases by an order of magnitude greater than the 5% assumed by Walsh & Warren. Further evidence that the quantity of water available, combined with

adequate sanitation, is crucial was reported by Esrey *et al.* (1985) who found that diarrhoeal disease morbidity rates were significantly reduced as evidenced in Table I. Esrey *et al.* anticipated even larger reductions on diarrhoeal disease mortality.

The World Health Organization (WHO) estimates that 500 million diarrhoeal episodes occur each year in children under five in Asia, Africa and Latin America, and three to four per cent of these end in death (WHO, 1979). Short-term curative measures such as ORT (Oral Rehydration Therapy) are indeed a moral imperative in reducing mortality in diarrhoea, and are amenable to instant quantification, but what of the 96 to 97% who do not die but who suffer repeated episodes of diarrhoea and who may subsequently be at risk of death again? ORT has no role in chronic diarrhoeal morbidity which is ubiquitous in many developing countries (in Zimbabwe, 80% of all diarrhoeas are chronic, Powell 1987, personal communication) and which further compromises poor nutritional status.

Inadequate environmental sanitation is a critical link in the chain of diarrhoeal disease that entraps young children of developing countries and claims a large percentage of deaths in the under five-year-old age group. Contributing factors are unsafe and insufficient water supplies, the lack of safe means of human waste disposal and poor personal and household hygiene, including unsanitary food-handling practices.

As Okun (1988) noted: "Unfortunately, the recent focus on ORT has diverted attention and funds from assessing the causes of diarrhoeal disease and from the broader preventive health aspects of water supply and sanitation." Okun (1988) lists the crucial economic, social and political impacts of water supply and sanitation projects as follows: "Prevention of disease, improvement in nutritional status, services to health centres, clinics, schools; time released for women; household irrigation and animal watering; promotion of commercial activities; strengthening community organizations; and, perhaps most important of all, an improved quality of life."

Prevention of disease goes beyond the prevention of diarrhoeal morbidity and deaths. Improvements in water supply and sanitation are effective in controlling cholera, typhoid, amoebiasis, giardiasis and a variety of helminthic diseases that drain limited food supplies and heighten malnutrition as well as many skin and eye diseases (notably trachoma). When water provides the only transmission route, as is the case with guinea worm, safe water supply is the single solution to combating the disease. However, most diseases spread through multiple faecal-oral transmission routes, necessitating improvements in sanitation, food hygiene and knowledge, as evidenced by Table 2.

In addition to soap and education, the above table does not indicate how much water is required to achieve the reduction of 84%, for example, in the case of Bangladesh. Furthermore, this figure is at variance with that given by Walsh & Warren (1979), who indicate that large quantities of water (plus soap and education) are required to reduce the incidence of shigellosis by 50%.

Table 2. *Effectiveness of public health interventions in various settings.*

Country	Setting	Intervention	Outcome Measure	Result
Bangladesh	Households with index case of shigellosis.	Soap and water and education vs. nothing	Secondary shigella cases	Reduction of 84%
USA	Day-care centres, children under 3.	Handwashing of staff and education of children vs. nothing	Incidence of diarrhoea	Reduction of 48%
Guatemala	Lowland villages children under 6.	Hygiene education vs. nothing	Incidence of diarrhoea	Reduction of 14%

Certainly the benefits of water supply and sanitation far exceed the impact on communicable diseases. Even seemingly peripheral socio-economic community benefits have a direct bearing on health. An accessible water supply can eliminate the wearisome labour of women and children who must fetch water over long distances – typically a walk of two to three hours each day. A trek of this length can consume 600 calories or more, using up to one-third of the daily nutritional intake. The impact of saving so many calories directly benefits the health of the woman, facilitates breast-feeding and aids the development of her children. In releasing the time of women for more productive activities, the introduction of accessible water supply is often the first step in women's advancement to full participation in the development process.

Significantly, the pendulum can be said to be swinging back. There is a perception that, once again, water and sanitation programmes are receiving greater emphasis.

GLOBAL ISSUES

The cardinal question of how soon water globally will become a limiting factor in the headlong increase in population expansion is extremely important from the point of view of water supply and health. One of many interesting calculations of global water supplies in the future is that by Keller (1984). He suggests that a realistic figure to use is 25 000 km^3 of available fresh water globally. Based on a projected population of 6.5 billion in the year 2000, it leads to a per capita figure of 3,800 m^3 per annum. If the water required for agriculture, industry and energy supply is also taken into account, 1,000 m^3 is needed each year, which brings the total per capita for general living to 2,800 m^3 per annum. From an absolute volume viewpoint, the figures look good, *i.e.,* in the year 2000

the *available* water is still eight m^3 per day per capita, *i.e.,* three orders of magnitude greater than the minimum WHO water requirement per capita. Theoretically, that is a comforting figure, especially in the humid tropics, but in practical terms quite unrealistic and misleading. Water is ill distributed in time and space, possibly polluted, and always decreasing with time.

As we enter the Twenty-First Century, the population, the gross national product and the resource projects in the humid tropics all imply rapidly increasing demands for fresh water. Projections by Unvala (1988) of per capita, all-purpose water availability, show a decrease from 5,000 m^3 per capita per year in 1980 to less than 1,000 m^3 per year by 2030 in Kenya, Uganda, Tanzania, Mozambique, Algeria, Tunisia, Morocco, Mauritania, Western Sahara, Bangladesh and Egypt.

Faced with such figures, the water and sanitation sector in the developing world must therefore urgently address the issues, among which the following are identified by Danida (1988):

(a) Institutions responsible for water resources, water supply and sanitation are frequently inefficient, lack appropriate skilled manpower and equipment, are financially weak, and often lack an efficient legal framework within which to operate – and an effective enforcement apparatus;

(b) Coordination and cooperation among external support agencies, between these agencies and national water agencies, and among water sector agencies themselves, is essential but often inadequate;

(c) The operation, maintenance, rehabilitation and training aspects of ongoing and completed projects often receive insufficient attention. These drawbacks are often aggravated by the use of inappropriate technologies which are neither affordable nor manageable;

(d) Public awareness, participation and support are often inadequate, as are public health promotion activities;

(e) There is a lack of scientific data and information on environmental processes and few ways to collect, process and communicate this information to decision-makers;

(f) Immediate attention is required to address the critical problems of desertification, deforestation, rehabilitation of degraded water-catchments, salinity and environmental health aspects of irrigated lands, protection of sensitive ecosystems with special reference to biological diversity, pesticide use, urban waste disposal, pollution and public health, and the effects of global climatic changes.

GROUNDWATER – THE PRE-EMINENT RESOURCE FOR THE FUTURE

In the humid areas of the world, knowledge about groundwater lagged for millenia with development being applied chiefly to natural springs. Even the origin of groundwater was controversial until late in the Eighteenth Century, yet groundwater is the largest source of fresh water in the world.

At most places on the earth's surface small quantities of groundwater, sufficient for domestic use, are available and eminently suitable for rural water supplies. Most authors divide groundwater into "shallow" (less than 300 m in depth) and "deep" (greater than 300 m in depth). There is little doubt that, in time, people will attempt to tap the deeper groundwater – Mexico City, the world's largest metropolis with a population of 18 million (expected to rise to 25 million by the year 2000), being a case in point. Generally, the deeper the groundwater, the older it is (some of the groundwater below the Sahara Desert is 50,000 years old) and the greater the dissolved solid content is within such bodies. Nevertheless, the depth of the source can ensure that the water is better protected from industrial pollution.

Tens of thousands of hand-pumped wells have been constructed in recent years and hundreds of thousands more are planned in an effort to meet the large and growing demand for safe water. The implementation of such water supply schemes needs to take into account the existing economic and technical constraints and opportunities which apply to groundwater development activities. Ideally, all groundwater development should be preceded by proper hydrogeological exploration in order to locate the optimum amount of groundwater. In many areas the construction of wells has proceeded without any detailed insight into the hydrogeological parameters which determine the presence and location of groundwater. Such construction has mainly been based on user convenience of distance to site, ownership of plot, etc. This approach will work well in certain terrain. However, expanding water demand, especially in marginal areas, now requires the application and proper use of groundwater investigation techniques. Geophysics (the electrical resistivity method in particular) is now commonly used in rural water supply exploration for shallow underground water.

Groundwater abstraction must also be carefully monitored and controlled. Over-exploitation (see Foster and Chilton, this volume), in addition to lowering the water table, can result in increased concentrations of pollutants in the aquifer due to reduced flow rates, increased risks of salt-water intrusion and land subsidence. This phenomenon has already occurred in many coastal areas and islands throughout the world (see Falkland and Brunel, this volume), where the available reserves of fresh water are so small that any continuous large-scale withdrawal of fresh water would result in saline intrusion and consequent pollution of limited fresh water resources. In Bangkok, Thailand, groundwater abstraction over decades for urban water supply has resulted in land subsidence, in some places more than 10 cm a year, and has contributed to serious flooding.

Groundwater quality

Groundwater is generally less polluted than surface water supplies. In contradistinction to surface water schemes where turnover times are in the order of days and weeks, groundwater reservoirs have turnover times in the order of hundreds to thousands of years so that once a groundwater reservoir becomes polluted, it is "forever."

Pollution and salinization of groundwater can occur as a result of irrigation schemes which often involve the excessive application of water to land with poor or non-existent drainage facilities. This may saturate the soil, impede aeration, leach nutrients and increase evaporation and soil salinization. The groundwater level rises and the land finally becomes waterlogged with increasingly saline water. At least 50% of the world's irrigated land now suffers from salinization. The effect on underlying groundwater can be devastating. In a country like Pakistan (although located just outside the humid tropics), large volumes of groundwater now show high salinity levels and cannot be used for drinking purposes. In addition, fertilizer (in the form of nitrates) and pesticides that have been used freely and abundantly over the past two to three decades are only now finding their way into the major groundwater aquifers.

The risk of contamination of groundwater is particularly great in areas with high water tables or areas which are subjected to periodic flooding. In such situations, a direct link can be established between the surface water and the groundwater. Therefore, the location of potential sources of pollution, e.g., sewerage and industrial wastewater outfalls, solid waste disposal sites, drainage channels, septic tanks and pit latrines should preferably avoid such areas.

Water quality problems in water supply schemes based on groundwater are reported by numerous countries. For example, in Sri Lanka, only 25% of the wells tested in 20 UNICEF schemes in the mid 1980s were bacteriologically safe. Since these wells are often multi-purpose (including bathing), their pollution is hardly surprising.

Groundwater chemistry can show wide variations regionally. High fluoride, iron, boron and salinity levels can cause significant health problems. For example, fluoride in drinking water in excess of 1.5 mg l^{-1} leads to dental fluorosis, a major dental public health problem in Kenya. Yet it is surprising that many communities appear to show few ill-effects despite ingesting water far in excess of WHO's maximum salinity limits. Levels above 2,500 mmols seem to have been well-tolerated in some areas for years (Tharparkar Desert, Sind Province, Pakistan) and, surprisingly, the cumulative effect is not manifest.

While major groundwater pollution is thus far not a feature of the developing countries of the humid tropics, a troubling unwillingness to face the reality of growing pollution problems has emerged in discussions between the author and relevant government officials in several humid tropic countries. In the face of literature which paints a far bleaker picture, personal perceptions seem limited by what is politically expedient.

RURAL WATER SUPPLY

Only in Asia has the pace of providing rural communities with safe water supplies been rapid enough to allow hope that the Water & Sanitation Decade targets will possibly be reached. In Africa, progress rates would leave half the population still without safe water in the year 2000, while in Latin America, unless progress improves dramatically, it may be 10 years into the next century before full-scale coverage is achieved (Arlosoroff *et al.*,1987).

Attempts to increase the pace of providing improved community water supplies have often been frustrated because the technology used has proved impossible *to sustain* in the village setting. Thus, to make a lasting impact, community water supply strategies must be based on sustainable and replicable programmes and must take account of the pace at which resource constraints can be overcome. A successful Community Water Supply (CWS) programme must involve a combination of hardware and software technology and institutional/organizational support elements matched in such a way that each community recognizes the benefits of the improved supply. The community also must be able to afford at least the cost of operating and maintaining the system, and must have the skills, spare parts, materials and tools available to sustain it. This "integrated approach" to CWS planning, involves consideration of a number of key issues, each individually important and together forming a complete package. Arlosoroff *et al.* (1987), stressed the importance of a package with:

(a) Effective involvement of the community in the design, implementation, maintenance and financing of planned improvements of the promoting agencies, providing technical assistance and support services as needed. The community's needs and wishes have to be reconciled with its capacity and willingness to pay for the level of service planned, and also with the technical aspects of the water facility.

(b) Provision for recurrent cost recovery for the support of capital (construction) costs for poorer communities. This should be offset by provision for full recovery from higher service levels.

(c) Maximum involvement of in-country industry in the supply of services and material, project construction and maintenance (*e.g.*, supply of pumps and spare parts, servicing and repairs) with the important proviso that quality control and reliability should be assured and that costs are competitive.

(d) Technology chosen to match the resources available to sustain it.

(e) Institutional and manpower development programmes matching the needs of the planned water supply system.

(f) Parallel programmes in health education and sanitation improvement.

RURAL WATER SUPPLY TECHNOLOGIES

The following alternative technologies are evaluated:

Rainwater harvesting
Desalination
Groundwater dams
Gravity flow schemes
The hand pump option

Particular attention will focus on the hand pump option which is considered to be the pre-eminent rural water supply technology of the future. The hand pump option includes consideration of shallow groundwater (shallow dug wells, shallow hand-drilled boreholes) and the tapping of deeper groundwater via machine-drilled boreholes.

Rainwater harvesting

Rainwater harvesting has been practised for more than 4,000 years and is necessary in areas having significant rainfall but lacking any kind of conventional centralized governmental supply system, as well as those where good quality fresh surface or groundwater is lacking. According to Hadwen (1985), for 2,000 million or more people living in the interior of continents, rainwater harvesting represents the only viable form of water supply. There are various levels of efficiency and cost which need to be considered. The review below is based essentially on Hadwen (1985).

Three main types of rainwater catchment systems are used: roof catchments, rock catchments and rainwater harvesting. In the latter case, the ground surfaces are modified in some way so as to collect the water, which then is usually fed into a storage tank.

Runoff from any surface depends mainly on rainfall intensity. Unless rainfall of 20 mm day^{-1} over several days occurs, runoff is not generated. To make effective and beneficial use

Table 3. *Costs of rural water supply options in Thailand (US $).*

Type of System	Capital Cost	Main- tenance Cost	Total Annual Cost	House- holds Served	Annual Cost/ H/holds
Rainwater jar	19	0.00	2.50	0.31	8.15
Rainwater tank	245	1.85	29.00	1.78	16.00
Shallow dugwell w/out HP	93	0.00	15.00	20.00	0.75
Shallow dugwell w/HP	540	93.00	165.00	20.00	8.15
Drilled deepwell w/HP	2660	93.00	440.00	20.00	22.00
Pond	1850	3.70	305.00	20.00	15.00
Weir	10000	18.50	1625.00	100.00	16.25

of the water, the natural surface needs modifying in order to reduce or eliminate its permeability. Compared with surface streamflow or groundwater levels, recovery from drought can be extremely rapid if a ground level or roof catchment has a high runoff efficiency and the water is adequately stored.

The most important public health consideration is to keep a water system working while, at the same time, ensuring that quality of the water is as high as possible. Properly designed roof catchment systems produce good quality water (see Falkland and Brunel, this volume, on this topic). However, groundwater catchments do not always achieve such high-quality standards owing to the difficulty of properly protecting them.

In the case of domestic supply storage tanks, it is essential to make the best use of the available runoff at the lowest cost. The cost is directly proportional to tank size; the problem becomes one of maximizing the supply while, at the same time, minimizing storage tank volume. A common container for many people is the 200 litre oil drum. These are cheap and durable, but too small to be for more than temporary use. One traditional method for a permanent, low-cost, easy-to-excavate tank is to excavate a hole and to line it with concrete.

For communal, as distinct from individual systems, localized storms may cause some tanks to overflow while others remain empty. This can be a particular problem on long, narrow islands. A relatively low-cost solution to this problem is to inter-link the reservoirs by a gravity-fed or pumped waterline.

Various innovations have been developed to solve the problem of cost, including above-ground sectional or corrugated galvanized iron tanks and all kinds of shapes and sizes of fiberglass, plastic, even wood tanks.

Studies in Thailand have shown that bamboo-reinforced water tanks are less cost-effective than water jars and construction of dug wells. However, wells are easily contaminated and jars are too small for many households. Therefore, the Population and Community Development Association selected water tanks as the optimal solution. In several African and Asian countries, combined roof catchments and storage tanks

have proved to be financially, socially and economically the most appropriate form of water supply.

An interesting cost analysis of various technologies from Thailand is given in Table 3 (Arlosoroff *et al.,* 1987).

Desalination

Approximately 50 million people currently benefit from desalination techniques. Buros (1985) is the primary source for the following review.

The arid zone was originally the centre of interest in desalination, but with the dramatic increase in groundwater pollution, particularly in areas of irrigation salinity, desalination has become an important technique in the humid tropics. Desalination is a separation process that treats saline water to reduce the dissolved salt content to a usable level. The use of desalination overcomes the paradox that many coastal communities face of having lots of water (saline) but no way to use it. Although some substances dissolved in water, such as calcium carbonate, can be removed by straightforward chemical methods, other common constituents like sodium chloride require more technically sophisticated methods such as desalination. In the past, the difficulty and expense of removing various dissolved salts from water made saline waters an impractical source of potable water. Early desalination methods employed variants of the distillation process with heat being supplied by fossil fuels or solar energy. This type of desalination was known and used at various locations through the Nineteenth and Twentieth Centuries.

During the 1950s, however, which was a decade of discovery for desalination, work began on developing various new methods, including distillation, freezing, electrodialysis and reverse osmosis.

The development of commercially viable desalination processes, begun in the 1950s, is continuing today with the four major processes having achieved different levels of commercial acceptance and success. By 1950, an additional 115 000 $m^3 day^{-1}$ (30 mgd), of desalination capacity was installed world-wide. In the next 10 years this was increased by approximately 720 000 $m^3 day^{-1}$ (190 mgd). By the beginning of

1984, the total installed desalination capacity was about 7.6 million m^3 day^{-1} (2 billion gpd). The majority of this capacity is in multiple-stage flash distillation units, many of which are installed as part of dual-purpose (power-water) generating stations in the Middle East and North Africa.

Aside from distillation, the other desalination processes that have become commercially viable are electrodialysis and reverse osmosis.

Freezing, which in the l950s seemed to be a viable process, has not attained commercial success. Membrane distillation, a fifth process for desalting saline water, is still under development. Only a limited number of commercial installations exist.

There are numerous areas in the world where society could benefit from the production of fresh water from saline or brackish water sources. Unfortunately, the economics involved seldom permit the use of desalination techniques. The solar distillation method is the only feasible (and remotely) appropriate low-cost technology suitable for poor rural communities. Small, relatively cheap solar stills employing glass covers are fragile and inefficient but may well be found appropriate in areas of brackish groundwater. Research is continuing into ways of maximizing efficiency and decreasing costs of solar family-type stills. Desalination, therefore, should only be undertaken after thorough exploitation of all available fresh water collection means have been exhausted.

Groundwater dams

The use of surface reservoirs to store water in areas with dry climates has several serious disadvantages, such as pollution risks, reservoir siltation and evaporation losses. Using groundwater is one way of overcoming these problems, but in some areas good aquifers are not available or they may only yield sufficient quantities of water seasonally. Experience has shown that conventional development of groundwater in developing countries involves serious problems related to operation and maintenance of drilling equipment, wells and pumps.

During the last few years, considerable attention has been given to the use of groundwater dams as a method of overcoming water shortages in regions with arid and tropical climates (Hanson & Nilsson, 1984). Damming groundwater for conservation purposes is certainly not a new concept. Groundwater dams were constructed in Sardinia in Roman times and damming of groundwater was practised by ancient civilizations in North Africa. More recently, various small-scale groundwater damming techniques have been developed and applied in many parts of the world, notably in southern and eastern Africa, and in India.

The rationale for damming groundwater is the irregular availability of surface and groundwater. Groundwater damming techniques may thus be applied in arid and semiarid areas where there is a need to conserve as much as possible of the scanty rainfall. They are also cost-effective and suitable in areas with a monsoon climate where there is a need to store surplus water from rainy seasons for use during dry periods. In such areas, groundwater dams are applicable in the rural areas, provided that they can be constructed at low-cost.

There are two types of groundwater dams: subsurface dams constructed below ground level to arrest the flow in a natural aquifer, and sand storage dams, structures which impound water in the sediments caused to accumulate by the dam itself.

These dams have the advantages over conventional surface dams in that they have:

(a) Low construction costs compared to a surface dam;
(b) Reduced evaporation and silting, as they are already filled with sediment;
(c) Better water quality than a surface dam.

A subsurface dam should be considered in valley situations where an aquifer of fairly permeable alluvial sediment supplies water to a village by a shallow well. Due to consumption and the natural groundwater flow, the aquifer is drained during the dry season. Consequently, the well runs dry. To prevent this, a trench is dug across the valley that reaches down to bedrock or some other solid, impervious layer. An impervious wall is constructed in the trench and, when the dam is completed, the trench is refilled with the excavated material. A reservoir built in this way will not be drained and may be used throughout the dry season, provided, of course, that the storage volume is sufficient to meet the water demand.

A sand dam may be considered where villagers collect their water from a small, non-perennial stream at the times when it carries water, or from holes dug in the shallow river bed for a short period after the rains. The quantity of water stored is not sufficient, however, to supply water to the village during the entire dry period. By constructing a weir of suitable height across the stream bed, coarse particles carried by heavy flows during the rains are caused to settle. Eventually the reservoir will be filled with sand. This artificial aquifer will be replenished each year during the rains and, if the dam is properly sited and constructed, water will be kept in the reservoir for use during the dry season.

Quite often, a groundwater dam is actually a combination of the two types. When constructing a subsurface dam in a river bed, the storage volume may be increased by letting the dam wall rise above ground level, thus creating an accumulation of sediment. Similarly, when a sand storage dam is constructed, it is necessary to excavate a trench in the sand bed in order to reach bedrock or a stable, impervious layer.

The topographical conditions govern to a large extent the technical possibilities of constructing the dams as well as achieving sufficiently large storage reservoirs with suitable recharge conditions and low seepage losses. The basin in which water is to be stored may be underlain by bedrock or unconsolidated formations of low permeability. It is generally preferable to site groundwater dams in well-defined and narrow valleys or river courses. This reduces construction costs and makes it possible to assess storage volumes and to control possible seepage losses. This is obviously difficult in areas of flat topography. On the other hand, efforts must be made to maximize storage volumes without the construction of unnecessarily high dams. In mountainous areas with very high gradients, it may be difficult to find an acceptable relation between storage volume and dam height. An optimum composition of river-bed material is generally found in the transition zones between mountains and plains.

One of the basic conditions justifying the construction of a subsurface dam is the depletion of groundwater storage through natural groundwater flow. The gradient of the groundwater table, and thus the extent of groundwater flow, is generally a function of the topographic gradient. This fact indicates that the construction of subsurface dams is feasible only at a certain minimum topographical gradient, which naturally varies according to local hydrogeological conditions. Most groundwater dams in existence today have been constructed in areas with a 1–5% slope, but there are examples of dams constructed where gradients are 10–15%.

The storage capacity of sand storage dams and subsurface dams constructed in river beds is a function of the specific yield of river sediment. Such dams, therefore, are preferably constructed in geological environments where the weathering products contain a substantial amount of sand and gravel. In consequence, cases where dam construction has met with success are mostly to be found where the bedrock consists of granite, gneiss and quartzite, whereas areas underlain by rocks such as basalt and rhyolite tend to be less favourable.

Gravity flow water supply systems

In this type of system, water is taken from the source, usually a spring or a stream, from whence it flows under pressure through pipes to an outlet or outlets – which must be below the level of the source. Gravity flow systems, therefore, tend to be constructed in mountainous areas. There are two main types of systems: Open and Closed, each having its own variations.

An open system is one in which the source can supply enough water to provide a continual flow on a 24-hour basis. Reservoir and break pressure tanks (BPT), faucets and many other valves which can fail, therefore, are unnecessary. The closed system is the more common type as the safe yield (safe yield: the lowest annual discharge from the source – usually occurring at the end of the dry season) (Jordan, 1984) – of a source cannot supply enough water to provide continual flow to all taps. Each tapstand must have a functioning faucet.

When the supply source cannot meet the peak water demands, a reservoir is required to store water during low demand periods. Closed systems often have break pressure tanks which are of two types – those with self-closing floatvalves and those with none. A self-closing floatvalve automatically shuts the inlet to the BPT and enables an upstream reservoir tank to fill. BPTs without floatvalves are used on intermittent supplies or open systems so as to reduce the static head.

The type of system to be built is determined by the projected water demands, the safe yield of the source and the topography of the design area. Other factors which should be considered are increased water demand, or increased per capita consumption, and system expansion. As with all other technologies, gravity flow scheme technology has to be appropriate, cost-effective and culturally acceptable.

In mountain terrain, gravity flow schemes are the technology of choice: capital and maintenance costs are low, there can be high community involvement, schemes can be easily extended, etc. In the foothills of mountainous regions, on the other hand, both gravity flow and groundwater may be feasible and a choice has to be made. The economic decision over which source to utilize should be based not only on construction costs but also operation and maintenance costs (Glennie, 1985).

The hand pump option

By far the most important rural water supply technology is the hand pump option which has been admirably expounded by Arlosoroff *et al.* (1987). Most of the following is based on their work.

Once a community has decided to avail itself of groundwater supplies, choices have to be made between hand pumps, public standpipes and yardtaps to the ultimate house connection, assuming that equal system reliability can be achieved. The dilemma is that, while for improved health, people should be encouraged to use copious quantities of water, *i.e.*, through house connections, in terms of water mining and ultimate budgets, they should be encouraged to conserve supplies by the use of low service levels.

The three main technology options, hand pumps, standpipes and yardtaps generally represent progressively increased service levels and call for increasing financial and technical resources for their implementation and maintenance. Table 4 illustrates this and the stepwise rise up the service level "ladder."

The aim is that the technology chosen should give a community the highest service level that it is willing to pay for and has the institutional capacity to sustain. The choice of appropriate technology for a particular project can only be made when resource constraints have been taken into account, including the capability of the users to operate and maintain the alternative systems.

In community water supply (CWS), one of the most important influences on system reliability is the length of time during which pumps stand idle when they break down. If an improved water supply system breaks down, people resort to traditional sources such as ponds, water holes and rivers which are often polluted. In such instances, any health benefits resulting from an improved supply are lost during such breakdown periods. The response times of centralized maintenance organizations covering dispersed communities can stretch to several months. Hand pump maintenance by an area mechanic within a week of breakdown makes a pump which breaks down on the average of once every eight months more "reli-

Table 4. *Options for community water supply.*

Step	Type of Service	Water Source	Quality Protection	Water Use LPCD*	Energy Source	Operation and Maintenance Needs	Costs	General Remarks
5	House connections	Groundwater Surface water Spring	Good, no treatment May need treatment Good, no treatment	100 to 150	Gravity Electric Diesel	Well-trained operator; reliable fuel and chemical supplies; many spare parts; waste-water disposal	High capital and O&M costs, except for gravity schemes	Most desirable service level but high resource needs
4	Yardtaps	Groundwater Surface water Spring	Good, no treatment May need treatment Good, no treatment	50 to 100	Gravity Electric Diesel	Well trained operator; reliable fuel and chemical supplies; many spare parts	High capital and O&M costs, except gravity schemes	Very good access to safe water; fuel and institutional support critical
3	Standpipes	Groundwater Surface water Spring	Good, no treatment May need treatment Good, no treatment	10 to 40	Gravity Electric Diesel Wind Solar	Well trained operator; reliable fuel and chemical Supplies; many spare parts	Moderate capital and O&M costs, except gravity schemes; collection time	Good access to safe water, cost competitive with handpumps at high pumping lifts
2	Handpumps	Groundwater	Good, no treatment	10 to 40	Manual	Trained repairer; few spare parts	Low capital & O&M costs; collection time	Good access to safe water; sustainable by villagers
1	Improved traditional sources (partially protected)	Groundwater Surface water Spring Rainwater	Variable poor Variable Good, if protected	10 to 40	Manual	General upkeep	Very low capital and O&M costs; collection time	Improvement if traditional source was badly contaminated
0	Traditional Sources (unprotected)	Surface water Groundwater Spring Rainwater	Poor Poor Variable Variable	10 to 40	Manual	General upkeep	Low O&M Costs (buckets, etc); collection time	Starting point for supply improvements

*LPCD = Litres Per Capita Per Day

After Arlosoroff, *et al*, 1987

Table 5. *Community water supply technology costs (for a community of 400 people).*

Technology	Low	High				
	Handpumps	Standpipes	Yardtaps	Handpumps	Standpipes	Yardtaps
Capital Cost (US $)						
Wells*	4000	2000	2500	10000	5000	6000
Pumps (hand/motor)	1300	4000	4500	2500	8000	9000
Distribution**	None	4500	16000	None	10000	30000
Sub Total	5300	10500	23000	12500	23000	45000
Cost per capita	13.3	26.3	57.5	31.2	57.5	112.5
Annual Cost (US $/year)						
Annualized capital ***	700	1500	3200	1400	3000	6000
Maintenance	200	600	1000	400	1200	2000
Operation (Fuel)	None	150	450	None	300	900
Sub Total (Cash)	900	2250	4650	1800	4700	8900
Haul costs (Labor)****	1400	1100	None	300	2200	None
TOTAL (including labor)	2300	3350	4650	4800	6900	8900
Total Annualized Cost Per Capita						
Cash only	2.3	5.6	11.6	4.5	11.8	22.3
Cash + Labor	5.89	8.4	11.6	12.0	17.3	22.3

* Pumping water level assumed to be 20 metres. Two wells assumed for handpump system (200 persons per handpump).

** Distribution system includes storage, piping, and tap with soakaway pits.

*** Capital costs with replacement of mechanical equipment after 10 years annualized at a discount rate of 10% over f 20 years.

**** Labor costs for walking to the water point, queuing, filling the container, and carrying the water back to the house. Time valued at US $ 0.125/h.

able" than one which lasts for an average of 18 months before it breaks down but then must wait two months for the mobile maintenance team to arrive. It is evident that a reliable hand pump supplying 30 litres per head per day for 95% of the year will provide a higher level of service than yardtaps designed for 150 litres per head per day but working for an average of only two hours a day because of leakage, breakdowns, fuel shortages or limited water available at the intake.

Financial implications

Arlosoroff, *et al.* (1987) also consider the financial implications, and provide some interesting and determinative figures when they compare the three types of technology based on groundwater (see Table 5).

Capital costs of these three technologies generally range from US$10 to $30 per capita for wells equipped with hand pumps to US$30 to $60 for motorized pumping and standpipes and US$60 to $100 or more for yardtap services. In global terms, that means cost estimates for meeting rural water supply needs to the year 2000 range from US$50,000 million to US$150 000 million, depending on the choice of technology. With the obvious difficulties of mobilizing financial resources for this scale of investment, rapid progress in meeting basic needs receives services at the lower end of the cost range. Upgrading to a higher service level may then be financed by the community at a later time as benefits from the initial investment and from other sources increase the available resources.

Analysis of data from a wide range of community water supply projects indicates a similar divergence in recurrent operation and maintenance costs of the three options to that already noted with the capital costs. With a centralized maintenance system, the annual per capita cost of maintenance for a hand pump-based community water supply system can range from US$0.50 to US$2.0. Well-planned community-level maintenance can bring that figure down to as low as US$0.50 per capita per year. By comparison, centralized maintenance of a standpipe system, with motorized pumping, costs from US$2 to $4 per capita per year and for yardtap maintenance, the range is US$4 to $8. There are circumstances where communities may value so highly the time saved due to the extra convenience of yardtaps that they are willing and able to pay the extra price.

More frequently, however, the serious shortage of readily available cash resources will mean that recurrent costs must be kept to a minimum and hand pumps will be the indicated choice.

There are many other valid reasons for promotion of a hand pump-based system. The most significant difference between hand pump projects and those based on standpipes or yardtaps is the switch to motorized pumping and the consequent need for dependable energy supplies and skilled pump mechanics when a piped distribution system is provided (Arlosoroff *et al.,* 1987).

However, in cases, where reliable low-cost electric power is available from a central grid, electric pumps can be a relatively inexpensive and operationally simple means of lifting water. Communities which have the financial and technical means available to implement and sustain projects based on electric pumping should be given every encouragement to do so, as this frees scarce public sector funds and external aid projects serving poorer communities, always assuming water availability. However, the number of communities in the humid tropics with a dependable electricity supply is presently small – well below 10% of the total rural population in Africa, only a little higher in most countries in Asia and reaching 40–50% in the more developed countries of Latin America.

The obvious conclusion from the work of Arlosoroff *et al.* (1987) is that hand pump technology should be the technology of choice for the *majority* of the global rural communities.

The hand pump

One of the major outcomes of the work of Arlosoroff *et al.* (1987), who headed a WB/UNDP hand pump project, is the concept of VLOM (a much-used acronym coined by the project) that stands for Village Level Operation and Maintenance. The VLOM concept seeks to avoid high costs, long response times, unreliable service and other operational difficulties in the repair of hand pumps through central maintenance systems. The VLOM concept is the cornerstone of the hand pump option approach.

It already has been highlighted that the main task in planning CWS improvements is to select the technology which will give the highest service level commensurate with the available resources, namely:

(a)   Financial resources for capital and recurrent costs.
(b)   Physical resources (water and energy).
(c)   Organizational resources (manpower institutions).

One of the primary criteria in the design of the pump is ease of maintenance. This is made possible by the use of new plastics for bearings and down-the-hole components. The bushes (allowing movement of the fulcrum) are now made of two different materials: polyester and polyamide. Field trials reveal minimum wear and the two bushes snap together to form one

self-contained unit that can be fitted into the pump without special tools. The down-the-hole components, plunger and floatvalve are also made of new third-generation plastic which can be replaced by two villagers within one hour, even at depths of 40 meters. One of the reasons for such rapid and easy replacement is the advance in the design of the pump rods. The WB/UNDP Project replaced conventional pump rods (using threaded connections) with lightweight rods joined by a system of hooks and eyes which allow the rods to be joined quickly and easily without tools.

An important aspect of any hand pump is the discharge rate versus pumping effort. In the case of the WB/UNDP pump (called "Afridev" in East Africa and elsewhere until a local name is coined – once the pump is being produced locally), the decision was taken that the "Afridev" would use a standard 50 mm diameter cylinder. The operative force needed to pump from different depths is varied by offering different mechanical advantages in the hand pump.

Another important concept in regard to the WB/UNDP pump is standardization. In developing countries, the scarcity of skilled mechanics and an unavailability of genuine spare parts have been prime reasons for rural water supply failures. Therefore, the major geometrical parameters of the pumps have been made standard and now apply to all global manufacturers. This will ensure that spare parts will be readily available and pumps can be readily interchanged.

In summary, the guidelines for the VLOM design are:

(a)   Ease of Maintenance – The pump design takes care of this.
(b)   Robustness – The pump design and rigorous quality control procedures ensure this.
(c)   Local Manufacture – In-country manufacture is promoted.
(d)   Standardization – At present four pumping manufacturers in U.K., Kenya, Pakistan and India have agreed to manufacture to a set of standards issued by the Swiss Appropriate Technology Centre.
(e)   Costs – Target costs should be US$300 to $400 for a pump complete with rods and a rising main to 25 metres.

In terms of South Asia, these costs seem high. In Pakistan, for example, a suction mode pump can be purchased for US$30. A force model can be secured for approximately US$200 that can pump from depths of 20 to 30 metres. These are, however, family pumps that are inappropriate for communities of 200 to 300 users. Their breakdown rate is high but, surprisingly, the reliability is also high since most members of a family are able to repair the pump with relative ease.

Whilst the emphasis in this section has focused on the "Afridev" hand pump, the author has attempted to highlight some of the peripheral issues that are so essential to the success of a rural water supply scheme based on hand pump technology. Many research and commercial institutions are carrying out excellent work on hand pumps and several outstanding hand pumps incorporating the VLOM concept are now available globally.

SANITATION AND ITS RELATIONSHIP TO HEALTH

There is a remarkable correlation between low service levels of sanitation and high infant mortality rates (IMR). It is anticipated that public acceptance of population control measures, amongst other things, will largely depend on the extent to which the IMR is reduced. Correspondingly, an improvement in the service level of sanitation can be expected to have a positive effect in curbing the high population growth rates in the humid tropics.

Although it is universally accepted that improving sanitation within a community should lead to an improvement in health, it is difficult to ascertain whether the impact is direct or indirect. While water and sanitation theoretically should be linked inextricably, they frequently are not. Witness the respective coverage figures in the *rural areas* of the humid tropics: 40% for water supplies and only 15% for adequate sanitation (several authors believe that 15% is an optimistic assessment). This disproportionate lag in sanitation has prompted new questions in the engineering sector that focus on what kind of technology would be the most appropriate to the communities to be served and how best to introduce this technology.

The need for technical specialists to be aware of the social and cultural context of engineering interventions has already been stressed, together with the need for popular participation in project design and implementation. Concepts such as grassroots development based on an approach which builds from below have offered a challenge to the supremacy of the topdown approach which is based on decisions made at high management levels. The grass-roots approach is critical in sanitation programmes (indeed, in rural programmes generally) since the effectiveness of these programmes depend not merely on community support, but more particularly on the consent and commitment of household and individual users. In sanitation, as with water programmes, the technical and social decisions are interrelated.

It is well-recognized that communities do not respond favourably to those ideas which have been initiated from the outside by governmental entities, by bilateral aid agencies, and by non-governmental aid units in the interest of progress. This is particularly evident in sanitation projects. The many constraints to improving health through better sanitation have to do with the political, economic, social and cultural context of health and disease. Each of these is brought into sharp relief within communities in what is frequently a working misunderstanding between those who promote sanitation improvements, *i.e.*, the outsiders, and the beneficiaries.

The professional promoters bring to the villagers those attitudes and values based on their training and experience (in the case of aid-agency personnel often on another continent and probably in another culture) which may not be acceptable to the villagers. Furthermore, the outsiders are seen as representatives of government. They are amongst the privileged, articulate minority who may be perceived as unable to identify easily with the problems of poor households. In such circumstances, it can be exceedingly difficult to persuade, encourage, cajole and advise rural people to change their age-old, deeplyingrained patterns and exchange traditional excretion areas for a dark (in the case of the improved ventilated pit-latrines), confined space.

Frames of reference in the developing world may differ substantially from those which apply in the developed world. Simpson-Gerbert (1984) points out that when people are told that new sanitation facilities will make their environment "cleaner," it is their own interpretation of this concept which will be used. "Clean" may have different meanings to project promoters and recipients. Simpson-Gerbert argues that it is essential, therefore, to look into the traditional categories of clean and dirty, purity, and pollution before embarking on a campaign to motivate people to accept a project in improved sanitation which, hopefully, will change their behaviour to comply with new standards of cleanliness.

The evidence for the value attached to cleanliness (and, by implication, environmental sanitation) by communities is gathered from studies of diarrhoea. Researchers frequently divide people's perception of its causes into two categories: physical and social or spiritual. De Zoysa *et al.* (1984) found in Zimbabwe that the physical causes were widely identified and although the germ theory was not explicitly stated, discussion of the faecal-oral transmission route of diarrhoea appeared to be understood. Some 55% of the households associated diarrhoea with a polluted environment, including uncovered food, dirty water and flies. The study also showed that social/spiritual causes of diarrhoea were perceived to be important. In conclusion, the authors stressed that this two-fold classification as to causes of diarrhoea should not be interpreted as mutually exclusive or divergent approaches to disease. The cultural practices of beliefs of any community can have a major bearing on the method chosen for excreta disposal and cleansing may involve the use of wet or dry material. If water is used and latrines are constructed, safe disposal of the water is essential to avoid problems of both volume over-loading of the latrine and of water ponding inside or outside the superstructure. The latter could lead to the inadvertent provision of possible breeding sites for disease-bearing organisms such as mosquitoes.

There may be a requirement for separate toilets for both sexes, or for toilets combined with washing facilities, or there may be taboos associated with the use of toilets. In addition, the hygiene practices of the community might negate any positive effect on health unless they are determined and remedied as part of the implementation programme. For example, the commonly observed neglect of thorough handwashing after defecation, and the mistaken belief that children's faeces are harmless, can be remedied through child and adult education programmes. Provision of water specifically for hand washing

Table 6. *Comparison of several sanitation technologies.*

Sanitation System	Rural Application	Urban Application	Construc-tion Cost	Operation Cost	Ease of Construction	Water Requirement	Hygiene
Pit latrines	Suitable in all areas	Not in high density suburbs	Low	Low	Very easy except in wet or rocky ground	None	Moderate
Bucket and cartage	Suitable	Suitable	Low	High	Easy	None	Bad
Vault and vacuum truck	Not suitable	Suitable where vehicle mainte-nance available	Medium	High	Requires skilled builder	None	Moderate
Aqua privies	Suitable	Suitable	High	Low	Requires skilled builder	Wat/source near privy	Good
Septic tanks	Suitable	Suitable for low density suburbs	V/high	Low	Requires skilled builder	Water piped to privy	Excellent
Pour flush and soakaway	Suitable	Not suitable	High	Low	Requires skilled builder	Wat/source near privy	Good
Sewerage	Not suitable	Suitable where it can be afforded	V/high	Medium	Requires experienced engineer	Water piped to privy	Excellent

closer to latrines, together with soap (or ash), multiplies the positive effects of sanitation on health.

While sanitation projects generally involve simple engineering, what complicates their design and implementation are that they are also projects in radical social intervention. Obviously, for many reasons, it is far simpler for the government official or aid agency worker to get the rural community to accept a water supply point than a sanitation facility.

Part of the rationale behind the "integrated approach," *i.e.,* the promotion of water, sanitation and health education as a package, rests upon the realization that communities are more likely to at least consider the question of sanitation facilities *if* the water supply is made conditional upon their acceptance of change in sanitation habits. But there are dangers in such an approach, and it is by no means universally accepted. Another approach which has found support advocates "water as an entry point," to stimulate demand for other services such as sanitation and health education.

The technology at least is well-established: simple, low-cost and appropriate latrines have been designed for different cultures and different topographies (as shown in Table 6) with the following guiding principles:

(a) Excreta disposal is a sanitation matter about which people have their own preferences.
(b) Rural people often require a reason or motivation for using a latrine.
(c) Any type of latrine needs cleaning and maintenance.
(d) The latrine must be suitable in terms of technical, social and economic criteria.

BEYOND THE DECADE – HELPING PEOPLE TO HELP THEMSELVES

While it is true that the United Nations Drinking Water and Sanitation Decade has focused attention on two billion people who have no potable water supply or sanitation facilities, it is a sobering thought that the number of functioning drinking water and sanitation facilities installed in developing countries during the first half of the Decade has hardly kept pace with population growth and the equipment breakdowns. The sad paradox is that it is the most affluent and accessible populations which have been the first to receive attention. Those remaining will be progressively more difficult to reach. Nevertheless, many positive aspects have emerged from the current Decade, among which can be counted the new thinking on the human component of the equation and the major advances in the development of technology which can bring water supplies and sanitation to the rural communities of the developing world. The successful technologies exhibit several common characteristics: they are affordable, socially acceptable, reliable and locally manufactured. A crucially important factor is that technology which is owned and managed by individuals, families or, indeed, closely-knit communities have a far higher probability of sustainability.

McGarry (1987) suggests ways of improving technology delivery beyond the decade by improving management, efficiency, and spreading the cost burden from government to the consumer via the community development approach, viz*:*

(a) Assisting the formation of an organization or committee within the community which becomes responsible for its participation on projects, for liaising with the implementing agency and, ultimately, for maintaining and, later, expanding the facilities.
(b) Training members of the community, including women, in technical operation, maintenance and repair; in management and sometimes in basic revenue collection and account procedures.
(c) Improving communications between the government and the community and often between sections of the community itself.
(d) Establishing a hygiene education programme which will reach the women and children and which should provide permanent behavioural changes leading to improved sanitation habits and greater awareness of the benefits accruing from sustained use of the facilities.
(e) Ensuring full and active participation of the communities in their own development – which must go well beyond the usual passive receipt of project benefits. The aim is to enable the community to effect its own self-development without undue dependence on governmental or other outside agencies.

Alas, a Sword of Damocles hangs over such an enlightened and forward-thinking strategy. To many politicians such ideas are heretical. Whether the politicians will ever see the irrefutable wisdom of many of the ideas quoted in this review remains to be seen, but certainly, the success or failure of the rural water supply and sanitation revolution is very much in political hands.

REFERENCES

Arlosoroff, S.; Tschannerl, G.; Grey, D.; Journey, W.; Karp, A.; Langeneffer, O. & Roche, R. (1987) *Community Water: The Hand Pump Option.* The World Bank, Washington, D.C., USA.

Briscoe, J. (1983) Selective Primary Health Care Revisited: Water Supply and Health in Developing Countries. Prepared for the Office of Health U.S. AID.

Buros, O. K. (1985) *An Introduction to Desalination. Non-Conventional Water Resources Use in Developing Countries.* United .Nations.

Danida. (1988). A Strategy for Water Resources Management. Danida Dept. of International Development Cooperation.

De Zoysa *et al.* (1984) Perceptions of Childhood Diarrhoea and its Treatment in Rural Zimbabwe. *Social Science and Medicine.*

Esrey, S. A. ,Feachem,R. J. & Hughes, J. M. (1985) Interventions for the control of diarrhoeal diseases among young children: Improving water supplies and excreta disposal facilities. *Bull. WHO 63(4).* 757–772.

Glennie, C. (1983) *Village Water Supply in the Decade: Lessons from the Field.* J. Wiley and Sons.

Hadwen, P. (1985). *Rainwater Harvesting: An Overview. Non-Conventional Water Resources Use In Developing Countries.* United Nations.

Hanson, G. & Nilsson, A. (In Draft 1984) Groundwater Dams for Rural Water Supplies in Developing Countries.

Jordan, D. J. (1984) *A Handbook of Gravity-Flow Water Systems.* Intermediate Tech. Pub.

Keller, R. (1984) The World's Freshwater: Yesterday–Today–Tomorrow. *Applied Geography and Development. 17.*

McGarry, M. G. (1987) Matching Water Supply Technology to the Needs and Resources of Developing Countries. *Natural Resources Forum.* United Nations.

Okun, D. A. (1988) The Value of Water Supply and Sanitation in Development: An Assessment. *American Journal of Public Health. 78,* (ll).

Shuval, H. I., Tilden, R. L., Perry, B. H. & Grosse,.R. N. (1981) Effect of investments in water supply and sanitation on health status: a threshold-saturation theory. *Bull. WHO 49 (2)* 248.

Simpson-Gerbert, M. (1984) Water and Sanitation: Cultural Considerations. In: Peter G. Bourne (ed.) *Water and Sanitation: Economic and Sociological Perspectives.*Academic Press, Orlando, Florida, USA.

Unvala, S. P. (1988) Will We Run Out of Water? *Water and Wastewater Int..* April.

Walsh, J. A. & Warren, K. S. (1979) Selective Primary Health Care: An interim strategy for developing countries. *New England Journal of Medicine 301(18).* 967–974.

World Health Organization.(1979) The WHO Diarrhoeal Disease Control Programme. *Weekly Epidemiological Record. 54*: 121–128.

VI. Management Issues

22: Water Resource Management

M. M. Hufschmidt

Program on Environment, East-West Center, 1777 East-West Road, Honolulu, Hawaii 96848

ABSTRACT

In the first major section of the chapter, water resource problems and issues are examined in two main contexts: *Urban and urbanizing* and *rural resource-related*. In the urban context, problems and management implications are summarized for *urban water supply, urban sanitation and water pollution, flooding and storm drainage, groundwater depletion and land subsidence*, and *integrated urban water and land management*. In the rural context, problems and management strategies are discussed for *lowland agriculture*, and *upland watersheds*. In the second major section of the chapter seven generic management issues are analyzed: *water resource assessment, integrated planning, the river basin as a management region, demand management, local involvement, implementation, and organizations and institutions*. For each issue, the major elements or problems are identified and some management solutions are proposed. The third major section of the chapter presents the major elements of a water resource-management strategy. Eight such elements are proposed for the planning stage, five for the implementation stage, and two relating to institutions and organizations. For each element, the presentation is illustrated with examples drawn from water resource management case studies in the humid tropics.

INTRODUCTION

The humid tropics are especially challenging for water resource management because of the extreme conditions encountered. High temperatures and rainfall, often occurring as severe storms, along with high rates of erosion, cause flooding and sedimentation of streams, lakes, reservoirs and estuaries. High rates of evapotranspiration impose stresses on vegetation in all but the wettest areas, such that irrigation is often required for crops during the dry season in these subhumid and wet-dry tropical areas. Management problems are compounded by high population densities and increasing urbanization, especially in South East Asia and parts of Africa and South and Central America. These special problems and challenges of the humid tropics add to the management difficulties

that are commonly encountered in most developing countries regardless of climate.

Management defined

Elsewhere we have defined water resource management as "a set of actions taken to use and control natural resources inputs such as water in order to obtain outputs and natural system conditions useful to society" (Hufschmidt & Kindler, 1991). Through management, water resources are put to beneficial use of humans and actions are taken to reduce detrimental effects of human-induced pollution and natural hazards such as floods and droughts on humans and natural systems. Management can be viewed as a process, involving the key stages of *assessment*, *planning* and *implementation*. It also can be considered as a set of linked activities and tasks, such as storing, transmitting, distributing and allocating water for irrigation or urban uses. Also, in its broadest sense, water resource management includes not only building and operating physical facilities, such as dams, canals, pumps and waterworks, but also devising and using the means for carrying out management plans (e.g, regulations and economic incentives) and creating or modifying the institutions and organizations needed for both planning and implementation. Integrated water resource management takes account of the important physical, economic, social, cultural and political linkages between land and water, and between surface water and groundwater; economic linkages between specific water uses such as irrigation and domestic supplies; and social, cultural and political linkages between water development, other sectors such as agriculture, industry, transportation and energy and the people who are benefited or adversely affected.

Scope of this chapter

Water resource management occurs in a variety of different settings in the humid tropics, as shown in the chapters in this volume on urban and urbanizing areas by Gladwell, Lee, Lind and Niemczynowicz and Low; on agricultural and forested areas by Tejwani; on water resource regions generally by Le Moigne and Kuffner; and on islands by Falkland and Brunel. Our analysis of water resource management in these settings will be summarized in two broad contexts which reflect different combinations of natural resources and human factors as they relate to water:

(a) The urban context, consisting of the urban or rapidly urbanizing areas, typified by Jakarta, Bangkok and Dhaka in Asia, Sao Paolo and Mexico City in Latin America, and Lagos in Africa.

(b) The rural, resource-related context, consisting of lowland agricultural areas, upland watersheds, tropical rain forests, coastal zones, including river deltas, and small islands which present special problems of management because of their size and geological origin.

Each of these contexts has distinctive water problems and management issues which warrant special attention here. However, both the urban and rural-resource settings share certain generic management issues which will be discussed in this chapter along with major elements of a management strategy for the humid tropics applicable to both urban and rural-resource settings.

MAJOR PROBLEM CONTEXTS

Urban and urbanizing

The major problems of large and rapidly growing cities in the humid tropics are four-fold:

(a) Need for dependable supplies of potable water

(b) Severe and growing sanitation problems, including pollution of streams, lakes, estuaries and groundwater from domestic and industrial sewage and solid wastes

(c) Large and increasing flood damages from urban occupancy of flood plains and high peak flows arising from urbanization of the watershed

(d) Depletion of groundwater aquifers, with associated land subsidence, caused by reduction of infiltration and over-pumping of the aquifers.

These typical water problems of urban occupancy are accentuated by the extremely rapid growth of many urban areas in developing countries and by the intense rainfall and consequent high water levels often experienced in these areas in the humid tropics. Solution of these problems calls for an integrated approach to water resource management that is specifically adapted to the urban setting in the humid tropics

Specific examples of many of these water resource problems and related management issues discussed in this section are given in the following chapters by Low, Lindh and Niemczynowicz, and Lee.

Urban water supply As Lee points out, a major constraint to providing a potable water supply to rapidly growing urban areas is shortage of investment funds. Historically, provision of infrastructure such as water supply, sanitation, transport, utilities and basic housing, has lagged behind population growth and economic activity in these cities, and recent rapid increases in population have accentuated the problem.

The capital investment needs for infrastructure including water supply are enormous in relation to need; and it may take decades to attain satisfactory levels of such urban services.

A special problem is the provision of minimal levels of water service for squatter settlements around major cities such as Calcutta, Dhaka, and Manila. At least in the short term, these settlements will have to make do with low-cost facilities such as standpipes, with upgrading of facilities to higher standards as a long-term objective.

As urbanization proceeds, readily available surface water and groundwater supplies are over-exploited; groundwater aquifers show rapid declines, as in Bangkok and Jakarta, and surface water and groundwater supplies become increasingly polluted. New and more remote sources of supply are sought, which increase the cost of supplying water to cities and bring the urban water user increasingly into conflict with other water uses, principally irrigation, as for example, in Bangkok, Jakarta, Surabaya and Madras. Serious problems of water allocation among competing uses are the result.

The urban water supply problem is further aggravated by the deterioration of existing systems arising from inadequate operation and maintenance. In some cases, over 50% of the available supply is lost through leakage of water mains or is otherwise unaccounted for.

Major water supply problems and implications for management are summarized in Table 1.

Urban sanitation and water pollution Most of the streams and estuaries in or adjacent to cities in the humid tropics are grossly polluted because centralized collection and disposal of urban-generated sewage is inadequate and continues to lag seriously behind urban growth. Further, much centrally collected sewage continues to be discharged into streams and estuaries without treatment. Even in the few instances, as in Singapore, where point-source pollution is adequately collected and treated, non-point sources of pollution including litter, solid wastes and sediment from the land and streets contribute to gross water pollution, especially during periods of high rainfall and associated runoff. In some cases, groundwater is also at risk of pollution (Gladwell, this volume).

As with urban water supply, the capital investments required for urban sanitation, including solid waste collection and sewage collection and treatment, are so great as to be beyond the immediate financial capacity of many cities, especially the fast-growing ones. With few exceptions, such as Singapore, wastewater treatment to the secondary level is not a realistic option, at least over the next decade or two.

In some cities, such as Jakarta and Surabaya, in the humid tropics, water pollution can be alleviated during the rainy season by using large volumes of water for flushing the streams and estuaries. However, during the dry periods, water used for other purposes such as irrigation or expensive reservoir storage would be required for this purpose.

Table 1. *Urban Water Supply*

Problems	Management Implications
(a) Many urban zones not served especially in poor sections: major causes – bias toward high-cost technology; shortage of investment funds; unequal distribution of infrastructure and services	(a) (i) Increased cost recovery from water users (ii) Adopt low-cost appropriate technology (iii) Promote community involvement in water supply (Lee, this volume)
(b) Unreliable performance of water system, including high rate of "unaccounted for" water: major causes: poor operation and maintenance; faulty metering and billing procedures.	(b) (i) Increased cost recovery to finance adequate operation and maintenance. (ii) Improved organization and staff training for operation and maintenance. (iii) Community involvement in managing local water supply
(c) Conflict with other water users (e.g., irrigation) for diminishing available water supplies.	(c) (i) Reduce water withdrawals, by demand management, recycling and waste water reuse. (ii) Integrated system management of surface water and groundwater to make best use of available water for various competing purposes.

Again, as with urban water supply, operation and mainte-nance of sewage collection and treatment systems and solid waste and litter control programs are often weak and ineffec-tive. These low-status activities are seriously underfunded and effective management is often given low priority by urban managers and policy-makers.

In coastal cities, waste water can often be disposed of satis-factorily, even with little or no treatment, via long outfalls to the sea. However, in inland cities, such as Bandung, this option is not available and some degree of treatment is required if gross pollution downstream is to be avoided.

A summary of major water pollution problems, causes and management implications is given in Table 2.

Flooding and storm drainage Human occupancy of urban flood plains in the humid tropics has greatly magnified the risk of flood damage, not only of the occupied areas, but also of downstream areas because of sharp increases of storm runoff from roads, streets, housing, parking lots and other impervious areas. Given existing urban occupancy of flood plains, struc-tural flood control works – reservoirs, levees, storm drains and channelization – are extremely costly and usually beyond the financial capacity of cities.

In practice, some cities such as Bangkok and Jakarta have accommodated to flooding, as the flood rises are gradual rather than torrential. In other cases, such as Dhaka, sudden flooding can cause severe damage, as in 1987 and 1988.

Cost-effective flood management for most cities in the humid tropics will involve a mixed strategy of inexpensive structural means such as drainage channels plus flood warning, flood plain zoning, flood proofing and flood evacuation mea-sures.

Installing and operating an effective program of this type requires a high level of management, usually not available to these cities. Few cities anywhere do a good job of urban flood management.

Urban storm drainage for flood control is closely related to non-point source water pollution as much of this pollution reaches the streams and estuaries via these drainage works.

A summary of major flooding and storm drainage problems and management implications is given in Table 3.

Groundwater depletion and land subsidence In some cities, such as Bangkok and Jakarta, over-pumping of groundwater, along with high concentrations of buildings, roads and streets, has led to serious subsidence of land with consequent flooding

Table 2. *Urban Water Pollution*

Problems and Causes	Management Implications
Gross pollution of urban streams, lakes, estuaries and groundwater is increasing, because:	
(a) Much domestic sewage is uncollected and flows directly into water bodies.	(a) (i) Low-cost appropriate sanitation technologies (ii) Community involvement in building and operating sanitation systems
(b) Most domestic sewage that is collected is discharged into water bodies without any treatment.	(b) (i) Appropriate low-cost treatment technologies e.g., detention basins (ii) Wastewater reuse (iii) Cost-sharing
(c) Much industrial waste is discharged untreated on land and into water bodies	(c) (i) Appropriate incentives (Subsidies for pollution control activities; effluent charges) (ii) Recycling
(d) Much pollution of surface water bodies originates from "non-point" sources – urban land runoff-litter, solid wastes, sediment.	(d) (i) Community involvement in sanitation & cleanup. (ii) Watershed management program
(e) Centralized sewage collection and treatment systems are costly and beyond immediate capacity of urban governments to finance.	(e) (i) Appropriate low-cost technology for wastewater collection & treatment (ii) Community involvement (iii) Cost sharing, via sewer charges
(f) Existing sewage collection and treatment systems are poorly maintained and operated, leading to sewage spills and failure of treatment systems, especially during heavy rainfall periods.	(f) (i) Community involvement in operation & maintenance of systems (ii) Adequate financing of operation and maintenance via sewer charges and property taxes

and damage to structures. Centralized control of groundwater pumping is required to reduce or eliminate land subsidence. This involves an advanced management program, with groundwater pumping permits, metering, monitoring and regulation of pumping rates.

In some cases, such as in Bangkok, conjunctive surface water-groundwater management is a preferred strategy. This could include artificial groundwater recharge via water-spreading fields and injection wells to raise groundwater levels. A possible problem is pollution of groundwater arising from these recharge activities.

A summary of groundwater depletion and land subsidence problems and management implications is given in Table 4.

Integrated urban water and land management These problems of water supply, sanitation, water and land pollution, flooding and storm drainage and groundwater deterioration are all inter-related. Nothing less than a fully integrated approach to water and land management at the urban scale will suffice to deal adequately with them. The reality is that in most cities in the humid tropics, land-use and water management decisions are each taken without regard for the other. Even within the water

Table 3. *Urban Flooding and Storm Drainage*

Problems		Management Implications	
(a)	Rapid urbanization, including human occupancy of flood plains has magnified flood risks and consequent damages.	(a)	Effective land use planning, including flood plain zoning, flood proofing, evacuation programs and redevelopment of flood-prone areas.
(b)	Costly structural flood control measures are beyond financial capacity of cities.	(b)	Emphasize non-structural measures and low-cost physical measures, e.g., using streets for temporary flood storage
(c)	Planning effective urban storm drainage systems requires rainfall and runoff data with very fine time and space resolutions; these are generally not available for humid tropics cities.	(c)	Establishing and maintaining hydrologic data systems suited to urban areas in humid tropics
(d)	Effective flood control program, combining structural and non-structural measures requires a high level of management, not available in most cities.	(d)	Building management capacity and appropriate institutions for integrated flood management

Table 4. *Groundwater Depletion and Land Subsidence*

Problems		Management Implications	
(a)	Excessive pumping of groundwater causes land subsidence in urban areas with consequent flood losses and damage to structures.	(a)	Aquifer-wide regulation of pumping rates, involving permits, metering, and monitoring.
(b)	Unsustainable groundwater pumping rates cause aquifer depletion, with increasing pumping costs, and saline water intrusion to coastal aquifers.	(b)	Conjunctive groundwater-surface water management, involving appropriate timing of groundwater pumping and artificial recharge of aquifers.

resource sector, urban water supply, sanitation and water pollution control and flood control and urban storm drainage are typically handled by different agencies, with little or no coordination among them. And all of these activities are undertaken typically with grossly inadequate basic data and scientific analysis of urban hydrology. Some of the detailed elements of an integrated approach to water management are presented in the sections to follow on generic management issues and management strategy.

Rural resources-related context

In contrast to the urban and urbanizing areas, human occupancy and economic development are typically of low density in rural, resource-related regions. However, human-natural system interactions can cause serious environmental problems, especially in heavily populated lowland and upland rural areas as in Java, Bangladesh, and the Philippines. Some of these problems are discussed in the context of lowland agriculture and upland watersheds.

Lowland agriculture Rice cultivation is dominant in lowland agriculture in the humid tropics, especially in South East Asia. According to Le Moigne and Kuffner (this volume), agricultural area under irrigation in 1980 (mainly devoted to rice) consists of almost 5.5 million hectares in Indonesia, 2.6 million hectares in Thailand and 1.2 million hectares in the Philippines. More recent data for 1984–86 reported by the World Resources Institute (1988) reveal that the proportions of arable and permanent cropland that are irrigated are 33% for Indonesia, 26% for Sri Lanka and Vietnam, 22% for Bangladesh, 19% for Thailand and 18% for the Philippines.

Over the past 30 years there has been a rapid expansion of irrigation, especially in South East Asia and South Asia. Much of this expansion has been in large-scale projects involving dams and reservoirs, planned and built by national irrigation agencies. Considerable potential for further irrigation development exists in the Niger Basin in Africa, along the Brahmaputra-Ganges in Bangladesh, the lower Mekong River, swamps and coastal lowlands of Indonesia and the floodlands (varzeas) of Amazonia (Barrow, 1987). Such development would be at high economic cost and ecological damage to terrestrial and aquatic ecosystems.

Management problems and strategies In recent years, it has become evident that the performance of many large-scale irrigation systems in developing countries is very poor. Studies by the World Bank (1981) and USAID (Steinberg *et al.*, 1983) revealed many deficiencies in donor-supported large-scale systems, including low productivity; insufficient provision of operation and maintenance; inadequate, unreliable, and inequitable water supplies; excessive water losses and limited productive life of irrigation infrastructure. In a study of irrigation development in Asia, Barker *et al.* (1984) point to the two broad types of irrigation systems: large-scale centrally-managed systems with entrenched bureaucracies, and small-scale community-managed systems. The increasing trend toward converting small community-based systems into the large centrally-managed systems is viewed as unfortunate, given the poor performance of the large-scale systems. Required is an approach by the central irrigation agency toward small systems that is different from that used for large systems, where community investment and control is often weak and ineffective. Part of the failure of large-scale systems lies at the planning and design stages where arrangements to encourage farmer participation and to provide appropriate incentives are not developed. All of these factors lead to inadequate performance, including breakdowns in the irrigation system at the main and secondary canal levels as well as at the tertiary or farm level (Barker *et al.*, 1984).

The perceived need for improvements in performance, especially at the tertiary or farm level, and involvement of the local irrigation community in management led to development in India of the Command Area Development approach. Under this approach, Command Area Development Agencies

(CADAs) were established by the government for specific irrigation sub-areas, independent of the central irrigation agency, to provide technical and financial help and management guidance to the local irrigation community. Experience of the CADAs over the past decade has been mixed. Although some improvements in irrigation performance were achieved, Sinha (1983) reports that targeted levels of crop production have not been attained because the CADAs have minimal control over key resource inputs, including water and infrastructure such as field channels and field drains. Co-ordination between the CADA and the irrigation department, which operates and maintains the main system from headworks to farm outlets, is weak. According to Sinha (1983), CADAs need to be strengthened and incorporated within the irrigation department in order to provide unified management of the entire system from headworks to the on-farm channels and drains. These revised and strengthened CADAs would then be able to enlist the active cooperation of farmers in constructing, operating and maintaining all irrigation works below the outlets to the farms.

Another major attempt at improving irrigation management focused on increasing farmer and local community involvement in water management beginning at the planning stage. This approach was begun in the Philippines in the 1970s by the National Irrigation Administration (NIA) and focused initially on developing water-user associations for small-scale systems (Korten, 1982; Korten & Siy, 1988). Beginning with a pilot project in 1976, the NIA soon moved ahead with a program for participatory small-scale irrigation development.

Experience with this approach in the Philippines and with similar attempts in Indonesia, Sri Lanka and Thailand has been mixed. However, the basic strategy of involving local communities in more and more aspects of irrigation management on large-scale systems as well as on community-based projects has proven to be sound. A 1982 irrigation management conference held in Manila made the following recommendations to national irrigation agencies:

(a) Agencies should recruit and train personnel who can work effectively with farmers and existing farmer-irrigation groups.

(b) Project planning, implementation, and evaluation at regional and central office levels should include perspectives beyond engineering and physical structure development.

(c) Legal status of farmer-irrigator organizations should be clarified and strengthened.

(d) Agency programs should build on, support and enhance existing irrigation associations wherever possible (Coward *et al.*, 1982).

An important example of an irrigation management problem relates to cost-sharing by and fee-collection from irrigation beneficiaries. Important questions are: How can alternative methods for collecting fees from beneficiaries be considered at the planning stage so that irrigation project plans will

Table 5. *Definitions of watershed-related terms.*

A *watershed* is a topographically delineated area that is drained by a stream system. The watershed is a hydrologic unit that has been described and used both as a physical-biological unit and as a socio-economic and socio-political unit for planning and implementing resource management activities. When the term *watershed* is used in this book, it refers to a subdrainage area of a major river basin.

A river *basin* is similarly defined but is of a larger scale (for example the Mekong River Basin, the Amazon River Basin, and the Mississippi River Basin).

Integrated watershed management is the process of formulating and implementing a course of action involving natural and human resources in a watershed, taking into account the social, political, economic, and institutional factors operating within the watershed and the surrounding river basin and other relevant regions to achieve specific social objectives. Typically, this process would include (1) establishing watershed management objectives, (2) formulating and evaluating alternative resource management actions involving various implementation tools and institutional arrangements, (3) choosing and implementing a preferred course of action, and (4) thorough monitoring of activities and outcomes, evaluating performance in terms of degrees of achievement of the specified objectives.

The *watershed approach* is the application of integrated watershed management in the planning and implementation of resource management and rural development projects or as part of planning for specific resource sectors such as agricultural, forestry, or mining. Imbedded in this approach is the linkage between uplands and lowlands in both biophysical and socioeconomic contexts.

Source: Easter, Dixon, & Hufschmidt 1986.

incorporate fee-collection systems? What measures are needed during the implementation stage to install an effective fee-collection program? What fee-collection strategies have worked and under what conditions? How does irrigation project design affect fee-collection strategies (Small and Carruthers 1991)?

According to Le Moigne and Kuffner (this volume), the rice-based agricultural economies in Asia were adversely affected in the 1980s by the sharp drop in rice prices. Attempts to shift to other crops were hampered by inflexible irrigation and farming systems. Not only have major investments in rice schemes become uneconomic, even improvements in distribution systems may not be warranted unless substantial increases in rice production can be achieved. There is need for research on design and management of irrigation systems to allow for wider crop choices for farmers.

Emerging water conflicts Rapidly increasing urban populations are pressing on available agricultural land and water supplies, especially in coastal and lowland areas of South East Asia and South Asia (Gladwell, Lee,this volume). Urban complexes such as Manila, Jakarta, Bangkok and Madras pose threats to existing agricultural water uses in their hinterlands and place severe constraints on new irrigation developments. Given the inherent inflexibility of irrigation institutions, the potential for conflict between urban, industrial and agricultural interests is great, leading to high transactions costs and unsustainable development of new surface and groundwater resources. The challenge for water management is to achieve cost-effective, efficient solutions to potential water conflicts through integrated planning on a collaborative basis with all contending interests.

Upland watersheds The major natural resource management problem of the upland watersheds in the humid tropics relates to land use. Under increasing population pressure, large sections of the uplands are being deforested and converted to agricultural and grazing uses. The severe impact of upland forest use or conversion on soil erosion rates, sediment in streams and reservoirs, water quantity in streams, peak stream discharges and nutrient input in streams has been discussed in this volume (Bonell with Balek, Fleming, Lal, Rose, Roche and Tejwani). To counteract these adverse impacts of faulty land-use in the uplands, watershed management approaches have been adopted by many countries and promoted by FAO and other international donor agencies.

Definitions of watershed and watershed-related terms are given in Table 5. As defined there, integrated watershed management gives promise of including often-ignored land and water interactions in planning and implementing development projects. These interactions are of utmost importance as shown, for example, in several river basins of the humid tropics, where high levels of soil erosion in the upper watersheds not only reduce forest and agricultural productivity but also cause sedimentation and water pollution problems downstream (Magrath and Arens 1989).

Many of the detailed aspects of watershed management for upland areas, including public awareness, population pressure, treatment of reservoir catchment areas, research and demonstration, training, cost-sharing and fiscal management, people participation and legislation, are discussed by Tejwani else-where in this volume. In the discussion to follow, we empha-size two important management issues: *implementation* and *institutions and organizations*.

Implementation Watershed management programs in the humid tropics have largely been the concern of forestry min-istries or departments or, in some cases, water development agencies concerned with construction and management of dams and reservoirs. In most cases, these agencies have encountered serious problems of implementing soil conserva-tion and watershed management programs in the rural upland areas. These implementation problems arise from a number of factors, including:

(a) little or no local participation,
(b) inadequate extension and technical assistance programs,
(c) inadequate testing and development of management practices,
(d) delays in delivery of key inputs, including financial resources,
(e) fragmented governmental management structure,
(f) exclusion of downstream interests,
(g) inappropriate institutional arrangements, and
(h) political boundaries unrelated to watershed boundaries (Easter *et al.,* 1985; Easter 1986a).

Many of these factors are closely related to the reality that actual land-use decisions are in the hands of watershed resi-dents – typically, many small farmers living in small rural communities. Implementation of watershed management pro-grams must largely be done by them or through their collabo-ration. It follows that undertaking watershed management practices must be shown to be in their interest and appropriate incentives must be provided to them. Crucial to this approach is a thorough understanding of the behavioral and social dimensions of rural watershed management. This involves (a) collection of baseline data on existing social and behavioral conditions and human-environmental relations, (b) identifica-tion of different social and ethnic groups in the watershed, (c) recognition of the cultural basis of different land use patterns, (d) giving special attention to ethnic minorities and the rural poor and (e) learning from other areas (Lovelace & Rambo, 1986).

Institutions and organizations Appropriate institutional and organizational arrangements are especially important to the success of watershed management programs (Gibbs, 1986). Perhaps the most important institutional factor affecting the watershed is concern with *ownership, rights and tenure*. At one extreme, land ownership is entirely in the hands of the government, and rights to use the land or its renewable resources such as trees and forage are subject to governmental regulation. At the other extreme, land ownership is in private hands, with private owners having great leeway in the use of their lands, subject only to general land-use restrictions on

some types of use that could create adverse off-site effects. In some cases, private land ownership – typically of agricultural and grazing lands – is concentrated in a relatively few owners, who, in turn, lease the right to use the land to tenants.

Between the extremes of government ownership and private ownership of land are "common property" rights to land, where rural groups have acquired by custom the right to use certain renewable resources such as forage, shrubs, fallen wood, fruits and nuts and even certain types of trees. The underlying land ownership may be in national, provincial or local governmental hands, but the customary use rights are handled by the local community.

In most upper watersheds, the institutional arrangements for land-use are a mix of these various types. From the standpoint of effective watershed management, the major issue is the extent to which the specific institutional arrangements are well adapted to this purpose. As Gibbs (1986) points out, perhaps the most important attribute is security of tenure: "If tenure is insecure, watershed occupants are unlikely to invest in long-term or low-maturing activities, or to be innovative or risk tak-ing .... The tenant who expects to be shifted or evicted by the landlord to prevent the establishment of right by prescription cannot plan for slow-growing perennial crops or soil-conserv-ing infrastructure that pay off years in the future." In terms of this criterion, outright land ownership and long-term leases by operating farmers are superior institutional arrangements.

Effective management of "common property" lands also requires certainty in the application of rules in order to avoid the "free-rider" problem in which some persons reap the bene-fits of the land resource, such as fuelwood and forage, without making any contribution to its management. Gibbs (1986) states that "The free-rider problem poses a major obstacle in expanding and maintaining protective land uses in watersheds that may be promoted with collective action by users." Community-based control of such common property resources offers a better opportunity to overcome this obstacle than sole reliance on the regulatory power of national or provincial gov-ernments.

To what extent can existing institutional arrangements be changed to contribute to improved water and land manage-ment? There are no easy answers to this question. At one extreme, these arrangements may be quite fixed and difficult to change, at least in the short-run. In many cases, however, there is some flexibility and it should be possible to change some institutional arrangements such as types of land, forestry, graz-ing or water rights and duties, once the need for such changes is demonstrated.

Closely related to institutional and organizational arrange-ments are the implementation tools [e.g., ways of installing and operating watershed management measures (Dixon *et al.,* 1989, Chapter 10)]. According to Easter (1986a), implementation tools can be grouped in four general classes: legal arrange-ments; monetary incentives or disincentives; technical assis-tance, education and research; and direct public installation or investment. A watershed management plan and program will generally include a combination of these implementation tools. A wide range of potential legal arrangements or regulations can be used to implement a watershed management plan. Regulation of activities on rural lands, especially those in private ownership, is often fraught with difficulties, particularly in developing countries where enforcement capabilities are weak. Economic incentives would appear to be appropriate tools to achieve environmental objectives of watershed management. Governmental cash subsidies have been used as incentives for farmers to install various watershed management practices but their effectiveness has been limited. As Easter (1986a) points out, many governments may find it hard to implement cost-sharing or low-interest-rate loan incentives because of budgetary constraints and the difficulties of distributing such incentives to the appropriate target population.

Technical assistance to farmers through extension services and demonstration projects is often a successful means of promoting watershed management on rural lands. According to Easter (1986a), "Once an effective set of crops, trees, and management practices has been developed through research and testing under local conditions, getting the information to users is the next step ... extension can play an important role in identifying the watershed management problems through meetings with local people. The information thus obtained should be an important input toward the development of workable resource management practices. It should also make it easier to obtain rapid adoption of practices once they have proven effective. Thus, the extension services can act as a conveyor of information and a source of information concerning target group characteristics, location, and research needs." Direct installation and maintenance of watershed management practices by government is appropriate for publicly-owned land such as upland forest and for off-site management measures such as check dams on small streams. Direct installation of on-site measures on private lands is more problematical. Such investments over extensive areas can be costly and will be limited by budgetary considerations. Also, if farmers have not been involved in installing the structures, they will not be inclined to maintain them (Easter 1986a).

In Indonesia, population densities commonly in excess of 500 persons per square kilometre make it very difficult to implement land-use regulations under the conditions found in Java's upper watersheds. Even on state forest lands and agricultural plantations, there has been little success with instituting controls on harvesting and other land management practices. For private lands, the policy of providing subsidies to farmers to encourage their adoption of a highly standardized soil conservation package has been followed. In the Citanduy Watershed, these subsidies and the formal demonstration farm approach were supplemented with a special upland credit program and a local program of small village grants as further economic incentives for adoption of the package. The Indonesian State Forest Corporation, which manages state for-est lands, has traditionally adopted a policing stance toward communities living in or adjacent to these lands. A pilot social forestry program, which has been underway for several years, is offering some alternatives that may be of use elsewhere. In selected locations throughout Java, forest officers have been encouraged to negotiate reforestation contracts with local residents, acknowledging their rights to plant seasonal and perennial crops for their own benefit (Hufschmidt & McCauley, 1989).

In summary, there is an increasing understanding of the important factors that influence the success or failure of watershed management projects in upland areas. Although programs in some countries are attempting to apply this knowledge to improve implementation success (Chanphaka, 1986), there is as yet no single management model or approach that has been demonstrated to be fully successful.

GENERIC MANAGEMENT ISSUES

It is clear from the preceding discussion of the various problem settings that many management issues are generic, regardless of the specific problem setting. Several of the major generic issues will be discussed here as they relate to the management tasks of assessment, planning and implementation. These management issues are: (a) water resource assessment, (b) integrated planning, (c) river basin as a management region, (d) demand management, (e) local involvement, (f) implementation, and (g) organizations and institutions.

Water resource assessment

This task is concerned with obtaining a thorough understanding of the natural water resource system, including the hydrologic cycle, with its components of precipitation, evaporation, surface runoff, and groundwater flows, as well as the broader "hydrologic continuum," defined by Leopold (1990) to include the soil, biota, and atmosphere as well as water.

The primary objective of water resource assessment is to expand and deepen the physical information base for effective water resource management. Ideally, the complete river basin should be used as the basic areal unit for data collection and interpretation, with appropriate subdivisions based on geology, climate, soil type, land use and land cover. All components of the hydrologic cycle including precipitation, evapotranspiration, stream flow, soil moisture, and groundwater flows and stocks should be measured. Quantitative assessments should use a complete water-budget approach, with emphasis on interactions between surface water and groundwater, and water quantity and water quality.

According to Manley & Askew, this volume, water resource assessment in the humid tropics is hampered by the fragmented nature of the hydrologic services and the low priority given to the task by governments, with consequent shortages of funds and trained staff. This situation has worsened over the past decade, especially in Africa, with a dramatic drop in collection of basic hydrologic data. The problem is especially acute because, for adequate reliability, 50 to 60 years of consistent hydrologic data are needed for much of the humid tropics.

Many groundwater systems in the humid tropics lack detailed studies (Foster & Chilton,this volume), in spite of the fact that this valuable and fragile resource is subject to pollution from liquid and solid wastes and saline water intrusion from overpumping. In addition, according to Bullock, this volume, hydrologic research on tropical wetlands is meagre. In spite of high potential wetland values and growing impacts of agricultural, industrial and urban development on wetlands, there is little or no understanding of wetland processes and values, and even worse, misconception of such values.

Improvement of water resource assessment in the humid tropics is a challenging task. It will involve the co-ordination of meteorological and hydrologic data collection and interpretation activities, which are often scattered throughout many government agencies. Successful assessments require: (a) common standards for water resource assessment and (b) unified data collection and measurement systems to record hydrologic, meteorologic and water-use measurements and other water-related data.

Integrated planning

A major issue is how to achieve integration in planning, given the complex physical, ecological, economic and social linkages in natural water and land resource systems and human-made urban and rural agricultural systems. Integrated planning is a challenging task, requiring the blending of many important related elements into a unified whole. In the context of the humid tropics, integrated water resource planning means incorporating *multiple objectives, multiple purposes and multiple means* in a systems context, as well as relating water resource plans to "outside factors" – plans and policies for other sectors of the economy, such as agriculture, energy, human settlements and transport.

Multiple objectives It is generally agreed that social and environmental objectives should be included along with economic development objectives in any water resource plan. The issue is precisely how these objectives are to be incorporated into planning. Most current approaches emphasize the economic development objective and bring in social and environmental aspects only as secondary elements, often by means of environmental impact statements. Clearly, this is not an adequate approach towards treating these objectives equally. However, the issue of precisely how to incorporate economic, equity, sustainability and environmental objectives in a water resource management plan remains problematic.

Multiple purposes An integrated plan requires the balanced consideration of a wide range of water uses and management purposes, including withdrawal uses such as domestic and industrial supply and irrigation, in-stream uses such as navigation, hydroelectric power, fish and wildlife and recreation and water problems such as flooding and pollution. Often, such balanced consideration is difficult to achieve, given the dominance of single-sector water development agencies who take a narrow single-purpose approach to planning. As a result, beneficial complementaries among purposes are not captured and tradeoffs among conflicting uses are not critically examined.

Environmental purposes, such as the maintenance of fish and wildlife habitat and biologically diverse ecosystems, are most often neglected under this approach. In urban areas, primary emphasis on domestic water supply often leads to relative neglect of point and non-point water pollution control, as in Bangkok and Jakarta.

Multiple means Fully integrated water resource planning involves using a multiplicity of means for achieving the plan's objectives. In most planning situations, the means are defined narrowly as comprising only physical facilities to meet water demands or solve water problems – dams, canals, pumping plants, hydroelectric power plants. Other management options, involving human adjustments to water shortages or to floods, are not given much consideration. The results are inefficient and wasteful uses of scarce water supplies and costly and ineffective flood control projects.

Outside factors To be fully integrated, a water resource management plan should be developed and carried out within a much larger context of regional and national goals, policies, plans and strategies (involving population growth, economic and social development), international trade, and sectors such as energy, agriculture, transport, and urban infrastructure. Often, these policies and plans act as constraints on the water resource management plan. For example, national policy on agricultural pricing for factor inputs (such as fertilizers, pesticides and water) and for crop outputs have significant effects on irrigation water use and water pollution (Repetto, 1986). In some cases, the drive to reduce foreign exchange demands will act as a constraint on water resource development. In other cases (such as subsidies for construction of water projects), national policies may speed developments but often at costs of economic efficiency and environmental and social disruption. It is clear that integrated planning requires careful assessment of such outside factors in terms of their impact on development of sound water management plans.

The river basin as a management region

A major unresolved issue is the appropriate region to be adopted for preparing and implementing water resource plans. Effective management of water resources must address the physical realities of the hydrologic cycle, including the biophysical linkages of the river basin. In spite of the theoretical advantages of managing water resource developments on a river basin or watershed basis, for example, as reported in

Easter *et al.*, (1986), cases of successful application of basin-wide management are relatively few. Most water resource developments are undertaken on a sectoral basis – for irrigation, hydroelectric power, domestic and industrial water supply or other purposes – with the individual project as the dominant element. Although the river basin or watershed is often the spatial unit used by public agencies for collecting and analysing physical data and preparing development plans with a broad perspective, there are few examples of detailed basin-wide plans. Only rarely do we encounter examples of detailed, integrated basin-wide management programs which involve both planning and implementation.

Part of the problem is the question of scale. At one extreme, there are the huge river basins typified by the Mekong in South East Asia, the Ganges in South Asia and the Amazon in South America. These large river systems have many tributary watersheds, each of which may be a sizeable management unit. Furthermore, some of them are international in scope. Detailed integrated management of such large river basins is probably administratively infeasible and may not be desirable. The leading example of detailed river basin management is the Tennessee Valley Authority (TVA) in the United States, which has purview over a relatively small sub-tributary area of the Mississippi River, with a drainage area of only 3% of the 3.22 million km^2 of the Mississippi River Basin. In South East Asia, the Lower Mekong Commission represents a special case of management of an international river basin, but up to now its activities have been largely limited to collecting and analysing basic data and to framework planning (Committee for Co-ordination of Investigations of the Lower Mekong Basin, 1988).

At the other extreme are many small watersheds composed of tributaries and sub-tributaries of larger river basins. Although detailed management of such watersheds is administratively feasible, in most cases such management has been limited to basic data collection and planning, with implementation continuing to be done by public or private agencies on a project or sectoral basis.

Of course, there are many examples of river basins and watersheds between these extremes, including the relatively small but populous river basins in the Philippines and the island of Java in Indonesia. In many of these cases, detailed water resource management on the river basin basis would be desirable and feasible.

Another major factor limiting the use of the river basin or watershed as the management area is that political and administrative boundaries rarely coincide with those of a river basin. Implementation of projects or programs is usually carried out either by provincial or district (city or village) governments or by regional offices of national sectoral agencies, usually organized on a provincial basis. Even when water resource management plans are prepared using river basin or watershed boundaries, their implementation is not likely to proceed on this basis. Also, urban water resource plans, involving water

supply, groundwater management, pollution control, flood control and storm drainage, typically do not conform with river basin boundaries.

Precisely how the river basin can be used most effectively as a spatial unit for management remains an unresolved issue which must be seriously considered in developing a water resource management strategy for a specific country, province, urban area or other region.

Demand management

One of the most serious problems facing water managers is the effective allocation of scarce water resources among rapidly growing and competing demands. In many cases, even in humid areas, supply-side solutions to meeting projected rising "demands" are no longer adequate. Demand management strategies that involve policies and activities to reduce per capita or per-unit-of-activity use rates are being increasingly advocated, especially for urban areas and irrigation projects. Adopting demand management as an integral part of water management would include formulating and evaluating demand reduction methods and strategies as complements to and substitutes for supply augmentation projects in order to bring projected demands for and supplies of water into balance. It would include using prices as a management tool by applying volumetric pricing of urban supply or wastewater on the basis of marginal supply or disposal costs, along with increasing block rates. Where appropriate, seasonal pricing and temporary drought surcharges would be imposed. Demand management would also use efficient technical means to reduce urban water use and transmission losses in the supply system, including changes in plumbing codes to require water-saving plumbing fixtures, programs of leak detection and control and sustained operation and maintenance. In addition, recycling and other technical means would be used to reduce withdrawal rates for water for industry, especially for cooling. Also, technical means that are economically feasible would be used to reduce irrigation water use, including drip irrigation, sprinkler irrigation, land levelling, canal lining, along with institutional means such as modifying water rights systems to encourage efficient use of irrigation water. Finally, lower-grade water would be used for certain domestic, commercial, industrial and agricultural purposes. The use of brackish water and treated wastewater for non-potable purposes may involve installation of dual water supply systems.

Local involvement

Most water resource management plans and programs will affect the interests or involve the activities of diverse private individuals and groups – such as village and urban dwellers, upland graziers and agroforesters, lowland farmers, fishermen and mining, energy and industrial firms. In most cases, the success of any plan is highly dependent on the effective participation, or at least the co-operation, of such individuals and groups. The issue of local participation is often defined in terms of a "top-down" strategy versus a "bottom-up" strategy. In reality, both "top down" and "bottom up" approaches are needed. As Lee points out in this volume, the benefits of community/user participation in water supply and sanitation projects are significant and the current trend is to emphasize community management of water supply and sanitation, which goes beyond participation and involves actual community control of their systems. The issue is how effectively to build in local involvement in a management program to fit the specific site and country situation, with the particular organizational and institutional milieu at hand.

Implementation

The weakest link in water resource management, especially in developing countries, is *implementation*. The literature is replete with examples of failure of project or program plans to be successfully implemented (see references in Chambers, 1988). Although the problems are generally well understood by donor and funding agencies, governments, professionals and scholars, attempts to solve the problems have met with limited success. Suggested approaches toward improvement include incorporating implementation strategies (including a set of implementation tools) into management plans from the very start of planning, building public involvement into both planning and implementation and critically examining institutional and organizational arrangements in terms of their effectiveness in implementing management plans. Some of these approaches are discussed by McCauley (1988) in the context of watershed management on the island of Java where lack of local participation in planning and highly fragmented or unclear responsibilities among national and regional agencies have led to serious failures in implementation.

A special problem of implementation relates to conflict management. In many situations, the need for allocation of scarce water resources and water-related services among competing claimants gives rise to conflicts among users. In other cases, conflicts arise because of damaging effects of water pollution on downstream areas and forced relocation of people from sites of water development projects and reservoir areas.

Such conflicts complicate the already difficult problems of achieving public participation in water management. Issues of equity arise in attempts to deal with such conflicts (e.g., conflicts between large-scale fish-pen operations and small-scale open-water fishermen on Laguna de Bay, Philippines [Fellizar *et al.*, 1989]). Another common conflict involves "head end" irrigators versus "tail end" irrigators served by surface water canals, where equity requires that "head enders" forego some water use to enable "tail enders" to avoid disastrous shortages (Chambers, 1988).

Conflict management can be viewed as closely related to public participation, where the realities of political forces, local elites and cliques must be recognized. However, increasing experience with application of mediation and conflict resolution techniques in a number of diverse situations provides a body of information that can be of use in including conflict management as a separate element in a water resource management program (McKinney, 1988). Issues of conflict management are discussed further in Nickum & Easter (1989).

Organizations and institutions for management

With respect to organization for management, one can postulate some ideal form, such as a river valley authority, that takes proper account of the spatial dimension (e.g., river basin), management objectives and need for an entity to speak for the river basin. However, in most cases the organizational arrangements for management are already established and must be accommodated. Usually this means that one is operating in the realm of the "second best" where certain desirable attributes of management (such as unified programming and budgeting) must be sacrificed to accommodate organizational realities (e.g., extreme division of water resource activities among sectors of government). The major organizational issue affecting development of a management plan is how to tailor the plan to work effectively either within the existing organizational structure, or within a modified structure where this is feasible and desirable. The particular means of accommodating management and organizational structure will be site- and country-specific but the goal of developing and implementing an effective management plan is common to all situations.

The term institutions for management is used here to mean those formal and informal rules which define property rights to land and water, and rights and obligations of individuals and groups. Any water resource management program must either accommodate to the existing set of institutional arrangements for a specific region or seek to change them as part of the management plan. Rarely are such institutional arrangements ideally suited for effective water resource management. For example, in most cases, rules for extraction of groundwater are atomistic and do not deal with the groundwater aquifer as a single management unit. A major issue in developing and implementing a management plan is fitting the plan into the existing institutional arrangements, where appropriate, and devising and obtaining necessary changes in the arrangements as part of the management plan.

MAJOR ELEMENTS OF A WATER RESOURCE MANAGEMENT STRATEGY FOR THE HUMID TROPICS

In this chapter we define *management strategy* as a plan or method of establishing, maintaining and operating a set of water resource management activities designed to achieve social-economic-environmental objectives embodying sustainable development. As so defined, a management strategy is broader than establishing an organizational framework, preparing a water resource development plan or implementing a management program. In this sense, the term *strategy* means the total approach to carrying out management, as distinguished from individual steps to organize for, plan or implement specific management activities or tasks. Developing a management strategy is a task for national, provincial and urban governments, although, in the case of developing countries, guidance may be provided by outside sources, such as specialized agencies of the United Nations.

The major elements of a proposed management strategy are presented in three parts: (a) the planning stage of management, (b) the implementation stage of management and (c) institutional and organizational factors in management. For each of these parts, major elements of the strategy are presented and illustrated with examples drawn from case studies of water resource management in the humid tropics. In most management situations, planning and implementation are not discrete stages but often are undertaken simultaneously. However, it is useful to separate them in this discussion of management strategy, if only because the specific activities carried out in planning are quite different from those in implementation. Institutional and organizational factors are given separate treatment because of the crucial role they play in the success or failure of a management strategy.

Elements of strategy at the planning stage

Eight major elements of strategy are presented as of particular importance at the planning stage. These deal with: (a) selecting an appropriate management area, (b) the multiple objective approach, (c) multiple purposes, (d) the economic-analytical approach, (e) multiple means, (f) planning for implementation, (g) local participation and (h) multidisciplinary planning team. Taken together, they provide a sound basis for building an effective management strategy at the planning stage.

Appropriate management area

Perhaps the first step in devising a management strategy for water resources is selection of the geographical area to be covered. In some situations, as in preparing a master plan, the nation is the appropriate spatial unit. Even here, however, as in the case of the national water plan for Bangladesh (Bangladesh Ministry of Irrigation, Water Development and Flood Control, 1986), it is necessary to consider the links with the upstream Ganges and Brahmaputra River Basins which lie in India, Nepal, Bhutan and China. And, of course, as in Malaysia, major river basins or watersheds within the nation are important sub-areas of a national water resource plan. In other cases, because of the nature of the problem, e.g., flood damages, water pollution or urban water supply, the political jurisdiction such as the city or province may be selected as the appropriate unit. Here again, the links between the selected management area and tributary watersheds must be taken into account. For example, when the water supply and water quality problems of Bangkok are studied, it is recognized that effects of activities in the Chao Phraya River Basin upstream of Bangkok have to be considered (Srivardhana, 1984, 1991).

Where management areas based on political or administrative jurisdictions are not co-terminous with hydrologic or other natural system boundaries such as river basins, watersheds or coastal zones, special attention must be paid to the linkages between portions of the natural system region (e.g., river basin, groundwater aquifer) that are outside the management region and those portions inside the region (Dixon & Easter, 1986). Where the choice of management region is still open, a general rule for choice is: (a) when administratively and politically feasible, the management region should be co-terminous with the surface water and groundwater hydrologic region or regions; or (b) where this is not feasible, at least the planning phase of management should be undertaken on the basis of hydrologic regions, even though implementation may be carried out on the basis of administrative or political regions.

In the special case of small islands, often with ephemeral streams, the entire island is obviously the appropriate management region.

Multiple objectives approach

One of the very first steps in the management process is the selection of objectives. These objectives vary from case to case and are a function of the specific political-institutional situation in the country involved, including its relationships with outside agencies such as the World Bank and other donors or technical aid organizations. Accordingly, water resource managers or analysts are not free to specify these objectives *a priori*, but must seek to identify them from the political process. For example, in the Bangladesh National Water Plan (Bangladesh Ministry of Irrigation, Water Development and Flood Control, 1986), achieving national foodgrain self-sufficiency was identified at the outset as an objective of prime importance to the government. In the Bangladesh case, other objectives such as economic efficiency and equity were implicitly considered to be of lesser importance.

Almost always, there will be more than one relevant objective. The most common objectives are economic development (technically defined as economic efficiency in the benefit-cost literature), equity (sometimes measured in terms of income distribution), regional or national self-sufficiency and environmental quality. Long-run sustainability has emerged recently as an additional objective of relevance to water resource management.

Identifying the relevant objectives is only the first step in incorporating multiple objectives in water resource planning. These objectives must be reflected in the water resource plans in one way or another. One approach is to formulate alternative plans which show different mixes of achievement of objectives. For example, one plan would emphasize economic development with less emphasis given to equity and environmental quality. Another plan would represent a more balanced emphasis on equity, environmental quality and economic development. The important point is that all relevant objectives would be seriously considered from the very start of plan

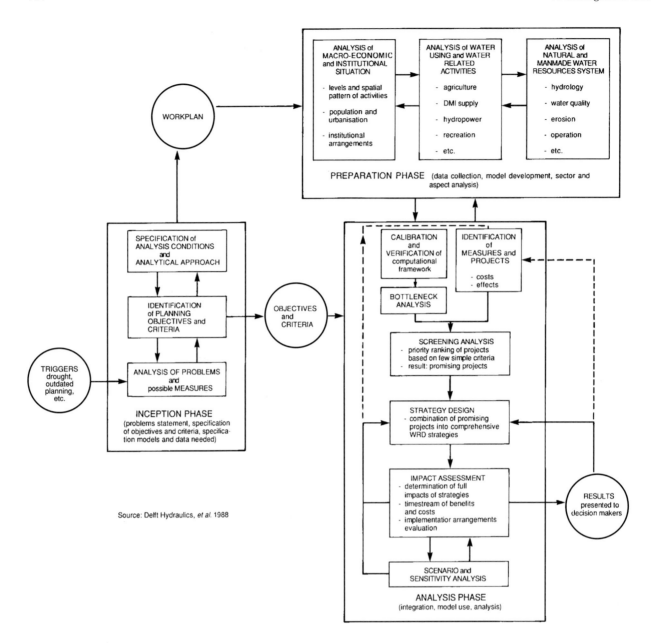

Figure 1: Framework of analysis for water resource management studies.

formulation. At least at the outset of planning, each objective would be given equal consideration in plan formulation so that information on trade-offs among objectives would be available to planners and decision-makers.

Much has been written over the past two decades about applying the multiple objective approach to water resource planning (Major, 1977; UNDTCD, 1988). A major problem has been incommensurability; measurements of achievement of the various objectives cannot be made in common units such as dollars or rupiah. Although some ingenious solutions have been proposed (Haimes *et al.*, 1975), the most practical approach is to formulate alternative plans with different levels of achievement of objectives, and compute the trade-offs among objectives. Under this approach, one can compute the loss (or opportunity cost) to one objective caused by achieving

a second objective. The effects on each objective would be measured in appropriate monetary or physical units. This information on trade-offs would be used in the decision-making process to rank alternative plans and ultimately to select a preferred plan. Although this approach sometimes may be cumbersome and is far from elegant, it appears to be the most practical at this time.

Several recent case studies in the humid tropics involve explicit or implicit concern for multiple objectives. In the Songkla Lake, Thailand, Laguna Bay, Philippines, and Saguling Reservoir, Indonesia cases three objectives; economic development, equity and environmental quality were identified as relevant [Hashimoto & Barrett (eds.) 1991].

An important related task is that of establishing the linkages between a water resource management strategy and broader societal objectives, including economic development. This can be done, in part, by making better quantitative assessments of the contribution that water services make to sustainable agricultural, industrial and urban development, and to the constraints placed on such development by water shortages or other water-related problems which decrease the level or increase the cost of water services.

Multiple purposes

Multiple purpose planning of water resources has been recognized for almost a century. By the 1930s in the United States, five major purposes for development of water resources were considered to be relevant: navigation, flood control, irrigation, hydroelectric power and domestic and industrial water supply. By the early 1950s, this list had expanded to include purposes such as fish and wildlife, water-based recreation, watershed management and water pollution control. In most water management situations in the humid tropics, more than one purpose is involved, although in only a few cases would all of the purposes listed above be present. At least in principle, a water resource management strategy should be broad in scope so as to accommodate all relevant purposes. Does that mean that there should be no single-purpose or dual-purpose projects? No, but it does mean that planning for a single-purpose project should take account of the relationship of that single-purpose project (such as an irrigation or urban water development) to other purposes in the river basin or region involved.

Problems arise when different purposes are the responsibility of different management agencies. For example, in the case of the Wonogiri Dam and Reservoir in Central Java, differences arose among several government agencies as to priorities for management of the upper catchment area. The Forestry Ministry proposed that priority for watershed management be given to the sub-catchment with the highest soil erosion rate. The Public Works Ministry, responsible for dam and reservoir management, proposed that priority be given to erosion control activities for the sub-catchment closest to the dam. The Agricultural Ministry favored work in the sub-catchment with greatest prospects for increased agricultural productivity while the Home Affairs Ministry and local government emphasized improving livelihood of farmers in the poorest sub-catchments. The issue was eventually resolved by adopting a more integrated and locationally balanced approach which forms the basis for a World Bank-funded project to rehabilitate the catchment and develop its agricultural and forestry base (McCauley, 1988).

A brief review of a number of humid tropics case studies discussed by Hufschmidt & McCauley (1991) reveals that all of them deal with several water resource management purposes. For example, Table 6 shows that the Lake Victoria Basin case involves 12 of the 13 listed purposes, followed by the Songkla Lake and Saguling Reservoir cases with 7 purposes. For all 5 cases, the average number of purposes is 7. A desirable water resource management strategy should have an open stance toward incorporating multiple purposes in planning. Given such stance, the task becomes one of identifying relevant purposes and incorporating them effectively into specific water management plans.

Economic-analytical approach

This approach to water resource planning is sometimes termed the systems approach or defined as the application of systems analysis methods to planning. One of its major features is the application of social benefit-cost analysis to the formulation and evaluation of alternative development and management plans (Dixon, in Hashimoto & Barrett, eds., 1991). In recent years, methods for valuation of benefits and costs in benefit-cost analysis have been extended to include many environmental aspects such as erosion and sedimentation, water pollution and aquatic and terrestrial ecosystem effects (Hufschmidt *et al.*, 1983; Dixon & Hufschmidt, 1986; Dixon *et al.*, 1988, 1989).

More generally, the approach is characterized by a comprehensive view of the multiple purposes of management in the context of the physical water and related land-resource system, to meet multiple objectives of society. The basic elements of strategy already discussed above – management area, multiple objectives and multiple purposes – are all a part of this comprehensive view. Examining the physical water and land complex of the river basin or watershed as an ecological-social system is also a part of this comprehensive view.

A recent example of the application of this approach is the Cisadene-Cimanuk Integrated Water Resource Development study in Indonesia (Delft Hydraulics *et al.*, 1988). In this study, involving the Jakarta urban and urbanizing region, systems analysis is defined as a systematic process of generating, analysing and evaluating alternative courses of action. Systems analysis includes a framework for analysis consisting of a co-ordinated regime of steps and a computational framework consisting of a coherent set of computational techniques, including computer models.

The details of this framework of analysis are shown in Fig. 1 taken from the Delft Hydraulics *et al.*, (1988) report. The precise form and content of the systems analysis approach to be used in a water resource management strategy may differ depending on the specific situation. But the main elements of the approach, as illustrated in this example, will be present in any strategy that embodies the economic-analytical or systems analytical methods.

Multiple means

Another attribute of the integrated approach to planning is a broad set of means for achieving water resource management objectives. Thus, in addition to considering supply-side projects (e.g., storage reservoirs, diversion dams, canals, groundwater pumping) for meeting assumed or projected water-resource demands, demand-side projects are considered as an integral part of a management plan. These include technical,

Table 6. *Water management purposes in humid tropics cases of water management.*

Case	Flood Control	Irrigation	Hydro Power	Navigation	Water Supply	Watershed Management	Recreation	Fish & Wildlife	Pollution Control	Insect Control	Drainage	Sediment Control	Salinity Control	Number of Purposes:
1. Lake Victoria basin, Kenya	X	X	X		X	X	X	X	X	X	X	X	X	12
2. Laguna Lake, Philippines	X				X	X		X	X					5
3. Songkla lake, Thailand	X			X			X	X	X			X	X	7
4. Brazil Amazon Reservoir			X							X				2
5. Saguling Reservoir, Indonesia		X	X			X		X	X	X		X		7
Number of Cases	3	2	3	1	2	3	2	4	4	3	1	3	2	–

Source: Hufschmidt and McCauley. 1991.

regulatory or economic ways of reducing per capita or other unit demands for water. Recycling of industrial water, waste water reuse including dual water supply systems and water-saving household equipment are examples of technical measures which can often be achieved via regulation or economic incentives. Metering accompanied by increasing block rate volumetric pricing is an example of combining technical and economic incentives to reduce water use. Examples of demand management options are shown in Millikan & Taylor (1981) and Hufschmidt *et al.*, (1988).

Multiple means can also be used to deal with problems such as flood damages and pollution. Flood plain zoning, flood insurance and flood warning systems can be used along with physical measures such as flood proofing, dams and levees to reduce flood hazards. Similarly, regulatory and economic incentives can be used to reduce generation of gaseous, liquid and solid residuals.

Planning for implementation

One of the most pervasive problems in water resource management, especially in developing countries, is achieving successful implementation. This is especially true for irrigation, fisheries, rural water supply, and upstream watershed management projects where the co-operative participation of many water and land users is often required for project success. This is also true in cases where large numbers of people are to be displaced by water developments and the successful relocation of people becomes an important task. The root of the problem often is at the planning stage where inappropriate project design or technology leads to difficulties in operation and application. For example, Indonesia watershed planners adopted a standardized package of check dams and agricultural practices to increase production while decreasing soil erosion. Because this package fits poorly with the diverse physical and socio-economic conditions of the watersheds, serious implementation difficulties were encountered (McCauley, 1988).

Because of the close relationships between project design and water and land management technology on the one hand and the ease or difficulty of implementing projects on the other, it is necessary to plan for the implementation of projects at the same time as for their design. Typically, such an implementation plan will be a package of implementation tools, consisting of regulatory measures, economic incentives, technical assistance and extension education, to ensure that the land and water users and managers take appropriate actions to achieve project success (Easter, 1986a). As discussed in more detail later in this chapter, the water users and land managers who will play key roles in implementing the project should participate in the combined project design and implementation planning. Of course, any initial implementation plan can only be a first draft, subject to change as experience is gained during implementation. The Laguna Bay, Philippines case study illustrates the problems of changing implementation plans as experience is gained in applying a fisheries management program to the lake (Fellizar *et al.*, 1989).

Planning for implementation can be illustrated with the case of irrigation in Indonesia. Irrigation management issues centering on water pricing, cost recovery and local financing of operation and maintenance are important concerns. Irrigation development has traditionally been the central concern of the main national-level water resource management agency. Experiments with a variety of approaches for the establishment of irrigation service fees are continuing. The need to improve the efficiency of irrigation water use is increasingly justified on the grounds of competing urban and industrial uses. There are ongoing efforts to turn over management of (especially smaller) irrigation systems from central government control to regional or local entities in an effort to improve management incentives.

Any implementation plan will be developed initially in terms of an existing set of institutional and organizational arrangements. Where these arrangements are found to be inadequate or constraining, the implementation plan may propose changes in these arrangements. This issue is discussed in more detail later in this chapter.

Local participation

Local participation enters into planning for implementation in two ways: First, farmers, fishermen, villagers, urban dwellers and others who will use the water resources or will be affected by development projects should be involved in the planning and design of projects. Second, the implementation plans that are developed should have specific roles or tasks for the water users and affected people. For example, an implementation plan for an irrigation project would define the roles and responsibilities of water users in canal management, water distribution and operation and maintenance. According to a recent study of medium-scale irrigation systems in north-west Thailand, "The (government) should adopt the concept of farmer involvement at all stages of irrigation development, from site design through 0 & M. The concept should be institutionalized in a Standard Operating Procedure, which establishes clear rights and responsibilities for farmers at all phases of system rehabilitation and operation," (Johnson *et al.*, 1989). An extensive discussion of research on farmer-managed irrigation system is contained in a paper by Coward & Levine (1987).

In the Philippines, local involvement in irrigation project planning and implementation is well advanced (Korten, 1982; Korten & Siy, 1988). On the other hand, community participation in planning and decision-making for Laguna de Bay is still in an embryonic stage. Its weakness is seen as a challenging problem facing the Laguna Lake Development Authority (Fellizar *et al.*, 1989).

Multidisciplinary planning team

Formulating water resource management plans is far more than a technical exercise involving the services of technical specialists such as hydrologists and civil engineers. In fact, *multiple objectives*: economic, social and environmental; *multiple purposes:* irrigation, hydroelectric power, domestic water supply; *multiple options*: demand-side and supply-side management; and *multiple means:* physical measures, implementation tools and institutional arrangements are all involved in planning. To perform the planning tasks effectively requires the services of people with training and experience in the social and biological sciences as well as in engineering and the physical sciences. Perhaps as important as having available the services of people covering the full range of these fields of knowledge and experience is the interdisciplinary composition of the leadership of the planning activity. The leadership should not be dominated by any one discipline, such as engineering or physical sciences. Rather, it should be a balanced composite of social science (economics, for example); biological science (ecology, for example); and physical/engineering science (civil and environmental engineering/hydrology, for example). Supporting such leadership team could be other specialists such as geologists, geographic information system specialists, soil and environmental scientists, agricultural crop specialists, foresters, fisheries scientists, anthropologists, rural sociologists, public health scientists, agricultural marketing specialists, management scientists and legal specialists, as required for the planning tasks in the specific case.

The Cisadene-Cimanuk Integrated Water resource Development study is a good example of effective use of a multidisciplinary planning team (Delft Hydraulics *et al.,* 1988). The Delft Hydraulics Laboratory team leader was a multidisciplinary-trained water resource planner. His team was composed of a broad spectrum of engineer/planners and social and natural scientists, including a systems analyst, computer modeller, civil engineer/planners, hydrologist, economist, sociologist, ecologist, water quality scientist, fisheries scientist, agronomist and irrigation engineer.

This multidisciplinary approach to water management planning is illustrated in Easter *et al.,* (1986a), where an integrated watershed management approach is described as embodying biophysical, economic, social, cultural and institutional concerns requiring the services of experts in all of these fields. The challenge in each specific planning situation is to establish a truly multidisciplinary leadership team and to keep a multidisciplinary perspective throughout the planning exercise so that the services of the experts from the many disciplines involved can contribute effectively in developing an integrated management plan.

Elements of strategy at the implementation stage

The major elements at the planning stage discussed above are directed toward assuring that an integrated water resource plan is prepared that meets the economic, social, environmental and sustainability objectives of the society. There remains, however, the challenging task of implementing this plan. For this purpose, five major elements of management strategy are advanced as of special importance at the implementation stage. These are: (a) an integrated program and budget, (b) local involvement in implementation, (c) monitoring and evaluation, (d) operation and maintenance, and (e) implementation tools. These five elements, when combined with appropriate institutional and organizational arrangements, form a sound basis for an effective management strategy for implementation.

Integrated program and budget

The integrated water resource plan which is the output of the planning stage typically will consist of a number of individual physical projects to be built or installed and individual management activities to be carried out by different agencies. For example, in Thailand at least 32 public and private agencies were found to be involved in an irrigation development (Bower & Hufschmidt, 1984). If the integrity of the plan is to be preserved during implementation, it is necessary to prepare and maintain an integrated multi-year program, along with an integrated budget for the first year of the program. The format for the integrated program consists of year-by-year sequencing of construction and operation of individual projects and activities, along with estimates of costs, outputs and benefits, and designation of responsible agencies. The integrated budget, typically, would consist of the first year of this multi-year integrated program.

Whether this integrated program and budget is to be funded by appropriating lump sums to a single ministry or department for allocation among individual agencies, or is to be provided as separate appropriations to individual ministries or departments, it is necessary that the integrity of the programs and budgets be maintained throughout the budgetary and appropriation process. This can best be done if (a) the multi-year programs and annual budget are prepared jointly by all relevant agencies, (b) if the basin-wide or region-wide programs and budget are kept intact as identifiable categories throughout the budgetary and appropriation process, and (c) if administering the program and budget during the implementation stage is a shared task of the relevant agencies. The details of how this could be done are contained in a paper by Hufschmidt (1986), developed for the case of Indonesia.

Because most water resource budgets are prepared, reviewed and approved on an individual agency or ministry basis, it is extremely difficult to obtain acceptance of this three-step procedure for achieving integration. However, where more than one agency implements a water resource plan, an integrated program and budget is essential to preserve the integrity of the plan.

Local involvement in implementation

The crucial need for local involvement in implementation has already been discussed under *Elements of Strategy at the Planning Stage*. However, the scope, form and method of this involvement will vary from case to case, and these factors, in turn, will affect the required strategy for public involvement.

In some cases, such as irrigation, rural water supply, fisheries and watershed management, the water or land users play key roles in project management. In other cases, where project construction and operation involve temporary or permanent displacement of people from the project site, or otherwise affect local people, these people have an obvious stake in how the project is managed. Also, in some river basins, the general public, although not directly affected by project activities, will have an interest in how the water resources are managed because of indirect or secondary effects on environmental, social and economic conditions in the region.

Appropriate strategies for public involvement for these cases range from direct involvement in project construction and operation, as for irrigation water distribution and on-farm water spreading facilities and close consultation and collaboration, as in the relocation of displaced people, to a strong advisory role on implementation in general, as in the case of citizen groups organized on a river basin basis.

As summarized by Howe (1986), water resource management should involve sufficient public participation at regional and local levels so that "(a) valuable local information and technical inputs are identified and utilized, (b) the interests of all affected groups are identified and taken into account, (and) (c) local water users feel a part of the river basin program and accept their responsibilities for operation and maintenance."

The key role that public participation plays in implementation of watershed management is discussed by Easter (1986a). Important issues of concern identified by Easter, Hufschmidt & McCauley (1985) are: (a) how to build local capacities for implementation; (b) need to pinpoint the obstacles to effective participation in water management at the policy, agency and local levels; (c) importance of examining the lessons of successful participatory approaches used in other programs of natural resource management and rural development and (d) finding out what can be learned from traditional systems of resource management that could be applicable to water resource management.

The institutional and organizational factors important for achieving successful local involvement are discussed below in the section dealing with institutions and organizations.

Monitoring and evaluation

As implementation proceeds, monitoring and evaluation of the performance of the water resources plan and program is an essential management activity. As pointed out by Easter (1986a), monitoring and evaluation is a continuous process designed to inform project or program managers what has happened or is happening while a project is being implemented. Accordingly, an effective monitoring and evaluation system will provide the kind of information that managers need in time to enable them to make necessary changes in project operations.

However, monitoring and evaluation is not merely a mechanistic exercise of learning to what extent the specifics of a pre-determined plan have been carried out in physical or financial terms. Rather, it is a broad management tool for identifying and assessing project and program effectiveness both in terms of meeting specific program objectives and of dealing with unanticipated social, economic and environmental consequences. As such, measures of effectiveness will emphasize socio-economic well-being of people, especially of lower-income groups and displaced people, and changes in environmental quality in terms of erosion, sedimentation and water pollution.

According to Easter (1986b), important components of an effective monitoring and evaluation systems are: (a) carefully designed baseline studies of physical and socio-economic conditions; (b) strong support of the system by project and program managers; (c) realistic targets for project inputs, activities and outputs; (d) local participation in the monitoring and evaluation, especially of outputs and impacts affecting local people; (e) appropriate measures of effectiveness; (f) timeliness and accuracy of information generated by monitoring and evaluation and (g) effective communication of results to planners, managers and decision-makers.

Monitoring and evaluation will need to be carried out not only at the individual project level (such as for a single dam and reservoir irrigation development, or urban water supply) but also at the river basin level and, in the case of large river basins, at the sub-basin or watershed level as well. This is especially necessary when major physical, economic and social linkages exist within watersheds, such as upstream-downstream effects involving erosion, sedimentation, water flow regimes and water yields.

Organization for monitoring and evaluation will have to be designed to fulfill the requirements listed above. Monitoring and evaluation of individual projects could be carried out by a unit of the implementing agency, while evaluations at the watershed and river basin levels could be the responsibility of an external agency, perhaps one with co-ordinating responsibilities for all water resource activities in the watershed or river basins.

In the case of the Laguna Lake development, monitoring and evaluation was shown to be weak, primarily because of inadequate staff and absence of clear and unambiguous evaluation criteria for projects and programs. As a result, new projects have been undertaken without the benefit of information on the success or failure of earlier projects (Fellizar, *et al.*, 1989).

Operation and maintenance

A major weakness in implementation of many water resource plans in the humid tropics is the failure to provide adequately for the operation and maintenance of projects once construction or installation is completed. Where projects are funded by loans or grants from international development banks or bilateral aid agencies, funds are provided only for construction, with the project sponsor – usually a national water resource agency – or the water users expected to provide for ongoing operation and maintenance. With few exceptions, water user fees and assessments are not adequate to pay for operation and maintenance of projects; in many cases, water services are provided free to users; in other cases, where fees are levied, as for irrigation water, the rate of collection is very low. As a result, operation and maintenance activities must seek funds in national water resource agency budgets – a difficult task when competing against more politically appealing projects.

In addition to being underfunded, operation and maintenance activities suffer from the low professional prestige accorded to those who undertake what are considered to be mundane tasks. In some cases, unprofessional performance in operation of irrigation systems, especially in the allocation of water, is prevalent (Chambers, 1988).

According to Howe & Dixon (1985), operation and maintenance problems stem from a series of distortions that start with project formulation and selection during the planning stage and continue through the project design, construction and operation stages. These distortions have their roots in decisions by donor/lenders and host countries and include adoption of inappropriate technology, capital intensive systems that are difficult to operate and maintain, little or no involvement of affected local people in project planning and inadequate funding and lack of staff incentives for good operation and maintenance.

To overcome these deficiencies it is necessary to develop and carry out an operations and maintenance plan which has for its objective the continued production of the outputs of the project on a sustainable basis. This requires that adequate funds be provided on a sustained basis in the integrated multi-year financial program and annual budget. To the extent practicable, funding should be provided from user charges and assessments on project beneficiaries. In addition, water users should participate in project operation and maintenance, especially for irrigation, rural water supply and watershed management activities. The monitoring and evaluation system should give special attention to the effectiveness of project operation and to the adequacy of maintenance of project facilities, and provide for prompt reporting of any problems to the project managers and decision-makers. Appropriate incentives and help should be provided to operation and maintenance staffs, including those involved in monitoring and evaluation, and to participating water users. These could be in the form of professional training, technical guidance, pay increases, professional recognition, higher status in the management hierarchy and security of tenure.

Implementation tools

The importance of appropriate implementation tools in an implementation strategy has been discussed above under *Planning for Implementation*. What is appropriate depends upon the individual situation or case and is often a function of (a) the kinds of physical management measures of concern, e.g., farm terraces, irrigation canals, water tubewells and (b) the institutional setting, e.g., land tenure, water rights. In the-

ory, there is an appropriate set of implementation tools for each specific combination of management measures and institutional arrangements. Sometimes, in the process of implementation, it is necessary to change the mix of management measures or the institutional arrangements to attain a workable combination of these with implementation tools. For example, soil conservation methods of low capital investment accompanied by security of land tenure may be combined with technical assistance to develop a workable implementation strategy for upland watersheds. Actual examples are found in Indonesia where package programs of management measures (bench terraces, agroforestry) and implementation tools (cash subsidies, technical aid, extension services) were applied to upstream watersheds in the Solo, Citanduy and Brantas River Basins where farmers had secure tenure (McCauley, 1988).

There are four general classes of implementation tools: Regulatory "command and control" arrangements, monetary incentives or disincentives, technical help, education and research and direct public installation and operation (Easter, 1986a). Most implementation strategies will use a combination of these classes. Each has its advantages and disadvantages. Direct "command and control" methods, involving regulation of land uses, forest harvesting, water withdrawals or polluted effluents, although potentially effective, often do not work well, especially on privately owned or operated lands in rural areas. Economic incentives, in the form of grants or subsidized loans, are often expensive and, once incentives are stopped, the desired management actions may also stop. Technical help via extension services, supported by education and research, for water, land-use and control practices have great potential for long-range effectiveness but the immediate pay-off is usually disappointing. Direct construction and operation by public agencies are most appropriate for large-scale multiple-purpose projects such as dams, reservoirs, levees and canals. But such projects are often very costly, and in many cases still require the participation of irrigators and other water users, such as fishermen, to achieve project outputs.

Elements of strategy: institutions and organizations

In this chapter we distinguish between *institutions* and *organizations*. Institutions as used here are the formal and informal rules of a society which define property rights to land, water and other natural resources, and spell out the rights and obligations of individuals and groups. As such, the definition also includes the rules under which organizations operate. Organizations are ordered groups of people in administrative or functional structures, such as private firms, non-profit public entities and government agencies (Gibbs, 1986). In almost all water resource management situations, more or less well-developed institutional arrangements and organizational patterns are already in place.

The major question to be addressed is the extent to which these existing arrangements and patterns provide adequate bases for water resource planning and implementation. In many cases, existing institutional and organizational arrangements are not ideally suited for effective management. However, usually such arrangements are difficult to change, at least in the short-run. Therefore, an important strategic issue is whether and how to adapt a water resource management activity to existing institutions and organizations, or alternatively, whether to seek immediate changes in existing arrangements as a part of implementing a water resource management plan. This issue must be taken up during the planning stage and any proposed changes in institutions and organizations would be included in the water resources plan and ideally put into effect at the first stages of implementation. In the following, we set forth certain desirable characteristics of institutions and organizations in an effective water resource management strategy.

Institutions: desirable characteristics

In general terms, water resource management institutions should promote or, as a minimum, not constrain the achievement of the multiple objectives of the water resource management plan – in particular, the development, equity, environmental quality and sustainability objectives. More specifically, the rights and obligations of water, land and related resource users should be clearly defined and consistently administered so as to allow efficient and equitable use of resources on a sustainable basis, with appropriate accountability for unavoidable adverse environmental and social consequences. These essential characteristics of clarity of definition and consistency in administration may take somewhat different forms for urban and rural systems as well as for surface water, groundwater, fisheries, forests, grasslands, rainfed and irrigated agricultural lands and aquatic ecosystems.

Turning first to agricultural lands: In general, security of tenure is the prime institutional characteristic contributing to productivity of agricultural lands on a sustainable basis. High security can take the form of outright private ownership of land as in Japan or the United States, or long-term leasehold tenancy arrangements with rights of renewal, either with private owners or with the government, as in many developing countries. Less secure are short-term tenancies with or without renewal rights, and customary land-use rights, with no legal ownership or tenancy status. For rainfed agricultural lands, secure rights to land is the important requirement but for irrigated lands, security of water rights is needed as well.

As pointed out by Nickum & Easter (1989), difficulties arise in achieving security and transferability of water rights because of: (a) interdependencies among water users, (b) high degree of variability of supply, (c) economies of scale of large water projects, (d) high social value of agricultural water supply relative to economic value, (e) conflicting social values concerning use, and (f) high transaction costs where many water users are involved. Given these difficulties, water markets for irrigation will be severely constrained by governmental action. One type of government-issued water-use right is a permit to withdraw a specific quantity or proportion of available water. Most secure and transferable are permits to a landowner in perpetuity or for very long terms, which can be bought and sold separate from the land. Other types with less security and transferability are permits for very short periods with no right to buy or sell. In some cases, as in canal irrigation, the quality of the use right depends on location of the user along the canal; "tailenders" are often at the mercy of "headenders" who divert more than proportionate shares of available water. Also, irrigation water users pumping from a common groundwater pool may have restrictions placed on their pumping rates which depend upon overall groundwater pumping and recharge rates.

In contrast to agricultural lands which are commonly in private ownership, grasslands and forests are often under communal (village) or governmental ownership and control. Use rights by individuals of forest and grassland products are typically controlled by communal groups or government. Unrestricted use of such open access resources leads to the "tragedy of the commons" and consequent loss of sustainability. In the case of restricted use, even under the best of circumstances, a common problem in developing countries is "government failure" of regulation, leading to inequities, widespread evasion of restrictions by "free riders," deterioration of the resources and often severe erosion, especially on steeply sloping lands. A clear definition of resource use rights and obligations and strict, consistent enforcement are required characteristics of institutional arrangements for such grazing and forest lands as well as other resources held in common.

Fisheries in freshwater lakes present a special type of open access problem. As shown in the Laguna Lake case (Fellizar *et al.*, 1989), some form of government regulation of fishery sites and rates of use was required to control adverse externalities and limit overall fish takes to sustainable levels. Here, too, the problem of "government failure" to regulate effectively an open-access resource becomes an important issue. The example of fishing rights in Tokyo Bay, described in Dixon & Hufschmidt (1986, Chapter 6), is a case where a government acted effectively to: (a) provide rights that were specific and transferable, (b) restrict access to members of user organizations (the fishery unions) and (c) vest decision-making rights to the fishery unions, with full participation by the users. In contrast, the Laguna Lake case is a less fortunate example of ineffective government and user performance (Fellizar *et al.*, 1989).

In cases of market failure involving spillover effects, such as water pollution, erosion and sedimentation, salinization and flooding, government will usually intervene with regulations or economic incentives to correct such distortions. As pointed out by Nickum & Easter (1989), bargaining among affected private parties could, in theory, lead to a socially optimal level of pollution. However, when, as is usually the case, many actors are involved or information on technical relationships is faulty, the transaction costs of bargaining are so high that this method becomes infeasible and government intervention is resorted to.

In the case of water pollution from urban areas, industry and agriculture, regulation can take various forms: effluent standards and ambient stream or lake standards. Economists advocate the use of effluent charges or tradeable permits as means of achieving economically efficient outcomes. All of these methods are most effective for large, clearly-defined point sources of pollution such as from industrial plants and urban sewer systems. They are least effective for dispersed pollution sources typically found in rural agricultural settings. In these cases the use of economic disincentives (taxes) to reduce application of potential pollutants such as fertilizers and pesticides at the source may be more effective, although effectiveness depends on how responsive farmers are to prices of these inputs. Banning the use of chemicals in areas where water leaches quickly to the groundwater may be necessary.

Control of erosion and sedimentation from watersheds composed largely of agricultural, grazing and forest lands presents special problems. Except in special situations, government regulation is not an effective approach, in part because of high transaction costs. Promotion of institutions that foster collective action and internalize off-site impacts, along with strong tenure arrangements, are more appropriate actions. These, along with technical help and economic incentives to landowners and tenants, can lead to successful watershed management programs (Easter, 1986a; Gibbs, 1986).

Organizations: desirable characteristics

The United Nations report on Integrated River Basin Development (1970) concluded:

There certainly is no single correct way to organize and administer a river basin programme. The plan of organization must in each case be fitted into the general governmental structure and into the cultural patterns and political traditions of the countries and regions which are involved.

This is still the case today. Accordingly, in developing the organizational dimension of a water resource management strategy, our approach is to look at a number of desirable characteristics that any organizational arrangement should have. How to fit these characteristics into an existing or proposed organization for a specific river basin or other management region is a separate and challenging task that is beyond the scope of this chapter.

Following are four important desirable characteristics for organization.

Regional or river basin focus The organizational structure should have an appropriate regional focus. This is obviously the case with a river basin authority such as the Tennessee Valley Authority in the United States, the Damodar Valley Authority in India and the Lake Victoria Basin Development Authority in Kenya. Regional focus can also be achieved by establishing a strong regional or river basin coordinating mechanism such as an inter-agency committee, preferably with independent leadership, or alternatively with leadership by a sectoral agency. Some attempts were made to achieve regional co-ordination by means of inter-agency committees for several major Javanese river basins, including the Upper Solo, Citanduy and Brantas River Basins (McCauley, 1988). Whatever form it takes, some regional entity or arrangement for co-ordination is required to provide the regional focus and to act as spokesman for the region.

Integration of objectives, purposes and means Along with a strong regional focus, organization for water resource management must achieve effective integration of the various multiple objectives, multiple purposes and multiple means that are part of management. Even where a river basin or regional authority exists, achievement of such integration is far from a routine exercise and requires sophisticated administrative skills. The problem is especially difficult in the usual case where responsibilities for water resource activities are divided among many sectoral agencies. As an example, Bower & Hufschmldt (1984) reported that in Thailand six national ministries and eight national enterprises are involved in such activities relating to irrigated agriculture.

In such cases of sectoral disaggregation of responsibility, strong inter-agency co-ordination mechanisms or special task forces are necessary. Howe (1986) gives several examples, including the Rio Fuerte Basin, Mexico and the Tana River Basin, Kenya. Leadership for such inter-agency co-ordination can be independent of any sectoral agency or can be the responsibility of one of the sectoral agencies. As Howe (1986) points out, the leadership must have power to obtain appropriate responses from the other member agencies.

Whatever the particular form of mechanism for integration, it is important that it be in place at the early stages of planning when some of the crucial decisions concerning scope, objectives and purposes of the water resource plan will be made. Equally important is to meet the need for integration throughout the implementation stage. In some cases, it may be desirable to set up a special co-ordinating agency for the implementation stage.

Management effectiveness Regardless of the particular organizational structure for management, the basic components, such as sectoral agencies, must have the legal authority and administrative capability to perform effectively the management tasks within their areas of responsibility. This is particularly important during the implementation phase of management. For example, the governmental organizations must provide fair and consistent administration of private rights and obligations to land, water, fisheries, forestry and grassland resources. Rights to surface and groundwater irrigation water must be fairly administered, preferably in collaboration with the water users. Management of open access resources such as lake fisheries, as in the Lake Laguna and Saguling Reservoir cases, groundwater aquifers and communal forestlands and grasslands requires rigorously fair and consistent enforcement of restrictions on use. Otherwise, regulatory failure can lead to inequities and conflicts among resource users, widespread evasion, and deterioration of the resources. Government administration of regulations and economic incentive systems for control of water pollution, erosion, sedimentation, salinization and flooding must also be equitable and consistent. Almost always, effectiveness of administration is enhanced if the local resource users are closely involved in the administration, often through water users organizations.

Decentralized management via local user organizations Although a considerable degree of centralization of activity is appropriate at the planning stage of management, a high degree of decentralization is usually necessary during implementation, especially when many water and land resource users are involved. Such decentralization of sectoral agency staffs to the project or sub-watershed levels is most effective when combined with devolution of responsibility to water user organizations, as in the case of irrigation projects, rural water supply schemes, watershed management and lake fisheries. Administration of water rights and economic incentives or regulations on pollution, sedimentation and groundwater withdrawals are often best accomplished through such water or land-user organizations.

In general, an effective organizational structure would provide a sound basis for initial planning as well as administration of the institutional arrangements that are appropriate for good water resource management and for specific application of the implementation tools, such as regulations, economic incentives, technical help and direct investments, needed to carry out the water resource plan.

SUMMARY

Water resource management is manifest in the humid tropics within two major problem contexts: *Urban and urbanizing*, characterized by: large and rapidly growing urban complexes with associated problems of potable water supply, sanitation, water pollution, flooding and depletion of groundwater aquifers. *Rural resources-related*, involving problems of maintaining and enhancing the productivity of lowland agriculture, including rice irrigation, and controlling the deterioration of upland watersheds through innovative programmes of watershed management.

Given the problem contexts, there are a number of generic or cross-cutting management issues that are of importance in achieving improvements in management. These issues include water resource assessment, integrated planning, the river basin as a management region, demand management, local involvement, implementation, and organizations and institutions for water management.

In order to deal effectively with these and related issues, a water resource management strategy appropriate for conditions in the humid tropics is required. The major elements of such a strategy are:

For the planning stage: selecting an appropriate management area, adopting a multiple objective approach, planning for multiple purposes, applying an economic-analytical approach, using multiple means, planning for implementation, obtaining local involvement, and using a multidisciplinary planning team.

For the implemention stage: adopting an integrated programme and budget, local involvement in implementation, effective monitoring and evaluation, improved operation and maintenance, and adopting effective implementation tools.

When combined with institutions and organizations with certain desirable characteristics, these elements of strategy, if adopted and rigorously applied, would go far toward assuring that water resource management would contribute to the sustainability of natural resources while meeting human needs for water and water-related services in the humid tropics.

REFERENCES

Bangladesh Ministry of Irrigation, Water Development and Flood Control (1986) *National Water Plan.* Dhaka, Bangladesh.

Barker, R., Coward, E. W. Jr., Levine, G. & Small, L. E. (1984) *Irrigation Development in Asia: Past Trends and Future Directions.* Cornell Univ., Studies in Irrigation, No. 1, Ithaca, New York.

Barrow, C. (1987) *Water Resources and Agricultural Development in the Tropics.* Longman & Sci. Tech., New York.

Bower, B. T., & Hufschmidt, M. M. (1984) A conceptual framework for water resources management in Asia. *Natural Resources Forum* 8(4): 343–356.

Chambers, R. (1988) *Managing Canal Irrigation.* Oxford and IBH Publ. Co., New Delhi. Chap. *6.*

Chanphaka, U. (1986) Water management and shifting cultivation: Three Asian approaches. *Unasylva 38*(1):21–27.

Committee for Coordination of Investigations of the Lower Mekong Basin. (1988) Perspectives for Mekong Development: *Revised Indicative Plan for the Development of Land, Water, and Related Resources of the Lower Mekong Basin.* Interim Comm. for Co-ordination of Investigations of the Lower Mekong Basin, Bangkok. Summary Report.

Coward, E. W., Jr., Koppel, B. & Siy, R. (1982) *Organization as a Strategic Resource In Irrigation Development: A Conference Report.* Asian Development Bank, Manila.

Coward, E. W., Jr. & Levine, G. (1987) Studies of farmer-managed irrigation systems: Ten years of cumulative knowledge and changing research priorities. *In: Public Intervention in Farmer Managed Irrigation Systems.* Digana Village via Kandy, Sri Lanka, Int'l. Irrigation Management Inst. 1–31.

Delft Hydraulics and Indonesian Ministry of Public Works. (1988) Cisadene-Cimanuk Integrated Water Resources Development. Jakarta, Ministry of Public Works. *Exec. Summary*, Vol. *1.*

Dixon, J. A., Carpenter, R. A., Fallon, L. A., Sherman, P. B. & Manopimoke, S. (1988) *Economic Analysis of Environmental Impacts of Development Projects.* Earthscan Publ., Ltd., London.

Dixon, J. A. & Easter, K. W. (1986) Integrated Watershed Management:

An Approach to Resource Management. *In*: K. W. Easter, J. A. Dixon & M. Hufschmidt (eds.) *Watershed Resources Management: An Integrated Framework with Studies from Asia and the Pacific.* Westview Press, Boulder, Colorado. 3–15.

Dixon, J. A. & Hufschmidt, M. M. (1986) *Economic Valuation Techniques for the Environment: A Case Study Workbook.* Johns Hopkins Univ. Press, Baltimore, Maryland.

Dixon, J. A., James D. E. & Sherman, P. B. (1989) *The Economics of Dryland Management.* Earthscan, London.

Easter, K. W. (1986a) Program implementation. *In*: K. W. Easter, J. A. Dixon & M.M. Hufschmidt (eds.) *Watershed Resources Management: An Integrated Framework with Studies from Asia and the Pacific.* Westview Press, Boulder, Colorado. 103–118.

Easter, K. W. (1986b) Monitoring and evaluation for integrated river basin development and watershed management. *In*: *Proc. of the Workshop on Integrated River Basin Development and Watershed Management.* Ministry of Public Works, Govt. of Indonesia, in Co-operation with Environ. and Policy Inst., East-West Center, Honolulu, Hawaii., Vol. 2:301–321.

Easter, K. W., M. M. Hufschmidt & McCauley, D. S. (1985) *Integrated Watershed Management Research for Developing Countries.* Report of workshop, Environ. and Policy Inst., East-West Center, Honolulu, Hawaii, 7–11 Jan.

Easter, K. W., Dixon, J. A. & Hufschmidt, M. M. (eds.) (1986c) *Watershed Resources Management: An Integrated Framework with Studies from Asia and the Pacific.* Westview Press, Boulder Colorado.

Fellizar, F. P., Jr., Pacardo, P., Francisco, F., Espaldon, M. O., Nepomuceno, D.N. & Espaldon, C. F. (1989) *Policy Responses to Fishery Conflict in Laguna Lake.* Paper presented at the Second Expert Group Workshop on River/Lake Basin Approaches to Environmentally Sound Management of Water Resources: Focus on Policy Responses to Water Resources Management Issues and Problems, 16–25 Jan. 1989, Bangkok and Hat Yai, Thailand.

Gibbs, C. J. N. (1986) Institutional and organizational concerns in upper watershed management. *In*: K. W. Easter, J. A. Dixon & M. M. Hufschmidt (eds.) *Watershed Resources Management: An Integrated Framework with Studies from Asia and the Pacific.* Westview Press, Boulder, Colorado. 91–102.

Haimes, Y. Y., Hall, W. A. & Freedman, H. T. (1975) *Multiobjective Optimization in Water Resources Systems.* Elsevier Sci. Publ. Co., Amsterdam.

Hashimoto, M. (ed.) (1991) *Guidelines of Lake Management*, Vol. 2. *Socio-Economic Aspects of Lake/Reservoir Management.* Int'l. Lake Environ. Committee Foundation, Otsu, Japan.

Howe, C. W. (1986) The institutional framework for river basin planning and management: Principles and case studies. *In*: *Proc. of the Workshop on Integrated River Basin Development and Watershed Management.* Ministry of Publ. Works, Govt. of Indonesia, in Co-operation with Environ. and Policy Inst., East-West Center, Honolulu, Hawaii. Vol. 2:251.

Hufschmidt, M. M. (1986) Planning, programming, and budgeting for integrated water resources development and watershed management. *In*: *Proc. of the Workshop on Integrated River Basin Development and Watershed Management.* Ministry of Public Works, Govt. of Indonesia, in Co-operation with Environ. and Policy Inst., East-West Center, Honolulu, Hawaii. Vol. 2:214–235.

Hufschmidt, M. M., James, D. E., Meister, A. D., Bower, B. T. & Dixon, J. A. (1983) *Environment, Natural Systems, and Development: An Economic Valuation Guide.* Johns Hopkins Univ. Press, Baltimore, Maryland.

Hufschmidt, M., Dixon, J., Fallon, L. & Zhu, Z. (1988) *Water Resources Policy and Management for the Beijing-Tianjin Region.* A joint summary report of the State Sci. and Tech. Commission of Beijing, China and the East-West Center, Honolulu, Hawaii.

Hufschmidt, M. M. & Kindler, J. (1991) *Approaches to Integrated Water Resources Management in Humid Tropical and Arid and Semiarid Zones in Developing Countries.* IHP Tech. Doc. in Hydrol., Paris, UNESCO.

Hufschmidt, M. M. & McCauley, D. S. (1989) *Institutional and Organizational Mechanisms for Integrating Land Use Decisions with Water Resources Management.* Paper presented at Second Expert Group Workshop on River/Lake Basin Approaches to Environmentally Sound Management of Water Resources, 16–21 Jan., Bangkok, Thailand.

Hufschmidt, M. M. & McCauley, D. S. (1991) *Strategies for Integrated Water Resources Management in a River/Lake Basin Context.* Environ. and Policy Inst., Honolulu, Hawaii. Working Paper No. 29.

Johnson, S. H., III et al. ;Patamatamkul, S; Apinantara, A.; Charoenwatana, T.; Issariyanukula, A.; Paranakian, K.; & Reiss, P.; (1989) *Medium Scale Irrigation System in Northeast Thailand: Future Directions.* ISPAN-Irrigation Support Proj. for Asia and the Near East, Arlington, Virginia.

Korten, F. F. (1982) *Building National Capacity to Develop Water Users' Associations: Experience from the Philippines.* World Bank Staff Working Paper No. *528*, Wash., D.C.

Korten, F. F. & R. Y. Siy, Jr. (1988) Transforming a Bureaucracy: The Experience of the Philippine National Irrigation Administration. Ateneo de Manila Univ. Press, Manila.

Leopold, L. B. (1990) Ethos, equity and the water resources. *Environ.* 32:(2)17–20; 37–42.

Lovelace, G. W. & Rambo, A. T. (1986) Behavioral and social dimensions. *In*: K. W. Easter, J. A. Dixon & M. M. Hufschmidt (eds.) *Watershed Resources Management: An Integrated Framework with Studies from Asia and the Pacific.* Westview Press, Boulder, Colorado. 81–90.

Magrath, W. & Arens, P. (1989) *The Costs of Soil Erosion on Java: A Natural Resource Accounting Approach.* World Bank Environ. Dept. Working Paper No. *18*, Washington, D.C.

Major, D. C. (1977) Multiobjective Water Resources Planning. Am. Geophys. Union, Washington, D.C. Water Resources Monograph *4*.

McCauley, D. S. (1988) *Overcoming Institutional and Organizational Constraints to Watershed Management for the Densely Populated Island of Java.* Paper presented at the Fifth Int'l. Soil Conservation Conf., 18–23 Jan., Bangkok, Thailand.

McKinney, M. J. (1988) Water resources planning: A collaborative consensus-building approach. Soc. and Nat. Resour. *1*(4):335–349.

Milliken, J. & Taylor, G. C. (1981) Metropolitan water management. Am. Geophys. Union, Wash., D.C. Water Resour. Monograph Series No. *6*.

Nickum, J. E. & Easter, K. W. (1989) *Institutional Arrangements for River/Lake Basin Management with Emphasis on Managing Conflicts.* Paper presented at the Second Expert Group Workshop on River/Lake Basin Approaches to Environ. Sound Management of Water Resources: Focus on Policy Responses to Water Resour. Management Issues and Problems, 16–25 Jan. 1989, Bangkok and Hat Yai, Thailand.

Repetto, R. (1986) *Skimming the Water: Rent-Seeking and the Performance of Public Irrigation Systems.* World Resources Institute, Wash., D.C.

Sinha, B. (1983) *Water Resources Management in Asia: Problems and Perspectives.* Environ. and Policy Inst., East-West Center, Honolulu, Hawaii.

Small, L. E. & Carruthers, I. (1991) *Farmer-financed Irrigation: The Economics of Reform.* Cambridge Univ. Press, Cambridge, U.K.

Srivardhana, R. (1984) No easy management: Irrigation development in the Chao Phya Basin, Thailand. *Natural Resources Forum 8*(2):135–145.

Srivardhana, R. (1991) A transaction cost economics approach to a study of industrial water in Samut Prakarn Province, Thailand. *Regional Development Dialogue 12*(4):73–81.

Steinberg, D. I., Clapp-Wincek, C. & Turner, A. G. (1983) Irrigation and AID's Experience: A Consideration Based on Evaluations. U.S. Agency for Int'l. Development, Wash., D.C. AID Program Eval. Report No. *8.*

United Nations (1970) *Integrated River Basin Development.* United Nations, New York.

UNDTCD (1988) *Assessment of Multiple Objective Water Resources Projects: Approaches for Developing Countries.* United Nations Dept. of Tech. Co-operation for Development and United Nations Environ. Programme, New York.

World Bank (1981) *Water Management in Bank-Supported Irrigation Project Systems: An Analysis of Past Experience.* Washington, D.C. Report No. *3421.*

World Resources Institute *et al.* (1988) *World Resources, 1988–89.* Basic Books, New York.

23: Water Management Issues: Population, Agriculture and Forests – a Focus on Watershed Management

K.G. Tejwani

Land-Use Consultants (International), 25/31 Old Najinder, Nagar, New Delhi, 110060, India

ABSTRACT

In developing countries of the humid tropics, the great pressure of human and livestock population on the land, forest and water resources leads to misuse and mismanagement of land, over-exploitation or clear felling of forests and overgrazing/destruction of grazing lands. Since rainfall is unevenly distributed in space and time, making rainfed agriculture uncertain, humans have turned to irrigated agriculture. Modern large-scale irrigation calls for construction of large dams and reservoirs for storage of water, and networks of canals to deliver water to the farm sites. When water flows overland, it carries varying quantities of suspended sediment (and bed load) which is deposited in the streams, reservoirs, canals and their distributaries, thus reducing the efficient functioning of the reservoirs, turbines and canals.

At the macroscale, the rivers of the world deliver to the ocean 25.7 billion tons of top soil from crop lands in excess of new soil formation. The rivers of Asia – which is very densely populated – are the most burdened with sediment. African rivers, which do not have very dense populations in their catchments, carry the least sediment. The chapter reviews the impact of forest land use on hydrologic disturbance and sediment generation at the microlevel. As long as the forest land use is sustainable and the forest remains intact, the rates of soil erosion and sediment load in streams, and flood peaks remain unchanged or only slightly changed. However, as soon as there is a significant change in any characteristic of the forest, adverse impacts arise from the hydrological disturbances, the erosion and the sedimentation. The intensity of the adverse impacts increases as the forest is converted to grassland and the grassland to cropland.

Of the biophysical, social and economic factors which lead to hydrologic disturbances, soil erosion and water mismanagement, some factors such as geology, geomorphology and rainfall are beyond the control of humans. Others are directly attributable to human intervention. Among these the most important is population pressure – both human and livestock – which leads to misuse and mismanagement of land, small farm sizes and landless people. Development activities – such as road construction, agriculture, and mining – carried out improperly lead to degradation of forest, soil erosion and mismanagement of water resources. The chapter compares the land use conditions and hydrological balance in densely populated areas of South Asia with sparsely populated areas in Africa's Sierra Leone and Tanzania.

Water mismanagement and soil erosion lead to watershed degradation with adverse ecological, economic and social impacts. Accelerated rates of runoff lead to reduction in the base flow of streams, a drop in groundwater tables and flashfloods. In densely populated areas, even the flood plain is populated, leading to frequent losses of life and property from flooding. The runoff from hydrologically-disturbed watersheds carrying high charges of sediment reduces the effective life of reservoirs, hydro-power generation, and water available for irrigation, with consequent loss of food production and industrial output. Silt-laden waters choke up the canals and distributaries, resulting in high cleaning and maintenance costs and consequent reduction of irrigated area and agricultural production. Sedimentation of streams leads to loss of inland water transport and loss of fish and fish breeding grounds. Examples from Asia and Africa are discussed.

The chapter looks at the programmes and policies promoting water management from a historical perspective. After World War II, the impetus to development and conservation of land and water resources was provided by FAO and UNESCO. In the Asia-Pacific region, China, India, Indonesia, Pakistan and Philippines have extensive programmes of soil conservation and drainage basin management. Currently there are no large-scale soil conservation programmes in Africa. At present, the governments of developing countries have varying degrees of concern about the relationship of water resource development to the sustainability of land-water-vegetation systems. Subject to the availability of funds, some are willing to invest in programmes related to soil conservation and reforestation. Realizing this, FAO/UNDP/WB/WRI have developed a major project on tropical forests which envisages investment of US$1,231 million in land use and drainage basin management programmes over a period of five years. About 20 per cent of this will be allocated to research, training, education and extension. The chapter discusses success stories in India, Nepal, China, Uganda, and Ethiopia and the demonstrated benefits of drainage basin management at the river basin, large drainage basin and small drainage basin scales. In the developing countries of Africa, with their low population pressure and where settled agriculture is yet to develop, the focus would be on increases in irrigated area, and increases in food production by good farming methods combined with conservation practices.

The chapter concludes with discussion of selected management issues. Population pressure is recognized as a key issue which can be ignored only at the cost of failure of development efforts. Another important issue is the simultaneous treatment of the reservoir catchment and development of the irrigation command area (the area to be irrigated) along with design and construction of the dam and reservoir. Other issues are the availability of research facilities for watershed management and of technically trained manpower. For implementing watershed management programmes, equitable fiscal policies are needed to determine who pays (the farmers, the community, the government), for what works and how much for each work. The fiscal arrangements must take into account the cost of operation and maintenance of the works, and should also include monitoring and evaluation of the project. In watershed management and soil conservation, participation of the people must be ensured and encouraged and the activities of all sectoral departments which impact on land and water uses must be coordinated.

INTRODUCTION

It is recognized that no life is possible on Earth without water. All of the ancient cultures – African, Pacific, Indian – recognized this fact in their legends as to how life started on the Earth. Humans, responding to this fact of life, have built their settlements and civilizations near the fresh water courses and rivers – such as the Nile, the Tigris, the Euphrates, the Indus and the river valleys of Peru. In the beginning, humans needed fresh water for drinking, washing and cooking. As they evolved from hunters and gatherers of food into agriculturists, they were dependent on rainfall for crop production. They must have soon realized that rainfall was too uncertain in time, space and quantity to assure successful crop growth, year after year. To ensure this they developed the techniques of water resource development, transport and use. Mention may be made of the ancient irrigation canals and tanks made in Egypt (3200 BC), China, Mesopotamia (modern Iraq), and the Indus Valley (2300 BC). Apart from transporting water by gravity in canals and qanats, humans developed devices to lift water, e.g., paddle wheels (the Summerians of Mesopotamia, 1900 BC); Persian wheels; shadufs – poles with a weight on one end and a scoop on the other – (Egypt, about 1430 BC); and the Archimedes screw (285 to 212 BC).

The search for water, which started with the dawn of civilization, continues as humans have multiplied in numbers and their need for water has increased, not only for agriculture, drinking, washing and cooking, but in modern times for industry, power, transport and many other uses. Fresh water, though a renewable natural resource, is limited in quantity in absolute terms. The distribution of this limited quantity of fresh water is further influenced by many natural factors and biotic influences. The unrestricted population growth during the last four decades, mostly in the developing countries, has impacted the availability of fresh water of adequate quality.

This chapter looks at fresh water management issues, with special reference to population, lands, forests and agriculture.

FRESH WATER, LAND AND POPULATION

While water is needed by humans for many varied uses, land is also essential to humans in order to grow food, fiber, fodder and fuel. In fact, humans cannot survive without using the land for agriculture and forests. It follows that there is a very close relationship between the availability of fresh water for use by humans and land-use management.

Fresh water: A limited, renewable resource

About three-quarters of the Earth's surface is covered with water. The volume of water is enormous – between 1.33 to 1.50 billion cubic kilometres. It is probable this quantity has always been the same. This water passes through an unending global cycle operating in and on the land, the oceans, and the Earth's atmosphere. The water cycle is described elsewhere in this volume (Fleming, Lal).

Of the 1.5 billion km^3 of total water present in the world, only 0.0085 per cent is available as fresh liquid water for human use. The remainder is distributed as salt-water in oceans and lakes (97.2 per cent), fresh water in ice and glaciers (2.15 per cent), fresh groundwater (0.64 per cent) and fresh water in plants, animals and the atmosphere (0.0015 per cent) (FAO, 1987a). Even though the liquid fresh water available to humans is a minute fraction of total water of the world, yet it would be sufficient to meet human needs, were it uniformly available in space and time.

Fresh water – unevenly distributed

In a year, the world's land surface receives an unimaginable amount of rain: 125 000 km^3 of fresh liquid water. Of this, 83 000 km^3 (66.6%) evaporates, 14 000 km^3 (11.1%) infiltrates into the ground, and 28 000 km^3 runs into lakes, rivers and then into the sea. Theoretically, the water in rivers, lakes and the ground is available for human use.

If the total rainfall were equally available to the present five billion people of the world, each person would receive 25 million litres of fresh water. However, with such an abundance of fresh water we face the paradox of floods in some parts of the world and severe droughts in others at the same time. In parts of the humid tropics, we can experience floods in the rainy season and drought in the dry season in the same place in the

Table 1. *Arable land and population in the developing market economies.*

Geographical Identity	Arable Land (km²)		Population (Millions)		Arable Land Per Capita (ha)		
	1970	1985	1970	1985	1970	1985	
World	13,190,360	13,757,360	3694.4	4836.8	0.35	0.28	(20)*
Africa	1,247,480	1,374,770	289.1	450.5	0.43	0.31	(28)
Latin America	1,205,810	1,482,920	283.4	404.8	0.43	0.37	(14)
Near East	775,590	769,310	161.6	241.9	0.48	0.32	(33)
Far East	2,355,190	2,390,250	986.5	1374.6	0.24	0.18	(25)

Source: FAO. 1987a.

*Figures in parentheses refer to percentage reduction in per capita arable land in 1985 with respect to 1970.

same year. These conditions arise due to the uneven distribution of rain in space and time on the one hand and land-use mismanagement on the other.

Humans do not exclusively depend on the use of rainfall directly, except where used for rainfed agriculture and forests. Humans depend also on stream runoff, primarily for irrigated agriculture, hydro-power generation and industry in all of the hydrologic regions of the world. However, given that fresh water supplies and runoff water supplies are limited, it is obvious that for any given region, the quantity of fresh water runoff available per capita will be highly influenced by the human and livestock population.

Land: a limited, non-renewable resource

It is well-recognized that land is a limited, non-renewable source. Currently 10.5 per cent of the land area in the world is under arable crops, 31.3 per cent under forest and woodland and 24.3 per cent under permanent pasture. In its natural state, land in the humid and subhumid tropics is covered with vegetation – usually forest. If the annual rainfall is between 280–750 mm per year, the vegetation will be woodland and/or savanna.

As soon as humans learned that food crops could be cultivated, they started clearing forests in their search for arable lands. They also needed firewood to cook their food, which also called for cutting the forests. Once humans learned to domesticate cattle and use them for milk and power, they started using the grasslands and lopping the trees for fodder. From that time on, the cutting of trees and clearing of forested lands has been a continuous process, conducted at an ever-increasing pace to satisfy the needs of an exponentially-expanding human population.

If one looks at the land-use figures of the world, one might take great comfort that currently over 55 per cent of the land is under forest or permanent pasture. However, this situation is not as rosy as it appears. First, the human population is not uniformly distributed over the landscape. Second, the population is settled on all types of lands – from very good to very poor.

In their search for arable lands, as long as humans did not greatly disturb the natural functioning of the land-water-plant system, there was little difficulty. However, as the human population increased, the compulsion to clear lands for habitation and cultivation extended to the upland drainage basins and to lands unsuited for agriculture. In many parts of the world, natural processes thus have been seriously disturbed, resulting in the disastrous consequences of soil erosion, degradation of the production base and loss of water resources.

Population: an unlimited, exponentially-expanding resource

When analyzing the availability and use of renewable fresh water resources and non-renewable land resources for human beings, the population of humans and their livestock is a critical factor. For centuries, the human population increased at a very slow pace. The world reached the first billion in the year 1800; it was inhabited by 2 billion in the year 1900, by 3 billion in 1950, by 4 billion in 1975, by 5 billion in 1989 and is expected to be over 6 billion by the year 2000. More specifically, population pressures are very heavy in the humid and subhumid tropical countries, including India, Sri Lanka, Bangladesh, Thailand, the Philippines and Indonesia. Such pressures arise not just from extra consumers of water, food and fuel or from users of land, but also from an increasing per capita use of these resources.

In 1970, the per capita arable land in the four large developing regions of the world varied from 0.24 ha in the Far East to 0.48 ha in the Near East. With the subsequent increase in population it decreased to 0.18 ha in the Far East and to 0.37 ha in the Near East by 1985 (Table 1). The Far East region includes southern Asia which has some of the most densely populated countries in the world, e.g., Bangladesh, India, Nepal, Pakistan and Sri Lanka. While conditions are not good even at present, one feels concerned and staggered by the terrific pressure which will be exerted by the fast-increasing human population in these countries. For example, the human population in India is estimated to become 931 million by the year 2000; the net

Table 2. *Livestock (expressed as millions of cattle units) in the developing market economies.*

	World		Africa		Latin America		Near East		Far East	
	1980	1985	1980	1985	1980	1985	1980	1985	1980	1985
Cattle Units	1872	1948 (4.1)	224	225 (0.5)	371	388 (4.7)	122	128 (4.5)	458	496 (8.4)
Cattle units per capita			0.8	0.5 (−38)	1.3	1.0 (−23)	0.8	0.5 (−38)	0.5	0.4 (−20)

Source: FAO. 1987a.

1 cattle unit = 0.75 buffalo/horse/camel = 1 mule = 2 asses = 5 sheep/goat

Figures in parentheses refer to percentage change in 1985 over 1980 values

land area per capita which was 0.9 ha in 1951 is expected to decline to 0.35 ha by the year 2000. Furthermore, although the cultivated area in India increased from 119 million ha in 1951 to 140 million ha in 1971, the per capita availability of land for production of food, fibre and other needs shrank from 0.33 ha to 0.29 ha during the same period; this will decrease to 0.175 ha by the year 2000 (DST, 1976; NCA, 1976). In Java, Indonesia, the population increased from 63 million in 1961 to 100 million in 1985 (CBS, 1987). As a consequence, subsistence farming spread into the marginal lands located in the uplands, leading to serious watershed problems (Pereira, 1989). (For definitions of the terms, "watershed" and "drainage basin," see Table l, Hufschmidt, this volume.)

These pressures are exerted not only on good arable land but also on non-arable land in the fragile mountain eco-systems.

Apart from humans, large numbers of livestock also depend on land. In 1980, 1.3, 0.8, 0.8 and 0.5 cattle units per capita were being maintained, respectively, in the developing market economies of Latin America, Africa, Near East and the Far East (as defined by FAO, 1987a). However, by 1985, as the population increased, the total number of cattle units increased only marginally so that the per capita cattle units decreased by 38% (Africa), 23% (Latin America), 38% (Near East), and 20% (Far East). See Table 2. Thus it is obvious that, as the population has been increasing, the per capita total land and arable land has been decreasing. At the same time, the livestock population has been increasing. Both of these trends indicate increasing pressure on land resources. It is also clear that the people realize that they cannot maintain the same numbers of livestock per capita.

Irrigation

There are many ways of increasing crop production, primarily irrigation, high-yielding plant varieties, fertilizers, plant protection and soil conservation. Human beings started irrigating crops quite early in their civilization – as indicated above.

Among the developing market economies, the Far East irrigated 24.2 per cent of arable land in 1985, followed by the Near East with 23.9 per cent, Latin American 9.9 per cent and Africa only 3.7 per cent (Table 3). The Near East and the Far East have centuries-old histories of developing reservoirs and tanks for storage of water and water conveyance systems. Although in arid and semi-arid areas, irrigation is the major means of making the land produce crops regularly and bountifully, even in humid and subhumid areas irrigation is often needed to increase the intensity of cropping as well as to provide supplemental water during periods of low rainfall. In some large river basins, such as the Indus, rainfall in the humid and subhumid tropics may be used for irrigation in the arid and semiarid areas of the basins. Africa, on the other hand, has a very low percentage of area under irrigation since its agriculture is not yet well-developed.

Modern large-scale irrigation works call for construction of large dams and reservoirs for water storage and a network of canals of varying capacities to deliver water to the farm site. The water stored in large reservoirs is often used simultaneously for hydro-power generation. When water flows over land, depending upon the catchment conditions, it carries varying quantities of suspended sediment (and bed load) which is deposited in the streams, reservoirs, canals and their distributaries. For the efficient functioning of the reservoirs, turbines and canals, it is essential that the water carry as little sediment as possible. Watershed management for reduction of erosion and sedimentation, therefore, is an essential element in integrated water management for irrigation, hydro-power or other water uses.

MANAGING THE LAND-WATER-VEGETATION SYSTEM

The streams, rivers and lakes receive the fresh liquid water as runoff during the rainy season and, wherever applicable, from

Table 3. *Irrigated area in the developing market economies.*

Geographical Identity	Irrigated Area (km²)		Irrigated Area as Percentage of Arable Land*		Irrigated Area (ha) Per Capita*		Percent Change in 1985 Over 1970	
							Per Capita Irrigated	Popu- lation
	1970	1985	1970	1985	1970	1985		
Africa	18,520	50,780	1.5	3.7	0.006	0.011	83	56
Latin America	102,150	147,550	8.5	9.9	0.036	0.036	0	43
Near East	168,980	183,670	21.8	23.9	0.105	0.076	−28	50
Far East	543,120	601,850	23.1	24.2	0.055	0.043	−22	39

Source: FAO. 1987.

*Arable land and population figures as in Table 1.

snowmelt and springs during the other seasons. Runoff, as is well-known, is determined by a large number of factors such as rainfall (intensity and duration), soil characteristics and conditions, topography (length and degree of slope), land-use, land management and cover (vegetation) management practices. (Lal & Fleming, this volume). Keeping the land-use and population situations in view, FAO (1986) has identified eight types of watersheds in the Asia-Pacific region (Table 4). All but the last type (arid and semi-arid) are present in the humid tropics. This classification is sufficiently broad and comprehensive to describe the types and behaviour of watersheds anywhere in the developing countries. It can also be observed that the rates of soil erosion and hydrologic disturbance are closely associated with population pressures.

The first attempt was made in 1938 to evaluate the soil erosion conditions of a large number of countries. The status of soil erosion, its causes and consequences, and the efforts being made to conserve the soil in 36 countries were described by Jacks & Whyte, (1938). This project was all the more remarkable because the effort was made in a period when the world was passing through the severe economic depression of the 1930s. It is also noteworthy that at the time, a concept of watershed conditions and the interrelationship of hydrologic conditions and soil erosion had not been clearly articulated. The next comprehensive attempt was made in 1982 when soil erosion by water was described in many tropical countries of Africa, Asia, South America, Central America, the Pacific and Caribbean islands and Australia (El-Swaify *et al.*, 1982).

Soil erosion and hydrologic disturbance: the macro level

One of the first scientists to assess the dimensions of world soil erosion was a geologist, Sheldon Judson, who estimated in 1968 that the amount of riverborne soil carried into the oceans had increased from 9.9 billion tons a year before the introduction of agriculture, grazing and related activities to 26.5 billion tonnes a year (Brown, 1984). It was estimated in 1984 that the world was losing 25.7 billions tons of topsoil annually from croplands in excess of new soil formation (Brown, 1984). El-Swaify *et al.*, (1982) have calculated the annual soil erosion within the drainage basins of selected major rivers of the humid tropics (Table 5). They reported that the three rivers having the maximum annual soil erosion in their drainage basins were in the Indian subcontinent, namely the Kosi, Damodar, and Ganges Rivers. The fourth river in order was in China and in Vietnam (the Red River); with the fifth being in South America (the Caroni River in Venezuela). Excepting the Caroni, the first nine rivers having the greatest soil erosion rates in their drainage basins were in Asia. The rivers having the lowest soil erosion rates in their drainage basins were from Africa (the Nile, Congo and Niger Rivers).

Although the geologic causes of erosion originating in the Himalayas are important, the Asian rivers with the highest rates of erosion also have high population pressures – both human and livestock – with consequent high rates of deforestation, forest degradation, and cultivation of marginal lands in their drainage basins. In Indonesia, the densely populated drainage basins in Java are much more disturbed than those having low population densities, as in Kalimantan. Although the low values of annual soil erosion in the African rivers may be of comfort at the river basin level, it is possible that in some sub-basin areas with a dense population there are high rates of erosion. Similarly, while the Asian rivers have very high average rates of soil erosion in their drainage basins, it is also possible that there are still considerably higher localized rates (due to roads, gullies and landslides) and also due to geologically young, unstable and highly erodible rocks, particularly in

Table 4. *Types of watershed in the Asia-Pacific Region.*

Watershed Description	Population Pressure	Current Rating for Hydrologic Disturbance and Soil Erosion Due to Biotic Interference
Predominantly pastoral	Low	Low
Predominantly forested	Low, but increasing	Low, but increasing
Predominantly farming	High	High
Predominantly intensive shifting cultivation	High	Medium to high
Predominantly extensive shifting cultivation	Low	Low
Predominantly plantation cash crops	High	Medium
Barren and degraded	Low	Medium to high
Arid and semi-arid	Low, but concentrated in pockets	Low, but medium to high in pockets where population is concentrated

Source: FAO. 1986.

the Himalayan region. For example, the first gross national estimate of erosion made in India in the 1950s reported that 6,000 million tonnes of soil were eroded by water every year (Kanwar: *vide* Vohra, 1981). This was subsequently verified by Tejwani & Rambabu (1981) and Narayana & Rambabu (1983), who concluded that in India 5,534 million tonnes of soil (16.4 tonnes per ha) are eroded annually. The country's rivers carry an approximate quantity of 2,052 million tonnes (6.26 tonnes ha^{-1}). Of this total, 480 million tonnes are deposited in various reservoirs and 1,572 million tonnes are carried out to sea.

Soil erosion and hydrologic disturbance: the micro level

The impact of forest use or conversion in watersheds on soil erosion rates, sediment in streams, groundwater, annual water quantity in streams, timing of stream discharge and nutrient input in streams has been reviewed by Hamilton with King (1983) and discussed in this volume (Bonell with Balek, Fleming, Lal, Roche and Rose). The forest use activities considered were of two categories, namely, the uses to which the forest is put and the uses to which forest is converted (i.e., forest replaced by grassland or cultivation). The impacts in all

of the cases were compared with a natural forest in good condition. The impacts of reforestation or afforestation of degraded lands were also considered in relation to the degraded land. The conclusions, based on authoritative research references and professional observations and judgment drawn by Hamilton with King (1983), and as adapted by the present author, are reported in Table 6. The impacts are described in simplified terms and are only indicative of general cases.

To summarize the information in Table 6, as long as the land-use is sustainable and remains in forest, the rates of soil erosion, the sediment load in streams, the groundwater level/springs, the water quantity in the streams, and the time to peak rate of runoff remain unchanged or only slightly changed. This observation is applicable also to long-cycle shifting cultivation. However, once there is a significant change of any characteristic of the forest, e.g., removal or burning of leaf litter, excessive burning of trees or the forest floor, or disturbance of the soil by commercial harvesting, there are significant adverse impacts. Further, if the forest is converted to other land-uses, then the adverse impacts are more severe and significant. Their degree will be determined by the amount of

Table 5. *Estimated annual soil erosion within drainages of selected rivers of the humid tropics.*

River	Countries Within Drainage Basin	Drainage Basin (sq. km 000)	Average Annual Suspended Load		Estimated Annual Soil Erosion from Fields		
			(million tonnes)	(tonnes/sq.km.)	(tonnes/sq.km)	(tonnes/ha)	Rank
Congo	Angola, Congo, Zaire, Cameroon, Central African Republic	4,014	65	16	320	3	13
Niger	Cameroon, Guinea, Dahoney, Chad Ivory Coast, Nigeria, Niger, Mali	1,114	5	4	80	0.8	14
Nile	Uganda, Kenya, Zaire, Ethiopia Tanzania, Sudan, Egypt, Rwanda, Burundi	2,978	111	37	740	8	12
Chao Phraya	Thailand	106	11	107	2,140	21	9
Ganges	India, Bangladesh Nepal, Tibet	1,076	1,455	1,352	27,040	270	3
Damodar	India	20	28	1,420	28,400	284	2
Irrawaddy	Burma	430	299	695	13,900	139	5
Kosi	India	62	172	2,774	55,480	555	1
Mahanadi	India	132	62	466	9,320	93	7
Mekong	China, Thailand, Laos, Tibet, Kampuchea, Vietnam, Burma	795	170	214	4,280	43	8
Red	China, Vietnam	120	130	1,083	21,660	217	4
Caroni	Venezuela	91	48	523	10,460	105	6
Amazon	Bolivia, Brazil Ecuador, Colombia, Peru, Venezuela	5,776	363	63	1,260	13	11
Orinoco	Venezuela, Colombia	950	87	91	1,820	18	10

Source: El-Swaify et al. (1982).

soil and water conservation in the new use. In case the natural forest is clear/felled and converted to a forest tree plantation or a commercial tree crop (coffee, tea, rubber, oil palm), the adverse impacts of clear felling will be observed in the initial stages until the commercial tree crop establishes itself. When the tree crop is established, the nature and degree of impacts will be determined by the methods of managing these plantations. If soil and water conservation practices are adopted and the roads and paths are properly aligned and maintained, the adverse impacts will be minimized.

Conversion of good forest to grassland leads to few adverse impacts. This is also true for the alang-alang *(Imperata cylin-* *drica)* dominated grasslands in Indonesia. But, if the grassland is grazed beyond its capacity, then the adverse impacts will be as serious as those of a degraded land. When the forest is converted to agricultural land, the adverse effects are potentially the most serious. Where the land is intrinsically non-suited to agriculture, the adverse effects are even more serious and the land will degrade very quickly. The most common impacts of deforestation are shown in diagrammatic form in Fig. 1. If soil and water conservation measures are adopted in any land-use, the adverse impacts of that particular land-use will be mitigated to some extent. However, if deforested and degraded land is successfully reforested, the impacts are reversed, and

Table 6. *Impact of watershed forest use and conversions.*

Forest Use or Conversion Activity	Erosion Rate at Site	Sediment in Stream	Ground-water	Water Quantity in Stream Over Year	Time to Peak Rate of Runoff	Nutrient Input to Stream
I Various forest uses						
Minor forest product gathering tapping but no tree cutting	No change	No change	NI	Increase due to removal or burning of leaf litter	Decrease due to removal or burning of leaf litter	NI *
Shifting cultivation – Fallow sufficiently long to sustain the system	Low	Low	No change	No change	No change	Low *
Shifting cultivation short fallow	High	High	NI	Increase	Decrease	High
Sustainable harvesting fuelwood & lopping fodder (no roads or vehicles)**	Low *	Low *	May * rise	May * increase	NI	Low *
Commercial harvesting of wood						
- Commonly used logging practices	Increase	Increase	Rise due to reduction in canopy till the full canopy is restored	Increase due to reduction in canopy & till the full canopy is restored	Decrease until the forest regrows	Medium in short term
- Improved watershed protection logging	Low *	Low				Medium in short term
Sustainable grazing ***	Little effect	Little effect	Rise	Little * effect	Little * effect	NI
Burning Forest Lands						
Controlled	Little * effect	Little * effect	NI	None to * slight	None to * slight	NI
Wild/uncontrolled	Increase	Increase	Fall	Increase	Decrease	Increase
II Conversion of forest to other land uses						
Conversion of natural forest to tree plantation	In the initial stages when the natural forest is clear felled, the impacts will be similar to deforestation, until the plantation is established, such as:					
	Increase *	Increase *	Rise *	Increase *	Decrease *	Increase *

Table 6 (*cont.*)

Forest Use or Conversion Activity	Erosion Rate at Site	Sediment in Stream	Ground-water	Water Quantity in Stream Over Year	Time to Peak Rate of Runoff	Nutrient Input to Stream
Conversion to grassland/savanna for grazing (and/ or burned to maintain). Includes overgrazing	Increase	Increase	Fall	Increase	Decrease	Increase
Conversion to commercial tree crop plantations (e.g. tea, coffee, rubber, oil palm, banana)	In the initial stages when the forest is clear cut the impacts will be similar to those of commercial harvesting with commonly used logging practices. When the tree crops grow, their impacts will be determined by management practices as under:					
-Scientific management concern for conserving soil	Slight increase	Slight increase	No change	Little effect	No change	Increase due to fertilizer
-Unscientific management with no concern for soil conservation	Increase	Increase	Fall	Increase	Decrease	Increase due to fertilizer
Conversion to annual rainfed cropping:						
-Without conservation and with poor farming	High increase	High increase	Fall	Increase	Decrease	Increase
-With conservation and good farming	Modest increase	Modest increase	Rise	Increase	Decrease	Increase
-Reforestation or afforestation of degraded areas	Decrease	Decrease	Rise	Decrease	Increase	NI

Source: Adapted from Hamilton with King (1983), generalized and simplified by the author.
NI: No information.

*   No research information.
**   Where roads are constructed and machinery used, the effects of such activity are treated as "Commercial harvesting of wood."
*** Unsustainable grazing involves annual or periodic burning and gradual removal of trees leading to conversion of forest to grassland. For effects see "Conversion to grassland. . . " below.

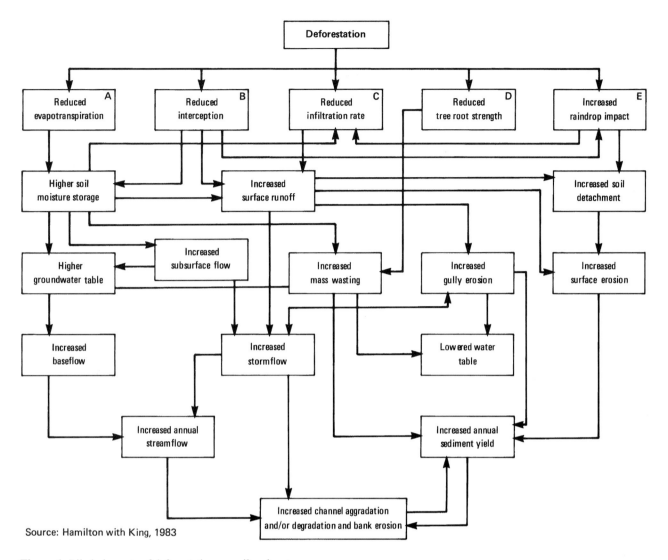

Source: Hamilton with King, 1983

Figure 1: Likely impacts of deforestation on soil and water

the original forested hydrological cycle and soil-conservation/formation processes are re-established over a period of time. The impacts of forest use or conversion, which are shown in qualitative terms in Table 6, are over-simplified in order to be presented in tabular form. Some of the quantitative data will be presented later in this chapter.

Factors and policies intensifying hydrologic disturbance
A number of biophysical, social and economic factors can lead to hydrologic disturbances, soil erosion and water mismanagement. Some of the factors such as geology, geomorphology and rainfall erosivity are beyond human control. Others are directly attributable to human activity. Of these, population pressure – both human and livestock – is the most important. High population numbers can lead to misuse and mismanagement of land, small farm sizes and landless people. In some developing countries, development activities carried out improperly can lead to degradation of forests, soil erosion and mismanagement of water resources. Some of the important factors are discussed below:

Geology and geomorphology Gupta (1975) reported that the Himalayan rivers carry far more sediment and much coarser material than the non-Himalayan rivers. This is partly because the Himalayas are young, still rising, and have very steep and long slopes. The shales, phyllites and other sedimentary rocks are more erodible than the igneous rocks. In addition, the shales and phyllites found there are more subject to landslips and slides. Laban (1979) reported that only 26 per cent of the landslides in the Nepalese Himalayas were due to human activities. Brundsen *et al.,* (1981) while working in the middle Himalayas of eastern Nepal, stated that landslides should be considered as a normal rather than exceptional process in that area. The phyllites, shales and schists found there showed no significant differences in degree or type of mass wasting, while the deeply weathered gneisses seemed considerably less subject to landslides but much more prone to piping and gully formation.

Extreme precipitation events, sometimes concurrent with earthquakes, often trigger large-scale, very extensive slope failures. The most recent large landslides occurred in the

Table 7. *Area of shifting cultivation and number of shifting cultivator households, Indonesia.*

Island Group	Area of Shifting Cultivation (ha)*	No. of Shifting Cultivator households**	Area (ha) Available per Household***	Shifting Cultivation Cycle (Years)# (Potential)
Sumatra	3,428,600	313,593	11	22
W. Nusa Tenggara	63,600	12,960	5	10
NIT & TT	422,100	273,033	1.5	3
Kalimantan	5,457,400	228,470	24	48
Sulawesi	526,900	243,570	2	4
Maluku	216,400	18,704	11.5	23
Irian Java	1,287,300	119,640	11	22
Total	11,402,300	1,209,970	9	18

Source: Ingram *et al.*, 1989.

* There is no shifting cultivation in Java and Bali.

** Five persons per household.

*** At the rate of 1 ha per household.

\# Assuming that land is cultivated for 2 years.

Darjeeling district of West Bengal in 1969. Tejwani *et al.,* (1969) investigated this phenomenon and attributed it to very heavy continuous rainfall in September when the soil was already fully saturated with water, as well as to the very long and steep slopes. There was no difference in the failure of slopes between the forested and non-forested areas. Starkel (1972), investigating the same phenomenon three years later, found that many new slides had been initiated and old slides re-activated, resulting in ten times more than the average yearly rates of erosion. The same phenomenon has been observed on Ambon Island in Indonesia. Geologically, the African mountains are more stable and have gentler and shorter slopes as compared to the Himalayas, hence erosion processes are also less active there.

Misuse of land Misuse and mismanagement of land in developing countries are generally directly related to human and livestock pressures. Traditional farming systems (e.g., shifting cultivation, home gardens and water harvesting techniques) can be sustainable and conservation-oriented, given a reasonable balance between the available natural resources and the population densities. However, when population pressures upset this balance, land degradation is the result.

Shifting cultivation Shifting cultivation is perhaps the most misunderstood land-use practice. However, a closer scrutiny reveals that, considering the hilly landscape, the high prevalent rainfall, the low inherent soil fertility, and with the most of the plant nutrients being tied up in the forest, shifting cultivation (or slash and burn) has been the most appropriate response of the humid tropic farmers of Africa, South and South East Asia. As long as the fallow period is long enough to allow the woody vegetation to return, thus restoring productivity, and killing the weeds, the system is sustainable, the soil is conserved and the hydrological behaviour of watersheds is normal (Nye & Greenland, 1960; Hamilton with King, 1983; Tejwani, 1988a, b). However, with short fallow periods in which the woody vegetation is not allowed to return, the same system completely breaks down in terms of decline in crop yields, degradation of the production base, accelerated surface runoff and soil erosion. It is reported from humid tropical northeast India that if shifting cultivation is practiced over a cycle of 10–15 years, it has both protective and productive benefits. Herbaceous species dominate during the first five years of the fallow; these are then replaced by bamboo (*Dendrocalamus hamiltonii*) which progressively increases in a 20-year fallow period (Ramakrishnan & Toky, 1978). However, in this area, where as much as 2.7 million ha are under shifting cultivation – 16.8 per cent being cultivated at one point in time – the cycle had shrunk to five to seven years by 1974. In earlier decades, the cycle had been 30–40 years (NEC, 1982).

The early stage of shifting cultivation is the most critical time for accelerated runoff and soil erosion; annual soil losses are reported to vary from 3.3 to 201 tonnes ha^{-1} (Singh &

Table 8. *Distribution of land in shifting cultivation in Indonesia, by land slope class.*

Land slope, class (per cent)	Per cent of all land in shifting cultivation	
	In Class	Cumulative
0 – 8	31.0	31.0
9 – 15	17.4	48.4
16 – 25	23.1	71.5
26 – 40	4.1	75.6
41 – 60	15.4	91.0
over 60	9.0	100.0

Source: Ingram *et al.*, 1989.

Singh, 1981). An analysis of the data on shifting cultivation in Indonesia (Table 7) indicates that the average potential shifting cultivation cycle in Kalimantan is 48 years; Sumatra, Maluku and Irian Jaya have a potential cycle of 22 years; East Nusa Tenggara and Sulawesi have the shortest cycles (three to four years). If the statistics are correct, the adverse impacts of short cycles in Sulawesi and East Nusa Tenggara could be very serious (Tejwani, 1990). The distribution of shifting cultivation among various classes of slope in Indonesia is shown in Table 8 (Ingram *et al.,* 1989). It is obvious that the shifting cultivators exercise judgment in selecting their sites. As much as 71.5 per cent of the area is accounted for by the three slope classes, up to 25 per cent. It should be remembered that shifting cultivation degrades the forest but settled agriculture destroys the forest. Therefore, the latter causes the maximum hydrologic and erosional disturbance.

Settled agriculture Strictly speaking, agricultural land-use should be based on the scientifically determined capability class of the land. However, when there is population pressure, there is great incentive for people to misuse the land. To take an example from India, it is generally believed that hills are less densely populated than the plains. However, the estimated population density per square kilometre of cultivated land is 1432 persons in the Himalayas as compared to 483 for whole of India. At the regional level, the population density in the Uttar Pradesh Himalayas is 94, which becomes 547 persons per square kilometre of cultivated land (Tejwani,1984a). The Nizam Sagar catchment on the Deccan Plateau, with a catchment of 20 950 ha, had a population density of 191 for the total area and 254 for the cultivated area.

In Indonesia, Java and Bali Islands carry a population density of 758 and 478 persons per sq. km. While the farmers have developed excellent farming systems on sloping uplands and valleys, severe watershed degradation still occurs on rainfed agricultural lands in the uplands. Even with a relatively low population density (69 persons per sq. km.), Sumatra experiences intense localized pressures as in the catchment of Lake Toba on Samosir Island. Here the people cultivate very steep slopes for arable crops. In fact, they scrape the topsoil from the uphill side to make beds for vegetables on the immediate downhill side. The whole catchment is full of severely degraded landscapes which have been abandoned.

A comparative study of land capability classes and actual land-use in two small watersheds in Himalaya showed, (a), that there was no land with capability classes I or II in either of the drainage basins and, (b), that there was only 2.0 per cent and 25.0 per cent of land in the combined capability classes III & IV in the Fakot and Kafra Bhaura watersheds, respectively. However, as much as 21.4 per cent and 50.8 per cent of land was actually under cultivation in these watersheds, respectively. Thus a large amount of unsuitable land was being used for agriculture. It is obvious that once crops are grown on land unsuited for agriculture, the land available for forests and other non-agricultural uses is reduced. In the case of the Fakot and Kafra Bhaura watersheds, 98 per cent and 75 per cent of the land, respectively should be under forest, pasture and other non-agricultural uses. However, only 35.8 per cent and 15.1 per cent, respectively, is under forest (CSWCRT, 1980; GBUAT, 1982).

This situation is true of all densely-populated hilly regions, especially in India, Nepal, Sri Lanka, and Indonesia. In Nepal, forested land decreased from 34.2 per cent to 29.1 per cent, between 1975 and 1980. Agricultural land has increased from 16.5 per cent to 22.2 per cent during the same period. Land-use statistics in some large watersheds (20 950 ha to 681 000 ha) in India indicated that the proportion of cultivated land in the watersheds located on the Deccan Plateau and the Vindhyan Plateau was higher (53.4 to 80.2 per cent) than that in the Himalayan watersheds (21.7 to 35.0 per cent). Also, forested land-use is less in the Deccan Plateau watersheds (3.8 to 21.3 per cent) than in the Himalayan drainage basins (48.5 per cent). To some extent this reflects the relatively greater availability of land suited for agriculture on the Deccan Plateau (AFC, 1988a, b and c; ASCI, 1987).

The 31 river valley projects in India, having a combined area of 78 574 000 ha, have an average of 62 per cent of land under agriculture, 20 per cent under forest and 18 per cent in other land uses – including grassland habitat and water bodies (Das *et al.,* 1981).

Costa Rica lost 75 per cent of its primary forest between 1940 and 1983. Eighty-two per cent had been tropical forest, 68 per cent had been premontane, 36 per cent had been lower montane and 30 per cent had been mountain forest (Sader & Joyce, 1988).

Mismanagement of land Mismanagement of land occurs either because of a generally high population of humans and livestock in the entire area or because of local pressure of high human population (e.g., urban areas or large villages) or a high cattle population with a low human population.

Small farm sizes When a high population exists in an area, farm holdings will be small. Such small farms are handicapped from the start, since they are subsistent in nature and their economically-weak owners cannot afford to use many resource inputs. In addition, the originally small holdings often are further fragmented. For example in the Himalayas, 70 per cent of the farm holdings are less than one hectare. Only 13 per cent are larger than two hectares. These fragmented small holdings are at many locations. At each location there is more than one field. Each location can have as many as four fields (Shah, 1981a; Ghildyal, 1981).

Because of the very high pressure of its human population, India has a unique problem of landless labour. This leads to a tendency to encroach on land. Also, it has been the policy of the government to distribute land – either belonging to the government or taken from land holders under land reform – at the rate of one ha per family. This ostensibly desirable policy is extremely harmful in practice, since the land which is distributed is invariably unfit for cultivation and already degraded. It is not clear how much more erosion occurs due to cultivation of this type of land but presumably, when it is cleared of the bush and grass which normally covers it, the runoff rates are considerably increased.

Forest land. Forest land in much of the humid tropics is also grossly mismanaged. However, it should be noted that not all land classed as forest is necessarily forested. For example, in the Uttar Pradesh Himalayas, only 60 per cent of the land under the control of the forest department in the hills is forested while forty per cent of this forested area is inferior or degraded. It is reported that the small Fakot watershed (369 ha) in the Uttar Pradesh Himalayas had 133 ha of forest land, of which not a single ha was under dense forest As many as 34 ha (26 per cent) had no canopy, and 16 ha (12 per cent) had thin forest (CSWCRT, 1980). At the macroscale, about 40 per cent of the forest in the entire Uttar Pradesh Himalayas was degraded.

Similar situations exist in parts of Indonesia and other tropical developing countries. For example, Indonesia's Java Island, with the greatest human population density (758 persons per sq. km.), has only 15.9 per cent of land in forest, while Irian Jaya and Kalimantan, which have low population densities (three and 14 persons per sq. km., respectively) have forest land cover of 98.8 per cent and 82.0 per cent, respectively (MOF, 1989).

It is generally believed that people encroach on government forests and degrade them. This is only partly true. The government itself is a party to the over-exploitation of forests. Gadgil (1989) reported that in India's Karnataka State, the total harvest of bamboo (160 000 tonnes) exceeded the annual increment to the stock by 30 000 tonnes. Field surveys showed that the government's forest resource survey consistently overestimated the existing bamboo stocks. It was also reported that the contractors supplying bamboo to the paper mills rarely adhered to the silvicultural prescriptions of the government. In fact, the suppliers of the public sector paper mill were in greater violation of the silvicultural norms than the suppliers of the private paper mills. Such examples of governmental mismanagement can be cited in many countries in Asia and Africa.

Agroforestry practices Agroforestry practices are old, traditional and very extensively practiced all over the subhumid tropics, e.g., shifting cultivation, taungya, growing tea and coffee under the shade of trees, intercropping under coconut, forest or Kandy gardens, multipurpose tree species on farmlands, and many silvo-pastoral systems. The only new element is the use of the term agroforestry to describe these practices. Not only have these systems been widely practiced for a long time, but some have also been widely studied, e.g., the growing of tea under shade since the latter part of the Nineteenth Century using groundwater, and the fodder-fuel plantations since the 1950s (Tejwani, 1987).

However, information on the impact of these practices on water disposal and soil erosion is very scanty. Wiersum (1984) has reviewed the impact of various tropical agroforestry systems on surface soil erosion in Indonesia. In advocating the use of agroforestry for erosion control, reference is usually made to the well-known positive effects of trees in controlling erosion. However, it is dangerous to generalize the erosion control potential of agroforestry because (i) many different agroforestry systems are possible within a continuum from almost pure forest at one extreme and almost pure agriculture at the other, and (ii) several management practices favoring erosion, such as tillage, harvesting, grazing, and road construction may take place in a specific agroforestry management system. A study of erosion under various tropical moist forest and tree crop systems led to the conclusion that individual trees cannot be expected to exert the same positive influence as a well-functioning undisturbed forest ecosystem (Wiersum, 1984). However, by virtue of the presence of trees, any agroforestry practice/system would have more protective function than any purely agricultural system. With respect to silvo-pastoral systems, results are available from two almost identical in size watersheds, each 32 ha in area. The first watershed consisted of grass (66 per cent) and shola forest (33 per cent). The second included a plantation of blue gum (leaving 33 per cent area under grass and 33 per cent under shola forest). The total water yield (runoff and base flow) reduced by 16 per cent in the latter watershed (an average 87 mm year^{-1}) during a period of ten years of plantation (i.e., one rotation cycle). The same plantation of blue gum (*Eucalyptus globuluus*) did not generate any extra sediment. This confirms the view that trees reduce runoff as compared to grass (CSWCRTI, 1987). Conversely, if the forest cover is removed and replaced by agricultural or tree crops, the water yield increases. It is reported that in the humid tropical climate of Malaysia, when the forest was removed and the land planted with cocoa and oil palm separately in two previously calibrated catchments, there was a significant increase in the yield of water in both catchments (Nik, 1988). Similar results are reported from Tanzania (Edwards, 1979) and Java in Indonesia (Bruijnzeel, 1986).

Common lands Common lands are provided to communities to meet their grazing, timber and fuelwood needs. The condition of these common lands in developing countries with high population densities – for example India and parts of Indonesia – almost without exception is deplorable. The degree of degradation is determined by the pressure of population, both human and livestock. Any forest land allotted either to civil authorities or village communities contain little if any forest. These as well as other grazing lands do not grow sufficient grass to support the livestock in India, Pakistan, Bangladesh, Nepal and parts of Indonesia (Sumatra). These lands are highly degraded and generally infested with obnoxious shrubs (for example, *Lantana camara*, *Berberis lysium*, *Rhuscotinus*, *Euphorbia royleana* at an elevation of 15 meters in the Tehri district of Uttar Pradesh in India). The grass cover also contains a much greater proportion of annuals than perennials. For example, in the Fakot watershed in Uttar Pradesh in India, *Lantana camara* was recorded as covering 33 per cent of the grazing lands. Grass yields were obviously low. In the watershed, yield of air-dry grass from one square metre quadrants was 24.2 gm in June (i.e., before the monsoon) and 312.0 gm in September (i.e., after the monsoon). An area of 157 ha of poor quality grassland with scrub is expected to support 529 cattle units (CSWCRTI, 1980). Being open access lands, they are no one's responsibility. The people tend to encroach on them and appropriate as much as they can. The encroachment itself creates additional pressures and intensifies the degradation. On the other hand, there are some examples of true communally-owned and operated lands, where rules of management are enforced in a social contract. Here the resource may be a well-managed common property resource.

Development activities Reservoirs and road construction and quarrying and mining activities are essential for economic development. However, if these are undertaken in an unecological manner, as is being currently done in most developing countries, they lead to deforestation, soil erosion and land degradation on an extensive scale. Thus, what are supposed to be development opportunities also turn out to be "development disasters." For example, the extensive network of roads in India's Himalayan region is estimated to generate 1.99 tonnes of sediment per metre length of road per annum (Narvana & Rambabu, 1983). Henderson & Rouysungnern (1984) reported that in Thailand, as a result of improper road construction, the sediment increased by 290 per cent per year. Paths, roads and house construction in a village settlement (40 huts, 240 people) accelerated the total sediment load 36 times over that from an evergreen hill forest in Thailand (Chunkao et al.,.1983). It is observed that dirt and gravel roads, which constitute more

than one-half the length of the whole road system in Indonesia, cause very severe erosion, damaging the land and property and creating gullies. The runoff and sediment generated by roadside erosion generally immediately ends up in a nearby stream (Tejwani, 1990).

In developing countries, no measures are taken to stabilize and protect mined areas. Chunkao et al., (1983) reported that tin mining in three watersheds in southern Thailand caused a high annual erosion rate of 98 tonnes ha^{-1}. Van der Meer (1981) observed that in Thailand, mining may cause more erosion than any other form of land-use.

Even in shifting cultivated areas – either with a high population density (e.g., in northeast India) or with low population density (e.g., in Sierra Leone, Tanzania, and Indonesia) – the sediment generated and the damage done to the landscape by gullying associated with roads in the upland areas is much greater than what the shifting cultivation itself creates (Tejwani, 1988 a, b; 1990).

The issues are not primarily technological but socio-economic and political. Also, the efficient functioning of roads (or mines or reservoirs), however scientifically and efficiently constructed and maintained, is related to how the surrounding lands are used. The best of roads can be easily damaged or destroyed with consequent cyclical soil erosion, high runoff and ecological disturbances, if the surrounding forest/grazing/agricultural land is not properly used.

Contrasting examples The preceding paragraphs dealing with misuse and mismanagement of land and the associated development activities, emphasize the accelerated soil erosion and hydrological disturbances due primarily to high population pressures. This section presents the contrasting case of how the land-water-vegetation system functions under low population pressure. Examples are cited from the Kigoma region of Tanzania in East Africa (south of the equator), Sierra Leone in West Africa, (north of the equator) and Kalimantan in Indonesia. Sierra Leone and the Kigoma region have a tropical rain climate, with one assured long wet season and one dry season. The annual rainfall varies from 2,000 to 3,000 mm in Sierra Leone and from 800 to 1,600 mm in the Kigoma region. The climate of Kalimantan, which straddles the equator, is tropical and typically monsoon in type. The average annual temperature is 27.5°C, with a mean daily maximum of 33°C and a mean daily minimum of 22.8°C. The average humidity is over 80 per cent, with the average annual rainfall being 2,694 mm at Banjarmann. In each of these regions, the assured rainfall, high humidity and moderately high to moderate temperatures provide excellent conditions for the growth of crops, trees and other vegetation.

Soils in Sierra Leone, the Kigoma region and Kalimantan are inherently of low erodibility. However, when associated with steep slopes and when devoid of vegetative cover, these soils also will be subject to high erosion under the given rainfall conditions. Topography is variable. The original vegeta-

Table 9. *Population and land use in Sierra Leone, the Kigoma region (Tanzania), and Kalimantan (Indonesia).*

	Sierra Leone	Kigoma Region (Tanzania)	Kalimantan (Indonesia)
Land area (sq. km.)	72,300	37,037	539,460
Water-Lake (sq. km.)	–	8,029	–
Population			
1974 (SIL)/1978 (Kigoma) 1961 (Kalimantan)	2,735,159	648,450	4,100,000
1985 (SIL)/1988 (Kigoma) 1985 (Kalimantan)	3,515,812	913,811	7,740,000
Population density (persons/sq. km.) 1985 (SIL)/1988 (Kigoma) 1988 (Kalimantan)	49	25	14
Land use (sq. km.)			
Cultivated area	5,440 (7.5)	1,336 (3.6)	14,707 [1] (2.7)
Reserve forest	2,850 (4.0)	8,839 (23.9)	508,556 [2] (94.3)
Other Land	63,724 [3] (88.1)	26,862 [4] (72.5)	16,197 [5] (3.0)
Mangrove and Swamps	286 (0.4)	–	–
Livestock (cattle units)	308,600	103,200	380,600

Source: Tejwani: 1988 a; 1988 b; 1990.

Figures in parentheses refer to percentage values.

[1]   Includes tree crops.

[2]   Includes shifting cultivation (54,574 sq. km.), shrubs (43,416 sq. km.) and grassland (14,368 sq. km.); the remaining (396,198 sq. km.) is in forest cover.

[3]   Land use statistics in Sierra Leone are confusing. This includes mostly forest land in private ownership and subject to shifting cultivation which is 80 per cent of agriculture.

[4]   This includes forest and pasture land kept apart for public use, as well as water bodies but excluding lake Tanganykia.

[5]   Includes human settlements, water bodies, unvegetated areas and unclassified.

tion of Sierra Leone was probably moist high forest except in the extreme north where drier and less humid conditions would have given rise to deciduous woodland. Due to human activities, this landscape has now been changed to a secondary high forest and derived savanna woodland resulting from shifting cultivation (called bush fallow) in the upland areas of the country. In the Kigoma region, the forest is the *miombo* woodland of Africa. The vegetation in Kalimantan is moist tropical forest, and mangrove where conditions permit.

As shown in Table 9, the area cultivated is very small (7.5 per cent in Sierra Leone, 3.6 per cent in the Kigoma region, and 2.7 per cent in Kalimantan). Most of the land is covered with forests, and the population pressure (48, 25 and 14 per-

sons per sq. kilometre in Sierra Leone, the Kigoma region, and Kalimantan, respectively) is low by Asian standards.

About 80 per cent of the agricultural land in Sierra Leone is in the uplands – under traditional shifting (bush fallow) cultivation. Operations are manual, using unsophisticated tools. When clearing the land, the farmers leave the naturally-growing oil palm trees untouched. They do not burn the trees but harvest the fuelwood and sell it. Although by African standards the pressure of population is high, the bush fallow cycle still averages 7.8 years. Because the land is not fully cleared of vegetative cover – leaving the valuable oil palm trees – and because of very fast regeneration of vegetation, the runoff and erosion losses are not serious. However, there is no doubt that the forest is degraded.

In the Kigoma region of Tanzania, there are two farming systems: (1) shifting cultivation with or without livestock and (2) subsistence agriculture, including cash crop farming (mainly cotton, and to a very limited extent tobacco) and agroforestry. It is reported that shifting cultivation is on its way out. Even where practiced, it is limited by the capability of the farmer to cultivate a piece of land. Also, the grass growth is so dominant and vigorous that uncontrolled runoff or accelerated erosion occur only during the early stages of land preparation and crop growth. Subsistence agriculture is the main occupation of the people in the Kigoma region, largely in the form of agroforestry. The main food crops are cassava, maize, beans and vegetables. The cash crops of cotton and tobacco are grown in the lowlands, and coffee in the highlands. However, all farms have mango trees and bananas where climatic conditions permit. Oil palm and coconut are also planted on the farms, especially near the lake zone. *Cassia seamia* trees are frequent and eucalyptus trees are also occasionally grown.

Given this situation, the potential for accelerated erosion and hydrologic disturbance due to agricultural practices is currently low. The extensive forest on the "other land" category is used primarily for fuelwood and charcoal production as well as to a limited extent in the salt industry, tobacco and fish curing, lime making and house construction. With 40 to 50 per cent of area under forest/woodland and with the very low population density in the Kigoma region, there would be no cause to worry, if the woodland were well managed. Unfortunately this is not the case. The forest reserve has been invaded for establishment of villages; and to that extent, it is exposed to degradation. The main cause of forest degradation, both in the forest reserve, as well as in the public lands, is fire. Reported reasons for setting fire to woodlands include: (a) hunting for wildlife and gathering of honey, (b) burning to induce grass growth for livestock fodder, (c) shifting cultivation, and (d) just for the "fun" of it or "pyromania." Since livestock pressure is low and shifting cultivation is reportedly on the decline, the remaining two causes are the apparent source of the very prevalent, wide-scale forest fires. Charcoal makers also are a contributing cause – primarily due to their negligence.

The adverse effects of human-made uncontrolled forest fires are (a) destruction of fauna and flora, (b) damaged and degraded forest, and (c) accelerated soil erosion and hydrologic disturbance. While the first two impacts invariably arise, the third effect does not necessarily occur as the land is quickly covered with grass and bushes.

In Kalimantan, Indonesia, the population pressure is very low (14 persons per sq. mi.). The total cultivated area is only 2.7 per cent of the land area. Of the cultivated area, 1.7 per cent is in wetland agriculture and 1 per cent is in tree crops. There is no settled upland agriculture. The potential shifting cultivation cycle (assuming five persons per household and cropping for two years) is 48 years (Tejwani, 1990). With only 380 600 cattle units, the cattle density is only 0.7 cattle units per sq. km. Almost 95 per cent of the area is under forest and other natural vegetative cover. Extensive fires occur in the *Imperata cylindrica* dominated grasslands.

The road networks are rather small in Sierra Leone, the Kigoma region of Tanzania, and Kalimantan (Indonesia). Small gullies develop where there is slope. However, since there is little population pressure on land, there is excellent vegetative/grass cover which comes up to the edges of the roads. Accordingly, road drainage systems need only to handle the runoff generated by the roads; consequently, the roads do not generate much runoff or sediment.

Impacts of water mismanagement and soil erosion
The impacts of watershed degradation are very pervasive and pernicious. Watershed degradation adversely affects the functioning of the natural ecosystem (ecological impacts), the production base, goods and services (economic impacts) and the well-being of the people (social impacts). In evaluating the impacts of watershed degradation in this chapter, all of these aspects are taken into account.

Loss of the water resource With the accelerated rate of runoff due to soil compaction, loss of topsoil and organic matter, and overall degradation of the watershed, there is less opportunity for the rainwater to infiltrate in an eroded soil and recharge the groundwater resources. Consequently, two situations arise. First, the period of base flow of streams is reduced, and second, the groundwater table falls. In the first situation, large amounts of runoff are available over a relatively shorter period of the total time of the flow of the stream. In the second case, water either is available for a shorter period or greater effort is needed to pump or fetch it. This phenomenon is a fact of life all over the world – in India, Nepal, and Africa. This also explains that when a watershed is rehabilitated by providing vegetative cover and soil conservation measures, the dry streams and springs are often regenerated, the water appears further down the stream and therefore flows for a longer period(see Bonell with Balek, this volume, for scientific issues related to this problem).

Floods At present, the statement that "deforestation causes flooding" is questionable as this conclusion is not supported by the hydrological behavior of forested and deforested watersheds. However, in developing countries, when a statement is made that "deforestation causes floods," what, in fact, is implied is a whole chain of activities, i.e., forest degradation and/or deforestation, overgrazing, and/or opening up of the land for agriculture contributes to flooding. Even forest hydrologists agree that deforestation will contribute to local flash-flooding (Hamilton with King, 1983). So one is certain that if a watershed degrades it must contribute to higher runoff in a shorter period and, therefore, to a local flash-flood. On the other hand, it is recognized that flood regulation by forests has physical limits that are from time to time overwhelmed by storms of sufficient size and duration.

Viewed from a macro-catchment level, floods are a part of the hydrologic behavior of the landscape. With abundant rainfall, they occur even in the least disturbed catchments. Other contributing factors for the intensification of floods are drainage congestion, heavy local rainfall, synchronization of upland floods with high tides, a backing-up of water in the tributaries at their outfalls into the main rivers and cyclones. In recent times, the degradation of the upland watersheds leading to gullying and the high rates of peak runoff, as well as raising of the beds of streams from sedimentation also are considered to be contributing factors. For example, interpretation of satellite imagery has clearly shown that some of the Himalayan torrents and streams have widened by 106 per cent and rivers by over 36 per cent over a period of seven years (Gupta, 1981). Another classical example is in the Hoshiarpur district of Punjab where areas under hill torrents increased from 19 282 ha in 1852 to 60 000 ha in 1936 (Jacks & Whyte, 1938; Tejwani, 1980).

River sedimentation problems are acute in Asia. The rivers reaching Bangladesh carry 2.4 billion tonnes of sediments per year, raising the beds of rivers, changing their courses and creating serious navigation problems (Ahmed, 1982). This is a result of natural processes aggravated by land degradation. In Nepal, the bed levels of the Terai Rivers are rising 15–30 cm annually. The Kosi River carries an annual load of 119 million tonnes of silt which is equivalent to two mm of topsoil depth over its entire catchment (FAO, 1982, 1986; Tejwani, 1984). Under high population pressure in the plains, even those floods occurring because of natural causes receive considerable attention due to the incursion of people in the flood zones (e.g., India, Bangladesh, and Ambon in Indonesia). However, in sparsely populated areas, e.g., Sierra Leone, the Kigoma region of Tanzania or Kalimantan in Indonesia, even the high floods pass unnoticed. The rivers and streams are covered with grass and trees up to the river banks and there is no hydrological disturbance, either in the catchments or along the stream banks.

Sedimentation If there is any single environmental, economic and social problem which affects the life of the people directly and constantly it is the quality and quantity of water. In the developing countries with a high population density, sediment is the biggest pollutant of water. Further, the quality and quantity of drinking water is poor for a vast majority of the people, whether or not a country is sparsely or densely populated.

By the year 1984–85, India had constructed 117 major and 869 medium-sized irrigation projects with an investment of Indian Rs 150.3 billion, creating a potential to irrigate 30.5 million ha and a hydro-power capacity of 19 855 MW. All the reservoirs are designed to receive sediments and to lose capacity. However, the reservoir sedimentation situation in case of India (and many other developing countries of the world) is very disturbing. For example, sedimentation studies of 21 major reservoirs in India (Gupta, 1980) have shown that the annual rates of siltation from a unit catchment is 40 to 2,166 per cent more than was assumed at the time of the project was designed. (It was lower only in the case of one reservoir). Using the average of 21 reservoirs, the actual sediment inflow has been about 200 per cent more than the design inflow. Nizamsagar Reservoir, which is the oldest (1931) of those studied, had lost 52.1 per cent of its capacity by 1967 (CBIP, 1981). In case of Sukhna Lake, 60 per cent of its storage was lost in a period of 15 years. The sediment was deposited at the rate of 150 tonnes year ha^{-1} of the catchment area (Mishra *et al.*, 1980).

Most of the existing reservoirs were planned with a provision for dead storage, with the designed live storage expected to be available for utilization throughout the projected life of the reservoirs. These assumptions have not been realized. Observations have shown that (i) the siltation is not confined to the dead storage pool only; and (ii) the quantum of siltation in the live storage pool is equal to or more than that in the dead storage (CBIP, 1981; Sinha, 1984a). The encroachment on the live storage has many serious consequences. The life of the reservoir is reduced and the site of the dam is lost. It also may not be possible to have another site; and the current irrigation potential and power capacity are reduced, resulting in a direct loss of food and industrial production. Encroachment on live storage also affects some important aspects of design such as the economic aspects of fixing the dead storage level, the outlet sill level, and the opening of the penstock (Murthy, 1981).

In Burma, siltation has reduced the availability of water behind Lowpita Dam (FAO, 1982); Kinda Dam is also suffering from increased siltation (FAO, 1986). Similar problems are reported from reservoirs all over the world. The life expectancy of Ambuklao Reservoir in the Philippines is reported to have been reduced from 62 years to 32 years (Weidelt, 1975). Pantabagan Reservoir in the Philippines, which impounds one of the largest artificial lakes in southeastern Asia, is threatened with a shortened life (PCARR, 1977). Kisongo Reservoir, which was built in Tanzania in 1960, had already reached its useful life in 1975 and was expected to be completely silted in by 1983 (Murray-Rust, 1972). The Anchicaya Reservoir in Colombia, which was predicted to have a long life, had already lost 25 per cent of its capacity two years after its completion in 1957 (Eckholm, 1976). The actual sedimentation rates observed for Sutami and Wonogiri Reservoirs in Indonesia were reported to be an average of 577 per cent greater than the design inflow (Sukaritko, 1986). Similar problems are reported for Bhumipol Reservoir in Thailand (FAO, 1986).

Sediment-laden waters and water uses:

Irrigation As stated above, by reducing reservoir capacity, sedimentation reduces the area which could otherwise be irrigated and impairs the functioning of existing irrigation facilities. The canals, their distributaries and water courses become silted up, making it essential to desilt them frequently. As a result, the economy and efficiency of the whole irrigation system are adversely affected. The malfunctioning of the Kosi canal system in India is a good example of this problem (Sinha, 1984 b). One of the tributaries of the Kosi River carries a very high sediment load of 60.76 ha m per 100 sq. km (Editor's Note: 1 hectare metre = 10 000 m³). A diversion barrage was constructed on the Kosi River, along with embankments on both sides of the river. A canal system was added to provide irrigation water to the area thus protected from flood damage. The scheme has been a success to a great extent as far as flood protection is concerned, but the development of the irrigation potential has remained low. A principal reason for the low utilization of the irrigation potential is the very heavy sediment load being transported to the Kosi canal system. Observations on sediment deposited in the main Kosi canal have shown that in the early years of the operation of this project (during the 1960s) the portion of the canal area encroached by sediment was as high as 20 per cent, resulting in a reduction of water conveyance capacity. Huge expenditures were incurred during the desilting of the canals. Thereafter, desilting of the canals became an annual activity, entailing heavy recurring costs. During the 20 years of canal operation, a number of canal siphons and ditches were filled up. Frequent closure of the canals has proven to be a serious constraint on the supply of irrigation water at times when it is most needed.

Power generation The adverse impacts of silt-laden reservoir waters on the generation of hydro-power show up in two ways: (i) the total volume of water available for power generation is reduced, hence less power is generated; and (ii) the silt-laden waters abrade the turbines, causing faster rate of wear and tear, frequent shut-downs, and higher maintenance costs.

Inland water transport Reduction in the depth of water or time of flow in rivers due to sedimentation results in restrictions or a complete loss of capacity to support inland water transport systems. Loss of such a system on the Ganges River in India is a classic example.

Fish production Inland water resources often have a considerable potential for fish production. No direct evidence of an adverse impact of heavy load of sediment in water on inland fish and fish breeding grounds has been reported. However, it is reasonable to conclude that siltation of tanks, streams and rivers creates conditions which adversely influence the breeding and multiplication of fresh water fish. Such conditions include reductions in depth and period of water flow, degradation of the quality of water with the high silt content and consequent variation in the oxygen content and the temperature of water. It is reported that sedimentation has impaired fish production in Laguna Lake in the Philippines (Gulcur, 1964).

Loss of soil fertility, rooting depth and land resource
It is now universally accepted that water erosion removes the most fertile part of the soil, which is rich in plant nutrients and organic matter. Not only does the soil fertility decline, but the condition of the soil also deteriorates. For example, a loss of organic matter leads to the breakdown of soil aggregates and a reduction in the water-holding capacity of the soil. Furthermore, as the soil is eroded, the soil depth decreases with a consequent loss of rooting depth for the crops and a reduction in water profile depth for crop growth. It is reported that in the Sholapur district of Maharashtra State in India, nearly 17 per cent of the land having medium soil depth (more than 45 cm) had deteriorated into shallow soils (less than 45 cm) over a 75 year period – from 1870 to 1945. Similarly in the Akola, Buldana and Yeotmal districts, the number of fields having less than a 37.5 cm soil depth increased during the same period by 54, 16 and 8 per cent, respectively. This implies that the soil has now lost some of its capacity to retain water, which would now appear as runoff and cause more erosion (Tejwani, 1980). It is estimated that in India 5.37 to 8.4 million tonnes of plant nutrients are lost every year due to soil erosion (MOA, 1985). Experiments have shown that when 2.5 cm of topsoil were removed artificially, maize grain and stalk yields each decreased by 14 per cent. When 7.5 cm of topsoil were removed, the maize grain and stalk yields were reduced by 33 and 27 per cent, respectively (Khybri *et al.*, 1981). Such decline in crop yields has been reported for maize from Malaysia: (Huat, 1974; Siew & Fatt, 1976); from Nigeria: (Lal, 1976, 1983); and from the USA by many sources.

PROGRAMMES AND POLICIES PROMOTING WATER MANAGEMENT

Historical perspective
There is a historical record, running back to the times before Christ, of many civilizations – such as the Egyptian, Indus, Mesopotamian, Chinese, and Babylonian – developing irrigation systems by building dams, tanks, and canals. The system of construction of ponds and tanks for irrigation and drinking water in India and Sri Lanka is centuries-old. There was a time when it was obligatory on the part of local rulers, large land holders or important men in villages to promote the construction of tanks and ponds.

Problems of sedimentation of tanks and ponds also received attention. Maintenance, repair and desilting of tanks and ponds was made a socio-religious responsibility in India. This responsibility used to be cooperative. Arabs are reported to have been working on building dikes to control sedimentation in the Euphrates River system (FAO, 1982).

In modern times, the British colonials took active steps to develop water resources and to control sedimentation and soil erosion. Sukkar Barrage on the Indus River (now in Pakistan) was constructed and commissioned in the early 1930s when the world was passing through an economic depression. A ravine reclamation project was undertaken in the 1920s in the United Provinces (now Uttar Pradesh) in India (Smythies, 1920). Soil conservation programmes in the dry farming areas of India were initiated in the early 1930s. In southern Rhodesia, in a period of eight years (1929–1936), 15,970 ha of agricultural land were treated with contour ridges (Jacks & Whyte, 1938).

Two other points should be noted regarding this modern colonial period. First, even though there was an awareness in a large number of humid tropical countries of these water problems, their causes, and their consequences, there were very few ongoing conservation programmes, except those mentioned above. Second, the components identified for the programmes were quite similar to those which are being recommended today. In terms of strategy, the components were legislation, propaganda, written or verbal demonstrations, advisory services and scientific services. In terms of techniques the components were for:

(a) Agricultural lands: Soil conservation by terracing, draining, contour planting and bunding.

(b) Other land uses: Protection of major catchment areas and drainage basins by means of forest reserves; preservation of natural vegetation on the poorer soils of the country; preservation of belts of forest and natural vegetation on the better lands to form wind-breaks; protection of stream banks and steep hillslopes under powers of the forest laws; maintenance of village forest areas which should normally occupy land unsuitable for (or least suited to) agriculture (Jacks & Whyte, 1938).

It may be noted that the lists of technical measures do not include reforestation, since in the 1930s, forests were still plentiful everywhere.

Programmes after World War II

After World War II, the impetus to development and conservation of water and land resources in the humid tropics was provided mostly by FAO and UNESCO. In the Asia-Pacific region, India, Indonesia, and Philippines have had extensive projects of development and conservation of land and water resources. For example, it is unique that within one year of independence, India passed the Damodar Valley Corporation Act in 1948, establishing an interstate water resource development system to moderate floods, to undertake integrated watershed development in the catchment of the Damodar River, and provide for irrigation, power generation and navigation. India also embarked on an ambitious programme of developing its irrigation potential and conserving its production base with planned development, starting with the first five-year plan (1951–56). The figures for water resource development, the creation of irrigation potential and hydro-power generation have been reported earlier. With respect to soil conservation, reforestation, afforestation and watershed management, there is a great variety of programmes and practices. Apart from bunding and terracing of agricultural lands as done in earlier years, there is now an integrated multi-treatment on a watershed basis. On non-agricultural lands, tree plantation programmes are undertaken. Special attention is paid to pro-grammes of nala (stream) bunding (so as to detain sediment and develop water resources), checkdams, gully plugging, land shaping, percolation tanks, water harvesting (i.e., constructing ponds), and water conveyance. Starting with the third five-year plan (1961–66), soil conservation in the river valley projects was started with the specific objective of protecting the developed water resources from sediment pollution. This programme is being implemented now in 31 river valley projects, with a total area 78.5 million ha (Das et al., 1981).

As of 1986–87, 20.8 million ha of this total area have been identified as priority watersheds and 2.13 million ha have been treated (MOA, 1988). The term "priority watershed" means that the watershed is in urgent need of treatment, either to reduce sediment where water yield is important or to reduce peak runoff where flood control is important. This delineation was important as it was clear that whole catchments of all reservoirs could not be treated at once or even in the long-term. Specific techniques were developed for making these delineations (Bali & Karale, 1973; Karale et al., 1975). For India as a whole, as of 1987–88, a total land area of 32.09 million ha had been treated at a total cost of 17.4 billion Indian Rs.

Up to 1985, Indonesia and the Philippines had implemented watershed management programmes on 1.0 million ha each. Thailand had done the same on 0.5 million ha (FAO, 1986). However, compared to the extent of these problems in humid tropical countries of the Asian-Pacific region, these achievements are not very significant (Table 10). In other regions of the world, the proportion of water management programme implementation to their number of problems is even poorer. FAO (1985) has reviewed the status of soil conservation in Africa. In tropical Africa, some watershed management programmes are under way in Ethiopia, Lesotho and Kenya – countries with high human and livestock pressures.

Governments of developing countries have varying degrees of concern about water resource development and the sustainability of the land-water-vegetation systems. Subject to the availability of resources, some governments are willing to invest in and undertake programmes related to soil conservation, afforestation and flood control, in addition to pro-grammes of agricultural development, food production, and irrigation development. In addition to the resources invested by the governments themselves, some funds are made available as grants and "soft" loans by international agencies (e.g., UNDP, FAO, UNEP, IFAD and UNESCO), bilateral aid agencies (AIDAB, CIDA, SATA, SIDA, GTZ, JICA and USAID), banking institutions (IBRD and ADB) and non-government organizations. The grants and loans provided by the international agencies are very small in relation to the dimensions of the problem world-wide and within each country. However, as mentioned above, these grants and loans are related to the overall development programmes of each country and the priority each country attaches to such conservation programmes. However, this problem has been now recognized and a very

Table 10. *Area of degraded land - treated and planned for treatment - in selected humid tropical countries in Asia and the Pacific.*

| Country | Total | Land Area, ('000 ha) | | |
		Total Degraded	Treated up to 1985	Planned for Treatment (1986-1990)
Bangladesh	114,400	989	138	60
Burma	67,655	210	16	26
Cook Islands	241	NA	NA	100
Fiji	1,827	NA	45	55
India	328,759	173,084	30,000	1,000
Indonesia	190,457	43,000	1,000	3,100
Laos	23,680	8,130	11	NA
Malaysia	32,975	NA	26	96
Philippines	30,000	5,000	1,000	165
Sri Lanka	6,561	656	112	35
Thailand	51,400	10,000	560	675
Tonga	70	3	NA	1
W. Samoa	286	32	NA	1

NA = data not available

Source: Adapted from FAO, 1986.

ambitious programme to invest about US$5,320 million in tropical forestry in the next five years has been developed by FAO, UNDP, WB and WRI (FAO, 1987b). This action plan has four major components: fuelwood and agroforestry ($1,899 million); land-use on upland watersheds ($1,231 million); forest management for industrial uses ($1,640 million); and conservation of tropical forest eco-systems ($548 million). The plan recognizes Ethiopia, Kenya, Madagascar, Zimbabwe, India, Indonesia, the Philippines, Brazil, Colombia, Ecuador, Jamaica, Panama and Peru as needing large-scale investments in land-use management on upland watersheds. About 20 per cent of this investment would be allocated to research, training, education and extension. If this plan is well-implemented, it should pave the way for still higher investments and lead not only to the conservation and better management of tropical forests but also to conservation and better management of other lands and water resources.

Demonstrated benefits

Water resource development for irrigation receives the first attention of any developing country having a high population density and an actual or potential acute shortage of food. However, the funds are never enough to protect the watershed which yields the water. Therefore, if any country decides to invest in programmes such as afforestation, soil conservation and watershed management, which give only long-range benefits and social benefits, it is pertinent to understand the direct and indirect benefits which occur. Benefits of soil conservation and watershed management have been demonstrated for indi-

vidual practices and integrated management at various scales e.g. river basins, reservoirs, catchments and small watersheds.

The WRI, WB and UNDP (WRI, 1985) entities recognize a number of success stories in watershed management in high rainfall areas (e.g., India, Nepal and China) and medium to low rainfall areas (e.g., China, Uganda, and Ethiopia). The success story of the Damodar Valley Corporation (DVC) at the river basin scale in India, which has all the components of water resource development for irrigation, hydro-power generation, flood control, inland navigation and consequent activities of growth in agriculture, industry and mining, is worth recalling in the context of an experience of a developing country. DVC was modelled on an earlier success story of the Tennessee Valley Authority (TVA) in the USA. The DVC project has moderated the floods, provided power for the expansion of mining (coal, iron, copper and aluminum), and for growth of large complexes of manufacturing and support industries. The water resource development is used for irrigation (261 000 ha), domestic and industrial purposes. Thus the Damodar River Basin has been transformed into a valley of prosperity (WRI, 1985; Tejwani, 1985). It is important to note that since the inception of the project, DVC established a soil conservation department for watershed management which has achieved significant success in reducing the rate of reservoir sedimentation. It is obvious that a ha m of sedimentation prevented is a ha m of water resource saved. Consequently, the useful life of the reservoir is prolonged, and irrigation and hydro-power generation capacity is maintained at a particular point of time. It is reported that the sedimentation rates in the

Table 11. *Benefits of watershed management at catchment level: three reservoirs in India.*

Parameter evaluated	Name of Reservoir		
	Nizamsagar	Matatila	Ukai
Year of construction	1931	1956	1972
Year of evaluation	1988	1988	1988
Catchment area (sq. km.)	20,960	21,060	62,400
Land use (%)			
Agriculture	75.3	53.4	63.3
Forest	3.8	21.3	21.3
Area treated			
Area (sq. km.)	5,760	4,710	19,650
Percent of catchment	27.5	22.4	31.5
Sediment inflow (SIRQ)*			
Initial	0.71	3.99	0.70
(year)	(1931-61)	(1962-64)	(1972-79)
Treated	0.42	0.66	0.58
(year)	(1931-75)	(1975-85)	(1972-83)
Loss of storage % and (years)	57 (44)**	22.4 (23)	–
Direct benefits in catchment			
Agricultural Land			
Increase in cropping intensity (%) ***	11.4	30.0	12.4
B:C ratio ****	1.80	1.68	1.80
Internal rate of return (%)	39.3	40.7	–
Protective and production B:C ratio #	1.25	3.8	1.36
Water harvesting			
B:C ratio ##	1.36	1.84	1.35
IRR (%) ***	48.1	31.9	33.0
Afforestation			
B:C ratio***	3.0	4.5	4.6
Employment operation			
Person years, total	387,798	214,001	952,239
Person years per sq. km. of catchment area	18.5	10.2	15.3

*   SIRQ = Sediment Inflow Rate per Runoff unit = ha m/100 km^2/million ha m

**   The dam has been recently raised

***   For a small watershed

**** Discount rate 12% over a period of 12 years

\#   For whole catchment over a period of 12 years

\#\#   Over a period of 20 years. ### = Over a period of 12 years.

Bhakra, Machkund, Maithan, Panchet and Hirakud catchments have decreased significantly as the area treated in their catchments has increased (Das *et al.*, 1981).

In recent evaluation studies on three Indian reservoirs – Nizam Sagar, Matatila and Ukai (AFC, 1988a, b & c) – watershed management practices led to a reduction in sediment inflow in the reservoirs (Table 11). The watershed treatments increased the cropping intensity (from 11.4 to 30.0 per cent), gave a positive economic benefit-cost-ratio for cropping (1.68 to 1.8), a positive benefit-cost ratio (1.25 to 3.8) for protective and production functions on agricultural land, and a positive benefit-cost ratio for afforestation (3.0 to 4.6). It also generated employment for unskilled and skilled persons on both a temporary and permanent employment basis (the total employment generated was 10 to 19 human years per sq. km. of the catchment over the period of observation). There were additional unquantified benefits related to the area reclaimed for agriculture and irrigation, including a rise in the water table in wells, an increase in the number of wells and flood control.

An evaluation of sediment reduction structures in the Hirakud Reservoir catchment in India indicated they not only provided protective benefits (of reduction of sediment and runoff) but also resulted in reclamation of land below them and a consequent increase in crop production. The benefit-cost-ratio for production purposes was 1.66 over a period of 20 years and, for protective purposes, 3.14 over a period of 20 years (Das & Singh, 1981). In a small experimental watershed (370 ha) management project in the Himalayas, the benefit-cost-ratios for individual land uses were: irrigated agriculture 0.976; rainfed agriculture 3.6, orchards 2.9, and for fuel fodder plantations 0.92. For the whole watershed, the benefit-cost-ratio was 1.5 (Seckler, 1981). Tejwani & Rambabu (1982) have also reported positive benefit-cost-ratios for a number of soil and water conservation practices on agricultural and non-agricultural lands.

STRATEGIES AND POLICIES FOR BETTER WATER MANAGEMENT

Comparison of high and low population situations in developing countries

At no time in the process of development can the outcomes be perfect. Development and growth are dynamic processes. As they take place, new problems and challenges arise – economic, social, and cultural. In this context, the strategies and policies in any program of development need constant review and adjustment. Similarly, in prescribing for water resource management for developing countries, it is difficult to suggest strategies and policies which could be universally applicable for the many different economic, social, cultural and political situations. The best one can do is to raise some fundamental issues and hope that in working towards solution

of the many challenging problems associated with water and watershed management, the policy-makers, decision-makers, and planners will consider them, evaluate their relevance to the specific conditions in a region or a country and keep them in view when preparing the development plans.

For example, Sierra Leone and Kerala State are almost similar in broad terms with respect to tropical climate, rainfall, soils, and vegetation. Yet they are quite different with respect to population density, agriculture, forestry, irrigation and general development. Some statistics for Sierra Leone and Kerala are reported in Table 12. Sierra Leone is over twice the area of Kerala but supports only 48 persons per sq. km whereas Kerala supports a fantastic 655 persons per sq. km. Kerala, therefore, cultivates 56.1 per cent of its land as compared to 7.5 per cent in Sierra Leone. Kerala also irrigates 62.4 per cent of its cropped land while Sierra Leone does not irrigate any land. Kerala uses 45.2 kg of fertilizer per ha whereas Sierra Leone, with no significant use of fertilizer, believes it is over-populated since the agricultural production system does not produce sufficient food to feed its people.

In Sierra Leone, the relevant strategies and policies for managing the land and water systems will be to wean its people from shifting cultivation in the uplands, encourage settled agriculture and agroforestry in the uplands, utilize the vast unexploited potential of its inland valley swamps and riverine grasslands by expanding the area under cultivation and practising intensive management. There will be a need to introduce coconut palm, encourage oil palm (which already grows in Sierra Leone) and other cash crops. The lowlands will need drainage as well as flood control. It must also develop its hydro-power and irrigation potential.

In the case of Kerala, there will still be a need to extend the area under irrigation, make better use of the irrigation potential, practise intensive cropping management and develop industry. Watershed management practices will be needed in both countries, but in the case of Sierra Leone the forest will have to be managed to make available more agricultural land, while in Kerala the remaining forest will need to be zealously protected and afforestation practised to meet the needs of its very high population. When Kerala can support 1,265 per cent more population than Sierra Leone under almost similar agro-ecological conditions, there is no reason for Sierra Leone to be worried, provided it can use its land and water resources properly.

Indonesia demonstrates both high and low population situations. Java and Bali have very dense populations, with advanced and intensive agriculture involving irrigation and home gardens. However, in the less densely populated islands of Indonesia there is a need to wean people from shifting cultivation. Fortunately, people in these islands are quite familiar with home gardens and plantations of rubber, oil palm, tea, cacao, coconut and cloves.

To summarize: The following definition of watershed management adopted by India, emphasizes not only the conserva-

Table 12. *Comparative socio-economic statistics of Kerala State in India and Sierra Leone in West Africa.*

	Kerala	Sierra Leone
Area (sq. km.)	38,863	72,300
Population (1981)	25,454,000	3,520,000
Population density (Per/sq. km.)	655	48
Decennial growth rate (%)	19.2 (1971–81)	23.1 (1974–85)
Rural population (%)	81.3	NA
Climate	Tropical – Humid	Tropical – Humid
Annual Rainfall (mm)	2,807	3,000
Land use (sq. km.)		
Agricultural land	21,800	5,440
	(56.1)*	(7.5)*
Forest	10,810.0	63,050
	(27.8)	(87.2)*
Tree crops & groves	640	NA
	(1.6)*	
Area irrigated (sq. km.)	13,610	0
	(62.4)**	
Yield of rice (kg/ha)	1,678	–
Consumption of fertilizer (NKP/kg/ha)	45.2	–
Consumption of electricity (GWH)		
Agricultural sector	126	None
	(4.1)***	
No. of villages per cent electrified	1268	None
	(100)	

Source: MOA. 1985.

NA = data not available
* Figures in parentheses refer to percentage values of total land area in agriculture, forest or tree crops and groves.
** Figures in parentheses refer to percentage of agricultural land that is irrigated.
*** Figure in parentheses refers to percentage of total electricity consumption in the agricultural sector.

tion aspects of land, water, and forest but also emphasizes the proper uses of these resources for production purposes, particularly under high population pressures. It therefore is apt for developing countries:

> Watershed Management is the rational utilization of land and water resources for optimum production with minimum hazard to the natural resources. It essentially relates to soil and water conservation, protecting land against all forms of deterioration, building and maintaining soil fertility, conserving water for farm use, proper management of local water for drainage, flood control and sediment reduction and increased production from all land uses.

Management issues

The strategies and policies discussed in this section are applicable to both high and low density population situations, although the intensity and focus of these strategies and policies may vary.

Management Management as a concept has two aspects: implementation tools and institutional arrangements (for getting things done), and management practices (i.e., things to be done). Examples of the latter are afforestation, soil conservation, and on-farm water management practices. The focus of this discussion is more on the first aspect than on the second.

With respect to watershed management and soil conservation practices, presumably sufficient is known at present to allow launching some programmes (after suitable adaptations and demonstrations). However, this is less true for the implementation and institutional aspects of watershed management; these deserve further discussion. The approach here will be to raise an issue, review it and, if possible to indicate a future course of action in view of the experiences gained elsewhere. In some cases, it may be possible to identify the problem but it may not be possible to identify the remedy. Even if a remedy is identified, it may be very difficult to apply for various social, cultural and economic reasons.

Awareness A foremost requirement for resolution of a problem is achieving an awareness of the problem, i.e., that something needs to be done and that if something is to be done, it must be correctly implemented to achieve the set goals. Whether it is control of a population explosion or managing the land-water-plant systems, it is the government, including its line/sectoral departments, and the socio-political system which have to be aware of the problems. These entities must alert the people at large, the farmers and the users of land and water so they can become aware, active and responsive (Tejwani, 1987).

Most developing countries in the humid tropics with high population pressures and food problems are well aware of the importance of irrigation and water management in meeting the needs of food for the people and hydro-power for industrial growth. However, with respect to the interrelationship between the watershed conditions and the life of reservoirs, usually they do not act. This could be due to either a lack of appreciation of the problem or a lack of funds or both. This also is true of people. For example, the farmers in the Indian and Nepal Himalayas currently are very much aware of the degradation of their production base and the loss of water resources. Presumably, they wish to act. Perhaps they are enmeshed in a socio-economic environment which impedes any action. Sometimes the situation may be so desperate that it stimulates a movement of the people to rise and correct the situation, in spite of the government's policies or even because of them. An outstanding example is the Chipko Movement in the Himalayas (Kunwar, 1982).

Awareness of problems should in fact lead to development and implementation of relevant land, water, and forest use policies, keeping in view the needs of the people and prevailing socio-economic conditions.

Population pressure Control of population explosions is a key issue in the entire development strategy – more so when dealing with nonrenewable resources such as land and forests in the catchments of large reservoirs. Tackling this issue squarely is usually avoided due to various social, political and religious factors, even though there are programmes of family planning. In countries with low overall population pressures, it may be physically possible to relocate people, as in Indonesia.

However, in areas of high population, e.g. in India, Bangladesh, Sri Lanka, and Thailand, this option is ruled out because there is insufficient vacant land. If any single factor is to determine the success or failure of any development activity, it is population control. For the long-term good, governments must be prepared to share even a measure of unpopularity on this account (Khoshoo, 1986). The positive steps taken by Indonesia in this direction are noteworthy.

Management of the catchment-reservoir-irrigation command area Reservoir construction in developing countries has a very high political profile. Reservoirs are showpieces, certainly in the short run. Currently in developing countries many reservoirs are planned and designed in isolation. In the enthusiasm to build a reservoir, a number of irregular things happen. The sediment inflow rates are underestimated and the life of reservoirs is overestimated. Although all design specifications of the dam and reservoir are clearly determined and estimated, the catchment conditions receive only a passing reference; the same lack of attention is observed in the case of the area to be irrigated. Consequently, when a project is approved, funds are made available only for the dam construction, related structures and operational costs. Funds are seldom provided for catchment treatment or for development of the area to be irrigated (command area). This is done primarily to obtain a positive benefit-cost-ratio and the reservoir construction becomes a goal in itself.

Only subsequent to project implementation is it realized that the rates of sedimentation were underestimated and that the remedial measures of catchment treatment are important. This is a short-sighted policy with very dangerous consequences, some of which are:

(a) the designed benefits in quantity and quality are seldom attained,

(b) much greater costs are incurred in construction, operation, repair and maintenance than originally provided for,

(c) the most feasible dam sites (which are rare) are lost in a shorter life span than was estimated, and

(d) the planned net benefits are rarely attained. If the costs and benefits had been realistically estimated at the outset, the project, in fact, may have had a benefit-cost-ratio of less than one.

It is observed that while hydro-power may be generated in time, the irrigation potential generated is seldom utilized because no timely steps are taken to develop the irrigation command area. To that extent, the fruits of water resource development for irrigation are delayed or even denied to the people and the country.

The rationale given by the government for neglecting the catchment area and not developing the command area simultaneously is lack of funds. This reason is not valid, as funds always seem to become available for constructing another reservoir or for salvaging projects which run into trouble from the above consequences.

It is extremely urgent that a firm policy decision be taken that, hereafter, the catchment-reservoir-command area will be surveyed, planned, designed, constructed and managed simultaneously. In fact, it may be wise to plan and implement catchment area programmes in advance of reservoir construction (Tejwani, 1984b). It will be more economical and beneficial if this policy is implemented everywhere a dam is to be constructed. For example, it has been reported that if soil conservation and watershed management practices are included as a part of the project from the planning and design stages, and a 25 per cent silt charge is planned to be reduced, the percentage increase in the utilization of water can be 1.48 – 1.79 and 3.02 – 3.81 over a period of 50 and 100 years, respectively. In another case study, it was observed that a pre-planned 25 per cent reduction in sediment would reduce the height of the dam from 32.6 to 32.0 m, thereby cutting construction costs.(Sinha, 1984a,b). In addition there will be less submergence, reduced costs of rehabilitating displaced persons, reduction in loss of land (and forest) resource; and reduced evaporation losses due to a smaller reservoir surface area.

Research and demonstration Because watershed management is a mix of many applied field techniques, a strong technological package developed through research and demonstration is needed. The current status of research and demonstration in watershed management in the developing countries of Asia and Pacific has been reviewed by FAO (1982, 1986) and world-wide by Tejwani (1986). It is noted that with the exception of tropical Australia and China, India, Indonesia, Malaysia, East Africa, Zimbabwe, Nigeria and Venezuela, research facilities are either non-existent or grossly inadequate. Since research is expensive, it is essential that the relevance of research be critically scrutinized.

Organization of research in developing countries can follow three courses of action: (a) adaptation of research findings from elsewhere (this option offers the greatest opportunity of starting programmes when a country is in need of an immediate package of practices. It is now believed that sufficient research information is available for a country to adapt a package for its needs); (b) develop a package based on previously conducted research in a country (reported/not reported) for example, in Indonesia; and (c) initiate research in response to the need of development programmes (Tejwani, 1986). Paths (a) and (b) can be followed to launch the programmes, but only up to a point. Ultimately, locally-based research will be needed, because of unique biophysical and socio-economic conditions in each country. The results of research need to be expressed in practical ways through demonstrations, operational research projects and "lab to land" programmes.

Pilot watershed projects or operational research projects will not only produce biophysical responses to management practices but also will provide feedback from people in the watershed with respect to their priorities and acceptance of the practices. This will enable researchers to suitably modify either the individual practices or the mix and the priority of practices. A significant finding of operational research projects in India has been that the people living in upland watersheds put low priority on soil conservation and sediment reduction (as they perceive the former to be of long-range benefit and the latter as an off-site benefit). Thus, even though the people are short of fuel and fodder, they are not enthusiastic about afforestation and grassland development. Their first priority is local water resource development for surface irrigation and drinking because they perceive this activity as of immediate, on-site, direct benefit to them.

However, a pilot watershed project or an operational research drainage basin project serves the purpose of pretesting and refining a package before it is implemented and expanded to a large-scale watershed management programme.

Of the humid tropical countries of the Asia-Pacific region, only India, Thailand, Indonesia and the Philippines have established demonstrations or pilot watershed and operational research projects. Indonesia has undertaken demonstrations in watershed management – based on technology developed by FAO-sponsored projects in the 1970's – with varying degrees of success. However, no mechanisms are available to capture the learning from these experiences. India undertakes operational research watershed projects on the basis of technology packages developed within the country. It is imperative that all demonstration or pilot watershed and operational research projects be evaluated so that lessons are learned, packages are refined and then transferred to large-scale development programmes. So far, these tasks have not been accomplished. This approach also presupposes a very close and active linkage between the research set-up and the developmental agencies, which again is often lacking (Tejwani, 1986).

Manpower planning and training Manpower planning and training is another key issue which often determines the success of any development programme. Watershed management requires trained people in numbers commensurate with the size of the programmes. FAO (1982, 1986) has evaluated the training and education status in Asia and the Pacific, and Tejwani (1986) has reviewed it for the developing countries of the world. It is concluded that training and education in soil conservation and watershed management are either non-existent or grossly inadequate in the developing countries of the Asian-Pacific region, Africa and Latin America, with a few exceptions (e.g., China, India, Indonesia, the Philippines, Thailand, Malawi, Jamaica and Venezuela). Among the developed countries, the USA, UK, Australia, New Zealand, France and Japan have a large number of colleges and universities which offer courses and training opportunities in the water sciences.

For successful implementation of a watershed management programme, there is a need for well-trained and committed personnel at supervisory, middle, and field technician levels. To achieve this goal, various types of courses are required of varying durations, e.g., training for specific practices, regular integrated training, refresher training, orientation training, trainers' training and seminars/workshops. The roles and responsibilities vary with the institution using the trained personnel, the training institute and the trainees (Tejwani, 1986). In addition, there is a need for training programmes for farmers.

International training opportunities, under technical programmes, are inadequate and costly. The diversion of trained personnel from the national programmes, inappropriate training, and, paradoxically, too many international training opportunities in many small countries with limited manpower, are some of the main reasons for the inadequate training. In many developing countries, the international training is a goal in itself with no endeavour being made to build up the local institutions so that the manpower could be trained locally. Unless the latter is done, it is certain that no country will be able to attain sufficient numbers and provide a sustained inflow of trained persons to implement the programmes.

Some countries have good training facilities and manpower to implement the programmes but still lack in the area of personnel planning. If the programmes are continually expanding in size and content as in India and Indonesia, there is a constant need to strengthen and expand the training facilities and types. In spite of a massive training effort by India, the lack of sufficient number of trained personnel has been recognized as a major constraint in the successful implementation of the watershed management programmes (Vohra, 1974; Tejwani, 1979, 1981).

Cost-sharing and fiscal management Watershed management offers some short-term direct benefits to the farmers but mostly it offers many long-term off-site and indirect benefits to farmers and communities located downstream. In view of this it is only fair that the people living in the watersheds are not made to pay for the entire cost of the works. Therefore, there is a need to develop equitable fiscal policies to determine who pays (the farmer, the community, the government) and for what and how much? The Colombian example of transferring a proportion (4 per cent) of the hydro-power revenue to help defray watershed rehabilitation costs is one model. There is an Indian model, with built-in subsidies in the watershed management programme. For example, for the works carried out on farmers' fields, varying amounts of subsidies are given for different types of work (e.g., contour terracing 25–100 per cent, bench terracing 50–100 per cent, land reclamation 25–75 per cent, farm pond and water harvesting 25–100 per cent, land leveling 25–100 per cent, and farm forestry and pasture development 50 per cent). On non-agricultural lands the government pays the cost, e.g., afforestation by the department of agriculture or forestry 100 per cent, torrent control 75 per cent. (Mukherjee *et al.,* 1985).

In Indonesia, the government pays 100 per cent of the capital costs of checkdams, gully control, and stream bank protection. However, it provides no material help (even tree

seedlings) to farmers or to shifting cultivators in its regreening and reforestation programmes. There could be other models, e.g., transmitting hydro-power upstream and establishing small-scale industries and/or providing subsidized electricity for residential and small-scale industrial enterprises in the catchment area.

Fiscal arrangements must take into account the maintenance of works. The bunds and bench terraces, even under the best of circumstances, are weak and give way during the rains. It is the farmer who is expected to maintain them but he rarely does so. There is no institutional provision for maintenance. Appropriate fiscal and institutional mechanisms need to be developed.

Fiscal management should also include monitoring and evaluation of the implementation and benefits of the programmes, not only in terms of funds spent but of the quality, performance, maintenance and life expectancy of the system. Evaluation is a dynamic and ongoing process and the decision-makers and policy-makers should have adequate resources, both fiscal and technical, to be able to carry out midcourse corrections.

Legislation There is a strong need for legislation (and, of course, a need for people to participate) for planning, designing and implementing watershed management programmes, e.g.

(a) Since these programmes concern land treatment and water disposal, their implementation goes beyond the boundaries of individual farms;

(b) These programmes require investment of funds – both public and private – with and without incentives;

(c) There is need for maintaining structures and plantations over a long period of time;

(d) These programmes have an element of sharing of long-term benefits, e.g., tree plantations.

A legislative framework can take the form of land, water, and forest policies, along with land, forest, water, and mining laws, as well as soil and water conservation acts. They may be national, regional, and interstate in character; they may create institutions and departments, they may determine fiscal policies and procedures; and they may determine the sole rights and responsibilities of the government, the people and non-governmental organizations in managing these resources. All legislation has to take into account the current social, economic, cultural and religious environment. There may be major issues of centralization versus decentralization of management and of private versus public management of resources. The correct code of action in a particular setting may be difficult to specify. However, some things are clear. For example, in the development, use and conservation of land, water and plant resources, active involvement cannot be decreed by law; people must wish to become active participants. Laws can only facilitate. Laws can also become obsolete with the passing of time and with changes in socio-economic conditions. For example, the original forest policy and forest act of India was found to be inadequate to meet the situation of the current times and had to be changed.

Furthermore, some of the land and water management issues in densely populated countries are extremely complex and almost defy solution, as in the extremely small and fragmented farm holdings in India and in communally-held lands in Sierra Leone, Indonesia and many other countries. These issues cannot be resolved by legislation under present farming and socio-economic conditions. Perhaps they could be resolved by firm population control and by weaning people from shifting cultivation to settled agriculture or from agriculture to other occupations. However, in sparsely populated areas, e.g., in Africa, and some island groups in Indonesia, it may be possible to exercise many options of proper land-use.

People participation If one asks a development/extension worker, a researcher, a policy-maker, or a social worker, what is implied by, and what is the scope of "people participation," the invariable response is that it implies a participatory role by the beneficiaries/farmers and sometimes by women. However, this definition camouflages the issues and often results in disastrous consequences, since the focus is placed on the group of beneficiaries/farmers/women at the grass-roots level. To some extent it is a tactic provided by the service groups to divert attention from themselves. However, if the same question were asked of the beneficiaries/farmers/women, one would indeed be surprised to hear that in their opinion in addition to themselves, all others starting from the top (the policy/decision-maker) to the lowest rung on the ladder (the official functionary/social worker) are participants and by implication "people." The issue, in fact, is that "people participation" should encompass all those concerned with policy/decision-making processes and project formulation/implementation, as well as technocrats, managers, middle level/field level technicians, the socio-political system, the community at large, and, of course, the beneficiaries/farmers/women (Tejwani, 1987b).

This issue of the appropriate participatory role of all people is crucial to the successful initiation/design/financing/implementation of development programmes. Such examples, where a key person at the policy-making level provides firm direction to generate programmes or where the lowest functionary makes a programme succeed (even though the programme is faulty in design) are well-recognized.

At the lowest level, the failures often come to our notice yet the successes achieved by individuals under great odds are seldom recognized by the agencies, although the people who benefit do remember.

It is neither possible nor desirable to review the role or commitment of each participant in watershed programmes but it may be worthwhile to highlight a few examples.

Neither a beneficiary nor a farmer will change land-use patterns or practices on which he is subsisting unless he perceives that the proposed change will benefit him directly and immediately. For example, he will not adopt erosion control measures and pay for them because they do not benefit him directly and immediately. He will not plant or protect grasses and trees on village common land since he is not sure of his share. Another contributing factor is that usually he gets free grass, fodder and fuel from the forest. He does not consider his labour spent on collection from long distances as a loss, since in his world labour has very low cost. On the other hand, if there is a project for irrigation water, then not only is he willing to participate by providing labour and borrowing money to pay for part of the work, but he is also very willing to improve his own bench terraces (Tejwani, 1987).

It is important that suitable entry points which concern the farmers directly and immediately be identified and the watershed management programme be built around them. Usually, farmers and the people are "water related." Hence, water resource development is a good entry point. There are also examples where health care or schools have proved to be good entry points.

Project level technicians For transfer of the technological package, it is imperative that project staff be technically competent and sensitive to the needs of watershed management as well as to the needs, urges and motivations of the local people. The project staff must be committed and oriented to doing the job well and on time. A defective operation or installation (such as wrong disposal of surplus water or wrong alignment of gully plugs/checkdams) is likely to cause more damage to the whole programme than lack of publicity or funds. Keeping to schedule is also relevant since terracing and earthwork jobs must obviously be completed before the onset of rains or sowing of crops. A farmer cannot afford to lose a crop season. If seeds, seedlings and fertilizers have to be supplied, they must be on-site at the time of sowing/planting/application. Failure to perform work on time alienates the farmers. To claim later that the people did not participate is a travesty of truth. As far as the field technicians are concerned, the farmers are their "clients" and the staff must be motivated by a sense of trusteeship and commitment.

Supervisors/managers Apart from being professionally competent, committed, sensitive and sympathetic to the beneficiaries, the supervisors and managers need to provide leadership and guidance to the staff. Leadership is acquired and respected among professionals by professional competence. A professional supervisor should be competent to resolve technical issues. A supervisor at any level, has to be obviously sensitive to the needs and motivations of his/her staff. These two abilities are the *sine qua non* of a successful supervisor.

Project director/technocrats/decision-makers This group of people shares the responsibility for (i) project identification, formulation, implementation, and evaluation, (ii) convincing and selling the programmes at the policy and political levels, and (iii) dealing with a large number of staff, setting targets, providing/arranging funds and inputs on time and providing a congenial environment for the success of the staff.

Although this group needs specific management skills, the basic inputs of professional competence, commitment to the job/programme/cause, and sensitivity to the needs and motivation of their large staff are *sine qua non* for the success of the project and the programme. These prerequisites may appear to be obvious. However, often it is the obvious of which sight is lost.

Co-ordination All developing countries have an infrastructure of many sectoral departments such as agriculture, forestry, irrigation, roads, fisheries and mining. As each department has its own programmes, they tend to work in isolation, unaware of the interdependence of land, water, irrigation, hydro-power, roads and mines, and oblivious to the need for co-ordination of their activities. In fact, it is observed in south and southeastern Asian countries that the activities of some departments cause damage to the land-water-plant system. For example, activities of forestry and irrigation departments cause damage to roads; activities of roads and power departments cause damage to forest plantations, water courses, and bench terraces constructed by the forestry, irrigation, agriculture and soil conservation departments.

Watershed management for integrated and sustainable use of natural resources is a recent development, being not more than sixty years old. It, therefore, is understandable that various sectoral departments are insensitive to the need for co-ordination. For this reason an awareness of the issues is important. In some countries, at least there is realization that co-ordinated action is needed and attempts are being made to achieve it, often without success. It will take some time, but sight of the need.should not be lost.

A number of management models are being followed. For example, if a watershed has more than 50 per cent agricultural land-use, then it is handled by the agricultural department. If it has more than 50 per cent forest land-use it is handled by the forestry department. This model calls for each department having professionals in other disciplines. Other models are creation of a separate soil conservation department and creation of an independent authority. Each of these models can be either successful or a failure, depending upon the leadership provided. In Indonesia, all land-use plans or watershed management projects involve a number of different departments. Although this is a step forward, integrated action is not necessarily assured.

ACKNOWLEDGMENTS

I am grateful to Colin Rosser, Director, ICIMOD for the facilities provided for the preparation of this work. I am thankful to Lawrence Hamilton and Maynard Hufschmidt of the East-West Center for critical comments and suggestions. I am also thankful to John S. Gladwell and UNESCO for the opportunity provided to prepare this work on the humid tropics.

REFERENCES

AFC. (1988a). Report on Evaluation Study of Soil Conservation in the River Valley Project of Nizamsagar. Agricultural Finance Corporation. Bombay, India.

AFC. (1988b) Report on Evaluation Study of Soil Conservation in the River Valley Project of Matatila. Agricultural Finance Corporation. Bombay, India.

AFC. (1988c) Summary Report: Evaluation Study of Soil Conservation in the River Valley Projects of Matatila, Nizamsagar and Ukai. Agricultural Finance Corporation. Bombay, India.

Ahmed, L. (1982) Catchment Management in Bangladesh for Optimum Use of Water Resources. ESCAP. Bangkok, Thailand.

ASCI. (1987) A study in Machkund Sileru in Andhra Pradesh and Orissa and Pochampad in Maharashtra. Administrative Staff, College of India. Hyderabad, India.

Bali, Y. P. and Karale, R. L. (1973) Priority delineation of river valley project catchments. All India Soil and Land Use Survey Organization, Ministry of Agriculture, Government of India. New Delhi, India.

Brown, L. R. (1984) The global loss of topsoil. *J. Soil and Water Conserv.* 162–165.

Bruijnzeel, P. M. (1986) Environmental impacts of (de) forestation in the humid tropics – A drainage basin perspective. *Wallaceana W46*: 3–13.

Brunsden, D.; Doorkamp, J. C.; Fookes, P. G. *et al.,* (1975) Large scale geomorphological mapping and engineering design. *J. Eng. Geol. 8*: 227–253.

Carson, B. (1985) Erosion and sedimentation processes in the Nepalese Himalaya. ICIMOD Occasional Paper No. 1. International Center for Integrated Mountain Development. Kathmandu, Nepal.

CBIP. (1981) Sedimentation studies in reservoirs. Tech. Report No. 20, *II*. Central Board of Irrigation and Power. New Delhi, India.

CBS. (1987) Statistic Year Book. Central Bureau of Statistics. Government of Indonesia. Jakarta, Indonesia.

Chaturvedi, M. C. (1978) Proc. of Symp.on Indian national water perspective, Technological Policy Issues and Systems Planning. Physical Research Laboratory. Ahmedabad, India. 459–489.

Chunkao, K. *et al.,* (1983) Research on hydrological evaluation of land-use factors related to water yields in the highlands as a basis for selecting substituting crops for opium poppy. Highland Agriculture Project – Contract No. 53–32 R 6–0–49. Faculty of Forestry, Kasetsart University. Bangkok, Thailand.

CSWCRTI. (1980) Report on operational research project: watershed management – Fakot. Central Soil and Water Conservation and Research and Training Institute Dehradun, India.

CSWCRTI. (1987) Effect of bluegum plantation on water yield in Nilgiri Hills. Bull No. T–18/0–3. Central Soil and Water Conservation Research and Training Institute, Research Centre. Udhagamandalam, India.

Das, D. C., Bali, Y. P., and Kaul, R. N. (1981) Soil conservation in multi-purpose river valley catchments: problems, programmes approaches and effectiveness. *Indian J. Soil Conserv. 7(1)*: 8–24.

Das, D. C. and Singh, S. (1981) Small storage works for erosion control and catchment improvement: mini case studies. In: Morgan, R. P. C., (ed.) *Soil Conservation: Problems and Prospects*. Wiley. Chichester, England. 425–450.

DST. (1976) Habitat. The UN Conference on Human Settlements, Vancouver, Canada. Department of Science and Technology, Government of India. New Delhi, India.

Eckholm, E. P. (1976) *Losing Ground.* Norton. New York, USA..

Edward, K. A. (1979) The water balance of Mbeya experimental catchment. *East African Agricultural and Forestry J. 43:* 231–247.

El-Swaify, S. A.; Dangler, E. W. and Armstrong, C. L. (1982) Soil Erosion by Water in the Tropics. Res Extn Series 024. College of Tropical Agriculture and Human Resources, Univ. of Hawaii. Honolulu, USA.

FAO. (1982) Watershed management in Asia and the Pacific. FAO/UNDP Project RAS/81/053. Food and Agriculture Organization, Rome, Italy.

FAO. (1985) Regional Soil Conservation Project for Africa Phase 1. AG/GCP/RAF/1–81/NOR Tech Report. Food and Agriculture Organization. Rome, Italy.

FAO. (1986) Watershed management in Asia and the Pacific: Needs and Opportunities for Action. FAO: RAS: 85: 017 Technical Report. Food and Agriculture Organization. Rome, Italy.

FAO. (1987a) FAO Production Year Book, 1986. Vol. 40. Food and Agriculture Organization. Rome, Italy. 306.

FAO. (1987b) Tropical Forestry Action Plan. Food and Agriculture Organization, United Nations Development Programme, World Bank and World Resources Institute, Rome, Italy.

Gadgil, M. (1989) Deforestation: problems and prospects. Supplement to Wastelands News 4(4): 44.

Gadgil, M. and Vartak, V. D. (1974) The sacred groves of Western Ghats in India. *Economic Botany 30*: 152–160.

GBPUAT. (1982) Integrated Natural and Human Resources Planning and Management in the Hills of UP. Govind Ballab Pant University of Agriculture and Technology. Pantnagar, India. *I* and II. (Mimeographed).

Ghildyal, B. P. (1981) Soils of the Central and Kumaon Himalayas. In: J.S. Lal, (ed.) *The Himalaya: Aspects of Change.* Oxford University Press. Delhi, India. 120–151.

Gulcur, M. Y. (1966) Watershed management in the Philippines. Paper presented at the Foresters Conference, at GSIS, Arroceros, Philippines.

Gupta, G. P. (1975) Sediment production: status report on data collection and utilization. *Soil Conserv. Digest 3(2)*: 10–21.

Gupta, G. P. (1980) Soil and water conservation in the catchment of river valley projects: status report. *Indian J. Soil Conserv. 8(1)*: 1–7.

Gupta, P. N. (1981) Integrated watershed rehabilitation and development project. UP Forest Department. Lucknow, India. (Mimeographed).

Gupta, S. K., Tejwani, K. G., Mathur, H. N. and Srivastava, M. M. (1970) Land resource regions and areas of India. *J. Indian Soc. Soil Sci. 18:* 187–198.

Hamilton, L. S. & Henderson, D. R. (1987) Country Papers on Status of Watershed Management Research in Asia and the Pacific. East-West Center, Honolulu, U.S.A.. 401.

Hamilton, L. S. with King, P. N. (1983) *Tropical Forested Watersheds: Hydrologic and Soil Responses to Major Uses or Conversions.* Westview Press Inc. Boulder, USA..

Henderson, G. S. and Rouysungnern, S. (1984) Erosion and sedimentation in Thailand. In: C. L. O'Loughlin and A. J. Pearce (eds.) Symposium on *Effects of Forest Land Use on Erosion and Slope Stability*. East-West Center. Honolulu, U.S.A.. 31–40.

Huat, Tan Eow. (1974) Effects of simulated erosion on performance of maize (Zea mays) grown on Serdang coluvium. Soil Conservation and Reclamation Report No. 1. Ministry of Agriculture and Fisheries. Kuala Lumpur, Malaysia.

Ingram, C. D., Constantinn, L. F. and Munir Mansyur. (1989) Statistical information related to the Indonesian forestry sector. Working Paper No. 5, UTF/INS: Forestry Studies. Ministry of Forestry, Government of Indonesia. Jakarta, Indonesia.

Jacks, G. V. and Whyte, R. O. (1938) Erosion and Soil Conservation. Tech. Communication 36. Imperial Bureau of Soil Sciences. Harpenden, England.

Kerale, R. L., Bali, Y. P. and Singh, C. P. (1975) Photo interpretation for erosion assessment in the Beas Catchments. Photonirvachak *J. Indian Soc. Photo Interpretation 4(1 & 2)*: 29–30.

Khoshoo, T. N. (1986) Environmental priorities in India and sustainable development. Presidential Address to the Indian Science Congress Associations. New Delhi, India.

Kosambi, D. D. (1962) *Myth and Reality.* Popular Press, Bombay.

Kunwar, S. S. (1982) *Hugging the Himalayas.* Dasholi Garam Swarjya Mandal. Gopeshwar, India. 102.

Laban, P. (1979) Landslide occurrence in Nepal. Phewa Tal Project Report No. SP/13. HMG/FAO Integrated Watershed Management Project. Kathmandu, Nepal.

Lal, R. (1976) Soil erosion problems on an Alfisol in Western Nigeria and their control. Monograph no. 1. International Institute of Tropical Agriculture. Ibadan, Nigeria.

Lal, R. (1983) Soil erosion and crop productivity relationships for tropical soils. Paper presented at Malama Aina 83. Honolulu, USA.

Mishra, P. R., Grewal, S. S., Mittal, S. P. and Agnihotri, Y. (1980) Operational research project on watershed development for sediment, drought and flood control Sukhomajri. Bull. Central Soil and Water Conservation Research and Training Institute, Research Centre. Chandigarh, India.

MOA. (1985) Indian Agriculture in Brief. Ministry of Agriculture, Government of India. New Delhi, India.

MOA. (1988) Soil and Water Conservation Programme and Progress. T. S. Land Resources – 3/88. Ministry of Agriculture, Government of India. New Delhi, India.

MOF. 1989 Statistic Kehutanan (Forest Statistics), Indonesia. Ministry of Forestry. Government of Indonesia. Jakarta, Indonesia.

Mukherjee, B. K., Das, D. C., Singh, S., Prasad, C. S. and Samual, J. C. (1985) Statistics: Soil and Water Conservation in India. Ministry of Agriculture, Government of India. New Delhi, India.

Murray-Rust, D. H. (1972) Soil erosion and reservoir sedimentation in a grazing area west of Arusha, northern Tanzania. *Geografisca Annaler 54 A*: 325–343.

Murthy, B.N. (1981) Sedimentation studies in reservoirs. Tech. Report No. 20, Vol. I. Central Board of Irrigation and Power. New Delhi, India.

Narayana, V. V. D. and Rambabu. (1983) Estimates of soil erosion in India. *J. Irrigation and Drainage Engineering 109(4)*: 419–143.

NCA. (1976) Report of the National Commission on Agriculture. Government of India. New Delhi, India.

NEC. (1982) Shifting cultivation in North Eastern Region. Pub. No. 17. North Eastern Council. Shillong, India.

Nik, A. R. (1988) Water yield changes after forest conversion to agricultural land-use in Peninsular Malaysia. *J. Trop. Forest Sci., 1(1)*: 67–84.

Nye, P. H. and Greenland, D. J. (1960) The Soil under Shifting Cultivation. Tech Communication 51. Commonwealth Agricultural Bureau. Farnham Royal, England.

PCARR. (1977) Annual Report Watershed and Range Research Division, Philippine Council for Agriculture and Resources Research. Philippines.

Pereira, H. C. (1989) *Policy and Practice in the Management of Tropical Watersheds.* Westview Press, Boulder & San Francisco, U.S.A..

Ram Boojh and Ramakrishnan, P. S. (1983) Strategies for Environmental Management. Department of Science and Environment, Government of

Uttar Pradesh. Lucknow, India. 6–8.

Ramakrishnan, P. S. and Toky, O. P. (1978) Preliminary observations on the impact of shifting agriculture on the forested eco-systems. In: National Seminar on *Resources Development and Environment in Himalayan Region*.Government of India. New Delhi, India, 343–354.

Rambabu; Tejwani, K. G.; Agarwal, M. C. and Bhushan, L. S. (1978) Distribution of erosion index and iso-erodent map of India. *Indian J. Soil Conserv. 6(1)*: 1–14.

Rao, K. L. (1975) *India's Water Wealth: Its Assessment, Uses and Projections*. Orient Longman Ltd. New Delhi, India.

Sader, S. A. and Joyce, A. T. (1988) Deforestation rates and trends in Costa Rica, 1940 to 1983. *Biotropica 20(1)*: 11–19 (cited from ISTF News 9(2), 7.

Seckler, D. W. (1981) Economic valuation of the Fakot project. In: Report on Operational Project – Watershed Management. Central Soil and Water Conservation Research and Training Institute. Dehradun, India.

Shah, S. L. (1981) Agricultural planning and development in the Northwestern Himalayas, India. In: *Nepal's Experience in Hill Agricultural Development*. Ministry of Food and Agriculture, HMG. Kathmandu, Nepal, 160–168.

Shalash, M. S. E. (1977) Erosion and soil matter transport in inland waters with reference to Nile basin. *IAHS-AISH Publ. No. 122*: 278–288.

Siew, T. K. and Chin, F. (1976) Effect of simulated erosion on performance of maize (Zea mays) grown on Durian series. Soil Conservation and Reclamation Report No. 3. Ministry of Agriculture. Kuala Lumpur, Malaysia.

Singh, A. and Singh, M. D. (1981) Soil erosion hazards in northeastern hill region. Bull No. 10. Indian Council Agric. Res., Res Complex, Northeastern Hill Region. Shillong, India.

Sinha, B. (1984a) Need of integrated approach to development of catchment and command of irrigation projects. *Water International 9(4)*: 158–160.

Sinha, B. (1984b) Role of drainage basin management in water resources development planning. Paper presented at the Workshop on the Management of River and Reservoir in Sedimentation in Asian Countries. East-West Center. Honolulu, USA (Mimeographed).

Smythies, E. A. (1920) Afforestation of ravine lands in the Etawah district, United Provinces. *Indian For Rec. 7*: 217–249.

Starkel, L. (1972) The role of catastrophic rainfall in the shaping of the relief of the lower Himalaya (Darjeeling Hills). *Geographic Polinica 21*: 103–147.

Stocking, M. (1984) Erosion and soil productivity – A review, Consultants Working Paper No. 1. AGLS, Food and Agriculture Organization. Rome, Italy.

Sukartiko, B. (1986) The Status and Needs of Watershed Management. Country Brief – Indonesia. RAS/85/017. Food and Agriculture Organization. Rome, Italy.

Tejwani, K. G. (1979) Soil and water conservation-promise and performance in the 1970s. *Indian J. Soil Conser. 7(2)*: 80–86.

Tejwani, K. G. (1980) Soil and water conservation. In: Handbook of Agriculture. Indian Council of Agricultural Research. New Delhi, India, 120–157.

Tejwani, K. G. (1981) Manpower needs of watershed management in India. Proceedings of the National Seminar on Watershed Management. Forest Research Institute & Colleges. Dehradun, India, 193–202.

Tejwani, K. G. (1982) Evaluation of the environmental benefits of soil and water conservation programmes. *Indian J. Soil Conserv. 10(2&3)*: 80–90.

Tejwani, K. G. (1984a) Biophysical and socioeconomic causes of land deforestation and a strategy to foster watershed rehabilitation in the Himalayas. In: O'Loughlin, C. L. and Pearce, A. J. (eds.) IUFRO Symposium on *Effects of Forest Land Use on Erosion and Slope Stability*. East-West Center, Honolulu, USA. 55–60.

Tejwani, K. G. (1984b) Reservoir sedimentation in India. Its causes, control and future course of action. *Water International 9(4)*: 150–154.

Tejwani, K. G. (1986) Training, research and demonstration in watershed management. In: *Strategies, Approaches and Systems in Integrated Drainage basin Management*. FAO Conserv. Guide 14, Food and Agriculture Organization. Rome, Italy. 201–219.

Tejwani, K. G. (1987a) Agroforestry practices and research in India. In: Gholz, H.L. (ed). *Agroforestry: Relatives, Possibilities and Potentials*. Martinus Nijhoff . Dordrecht, Netherlands, 109–136.

Tejwani, K. G. (1987b) Watershed management in the Indian Himalaya. In: Khoshoo, T. N. (ed.) *Perspective in Environmental Management*. Oxford and IBH Publishing House Co. Pvt. Ltd. New Delhi, India. 203–227.

Tejwani, K. G. (1988a) Watershed Management and Soil Conservation in Kigoma District, Tanzania. TCP/URT/8851(F) – Assessment of Wood Supply Situation and Environment Degradation in Kigoma Region-Tanzania. Food and Agriculture Organization. Rome, Italy (Unpublished).

Tejwani, K. G. (1988b) Watershed Management in Sierra Leone. SIL/87/010/12/A – Joint Interagency Forestry Sector Review. Food and Agriculture Organization. Rome, Italy. (Unpublished).

Tejwani, K. G. (1990) Technical and institutional aspects of watershed technology centers in Indonesia. Field Document 8, FD:DP/INS/86/024. Food and Agriculture Organization. Jakarta, Indonesia.

Tejwani, K. G., Gupta, S. K. and Singh, G. (1969) Report on Landslide Erosion in Darjeeling District, West Bengal, Ministry of Agriculture, Government of India. New Delhi, India. (Unpublished).

Tejwani, K. G. and Rambabu (1981) Unpublished data. Central Soil and Water Conservation Research and Training Institute. Dehradun, India.

USDA. (1980) Soils depletion study. Reference Report-Southern Iowan Rivers Basin. U.S. Dept. Agric-SCS-Economics, Statistics and Cooperative Services. Washington D.C., USA

Van der Meer, C. L. J. (1981) Rural Development in Northern Thailand. Dissertation-Gromingen University, Netherlands.

Vohra, B. B. (1974) A charter for the land. *Soil Conserv. Digest 2(2)*, 1–25.

Vohra, B. B. (1981). A policy for land and water. Sardar Patel Memorial Lectures, 1980. Department of Environment, Government of India, New Delhi, India.

Weidelt, H. J. (1975) (Compiler). Manual of Reforestation and Erosion Control for Philippines. German Agency for Technical Cooperation, Ltd. Eshborn, West Germany.

Whitney, M. D. (1905) Atharva Veda Samhita. Harvard Oriental Series. Authorized Indian Reprint. Motilal Banarsidas, 1984. Delhi, India.

Wiersum, K. F. (1984) Surface erosion under various tropical agroforestry systems. In: O'Loughlin, C. L., and Pearce, A. J. (eds.) Symposium on *Effects of Forest Land Use on Erosion and Slope Stability*. East-West Center. Honolulu, USA, 231–240.

World Resources Institute (WRI). (1985) Tropical Forests: A Call for Action. Report of an International Task Force Convened by the World Resources Institute, the World Bank and the United Nations Development Organization. Washington D.C., USA.

24: Urban Water Resources in the Humid Tropics: An Overview of the ASEAN Region

K. S. Low

Institute for Advanced Studies, University of Malaya, 59100 Kuala Lumpur, Malaysia

ABSTRACT

South East Asia is one of the most populated regions within the humid tropics and is fast becoming an enclave of industrial development. Geographically, it is endowed with a good source of water, due to its copious rainfall. However water resource problems often arise because of variations in distribution, timing and the small sizes of the catchments. The last few decades also witnessed the rapid deterioration of the available water due to anthropogenic activities causing pollution, reducing further the available water. Demands, on the other hand, have increased with rising urbanization and industrial development. In many countries, balancing a good water supply and demands is a major issue requiring more than environmental controls. It may require very stringent legislation to control the water sources. It is the aim of this chapter to examine some of those inherent problems so that scientists, mitigation planners and engineers can address them in their detailed resource planning.

INTRODUCTION

The humid tropics is a geographical region recognized by its climatic similarities. Quantitative definitions have been given by Chang & Lau (1983) as a region having a minimum of 4.5 wet months per year and an average minimum temperature of l8°C (See Appendix A, this volume). Vitousek & Sanford (1986) have defined it as an area lying 23° N and 23° S and receiving at least 1,600mm of annual rainfall. Within this broad region astride the equator lies an assortment of countries which could be categorized further, based either on economic status, geographical size or political alignment. The first is used by macro-economists to show the levels of development within and between regions, and will also be used in this chapter. Thus, for the purpose of this chapter, the countries in the humid tropics will be grouped into low income (underdeveloped), middle income (developing), newly industrialized and high income (developed) economies. This grouping is significant because there is a great deal of resource diversity between the countries in the region and hence the stages or levels of development are different.

As shown in Table I, almost all of the countries are in the low and middle income categories. The only exceptions are Singapore with a per capita GNP of US$7,940 in 1987 (World Bank Report, 1989) and oil-rich Brunei Darussalam which, although having one of the highest per capita incomes in the world, is economically unindustrialized.

Although basically underdeveloped, these humid tropical countries are major players in resource exploitation and are undergoing very rapid environmental transformation. The pace of change is more rapid than was the case for developed countries at a similar stage of economic development. A major feature of this transformation has been the conversion of forests to agricultural and urban lands on a scale which constitutes almost a new "great age of land clearance" (Brookfield, 1988). The forces behind such transformations have been the rapid growth of population, along with policies and strategies to develop resources for economic gains, partly underpinned by strong socio-economic objectives of equitable distribution of wealth which accompanies economic growth.

Today, the countries shown in Table I have a total population of 1960 million, densities which range from 8.7 to 3,000 persons per sq. km., and a total of lll cities with populations above 500 000 (1980). Gladwell & Bonell (1990) estimate that the population in the humid tropics will double between 1980 and the year 2000 by which time it will constitute 35% of the world's forecasted 6.25 billion population.

Throughout the whole region, population growth is rapid and accelerating. Between 1980–1990, it averaged 2.8% compared to an average global growth rate of 1.74% (World Resources Institute, 1990) and 0.6–0.7% in the developed countries (Maione, 1988).

The urban population growth rates are very much higher. In the ASEAN countries the average is 4.0% (excluding Singapore and Brunei), a value which is substantially affected by rural-urban migration. Other values (1980–1987) range from a high of 8.6% in Kenya to lows of 1.2% in Sri Lanka and 1.1% in Singapore, as shown in Table 1. McGee & Yeung (1990) indicate a perceptible change in the economies of developing countries between 1965–1980 and 1980–1988 which accelerated the flow to the urban areas. This is certainly the case in the ASEAN

Table 1. *Population and basic urban/economic indicators in selected humid tropical countries*

Countries	Population (mil) 1990	Density (person/sq.km)	Urban Population Growth Rates (%) 1980–87	Per Capita GNP (US$) 1987	No.Cities Over 500,000 1980
Low income economies					
Ethiopia	48.0	39.3	4.6	130	1
Zaire	37.0	15.8	4.6	150	2
Bangladesh	114.0	791.7	5.8	160	3
Laos	6.0	25.3	6.1	170	0
Zambia	8.0	10.6	6.6	250	1
Burma	43.0	63.5	2.3		2
Uganda	18.0	76.3	5.0	260	1
India	844.0	256.7	4.1	300	36
Central Africa					
Republic	3.0	4.8	4.7	330	0
Kenya	25.0	42.9	8.6	330	1
Pakistan	106.0	131.8	4.5	350	7
Middle income economies					
Nigeria	118.0	127.7	6.3	370	9
Indonesia	179.0	93.3	5.0	450	9
Philippines	61.0	203.3	3.8	590	9
Papua New					
Guinea	4.0	8.7	4.8	700	0
Thailand	56.0	108.9	4.9	850	1
Colombia	31.0	27.2	2.9	1240	4
Malaysia	17.0	51.5	5.0	1810	1
Mexico	89.0	45.1	3.2	1830	7
Brazil	150.0	17.6	3.7	2020	14
High income economies					
Singapore	3.0	3000.0	1.1	7940	1
Brunei					
Darussalam	0.2	39.2	=	15390	1
TOTAL	1960.2				111

Source: World Development Report 1985
World Bank Annual Report 1989

countries where the prolonged economic boom since the early 1960s had caused large-scale movements of the rural population to the urban areas in search of economic opportunities in the many factories that had sprung up. The latter were encouraged by the industrialization policies. In some of the African countries, particularly Ethiopia and Nigeria, the continuing drought has caused a shift of the population to the urban centres located further south. In Uganda, the case was complicated by a change in government that had forced out-migration on a large-scale in the 1960s. The opening up of lands in the interior of Brazil and the construction of a new city (Brazilia) in the Amazon jungle also has caused a large clearance of land and a shift of population away from the coast.

Thus, in the countries shown in Table 1, there were 111 cities with populations of more than 0.5 million in 1980. Of the 17 cities with a 1980 population of 1 million and over as shown in Table 2, nine will have estimated populations in excess of 10 million by the year 2000. The two largest cities, Mexico City and Sao Paulo, are expected to have populations nearing 25 million by the year 2000.

The primacy of cities is well documented (McGee, 1967; Pryor, 1979). They have been described as keys to economic development. Most non-resource-oriented economic activities, whether commercial, financial or industrial, are located in these cities and their immediate hinterlands. Hence, they also consume the largest amount of resources brought in from the

Table 2. *Selected large urban areas in humid tropical countries.*

Cities	Population (in millions)		
	1950	1980	2000
Mexico City	3.1	14.5	25.8
Sao Paulo	2.8	12.8	24.0
Calcutta	4.5	9.5	16.5
Rio de Janeiro	3.5	9.2	13.3
Bombay	3.0	8.5	16.0
Jakarta	1.8	6.7	13.3
Manila	1.6	6.0	11.1
New Delhi	1.4	5.9	13.2
Bangkok	1.4	5.0	10.7
Madras	1.4	4.4	8.2
Dhaka	0.4	3.4	11.2
Lagos	0.4	2.8	8.3
Surabaya	0.7	2.3	5.0
Singapore	0.5	2.3	3.0
Kinshasa	0.2	2.2	5.0
Bandung	0.5	1.8	4.1
Kuala Lumpur	0.2	1.0	1.8

Source: Series from World Development Reports.

hinterlands or from abroad. For example, Bangkok, built on the alluvial deposits of the Chao Phraya River, with a 1980 population of 5 million people, is at least 50 times larger than Thailand's next largest city, Chiangmai. Bangkok also possesses almost 80% of all the telephones, 50% of the motor vehicles, and consumes over 80% of the electricity and potable water supply in Thailand (Phantumvanit & Liengcharensit, 1989). The metropolitan area of Manila has 8 million people, or 30% of the total urban population in the Philippines, within an area of 636 km^2. The urban population congestion has serious implications in terms of shortages in public services such as water supply and sanitation (Jimenez & Velasquez, 1989). Similarly, Mexico City, Dhaka and Kuala Lumpur have the largest share of commercial and institutional establishments in comparison with the second largest cities in their countries.

The relentless demands of cities – due to their sheer size – definitely impose an incredible strain on the resources of the cities and their hinterlands. Nowhere is this strain felt more than in water resources. It is the aim of this chapter to analyse the issues underlying the water resource problem to show how they determine the manner in which this resource is becoming unsustainable in some humid tropical countries. The paper focuses on four major countries in southeastern Asia: Indonesia, Malaysia, the Philippines and Thailand, where urbanization and water problems are inextricably connected. However, wherever possible, references will be made to other humid tropical countries.

OVERVIEW OF URBAN WATER RESOURCES

In these southeastern Asian countries, a combination of climatic variability and general *ad hoc* management of water resources has resulted in widespread water pollution and water scarcity. Basically, there is underdevelopment of the water resources which is exacerbated by the widespread contamination of what is available, the latter being a consequence of the misuse of water, caused directly or indirectly by the rapid pace of industrialization and urbanization. The simultaneous existence of these two contrasting phenomena raises many questions without immediate answers. The water resource situation is highly complex and, in the case of pollution, there have been outcries of "ecological disasters" by the international environmental communities which seek to conserve and preserve the water catchment areas upstream of the urban areas. Even if such outcries were heeded, the water problems could not be solved in the immediate future due to a combination of other factors.

Physical location of cities and rainfall variability

One important factor is the physical location of these cities in areas of the humid tropics which experience high rainfall variability. This variability arises from climatic conditions which are different in South East Asia from those in the humid South American and central African regions. The average annual rainfall in South East Asia ranges from 1,000 mm to as high as 4,000 mm with marked seasonal concentrations (ACST, 1982). Areas in the higher latitudes, such as northern Thailand which can be classified as sub-equatorial (though still within the humid tropics), experience marked seasonality with 50% of the annual rainfall occurring in two to three months, yet with water stress during the remainder of the year.

Furthermore, all of the areas in the region are affected by the northeast and southwest monsoons, the former occurring between late October and early January, and the latter between May and September. Depending on the location of the countries, heavy rainfall occurs during these months, with intervening months being relatively dry. For example, during the northeast monsoons, the northern monsoon shearline (see Manton & Bonell, this volume) lies just north of the equator between longitudes 60°E and 160°E. Cyclonic vortices often develop in this trough over the South China Sea because the interactions of air-sea mixing cause conditional instability. These disturbances migrate southwestwards, picking up additional moisture during their long travel across the South China Sea, and deposit heavy rainfall on the east coasts of Peninsular Malaysia, Sabah and Sarawak and the eastern regions of Thailand and Kalimantan. The southwest monsoons, on the other hand, occurring between May and September, bring heavy rainfall to the west coasts of Sumatra, Java and west Peninsular Malaysia. It is the strong influence of these two monsoonal seasons that imparts the variabilities which result in the two hydrological phenomena: floods and droughts.

Flood-causing rainstorms with a maximum 24-hour rainfall exceeding 600 mm have been recorded in all major cities in South East Asia. Similarly, maximum monthly rainfalls above 2,000 mm have also been recorded in all the major cities except Singapore. Floods from concentrations of heavy rainfall in a short period of time are more common than droughts, although prolonged droughts are known historically to occur regularly, especially when intensified by the reversal of the Walker circulation in the periodic El Niño/Southern Oscillation (ENSO) events. Taking the average evapotranspiration rates as constant at about 4–5 mm day^{-1} or 120–150 mm month^{-1}, many parts of the region receive rainfall less than this amount for 4 to 5 months in a year. Water-balance studies in Peninsular Malaysia showed that water deficits and stresses were quite prevalent unless adequate natural or artificial storages are available (Niewolt, 1965; Low & Goh, 1972). Kuala Lumpur, for example, has monthly rainfall of less than 150 mm for 40% to 50% of the time. The case is similar in the Philippines and Indonesia, while water deficits are even higher in central and northern Thailand during the intermonsoonal months. Climatic variability is a major causative factor of water problems in the humid tropical cities.

In the case of Bangkok, its water supply problem is compounded by the fact that the city is located at the downstream end of the Chao Phraya River while the water supply intake is upstream in the headwater region which is away from the humid tropics and experiences high rainfall variability. Moreover, much of the upstream water is diverted for irrigation and the return flow is highly contaminated by residues of pesticides and fertilizers. Thus it cannot be used for drinking purposes unless it heavily treated. As a result, up to 30% of Bangkok's daily water demands is met by groundwater. This unregulated form of supply has caused considerable land subsidence over a long period. According to Nutalaya (1990), most of the areas of Bangkok are only 1.5 m above mean sea-level, leaving the entire area very susceptible to flooding.

Sizes of river basins and locations of urban centres

Another major problem facing water resources development in these four countries is the relatively small size of their river basins. A large proportion of the highly-populated areas is either on small- or medium-sized islands or peninsulas, with associated small river basins. The largest basins in Peninsular Malaysia, Indonesia (excluding Kalimantan) and the Philippines are in the range of 20 000–50 000 km^2 but there are only six of these. Actually, most urban centres are located within very small river basins, whose sizes do not exceed 2,000 km^2. The small size of these basins poses several problems. First, the urban activities often generate pollutants and sediment loads on the river systems, which because of their small sizes, do not have the capacity to assimilate and dilute these wastes. Second, as the volume of water available is relatively small (because of small catchment areas with minimal natural storage), when the available water is polluted, it effectively reduces the net amount suitable for use without costly treatment. Furthermore, the relief-length ratios exhibited in the river basins are usually high and, under the humid tropical conditions with highly-weathered soils, the delivery of sediments into the river network is very efficient, thus polluting the water sources further.

Kuala Lumpur is a good example. The city is located in the heart of the Kelang River Basin which is about 1,200 km^2 in size. Due to heavy demands for potable water from two dams located upstream, very little water is released to augment the downstream low flows. In many ways, this has magnified the water pollution situation. Only a few kilometers downstream from the dams, the extremely low flows in the river, especially during the dry seasons in February and July, are unable to dilute and flush the incoming liquid and solid wastes generated by the urban population and its activities. In effect, the river has become an open sewer. In addition, housing development and road construction in the foothills bordering the river valley have brought an average of 500 tonnes km^{-2} year^{-1} of sediments downstream to the city of Kuala Lumpur . A large proportion of the potable water, therefore, must be brought into the city by transfer from the nearby Semenyih River. This supply will have to be augmented by the year 2000 from other river basins located even farther away from the city if the rapidly growing water demands of the urban population are to be met.

In contrast, in areas with ample water to meet demands such as on the island of Borneo, the demand is very low due to the low population density. The density of population in Sarawak or Kalimantan, for example, is hardly more than five persons per km^2, and hence does not justify the cost of developing large water resources projects unless hydroelectricity is also tapped. In fact, it is hydroelectricity rather than water supply that is the major rationale for developing the water resources in the upper reaches of such river basins in this area. This certainly is the case for the Batang Ai Dam in Sarawak where hydroelectricity is fed into the national electricity grid to supply the city of Kuching, the capital of Sarawak, while potable water for the city is obtained through direct river abstraction from a number of tributaries rather than from the reservoir created by the dam.

Thailand is an exception. The river basins in Thailand are large but receive the lowest annual rainfall compared to the other countries in the region. The largest river, the Chao Phraya, with a drainage area of about 120 000 km^2, has abundant water but the major usage is for agriculture. The withdrawal of water for irrigation, especially for wet-rice cultivation, and its subsequent release during harvesting has increased the fluctuations on the supply side of the water balance equation and contaminated the released water with fertilizers and pesticides. In either case, the problem is not simply a matter of deducting the consumptive use from the runoff. It is much more complex and is related to the land-use pattern in the river basins, where generally the urban enclaves receive the brunt of the problem because of their location downstream.

Table 3. *Methods of waste water disposal in Bangkok (adapted from Xoomsai, 1988).*

Method	(Percentage of population)		
	Bangkok Metropolis	City Core	Outer Districts
Dumping into public sewers	60.8	75.1	13.8
Drainage to vacant land	5.0	3.2	4.1
Septic tank	1.0	0.7	3.8
Cesspool latrine or left on ground	13.1	7.2	31.9
Disposal into rivers or canals	18.2	12.5	44.5
Other means	1.9	1.3	1.9

Table 4. *Population served by different types of sewage treatment systems in Malaysia (Pillai, 1987).*

Type of facility	% of population served		
	1970 census	1980 census	1985 (est.)
Central sewerage system	3.4	4.0	5.3
Flush toilets connected to septic tanks or communal treatment plants	16.0	21.8	30.6
Pour flush latrlne	2.6	30.3	39.2
Bucket latrine	17.1	7.7	3.4
Pit latrine	27.8	15.3	8.5
Hanging latrine		4.5	2.8
No facility	33.1*	16.4	10.2

*These two categories were grouped together in the 1970 census.

Thus, Bangkok relies more on groundwater than most other cities in the region, and although this has often been cited as contributing to land subsidence in the city, such continued reliance on groundwater seems to be necessary. In the last decade, Bangkok has resorted to some inter-basin transfer of water, especially from the hilly Kanchanaburi Province southwest of Bangkok. Here again, hydroelectricity takes priority over water supply but, nevertheless, the objective of augment-

ing potable water to Bangkok has been met to a certain extent. There have also been suggestions of transferring water from the Mekong River to the Chao Phraya. The former has an estimated excess of 950 MGD to contribute (Phantumvanit & Liengcharensit, 1989) which would be ample for part of Bangkok's future demands.

WATER PROBLEMS CAUSED BY URBAN ACTIVITIES

Water resource management now and in the next decade has to reconcile climatic variabilities with river basin development and with water contamination caused by human activities. Water pollution has been defined as "anything causing or inducing objectionable conditions in any water course and affecting adversely any use, or uses, to which the water may be put" (Klein, 1957). The word "uses" in this definition is particularly meaningful with respect to water pollution in the urban centres of the humid tropics where the uses of rivers are many and diverse. The waters in the urban rivers are a source of drinking water, a conveyance for sewage disposal, a mode of transportation, a recreational resource, a source of food and, more often than not, a rubbish dump. With ever-progressive urban development and population pressure, many of these rivers are increasingly polluted, becoming turbid, septic and emitting unbearable odors. Several rivers have been classified as "biologically dead" when their dissolved oxygen content in the water approached zero. The major sources of water pollution are sewage effluent and untreated faecal matter, industrial effluents and untreated industrial wastes, domestic and industrial garbage, agricultural chemicals, and sediment.

Sewage disposal and water pollution

Although water contamination by sewage is a universal problem, it is especially severe in large cities of the humid tropics. Sewage, if adequately treated and disposed of, poses very minimal pollution or health problems. However, sewage treatment and disposal require heavy capital financing which many cities in the humid tropics would rather spend on something yielding more tangible benefits. This is certainly the case in Sao Paulo, Calcutta, Bombay and Jakarta. In Bangkok in 1980, for example, only 61% of the sewage was discharged into public sewers. The remainder was disposed of by drainage onto vacant land or discharged into rivers and canals (Xoomsai, 1988). In the outer districts of Bangkok, almost 45% of the untreated sewage ends up in rivers and canals (Table 3). In urban areas in Malaysia, although the majority of the population had some form of decent sanitation in 1985, more than 10% of the population had no facilities at all. Fifteen per cent had only rudimentary latrines (Table 4). In Kuala Lumpur, the largest and best-kept city in Malaysia, more than 1% of its population in 1983 had latrines located directly over rivers, mainly over the heavily polluted Kelang River. In 1980, the faecal coliform counts in the Kelang River ranged from 113 000 to 160 000

Table 5. *Heavy metals in selected Malaysian rivers, 1985–1986 (DOE, 1986).*

Heavy Metals	Rivers				
	Perai	Perak	Kelang	Langat	Muar
(concentrations, mg/l)					
Lead	0.149	0.008	0.025	0.035	0.008
Mercury	0.028	0.002	0.001	–	–
Cadmium	0.025	–	0.010	–	0.010
Copper	–	–	0.040	–	0.030

per 100 ml (Low, 1980) which were extremely high compared to the WHO standard of 100 per ml for bathing water. The above situation is also true in Jakarta where less than 75% of the 1980 population was served with some form of sanitation. The remainder used pit privies or direct disposal to water courses (Douglas, 1983). In Calcutta, there is hardly any central sewage treatment. Individual pit toilets are prevalent, leading to contamination of the groundwater. The final disposal points of most sewage systems are the water courses where the waters are often used for bathing and drinking or for aquaculture.

There are no easy or simple answers to these pervasive problems of urban pollution. Constructing and operating adequate sewage collection and treatment facilities entail enormous expenditures to keep pace with the demands imposed by rapid urban growth. Funds for needed capital expenditure are severely constrained, and, under the best of circumstances, it will take many years for most large cities in the humid tropics to provide proper sanitation services for their inhabitants.

Industrial growth and water pollution

The recent rapid growth of the manufacturing and processing industries has brought with it major problems in terms of water contamination and pollution. Although the impacts of these industries are well-known, treatment is often belated. During the early years of industrial expansion in Malaysia and Thailand – the 1960s and early 1970s – environmental problems were ignored in the pursuit of economic enhancement. However, this has proved to be a costly policy. Many industries in Kuala Lumpur and Bangkok are now actually polluting the very rivers from which they abstract their waters for their own use. Even though legislation had been enacted to require industries to treat their wastes before discharge into rivers, such requirements were never adequately enforced in the early 1970s when industrialization gathered momentum. Only since the early 1980s, when the governments began to enforce stringent standards and to apply severe penalties, have factories started treating their wastes.

Processing of tropical agricultural products has long been identified as a major cause of water pollution. In Malaysia, Thailand, Indonesia and Nigeria, for example, the processors of rubber, palm oil and cocoa have been cited often as major water polluters. It is estimated that palm oil mills in Malaysia produce a total of 15 million tonnes of liquid wastes per year, of which a substantial amount is discharged into rivers. Over time, these wastes steadily diminish the oxygen content and hence slowly reduce the diversity and productivity of the aquatic species in the rivers, while at the same time making the water more expensive to treat.

The presence of heavy metals in the waters of urban rivers is common. In areas where medium-sized and large industries are located near rivers, heavy metal concentrations can be very high, indeed. Studies by the Department of Environment in Malaysia showed that, during 1985 and 1986, several major urban rivers had heavy metal concentrations regularly above the minimum recommended levels. Many of these rivers have industries along their banks and much of their effluent finds its way into the water courses. Law & Singh (1982) reported that the mercury content in the waters and sediment in the Kelang River estuary were 1,690 mg l^{-1} and 0.20 mg kg^{-1}, respectively, indicating significant levels of mercury pollution. Table 5 shows the heavy metal concentrations in some selected Malaysian rivers.

The industrial pollution problem was recognized earlier in the Philippines. In 1973, a total of 115 firms had water pollution cases pending, of which 79 were from Metro Manila (Marlay, 1977). At that time, it was also reported that more than 30 rivers in the country were grossly polluted. An interesting case concerning industrialization and water pollution was that of the city of Malaban, north of Manila. This once-quiet town was hit by an influx of factories which discharged their effluents into the clear-running Tinajeros River. The pollution devastated the fishing industries along the river which had been in continuous productive operation for at least 200 years. A series of bitter complaints, feuds, and legal actions followed between the fishermen, governmental agencies and factory owners. It ended with the arrest of several factory owners, but not before the Philippine government was jolted into environmental awareness.

The situation is similar in Thailand. In Bangkok and the adjacent province of Santprakarn, the dissolved oxygen levels in the tributaries of the Chao Phraya in 1986 were only 1 to 3 mg l^{-1}. Factories, both upstream and downstream of Bangkok, were responsible for up to 41% of total wastes discharged into the Chao Phraya. It was estimated that of the 267 530 kg of BOD wastes discharged into the Chao Phraya in 1979, 36% were from factories and 62% were from households situated along canals which flow into the river.

Deforestation, soil erosion and sedimentation

Sedimentation in the rivers of the humid tropics is also an important factor in reducing the quality of water available for use. Soil erosion can be very severe due to the high erosivity of rainstorms (high intensities, large drop sizes and strong accompanying winds) and generally highly-weathered soil profiles, as discussed elsewhere by Lal, this volume. Thus, when land is cleared of its vegetation, excessive amounts of soil are eroded. The rapid runoff carries the debris into the river systems and transports it downstream as sediment. Rapid urbanization in the past two decades also has resulted in the clearance of large areas of forest, particularly in the urban fringe areas. Increasing population pressure in terms of housing and food requirements has stimulated development in the fringes and environmentally-sensitive areas around the cities.

The suspended sediment loads of rivers in the humid tropics are very high. Intensively cultivated river basins in Java and the Philippines contribute large volumes of sediments to the rivers. The Tjatjaban and Tjeloetoeng Rivers in Java were reported to carry suspended loads of 2,500 and 1,350 m^3 km^{-2} year, respectively (Douglas, 1978). The tributaries of the Kelang River in Kuala Lumpur carry estimated suspended sediment loads of 250 to 550 m^3 km^{-2} year. The sediments are often deposited on flood plains near the city, resulting in raised river bed levels and a greater frequency and magnitude of flooding. Extremely high concentrations of sediment have been reported in the Kelang River, following land clearance. Douglas (1978) recorded a maximum concentration of 81 230 mg l^{-1} in the Air Batu River following a 75 mm rain which fell in 45 minutes where land was being cleared for housing construction. Suspended sediment concentrations above 10 000 mg l^{-1} following periods of heavy rainfall are common in many urban rivers in Malaysia. It is also common for urban rivers in the Philippines and Indonesia to carry suspended sediment concentrations of above 5,000 mg l^{-1} during high flows. In Manila and Jakarta, clearing of urban fringe areas has caused excessive amounts of sediment to be delivered into the rivers and canals. The canals in Jakarta, which are meant to carry sewage and flood waters, are now largely ineffective for flood protection.

The large concentration of sediments in the rivers has both negative and positive impacts. The large sediment plumes at the mouths of rivers are a clear indication of the high rates of sediment transport downstream. Such large quantities of sediment help contribute to the growth of deltas and the consequent increase of land area. They also provide soil nutrients to the flood plains which in turn raise the productivity of the land,

with beneficial effects to the agricultural communities. On the other hand, sediment can clog channels and harbours, thus requiring costly dredging of shipping lanes and harbours.

Solid waste and garbage disposal into rivers

Large volumes of solid wastes and garbage in rivers and on riparian land are a common sight in the tropics. Because public disposal services are inadequate for the task of removal and safe disposal, much of the garbage is handled privately by improper means such as open burning or dumping along roadsides or into rivers or canals. The situation is more critical in larger cities.

In Bangkok (population 5 million in 1980), at least 25% of the garbage produced is not disposed of properly (Tienchai, 1982 cited in Xoomsai,1988). Conservatively, an average person in an urban area generates an estimated 0.5 to 1.0 kg of garbage per day. Thus, the total garbage generated in the city is in the range of 2,500 to 5,000 tonnes day^{-1}. About 2.0% of the total produced is dumped into rivers or canals, which is equivalent to 50–100 tonnes, day^{-1}. In areas outside Bangkok, about 75% of the garbage is improperly discarded, with most of that amount being openly burned. In Kuala Lumpur, only 75% of the households are served by formal garbage collection and disposal. The remaining households, especially those in squatter settlements along the rivers, either rely on open burning or dumping into the rivers. The total amount of garbage collected at the mouth of the Kelang River by the drainage and irrigation department is about 10 000 tonnes month^{-1}. In addition, the garbage collected by the public authorities is often dumped onto former mining lands and swampy areas. The exposed garbage, in most cases, is in direct contact with the water table. Large amounts of leachates thus seep from these landfill sites and contaminate the groundwater. Sanitary landfills are uncommon in Kuala Lumpur, Manila, Jakarta and Bangkok. The dumping grounds in Calcutta and Manila are unique in that they provide a means of livelihood to a large group of urban poor through their scavenging activities.

Floods and drainage

The hydrology of many urban areas in the humid tropics is characterized by high coefficients of surface runoff, high drainage density and prominence of the dual water cycle (stormwater and lower-quality recycled sewage water). In addition, it is characterized by heavy sediment loads, choked drains, inadequate culverts, intensive human activities along rivers, and active river channel morphological processes. Thus the urban hydrology of the region is of special importance because of the strong interrelationships between human occupancy and the river, in a manner that is both symbiotic and parasitic.

Flash-floods are common in many cities of the humid tropics. Endowed with heavy rainfall but without adequate drainage, Bangkok, Kuala Lumpur and Manila often experience damaging flash-floods,following moderately intense rainfall. In Bangkok especially, severe flooding occurs almost every year during the months of October and November, with enormous economic damages. The 1975 flood in Bangkok caused damages of up to 1 billion bahts (US$ 50 million) while damages from the 1982 floods amounted to 6.6 billion bahts (US$ 330 million). Although the Bangkok flood damages are largely due to the heavy monsoon rains, improper drainage and inadequate flood mitigation measures compound the problem.

Occurrences of flash-floods are associated with several interrelated factors which need thorough scrutiny. First are the factors related to urbanization. Early urban development occurred without much consideration of the hydrological and geomorphological processes. Land surfaces were paved, thus increasing the peak runoff, without providing any increase in the capacity of the drainage network. Expansion of human occupancy into the urban fringes, with consequent soil erosion, aggravated the problem by delivering large amounts of sediment into the already clogged drainage system. Slums and squatter settlements, often located along canals and rivers, further obstructed the drainage networks. In addition, much of the garbage produced in these settlements is dumped into the water courses, further hindering the flow of water.

A second set of factors involves natural processes. The geomorphology of urban streams is particularly active because of human interference. This effect is most prominent when channels are forced to accommodate the incoming sediment. Small streams or drains change their paths to avoid choked pathways. Drains and culverts broken in the process of road and housing construction remain unrepaired. The hydrological and hydraulic processes in these urban areas are in constant turmoil – trying to accommodate the extraneous shock caused by urban activities. The prevalence of the dual water cycle further affects the river morphology (mostly on smaller rivers) due to the increased water transmission demands.

The severe damages from urban flash-floods in developing countries are basically a recurring process arising from a vicious socio-economic cycle: economic development, followed by the general inability of the infrastructure to accommodate the development. Often, the existing infrastructure is not adequate to support such rapid urbanization, and the programs for expansion of the infrastructure lag seriously behind needs. The results are degradation of the rivers and drainage networks due to sedimentation, garbage disposal and damages to drains and culverts. The inadequate potable water supply leads to over-abstraction of groundwater, often causing land subsidence which further aggravates the flood problem. In Bangkok, the groundwater abstraction is estimated to be over 1 million $m^3 \, day^{-1}$. As a result, the city is sinking at the rate of 4 cm per year, with some localities subsiding as much as 14 cm per year (See Gladwell, this volume, for recent management action directed toward reducing the rate of subsidence). Jakarta, for example, has practically no sewer system. The canal system – built during the Dutch occupation to receive waste water and to provide some flood control for a small city – is not able to function effectively as an outlet for the garbage and sewage of the present day population of almost seven million. Because of sedimentation and improper drainage, the canals no longer provide much flood control. Many roads and structures become flooded during the frequent heavy tropical rains.

An assessment of the water situation in humid tropical cities would not be complete without mentioning the perception of the people most affected by floods. Human adjustments and adaptations to floods are not merely a function of physical processes but are also a reflection of the different human perceptions and cultural attitudes towards the situation itself. The main question that arises here is "Why do people still remain in the cities?"

Several perception studies carried out in Peninsular Malaysia (e.g., Chua, 1972; Leigh & Low, 1984), particularly in relation to floods, have revealed the benign attitudes of the city dwellers who tended to view the adverse water situation as part of everyday living. The main reason for this attitude is that the people have no satisfactory alternatives. Much of the city's population consists of migrants from rural areas where conditions are even worse than those in the cities and where economic opportunities are limited. Hence, they tend to adapt to the situation as best they can.

WATER RESOURCE MANAGEMENT POLICIES AND CONSTRAINTS

At the Commonwealth Geographical Bureau Seminar held in Hong Kong, Douglas *et al.*, (1990) predicted that there will be more decades of mud in the river systems of the humid tropics. Moreover, McGee & Yeung (1990) indicated that the megalopolis cities are here to stay in the next century, and that effective urban water management will be exceptionally difficult to carry out. Even though there are strong lobbies for "sustainable development" to ensure that water resources are available in the quantity and quality that are required, and even with this management policy having been advocated strongly in many governments, the necessary steps to apply this policy still have not been taken. In many instances, *ad hoc* solutions are the cause of major water disasters. The short-term solutions often are not compatible with long-term aims. One of the major obstacles has been the financial constraints faced by many countries. For example, only in 1989 was a centralized sewage system completed in Kuala Lumpur, and this system covers only half of the city. Although it is unlikely that water shortages would impose limits on urban population growth and economic expansion in the humid tropics, nevertheless they will impose serious challenges to water management. Many areas of tropical Africa are experiencing extreme water shortages due to over-withdrawal and contamination of the water. In southeastern Asian countries the situation is not so severe, but there is a renewed interest by some cities (e.g.,

Bangkok, Jakarta) in developing new and more distant surface water sources, including inter-basin transfers, in anticipation of the increasing water demands.

With regard to pollution control in urban areas, the common factor seems to be inadequate enforcement of laws and regulations. In most countries, water pollution monitoring and enforcement seems to be the weakest links as they depend on adequate manpower and funding, as well as on the will to act, which many developing countries in the humid tropics lack. The financial and economic gains from unfettered development are too attractive to be hampered by enforcement of anti-pollution legislation.

The high population growth rates in the urban centres of developing countries constrain these countries from effectively providing essential water supply and pollution control services for the benefit of all. Water projects are expensive and in the absence of subsidies from central governments or the donor community, city governments would find it extremely difficult to finance needed projects from local sources, either through taxes or user fees. In all developing countries, urban dwellers are already burdened with water bills, conservancy fees, electricity bills and land rents. Hence the options to develop water management projects are often weighed against the costs, the difficulties of financing and the effects on economic development. The options chosen reflect the government's policies and aspirations concerning the range of outstanding issues which will influence decision-making on the scale of water development required and the types of water technologies adopted. In the final analysis, no single option is appropriate for water management for all countries in the humid tropics. Rather, there will be a number of relevant options, ranging from achieving economies of scale in the allocation of water supplies to management of water demands and savings of costs. There will be extensive trade-offs which urban governments must consider in choosing among the various water management options.

REFERENCES

ASEAN Committee on Science & Technology (1982) The ASEAN Compendium of Climatic Statistics, ASEAN Committee of Science and Technology, Jakarta.

Brookfield, H. C. (1988) The new great age of clearance and beyond: What sustainable development is possible? In: J. Denslow, & C. Padoch (eds.), *People of the Tropical Rain Forest*, Univ. of California Press, Berkeley & Los Angeles.

Chang, J. H. & L. S. Lau (1983) Definition of the Humid Tropics. Paper presented at the International Association of Hydrological Sciences in Hamburg, Germany.

Chua, L. (1972) Pekan Floodplain, Pahang: Attitude and behaviour of the floodplain occupants towards possible resettlement. *Graduation Exercise, University of Malaya.*

DOE (1986) Department of Environmental Annual Reports. Ministry of Science, Technology and Environment, Malaysia.

Douglas, I. (1978) The impact of urbanization on fluvial geomorphology in the tropics. *Geo-eco-Trop. 2*, 229–242.

Douglas, I. (1983) *The Urban Environment*. Edward Arnold Publ., London.

Douglas, I., Wong, W. M. & Greer, A. (1990) Rain forest logging, erosion, siltation and water quality: A Sabah experience. *Commonwealth Geographical Bureau Seminar*, Hong Kong.

Gladwell, J. S. & Bonell, M. (1990) *An International Programme for Environmentally Sound Hydrologic and Water Management Strategies in the Humid Tropics*. Paper presented at Symposium on Tropical Hydrology, San Juan, Puerto Rico, July 23–27, 1–10.

Jimenez, R. D. & Velasquez, A. (1989) Metropolitan Manila: A framework for its sustained development. *Environment and Urbanization, 1*(1), 51–58.

Klein, L. (1957) *Aspects of River Pollution*. Academic Press Inc., New York.

Law, A. T. (1980) Sewage pollution in Kelang River and its Estuary. *Pertanika, 3*(1), 13–19.

Law, A. T. & Singh, A. (1982) Mercury in the Kelang River Estuary. *Pertanika, 5.*

Leigh, C. H. & Low, K. S. (1984) Attitude and adjustments to the flood hazard in a mixed ethnic community in Malacca Town, Peninsular Malaysia. *Singapore J. of Tropical Geography, 4*, 40–52.

Low, K. S. & Goh, K. C. (1972) The water balance of five catchments in Selangor, West Malaysia. *J. of Trop. Geography, 35*, 60–66.

Maione, U. (1988) Present and future perspectives on water resources in developed countries. *J. Hydrol. Sc.* Vol. *33*, 87–102.

Marlay, R. (1977) *Pollution and Politics in the Philippines*. Paper in Int. Studies, Southeast Asia Series No. 43, Ohio University.

McGee, T. G. (1967) *The Southeast Asian City: A Social Geography of the Primate Cities of Southeast Asia*. G. Bells & Sons Ltd., London.

McGee, T. G. & Yeung, Y. M. (1990) Urban Futures for Pacific Asia: Towards the Twenty-First Century. *Commonwealth Geographical Bureau Seminar*, Hong Kong.

Niewolt, S. (1965) Evaporation and water balances in Malaya. *J. of Trop. Geography. 20*, 17–26.

Nutalaya, P. (1990) The Atlantic of the Orient. In: *IDRC Reports. 18*(4), 7.

Phantumvanit, D. & Liengcharensit, W. (1989) Coming to terms with Bangkok's environmental problems. *Environment and Urbanization. 1*(1), 31–39.

Pillai, M. S. (1987) An overview of sewerage systems development in Malaysia. Paper presented at *Seminar on Domestic Wastewater Treatment Alternatives*, Universiti Pertanian Malaysia, 9–11 Nov., Kuala Lumpur.

Pryor, R. J. (1979) *Migration and Developments in Southeast Asia: A Demographic Perspective*. Oxford University Press, Kuala Lumpur.

Vitousek, P. M. & Sanford, R. L. (1986) Nutrient cycling in moist tropical forest. *Annual Review of Ecology and Systematics. 17*, 137–167.

World Resources Institute (1990) *World Resources*. Oxford University Press, New York.

Xoomsai, T. N. (1988) Bangkok: Environmental quality in a primate city. In: P. Hills & J. Whitney (eds.), *Environmental Quality Issues in Asian Cities*, , Project Ecoville Working Paper 43, Univ. Hong Kong, 1–23.

25: Urban Water Problems in the Humid Tropics

G. Lindh and J. Niemczynowicz

Water Resources Engineering, University of Lund, Sweden

ABSTRACT

A continuous increase in urban populations and a further growth of urban agglomerations are to be expected in the world. By the year 2010, about 33 per cent of the world's population will live in urban areas. More than 50 per cent will live within the humid tropics by the year 2000. In humid tropics there are many very difficult water-related problems which urgently need solving if living conditions there are to be improved and sustainable development maintained. Most of the existing water-related problems are being intensified by the very fast development of large metropolitan areas. Problems connected to water shortages, pollution, solid waste management and the general lack of adequate data are described and exemplified with several cases. The influence of the existing social and economic structure of society on the possibilities of solving water-related problems is strongly emphasized. It is also stressed that the water sector in the urban society cannot be considered in isolation from other urban sectors, as the management of liquid and solid wastes is closely related to the management of water and other resources.

The region of the humid tropics is rather heterogeneous with regard to the various stages of development, urbanization and industrialization. The magnitude and kind of water-related problems bear some relation to the stage of development. Many problems are highly site-specific. However, several obstacles preventing reasonable water management and implementation of environmentally-sound solutions are common in many countries. These nations can be characterized as lacking sufficient economic funds, information and the data necessary to plan countermeasures. Inefficient communication between the institutions and between the users, planners and designers and a lack of adequately educated personnel also are their hallmarks.

Available solutions include an increase of water reuse for irrigation, groundwater recharging and industrial activity. Municipal waste water is a resource that should be utilized but with the provision for health safeguards. It should be included in water resources planning. Governments should establish standards and guidelines. Stabilization ponds, plant filters and wetlands are preferable, in some cases, to conventional treatment systems. Health protection measures must be evaluated and monitored. Another option is waste water disposal through submarine outfalls. Waste water volumes also may be reduced by application of water-saving skims. The volume of stormwater and runoff from urban surfaces may be reduced by the use of different local disposal, source control and reuse options. The best management of solid wastes must contain elements of source control and reuse.

All these techniques may be applied successfully without damaging the environment or the population, provided they are done in a controllable manner with certain rules being observed. These rules must be based on a general understanding of the processes, interactions and feedback mechanisms involved. However, the application of present technical solutions is not sufficient. In order to achieve sustainability of the environment in the humid tropics, new, radical, economically efficient and environmentally sound technologies that provide solutions to the widespread pollution must be developed. These solutions need not necessarily follow the traditional ways of constructing sewerage and stormwater systems and water purification and treatment plants. Thus, scientific work must continue on new solutions that are based on ecological approaches and environmental concerns. Parallel cooperation leading to the transfer of a modern approach to pollution problems and up-to-date knowledge must continue. Technologies which are appropriate for the humid tropics may emerge from the application of this modern ecological approach.

WATER AVAILABILITY

As background for understanding water problems in the humid tropics, it is of paramount importance to know what water resources are available at a certain location and point in time. Contrary to what may be expected, it is necessary in a region like the humid tropics to carefully inventory the water resources (Ayoade, 1987). One practical reason for studying the components of the hydrological cycle and making water budgets is that precipitation is the most variable element of tropical climates. The annual rainfall total varies considerably

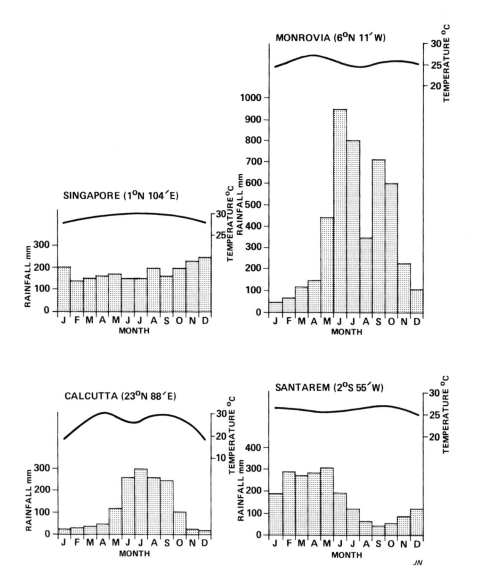

Figure 1: Different rain regimes in the humid tropics.

from year to year and from place to place. Furthermore, other characteristic features of precipitation, such as the seasonal and diurnal distribution, and the intensity and duration of rainy days, show great local variations. The density of recording raingauges is often insufficient to allow an accurate investigation of these variations. Moreover, in the absence of a long series of records, one will not be able to make reliable prognoses of the probability of rainfall. Such a situation may lead to generalized judgments, the reliability of which may be questionable. This is a severe problem because the availability of water is a critical factor for socio-economic development.

When discussing precipitation, something should be said about the differences in the origin of rainfall in tropical and extra-tropical regions. (However, it must be kept in mind that for this subject and others in the tropics there are still large uncertainties because of information shortages.) There mainly seems to be two such characteristic differences. The first one has to do with how the precipitation is generated. In tropical regions, it may be that the uplift of air masses is caused by

convection, whereby in the mid-latitudes, the forcing mechanism is connected with cyclonic activities. One consequence of this difference is that the ascent of air masses in tropical areas occurs at higher speeds but will be of shorter duration. The second major difference originates from the fact that tropical air masses are generally warmer and more humid, compared with those of the mid-latitudes. Such air masses, after having reached condensation at a rather high temperature, produce what sometimes is called "warm rain." The different rainfall regimes which occur in different parts of the humid tropics are shown in Fig. 1.

THE IMPORTANCE OF MAKING WATER BUDGETS

The most fundamental tool for calculating water availability is the water budget. Making such a budget for an urban area is a complicated task. It is, of course, based on estimations of the hydrological components that are needed for a water balance computation (Sokolov & Chapman, 1974). For an urban area, this task is complicated because one has to consider not only what is called the internal system which is made up of the flows within the city (Lindh, 1979), but also the outer system. The internal system shows the water that is withdrawn from nearby sources or imported from more or less remote locations, then used for different purposes. The outer system depicts a water balance based on the natural hydrological cycle, that is, it shows the precipitation, evaporation, infiltration and runoff as they occur in the urban area. Both the inner and outer systems are combined in a joint runoff process where water is discharged to a water course, treated or not. The difficulties that may occur in making such a water budget, based on available data, are evident when we consider the complexity which the urban area itself reflects, being the composition not only of a city core but of unstructured slum quarters. However, even if the difficulties are obvious, such water balances have been made for many cities situated in the developed as well as in the developing countries. See, for example, Chagnon *et al.* 1977; McPherson 1979; Niemczynowicz & Falk 1981. Such water budgets, when used in conjunction with budgets illustrating other urban processes – for instance, the flows of energy, people, goods, etc. – have been excellent when used in planning and decision-making.

Determining the hydrology of urban areas, where one is dealing with relatively small catchments compared with those of rural areas, requires data with very fine time-and-space resolution. Such information is usually not available from national meteorological services. For example, rainfall data necessary for meaningful runoff calculations must have a time resolution of the order of single minutes (Niemczynowicz, 1990; Krejci & Shilling, 1989). A dense network of gauges also is necessary to achieve the required spatial representativity. Such data are usually not available without special measurements. This situation is especially critical for countries situated in the humid tropics. The development of modern urban water management in developed countries situated mainly in the temperate climate zones was preceded by a period of learning about basic hydrological and hydraulic urban processes. Gathering this basic knowledge was facilitated by establishment of several urban experimental basins where different hydrological elements were accurately measured on appropriate time and space scales. There are no such experimental basins, fully equipped with instrumentation, situated in the humid tropics region.

URBAN WATER-RELATED ISSUES IN THE TROPICS

The concept of carrying capacity

Before discussing some specific water-related problems in the humid tropic urban areas, it may be useful to dwell upon the consequences of urban growth. We would like to include not only the problems that are directly connected to the water sector but also all the other related sectors – waste water, urban storm drainage, solid wastes, water treatment, energy, etc. Water problems cannot be considered in isolation from other urban sector problems if we really intend to solve the urban problem from a societal point of view. This is the view taken by the modern approach to city planning. A Man and the Biosphere (MAB) study of decision-making process in Frankfurt am Main is an example (Vester & von Hesler, 1980). Closely related to such an approach is also the idea of a decision-support system (Carlsen, 1987). Another complication in water-budgeting is the interplay of the nearby rural areas with the urban.

As is well known from the development of urbanized areas, the all-pervading problems of managing such development are overpopulation and carrying capacity. The latter concept refers to the capacity of land to support a certain population. The two problems thus are interrelated. If an area, because of in-migration, cannot adequately support its population with the necessities of life, the carrying capacity of this area has been exceeded. However, the concept of carrying capacity only may be a guide for managing urban issues as we run into difficulties if we try to define this concept. Carrying capacity depends on what area it is applied to, since it will be modified by a wide range of factors, including technical change and extreme physical events (Barke & O'Hare, 1986). It has been pointed out by Brown (1987) that the concept of carrying capacity was originally a term used by biologists but nowadays it is applied to entire ecosystems or even a country. Carrying capacity is also a concept of "sustainable development" (Bruntland, 1987). The carrying capacity concept may be useful as a tool for understanding the interplay between different parts of an urbanized area as well as between a city and rural areas or other parts of a country. The concept, ultimately, can be applied to the whole world.

Water management problems

Some urban areas situated in humid tropics, such as Singapore and Hong Kong, have reached a relatively high level of economic and social development with well-organized water services and infrastructure. However, most of the countries in this region belong to the developing world. Economic, political, social and logistic problems which make it difficult to solve water-related problems are common. The origin and expression of these problems, however, may be quite different in the humid tropics than in arid regions. The factors which create these differences include climatic characteristics, geography and topography, the type of agriculture, a social structure specific for an agricultural society, recent and very rapid urbanization and a connected underdevelopment of the urban infrastructure. These are the factors which create site-specific problems and which need the development of special methods and technology.

One typical, though unexpected, problem in this region is the shortage of drinking water of good quality. This deficit is caused mainly by the environmental pressures on the water

Table 1. *Comparison of average annual loadings of rainwater quality parameters (partially after De Luca, 1990) and per cent contribution to the receiving water body of Porto Alegre.*

Parameters	Loadings (kg/ha/year)			% Contribution to Receiving Body
	1	2	3	
SO_4^{2-}	12.85	25.00	116.30	83
Cl^-	3.45	12.30	81.26	79
NO_3^-	3.64	8.00	8.66	99
Ca^{2+}	3.83	14.95	18.29	–
Mg^{2+}	1.15	5.77	15.60	–
K^+	0.77	1.22	12.70	–
Zn	0.68*	24.66	0.19	50
Pb	0.16*	–	1.16	99
Cu	0.23*	2.26	0.19	98

1 – Non-industrial region (USA, Europe),
2 – Highly industrialized region,
3 – Metropolitan Area of Porto Alegre.
* – Data from Sweden (Hogland & Niemczynowicz, 1980).

bodies. Such pressures originated with the underdevelopment of sewerage systems and water treatment facilities. The development of these facilities did not keep up with the industrial and urban development. Thus, the water quality became a crucial issue.

A number of water quality problems, characteristic and common for most of the countries in the region of the humid tropics have been pointed out (Bartone, 1990):

(a) Deterioration of the water supply sources. The rivers serve as receiving water bodies for the major portion of the industrial and urban effluents. The same rivers that serve as major sources of drinking water are also used for personal hygiene and recreational purposes.

(b) Inadequate disposal of waste water. Domestic sewage creates eutrophication and microbiological contamination. Industrial effluents can be shock loads of toxic elements, capable of destroying the biological self-purification capacity of water bodies.

(c) Increasing and unregulated use of raw sewage for irrigation and the mixing of sewage with river water is contaminating the soil and decreasing the soil productivity, while at the same time increasing the potential risks for public health and ecological degradation.

(d) Increasing flooding problems due to decreases in the storage capacity of the soil. Urbanization is the culprit here. Flooding, in turn, brings further environmental stress from non-point pollution sources. Runoff from densely populated areas – in connection with often inadequate management of solid wastes and insufficient street cleaning – constitutes an important pollution source.

(e) Soil erosion from the increased frequency of flooding heightens problems with sediment transport in rivers and siltation in reservoirs.

Besides their obvious regional and global impact on the environment, pollution problems connected with urbanization have a major influence on the health of the population. About 80 per cent of tropical diseases can be attributed to poor or non-existent sewage treatment and the lack of safe drinking water (Gladwell, 1990). The situation may be best enlightened by a few examples.

The first is the city of Hong Kong which has progressed recently in economic activity and urban development but has contributed significantly to environmental pollution at the same time, (Iayawardena, 1987). Most of rivers within the New Territories carry increasing amounts of urban and industrial effluents. During the early 1970s, it was reported that of the 400 km of streams, 165 km were polluted and 240 km could be classified as clean. By 1981, the situation had changed significantly; the percentage of deteriorated stream mileage had increased from 40 in 1972 to 64 in 1981 (Holmes, 1983). Hong Kong reached the practical limit of its fresh water resources in the late l980s. Further increases in demand are being met by importation from China.

Low-lying regional areas also are subjected to increased flooding, thus causing damage to the farmland. Coastal waters are significantly affected by pollution brought by rivers and runoff from the land areas. The so-called Hong Kong Epidemiological Study (Cheung *et al.*, 1990), performed over a period of two months in 1987, revealed that pollution of coastal waters resulted in 60 000 people per year suffering from swimming-associated gastrointestinal and skin maladies.

The second example, the Porto Alegre metropolitan area of Brazil, with three million inhabitants, shows that the type of water-related problems is highly site-specific. In this region, due to rapid industrialization and the increased burning of fossil fuels, significant changes in the quality of the air, rain, and the surface and coastal waters have been observed. High concentrations of heavy metals, associated with fly ash from mineral and vegetal coal burning and other industrial sources, have been measured in the air. In this situation, the relative importance of pollution released by surface runoff and by stormwater is very high compared to the pollution loading from, in this case, well-organized sewerage and treatment plant systems. It was found that about 90% of the Pb, 50% of the Zn, 96% of the Cr, 52% of the BOD and 24% of the ammonia was being released via the stormwater system and surface runoff. Further treatment of domestic sewage would not do much to improve the quality of receiving waters (De Luca *et al.*, 1990). It is perhaps instructive to compare the average annual loadings of rainwater washed off from areas in developed countries (Europe, USA) and in Porto Alegre. Values from Porto Alegre may be seen as representative for several other highly urbanized locations in the developing countries. Table 1 gives such a comparison.

Another example is from Bangkok. One-third of that city's population has no access to public water and must obtain water from vendors (Sivaramakrishnan & Green, 1986). According to Low (this volume), much of the human wastes are discharged onto vacant land or into rivers and canals. The stormwater system is inefficient causing flooding of the city several times a year. Stormwater polluted by waste water, as well as runoff from the urban area seriously pollutes the Chao Phraya River.

It is clear that the use of septic tanks is by no means a good solution for taking care of human waste, especially not in growing urban areas (Feacham et al., 1977). According to the authors, during the early stage of urban growth, a municipality is tempted to rely upon privately owned systems. Later on, it may be very costly to replace inadequate systems.

As a consequence of insufficient methods of garbage disposal, about half of Bangkok's 2,500 tons of solid waste is deposited in the stormwater clongs. About 25% of the garbage is improperly disposed of – with 2% being deposited in the rivers.

Continuing land subsidence also is connected to the rapid urbanization and is one of the reasons for the increasing frequency of floods within the flat deltaic plain of the Chao Phraya River and the rising of groundwater that has been observed. In northern Thailand, more than 50% of the yearly rainfall occurs within a period of two to three months. Bangkok experiences severe floods several times a year, especially during October and November. Enormous economic losses occur.

In order to bring about a permanent solution which will be effective for the topographical changes and urban growth expected by the year 2000, a master plan has been prepared which incorporates a comprehensive flood protection and drainage system for eastern Bangkok (Noppun & Klankrong, 1990).

In spite of the ambitious target – over 95% of the Thai population (more than 50 million people) was to be served with clean water by 1990 – set up by the Thai government in the early 1980s, only 58% of the population was actually covered by water supply systems (Lowatanatrakul, 1990). The quality of water delivered by existing water supply systems also is steadily deteriorating.

In Malaysia, central sewerage systems in 1985 served only 5.3% of the population. One-third is served by septic tanks, the rest (61.6%) use pit and bucket latrines not connected to any treatment facility (See Table 4, Low, this volume). The fecal coliform content of the Kelang River was reported to be from 113–16 000 100^{-1} ml, compared with the WHO standard of 100 coliforms 100^{-1} ml for bathing waters. The Rangamm River which flows through the industrial area of Shah Alam shows an increase of BOD from 1 mg l^{-1} upstream to 34 mg l^{-1} downstream.

In northern Malaysia, 40 factories built close to the Juru River resulted in the following pollution levels: oil and grease 1,800 mg l^{-1}; mercury 2.3 m l^{-1}; lead 1.5 mg l^{-1}; chromium 1.4 mg l^{-1}; and, cadmium 0.3 mg l^{-1}. The mercury level was 400 times higher than what the USA's Environmental Protection Agency recommends. The cadmium level was 30 times higher. Fishing offshore from the Jahor River was severely affected and finally abandoned. A number of food poisoning cases have occurred.

The rubber and palm oil industries are a severe source of pollution. In the year 1986, oil palm mills disposed of 1,893 tonnes of waste per day, resulting in the devastation of prawn farms.

Urban development clears large areas of land and increases erosion damage. High rainfall intensities, large drop sizes and high winds characteristic of the climate cause erosion and very high levels of sediment transport in the rivers.

The Tjatjaban and Ttjeloetoeng Rivers in Java carry suspended loads of 2,500 and 1,350 m^3 km^{-1} per year with the highest recorded value being 81 230 mg l^{-1}. In Kuala Lumpur, only 75% of households are served by garbage collection services. The deposited garbage, in most cases, is in direct contact with the water table. Floating garbage is a common sight in rivers. The total garbage generated in the cities of Malaysia is in the range of 2,500–5,000 tons day^{-1} (Low, 1989).

Similar problems can be found in Jakarta. The water supply and disposal system was intended to serve a population of half a million (Sivaramakrishnan & Green, 1986). However, the city, with a population of more than 7.7 million in 1985 – projected to grow to 17 million by the year 2000 (UNESCO, MAB 1988) – is suffering from a continuous water shortage (For a lower projection for the year 2000, see Table 2, Low, this volume). Less than a quarter of the population has direct access to the water system, with about 30% of the households depending on water sold by vendors.

There are also other site-specific, water-related problems. One is the over-exploitation of groundwater which increases the natural salt-water intrusion in the aquifers. The total water demand has increased in Jakarta from 10 mln(million) m^3 $year^{-1}$ in 1950 to more than 500 mln m^3 $year^{-1}$ in 1985 and is expected to reach 1,200 mln m^3 $year^{-1}$ in 2005 (JICA, 1985). The water level in the originally artesian aquifer is now generally below sea-level (locally 30 m) as far as 10–15 km inland (Schmidt et al., 1988). The resulting sea-water encroachment is spoiling this source of drinking water. Planning of counter measures is difficult because adequate data are lacking.

The sewerage system in Jakarta is practically non-existent. Septic tanks are used by a quarter of the population. The remaining citizens use pot latrines, cesspools and ditches along the roadside. At the same time, the population uses the drainage canals for bathing and laundering. A World Bank sewerage and sanitation project was set up for Jakarta in order to demonstrate a comprehensive environmental clean-up in a pilot area with a population of about half a million (Zajac et al., 1984). We will return to the water-related problems of Jakarta and the World Bank project in the next section.

One positive example of a location where water supply and, to some degree, sewage disposal, are efficiently managed, is the island of Penang, located on the northwestern shore of Malaysia. A drastic increase in domestic and industrial water consumption, concomitant with the high rate of urbanization and industrialization that occurred over two decades, was met by efficient management that could be a model for several other cities located in the developing countries. The creation of the Penang Water Authority (PWA) in 1972 was the key to success. With the introduction of water metering and reduction of water losses, the annual income and expenditures showed that the PWA did more than break even, unlike the situation in many large cities. Without going into details about the solution, it may be stated that comprehensive planning, together with a well-balanced application of technical, institutional and infrastructural measures, resulted in 95% of the island's residents having access to a potable water supply by the year 1985 (Goh, 1988).

Some waste management problems

As was mentioned in a previous section, solid waste management in most of the countries in the humid tropics is not well-organized, which gives rise to specific environmental problems. These problems may be classified into three main groups: (a) problems related to the existing practices of disposing of excreta, (b) problems connected with bad management of municipal solid waste, and (c) problems with disposal of toxic industrial waste.

The development of water supply systems has advanced much faster than has waste management. According to McGarry (1977), the slow development of facilities for the disposal of human wastes may be explained by the following reasons: the collection of human wastes is aesthetically displeasing; governmental commitment and motivation is low; and social constraints on the design of the technology have led to capital-intensive systems.

Improvements in human excreta disposal practices, development of low-cost facilities and alternative technologies are crucial to increasing the level of public health. As a matter of fact, human waste possesses a high content of organic matter; it may be mixed together with other solid waste and composted into a useful product (Obeng & Wright, 1987).

Management of municipal solid wastes has been neglected for a long time because of low governmental and societal motivation, a lack of appropriate technologies and a lack of understanding as to how solid waste influences the water quality of receiving water bodies. As a result, solid waste management frequently suffers more than other municipal services when budget allocations and cuts are made (Cointreau, 1982). Even in resource-scarce societies, governments have not yet regarded waste as a potential asset capable of reuse (Pollock, 1987). In many humid tropic countries, a stormwater system is constructed as open ditches, channels and clongs. Solid wastes which are disposed of in the stormwater system create obsta-

cles, reducing the hydraulic capacity of the system. Floods can result. During dry spells, anaerobic conditions may develop in the channels, producing hydrogen sulfide odours.

Mountains of garbage accumulate along the roadside. A large portion of solid waste is burned, causing air pollution. Leachate from domestic and industrial waste deposits and landfills pollute the rivers and the groundwater. Waste management and disposal by composting, landfilling or incineration is often hindered by lack of basic information about the composition of the waste and its origin. In addition, economic constraints may require that the questions of construction, location, transportation and marketing be solved in a reasonable way.

As industrialization progresses, the volume of industrial solid and liquid waste grows. Also, as we know from experience in developed countries, industrial waste often contains highly toxic elements, such as heavy metals and various hydrocarbons. These kinds of deposits create problems which may be hidden for decades before the real effects are discovered. There are no data available which can illuminate the present extent of these problems in the humid tropics.

The need for a modern approach

By the year 2000, about 50% of the globe's population will live in the humid tropics. The large urban agglomerations also are growing the fastest in that region. But in spite of great efforts, the basic human needs of water supply, sanitation and waste disposal are still not satisfied in the humid tropics. It is fundamental that all actors who might influence this situation recognize the importance of increasing their efforts to effectively fight the environmental problems related to water and waste management in the large urban areas of the humid tropics. In order to keep pace with the urban population increase, water-related problems of large urban systems must be addressed faster and more efficiently than they are today. Such progress is necessary to move toward sustainable development in this region.

Let us exemplify with two cases what has gone wrong with water management in some urban areas of the humid tropics and what were the major hindrances against harmonic development of water-related infrastructure in these areas.

Let us once more take Jakarta as an example of a city where urban water problems are significant. An early failure in population growth estimation has resulted in an infrastructure with "built-in" errors. Although the system has been expanded, less than a quarter of the population has direct connection to the water system. Because of Jakarta's local situation, salt-water intrusion is causing deterioration of the groundwater. Furthermore, the demand for water from a still increasing population is resulting in over-exploitation of the salt-polluted groundwater.

As stated before, sewerage systems in Jakarta are almost non-existent. Siwaramakrishnan & Green (1986) comment that "much of the population, however, has no alternative other than to use the drainage canals for bathing, laundering and defecation." In light of these words, it is interesting to note the statement made by Zajac *et al.*, (1984) that "the majority of the people in developing countries do not enjoy the benefits of an adequate supply of safe water and facilities for sanitary disposal of their wastes. Progress in improving this situation has been slow."

With regard to this latter statement, it may be interesting to acquaint oneself with the Jakarta Sewerage and Sanitation Project which the above-mentioned authors discussed in a World Bank report. The main goal of the project was a comprehensive environmental clean-up in a pilot demonstration area of Jakarta that had a population of about half of million. Not only was this the first sewerage program for Jakarta but it also included another set of measures, especially in the poor Kampung zones. The project contained such additional measures as surface drainage improvements, public water tap installations, improvements in the system of individual leaching pits used by poor people to dispose of excreta, provision of public washing/bathing/toilet units and improvement in solid waste management. Thus, the World Bank project aimed beyond the traditional design of water systems. Its ambition was also to integrate several measures on different levels of social needs.

The main difficulty in carrying out this project was the lack of information on the adequacy of the existing sanitation facilities. It was recognized that there was not enough information on the existing system to enable an engineer to design a new, improved system. Thus, there was an apparent need for an engineering survey to quantify the sanitation gaps and their origins. The next step was to propose solutions based on the desired sanitation level. This, of course, included the contribution of engineers and an adequate and carefully prepared design of the system.

The authors of the World Bank report added that it may be desirable to include a minimum-cost program of continuing to monitor the community facilities. There is one very interesting sociological aspect of preparing such a program. People living in poor conditions may not always accept paying for maintenance and repair if they are not convinced of the effects of such measures. It may be difficult to convince people about meaningful returns on investment in such facilities. Thus there is an obvious connection between sanitary projects and social aspects. Gathering of socio-cultural data prior to building water supply and sanitation projects becomes important. The engineers may need to work together with sociologists, anthropologists and/or health educators.

With respect to the planned technical solutions, it also may be beneficial for the project to have the engineers work in an interdisciplinary team. It is of paramount importance to have excellent communications between all persons involved in the project. It also should be stressed that the success of changing to an alternative way of managing water supply, waste water or solid waste problems is closely related to communication

activities before the project starts. See Perrett (1984). A dialogue must be established between the users of the new system and the designer of the system. In fact, starting such a dialogue at an early stage of the project development may be useful as good ideas then may be utilized during the implementation process.

Another example of "what can go wrong" with water management in the humid tropics took place at Greater Sao Paulo, one of the world's largest metropolitan areas. Less than half of Sao Paulo's 14 million inhabitants are served by sewer systems. Only 5% of the sewage is properly treated. In 1975 the state government initiated creation of a master plan dealing with sewage treatment and disposal. The resulting SANE-GRAN plan contained several measures aimed at solving existing sewage problems in the city. About 50% of all sewage was to be treated in a two-stage treatment plant in order to remove 90% of the organic load by the year 2005 (Water International, 1987). The rest of sewage was to be discharged into the sea. In the 1970s, an integrated solution of Sao Paulo's sewerage problems was proposed. Stabilization lagoons were to be utilized to dispose of about 80% of the total sewage volume from the city. However, by 1983 the capacity of the treatment plants did not exceed 5% of the total sewage volume. Progress has continued to be slower than planned.

The above examples of Jakarta and Greater Sao Paulo can teach us some basic lessons, summarized as follows:

(a) The existing infrastructure for water supply and sanitation is a result of a social and technological development process that is combined with changing perceptions of societal needs. Mistakes in this perception, committed during the early stages of development, are "built into" the infrastructure during later stages.

(b) It is costly and difficult to change existing systems. It is more difficult to convince people to pay for improvements than for new systems.

(c) The social structure and traditional behavior influence the need and the type, extent and design of the water-related facilities. Disregard of the socio-economical aspects is a common reason for failure.

(d) Only integrated actions performed interdisciplinarily and cooperatively may be able to build upon existing socio-economic conditions to create viable (sustainable) solutions.

(e) Optimistic plans and financial schemes are seldom fully realized in time.

The question remains as to which technical solutions should be used to meet societal needs while simultaneously protecting the degrading environment. Perhaps one of the reasons for failure of ambitious schemes is the use of the traditional approach that promotes only the technical, large-scale, "end-of-pipe" solutions. However, the understanding that our past approaches to water-related problems in urban areas have not always been right is spreading among scientists and technicians in the developed countries. It is essential that this change in basic technical paradigm now be promoted also in the economically weak regions of the world. Application of the new, multidisciplinary approach may give substance to the concept of a new, ecologically sound and economically effective technology.

Summary of the problems

The humid tropics region is heterogeneous with regard to development, urbanization and industrialization. The magnitude and kind of water-related problems in the region bear some relation to the various stages of development. Many of the problems are highly site-specific. Each location has its own problems and requires its own solutions. Common for all countries in the humid tropics are the climatic conditions characterized by large rainfall volumes and high intensities. Another common factor is the rapid urban development of large metropolitan areas and the problems connected with them, such as deteriorating quality of the surface water bodies, the groundwater and the soil, and in some places, land subsidence and flooding.

Several obstacles preventing reasonable water management and implementation of environmentally sound solutions to the existing problems are common in many countries. They are:

Lack of sufficient funds Sanitation facilities and sewerage systems could be built and industrial wastes and garbage could be treated if sufficient financial resources could be obtained. Despite the questionable truth of such a statement, it is clearly unrealistic to count on enormous sums (that would be necessary) being raised in the near future. Therefore, the problem must be redefined: what are the realistic options for actions to improve the present situation? In other words, we must look for modern solutions which are financially realistic as well as environmentally sound and find logistic ways to implement them.

Insufficient information necessary to plan countermeasure Without appropriate data it is not possible to make accurate calculations, or model hydrological processes. Reasonable water management and design of drainage systems in cities is difficult, lacking the answers from such calculations. On the other hand, without experimental data it is not possible to acquire the basic knowledge about the hydrological processes governing the hydrological cycle in urban areas. Such knowledge is necessary in order to develop calculation methods and design procedures relevant for the local climatic, geological and socio-economic conditions of the humid tropics. Therefore, the first step must be the establishment of several urban experimental basins where different hydrological elements are accurately measured on appropriate time and space scales. There are no such experimental basins, fully equipped with instrumentation, situated in the humid tropics region. The next step would be an implementation of system solutions, i.e., a total environmental restoration of several carefully chosen urban catchments.

Lack of, or inefficient, communication between the institutions and between the users, the planners and the designers It was noted during several development projects that this was the main cause of failure. The positive example of Penang Island given above can teach us that a well-organized water authority may establish proper ways of communication and create viable solutions. It must be stressed also that the environmentally sound solutions can be found only if multidisciplinary cooperation is established and water management is considered together with management of other resources and urban development issues. The existing socio-economic and cultural conditions also must be taken into account (Lindh, 1983, 1985).

A lack of educated personnel This statement is not always valid. The question is rather whether the personnel have been adequately trained to solve the problem in its social context. Another important question is whether the responsible persons are really able to implement the multidisciplinary actions and apply the integrated water management necessary for finding environmentally sound and economically effective solutions.

The scarcity of positive examples due to failure of several pilot projects The reason is not only financial. The lack of adequate information and communication were identified as the main reasons for the failure.

Available technical solutions

Let us examine what types of options are available:

(a) Considering the existing shortage of water resources of good quality in the region, it is advisable that more waste water be reused. Waste water may, after adequate treatment, be used for irrigation and for recharge of groundwater aquifers (Bartone, 1990a). The following conclusions and recommendations were formulated by a scientific group dealing with agricultural waste water: (1) Municipal waste water is a resource that should be utilized (but) with health safeguards.(2) Reuse of waste water is preferable and should be included in water resources planning. (3) Reuse should be supported by an integrated set of measures including treatment, crop restriction and human exposure control. (4) Governments should establish standards and guidelines. (5) Stabilization ponds are preferable to conventional treatment systems. (6) Health protection must be evaluated and monitored. (WHO, 1989; Kreisel, 1990).

(b) Industrial waters should be recycled. This requires changes in industrial processes and development of new technologies.

(c) Another option for disposal of waste water that is available to coastal cities is submarine outfalls (Sharp 1990; Bartone 1990a). This technology is widely used in many countries. One example is the Ipamena outfall in Brazil. A 2.4 m diameter outfall pipe discharges 6 m^3 s^{-1} (will be expanded to l2 m^3 s^{-1}) of waste water from Rio de Janeiro.

(d) Yet another option is to apply some kind of "simplified sewage treatment." Waste stabilization ponds, plant filters and wetlands are options for the disposal of municipal waste water. The volume of waste water also may be reduced by the use of water-saving skims.

(e) Stormwater and runoff from urban surfaces may be reduced by the use of different local disposal, source control and reuse options. Specially designed infiltration and percolation facilities, detention ponds and retention reservoirs, and deep wells for recharge of groundwater, belong to this group of options.

(f) The best management of solid wastes must contain elements of source control and reuse. Solid wastes constitute a valuable resource, reuse of which may provide an economical solution. Another option is incineration which may bring economic benefits through energy production.

All these techniques may be applied without damaging the environment or the population, provided that they are done in a controllable manner with certain rules being observed. These rules must be based on general understanding of involved processes, interactions and feedback mechanisms. This procedure requires, in turn, an ecological approach that is based on understanding the cyclicity of material and energy flows in nature. Contrary to the above, the application of some techniques for waste and sewage disposal as they are used today, emerged from necessity and economic constraints rather than from scientific considerations. Irrigation with waste water may be performed safely, provided that the ways of pollution are known and controlled to a degree that ensures that it is safe to use the crops and that there is no accumulation in the soil or groundwater. It has been shown that properly designed ocean outfalls may be a beneficial alternative to land-based treatment by diluting domestic sewage waters to the levels that have no adverse ecological effects (Bartone, 1990; Margetta *et al.*, 1990). However, in a global sense, it is not enough if the land-based treatment plants or the ocean outfalls operate satisfactorily on a short time scale, say 10–20 years. It may be advisable to release waste water in the ocean, provided that the biological activity is able to disarm all the pollutants, depositing the residuals in a similar manner to what occurs in nature. But if the actions lead to the accumulation of pollutants that are not biodegradable, environmental problems will occur sooner or later.

It is doubtful whether this modern ecological approach is really being applied in the developing countries today or whether it will be applied in the near future. The process of changing approaches must be accelerated in these countries.

The future

In order to sustain the environment of the humid tropics, as well as the environment of the developing countries located elsewhere, new, radical, economically efficient and environmentally sound technologies that provide solutions to widespread pollution must be developed. These solutions need not necessarily follow traditional ways of constructing sewerage and stormwater systems, nor the new water purification and treatment plants.

More than 90 m³ sec of waste water is produced within Greater Sao Paulo. In order to treat this amount, 156 treatment plants with a capacity of 50 000 m³ day should be constructed. Such a program, of course, is possible, but few countries can really handle the costs of such an investment. We know now that several subsequent plans for construction of waste water treatment plants in Sao Paulo have not been realized, with the sanitation problems remaining enormous. As another example, 37 billion m³ sewage water per year is released in China without treatment (Zhixin, 1989). Taking 50 000 m³ day as the capacity of a medium- to large-sized treatment plant, more than 2,000 such plants would be needed. It is, of course, unrealistic to believe that their construction could be accomplished in the near future. Thus, European sewage treatment technology remains uninteresting to China. Nor is it the most optimal solution for Sao Paulo. Thus, new and less expensive solutions must be found. Scientific work must continue on new solutions that are based on an ecological approach and environmental concern. These solutions also should be economically effective, because only such solutions are affordable for these countries. At the same time, cooperation must continue that leads to a modern approach to pollution problems and the transfer of up-to-date knowledge.

World organizations, such as UNESCO, UNDP, The World Bank, UNEP, FAO, and WHO, have done considerable work for developing countries and are continuing to do so. The question is whether these efforts are optimal in relation to their possible impact. Many efforts have been devoted to the "transfer of knowledge and technology" and have made a real impact on the state of knowledge in many countries. However, are these actions what are needed the most? What technology and what knowledge is transferred? What should be transferred? The transfer of modern ideas and approaches to the environmental problems of today may be more important than the transfer of traditional knowledge and technology.

Perhaps the best way to develop answers to these questions is to establish carefully selected groups of top-ranked scientists and technicians from the developed and developing countries to discuss the problem in question until a consensus is reached. The group members should create a multidisciplinary team that covers the interests of all parties involved, with Nature being treated as a very important partner. Only through such actions may water management really be integrated with other human activities. Positive examples of such actions may be created by working on small, well-defined areas in pilot projects where the real goal is the transfer of ideas and approaches in order to promote the change of paradigm which must occur in those countries. Perhaps it is more important to allocate money for such actions instead of constructing gigantic dams and factories.

It may be enlightening to discuss in more detail what is the modern approach to water/environmental problems (they cannot be divided) and what the reasons are for the change of paradigm which has occurred recently in developed countries. There are certain key words representing the modern approach to how we should organize our activities in a more environmentally-friendly manner. These key-words also represent our present stage of knowledge; they are valid across all national, climatic and regional borders; and they should be applied independently to the present stage of development in the humid tropics and elsewhere. The technologies which are appropriate for different regions may emerge from the application of the following key-words:

Integrated system approaches, in contrast to narrow-minded technical solutions. Multidisciplinary cooperation as well as an integrated system approach are crucial for solving complex problems. Relevant and environmentally-sound technology can emerge as a result of such cooperation.

Small scale in contrast to technological monumentalism. It is less expensive to construct small-scale treatment units than large ones. The distances of transport are shorter, plants are less vulnerable, and it is possible to design plants adjusted to the local needs and conditions. Perhaps the clean-up of Sao Paulo would be smoother if several, smaller treatment plants, adjusted to the quality of the local sewage, were constructed instead of large plants. This alternative should be considered.

Source control instead of the "end of pipe" approach.

Local disposal and reuse instead of exploitation and wastefulness. The volume of sewage waters and wastes may be reduced at the source by changing the routines in the production stage and/or by local disposal. Examples are the infiltration of storm and waste water, reduction of water use by structural and non-structural measures, construction of small treatment facilities close to the polluting industry, biologically balanced infiltration of waste water to the ground and oxidation ponds. Water may be reused for industry after local treatment, wastes may be reused after separation at source, ashes may be used for road construction, heavy metals can be extracted from effluents, using bacterial uptake, etc. The traditional solutions, both for treating waste water and solids, mix together several pollution components which makes reuse difficult.

Pollution prevention instead of reacting to damage already done.

The use of biological systems in waste water and solid waste management. Biological systems may, in some cases, substitute for traditional treatment facilities. Biological activity in the soil may serve as the best treatment plant. Bacteria, algae, plants and animals may together constitute a system which selectively removes and concentrates all pollution components in any waste water. Waste water enters the system, clean water leaves the system, plus then we have reusable resources such as wood (energy), paper, cattle food, chemicals, and food. Such systems have yet to be developed, but parts of them may be used to-day. Literature on this topic is extensive (Mara & Caimcross, 1989; WHO, 1989; Zweig, 1985). Treatment plants using biological systems may be much cheaper than traditional plants. According to Widmer (1981), a traditional plant, with a trickling filter or using activated sludge, that is large enough to serve a population of 100 000, is five times more expensive than a waste stabilization pond system. A pond without bottom sealing costs between seven and ten times less than the same capacity conventional treatment plant (Arthur, 1983). However, the price of the land may be a limiting factor. Work is going on to reduce the land requirements (Yuki *et al.*,1991). This is a field of work for many scientists in various disciplines.

Ecological philosophy Existing or artificially-created ecological systems may be used in the treatment of water. Wetlands, ecotones close to the rivers and riparian zones with root uptake may take part in the purification of water. We must give them a chance to do this by increasing the time of contact. Water seeping through wetlands, the meandering courses of rivers, plant filters, etc., may be used. Ecological systems may be specially designed to do the work. In the hydrology of urban areas, "blue-green zones" touch upon this possibility (Ontario, 1987). In these zones, stormwater, waste water and industrial waste water are led through a water course with specially designed lakes/ponds which attenuate the flow, increase the residence time and make biological activity possible. Sediment is dredged periodically, concentrated and reused. Plants are collected and used.

The use of wetlands and plant filters for treatment of waste waters may become a major method that will be possible to apply in developing countries where construction of traditional treatment plants is economically impossible. The use of plant filters in conjunction with production of renewable biomass as a fuel is equally attractive for developed countries, where alternative energy sources must be developed in order to reduce CO_2 emissions. It is also an alternative to the development of new steps in water treatment plants to reduce the emission of nitrogen. It may also be an attractive method for cutting pollution from stormwater. These ecologically sound and economically attractive methods may soon substitute for traditional treatment methods. Thus, waste water may become a valuable resource which is used in the biological process of biomass production. This process is closer to the natural biological cycles of matter and energy.

Understanding this philosophical approach and its application to solving of water-related problems may lead to the development of technology which will be really appropriate in the humid tropics as well as in any other climatic region.

REFERENCES

Arthur, J. P. (1983) .Notes on the design and operation of waste stabilization ponds in warm climates of developing countries. World Bank Technical Paper No 7. Washington, D. C. USA.

Ayaode, J. O. (1987) Forecasting and managing the demand for water in Nigeria. *Water Resources Development*, *3*, No. 4.

Barke, M. & O'Hare, G. (1986) *The Third World*. Oliver and Boyd.

Bartone, B. C. (1990) Water Quality and Urbanization in Latin America. *Water Inter.*, *15* No 1, 3–14.

Bartone, B. C. (1990a) International perspective on water resources management and wastewater reuse – appropriate technology. Proc. of the Fifteenth Biennial Conference of the IAWPRC: *Water Pollution Research and Control*, Kyoto, Japan, 2039–2048.

Brown, L. R. (1987) Analyzing the demographic trap. In: L. Brown (ed) *State of the World, 1987*. W.W. Norton and Co., New York.

Bruntland, G. H. (1987) *Our Common Future*. Oxford University Press.

Carlsen, G. H. (1986) Decision-making in water resources planning. Proc. UNESCO Symp., Oslo, Norway.

Changon, S. A., Huff, F. A., Schickedanz, P. T. & Vogel, L. L. (1977) Summary of Metromex, Vol. 1, Weather anomalies and impacts. Illinois State Water Survey, Bulletin 62, Urbana, USA.

Cheung, W. H. S., Hung, R. O. S., Chang, K. C. K. & Kleevens, J. W. L. (1990) Epidemiological study of beach water pollution and health-related bathing water standards in Hong Kong. Proc. of the Fifteenth Biennial Conference of the IAWPRC: *Water Pollution Research and Control*, Kyoto, Japan, 243–252.

Cointreau, S. (1982) *Environmental management of urban solid wastes in developing countries*. The World Bank, Washington.

Feachem, R., Mc Garry, M. & Mara, D. (1977) *Water, Wastes and Health in Hot Climates*. John Wiley and Sons, Chichester.

Gladwell, J. (1990) Keeping the boiler stoked. In: UNESCO Sources. *Water: our most precious resource*. UNESCO, March 1990.

Goh, K. C. (1988) Challenges of water supply development in the urbanized island of Penang, Malaysia. Proc. Internat. Symp. *Hydrological Processes and Water Management in Urban Areas*, Duisburg, FRG. 657–665.

Hogland, W. & Niemczynowicz, J. (1980) Kvantitativ och kvalitativ vattenomsattningsbudget for Lund centralort. Department of Water Resources Engineering, University of Lund, Report No 3038.

Holmes, P. R. (1976) Water quality management in Hong Kong. *J. of the Hong Kong Institution of Engineers*, June 1983.

Jayawardena, A. W. (1987) Water supply and development in Hong Kong. In: *Rivers and Water Resources in East Asia*. (Proc. of the Symp. in celebration of Prof. Y.Takahasi's retirement), University of Tokyo, December 1987. 93–98.

JICA, Japan International Cooperation Agency (1983) Master plan for the Jakarta Water Supply Development Project. Internal Report, Ministry of Public Works, Jakarta.

Krejci, V. & Schilling, W. (1989) Urban hydrologists need meteorologists!. Proc. WMO/IAHS/ETH Workshop on Precipitation Measurement, St. Moritz, Switzerland, 3–7 December, 371–376.

Kreisel, W. (1990) Water quality and health. Proc. Fifth Internat. Conf. on Urban Storm Drainage, Osaka, Japan. 201–209.

Lindh, G. (1979) Socio-economic aspects of urban hydrology studies and reports in hydrology. No. 27, UNESCO.

Lindh G. (1985) The Planning and Management of Water Resources in Metropolitan Areas. Dep. of Water Res. Eng., University of Lund, Report No 3105.

Lindh G. (1983) *Water and the City*. UNESCO.

Lowatanatrakul, W. (1990) The Provincial Water Supply in Thailand and the Water Decade. Proc. of the Fifteenth Biennial Conference of the IAWPRC: *Water Pollution Research and Control*, Kyoto, Japan, 223–228.

De Luca, S. J., Milano, L. B. & Ide, C. N. (1990) Rain and urban stormwater quality. Proc. Fifth Internat. Conf. on *Urban Storm Drainage*, Osaka, Japan. 133–140.

Mara, D. & Cairncross, S. (1989) Guidelines for the safe use of waste water and excreta in agriculture and aquaculture: Measures for public health protection. World Health Organization, Geneva.

McPherson, M. B. (1979) International Symp. on Urban Hydrology. Tech. Memo. 38, ASCE, Urban Water Resources Research Programme, New York, N.Y.

Margetta, J., Fontane, D. G. & Seok-ku, (1990) Multicriteria ranking wastewater disposal for coastal towns. *Water Inter.*, *15*, No 2.

Niemczynowicz, J., & Falk. J. (1981) Water budget for the City of Lund. Proc. of the Internat. Conf. on *Urban Storm Drainage*, Urbana, Illinois. USA.

Niemczynowicz:, J., (1990) Necessary level of accuracy in rainfall input for runoff modelling. Fifth Int. *Conf. on Urban Storm Drainage*, July 23–27, Osaka, Japan.

Noppun, M. & Klankrong, B. (1990) Flood protection and drainage project in Bangkok. Proc. Fifth Internat. Conf. on *Urban Storm Drainage*, Osaka, Japan. 1549–1557.

Obeng, L. A. & Wright, F. W. (1987) Integrated resource recovery. The composting of domestic solid and human wastes. Technical Paper No 57, The World Bank, Washington.

Ontario Ministries of Natural Resources, Environment, Municipal Affairs and Transportation and Communication, Association of Conservation Authorities of Ontario, Municipal Engineering Association, Urban Development Institute, Ontario (1987) Urban Drainage Design Guidelines.

Pollock, C. (1987) Mining urban wastes: The potential for recycling. *Worldwatch Paper* No 76.

Schmidt, G., Soefner, B. & Soekardi, P. (1988) Possibilities for Groundwater Development for the City of Jakarta, Indonesia. Proc. Internat. Symposium *Hydrological Processes and Water Management in Urban Areas*, Duisburg, FRG 24–29 April. 505–516.

Sharp, J. J. (1990) The use of ocean outfalls for effluent disposal in small communities and developing countries. *Water Inter.*, *15*, No 1.

Low, K S. (1989) Urbanization and urban water problems: a case of unsustainable development in Asean countries. Internat. Colloquium, *Development of Hydrological and Water Resource Strategies in the Humid Tropics*, Townsville, Australia, 15–22 July.

Sivaramakrishnan, K. C. & Green, L. (1986) *Metropolitan Management: The Asian Experience*. Oxford University Press.

Sokolov, A. A. & Chapman, T. G. (1974) Methods for water balance computations. Studies and reports in hydrology No. 17, UNESCO.

UNESCO, MAB (1988) Man belongs to the earth. UNESCO'S Man and the Biosphere Programme. Paris.

Vester, F. & Von Hesler, A. (1980) Sensitivitetsmodell. Regionale Planungsgemeinschaft Untermain, Frankfurt am Main, FRG.

Widmer, W. J. (1981) Summary previews of waste stabilization ponds. In: Waste Stabilization Ponds: Design and Operation. WHO/EMPO Technical Publication No. 3:73–111.

World Health Organization (1989) Health guidelines for the use of wastewater in agriculture and aquaculture." Report of a WHO Scientific Group. WHO Technical Report Series 778. Geneva.

WQI, Water Quality International (1987) Sao Paulo faces seven options". No 3, p.21.

Yuki, Y., Takayanagi, E. & Abe, T. (1991) Design of multi-storey sewage treatment facilities. *Wat. Sci. Tech.* Vol. 23, Kyoto, Japan, pp 1733–1742.

Zajac, V., Mertodiningrat, S., Susanto, H. S. & Ludwig, H.F. (1984)

Appropriate technology for water supply and sanitation. Technical Paper No 18., The World Bank, Washington, D.C.

Zhixin, L. (1989) Report on urban hydrology in China. Proc. IHP Conf: Integrated Water Management and Conservation in Urban Areas, 28 Aug.–5 Sept. Nagoya, Japan.

Zweig, R. O. (1985) Freshwater aquaculture in China/ecosystem management for survival, *Ambio*: *l4 No.2*, 66–74.

26: Rethinking Urban Water Supply and Sanitation Strategy in Developing Countries in the Humid Tropics: Lessons from the International Water Decade

Yok-Shiu F. Lee

Program on Environment, East-West Center, 1777 East-West Road, Honolulu, Hawaii 96848 USA

ABSTRACT

A full assessment of the International Water Decade's experience requires us to look beyond water supply and sanitation coverage figures and to begin to identify new ways of thinking about the sectors that have emerged. Four major lessons can be derived from the Decade. First, technology alone is not enough; managerial, organizational and social issues must also be considered in designing efficient systems. Second, appropriate and low-cost technology maximizes the value of investments. Third, innovative cost recovery programs are essential to sustainable projects. Fourth, community participation in planning, operation and maintenance tasks is the key to successful overall system performance. Whereas the objectives were technology oriented and were defined in terms of "coverage" at the beginning of the Decade, the emerging approach goes beyond coverage to emphasize institutional aspects and the "human factor" as central to achieving "sustainable, effectively used services."

INTRODUCTION

The International Water Supply and Sanitation Decade (1980–1990) was the most notable international and national effort that sought to improve the provision of clean water supply and adequate sanitation facilities in the developing countries. Many of these countries are located within the humid tropics and other warm humid regions and will represent nearly 33% of the world's population by the end of this century. The continuing population shift from developed to developing countries and from rural to urban areas, and the concomitant need to sustain the urban environments present a formidable challenge to cope with the requirements and impacts of water-related activities that will accompany such transformations in the humid tropics and other warm humid regions. A critical review of the accomplishments and shortfalls of the International Water Decade will provide an important step towards rethinking the urban water and sanitation strategy for these regions.

The Decade called for 100% coverage of safe water and sanitary services in all rural and urban areas by 1990 (UNCHS, 1987). At the time this goal was pronounced, it did not seem overly ambitious. However, the launching of the Decade coincided with the beginning of a world recession. Three years into the Decade, a gloomy assessment by World Health Organization officials pointed to "the absence of strong popular and official support, weak institutions, shortage of trained personnel, doubts about technology and insufficient financial resources" (Urban Edge, 1983). Experience during the Decade's first half led many countries to trim their plans so that, by 1985, the proportion of countries aiming for 100% urban water supply coverage fell from 48% to 40% and those with similar hopes for urban sanitation dropped from 33% to 26% (WHO, 1987).

We all know by now that, despite the momentum generated by the Decade, the coverage of population in developing countries with water supply and sanitation services is far from satisfactory. High rates of population growth and low or even negative growth rates in GNP per capita are commonly cited as two major factors that have curtailed the coverage rate of water and sanitation (Christmas & de Rooy, 1991; Najlis & Edwards, 1991).

For instance, a recent WHO estimate of sector investments by governments and External Support Agencies (ESAs) in the 1981–1985 period indicates a total of US$70 billion (1985 dollars) – an average of US$14 billion a year. For the 1985–1989 period, the World Bank estimates sector investments by governments and ESAs at about US$9.3 billion (1985 dollars) a year. The estimated lower levels of investment during the last half of the Decade have been attributed to the generally lower rates of economic growth during this period in most developing countries (McGarry, 1991). But it has also been pointed out that, in many cases, developing countries have little choice on how to spend their money. Structural adjustment loans from the International Monetary Fund often come with requirements that include stiff cuts in public spending without regard, for example, for the negative impacts of such policies on a population's welfare (Mfutakamba, 1989).

Table 1. *Water supply and sanitation coverage by region, 1980-1990 for developing countries (population in millions).*

Region/ sector	1980				1990			
	Population	Percent coverage	Number served	Number unserved	Population	Percent coverage	Number served	Number unserved
Africa								
Urban Water	119.77	83	99.41	20.36	202.54	87	176.21	26.33
Rural Water	332.83	33	109.83	223.00	409.64	42	172.06	237.59
Urban Sanitation	119.77	65	77.85	41.92	202.54	78	160.01	42.53
Rural Sanitation	332.83	18	59.91	272.92	409.64	26	106.51	303.13
Latin America and the Caribbean								
Urban Water	236.72	82	194.11	42.61	324.08	87	281.95	42.13
Rural Water	124.91	47	58.71	66.20	123.87	62	76.80	47.07
Urban Sanitation	236.72	78	184.64	52.08	324.08	79	256.02	68.06
Rural Sanitation	124.91	22	27.48	97.43	123.87	37	45.83	78.04
Asia and the Pacific								
Urban Water	549.44	73	401.09	148.35	761.18	77	586.11	175.07
Rural Water	1823.30	28	510.52	1312.78	2099.40	67	1406.60	692.80
Urban Sanitation	549.44	65	357.14	192.30	761.18	65	494.77	266.41
Rural Sanitation	1823.30	42	765.79	1057.51	2099.40	54	1133.68	965.72
Western Asia (Middle East)								
Urban Water	27.54	95	26.16	1.38	44.42	100	44.25	0.17
Rural Water	21.95	51	11.19	10.76	25.60	56	14.34	11.26
Urban Sanitation	27.54	79	21.76	5.78	44.42	100	44.42	0.00
Rural Sanitation	21.95	34	7.46	14.49	25.60	34	8.70	16.90
Global totals								
Urban Water	933.47	77	720.77	212.70	1332.22	82	1088.52	243.70
Rural Water	2302.99	30	690.25	1612.74	2658.51	63	1669.79	988.72
Urban Sanitation	933.47	69	641.39	292.08	1332.23	72	955.22	377.00
Rural Sanitation	2302.99	37	860.64	1442.35	2658.51	49	1294.72	1363.79

Source: Christmas, J. and C. de Rooy. 1991. "The Decade and Beyond: At a Glance." *Water International*, 16(3), p. 129.

The changes in service coverage throughout the Decade in the developing world are summarized in Table 1. While the Decade has made tremendous gains in terms of extending the provision of water and sanitation services to *additional* numbers of people during the 1980s, in some parts of the developing world (notably Africa and South Asia), the total number of residents still without safe water supply and adequate sanitation has remained the same or has actually increased. According to the United Nations Secretary General's report on the achievement of the Decade:

... about 1,348 million more people were provided with safe drinking water supply in developing countries during the 1980s, 368 million in urban areas and 980 in rural areas. Similarly, 748 million more people, 314 million urban dwellers, and 434 million people in rural areas

were provided with suitable sanitation services. Overall, the number of people without safe water decreased from 1,825 million to 1,232 million, while the number of people without suitable sanitation remained virtually the same (Najlis & Edwards, 1991).

The most dramatic result of the Decade was recorded in the rural areas where the number of people without safe water supply decreased by 624 million, and those without adequate sanitation by 79 million (Najlis & Edwards, 1991). Urban areas, however, are still better served than rural areas with water supply and sanitation services throughout almost all of the developing world (Table 1). Thus, the problem of providing water and sanitation services is, at present, overwhelmingly rural (as discussed elsewhere in this volume by Wurzel) rather than urban in nature, if one is to judge the magnitude of the problem

Table 2. *Unaccounted-for water in municipal water supply systems in developing countries in humid tropics and other warm humid regions.*

City, country	Proportion of unaccounted-for water to total water supply	Year
	(in percent)	
Manila, Philippines	55–65	1984
Jakarta, Indonesia	50	1976
Mexico City, Mexico	50	1983
Bangkok, Thailand	32	1990

Sources: Manila: Richardson, J. 1988. "Non-revenue Water-a Lost Cause?" *The Proceedings of the 14th WEDC Conference*, p. 147.

Jakarta: Mehta, R. S. 1982. "Problems of Shelter, Water Supply and Sanitation in Large Urban Areas: in UNEP, *Environment and Development in Asia and the Pacific*, p. 242.

Mexico City: UNCHS. 1984. *Environmental Aspects of Water Management in Metropolitan Areas of Developing Countries: Issues and Guidelines*, Nairobi, p. 38.

Bangkok: Sethaputra, S. *et al.*, 1990. *Water Shortages: Managing Demand to Expand Supply*. Bangkok: TDRI, December, p. 97.

by the number of people affected. While there is no doubt that undue priority has been given in the past to urban areas, the urban fringe where the majority of urban poor have settled in slums and shanty towns has been largely overlooked. The limited information available indicates most urban infrastructure investments are directed to the well-off neighbourhoods. Residents in urban slums may have a lower level of services than their rural counterparts who often have at least the choice of using alternative sources of water (World Bank, 1988; UNCHS, 1987).

BEYOND NUMBERS: AN EMERGING NEW APPROACH

It is clear by now that the Decade's primary goal of full access to water supply and sanitation was not achieved by the target year of 1990. However, the Decade's experience cannot be summarized just in numbers. Despite the inability to extend 100% coverage, it has succeeded in crystallizing a growing awareness of the seriousness of the health problems associated with inadequate water supply and sanitation facilities. Moreover, the Decade's activities have led to concerted efforts in establishing international collaboration concerning water supply and sanitation services and in accelerating the growing attention on the sector's institutional and sociological aspects.

Furthermore, research and development work undertaken during the Decade has resulted in developing and introducing new, low-cost technologies, and in focusing attention on the user communities as active participants in the developmental process rather than merely passive recipients (Christmas & de Rooy, 1991; Grover & Howarth, 1991; Najlis & Edwards, 1991; Okun, 1991; Warner & Laugeri, 1991).

A full assessment of the Decade's experience thus requires us to look beyond the conventional measuring yardsticks of global and regional coverage figures and to begin to identify new ways of thinking about the sector that have emerged in the course of the Decade. In fact, the accumulation of small events over the last 10 years has produced a substantially new approach that has far more significance than the accumulated statistics.

The approach being taken contains a new thrust which has been shaped by past evaluations of water and sanitation projects in developing countries and is directing attention to the following issues: the mobilization of communities to manage their programs, including increased involvement of women and development of cost recovery mechanisms for sustainable, effectively used services; the development and utilization of affordable, appropriate technologies, including the development of appropriate delivery and operation and maintenance systems; and the continuing need for international co-ordination regarding sector inputs, particularly with regard to capacity building and human resources development. In short, whereas the objectives were technology oriented and were defined in terms of "coverage" at the beginning of the Decade, the emerging approach goes beyond coverage to emphasize the "human factor" as central to achieving "sustainable, effectively used services" (Melchior-Tellier, 1991).

LESSONS FROM THE DECADE

Lesson 1: Technology alone is not enough

One major realization from the experience of numerous projects and programs implemented during the International Decade is that technology alone is not enough. Far too much emphasis was given to the construction of new facilities by national and international agencies at the expense of developing appropriate provisions for the proper operation and maintenance of existing and new installations (Warner & Laugeri, 1991). Little attention was given in particular to the institutional aspects associated with the implementation of such functions (Najlis & Edwards, 1991).

An excellent example of the serious consequences of inadequate operation and maintenance is the large volume of unaccounted-for-water in many urban centres in developing countries. In many cities in developing countries located within the humid tropics and other warm humid regions, around 50% of the water that is treated and distributed at public expense is not accounted for by sales (Table 2). There is no record of it having been delivered to consumers and it does not earn revenue

Table 3. *Per capita unit costs (median) of construction of water supply and sanitation system in the Asia-Pacific region.* *(Current US$)*

| | Urban water supply | | | | Urban sanitation | | | |
| | Home connections | | Standposts | | Sewer connections | | Individual household systems | |
	1980	1985	1980	1985	1980	1985	1980	1985
South East Asia	55	60	–	35	63	81	15	20
Western Pacific	80	96	20	42	220	444	50	73

– Not available.

Source: World Health Organization, 1987. *The International Drinking Water Supply and Sanitation Decade, 1981–1990.*

for the water supply authorities. World Bank research suggests that, as a rule, if more than 25% of the treated water is not accounted for, a program to control the losses may prove to be cost-effective (Urban Edge, 1986).

Implementing a formal control policy to reduce both physical losses (through leakage detection and repair) and nonphysical losses (through improved management practices) typically costs US$ 5–10 per capita. Studies have shown that savings and increased revenue will repay this cost within one or two years (Richardson, 1988). Investment to improve the performance of existing assets is thus highly cost-effective. Reduction in unaccounted-for-water can allow investments in new works to be deferred or at least reduced in scope, with significant savings as a result. In addition, by improving the system of meter reading and billing or by detecting and charging for illegal connections, revenue can be greatly increased to pay for the costs of treating and distributing the water, as well as the costs of operation and maintenance of the system. Also, if illegal connections are found and charged for, willingness to pay by all may improve. For example, in urban areas in Thailand, each 10% of unaccounted-for-water saved would immediately generate an additional US$8 million per annum from the 3.5 million people served (Richardson, 1988).

Furthermore, where the distribution systems are corroded and broken, appreciable increases in supply do not reach the consumers but result in higher leakage losses. That is, implementation of augmentation projects without controlling leakages could become counter-productive (Kumer & Abhyankar, 1988). Passive control of water loss, such as repairing leaks only when they are noticed, is inadequate. Active control measures are needed such as zone metering to monitor for suspected leaks and systematic leakage detection (Richardson, 1988). Leakage from newly constructed systems, which can be as significant as that of old piping networks, can be minimized through careful review of design, materials and construction standards and tightened monitoring over construction (Urban Edge, 1986).

Minimizing leakage alone is not enough, however. Non-physical losses can account for a quarter (Bangkok) to one half

(Manila) of unaccounted-for-water and they can be reduced at less cost than leakage (Richardson, 1988). Major strategies for controlling non-physical losses include: the installation, prompt servicing and recalibration of meters; the updating and reviewing of consumer records to establish a sound basis for estimating consumption when meters are unserviceable; and the streamlining of bureaucratic procedures to make it easier for customers to make new legal connections to reduce the "theft" of water (Richardson, 1988).

Although these strategies and measures are common-sense, well-intentioned efforts to reduce high levels of unaccounted-for-water, they have met with little success, indicating the difficulty of solving this seemingly simple problem in the developing countries.

This situation also reflects the tension between the apparent need for investment in system expansion and the requirements for routine repair and maintenance of the existing system. Whereas new investment in system expansion usually receives enthusiastic support from managers, engineers and politicians, routine repair and maintenance work is somehow always accorded a low priority. This highlights the fact that high rates of unaccounted-for-water are linked not only to technical problems, but to broader managerial, organizational and social issues that must also be taken into consideration in designing efficient water supply systems.

Lesson 2: Appropriate technology maximizes the value of investments

The availability and quality of urban water supply and sanitation services depend to a great extent on the standards of physical infrastructure systems such as water piping and sewer networks. In many developing countries, there is a tendency to insist on standards higher than necessary, which sometimes doubles the cost of service delivery. The result is poor access to water supply and sanitation services (UNEP, 1982; Ridgley, 1989; Gakenheimer & Brando, 1987). Per capita unit costs of providing services have generally continued to increase despite the development of less expensive technologies (Table 3). Only a drastic revision of design standards to sharply reduce construction costs is likely to offer hope of providing even minimal levels of public water services to extensive low-income urban neighborhoods.

With few exceptions, the technologies currently in use in developing countries are the same as those employed in the developed countries: piped water, full internal plumbing, and conventional water-borne sewage (Ridgley, 1989). Services tend to be provided to those sectors of the population with developed country incomes. Even here there are problems because these systems require expensive equipment and materials, often imported, and trained manpower to operate them, none of which is easily available in the resource-poor developing world.

A full-scale attack on urban sanitation problems in developing countries in the humid tropics would require huge increases in investment only if Third World governments insist upon adopting water-borne sewage – the conventional form of urban sanitation. However, conventional sewerage is inappropriate for use in low-income communities because of its high costs which are in turn the result of the use of inappropriate construction standards (Pickford, 1990; Taylor, 1990). World Bank research has demonstrated that a wide range of household and community systems could greatly improve the sanitation conditions at affordable costs to the urban poor (Sinnatamby, 1990). The solutions involve low-cost, locally manufactured hardware (plumbing, sanitary sheds, concrete cans for pit latrines) that can be installed using labor-intensive techniques. The central technologies range from improved ventilated pit latrines to simple modifications of standard sewerage designs that reduce diameters, excavation, inspection chambers and other standard specifications (Campbell, 1989). The total annual cost per household of several of these options was only one-tenth to one-twentieth that of the conventional sewerage systems. Most demand far less water to allow for their efficient operations and it is possible to install one of the lowest-cost systems initially and then upgrade it gradually (Hardoy & Satterthwaite, 1989).

Estimates are that the current distribution of sector investments in developing countries to high-cost and low-cost technology is in the order of 80% and 20% respectively (Christmas & de Rooy, 1991; Kalbermatten, 1991). This essentially means that 70 to 80% of funds go to serve 20 to 30% of the population, mostly the higher income groups. Experience gathered from Decade programs, however, demonstrates that measures designed to make more effective use of existing resources play an important role in maximizing the impact of investments. Considerable progress has been made in the development of low-cost appropriate technologies with increased operational reliability. The major problem of the Decade, one which will remain for years to come, is to extend low-cost appropriate technologies to the low-income communities hitherto unserved.

Conventional wisdom suggests that aid-giving agencies and consultants from developed countries encourage the use of costly imported equipment and materials which are produced in their lands of origin. Gakenheimer & Brando (1987) argue, however, that there are strong influences within the developing countries themselves – an "unintentional conspiracy" – that insist on unnecessarily high standards. These include engineers who are most familiar with modern solutions, government agencies who pursue failure-proof and maintenance-free construction and politicians who wish to avoid being accused of "demodernizing" services. Taken together, these actions and inactions result in an unfortunate tendency toward unrealistically high standards.

Appropriate technologies and the methods of selecting them are, therefore, more institutional than technological. Because the tension between financial constraint and the adoption of high cost infrastructure is caused by numerous actors with different motivations, an overall restructuring of the major institutional relationships may be necessary to yield an effective solution. One of the future challenges for research in the sector is to delineate specific, feasible actions that could be effectively taken in different cities in the developing world to make such a rearrangement.

Lesson 3: Innovative cost recovery programs are essential to sustainable water and sanitation projects

The problem of cost recovery is to some extent linked to the difficulty in attracting skilled staff because it is cost recovery that makes it possible to pay salaries – from a secure revenue base – which attract and retain trained personnel. Therefore, in many ways, the key to improving the performance of urban water supply and sanitation services over the short- and medium-term lies in the ability of these public utilities to recover an increasing percentage of the cost of providing services from their customers. Nonetheless, the most recent data from WHO indicate that with the exception of Singapore and the Philippines, the average water tariffs in countries in the humid tropics and other warm humid regions do not cover the average operating costs of water production (Table 4).

At the beginning of the Decade in 1980, cost recovery issues were given little attention. But, by 1985, "an inadequate cost recovery framework" was cited by the WHO as the second most serious constraint facing the Decade (WHO, 1987). By the late 1980s, many External Support Agencies and governments in developing countries had reconsidered their cost recovery policies and had started looking seriously for means of implementing cost recovery programs (Franceys, 1990; Katko, 1990). By now the question is no longer whether to charge but how much (Melchior-Tellier, 1991). The debate is taking shape over whether water supply tariffs should cover only operation and maintenance costs or whether they should also generate resources for future investment. A closely related conflict is arising between the increasing demand for sector financial viability through full-cost recovery and the long-standing commitment by most urban governments to provide subsidized services to their constituencies (Warner & Laugeri,

Table 4. *Unit costs of water production (operation only) and water tariffs in countries in the humid tropics and other warm humid regions, circa 1985.*

Country	Average cost of water production (US\$/m^3)	Average water tariff (US\$/m^3)
India	0.08	0.05
Bangladesh	0.09	0.08
Thailand	0.21	0.21
Burma	0.25	0.20
Sri Lanka	0.25	0.20
Philippines	0.05	0.15
Papua New Guinea	0.55	0.55
Singapore	0.24	0.29
Samoa	0.09	0.03

Source: World Health Organization, 1987. *The International Drinking Water Supply and Sanitation Decade, 1981–1990.*

Table 5. *Prices charged by water vendors, mid 1970s–1980.*

City, country	Multiples of price charged by public water utility
Surabaya, Indonesia	20–60
Dacca, Bangladesh	12–15

Source: World Resources Institute, 1990. *World Resources 1990–1991.* New York: Oxford University Press, p. 77.

1991). So far, very little attention has been paid to cost recovery from sewage services.

In most developing countries, the conventional wisdom is that the poverty of the vast majority of the citizens may make cost recovery very difficult. Yet the supposition that the poor in developing countries cannot afford or will not pay for water services is belied by the widespread practice of water vending at market prices, indicating a high level of affordability and willingness to pay for water by the poor. Nonetheless, it is unclear whether governments have the political will or administrative capability to charge the poor.

Although water vending is an old tradition all over the world, little attention has been paid to it in studies of water supply. Recent investigations show that in the absence of access to a public water supply system, people spend substantial amounts of money on vended water (Katko, 1990; Okun, 1988). The evidence clearly shows that service from water vendors costs substantially more than is paid by customers served by piped water system in the same area (Table 5). The poor may pay as much as 30% of their income for water

whereas the well-to-do pay less than 2% (Okun, 1988). Supplying free or almost free water, therefore, often produces very inequitable results where only the better-off consumers with house connections reap the benefits.

Instances accumulated over the Decade indicate that the success of a revenue recovery program depends to a great extent on how the revenue is collected. Very often, the major barrier to securing financial involvement of low-income groups lies not with their absolute inability to contribute but with the lack of innovative cost recovery programs (Najlis & Edwards, 1991). There is an understandable reluctance of many low-income people in developing countries to pay money to a government department which they suspect of corruption or in which they have little or no confidence. With increasing demand to increase the rate of cost recovery, there is a greater need to understand not only how much the user is willing to pay for service but, from the user's perspective, how the money is to be collected and managed (McGarry, 1987). The need to implement cost recovery programs thus leads to a fundamental, but as yet unanswered, question: What roles should each party in urban water supply development and planning play? The major actors are the central government, the municipal administration, the water agencies, the consumers (household and institutional), and the private supply sector.

Lesson 4: Community management is key to successful overall system performance

The first half of the Decade was marked by the development of new, low-cost technologies appropriate to the needs of the developing countries. Yet, both the urban and rural landscapes of the developing world are littered with inoperative pumps that may have been well-conceived at the office of a donor agency and a country ministry but have fallen into disrepair because of the lack of commitment and participation of the local populations who were purportedly the beneficiaries of such projects (Okun, 1988). The need now is not so much for further technological innovation, but for the rethinking of the institutional framework to draw forth improved performance through alternative approaches to management and maintenance. For the full benefits of water supply and sanitation to be realized, more is required than the installation of the structures, pumps, and pipes.

One major reason for failure in water supply and sanitation projects has been that in the minds of international and national planners in this sector, health improvements were the greatest, if not the only, benefits of water supply and sanitation. But the population receiving these services, whether rural villagers or urban squatters, have additional concerns. For them, the reduction in labor spent in collecting water, the prestige of having water in or near the home, or privacy and comfort of a water closet could be the primary reasons in demanding improvements in water supply and sanitation. Improved health is often a distant third or fourth priority for many users (McGarry, 1987). Well-meaning low-tech alternatives such as

Table 6. *Developing country needs for sector services, 1990–2000.*

	Population Not Served in 1990	Expected Population Increase 1990-2000	Total Additional Population Requiring Service by 2000
	(millions)		
Water Supply			
Urban	243	570	813
Rural	989	312	1,301
Total	1,232	882	2,114
Sanitation			
Urban	377	570	947
Rural	1,364	312	1,676
Total	1,741	882	2,623

Source: Report A/45/327 of the Secretary General of the Economic and Social Council to the UN General Assembly, July 1990.

VIP latrines are unlikely to be accepted, for example, because they brand the recipient as socially inferior or their design does not provide a certain minimum degree of privacy for women. The success of introducing appropriate technologies requires a thorough understanding of the perceived needs of the user communities. In particular, the participation of women, who are the major system users, in the design, construction and sustained effective use and management of projects, has been increasingly recognized as critical to the success of sector initiatives (Ellis, 1990; Najlis & Edwards, 1991).

Benefits of community/user participation in all stages of water supply and sanitation projects could include lower costs, greater likelihood of acceptance of the technology, and greater user maintenance of the facilities. Studies have shown that those projects with strong community input are the most successful in terms of reaching the greatest number of the poor with long-lasting services (McGarry, 1987).

Reorienting project design and implementation methods to incorporate meaningful users' participation is not an easy task, however. Twenty years ago, "community development" was used to refer simply to the generation of local contributions. At the beginning of the Decade, the emphasis had shifted to the concept of "community participation," which calls for varying degrees of local involvement (Melchior-Tellier, 1991). Current terminology now centers around "community management," which "goes beyond simple participation. It aims to empower and equip communities to control their own systems" (Warner & Laugeri, 1991). It thus requires substantial structural and attitudinal changes within the implementing agencies. It also highlights the tension between the bias of many managers and administrators to adhere to conventional municipal systems in urban centres and the need to design and implement innovative, decentralized, low-cost communal systems for low-income settlements. Research is clearly needed to help address the question of how to motivate the centralized urban water and sanitation agencies to undertake effective outreach measures to attend to low-income communities where the needs for their services are the most acute but largely unmet.

CONCLUSION: CHALLENGES IN THE 1990s

The Decade has made tremendous gains in extending water and sanitation services to more than 1.3 billion people during the 1980s. However, roughly one person in three in developing countries still did not have a reliable supply of safe drinking water in 1990. About 44% of the population also lacked access to sanitation. United Nations estimates are that the population in these countries will increase by almost 900 million in the 1990s. Two-thirds of this increase will occur in already crowded urban centres located in the humid tropics and other warm humid regions (Table 6). That is, by the year 2000, combining those who are not yet served and the expected population increase in the coming decade, more than two billion additional persons will require sector services.

Recent estimates conducted by UNICEF in consultation with the World Bank and UNDP indicate that if sector services were to be provided to 90% of the population in developing countries, "the average annual level of investment required for new services alone, excluding operation and maintenance and rehabilitation costs, would need to be nearly three times higher than the average achieved during the 1980s" (Najlis & Edwards, 1991). This formidable task is compounded by the challenge of sustaining existing facilities which require substantial maintenance and repair expenses as well as the prospect of a rising per capita cost of water provision and its environmentally safe collection and disposal (Grover & Howarth, 1991; McGarry, 1991).

This task is further exacerbated by the shift of the burden of poverty from rural areas to the urban centres as the major population movement to the cities in the developing countries continue in the coming years. By the year 2000, the number of urban households living in absolute poverty is projected to increase by 76% to 72 million, whereas that of the poor rural households is expected to fall by 29% to 56 million (UNDP, 1990). As demand for sector services will become increasingly pressing in urban areas, particularly by the poor who generally remain unserved, there is a need to examine critically such distributive questions as who is to be served now, who is to be served later, and at what service level (McGarry, 1991). One major challenge in the coming decade is, therefore, posed by the increasing number of urban poor who lack basic water and sanitation services.

Although one might expect that the level of sector assistance from External Support Agencies (ESAs) would increase in the coming decade, the ESAs collectively have provided only a small portion of the roughly US$ 10 billion invested annually in the sector and will in all likelihood finance a declining share of overall future investments (Grover & Howarth, 1991). The common outlook for the sector is that the bulk of resources for future initiatives will have to be generated within the developing countries themselves. However, it is doubtful, even under optimistic assumptions about economic growth, that the share of national development financing in many developing countries allocated to water supply and sanitation could be dramatically increased (Najlis & Edwards, 1991).

In conclusion, the major challenge of the next decade is to develop appropriate institutional mechanisms in urban centres in developing countries whereby municipal and local resources could be mobilized and utilized in a sustainable, efficient and equitable fashion. Lessons from the Decade indicate that the "human factor" is particularly critical in the design of future strategy to successfully meet this challenge. The task is no longer a narrow and technologically-oriented one of installing facilities, but one of assuring their sustainable and effective utilization in interaction with the user communities.

Given that the proportion of unserved population will remain sizeable by the end of the century, public financing of water and sanitation sector activities will continue to be limited, and the per capita cost of provision of conventional sector facilities will rise dramatically, the principal courses of actions to approach the sector's future will have to include increasing attention to efficient operation and maintenance of existing systems, further development and installation of appropriate and low-cost alternative technologies, the development of innovative cost-recovery programs, and reorienting project design and implementation methods to support meaningful community management.

ACKNOWLEDGMENTS

Valuable suggestions and comments on earlier drafts of this manuscript were given by Maynard Hufschmidt and James Nickum. Editorial assistance was provided by Regina Gregory and Angelina Lau typed the manuscript with efficiency and accuracy.

REFERENCES

Campbell, T. (1989) Environmental dilemmas and the urban poor. In: J. Leonard (ed.) *Environment and the Poor: Development Strategies for a Common Approach.* Transaction Books, New Brunswick. 165–187.

Christmas, J. & de Rooy, C. (1991) The decade and beyond: at a glance. *Water Int.*, *16* (3): 127–134.

Ellis, K. V. (1990) Potable water for the developing world – some of the problems. *Aqua 39* (6): 368–375.

Franceys, R. (1990) Paying for water – urban water tariffs. *Waterlines*, 9 (1): 9–12.

Gakenheimer, R. & Brando, C. H. J. (1987) Infrastructure standards. In: L. Rodwin, (ed.) *Shelter, Settlement, and Development* Allen and Unwin, Boston. 133–150.

Grover, B. & Howarth, D. (1991) Evolving international collaborative arrangements for water supply and sanitation. *Water Int., 16* (3): 145–152.

Hardoy, J. & Satterthwaite, D. (1989) *Squatter Citizen.* Earthscan, London.

Kalbermatten, J. M. (1991) Become reality or remain a dream? *Water Int., 16 (3)*: 121–126.

Katko, T. S. (1990) Cost recovery in water supply in developing countries. *Int. J. of Water Resources Development, 6* (2): 86–94.

Kumar, A. & Abhyankar, G. V. (1988) Assessment of leakages and wastages. In: *Proc.of the 14th WEDC Conference-Water and Urban Services in Asia and the Pacific*, 23–26.

Lindh, G. (1983) *Water and the City.* UNESCO, Paris.

McGarry, M. G. (1987) Matching water supply technology to the needs and resources of developing countries. *Natural Resources Forum, 11* (2): 141–151.

McGarry, M. G. (1991) Water supply and sanitation in the 1990s. *Water Int., 16* (3): 153–160.

Mehta, R. S. (1982) Problems of shelter, water supply and sanitation in large urban Areas. In: UNEP, *Environment and Development in Asia and the Pacific.* 227–258.

Mekvichai, B., Foster, D. & Kritiporn, P. (1990) *Urbanization and Environment: Managing the Conflict,* Ambassador City Jomtien, Chon Buri.

Melchior-Tellier, S. (1991) Women, water and sanitation. *Water Int., 16* (3): 161–168.

Mfutakamba, A. R. (1989) Safe drinking water and sanitation for all: An 80s goal dashed by development policies. *Environmental News Digest, 7* (3): 6.

Munasinghe, M. (1990) Water supply policies and issues in developing countries. *Natural Resources Forum* Feb. 1990. 33–48.

Najlis, P. & Edwards, A. (1991) The International Drinking Water Supply and Sanitation Decade in retrospect and implications for the future. *Natural Resources Forum, 1* (2): 110–117.

Okun, D. A. (1988) The value of water supply and sanitation in development: An assessment. *American J. of Public Health, 78* (11): 1463–1467.

Okun, D. A. (1990) Water reuse in developing countries. *Water and Wastewater Int., 5* (1): 13–21.

Pickford, J. (1991) Training and human resource development in water supply and sanitation. *Water Int., 16* (3): 169–175.

Richardson, J. (1988) Non-revenue water-a lost cause? In: Proc.*of the 14th WEDC Conference.* 147–148.

Ridgley, M. A. (1989) Evaluation of water-supply and sanitation options in third world cities: An example from Cali, Colombia. *Geo J. 18* (2): 199–211.

Sethaputra, S. Panayotou, T. & Wangwacharakul, V. (1990) Water Shortages: *Managing Demand to Expand Supply.* Bangkok: TRDI, December.

Sinnatamby, G. (1990) Low cost sanitation. In: Cairncross, S., Hardoy, J.& Satterthwaite, D. (eds.) *The Poor Die Young.* Earthscan, London. 127–157.

Taylor, K. (1990) Sewerage for low-income communities in Pakistan. *Waterlines* 9 (1): 21–24.

United Nations Center for Human Settlements (UNCHS) (1984) *Environmental Aspects of Water Management in Metropolitan Areas of Developing Countries: Issues and Guidelines,* Nairobi.

UNCHS (1987) *Global Report on Human Settlements.* Oxford University Press, Oxford.

UNEP (1982) *Environment and Development in Asia and the Pacific* Nairobi.

Unvala, S. P. (1989) Bombay's water supply situation: Drought and migration wreak havoc on limited resource. *Water and Wastewater Int.,* 4 (1): 33–37.

Urban Edge (1983) Water's uncertain decade of development, 7, (8).

Urban Edge (1984) Participation in urban development, 8 (5).

Urban Edge (1986) Tackling the Problem of "Lost" Water, 10, (6).

van der Mandele, H. (1989) Resolving riddles of rice/demand. *Water Resources Journal.* June.

Warner, D. B. & Laugeri, L. (1991) Health for all: The legacy of the Water Decade. *Water Int., 16* (3): 135–141.

Wiseman, R (1990) Low-cost technology favored by UNDP/World Bank program. *Water and Wastewater Int., 5* (2): 11–15.

World Bank (1988) *Information and Training for Low-Cost Water Supply and Sanitation.* Washington, D.C.

World Health Organization (WHO) (1987) *The International Drinking Water Supply and Sanitation Decade*, 1981–1990.

WHO (1988a) *Urbanization and Its Implications for Child Health.* Geneva.

WHO (1988b) *Review of Progress of the International Drinking Water Supply and Sanitation Decade 1981–1990: Eight Years of Implementation.*

World Resources Institute (WRI) (1990) *World Resources 1990–1991* Oxford University Press. New York.

27: Water Resource Management Issues in the Humid Tropics

G. Le Moigne and U. Küffner

The World Bank, Washington, D.C., 20433 United States of America

ABSTRACT

The particular water management issues in the humid tropics arise from the hydrological features of this region (excess water, floods, greater variation of flows than in temperate zones), from the unique potential for hydro-power generation and from the traditional emphasis on growing irrigated rice. The central cause of water management problems is the rapid population growth with its increasing demand for water, food and energy, resulting in increasing pollution, and environmental and social problems. The capacity to deal with these problems is constrained by the limited availability of human and financial resources. Organizations at various levels, national and international, have been created to deal with the numerous issues, but the situation is far from satisfactory. Policy-makers and managers in the water sector should give special attention to the issues of institutional and water-sharing arrangements and make determined efforts to adopt modern tools and procedures to improve water development planning and implementation.

INTRODUCTION

Water utilization is on the increase as a result of rapid population growth. As rivers and lakes become more and more polluted, clean water becomes scarce; as public awareness of clean water dependency intensifies, so does the demand for efforts to protect this resource. But there is a disparity between official policies and plans for water resource utilization, and the actual implementation efforts, as is the case for resources in general. In view of these broad issues, there is a need to determine realistic water development policies for efficient, integrated and environmentally sound water resource utilization.

Water resource management has to address such diverse issues as comprehensive national and basin planning, competing demands of water users, international agreements on water uses, the implementation and operation of water development schemes, and increasing needs for training and support systems. Because of the accelerating technological development, population growth, and the urgent task to develop and use water resources more efficiently, it is necessary to broaden traditional planning, to consider the particular aspects of decision-making, the process of implementation, and operation and maintenance issues.

The changing social and technological conditions and the widening range of alternatives require more comprehensive and at the same time more flexible planning and implementation strategies. In addition to this focus on water management as such, increasing emphasis is now also placed on a sustainable development of water and land resources and the equitable sharing of these resources by society as a whole.

Water management in the humid tropics does not differ essentially from water management in other climatic regions, but there are some specific issues which are of greater importance in this region than elsewhere. They are most evident in the large irrigated rice schemes, in flood control development and in hydro-power generation.

CHARACTERISTICS OF THE HUMID TROPICS

For the purposes of this colloquium, the Chang and Lau definition of the humid tropics has been adopted. It determines the outer limits of humid tropics where the mean temperature of the coldest month is at least 18° C, with a minimum of 4 1/2 wet months[1]. The core of the humid tropics is the dense tropical rain forest in the equatorial regions. Some climatologists defined the humid tropics even on just this particular criterion: the extension of the contiguous tropical vegetation.

The tropical rain forest plays an important role in the hydrological cycle. The dense vegetation maintains a constantly saturated atmosphere in a layer of about 40 meters above the ground, in addition to the water retained in the plants and the soil. Investigations in the Amazon have shown that the water reservoir of the rain forest is partially replenished by moisture brought in over long distances, and partially recycled in the forest and between the forest and the atmosphere immediately above the rain forest. When the forest is cut down, this local reservoir of water is reduced (see Bonell with Balek, this volume). Typical soils of the humid tropics are *in situ* weathered rocks, commonly described as laterites which are easily erodible where not protected by dense vegetation or artificial protective measures, as discussed elsewhere by Lal and Rose, this volume. Because of the high rainfall and sudden heavy rainstorms, which frequently occur, laterites are particularly vulnerable.

The most significant recent changes in the humid tropics which affect water resource development and management are the population growth and urbanization, the reduction of the tropical rain forest with the associated increased erosion and, of course, construction works to utilize the water resources.

The population in the countries of the humid tropics increased between two and four per cent annually during the past decade. In most of these countries the population doubled during the last 25 to 30 years and a similar increase has to be expected in the near future. The urban population will increase even faster, leading to a rapid urbanization, particularly in the humid tropical regions. The Global Report on Human Settlements issued by Habitat (Habitat, 1987) indicated that in 1990, six of the world's twenty largest urban agglomerations would be located in the humid tropics; by the year 2000, their number is expected to rise to 10 of the largest agglomerations, and each one of them would then be inhabited by more than 10 million people (See Table 2, Low, this volume). The task of providing adequate services to all these people – especially water supply, sewerage, flood and storm drainage – will require billions of dollars. The expressed goal of providing adequate services to all urban inhabitants will be particularly difficult to achieve, since by the year 2000 the 20 largest agglomerations will be in developing countries, where low service levels prevail. The dramatic population increase is, of course, the main reason for the increasing water use, and for changes in the environment, deforestation and accelerating soil erosion. The rapid urbanization with its accompanying industrialization is one of the main reasons for the pollution of rivers, lakes, groundwater and parts of the oceans, which are, in turn, damaging the environment.

A closer review of some statistical data available for the main countries in the humid tropics reveals substantial differences, but also some similarities, particularly on a regional basis. In Asia and Latin America, the annual population growth from 1980-1987 was about 2%; in Africa, however, between 3 and 4%. Education, as measured by school enrollment, was generally higher in Asia and Latin America than in Africa, but an increased enrollment was reported for all countries.

The most significant differences appear in the data on GNP growth and the growth rate for GNP per capita. Modest gains in the average annual per capita GNP growth from 1980 to 1987 were reported for major Latin American countries (Brazil 1.0~% Colombia 0.6%), widely differing rates for Asian countries (Thailand 2.6%, Indonesia 1.9%, Philippines -3.3%) and generally negative rates in Africa (Cote d'Ivoire -3.0%, Zaire -2.8%).

Deforestation, particularly in the Amazon Basin, has been widely publicized and criticized in recent years. One of the early publications that predicted the serious consequences of deforestation of the Amazon Basin was the work of Goodland & Irvin with the title "Amazon jungle: Green hell to red desert?." Evidence collected during the years after its publication in 1975 confirmed the predictions of the authors about the effects of deforestation in general (Leopoldo *et al.,*1987). While traditional subsistence agriculture with its temporary small forest clearings does not harm the forest, large long-term clearings of 10 000 ha or more, as often carried out in the Amazon Basin in recent years, cause drastic changes in the ecosystem. The deforestation leads to a reduction of evaporation and water retention, to direct exposure of the soil to solar radiation, soil compaction, decreased infiltration into the soil and increased runoff. And, of course, the deforestation destroys the special ecosystem which supports the unique and rich biodiversity.

The exposure of the soil to heavy rainstorms and increased runoff causes in turn increased soil erosion. This erosion is most noticeable, sometimes with catastrophic consequences, in mountainous regions where landslides and avalanches can change the topography drastically – as documented for the southern range of the Himalayan Mountains. But deforestation in the plains of the Amazon Basin also is leading to increased erosion, as confirmed by recent investigations.

The increasing population, singled out as the most significant change in recent years is, of course, also the main reason for the rapid expansion of water development works, irrigation schemes, hydro-power facilities, water supply installations and flood control schemes, which together with the construction of reservoirs and canals, have altered river regimes and regional water balances. The increasing population also is causing a serious deterioration of the water quality, even in this region with plentiful water.

WATER DEVELOPMENT ISSUES IN THE HUMID TROPICS

Rice cultivation has been the traditional and typical main development of water resources in the humid tropics, primarily in southeast Asia and Madagascar. Terraced hillsides for rice cultivation are often shown as typical pictures of water development in the tropics. Beginning in the late last century, the vast potential of the region for hydro-power generation was developed, and flood control and drainage schemes were constructed to reduce the frequent flood damages and to deal with the problem of excess water in rural and urban areas. These three purposes of water development are typical and most important for the humid tropics.

Irrigation

The widespread traditional rice irrigation can be traced back to prehistoric times. The construction of terraced basins in diffi-

BOX 1 *Purposes of Water Development*

Irrigation
Hydro-power generation
Flood control and drainage
Domestic/Industrial water supply
Erosion and sedimentation control
Navigation
Recreation
Fishery and wildlife
Coastal protection

---

cult terrain and the elaborate organizational arrangements required to utilize these systems effectively show the high degree of past achievements in water development. The channelling of water into the rice fields, the distribution in an elaborate system where water has to be shared by numerous farmers, requires an organization that is functioning well and all participants cooperating fully.

The need to increase food production, the success of the traditional rice production, and the development of modern agricultural techniques and improved rice varieties led to a rapid expansion of irrigation for rice production in this century, and particularly during the past 30 years. In Indonesia, the area under irrigation, which is mainly devoted to rice, increased from about 4 million ha in 1968 to almost 5.5 million ha in 1980; in the Philippines, from 740 000 ha to 1.2 million ha; and in Thailand, from 1.8 to 2.6 million ha in the same time period.

The rice-based agricultural economies in Asia experienced unexpected problems in the 1980s, when increasing surpluses appeared which forced down rice prices in real terms in domestic and world markets to their lowest level since early this century. The affected countries tried to diversify out of rice, but these efforts were hampered by inflexible irrigation and farming systems and the general deterioration of agricultural commodity prices in world markets (World Bank, 1988).

In most rice cultivation systems, the dominant irrigation method is the flow of water from basin to basin on a continuous basis. The flexibility of individual farmers to adopt other crops is limited or impossible because of collective irrigation and drainage practices required for groups of farmers. Non-rice crops require larger delivery flows intermittently instead of the smaller continuous flow needed for rice cultivation. Also, traditional rice irrigation projects often lack sufficient storage capacity for cropping in the dry season.

Diversification from rice cultivation to non-rice crops is severely restricted by the existing irrigation facilities, soil types, drainage facilities and the land preparation for rice. Irrigated rice in Asia is grown predominantly on heavy clay soils which are ideal for rice cultivation because of their water-retention capability which facilitates the maintaining of a high degree of moisture which is needed for rice. However, most other crops do not grow well under such conditions.

The fall of rice prices in the 1980s has made major new investments in rice schemes uneconomic. The only viable investments in rice irrigation are those which are based on sunk costs in existing infrastructure. Even improvements of distribution systems may not be economic, unless substantial increases in rice production can be achieved. Therefore, research on design and management of irrigation systems to make crop choices wider should be undertaken to broaden the cropping opportunities of farmers.

Hydro-power generation

The potential for hydro-power generation in the humid tropics is very large due to the high river flows, and it is especially attractive in the mountainous regions. Some of the largest hydro-power plants, such as Guri in Venezuela and Itaipu in Brazil, have been built in these regions, but the potential is far from being fully exploited. A study carried out by the World Bank showed that only 2% of the potential was developed in the humid tropical countries of Africa and Asia, and 7% in Latin America (World Bank, 1984).

The abundance of hydro-power has led to the construction of numerous micro-hydroelectric plants supplying power to local consumers. The main problem of the development of the large hydro-power potential is, however, its geographic location relative to the power demand centers. In Indonesia, the greatest potential – over 25% – lies in Irian Jaya, where the demand is less than 1% of the total domestic power demand, while Java with its 80% of the present power demand has only 5% of the hydro-power potential.

The distance of the power generation plants to the demand centers requires the construction of long transmission lines, often through difficult mountainous terrain covered with tropical forest, with the danger of landslides and the need of expensive access roads, or the construction of power lines with the help of helicopters. Consequently, the cost of electrical energy delivered at the demand centers is often high and the economic justification of the power development doubtful or less attractive than anticipated, particularly in view of relatively low prices of alternative energies. Thus, planning and design have to be carried out very carefully to avoid uneconomic investments. While in some countries the development of the most economic sites has already taken place, hydro-power development continues to be an attractive investment, particularly in the humid tropical countries. From 1974 to 1988, total funding of power facilities by the World Bank amounted to US$ 25 480 million, of which 26% was for hydro-power. Hydro-power in the humid tropical areas accounted for about two-thirds of the hydro-power investments.

Flood control and drainage

Fertile flood plains and deltas such as the Ganges Delta have been settled and cultivated historically by farming communities which have often been threatened by devastating floods. Demands for flood protection started to rise when accelerated population growth led people to increasingly settle in flood plains. This applied especially to urban areas. Thus flood control has become an increasingly urgent task. While previously, people had generally a choice, opting for settling in flood plains because of the fertility of the soils, nowadays many are forced to take this risk as there are no alternatives. As a consequence, political pressures are developing to take measures against floods which threaten farmers and urban dwellers, who are often the poorest segment of the population.

Flood control and flood damage reduction can be achieved in several ways. A complete control usually requires the construction of flood retention reservoirs, embankments along the rivers which cause floods, and possibly bypasses to divert major flood flows. In the tropics where major river floods are often accompanied by heavy rainstorms, additional measures have to be taken to evacuate water from the areas behind the embankments which protect the land from river flooding but also prevent natural drainage. To evacuate this water, pumping installations are needed which are usually justified to protect residential areas, industrial installations and high-value crops.

Flood control works for extensive flood plains, particularly those which have been settled only recently and where the population is not very dense, can rarely be designed for complete flood control. Difficult choices have then to be made to determine which areas deserve full protection, partial protection or should be left without any protection. In the latter case, the installation of a flood warning system may be the only economic measure to save lives and to reduce property damages.

Drainage of residential areas is normally provided in connection with sewerage works. The large volume of rainwater in the humid tropics is often difficult to drain. A combination of buried pipes, surface drains and retention reservoirs has to be considered. Under certain conditions, temporary flooding of low-lying areas may be justifiable and acceptable to reduce the cost of a drainage system which could be overly expensive if designed to evacuate immediately on the occasion of every foreseeable rainstorm.

Agricultural drainage means the maintenance or rehabilitation of agricultural lands through lowering the groundwater table to such a level as required for efficient crop production. Structural measures include the installation of open or buried drain pipes and, in some cases, the construction of wells to lower the groundwater table. If drainage is installed too late, water-logging may develop, and lead to a loss of agricultural productivity until proper drainage is provided. Irrigated areas in tropical regions encounter few salinization problems, except in coastal areas, where salt-water intrusion can create production problems. Unique properties of tropical soils sometimes require special drainage systems to minimize problems. For example, acid sulfite soil conditions require controlled water table provisions to minimize toxic crop damage, and peat soils need water table control to minimize subsidence.

The draining of wetlands and swamps to develop agriculture or reclaim land for industrial and residential use, which was carried out indiscriminately only a short time ago, is today questioned not only by environmentalists who wish to preserve existing ecosystems, but also by advocates of a more sensitive approach to wetland development. The latter recommend exploiting the potential for fishery and livestock raising and using the capacity of the wetlands for retaining floods.

An ambitious program to develop the agricultural potential of the humid tropical region of Mexico was initiated over 10 years ago. It was one of the main recommendations of a national water plan which concluded that Mexico needed a reorientation towards the development of the humid tropics after developing most of the agricultural and irrigation potential in the arid and semiarid parts of the country. The main objective of the program, which is still ongoing, is an appropriate development and utilization of the potential of the region through the construction of drainage and infrastructure works, the provision of services including extension to the farmers and the strengthening of research on the unique conditions of the region.

Other water development

While irrigation, flood control and hydro-power generation are the predominant and typical purposes of water development in the humid tropics, other purposes can also play important roles.

Municipal and industrial water supply, although more easily developed than in arid or semiarid regions with their almost continuous shortages of water, may nevertheless face serious problems, especially when dependent on fluctuating surface water supplies. Even larger rivers may not supply sufficient water of adequate quality to growing urban-industrial areas, and the demand for high water quality may require the construction of costly water treatment plants. Cities located near the ocean may also confront the problem of salt-water intrusion in aquifers used for their water supply or in the rivers where strong tidal actions can force the occasional shut down of water intakes, as experienced by the city of Guayaquil, Ecuador.

BOX 2 *Water Development Problems and Their Effects*

Problems	Primary Effects	Secondary Effects
Population growth	Growing demand for water, food, energy	Increasing pollution, damage to environment
Inadequate data for planning and design	Design and evaluation problems	Misallocation of resources
Inadequate human and financial resources	Insufficient implementation and operation capacity	Production targets not reached; delays; low quality construction: deterioration of investments
Inadequate institutional and legal framework	Insufficient management, conflicts between users	Insufficient and inadequate use of resources
Inadequate planning and project preparation	Deficient plans and works	Problems of health; erosion, sedimentation, resettlement

Erosion and sedimentation problems have become more and more serious in the past decades, and increasingly steps are taken to address these problems. But such measures cannot arrest erosion and sedimentation completely. Both phenomena are critical elements of dynamic river systems. Ideally, sediment and bed loads should be in a relatively stable condition in relation to the river flows.

Occasionally, high flows in mountainous terrain carry such an amount of debris, rocks, gravel and sand that the streams must change their courses where the gradients of the river flatten and the material is deposited. Structural measures to stabilize such rivers will often be prohibitively expensive.

Coastal protection is of major importance where fertile soils, as in the case of deltas, are subject to serious erosion. The densely populated Ganges Delta is regularly threatened by major storms, but the magnitude of the task has not allowed the planners to find a comprehensive, economic solution.

Water requirements for navigation and fisheries may force planners into difficult compromises. In Bangladesh, it is expected that there will be a continuous loss of flood plain fishery habitat to other water users, mainly through flood control projects for agriculture. As water requirements for food production increase over the next 20 years, there will be an increasing need to use dry season flows from the major rivers for irrigation. The increased agricultural production will mean more use of agro-chemicals, and the development of barrages and regulators for more water control increases the intensity of disruption of the life cycle of the open water fishery.

Also, in Bangladesh, where navigation has always been a principal means of transport, the geologically young deltaic plain is continuously undergoing changes which create prob-

lems for the river traffic. Navigation will continue to be affected by shifting channels unless some special measures like river training or dredging are adopted. Other problems are the massive siltation and the reduction of flows which will gradually limit the movement of the river traffic.

WATER MANAGEMENT ISSUES AND STRATEGIES

Rapid population growth, with its increasing demand for water, food and energy, is the central cause of water development problems. The specific problems stemming from a particular region include technical, environmental and social problems, and the capacity to deal with these problems depends on the availability of human and financial resources. Failure to prepare and carry out projects adequately, to achieve targets, and to maintain facilities properly can often be traced to these problems. Furthermore, the lack of reliable data on the prevailing conditions – hydrologic, geologic, agronomic or social – can compound these problems. Not only developing countries, with their severely limited financial resources, have to face such problems, which may lead to failures or unsatisfactory results, but also industrialized countries and international donor agencies with much greater resources.

One of the most serious social issues related to population growth is the lack of water supply facilities in the developing world. In the urban areas, 25% of the people have no access to a reasonable water supply. In the rural areas, this percentage is over 60%. Even more people lack access to sanitary facilities. Other authors in this volume provide more details, e.g. Lee, Low, Prost, Roche, and Wurzel

BOX 3 *Outline of the Management Process*

Planning
Definition of objectives
Inventories and projections
Project identification, analysis and design
Program development
Implementation
Project authorization, financing arrangements
Establishment of project organization
Contract adjudication, construction, installation
Initial operation
Operation
Operation of facilities
Maintenance and repairs
Monitoring and controls
Improvements and modernization

It is not surprising that water pollution is increasing. Rapid urbanization, industrialization and increasing use of agro-chemicals on the one hand, and lacking or inadequate pollution control on the other, is leading to depressing levels of contamination of surface supplies and groundwater. The increasing prevalence of those tropical diseases which are predominantly water-related has been associated with the lack of safe drinking water and nonexistent sewage treatment facilities (see Prost and Wurzel, this volume). Furthermore, where nutrients from domestic, industrial and agricultural sources enter reservoirs or lakes, they promote the production of excessive amounts of algae and aquatic plants. When these die, they use oxygen faster than it can be replenished. This eutrophication process can kill fish populations and make the water less palatable as drinking water.

A systematic review of such problems and the relevant water management issues may follow several different paths to address all relevant issues comprehensively. One path is the review of the various water development purposes, as discussed previously, with special consideration of the humid tropical water development. A second set of issues can be derived by following the management process or project cycle from planning through implementation to operation. Thirdly, the consideration of the elements of what may be called the macro-context will identify issues important for water management and even more so for formulating water policies. The elements of the macro-context include the overall economy, the institutional framework, and economic, financial, social, environmental and technical aspects. The systematic and comprehensive review of the issues should determine those that are of particular importance to water management and, in this review, to water management in the humid tropics.

Water management strategies – The management process
The management process or project cycle leads from planning

through implementation to the operation of water development schemes.

Planning
Planning and project preparation should reveal and address relevant problems by focusing on the essential aspects. To the initial purely technical planning and economic analysis, several important aspects have been added during the past few decades: social aspects such as the consideration of poverty and basic needs, and the resettlement of people in connection with major water development projects; environmental aspects, not only the possible degradation of the environment but also the possible destruction of the environment for rare or endangered plants and animals; the aspect of providing for operation and maintenance of project facilities, and organizing executing agencies and beneficiaries, who themselves should also be involved in the preparation phase; and the aspect of developing the necessary human resources, technicians and workers as well as beneficiaries.

Also at issue emerges the question, how should planning be done? Although there are well-established and generally agreed upon procedures, problems have arisen again and again because of a failure to follow the proper steps. Any planning should at the very beginning define objectives and targets, or obtain them from the decision-makers, but this essential step has in the past often been neglected, and data collection activities have occupied the planners excessively. To meet the objectives and targets, projects should then be identified, analyzed and ranked so that alternative programs can be developed. The latter should be evaluated according to their costs and benefits and other criteria, such as employment generation and foreign exchange savings or earnings, which are of importance for the development program. These evaluations should lead to the selection of the preferred program which should, of course, also achieve the initially formulated targets to the extent possible.

Often plans have been little more than a collection or a catalogue of all known projects with incomplete economic analyses and without a clear relation to the country's overall economic development program and the availability of funds. A clear definition of objectives and targets helps the planners to focus their work on these essential aspects. The objectives should make it clear that an integrated plan has to be developed. In almost all cases, multiple objectives will have to be defined, including technical, social, economic and environmental aspects and consequences, at all stages of planning.

How detailed, how "final" should plans be? Planning is a dynamic process. Plans are prepared, approved, modified, changed. The database changes over time. Projects are built which fit into the plan or change essential aspects, requiring modifications. Consequently, plans should maintain flexibility to allow for such changes. To stress the element of flexibility in planning, some critics have suggested avoiding the term "master plan", which is often seen as too rigid a plan, and have therefore advocated such terms as "framework plan" to indicate that further adjustments and further planning are required.

Who should prepare plans and projects? Specialized agencies and consulting firms have developed the capabilities for this work, and the political decision-makers are often involved at various stages of the planning process. The involvement of beneficiaries of projects and those affected by them have been neglected for a long time, but it has gradually been accepted. Not only should the water users and other groups interested in or affected by the water development have a right to express their views at all stages, they are also often in the best position to assist the planners – to warn them about negative aspects and recommend more advantageous solutions which will ultimately assure the acceptance and long-term success of the plan.

The conventional water development plans are river basin plans defined by the drainage or catchment area. This natural definition is generally accepted, but it may be questioned whether a river basin is the appropriate planning area when important cities or industries with large water requirements are located just outside the basin. In such cases, regional plans which address the issues of a geographic or economic region may be more suitable.

The broadest plan is a multinational basin plan. Although attempts have been made to develop such plans, they have usually had only partial success. Most international water agreements have been reached on a bilateral basis, and do not include the formulation of a joint plan. They consist, essentially, of a water-sharing agreement which allows the planning and use of water on a national basis but does not formulate a joint plan. For international river basins, as for the Mekong, Nile and Zambezi Basins, plans have nevertheless been attempted with the assistance of UN and regional agencies, but experience has shown that decades of work are needed to make some progress.

Implementation

The implementation of water management activities includes the construction of the water development works and the installation of the necessary equipment. Furthermore, the final decision on a project, the raising of funds for the construction, and the organization of the project management unit are all part of the implementation. Usually, the early stage of project operation, the phasing-in period, is also considered part of the implementation, while operation, maintenance and repairs are separated from implementation by some authors, and included in the implementation by others. Controls and a comprehensive monitoring system are also essential elements of implementation.

Implementation of water development projects in the humid tropics does not differ substantially from implementation in other regions. The particular features of the humid tropics, the high rainfall, floods, high temperatures, erosion and sedimentation problems, however, can affect construction processes, endanger construction sites and thus ultimately lead to higher costs.

Implementation issues also are linked to institutional questions: Are the executing agencies adequately organized, staffed and financed? Do they have clear responsibilities for their tasks, or do they depend on other agencies that interfere frequently in the implementation? Do financial arrangements cover the total cost of projects, or do the executing agencies depend on yearly budget allocations which may affect or interrupt the progress of construction?

Operation and maintenance

The operation of water development projects is primarily geared towards the most efficient use of water for the purposes of the project as determined during the planning and design phase. Maintenance should assure the proper functioning of all parts of a project or system. The lack of adequate operation and maintenance in many developing countries caused a serious deterioration of valuable assets and created the need for numerous rehabilitation projects. This fact was one of the reasons why operation and maintenance have gained more and more attention and their preparation is now considered a necessary part of a comprehensive plan.

One of the most important and serious maintenance problems in the humid tropics is the removal of sediments from canals and, where feasible, from reservoirs through flushing procedures. Some large irrigation schemes are nowadays so overloaded with sediments that the managers see it as the main issue which they have to address. The large irrigation schemes in the Sudan are an example of this phenomenon. Although they are not located in the humid tropics, they are affected by this region from which they receive their water containing a high sediment load. In view of the operation and maintenance problems (O & M), it has been suggested that design features which reduce these costs be adopted, even if the initial investments are higher. Managers who have difficulties in obtaining

BOX 4 *Elements of the Macro-Context*

Overall economy
Institutional framework
Financing and cost recovery
Social framework
Environment
Technical conditions

adequate funds for O&M usually prefer this solution, which may not necessarily be the most economic when all aspects are analyzed.

The main cause of operation and maintenance problems is the difficult mobilization of funds for these activities. Although the principle of cost recovery from project beneficiaries for the payment of at least O&M costs (if feasible, also for a share of the investments) has been advocated for many years, even imposed as a condition on development credits, it is still not widely applied. The early involvement of beneficiaries in the preparation of a project, even in the review of funding arrangements, is essential to obtain their collaboration – their willingness to pay – as they will be more willing to contribute to a scheme for which they were able to influence the design and the maintenance procedures.

The macro-context

Like other sectors, such as transport, industry or construction, water management is directly affected by the elements of what may be called the macro-context (Box 4).

The elements of the macro-context are, of course, relevant for water resource planning, implementation and operations, but the systematic review of these elements should ensure that all relevant aspects are considered.

As indicated in the above discussion of water resource planning, it is necessary to link water development plans to overall economic development plans and develop integrated plans and investment programs.

The institutional framework affects water management in various ways, as it consists of laws, policies and water-related organizations. The countries in the humid tropical regions have not developed unique or distinctive institutional frameworks, but their laws, policies and organizations have been influenced by regional traditions, the specific natural conditions and by the legal systems of Asian, European and Islamic states.

Unfortunately, water policies and laws do not exist in many countries, or they are antiquated or virtually unknown by field personnel (Radosevich, 1987). Legislative enactments commonly deal separately with irrigation, hydro-power, flood control, navigation, and mining, and they grant executive powers, often conflicting, to the relevant organizations.

The need for a coherent, integrated legal and administrative system was well formulated for the Asian countries by the ECAFE Working Group of Experts in 1967, but their call for a water code, a basic, national water law (UN, 1981) met with virtually no response. In the absence of such legislation, and in view of the difficulties in overcoming well-entrenched administrative interests, it has been proposed to establish a central authority for the purpose of enhancing cooperation in planning and execution of water sector policies. Central agencies may be designed to delegate powers to sectoral and regional agencies which can then exercise them only within the prescribed limits.

There are numerous organizations, at many levels, dealing with water management tasks, from local groups operating small projects, to international organizations dealing with global water issues. An attempt has been made to identify the major organizations and their tasks in relation to each other (Box 5).

The United Nations and its specialized agencies have stimulated increasing world-wide collaboration in many fields. The water sector is of special interest to many of the specialized agencies, such as FAO, UNESCO, UNDP, WMO, WHO and the World Bank. With the help of international agencies, conflicts over water between nations have been mediated and eventually brought to agreement. Valuable assistance is being provided in planning, project preparation and human resource development.

Without any doubt, international cooperation in the water sector is widely accepted and increasing, but the central decision-making role rests with the sovereign national governments. In this connection, John Waterbury's statement is apt: ". . .the nation state with its national goals and foreign policy and domestic concerns remains the framework for dealing with transnational assets." (Waterbury, 1979). This is most evident when national interests are threatened by other governments which plan to develop projects in international river basins. Although general principles for the development of such projects exist, governments have often proceeded with projects which are not in the interest of other riparian states.

In some cases, governments ceded some authority to bilateral or multilateral agencies, such as international river basin commissions, but their record of successful operation is mixed, not only because of the difficult reconciliation of interests, but also because of the problems of financial contributions to such agencies. Poorer governments are often in arrears

BOX 5 *Organizational Context*

Context	Organization	Planning	Project Prepn.	Decision Making	Regu-lating	Funding	Project Implem.	O&M	Control/ Monitor	Conflict Solving	Human Res. Dev.
Global	Int'l Agencies	X	X			X				X	X
Multi-national	Governments, Multi- and Bi-lateral Agencies	X	X	X	X	X				X	X
National	Specialized Government Agencies	X	X	X			X	X	X		X
Regional-Basin	Regional Agencies, Basin Commissions	X	X	(X)	(X)		(X)		X	X	X
Project	Specialized Agencies, Project Authorities, User Groups	X	X				X	X	X		X

(X) Not always applicable.

with their contributions, and the agencies consequently see their capabilities and thus their authority eroding.

Economic and financial conditions determine the rate of investments in the water sector which places heavy demands on national budgets. Undoubtedly, this demand will increase in the future. Consequently, cost recovery questions will become more and more critical, but cost recovery from beneficiaries has been difficult in almost all sectors and countries.

The effectiveness of cost recovery mechanisms depends on a number of factors, including government pricing and taxation policies. Subsidies to achieve social goals – such as improved health through safe drinking water, food security through irrigation, or industrialization through low electricity tariffs for hydro-power generation – must be incorporated in water pricing policies. Taxes and fees imposed on water users, such as high taxes on irrigated land, must be taken into account. Water charges may be transferred to the government treasury, and, consequently, considered as a form of taxation but not as cost recovery related to water investments. External costs, such as damages created by irrigation, drainage and sewage disposal, also are rarely incorporated in water tariffs. Thus, mechanisms for cost recovery should provide incentives for appropriate water use and take into account the environmental effects and market distortions created by other governmental programs.

Social considerations affect water management decisions in various ways. As the investments in the water sector amount to a large share of public and private investments, equity considerations are very important. Who will benefit primarily from those investments? Are certain regions and income groups favored over others?

The cost of providing safe drinking water to all people, and to control water pollution, as postulated repeatedly, is enormous. While the goal is undisputed, the appropriate steps, the possible level of services to be provided to population groups, their contribution to the costs and the feasibility of imposing the cost of effluent treatment on financially weak industries raise serious political, social and financial questions.

Another serious impact of large water development projects is the displacement of individuals or communities who lived on the land inundated by reservoirs. While the laws of most countries provide for the right to fair compensation, frequently problems arise concerning the value and timing of compensation, the claims of people without legal titles, and the problems of re-establishing productive communities. The shortcomings of the compensation approach have led to the development of the concept of rehabilitation of the people who are affected by the construction of such projects. The main objective is to ensure that all displaced people should be given opportunities to become established in the shortest possible period at living standards that at least match and, if possible, improve on those existing before resettlement.

Environmental degradation was referred to earlier. Water development works, if inadequately designed and constructed, may cause environmental and social/health problems. The construction of water impoundments in warm climates often increases the prevalence of schistosomiasis and malaria and occasionally other diseases, such as onchocerciasis and Japanese B. encephalitis. Malaria and schistosomiasis are most often associated with irrigation systems and small, shallow reservoirs which provide a favorable habitat for the disease vectors. Methods of control include environmental sanitation, application of pesticides and biological controls, and chemotherapy of infected people.

Furthermore, the destruction of vegetative cover leads to accelerated erosion resulting in the loss of fertile topsoil and sedimentation in river beds and reservoirs. Thus, environmental protection will necessarily have to play a growing role in water planning and management.

Many water development problems can be solved by adopting appropriate technical solutions. Technical measures were usually the first steps to tap water sources, and to convey and distribute water to many users. Technical measures have also been developed to deal with water shortages, to store water, to conserve water and to apply it in exactly the quantities needed.

During the past decades, many new technical solutions have been found for treating water, and for desalinating and recycling it. Advanced irrigation methods and efficient drainage technologies have been developed. Computers and satellite monitoring also have added new tools for engineers and managers.

Given the importance of technical change in ameliorating water scarcity and water quality degradation, several issues arise. The integration of technical change into water planning and management decisions is one critical element. The lack of technical infrastructure in many countries to develop appropriate water conservation and pollution control technologies and to adapt them to local conditions is another important issue. Mechanisms for technology transfer need to be improved. In this context, the impact of pricing and subsidy programs on technical change has to be considered.

The changing focus of water resource management

The past three decades have broadened rapidly the horizon of water resource planning and management, an experience which has taken place in many other sectors. But there still remains the first and elementary task of understanding and studying water resources and their evolution in the natural setting. Engineers have been accused of a narrow focus on technical aspects of water development, but their attention to engineering problems and possible solutions remains the elementary task. Engineering has a long history, and lessons can be learned from records which go back for centuries. Good engineering has traditionally tried to determine future scenarios. Perhaps these attempts will be found to be inadequate or wrong. It remains to be seen how our work will be judged in the future.

Economic and financial aspects were always included in the work of planners, and the methods of economic and financial analysis became more elaborate and sophisticated. But the future-oriented element of this analysis, the prediction of costs and benefits, remains, of course, as uncertain as the prediction of other future developments, and surely even more so than the engineers' prediction of the future behaviour of a water source. Nevertheless, careful economic and financial analyses of projects and programs are indispensable, as they are the tools for comparing alternative investments in order that financial resources are utilized in the best interest of society.

Legal provisions have also traditionally affected the work of planners. In most societies, customary laws were originally observed. They gradually were replaced by a system of judicial apportionment of water. In many countries, the executive branch of the government eventually assumed the authority to grant water rights. The laws determining or affecting water use are, in many countries, antiquated. The difficulty of changing legal provisions has the result that they are often constraints rather than facilitators to water development. But the growing awareness of environmental issues has put pressure on legislators to introduce more appropriate laws dealing with the environment, and many countries have passed new laws in view of environmental concerns.

The first step to a more comprehensive concept of water resource management was the development of river basin plans, the consideration of water use in this hydrographically defined area. As a logical consequence of plans for river basins in a country, plans for international river basins have been promoted. While the record of these planning activities is mixed, such international, bilateral and multilateral attempts have shown the way to solve the inherent conflicts between riparian countries for the use of a shared water source.

River basins are the basic geographic units for which water resource plans are prepared. In recent years, many new aspects have been added to the traditional planning work, which focused on technical and economic factors. The distinctive elements of the river basin, the special features of upstream and downstream development, found more attention. Upstream water users normally appeared to be in an advantageous position because of their early access to the water. But when the principle of equitable sharing of resources was applied, the upstream users were the ones who had to guarantee a certain flow to the downstream users, to treat waste waters before discharging them into a common river, while downstream users could discharge them into the ocean without major treatment costs, and the upstream users faced additional disadvantages of longer distances to ports and market centers, usually located near the coast (Goulter, 1985).

Water resource management was also strongly influenced by the growing concern for the environment and the socio-cultural conditions of the societies, as discussed in the previous sections. Together with the attention given to institutional and organizational issues and to the whole process from planning through implementation to the operation of water development schemes, water management calls for an increasingly compre-

hensive approach (Ali *et al.,* 1987). In addition, it has been advocated to treat water and land development together through environmental management because of the interdependence and interactions between land and water management (Lundqvist *et al.,* 1985).

The disappointing results of projects in the water sector and in other sectors, particularly in the poorer countries, led to repeated critical reviews and a search for the reasons for the failures. Lately, it has been recognized that more and more projects did not show a sustained viability. Therefore, a special emphasis is now placed on an analysis of the sustainability of the investments – the technical, financial and environmental sustainability. The disappointment with the performance of public agencies gave arguments to the proponents of a stronger participation of the private sector. Critics may point out that good planners and managers would consider all these aspects. This may be correct. But the decisive point is the need to place a special emphasis on some aspects at a given time. There may be some fashionable trends, even in water resource management, but the changing focus is a response to changing problems, needs and opportunities.

The comprehensive approach to water management has been aided greatly by the development of modern tools and processes which in turn have influenced water management. Computers made it possible to process, analyse, and thus efficiently use the large amount of data which are needed to evaluate the numerous aspects of water development. Mathematical modelling enabled planners to develop and compare many more alternatives than previously. Systems analyses facilitated multi-objective investigations and the development of comprehensive plans. Satellite remote sensing also has provided not only a completely new perspective, but new data and information which are only beginning to affect water management.

The overall framework of water resource management is determined by policy statements dealing directly or indirectly with water. While policy-makers are usually interested primarily in political and economic issues, planners and managers focus on technical, economic and other related-sector aspects. Major decisions are, therefore, often made without sufficient interaction between these parties. The decision-making process itself, determining policies and the support for major projects, may even be unclear, which can create an atmosphere of suspicion and mistrust. Undoubtedly, it is necessary to establish clear and open decision-making processes with the active participation of the principal interested parties.

Unfortunately, water policies are inadequate in many countries. Even where policies have been formulated, the degradation of water quality and the rapidly increasing water demand requires the adjustment or reformulation of water policies.

Recently, the World Bank initiated work on integrated water resource management policy because in many countries impending water shortages threaten economic growth and call for government review of the relevant policies. An interdepartmental work program of the World Bank is aimed at achieving such an integrated approach to water resource management. The work on water resource management will examine and develop policy guidelines for some of the main issues affecting the management of, and investments in, the water sector. It will cover intersectoral issues pertaining to economic, legal, institutional, environmental and technological considerations. With the growing demand for water, competition among water users over existing supplies will intensify, and, therefore, will become of prime importance. As water allocation issues become more contentious, the water resources sector will place increasingly large demands on national budgets. Consequently, cost recovery questions will become critical. Institutional issues include organizational reforms, the strengthening of agencies, training and educational programs, and changes in the legal and regulatory systems. The issue of degradation and depletion of water resources will necessarily play a growing role in water planning and management. One of the challenges countries have to face is to define an appropriate balance between environmental protection and income growth. The comprehensive approach to water resource development requires an addressing of the technical issues that are common to all subsectors, i.e., the comprehensive assessment of all water resources, their quantity and quality; the assessment of technical solutions to optimize overall water use, such as the reuse of urban waste water for irrigation; water quality monitoring to safeguard against public health and environmental hazards; and the promotion of technology transfer.

In addition to proposing policy guidelines for the Bank's project work, the study will address issues relevant to the policy dialogue between the Bank and its borrowers and will analyse water resource issues for the Twenty-first Century.

CONCLUSIONS

Water resource management has to find responses to the recurrent problems of floods, water shortages and the deterioration of water quality; it has to find solutions to satisfy the growing and often conflicting demands for water and it must determine ways and means to develop, use and manage this resource. Most governments have responded to these challenges and have taken steps to determine policies for the utilization of water and other natural resources. However, the institutional framework, laws and organizational structures are frequently constraints instead of means for the necessary development.

Water resource policies and strategies must try to combine such diverse aspects as increasing and competing demands by irrigated agriculture and growing urban centers and industries, the disposal of waste water, the control of floods, environmental protection, resettlement, river basin planning, and national and international legal agreements. The answer to this complex question is provided by comprehensive and coherent planning efforts. These efforts to address multiple objectives within a planning system must respond to the needs and demands of the public. Therefore, if the planning process is

properly carried out in consultation with beneficiaries and all parties affected and interested, it should produce results which are socially and politically acceptable.

The unprecedented population growth, the rapid pace of technological change and constantly changing needs of the society must be accommodated not only in a comprehensive planning framework, but also through a flexibility of water resource plans which leave sufficient room for adjustments and modifications.

Unfortunately, the interaction between policy-makers and planners is rarely adequate. Policy-makers are usually preoccupied with political and economic criteria, while planners concentrate on technological, economic and other aspects and try to distance themselves from politics. As a result, the decision-making process is often unclear and unsatisfactory to the actors involved. A recognition of the problems and the necessary interaction may lead to a better mutual understanding.

There is an interdependence between water policy and water management, a process moving from policy formulation to planning, construction and operation in which many actors are involved at different levels representing local, regional, sectoral, national and international interests.

In view of the issues in the humid tropical regions, it is suggested that policy-makers and managers responsible for the water sector focus on the particular issues of irrigation, flood control and hydro-power development, and give special attention to the increasing environmental problems, including water quality, erosion and sedimentation; examine the adequacy of organizational, institutional, legal and water-sharing arrangements and make the necessary adjustments; and make determined efforts to adopt modern tools and procedures to improve the planning and implementation of water development.

NOTE

The views expressed in this paper are those of the authors, and are not presented as official or unofficial views of the World Bank.

END NOTE

1. A wet month is one having more than 100 mm of rainfall. When the monthly rainfall is between 60 and 100 mm, it is considered that the water supply for plant growth is inadequate for at least half of the month, and this is credited as a half wet month (Chang & Lau, 1987 IHP Project 4.2 Report)

REFERENCES

Ali, M., Rhan, A. A. & Radosevich, G. E. (1987) National water goals, policies and laws: The institutional framework. *Water Resources Policy for Asia.* A.A. Balkema, Rotterdam, Netherlands.

Goodland, R. J. A. & Irvin, H. S. (1975) Amazon jungle: green hell to red desert? *Elsevier Scientific.* Amsterdam, Netherlands.

Goulter, I. C. (1985) Equity issues in the implementation of river basin planning. *Strategies for River Basin Management.* D. Reidel, Dordrecht, Netherlands.

Habitat (United Nations Centre for Human Settlements) (1987) *Global Report on Human Settlements.* Oxford University Press, Oxford, England.

Leopoldo, P. R., Franken, W., Ribeiro, M. N. & Salati, E. (1987) Towards a water balance in the central Amazonian region. *Experientia 43* Basel, Switzerland.

Lundqvist, J., Falkenmark, M. & Lohm, U. (1985) River basin strategy for coordinated land and water conservation: Synthesis and conclusion. *Strategies for River Basin Management.* D. Reidel, Dordrecht, Netherlands.

Radosevich, G. E., Ali, M. & Rhan, A. A. (1987) Approaching water resources development. *Water Resources Policy for Asia.* A.A. Balkema, Rotterdam, Netherlands.

United Nations (1981) *Third Report on the Law of the NonNavigational Uses of International Watercourses.* United Nations International Law Commission (UN doc A/CN 4/348).

Waterbury, J. (1979) *Hydropolitics of the Nile Valley.* Syracuse University Press, Syracuse, NY, USA.

World Bank (1984) *A Survey of the Future Role of Hydroelectric Power in 100 Developing Countries.* Washington DC, USA.

World Bank (1988) *Diversification in Rural Asia.* Washington, DC, USA

VII. Appendices

Appendix A: Definition of the Humid Tropics[1]

J-H. Chang
Chinese Culture University, HWA Kang, Yang Ming Shan, Taiwan

L. S. Lau
Water Resources Research Center, University of Hawaii at Manoa, 2540 Dole Street, Honolulu, Hawaii 96822 USA

The humid tropics can be defined in different ways, depending on the objectives. The three best known approaches may be briefly summarized.

Köppen classification. Wladimir Köppen, German biologist, was the pioneer of climatic classification with his original work being published in 1900 and revised in 1936.

Köppen presented formulas, which included monthly temperature and rainfall data, to delineate arid (desert) and semiarid (steppe) climates. Outside the arid and semiarid zones, all climates are forest climates, except in polar regions and high mountains where low temperatures prohibit the growth of trees.

Within the forest climates, three major types are recognized according to temperature. Köppen's tropical climate (**A** climate) includes areas where the mean temperature of the coldest month exceeds 18°C, the optimum temperature for human comfort. The 18°C isotherm also agrees well with the distribution of palm trees, a typical tropical plant.

The **A** climate has three sub-types: **Af** (tropical rain forest), **Am** (monsoon forest), and **Aw** or **As** (savanna). In the **Af** climate, monthly rainfall exceeds 60 mm throughout the year. In the **Am** climate, the dry season is either very short or the rainfall during the wet season is copious. Both the **Aw** and **As** climates have a distinct dry and wet season, with the dry season occurring during the winter and summer, respectively. However, the **As** climate has a limited distribution.

Some climatologists consider all **A** climates to be humid tropical since they occur outside the arid and semiarid zone. Köppen's **Aw** climate, however, is too extensive and consequently includes a wide range of ecological and agroclimatic zones.

Thornthwaite classification. In his climatic classification – first published in 1931 and later revised in 1948 – Thornthwaite introduced and elaborated on the concept of potentievapotranspiration. He derived a formula for the estimation of potential evapotranspiration as a function of temperature, but adjusted it according to day length and latitude. Thornthwaite defined megathermal (or tropical) climates as areas where the annual potential evapotranspiration exceeds 1,140 mm, which roughly corresponds to a mean annual temperature of 23°C.

Because he contended that the moisture status of a place should be determined by a balance between rainfall and evapotranspiration, Thornthwaite devised a water budget method for its computation. This was an improvement over Köppen's aridity criterion. The Thornthwaite classification was adopted by Meigs (1953) in the preparation of an arid zone world map which was published by UNESCO.

Studies in recent years have shown that the Thornthwaite estimates depart appreciably from field observations in many parts of the world. Evapotranspiration is a physical process that requires a supply of energy as well as a means for turbulent transfer. The Penman equation (1948), which incorporates net radiation, wind, humidity, and temperature, is more accurate than the Thornthwaite formula. In a recent map showing the distribution of arid zones in the world, also published by UNESCO (1979), the Penman approach was used.

Garnier's map. In 1961 the UNESCO Humid Tropics Research Program presented two maps showing the extent of the humid tropics (Fosberg *et al.*, 1961). The map prepared by Garnier used climatic parameters; the other by Küchler was a vegetation map. The latter author stated that "a generally mesophytic vegetation, showing no great predominance of xeromorphic characteristics or putative adaptations to reduce transpiration, indicates an excess of available moisture over potential evapotranspiration (of its habitat)."

Taking advantage of his experience in West Africa, Garnier listed three criteria for humid tropicality: (1) the mean monthly temperature for at least eight months of the year equals or exceeds 20°C; (2) the vapour pressure and relative humidity for at least six months of the year average respectively at least 20 millibars and 65%; and (3) the mean annual rainfall totals at least 1,000 mm, and, for at least six months, precipitation exceeds 75 mm each month.

Garnier's map includes several areas that are atmospherically moist as regards humidity, but ecologically dry as regards rainfall – an approach clearly in the vein of human autecology but less suitable for land-use planning.

PROPOSED CRITERIA

It is apparent that the three major approaches differ considerably. After reviewing these and other studies, we propose the following criteria.

Thermal criterion. Although some climatologists have used the annual cumulative potential evapotranspiration or total annual net radiation (Budyko, 1948) to define the tropics, the use of temperature as a criterion is preferred. There has been a convergence on 20°C where mean annual temperature have been used and on 18°C for the mean coldest month (Oliver, 1979). The latter value is adopted not only because of its physiological and botanical implications, but also because it includes approximately 95% of the lowlands between the 23°30' parallels, the conventional definition of the tropics. On the other hand, the mean annual isotherm of 20°C extends well beyond the Tropic of Cancer and the Tropic of Capricorn in several parts of the world.

The use of temperature as a criterion necessarily excludes tropical uplands. The generally accepted average lapse rate ranges from 0.6°C to 0.65°C per 100 m. Yoshino (1980) has analyzed temperature data for 14 stations in northwestern Thailand, and has found that in some cases the lapse rate greatly exceeds 0.65°C per 100 m.

In order to delineate the average elevation limits of the coldest month isotherm of 18°C in different latitudinal belts of the tropics, temperature data of 207 stations have been analyzed. The following equation has been established:

$$T - 0.0065\,A = 31.324833 - 0.457172\,L$$

where T is temperature in °C, A is elevation in meters and L is latitude in degree. The correlation coefficient is -0.741 and r^2 = 0.549.

According to this equation, the average elevation where the coldest month isotherm of 18°C is observed varies with latitude as follows:

Latitude	0°	5°	10°	15°	20°	25°
Elevation (meters)	1,979	1,698	1,347	995	643	330

Wet-month definition. From the standpoint of agriculture and vegetation ecology, a wet month may be defined as one in which rainfall is equal to or exceeds potential evapotranspiration. Köppen selected a monthly rainfall of 60 mm as the minimum requirement for his **Af** climate, whereas Garnier adopted a value of 75 mm. Recent studies indicate that both values are too low. Observed evapotranspiration rates from tropical forests in the Ivory Coast (Huttel, 1962), Java (Coster, 1937), Zaire (Malaisse, 1973), Thailand (Sabhasri *et al.*, 1970), and Bengal (Banerjee, 1972) ranged from 948 to 1,150 mm per year, or slightly over 90 mm per month. Alexandre (1977) has found that during the rainy season the evapotranspiration rate from an evergreen forest in Zaire averaged 100 mm per month.

Penman (1970) has estimated the average annual evapotranspiration rate from humid tropical forests to be 1,200 to 1,500 mm.

Monthly evapotranspiration rates for crops in the humid tropics often exceed 100 mm, presumably because advected energy is greater over small crop fields than it is over extensive forest canopy. In the absence of advection, 100 mm, however, is a widely accepted monthly average value. For instance, in an attempt to evaluate climatic potential for cropping systems in South East Asia, scientists at the International Rice Research Institute (IRRI) used 100 mm as the average monthly water needed for a rice crop (IRRI, 1974).

In an analysis of agricultural drought in East Africa, Nieuwold (1980) stated that any month with rainfall in excess of 100 mm cannot be considered a drought month even though that month has less than half of the long-term average. In West Africa, 100 mm per month also appears to be the average value of potential evapotranspiration (Oladipo, 1981).

After analyzing extensive rainfall and soil moisture data from Indonesia, Mohr (1933) divided the various months of the year into three categories: (1) a wet month with more than 100 mm of rainfall, (2) a moist month with 60 to 100 mm of rainfall, and (3) a dry month with less than 60 mm of rainfall. It has been demonstrated that Mohr's system has a broader application to agriculture on the island of Java than the Köppen system (Sukanto, 1969).

From the above discussion, it seems reasonable that we define a wet month as one having more than 100 mm of rainfall. When the monthly rainfall is between 60 and 100 mm, it is highly probable that the water supply for plant growth is adequate for at least half of the month, and thus can be credited as a half wet-month.

Hydrologic growth season. One of the central issues in defining the humid tropics is how much seasonality of rainfall can be allowed. In the humid tropics, the length of the growing season is primarily determined by the water balance. The length of the wet season is often referred to as the hydrologic growing season.

Fuson (1963) has stated that "the most critical factor determining vegetation distribution is not total rainfall but the length and intensity of the dry season." Bunting (1961) considers the mean annual rainfall of "little significance for the agricultural assessment of climate in tropical Africa." Phillips (1959) has demonstrated a close association between the length of the wet season and the proportion of woody plants in the vegetation.

In Panama, the boundary between rain forest (three levels of growth with the top one usually being continuous) and the semi-evergreen forest (two levels of growth with the top forming a closed canopy) coincide with the one-month isoxeromene, or isoline of one-month dry season. The climate that has a monthly rainfall in excess of 100 mm is so unique that Serebrenick (1945) has designated it as super humid (tropico iso-superumido). In the Amazon, it occurs in an area west of the mouth of the Purus River. In Sumatra, it is found in the Medan area. Serebrenick considers these two areas to be homoclimatic and suitable for rubber trees. In Africa, super humid climate is found only in a very small area in the Congo – represented by Boende (O°13'S, 20°51'E) and Yalusake (1°02'S, 22°50'E).

In Indonesia, Schmidt & Ferguson (1951) have classified rainfall types based on wet and dry period ratios, following the definition by Hohr. Whitmore (1975) has found that such a classification based on wet and dry seasons is more useful in reflecting vegetation types in South East Asia than either the Köppen or Thornthwaite classifications. A similar approach has also been used by Aubréville (1949) in the study of vegetation in Africa.

Lauer (1952) has shown that in Africa, the boundary between the wet and dry savannas conform to the five-month isozeromene, or lines connecting points of equal number of dry months. All major oil palm-producing areas have less than five dry months (Benneh, 1972). The concept of the hydrologic growing season has also been used in assessing agricultural potential in Thailand by Eelaart (1973) and in Kenya by Jatzold (1976).

The most elaborate system that makes use of the concept of the hydrologic growing season in assessing agricultural potential, however, is in the work by IRRI mentioned above. In that study, it was assumed that in paddy fields there is an additional water loss of 100 mm per month due to percolation and seepage which varies greatly according to the layout of the field, groundwater level, topography, soil texture, and the like (Zamstra et al., 1982). Consequently, the water requirements for wetland rice and rainfed rice are different. For wetland rice, a wet month requires 200 mm of rainfall while a dry month is defined as having less than 100 mm. A detailed map for South East Asia showing areas with different combinations of dry and wet months has been prepared. In areas with more than nine consecutive wet months, double cropping of a paddy crop is possible. In areas with less than five consecutive wet months, only a single crop can be grown unless irrigation is practiced.

The USA President's Scientific Advisory Committee (1967) classified tropical climates according to the following criteria: (1) rainy climate with 9–1/2 to 12 wet months, (2) humid seasonal climate with 7 to 9–1/2 wet months, (3) wet-dry climate with 4–1/2 to 7 wet months, and (4) dry climate with less than 4–1/2 wet months. The Committee did not, however, present a map showing the distribution of these climatic types. The selection of the threshold number of wet months by the Committee is reasonable. Their criteria will be adopted in our classification. The term "rainy climate" will be changed, however, to "humid tropical" and "humid seasonal climate" to "subhumid tropical."

PREPARATION OF THE MAP

A map showing the distribution of the four types of climate has been prepared using the data from various sources (see Figure 1, Chapter 1). For some mountainous areas detailed patterns cannot be shown on such a small-scale world map. The problem, however, is not too serious because isolines of equal number of wet or dry months have a tendency to show abrupt changes than do isohyets. In this regard, isolines of numbers of dry and wet months are a useful guide for the delineation of agroclimatic regions.

ENDNOTE

1 Paper originally delivered orally at the International Association of Hydrological Sciences, Hamburg Symposium, (August, 1983) entitled Hydrology of Humid Tropical Regions with Particular Reference to the Hydrological Effects of Agriculture and Forestry Practices and later presented in the UNESCO report entitled Hydrology of Humid Tropical Regions (May 1986) based on IHP-II Project A.l.10 and IHP III Project 4.2 (b).– available from UNESCO, Division of Water Sciences, Paris, France.

REFERENCES

Alexandre, J. (1977) Le bilan de l'eau dans le miombo (forêt claire tropicale). Société Geographique de Liège Bulletin, 13: 107–126.

Aubréville, A. (1949) Climats, forêts et desertification de l'Afrique tropicale. Paris, Soc. Ed. Geographiques, Maritimes et Coloniales.

Banerjee, A. K. (1972) Evapotranspiration from a young Eucalyptus hybrid plantation of West Bengal. Proc. and Technical Papers, Symp.of Man-Made Forest in India, Soc. of Indian Foresters, Dehra Dun.

Benneh, G. (1972) The importance of rainfall and soil moisture in the agricultural development of the closed forest zone of Ghana. Proc. of Symp. on World Water Balance, Reading, 15–23 July 1970. IASH-UNESCO-WMO 467–478.

Budyko, M. I. (1948) Heat balance of the earth. Trans. by N. A. Stepanova. Washington, D.C., USA.

Bunting, A. H. (1961) Some problems of agricultural climatology in tropical Africa. Geography, 46: 283–294.

Coster, C. D. (1937) Verdamping van verschillende vegetatie vormen of Java. Tectona 30: 1–102.

Eelaart, A. T. J. van der. (1973) Climate and crop in Thailand. FAO Bangkok Report SSR-96: 23

Fosberg, F. R., Garnier, B. J., & Küchler, A. W. (1961) Delimitation of the humid tropics. Geograph. Rev., 51: 333–347.

Fuson, R. H. (1963) The isoxeromene as a tropical climatic boundary. Prof. Geogr., 15 (3): 4–7.

Huttel, C. (1962) Estimation du bilan hydrique dans une forêt sempervirente de basse Côte-d'Ivoire. Radioisotopes in Soil-Plant Nutrition Studies, Vienna, AIEA, 461.

International Rice Research Institute (1974) An agroclimatic classification for evaluating cropping systems potential in South East Asia rice growing regions. Los Banos, Philippines, 12.

Jatzold, R. (1976) Isolinien humider Monate als agrarplanerisches Hilfsmittel am Biespiel von Kenia. Zeitschrift für Auslandische Landwirtschaft, 15: 330–350.

Köppen, W. (1900) Versuch einer Klassifikation der Klimate. *Geogr. Zeit.*, *6*: 593–611.

Köppen, W. (1936) Das Geographische System de Klimate. *Handbuch der Klimatologie*, (eds). W. Koppen and R. Geiger, *I* (C).

Lauer, W. (1952) Humide und aride Jahrezeiten in Afrika und Südamerika und ihre Beziehung zu den Vegetations-gürteln. *Bonner Geographische Abhandlungen*, *9*: 15–98.

Malaisse, F. (1973) Contribution à l'étude de l'écosystème fôret claire (miombo). Note 8, le Project Miombo, Ann., Univ. Abidjan, E, *6* (2): 227–250.

Meigs, P. (1953) World distribution of arid and semi-arid homoclimates. *Review of Research on Arid Zone Hydrology*, I, Paris, UNESCO, 203–209.

Mohr, E. C. J. (1933) De bodem der tropen in het algemeen en die van Nederlandsch India in het bijzonder. Koninkl Ver Koloniaal Inst., Amsterdam, Mededeel, *31*: deel 1.

Nieuwolt, S. (1980) Seasonal droughts in East Africa. *East Afr. Agric. For. J.*, *43*: 208–222.

Oladipo, E. O. (1980) An analysis of heat and water balances in West Africa. *Geograph. Rev.*: *70*, 194–209.

Oliver, J. (1979) A study of geographical imprecision: The tropics. *Australian Geographical Studies*, *17*: 3–17.

Penman, H. L. (1948) Natural evaporation from open water bare soil and grass. *Proc. Royal Soc.*, ser. A, *193*: 120–145.

Penman, L. (1970) The water cycle. *Sci. Am.*, *223*: 99–108.

Phillips, J. (1959) *Agriculture and ecology in Africa*. London: Faber & Faber.

Sabhasri, S., Chunkao, K. & Ngampongsai, C. (1970) The estimation of evapotranspiration of the old clearing and the dry evergreen forest. Sakaerat, Nakorn Rachasima, Bangkok, Faculty of Forestry, Kasetsart University, 6.

Schmidt, F. H.& Ferguson, J. H. A. (1951) Rainfall types based on wet and dry period ratios for Indonesia. *Verhandel Djawatan Meteorol. dan Goefis Djakarta*, *42*: 77.

Serebrenick, S. (1945) Notas sôbre o clima do Brasil. Bol. Minist. Agric., off-print.

Sukanto, M. (1969) Climate of Indonesia. *Climates of Northern and Eastern Asia, World Survey of Climatology*, (ed.) H. E. Landsbery, *8*: 215–229.

Thornthwaite, C. W. (1931) The climates of North America according to a new classification. *Geograph. Rev.*, *21*: 633–655.

Thornthwaite, C. W. (1948) An approach toward a rational classification of climate. *Geograph. Rev.*, *38*: 55–95.

UNESCO (1979) Map of the world distribution of arid regions. MAB Technical Notes 7, Paris, France.

USA. President's Science Advisory Committee (1967) The world food problem: Report of the panel on the world food supply, *1–2* Washington, D.C.: Government Printing Office.

Whitmore, T. C. (1975) *Tropical rain forests of the Far East*. Oxford, Clarendon Press.

Yoshino, M. M. (1980) Local climatological differences between highlands and lowlands in Thailand. *Conservation and Development in Northern Thailand*, (ed.) J. D. Ives, S. Sabhasri, and P. Voraurai, The United Nation University, 63–74.

Zamstra, H. G., Samarita, D. E. & Pontipedra, A. N. (1982) Growing season analyses for rainfed wetland fields. IRRI Research Paper Series, *73*: 14.

Appendix B: Workshop Reports

Hydrological Processes Workshop

Rapporteur:

A. BULLOCK

Institute of Hydrology, Wallingford, Oxfordshire OX10 8BB United Kingdom

RESEARCH NEEDS AND STRATEGIES

Identification of hydrologically homogeneous domains

To evaluate, understand and define the spatial and temporal variability of key hydrological measures to enable the delimitation and characterization of homogeneous domains.

Recognition of processes in urban areas

To evaluate specifics of hydrological and hydraulic processes taking place in urban and urbanizing areas.

Focus process studies within reference catchments with the objectives of:

(a) defining the vegetation/soil/climate complex and its dynamic role in the energy and water balances.

(b) the development and refinement of empirical and deterministic models, stressing domains of applicability.

(c) the establishment of techniques and framework for model parameterization at multiple scales, emphasizing point to areal extrapolation, vertical scales and macrohydrology.

(d) investigating the recovery of degraded catchments

(e) to establish urban experimental basins in order to evaluate urban water budgets, including water flow in manmade structures such as conduits, constructions and impermeable surfaces with objectives similar to a, b and c.

Impact studies through hydrological models

(a) To use models developed on research basins of various scales, and where appropriate, network data for evaluating the impact of human and climate change, giving due consideration to the:

    (i) magnitude of induced change in the context of natural variability.

    (ii) nature, magnitude and scale of change.

    (iii) geographical variability of change.

    (iv) temporal variability of change.

(b) To evaluate and adopt existing calculation methods and models to the special needs of hydrology of urban areas.

Environmental protection and mitigation

(a) To define and delimit a hierarchy of environmental susceptibility to change as a basis for the construction and enforcement of policies for protection, mitigation and recovery.

(b) To define environmental impacts from large urban agglomerations.

(c) To define appropriate water management methods applicable to urban areas of humid tropics – with respect to environmental impact.

(d) To define appropriate methods for designing sewerage systems.

TECHNOLOGY ADAPTION

(a) To apply consistent methods of collecting, archiving and processing user-friendly, menu-driven multi-language software systems for the quality control and analysis of network hydrological and hydrogeological data.

(b) To apply consistent methods of analysis to a number of research basins of various scales in order to support the development of an international standard for the collection, archiving and analysis of research basin data.

(c) To develop guidelines for the appropriate generic classification of existing models and for the application of hydrological models in engineering design and environmental impact and management studies.

(d) To promote the appropriate application of water science through training programs and dissemination

(e) To define means of elimination of economical and other losses due to flooding of urban areas.

(f) To investigate and adopt existing methods of solid waste disposal with respect to needs of humid tropical countries.

(g) To evaluate environmental impacts of solid waste disposal using different methods (disposal, co-composting, burning).

(h) To enhance creation of databanks containing information from urban catchments reflecting quantitative and qualitative characteristics of various elements of the urban hydrological cycle.

TECHNICAL REQUIREMENTS

(a) To coordinate the dissemination of information for the use of hydrologically relevant aspects of database systems, GIS and remote sensing techniques, and to develop new methodologies for their application.

(b) To establish a network of reference basins representing complexes of hydrological domains and aquatic ecosystems.

(c) To assess and, if appropriate, define the requirement for minimum data observation standards for specific management objectives.

(d) To establish and maintain long-term experiments to assess impact of land-use change and natural variability.

(e) To arrest the decline in the operational collection and storage of hydrological and related data in the humid tropics which are essential for the monitoring of the water environment and for the development of water resources; and to reestablish and augment these programs to meet the increasing demands.

(f) To promote national/international hydrological databases – specifically, identification of source data and responsible organizations, establish institutional framework and define terms of access.

EDUCATIONAL/TECHNICAL SUPPORT

(a) To promote improved training for the maintenance of instrumentation and provision of operational support for an agreed minimum period

(b) To initiate core curricula within hydrological training courses to ensure consistency within agency training programs and within university/postgraduate courses

(c) To ensure that training programmes and curricula stress the specific research and water resource management requirements of the humid tropics

(d) To improve coordination of education and training elements amongst existing organizations

(e) To enhance the awareness of policy-makers as to the increasing significance of good water management in rapidly growing urban areas.

INSTITUTIONAL ARRANGEMENTS

(a) To initiate research investments from resource development schemes

(b) To promote national, international and interregional collaboration

(c) To co-ordinate interdisciplinary research and implementation programs amongst agencies of the United Nations, NGO's and educational establishments

(d) To review the successes/failures of major United Nations agencies' programs

GENERAL RECOMMENDATIONS

This Workshop recognizes the importance of hydrological processes amongst urban and rural surface water, groundwater and climate-related systems in an environrment subject to anthropogenic and natural change and increasing pressure on water and land resources.

The Workshop recognizes the growing importance of understanding appropriate management of water resources, as well as effluents from urban and urbanizing areas.

Erosion and Sedimentation Workshop
Rapporteur:

V. NOGUEIRA
Centro de Ciencias do Ambiente, Universidade do Amazonas, 6900 Manaus-Amazonas, Brazil

THREE MAJOR CLASSES OF EROSION/SEDIMENTATION PROBLEMS EXIST:

(a) On-site erosion which threatens sustainability of rural land uses

(b) Off-site problems caused by sediment transport and deposition

(c) Urban erosion/sedimentation

For Type (a) problems:

(i) Active participation by farmers and other land users is essential for successful changes in management systems.

(ii) Surface contact cover or mulch is essential to minimize surface erosion.

(iii) Land protection via changes in land management must be tied to direct economic benefits for land users.

(iv) We know (in principle at least) how to manage on-site erosion. Securing active participation and cost-effective implementation is the problem.

For Types (b) and (c) problems:

(i) We know much less that is universally applicable although there is some understanding of the interactions between river flow and sediment transport.

(ii) Other studies ought to include the interaction between erosion and fertility, calibration of USLE and alternatives, data on soil loss, bedload transport, runoff rates, rill and gully processes, and the relation of bed to suspended load.

COLLATE AND SYSTEMATISE EXISTING SMALL CATCHMENT DATA ON EROSION AND SEDIMENTATION INTO PREDICTIVE FRAMEWORKS OR MODEL(S)

Purpose:

(a) To improve spatial and temporal transferability of data for predicting management impacts in areas with limited or no data.

SET UP *REFERENCE* BASINS AS ADJUNCTS TO SYSTEMATISING EXISTING DATA, EMPHASIZING CONSISTENT DATA COLLECTION AND ANALYSIS METHODS

Purposes:

(a) To monitor long-term changes driven by climate change and human activities.

(b) To improve and expand predictive framework or models derived from existing data.

CALIBRATE REMOTE SENSING DATA TO SEDIMENT YIELD OR CONCENTRATION DATA IF POSSIBLE. COMBINE WITH GIS + LAND RESOURCE DATA + HYDROLOGICAL DATA TO MAP AND PREDICT CATCHMENT AND REGIONAL SEDIMENT YIELDS

Purpose:

(a) Potentially the most cost-effective tool for space-time integration of sediment yield data. Needs developing.

EDUCATION AND TRAINING

(a) Explore the desirability of developing a first-degree (BS) course in water sciences (hydrological sciences/hydrology).

(b) Increase erosion and sedimentation/fluvial geomorphology components of civil engineering and other technical professional training (forestry, agronomy, etc.).

(c) Develop specific erosion and sedimentation in-service training for existing technical professionals.

(d) Develop interpretation/communication skills of technical professionals and planners to communicate information and concepts to policy-makers, politicians and general public. Develop awareness and ability to exploit opportunities to communicate concepts for policy development and public education.

(e) Develop "twinning" of government agencies, research organizations and educational institutions concerned with

erosion and sedimentation.

(f) Provide training in GIS, land resource and hydrologic data collection and integration in order to take advantage of new or emerging technolgies.

POLICY CHANGES NEEDED

(a) Policies should be set to match local characteristics with due consideration to both biophysical and socio-economic settings.

(b) Environmental standards should not be unrealistically stringent and should not exceed natural background levels.

(c) Modify legislation to match physical and financial capacities for implementation.

(d) Provide statutory disincentives for forest clearance and development of areas at risk from serious erosion and/or sedimentation.

(e) Use non-structural measures wherever possible for flood protection and sediment management, e.g. linking insurance premiums to risk.

(f) Management should follow the "beneficiaries pay" principle, e.g.downstream beneficiaries should pay for upstream costs of achieving benefits.

POINTS FROM WORKSHOP DISCUSSION

(a) Logging and clearing practices are a major cause of erosion and sedimentation, therefore, there is a need for a change in these practices.

(b) The humid tropics are different from other regions of the world because of:
    (i) rainfall intensity,
    (ii) irreversible dehydration of the soil,
    (iii) some of the major sedimentation that reaches the ocean is from humid tropical areas.

(c) A two-tiered approach is required for training technicians

(d) Policy changes are needed – such as linking insurance premiums to risk and public participation in the decision-making process

Water Quality Control Workshop

Rapporteur:

B. J. STEWART
Hydrology Branch, Bureau of Meteorology, Melbourne, Australia, 3000

OBJECTIVE

To focus on the most important problems in water quality in the humid tropics, identify them, and recommend strategies for their solution.

MANAGEMENT PROBLEMS AND ISSUES

National and international cooperation
Problem:
A lack of coordination and cooperation between agencies at all levels of operation from local to international.
Recommendation:
That countries establish regular meetings between all agencies involved in water quality issues. International organizations should expand on their avenues of coordination and cooperation.

Problem:
Areas of responsibility at all levels need to be clearly defined.
Recommendation:
That an inventory of all agencies involved in aspects of water quality be compiled for each country with a statement on roles of each organization. A similar inventory for international organizations should be developed. International organizations should provide quality assurance and quality control guidelines (e.g. for laboratories).

Education, training and information exchange
Problem:
Decision-makers, peers and the community at large must be supplied with adequate information by which to evaluate issues.
Recommendation:
The media should be used to raise awareness and influence public opinion on water quality issues. The education of school children about water quality issues is seen as a viable and longer-term strategy.

Problem:
Presentation of knowledge, information and data must be aimed at the level of the receiver.
Recommendation:
Education institutions, such as universities, be encouraged to act as information dissemination points, but should ensure that information is delivered at the right level. Distributors of information (e.g. librarians) should be knowledgeable in the topic.

Problem:
There appears to be an abundance of courses but the content with respect to water quality issues in the humid tropics needs to be established.
Recommendation:
That a course must be encouraged to keep content in line with the most recent developments and new practices. Courses should be developed that enable practitioners to easily bridge these gaps. Courses should also be developed to train people to teach the subject matter back in their own country. The medium of instruction should be variable.

Problem:
There is a lack of information exchange between agencies at all levels, both nationally and internationally.
Recommendation:
International agencies should continue to provide information systems, and facilitate access to computer-based information systems.

Mathematical models
Problem:
Models are powerful and yet, if used unwisely, dangerous tools. Good quality data are also essential for their proper application.
Recommendation:
Training courses be organized to ensure that users of models are made aware of requirements and the limitations of various models.

Problem:
Models (water quality) for the humid tropics are not readily available, nor are models developed in other areas necessarily suitable for transfer.
Recommendation:
International agencies aid the development of water quality models for the humid tropics and educate the humid tropic countries in their use. Research funding should be made available in this area.

Inventory/diagnosis of national status of water quality
Problem:
There is a need for national assessments on a regular basis of water quality and quantity .
Recommendation:
That agencies continue to conduct or instigate national assessments of water quality and quantity using guidelines appropriate for the humid tropics – such as those produced by international agencies for Africa, Asia, and Latin America.

Long-term monitoring networks
Problem:
Network design strategies for water quality networks are inadequate.
Recommendation:
An international working group be established to examine the current network design philosophy and to develop guidelines in cooperation with the WMO Guide to Hydrological Practices.

Problem:
There have been significant advances in water quality data collection techniques. However, they are expensive and relatively new to the humid tropics.
Recommendation:
Data collection techniques be implemented on the basis of the purpose for collecting the information and the accuracy

of the data required. Agencies should not introduce new technology until it has been tested in the humid tropics and unless it is more cost effective than the current collection techniques.

Problem:

Information transfer on data collection techniques.

Recommendation:

Humid tropic countries should make use of the WMO HOMS system to enable technology transfer of data collection and processing systems.

Problem:

Data collection at remote localities.

Recommendation:

Use be made of satellite data collection techniques when they have been implemented successfully in the humid tropics.

Problem:

Urban data requirements are different from rural data in magnitude – in both time and space.

Recommendation:

That a conference/workshop be held on urban water quality which includes all aspects of data collection as well as analysis and interpretation.

Problem:

High technology water quality data collection techniques are expensive to operate and maintain.

Recommendation:

When the situation and conditions warrant, use be made of water quality field test kits which furnish broad indices of water quality instead of high technology.

Vector, parasites and pesticide control

Problem:

The major water quality pollutants in this area are fertilizers and pesticides.

Recommendation:

Companies which produce fertilizers be required to have labeling that indicates which countries have banned the use of the product. They should be held responsible for any damage caused by their product.

Problem:

The hydrological processes which transfer fertilizers and pesticides through the hydrological cycle (including groundwater) and aid degradation into by-products are inadequately understood.

Recommendation:

The World Health Organization, in conjunction with the Food and Agriculture Organization, investigate the above processes through the establishment of a joint working party to determine suitable research projects in this area.

Problem:

The production of genetically engineered micro-organisms in the humid tropics is a potential source of pollution.

Recommendation:

The tropical countries pass legislation to ensure the protection of all water sources from the possible introduction of genetically engineered micro-organisms.

Micro-organisms, biodegradability and fecal contamination

Problem:

Current research into fecal coliforms may have indicated that existing WHO guidelines for water quality are misleading.

Recommendation:

Current research be extended to other parts of the humid tropics and another indicator of disease risk be established for the humid tropics, if required.

Problem:

The WHO guidelines for water quality for drinking purposes are being adopted as standards for all sites, and all uses of water, and are thus restrictive to development.

Recommendation:

The World Health Organization reemphasize in the humid tropic countries their statement that they are guidelines for the safeguard of human health and should be adjusted for local conditions and the proposed use of the water resources.

Domestic and industrial organic pollution

Problem:

Release of untreated sewage into hydrological systems causes pollution in water systems – from streams to estuaries.

Recommendation:

Raw waste water should not be discharged into hydrologic systems.

Problem:

Development of water supply systems, sanitation and waste disposal systems is fragmented and uncoordinated.

Recommendation:

Water supply systems, sanitation and waste disposal systems be designed and constructed in a coordinated fashion. Guidelines should be prepared by the WHO.

Problem:

Solid waste materials are potential sources of pollution.

Recommendation:

The onus for treatment of industrial wastes be placed on the industry and their liability for what they release into the environment established legally. Municipal wastes should be treated and where possible, reused. If reuse is not possible, landfills should be constructed to maximise the containment of the pollutant.

Problem:

Water is a scarce and valuable resource in the humid tropics.

Recommendation:

The reuse/recycling of water in the humid tropics be strongly supported and appropriate uses of the water be determined on the basis of water quality requirements provided by the WHO. Regional workshops to promote reuse and adjust techniques to regional conditions in the humid tropics should be supported by WHO and UNESCO.

Industrial toxic waste and other chemical control

Problem:

The release of industrial microtoxins is a potentially harmful source of pollution.

Recommendation:

Funding resources be allocated to the establishment of international laboratories and to research into the hydrological, chemical and environmental processes associated with microtoxins. Preventative techniques which control the collection and disposal of toxic wastes be researched and strongly supported by international organizations. Legislation be introduced in all countries to ensure that the polluters are held financially responsible for all polution they cause. International agencies should prepare guidelines for the control and handling of microtoxins.

Problem:

Information transfer is lacking in this area.

Recommendation:

UNEP support international cooperation between all developing and developed countries with respect into microtoxins, and their impact on the water quality of the humid tropics.

Problem:

Potential water quality problems exist from polluted air.

Recommendation:

Humid tropical countries be made aware that air pollution can lead to water quality problems. Integrated air and water pollution control be coordinated by the relative agencies. There are particular problems in urban areas which are especially susceptible.

Integrated watershed management

Problem:

Knowledge of the hydrological and environmental processes in the humid tropics is lacking.

Recommendation:

Research into these processes be supported and encouraged for both urban and rural areas. Reduction of surface runoff in humid tropic urban areas is a major issue requiring attention.

Problem:

Urban drainage systems are being subjected to over-use by increasing populations.

Recommendation

Countries strive to improve drainage systems and reconstruct sewerage and other waste disposal systems in order to reduce pollution. Use should be made of current technology.

Problem:

Waste disposal from feedlots and pig farms are sources of pollution.

Recommendation:

Countries be encouraged to make use of the variety of disposal waste techniques (e.g. biogas digesters). However they should be made aware of retention times and potential health risks.

Problem:

Erosion and sedimentation cause pollution

Recommendation:

An inventory of farming technology and practices that reduce erosion be prepared as part of an international program.

GENERAL RECOMMENDATION FOR FUTURE RESEARCH ON URBAN HYDROLOGY IN THE HUMID TROPICS

The workshop considered that the specific needs of urban water resources management have not been fully addressed by this Colloquium and recommended that the International Hydrological Programme develop a strong theme of urban hydrology in its future activities. As indicated previously, urban hydrological problems are different in terms of magnitude in temporal and spatial dimensions, and require specialized attention.

Water Resource Management Workshop

Rapporteurs:

D. S. McCAULEY
U.S. Agency For International Development, Washington, DC

E. VAN BEEK
Delft Hydraulics Laboratory, Delft, The Netherlands

INTRODUCTION

Water resource management (WRM) is defined as a set of actions taken to use and control water resource inputs in order to obtain outputs and natural system conditions useful to society. Hence, the workshop mainly addressed issues related to how problems should be solved and by whom, rather than the technical problems themselves. Reference is made to the results of the other three workshops for those technical problems.

Specific characteristics of WRM in the humid tropics were defined as management under highly variable natural conditions (rainfall, river flow), excess rainfall, extreme weather conditions (cyclones), rapidly changing socio-economic conditions (population, economic development, urbanization) and the fact that WRM mainly pertains to developing countries with a lack of good data. The fast growing population causes water-related issues such as demand for food (irrigation, utilization of unsuitable land leading to erosion), drinking and industrial water demand, and water pollution. Wetland areas and their management also require special attention in the humid tropics. Water resource managers are facing very challenging tasks in the humid tropics to solve these problems. In fact, the majority of the workshop participants felt that, given the rapid population growth in many parts of the humid tropics, the possibilities for adequate WRM are indeed quite low in these regions. Some participants also felt that the rapid economic development was the main cause of stress on water resources in many regions of the humid tropics.

The workshop participants attempted to confine the discussion to topics of direct relevance to the humid tropics. It became clear, however, that the relevant management problems and solutions are not confined to the humid tropics but apply to other regions as well. The special nature of the humid tropical region lies in the high variability of dominant physical processes and socio-economic situations which causes these management problems to be more intense. In addition, severe deficiencies in data, knowledge, and manpower in many countries make these problems more difficult to handle.

MANAGEMENT PROBLEMS AND ISSUES IN THE HUMID TROPICS

Development issues

The humid tropics include a wide variety of conditions. From a WRM point of view, it is important to distinguish between densely populated areas (e.g. major parts of South and South East Asia) and sparsely populated areas (much of South America and Africa). Another important distinction can be made between urban and rural areas. Small tropical islands are a further specific category. Each area has its own water-related problems requiring specific WRM approaches, making it difficult to draw general conclusions.

Session XII on water management identified irrigation, flood damage reduction and hydropower generation as the three most characteristic and important purposes of water resource development in the humid tropics. Only the first two topics were discussed in any detail during the workshop. Other important water management opportunities include urban and rural water supply, sanitation and water quality management, as well as control of erosion and sedimentation.

Irrigation In the Asian section of the humid tropics, rice dominates agricultural activity. Low rice prices in the 1980s have made major new investments in rice schemes uneconomic and have led some countries to try to diversify out of rice. These efforts were hampered by inflexible irrigation and farming systems. Small-scale irrigation and non-irrigated agriculture deserve more attention under these conditions. Integration of various components of water resource management systems and inclusion of non-water resource sectors (e.g. construction

of village roads) should be encouraged for the improvement of social and economic conditions in rural areas. It was stated that the national objective of self-sufficiency in rice production in Indonesia means that rice production continues to be the most important water-related development issue in that country. Also, African countries continue their efforts in irrigation development in view of the need to increase food production. The development of irrigation projects in the past has led in many cases to disappointing results (lagging project completion, cost overruns and planned outputs not achieved). Some of these problems relate to weak project preparation, insufficient training of farmers and an inadequate institutional structure. It is noted that aquaculture development, especially intensive brackish water shrimp culture is becoming an important sector, competing with irrigation for available water supplies.

Flood damage reduction The main item discussed with regard to flood damage reduction was urban drainage. Urban floods and drainage are of particular importance in the humid tropics because of the links to pollution, sewerage and related health hazards. The intense rainfall and high frequency of flooding makes this a special problem in the humid tropics. In several cases, it appears that an adequate knowledge about urban hydrologic processes and flood damage aspects are lacking. Flooding problems in rural areas often relate to land-use planning issues. For both types of flooding there seems to be a tendency to shift management approaches from structural to non-structural (e.g. zoning, relocation, flood warning systems) and from upstream solutions (such as reservoirs and flood diversions) to local solutions. Upper watershed management (an upstream solution) continues to be of high importance. It was noted that the preferred solution depends very much on the development stage (availability of funds) of the area involved, with Singapore as extreme example of a full structural solution with very high protection.

Management problems

An evaluation of water related problems and the results of projects carried out in this field reveals a number of management problems, i.e., deficiencies in the performance of these projects, which includes the physical structures and their management. Often:

- projects are not completed on time, and there are severe cost overruns;
- planned outputs are not achieved;
- projects lack economic feasibility;
- projects are only physically oriented and lack required related managerial and incentive measures;
- projects have severe environmental and social (e.g. forced relocation) impacts;
- reservoirs suffer rapid sedimentation; and
- the physical infrastructure suffers rapid deterioration because of faulty operation and inadequate maintenance.

Causes of deficiencies

The management problems and their causes can be grouped in the following terms:

- the upstream-downstream problem, where downstream water development and upstream watershed management proceed independently or without adequate coordination.
- the environmental impact problem, where there is failure to adequately identify and deal with important environmental and social consequences in the planning and implementation of water resource activities.
- the water misallocation problem, where there is failure to allocate scarce resources to the highest and best economic and social uses.
- the implementation problems as reflected by low project outputs, high costs, adverse environmental and social impacts, and project deterioration and failure.

In terms of activities, these causes relate to inadequate planning and implementation. Reasons for this situation are:

- inadequate data;
- too narrow a scope of water resource planning in temporal, geographic, objectives, and purpose terms;
- inadequately trained and too narrow professional composition of planning and implementation staff;
- fractionated organizational and administrative structure, inhibiting the collaboration necessary for integrated WRM;
- budgetary/financing and professional reward systems that overemphasize construction at the expense of operation and maintenance;
- weak and inappropriate institutional structures;
- exogenous national and donor programs and policies that work against effective and efficient WRM; and
- low priority for the funding of planning and implementation.

SUGGESTED MANAGEMENT STRATEGIES

Planning

It is recommended that international development agencies and national water management organizations encourage the use of strategies to specifically address the most common water resource management problems of the humid tropics. Six approaches were suggested by the workshop participants:

Integrated watershed/river basin planning Where there are strong upstream-downstream linkages affecting water resource planning, a watershed/basin perspective (including the acknowledgement of inter-basin transfers) offers the best understanding of natural flows. Where these linkages are weaker – such as when there are limited downstream negative effects of upstream land and water use – there is less need for integration on this basis (leaving emphasis on on-site concerns). Regional and national priorities must also be balanced.

Multiple-objective planning approach Planning of water resource management activities can define its objectives in more than the traditional economic development terms to explicitly include environmental sustainability and social goals. When properly applied this approach reduces the need for separate environmental impact assessments.

Planning for implementation Especially where the operation and maintenance stage of a water resource management plan is heavily dependent on human organization, it is important to consider the implementation tools and the institutional and organizational arrangements at the planning phase. Involvement of local communities affected by the water resource management plan is an important part of this strategy which also can improve project design in other ways.

Demand management Especially where seasonal water shortages or water quality problems are leading to increasing competition among water uses and users, attention to the more efficient use of existing water supplies (including recycling, adjusting water pricing policy, water reuse and reduction in per capita and per unit water use) can lead to cost effective.

Judicious use of limited and uncertain information Planning approaches can recognize the often severe data limitations by making use of simple system analytical tools. Data uncertainty and environmental variability – combined with often rapid shifts in development patterns – can be accommodated through robust and flexible water resource management strategies.

International cooperation Where water resource systems cross international boundaries, bilateral and multilateral governmental arrangements can lead to more efficient allocation of outputs from water resource management.

Implementation

There are also approaches which should be encouraged for dealing with problems commonly encountered at the implementation stage of water resource management. Two strategies were discussed and recommended:

Monitoring and evaluation During both the installation/construction as well as the operation/maintenance phases of implementation, information feed-back mechanisms can be used to adjust water resource management activities to changing circumstances or improved understanding of the water resource system. As data are relatively limited in the humid tropics and the environment is more variable, this flexibility is especially important.

Appropriate institutional and organizational arrangements Since the "local" context is so important to determining the most desirable and feasible institutional and organizational patterns for water resource management, these cannot be generally prescribed. Except where there is a congruence of hydrological and administrative boundaries, it is better to organize implementation according to the latter and to decentralize such authority wherever feasible. Opportunities should also be sought for private sector involvement in water resource management.

RESEARCH, TRAINING, EDUCATION AND PUBLIC AWARENESS

Research on water resource management problems of the humid tropics

Water management professionals need better information about the unique challenges of the humid tropical zone. Some factors include different dominating hydrological processes, rapid weathering processes, high rainfall variability and frequency of extreme events, rich and still weakly understood biological resources and ecological processes, and often rapid social and economic change in a developing country context. The study could lead to improved design criteria and management strategies plus research priorities.

Training and education for improved water management in the humid tropics

Based on improved understanding of the special characteristics and management challenges of the humid tropics, complementary education and training programs are needed for all levels of professionals involved in water management. High priority should be given to improving the understanding of decision-makers including mid-level and senior water agency officials regarding the special tasks and problems of water management and policy in the humid tropics.

Public awareness of water resource management issues and strategies in the humid tropics

Public awareness should be improved, since it is crucial to popular understanding of and support for water resource management efforts in the humid tropics. Such awareness is likely to enhance the political will to give a higher priority to water resource management, resulting in an increased allocation of financial resources to such efforts. This may also reduce the likelihood of problems reaching a crisis stage before public action is taken.

Appendix C: Committees for the 'International Colloquium on the Development of Hydrologic and Water Management Strategies in the Humid Tropics'

ORGANIZING COMMITTEE

Dr. Michael Bonell (Chairman), Department of Geography, Institute for Tropical Rainforest Studies, James Cook University of North Queensland, Townsville, QLD 4811, Australia. Present address: UNESCO—SC/HYD, 1, rue Miollis,75732 Paris CEDEX 15, France.

Dr. Jaroslav Balek (Representing IAHS), Senior Programme Officer ME, UNEP, P. O. Box 30552, Nairobi, Kenya. Present address: ENEX – Environmental Engineering Consultancy, Kopecko 8, Prague 6, 16900 Czech Republic.

Dr. Pierre Dubreuil, CIRAD, 42 rue Scheffer, 75116 Paris, France.

Mr. David C. Flaherty (Text Editor), N.W. 1000 Bryant, Pullman, WA 99163, USA.

Dr. John S. Gladwell. Hydro Tech International, Vancouver, B.C. Canada (formerly UNESCO-SC/HYD, 1 rue Miollis, 75732 Paris CEDEX 15, France).

Dr. Lawrence S. Hamilton, Environmental and Policy Institute, East-West Center, 1777 East-West Road, Honolulu, Hawaii 96848, USA.

Dr. Maynard Hufschmidt, Environmental and Policy Institute, East-West Center, 1777 East-West Road, Honolulu, Hawaii 96848, USA.

Dr. L. Stephen Lau, Water Resources Research Center, University of Hawaii at Manoa, 2540 Dole Street, Holmes Hall 283, Honolulu, Hawaii 96822, USA.

Dr. David S. McCauley, Environmental Advisor, USAID, c/o US Embassy, Jakarta, Indonesia. Present address: Office of Environmental and Natural Resources, U.S. Agency for International Development, Washington, D.C. 20523–1812, USA.

Ir. Mardjono Notodihardjo, President Director, Pt Bina Karya (Persero), Architects and Consulting Engineers, Jalan DI Panjaitan Kav No. 2 Cawang, Jakarta 13001, Indonesia.

Dr. Glenn E. Stout, Executive Director, IWRA, 205 North Mathews Avenue, Urbana, IL 61801, USA.

COORDINATING COMMITTEE

Dr. Michael Bonell (Chairman)
Dr. Jaroslav Balek (Representing lAHS)
Dr. Pierre Dubreuil
Dr. John S. Gladwell (Project Officer)

The Organising Committee

International Colloquium on the Development of Hydrologic and Water Management Strategies in the Humid Tropics, James Cook University of North Queensland, Townsville, Australia. July 15–22, 1989.

Front row: (L to R) Dr Jack Gladwell, Dr Stephen Lau, Ir Mardjono Notodihardjo, Dr Mike Bonell, Dr Maynard Hufschmidt.

Back row: Dr Jaroslav (Jerry) Balek, Dr Glenn Stout, Mr David Flaherty, Dr Lawrence Hamilton, Dr David McCauley. (Absent: Dr Pierre Dubreuil)

The Participants

International Colloquium on the Development of Hydrologic and Water Management Strategies in the Humid Tropics, James Cook University of North Queensland, Townsville, Australia. July 15–22, 1989.

Front row: (L to R) S.T. Malling, G.E. Stout, P. Wurzel, U. Kuffner, F.R. Francisco, L. Stephen Lau, Maynard M. Hufschmidt, David S. McCauley, D.N. Body, J.S. Gladwell, M. Bonell, L.A. Mandalia, Femi Odumosu, L.S. Hamilton, Liu Chang Ming.

Second row: Law Kong Fook, Lam Kin Che, Janusz Niemczynowicz, Low Kwai Sim, K.G. Tejwani, Abdul Rahim Nik, Nitaya Mahabhol, Bernard Griesinger, Tanyileke Gregory, Pierre Dubreuil, Ian Gordon, Kate Duggan.

Third row: Kenji Oya, Juan G. Limon, D.C. Goswami, Michel-Alain Roche, Jean-Pierre Brunel, John Kilani, Carlos Fernandez-Jauregui, Keith L. Bristow, Ray Voker, Gaddi G. Ngirane-Katashya.

Fourth row: M. Manton, Calvin Rose, Vicente Nogueira, Ben Braga, Kuniyoshi Takeuchi, Brian Prove, Mardjono Notodihardjo, Charles G. Birigenda, A. Prost, Roger S. Fujioka, Mick Fleming, Andrew Pearce, Geoff Heatherwick.

Back row: Eello Van Beek, Terry C. Hazen, V. Klemes, J. Balek, G. Strigel, K.C. Goh, Andrew Bullock, Ausaf-Ur Rahman, David Flaherty, Thomas Stadtmuller, Bruce Stewart, Mamadou Sakho, Dirk Libbrecht, Tony Falkland.

Appendix D: Organizations Supporting UNESCO and UNEP in the Presentation of the Colloquium

CO-SPONSORING ORGANIZATIONS

Australian International Development Assistance Bureau (AIDAB)
Australian National Commission for UNESCO
Australian National Committee of the IHP
CSIRO, Division of Water Resources
East-West Center, Environment and Policy Institute
Federal Republic of Germany National Committee for the IHP
Institut Français de Recherche Scientifique pour le Développement en Cooperation (ORSTOM)
International Association of Hydrogeologists (IAH)
International Association of Hydrological Sciences (IAHS)
International Water Resources Association (IWRA)
James Cook University of North Queensland
Organization of American States
USA National Committee on Scientific Hydrology

In cooperation and with the support of:
Commission of European Communities (CEC)
Department of Technical Co-operation for Development (UNDTCD)
Economic and Social Commission for Asia and the Pacific(ESCAP)
Economic Commission for Latin America and the Caribbean (ECLAC)
Japan National Committee for IHP
United Nations Children's Fund (UNICEF)
World Bank
World Health Organization (WHO)
World Meteorological Organization (WMO)

The International Hydrological Programme (IHP) wishes also to acknowledge the support of UNESCO's Man and the Biosphere (MAB) Programme.

Index